31st EDITION

CRC
STANDARD
MATHEMATICAL
TABLES AND
FORMULAE

31st EDITION

CRC
STANDARD
MATHEMATICAL
TABLES AND
FORMULAE

DANIEL ZWILLINGER

CHAPMAN & HALL/CRC

A CRC Press Company

Boca Raton   London   New York   Washington, D.C.

Visit the CRC Press Web site at www.crcpress.com

Preface

It has long been the established policy of CRC Press to publish, in handbook form, the most up-to-date, authoritative, logically arranged, and readily usable reference material available. Prior to the preparation of this *31ˢᵗ Edition* of the *CRC Standard Mathematical Tables and Formulae*, the content of such a book was reconsidered. The previous edition was carefully analyzed, and input was obtained from practitioners in the many branches of mathematics, engineering, and the physical sciences. The consensus was that numerous small additions were required in several sections, and several new areas needed to be added.

Some of the new materials included in this edition are: game theory and voting power, heuristic search techniques, quadratic fields, reliability, risk analysis and decision rules, a table of solutions to Pell's equation, a table of irreducible polynomials in $Z_2[x]$, a longer table of prime numbers, an interpretation of powers of 10, a collection of "proofs without words", and representations of groups of small order. In total, there are more than 30 completely new sections, more than 50 new and modified entries in the sections, more than 90 distinguished examples, and more than a dozen new tables and figures. This brings the total number of sections, sub-sections, and sub-sub-sections to more than 1,000. Within those sections are now more than 3,000 separate items (a definition, a fact, a table, or a property). The index has also been extensively re-worked and expanded to make finding results faster and easier; there are now more than 6,500 index references (with 75 cross-references of terms) and more than 750 notation references.

The same successful format which has characterized earlier editions of the *Handbook* is retained, while its presentation has been updated and made more consistent from page to page. Material is presented in a multi-sectional format, with each section containing a valuable collection of fundamental reference material—tabular and expository.

In line with the established policy of CRC Press, the *Handbook* will be kept as current and timely as is possible. Revisions and anticipated uses of newer materials and tables will be introduced as the need arises. Suggestions for the inclusion of new material in subsequent editions and comments regarding the present edition are welcomed. The home page for this book, which will include errata, will be maintained at http://smtf.mathtable.com.

The major material in this new edition is as follows:

Chapter 1: *Analysis* begins with numbers and then combines them into series and products. Series lead naturally into Fourier series. Numbers also lead to functions which results in coverage of real analysis, complex analysis, and generalized functions.

Chapter 2: *Algebra* covers the different types of algebra studied: elementary algebra, vector algebra, linear algebra, and abstract algebra. Also included are details on polynomials and a separate section on number theory. This chapter includes many new tables.

Chapter 3: *Discrete Mathematics* covers traditional discrete topics such as combinatorics, graph theory, coding theory and information theory, operations research, and game theory. Also included in this chapter are logic, set theory, and chaos.

Chapter 4: *Geometry* covers all aspects of geometry: points, lines, planes, surfaces, polyhedra, coordinate systems, and differential geometry.

Chapter 5: *Continuous Mathematics* covers calculus material: differentiation, integration, differential and integral equations, and tensor analysis. A large table of integrals is included. This chapter also includes differential forms and orthogonal coordinate systems.

Chapter 6: *Special Functions* contains a sequence of functions starting with the trigonometric, exponential, and hyperbolic functions, and leading to many of the common functions encountered in applications: orthogonal polynomials, gamma and beta functions, hypergeometric functions, Bessel and elliptic functions, and several others. This chapter also contains sections on Fourier and Laplace transforms, and includes tables of these transforms.

Chapter 7: *Probability and Statistics* begins with basic probability information (defining several common distributions) and leads to common statistical needs (point estimates, confidence intervals, hypothesis testing, and ANOVA). Tables of the normal distribution, and other distributions, are included. Also included in this chapter are queuing theory, Markov chains, and random number generation.

Chapter 8: *Scientific Computing* explores numerical solutions of linear and non-linear algebraic systems, numerical algorithms for linear algebra, and how to numerically solve ordinary and partial differential equations.

Chapter 9: *Financial Analysis* contains the formulae needed to determine the return on an investment and how to determine an annuity (i.e., the cost of a mortgage). Numerical tables covering common values are included.

Chapter 10: *Miscellaneous* contains details on physical units (definitions and conversions), formulae for date computations, lists of mathematical and electronic resources, and biographies of famous mathematicians.

It has been exciting updating this edition and making it as useful as possible. But it would not have been possible without the loving support of my family, Janet Taylor and Kent Taylor Zwillinger.

Daniel Zwillinger
zwillinger@alum.mit.edu
15 October 2002

Contributors

Karen Bolinger
Clarion University
Clarion, Pennsylvania

Patrick J. Driscoll
U.S. Military Academy
West Point, New York

M. Lawrence Glasser
Clarkson University
Potsdam, New York

Jeff Goldberg
University of Arizona
Tucson, Arizona

Rob Gross
Boston College
Chestnut Hill, Massachusetts

George W. Hart
SUNY Stony Brook
Stony Brook, New York

Melvin Hausner
Courant Institute (NYU)
New York, New York

Victor J. Katz
MAA
Washington, DC

Silvio Levy
MSRI
Berkeley, California

Michael Mascagni
Florida State University
Tallahassee, Florida

Ray McLenaghan
University of Waterloo
Waterloo, Ontario, Canada

John Michaels
SUNY Brockport
Brockport, New York

Roger B. Nelsen
Lewis & Clark College
Portland, Oregon

William C. Rinaman
LeMoyne College
Syracuse, New York

Catherine Roberts
College of the Holy Cross
Worcester, Massachusetts

Joseph J. Rushanan
MITRE Corporation
Bedford, Massachusetts

Les Servi
MIT Lincoln Laboratory
Lexington, Massachusetts

Peter Sherwood
Interactive Technology, Inc.
Newton, Massachusetts

Neil J. A. Sloane
AT&T Bell Labs
Murray Hill, New Jersey

Cole Smith
University of Arizona
Tucson, Arizona

Mike Sousa
Veridian
Ann Arbor, Michigan

Gary L. Stanek
Youngstown State University
Youngstown, Ohio

Michael T. Strauss
HME
Newburyport, Massachusetts

Nico M. Temme
CWI
Amsterdam, The Netherlands

Ahmed I. Zayed
DePaul University
Chicago, Illinois

Table of Contents

Chapter 1
Analysis **1**
> Karen Bolinger, M. Lawrence Glasser, Rob Gross, and
> Neil J. A. Sloane

Chapter 2
Algebra **79**
> Patrick J. Driscoll, Rob Gross, John Michaels, Roger B.
> Nelsen, and Brad Wilson

Chapter 3
Discrete Mathematics **197**
> Jeff Goldberg, Melvin Hausner, Joseph J. Rushanan, Les
> Servi, and Cole Smith

Chapter 4
Geometry **297**
> George W. Hart, Silvio Levy, and Ray McLenaghan

Chapter 5
Continuous Mathematics **383**
> Ray McLenaghan and Catherine Roberts

Chapter 6
Special Functions **499**
> Nico M. Temme and Ahmed I. Zayed

Chapter 7
Probability and Statistics **615**
> Michael Mascagni, William C. Rinaman, Mike Sousa, and
> Michael T. Strauss

Chapter 8
Scientific Computing **727**
> Gary Stanek

Chapter 9
Financial Analysis **779**
> Daniel Zwillinger

Chapter 10
Miscellaneous **791**
> Rob Gross, Victor J. Katz, and Michael T. Strauss

Table of Contents

Chapter 1
Analysis **1**

   1.1   Constants 3
   1.2   Special numbers 10
   1.3   Series and products 31
   1.4   Fourier series 48
   1.5   Complex analysis 53
   1.6   Interval analysis 65
   1.7   Real analysis 66
   1.8   Generalized functions 76

Chapter 2
Algebra **79**

   2.1   Proofs without words 81
   2.2   Elementary algebra 83
   2.3   Polynomials 89
   2.4   Number theory 93
   2.5   Vector algebra 131
   2.6   Linear and matrix algebra 137
   2.7   Abstract algebra 160

Chapter 3
Discrete Mathematics **197**

   3.1   Symbolic logic 199
   3.2   Set theory 202
   3.3   Combinatorics 206
   3.4   Graphs 219
   3.5   Combinatorial design theory 241
   3.6   Communication theory 253
   3.7   Difference equations 265
   3.8   Discrete dynamical systems and chaos 272
   3.9   Game theory 274
   3.10 Operations research 280

Chapter 4
Geometry **297**

   4.1   Coordinate systems in the plane 299
   4.2   Plane symmetries or isometries 305
   4.3   Other transformations of the plane 312
   4.4   Lines 314

4.5 Polygons 317
4.6 Conics 325
4.7 Special plane curves 336
4.8 Coordinate systems in space 345
4.9 Space symmetries or isometries 348
4.10 Other transformations of space 352
4.11 Direction angles and direction cosines 353
4.12 Planes 354
4.13 Lines in space 355
4.14 Polyhedra 357
4.15 Cylinders 361
4.16 Cones 361
4.17 Surfaces of revolution: the torus 363
4.18 Quadrics 364
4.19 Spherical geometry & trigonometry 368
4.20 Differential geometry 373
4.21 Angle conversion 381
4.22 Knots up to eight crossings 382

Chapter 5
Continuous Mathematics **383**

5.1 Differential calculus 385
5.2 Differential forms 395
5.3 Integration 398
5.4 Table of indefinite integrals 412
5.5 Table of definite integrals 448
5.6 Ordinary differential equations 456
5.7 Partial differential equations 468
5.8 Eigenvalues 477
5.9 Integral equations 478
5.10 Tensor analysis 482
5.11 Orthogonal coordinate systems 492
5.12 Control theory 497

Chapter 6
Special Functions **499**

6.1 Trigonometric or circular functions 503
6.2 Circular functions and planar triangles 512
6.3 Inverse circular functions 518
6.4 Ceiling and floor functions 520
6.5 Exponential function 520
6.6 Logarithmic functions 522
6.7 Hyperbolic functions 523
6.8 Inverse hyperbolic functions 527
6.9 Gudermannian function 530
6.10 Orthogonal polynomials 532

6.11 Gamma function 540
6.12 Beta function 544
6.13 Error functions 545
6.14 Fresnel integrals 547
6.15 Sine, cosine, and exponential integrals 549
6.16 Polylogarithms 551
6.17 Hypergeometric functions 552
6.18 Legendre functions 554
6.19 Bessel functions 559
6.20 Elliptic integrals 568
6.21 Jacobian elliptic functions 572
6.22 Clebsch–Gordan coefficients 574
6.23 Integral transforms: Preliminaries 576
6.24 Fourier transform 576
6.25 Discrete Fourier transform (DFT) 582
6.26 Fast Fourier transform (FFT) 584
6.27 Multidimensional Fourier transform 585
6.28 Laplace transform 585
6.29 Hankel transform 589
6.30 Hartley transform 591
6.31 Hilbert transform 591
6.32 Z-Transform 594
6.33 Tables of transforms 599

Chapter 7
Probability and Statistics **615**

7.1 Probability theory 617
7.2 Classical probability problems 627
7.3 Probability distributions 630
7.4 Queuing theory 637
7.5 Markov chains 640
7.6 Random number generation 644
7.7 Control charts and reliability 650
7.8 Risk analysis and decision rules 656
7.9 Statistics 658
7.10 Confidence intervals 666
7.11 Tests of hypotheses 669
7.12 Linear regression 682
7.13 Analysis of variance (ANOVA) 686
7.14 Probability tables 695
7.15 Signal processing 718

Chapter 8
Scientific Computing **727**

8.1 Basic numerical analysis 728
8.2 Numerical linear algebra 740

8.3   Numerical integration and differentiation 750
8.4   Programming techniques 777

Chapter 9
Financial Analysis **779**

9.1   Financial formulae 779
9.2   Financial tables 783

Chapter 10
Miscellaneous **791**

10.1   Units 792
10.2   Interpretations of powers of 10 798
10.3   Calendar computations 799
10.4   AMS classification scheme 801
10.5   Fields medals 802
10.6   Greek alphabet 803
10.7   Computer languages 803
10.8   Professional mathematical organizations 804
10.9   Electronic mathematical resources 807
10.10   Biographies of mathematicians 810

List of references **817**

List of figures **821**

List of notation **823**

Index **835**

Chapter 1

Analysis

1.1 CONSTANTS **3**
1.1.1   Types of numbers 3
1.1.2   Roman numerals 4
1.1.3   Arrow notation 4
1.1.4   Representation of numbers 5
1.1.5   Binary prefixes 6
1.1.6   Decimal multiples and prefixes 6
1.1.7   Decimal equivalents of common fractions 7
1.1.8   Hexadecimal addition and subtraction table 8
1.1.9   Hexadecimal multiplication table 8
1.1.10   Hexadecimal–decimal fraction conversion table 9

1.2 SPECIAL NUMBERS **10**
1.2.1   Powers of 2 10
1.2.2   Powers of 16 in decimal scale 12
1.2.3   Powers of 10 in hexadecimal scale 13
1.2.4   Special constants 13
1.2.5   Constants in different bases 16
1.2.6   Factorials 17
1.2.7   Bernoulli polynomials and numbers 19
1.2.8   Euler polynomials and numbers 20
1.2.9   Fibonacci numbers 21
1.2.10   Powers of integers 21
1.2.11   Sums of powers of integers 22
1.2.12   Negative integer powers 23
1.2.13   de Bruijn sequences 24
1.2.14   Integer sequences 25

1.3 SERIES AND PRODUCTS **31**
1.3.1   Definitions 31
1.3.2   General properties 32
1.3.3   Convergence tests 33
1.3.4   Types of series 34
1.3.5   Summation formulae 40
1.3.6   Improving convergence: Shanks transformation 40
1.3.7   Summability methods 41
1.3.8   Operations with power series 41
1.3.9   Miscellaneous sums and series 41
1.3.10   Infinite series 42
1.3.11   Infinite products 47

1-58488-291-3/02/$0.00+$1.50
© 2003 CRC Press, Inc.

1.3.12   *Infinite products and infinite series* *47*

1.4   **FOURIER SERIES** **48**
1.4.1   *Special cases* *49*
1.4.2   *Alternate forms* *50*
1.4.3   *Useful series* *50*
1.4.4   *Expansions of basic periodic functions* *51*

1.5   **COMPLEX ANALYSIS** **53**
1.5.1   *Definitions* *53*
1.5.2   *Operations on complex numbers* *54*
1.5.3   *Functions of a complex variable* *54*
1.5.4   *Cauchy–Riemann equations* *55*
1.5.5   *Cauchy integral theorem* *55*
1.5.6   *Cauchy integral formula* *55*
1.5.7   *Taylor series expansions* *55*
1.5.8   *Laurent series expansions* *56*
1.5.9   *Zeros and singularities* *56*
1.5.10   *Residues* *57*
1.5.11   *The argument principle* *58*
1.5.12   *Transformations and mappings* *58*

1.6   **INTERVAL ANALYSIS** **65**
1.6.1   *Interval arithmetic rules* *65*
1.6.2   *Interval arithmetic properties* *65*

1.7   **REAL ANALYSIS** **66**
1.7.1   *Relations* *66*
1.7.2   *Functions (mappings)* *66*
1.7.3   *Sets of real numbers* *67*
1.7.4   *Topological space* *69*
1.7.5   *Metric space* *69*
1.7.6   *Convergence in $\mathbb{R}$ with metric $|x - y|$* *70*
1.7.7   *Continuity in $\mathbb{R}$ with metric $|x - y|$* *71*
1.7.8   *Banach space* *72*
1.7.9   *Hilbert space* *74*
1.7.10   *Asymptotic relationships* *75*

1.8   **GENERALIZED FUNCTIONS** **76**
1.8.1   *Delta function* *76*
1.8.2   *Other generalized functions* *77*

1.1   CONSTANTS

1.1.1   TYPES OF NUMBERS

1.1.1.1   Natural numbers

The set of *natural numbers*, $\{0, 1, 2, \ldots\}$, is customarily denoted by $\mathbb{N}$. Many authors do not consider 0 to be a natural number.

1.1.1.2   Integers

The set of *integers*, $\{0, \pm 1, \pm 2, \ldots\}$, is customarily denoted by $\mathbb{Z}$. The *positive integers* are $\{1, 2, 3, \ldots\}$.

1.1.1.3   Rational numbers

The set of *rational numbers*, $\{\frac{p}{q} \mid p, q \in \mathbb{Z}, q \neq 0\}$, is customarily denoted by $\mathbb{Q}$. Two fractions $\frac{p}{q}$ and $\frac{r}{s}$ are equal if and only if $ps = qr$.

Addition of fractions is defined by $\frac{p}{q} + \frac{r}{s} = \frac{ps+qr}{qs}$. Multiplication of fractions is defined by $\frac{p}{q} \cdot \frac{r}{s} = \frac{pr}{qs}$.

1.1.1.4   Real numbers

The set of *real numbers* is customarily denoted by $\mathbb{R}$. Real numbers are defined to be converging sequences of rational numbers or as decimals that might or might not repeat.

Real numbers are often divided into two subsets. One subset, the *algebraic numbers*, are real numbers which solve a polynomial equation in one variable with integer coefficients. For example; $\frac{1}{\sqrt{2}}$ is an algebraic number because it solves the polynomial equation $2x^2 - 1 = 0$; and all rational numbers are algebraic. Real numbers that are not algebraic numbers are called *transcendental numbers*. Examples of transcendental numbers include π and e.

1.1.1.5   Complex numbers

The set of *complex numbers* is customarily denoted by $\mathbb{C}$. They are numbers of the form $a + bi$, where $i^2 = -1$, and a and b are real numbers. See page 53.

Operation	computation	result
addition	$(a + bi) + (c + di)$	$(a + c) + i(b + d)$
multiplication	$(a + bi)(c + di)$	$(ac - bd) + (ad + bc)i$
reciprocal	$\dfrac{1}{a + bi}$	$\dfrac{a}{a^2 + b^2} - \dfrac{b}{a^2 + b^2}i$
complex conjugate	$z = a + bi$	$\bar{z} = a - bi$

Properties include: $\overline{z + w} = \bar{z} + \bar{w}$ and $\overline{zw} = \bar{z}\,\bar{w}$.

1.1.2   ROMAN NUMERALS

The major symbols in Roman numerals are $I = 1, V = 5, X = 10, L = 50, C = 100$, $D = 500$, and $M = 1,000$. The rules for constructing Roman numerals are:

1. A symbol following one of equal or greater value adds its value. (For example, $II = 2, XI = 11$, and $DV = 505$.)

2. A symbol following one of lesser value has the lesser value subtracted from the larger value. An I is only allowed to precede a V or an X, an X is only allowed to precede an L or a C, and a C is only allowed to precede a D or an M. (For example $IV = 4, IX = 9$, and $XL = 40$.)

3. When a symbol stands between two of greater value, its value is subtracted from the second and the result is added to the first (for example, $XIV = 10 + (5 - 1) = 14, CIX = 100 + (10 - 1) = 109, DXL = 500 + (50 - 10) = 540$).

4. When two ways exist for representing a number, the one in which the symbol of larger value occurs earlier in the string is preferred. (For example, 14 is represented as XIV, not as VIX.)

Decimal number	1	2	3	4	5	6	7	8	9
Roman numeral	I	II	III	IV	V	VI	VII	VIII	IX

10	14	50	200	400	500	600	999	1000
X	XIV	L	CC	CD	D	DC	CMXCIX	M

1950	1960	1970	1980	1990
MCML	MCMLX	MCMLXX	MCMLXXX	MCMXC

1995	1999	2000	2001	2004	2010
MCMXCV	MCMXCIX	MM	MMI	MMIV	MMX

1.1.3   ARROW NOTATION

Arrow notation is a way to represent large numbers in which evaluation proceeds from the right:

$$m \uparrow n = \underbrace{m \cdot m \cdots m}_{n}$$

$$m \uparrow\uparrow n = \underbrace{m \uparrow m \uparrow \cdots \uparrow m}_{n} \qquad (1.1.1)$$

$$m \uparrow\uparrow\uparrow n = \underbrace{m \uparrow\uparrow m \uparrow\uparrow \cdots \uparrow\uparrow m}_{n}$$

For example, $m \uparrow n = m^n, m \uparrow\uparrow 2 = m^m$, and $m \uparrow\uparrow 3 = m^{(m^m)}$.

1.1.4 REPRESENTATION OF NUMBERS

Numerals as usually written have radix or base 10, so that the numeral $a_n a_{n-1} \ldots a_1 a_0$ represents the number $a_n 10^n + a_{n-1} 10^{n-1} + \cdots + a_2 10^2 + a_1 10 + a_0$. However, other bases can be used, particularly bases 2, 8, and 16. When a number is written in base 2, the number is said to be in binary notation. The names of other bases are:

2	binary	9	nonary
3	ternary	10	decimal
4	quaternary	11	undenary
5	quinary	12	duodecimal
6	senary	16	hexadecimal
7	septenary	20	vigesimal
8	octal	60	sexagesimal

When writing a number in base b, the digits used range from 0 to $b - 1$. If $b > 10$, then the digit A stands for 10, B for 11, etc. When a base other than 10 is used, it is indicated by a subscript:

$$10111_2 = 1 \times 2^4 + 0 \times 2^3 + 1 \times 2^2 + 1 \times 2 + 1 = 23,$$
$$A3_{16} = 10 \times 16 + 3 = 163, \qquad (1.1.2)$$
$$543_7 = 5 \times 7^2 + 4 \times 7 + 3 = 276.$$

To convert a number from base 10 to base b, divide the number by b, and the remainder will be the last digit. Then divide the quotient by b, using the remainder as the previous digit. Continue dividing the quotient by b until a quotient of 0 is arrived at.

EXAMPLE To convert 573 to base 12, divide 573 by 12, yielding a quotient of 47 and a remainder of 9; hence, "9" is the last digit. Divide 47 by 12, yielding a quotient of 3 and a remainder of 11 (which we represent with a "B"). Divide 3 by 12 yielding a quotient of 0 and a remainder of 3. Therefore, $573_{10} = 3B9_{12}$.

In general, to convert from base b to base r, it is simplest to convert to base 10 as an intermediate step. However, it is simple to convert from base b to base b^n. For example, to convert 110111101_2 to base 16, group the digits in fours (because 16 is 2^4), yielding 1 1011 1101_2, and then convert each group of 4 to base 16 directly, yielding $1BD_{16}$.

1.1.5 BINARY PREFIXES

A byte is 8 bits. A kibibyte is $2^{10} = 1024$ bytes. Other prefixes for power of 2 are:

Factor	Prefix	Symbol
2^{10}	kibi	Ki
2^{20}	mebi	Mi
2^{30}	gibi	Gi
2^{40}	tebi	Ti
2^{50}	pebi	Pi
2^{60}	exbi	Ei

1.1.6 DECIMAL MULTIPLES AND PREFIXES

The prefix names and symbols below are taken from Conference Générale des Poids et Mesures, 1991. The common names are for the U.S.

Factor	Prefix	Symbol	Common name
$10^{(10^{100})}$			googolplex
10^{100}			googol
10^{24}	yotta	Y	heptillion
10^{21}	zetta	Z	hexillion
10^{18}	exa	E	quintillion
10^{15}	peta	P	quadrillion
10^{12}	tera	T	trillion
10^{9}	giga	G	billion
10^{6}	mega	M	million
10^{3}	kilo	k	thousand
10^{2}	hecto	H	hundred
10^{1}	deka	da	ten
10^{-1}	deci	d	tenth
10^{-2}	centi	c	hundreth
10^{-3}	milli	m	thousandth
10^{-6}	micro	μ (Greek mu)	millionth
10^{-9}	nano	n	billionth
10^{-12}	pico	p	trillionth
10^{-15}	femto	f	quadrillionth
10^{-18}	atto	a	quintillionth
10^{-21}	zepto	z	hexillionth
10^{-24}	yocto	y	heptillionth

1.1.7 DECIMAL EQUIVALENTS OF COMMON FRACTIONS

		1/64	0.015625			33/64	0.515625
	1/32	2/64	0.03125		17/32	34/64	0.53125
		3/64	0.046875			35/64	0.546875
1/16	2/32	4/64	0.0625	9/16	18/32	36/64	0.5625
		5/64	0.078125			37/64	0.578125
	3/32	6/64	0.09375		19/32	38/64	0.59375
		7/64	0.109375			39/64	0.609375
1/8	4/32	8/64	0.125	5/8	20/32	40/64	0.625
		9/64	0.140625			41/64	0.640625
	5/32	10/64	0.15625		21/32	42/64	0.65625
		11/64	0.171875			43/64	0.671875
3/16	6/32	12/64	0.1875	11/16	22/32	44/64	0.6875
		13/64	0.203125			45/64	0.703125
	7/32	14/64	0.21875		23/32	46/64	0.71875
		15/64	0.234375			47/64	0.734375
1/4	8/32	16/64	0.25	3/4	24/32	48/64	0.75
		17/64	0.265625			49/64	0.765625
	9/32	18/64	0.28125		25/32	50/64	0.78125
		19/64	0.296875			51/64	0.796875
5/16	10/32	20/64	0.3125	13/16	26/32	52/64	0.8125
		21/64	0.328125			53/64	0.828125
	11/32	22/64	0.34375		27/32	54/64	0.84375
		23/64	0.359375			55/64	0.859375
3/8	12/32	24/64	0.375	7/8	28/32	56/64	0.875
		25/64	0.390625			57/64	0.890625
	13/32	26/64	0.40625		29/32	58/64	0.90625
		27/64	0.421875			59/64	0.921875
7/16	14/32	28/64	0.4375	15/16	30/32	60/64	0.9375
		29/64	0.453125			61/64	0.953125
	15/32	30/64	0.46875		31/32	62/64	0.96875
		31/64	0.484375			63/64	0.984375
1/2	16/32	32/64	0.5	1/1	32/32	64/64	1

1.1.8   HEXADECIMAL ADDITION AND SUBTRACTION TABLE

$A = 10, B = 11, C = 12, D = 13, E = 14, F = 15.$
Example: $6 + 2 = 8$; hence $8 - 6 = 2$ and $8 - 2 = 6$.
Example: $4 + E = 12$; hence $12 - 4 = E$ and $12 - E = 4$.

	1	2	3	4	5	6	7	8	9	A	B	C	D	E	F
1	02	03	04	05	06	07	08	09	0A	0B	0C	0D	0E	0F	10
2	03	04	05	06	07	08	09	0A	0B	0C	0D	0E	0F	10	11
3	04	05	06	07	08	09	0A	0B	0C	0D	0E	0F	10	11	12
4	05	06	07	08	09	0A	0B	0C	0D	0E	0F	10	11	12	13
5	06	07	08	09	0A	0B	0C	0D	0E	0F	10	11	12	13	14
6	07	08	09	0A	0B	0C	0D	0E	0F	10	11	12	13	14	15
7	08	09	0A	0B	0C	0D	0E	0F	10	11	12	13	14	15	16
8	09	0A	0B	0C	0D	0E	0F	10	11	12	13	14	15	16	17
9	0A	0B	0C	0D	0E	0F	10	11	12	13	14	15	16	17	18
A	0B	0C	0D	0E	0F	10	11	12	13	14	15	16	17	18	19
B	0C	0D	0E	0F	10	11	12	13	14	15	16	17	18	19	1A
C	0D	0E	0F	10	11	12	13	14	15	16	17	18	19	1A	1B
D	0E	0F	10	11	12	13	14	15	16	17	18	19	1A	1B	1C
E	0F	10	11	12	13	14	15	16	17	18	19	1A	1B	1C	1D
F	10	11	12	13	14	15	16	17	18	19	1A	1B	1C	1D	1E

1.1.9   HEXADECIMAL MULTIPLICATION TABLE

Example: $2 \times 4 = 8$.
Example: $2 \times F = 1E$.

	1	2	3	4	5	6	7	8	9	A	B	C	D	E	F
1	01	02	03	04	05	06	07	08	09	0A	0B	0C	0D	0E	0F
2	02	04	06	08	0A	0C	0E	10	12	14	16	18	1A	1C	1E
3	03	06	09	0C	0F	12	15	18	1B	1E	21	24	27	2A	2D
4	04	08	0C	10	14	18	1C	20	24	28	2C	30	34	38	3C
5	05	0A	0F	14	19	1E	23	28	2D	32	37	3C	41	46	4B
6	06	0C	12	18	1E	24	2A	30	36	3C	42	48	4E	54	5A
7	07	0E	15	1C	23	2A	31	38	3F	46	4D	54	5B	62	69
8	08	10	18	20	28	30	38	40	48	50	58	60	68	70	78
9	09	12	1B	24	2D	36	3F	48	51	5A	63	6C	75	7E	87
A	0A	14	1E	28	32	3C	46	50	5A	64	6E	78	82	8C	96
B	0B	16	21	2C	37	42	4D	58	63	6E	79	84	8F	9A	A5
C	0C	18	24	30	3C	48	54	60	6C	78	84	90	9C	A8	B4
D	0D	1A	27	34	41	4E	5B	68	75	82	8F	9C	A9	B6	C3
E	0E	1C	2A	38	46	54	62	70	7E	8C	9A	A8	B6	C4	D2
F	0F	1E	2D	3C	4B	5A	69	78	87	96	A5	B4	C3	D2	E1

1.1.10   HEXADECIMAL–DECIMAL FRACTION CONVERSION TABLE

The values below are correct to all digits shown.

Hex	Decimal	Hex	Decimal	Hex	Decimal	Hex	Decimal
.00	0	.40	0.250000	.80	0.500000	.C0	0.750000
.01	0.003906	.41	0.253906	.81	0.503906	.C1	0.753906
.02	0.007812	.42	0.257812	.82	0.507812	.C2	0.757812
.03	0.011718	.43	0.261718	.83	0.511718	.C3	0.761718
.04	0.015625	.44	0.265625	.84	0.515625	.C4	0.765625
.05	0.019531	.45	0.269531	.85	0.519531	.C5	0.769531
.06	0.023437	.46	0.273437	.86	0.523437	.C6	0.773437
.07	0.027343	.47	0.277343	.87	0.527343	.C7	0.777343
.08	0.031250	.48	0.281250	.88	0.531250	.C8	0.781250
.09	0.035156	.49	0.285156	.89	0.535156	.C9	0.785156
.0A	0.039062	.4A	0.289062	.8A	0.539062	.CA	0.789062
.0B	0.042968	.4B	0.292968	.8B	0.542968	.CB	0.792968
.0C	0.046875	.4C	0.296875	.8C	0.546875	.CC	0.796875
.0D	0.050781	.4D	0.300781	.8D	0.550781	.CD	0.800781
.0E	0.054687	.4E	0.304687	.8E	0.554687	.CE	0.804687
.0F	0.058593	.4F	0.308593	.8F	0.558593	.CF	0.808593
.10	0.062500	.50	0.312500	.90	0.562500	.D0	0.812500
.11	0.066406	.51	0.316406	.91	0.566406	.D1	0.816406
.12	0.070312	.52	0.320312	.92	0.570312	.D2	0.820312
.13	0.074218	.53	0.324218	.93	0.574218	.D3	0.824218
.14	0.078125	.54	0.328125	.94	0.578125	.D4	0.828125
.15	0.082031	.55	0.332031	.95	0.582031	.D5	0.832031
.16	0.085937	.56	0.335937	.96	0.585937	.D6	0.835937
.17	0.089843	.57	0.339843	.97	0.589843	.D7	0.839843
.18	0.093750	.58	0.343750	.98	0.593750	.D8	0.843750
.19	0.097656	.59	0.347656	.99	0.597656	.D9	0.847656
.1A	0.101562	.5A	0.351562	.9A	0.601562	.DA	0.851562
.1B	0.105468	.5B	0.355468	.9B	0.605468	.DB	0.855468
.1C	0.109375	.5C	0.359375	.9C	0.609375	.DC	0.859375
.1D	0.113281	.5D	0.363281	.9D	0.613281	.DD	0.863281
.1E	0.117187	.5E	0.367187	.9E	0.617187	.DE	0.867187
.1F	0.121093	.5F	0.371093	.9F	0.621093	.DF	0.871093
.20	0.125000	.60	0.375000	.A0	0.625000	.E0	0.875000
.21	0.128906	.61	0.378906	.A1	0.628906	.E1	0.878906
.22	0.132812	.62	0.382812	.A2	0.632812	.E2	0.882812
.23	0.136718	.63	0.386718	.A3	0.636718	.E3	0.886718
.24	0.140625	.64	0.390625	.A4	0.640625	.E4	0.890625
.25	0.144531	.65	0.394531	.A5	0.644531	.E5	0.894531
.26	0.148437	.66	0.398437	.A6	0.648437	.E6	0.898437
.27	0.152343	.67	0.402343	.A7	0.652343	.E7	0.902343

Hex	Decimal	Hex	Decimal	Hex	Decimal	Hex	Decimal
.28	0.156250	.68	0.406250	.A8	0.656250	.E8	0.906250
.29	0.160156	.69	0.410156	.A9	0.660156	.E9	0.910156
.2A	0.164062	.6A	0.414062	.AA	0.664062	.EA	0.914062
.2B	0.167968	.6B	0.417968	.AB	0.667968	.EB	0.917968
.2C	0.171875	.6C	0.421875	.AC	0.671875	.EC	0.921875
.2D	0.175781	.6D	0.425781	.AD	0.675781	.ED	0.925781
.2E	0.179687	.6E	0.429687	.AE	0.679687	.EE	0.929687
.2F	0.183593	.6F	0.433593	.AF	0.683593	.EF	0.933593
.30	0.187500	.70	0.437500	.B0	0.687500	.F0	0.937500
.31	0.191406	.71	0.441406	.B1	0.691406	.F1	0.941406
.32	0.195312	.72	0.445312	.B2	0.695312	.F2	0.945312
.33	0.199218	.73	0.449218	.B3	0.699218	.F3	0.949218
.34	0.203125	.74	0.453125	.B4	0.703125	.F4	0.953125
.35	0.207031	.75	0.457031	.B5	0.707031	.F5	0.957031
.36	0.210937	.76	0.460937	.B6	0.710937	.F6	0.960937
.37	0.214843	.77	0.464843	.B7	0.714843	.F7	0.964843
.38	0.218750	.78	0.468750	.B8	0.718750	.F8	0.968750
.39	0.222656	.79	0.472656	.B9	0.722656	.F9	0.972656
.3A	0.226562	.7A	0.476562	.BA	0.726562	.FA	0.976562
.3B	0.230468	.7B	0.480468	.BB	0.730468	.FB	0.980468
.3C	0.234375	.7C	0.484375	.BC	0.734375	.FC	0.984375
.3D	0.238281	.7D	0.488281	.BD	0.738281	.FD	0.988281
.3E	0.242187	.7E	0.492187	.BE	0.742187	.FE	0.992187
.3F	0.246093	.7F	0.496093	.BF	0.746093	.FF	0.996093

1.2   SPECIAL NUMBERS

1.2.1   POWERS OF 2

n	2^n	2^{-n}
1	2	0.5
2	4	0.25
3	8	0.125
4	16	0.0625
5	32	0.03125
6	64	0.015625
7	128	0.0078125
8	256	0.00390625
9	512	0.001953125
10	1024	0.0009765625

n	2^n	2^{-n}
11	2048	0.00048828125
12	4096	0.000244140625
13	8192	0.0001220703125
14	16384	0.00006103515625
15	32768	0.000030517578125
16	65536	0.0000152587890625
17	131072	0.00000762939453125
18	262144	0.000003814697265625
19	524288	0.0000019073486328125
20	1048576	0.00000095367431640625
21	2097152	0.000000476837158203125
22	4194304	0.0000002384185791015625
23	8388608	0.00000011920928955078125
24	16777216	0.000000059604644775390625
25	33554432	0.0000000298023223876953125
26	67108864	0.00000001490116119384765625
27	134217728	0.000000007450580596923828125
28	268435456	0.0000000037252902984619140625
29	536870912	0.00000000186264514923095703125
30	1073741824	0.000000000931322574615478515625
31	2147483648	0.0000000004656612873077392578125
32	4294967296	0.00000000023283064365386962890625
33	8589934592	0.000000000116415321826934814453125
34	17179869184	0.0000000000582076609134674072265625
35	34359738368	0.00000000002910383045673370361328125
36	68719476736	0.000000000014551915228366851806640625
37	137438953472	0.0000000000072759576141834259033203125
38	274877906944	0.00000000000363797880709171295166015625
39	549755813888	0.000000000001818989403545856475830078125
40	1099511627776	0.0000000000009094947017729282379150390625

n	2^n	n	2^n
41	2199023255552	42	4398046511104
43	8796093022208	44	17592186044416
45	35184372088832	46	70368744177664
47	140737488355328	48	281474976710656
49	562949953421312	50	1125899906842624
51	2251799813685248	52	4503599627370496
53	9007199254740992	54	18014398509481984
55	36028797018963968	56	72057594037927936
57	144115188075855872	58	288230376151711744
59	576460752303423488	60	1152921504606846976
61	2305843009213693952	62	4611686018427387904
63	9223372036854775808	64	18446744073709551616

n	2^n	n	2^n
65	36893488147419103232	66	73786976294838206464
67	147573952589676412928	68	295147905179352825856
69	590295810358705651712	70	1180591620717411303424
71	2361183241434822606848	72	4722366482869645213696
73	9444732965739290427392	74	18889465931478580854784
75	37778931862957161709568	76	75557863725914323419136
77	151115727451828646838272	78	302231454903657293676544
79	604462909807314587353088	80	1208925819614629174706176
81	2417851639229258349412352	82	4835703278458516698824704
83	9671406556917033397649408	84	19342813113834066795298816
85	38685626227668133590597632	86	77371252455336267181195264
87	154742504910672534362390528	88	309485009821345068724781056
89	618970019642690137449562112	90	1237940039285380274899124224

1.2.2 POWERS OF 16 IN DECIMAL SCALE

n	16^n	16^{-n}
0	1	1
1	16	0.0625
2	256	0.00390625
3	4096	0.000244140625
4	65536	0.0000152587890625
5	1048576	0.00000095367431640625
6	16777216	0.0000000059604644775390625
7	268435456	0.00000000037252902984619140625
8	4294967296	0.0000000000232830643653869628906.25
9	68719476736	0.00000000000145519152283668518066.40625
10	1099511627776	0.00000000000009094947017729282379150390625
11	17592186044416	$5.6843418860808014869689941 40625 \times 10^{-14}$
12	281474976710656	$3.5527136788005009293556213 37890 \cdots \times 10^{-15}$
13	4503599627370496	$2.2204460492503130808472633 36181 \cdots \times 10^{-16}$
14	72057594037927936	$1.3877787807814456755295395 85113 \cdots \times 10^{-17}$
15	1152921504606846976	$8.6736173798840354720596224 06959 \cdots \times 10^{-19}$
16	18446744073709551616	$5.4210108624275221700372640 04349 \cdots \times 10^{-20}$
17	295147905179352825856	$3.3881317890172013562732900 02718 \cdots \times 10^{-21}$
18	4722366482869645213696	$2.1175823681357508476708062 51699 \cdots \times 10^{-22}$
19	75557863725914323419136	$1.3234889800848442797942539 07311 \cdots \times 10^{-23}$
20	1208925819614629174706176	$8.2718061255302767487140869 20699 \cdots \times 10^{-25}$

1.2.3   POWERS OF 10 IN HEXADECIMAL SCALE

n	10^n	10^{-n}
0	1_{16}	1_{16}
1	A_{16}	$0.19999999999999999999\ldots_{16}$
2	64_{16}	$0.028F5C28F5C28F5C28F5\ldots_{16}$
3	$3E8_{16}$	$0.004189374BC6A7EF9DB2\ldots_{16}$
4	2710_{16}	$0.00068DB8BAC710CB295E\ldots_{16}$
5	$186A0_{16}$	$0.0000A7C5AC471B478423\ldots_{16}$
6	$F4240_{16}$	$0.000010C6F7A0B5ED8D36\ldots_{16}$
7	989680_{16}	$0.000001AD7F29ABCAF485\ldots_{16}$
8	$5F5E100_{16}$	$0.0000002AF31DC4611873\ldots_{16}$
9	$3B9ACA00_{16}$	$0.000000044B82FA09B5A5\ldots_{16}$
10	$2540BE400_{16}$	$0.000000006DF37F675EF6\ldots_{16}$
11	$174876E800_{16}$	$0.000000000AFEBFF0BCB2\ldots_{16}$
12	$E8D4A51000_{16}$	$0.00000000119799812DE\ldots_{16}$
13	$9184E72A000_{16}$	$0.00000000001C25C26849\ldots_{16}$
14	$5AF3107A4000_{16}$	$0.000000000002D09370D4\ldots_{16}$
15	$38D7EA4C68000_{16}$	$0.000000000000480EBE7B\ldots_{16}$
16	$2386F26FC10000_{16}$	$0.0000000000000734ACA5\ldots_{16}$

1.2.4   SPECIAL CONSTANTS

1.2.4.1   The constant π

The transcendental number π is defined as the ratio of the circumference of a circle to the diameter. It is also the ratio of the area of a circle to the square of the radius (r) and appears in several formulae in geometry and trigonometry (see Section 6.1)

$$\text{circumference of a circle} = 2\pi r, \qquad \text{volume of a sphere} = \frac{4}{3}\pi r^3,$$

$$\text{area of a circle} = \pi r^2, \qquad \text{surface area of a sphere} = 4\pi r^2.$$

One method of computing π is to use the infinite series for the function $\tan^{-1} x$ and one of the identities

$$\begin{aligned}
\pi &= 4\tan^{-1} 1 = 6\tan^{-1}\frac{1}{\sqrt{3}} \\
&= 2\tan^{-1}\frac{1}{2} + 2\tan^{-1}\frac{1}{3} + 8\tan^{-1}\frac{1}{5} - 2\tan^{-1}\frac{1}{239} \\
&= 24\tan^{-1}\frac{1}{8} + 8\tan^{-1}\frac{1}{57} + 4\tan^{-1}\frac{1}{239} \\
&= 48\tan^{-1}\frac{1}{18} + 32\tan^{-1}\frac{1}{57} - 20\tan^{-1}\frac{1}{239}
\end{aligned} \qquad (1.2.1)$$

There are many other identities involving π. See Section 1.4.3. For example:

$$\pi = \sum_{i=0}^{\infty} \frac{1}{16^i} \left(\frac{4}{8i+1} - \frac{2}{8i+4} - \frac{1}{8i+5} - \frac{1}{8i+6} \right)$$

$$\pi = \lim_{k\to\infty} \left[2^k \underbrace{\sqrt{2 - \sqrt{2 + \sqrt{2 + \sqrt{2 + \sqrt{2 + \sqrt{2 + \cdots + \sqrt{2 + \sqrt{2}}}}}}}}}_{k \text{ square roots}} \right]$$

$$\frac{2}{\pi} = \prod_{k=1}^{\infty} \left[\frac{\overbrace{\sqrt{2 + \sqrt{2 + \cdots + \sqrt{2 + \sqrt{2}}}}}^{k \text{ square roots}}}{2} \right]$$

$$\frac{\pi^3}{32} = \sum_{n=0}^{\infty} \frac{(-1)^n}{(2n+1)^3} = 1 - \frac{1}{27} + \frac{1}{125} - \frac{1}{343} + \cdots$$

$$(1.2.2)$$

To 200 decimal places:

$\pi \approx 3.$ 14159 26535 89793 23846 26433 83279 50288 41971 69399 37510
 58209 74944 59230 78164 06286 20899 86280 34825 34211 70679
 82148 08651 32823 06647 09384 46095 50582 23172 53594 08128
 48111 74502 84102 70193 85211 05559 64462 29489 54930 38196

To 50 decimal places:

$\pi/20 \approx$ 0.15707 96326 79489 66192 31321 69163 97514 42098 58469 96876
$\pi/15 \approx$ 0.20943 95102 39319 54923 08428 92218 63352 56131 44626 62501
$\pi/12 \approx$ 0.26179 93877 99149 43653 85536 15273 29190 70164 30783 28126
$\pi/11 \approx$ 0.28559 93321 44526 65804 20584 89389 04571 67451 97218 12501
$\pi/10 \approx$ 0.31415 92653 58979 32384 62643 38327 95028 84197 16939 93751
$\pi/9 \approx$ 0.34906 58503 98865 91538 47381 53697 72254 26885 74377 70835
$\pi/8 \approx$ 0.39269 90816 98724 15480 78304 22909 93786 05246 46174 92189
$\pi/7 \approx$ 0.44879 89505 12827 60549 46633 40468 50041 20281 67057 05359
$\pi/6 \approx$ 0.52359 87755 98298 87307 71072 30546 58381 40328 61566 56252
$\pi/5 \approx$ 0.62831 85307 17958 64769 25286 76655 90057 68394 33879 87502
$\pi/4 \approx$ 0.78539 81633 97448 30961 56608 45819 87572 10492 92349 84378
$\pi/3 \approx$ 1.04719 75511 96597 74615 42144 61093 16762 80657 23133 12504
$\pi/2 \approx$ 1.57079 63267 94896 61923 13216 91639 75144 20985 84699 68755
$2\pi/3 \approx$ 2.09439 51023 93195 49230 84289 22186 33525 61314 46266 25007
$3\pi/2 \approx$ 4.71238 89803 84689 85769 39650 74919 25432 62957 54099 06266

$5\pi/2 \approx$ 7.85398 16339 74483 09615 66084 58198 75721 04929 23498 43776
$\sqrt{\pi} \approx$ 1.77245 38509 05516 02729 81674 83341 14518 27975 49456 12239

In 1999 π was computed to $206,158,430,208 = 3 \cdot 2^{36}$ decimal digits. The frequency distribution of the digits for $\pi - 3$, up to 200,000,000,000 decimal places, is:

digit 0: 20000030841	digit 5: 19999917053
digit 1: 19999914711	digit 6: 19999881515
digit 2: 20000136978	digit 7: 19999967594
digit 3: 20000069393	digit 8: 20000291044
digit 4: 19999921691	digit 9: 19999869180

1.2.4.2 The constant e

The transcendental number e is the base of natural logarithms. It is given by

$$e = \lim_{n\to\infty} \left(1 + \frac{1}{n}\right)^n = \sum_{n=0}^{\infty} \frac{1}{n!}. \tag{1.2.3}$$

To 200 decimal places:

$e \approx 2.$ 71828 18284 59045 23536 02874 71352 66249 77572 47093 69995
95749 66967 62772 40766 30353 54759 45713 82178 52516 64274
27466 39193 20030 59921 81741 35966 29043 57290 03342 95260
59563 07381 32328 62794 34907 63233 82988 07531 95251 01901

To 50 decimal places:

$e/8 \approx$ 0.33978 52285 57380 65442 00359 33919 08281 22196 55886 71249
$e/7 \approx$ 0.38832 59754 94149 31933 71839 24478 95178 53938 92441 95714
$e/6 \approx$ 0.45304 69714 09840 87256 00479 11892 11041 62928 74515 61666
$e/5 \approx$ 0.54365 63656 91809 04707 20574 94270 53249 95514 49418 73999
$e/4 \approx$ 0.67957 04571 14761 30884 00718 67838 16562 44393 11773 42499
$e/3 \approx$ 0.90609 39428 19681 74512 00958 23784 22083 25857 49031 23332
$e/2 \approx$ 1.35914 09142 29522 61768 01437 35676 33124 88786 23546 84998
$2e/3 \approx$ 1.81218 78856 39363 49024 01916 47568 44166 51714 98062 46664
$e^{\pi} \approx$ 23.14069 26327 79269 00572 90863 67948 54738 02661 06242 60021
$\pi^e \approx$ 22.45915 77183 61045 47342 71522 04543 73502 75893 15133 99669

The function e^x is defined by $e^x = \sum_{n=0}^{\infty} \frac{x^n}{n!}$ (see page 521). The numbers e and π are related by the formula

$$e^{\pi i} = -1 \tag{1.2.4}$$

1.2.4.3 The constant γ

Euler's constant γ is defined by

$$\gamma = \lim_{n\to\infty} \left(\sum_{k=1}^{n} \frac{1}{k} - \log n\right). \tag{1.2.5}$$

It is not known whether γ is rational or irrational. To 200 decimal places:

$\gamma \approx 0.$ 57721 56649 01532 86060 65120 90082 40243 10421 59335 93992
35988 05767 23488 48677 26777 66467 09369 47063 29174 67495
14631 44724 98070 82480 96050 40144 86542 83622 41739 97644
92353 62535 00333 74293 73377 37673 94279 25952 58247 09492

1.2.4.4   The constant ϕ

The golden ratio, ϕ, is defined as the positive root of the equation $\frac{\phi}{1} = \frac{1+\phi}{\phi}$; that is $\phi = \frac{1+\sqrt{5}}{2}$. There is the continued fraction representation $\phi = \overline{[1]}$ (see Section 2.4.4) and the representation in square roots

$$\phi = \sqrt{1 + \sqrt{1 + \sqrt{1 + \sqrt{1 + \ldots}}}}$$

To 200 decimal places:

$\phi \approx 1.$ 61803 39887 49894 84820 45868 34365 63811 77203 09179 80576
28621 35448 62270 52604 62818 90244 97072 07204 18939 11374
84754 08807 53868 91752 12663 38622 23536 93179 31800 60766
72635 44333 89086 59593 95829 05638 32266 13199 28290 26788

1.2.4.5   Other constants

To 50 decimal places

$\sqrt{2} \approx$ 1.41421 35623 73095 04880 16887 24209 69807 85696 71875 37695
$\sqrt{3} \approx$ 1.73205 08075 68877 29352 74463 41505 87236 69428 05253 81038
$\sqrt{5} \approx$ 2.23606 79774 99789 69640 91736 68731 27623 54406 18359 61153
$\sqrt{6} \approx$ 2.44948 97427 83178 09819 72840 74705 89139 19659 47480 65667
$\sqrt{7} \approx$ 2.64575 13110 64590 59050 16157 53639 26042 57102 59183 08245
$\sqrt{8} \approx$ 2.82842 71247 46190 09760 33774 48419 39615 71393 43750 75390
$\ln 2 \approx$ 0.69314 71805 59945 30941 72321 21458 17656 80755 00134 36026
$\ln 3 \approx$ 1.09861 22886 68109 69139 52452 36922 52570 46474 90557 82275
$\ln 5 \approx$ 1.60943 79124 34100 37460 07593 33226 18763 95256 01354 26852
$\log 2 \approx$ 0.30102 99956 63981 19521 37388 94724 49302 67681 89881 46211
$\log 3 \approx$ 0.47712 12547 19662 43729 50279 03255 11530 92001 28864 19070
$\log 5 \approx$ 0.69897 00043 36018 80478 62611 05275 50697 32318 10118 53789

1.2.5   CONSTANTS IN DIFFERENT BASES

Base 2

$\pi \approx$ `11.001001000011111101101010100010001000010110100001...`$_2$
$e \approx$ `10.10110111111000010101000101100010100010101110110...`$_2$
$\gamma \approx$ `0.10010011110001000110011111100011011111011011000...`$_2$
$\sqrt{2} \approx$ `1.0110101000001001111001100110011111111001110111110...`$_2$
$\ln 2 \approx$ `0.1011000101110010000010111111101111110100011100111...`$_2$

Base 8

$$\pi \approx 3.11037552421026430215142306305056006701632112201\ldots_8$$
$$e \approx 2.55760521305053551246527734254200471723636166134\ldots_8$$
$$\gamma \approx 0.44742147706766606172232157437601002513132552071\ldots_8$$
$$\sqrt{2} \approx 1.32404746317716746220426276611546725125751743533\ldots_8$$
$$\ln 2 \approx 0.54271027757507173632571170731630007713665364036\ldots_8$$

Base 12

$$\pi \approx 3.184809493B918664573A6211BB151551A05729290A78\ldots_{12}$$
$$e \approx 2.8752360698219BA71971009B388AA876676025642727\ldots_{12}$$
$$\gamma \approx 0.6B15188A6760B381B754334520434A22560A590A6A5\ldots_{12}$$
$$\sqrt{2} \approx 1.4B79170A07B85737704B085486853504563650B559B8\ldots_{12}$$
$$\ln 2 \approx 0.839912483369AB213742A34679253788658A1402A540\ldots_{12}$$

Base 16

$$\pi \approx 3.243F6A8885A308D313198A2E03707344A4093822299F\ldots_{16}$$
$$e \approx 2.B7E151628AED2A6ABF7158809CF4F3C762E7160F3\ldots_{16}$$
$$\gamma \approx 0.93C467E37DB0C7A4D1BE3F810152CB56A1CECC3A\ldots_{16}$$
$$\sqrt{2} \approx 1.6A09E667F3BCC908B2FB1366EA957D3E3ADEC175\ldots_{16}$$
$$\ln 2 \approx 0.B17217F7D1CF79ABC9E3B39803F2F6AF40F3432672\ldots_{16}$$

1.2.6 FACTORIALS

For non-negative integers n, the factorial of n, denoted $n!$, is the product of all positive integers less than or equal to n; $n! = n \cdot (n-1) \cdot (n-2) \cdots 2 \cdot 1$. If n is a negative integer ($n = -1, -2, \ldots$) then $n! = \pm\infty$. Note that, since the empty product is 1, it follows that $0! = 1$. The generalization of the factorial function to non-integer arguments is the gamma function (see page 540). When n is an integer, $\Gamma(n) = (n-1)!$.

The double factorial of n, denoted $n!!$, is the product of every other integer: $n!! = n \cdot (n-2) \cdot (n-4) \cdots$, where the last element in the product is either 2 or 1, depending on whether n is even or odd. The *shifted factorial* (also called the *rising factorial* and *Pochhammer's symbol*) is denoted by $(a)_n$ (sometimes $a^{\overline{n}}$) and is defined as

$$(a)_n = \underbrace{a \cdot (a+1) \cdot (a+2) \cdots (a+n-1)}_{n \text{ terms}} = \frac{(a+n-1)!}{(a-1)!} = \frac{\Gamma(a+n)}{\Gamma(a)}. \quad (1.2.6)$$

Approximations to $n!$ for large n include Stirling's formula

$$n! \approx \sqrt{2\pi e} \left(\frac{n}{e}\right)^{n+\frac{1}{2}}, \quad (1.2.7)$$

and Burnsides's formula

$$n! \approx \sqrt{2\pi} \left(\frac{n+\frac{1}{2}}{e}\right)^{n+\frac{1}{2}}. \quad (1.2.8)$$

n	$n!$	$\log_{10} n!$	$n!!$	$\log_{10} n!!$
0	1	0.00000	1	0.00000
1	1	0.00000	1	0.00000
2	2	0.30103	2	0.30103
3	6	0.77815	3	0.47712
4	24	1.38021	8	0.90309
5	120	2.07918	15	1.17609
6	720	2.85733	48	1.68124
7	5040	3.70243	105	2.02119
8	40320	4.60552	384	2.58433
9	3.6288×10^5	5.55976	945	2.97543
10	3.6288×10^6	6.55976	3840	3.58433
11	3.9917×10^7	7.60116	10395	4.01682
12	4.7900×10^8	8.68034	46080	4.66351
13	6.2270×10^9	9.79428	1.3514×10^5	5.13077
14	8.7178×10^{10}	10.94041	6.4512×10^5	5.80964
15	1.3077×10^{12}	12.11650	2.0270×10^6	6.30686
16	2.0923×10^{13}	13.32062	1.0322×10^7	7.01376
17	3.5569×10^{14}	14.55107	3.4459×10^7	7.53731
18	6.4024×10^{15}	15.80634	1.8579×10^8	8.26903
19	1.2165×10^{17}	17.08509	6.5473×10^8	8.81606
20	2.4329×10^{18}	18.38612	3.7159×10^9	9.57006
21	5.1091×10^{19}	19.70834	1.3749×10^{10}	10.13828
22	1.1240×10^{21}	21.05077	8.1750×10^{10}	10.91249
23	2.5852×10^{22}	22.41249	3.1623×10^{11}	11.50001
24	6.2045×10^{23}	23.79271	1.9620×10^{12}	12.29270
25	1.5511×10^{25}	25.19065	7.9059×10^{12}	12.89795
30	2.6525×10^{32}	32.42366	4.2850×10^{16}	16.63195
40	8.1592×10^{47}	47.91165	2.5511×10^{24}	24.40672
50	3.0414×10^{64}	64.48307	5.2047×10^{32}	32.71640
60	8.3210×10^{81}	81.92017	2.8481×10^{41}	41.45456
70	1.1979×10^{100}	100.07841	3.5504×10^{50}	50.55028
80	7.1569×10^{118}	118.85473	8.9711×10^{59}	59.95284
90	1.4857×10^{138}	138.17194	4.2088×10^{69}	69.62416
100	9.3326×10^{157}	157.97000	3.4243×10^{79}	79.53457
110	1.5882×10^{178}	178.20092	4.5744×10^{89}	89.66033
120	6.6895×10^{198}	198.82539	9.5934×10^{99}	99.98197
130	6.4669×10^{219}	219.81069	3.0428×10^{110}	110.48328
140	1.3462×10^{241}	241.12911	1.4142×10^{121}	121.15050
150	5.7134×10^{262}	262.75689	9.3726×10^{131}	131.97186
500	1.2201×10^{1134}	1134.0864	5.8490×10^{567}	567.76709
1000	4.0239×10^{2567}	2567.6046	3.9940×10^{1284}	1284.6014

1.2.7 BERNOULLI POLYNOMIALS AND NUMBERS

The *Bernoulli polynomials* $B_n(x)$ are defined by the generating function

$$\frac{te^{xt}}{e^t - 1} = \sum_{n=0}^{\infty} B_n(x)\frac{t^n}{n!}. \qquad (1.2.9)$$

These polynomials can also be defined recursively by means of $B_0(x) = 1$, $B'_n(x) = nB_{n-1}(x)$, and $\int_0^1 B_n(x)\,dx = 0$ for $n \geq 1$. The identity $B_{k+1}(x+1) - B_{k+1}(x) = (k+1)x^k$ means that sums of powers can be computed in terms of Bernoulli polynomials

$$1^k + 2^k + \cdots + n^k = \frac{B_{k+1}(n+1) - B_{k+1}(0)}{k+1}. \qquad (1.2.10)$$

n	$B_n(x)$
0	1
1	$(2x - 1)/2$
2	$(6x^2 - 6x + 1)/6$
3	$(2x^3 - 3x^2 + x)/2$
4	$(30x^4 - 60x^3 + 30x^2 - 1)/30$
5	$(6x^5 - 15x^4 + 10x^3 - x)/6$

The Bernoulli numbers are the Bernoulli polynomials evaluated at 0: $B_n = B_n(0)$. A generating function for the Bernoulli numbers is $\displaystyle\sum_{n=0}^{\infty} B_n\frac{t^n}{n!} = \frac{t}{e^t - 1}$.
In the following table each Bernoulli number is written as a fraction of integers: $B_n = N_n/D_n$. Note that $B_{2m+1} = 0$ for $m \geq 1$.

n	N_n	D_n	B_n
0	1	1	1.000000000×10^0
1	-1	2	$-5.000000000 \times 10^{-1}$
2	1	6	$1.666666667 \times 10^{-1}$
4	-1	30	$-3.333333333 \times 10^{-2}$
6	1	42	$2.380952381 \times 10^{-2}$
8	-1	30	$-3.333333333 \times 10^{-2}$
10	5	66	$7.575757576 \times 10^{-2}$
12	-691	2730	$-2.531135531 \times 10^{-1}$
14	7	6	1.166666667×10^0
16	-3617	510	-7.092156863×10^0
18	43867	798	5.497117794×10^1
20	-174611	330	-5.291242424×10^2
22	854513	138	6.192123188×10^3
24	-236364091	2730.	-8.658025311×10^4
26	8553103	6	1.425517167×10^6
28	-23749461029	870	-2.729823107×10^7
30	8615841276005	14322	6.015808739×10^8
32	-7709321041217	510	$-1.511631577 \times 10^{10}$
34	2577687858367	6	$4.296146431 \times 10^{11}$

1.2.8   EULER POLYNOMIALS AND NUMBERS

The *Euler polynomials* $E_n(x)$ are defined by the generating function

$$\frac{2e^{xt}}{e^t + 1} = \sum_{n=0}^{\infty} E_n(x)\frac{t^n}{n!}. \tag{1.2.11}$$

n	$E_n(x)$
0	1
1	$(2x - 1)/2$
2	$x^2 - x$
3	$(4x^3 - 6x^2 + 1)/4$
4	$x^4 - 2x^3 + x$
5	$(2x^5 - 5x^4 + 5x^2 - 1)/2$

Alternating sums of powers can be computed in terms of Euler polynomials

$$\sum_{i=1}^{n}(-1)^{n-i}i^k = n^k - (n-1)^k + \cdots \mp 2^k \pm 1^k = \frac{E_k(n+1) + (-1)^n E_k(0)}{2}.$$
$$\tag{1.2.12}$$

The Euler numbers are the Euler polynomials evaluated at $1/2$, and scaled: $E_n = 2^n E_n(\frac{1}{2})$. A generating function for the Euler numbers is

$$\sum_{n=0}^{\infty} E_n \frac{t^n}{n!} = \frac{2e^t}{e^{2t} + 1} \tag{1.2.13}$$

n	E_n
2	-1
4	5
6	-61
8	1385
10	-50521
12	2702765
14	-199360981
16	19391512145
18	-2404879675441
20	370371188237525
22	-69348874393137901
24	15514534163557086905
26	-4087072509293123892361
28	$1252259641403629865468285$
30	$-441543893249023104553682821$
32	$177519391579539289436664789665$
34	$-80723299235887898062168247453281$
36	$41222060339517702122347079671259045$
38	$-23489580527043108252017828576198947741$
40	$14851150718114980017877156781405826684425$

1.2.9 FIBONACCI NUMBERS

The Fibonacci numbers $\{F_n\}$ are defined by the recurrence:

$$F_1 = 1, \qquad F_2 = 1, \qquad F_{n+2} = F_n + F_{n+1}. \qquad (1.2.14)$$

An exact formula is available:

$$F_n = \frac{1}{\sqrt{5}} \left[\left(\frac{1 + \sqrt{5}}{2} \right)^n - \left(\frac{1 - \sqrt{5}}{2} \right)^n \right]. \qquad (1.2.15)$$

Note that $\lim\limits_{n \to \infty} \dfrac{F_{n+1}}{F_n} = \phi$, the golden ratio. Also, $F_n \sim \phi^n / \sqrt{5}$ as $n \to \infty$.

n	F_n	n	F_n	n	F_n	n	F_n
1	1	14	377	27	196418	40	102334155
2	1	15	610	28	317811	41	165580141
3	2	16	987	29	514229	42	267914296
4	3	17	1597	30	832040	43	433494437
5	5	18	2584	31	1346269	44	701408733
6	8	19	4181	32	2178309	45	1134903170
7	13	20	6765	33	3524578	46	1836311903
8	21	21	10946	34	5702887	47	2971215073
9	34	22	17711	35	9227465	48	4807526976
10	55	23	28657	36	14930352	49	7778742049
11	89	24	46368	37	24157817	50	12586269025
12	144	25	75025	38	39088169	51	20365011074
13	233	26	121393	39	63245986	52	32951280099

1.2.10 POWERS OF INTEGERS

n	n^3	n^4	n^5	n^6	n^7	n^8	n^{10}
1	1	1	1	1	1	1	1
2	8	16	32	64	128	256	1024
3	27	81	243	729	2187	6561	59049
4	64	256	1024	4096	16384	65536	1048576
5	125	625	3125	15625	78125	390625	9765625
6	216	1296	7776	46656	279936	1679616	60466176
7	343	2401	16807	117649	823543	5764801	282475249
8	512	4096	32768	262144	2097152	16777216	1073741824
9	729	6561	59049	531441	4782969	43046721	3486784401
10	1000	10000	100000	1000000	10000000	100000000	10000000000
11	1331	14641	161051	1771561	19487171	214358881	25937424601
12	1728	20736	248832	2985984	35831808	429981696	61917364224

1.2.11   SUMS OF POWERS OF INTEGERS

1. Define

$$s_k(n) = 1^k + 2^k + \cdots + n^k = \sum_{m=1}^{n} m^k. \tag{1.2.16}$$

Properties include:

(a) $s_k(n) = (k+1)^{-1}[B_{k+1}(n+1) - B_{k+1}(0)]$
(where the B_k are Bernoulli polynomials, see Section 1.2.7).

(b) If $s_k(n) = \sum_{m=1}^{k+1} a_m n^{k-m+2}$, then

$$s_{k+1}(n) = \left(\frac{k+1}{k+2}\right) a_1 n^{k+2} + \cdots + \left(\frac{k+1}{k}\right) a_3 n^k$$

$$+ \cdots + \left(\frac{k+1}{2}\right) a_{k+1} n^2 + \left[1 - (k+1)\sum_{m=1}^{k+1} \frac{a_m}{k+3-m}\right] n.$$

$$s_1(n) = 1 + 2 + 3 + \cdots + n = \frac{1}{2}n(n+1)$$

$$s_2(n) = 1^2 + 2^2 + 3^2 + \cdots + n^2 = \frac{1}{6}n(n+1)(2n+1)$$

$$s_3(n) = 1^3 + 2^3 + 3^3 + \cdots + n^3 = \frac{1}{4}n^2(n+1)^2 = [s_1(n)]^2$$

(c)

$$s_4(n) = 1^4 + 2^4 + 3^4 + \cdots + n^4 = \frac{1}{5}(3n^2 + 3n - 1)s_2(n)$$

$$s_5(n) = 1^5 + 2^5 + 3^5 + \cdots + n^5 = \frac{1}{12}n^2(n+1)^2(2n^2 + 2n - 1)$$

$$s_6(n) = 1^6 + 2^6 + 3^6 + \cdots + n^6$$

$$= \frac{n}{42}(n+1)(2n+1)(3n^4 + 6n^3 - 3n + 1).$$

2. $\displaystyle\sum_{k=1}^{n}(km - 1) = \frac{1}{2}mn(n+1) - n$

3. $\displaystyle\sum_{k=1}^{n}(km - 1)^2 = \frac{n}{6}\left[m^2(n+1)(2n+1) - 6m(n+1) + 6\right]$

4. $\displaystyle\sum_{k=1}^{n}(km-1)^3 = \frac{n}{4}\left[m^3 n(n+1)^2 - 2m^2(n+1)(2n+1) + 6m(n+1) - 4\right]$

5. $\displaystyle\sum_{k=1}^{n}(-1)^{k+1}(km - 1) = \frac{(-1)^n}{4}[2 - (2n+1)m] + \frac{m-2}{4}$

6. $\displaystyle\sum_{k=1}^{n}(-1)^{k+1}(km-1)^2 = \frac{(-1)^{n+1}}{2}\left[n(n+1)m^2 - (2n+1)m + 1\right] + \frac{1-m}{2}$

n	$\sum_{k=1}^{n} k$	$\sum_{k=1}^{n} k^2$	$\sum_{k=1}^{n} k^3$	$\sum_{k=1}^{n} k^4$	$\sum_{k=1}^{n} k^5$
1	1	1	1	1	1
2	3	5	9	17	33
3	6	14	36	98	276
4	10	30	100	354	1300
5	15	55	225	979	4425
6	21	91	441	2275	12201
7	28	140	784	4676	29008
8	36	204	1296	8772	61776
9	45	285	2025	15333	120825
10	55	385	3025	25333	220825
11	66	506	4356	39974	381876
12	78	650	6084	60710	630708
13	91	819	8281	89271	1002001
14	105	1015	11025	127687	1539825
15	120	1240	14400	178312	2299200
16	136	1496	18496	243848	3347776
17	153	1785	23409	327369	4767633
18	171	2109	29241	432345	6657201
19	190	2470	36100	562666	9133300
20	210	2870	44100	722666	12333300
21	231	3311	53361	917147	16417401
22	253	3795	64009	1151403	21571033
23	276	4324	76176	1431244	28007376
24	300	4900	90000	1763020	35970000
25	325	5525	105625	2153645	45735625

1.2.12 NEGATIVE INTEGER POWERS

Riemann's zeta function is $\zeta(n) = \sum_{k=1}^{\infty} \frac{1}{k^n}$ (it is defined for Re $k > 1$ and extended to $\mathbb{C}$). Related functions are

$$\alpha(n) = \sum_{k=1}^{\infty} \frac{(-1)^{k+1}}{k^n}, \qquad \beta(n) = \sum_{k=0}^{\infty} \frac{(-1)^k}{(2k+1)^n}, \qquad \gamma(n) = \sum_{k=0}^{\infty} \frac{1}{(2k+1)^n}.$$

Properties include:

1. $\alpha(n) = (1 - 2^{1-n})\zeta(n)$

2. $\zeta(2k) = \dfrac{(2\pi)^{2k}}{2(2k)!} |B_{2k}|$

3. $\gamma(n) = (1 - 2^{-n})\zeta(n)$

4. $\beta(2k+1) = \dfrac{(\pi/2)^{2k+1}}{2(2k)!} |E_{2k}|$

5. The series $\beta(1) = 1 - \frac{1}{3} + \frac{1}{5} - \cdots = \pi/4$ is known as Gregory's series.

6. Catalan's constant is $\mathbf{G} = \beta(2) \approx 0.915966$.

7. *Riemann hypothesis*: The non-trivial zeros of the Riemann zeta function (i.e., the $\{z_i\}$ that satisfy $\zeta(z_i) = 0$) lie on the *critical line* given by $\mathrm{Re}\; z_i = \frac{1}{2}$. (The trivial zeros are $z = -2, -4, -6, \dots$.)

n	$\zeta(n) = \displaystyle\sum_{k=1}^{\infty} \frac{1}{k^n}$	$\displaystyle\sum_{k=1}^{\infty} \frac{(-1)^{k+1}}{k^n}$	$\displaystyle\sum_{k=0}^{\infty} \frac{(-1)^k}{(2k+1)^n}$	$\displaystyle\sum_{k=0}^{\infty} \frac{1}{(2k+1)^n}$
1	∞	0.6931471805	0.7853981633	∞
2	1.6449340669	0.8224670334	0.9159655941	1.2337005501
3	1.2020569032	0.9015426773	0.9689461463	1.0517997903
4	1.0823232337	0.9470328294	0.9889445517	1.0146780316
5	1.0369277551	0.9721197705	0.9961578281	1.0045237628
6	1.0173430620	0.9855510912	0.9986852222	1.0014470766
7	1.0083492774	0.9925938199	0.9995545079	1.0004715487
8	1.0040773562	0.9962330018	0.9998499902	1.0001551790
9	1.0020083928	0.9980942975	0.9999496842	1.0000513452
10	1.0009945752	0.9990395075	0.9999831640	1.0000170414
11	1.0004941886	0.9995171435	0.9999943749	1.0000056661
12	1.0002460866	0.9997576851	0.9999981224	1.0000018858
13	1.0001227133	0.9998785428	0.9999993736	1.0000006281
14	1.0000612482	0.9999391703	0.9999997911	1.0000002092
15	1.0000305882	0.9999695512	0.9999999303	1.0000000697
16	1.0000152823	0.9999847642	0.9999999768	1.0000000232
17	1.0000076372	0.9999923783	0.9999999923	1.0000000077
18	1.0000038173	0.9999961879	0.9999999974	1.0000000026
19	1.0000019082	0.9999980935	0.9999999991	1.0000000009
20	1.0000009540	0.9999990466	0.9999999997	1.0000000003

$$\beta(1) = \pi/4 \qquad\qquad \zeta(2) = \pi^2/6$$
$$\beta(3) = \pi^3/32 \qquad\qquad \zeta(4) = \pi^4/90$$
$$\beta(5) = 5\pi^5/1536 \qquad\quad \zeta(6) = \pi^6/945$$
$$\beta(7) = 61\pi^7/184320 \qquad \zeta(8) = \pi^8/9450$$
$$\beta(9) = 277\pi^9/8257536 \quad \zeta(10) = \pi^{10}/93555$$

1.2.13   DE BRUIJN SEQUENCES

A sequence of length q^n over an alphabet of size q is a *de Bruijn sequence* if every possible n-tuple occurs in the sequence (allowing wraparound to the start of the sequence). There are de Bruijn sequences for any q and n. The table below gives some small examples.

q	n	Length	Sequence
2	1	2	01
2	2	4	0110
2	3	8	01110100
2	4	16	0101001101111000
3	2	9	001220211
4	2	16	0011310221203323

1.2.14 INTEGER SEQUENCES

These sequences are arranged in numerical order (disregarding any leading zeros or ones). Note that $C(n, k) = \binom{n}{k}$; see page 206.

1. $1, -1, -1, 0, -1, 1, -1, 0, 0, 1, -1, 0, -1, 1, 1, 0, -1, 0, -1, 0, 1, 1, -1, 0, 0, 1, 0,$
   $0, -1, -1, -1, 0, 1, 1, 1, 0, -1, 1, 1, 0, -1, -1, -1, 0, 0, 1, -1, 0, 0, 0, 1, 0, -1, 0,$
   $1, 0$
   \hfill **Möbius function $\mu(n)$, $n \geq 1$**

2. $1, 1, 0, 1, 1, 0, 0, 1, 1, 1, 0, 0, 1, 0, 0, 1, 1, 1, 0, 1, 0, 0, 0, 0, 2, 1, 0, 0, 1, 0, 0, 1, 0, 1,$
   $0, 1, 1, 0, 0, 1, 1, 0, 0, 0, 1, 0, 0, 0, 1, 2, 0, 1, 1, 0, 0, 0, 0, 1, 0, 0, 1, 0, 0, 1, 2, 0, 0, 1,$
   0
   \hfill **Number of ways of writing n as a sum of 2 squares, $n \geq 0$**

3. $0, 1, 1, 1, 1, 2, 1, 1, 1, 2, 1, 2, 1, 2, 2, 1, 1, 2, 1, 2, 2, 2, 1, 2, 1, 2, 1, 2, 1, 3, 1, 1, 2, 2,$
   $2, 2, 1, 2, 2, 2, 1, 3, 1, 2, 2, 2, 1, 2, 1, 2, 2, 2, 1, 2, 2, 2, 2, 2, 1, 3, 1, 2, 2, 1, 2, 3, 1, 2,$
   2
   \hfill **Number of distinct primes dividing n, $n \geq 1$**

4. $1, 1, 1, 2, 1, 1, 1, 3, 2, 1, 1, 2, 1, 1, 1, 5, 1, 2, 1, 2, 1, 1, 1, 3, 2, 1, 3, 2, 1, 1, 1, 7, 1, 1,$
   $1, 4, 1, 1, 1, 3, 1, 1, 1, 2, 2, 1, 1, 5, 2, 2, 1, 2, 1, 3, 1, 3, 1, 1, 1, 2, 1, 1, 2, 11, 1, 1, 1,$
   2
   \hfill **Number of abelian groups of order n, $n \geq 1$**

5. $1, 1, 1, 2, 1, 2, 1, 5, 2, 2, 1, 5, 1, 2, 1, 14, 1, 5, 1, 5, 2, 2, 1, 15, 2, 2, 5, 4, 1, 4, 1, 51,$
   $1, 2, 1, 14, 1, 2, 2, 14, 1, 6, 1, 4, 2, 2, 1, 52, 2, 5, 1, 5, 1, 15, 2, 13, 2, 2, 1, 13, 1, 2, 4,$
   267
   \hfill **Number of groups of order n, $n \geq 1$**

6. $0, 1, 1, 2, 1, 2, 2, 3, 1, 2, 2, 3, 2, 3, 3, 4, 1, 2, 2, 3, 2, 3, 3, 4, 2, 3, 3, 4, 3, 4, 4, 5, 1, 2,$
   $2, 3, 2, 3, 3, 4, 2, 3, 3, 4, 3, 4, 4, 5, 2, 3, 3, 4, 3, 4, 4, 5, 3, 4, 4, 5, 4, 5, 5, 6, 1, 2, 2, 3,$
   2
   \hfill **Number of 1's in binary expansion of n, $n \geq 0$**

7. $1, 2, 1, 2, 3, 6, 9, 18, 30, 56, 99, 186, 335, 630, 1161, 2182, 4080, 7710, 14532, 27594,$
   $52377, 99858, 190557, 364722, 698870, 1342176, 2580795, 4971008$
   Number of binary irreducible polynomials of degree n, or n-bead necklaces, $n \geq 0$

8. $1, 1, 1, 2, 1, 3, 1, 4, 2, 3, 1, 8, 1, 3, 3, 8, 1, 8, 1, 8, 3, 3, 1, 20, 2, 3, 4, 8, 1, 13, 1, 16, 3,$
   $3, 3, 26, 1, 3, 3, 20, 1, 13, 1, 8, 8, 3, 1, 48, 2, 8, 3, 8, 1, 20, 3, 20, 3, 3, 113$
   Number of perfect partitions of n, or ordered factorizations of $n + 1$, $n \geq 0$

9. $1, 2, 2, 1, 2, 1, 1, 2, 2, 1, 1, 2, 1, 2, 2, 1, 2, 1, 1, 2, 1, 2, 2, 1, 1, 2, 2, 1, 2, 1, 1, 2, 2, 1,$
   $1, 2, 1, 2, 2, 1, 1, 2, 2, 1, 2, 1, 1, 2, 1, 2, 2, 1, 2, 1, 1, 2, 2, 1, 1, 2, 1, 2, 2, 1, 2, 1, 1, 2,$
   1
   \hfill **Thue–Morse non-repeating sequence**

10. $1, 2, 1, 4, 1, 2, 1, 8, 1, 2, 1, 4, 1, 2, 1, 9, 1, 2, 1, 4, 1, 2, 1, 8, 1, 2, 1, 4, 1, 2, 1, 10, 1, 2,$
    $1, 4, 1, 2, 1, 8, 1, 2, 1, 4, 1, 2, 1, 9, 1, 2, 1, 4, 1, 2, 1, 8, 1, 2, 1, 4, 1, 2, 1, 12, 1, 2, 1,$
    4
    \hfill **Hurwitz–Radon numbers**

11. $1, 2, 2, 3, 2, 4, 2, 4, 3, 4, 2, 6, 2, 4, 4, 5, 2, 6, 2, 6, 4, 4, 2, 8, 3, 4, 4, 6, 2, 8, 2, 6, 4, 4,$
    $4, 9, 2, 4, 4, 8, 2, 8, 2, 6, 6, 4, 2, 10, 3, 6, 4, 6, 2, 8, 4, 8, 4, 4, 2, 12, 2, 4, 6, 7, 4, 8, 2,$
    6
    \hfill **$d(n)$, the number of divisors of n, $n \geq 1$**

12. $0, 1, 2, 2, 3, 3, 4, 4, 4, 4, 5, 5, 6, 6, 6, 6, 7, 7, 8, 8, 8, 8, 9, 9, 9, 9, 9, 9, 10, 10, 11, 11,$
    $11, 11, 11, 11, 12, 12, 12, 12, 13, 13, 14, 14, 14, 14, 15, 15, 15, 15, 15, 15, 16, 16, 16,$
    16
    \hfill **$\pi(n)$, the number of primes $\leq n$, for $n \geq 1$**

13. $1, 1, 2, 2, 3, 4, 5, 6, 8, 10, 12, 15, 18, 22, 27, 32, 38, 46, 54, 64, 76, 89, 104, 122,$
    $142, 165, 192, 222, 256, 296, 340, 390, 448, 512, 585, 668, 760, 864, 982, 1113, 1260,$
    1426
    \hfill **Number of partitions of n into distinct parts, $n \geq 1$**

14. $1, 1, 2, 2, 4, 2, 6, 4, 6, 4, 10, 4, 12, 6, 8, 8, 16, 6, 18, 8, 12, 10, 22, 8, 20, 12, 18, 12, 28,$
    $8, 30, 16, 20, 16, 24, 12, 36, 18, 24, 16, 40, 12, 42, 20, 24, 22, 46, 16, 42$
    Euler totient function $\phi(n)$: count numbers $\leq n$ and prime to n, for $n \geq 1$

15. 1, 1, 1, 0, 1, 1, 2, 2, 4, 5, 10, 14, 26, 42, 78, 132, 249, 445, 842, 1561, 2988, 5671, 10981, 21209, 41472, 81181, 160176, 316749, 629933, 1256070, 2515169, 5049816
Number of series-reduced trees with n unlabeled nodes, $n \geq 0$

16. 1, 2, 3, 4, 5, 7, 8, 9, 11, 13, 16, 17, 19, 23, 25, 27, 29, 31, 32, 37, 41, 43, 47, 49, 53, 59, 61, 64, 67, 71, 73, 79, 81, 83, 89, 97, 101, 103, 107, 109, 113, 121, 125, 127, 128, 131
Powers of prime numbers

17. 1, 2, 3, 4, 6, 8, 10, 12, 16, 18, 20, 24, 30, 36, 42, 48, 60, 72, 84, 90, 96, 108, 120, 144, 168, 180, 210, 216, 240, 288, 300, 336, 360, 420, 480, 504, 540, 600, 630, 660
Highly abundant numbers: where sum-of-divisors function increases

18. 1, 2, 3, 4, 6, 8, 11, 13, 16, 18, 26, 28, 36, 38, 47, 48, 53, 57, 62, 69, 72, 77, 82, 87, 97, 99, 102, 106, 114, 126, 131, 138, 145, 148, 155, 175, 177, 180, 182, 189, 197, 206, 209
Ulam numbers: next is uniquely the sum of 2 earlier terms

19. 2, 3, 5, 7, 11, 13, 17, 19, 23, 29, 31, 37, 41, 43, 47, 53, 59, 60, 61, 67, 71, 73, 79, 83, 89, 97, 101, 103, 107, 109, 113, 127, 131, 137, 139, 149, 151, 157, 163, 167, 168, 173
Orders of simple groups

20. 2, 3, 5, 7, 11, 13, 17, 19, 23, 29, 31, 37, 41, 43, 47, 53, 59, 61, 67, 71, 73, 79, 83, 89, 97, 101, 103, 107, 109, 113, 127, 131, 137, 139, 149, 151, 157, 163, 167, 173, 179, 181
Prime numbers

21. 1, 2, 3, 5, 7, 11, 15, 22, 30, 42, 56, 77, 101, 135, 176, 231, 297, 385, 490, 627, 792, 1002, 1255, 1575, 1958, 2436, 3010, 3718, 4565, 5604, 6842, 8349, 10143, 12310, 14883
Number of partitions of n, $n \geq 1$

22. 2, 3, 5, 7, 13, 17, 19, 31, 61, 89, 107, 127, 521, 607, 1279, 2203, 2281, 3217, 4253, 4423, 9689, 9941, 11213, 19937, 21701, 23209, 44497, 86243, 110503, 132049, 216091, 756839, 859433
Mersenne primes: n such that $2^n - 1$ is prime

23. 1, 1, 2, 3, 5, 8, 13, 21, 34, 55, 89, 144, 233, 377, 610, 987, 1597, 2584, 4181, 6765, 10946, 17711, 28657, 46368, 75025, 121393, 196418, 317811, 514229, 832040, 1346269
Fibonacci numbers: $F(n) = F(n-1) + F(n-2)$

24. 1, 2, 3, 6, 10, 20, 35, 70, 126, 252, 462, 924, 1716, 3432, 6435, 12870, 24310, 48620, 92378, 184756, 352716, 705432, 1352078, 2704156, 5200300, 10400600, 20058300
Central binomial coefficients: $C(n, \lfloor n/2 \rfloor)$, $n \geq 1$

25. 1, 1, 2, 3, 6, 11, 20, 40, 77, 148, 285, 570, 1120, 2200, 4323, 8498, 16996, 33707, 66844, 132568, 262936, 521549, 1043098, 2077698, 4138400, 8243093
Stern's sequence: $a(n+1)$ is sum of preceding $\left\lceil \frac{\sqrt{8n+1}-1}{2} \right\rceil$ terms, $n \geq 1$

26. 1, 1, 2, 3, 6, 11, 22, 42, 84, 165, 330, 654, 1308, 2605, 5210, 10398, 20796, 41550, 83100, 166116, 332232, 664299, 1328598, 2656866, 5313732, 10626810
Narayana–Zidek–Capell numbers: $a(2n) = 2a(2n-1)$, $a(2n+1) = 2a(2n) - a(n)$

27. 1, 1, 1, 2, 3, 6, 11, 23, 46, 98, 207, 451, 983, 2179, 4850, 10905, 24631, 56011, 127912, 293547, 676157, 1563372, 3626149, 8436379, 19680277, 46026618, 107890609
Wedderburn–Etherington numbers: interpretations of X^n, $n \geq 1$

28. 1, 1, 1, 2, 3, 6, 11, 23, 47, 106, 235, 551, 1301, 3159, 7741, 19320, 48629, 123867, 317955, 823065, 2144505, 5623756, 14828074, 39299897, 104636890, 279793450
Number of trees with n unlabeled nodes, $n \geq 1$

29. 2, 3, 6, 20, 168, 7581, 7828354, 2414682040998, 56130437228687557907788
Dedekind numbers: number of monotone Boolean functions of n variables, $n \geq 0$

30. 1, 1, 2, 3, 7, 16, 54, 243, 2038, 33120, 1182004, 87723296, 12886193064, 3633057074584, 1944000150734320, 1967881448329407496
    Number of Euler graphs or 2-graphs with n nodes, $n \geq 1$

31. 0, 0, 1, 1, 2, 3, 7, 18, 41, 123, 367, 1288, 4878
    Number of alternating prime knots with n crossings, $n \geq 1$

32. 0, 0, 1, 1, 2, 3, 7, 21, 49, 165, 552, 2176, 9988
    Number of prime knots with n crossings, $n \geq 1$

33. 1, 1, 2, 3, 8, 14, 42, 81, 262, 538, 1828, 3926, 13820, 30694, 110954, 252939, 933458, 2172830, 8152860, 19304190, 73424650, 176343390, 678390116, 1649008456
    Meandric numbers: ways a river can cross a road n times, $n \geq 1$

34. 0, 1, 2, 4, 5, 8, 9, 10, 13, 16, 17, 18, 20, 25, 26, 29, 32, 34, 36, 37, 40, 41, 45, 49, 50, 52, 53, 58, 61, 64, 65, 68, 72, 73, 74, 80, 81, 82, 85, 89, 90, 97, 98, 100, 101, 104, 106
    Numbers that are sums of 2 squares

35. 1, 2, 4, 5, 8, 10, 14, 15, 16, 21, 22, 25, 26, 28, 33, 34, 35, 36, 38, 40, 42, 46, 48, 49, 50, 53, 57, 60, 62, 64, 65, 70, 77, 80, 81, 83, 85, 86, 90, 91, 92, 100, 104, 107
    MacMahon's prime numbers of measurement, or segmented numbers

36. 1, 2, 4, 6, 10, 14, 20, 26, 36, 46, 60, 74, 94, 114, 140, 166, 202, 238, 284, 330, 390, 450, 524, 598, 692, 786, 900, 1014, 1154, 1294, 1460, 1626, 1828, 2030, 2268, 2506
    Binary partitions (partitions of $2n$ into powers of 2), $n \geq 0$

37. 1, 2, 4, 8, 16, 32, 64, 128, 256, 512, 1024, 2048, 4096, 8192, 16384, 32768, 65536, 131072, 262144, 524288, 1048576, 2097152, 4194304, 8388608, 16777216, 33554432, 67108864, 134217728, 268435456, 536870912   **Powers of 2**

38. 1, 1, 2, 4, 9, 20, 48, 115, 286, 719, 1842, 4766, 12486, 32973, 87811, 235381, 634847, 1721159, 4688676, 12826228, 35221832, 97055181, 268282855, 743724984, 2067174645   **Number of rooted trees with n unlabeled nodes, $n \geq 1$**

39. 1, 1, 2, 4, 9, 21, 51, 127, 323, 835, 2188, 5798, 15511, 41835, 113634, 310572, 853467, 2356779, 6536382, 18199284, 50852019, 142547559, 400763223, 1129760415
    Motzkin numbers: ways to join n points on a circle by chords

40. 1, 1, 2, 4, 9, 22, 59, 167, 490, 1486, 4639, 14805, 48107, 158808, 531469, 1799659, 6157068, 21258104, 73996100, 259451116, 951695102, 3251073303
    Number of different scores in n-team round-robin tournament, $n \geq 1$

41. 1, 1, 2, 4, 11, 34, 156, 1044, 12346, 274668, 12005168, 1018997864, 165091172592, 50502031367952, 29054155657235488, 31426485969804308768
    Number of graphs with n unlabeled nodes, $n \geq 0$

42. 0, 1, 2, 5, 12, 29, 70, 169, 408, 985, 2378, 5741, 13860, 33461, 80782, 195025, 470832, 1136689, 2744210, 6625109, 15994428, 38613965, 93222358, 225058681, 543339720   **Pell numbers: $a(n) = 2a(n-1) + a(n-2)$**

43. 1, 1, 2, 5, 12, 35, 108, 369, 1285, 4655, 17073, 63600, 238591, 901971, 3426576, 13079255, 50107909, 192622052, 742624232, 2870671950, 11123060678, 43191857688, 168047007728, 654999700403   **Polyominoes with n cells, $n \geq 1$**

44. 1, 1, 2, 4, 12, 56, 456, 6880, 191536, 9733056, 903753248, 154108311168, 48542114686912, 28401423719122304, 31021002160355166848
    Number of outcomes of n-team round-robin tournament, $n \geq 1$

45. 1, 1, 2, 5, 14, 38, 120, 353, 1148, 3527, 11622, 36627, 121622, 389560, 1301140, 4215748, 13976335, 46235800, 155741571, 512559185, 1732007938, 5732533570   **Number of ways to fold a strip of n blank stamps, $n \geq 1$**

46. 1, 1, 2, 5, 14, 42, 132, 429, 1430, 4862, 16796, 58786, 208012, 742900, 2674440, 9694845, 35357670, 129644790, 477638700, 1767263190, 6564120420, 24466267020   **Catalan numbers:** $C(2n, n)/(n + 1), n \geq 0$

47. 1, 1, 2, 5, 15, 52, 203, 877, 4140, 21147, 115975, 678570, 4213597, 27644437, 190899322, 1382958545, 10480142147, 82864869804, 682076806159, 5832742205057   **Bell or exponential numbers: expansion of** $e^{(e^x - 1)}$

48. 1, 1, 1, 2, 5, 16, 61, 272, 1385, 7936, 50521, 353792, 2702765, 22368256, 199360981, 1903757312, 19391512145, 209865342976, 2404879675441, 29088885112832   **Euler numbers: expansion of** $\sec x + \tan x$

49. 0, 2, 6, 12, 20, 30, 42, 56, 72, 90, 110, 132, 156, 182, 210, 240, 272, 306, 342, 380, 420, 462, 506, 552, 600, 650, 702, 756, 812, 870, 930, 992, 1056, 1122, 1190, 1260, 1332   **Pronic numbers:** $n(n + 1), n \geq 0$

50. 1, 2, 6, 20, 70, 252, 924, 3432, 12870, 48620, 184756, 705432, 2704156, 10400600, 40116600, 155117520, 601080390, 2333606220, 9075135300, 35345263800   **Central binomial coefficients:** $C(2n, n), n \geq 0$

51. 1, 1, 1, 2, 6, 21, 112, 853, 11117, 261080, 11716571, 1006700565, 164059830476, 50335907869219, 29003487462848061, 31397381142761241960   **Number of connected graphs with** n **unlabeled nodes,** $n \geq 0$

52. 1, 2, 6, 22, 101, 573, 3836, 29228, 250749, 2409581, 25598186, 296643390, 3727542188, 50626553988, 738680521142   **Kendall–Mann numbers: maximal inversions in permutation of** n **letters,** $n \geq 1$

53. 1, 1, 2, 6, 24, 120, 720, 5040, 40320, 362880, 3628800, 39916800, 479001600, 6227020800, 87178291200, 1307674368000, 20922789888000, 355687428096000, 6402373705728000   **Factorial numbers:** $n!, n \geq 0$

54. 1, 2, 7, 42, 429, 7436, 218348, 10850216, 911835460, 129534272700, 31095744852375, 12611311859677500, 8639383518297652500   **Robbins numbers:** $\prod_{k=0}^{n-1}(3k + 1)!/(n + k)!, n \geq 1$

55. 1, 2, 8, 42, 262, 1828, 13820, 110954, 933458, 8152860, 73424650, 678390116, 6405031050, 61606881612, 602188541928, 5969806669034, 59923200729046   **Closed meandric numbers: ways a loop can cross a road** $2n$ **times,** $n \geq 1$

56. 1, 2, 8, 48, 384, 3840, 46080, 645120, 10321920, 185794560, 3715891200, 81749606400, 1961990553600, 51011754393600, 1428329123020800, 42849873690624000   **Double factorial numbers:** $(2n)!! = 2^n n!, n \geq 0$

57. 0, 1, 2, 9, 44, 265, 1854, 14833, 133496, 1334961, 14684570, 176214841, 2290792932, 32071101049, 481066515734, 7697064251745, 130850092279664   **Derangements: permutations of** n **elements with no fixed points,** $n \geq 1$

58. 1, 2, 16, 272, 7936, 353792, 22368256, 1903757312, 209865342976, 29088885112832, 4951498053124096, 1015423886506852352, 246921480190207983616   **Tangent numbers: expansion of** $\tan x$

59. 1, 3, 4, 7, 6, 12, 8, 15, 13, 18, 12, 28, 14, 24, 24, 31, 18, 39, 20, 42, 32, 36, 24, 60, 31, 42, 40, 56, 30, 72, 32, 63, 48, 54, 48, 91, 38, 60, 56, 90, 42, 96, 44, 84, 78, 72, 48, 124 $\sigma(n)$, **sum of the divisors of** $n, n \geq 1$

60. 1, 3, 4, 7, 9, 12, 13, 16, 19, 21, 25, 27, 28, 31, 36, 37, 39, 43, 48, 49, 52, 57, 61, 63, 64, 67, 73, 75, 76, 79, 81, 84, 91, 93, 97, 100, 103, 108, 109, 111, 112, 117, 121, 124, 127   **Numbers of the form** $x^2 + xy + y^2$

61. 1, 3, 4, 7, 11, 18, 29, 47, 76, 123, 199, 322, 521, 843, 1364, 2207, 3571, 5778, 9349, 15127, 24476, 39603, 64079, 103682, 167761, 271443, 439204, 710647, 1149851, 1860498   **Lucas numbers:** $L(n) = L(n-1) + L(n-2)$

62. 1, 1, 1, 3, 4, 12, 27, 82, 228, 733, 2282, 7528, 24834, 83898, 285357, 983244, 3412420, 11944614, 42080170, 149197152, 531883768, 1905930975, 6861221666, 24806004996   **Number of ways to cut an n-sided polygon into triangles,** $n \geq 1$

63. 1, 3, 6, 10, 15, 21, 28, 36, 45, 55, 66, 78, 91, 105, 120, 136, 153, 171, 190, 210, 231, 253, 276, 300, 325, 351, 378, 406, 435, 465, 496, 528, 561, 595, 630, 666, 703, 741, 780   **Triangular numbers:** $n(n+1)/2, n \geq 1$

64. 1, 3, 6, 11, 17, 25, 34, 44, 55, 72, 85, 106, 127, 151
       Shortest Golomb ruler with n marks, $n \geq 2$

65. 1, 3, 6, 13, 24, 48, 86, 160, 282, 500, 859, 1479, 2485, 4167, 6879, 11297, 18334, 29601, 47330, 75278, 118794, 186475, 290783, 451194, 696033, 1068745, 1632658
       Number of planar partitions of n, $n \geq 1$

66. 1, 3, 7, 9, 13, 15, 21, 25, 31, 33, 37, 43, 49, 51, 63, 67, 69, 73, 75, 79, 87, 93, 99, 105, 111, 115, 127, 129, 133, 135, 141, 151, 159, 163, 169, 171, 189, 193, 195, 201, 205   **Lucky numbers (defined by sieve similar to prime numbers)**

67. 1, 3, 7, 19, 47, 130, 343, 951, 2615, 7318, 20491, 57903, 163898, 466199, 1328993, 3799624, 10884049, 31241170, 89814958, 258604642
       Number of mappings from n unlabeled points to themselves, $n \geq 1$

68. 1, 3, 9, 25, 65, 161, 385, 897, 2049, 4609, 10241, 22529, 49153, 106497, 229377, 491521, 1048577, 2228225, 4718593, 9961473, 20971521, 44040193, 92274689   **Cullen numbers:** $n \cdot 2^n + 1, n \geq 0$

69. 1, 3, 9, 27, 81, 243, 729, 2187, 6561, 19683, 59049, 177147, 531441, 1594323, 4782969, 14348907, 43046721, 129140163, 387420489, 1162261467, 3486784401, 10460353203   **Powers of 3**

70. 1, 3, 9, 33, 139, 718, 4535
Number of topologies or transitive-directed graphs with n unlabeled nodes, $n \geq 1$

71. 1, 1, 3, 11, 45, 197, 903, 4279, 20793, 103049, 518859, 2646723, 13648869, 71039373, 372693519, 1968801519, 10463578353, 55909013009, 300159426963
       Schroeder's second problem: ways to interpret $X_1 X_2 \ldots X_n, n \geq 1$

72. 1, 3, 11, 50, 274, 1764, 13068, 109584, 1026576, 10628640, 120543840, 1486442880, 19802759040, 283465647360, 4339163001600, 70734282393600, 1223405590579200   **Stirling cycle numbers:** $\left[{n \atop 2}\right], n \geq 2$.

73. 1, 3, 13, 75, 541, 4683, 47293, 545835, 7087261, 102247563, 1622632573, 28091567595, 526858348381, 10641342970443, 230283190977853, 5315654681981355   **Preferential arrangements of n things,** $n \geq 1$

74. 1, 3, 15, 105, 945, 10395, 135135, 2027025, 34459425, 654729075, 13749310575, 316234143225, 7905853580625, 213458046676875, 6190283353629375
       Double factorial numbers: $(2n+1)!! = 1 \cdot 3 \cdot 5 \cdots (2n+1), n \geq 1$

75. 1, 3, 16, 125, 1296, 16807, 262144, 4782969, 100000000, 2357947691, 61917364224, 1792160394037, 56693912375296, 1946195068359375, 72057594037927936   **Number of trees with n labeled nodes:** $n^{n-2}, n \geq 2$

76. 1, 3, 16, 218, 9608, 1540944, 882033440, 1793359192848, 13027956824399552, 341260431952972580352, 325229093850558861111197440
       Directed graphs with n unlabeled nodes, $n \geq 1$

77. 1, 1, 3, 17, 155, 2073, 38227, 929569, 28820619, 1109652905, 51943281731, 2905151042481, 191329672483963, 14655626154768697, 1291885088448017715
Genocchi numbers: expansion of $\tan(x/2)$

78. 0, 1, 4, 5, 16, 17, 20, 21, 64, 65, 68, 69, 80, 81, 84, 85, 256, 257, 260, 261, 272, 273, 276, 277, 320, 321, 324, 325, 336, 337, 340, 341, 1024, 1025, 1028, 1029, 1040, 1041
Moser–de Bruijn sequence: sums of distinct powers of 4

79. 4, 7, 8, 9, 10, 11, 12, 12, 13, 13, 14, 15, 15, 16, 16, 16, 17, 17, 18, 18, 19, 19, 19, 20, 20, 20, 21, 21, 21, 22, 22, 22, 23, 23, 23, 24, 24, 24, 24, 25, 25, 25, 25, 26, 26, 26
Chromatic number of surface of genus $n, n \geq 0$

80. 1, 4, 9, 16, 25, 36, 49, 64, 81, 100, 121, 144, 169, 196, 225, 256, 289, 324, 361, 400, 441, 484, 529, 576, 625, 676, 729, 784, 841, 900, 961, 1024, 1089, 1156, 1225, 1296
The squares

81. 1, 4, 10, 19, 31, 46, 64, 85, 109, 136, 166, 199, 235, 274, 316, 361, 409, 460, 514, 571, 631, 694, 760, 829, 901, 976, 1054, 1135, 1219, 1306, 1396, 1489, 1585, 1684, 1786
Centered triangular numbers: $(3n^2 + 3n + 2)/2, n \geq 0$

82. 1, 4, 10, 20, 35, 56, 84, 120, 165, 220, 286, 364, 455, 560, 680, 816, 969, 1140, 1330, 1540, 1771, 2024, 2300, 2600, 2925, 3276, 3654, 4060, 4495, 4960, 5456, 5984
Tetrahedral numbers: $C(n + 3, 3), n \geq 0$

83. 1, 1, 4, 26, 236, 2752, 39208, 660032, 12818912, 282137824, 6939897856, 188666182784, 5617349020544, 181790703209728, 6353726042486272
Schroeder's fourth problem: families of subsets of an n **set,** $n \geq 1$

84. 1, 4, 29, 355, 6942, 209527, 9535241, 642779354, 63260289423, 8977053873043, 1816846038736192, 519355571065774021
Number of transitive-directed graphs with n **labeled nodes,** $n \geq 1$

85. 1, 5, 12, 22, 35, 51, 70, 92, 117, 145, 176, 210, 247, 287, 330, 376, 425, 477, 532, 590, 651, 715, 782, 852, 925, 1001, 1080, 1162, 1247, 1335, 1426, 1520, 1617, 1717, 1820
Pentagonal numbers: $n(3n - 1)/2, n \geq 1$

86. 1, 5, 13, 25, 41, 61, 85, 113, 145, 181, 221, 265, 313, 365, 421, 481, 545, 613, 685, 761, 841, 925, 1013, 1105, 1201, 1301, 1405, 1513, 1625, 1741, 1861, 1985, 2113, 2245
Centered square numbers: $n^2 + (n - 1)^2, n \geq 1$

87. 1, 5, 14, 30, 55, 91, 140, 204, 285, 385, 506, 650, 819, 1015, 1240, 1496, 1785, 2109, 2470, 2870, 3311, 3795, 4324, 4900, 5525, 6201, 6930, 7714, 8555, 9455, 10416
Square pyramidal numbers: $n(n + 1)(2n + 1)/6, n \geq 1$

88. 1, 5, 25, 125, 625, 3125, 15625, 78125, 390625, 1953125, 9765625, 48828125, 244140625, 1220703125, 6103515625, 30517578125, 152587890625, 762939453125, 3814697265625
Powers of 5

89. 1, 5, 52, 1522, 145984, 48464496, 56141454464, 229148550030864, 3333310786076963968, 17469527274674991958 0928
Number of possible relations on n **unlabeled points,** $n \geq 1$

90. 1, 1, 5, 61, 1385, 50521, 2702765, 199360981, 19391512145, 2404879675441, 370371188237525, 69348874393137901, 15514534163557086905, 4087072509293123892361
Euler numbers: expansion of $\sec x$

91. 1, 5, 109, 32297, 2147321017, 9223372023970362989, 170141183460469231667123699502996689125
Number of ways to cover an n **set,** $n \geq 1$

92. 1, 6, 15, 28, 45, 66, 91, 120, 153, 190, 231, 276, 325, 378, 435, 496, 561, 630, 703, 780, 861, 946, 1035, 1128, 1225, 1326, 1431, 1540, 1653, 1770, 1891, 2016, 2145, 2278   **Hexagonal numbers:** $n(2n-1), n \geq 1$

93. 1, 6, 25, 90, 301, 966, 3025, 9330, 28501, 86526, 261625, 788970, 2375101, 7141686, 21457825, 64439010, 193448101, 580606446, 1742343625, 5228079450, 15686335501   **Stirling subset numbers:** $\left\{ {n \atop 3} \right\}, n \geq 3$

94. 6, 28, 496, 8128, 33550336, 8589869056, 137438691328, 2305843008139952128, 2658455991569831744654692615953842176
             Perfect numbers: equal to the sum of their proper divisors

95. 1, 8, 21, 40, 65, 96, 133, 176, 225, 280, 341, 408, 481, 560, 645, 736, 833, 936, 1045, 1160, 1281, 1408, 1541, 1680, 1825, 1976, 2133, 2296, 2465, 2640, 2821, 3008, 3201
             Octagonal numbers: $n(3n-2), n \geq 1$

96. 1, 8, 27, 64, 125, 216, 343, 512, 729, 1000, 1331, 1728, 2197, 2744, 3375, 4096, 4913, 5832, 6859, 8000, 9261, 10648, 12167, 13824, 15625, 17576, 19683, 21952, 24389
             The cubes

97. 1, -24, 252, -1472, 4830, -6048, -16744, 84480, -113643, -115920, 534612, -370944, -577738, 401856, 1217160, 987136, -6905934, 2727432, 10661420   **Ramanujan τ function**

98. 341, 561, 645, 1105, 1387, 1729, 1905, 2047, 2465, 2701, 2821, 3277, 4033, 4369, 4371, 4681, 5461, 6601, 7957, 8321, 8481, 8911, 10261, 10585, 11305, 12801, 13741, 13747   **Sarrus numbers: pseudo-primes to base 2**

99. 561, 1105, 1729, 2465, 2821, 6601, 8911, 10585, 15841, 29341, 41041, 46657, 52633, 62745, 63973, 75361, 101101, 115921, 126217, 162401, 172081, 188461, 252601, 278545   **Carmichael numbers**

100. 1, 744, 196884, 21493760, 864299970, 20245856256, 333202640600, 4252023300096, 44656994071935, 401490886656000, 3176440229784420, 22567393309593600   **Coefficients of the modular function j**

For more information about these sequences and tens of thousands of others, including formulae and references, see "The On-Line Encyclopedia of Integer Sequences", published electronically at `www.research.att.com/~njas/sequences/`.

1.3 SERIES AND PRODUCTS

1.3.1 DEFINITIONS

If $\{a_n\}$ is a sequence of numbers or functions, then

1. $S_N = \sum_{n=1}^{N} a_n = a_1 + a_2 + ... + a_N$ is the Nth *partial sum*.

2. For an infinite series: $S = \lim_{N \to \infty} S_N = \sum_{n=1}^{\infty} a_n$ (when the limit exists). Then S is called the *sum* of the series.

3. The series is said to *converge* if the limit exists and *diverge* if it does not.

4. If $a_n = b_n x^n$, where b_n is independent of x, then S is called a *power series*.

5. If $a_n = (-1)^n |a_n|$, then S is called an *alternating series*.

6. If $\sum |a_n|$ converges, then the series *converges absolutely*.

7. If S_N converges, but not absolutely, then it *converges conditionally*.

EXAMPLES

1. The *harmonic series* $S = 1 + \frac{1}{2} + \frac{1}{3} + \ldots$ diverges. The corresponding alternating series (called the *alternating harmonic series*) $S = 1 - \frac{1}{2} + \frac{1}{3} - \cdots + (-1)^{n-1} \frac{1}{n} + \ldots$ converges (conditionally) to $\log 2$.

2. The harmonic numbers are $H_n = \sum_{k=1}^{n} \frac{1}{k}$. The first few values are $\{1, \frac{3}{2}, \frac{11}{6}, \frac{25}{12}, \ldots\}$. Asymptotically, $H_n \sim \ln n + \gamma + \frac{1}{2n}$.

3. $S_N = \sum_{n=0}^{N} x^n = \frac{1 - x^{N+1}}{1 - x}$ if $x \neq 1$.

1.3.2   GENERAL PROPERTIES

1. Adding or removing a finite number of terms does not affect the convergence or divergence of an infinite series.

2. The terms of an absolutely convergent series may be rearranged in any manner without affecting its value.

3. A conditionally convergent series can be made to converge to any value by suitably rearranging its terms.

4. If the component series are convergent, then

$$\sum (\alpha a_n + \beta b_n) = \alpha \sum a_n + \beta \sum b_n.$$

5. $\left(\sum_{n=0}^{\infty} a_n \right) \left(\sum_{n=0}^{\infty} b_n \right) = \sum_{n=0}^{\infty} c_n$ where $c_n = a_0 b_n + a_1 b_{n-1} + \cdots + a_n b_0$.

6. Summation by parts: let $\sum a_n$ and $\sum b_n$ converge. Then

$$\sum a_n b_n = \sum S_n (b_n - b_{n+1})$$

where S_n is the nth partial sum of $\sum a_n$.

7. A power series may be integrated and differentiated term-by-term within its interval of convergence.

8. *Schwarz inequality*:

$$\sum |a_n| \, |b_n| \leq \left(\sum |a_n|^2 \right)^{1/2} \left(\sum |b_n|^2 \right)^{1/2}$$

9. *Holder's inequality*: when $1/p + 1/q = 1$ and $p, q > 1$

$$\sum |a_n b_n| \le \left(\sum |a_n|^p\right)^{1/p} \left(\sum |b_n|^q\right)^{1/q}$$

10. *Minkowski's inequality*: when $p \ge 1$

$$\left(\sum |a_n + b_n|^p\right)^{1/p} \le \left(\sum |a_n|^p\right)^{1/p} + \left(\sum |b_n|^p\right)^{1/p}$$

11. *Arithmetic mean–geometric mean inequality*: If $a_i > 0$ then

$$\frac{a_1 + a_2 + \ldots + a_n}{n} \ge (a_1 a_2 \cdots a_n)^{1/n}$$

12. *Kantorovich inequality*: Suppose that $0 < x_1 < x_2 < \ldots < x_n$. If $\lambda_i \ge 0$ and $\sum_{i=1}^{n} \lambda_i = 1$ then

$$\left(\sum \lambda_i x_i\right) \left(\sum \frac{\lambda_i}{x_i}\right) \le A^2 G^{-2}$$

where $A = \frac{1}{2}(x_1 + x_n)$ and $G = \sqrt{x_1 x_n}$.

EXAMPLES

1. Let T be the alternating harmonic series S rearranged so that each positive term is followed by the next two negative terms. By combining each positive term of T with the succeeding negative term, we find that $T_{3N} = \frac{1}{2} S_{2N}$. Hence, $T = \frac{1}{2} \log 2$.

2. The series $1 + \frac{1}{2} - \frac{1}{3} + \frac{1}{4} + \frac{1}{5} - \frac{1}{6} + \frac{1}{7} + \frac{1}{8} - \frac{1}{9} + \ldots$ diverges, whereas

$$\left(1 + \frac{1}{3} - \frac{1}{2}\right) + \left(\frac{1}{5} + \frac{1}{7} - \frac{1}{4}\right) + \cdots + \left(\frac{1}{4n-3} + \frac{1}{4n-1} - \frac{1}{2n}\right) + \ldots$$

converges to $\log(2\sqrt{2})$.

1.3.3 CONVERGENCE TESTS

1. *Comparison test*: If $|a_n| \le b_n$ and $\sum b_n$ converges, then $\sum a_n$ converges.

2. *Limit test*: If $\lim_{n \to \infty} a_n \ne 0$, or the limit does not exist, then $\sum a_n$ is divergent.

3. *Ratio test*: Let $\rho = \lim_{n \to \infty} |\frac{a_{n+1}}{a_n}|$. If $\rho < 1$, the series converges absolutely. If $\rho > 1$, the series diverges.

4. *Cauchy root test*: Let $\sigma = \lim_{n \to \infty} |a_n|^{1/n}$. If $\sigma < 1$, the series converges. If $\sigma > 1$, it diverges.

5. *Integral test*: Let $|a_n| = f(n)$ with $f(x)$ being monotone decreasing, and $\lim_{x \to \infty} f(x) = 0$. Then $\int_A^\infty f(x)\,dx$ and $\sum a_n$ both converge or both diverge for any $A > 0$.

6. *Gauss's test*: If $\left|\dfrac{a_{n+1}}{a_n}\right| = 1 - \dfrac{p}{n} + \dfrac{A_n}{n^q}$ where $q > 1$ and the sequence $\{A_n\}$ is bounded, then the series is absolutely convergent if and only if $p > 1$.

7. *Alternating series test*: If $|a_n|$ tends monotonically to 0, then $\sum (-1)^n |a_n|$ converges.

EXAMPLES

1. For $S = \sum_{n=1}^{\infty} n^c x^n$, $\rho = \lim_{n \to \infty} (1 + \frac{1}{n})^c x = x$. Hence, using the ratio test, S converges for $0 < x < 1$ and any value of c.

2. For $S = \sum_{n=1}^{\infty} \frac{5^n}{n^{20}}$, $\sigma = \lim_{n \to \infty} (\frac{5^n}{n^{20}})^{1/n} = 5$. Therefore the series diverges.

3. For $S = \sum_{n=1}^{\infty} n^{-t}$, consider $f(x) = x^{-t}$. Then

$$\int_1^{\infty} f(x)\, dx = \int_1^{\infty} \frac{dx}{x^t} = \begin{cases} \dfrac{1}{t-1} & \text{for } t > 1 \\ \text{diverges} & \text{for } t \leq 1 \end{cases}$$

Hence, S converges for $t > 1$.

4. The sum $\sum_{n=2}^{\infty} \frac{1}{n(\log n)^s}$ converges for $s > 1$ by the integral test.

5. Let $a_n = \dfrac{(c)_n}{n!} = \dfrac{c(c+1)\ldots(c+n-1)}{n!}$ where c is not 0 or a negative integer. Then $|a_{n+1}/a_n| = 1 - (c+1)/n + (c+1)/n^2(1+1/n)$. By Gauss's test, the series converges absolutely if and only if $c > 0$.

1.3.4   TYPES OF SERIES

1.3.4.1   Bessel series

1. *Fourier–Bessel series*: $\displaystyle\sum_{n=0}^{\infty} a_n J_\nu(j_{\nu,n} z)$ ($j_{\nu,k}$ is a zero of $J_\nu(x)$)

2. *Neumann series*: $\displaystyle\sum_{n=0}^{\infty} a_n J_{\nu+n}(z)$

3. *Kapteyn series*: $\displaystyle\sum_{n=0}^{\infty} a_n J_{\nu+n}[(\nu + n)z]$

4. *Schlömilch series*: $\displaystyle\sum_{n=1}^{\infty} a_n J_\nu(nz)$

EXAMPLES

1. $\sum_{n=0}^{\infty} \frac{1}{n!} J_{\nu+n}(2) = \frac{1}{\Gamma(\nu+1)}$

2. $\sum_{n=1}^{\infty} J_n(nz) = \frac{1}{2}\frac{z}{1-z}$ for $0 < z < 1$

3. $\sum_{n=1}^{\infty} (-1)^{n+1} J_0(nz) = \frac{1}{2}$ for $0 < z < \pi$

1.3.4.2 Dirichlet series

These are series of the form $\sum_{n=1}^{\infty} \frac{a_n}{n^x}$. They converge for $x > x_0$, where x_0 is the *abscissa of convergence*. Assuming the limits exist:

1. If $\sum a_n$ diverges, then $x_0 = \lim_{n \to \infty} \dfrac{\log |a_1 + \cdots + a_n|}{\log n}$.

2. If $\sum a_n$ converges, then $x_0 = \lim_{n \to \infty} \dfrac{\log |a_{n+1} + a_{n+2} + \ldots|}{\log n}$.

EXAMPLES

1. *Riemann zeta function:* $\zeta(x) = \sum_{n=1}^{\infty} \frac{1}{n^x}$, $x_0 = 1$

2. $\sum_{n=1}^{\infty} \frac{\mu(n)}{n^x} = \frac{1}{\zeta(x)}$, $x_0 = 1$ ($\mu(n)$ denotes the Möbius function; see Section 2.4.9)

3. $\sum_{n=1}^{\infty} \frac{d(n)}{n^x} = \zeta^2(x)$, $x_0 = 1$ ($d(n)$ is the number of divisors of n; see page 128)

1.3.4.3 Fourier series

If $f(x)$ satisfies certain properties, then (see page 48)

$$f(x) = \frac{a_0}{2} + \sum_{n=1}^{\infty} \left(a_n \cos \frac{n\pi x}{L} + b_n \sin \frac{n\pi x}{L} \right) \qquad (1.3.1)$$

1. If $f(x)$ has the Laplace transform $F(k) = \int_0^{\infty} e^{-xk} f(x)\, dx$, then

$$\sum_{k=1}^{\infty} F(k) \cos(kt) = \frac{1}{2} \int_0^{\infty} \frac{\cos(t) - e^{-x}}{\cosh(x) - \cos(t)} f(x)\, dx,$$

$$\sum_{k=1}^{\infty} F(k) \sin(kt) = \frac{1}{2} \int_0^{\infty} \frac{f(x)}{\cosh(x) - \cos(t)}\, dx. \qquad (1.3.2)$$

2. Since the cosine transform of $(\cosh(x) - \cos(t))^{-1}$ with respect to x is $\pi \csc(t)$ $\operatorname{csch}(\pi y) \sinh(\pi - t)y$, we find that

$$\sum_{k=1}^{\infty} \frac{k \sin(kt)}{k^2 + y^2} = \frac{\pi}{2} \frac{\sinh(\pi - t)y}{\sinh(\pi y)}.$$

3. $\sum_{n=1}^{\infty} \frac{\sin(2n\pi x)}{n^{2k+1}} = \frac{(-1)^{k-1}}{2} \frac{(2\pi)^{2k+1}}{2k+1} B_{2k+1}(x)$, for $0 < x < \frac{1}{2}$

4. $\sum_{n=1}^{\infty} \frac{\cos(2n\pi x)}{n^{2k}} = \frac{(-1)^{k-1}}{2} \frac{(2\pi)^{2k}}{(2k)!} B_{2k}(x)$ for $0 < x < \frac{1}{2}$

5. $\sum_{n=1}^{\infty} \frac{\sin((2n+1)\pi x - \pi k/2)}{(2n+1)^{k+1}} = \frac{\pi^{k+1}}{4k!} E_k(x)$

6. $\sum_{n=1}^{\infty} a^n \sin(nx) = \frac{a \sin(x)}{1 - 2a \cos(x) + a^2}$ for $|a| < 1$

7. $\sum_{n=0}^{\infty} a^n \cos(nx) = \frac{1 - a \cos(x)}{1 - 2a \cos(x) + a^2}$ for $|a| < 1$

1.3.4.4 Hypergeometric series

The hypergeometric function is

$$
{}_pF_q \left(\begin{array}{cccc} a_1 & a_2 & \cdots & a_p \\ b_1 & b_2 & \cdots & b_q \end{array} \middle| x \right) = \sum_{n=0}^{\infty} \frac{(a_1)_n (a_2)_n \cdots (a_p)_n}{(b_1)_n (b_2)_n \cdots (b_q)_n} \frac{x^n}{n!} \tag{1.3.3}
$$

where $(a)_n = \Gamma(a+n)/\Gamma(a)$ is the shifted factorial. Any infinite series $\sum A_n$ with A_{n+1}/A_n a rational function of n is a hypergeometric series. These include series of products and quotients of binomial coefficients.

EXAMPLES

1. $\displaystyle {}_2F_1 \left(\begin{array}{cc} a, & b \\ & c \end{array} \middle| 1 \right) = \frac{\Gamma(c)\Gamma(c-a-b)}{\Gamma(c-a)\Gamma(c-b)}$ (Gauss)

2. $\displaystyle {}_3F_2 \left(\begin{array}{ccc} -n, & a, & b \\ c, & 1+a+b-c-n & \end{array} \middle| 1 \right) = \frac{(c-a)_n(c-b)_n}{(c)_n(c-a-b)_n}$ (Saalschutz)

3. $\displaystyle {}_4F_3 \left(\begin{array}{cccc} a, & 1+a/2, & b, & -n \\ a/2, & 1+a-b, & 1+2b-n & \end{array} \middle| 1 \right) = \frac{(a-2b)_n(-b)_n}{(1+a-b)_n(-2b)_n}$ (Bailey)

4. $\displaystyle \sum_{m=0}^{2n} (-1)^m \frac{\binom{2n}{m}\binom{2n+m+1}{m}}{\binom{4n+2m+2}{2m}} (3+2\sqrt{2})^m = \frac{(3/4)_n(5/4)_n}{(7/8)_n(9/8)_n}$

1.3.4.5 Power series

1. The values of x, for which the power series $\sum_{n=0}^{\infty} a_n x^n$ converges, form an interval (*interval of convergence*) which may or may not include one or both endpoints.

2. A power series may be integrated and differentiated term-by-term within its interval of convergence.

3. Note that $[1 + \sum_{n=1}^{\infty} a_n x^n]^{-1} = 1 - \sum_{n=1}^{\infty} b_n x^n$, where $b_1 = a_1$ and $b_n = a_n + \sum_{k=1}^{n-1} b_{n-k} a_k$ for $n \geq 2$.

4. *Inversion of power series:* If $s = \sum_{n=1}^{\infty} a_n x^n$, then $x = \sum_{n=1}^{\infty} A_n s^n$, where $A_1 = 1/a_1$, $A_2 = -a_2/a_1^3$, $A_3 = (2a_2^2 - a_1 a_3)/a_1^5$, $A_4 = (5a_1 a_2 a_3 - a_1^2 a_4 - 5a_2^3)/a_1^7$, $A_5 = (6a_1^2 a_2 a_4 + 3a_1^2 a_3^2 + 14a_2^4 - a_1^3 a_5 - 21a_1 a_2^2 a_3)/a_1^9$.

1.3.4.6 Taylor series

1. Taylor series in 1 variable:

$$
f(a+x) = \sum_{n=0}^{N} \frac{f^{(n)}(a)}{n!} x^n + R_N
$$

or

$$f(x) = f(a) + f'(a)(x-a) + \frac{f''(a)}{2!}(x-a)^2 + \frac{f'''(a)}{3!}(x-a)^3 + \ldots$$

or, specializing to $a = 0$, results in the MacLaurin series

$$f(x) = f(0) + f'(0)x + \frac{f''(0)}{2!}x^2 + \frac{f'''(0)}{3!}x^3 + \ldots.$$

2. Lagrange's form of the remainder:

$$R_N = \frac{x^{N+1}}{(N+1)!}f^{(N+1)}(a + \theta x), \qquad \text{for some } 0 < \theta < 1.$$

3. Taylor series in 2 variables:

$$f(a+x, b+y) = f(a,b) + xf_x(a,b) + yf_y(a,b) +$$
$$\frac{1}{2!}\left[x^2 f_{xx}(a,b) + 2xy f_{xy}(a,b) + y^2 f_{yy}(a,b)\right] + \ldots.$$

4. Taylor series for vectors:

$$f(\mathbf{a} + \mathbf{x}) = \sum_{n=0}^{N} \frac{[(\mathbf{x} \cdot \nabla)^n f](\mathbf{a})}{n!} + R_N(\mathbf{a}) = f(\mathbf{a}) + \mathbf{x} \cdot \nabla f(\mathbf{a}) + \ldots.$$

EXAMPLES

1. *Binomial series*:

$$(x+y)^\nu = \sum_{n=0}^{\infty} \frac{\Gamma(\nu+1)}{\Gamma(\nu-n+1)} \frac{x^n y^{\nu-n}}{n!}.$$

When ν is a positive integer, this series terminates at $n = \nu$.

2. $\frac{1}{\sqrt{1-4x}} = \sum_{k=0}^{\infty} \frac{(2k)!}{(k!)^2} x^k$ for $|x| < 1/4$

3. $\frac{xe^{xt}}{e^x - 1} = \sum_{n=0}^{\infty} B_n(t) \frac{x^n}{n!}$

4. $\frac{2e^{xt}}{e^x + 1} = \sum_{n=0}^{\infty} E_n(t) \frac{x^n}{n!}$

5. $\sum_{n=1}^{\infty} \frac{x^n}{(n+1)(n+3)} = \frac{1}{x^3} \int_0^x u\, du \int_0^u dt \sum_0^{\infty} t^n$
   $= \frac{1}{2x^3}\left[x + \frac{1}{2}x^2 + (1-x^2)\log(1-x)\right]$ for $|x| < 1$

6. $\sum_{k=1}^{\infty} \frac{x^k}{k^n} = \text{Li}_n(x)$ \qquad (polylogarithm)

1.3.4.7 Telescoping series

If $\lim_{n\to\infty} F(n) = 0$, then $\sum_{n=1}^{\infty}[F(n) - F(n+1)] = F(1)$. For example,

$$\sum_{n=1}^{\infty} \frac{1}{(n+1)(n+2)} = \sum_{n=1}^{\infty} \left[\frac{1}{n+1} - \frac{1}{n+2}\right] = \frac{1}{2}.$$

The GWZ algorithm expresses a proposed identity in the form of a telescoping series $\sum_k [F(n+1,k) - F(n,k)] = 0$, then searches for a $G(n,k)$ that satisfies $F(n+1,k) - F(n,k) = G(n,k+1) - G(n,k)$ and $G(n, \pm\infty) = 0$. The search assumes that $G(n,k) = R(n,k)F(n,k-1)$ where $R(n,k)$ is a rational expression in n and k. When R is found, the proposed identity is verified. For example, the Pfaff–Saalschutz identity has the following proof:

$$\sum_{k=-\infty}^{\infty} \frac{(a+k)!(b+k)!(c-a-b+n-1-k)!}{(k+1)!(n-k)!(c+k)!} = \frac{(c-a+n)!(c-b+n)!}{(n+1)!(c+n)!},$$

$$R(n,k) = -\frac{(b+k)(a+k)}{(c-b+n+1)(c-a+n+1)}.$$

1.3.4.8   Other types of series

1. *Arithmetic series:*

$$\sum_{n=1}^{N}(a+nd) = Na + \frac{1}{2}N(N+1)d.$$

2. *Arithmetic power series:*

$$\sum_{n=0}^{N}(a+nb)x^n = \frac{a - (a+bN)x^{N+1}}{1-x} + \frac{bx(1-x^N)}{(1-x)^2}, \qquad (x \neq 1).$$

3. *Geometric series:*

$$1 + x + x^2 + x^3 + \cdots = \frac{1}{1-x}, \qquad (|x| < 1).$$

4. *Arithmetic–geometric series:*

$$a + (a+b)x + (a+2b)x^2 + (a+3b)x^3 + \cdots = \frac{a}{1-x} + \frac{bx}{(1-x)^2},$$
$$(|x| < 1).$$

5. *Combinatorial sums:*

   (a) $\sum_{k=0}^{n} \binom{x-k}{n-k} = \binom{x+1}{n}$

   (b) $\sum_{k=-\infty}^{m} (-1)^k \binom{x}{k} = (-1)^m \binom{x-1}{m}$

   (c) $\sum_{k=0}^{n} \binom{k+m}{k} = \binom{m+n+1}{n}$

   (d) $\sum_{k=-\infty}^{m} (-1)^k \binom{x+m}{k} = \binom{-x}{m}$

   (e) $\sum_{k=-\infty}^{\infty} \binom{x}{m+k}\binom{y}{n-k} = \binom{x+y}{m+n}$

   (f) $\sum_{k=-\infty}^{\infty} \binom{l}{m+k}\binom{x}{n+k} = \binom{l+x}{l-m+n}$

(g) $\sum_{k=-\infty}^{\infty} (-1)^k \binom{l}{m+k}\binom{x+k}{n} = (-1)^{l+m}\binom{x-m}{n-l}$

(h) $\sum_{k=-\infty}^{l} (-1)^k \binom{l-k}{m}\binom{x}{k-n} = (-1)^{l+m}\binom{x-m-l}{l-m-n}$

(i) $\sum_{k=0}^{l} \binom{l-k}{m}\binom{q+k}{n} = \binom{l+q+1}{m+n+1}$ (for $m \geq q$)

6. *Generating functions*:

    (a) Bessel functions: $\sum_{k=-\infty}^{\infty} J_k(x)z^k = \exp(\frac{1}{2}x\frac{z^2-1}{z})$

    (b) Chebyshev polynomials: $\sum_{n=1}^{\infty} T_n(x)z^n = \frac{z(z+2x)}{2xz-z^2-1}$

    (c) Hermite polynomials: $\sum_{n=0}^{\infty} \frac{H_n(x)}{n!}z^n = \exp(2xz - z^2)$

    (d) Laguerre polynomials: $\sum_{n=0}^{\infty} L_n^{(\alpha)}(x)z^n = (1-z)^{-\alpha-1}\exp[\frac{xz}{z-1}]$

    (e) Legendre polynomials: $\sum_{n=0}^{\infty} P_n(z)x^n = \frac{1}{\sqrt{1-xz+x^2}}$, for $|x| < 1$

7. *Multiple series*:

    (a) $\sum \frac{(-1)^{l+m+n}}{\sqrt{(l+1/6)^2+(m+1/6)^2+(n+1/6)^2}} = \sqrt{3}$
    where $-\infty < l, m, n < \infty$ and they are not all zero

    (b) $\sum \frac{1}{(m^2+n^2)^z} = 4\beta(z)\zeta(z)$ for $-\infty < m, n < \infty$ not both zero

    (c) $\sum_{m,n=0}^{\infty} \frac{(-1)^n}{n!}\frac{\Gamma(n+1/2)}{\Gamma(m+n+1/2)}z^{m+n} = \sqrt{\pi}e^z\frac{\operatorname{erf}(\sqrt{z-z})}{\sqrt{z-z}}$ for $z > 0$

    (d) $\sum_{m,n=1}^{\infty} \frac{m^2-n^2}{(m^2+n^2)^2} = \frac{\pi}{4}$

    (e) $\sum \frac{1}{k_1^2 k_2^2 \dots k_n^2} = \frac{\pi^{2n}}{(2n+1)!}$ for $1 \leq k_1 < \cdots < k_n < \infty$

8. *Theta series*:

$$\sum_{n=-\infty}^{\infty} e^{n^2\pi it+2niz} = \sqrt{\frac{i}{t}}\sum_{n=-\infty}^{\infty} e^{(z-n\pi)^2/\pi it}$$

9. *Lagrange series*: If $f(z)$ is analytic at $z = z_0$, $f(z_0) = w_0$, and $f'(z_0) \neq 0$, then the equation $w = f(z)$ has the unique solution $z = F(w)$. If both functions are expanded

$$f(z) = f_0 + f_1(z - z_0) + f_2(z - z_0)^2 + \dots$$
$$F(w) = F_0 + F_1(w - w_0) + F_2(w - w_0)^2 + \dots$$

(1.3.5)

with $F_0 = F(w_0) = z$, then

$$F_j = \frac{1}{j!}\left[\frac{d^{j-1}}{dz^{j-1}}\left\{\frac{z-z_0}{f(z)-f_0}\right\}^j\right]_{z=z_0}$$

(1.3.6)

For example: $F_1 = \frac{1}{f_1}$, $F_2 = -\frac{f_2}{f_1^3}$, $F_3 = \frac{2f_2^2-f_1f_3}{f_1^5}, \dots$

1.3.5 SUMMATION FORMULAE

1. *Euler–MacLaurin summation formula:* As $n \to \infty$,

$$\sum_{k=0}^{n} f(k) \sim \frac{1}{2} f(n) + \int_{0}^{n} f(x)\, dx + C + \sum_{j=1}^{\infty} (-1)^{j+1} B_{j+1} \frac{f^{(j)}(n)}{(j+1)!}$$

where B_j is the jth Bernoulli number and

$$C = \lim_{m \to \infty} \left[\sum_{j=1}^{m} (-1)^j B_{j+1} \frac{f^{(j)}(0)}{(j+1)!} + \frac{1}{2} f(0) \right.$$
$$\left. + \frac{(-1)^m}{(m+1)!} \int_{0}^{\infty} B_{m+1}(x - \lfloor x \rfloor) f^{(m+1)}(x)\, dx \right].$$

2. *Poisson summation formula:* If f is continuous,

$$\frac{1}{2} f(0) + \sum_{n=1}^{\infty} f(n) = \int_{0}^{\infty} f(x)\, dx + 2 \sum_{n=1}^{\infty} \left[\int_{0}^{\infty} f(x) \cos(2n\pi x) dx \right].$$

3. *Plana's formula:*

$$\sum_{k=1}^{n} f(k) = \frac{1}{2} f(n) + \int_{a}^{n} f(x)\, dx + \sum_{j=1}^{\infty} (-1)^{j-1} B_j \frac{f^{(2j+1)}(n)}{2j!}$$

where f is analytic, a is a constant dependent on f, and B_j is the jth Bernoulli number.

EXAMPLES

1. $\displaystyle\sum_{k=1}^{n} \frac{1}{k} \sim \log n + \gamma + \frac{1}{2n} - \frac{B_2}{2n^2} - \dots$ where γ is Euler's constant.

2. $\displaystyle 1 + 2 \sum_{n=1}^{\infty} e^{-n^2 x} = \sqrt{\frac{\pi}{x}} \left[1 + 2 \sum_{n=1}^{\infty} e^{-\pi^2 n^2 / x} \right]$ (Jacobi)

1.3.6 IMPROVING CONVERGENCE: SHANKS TRANSFORMATION

Let s_n be the nth partial sum. The sequences $\{S(s_n)\}$, $\{S(S(s_n))\}, \dots$ often converge successively more rapidly to the same limit as $\{s_n\}$, where

$$S(s_n) = \frac{s_{n+1} s_{n-1} - s_n^2}{s_{n+1} + s_{n-1} - 2s_n}. \tag{1.3.7}$$

EXAMPLE   For $s_n = \sum_{k=0}^n (-1)^k z^k$, we find $S(s_n) = \frac{1}{1+z}$ for all n.

1.3.7 SUMMABILITY METHODS

Unique values can be assigned to divergent series in a variety of ways which preserve the values of convergent series.

1. Abel summation: $\displaystyle\sum_{n=0}^\infty a_n = \lim_{r\to 1^-} \sum_{n=0}^\infty a_n r^n$.

2. Cesaro $(C,1)$-summation: $\displaystyle\sum_{n=0}^\infty a_n = \lim_{N\to\infty} \frac{s_0 + s_1 + \ldots s_N}{N+1}$

   where $s_n = \sum_{m=0}^n a_m$.

EXAMPLES

(a) $1 - 1 + 1 - 1 + \cdots = \frac{1}{2}$ (in the sense of Abel summation)

(b) $1 - 1 + 0 + 1 - 1 + 0 + 1 - \cdots = \frac{1}{3}$ (in the sense of Cesaro summation)

1.3.8 OPERATIONS WITH POWER SERIES

Let $y = a_1 x + a_2 x^2 + a_3 x^3 + \ldots$, and let $z = z(y) = b_1 x + b_2 x^2 + b_3 x^3 + \ldots$.

$z(y)$	b_0	b_1	b_2	b_3
$1/(1-y)$	1	a_1	$a_1^2 + a_2$	$a_1^3 + 2a_1 a_2 + a_3$
$\sqrt{1+y}$	1	$\frac{1}{2}a_1$	$-\frac{1}{8}a_1^2 + \frac{1}{2}a_2$	$\frac{1}{16}a_1^3 - \frac{1}{4}a_1 a_2 + \frac{1}{2}a_3$
$(1+y)^{-1/2}$	1	$-\frac{1}{2}a_1$	$\frac{3}{8}a_1^2 - \frac{1}{2}a_2$	$-\frac{5}{16}a_1^3 + \frac{3}{4}a_1 a_2 - \frac{1}{2}a_3$
e^y	1	a_1	$\frac{1}{2}a_1^2 + a_2$	$\frac{1}{6}a_1^3 + a_1 a_2 + a_3$
$\log(1+y)$	0	a_1	$a_2 - \frac{1}{2}a_1^2$	$a_3 - a_1 a_2 + \frac{1}{3}a_1^3$
$\sin y$	0	a_1	a_2	$-\frac{1}{6}a_1^3 + a_3$
$\cos y$	1	0	$-\frac{1}{2}a_1^2$	$-a_1 a_2$
$\tan y$	0	a_1	a_2	$\frac{1}{3}a_1^3 + a_3$

1.3.9 MISCELLANEOUS SUMS AND SERIES

1. $\sum_{k=1}^n (-1)^{k+1} k^j = \frac{1}{2}(-1)^{n+1}\left[E_j(n+1) + (-1)^n E_j\right]$

2. $\sum_{k=1}^n \frac{1}{k(k+1)(k+2)} = \frac{1}{4} - \frac{1}{2(n+1)(n+2)}$

3. $\sum_{k=1}^n \frac{km^k}{(k+m)!} = \frac{1}{(m-1)!} - \frac{m^{n+1}}{(m+n)!}$

4. $\sum_{k=0}^{2n}(-1)^k\binom{2n}{k}^{-1} = \frac{2n+1}{n+1}$

5. $\sum_{k=0}^{n}\binom{m}{k}\binom{n}{k}\binom{m+n+k}{m+n} = \binom{m+n}{m}^2$

6. $\sum_{k=0}^{n-1}\frac{1}{4^k}\operatorname{sech}^2\frac{x}{2^k} = \frac{1}{4^n}\operatorname{csch}^2\frac{x}{2^n} - \operatorname{csch}^2 x$

7. $\sum_{k=0}^{n-1}(-1)^k\cos^n\frac{\pi k}{n} = \frac{n}{2^{n-1}}, \qquad \sum_{k=0}^{n}\sec\frac{\pi k}{n} = 0, \qquad (k \neq n/2)$

8. $\sum_{k=-\infty}^{\infty}\frac{1}{(k+a)(k+b)} = \frac{\pi}{b-a}\left[\cot(\pi a) - \cot(\pi b)\right]$

9. $\sum_{k=0}^{\infty}\frac{1}{k^4+a^4} = \frac{1}{2a^4} + \frac{\pi}{2a^3\sqrt{2}}\frac{\sinh(a\pi\sqrt{2})+\sin(a\pi\sqrt{2})}{\cosh(a\pi\sqrt{2})-\cos(a\pi\sqrt{2})}$

10. $\sum_{k=1}^{\infty}\frac{k}{e^{2\pi k}-1} = \frac{1}{24} - \frac{1}{8\pi}$

11. $\sum_{k=0}^{\infty}\frac{1}{k(k+1)} = 1$

12. $\sum_{k=0}^{\infty}\frac{1}{k(k+1)(k+2)} = \frac{1}{4}$

13. $\sum_{k=0}^{\infty}\frac{1}{k(k+1)(k+2)(k+3)} = \frac{1}{18}$

14. $\sum_{k=0}^{\infty}\frac{1}{k(k+1)\cdots(k+p)} = \frac{1}{p\cdot p!}$

15. The series $\displaystyle\sum_{k=3}^{\infty}\frac{1}{k\log k(\log\log k)^2}$ converges to 38.43... so slowly that it requires $10^{3.14\cdot10^{86}}$ terms to give two-decimal accuracy

16. The series $\displaystyle\sum_{k=3}^{\infty}\frac{1}{k\log k(\log\log k)}$ diverges, but the partial sums exceed 10 only after a googolplex of terms have appeared

17. $\sum_{k=1}^{\infty}\frac{(-1)^k k^2}{k^3+1} = \frac{1}{3}[1 - \ln 2 + \pi\operatorname{sech}(\pi\sqrt{3}/2)]$

1.3.10 INFINITE SERIES

1.3.10.1 Algebraic functions

$$(x + y)^n = x^n + \binom{n}{1}x^{n-1}y + \binom{n}{2}x^{n-2}y^2 + \ldots.$$

$$(1 \pm x)^n = 1 \pm \binom{n}{1}x + \binom{n}{2}x^2 \pm \binom{n}{3}x^3 + \ldots, \qquad (x^2 < 1).$$

$$(1 \pm x)^{-n} = 1 \mp \binom{n}{1}x + \binom{n+1}{2}x^2 \mp \binom{n+2}{3}x^3 + \ldots, \qquad (x^2 < 1).$$

$$\sqrt{1 + x} = 1 + \frac{1}{2}x - \frac{1}{8}x^2 + \frac{1}{16}x^3 - \frac{5}{128}x^4 + \ldots, \qquad (x^2 < 1).$$

$$(1+x)^{-1/2} = 1 - \frac{1}{2}x + \frac{3}{8}x^2 - \frac{5}{16}x^3 + \frac{35}{128}x^4 + \ldots, \qquad (x^2 < 1).$$

$$(1 \pm x)^{-1} = 1 \mp x + x^2 \mp x^3 + x^4 \mp x^5 + \ldots, \qquad (x^2 < 1).$$

$$(1 \pm x)^{-2} = 1 \mp 2x + 3x^2 \mp 4x^3 + 5x^4 \mp 6x^5 + \ldots, \qquad (x^2 < 1).$$

1.3.10.2 Exponential functions

$$e = 1 + \frac{1}{1!} + \frac{1}{2!} + \cdots + \frac{1}{n!} + \cdots .$$

$$e^x = 1 + \frac{x}{1!} + \frac{x^2}{2!} + \cdots + \frac{x^n}{n!} + \cdots , \qquad \text{(all real values of } x\text{)}$$

$$= \frac{1}{1-x} + \sum_{n=1}^{\infty} \frac{x^{n+1}}{n!(x-n)(n+1-x)} \qquad (x \text{ not a positive integer})$$

$$= e^a \left[1 + (x-a) + \frac{(x-a)^2}{2!} + \cdots + \frac{(x-a)^n}{n!} + \ldots \right] .$$

$$a^x = 1 + x \log_e a + \frac{(x \log_e a)^2}{2!} + \cdots + \frac{(x \log_e a)^n}{n!} + \cdots .$$

$$\text{(all real values of } x\text{)}$$

1.3.10.3 Logarithmic functions

$$\log x = \frac{x-1}{x} + \frac{1}{2}\left(\frac{x-1}{x}\right)^2 + \cdots + \frac{1}{n}\left(\frac{x-1}{x}\right)^n + \cdots , \qquad (x > 1/2),$$

$$= (x-1) - \frac{1}{2}(x-1)^2 + \frac{1}{3}(x-1)^3 - \ldots, \qquad (2 \ge x > 0),$$

$$= 2\left[\frac{x-1}{x+1} + \frac{1}{3}\left(\frac{x-1}{x+1}\right)^3 + \frac{1}{5}\left(\frac{x-1}{x+1}\right)^5 + \ldots\right], \qquad (x > 0).$$

$$= \log a + \frac{(x-a)}{a} - \frac{(x-a)^2}{2a^2} + \frac{(x-a)^3}{3a^2} - \ldots, \qquad (0 < x \le 2a).$$

$$\log(1+x) = x - \frac{x^2}{2} + \frac{x^3}{3} - \frac{x^4}{4} + \ldots, \qquad -1 < x \le 1.$$

$$\log(n+1) = \log(n-1) + 2\left[\frac{1}{n} + \frac{1}{3n^3} + \frac{1}{5n^5} + \ldots\right].$$

$$\log(a+x) = \log a + 2\left[\frac{x}{2a+x} + \frac{1}{3}\left(\frac{x}{2a+x}\right)^3 + \frac{1}{5}\left(\frac{x}{2a+x}\right)^5 + \ldots\right],$$

$$(a > 0, -a < x).$$

$$\log\frac{1+x}{1-x} = 2\left[x + \frac{x^3}{3} + \cdots + \frac{x^{2n-1}}{2n-1} + \ldots\right], \qquad (-1 < x < 1).$$

1.3.10.4 Trigonometric functions

$$\sin x = x - \frac{x^3}{3!} + \frac{x^5}{5!} - \frac{x^7}{7!} + \ldots \qquad \text{(all real values of } x\text{)}.$$

$$\cos x = 1 - \frac{x^2}{2!} + \frac{x^4}{4!} - \frac{x^6}{6!} + \ldots \qquad \text{(all real values of } x\text{)}.$$

$$\tan x = x + \frac{x^3}{3} + \frac{2x^5}{15} + \ldots 5! + \frac{(-1)^{n-1}2^{2n}(2^{2n}-1)B_{2n}}{(2n)!}x^{2n-1} + \ldots$$

$$(x^2 < \pi^2/4, \, B_n \text{ is the } n^{\text{th}} \text{ Bernoulli number}).$$

$$\cot x = \frac{1}{x} - \frac{x}{3} - \frac{x^3}{45} - \frac{2x^5}{945} - \frac{x^7}{4725} - \ldots + \frac{(-1)^{n+1}2^{2n}B_{2n}}{(2n)!}x^{2n-1} + \ldots$$

$$(x^2 < \pi^2, \, B_n \text{ is the } n^{\text{th}} \text{ Bernoulli number}).$$

$$\sec x = 1 + \frac{x^2}{2} + \frac{5}{24}x^4 + \frac{61}{720}x^6 + \frac{277}{8064}x^8 + \ldots + \frac{(-1)^n E_{2n}}{(2n)!}x^{2n} + \ldots$$

$$(x^2 < \pi^2/4, \, E_n \text{ is the } n^{\text{th}} \text{ Euler number}).$$

$$\csc x = \frac{1}{x} + \frac{x}{6} + \frac{7x^3}{360} + \frac{31x^5}{15120} + \ldots + \frac{(-1)^{n+1}2(2^{2n-1}-1)B_{2n}}{(2n)!}x^{2n-1} + \ldots$$

$$(|x| < \pi, \, B_n \text{ is the } n^{\text{th}} \text{ Bernoulli number}).$$

$$\log \sin x = \log x - \frac{x^2}{6} - \frac{x^4}{180} - \frac{x^6}{2835} - \ldots \qquad (x^2 < \pi^2).$$

$$\log \cos x = -\frac{x^2}{2} - \frac{x^4}{12} - \frac{x^6}{45} - \frac{17x^6}{2520} - \ldots \qquad (x^2 < \pi^2/4).$$

$$\log \tan x = \log x + \frac{x^2}{3} + \frac{7x^4}{90} + \frac{62x^6}{2835} + \ldots \qquad (x^2 < \pi^2/4).$$

$$e^{\sin x} = 1 + x + \frac{x^2}{2!} - \frac{3x^4}{4!} - \frac{8x^5}{5!} - \frac{3x^6}{6!} + \frac{56x^7}{7!} + \ldots$$

$$e^{\cos x} = e\left(1 - \frac{x^2}{2!} + \frac{4x^4}{4!} - \frac{31x^6}{6!} + \ldots\right).$$

$$e^{\tan x} = 1 + x + \frac{x^2}{2!} + \frac{3x^3}{3!} + \frac{9x^4}{4!} + \frac{37x^5}{5!} + \ldots \qquad (x^2 < \pi^2/4).$$

$$\sin x = \sin a + (x-a)\cos a - \frac{(x-a)^2}{2!}\sin a - \frac{(x-a)^3}{3!}\cos a + \ldots.$$

1.3.10.5 Inverse trigonometric functions

$$\sin^{-1} x = x + \frac{1}{2 \cdot 3} x^3 + \frac{1 \cdot 3}{2 \cdot 4 \cdot 5} x^5 + \frac{1 \cdot 3 \cdot 5}{2 \cdot 4 \cdot 6 \cdot 7} x^7 + \dots$$

$$(x^2 < 1, -\tfrac{\pi}{2} < \sin^{-1} x < \tfrac{\pi}{2}).$$

$$\cos^{-1} x = \frac{\pi}{2} - \left(x + \frac{1}{2 \cdot 3} x^3 + \frac{1 \cdot 3}{2 \cdot 4 \cdot 5} x^5 + \frac{1 \cdot 3 \cdot 5}{2 \cdot 4 \cdot 6 \cdot 7} x^7 + \dots \right)$$

$$(x^2 < 1, 0 < \cos^{-1} x < \pi).$$

$$\tan^{-1} x = x - \frac{x^3}{3} + \frac{x^5}{5} - \frac{x^7}{7} + \dots \qquad (x^2 < 1),$$

$$= \frac{\pi}{2} - \frac{1}{x} + \frac{1}{3x^3} - \frac{1}{5x^5} + \frac{1}{7x^7} - \dots \qquad (x > 1),$$

$$= -\frac{\pi}{2} - \frac{1}{x} + \frac{1}{3x^3} - \frac{1}{5x^5} + \frac{1}{7x^7} - \dots \qquad (x < -1).$$

$$\cot^{-1} x = \frac{\pi}{2} - x + \frac{x^3}{3} - \frac{x^5}{5} + \frac{x^7}{7} - \dots \qquad (x^2 < 1).$$

1.3.10.6 Hyperbolic functions

$$\sinh x = x + \frac{x^3}{3!} + \frac{x^5}{5!} + \frac{x^7}{7!} + \dots + \frac{x^{(2n+1)}}{(2n+1)!} + \dots .$$

$$\sinh ax = \frac{2}{\pi} \sinh \pi a \left[\frac{\sin x}{a^2 + 1^2} - \frac{2 \sin 2x}{a^2 + 2^2} + \frac{3 \sin 3x}{a^2 + 3^2} + \dots \right] \qquad (|x| < \pi).$$

$$\cosh x = 1 + \frac{x^2}{2!} + \frac{x^4}{4!} + \frac{x^6}{6!} + \dots + \frac{x^{2n}}{(2n)!} + \dots .$$

$$\cosh ax = \frac{2a}{\pi} \sinh \pi a \left[\frac{1}{2a^2} - \frac{\cos x}{a^2 + 1^2} + \frac{\cos 2x}{a^2 + 2^2} - \frac{\cos 3x}{a^2 + 3^2} + \dots \right] \qquad (|x| < \pi).$$

$$\tanh x = x - \frac{1}{3} x^3 + \frac{2}{15} x^5 - \dots + \frac{2^{2n}(2^{2n} - 1)B_{2n}}{(2n)!} x^{2n-1} + \dots \qquad (|x| < \pi/2),$$

$$= 1 - 2e^{-2x} + 2e^{-4x} - 2e^{-6x} + \dots \qquad (\text{Re } x > 0),$$

$$= 2x \left[\frac{1}{\left(\frac{\pi}{2}\right)^2 + x^2} + \frac{1}{\left(\frac{3\pi}{2}\right)^2 + x^2} + \frac{1}{\left(\frac{5\pi}{2}\right)^2 + x^2} + \dots \right].$$

$$\coth x = \frac{1}{x} + \frac{x}{3} - \frac{x^3}{45} + \frac{2x^5}{945} + \dots + \frac{2^{2n} B_{2n}}{(2n)!} x^{2n-1} + \dots \qquad (0 < |x| < \pi),$$

$$= 1 + 2e^{-2x} + 2e^{-4x} + 2e^{-6x} + \dots \qquad (\text{Re } x > 0),$$

$$= \frac{1}{x} + 2x \left[\frac{1}{\pi^2 + x^2} + \frac{1}{(2\pi)^2 + x^2} + \frac{1}{(3\pi)^2 + x^2} + \dots \right] \qquad (\text{Re } x > 0).$$

$$\operatorname{sech} x = 1 - \frac{1}{2!}x^2 + \frac{5}{4!}x^4 - \frac{61}{6!}x^6 + \cdots + \frac{E_{2n}}{(2n)!}x^{2n} + \ldots$$

$(|x| < \pi/2, E_n$ is the nth Euler number$)$,

$$= 2\left(e^{-x} - e^{-3x} + e^{-5x} - e^{-7x} + \ldots\right) \qquad (\operatorname{Re} x > 0),$$

$$= 4\pi\left[\frac{1}{\pi^2 + 4x^2} - \frac{3}{(3\pi)^2 + 4x^2} + \frac{5}{(5\pi)^2 + 4x^2} + \ldots\right].$$

$$\operatorname{csch} x = \frac{1}{x} - \frac{x}{6} + \frac{7x^3}{360} + \cdots + \frac{2(2^{2n-1}-1)B_{2n}}{(2n)!}x^{2n-1} + \ldots \qquad (0 < |x| < \pi),$$

$$= 2\left(e^{-x} + e^{-3x} + e^{-5x} + e^{-7x} + \ldots\right) \qquad (\operatorname{Re} x > 0),$$

$$= \frac{1}{x} - \frac{2x}{\pi^2 + x^2} + \frac{2x}{(2\pi)^2 + x^2} - \frac{2x}{(3\pi)^2 + x^2} + \ldots.$$

$$\sinh nu = \sinh u\left[(2\cosh u)^{n-1} - \frac{(n-2)}{1!}(2\cosh u)^{n-3}\right.$$

$$+ \frac{(n-3)(n-4)}{2!}(2\cosh u)^{n-5}$$

$$\left. - \frac{(n-4)(n-5)(n-6)}{3!}(2\cosh u)^{n-7} + \ldots\right].$$

$$\cosh nu = \frac{1}{2}\left[(2\cosh u)^n - \frac{n}{1!}(2\cosh u)^{n-2} + \frac{n(n-3)}{2!}(2\cosh u)^{n-4}\right.$$

$$\left. - \frac{n(n-4)(n-5)}{3!}(2\cosh u)^{n-6} + \ldots\right].$$

1.3.10.7 Inverse hyperbolic functions

$$\sinh^{-1} x = x - \frac{1}{2 \cdot 3}x^3 + \frac{1 \cdot 3}{2 \cdot 4 \cdot 5}x^5 - \frac{1 \cdot 3 \cdot 5}{2 \cdot 4 \cdot 6 \cdot 7}x^7 + \ldots \qquad (|x| < 1),$$

$$= \log(2x) + \frac{1}{2} \cdot \frac{1}{2x^2} + \frac{1 \cdot 3}{2 \cdot 4} \cdot \frac{1}{4x^4} + \frac{1 \cdot 3 \cdot 5}{2 \cdot 4 \cdot 6} \cdot \frac{1}{6x^6} + \ldots \qquad (|x| > 1).$$

$$\cosh^{-1} x = \pm\left[\log(2x) - \frac{1}{2} \cdot \frac{1}{2x^2} - \frac{1 \cdot 3}{2 \cdot 4} \cdot \frac{1}{4x^4} + \ldots\right] \qquad (x > 1).$$

$$\operatorname{csch}^{-1} x = \frac{1}{x} - \frac{1}{2} \cdot \frac{1}{3x^3} - \frac{1 \cdot 3}{2 \cdot 4} \cdot \frac{1}{5x^5} - \frac{1 \cdot 3 \cdot 5}{2 \cdot 4 \cdot 6} \cdot \frac{1}{7x^7} + \ldots \qquad (|x| > 1),$$

$$= \log\frac{2}{x} + \frac{1}{2} \cdot \frac{x^2}{2} - \frac{1 \cdot 3}{2 \cdot 4} \cdot \frac{x^4}{4} + \frac{1 \cdot 3 \cdot 5}{2 \cdot 4 \cdot 6} \cdot \frac{x^6}{6} - \ldots \qquad (0 < x < 1).$$

$$\operatorname{sech}^{-1} x = \log\frac{2}{x} - \frac{1}{2} \cdot \frac{x^2}{2} - \frac{1 \cdot 3}{2 \cdot 4} \cdot \frac{x^4}{4} - \frac{1 \cdot 3 \cdot 5}{2 \cdot 4 \cdot 6} \cdot \frac{x^6}{6} - \ldots \qquad (0 < x < 1).$$

$$\tanh^{-1} x = x + \frac{x^3}{3} + \frac{x^5}{5} + \frac{x^7}{7} + \cdots + \frac{x^{2n+1}}{2n+1} + \ldots \qquad (|x| < 1).$$

$$\coth^{-1} x = \frac{1}{x} + \frac{1}{3x^3} + \frac{1}{5x^5} + \frac{1}{7x^7} + \cdots + \frac{1}{(2n+1)x^{2n+1}} + \ldots \qquad (|x| > 1).$$

$$\operatorname{gd} x = x - \frac{1}{6}x^3 + \frac{1}{24}x^5 + \cdots + \frac{E_{2n}}{(2n+1)!}x^{2n+1} + \ldots \qquad (|x| < 1).$$

1.3.11 INFINITE PRODUCTS

For the sequence of complex numbers $\{a_k\}$, an infinite product is $\prod_{k=1}^{\infty}(1 + a_k)$. A necessary condition for convergence is that $\lim_{n\to\infty} a_n = 0$. A necessary and sufficient condition for convergence is that $\sum_{k=1}^{\infty} \log(1 + a_k)$ converges. Examples:

(a) $z! = \prod_{k=1}^{\infty} \dfrac{\left(1 + \frac{1}{k}\right)^z}{1 + \frac{z}{k}}$

(b) $\sin z = z \prod_{k=1}^{\infty} \cos \dfrac{z}{2^k}$

(c) $\sin(a + z) = (\sin a) \prod_{k=0,\pm 1,\pm 2,\dots}^{\infty} \left(1 + \dfrac{z}{a + k\pi}\right)$

(d) $\cos(a + z) = (\cos a) \prod_{k=\pm 1,\pm 3,\pm 5,\dots}^{\infty} \left(1 + \dfrac{2z}{2a + k\pi}\right)$

(e) $\sin \pi z = \pi z \prod_{k=1}^{\infty} \left(1 - \dfrac{z^2}{k^2}\right)$

(f) $\cos \pi z = \prod_{k=1}^{\infty} \left(1 - \dfrac{4z^2}{(2k-1)^2}\right)$

(g) $\sinh z = z \prod_{k=1}^{\infty} \left(1 + \dfrac{z^2}{k^2\pi^2}\right)$

(h) $\cosh z = \prod_{k=0}^{\infty} \left(1 + \dfrac{4z^2}{(2k+1)^2\pi^2}\right)$

1.3.11.1 Weierstrass theorem

Define $E(w, m) = (1 - w) \exp\left(w + \dfrac{w^2}{2} + \cdots + \dfrac{w^m}{m}\right)$. For $k = 1, 2, \dots$ let $\{b_k\}$ be a sequence of complex numbers such that $|b_k| \to \infty$. Then the infinite product

$$P(z) = \prod_{k=1}^{\infty} E\left(\dfrac{z}{b_k}, k\right)$$

is an entire function with zeros at b_k and at these points only. The multiplicity of the root at b_n is equal to the number of indices j such that $b_j = b_n$.

1.3.12 INFINITE PRODUCTS AND INFINITE SERIES

1. The Rogers–Ramanujan identities (for $a = 0$ or $a = 1$) are

$$1 + \sum_{k=1}^{\infty} \frac{q^{k^2+ak}}{(1-q)(1-q^2)\cdots(1-q^k)}$$

$$= \prod_{j=0}^{\infty} \frac{1}{(1 - q^{5j+a+1})(1 - q^{5j-a+4})}. \tag{1.3.8}$$

2. Jacobi's triple product identity is

$$\sum_{k=-\infty}^{\infty} q^{\binom{k}{2}} x^k = \prod_{j=1}^{\infty} (1 - q^j)(1 + x^{-1}q^j)(1 + xq^{j-1}). \qquad (1.3.9)$$

3. The quintuple product identity is

$$\sum_{k=-\infty}^{\infty} (-1)^k q^{(3k^2-k)/2} x^{3k} (1 + xq^k)$$

$$(1.3.10)$$

$$= \prod_{j=1}^{\infty} (1 - q^j)(1 + x^{-1}q^j)(1 + xq^{j-1})(1 + x^{-2}q^{2j-1})(1 + x^2 q^{2j-1}).$$

1.4 FOURIER SERIES

If $f(x)$ is a bounded periodic function of period $2L$ (that is, $f(x + 2L) = f(x)$) and satisfies the Dirichlet conditions,

1. In any period, $f(x)$ is continuous, except possibly for a finite number of jump discontinuities.
2. In any period $f(x)$ has only a finite number of maxima and minima.

Then $f(x)$ may be represented by the Fourier series,

$$f(x) = \frac{a_0}{2} + \sum_{n=1}^{\infty} \left(a_n \cos \frac{n\pi x}{L} + b_n \sin \frac{n\pi x}{L} \right), \qquad (1.4.1)$$

where $\{a_n\}$ and $\{b_n\}$ are determined as follows:

$$a_n = \frac{1}{L} \int_{\alpha}^{\alpha+2L} f(x) \cos \frac{n\pi x}{L} \, dx \qquad \text{for } n = 0, 1, 2, \ldots,$$

$$= \frac{1}{L} \int_{0}^{2L} f(x) \cos \frac{n\pi x}{L} \, dx, \qquad (1.4.2)$$

$$= \frac{1}{L} \int_{-L}^{L} f(x) \cos \frac{n\pi x}{L} \, dx;$$

$$b_n = \frac{1}{L} \int_{\alpha}^{\alpha+2L} f(x) \sin \frac{n\pi x}{L} \, dx \qquad \text{for } n = 1, 2, 3, \ldots,$$

$$= \frac{1}{L} \int_{0}^{2L} f(x) \sin \frac{n\pi x}{L} \, dx, \qquad (1.4.3)$$

$$= \frac{1}{L} \int_{-L}^{L} f(x) \sin \frac{n\pi x}{L} \, dx,$$

where α is any real number (the second and third lines of each formula represent $\alpha = 0$ and $\alpha = -L$ respectively).

The series in Equation (1.4.1) will converge (in the Cesaro sense) to every point where $f(x)$ is continuous, and to $\dfrac{f(x^+) + f(x^-)}{2}$ (i.e., the average of the left hand and right hand limits) at every point where $f(x)$ has a jump discontinuity.

1.4.1 SPECIAL CASES

1. If, in addition to the Dirichlet conditions in Section 1.4, $f(x)$ is an even function (i.e., $f(x) = f(-x)$), then the Fourier series becomes

$$f(x) = \frac{a_0}{2} + \sum_{n=1}^{\infty} a_n \cos \frac{n\pi x}{L}. \qquad (1.4.4)$$

That is, every $b_n = 0$. In this case, the $\{a_n\}$ may be determined from

$$a_n = \frac{2}{L} \int_0^L f(x) \cos \frac{n\pi x}{L} \, dx \qquad n = 0, 1, 2, \ldots. \qquad (1.4.5)$$

If, in addition to the above requirements, $f(x) = -f(L - x)$, then a_n will be zero for all even values of n. In this case the expansion becomes

$$f(x) = \sum_{m=1}^{\infty} a_{2m-1} \cos \frac{(2m-1)\pi x}{L}. \qquad (1.4.6)$$

2. If, in addition to the Dirichlet conditions in Section 1.4, $f(x)$ is an odd function (i.e., $f(x) = -f(-x)$), then the Fourier series becomes

$$f(x) = \sum_{n=1}^{\infty} b_n \sin \frac{n\pi x}{L}. \qquad (1.4.7)$$

That is, every $a_n = 0$. In this case, the $\{b_n\}$ may be determined from

$$b_n = \frac{2}{L} \int_0^L f(x) \sin \frac{n\pi x}{L} \, dx \qquad n = 1, 2, 3, \ldots. \qquad (1.4.8)$$

If, in addition to the above requirements, $f(x) = f(L - x)$, then b_n will be zero for all even values of n. In this case the expansion becomes

$$f(x) = \sum_{m=1}^{\infty} b_{2m-1} \sin \frac{(2m-1)\pi x}{L}. \qquad (1.4.9)$$

The series in Equation (1.4.6) and Equation (1.4.9) are known as odd harmonic series, since only the odd harmonics appear. Similar rules may be stated for even harmonic series, but when a series appears in even harmonic form, it means that $2L$

has not been taken to be the smallest period of $f(x)$. Since any integral multiple of a period is also a period, series obtained in this way will also work, but, in general, computation is simplified if $2L$ is taken as the least period.

Writing the trigonometric functions in terms of complex exponentials, we obtain the complex form of the Fourier series known as the *complex Fourier series* or as the *exponential Fourier series*. It is represented by

$$f(x) = \sum_{n=-\infty}^{\infty} c_n e^{i\omega_n x} \tag{1.4.10}$$

where $\omega_n = \dfrac{n\pi}{L}$ for $n = 0, \pm 1, \pm 2, \ldots$ and the $\{c_n\}$ are determined from

$$c_n = \frac{1}{2L} \int_{-L}^{L} f(x) e^{-i\omega_n x} \, dx. \tag{1.4.11}$$

The set of coefficients $\{c_n\}$ is often referred to as the *Fourier spectrum*.

1.4.2 ALTERNATE FORMS

The Fourier series in Equation (1.4.1) may be represented in the alternate forms:

1. When $\phi_n = \tan^{-1}(-a_n/b_n)$, $a_n = c_n \sin \phi_n$, $b_n = -c_n \cos \phi_n$, and $c_n = \sqrt{a_n^2 + b_n^2}$, then

$$f(x) = \frac{a_0}{2} + \sum_{n=1}^{\infty} c_n \sin \left(\frac{n\pi x}{L} + \phi_n \right). \tag{1.4.12}$$

2. When $\phi_n = \tan^{-1}(a_n/b_n)$, $a_n = c_n \sin \phi_n$, $b_n = c_n \cos \phi_n$, and $c_n = \sqrt{a_n^2 + b_n^2}$, then

$$f(x) = \frac{a_0}{2} + \sum_{n=1}^{\infty} c_n \cos \left(\frac{n\pi x}{L} + \phi_n \right). \tag{1.4.13}$$

1.4.3 USEFUL SERIES

(a) $1 = \dfrac{4}{\pi} \left[\sin \dfrac{\pi x}{L} + \dfrac{1}{3} \sin \dfrac{3\pi x}{L} + \dfrac{1}{5} \sin \dfrac{5\pi x}{L} + \ldots \right]$ $\qquad (0 < x < L)$

(b) $x = \dfrac{2L}{\pi} \left[\sin \dfrac{\pi x}{L} - \dfrac{1}{2} \sin \dfrac{2\pi x}{L} + \dfrac{1}{3} \sin \dfrac{3\pi x}{L} + \ldots \right]$ $\qquad (-L < x < L)$

(c) $x = \dfrac{L}{2} - \dfrac{4L}{\pi^2} \left[\cos \dfrac{\pi x}{L} + \dfrac{1}{3^2} \cos \dfrac{3\pi x}{L} + \dfrac{1}{5^2} \cos \dfrac{5\pi x}{L} + \ldots \right]$ $\qquad (0 < x < L)$

(d) $x^2 = \dfrac{2L^2}{\pi^3}\left[\left(\dfrac{\pi^2}{1}-\dfrac{4}{1}\right)\sin\dfrac{\pi x}{L}-\dfrac{\pi^2}{2}\sin\dfrac{2\pi x}{L}+\left(\dfrac{\pi^2}{3}-\dfrac{4}{3^3}\right)\sin\dfrac{3\pi x}{L}\right.$

$\left.-\dfrac{\pi^2}{4}\sin\dfrac{4\pi x}{L}+\left(\dfrac{\pi^2}{5}-\dfrac{4}{5^3}\right)\sin\dfrac{5\pi x}{L}\cdots\right]\quad(0<x<L)$

(e) $x^2 = \dfrac{L^2}{3}-\dfrac{4L^2}{\pi^2}\left[\cos\dfrac{\pi x}{L}-\dfrac{1}{2^2}\cos\dfrac{2\pi x}{L}+\dfrac{1}{3^2}\cos\dfrac{3\pi x}{L}-\dfrac{1}{4^2}\cos\dfrac{4\pi x}{L}+\cdots\right]$

$(-L<x<L)$

(f) $\dfrac{\pi}{4}=1-\dfrac{1}{3}+\dfrac{1}{5}-\dfrac{1}{7}+\cdots$

(g) $\dfrac{\pi^2}{6}=1+\dfrac{1}{2^2}+\dfrac{1}{3^2}+\dfrac{1}{4^2}+\cdots=2\left(1-\dfrac{1}{2^2}+\dfrac{1}{3^2}-\dfrac{1}{4^2}+\cdots\right)$

(h) $\dfrac{\pi^2}{8}=1+\dfrac{1}{3^2}+\dfrac{1}{5^2}+\dfrac{1}{7^2}+\cdots=3\left(\dfrac{1}{2^2}+\dfrac{1}{4^2}+\dfrac{1}{6^2}+\dfrac{1}{8^2}+\cdots\right)$

1.4.4 EXPANSIONS OF BASIC PERIODIC FUNCTIONS

(a) $f(x)=\dfrac{c}{L}+\dfrac{2}{\pi}\displaystyle\sum_{n=1}^{\infty}\dfrac{(-1)^n}{n}\sin\dfrac{n\pi c}{L}\cos\dfrac{n\pi x}{L}$

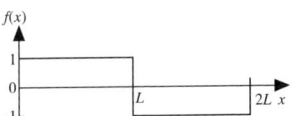

(b) $f(x)=\dfrac{4}{\pi}\displaystyle\sum_{n=1,3,5,\dots}\dfrac{1}{n}\sin\dfrac{n\pi x}{L}$

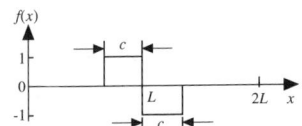

(c) $f(x)=\dfrac{2}{\pi}\displaystyle\sum_{n=1}^{\infty}\dfrac{(-1)^n}{n}\left(\cos\dfrac{n\pi c}{L}-1\right)\sin\dfrac{n\pi x}{L}$

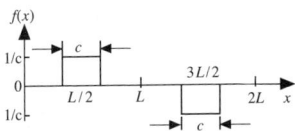

(d) $f(x)=\dfrac{2}{L}\displaystyle\sum_{n=1}^{\infty}\sin\dfrac{n\pi}{2}\dfrac{\sin(n\pi c/2L)}{n\pi c/2L}\sin\dfrac{n\pi x}{L}$

(e) $f(x)=\dfrac{4}{\pi}\displaystyle\sum_{n=1}^{\infty}\dfrac{1}{n}\sin\dfrac{n\pi}{4}\sin n\pi a\sin\dfrac{n\pi x}{L}\qquad\left(a=\dfrac{c}{2L}\right)$

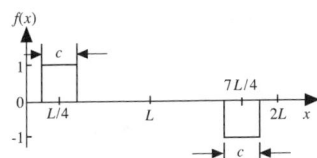

(f) $f(x) = \dfrac{1}{2} - \dfrac{1}{\pi} \displaystyle\sum_{n=1}^{\infty} \dfrac{1}{n} \sin \dfrac{n\pi x}{L}$

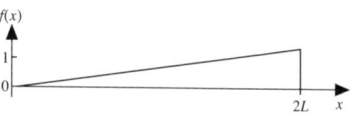

(g) $f(x) = \dfrac{1}{2} - \dfrac{4}{\pi^2} \displaystyle\sum_{n=1,3,5,\ldots} \dfrac{1}{n^2} \cos \dfrac{n\pi x}{L}$

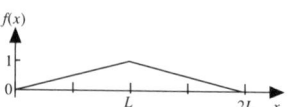

(h) $f(x) = \dfrac{1+a}{2} + \dfrac{2}{\pi^2(1-a)} \displaystyle\sum_{n=1}^{\infty} \dfrac{1}{n^2} \left[(-1)^n \cos n\pi a - 1\right] \cos \dfrac{n\pi x}{L}$ $\left(a = \dfrac{c}{2L}\right)$

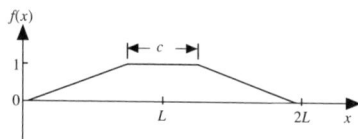

(i) $f(x) = \dfrac{1}{2} - \dfrac{4}{\pi^2(1-2a)} \displaystyle\sum_{n=1,3,5,\ldots} \dfrac{1}{n^2} \cos n\pi a \cos \dfrac{n\pi x}{L}$ $\left(a = \dfrac{c}{2L}\right)$

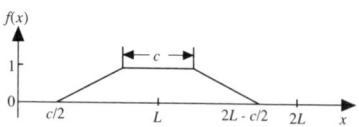

(j) $f(x) = \dfrac{2}{\pi} \displaystyle\sum_{n=1}^{\infty} \dfrac{(-1)^{n-1}}{n} \left[1 + \dfrac{\sin n\pi a}{n\pi(1-a)}\right] \sin \dfrac{n\pi x}{L}$ $\left(a = \dfrac{c}{2L}\right)$

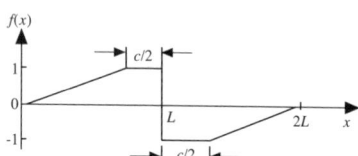

(k) $f(x) = \dfrac{2}{\pi} \displaystyle\sum_{n=1}^{\infty} \dfrac{(-1)^n}{n} \left[1 + \dfrac{1+(-1)^n}{n\pi(1-2a)} \sin n\pi a\right] \sin \dfrac{n\pi x}{L}$ $\left(a = \dfrac{c}{2L}\right)$

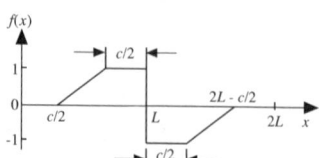

(l) $f(x) = \dfrac{2}{\pi} \displaystyle\sum_{n=1}^{\infty} \dfrac{(-1)^{n+1}}{n} \sin \dfrac{n\pi x}{L}$

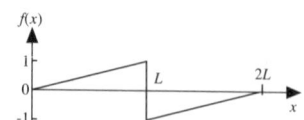

(m) $f(x) = \dfrac{8}{\pi^2} \displaystyle\sum_{n=1,3,5,\ldots} \dfrac{(-1)^{(n-1)/2}}{n^2} \sin \dfrac{n\pi x}{L}$

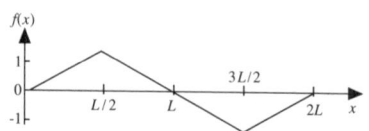

(n) $f(x) = \dfrac{9}{\pi^2} \displaystyle\sum_{n=1}^{\infty} \dfrac{1}{n^2} \sin \dfrac{n\pi}{3} \sin \dfrac{n\pi x}{L}$

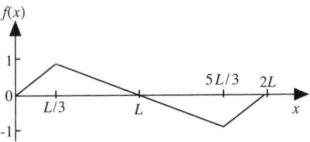

(o) $f(x) = \dfrac{32}{3\pi^2} \displaystyle\sum_{n=1}^{\infty} \dfrac{1}{n^2} \sin \dfrac{n\pi}{4} \sin \dfrac{n\pi x}{L}$

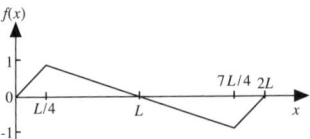

(p) $f(x) = \dfrac{1}{\pi} + \dfrac{1}{2} \sin \omega t - \dfrac{2}{\pi} \displaystyle\sum_{n=2,4,6,\ldots} \dfrac{1}{n^2 - 1} \cos n\omega t$

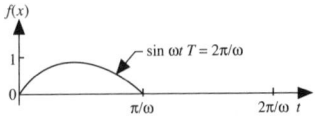

1.5 COMPLEX ANALYSIS

1.5.1 DEFINITIONS

A complex number z has the form $z = x + iy$ where x and y are real numbers, and $i = \sqrt{-1}$; the number i is sometimes called the imaginary unit. We write $x = \operatorname{Re} z$ and $y = \operatorname{Im} z$. The number x is called the *real* part of z and y is called the *imaginary* part of z. This form is also called the *Cartesian form* of the complex number.

Complex numbers can also be written in *polar form*, $z = r e^{i\theta}$, where r, called the *modulus*, is given by $r = |z| = \sqrt{x^2 + y^2}$, and θ is called the *argument*: $\theta = \arg z = \tan^{-1} \frac{y}{x}$ (when $x > 0$). The geometric relationship between Cartesian and polar forms is shown below

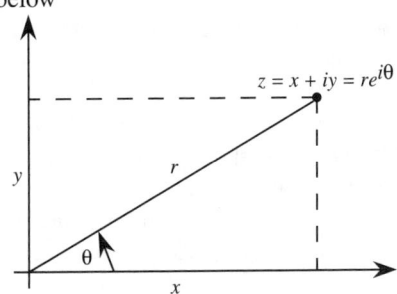

The *complex conjugate* of z, denoted $\bar{z}$, is defined as $\bar{z} = x - iy = re^{-i\theta}$. Note that $|z| = |\bar{z}|$, $\arg \bar{z} = -\arg z$, and $|z| = \sqrt{z\bar{z}}$. In addition, $\bar{\bar{z}} = z$, $\overline{z_1 + z_2} = \bar{z}_1 + \bar{z}_2$, and $\overline{z_1 z_2} = \bar{z}_1 \bar{z}_2$.

1.5.2 OPERATIONS ON COMPLEX NUMBERS

1. Addition and subtraction:

$$z_1 \pm z_2 = (x_1 + iy_1) \pm (x_2 + iy_2) = (x_1 \pm x_2) + i(y_1 \pm y_2).$$

2. Multiplication:

$$z_1 z_2 = (x_1 + iy_1)(x_2 + iy_2) = (x_1 x_2 - y_1 y_2) + i(x_1 y_2 + x_2 y_1) = r_1 r_2 e^{i(\theta_1 + \theta_2)}.$$
$$|z_1 z_2| = |z_1||z_2|, \qquad \arg(z_1 z_2) = \arg z_1 + \arg z_2 = \theta_1 + \theta_2.$$

3. Division:

$$\frac{z_1}{z_2} = \frac{z_1 \bar{z}_2}{z_2 \bar{z}_2} = \frac{(x_1 x_2 + y_1 y_2) + i(x_2 y_1 - x_1 y_2)}{x_2^2 + y_2^2} = \frac{r_1}{r_2} e^{i(\theta_1 - \theta_2)}.$$
$$\left|\frac{z_1}{z_2}\right| = \frac{|z_1|}{|z_2|}, \qquad \arg\left(\frac{z_1}{z_2}\right) = \arg z_1 - \arg z_2 = \theta_1 - \theta_2.$$

4. Powers:
$$z^n = r^n e^{in\theta} = r^n(\cos n\theta + i \sin n\theta) \qquad \textit{DeMoivre's Theorem.}$$

5. Roots:

$$z^{1/n} = r^{1/n} e^{i(\theta + 2k\pi)/n} = r^{1/n}\left(\cos\frac{\theta + 2k\pi}{n} + i\sin\frac{\theta + 2k\pi}{n}\right),$$

for $k = 0, 1, \ldots, n - 1$. The *principal root* has $-\pi < \theta \le \pi$ and $k = 0$.

1.5.3 FUNCTIONS OF A COMPLEX VARIABLE

A complex function

$$w = f(z) = u(x, y) + iv(x, y) = |w|e^{i\phi},$$

where $z = x + iy$, associates one or more values of the complex dependent variable w with each value of the complex independent variable z for those values of z in a given domain.

A function $f(z)$ is said to be *analytic* (or *holomorphic*) at a point z_0 if $f(z)$ is defined in each point z of a disc with positive radius R around z_0, h is any complex number with $|h| < R$, and the limit of $[f(z_0+h)-f(z_0)]/h$ exists and is independent of the mode in which h tends to zero. This limiting value is the *derivative* of $f(z)$ at z_0 denoted by $f'(z_0)$. A function is called *analytic* in a connected domain if it is analytic at every point in that domain.

A function is called *entire* if it is analytic in $\mathbb{C}$.

Liouville's theorem: A bounded entire function is constant.

EXAMPLES

1. $f(z) = z^n$ is analytic everywhere when n is a non-negative integer. If n is a negative integer, then $f(z)$ is analytic except at the origin.

2. $f(z) = \bar{z}$ is nowhere analytic.

3. $f(z) = e^z$ is analytic everywhere.

1.5.4 CAUCHY–RIEMANN EQUATIONS

A necessary and sufficient condition for $f(z) = u(x, y) + iv(x, y)$ to be analytic is that it satisfies the Cauchy–Riemann equations,

$$\frac{\partial u}{\partial x} = \frac{\partial v}{\partial y}, \text{ and } \frac{\partial u}{\partial y} = -\frac{\partial v}{\partial x}. \tag{1.5.1}$$

1.5.5 CAUCHY INTEGRAL THEOREM

If $f(z)$ is analytic at all points within and on a simple closed curve C, then

$$\int_C f(z) \, dz = 0. \tag{1.5.2}$$

1.5.6 CAUCHY INTEGRAL FORMULA

If $f(z)$ is analytic inside and on a simple closed contour C and if z_0 is interior to C, then

$$f(z_0) = \frac{1}{2\pi i} \int_C \frac{f(z)}{z - z_0} \, dz. \tag{1.5.3}$$

Moreover, since the derivatives $f'(z_0), f''(z_0), \ldots$ of all orders exist, then

$$f^{(n)}(z_0) = \frac{n!}{2\pi i} \int_C \frac{f(z)}{(z - z_0)^{n+1}} \, dz. \tag{1.5.4}$$

1.5.7 TAYLOR SERIES EXPANSIONS

If $f(z)$ is analytic inside of and on a circle C of radius r centered at the point z_0, then a unique and uniformly convergent series expansion exists in powers of $(z - z_0)$ of the form

$$f(z) = \sum_{n=0}^{\infty} a_n (z - z_0)^n, \quad |z - z_0| < r, \quad z_0 \neq \infty, \tag{1.5.5}$$

where

$$a_n = \frac{1}{n!} f^{(n)}(z_0) = \frac{1}{2\pi i} \int_C \frac{f(z)}{(z - z_0)^{n+1}} \, dz. \tag{1.5.6}$$

If $M(r)$ is an upper bound of $|f(z)|$ on C, then

$$|a_n| = \frac{1}{n!}|f^{(n)}(z_0)| \leq \frac{M(r)}{r^n} \quad \text{(Cauchy's inequality)}. \quad (1.5.7)$$

If the series is truncated with the term $a_n(z - z_0)^n$, the remainder $R_n(z)$ is given by

$$R_n(z) = \frac{(z - z_0)^{n+1}}{2\pi i} \int_C \frac{f(s)}{(s - z)(s - z_0)^{n+1}} \, ds, \quad (1.5.8)$$

and

$$|R_n(z)| \leq \left(\frac{|z - z_0|}{r}\right)^n \frac{rM(r)}{r - |z - z_0|}. \quad (1.5.9)$$

1.5.8 LAURENT SERIES EXPANSIONS

If $f(z)$ is analytic inside the annulus between the concentric circles C_1 and C_2 centered at z_0 with radii r_1 and r_2 ($r_1 < r_2$), respectively, then a unique series expansion exists in terms of positive and negative powers of $z - z_0$ of the following form:

$$f(z) = \sum_{n=1}^{\infty} b_n(z - z_0)^{-n} + \sum_{n=0}^{\infty} a_n(z - z_0)^n \quad (1.5.10)$$

$$= \cdots + \frac{b_2}{(z - z_0)^2} + \frac{b_1}{z - z_0} + a_0 + a_1(z - z_0) + a_2(z - z_0)^2 + \ldots$$

with (here C is a contour between C_1 and C_2)

$$a_n = \frac{1}{2\pi i} \int_C \frac{f(s)}{(s - z_0)^{n+1}} \, ds, \quad n = 0, 1, 2, \ldots,$$

$$b_n = \frac{1}{2\pi i} \int_C f(s)(s - z_0)^{n-1} \, ds, \quad n = 1, 2, 3, \ldots. \quad (1.5.11)$$

Equation (1.5.10) is often written in the form

$$f(z) = \sum_{n=-\infty}^{\infty} c_n(z - z_0)^n \quad \text{for } r_1 < |z - z_0| < r_2 \quad (1.5.12)$$

with

$$c_n = \frac{1}{2\pi i} \int_C \frac{f(z)}{(z - z_0)^{n+1}} \, dz \quad \text{for } n = 0, \pm 1, \pm 2, \ldots. \quad (1.5.13)$$

1.5.9 ZEROS AND SINGULARITIES

The points z for which $f(z) = 0$ are called *zeros* of $f(z)$. A function $f(z)$ which is analytic at z_0 has a zero of *order* m there, where m is a positive integer, if and only if the first m coefficients $a_0, a_1, \ldots, a_{m-1}$ in the Taylor expansion at z_0 vanish.

A *singular point* or *singularity* of the function $f(z)$ is any point at which $f(z)$ is not analytic. An *isolated singularity* of $f(z)$ at z_0 may be classified in one of three ways:

1. A *removable* singularity if and only if all coefficients b_n in the Laurent series expansion of $f(z)$ at z_0 vanish. This implies that $f(z)$ can be analytically extended to z_0.

2. A *pole* of order m if and only if $(z - z_0)^m f(z)$, but not $(z - z_0)^{m-1} f(z)$, is analytic at z_0 (i.e., if and only if $b_m \neq 0$ and $0 = b_{m+1} = b_{m+2} = \ldots$ in the Laurent series expansion of $f(z)$ at z_0). Equivalently, $f(z)$ has a pole of order m if $1/f(z)$ is analytic at z_0 and has a zero of order m there.

3. An isolated *essential singularity* if and only if the Laurent series expansion of $f(z)$ at z_0 has an infinite number of terms involving negative powers of $z - z_0$.

Theorems:

Riemann removable singularity theorem   Suppose that a function f is analytic and bounded in some deleted neighborhood $0 < |z - z_0| < \epsilon$ of a point z_0. If f is not analytic at z_0, then it has a removable singularity there.

Casorati–Weierstrass theorem   Suppose that z_0 is an essential singularity of a function f, and let w be an arbitrary complex number. Then, for any $\epsilon > 0$, the inequality $|f(z) - w| < \epsilon$ is satisified at some point z in each deleted neighborhood $0 < |z - z_0| < \delta$ of z_0.

1.5.10   RESIDUES

Given a point z_0 where $f(z)$ is either analytic or has an isolated singularity, the *residue* of $f(z)$ is the coefficient of $(z - z_0)^{-1}$ in the Laurent series expansion of $f(z)$ at z_0, or

$$\text{Res}(z_0) = b_1 = \frac{1}{2\pi i} \int_C f(z)\,dz. \tag{1.5.14}$$

If $f(z)$ is either analytic or has a removable singularity at z_0, then $b_1 = 0$ there. If z_0 is a pole of order m, then

$$b_1 = \frac{1}{(m-1)!} \frac{d^{m-1}}{dz^{m-1}} [(z - z_0)^m f(z)] \Big|_{z=z_0}. \tag{1.5.15}$$

For every simple closed contour C enclosing at most a finite number of singularities $z_1, z_2, \ldots, z_n$ of a function analytic in a neighborhood of C,

$$\int_C f(z)\,dz = 2\pi i \sum_{k=1}^{n} \text{Res}(z_k), \tag{1.5.16}$$

where $\text{Res}(z_k)$ is the residue of $f(z)$ at z_k.

1.5.11   THE ARGUMENT PRINCIPLE

Let $f(z)$ be analytic on a simple closed curve C with no zeros on C and analytic everywhere inside C except possibly at a finite number of poles. Let $\Delta_C \arg f(z)$ denote the change in the argument of $f(z)$ (that is, final value $-$ initial value) as z transverses the curve once in the positive sense. Then

$$\frac{1}{2\pi} \Delta_C \arg f(z) = \frac{1}{2\pi i} \int_C \frac{f'(z)}{f(z)} \, dz = N - P, \qquad (1.5.17)$$

where N is number of zeros of $f(z)$ inside C, and P is the number of poles inside C. The zeros and poles are counted according to their multiplicities.

1.5.12   TRANSFORMATIONS AND MAPPINGS

A function $w = f(z) = u(z) + iv(z)$ maps points of the z-plane into corresponding points of the w-plane. At every point z such that $f(z)$ is analytic and $f'(z) \neq 0$, the mapping is *conformal*, i.e., the angle between two curves in the z-plane through such a point is equal in magnitude and sense to the angle between the corresponding curves in the w-plane. A table giving real and imaginary parts, zeros, and singularities for frequently used functions of a complex variable and a table illustrating a number of special transformations of interest are at the end of this section.

A function is said to be *simple* in a domain D if it is analytic in D and assumes no value more than once in D. *Riemann's mapping theorem* states:

> If D is a simply-connected domain in the complex z plane, which is neither the z plane nor the extended z plane, then there is a simple function $f(z)$ such that $w = f(z)$ maps D onto the disc $|w| < 1$.

1.5.12.1   Bilinear transformations

The *bilinear* transformation is defined by $w = \dfrac{az + b}{cz + d}$, where a, b, c, and d are complex numbers and $ad \neq bc$. It is also known as the *linear fractional* transformation. The bilinear transformation is defined for all $z \neq -d/c$. The bilinear transformation is conformal and maps circles and lines onto circles and lines.

The inverse transformation is given by $z = \dfrac{-dw + b}{cw - a}$, which is also a bilinear transformation. Note that $w \neq a/c$.

The *cross ratio* of four distinct complex numbers z_k (for $k = 1, 2, 3, 4$) is given by

$$(z_1, z_2, z_3, z_4) = \frac{(z_1 - z_2)(z_3 - z_4)}{(z_1 - z_4)(z_3 - z_2)}.$$

If any of the z_k's is complex infinity, the cross ratio is redefined so that the quotient of the two terms on the right containing z_k is equal to 1. Under the bilinear transformation, the cross ratio of four points is invariant: $(w_1, w_2, w_3, w_4) = (z_1, z_2, z_3, z_4)$.

The *Möbius* transformation is special case of the bilinear transformation; it is defined by $w(z) = \frac{z - a}{1 - \bar{a}z}$ where a is a complex constant of modulus less than 1. It maps the unit disk onto itself.

1.5.12.2 Table of transformations

$f(z) = w(x,y)$	$u(x,y) = \operatorname{Re} w(x,y)$	$v(x,y) = \operatorname{Im} w(x,y)$	Zeros (and order m)	Singularities (and order m)
z	x	y	$z = 0,\, m = 1$	Pole ($m = 1$) at $z = \infty$
z^2	$x^2 - y^2$	$2xy$	$z = 0,\, m = 2$	Pole ($m = 2$) at $z = \infty$
$\dfrac{1}{z}$	$\dfrac{x}{x^2 + y^2}$	$\dfrac{-y}{x^2 + y^2}$	$z = \infty,\, m = 1$	Pole ($m = 1$) at $z = 0$
$\dfrac{1}{z^2}$	$\dfrac{x^2 - y^2}{(x^2 + y^2)^2}$	$\dfrac{-2xy}{(x^2 + y^2)^2}$	$z = \infty,\, m = 2$	Pole ($m = 2$) at $z = 0$
$z - (a + ib)$ a, b real	$\dfrac{x - a}{(x - a)^2 + (y - b)^2}$	$\dfrac{-(y - b)}{(x - a)^2 + (y - b)^2}$	$z = \infty,\, m = 1$	Pole ($m = 1$) at $z = a + ib$
$\sqrt{z}$	$\pm\left(\dfrac{x + \sqrt{x^2 + y^2}}{2}\right)^{1/2}$	$\pm\left(\dfrac{-x + \sqrt{x^2 + y^2}}{2}\right)^{1/2}$	$z = 0,\, m = 1$	Branch point ($m = 1$) at $z = 0$ Branch point ($m = 1$) at $z = \infty$
e^z	$e^x \cos y$	$e^x \sin y$	None	Essential singularity at $z = \infty$
$\sin z$	$\sin x \cosh y$	$\cos x \sinh y$	$z = k\pi,\, m = 1$ $(k = 0, \pm 1, \pm 2, \ldots)$	Essential singularity at $z = \infty$
$\cos z$	$\cos x \cosh y$	$-\sin x \sinh y$	$z = (k + 1/2)\pi,\, m = 1$ $(k = 0, \pm 1, \pm 2, \ldots)$	Essential singularity at $z = \infty$
$\sinh z$	$\sinh x \cos y$	$\cosh x \sin y$	$z = k\pi i,\, m = 1$ $(k = 0, \pm 1, \pm 2, \ldots)$	Essential singularity at $z = \infty$
$\cosh z$	$\cosh x \cos y$	$\sinh x \sin y$	$z = (k + 1/2)\pi i,\, m = 1$ $(k = 0, \pm 1, \pm 2, \ldots)$	Essential singularity at $z = \infty$
$\tan z$	$\dfrac{\sin 2x}{\cos 2x + \cosh 2y}$	$\dfrac{\sinh 2y}{\cos 2x + \cosh 2y}$	$z = k\pi,\, m = 1$ $(k = 0, \pm 1, \pm 2, \ldots)$	Essential singularity at $z = \infty$ Poles ($m = 1$) at $z = (k + 1/2)\pi$ $(k = 0, \pm 1, \pm 2, \ldots)$
$\tanh z$	$\dfrac{\sinh 2x}{\cosh 2x + \cos 2y}$	$\dfrac{\sin 2y}{\cosh 2x + \cos 2y}$	$z = k\pi i,\, m = 1$ $(k = 0, \pm 1, \pm 2, \ldots)$	Essential singularity at $z = \infty$ Poles ($m = 1$) at $z = (k + 1/2)\pi i$ $(k = 0, \pm 1, \pm 2, \ldots)$
$\log z$	$\tfrac{1}{2}\log(x^2 + y^2)$	$\tan^{-1}\dfrac{x}{y} + 2k\pi$ $(k = 0, \pm 1, \pm 2, \ldots)$	$z = 1,\, m = 1$	Branch points at $z = 0,\, z = \infty$

1.5.12.3   Table of conformal mappings

In the following functions $z = x + iy$ and $w = u + iv = \rho e^{i\phi}$.

1. 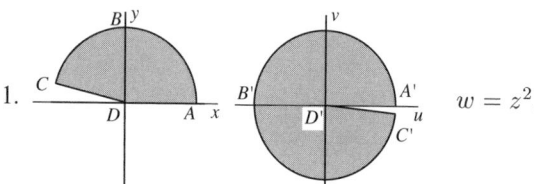 $w = z^2$.

2. 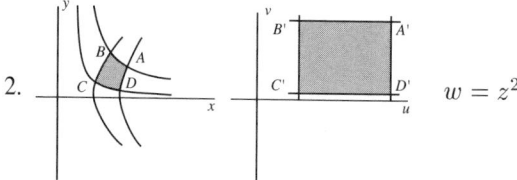 $w = z^2$.

3. 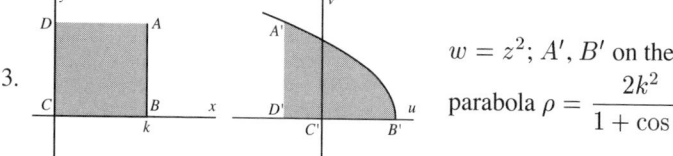 $w = z^2$; A', B' on the parabola $\rho = \dfrac{2k^2}{1 + \cos \phi}$.

4. 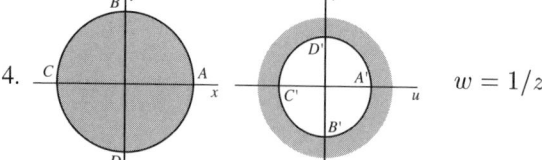 $w = 1/z$.

5. 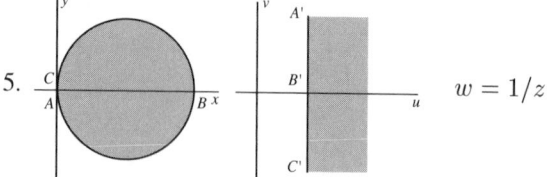 $w = 1/z$.

6. 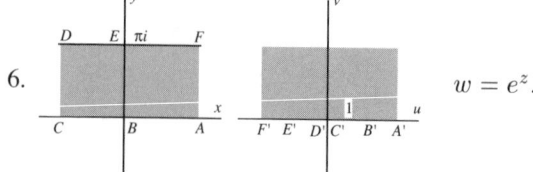 $w = e^z$.

7. 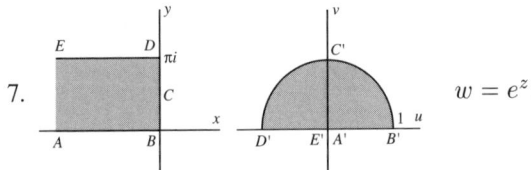 $w = e^z$.

8. 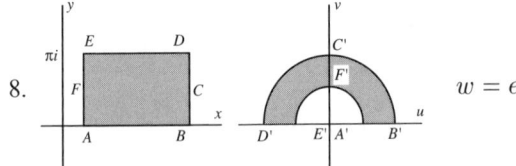 $w = e^z$.

9. 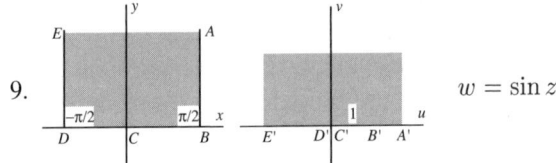 $w = \sin z$.

10. 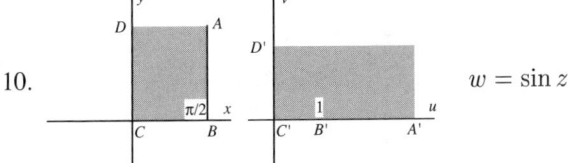 $w = \sin z$.

11. 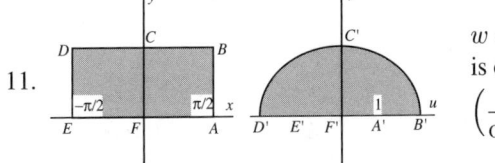 $w = \sin z$; BCD: $y = k$, $B'C'D'$ is on the ellipse
$$\left(\frac{u}{\cosh k}\right)^2 + \left(\frac{v}{\sinh k}\right)^2 = 1.$$

12. 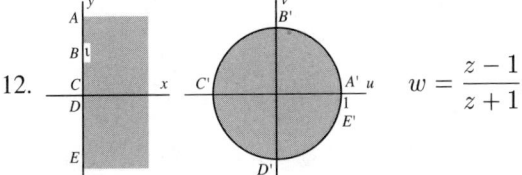 $w = \dfrac{z-1}{z+1}$.

13. 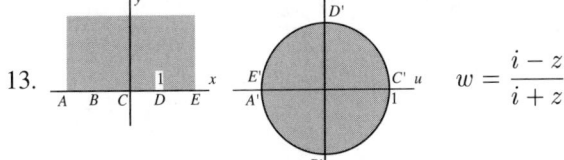 $w = \dfrac{i-z}{i+z}$.

14.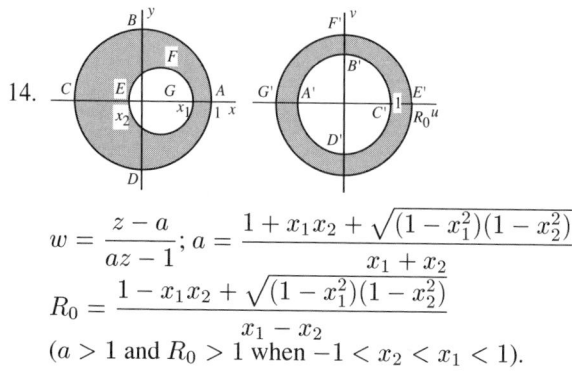

$$w = \frac{z - a}{az - 1}; a = \frac{1 + x_1 x_2 + \sqrt{(1 - x_1^2)(1 - x_2^2)}}{x_1 + x_2};$$
$$R_0 = \frac{1 - x_1 x_2 + \sqrt{(1 - x_1^2)(1 - x_2^2)}}{x_1 - x_2}$$
($a > 1$ and $R_0 > 1$ when $-1 < x_2 < x_1 < 1$).

15.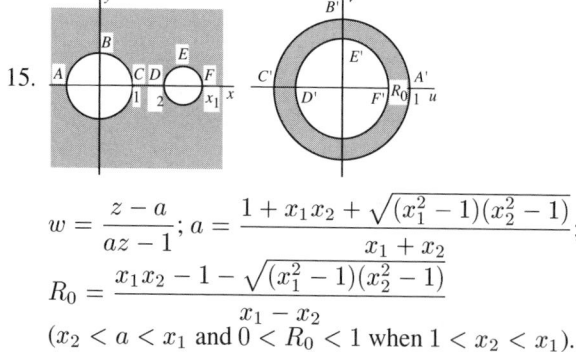

$$w = \frac{z - a}{az - 1}; a = \frac{1 + x_1 x_2 + \sqrt{(x_1^2 - 1)(x_2^2 - 1)}}{x_1 + x_2};$$
$$R_0 = \frac{x_1 x_2 - 1 - \sqrt{(x_1^2 - 1)(x_2^2 - 1)}}{x_1 - x_2}$$
($x_2 < a < x_1$ and $0 < R_0 < 1$ when $1 < x_2 < x_1$).

16.   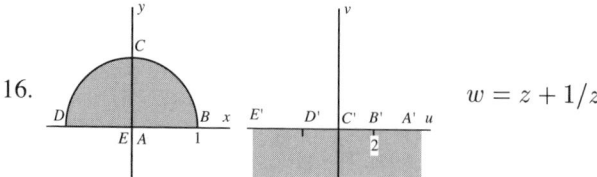 $w = z + 1/z.$

17.   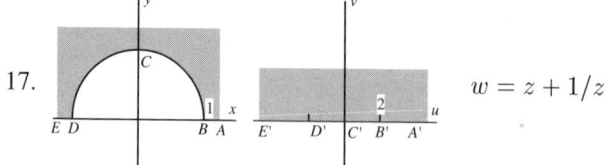 $w = z + 1/z.$

18.   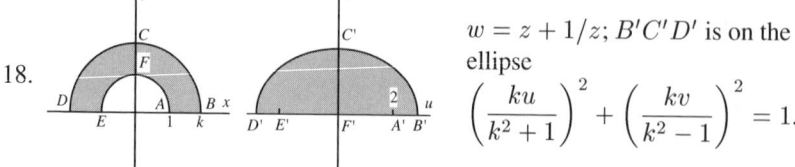 $w = z + 1/z;$ $B'C'D'$ is on the ellipse
$$\left(\frac{ku}{k^2 + 1}\right)^2 + \left(\frac{kv}{k^2 - 1}\right)^2 = 1.$$

19. 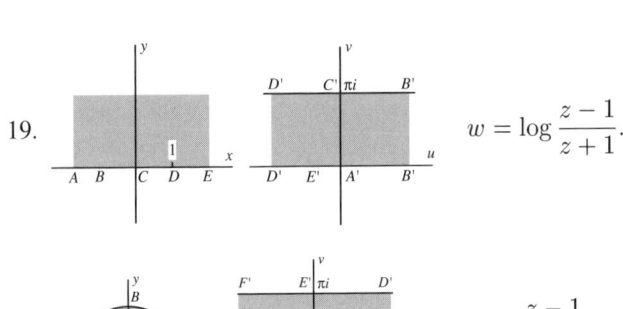 $w = \log \dfrac{z-1}{z+1}.$

20. 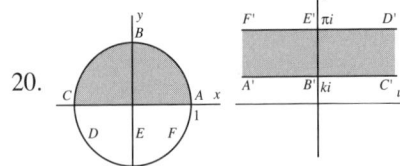 $w = \log \dfrac{z-1}{z+1}$; ABC is on the circle $x^2 + y^2 - 2y \cot k = 1$.

21. 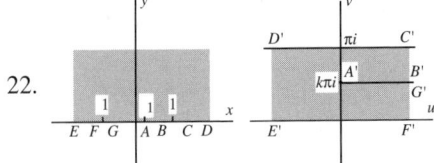 $w = \log \dfrac{z+1}{z-1}$; relationship between centers and radii: centers of circles at $z_n = \coth c_n$, radii are $\operatorname{csch} c_n$, $n = 1, 2$.

22. 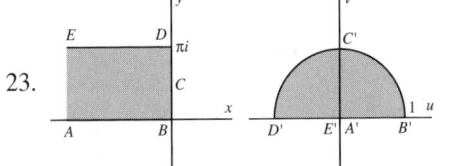 $w = k \log \dfrac{k}{1-k} + \log 2(1-k) + i\pi - k \log(z+1) - (1-k)\log(z-1)$; $x_1 = 2k - 1$.

23. 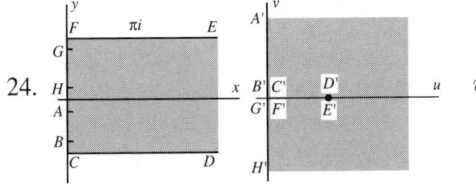 $w = \tan^2(z/2).$

24. 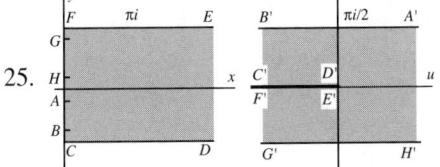 $w = \coth(z/2).$

25. 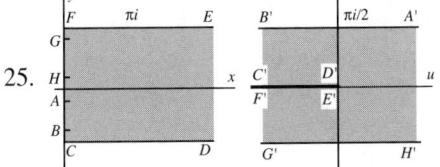 $w = \log \coth(z/2).$

26. 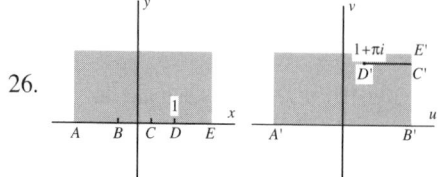 $w = \pi i + z - \log z.$

27.

$$w = 2(z + 1)^{1/2} + \log \frac{(z + 1)^{1/2} - 1}{(z + 1)^{1/2} + 1}.$$

28.

$$w = \frac{i}{k} \log \frac{1 + ikt}{1 - ikt} + \log \frac{1 + t}{1 - t};$$

$$t = \left(\frac{z - 1}{z + k^2} \right)^{1/2}.$$

29.

$$w = \frac{h}{\pi} \left[(z^2 - 1)^{1/2} + \cosh^{-1} z \right].$$

30.

$$w = \cosh^{-1} \left(\frac{2z - k - 1}{k - 1} \right) - \frac{1}{k} \cosh^{-1} \left[\frac{(k + 1)z - 2k}{(k - 1)z} \right].$$

1.6 INTERVAL ANALYSIS

1. An *interval* x is a subset of the real line: $x = [\underline{x}, \overline{x}] = \{z \in \mathbb{R} \mid \underline{x} \le z \le \overline{x}\}$.
2. A *thin interval* is a real number: x is thin if $\underline{x} = \overline{x}$

3. $\text{mid}(x) = \frac{\overline{x} + \underline{x}}{2}$

4. $\text{rad}(x) = \frac{\overline{x} - \underline{x}}{2}$

5. $|x| = \text{mag}(x) = \max\limits_{z \in x} |z|$

6. $\langle x \rangle = \text{mig}(x) = \min\limits_{z \in x} |z|$

1.6.1 INTERVAL ARITHMETIC RULES

Operation	Rule
$x + y$	$[\underline{x} + \underline{y}, \overline{x} + \overline{y}]$
$x - y$	$[\underline{x} - \overline{y}, \overline{x} - \underline{y}]$
xy	$[\min(\underline{x}\underline{y}, \underline{x}\overline{y}, \overline{x}\underline{y}, \overline{x}\overline{y}), \max(\underline{x}\underline{y}, \underline{x}\overline{y}, \overline{x}\underline{y}, \overline{x}\overline{y})]$
$\frac{x}{y}$	$\left[\min\left(\frac{\underline{x}}{\underline{y}}, \frac{\underline{x}}{\overline{y}}, \frac{\overline{x}}{\underline{y}}, \frac{\overline{x}}{\overline{y}}\right), \max\left(\frac{\underline{x}}{\underline{y}}, \frac{\underline{x}}{\overline{y}}, \frac{\overline{x}}{\underline{y}}, \frac{\overline{x}}{\overline{y}}\right)\right]$ if $0 \notin y$

EXAMPLES

1. $[1, 2] + [-2, 1] = [-1, 3]$

2. $[1, 2] - [1, 2] = [-1, 1]$

3. $[1, 2] * [-2, 1] = [-4, 2]$

4. $[1, 2]/[1, 2] = [\frac{1}{2}, 2]$

1.6.2 INTERVAL ARITHMETIC PROPERTIES

Property	$+$ and $-$	$*$ and $/$
commutative	$x + y = y + x$	$xy = yx$
associative	$x + (y + z) = (x + y) + z$	$x(yz) = (xy)z$
identity elements	$0 + x = x + 0 = x$	$1 * y = y * 1 = y$
sub-distributivity	$x(y \pm z) \subseteq xy \pm xz$ (equality holds if x is thin)	
sub-cancellation	$x - y \subseteq (x + z) - (y + z)$	$\frac{x}{y} \subseteq \frac{xz}{yz}$
	$0 \in x - x$	$1 \in \frac{y}{y}$

1.7   REAL ANALYSIS

1.7.1   RELATIONS

For two sets A and B, the *product* $A \times B$ is the set of all ordered pairs (a, b) where a is in A and b is in B. Any subset of the product $A \times B$ is called a *relation*. A relation R on a product $A \times A$ is called an *equivalence relation* if the following three properties hold:

1. *Reflexive*: (a, a) is in R for every a in A.
2. *Symmetric*: If (a, b) is in R, then (b, a) is in R.
3. *Transitive*: If (a, b) and (b, c) are in R, then (a, c) is in R.

When R is an equivalence relation then the *equivalence class* of an element a in A is the set of all b in A such that (a, b) is in R.

1. If $|A| = n$, there are 2^{n^2} relations on A.

2. If $|A| = n$, the number of equivalence relations on A is given by the Bell number B_n.

EXAMPLE   The set of rational numbers has an equivalence relation "=" defined by the requirement that an ordered pair $(\frac{a}{b}, \frac{c}{d})$ belongs in the relation if and only if $ad = bc$. The equivalence class of $\frac{1}{2}$ is the set $\{\frac{1}{2}, \frac{2}{4}, \frac{3}{6}, \ldots, \frac{-1}{-2}, \frac{-2}{-4}, \ldots\}$.

1.7.2   FUNCTIONS (MAPPINGS)

A relation f on a set $X \times Y$ is a *function* (or *mapping*) from X into Y if (x, y) and (x, z) in the relation implies that $y = z$, and each $x \in X$ has a $y \in Y$ such that (x, y) is in the relation. The last condition means that there is a unique pair in f whose first element is x. We write $f(x) = y$ to mean that (x, y) is in the relation f, and emphasize the idea of mapping by the notation $f: X \to Y$. The *domain* of f is the set X. The *range* of a function f is a set containing all the y for which there is a pair (x, y) in the relation. The *image* of a set A in the domain of a function f is the set of y in Y such that $y = f(x)$ for some x in A. The notation for the image of A under f is $f[A]$. The *inverse image* of a set B in the range of a function f is the set of all x in X such that $f(x) = y$ for some y in B. The notation is $f^{-1}[B]$.

A function f is *one-to-one* (or *univalent*, or *injective*) if $f(x_1) = f(x_2)$ implies $x_1 = x_2$. A function $f: X \to Y$ is *onto* (or *surjective*) if for every y in Y there is some x in X such that $f(x) = y$. A function is *bijective* if it is both one-to-one and onto.

EXAMPLES

1. $f(x) = e^x$, as a mapping from $\mathbb{R}$ to $\mathbb{R}$, is one-to-one because $e^{x_1} = e^{x_2}$ implies $x_1 = x_2$ (by taking the natural logarithm). It is not onto because -1 is not the value of e^x for any x in $\mathbb{R}$.

2. $g(x) = x^3 - x$, as a mapping from $\mathbb{R}$ to $\mathbb{R}$, is onto because every real number is attained as a value of $g(x)$, for some x. It is not one-to-one because $g(-1) = g(0) = g(1)$.

3. $h(x) = x^3$, as a mapping from $\mathbb{R}$ to $\mathbb{R}$, is bijective.

For an injective function f mapping X into Y, there is an *inverse function* f^{-1} mapping the range of f into X which is defined by: $f^{-1}(y) = x$ if and only if $f(x) = y$.

EXAMPLE   The function $f(x) = e^x$ mapping $\mathbb{R}$ into $\mathbb{R}^+$ (the set of positive reals) is bijective. Its inverse is $f^{-1}(x) = \ln(x)$ which maps $\mathbb{R}^+$ into $\mathbb{R}$.

For functions $f\colon X \to Y$ and $g\colon Y \to Z$, with the range of f contained in the domain of g, the *composition* $(g \circ f)\colon X \to Z$ is a function defined by $(g \circ f)(x) = g(f(x))$ for all x in the domain of f.

1. Note that $g \circ f$ may not be the same as $f \circ g$. For example, for $f(x) = x + 1$, and $g(x) = 2x$, we have $(g \circ f)(x) = g(f(x)) = 2f(x) = 2(x+1) = 2x+2$. However $(f \circ g)(x) = f(g(x)) = g(x) + 1 = 2x + 1$.

2. For every function f and its inverse f^{-1}, we have $(f \circ f^{-1})(x) = x$, for all x in the domain of f^{-1}, and $(f^{-1} \circ f)(x) = x$ for all x in the domain of f. (Note that the inverse function, f^{-1}, does not mean $\frac{1}{f}$).

1.7.3   SETS OF REAL NUMBERS

A *sequence* is the range of a function having the natural numbers as its domain. It can be denoted by $\{x_n \mid n \text{ is a natural number}\}$ or simply $\{x_n\}$. For a chosen natural number N, a *finite* sequence is the range of a function having natural numbers less than N as its domain. Sets A and B are in a *one-to-one correspondence* if there is a bijective function from A into B. Two sets A and B have the same cardinality if there is a one-to-one correspondence between them. A set which is equivalent to the set of natural numbers is *denumerable* (or *countably infinite*). A set which is empty or is equivalent to a finite sequence is *finite* (or *finite countable*).

EXAMPLES   The set of letters in the English alphabet is finite. The set of rational numbers is denumerable. The set of real numbers is uncountable.

1.7.3.1   Axioms of order

1. There is a subset P (positive numbers) of $\mathbb{R}$ for which $x + y$ and xy are in P for every x and y in P.

2. Exactly one of the following conditions can be satisfied by a number x in $\mathbb{R}$ (*trichotomy*): $x \in P$, $-x \in P$, or $x = 0$.

1.7.3.2 Definitions

A number b is an *upper* (or *lower*) *bound* of a subset S in $\mathbb{R}$ if $x \le b$ (or $x \ge b$) for every x in S. A number c is a *least upper bound* (*lub*, *supremum*, or *sup*) of a subset S in $\mathbb{R}$ if c is an upper bound of S and $b \ge c$ for every upper bound b of S. A number c is a *greatest lower bound* (*glb*, *infimum*, or *inf*) if c is a lower bound of S and $c \ge b$ for every lower bound b of S.

1.7.3.3 Completeness (or least upper bound) axiom

If a non-empty set of real numbers has an upper bound, then it has a least upper bound.

1.7.3.4 Characterization of the real numbers

The set of real numbers is the smallest complete ordered field that contains the rationals. Alternatively, the properties of a field, the order properties, and the least upper bound axiom characterize the set of real numbers. The least upper bound axiom distinguishes the set of real numbers from other ordered fields.

Archimedean property of $\mathbb{R}$: For every real number x, there is an integer N such that $x < N$. For every pair of real numbers x and y with $x < y$, there is a rational number r such that $x < r < y$. This is sometimes stated: The set of rational numbers is dense in $\mathbb{R}$.

1.7.3.5 Definition of infinity

The *extension* of $\mathbb{R}$ by ∞ is accomplished by including the symbols $+\infty$ and $-\infty$ with the following definitions (for all $x \in \mathbb{R}$)

1.	for all x in $\mathbb{R}$: $-\infty < x < \infty$	6.	if $x > 0$ then $x \cdot (-\infty) = -\infty$
2.	for all x in $\mathbb{R}$: $x + \infty = \infty$	7.	$\infty + \infty = \infty$
3.	for all x in $\mathbb{R}$: $x - \infty = -\infty$	8.	$-\infty - \infty = -\infty$
4.	for all x in $\mathbb{R}$: $\dfrac{x}{\infty} = \dfrac{x}{-\infty} = 0$	9.	$\infty \cdot \infty = \infty$
5.	if $x > 0$ then $x \cdot \infty = \infty$	10.	$-\infty \cdot (-\infty) = \infty$

1.7.3.6 Inequalities among real numbers

The expression $a > b$ means that $a - b$ is a positive real number.

1. If $a < b$ and $b < c$ then $a < c$.
2. If $a < b$ then $a \pm c < b \pm c$ for any real number c.
3. If $a < b$ and $\begin{cases} \text{if } c > 0 \text{ then } ac < bc \\ \text{if } c < 0 \text{ then } ac > bc \end{cases}$
4. If $a < b$ and $c < d$ then $a + c < b + d$.
5. If $0 < a < b$ and $0 < c < d$ then $ac < bd$.
6. If $a < b$ and $\left\{ \begin{matrix} ab > 0 \\ ab < 0 \end{matrix} \right\}$ then $\left\{ \begin{matrix} \dfrac{1}{a} > \dfrac{1}{b} \\ \dfrac{1}{a} < \dfrac{1}{b} \end{matrix} \right\}$

1.7.4 TOPOLOGICAL SPACE

A *topology* on a set X is a collection T of subsets of X (called *open sets*) having the following properties:

1. The empty set and X are in T.
2. The union of elements in an arbitrary subcollection of T is in T.
3. The intersection of elements in a finite subcollection of T is in T.

The complement of an open set is a *closed set*. A set is *compact* if every open cover has a finite subcover. The set X together with a topology T is a *topological space*.

1.7.4.1 Notes

1. A subset E of X is closed if and only if E contains all its limit points.
2. The union of finitely many closed sets is closed.
3. The intersection of an arbitrary collection of closed sets is closed.
4. The image of a compact set under a continuous function is compact.

1.7.5 METRIC SPACE

A *metric* (or *distance function*) on a set E is a function $\rho \colon E \times E \to \mathbb{R}$ that satisfies the following conditions:

1. *Positive definiteness*: $\rho(x,y) \geq 0$ for all x, y in E, and $\rho(x,y) = 0$ if and only if $x = y$.
2. *Symmetry*: $\rho(x,y) = \rho(y,x)$ for all x, y in E.
3. *Triangle inequality*: $\rho(x,y) \leq \rho(x,z) + \rho(z,y)$ for all x, y, z in E.

EXAMPLE The set of real numbers with distance defined by $d(x,y) = |x-y|$ is a metric space.

A δ *neighborhood* of a point x in a metric space E is the set of all y in E such that $d(x,y) < \delta$. For example, a δ neighborhood of x in $\mathbb{R}$ is the interval centered at x with radius δ, $(x - \delta, x + \delta)$. In a metric space the topology is generated by the δ neighborhoods.

1. A subset G of $\mathbb{R}$ is *open* if, for every x in G, there is a δ neighborhood of x which is a subset of G. For example, intervals (a,b), (a,∞), $(-\infty,b)$ are open in $\mathbb{R}$.

2. A number x is a *limit point* (or a *point of closure,* or an *accumulation point*) of a set F if, for every $\delta > 0$, there is a point y in F, with $y \neq x$, such that $|x - y| < \delta$.

3. A subset F of $\mathbb{R}$ is *closed* if it contains all of its limit points. For example, intervals $[a, b]$, $(-\infty, b]$, and $[a, \infty)$ are closed in $\mathbb{R}$.

4. A subset F is *dense* in $\mathbb{R}$ if every element of $\mathbb{R}$ is a limit point of F.

5. A metric space is *separable* if it contains a denumerable dense set. For example, $\mathbb{R}$ is separable because the subset of rationals is a denumerable dense set.

6. Theorems:

Bolzano–Weierstrass theorem   Any bounded infinite set of real numbers has a limit point in $\mathbb{R}$.

Heine–Borel theorem   A subset of R is compact if and only if it is closed and bounded.

1.7.6 CONVERGENCE IN $\mathbb{R}$ WITH METRIC $|x - y|$

1.7.6.1 Limit of a sequence

A number L is a *limit point* of a sequence $\{x_n\}$ if, for every $\epsilon > 0$, there is a natural number N such that $|x_n - L| < \epsilon$ for all $n > N$. If it exists, a limit of a sequence is unique. A sequence is said to *converge* if it has a limit. A number L is a *cluster point* of a sequence $\{x_n\}$ if, for every $\epsilon > 0$ and every index N, there is an $n > N$ such that $|x_n - L| < \epsilon$.

EXAMPLE   The limit of a sequence is a cluster point, as in $\{\frac{1}{n}\}$, which converges to 0. However, cluster points are not necessarily limits, as in $\{(-1)^n\}$, which has cluster points $+1$ and -1 but no limit.

Let $\{x_n\}$ be a sequence. A number L is the *limit superior (limsup)* if, for every $\epsilon > 0$, there is a natural number N such that $x_n > L - \epsilon$ for infinitely many $n \geq N$, and $x_n > L + \epsilon$ for only finitely many terms. An equivalent definition of the limit superior is given by

$$\limsup x_n = \inf_N \sup_{k \geq N} x_k. \tag{1.7.1}$$

The *limit inferior (liminf)* is defined in a similar way by

$$\liminf x_n = \sup_N \inf_{k \geq N} x_k. \tag{1.7.2}$$

For example, the sequence $\{x_n\}$ with $x_n = 1 + (-1)^n + \frac{1}{2^n}$ has $\limsup x_n = 2$, and $\liminf x_n = 0$.

Theorem   Every bounded sequence $\{x_n\}$ in $\mathbb{R}$ has a lim sup and a lim inf. In addition, if $\limsup x_n = \liminf x_n$, then the sequence converges to their common value.

A sequence $\{x_n\}$ is a *Cauchy* sequence if, for any $\epsilon > 0$, there exists a positive integer N such that $|x_n - x_m| < \epsilon$ for every $n > N$ and $m > N$.

Theorem   A sequence $\{x_n\}$ in $\mathbb{R}$ converges if and only if it is a Cauchy sequence.

A metric space in which every Cauchy sequence converges to a point in the space is called *complete*. For example, $\mathbb{R}$ with the metric $d(x, y) = |x - y|$ is complete.

1.7.6.2 Limit of a function

A number L is a *limit* of a function f as x approaches a number a if, for every $\epsilon > 0$, there is a $\delta > 0$ such that $|f(x) - L| < \epsilon$ for all x with $0 < |x - a| < \delta$. This is represented by the notation $\lim_{x \to a} f(x) = L$. The symbol ∞ is the limit of a function f as x approaches a number a if, for every positive number M, there is a $\delta > 0$ such that $f(x) > M$ for all x with $0 < |x - a| < \delta$. The notation is $\lim_{x \to a} f(x) = \infty$. A number L is a limit of a function f as x approaches ∞ if, for every $\epsilon > 0$, there is a positive number M such that $|f(x) - L| < \epsilon$ for all $x > M$; this is written $\lim_{x \to \infty} f(x) = L$. The number L is said to be the *limit at infinity*.

EXAMPLES $\displaystyle\lim_{x \to 2} 3x - 1 = 5,$ $\displaystyle\lim_{x \to 0} \frac{1}{x^2} = \infty,$ $\displaystyle\lim_{x \to \infty} \frac{1}{x} = 0.$

1.7.6.3 Limit of a sequence of functions

A sequence of functions $\{f_n(x)\}$ is said to *converge pointwise* to the function $f(x)$ on a set E if for every $\epsilon > 0$ and $x \in E$ there is a positive integer N such that $|f(x) - f_n(x)| < \epsilon$ for every $n \geq N$. A sequence of functions $\{f_n(x)\}$ is said to *converge uniformly* to the function f on a set E if, for every $\epsilon > 0$, there exists a positive integer N such that $|f(x) - f_n(x)| < \epsilon$ for all x in E and $n \geq N$.

Note that these formulations of convergence are not equivalent. For example, the functions $f_n(x) = x^n$ on the interval $[0, 1]$ converge pointwise to the function $f(x) = 0$ for $0 \leq x < 1$, $f(1) = 1$. They do not converge uniformly because, for $\epsilon = 1/2$, there is no N such that $|f_n(x) - f(x)| < 1/2$ for all x in $[0, 1]$ and every $n \geq N$.

A function f is *Lipschitz* if there exists $k > 0$ in $\mathbb{R}$ such that $|f(x) - f(y)| \leq k|x - y|$ for all x and y in its domain. The function is a *contraction* if $0 < k < 1$.

> **Fixed point or contraction mapping theorem** Let E be a complete metric space. If the function $f \colon E \to E$ is a contraction, then there is a unique point x in E such that $f(x) = x$. The point x is called a *fixed point* of f.

EXAMPLE Newton's method for finding a zero of $f(x) = (x + 1)^2 - 2$ on the interval $[0, 1]$ produces $x_{n+1} = g(x_n)$ with the contraction $g(x) = \frac{x}{2} - \frac{1}{2} + \frac{1}{x+1}$. This has the unique fixed point $\sqrt{2} - 1$ in $[0, 1]$.

1.7.7 CONTINUITY IN $\mathbb{R}$ WITH METRIC $|x - y|$

A function $f \colon \mathbb{R} \to \mathbb{R}$ is *continuous at a point* a if f is defined at a and

$$\lim_{x \to a} f(x) = f(a). \tag{1.7.3}$$

The function f is *continuous on a set* E if it is continuous at every point of E. A function f is *uniformly continuous* on a set E if, for every $\epsilon > 0$, there exists a $\delta > 0$ such that $|f(x) - f(y)| < \epsilon$ for every x and y in E with $|x - y| < \delta$. A sequence $\{f_n(x)\}$ of continuous functions on the interval $[a, b]$ is *equicontinuous* if, for every $\epsilon > 0$, there exists a $\delta > 0$ such that $|f_n(x) - f_n(y)| < \epsilon$ for every n and for all x and y in $[a, b]$ with $|x - y| < \delta$.

1. A function can be continuous without being uniformly continuous. For example, the function $g(x) = \frac{1}{x}$ is continuous but not uniformly continuous on the open interval $(0, 1)$.

2. A collection of continuous functions can be bounded on a closed interval without having a uniformly convergent sub-sequence. For example, the continuous functions $f_n(x) = \frac{x^2}{x^2 + (1 - nx)^2}$ are each bounded by 1 in the closed interval $[0, 1]$ and for every x there is the limit: $\lim_{n \to \infty} f_n(x) = 0$. However, $f_n\left(\frac{1}{n}\right) = 1$ for every n, so that no sub-sequence can converge uniformly to 0 everywhere on $[0, 1]$. This sequence is not equicontinuous.

Theorems:

Theorem   Let $\{f_n(x)\}$ be a sequence of functions mapping $\mathbb{R}$ into $\mathbb{R}$ which converges uniformly to a function f. If each $f_n(x)$ is continuous at a point a, then $f(x)$ is also continuous at a.

Theorem   If a function f is continuous on a closed bounded set E, then it is uniformly continuous on E.

Ascoli–Arzela theorem   Let K be a compact set in $\mathbb{R}$. If $\{f_n(x)\}$ is uniformly bounded and equicontinuous on K, then $\{f_n(x)\}$ contains a uniformly convergent sub-sequence on K.

Weierstrass polynomial approximation theorem   Let K be a compact set in $\mathbb{R}$. If f is a continuous function on K then there exists a sequence of polynomials that converges uniformly to f on K.

1.7.8 BANACH SPACE

A *norm* on a vector space E with scalar field $\mathbb{R}$ is a function $|| \cdot ||$ from E into $\mathbb{R}$ that satisfies the following conditions:

1. *Positive definiteness*: $||x|| \geq 0$ for all x in E, and $||x|| = 0$ if and only if $x = 0$.
2. *Scalar homogeneity*: For every x in E and a in $\mathbb{R}$, $||ax|| = |a| \, ||x||$.
3. *Triangle inequality*: $||x + y|| \leq ||x|| + ||y||$ for all x, y in E.

Every norm $|| \cdot ||$ gives rise to a metric ρ by defining: $\rho(x, y) = ||x - y||$.

EXAMPLES

1. $\mathbb{R}$ with absolute value as the norm has the metric $\rho(x, y) = |x - y|$.
2. $\mathbb{R} \times \mathbb{R}$ (denoted $\mathbb{R}^2$) with the Euclidean norm $||(x, y)|| = \sqrt{x^2 + y^2}$ has the metric $\rho((x_1, y_1), (x_2, y_2)) = \sqrt{(x_1 - x_2)^2 + (y_1 - y_2)^2}$.

A *Banach space* is a complete normed space.

A widely studied example of a Banach space is the (vector) space of measurable functions f on $[a, b]$ for which $\int_a^b |f(x)|^p \, dx < \infty$ with $1 < p < \infty$. This is denoted by $L_p[a, b]$ or simply L_p. The space of essentially bounded measurable functions on $[a, b]$ is denoted by $L_\infty[a, b]$.

The L_p *norm* for $0 < p < \infty$ is defined by $\|f\|_p = \left(\int_a^b |f(x)|^p \, dx \right)^{1/p}$. The L_∞ norm is defined by

$$\|f\|_\infty = \operatorname*{ess\ sup}_{a \leq x \leq b} |f(x)|, \qquad (1.7.4)$$

where

$$\operatorname*{ess\ sup}_{a \leq x \leq b} |f(x)| = \inf\{M | m\{t : f(t) > M\} = 0\}. \qquad (1.7.5)$$

Let $\{f_n(x)\}$ be a sequence of functions in L_p (with $1 \leq p < \infty$) and f be some function in L_p. We say that $\{f_n\}$ *converges in the mean of order p* (or simply in L_p-norm) to f if $\lim_{n \to \infty} \|f_n - f\|_p = 0$.

Riesz–Fischer theorem   The L_p spaces are complete.

1.7.8.1   Inequalities

1. *Minkowski inequality*   If f and g are in L_p with $1 \leq p \leq \infty$, then $\|f + g\|_p \leq \|f\|_p + \|g\|_p$. That is,

$$\left(\int_a^b |f + g|^p \right)^{1/p} \leq \left(\int_a^b |f|^p \right)^{1/p} + \left(\int_a^b |g|^p \right)^{1/p} \quad \text{for } 1 \leq p < \infty,$$

(1.7.6)

$$\operatorname{ess\ sup} |f + g| \leq \operatorname{ess\ sup} |f| + \operatorname{ess\ sup} |g|.$$

2. *Hölder inequality*   If p and q are non-negative extended real numbers such that $1/p + 1/q = 1$ and $f \in L_p$ and $g \in L_q$, then $\|fg\|_1 \leq \|f\|_p \|g\|_q$. That is

$$\int_a^b |fg| \leq \left(\int_a^b |f|^p \right)^{1/p} \left(\int_a^b |g|^q \right)^{1/q} \quad \text{for } 1 \leq p < \infty, \qquad (1.7.7)$$

$$\int_a^b |fg| \leq (\operatorname{ess\ sup} |f|) \int_a^b |g|. \qquad (1.7.8)$$

3. *Schwartz* (or *Cauchy–Schwartz*) *inequality*   If f and g are in L_2, then $\|fg\|_1 \leq \|f\|_2 \|g\|_2$. This is the special case of Hölder's inequality with $p = q = 2$.

4. *Arithmetic mean–geometric mean inequality*   If A_n and G_n are the arithmetic and geometric means of the set of positive numbers $\{a_1, a_2, \ldots, a_n\}$ then $A_n \geq G_n$. That is

$$\frac{a_1 + a_2 + \ldots + a_n}{n} \geq (a_1 a_2 \cdots a_n)^{1/n}$$

5. *Carleman's inequality*   If A_n and G_n are the arithmetic and geometric means of the set of positive numbers $\{a_1, a_2, \ldots, a_n\}$ then

$$\sum_{r=1}^{n} G_r \leq n e A_n.$$

1.7.9 HILBERT SPACE

An *inner product* on a vector space E with scalar field $\mathbb{C}$ (complex numbers) is a function from $E \times E$ into $\mathbb{C}$ that satisfies the following conditions:

1. $\langle x, x \rangle \geq 0$, and $\langle x, x \rangle = 0$ if and only if $x = 0$.
2. $\langle x + y, z \rangle = \langle x, z \rangle + \langle y, z \rangle$
3. $\langle cx, y \rangle = c \langle x, y \rangle$
4. $\langle x, y \rangle = \overline{\langle y, x \rangle}$

Every inner product $\langle x, y \rangle$ gives rise to a norm $\|x\|$ by defining $\|x\| = \langle x, x \rangle^{1/2}$.

A *Hilbert space* is a complete inner product space. A widely studied Hilbert space is $L_2[a, b]$ with the inner product $\langle f, g \rangle = \int_a^b f(t)\overline{g(t)}\, dt$.

Two functions f and g in $L_2[a, b]$ are *orthogonal* if $\int_a^b fg = 0$. A set of L_2 functions $\{\phi_n\}$ is *orthogonal* if $\int_a^b \phi_m \phi_n = 0$ for $m \neq n$. The set is *orthonormal* if, in addition, each member has norm 1. That is, $\|\phi_n\|_2 = 1$. For example, the functions $\{\sin nx\}$ are mutually orthogonal on $(-\pi, \pi)$. The functions $\{\frac{\sin nx}{\sqrt{\pi}}\}$ form an orthonormal set on $(-\pi, \pi)$.

Let $\{\phi_n\}$ be an orthonormal set in L_2 and f be in L_2. The numbers $c_n = \int_a^b f\phi_n\, dx$ are the *generalized Fourier coefficients* of f with respect to $\{\phi_n\}$, and the series $\sum_{n=1}^{\infty} c_n \phi_n(x)$ is called the *generalized Fourier series* of f with respect to $\{\phi_n\}$.

For a function f in L_2, the *mean square error* of approximating f by the sum $\sum_{n=1}^{N} a_n \phi_n$ is $\frac{1}{b-a} \int_a^b |f(x) - \sum_{n=1}^{N} a_n \phi_n(x)|^2\, dx$. An orthonormal set $\{\phi_n\}$ is *complete* if the only measurable function f that is orthogonal to every ϕ_n is zero. That is, $f = 0$ a.e. (In the context of elementary measure theory, two measurable functions f and g are *equivalent* if they are equal except on a set of measure zero. They are said to be equal *almost everywhere*. This is denoted by $f = g$ a.e.)

Bessel's inequality: For a function f in L_2 having generalized Fourier coefficients $\{c_n\}$, $\sum_{n=1}^{\infty} |c_n^2| \leq \int_a^b |f(x)|^2\, dx$.

Theorems:

Riesz–Fischer theorem   Let $\{\phi_n\}$ be an orthonormal set in L_2 and let $\{c_n\}$ be constants such that $\sum_{n=1}^{\infty} |c_n^2|$ converges. Then a unique function f in L_2 exists such that the c_n are the Fourier coefficients of f with respect to $\{\phi_n\}$ and $\sum_{n=1}^{\infty} c_n \phi_n$ converges in the mean (or order 2) to f.

Theorem   The generalized Fourier series of f in L_2 converges in the mean (of order 2) to f.

Theorem   Parseval's identity holds: $\displaystyle\int_a^b |f(x)|^2\, dx = \sum_{n=1}^{\infty} |c_n^2|$.

Theorem   The mean square error of approximating f by the series $\sum_{n=1}^{\infty} a_n \phi_n$ is minimum when all coefficients a_n are the Fourier coefficients of f with respect to $\{\phi_n\}$.

EXAMPLE   Suppose that the series $\frac{a_o}{2} + \sum_{n=1}^{\infty} (|a_n|^2 + |b_n|^2)$ converges. Then the trigonometric series $\frac{a_o}{2} + \sum_{n=1}^{\infty} (a_n \cos nx + b_n \sin nx)$ is the Fourier series of some function in L_2.

1.7.10 ASYMPTOTIC RELATIONSHIPS

Asymptotic relationships are indicated by the symbols O, Ω, Θ, o, and $\sim$.

1. The symbol O (pronounced "big-oh"): $f(x) \in O(g(x))$ as $x \to x_0$ if a positive constant C exists such that $|f(x)| \le C\,|g(x)|$ for all x sufficiently close to x_0. Note that $O(g(x))$ is a class of functions. Sometimes the statement $f(x) \in O(g(x))$ is written (imprecisely) as $f = O(g)$.

2. The symbol Ω : $f(x) \in \Omega(g(x))$ as $x \to x_0$ if a positive constant C exists such that $|g(x)| \le C\,|f(x)|$ for all x sufficiently close to x_0.

3. The symbol Θ : $f(x) \in \Theta(g(x))$ as $x \to x_0$ if positive constants c_1 and c_2 exist such that $c_1 g(x) \le f(x) \le c_2 g(x)$ for all x sufficiently close to x_0. This is equivalent to: $f(x) = O(g(x))$ and $g(x) = O(f(x))$. The symbol $\approx$ is often used for Θ (i.e., $f(x) \approx g(x)$).

4. The symbol o (pronounced "little-oh"): $f(x) \in o(g(x))$ as $x \to x_0$ if, given any $\mu > 0$, we have $|f(x)| < \mu|g(x)|$ for all x sufficiently close to x_0.

5. The symbol $\sim$ (pronounced "asymptotic to"): $f(x) \sim (g(x))$ as $x \to x_0$ if $f(x) = g(x)\,[1 + o(1)]$ as $x \to x_0$.

6. Two functions, $f(x)$ and $g(x)$, are *asymptotically equivalent* as $x \to x_0$ if $f(x)/g(x) \sim 1$ as $x \to x_0$.

7. A sequence of functions, $\{g_k(x)\}$, forms an *asymptotic series* at x_0 if $g_{k+1}(x) = o(g_k(x))$ as $x \to x_0$.

8. Given a function $f(x)$ and an asymptotic series $\{g_k(x)\}$ at x_0, the formal series $\sum_{k=0}^{\infty} a_k g_k(x)$, where the $\{a_k\}$ are given constants, is an *asymptotic expansion* of $f(x)$ if $f(x) - \sum_{k=0}^{n} a_k g_k(x) = o(g_n(x))$ as $x \to x_0$ for every n; this is expressed as $f(x) \sim \sum_{k=0}^{\infty} a_k g_k(x)$. Partial sums of this formal series are called *asymptotic approximations* to $f(x)$. This formal series need not converge.

Think of O being an upper bound on a function, Ω being a lower bound, and Θ being both an upper and lower bound. For example: $\sin x \in O(x)$ as $x \to 0, \log n \in o(n)$ as $n \to \infty$, and $n^9 \in \Omega(n^9 + n^2)$ as $n \to \infty$.

The statements: $n^2 = o(n^5)$, $n^5 = o(2^n)$, $2^n = o(n!)$, and $n! = o(n^n)$ as $n \to \infty$ can be illustrated as follows. If a computer can perform 10^9 operations per second, and a procedure takes $f(n)$ operations, then the following table indicates approximately how long it will take a computer to perform the procedure, for various $f(n)$ functions and values of n.

complexity	$n = 10$	$n = 20$	$n = 50$	$n = 100$	$n = 300$
$f(n) = n^2$	10^{-7} sec	10^{-7} sec	10^{-6} sec	10^{-5} sec	10^{-4} sec
$f(n) = n^5$	10^{-4} sec	10^{-3} sec	0.3 sec	10 sec	41 minutes
$f(n) = 2^n$	10^{-6} sec	10^{-3} sec	2 weeks	10^{11} centuries	10^{72} centuries
$f(n) = n!$	10^{-3} sec	77 years	10^{46} centuries	10^{139} centuries	10^{596} centuries
$f(n) = n^n$	10 sec	10^7 centuries	10^{66} centuries	10^{181} centuries	10^{724} centuries

1.8   GENERALIZED FUNCTIONS

1.8.1   DELTA FUNCTION

Dirac's delta function is a distribution defined by $\delta(x) = \begin{cases} 0 & x \neq 0 \\ \infty & x = 0 \end{cases}$, and is normalized so that $\int_{-\infty}^{\infty} \delta(x)\,dx = 1$. Properties include (assuming that $f(x)$ is continuous):

1. $\int_{-\infty}^{\infty} f(x)\delta(x - a)\,dx = f(a)$.

2. $\int_{-\infty}^{\infty} f(x)\frac{d^m \delta(x)}{dx^m}\,dx = (-1)^m \frac{d^m f(0)}{dx^m}$.

3. $x\delta(x)$, as a distribution, equals zero.

4. $\delta(ax) = \frac{1}{|a|}\delta(x)$ when $a \neq 0$.

5. $\delta(x^2 - a^2) = \frac{1}{2a}[\delta(x + a) + \delta(x - a)]$.

6. $\delta(x) = \frac{1}{2L} + \frac{1}{L}\sum_{n=1}^{\infty} \cos \frac{n\pi x}{L}$ (Fourier series).

7. $\delta(x - \xi) = \frac{2}{L}\sum_{n=1}^{\infty} \sin \frac{n\pi \xi}{L} \sin \frac{n\pi x}{L}$ for $0 < \xi < L$ (Fourier sine series).

8. $\delta(x) = \frac{1}{2\pi} \int_{-\infty}^{\infty} e^{ikx}\,dk$ (Fourier transform).

9. $\delta(\rho - \rho') = \rho \int_0^\infty k J_m(k\rho) J_m(k\rho') \, dk.$

Sequences of functions $\{\phi_n\}$ that approximate the delta function as $n \to \infty$ are known as delta sequences.

EXAMPLES

1. $\phi_n(x) = \frac{n}{\pi} \frac{1}{1+n^2 x^2}$

2. $\phi_n(x) = \frac{n}{\sqrt{\pi}} e^{-n^2 x^2}$

3. $\phi_n(x) = \frac{1}{n\pi} \frac{\sin^2 nx}{x^2}$

4. $\phi_n(x) = \begin{cases} 0 & |x| \geq 1/n \\ n/2 & |x| < 1/n \end{cases}$

The delta function $\delta(\mathbf{x} - \mathbf{x}') = \delta(x_1 - x_1')\delta(x_2 - x_2')\delta(x_3 - x_3')$ in terms of the coordinates (ξ_1, ξ_2, ξ_3), related to (x_1, x_2, x_3), via the Jacobian $J(x_i, \xi_j)$, is written

$$\delta(\mathbf{x} - \mathbf{x}') = \frac{1}{|J(x_i, \xi_j)|} \delta(\xi_1 - \xi_1')\delta(\xi_2 - \xi_2')\delta(\xi_3 - \xi_3'). \qquad (1.8.1)$$

For example, in spherical polar coordinates

$$\delta(\mathbf{x} - \mathbf{x}') = \frac{1}{r^2} \delta(r - r')\delta(\phi - \phi')\delta(\cos\theta - \cos\theta'). \qquad (1.8.2)$$

The solutions to differential equations involving delta functions are called Green's functions (see pages 463 and 471).

1.8.2   OTHER GENERALIZED FUNCTIONS

The Heaviside function, or step function, is defined as

$$H(x) = \int_{-\infty}^{x} \delta(t) \, dt = \begin{cases} 0 & x < 0 \\ 1 & x > 0. \end{cases} \qquad (1.8.3)$$

Sometimes $H(0)$ is stated to be $1/2$. This function has the representations:

1. $H(x) = \frac{1}{2} + \frac{2}{\pi} \sum_{n=\text{odd}}^{\infty} \frac{1}{n} \sin \frac{n\pi x}{L}$

2. $H(x) = \frac{1}{2\pi i} \int_{-\infty}^{\infty} \frac{e^{ikx}}{k} \, dk$

The related signum function gives the sign of its argument:

$$\text{sgn}(x) = 2H(x) - 1 = \begin{cases} -1 & \text{if } x < 0 \\ 1 & \text{if } x > 0. \end{cases} \qquad (1.8.4)$$

Chapter 2

Algebra

2.1   **PROOFS WITHOUT WORDS** 81

2.2   **ELEMENTARY ALGEBRA** 83
    2.2.1   *Basic algebra* *83*
    2.2.2   *Progressions* *86*
    2.2.3   *DeMoivre's theorem* *87*
    2.2.4   *Partial fractions* *87*

2.3   **POLYNOMIALS** 89
    2.3.1   *Quadratic polynomials* *89*
    2.3.2   *Cubic polynomials* *89*
    2.3.3   *Quartic polynomials* *90*
    2.3.4   *Quartic curves* *90*
    2.3.5   *Quintic polynomials* *90*
    2.3.6   *Tschirnhaus' transformation* *91*
    2.3.7   *Polynomial norms* *91*
    2.3.8   *Cyclotomic polynomials* *91*
    2.3.9   *Other polynomial properties* *93*

2.4   **NUMBER THEORY** 93
    2.4.1   *Divisibility* *93*
    2.4.2   *Congruences* *94*
    2.4.3   *Chinese remainder theorem* *96*
    2.4.4   *Continued fractions* *96*
    2.4.5   *Diophantine equations* *98*
    2.4.6   *Greatest common divisor* *101*
    2.4.7   *Least common multiple* *101*
    2.4.8   *Magic squares* *101*
    2.4.9   *Möbius function* *102*
    2.4.10   *Prime numbers* *103*
    2.4.11   *Prime numbers of special forms* *105*
    2.4.12   *Prime numbers less than 100,000* *108*
    2.4.13   *Factorization table* *125*
    2.4.14   *Factorization of $2^m - 1$* *128*
    2.4.15   *Euler Totient function* *128*

2.5   **VECTOR ALGEBRA** 131
    2.5.1   *Notation for vectors and scalars* *131*
    2.5.2   *Physical vectors* *131*
    2.5.3   *Fundamental definitions* *131*

1-58488-291-3/02/$0.00+$1.50
© 2003 CRC Press, Inc.

2.5.4   *Laws of vector algebra* *132*
2.5.5   *Vector norms* *133*
2.5.6   *Dot, scalar, or inner product* *133*
2.5.7   *Vector or cross product* *135*
2.5.8   *Scalar and vector triple products* *136*

2.6   LINEAR AND MATRIX ALGEBRA **137**
2.6.1   *Definitions* *137*
2.6.2   *Types of matrices* *138*
2.6.3   *Conformability for addition and multiplication* *142*
2.6.4   *Determinants and permanents* *144*
2.6.5   *Matrix norms* *146*
2.6.6   *Singularity, rank, and inverses* *147*
2.6.7   *Systems of linear equations* *148*
2.6.8   *Linear spaces and linear mappings* *149*
2.6.9   *Traces* *150*
2.6.10   *Generalized inverses* *151*
2.6.11   *Eigenstructure* *152*
2.6.12   *Matrix diagonalization* *154*
2.6.13   *Matrix exponentials* *155*
2.6.14   *Quadratic forms* *155*
2.6.15   *Matrix factorizations* *156*
2.6.16   *Theorems* *157*
2.6.17   *The vector operation* *158*
2.6.18   *Kronecker products* *159*
2.6.19   *Kronecker sums* *160*

2.7   ABSTRACT ALGEBRA **160**
2.7.1   *Definitions* *160*
2.7.2   *Groups* *161*
2.7.3   *Rings* *164*
2.7.4   *Fields* *167*
2.7.5   *Quadratic fields* *167*
2.7.6   *Finite fields* *169*
2.7.7   *Homomorphisms and isomorphisms* *170*
2.7.8   *Matrix classes that are groups* *171*
2.7.9   *Permutation groups* *172*
2.7.10   *Tables* *176*

2.1 PROOFS WITHOUT WORDS

A Property of the Sequence of Odd Integers (Galileo, 1615)

The Pythagorean Theorem

$$\frac{1}{3} = \frac{1+3}{5+7} = \frac{1+3+5}{7+9+11} = \cdots$$

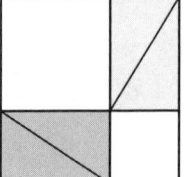 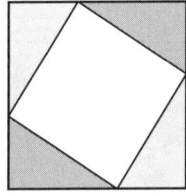

—the *Chou pei suan ching*
(author unknown, circa B.C. 200?)

$$\frac{1+3+\cdots+(2n-1)}{(2n+1)+(2n+3)+\cdots+(4n-1)} = \frac{1}{3}$$

$$1+2+\cdots+n = \frac{n(n+1)}{2}$$

$$1+3+5+\cdots+(2n-1) = n^2$$

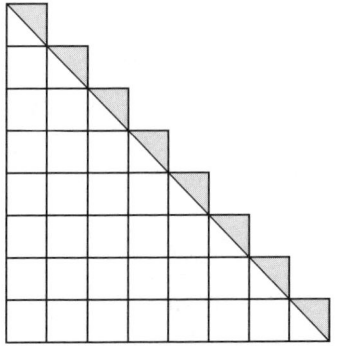

$$1+2+\cdots+n = \frac{1}{2}\cdot n^2 + n\cdot\frac{1}{2} = \frac{n(n+1)}{2}$$

—Ian Richards

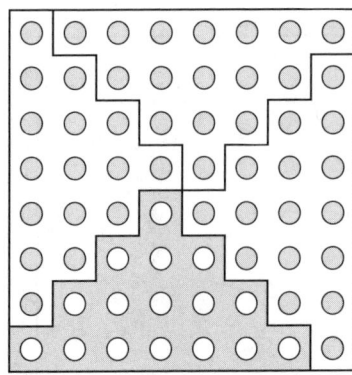

$$1+3+\cdots+(2n-1) = \frac{1}{4}(2n)^2 = n^2$$

Geometric Series

$$\frac{1}{4} + \left(\frac{1}{4}\right)^2 + \left(\frac{1}{4}\right)^3 + \cdots = \frac{1}{3}$$

—Rick Mabry

Geometric Series

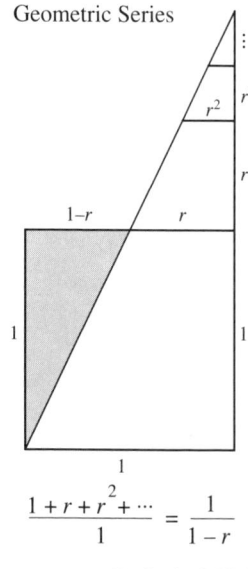

$$\frac{1 + r + r^2 + \cdots}{1} = \frac{1}{1 - r}$$

—Benjamin G. Klein
and Irl C. Bivens

Addition Formulae for the Sine
and Cosine

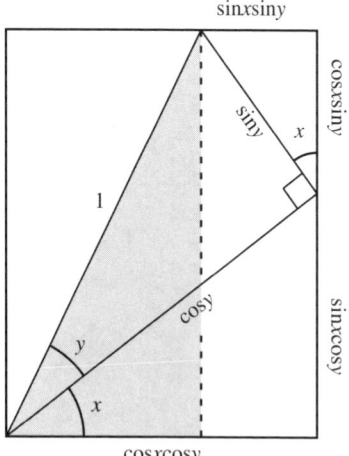

$\sin(x + y) = \sin x \cos y + \cos x \sin y$
$\cos(x + y) = \cos x \cos y - \sin x \sin y$

The Distance Between a Point and a Line

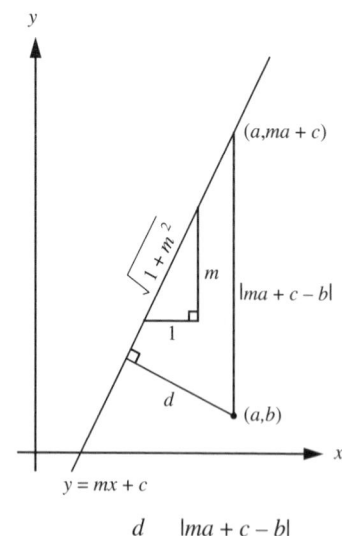

$$\frac{d}{1} = \frac{|ma + c - b|}{\sqrt{1 + m^2}}$$

—R. L. Eisenman

The Arithmetic Mean-Geometric Mean Inequality

$$a,b > 0 \Rightarrow \frac{a+b}{2} \geq \sqrt{ab}$$

The Mediant Property

$$\frac{a}{b} < \frac{c}{d} \Rightarrow \frac{a}{b} < \frac{a+c}{b+d} < \frac{c}{d}$$

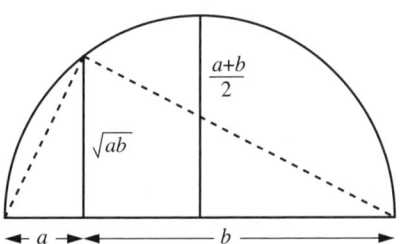

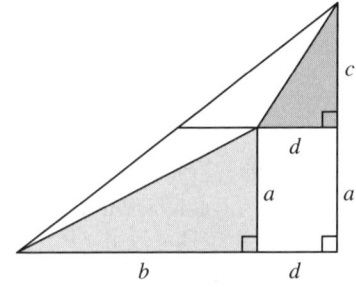

—Charles D. Gallant

—Richard A. Gibbs

Reprinted from "Proofs Without Words: Exercises in Visual Thinking", by Roger B. Nelsen. Copyright 1993 by The Mathematical Association of America, pages: 3, 40, 49, 60, 70, 72, 115, 120.

Reprinted from "Proofs Without Words II: More Exercises in Visual Thinking", by Roger B. Nelsen. Copyright 2000 by The Mathematical Association of America, pages 46, 111.

2.2 ELEMENTARY ALGEBRA

2.2.1 BASIC ALGEBRA

2.2.1.1 Algebraic equations

A *polynomial equation in one variable* has the form $f(x) = 0$ where $f(x)$ is a polynomial of degree n

$$f(x) = a_n x^n + a_{n-1} x^{n-1} + \cdots + a_1 x + a_0 \tag{2.2.1}$$

and $a_n \neq 0$.

A complex number z is a *root* of the polynomial $f(x)$ if $f(z) = 0$. A complex number z is a *root of multiplicity* k if $f(z) = f'(z) = f''(z) = \cdots = f^{(k-1)}(z) = 0$, but $f^{(k)}(z) \neq 0$. A root of multiplicity 1 is called a *simple root*. A root of multiplicity 2 is called a *double root*, and a root of multiplicity 3 is called a *triple root*.

2.2.1.2   Roots of polynomials

1. *Fundamental theorem of algebra*

   A polynomial equation of degree n has exactly n complex roots, where a double root is counted twice, a triple root three times, and so on. If the n roots of the polynomial $f(x)$ are $z_1, z_2, \ldots, z_n$ (where a double root is listed twice, a triple root three times, and so on), then the polynomial can be written as

   $$f(x) = a_n(x - z_1)(x - z_2) \ldots (x - z_n). \tag{2.2.2}$$

2. If the coefficients of the polynomial, $\{a_0, a_1, \ldots, a_n\}$, are real numbers, then the polynomial will always have an even number of complex roots occurring in pairs. That is, if z is a complex root, then so is $\bar{z}$. If the polynomial has an odd degree and the coefficients are real, then it must have at least one real root.

3. The coefficients of the polynomial may be expressed as symmetric functions of the roots. For example, the elementary symmetric functions $\{s_i\}$, and their values for a polynomial of degree n (known as *Viete's formulae*), are:

   $$s_1 = z_1 + z_2 + \cdots + z_n = -\frac{a_{n-1}}{a_n},$$

   $$s_2 = z_1 z_2 + z_1 z_3 + z_2 z_3 + \cdots = \sum_{i>j} z_i z_j = \frac{a_{n-2}}{a_n},$$

   $$\tag{2.2.3}$$

   $$\vdots$$

   $$s_n = z_1 z_2 z_3 \ldots z_n = (-1)^n \frac{a_0}{a_n},$$

   where s_k is the sum of $\binom{n}{k}$ products, each product combining k factors without repetition.

4. The *discriminant* of the polynomial is defined by $\prod_{i>j}(z_i - z_j)^2$, where the ordering of the roots is irrelevant. The discriminant can always be written as a polynomial combination of $a_0, a_1, \ldots, a_n$, divided by a_n.

   (a) For the quadratic equation $ax^2 + bx + c = 0$ the discriminant is $\frac{b^2 - 4ac}{a^2}$.

   (b) For the cubic equation $ax^3 + bx^2 + cx + d = 0$ the discriminant is

   $$\frac{b^2 c^2 - 4b^3 d - 4ac^3 + 18abcd - 27a^2 d^2}{a^4}.$$

5. The number of roots of a polynomial in modular arithmetic is difficult to predict. For example

   (a) $y^4 + y + 1 = 0$ has one root modulo 51: $y = 37$

   (b) $y^4 + y + 2 = 0$ has no roots modulo 51

   (c) $y^4 + y + 3 = 0$ has six roots modulo 51: $y = \{15, 27, 30, 32, 44, 47\}$

2.2.1.3 Resultants

Let $f(x) = a_n x^n + a_{n-1} x^{n-1} + \cdots + a_1 x + a_0$ and $g(x) = b_m x^m + b_{m-1} x^{m-1} + \cdots + b_1 x + b_0$, where $a_n \neq 0$ and $b_m \neq 0$. The *resultant* of f and g is the determinant of the $(m+n) \times (m+n)$ matrix

$$
\det \begin{bmatrix}
a_n & a_{n-1} & \cdots & \cdots & a_0 & 0 & \cdots & 0 \\
0 & a_n & a_{n-1} & \cdots & a_1 & a_0 & & 0 \\
\vdots & & \ddots & \ddots & \ddots & \ddots & \ddots & \ddots \\
0 & \cdots & 0 & a_n & a_{n-1} & \cdots & a_1 & a_0 \\
\hdashline
0 & \cdots & \cdots & 0 & b_m & \cdots & b_1 & b_0 \\
\vdots & & & & \ddots & & \ddots & \\
0 & b_m & b_{m-1} & \cdots & b_0 & 0 & \cdots & 0 \\
b_m & b_{m-1} & \cdots & & b_0 & 0 & \cdots & 0
\end{bmatrix} \tag{2.2.4}
$$

The resultant of $f(x)$ and $g(x)$ is 0 if and only if $f(x)$ and $g(x)$ have a common root. Hence, the resultant of $f(x)$ and $f'(x)$ is zero if and only if $f(x)$ has a multiple root.

EXAMPLES

1. If $f(x) = x^2 + 2x + 3$ and $g(x; \alpha) = 4x^3 + 5x^2 + 6x + (7 + \alpha)$, then the resultant of $f(x)$ and $g(x; \alpha)$ is

$$
\det \begin{bmatrix}
1 & 2 & 3 & 0 & 0 \\
0 & 1 & 2 & 3 & 0 \\
0 & 0 & 1 & 2 & 3 \\
0 & 4 & 5 & 6 & 7+\alpha \\
4 & 5 & 6 & 7+\alpha & 0
\end{bmatrix} = (16 + \alpha)^2
$$

Note that $g(x, -16) = (4x - 3)(x^2 + 2x + 3) = (4x - 3)f(x)$.

2. The resultant of $ax + b$ and $cx + d$ is $da - bc$.
3. The resultant of $(x + a)^5$ and $(x + b)^5$ is $(b - a)^{25}$.

2.2.1.4 Algebraic identities

$$(a \pm b)^2 = a^2 \pm 2ab + b^2.$$

$$(a \pm b)^3 = a^3 \pm 3a^2 b + 3ab^2 \pm b^3.$$

$$(a \pm b)^4 = a^4 \pm 4a^3 b + 6a^2 b^2 \pm 4ab^3 + b^4.$$

$$(a \pm b)^n = \sum_{k=0}^{n} \binom{n}{k} a^k (\pm b)^{n-k} \quad \text{where} \quad \binom{n}{k} = \frac{n!}{k!(n-k)!}.$$

$$a^2 + b^2 = (a + bi)(a - bi).$$

$$a^4 + b^4 = (a^2 + \sqrt{2}ab + b^2)(a^2 - \sqrt{2}ab + b^2).$$

$$a^2 - b^2 = (a - b)(a + b).$$

$$a^3 - b^3 = (a - b)(a^2 + ab + b^2).$$

$$a^n - b^n = (a - b)(a^{n-1} + a^{n-2}b + \cdots + ab^{n-2} + b^{n-1}).$$

$(a + b + c)^2 = a^2 + b^2 + c^2 + 2ab + 2ac + 2bc.$

$(a + b + c)^3 = a^3 + b^3 + c^3 + 3(a^2b + ab^2 + a^2c + ac^2 + b^2c + bc^2) + 6abc.$

$a^n + b^n = (a + b)\left(a^{n-1} + (-b)a^{n-2} + (-b)^2a^{n-3} + \cdots + (-b)^{n-1}\right)$ n odd

2.2.1.5   Laws of exponents

Assuming all quantities are real, a and b positive, and no denominators are zero, then

$$a^x a^y = a^{x+y}, \qquad \frac{a^x}{a^y} = a^{x-y}, \qquad (ab)^x = a^x b^x,$$

$$a^0 = 1 \text{ if } a \neq 0, \qquad a^{-x} = \frac{1}{a^x}, \qquad \left(\frac{a}{b}\right)^x = \frac{a^x}{b^x},$$

$$(a^x)^y = a^{xy}, \qquad a^{\frac{1}{x}} = \sqrt[x]{a}, \qquad \sqrt[x]{ab} = \sqrt[x]{a}\,\sqrt[x]{b},$$

$$\sqrt[x]{\sqrt[y]{a}} = \sqrt[xy]{a}, \qquad a^{\frac{x}{y}} = \sqrt[y]{a^x} = \left(\sqrt[y]{a}\right)^x, \qquad \sqrt[x]{\frac{a}{b}} = \frac{\sqrt[x]{a}}{\sqrt[x]{b}}.$$

2.2.1.6   Proportion

If $\frac{a}{b} = \frac{c}{d}$, then $\frac{a}{c} = \frac{b}{d}$, $ad = bc$, $\frac{a+b}{b} = \frac{c+d}{d}$, $\frac{a-b}{b} = \frac{c-d}{d}$, and $\frac{a-b}{a+b} = \frac{c-d}{c+d}$.

   If $\frac{a}{b} = \frac{c}{d}$, where a, b, c, and d are all positive numbers and a is the largest of the four numbers, then $a + d > b + c$.

2.2.2   PROGRESSIONS

2.2.2.1   Arithmetic progression

An *arithmetic progression* is a sequence of numbers such that the difference of any two consecutive numbers is constant. If the sequence is $a_1, a_2, \ldots a_n$, where $a_{i+1} - a_i = d$, then $a_k = a_1 + (k - 1)d$ and

$$a_1 + a_2 + \cdots + a_n = \frac{n}{2}\left(2a_1 + (n - 1)d\right). \tag{2.2.5}$$

In particular, the arithmetic progression $1, 2, \ldots, n$ has the sum $n(n + 1)/2$. (See page 22.)

2.2.2.2   Geometric progression

A *geometric progression* is a sequence of numbers such that the ratio of any two consecutive numbers is constant. If the sequence is $a_1, a_2, \ldots, a_n$, where $a_{i+1}/a_i = r$, then $a_k = a_1 r^{k-1}$.

$$a_1 + a_2 + \cdots + a_n = \begin{cases} a_1 \frac{1-r^n}{1-r} & r \neq 1 \\ na_1 & r = 1. \end{cases} \tag{2.2.6}$$

If $|r| < 1$, then the infinite geometric series $a_1(1 + r + r^2 + r^3 + \cdots)$ converges to $\frac{a_1}{1-r}$. For example, $1 + \frac{1}{2} + \frac{1}{4} + \frac{1}{8} + \cdots = 2$.

2.2.2.3 Means

1. The *arithmetic mean* of a and b is given by $\frac{a+b}{2}$. More generally, the arithmetic mean of $a_1, a_2, \ldots, a_n$ is given by $(a_1 + a_2 + \cdots + a_n)/n$.
2. The *geometric mean* of a and b is given by $\sqrt{ab}$. More generally, the geometric mean of $a_1, a_2, \ldots, a_n$ is given by $\sqrt[n]{a_1 a_2 \ldots a_n}$. The geometric mean of n positive numbers is less than the corresponding arithmetic mean, unless all of the numbers are equal.
3. The *harmonic mean* of a and b is given by $\dfrac{1}{\frac{1}{2}\left(\frac{1}{a} + \frac{1}{b}\right)} = \dfrac{2ab}{a+b}$.

If A, G, and H represent the arithmetic, geometric, and harmonic means of a and b, then $AH = G^2$.

2.2.3 DEMOIVRE'S THEOREM

A complex number $a + bi$ can be written in the form $re^{i\theta}$, where $r^2 = a^2 + b^2$ and $\tan\theta = b/a$. Because $e^{i\theta} = \cos\theta + i\sin\theta$,

$$(a + bi)^n = r^n(\cos n\theta + i\sin n\theta),$$

$$\sqrt[n]{1} = \cos\frac{2k\pi}{n} + i\sin\frac{2k\pi}{n}, \qquad\qquad k = 0, 1, \ldots, n - 1. \quad (2.2.7)$$

$$\sqrt[n]{-1} = \cos\frac{(2k+1)\pi}{n} + i\sin\frac{(2k+1)\pi}{n}, \qquad k = 0, 1, \ldots, n - 1.$$

2.2.4 PARTIAL FRACTIONS

The technique of partial fractions allows a quotient of two polynomials to be written as a sum of simpler terms.

Let $f(x)$ and $g(x)$ be polynomials and let the fraction be $\frac{f(x)}{g(x)}$. If the degree of $f(x)$ is greater than the degree of $g(x)$ then divide $f(x)$ by $g(x)$ to produce a quotient $q(x)$ and a remainder $r(x)$, where the degree of $r(x)$ is less than the degree of $g(x)$. That is, $\frac{f(x)}{g(x)} = q(x) + \frac{r(x)}{g(x)}$. Therefore, assume that the fraction has the form $\frac{r(x)}{g(x)}$, where the degree of the numerator is less than the degree of the denominator. The techniques used to find the partial fraction decomposition of $\frac{r(x)}{g(x)}$ depend on the factorization of $g(x)$.

2.2.4.1 Single linear factor

Suppose that $g(x) = (x - a)h(x)$, where $h(a) \neq 0$. Then

$$\frac{r(x)}{g(x)} = \frac{A}{x - a} + \frac{s(x)}{h(x)}, \quad (2.2.8)$$

where $s(x)$ can be computed and the number A is given by $r(a)/h(a)$. For example (here $r(x) = 2x$, $g(x) = x^2 - 1 = (x - 1)(x + 1)$, $h(x) = x + 1$, and $a = 1$):

$$\frac{2x}{x^2 - 1} = \frac{1}{x - 1} + \frac{1}{x + 1}. \quad (2.2.9)$$

2.2.4.2   Repeated linear factor

Suppose that $g(x) = (x - a)^k h(x)$, where $h(a) \neq 0$. Then

$$\frac{r(x)}{g(x)} = \frac{A_1}{x - a} + \frac{A_2}{(x - a)^2} + \cdots + \frac{A_k}{(x - a)^k} + \frac{s(x)}{h(x)}, \qquad (2.2.10)$$

for a computable $s(x)$ where

$$A_k = \frac{r(a)}{h(a)},$$

$$A_{k-1} = \frac{d}{dx}\left(\frac{r(x)}{h(x)}\right)\bigg|_{x=a},$$

$$A_{k-2} = \frac{1}{2!}\frac{d^2}{dx^2}\left(\frac{r(x)}{h(x)}\right)\bigg|_{x=a}, \qquad (2.2.11)$$

$$A_{k-j} = \frac{1}{j!}\frac{d^j}{dx^j}\left(\frac{r(x)}{h(x)}\right)\bigg|_{x=a}.$$

2.2.4.3   Single quadratic factor

Suppose that $g(x) = (x^2 + bx + c)h(x)$, where $b^2 - 4c < 0$ (so that $x^2 + bx + c$ does not factor into real linear factors) and $h(x)$ is relatively prime to $x^2 + bx + c$ (that is $h(x)$ and $x^2 + bx + c$ have no factors in common). Then

$$\frac{r(x)}{g(x)} = \frac{Ax + B}{x^2 + bx + c} + \frac{s(x)}{h(x)}. \qquad (2.2.12)$$

In order to determine A and B, multiply the equation by $g(x)$ so that there are no denominators remaining, and substitute any two values for x, yielding two equations for A and B.

When A and B are both real, if after multiplying the equation by $g(x)$ a root of $x^2 + bx + c$ is substituted for x, then the values of A and B can be inferred from this single complex equation by equating real and imaginary parts. (Since $x^2 + bx + c$ divides $g(x)$, there are no zeros in the denominator.) This technique can also be used for repeated quadratic factors (below).

2.2.4.4   Repeated quadratic factor

Suppose that $g(x) = (x^2 + bx + c)^k h(x)$, where $b^2 - 4c < 0$ (so that $x^2 + bx + c$ does not factor into real linear factors) and $h(x)$ is relatively prime to $x^2 + bx + c$. Then

$$\frac{r(x)}{g(x)} = \frac{A_1 x + B_1}{x^2 + bx + c} + \frac{A_2 x + B_2}{(x^2 + bx + c)^2} + \frac{A_3 x + B_3}{(x^2 + bx + c)^3}$$

$$+ \cdots + \frac{A_k x + B_k}{(x^2 + bx + c)^k} + \frac{s(x)}{h(x)}.$$

In order to determine A_i and B_i, multiply the equation by $g(x)$ so that there are no denominators remaining, and substitute any $2k$ values for x, yielding $2k$ equations for A_i and B_i.

2.3 POLYNOMIALS

All polynomials of degree 2, 3, or 4 are solvable by radicals. That is, their roots can be written in terms of a finite number of algebraic operations $(+, -, \times, \text{and} \div)$ and root-taking ($\sqrt[n]{}$). While some polynomials of higher degree can be solved by radicals (e.g., $x^{10} = 1$ is easy to solve), the general polynomial of degree 5 or higher cannot be solved by radicals. However, general polynomials of degree 5 and higher can be solved using hypergeometric functions.

2.3.1 QUADRATIC POLYNOMIALS

The solution of the equation $ax^2 + bx + c = 0$, where $a \neq 0$, is given by

$$x = \frac{-b \pm \sqrt{b^2 - 4ac}}{2a}. \tag{2.3.1}$$

The discriminant of the quadratic equation is $(b^2 - 4ac)/a^2$. Suppose that a, b, and c are all real. If the discriminant is negative, then the two roots are complex numbers which are conjugate. If the discriminant is positive, then the two roots are unequal real numbers. If the discriminant is 0, then the two roots are equal.

2.3.2 CUBIC POLYNOMIALS

To solve the equation $ax^3 + bx^2 + cx + d = 0$, where $a \neq 0$, begin by making the substitution $y = x + \frac{b}{3a}$. That gives the equation $y^3 + 3py + q = 0$, where $p = \frac{3ac - b^2}{9a^2}$ and $q = \frac{2b^3 - 9abc + 27a^2 d}{27a^3}$. The discriminant of this polynomial is $4p^3 + q^2$.

The solutions of $y^3 + 3py + q = 0$ are given by $\sqrt[3]{\alpha} + \sqrt[3]{\beta}$, $e^{\frac{2\pi i}{3}}\sqrt[3]{\alpha} + e^{\frac{4\pi i}{3}}\sqrt[3]{\beta}$, and $e^{\frac{4\pi i}{3}}\sqrt[3]{\alpha} + e^{\frac{2\pi i}{3}}\sqrt[3]{\beta}$, where

$$\alpha = \frac{-q + \sqrt{q^2 + 4p^3}}{2} \quad \text{and} \quad \beta = \frac{-q - \sqrt{q^2 + 4p^3}}{2}. \tag{2.3.2}$$

Suppose that p and q are real numbers. If the discriminant is positive, then one root is real, and two are complex conjugates. If the discriminant is 0, then there are three real roots, of which at least two are equal. If the discriminant is negative, then there are three unequal real roots.

2.3.2.1 Trigonometric solution of cubic polynomials

In the event that the roots of the polynomial $y^3 + 3py + q = 0$ are all real, meaning that $q^2 + 4p^3 \leq 0$, then the expressions above involve complex numbers. In that case one can also express the solution in terms of trigonometric functions. Define r and θ by

$$r = \sqrt{-p^3} \quad \text{and} \quad \theta = \cos^{-1}\frac{-q}{2r}. \tag{2.3.3}$$

Then the three roots are given by

$$2\sqrt[3]{r}\cos\frac{\theta}{3}, \qquad 2\sqrt[3]{r}\cos\frac{\theta+2\pi}{3}, \qquad \text{and} \qquad 2\sqrt[3]{r}\cos\frac{\theta+4\pi}{3}. \qquad (2.3.4)$$

2.3.3 QUARTIC POLYNOMIALS

To solve the equation $ax^4 + bx^3 + cx^2 + dx + e = 0$, where $a \neq 0$, start with the substitution $y = x + \frac{b}{4a}$. This gives $y^4 + py^2 + qy + r = 0$, where $p = \frac{8ac-3b^2}{8a^2}$, $q = \frac{b^3-4abc+8a^2d}{8a^3}$, and $r = \frac{16ab^2c+256a^3e-3b^4-64a^2bd}{256a^4}$.

The *cubic resolvent* of this polynomial is defined as $t^3 - pt^2 - 4rt + (4pr - q^2) = 0$. If u is a root of the cubic resolvent (see previous section), then the solutions of the original quartic are given by $x = y - \frac{b}{4a}$ where y is a solution of:

$$y^2 \pm \sqrt{u-p}\left(y - \frac{q}{2(u-p)}\right) + \frac{u}{2} = 0. \qquad (2.3.5)$$

2.3.4 QUARTIC CURVES

Any quartic curve of the form $y^2 = (x-\alpha)(x-\beta)(x-\gamma)(x-\delta)$ can be written as

$$\left(\frac{y}{(x-\alpha)^2}\right)^2 = \left(1 - \frac{\beta-\alpha}{x-\alpha}\right)\left(1 - \frac{\gamma-\alpha}{x-\alpha}\right)\left(1 - \frac{\delta-\alpha}{x-\alpha}\right) \qquad (2.3.6)$$

and hence it is cubic in the coordinates $X = \frac{1}{x-\alpha}$ and $Y = \frac{y}{x-\alpha^2}$. For example, $y^2 = 1-x^4$ is the cubic $Y^2 = 4X^3-6X^2+4X-1$ in the coordinates $X = 1/(1-x)$ and $Y = y/(1-x)^2$.

2.3.5 QUINTIC POLYNOMIALS

Some quintic equations are solvable by radicals. If the function $f(x) = x^5 + ax + b$ (with a and b rational) is irreducible, then $f(x) = 0$ is solvable by radicals if, and only if, numbers ϵ, c, and e exist (with $\epsilon = \pm 1$, $c \geq 0$, and $e \neq 0$) such that

$$a = \frac{5e^4(3-4\epsilon c)}{c^2+1} \quad \text{and} \quad b = \frac{-4e^5(11\epsilon+2c)}{c^2+1}. \qquad (2.3.7)$$

In this case, the roots are given by $x = e\left(\omega^j u_1 + \omega^{2j} u_2 + \omega^{3j} u_3 + \omega^{4j} u_4\right)$ for $j = 0,1,2,3,4$, where ω is a fifth root of unity ($\omega = \exp(2\pi i/5)$) and

$$u_1 = \left(\frac{v_1^2 v_3}{D^2}\right), \quad u_2 = \left(\frac{v_3^2 v_4}{D^2}\right), \quad u_3 = \left(\frac{v_2^2 v_1}{D^2}\right), \quad u_4 = \left(\frac{v_4^2 v_2}{D^2}\right),$$

$$v_1 = \sqrt{D} + \sqrt{D - \epsilon\sqrt{D}}, \qquad v_2 = -\sqrt{D} - \sqrt{D + \epsilon\sqrt{D}},$$

$$v_3 = -\sqrt{D} + \sqrt{D + \epsilon\sqrt{D}}, \qquad v_4 = \sqrt{D} - \sqrt{D - \epsilon\sqrt{D}}, \quad \text{and}$$

$$D = c^2 + 1.$$

EXAMPLE   The quintic $f(x) = x^5 + 15x + 12$ has the values $\epsilon = -1$, $c = 4/3$, and $e = 1$. Hence the unique real root is given by

$$x = \left(\frac{-75 + 21\sqrt{10}}{125}\right)^{1/5} + \left(\frac{-75 - 21\sqrt{10}}{125}\right)^{1/5}$$
$$+ \left(\frac{225 + 72\sqrt{10}}{125}\right)^{1/5} + \left(\frac{225 - 72\sqrt{10}}{125}\right)^{1/5}.$$

2.3.6   TSCHIRNHAUS' TRANSFORMATION

The nth degree polynomial equation

$$a_n x^n + a_{n-1} x^{n-1} + \cdots + a_1 x + a_0 = 0 \tag{2.3.8}$$

can be transformed to one with up to three fewer terms,

$$z^n + b_{n-4} z^{n-4} + \cdots + b_1 z + b_0 = 0 \tag{2.3.9}$$

by making a transformation of the form

$$z_j = \gamma_4 x_j^4 + \gamma_3 x_j^3 + \gamma_2 x_j^2 + \gamma_1 x_j + \gamma_0 \tag{2.3.10}$$

for $j = 1, \ldots, n$ where the $\{\gamma_i\}$ can be computed, in terms of radicals, from the $\{a_i\}$. Hence, a general quintic polynomial can be transformed to the form $z^5 + az + b = 0$.

2.3.7   POLYNOMIAL NORMS

The polynomial $P(x) = \displaystyle\sum_{j=0}^{n} a_j x^j$ has the norms:

$$\|P\|_1 = \int_0^{2\pi} |P(e^{i\theta})| \frac{d\theta}{2\pi} \qquad\qquad |P|_1 = \sum_{j=0}^{n} |a_j|. \tag{2.3.11}$$

$$\|P\|_2 = \left(\int_0^{2\pi} |P(e^{i\theta})|^2 \frac{d\theta}{2\pi}\right)^{1/2} \qquad |P|_2 = \left(\sum_{j=1}^{n} |a_j|^2\right)^{1/2}. \tag{2.3.12}$$

$$\|P\|_\infty = \max_{|z|=1} |P(z)| \qquad\qquad\quad |P|_\infty = \max_j |a_j|. \tag{2.3.13}$$

For the double bar norms, P is considered as a function on the unit circle; for the single bar norms, P is identified with its coefficients. These norms are comparable:

$$|P|_\infty \le \|P\|_1 \le |P|_2 = \|P\|_2 \le \|P\|_\infty \le |P|_1 \le n|P|_\infty. \tag{2.3.14}$$

2.3.8   CYCLOTOMIC POLYNOMIALS

The dth *cyclotomic polynomial*, $\Phi_d(x)$, is

$$\Phi_d(x) = \prod_{\substack{k=1,2,\ldots \\ (k,d)=1}}^{d} (x - \xi_k) = x^{\phi(d)} + \ldots \tag{2.3.15}$$

where the $\xi_k = e^{2\pi i k/d}$ are the primitive d^{th} roots of unity

$$\Phi_p(x) = \sum_{k=0}^{p-1} x^k = \frac{x^p - 1}{x - 1} \qquad \text{if } p \text{ is prime}$$

$$x^n - 1 = \prod_{d|n} \Phi_d(x) \tag{2.3.16}$$

$$x^n + 1 = \frac{x^{2n} - 1}{x^n - 1} = \frac{\prod_{d|2n} \Phi_d(x)}{\prod_{d|n} \Phi_d(x)} = \prod_{d|m} \Phi_{2^t d}(x)$$

where $n = 2^{t-1}m$ and m is odd.

n	cyclotomic polynomial of degree n
1	$-1 + x$
2	$1 + x$
3	$1 + x + x^2 = \frac{x^3 - 1}{x - 1}$
4	$1 + x^2$
5	$1 + x + x^2 + x^3 + x^4 = \frac{x^5 - 1}{x - 1}$
6	$1 - x + x^2$
7	$1 + x + x^2 + x^3 + x^4 + x^5 + x^6 = \frac{x^7 - 1}{x - 1}$
8	$1 + x^4$
9	$1 + x^3 + x^6$
10	$1 - x + x^2 - x^3 + x^4$
11	$1 + x + x^2 + x^3 + x^4 + x^5 + x^6 + x^7 + x^8 + x^9 + x^{10} = \frac{x^{11} - 1}{x - 1}$
12	$1 - x^2 + x^4$
13	$\frac{x^{13} - 1}{x - 1}$
14	$1 - x + x^2 - x^3 + x^4 - x^5 + x^6$
15	$1 - x + x^3 - x^4 + x^5 - x^7 + x^8$
16	$1 + x^8$
17	$\frac{x^{17} - 1}{x - 1}$
18	$1 - x^3 + x^6$
19	$\frac{x^{19} - 1}{x - 1}$
20	$1 - x^2 + x^4 - x^6 + x^8$
21	$1 - x + x^3 - x^4 + x^6 - x^8 + x^9 - x^{11} + x^{12}$
22	$1 - x + x^2 - x^3 + x^4 - x^5 + x^6 - x^7 + x^8 - x^9 + x^{10}$
23	$\frac{x^{23} - 1}{x - 1}$
24	$1 - x^4 + x^8$
25	$1 + x^5 + x^{10} + x^{15} + x^{20}$
26	$1 - x + x^2 - x^3 + x^4 - x^5 + x^6 - x^7 + x^8 - x^9 + x^{10} - x^{11} + x^{12}$
27	$1 + x^9 + x^{18}$
28	$1 - x^2 + x^4 - x^6 + x^8 - x^{10} + x^{12}$
29	$\frac{x^{29} - 1}{x - 1}$
30	$1 + x - x^3 - x^4 - x^5 + x^7 + x^8$

2.3.9 OTHER POLYNOMIAL PROPERTIES

1. *Jensen's inequality:* For the polynomial $P(x) = \sum\limits_{j=0}^{n} a_j x^j$, with $a_0 \neq 0$

$$\int_0^{2\pi} \log \left| P\left(e^{i\theta}\right) \right| \frac{d\theta}{2\pi} \geq \log |a_0| \, . \tag{2.3.17}$$

2. *Symmetric form:* The polynomial $P(x_1, \ldots, x_n) = \sum\limits_{|\alpha|=m} a_\alpha x_1^{\alpha_1} x_2^{\alpha_2} \ldots x_N^{\alpha_N}$,

   where $\alpha = (\alpha_1, \ldots, \alpha_N)$ can be written in the symmetric form

$$P(x_1, \ldots, x_n) = \sum\limits_{i_1, \ldots, i_m = 1}^{N} c_{i_1, \ldots, i_m} x_{i_1} x_{i_2} \ldots x_{i_m} \tag{2.3.18}$$

   with $c_{i_1, \ldots, i_m} = \frac{1}{m!} \frac{\partial^m P}{\partial x_{i_1} \ldots \partial x_{i_m}}$. This means that the $x_1 x_2$ term is written as $\frac{1}{2}(x_1 x_2 + x_2 x_1)$, the term $x_1 x_2^2$ becomes $\frac{1}{3}(x_1 x_2 x_2 + x_2 x_1 x_2 + x_2 x_2 x_1)$.

3. The *Mahler measure* (a valuation) of the polynomial $P(x) = a_n x^n + a_{n-1} x^{n-1} + \cdots + a_0 = a_n(x - z_1)(x - z_2) \ldots (x - z_n)$ is given by

$$M(P) = |a_n| \prod\limits_{i=1}^{n} \max(1, |z_i|) = \exp\left(\int_0^1 \log \left| P\left(e^{2\pi it}\right) \right| \, dt \right) \, .$$

   This valuation satisfies the properties:

   (a) $M(P) \, M(Q) = M(PQ)$

   (b) $M(P(x)) = M(P(x^k))$ for $k \geq 1$

   (c) $M(x^n P(x^{-1})) = M(P(x))$

2.4 NUMBER THEORY

2.4.1 DIVISIBILITY

The notation "$a|b$" means that the number a evenly divides the number b. That is, the ratio $\frac{b}{a}$ is an integer.

2.4.2 CONGRUENCES

1. If the integers a and b leave the same remainder when divided by the number n, then a and b are *congruent* modulo n. This is written $a \equiv b \pmod{n}$.

2. If the congruence $x^2 \equiv a \pmod{p}$ has a solution, then a is a *quadratic residue* of p. Otherwise, a is a *quadratic non-residue* of p.

   (a) Let p be an odd prime. *Legendre's symbol* $\left(\frac{a}{p}\right)$ has the value $+1$ if a is a quadratic residue of p, and the value -1 if a is a quadratic non-residue of p. This can be written $\left(\frac{a}{p}\right) \equiv a^{(p-1)/2} \pmod{p}$.

   (b) The *Jacobi symbol* generalizes the Legendre symbol to non-prime moduli. If $n = \prod_{i=1}^{k} p_i^{b_i}$ then the Jacobi symbol can be written in terms of the Legendre symbol as follows

$$\left(\frac{a}{n}\right) = \prod_{i=1}^{k} \left(\frac{a}{p_i}\right)^{b_i}. \tag{2.4.1}$$

3. An *exact covering* sequence is a set of non-negative ordered pairs $\{(a_i, b_i)\}_{i=1}^{k}$ such that every non-negative integer n satisfies $n \equiv a_i \pmod{b_i}$ for exactly one i. An exact covering sequence satisfies

$$\sum_{i=1}^{k} \frac{x^{a_i}}{1 - x^{b_i}} = \frac{1}{1 - x}. \tag{2.4.2}$$

For example, every positive integer n is either congruent to 1 mod 2, or 0 mod 4, or 2 mod 4. Hence, the three pairs $\{(1, 2), (0, 4), (2, 4)\}$ of residues and moduli *exactly cover* the positive integers. Note that

$$\frac{x}{1 - x^2} + \frac{1}{1 - x^4} + \frac{x^2}{1 - x^4} = \frac{1}{1 - x}. \tag{2.4.3}$$

4. Carmichael numbers are composite numbers n that satisfy $a^{n-1} \equiv 1 \pmod{n}$ for every a (with $1 < a < n$) that is relatively prime to n. There are infinitely many Carmichael numbers. Every Carmichael number has at least three prime factors. If $n = \prod_i p_i$ is a Carmichael number, then $(p_i - 1)$ divides $(n - 1)$ for each i.

There are 43 Carmichael numbers less than 10^6 and 105,212 less than 10^{15}. The Carmichael numbers less than ten thousand are 561, 1105, 1729, 2465, 2821, 6601, and 8911.

2.4.2.1 Properties of congruences

1. If $a \equiv b \pmod{n}$, then $b \equiv a \pmod{n}$.

2. If $a \equiv b \pmod{n}$, and $b \equiv c \pmod{n}$, then $a \equiv c \pmod{n}$.

3. If $a \equiv a' \pmod{n}$, and $b \equiv b' \pmod{n}$, then $a \pm b \equiv a' \pm b' \pmod{n}$.

4. If $a \equiv a' \pmod{n}$, then $a^2 \equiv (a')^2 \pmod{n}$, $a^3 \equiv (a')^3 \pmod{n}$, etc.

5. If $\text{GCD}(k, m) = d$, then the congruence $kx \equiv n \pmod{m}$ is solvable if and only if d divides n. It then has d solutions.

6. If p is a prime, then $a^p \equiv a \pmod{p}$.

7. If p is a prime, and p does not divide a, then $a^{p-1} \equiv 1 \pmod{p}$.

8. If $\text{GCD}(a, m) = 1$ then $a^{\phi(m)} \equiv 1 \pmod{m}$. (See Section 2.4.15 for $\phi(m)$.)

9. If p is an odd prime and a is not a multiple of p, then Wilson's theorem states $(p-1)! \equiv -\left(\frac{a}{p}\right) a^{(p-1)/2} \pmod{p}$.

10. If p and q are odd primes, then Gauss's *law of quadratic reciprocity* states that $\left(\frac{p}{q}\right)\left(\frac{q}{p}\right) = (-1)^{(p-1)(q-1)/4}$. Therefore, if a and b are relatively prime odd integers and $b \geq 3$, then $\left(\frac{a}{b}\right) = (-1)^{(a-1)(b-1)/4}\left(\frac{b}{a}\right)$.

11. The number -1 is a quadratic residue of primes of the form $4k + 1$ and a non-residue of primes of the form $4k + 3$. That is

$$\left(\frac{-1}{p}\right) = (-1)^{(p-1)/2} = \begin{cases} +1 & \text{when } p \equiv 1 \pmod{4} \\ -1 & \text{when } p \equiv 3 \pmod{4} \end{cases}$$

12. The number 2 is a quadratic residue of primes of the form $8k \pm 1$ and a non-residue of primes of the form $8k \pm 3$. That is

$$\left(\frac{2}{p}\right) = (-1)^{(p^2-1)/8} = \begin{cases} +1 & \text{when } p \equiv \pm 1 \pmod{8} \\ -1 & \text{when } p \equiv \pm 3 \pmod{8} \end{cases}$$

13. The number -3 is a quadratic residue of primes of the form $6k + 1$ and a non-residue of primes of the form $6k + 5$.

14. The number 3 is a quadratic residue of primes of the form $12k \pm 1$ and a non-residue of primes of the form $12k \pm 5$.

2.4.3 CHINESE REMAINDER THEOREM

Let $m_1, m_2, \ldots, m_r$ be pairwise relatively prime integers. Then the system of congruences

$$
\begin{aligned}
x &\equiv a_1 \quad (\mathrm{mod}\ m_1) \\
x &\equiv a_2 \quad (\mathrm{mod}\ m_2) \\
&\ \ \vdots \\
x &\equiv a_r \quad (\mathrm{mod}\ m_r)
\end{aligned}
\tag{2.4.4}
$$

has a unique solution modulo $M = m_1 m_2 \cdots m_r$. This unique solution can be written as

$$
x = a_1 M_1 y_1 + a_2 M_2 y_2 + \cdots + a_r M_r y_r
\tag{2.4.5}
$$

where $M_k = M/m_k$, and y_k is the inverse of M_k (modulo m_k).

EXAMPLE   For the system of congruences

$$
\begin{aligned}
x &\equiv 1 \quad (\mathrm{mod}\ 3) \\
x &\equiv 2 \quad (\mathrm{mod}\ 5) \\
x &\equiv 3 \quad (\mathrm{mod}\ 7)
\end{aligned}
$$

we have $M = 3 \cdot 5 \cdot 7 = 105$. Hence $M_1 = 35$, $M_2 = 21$, and $M_3 = 15$. The equation for y_1 is $M_1 y_1 = 35 y_1 \equiv 1 \pmod{3}$ with solution $y_1 \equiv 2 \pmod 3$. Likewise, $y_2 \equiv 1 \pmod 5$ and $y_3 \equiv 1 \pmod 7$. This results in $x = 1 \cdot 35 \cdot 2 + 2 \cdot 21 \cdot 1 + 3 \cdot 15 \cdot 1 \equiv 52 \pmod{105}$.

2.4.4 CONTINUED FRACTIONS

The symbol $[a_0, a_1, \ldots, a_N]$, with $a_i > 0$, represents the simple *continued fraction*,

$$
[a_0, a_1, \ldots, a_N] = a_0 + \cfrac{1}{a_1 + \cfrac{1}{a_2 + \cfrac{1}{a_3 + \cfrac{1}{a_4 + \cfrac{\cdots}{\cdots + \cfrac{1}{a_N}}}}}}.
\tag{2.4.6}
$$

The nth *convergent* (with $0 < n < N$) of $[a_0, a_1, \ldots, a_N]$ is defined to be $[a_0, a_1, \ldots, a_n]$. If $\{p_n\}$ and $\{q_n\}$ are defined by

$$
\begin{aligned}
p_0 &= a_0, & p_1 &= a_1 a_0 + 1, & p_n &= a_n p_{n-1} + p_{n-2} \quad (2 \leq n \leq N) \\
q_0 &= 1, & q_1 &= a_1, & q_n &= a_n q_{n-1} + q_{n-2} \quad (2 \leq n \leq N)
\end{aligned}
$$

then $[a_0, a_1, \ldots, a_n] = p_n/q_n$. The continued fraction is convergent if and only if the infinite series $\sum_i^\infty a_i$ is divergent.

If the positive rational number x can be represented by a simple continued fraction with an odd (even) number of terms, then it is also representable by one with an even (odd) number of terms. (Specifically, if $a_n = 1$ then $[a_0, \ldots, a_{n-1}, 1] = [a_0, \ldots, a_{n-1} + 1]$, and if $a_n \geq 2$, then $[a_0, \ldots, a_n] = [a_0, \ldots, a_n - 1, 1]$.) Aside from this indeterminacy, the simple continued fraction of x is unique. The error in approximating by a convergent is bounded by

$$\left| x - \frac{p_n}{q_n} \right| \leq \frac{1}{q_n q_{n+1}} < \frac{1}{q_n^2}. \tag{2.4.7}$$

The algorithm for finding a continued fraction expansion of a number is to remove the integer part of the number (this becomes a_i), take the reciprocal, and repeat. For example, for the number π:

$$\beta_0 = \pi \approx 3.14159 \qquad\qquad a_0 = \lfloor \beta_0 \rfloor = 3$$
$$\beta_1 = 1/(\beta_0 - a_0) \approx 7.062 \qquad a_1 = \lfloor \beta_1 \rfloor = 7$$
$$\beta_2 = 1/(\beta_1 - a_1) \approx 15.997 \qquad a_2 = \lfloor \beta_2 \rfloor = 15$$
$$\beta_3 = 1/(\beta_2 - a_2) \approx 1.0034 \qquad a_3 = \lfloor \beta_3 \rfloor = 1$$
$$\beta_4 = 1/(\beta_3 - a_3) \approx 292.6 \qquad a_4 = \lfloor \beta_4 \rfloor = 292$$

Approximations to π and e may be found from $\pi = [3, 7, 15, 1, 292, 1, 1, 1, 2, 1, 3, 1, 14, 2, \ldots]$ and $e = [2, 1, 2, 1, 1, 4, 1, 1, 6, \ldots, 1, 1, 2n, \ldots]$. The convergents for π are $\frac{22}{7} \approx 3.14\underline{2}$, $\frac{333}{106} \approx 3.1415\underline{0}$, $\frac{355}{113} \approx 3.141592\underline{9}$, $\frac{103993}{33102} \approx 3.141592653\underline{0}, \ldots$. The convergents for e are $\frac{8}{3} \approx 2.\underline{6}$, $\frac{11}{4} \approx 2.7\underline{5}$, $\frac{19}{7} \approx 2.71\underline{4}$, $\frac{87}{32} \approx 2.718\underline{7}, \ldots$.

A *periodic continued fraction* is an infinite continued fraction in which $a_l = a_{l+k}$ for all $l \geq L$. The set of partial quotients $a_L, a_{L+1}, \ldots, a_{L+k-1}$ is the *period*. A periodic continued fraction may be written as

$$\left[a_0, a_1, \ldots, a_{L-1}, \overline{a_L, a_{L+1}, \ldots, a_{L+k-1}} \right]. \tag{2.4.8}$$

For example,

$$\sqrt{2} = [1, \overline{2}] \qquad \sqrt{6} = [2, \overline{2, 4}] \qquad \sqrt{10} = [3, \overline{6}] \qquad \sqrt{14} = [3, \overline{1, 2, 1, 6}]$$
$$\sqrt{3} = [1, \overline{1, 2}] \qquad \sqrt{7} = [2, \overline{1, 1, 1, 4}] \qquad \sqrt{11} = [3, \overline{3, 6}] \qquad \sqrt{15} = [3, \overline{1, 6}]$$
$$\sqrt{4} = [2] \qquad \sqrt{8} = [2, \overline{1, 4}] \qquad \sqrt{12} = [3, \overline{2, 6}] \qquad \sqrt{16} = [4]$$
$$\sqrt{5} = [2, \overline{4}] \qquad \sqrt{9} = [3] \qquad \sqrt{13} = [3, \overline{1, 1, 1, 1, 6}] \qquad \sqrt{17} = [4, \overline{8}]$$

If $x = [\overline{b, a}]$ then $x = \frac{1}{2}(b + \sqrt{b^2 + \frac{4b}{a}})$. For example, $[\overline{1}] = [\overline{1, 1}] = (1 + \sqrt{5})/2$, $[\overline{2}] = [\overline{2, 2}] = 1 + \sqrt{2}$, and $[\overline{2, 1}] = 1 + \sqrt{3}$.

Functions can be represented as continued fractions. Using the notation

$$b_0 + \cfrac{a_1}{b_1 + \cfrac{a_2}{b_2 + \cfrac{a_3}{b_3 + \cfrac{a_4}{b_4 + \ldots}}}} \equiv b_0 + \frac{a_1}{b_1 +} \frac{a_2}{b_2 +} \frac{a_3}{b_3 +} \frac{a_4}{b_4 +} \cdots \tag{2.4.9}$$

we have (allowable values of z may be restricted in the following)

(a) $\ln(1+z) = \dfrac{z}{1+}\dfrac{z}{2+}\dfrac{z}{3+}\dfrac{4z}{4+}\dfrac{4z}{5+}\dfrac{9z}{6+}\cdots$

(b) $e^z = \dfrac{1}{1-}\dfrac{z}{1+}\dfrac{z}{2-}\dfrac{z}{3+}\dfrac{z}{2-}\dfrac{z}{5+}\dfrac{z}{2-}\cdots = 1 + \dfrac{z}{1-}\dfrac{z}{2+}\dfrac{z}{3-}\dfrac{z}{2+}\dfrac{z}{5-}\dfrac{z}{2+}\dfrac{z}{7-}\cdots$

(c) $\tan z = \dfrac{z}{1-}\dfrac{z^2}{3-}\dfrac{z^2}{5-}\dfrac{z^2}{7-}\cdots$

(d) $\tanh z = \dfrac{z}{1+}\dfrac{z^2}{3+}\dfrac{z^2}{5+}\dfrac{z^2}{7+}\cdots$

2.4.5 DIOPHANTINE EQUATIONS

A *diophantine equation* is one which requires the solutions to come from the set of integers.

1. Apart from the trivial solutions (with $x = y = 0$ or $x = u$), the general solution to the equation $x^3 + y^3 = u^3 + v^3$ is given by

$$x = \lambda\left[1 - (a - 3b)(a^2 + 3b^2)\right] \qquad y = \lambda\left[(a + 3b)(a^2 + 3b^2) - 1\right]$$
$$u = \lambda\left[(a + 3b) - (a^2 + 3b^2)^2\right] \qquad v = \lambda\left[(a^2 + 3b^2)^2 - (a - 3b)\right]$$
$$(2.4.10)$$

where $\{\lambda, a, b\}$ are any rational numbers except that $\lambda \neq 0$.

2. A parametric solution to $x^4 + y^4 = u^4 + v^4$ is given by

$$x = a^7 + a^5 b^2 - 2a^3 b^4 + 3a^2 b^5 + ab^6$$
$$y = a^6 b - 3a^5 b^2 - 2a^4 b^3 + a^2 b^5 + b^7$$
$$u = a^7 + a^5 b^2 - 2a^3 b^4 - 3a^2 b^5 + ab^6 \qquad (2.4.11)$$
$$v = a^6 b + 3a^5 b^2 - 2a^4 b^3 + a^2 b^5 + b^7$$

3. Fermat's last theorem states that there are no integer solutions to $x^n + y^n = z^n$, when $n > 2$. This was proved by Andrew Wiles in 1995.

4. Bachet's equation, $y^2 = x^3 + k$, has no solutions for k equal to any of the following: $-144, -105, -78, -69, -42, -34, -33, -31, -24, -14, -5, 7,$ $11, 23, 34, 45, 58, 70$.

2.4.5.1 Pythagorean triples

If the positive integers A, B, and C satisfy the relationship $A^2 + B^2 = C^2$, then the triplet (A, B, C) is a *Pythagorean triple*. It is possible to construct a right triangle with sides of length A and B and a hypotenuse of C.

There are infinitely many Pythagorean triples. The most general solution to $A^2 + B^2 = C^2$, with $\text{GCD}(A, B) = 1$ and A even, is given by

$$A = 2xy \qquad B = x^2 - y^2 \qquad C = x^2 + y^2, \qquad (2.4.12)$$

where x and y are relatively prime integers of opposite parity (i.e., one is even and the other is odd) with $x > y > 0$. The following table shows some Pythagorean triples with the associated (x, y) values.

x	y	$A = 2xy$	$B = x^2 - y^2$	$C = x^2 + y^2$
2	1	4	3	5
4	1	8	15	17
6	1	12	35	37
8	1	16	63	65
10	1	20	99	101
3	2	12	5	13
5	2	20	21	29
7	2	28	45	51
4	3	24	7	25

2.4.5.2 Pell's equation

Pell's equation is $x^2 - dy^2 = 1$. The solutions, integral values of (x, y), arise from continued fraction convergents of $\sqrt{d}$ (see page 96). If (x, y) is the least positive solution to Pell's equation (with d square-free), then every positive solution (x_k, y_k) is given by

$$x_k + y_k\sqrt{d} = (x + y\sqrt{d})^k. \qquad (2.4.13)$$

The following tables contain the least positive solutions to Pell's equation with d square-free and $d < 100$.

d	x	y	d	x	y	d	x	y
2	3	2	35	6	1	69	7,775	936
3	2	1	37	73	12	70	251	30
5	9	4	38	37	6	71	3,480	413
6	5	2	39	25	4	73	2,281,249	267,000
7	8	3	41	2,049	320	74	3,699	430
10	19	6	42	13	2	77	351	40
11	10	3	43	3,482	531	78	53	6
13	649	180	46	24,335	3,588	79	80	9
14	15	4	47	48	7	82	163	18
15	4	1	51	50	7	83	82	9
17	33	8	53	66,249	9,100	85	285,769	30,996
19	170	39	55	89	12	86	10,405	1,122
21	55	12	57	151	20	87	28	3
22	197	42	58	19,603	2,574	89	500,001	53,000
23	24	5	59	530	69	91	1,574	165
26	51	10	61	1,766,319,049	226,153,980	93	12,151	1,260
29	9,801	1,820	62	63	8	94	2,143,295	221,064
30	11	2	65	129	16	95	39	4
31	1,520	273	66	65	8	97	62,809,633	6,377,352
33	23	4	67	48,842	5,967			

EXAMPLES

1. The number $\sqrt{2}$ has the continued fraction expansion $[1, 2, 2, 2, 2, \ldots]$, with convergents $\frac{3}{2}, \frac{7}{5}, \frac{17}{12}, \frac{41}{29}, \frac{99}{70}, \ldots$. In this case, every second convergent represents a solution:

$$3^2 - 2 \cdot 2^2 = 1,$$
$$17^2 - 2 \cdot 12^2 = 1, \text{ and}$$
$$99^2 - 2 \cdot 70^2 = 1.$$

2. The least positive solution for $d = 11$ is $(x, y) = (10, 3)$. Since $(10 + 3\sqrt{11})^2 = 199 + 60\sqrt{11}$, another solution is given by $(x_2, y_2) = (199, 60)$.

2.4.5.3 Waring's problem

If each positive integer can be expressed as a sum of n k^{th} powers, then there is a least value of n for which this is true: this is the number $g(k)$. For all sufficiently large numbers, however, a smaller value of n may suffice: this is the number $G(k)$. Waring's problem is to determine $g(n)$ and $G(n)$.

1. *Lagrange's theorem* states: "Every positive integer is the sum of four squares"; this is equivalent to the statement $g(2) = 4$. The following identity shows how a product can be written as the sum of four squares:

$$(x_1^2 + x_2^2 + x_3^2 + x_4^2)(y_1^2 + y_2^2 + y_3^2 + y_4^2) = \qquad (2.4.14)$$
$$(x_1 y_1 + x_2 y_2 + x_3 y_3 + x_4 y_4)^2 + (x_1 y_2 - x_2 y_1 + x_3 y_4 - x_4 y_3)^2$$
$$+ (x_1 y_3 - x_3 y_1 + x_4 y_2 - x_2 y_4)^2 + (x_1 y_4 - x_4 y_1 + x_2 y_3 - x_3 y_2)^2$$

2. Consider $k = 3$; all numbers can be written as the sum of not more than 9 cubes, so that $g(3) = 9$. However, only the two numbers

$$23 = 2^3 + 2^3 + 1^3 + 1^3 + 1^3 + 1^3 + 1^3 + 1^3 + 1^3,$$
$$239 = 4^3 + 4^3 + 3^3 + 3^3 + 3^3 + 3^3 + 1^3 + 1^3 + 1^3,$$

require the use of 9 cubes; so $G(3) \leq 8$.

3. $G(k) \leq 6k \log k + \left(4 + 3 \log \left(3 + \frac{2}{k}\right)\right) k + 3$.

4. Known values include

$g(2) = 4$	$G(2) = 4$
$g(3) = 9$	$4 \leq G(3) \leq 7$
$g(4) = 19$	$G(4) = 16$
$g(5) = 37$	$6 \leq G(5) \leq 18$
$g(6) = 73$	$9 \leq G(6) \leq 27$
$143 \leq g(7) \leq 306$	$8 \leq G(7) \leq 36$
$279 \leq g(8) \leq 36,119$	$32 \leq G(8) \leq 42$
$548 \leq g(9)$	$13 \leq G(9) \leq 82$
$1,079 \leq g(10)$	$12 \leq G(10) \leq 102$

2.4.6 GREATEST COMMON DIVISOR

The *greatest common divisor* of the integers n and m is the largest integer that evenly divides both n and m; this is written as $\text{GCD}(n, m)$ or (n, m). The Euclidean algorithm is frequently used for computing the GCD of two numbers; it utilizes the fact that $m = \left\lfloor \frac{m}{n} \right\rfloor n + p$ where $0 \le p < n$.

Given the integers m and n, two integers a and b can always be found so that $am + bn = \text{GCD}(n, m)$.

Two numbers, m and n, are said to be *relatively prime* if they have no divisors in common; i.e., if $\text{GCD}(a, b) = 1$. The probability that two integers chosen randomly are relatively prime is $\pi/6$.

EXAMPLE Consider 78 and 21. Since $78 = 3 \cdot 21 + 15$, the largest integer that evenly divides both 78 and 21 is also the largest integer that evenly divides both 21 and 15. Iterating results in:

$$78 = 3 \cdot 21 + 15$$
$$21 = 1 \cdot 15 + 6$$
$$15 = 2 \cdot 6 + 3$$
$$6 = 2 \cdot 3 + 0$$

Hence $\text{GCD}(78, 21) = \text{GCD}(21, 15) = \text{GCD}(15, 6) = \text{GCD}(6, 3) = 3$. Note that $78 \cdot (-4) + 21 \cdot 15 = 3$.

2.4.7 LEAST COMMON MULTIPLE

The *least common multiple* of the integers a and b (denoted $\text{LCM}(a, b)$) is the smallest integer r that is divisible by both a and b. The simplest way to find the LCM of a and b is via the formula $\text{LCM}(a, b) = ab/\text{GCD}(a, b)$. For example, $\text{LCM}(10, 4) = \frac{10 \cdot 4}{\text{GCD}(10,4)} = \frac{10 \cdot 4}{2} = 20$.

2.4.8 MAGIC SQUARES

A *magic square* is a square array of integers with the property that the sum of the integers in each row or column is the same. If $(c, n) = (d, n) = (e, n) = (f, n) = (cf - en, n) = 1$, then the array $A = (a_{ij})$ will be magic (and use the n^2 numbers $0, 1, \ldots, n^2 - 1$) if $a_{ij} = k$ with

$$i \equiv ck + e \left\lfloor \frac{k}{n} \right\rfloor \quad (\text{mod } n) \qquad \text{and} \qquad j \equiv dk + f \left\lfloor \frac{k}{n} \right\rfloor \quad (\text{mod } n)$$

For example, with $c = 1$, $d = e = f = 2$, and $n = 3$, a magic square is

6	1	5
2	3	7
4	8	0

2.4.9 MÖBIUS FUNCTION

The *Möbius function* is defined by

1. $\mu(1) = 1$ 2. $\mu(n) = 0$ if n has a squared factor

3. $\mu(p_1 p_2 \ldots p_k) = (-1)^k$ if all the primes $\{p_1, \ldots, p_k\}$ are distinct

Its properties include:

1. If $\mathrm{GCD}(m, n) = 1$ then $\mu(mn) = \mu(m)\,\mu(n)$

2. $\displaystyle\sum_{d|n} \mu(d) = \begin{cases} 1 & \text{if } n = 1 \\ 0 & \text{if } n > 1 \end{cases}$

3. Generating function: $\displaystyle\sum_{n=0}^{\infty} \mu(n) n^{-s} = \dfrac{1}{\zeta(s)}$

The Möbius inversion formula states that, if $g(n) = \displaystyle\sum_{d|n} f(d)$, then

$$f(n) = \sum_{d|n} \mu\left(\frac{n}{d}\right) g(d) = \sum_{d|n} \mu(d) g\left(\frac{n}{d}\right). \qquad (2.4.15)$$

For example, the Möbius inversion of $n = \displaystyle\sum_{d|n} \phi(d)$ is $\phi(n) = n \displaystyle\sum_{d|n} \dfrac{\mu(d)}{d}$.

The table below can be derived from the table in Section 2.4.13. The value of $\mu(10n + k)$ is in row n_- and column $\_k$.

Möbius function values

(For example, $\mu(2) = -1$, $\mu(4) = 0$, and $\mu(6) = 1$.)

	_0	_1	_2	_3	_4	_5	_6	_7	_8	_9
0_		1	−1	−1	0	−1	1	−1	0	0
1_	1	−1	0	−1	1	1	0	−1	0	−1
2_	0	1	1	−1	0	0	1	0	0	−1
3_	−1	−1	0	1	1	1	0	−1	1	1
4_	0	−1	−1	−1	0	0	1	−1	0	0
5_	0	1	0	−1	0	1	0	1	1	−1
6_	0	−1	1	0	0	1	−1	−1	0	1
7_	−1	−1	0	−1	1	0	0	1	−1	−1
8_	0	0	1	−1	0	1	1	1	0	−1
9_	0	1	0	1	1	1	0	−1	0	0
10_	0	−1	−1	−1	0	−1	1	−1	0	−1
11_	−1	1	0	−1	−1	1	0	0	1	1

2.4.10   PRIME NUMBERS

1. A *prime number* is a positive integer greater than 1 with no positive, integral divisors other than 1 and itself. There are infinitely many prime numbers, $2, 3, 5, 7, \ldots$. The sum of the reciprocals of the prime numbers diverges: $\sum_n \frac{1}{p_n} = \frac{1}{3} + \frac{1}{5} + \frac{1}{7} + \ldots = \infty$.

2. *Twin primes* are prime numbers that differ by two: $(3,5)$, $(5,7)$, $(11,13)$, $(17,19), \ldots$. It is not known whether there are infinitely many twin primes. The sum of the reciprocals of the twin primes converges; the value

$$B = \left(\frac{1}{3} + \frac{1}{5}\right) + \left(\frac{1}{5} + \frac{1}{7}\right) + \left(\frac{1}{11} + \frac{1}{13}\right) + \ldots + \left(\frac{1}{p} + \frac{1}{p+2}\right) + \ldots$$

   known as Brun's constant is approximately $B \approx 1.90216054$.

3. For every integer $n \geq 2$, the numbers $\{n! + 2, n! + 3, \ldots, n! + n\}$ are a sequence of $n - 1$ consecutive composite (i.e., not prime) numbers.

4. *Dirichlet's theorem on primes in arithmetic progressions*: Let a and b be relatively prime positive integers. Then the arithmetic progression $an + b$ (for $n = 1, 2, \ldots$) contains infinitely many primes.

5. *Goldbach conjecture*: every even number is the sum of two prime numbers.

6. The function $\pi(x)$ represents the number of primes less than x. The prime number theorem states that $\pi(x) \sim x/\log x$ as $x \to \infty$. The exact number of primes less than a given number is:

x	100	1000	10,000	10^5	10^6	10^7	10^8
$\pi(x)$	25	168	1,229	9,592	78,498	664,579	5,761,455

x	10^{10}	10^{15}	10^{21}
$\pi(x)$	455,052,511	29,844,570,422,669	21,127,269,486,018,731,928

2.4.10.1   Prime formulae

The polynomial $x^2 - x + 41$ yields prime numbers when evaluated at $x = 0$, 1, 2, $\ldots$, 39.

The set of prime numbers is identical with the set of positive values taken on by the polynomial of degree 25 in the 26 variables $\{a, b, \ldots, z\}$:

$$
\begin{aligned}
&(k+2)\{1 - [wz + h + j - q]^2 - [(gk + 2g + k + 1)(h + j) + h - z]^2 - [2n + p + q + z - e]^2 \\
&- [16(k+1)^3(k+2)(n+1)^2 + 1 - f^2]^2 - [e^3(e+2)(a+1)^2 + 1 - o^2]^2 - [(a^2-1)y^2 + 1 - x^2]^2 \\
&- [16r^2y^4(a^2-1) + 1 - u^2]^2 - [((a + u^2(u^2 - a))^2 - 1)(n + 4dy)^2 + 1 - (x + cu)^2]^2 - [n + l + v - y]^2 \\
&- [(a^2-1)l^2 + 1 - m^2]^2 - [ai + k + 1 - l - i]^2 - [p + l(a - n - 1) + b(2an + 2a - n^2 - 2n - 2) - m]^2 \\
&- [q + y(a - p - 1) + s(2ap + 2a - p^2 - 2p - 2) - x]^2 - [z + pl(a - p) + t(2ap - p^2 - 1) - pm]^2\}.
\end{aligned}
$$
$$(2.4.16)$$

Although this polynomial appears to factor, the factors are improper, $P = P \cdot 1$. Note that this formula will also take on negative values, such as -76. There also exists a prime representing polynomial with 12 variables of degree 13697, and one of 10 variables and degree about 10^{45}.

2.4.10.2 Lucas–Lehmer primality test

Define the sequence $r_{m+1} = r_m^2 - 2$ with $r_1 = 3$. If p is a prime of the form $4n + 3$ and $M_p = 2^p - 1$, then M_p will be prime (called a *Mersenne prime*) if, and only if, M_p divides r_{p-1}.

This simple test is the reason that the largest known prime numbers are Mersenne primes. For example, consider $p = 7$ and $M_7 = 127$. The $\{r_n\}$ sequence is $\{3, 7, 47, 2207 \equiv 48, 2302 \equiv 16, 254 \equiv 0\}$; hence M_7 is prime.

2.4.10.3 Primality test certificates

A *primality certificate* is an easily verifiable statement (easier than it was to determine that it was prime in the first place) that proves that a specific number is prime. There are several types of certificates that can be given. The Atkin–Morain certificate uses elliptic curves.

To prove that the number p is prime, Pratt's certificate consists of a number a and the factorization of the number $p - 1$. The number p will be prime if there exists a primitive root a in the field GF[p]. This primitive root must satisfy the conditions $a^{p-1} = 1 \pmod{p}$ and $a^{(p-1)/q} \neq 1 \pmod{p}$ for any prime q that divides $p - 1$.

EXAMPLE   The number $p = 31$ has $p - 1 = 30 = 2 \cdot 3 \cdot 5$, and a primitive root is given by $a = 3$. Hence, to verify that $p = 31$ is prime, we compute

$$3^{(31-1)/2} = 3^{15} \equiv 14348907 \equiv -1 \neq 1 \qquad (\text{mod } 31),$$

$$3^{(31-1)/3} = 3^{10} \equiv 59049 \equiv 25 \neq 1 \qquad (\text{mod } 31),$$

$$3^{(31-1)/5} = 3^{6} \equiv 729 \equiv 16 \neq 1 \qquad (\text{mod } 31),$$

$$3^{(31-1)} = \left(3^{(31-1)/2}\right)^2 \equiv (-1)^2 = 1 \qquad (\text{mod } 31).$$

2.4.10.4 Probabilistic primality test

Let n be a number whose primality is to be determined. Probabilistic primality tests can return one of two results: either a proof that the number n is composite or a statement of the form, "The probability that the number n is not prime is less than ϵ", where ϵ can be specified by the user. Typically, we take $\epsilon = 2^{-200} < 10^{-60}$.

¿From Fermat's theorem, if $b \neq 0$, then $b^{n-1} = 1 \pmod{n}$ whenever n is prime. If this holds, then n *is a probable prime to the base* b. Given a value of n, if a value of b can be found such that this does not hold, then n cannot be prime. It can happen, however, that a probable prime is not prime.

Let $P(x)$ be the probability that n is composite under the hypotheses:

1. n is an odd integer chosen randomly from the range $[2, x]$;
2. b is an integer chosen randomly from the range $[2, n - 2]$;
3. n is a probable prime to the base b.

Then $P(x) \leq (\log x)^{-197}$ for $x \geq 10^{10000}$.

A different test can be obtained from the following theorem. Given the number n, find s and t with $n - 1 = 2^s t$, with t odd. Then choose a random integer b from the range $[2, n - 2]$. If either

$$b^t = 1 \pmod{n} \qquad \text{or} \qquad b^{2^i t} = -1 \pmod{n}, \quad \text{for some } i < s,$$

then n *is a strong probable prime to the base* b. Every odd prime must pass this test. If $n > 1$ is an odd composite, then the probability that it is a strong probable prime to the base b, when b is chosen randomly, is less than $1/4$.

A stronger test can be obtained by choosing k independent values for b in the range $[2, n - 2]$ and checking the above relation for each value of b. Let $P_k(x)$ be the probability that n is found to be a strong probable prime to each base b. Then $P_k(x) \leq 4^{-(k-1)} P(x)/(1 - P(x))$.

2.4.11 PRIME NUMBERS OF SPECIAL FORMS

1. The largest known prime numbers, in descending order, are

Prime number	Number of digits
$2^{13466917} - 1$	4,053,946
$2^{6972593} - 1$	2,098,960
$2^{3021377} - 1$	909,526
$2^{2976221} - 1$	895,932
$2^{1398269} - 1$	420,921
$1266062^{65536} + 1$	399,931
$5 \cdot 2^{1320487} + 1$	397,507
$857678^{65536} + 1$	388,847
$843832^{65536} + 1$	388,384
$5 \cdot 2^{1282755} + 1$	386,149

2. The largest known twin primes are: $318032361 \cdot 2^{107001} \pm 1$ (with 32,220 digits), $1807318575 \cdot 2^{98305} \pm 1$ (with 29,603 digits), and $665551035 \cdot 2^{80025} \pm 1$ (with 24,099 digits).

3. There exist constants $\theta \approx 1.30637788$ and $\omega \approx 1.9287800$ such that $\left\lfloor \theta^{3^n} \right\rfloor$ and $\left\lfloor \underbrace{2^{2^{2^{\cdots^{2^\omega}}}}}_{n} \right\rfloor$ are prime for every $n \geq 1$.

4. Primes with special properties

   (a) A *Sophie Germain prime* p has the property that $2p + 1$ is also prime. Sophie Germain primes include: 2, 3, 5, 11, 23, 29, 41, 53, 83, 89, 113, 131, ..., $3714089895285 \cdot 2^{60000} - 1$, $984798015 \cdot 2^{66444} - 1$, $109433307 \cdot 2^{66452} - 1, \ldots$.

(b) An odd prime p is called a *Wieferich prime* if $2^{p-1} \equiv 1 \pmod{p^2}$. Wieferich primes include 1093 and 3511.

(c) A *Wilson prime* satisfies $(p-1)! \equiv -1 \pmod{p^2}$. Wilson primes include 5, 13, and 563.

5. For each n shown below, the numbers $\{a + md \mid m = 0, 1, \ldots, n-1\}$ are an arithmetic sequence of n prime numbers:

n	a	$a + (n-1)d$	d
3	3	7	2
4	61	79	6
5	11	131	30
10	199	2089	210
22	11410337850553	108201410428753	4609098694200

6. Define $p\#$ to be the product of the prime numbers less than or equal to p.

Form	Values of n or p for which the form is prime
$2^{2^n} + 1$	0, 1, 2, 3, 4 ... (Fermat primes)
$2^n - 1$	2, 3, 5, 7, 13, 17, 19, 31, 61, 89, 107, 127, 521, 607, 1279, 2203, 2281, 3217, 4253, 4423, 9689, 9941, 11213, 19937, 21701, 23209, 44497, 86243, 110503, 132049, 216091, 756839, 859433, 1257787, 1398269, 2976221, 3021377, 6972593, . . . , 13466917 . . . (Mersenne primes)
$n! - 1$	3, 4, 6, 7, 12, 14, 30, 32, 33, 38, 94, 166, 324, 379, 469, 546, 974, 1963, 3507, 3610, 6917, . . . (factorial primes)
$n! + 1$	1, 2, 3, 11, 27, 37, 41, 73, 77, 116, 154, 320, 340, 399, 427, 872, 1477, 6380, . . . (factorial primes)
$p\# - 1$	3, 5, 11, 13, 41, 89, 317, 337, 991, 1873, 2053, 2377, 4093, 4297, 4583, 6569, 13033, 15877, . . . (primorial primes)
$p\# + 1$	2, 3, 5, 7, 11, 31, 379, 1019, 1021, 2657, 3229, 4547, 4787, 11549, 13649, 18523, 23801, 24029, 42209, . . . , 145823, 366439, 392113, . . . (primorial primes)
$n2^n + 1$	1, 141, 4713, 5795, 6611, 18496, 32292, 32469, 59656, 90825, 262419, 361275, . . . , 481899, . . . (Cullen primes)
$n2^n - 1$	2, 3, 6, 30, 75, 81, 115, 123, 249, 362, 384, 462, 512, 751, 822, 5312, 7755, 9531, 12379, 15822, 18885, . . . 143018, 151023, 667071, . . . (Woodall primes)

7. Prime numbers of the form $\dfrac{a^n - 1}{a - 1}$ (called repunits).

Form	Values of n for which the form is prime
$\dfrac{2^n - 1}{1}$	These are Mersenne primes; see the previous table.
$\dfrac{3^n - 1}{2}$	$3, 7, 13, 71, 103, 541, 1091, 1367, 1627, 4177, 9011, 9551, \ldots$
$\dfrac{5^n - 1}{4}$	$3, 7, 11, 13, 47, 127, 149, 181, 619, 929, 3407, 10949, \ldots$
$\dfrac{6^n - 1}{5}$	$2, 3, 7, 29, 71, 127, 271, 509, 1049, 6389, 6883, 10613, \ldots$
$\dfrac{7^n - 1}{6}$	$5, 13, 131, 149, 1699, \ldots$
$\dfrac{10^n - 1}{9}$	$2, 19, 23, 317, 1031, 49081, 86453, \ldots$
$\dfrac{11^n - 1}{10}$	$17, 19, 73, 139, 907, 1907, 2029, 4801, 5153, 10867, \ldots$
$\dfrac{12^n - 1}{11}$	$2, 3, 5, 19, 97, 109, 317, 353, 701, 9739, \ldots$

8. Prime numbers of the forms $2^n \pm a$, $10^n \pm b$, and $16^n \pm c$.

In the following table, for a given value of n, the quantities $a_\pm$, $b_\pm$, and $c_\pm$ are the least values such that $2^n + a_\pm$, $10^n + b_\pm$, and $16^n + c_\pm$ are probably primes. (A probabilistic primality test was used.) For example, for $n = 3$, the numbers $2^3 - 1 = 7$, $2^3 + 3 = 11$, $10^3 - 3 = 997$, $10^3 + 9 = 1009$, $16^3 - 3 = 4093$, and $16^3 + 3 = 4099$ are all prime.

	$2^n + a$		$10^n + b$		$16^n + c$	
n	a_-	a_+	b_-	b_+	c_-	c_+
2	-1	1	-3	1	-5	1
3	-1	3	-3	9	-3	3
4	-3	1	-27	7	-15	1
5	-1	5	-9	3	-3	7
6	-3	3	-17	3	-3	43
7	-1	3	-9	19	-57	3
8	-5	1	-11	7	-5	15
9	-3	9	-63	7	-5	31
10	-3	7	-33	19	-87	15
11	-9	5	-23	3	-17	7
12	-3	3	-11	39	-59	21
13	-1	17	-29	37	-47	21
14	-3	27	-27	31	-5	81

n	$2^n + a$		$10^n + b$		$16^n + c$	
	a_-	a_+	b_-	b_+	c_-	c_+
15	-19	3	-11	37	-93	33
16	-15	1	-63	61	-59	13
17	-1	29	-3	3	-23	33
18	-5	3	-11	3	-93	15
19	-1	21	-39	51	-15	15
20	-3	7	-11	39	-65	13
50	-27	55	-57	151	-75	235
100	-15	277	-797	267	-593	181
150	-3	147	-273	67	-95	187
200	-75	235	-189	357	-105	25
300	-153	157	-69	331	-305	1515
400	-593	181	-513	69	-2273	895
500	-863	55	-1037	961	-2217	841
600	-95	187	-1791	543	-5	255
700	-1113	535	-2313	7	-909	2823
800	-105	25	-1007	1537	-1683	751
900	-207	693	-773	1873	-1193	8767
1000	-1245	297	-1769	453	-2303	63

2.4.12   PRIME NUMBERS LESS THAN 100,000

The prime number p_{10n+k} is found by looking at row n_- and the column $_-k$.

	$_-0$	$_-1$	$_-2$	$_-3$	$_-4$	$_-5$	$_-6$	$_-7$	$_-8$	$_-9$
0_-		2	3	5	7	11	13	17	19	23
1_-	29	31	37	41	43	47	53	59	61	67
2_-	71	73	79	83	89	97	101	103	107	109
3_-	113	127	131	137	139	149	151	157	163	167
4_-	173	179	181	191	193	197	199	211	223	227
5_-	229	233	239	241	251	257	263	269	271	277
6_-	281	283	293	307	311	313	317	331	337	347
7_-	349	353	359	367	373	379	383	389	397	401
8_-	409	419	421	431	433	439	443	449	457	461
9_-	463	467	479	487	491	499	503	509	521	523
10_-	541	547	557	563	569	571	577	587	593	599
11_-	601	607	613	617	619	631	641	643	647	653
12_-	659	661	673	677	683	691	701	709	719	727
13_-	733	739	743	751	757	761	769	773	787	797
14_-	809	811	821	823	827	829	839	853	857	859
15_-	863	877	881	883	887	907	911	919	929	937
16_-	941	947	953	967	971	977	983	991	997	1009
17_-	1013	1019	1021	1031	1033	1039	1049	1051	1061	1063
18_-	1069	1087	1091	1093	1097	1103	1109	1117	1123	1129
19_-	1151	1153	1163	1171	1181	1187	1193	1201	1213	1217
20_-	1223	1229	1231	1237	1249	1259	1277	1279	1283	1289
21_-	1291	1297	1301	1303	1307	1319	1321	1327	1361	1367
22_-	1373	1381	1399	1409	1423	1427	1429	1433	1439	1447

	.0	.1	.2	.3	.4	.5	.6	.7	.8	.9
23_	1451	1453	1459	1471	1481	1483	1487	1489	1493	1499
24_	1511	1523	1531	1543	1549	1553	1559	1567	1571	1579
25_	1583	1597	1601	1607	1609	1613	1619	1621	1627	1637
26_	1657	1663	1667	1669	1693	1697	1699	1709	1721	1723
27_	1733	1741	1747	1753	1759	1777	1783	1787	1789	1801
28_	1811	1823	1831	1847	1861	1867	1871	1873	1877	1879
29_	1889	1901	1907	1913	1931	1933	1949	1951	1973	1979
30_	1987	1993	1997	1999	2003	2011	2017	2027	2029	2039
31_	2053	2063	2069	2081	2083	2087	2089	2099	2111	2113
32_	2129	2131	2137	2141	2143	2153	2161	2179	2203	2207
33_	2213	2221	2237	2239	2243	2251	2267	2269	2273	2281
34_	2287	2293	2297	2309	2311	2333	2339	2341	2347	2351
35_	2357	2371	2377	2381	2383	2389	2393	2399	2411	2417
36_	2423	2437	2441	2447	2459	2467	2473	2477	2503	2521
37_	2531	2539	2543	2549	2551	2557	2579	2591	2593	2609
38_	2617	2621	2633	2647	2657	2659	2663	2671	2677	2683
39_	2687	2689	2693	2699	2707	2711	2713	2719	2729	2731
40_	2741	2749	2753	2767	2777	2789	2791	2797	2801	2803
41_	2819	2833	2837	2843	2851	2857	2861	2879	2887	2897
42_	2903	2909	2917	2927	2939	2953	2957	2963	2969	2971
43_	2999	3001	3011	3019	3023	3037	3041	3049	3061	3067
44_	3079	3083	3089	3109	3119	3121	3137	3163	3167	3169
45_	3181	3187	3191	3203	3209	3217	3221	3229	3251	3253
46_	3257	3259	3271	3299	3301	3307	3313	3319	3323	3329
47_	3331	3343	3347	3359	3361	3371	3373	3389	3391	3407
48_	3413	3433	3449	3457	3461	3463	3467	3469	3491	3499
49_	3511	3517	3527	3529	3533	3539	3541	3547	3557	3559
50_	3571	3581	3583	3593	3607	3613	3617	3623	3631	3637
51_	3643	3659	3671	3673	3677	3691	3697	3701	3709	3719
52_	3727	3733	3739	3761	3767	3769	3779	3793	3797	3803
53_	3821	3823	3833	3847	3851	3853	3863	3877	3881	3889
54_	3907	3911	3917	3919	3923	3929	3931	3943	3947	3967
55_	3989	4001	4003	4007	4013	4019	4021	4027	4049	4051
56_	4057	4073	4079	4091	4093	4099	4111	4127	4129	4133
57_	4139	4153	4157	4159	4177	4201	4211	4217	4219	4229
58_	4231	4241	4243	4253	4259	4261	4271	4273	4283	4289
59_	4297	4327	4337	4339	4349	4357	4363	4373	4391	4397
60_	4409	4421	4423	4441	4447	4451	4457	4463	4481	4483
61_	4493	4507	4513	4517	4519	4523	4547	4549	4561	4567
62_	4583	4591	4597	4603	4621	4637	4639	4643	4649	4651
63_	4657	4663	4673	4679	4691	4703	4721	4723	4729	4733
64_	4751	4759	4783	4787	4789	4793	4799	4801	4813	4817
65_	4831	4861	4871	4877	4889	4903	4909	4919	4931	4933
66_	4937	4943	4951	4957	4967	4969	4973	4987	4993	4999
67_	5003	5009	5011	5021	5023	5039	5051	5059	5077	5081
68_	5087	5099	5101	5107	5113	5119	5147	5153	5167	5171
69_	5179	5189	5197	5209	5227	5231	5233	5237	5261	5273
70_	5279	5281	5297	5303	5309	5323	5333	5347	5351	5381
71_	5387	5393	5399	5407	5413	5417	5419	5431	5437	5441
72_	5443	5449	5471	5477	5479	5483	5501	5503	5507	5519
73_	5521	5527	5531	5557	5563	5569	5573	5581	5591	5623
74_	5639	5641	5647	5651	5653	5657	5659	5669	5683	5689
75_	5693	5701	5711	5717	5737	5741	5743	5749	5779	5783
76_	5791	5801	5807	5813	5821	5827	5839	5843	5849	5851
77_	5857	5861	5867	5869	5879	5881	5897	5903	5923	5927
78_	5939	5953	5981	5987	6007	6011	6029	6037	6043	6047

	.0	.1	.2	.3	.4	.5	.6	.7	.8	.9
79_	6053	6067	6073	6079	6089	6091	6101	6113	6121	6131
80_	6133	6143	6151	6163	6173	6197	6199	6203	6211	6217
81_	6221	6229	6247	6257	6263	6269	6271	6277	6287	6299
82_	6301	6311	6317	6323	6329	6337	6343	6353	6359	6361
83_	6367	6373	6379	6389	6397	6421	6427	6449	6451	6469
84_	6473	6481	6491	6521	6529	6547	6551	6553	6563	6569
85_	6571	6577	6581	6599	6607	6619	6637	6653	6659	6661
86_	6673	6679	6689	6691	6701	6703	6709	6719	6733	6737
87_	6761	6763	6779	6781	6791	6793	6803	6823	6827	6829
88_	6833	6841	6857	6863	6869	6871	6883	6899	6907	6911
89_	6917	6947	6949	6959	6961	6967	6971	6977	6983	6991
90_	6997	7001	7013	7019	7027	7039	7043	7057	7069	7079
91_	7103	7109	7121	7127	7129	7151	7159	7177	7187	7193
92_	7207	7211	7213	7219	7229	7237	7243	7247	7253	7283
93_	7297	7307	7309	7321	7331	7333	7349	7351	7369	7393
94_	7411	7417	7433	7451	7457	7459	7477	7481	7487	7489
95_	7499	7507	7517	7523	7529	7537	7541	7547	7549	7559
96_	7561	7573	7577	7583	7589	7591	7603	7607	7621	7639
97_	7643	7649	7669	7673	7681	7687	7691	7699	7703	7717
98_	7723	7727	7741	7753	7757	7759	7789	7793	7817	7823
99_	7829	7841	7853	7867	7873	7877	7879	7883	7901	7907
100_	7919	7927	7933	7937	7949	7951	7963	7993	8009	8011
101_	8017	8039	8053	8059	8069	8081	8087	8089	8093	8101
102_	8111	8117	8123	8147	8161	8167	8171	8179	8191	8209
103_	8219	8221	8231	8233	8237	8243	8263	8269	8273	8287
104_	8291	8293	8297	8311	8317	8329	8353	8363	8369	8377
105_	8387	8389	8419	8423	8429	8431	8443	8447	8461	8467
106_	8501	8513	8521	8527	8537	8539	8543	8563	8573	8581
107_	8597	8599	8609	8623	8627	8629	8641	8647	8663	8669
108_	8677	8681	8689	8693	8699	8707	8713	8719	8731	8737
109_	8741	8747	8753	8761	8779	8783	8803	8807	8819	8821
110_	8831	8837	8839	8849	8861	8863	8867	8887	8893	8923
111_	8929	8933	8941	8951	8963	8969	8971	8999	9001	9007
112_	9011	9013	9029	9041	9043	9049	9059	9067	9091	9103
113_	9109	9127	9133	9137	9151	9157	9161	9173	9181	9187
114_	9199	9203	9209	9221	9227	9239	9241	9257	9277	9281
115_	9283	9293	9311	9319	9323	9337	9341	9343	9349	9371
116_	9377	9391	9397	9403	9413	9419	9421	9431	9433	9437
117_	9439	9461	9463	9467	9473	9479	9491	9497	9511	9521
118_	9533	9539	9547	9551	9587	9601	9613	9619	9623	9629
119_	9631	9643	9649	9661	9677	9679	9689	9697	9719	9721
120_	9733	9739	9743	9749	9767	9769	9781	9787	9791	9803
121_	9811	9817	9829	9833	9839	9851	9857	9859	9871	9883
122_	9887	9901	9907	9923	9929	9931	9941	9949	9967	9973
123_	10007	10009	10037	10039	10061	10067	10069	10079	10091	10093
124_	10099	10103	10111	10133	10139	10141	10151	10159	10163	10169
125_	10177	10181	10193	10211	10223	10243	10247	10253	10259	10267
126_	10271	10273	10289	10301	10303	10313	10321	10331	10333	10337
127_	10343	10357	10369	10391	10399	10427	10429	10433	10453	10457
128_	10459	10463	10477	10487	10499	10501	10513	10529	10531	10559
129_	10567	10589	10597	10601	10607	10613	10627	10631	10639	10651
130_	10657	10663	10667	10687	10691	10709	10711	10723	10729	10733
131_	10739	10753	10771	10781	10789	10799	10831	10837	10847	10853
132_	10859	10861	10867	10883	10889	10891	10903	10909	10937	10939
133_	10949	10957	10973	10979	10987	10993	11003	11027	11047	11057
134_	11059	11069	11071	11083	11087	11093	11113	11117	11119	11131

	_0	_1	_2	_3	_4	_5	_6	_7	_8	_9
135_	11149	11159	11161	11171	11173	11177	11197	11213	11239	11243
136_	11251	11257	11261	11273	11279	11287	11299	11311	11317	11321
137_	11329	11351	11353	11369	11383	11393	11399	11411	11423	11437
138_	11443	11447	11467	11471	11483	11489	11491	11497	11503	11519
139_	11527	11549	11551	11579	11587	11593	11597	11617	11621	11633
140_	11657	11677	11681	11689	11699	11701	11717	11719	11731	11743
141_	11777	11779	11783	11789	11801	11807	11813	11821	11827	11831
142_	11833	11839	11863	11867	11887	11897	11903	11909	11923	11927
143_	11933	11939	11941	11953	11959	11969	11971	11981	11987	12007
144_	12011	12037	12041	12043	12049	12071	12073	12097	12101	12107
145_	12109	12113	12119	12143	12149	12157	12161	12163	12197	12203
146_	12211	12227	12239	12241	12251	12253	12263	12269	12277	12281
147_	12289	12301	12323	12329	12343	12347	12373	12377	12379	12391
148_	12401	12409	12413	12421	12433	12437	12451	12457	12473	12479
149_	12487	12491	12497	12503	12511	12517	12527	12539	12541	12547
150_	12553	12569	12577	12583	12589	12601	12611	12613	12619	12637
151_	12641	12647	12653	12659	12671	12689	12697	12703	12713	12721
152_	12739	12743	12757	12763	12781	12791	12799	12809	12821	12823
153_	12829	12841	12853	12889	12893	12899	12907	12911	12917	12919
154_	12923	12941	12953	12959	12967	12973	12979	12983	13001	13003
155_	13007	13009	13033	13037	13043	13049	13063	13093	13099	13103
156_	13109	13121	13127	13147	13151	13159	13163	13171	13177	13183
157_	13187	13217	13219	13229	13241	13249	13259	13267	13291	13297
158_	13309	13313	13327	13331	13337	13339	13367	13381	13397	13399
159_	13411	13417	13421	13441	13451	13457	13463	13469	13477	13487
160_	13499	13513	13523	13537	13553	13567	13577	13591	13597	13613
161_	13619	13627	13633	13649	13669	13679	13681	13687	13691	13693
162_	13697	13709	13711	13721	13723	13729	13751	13757	13759	13763
163_	13781	13789	13799	13807	13829	13831	13841	13859	13873	13877
164_	13879	13883	13901	13903	13907	13913	13921	13931	13933	13963
165_	13967	13997	13999	14009	14011	14029	14033	14051	14057	14071
166_	14081	14083	14087	14107	14143	14149	14153	14159	14173	14177
167_	14197	14207	14221	14243	14249	14251	14281	14293	14303	14321
168_	14323	14327	14341	14347	14369	14387	14389	14401	14407	14411
169_	14419	14423	14431	14437	14447	14449	14461	14479	14489	14503
170_	14519	14533	14537	14543	14549	14551	14557	14561	14563	14591
171_	14593	14621	14627	14629	14633	14639	14653	14657	14669	14683
172_	14699	14713	14717	14723	14731	14737	14741	14747	14753	14759
173_	14767	14771	14779	14783	14797	14813	14821	14827	14831	14843
174_	14851	14867	14869	14879	14887	14891	14897	14923	14929	14939
175_	14947	14951	14957	14969	14983	15013	15017	15031	15053	15061
176_	15073	15077	15083	15091	15101	15107	15121	15131	15137	15139
177_	15149	15161	15173	15187	15193	15199	15217	15227	15233	15241
178_	15259	15263	15269	15271	15277	15287	15289	15299	15307	15313
179_	15319	15329	15331	15349	15359	15361	15373	15377	15383	15391
180_	15401	15413	15427	15439	15443	15451	15461	15467	15473	15493
181_	15497	15511	15527	15541	15551	15559	15569	15581	15583	15601
182_	15607	15619	15629	15641	15643	15647	15649	15661	15667	15671
183_	15679	15683	15727	15731	15733	15737	15739	15749	15761	15767
184_	15773	15787	15791	15797	15803	15809	15817	15823	15859	15877
185_	15881	15887	15889	15901	15907	15913	15919	15923	15937	15959
186_	15971	15973	15991	16001	16007	16033	16057	16061	16063	16067
187_	16069	16073	16087	16091	16097	16103	16111	16127	16139	16141
188_	16183	16187	16189	16193	16217	16223	16229	16231	16249	16253
189_	16267	16273	16301	16319	16333	16339	16349	16361	16363	16369
190_	16381	16411	16417	16421	16427	16433	16447	16451	16453	16477

	.0	.1	.2	.3	.4	.5	.6	.7	.8	.9
191_	16481	16487	16493	16519	16529	16547	16553	16561	16567	16573
192_	16603	16607	16619	16631	16633	16649	16651	16657	16661	16673
193_	16691	16693	16699	16703	16729	16741	16747	16759	16763	16787
194_	16811	16823	16829	16831	16843	16871	16879	16883	16889	16901
195_	16903	16921	16927	16931	16937	16943	16963	16979	16981	16987
196_	16993	17011	17021	17027	17029	17033	17041	17047	17053	17077
197_	17093	17099	17107	17117	17123	17137	17159	17167	17183	17189
198_	17191	17203	17207	17209	17231	17239	17257	17291	17293	17299
199_	17317	17321	17327	17333	17341	17351	17359	17377	17383	17387
200_	17389	17393	17401	17417	17419	17431	17443	17449	17467	17471
201_	17477	17483	17489	17491	17497	17509	17519	17539	17551	17569
202_	17573	17579	17581	17597	17599	17609	17623	17627	17657	17659
203_	17669	17681	17683	17707	17713	17729	17737	17747	17749	17761
204_	17783	17789	17791	17807	17827	17837	17839	17851	17863	17881
205_	17891	17903	17909	17911	17921	17923	17929	17939	17957	17959
206_	17971	17977	17981	17987	17989	18013	18041	18043	18047	18049
207_	18059	18061	18077	18089	18097	18119	18121	18127	18131	18133
208_	18143	18149	18169	18181	18191	18199	18211	18217	18223	18229
209_	18233	18251	18253	18257	18269	18287	18289	18301	18307	18311
210_	18313	18329	18341	18353	18367	18371	18379	18397	18401	18413
211_	18427	18433	18439	18443	18451	18457	18461	18481	18493	18503
212_	18517	18521	18523	18539	18541	18553	18583	18587	18593	18617
213_	18637	18661	18671	18679	18691	18701	18713	18719	18731	18743
214_	18749	18757	18773	18787	18793	18797	18803	18839	18859	18869
215_	18899	18911	18913	18917	18919	18947	18959	18973	18979	19001
216_	19009	19013	19031	19037	19051	19069	19073	19079	19081	19087
217_	19121	19139	19141	19157	19163	19181	19183	19207	19211	19213
218_	19219	19231	19237	19249	19259	19267	19273	19289	19301	19309
219_	19319	19333	19373	19379	19381	19387	19391	19403	19417	19421
220_	19423	19427	19429	19433	19441	19447	19457	19463	19469	19471
221_	19477	19483	19489	19501	19507	19531	19541	19543	19553	19559
222_	19571	19577	19583	19597	19603	19609	19661	19681	19687	19697
223_	19699	19709	19717	19727	19739	19751	19753	19759	19763	19777
224_	19793	19801	19813	19819	19841	19843	19853	19861	19867	19889
225_	19891	19913	19919	19927	19937	19949	19961	19963	19973	19979
226_	19991	19993	19997	20011	20021	20023	20029	20047	20051	20063
227_	20071	20089	20101	20107	20113	20117	20123	20129	20143	20147
228_	20149	20161	20173	20177	20183	20201	20219	20231	20233	20249
229_	20261	20269	20287	20297	20323	20327	20333	20341	20347	20353
230_	20357	20359	20369	20389	20393	20399	20407	20411	20431	20441
231_	20443	20477	20479	20483	20507	20509	20521	20533	20543	20549
232_	20551	20563	20593	20599	20611	20627	20639	20641	20663	20681
233_	20693	20707	20717	20719	20731	20743	20747	20749	20753	20759
234_	20771	20773	20789	20807	20809	20849	20857	20873	20879	20887
235_	20897	20899	20903	20921	20929	20939	20947	20959	20963	20981
236_	20983	21001	21011	21013	21017	21019	21023	21031	21059	21061
237_	21067	21089	21101	21107	21121	21139	21143	21149	21157	21163
238_	21169	21179	21187	21191	21193	21211	21221	21227	21247	21269
239_	21277	21283	21313	21317	21319	21323	21341	21347	21377	21379
240_	21383	21391	21397	21401	21407	21419	21433	21467	21481	21487
241_	21491	21493	21499	21503	21517	21521	21523	21529	21557	21559
242_	21563	21569	21577	21587	21589	21599	21601	21611	21613	21617
243_	21647	21649	21661	21673	21683	21701	21713	21727	21737	21739
244_	21751	21757	21767	21773	21787	21799	21803	21817	21821	21839
245_	21841	21851	21859	21863	21871	21881	21893	21911	21929	21937
246_	21943	21961	21977	21991	21997	22003	22013	22027	22031	22037

	_0	_1	_2	_3	_4	_5	_6	_7	_8	_9
247_	22039	22051	22063	22067	22073	22079	22091	22093	22109	22111
248_	22123	22129	22133	22147	22153	22157	22159	22171	22189	22193
249_	22229	22247	22259	22271	22273	22277	22279	22283	22291	22303
250_	22307	22343	22349	22367	22369	22381	22391	22397	22409	22433
251_	22441	22447	22453	22469	22481	22483	22501	22511	22531	22541
252_	22543	22549	22567	22571	22573	22613	22619	22621	22637	22639
253_	22643	22651	22669	22679	22691	22697	22699	22709	22717	22721
254_	22727	22739	22741	22751	22769	22777	22783	22787	22807	22811
255_	22817	22853	22859	22861	22871	22877	22901	22907	22921	22937
256_	22943	22961	22963	22973	22993	23003	23011	23017	23021	23027
257_	23029	23039	23041	23053	23057	23059	23063	23071	23081	23087
258_	23099	23117	23131	23143	23159	23167	23173	23189	23197	23201
259_	23203	23209	23227	23251	23269	23279	23291	23293	23297	23311
260_	23321	23327	23333	23339	23357	23369	23371	23399	23417	23431
261_	23447	23459	23473	23497	23509	23531	23537	23539	23549	23557
262_	23561	23563	23567	23581	23593	23599	23603	23609	23623	23627
263_	23629	23633	23663	23669	23671	23677	23687	23689	23719	23741
264_	23743	23747	23753	23761	23767	23773	23789	23801	23813	23819
265_	23827	23831	23833	23857	23869	23873	23879	23887	23893	23899
266_	23909	23911	23917	23929	23957	23971	23977	23981	23993	24001
267_	24007	24019	24023	24029	24043	24049	24061	24071	24077	24083
268_	24091	24097	24103	24107	24109	24113	24121	24133	24137	24151
269_	24169	24179	24181	24197	24203	24223	24229	24239	24247	24251
270_	24281	24317	24329	24337	24359	24371	24373	24379	24391	24407
271_	24413	24419	24421	24439	24443	24469	24473	24481	24499	24509
272_	24517	24527	24533	24547	24551	24571	24593	24611	24623	24631
273_	24659	24671	24677	24683	24691	24697	24709	24733	24749	24763
274_	24767	24781	24793	24799	24809	24821	24841	24847	24851	24859
275_	24877	24889	24907	24917	24919	24923	24943	24953	24967	24971
276_	24977	24979	24989	25013	25031	25033	25037	25057	25073	25087
277_	25097	25111	25117	25121	25127	25147	25153	25163	25169	25171
278_	25183	25189	25219	25229	25237	25243	25247	25253	25261	25301
279_	25303	25307	25309	25321	25339	25343	25349	25357	25367	25373
280_	25391	25409	25411	25423	25439	25447	25453	25457	25463	25469
281_	25471	25523	25537	25541	25561	25577	25579	25583	25589	25601
282_	25603	25609	25621	25633	25639	25643	25657	25667	25673	25679
283_	25693	25703	25717	25733	25741	25747	25759	25763	25771	25793
284_	25799	25801	25819	25841	25847	25849	25867	25873	25889	25903
285_	25913	25919	25931	25933	25939	25943	25951	25969	25981	25997
286_	25999	26003	26017	26021	26029	26041	26053	26083	26099	26107
287_	26111	26113	26119	26141	26153	26161	26171	26177	26183	26189
288_	26203	26209	26227	26237	26249	26251	26261	26263	26267	26293
289_	26297	26309	26317	26321	26339	26347	26357	26371	26387	26393
290_	26399	26407	26417	26423	26431	26437	26449	26459	26479	26489
291_	26497	26501	26513	26539	26557	26561	26573	26591	26597	26627
292_	26633	26641	26647	26669	26681	26683	26687	26693	26699	26701
293_	26711	26713	26717	26723	26729	26731	26737	26759	26777	26783
294_	26801	26813	26821	26833	26839	26849	26861	26863	26879	26881
295_	26891	26893	26903	26921	26927	26947	26951	26953	26959	26981
296_	26987	26993	27011	27017	27031	27043	27059	27061	27067	27073
297_	27077	27091	27103	27107	27109	27127	27143	27179	27191	27197
298_	27211	27239	27241	27253	27259	27271	27277	27281	27283	27299
299_	27329	27337	27361	27367	27397	27407	27409	27427	27431	27437
300_	27449	27457	27479	27481	27487	27509	27527	27529	27539	27541
301_	27551	27581	27583	27611	27617	27631	27647	27653	27673	27689
302_	27691	27697	27701	27733	27737	27739	27743	27749	27751	27763

	_0	_1	_2	_3	_4	_5	_6	_7	_8	_9
303_	27767	27773	27779	27791	27793	27799	27803	27809	27817	27823
304_	27827	27847	27851	27883	27893	27901	27917	27919	27941	27943
305_	27947	27953	27961	27967	27983	27997	28001	28019	28027	28031
306_	28051	28057	28069	28081	28087	28097	28099	28109	28111	28123
307_	28151	28163	28181	28183	28201	28211	28219	28229	28277	28279
308_	28283	28289	28297	28307	28309	28319	28349	28351	28387	28393
309_	28403	28409	28411	28429	28433	28439	28447	28463	28477	28493
310_	28499	28513	28517	28537	28541	28547	28549	28559	28571	28573
311_	28579	28591	28597	28603	28607	28619	28621	28627	28631	28643
312_	28649	28657	28661	28663	28669	28687	28697	28703	28711	28723
313_	28729	28751	28753	28759	28771	28789	28793	28807	28813	28817
314_	28837	28843	28859	28867	28871	28879	28901	28909	28921	28927
315_	28933	28949	28961	28979	29009	29017	29021	29023	29027	29033
316_	29059	29063	29077	29101	29123	29129	29131	29137	29147	29153
317_	29167	29173	29179	29191	29201	29207	29209	29221	29231	29243
318_	29251	29269	29287	29297	29303	29311	29327	29333	29339	29347
319_	29363	29383	29387	29389	29399	29401	29411	29423	29429	29437
320_	29443	29453	29473	29483	29501	29527	29531	29537	29567	29569
321_	29573	29581	29587	29599	29611	29629	29633	29641	29663	29669
322_	29671	29683	29717	29723	29741	29753	29759	29761	29789	29803
323_	29819	29833	29837	29851	29863	29867	29873	29879	29881	29917
324_	29921	29927	29947	29959	29983	29989	30011	30013	30029	30047
325_	30059	30071	30089	30091	30097	30103	30109	30113	30119	30133
326_	30137	30139	30161	30169	30181	30187	30197	30203	30211	30223
327_	30241	30253	30259	30269	30271	30293	30307	30313	30319	30323
328_	30341	30347	30367	30389	30391	30403	30427	30431	30449	30467
329_	30469	30491	30493	30497	30509	30517	30529	30539	30553	30557
330_	30559	30577	30593	30631	30637	30643	30649	30661	30671	30677
331_	30689	30697	30703	30707	30713	30727	30757	30763	30773	30781
332_	30803	30809	30817	30829	30839	30841	30851	30853	30859	30869
333_	30871	30881	30893	30911	30931	30937	30941	30949	30971	30977
334_	30983	31013	31019	31033	31039	31051	31063	31069	31079	31081
335_	31091	31121	31123	31139	31147	31151	31153	31159	31177	31181
336_	31183	31189	31193	31219	31223	31231	31237	31247	31249	31253
337_	31259	31267	31271	31277	31307	31319	31321	31327	31333	31337
338_	31357	31379	31387	31391	31393	31397	31469	31477	31481	31489
339_	31511	31513	31517	31531	31541	31543	31547	31567	31573	31583
340_	31601	31607	31627	31643	31649	31657	31663	31667	31687	31699
341_	31721	31723	31727	31729	31741	31751	31769	31771	31793	31799
342_	31817	31847	31849	31859	31873	31883	31891	31907	31957	31963
343_	31973	31981	31991	32003	32009	32027	32029	32051	32057	32059
344_	32063	32069	32077	32083	32089	32099	32117	32119	32141	32143
345_	32159	32173	32183	32189	32191	32203	32213	32233	32237	32251
346_	32257	32261	32297	32299	32303	32309	32321	32323	32327	32341
347_	32353	32359	32363	32369	32371	32377	32381	32401	32411	32413
348_	32423	32429	32441	32443	32467	32479	32491	32497	32503	32507
349_	32531	32533	32537	32561	32563	32569	32573	32579	32587	32603
350_	32609	32611	32621	32633	32647	32653	32687	32693	32707	32713
351_	32717	32719	32749	32771	32779	32783	32789	32797	32801	32803
352_	32831	32833	32839	32843	32869	32887	32909	32911	32917	32933
353_	32939	32941	32957	32969	32971	32983	32987	32993	32999	33013
354_	33023	33029	33037	33049	33053	33071	33073	33083	33091	33107
355_	33113	33119	33149	33151	33161	33179	33181	33191	33199	33203
356_	33211	33223	33247	33287	33289	33301	33311	33317	33329	33331
357_	33343	33347	33349	33353	33359	33377	33391	33403	33409	33413
358_	33427	33457	33461	33469	33479	33487	33493	33503	33521	33529

	.0	.1	.2	.3	.4	.5	.6	.7	.8	.9
359_	33533	33547	33563	33569	33577	33581	33587	33589	33599	33601
360_	33613	33617	33619	33623	33629	33637	33641	33647	33679	33703
361_	33713	33721	33739	33749	33751	33757	33767	33769	33773	33791
362_	33797	33809	33811	33827	33829	33851	33857	33863	33871	33889
363_	33893	33911	33923	33931	33937	33941	33961	33967	33997	34019
364_	34031	34033	34039	34057	34061	34123	34127	34129	34141	34147
365_	34157	34159	34171	34183	34211	34213	34217	34231	34253	34259
366_	34261	34267	34273	34283	34297	34301	34303	34313	34319	34327
367_	34337	34351	34361	34367	34369	34381	34403	34421	34429	34439
368_	34457	34469	34471	34483	34487	34499	34501	34511	34513	34519
369_	34537	34543	34549	34583	34589	34591	34603	34607	34613	34631
370_	34649	34651	34667	34673	34679	34687	34693	34703	34721	34729
371_	34739	34747	34757	34759	34763	34781	34807	34819	34841	34843
372_	34847	34849	34871	34877	34883	34897	34913	34919	34939	34949
373_	34961	34963	34981	35023	35027	35051	35053	35059	35069	35081
374_	35083	35089	35099	35107	35111	35117	35129	35141	35149	35153
375_	35159	35171	35201	35221	35227	35251	35257	35267	35279	35281
376_	35291	35311	35317	35323	35327	35339	35353	35363	35381	35393
377_	35401	35407	35419	35423	35437	35447	35449	35461	35491	35507
378_	35509	35521	35527	35531	35533	35537	35543	35569	35573	35591
379_	35593	35597	35603	35617	35671	35677	35729	35731	35747	35753
380_	35759	35771	35797	35801	35803	35809	35831	35837	35839	35851
381_	35863	35869	35879	35897	35899	35911	35923	35933	35951	35963
382_	35969	35977	35983	35993	35999	36007	36011	36013	36017	36037
383_	36061	36067	36073	36083	36097	36107	36109	36131	36137	36151
384_	36161	36187	36191	36209	36217	36229	36241	36251	36263	36269
385_	36277	36293	36299	36307	36313	36319	36341	36343	36353	36373
386_	36383	36389	36433	36451	36457	36467	36469	36473	36479	36493
387_	36497	36523	36527	36529	36541	36551	36559	36563	36571	36583
388_	36587	36599	36607	36629	36637	36643	36653	36671	36677	36683
389_	36691	36697	36709	36713	36721	36739	36749	36761	36767	36779
390_	36781	36787	36791	36793	36809	36821	36833	36847	36857	36871
391_	36877	36887	36899	36901	36913	36919	36923	36929	36931	36943
392_	36947	36973	36979	36997	37003	37013	37019	37021	37039	37049
393_	37057	37061	37087	37097	37117	37123	37139	37159	37171	37181
394_	37189	37199	37201	37217	37223	37243	37253	37273	37277	37307
395_	37309	37313	37321	37337	37339	37357	37361	37363	37369	37379
396_	37397	37409	37423	37441	37447	37463	37483	37489	37493	37501
397_	37507	37511	37517	37529	37537	37547	37549	37561	37567	37571
398_	37573	37579	37589	37591	37607	37619	37633	37643	37649	37657
399_	37663	37691	37693	37699	37717	37747	37781	37783	37799	37811
400_	37813	37831	37847	37853	37861	37871	37879	37889	37897	37907
401_	37951	37957	37963	37967	37987	37991	37993	37997	38011	38039
402_	38047	38053	38069	38083	38113	38119	38149	38153	38167	38177
403_	38183	38189	38197	38201	38219	38231	38237	38239	38261	38273
404_	38281	38287	38299	38303	38317	38321	38327	38329	38333	38351
405_	38371	38377	38393	38431	38447	38449	38453	38459	38461	38501
406_	38543	38557	38561	38567	38569	38593	38603	38609	38611	38629
407_	38639	38651	38653	38669	38671	38677	38693	38699	38707	38711
408_	38713	38723	38729	38737	38747	38749	38767	38783	38791	38803
409_	38821	38833	38839	38851	38861	38867	38873	38891	38903	38917
410_	38921	38923	38933	38953	38959	38971	38977	38993	39019	39023
411_	39041	39043	39047	39079	39089	39097	39103	39107	39113	39119
412_	39133	39139	39157	39161	39163	39181	39191	39199	39209	39217
413_	39227	39229	39233	39239	39241	39251	39293	39301	39313	39317
414_	39323	39341	39343	39359	39367	39371	39373	39383	39397	39409

	_0	_1	_2	_3	_4	_5	_6	_7	_8	_9
415_	39419	39439	39443	39451	39461	39499	39503	39509	39511	39521
416_	39541	39551	39563	39569	39581	39607	39619	39623	39631	39659
417_	39667	39671	39679	39703	39709	39719	39727	39733	39749	39761
418_	39769	39779	39791	39799	39821	39827	39829	39839	39841	39847
419_	39857	39863	39869	39877	39883	39887	39901	39929	39937	39953
420_	39971	39979	39983	39989	40009	40013	40031	40037	40039	40063
421_	40087	40093	40099	40111	40123	40127	40129	40151	40153	40163
422_	40169	40177	40189	40193	40213	40231	40237	40241	40253	40277
423_	40283	40289	40343	40351	40357	40361	40387	40423	40427	40429
424_	40433	40459	40471	40483	40487	40493	40499	40507	40519	40529
425_	40531	40543	40559	40577	40583	40591	40597	40609	40627	40637
426_	40639	40693	40697	40699	40709	40739	40751	40759	40763	40771
427_	40787	40801	40813	40819	40823	40829	40841	40847	40849	40853
428_	40867	40879	40883	40897	40903	40927	40933	40939	40949	40961
429_	40973	40993	41011	41017	41023	41039	41047	41051	41057	41077
430_	41081	41113	41117	41131	41141	41143	41149	41161	41177	41179
431_	41183	41189	41201	41203	41213	41221	41227	41231	41233	41243
432_	41257	41263	41269	41281	41299	41333	41341	41351	41357	41381
433_	41387	41389	41399	41411	41413	41443	41453	41467	41479	41491
434_	41507	41513	41519	41521	41539	41543	41549	41579	41593	41597
435_	41603	41609	41611	41617	41621	41627	41641	41647	41651	41659
436_	41669	41681	41687	41719	41729	41737	41759	41761	41771	41777
437_	41801	41809	41813	41843	41849	41851	41863	41879	41887	41893
438_	41897	41903	41911	41927	41941	41947	41953	41957	41959	41969
439_	41981	41983	41999	42013	42017	42019	42023	42043	42061	42071
440_	42073	42083	42089	42101	42131	42139	42157	42169	42179	42181
441_	42187	42193	42197	42209	42221	42223	42227	42239	42257	42281
442_	42283	42293	42299	42307	42323	42331	42337	42349	42359	42373
443_	42379	42391	42397	42403	42407	42409	42433	42437	42443	42451
444_	42457	42461	42463	42467	42473	42487	42491	42499	42509	42533
445_	42557	42569	42571	42577	42589	42611	42641	42643	42649	42667
446_	42677	42683	42689	42697	42701	42703	42709	42719	42727	42737
447_	42743	42751	42767	42773	42787	42793	42797	42821	42829	42839
448_	42841	42853	42859	42863	42899	42901	42923	42929	42937	42943
449_	42953	42961	42967	42979	42989	43003	43013	43019	43037	43049
450_	43051	43063	43067	43093	43103	43117	43133	43151	43159	43177
451_	43189	43201	43207	43223	43237	43261	43271	43283	43291	43313
452_	43319	43321	43331	43391	43397	43399	43403	43411	43427	43441
453_	43451	43457	43481	43487	43499	43517	43541	43543	43573	43577
454_	43579	43591	43597	43607	43609	43613	43627	43633	43649	43651
455_	43661	43669	43691	43711	43717	43721	43753	43759	43777	43781
456_	43783	43787	43789	43793	43801	43853	43867	43889	43891	43913
457_	43933	43943	43951	43961	43963	43969	43973	43987	43991	43997
458_	44017	44021	44027	44029	44041	44053	44059	44071	44087	44089
459_	44101	44111	44119	44123	44129	44131	44159	44171	44179	44189
460_	44201	44203	44207	44221	44249	44257	44263	44267	44269	44273
461_	44279	44281	44293	44351	44357	44371	44381	44383	44389	44417
462_	44449	44453	44483	44491	44497	44501	44507	44519	44531	44533
463_	44537	44543	44549	44563	44579	44587	44617	44621	44623	44633
464_	44641	44647	44651	44657	44683	44687	44699	44701	44711	44729
465_	44741	44753	44771	44773	44777	44789	44797	44809	44819	44839
466_	44843	44851	44867	44879	44887	44893	44909	44917	44927	44939
467_	44953	44959	44963	44971	44983	44987	45007	45013	45053	45061
468_	45077	45083	45119	45121	45127	45131	45137	45139	45161	45179
469_	45181	45191	45197	45233	45247	45259	45263	45281	45289	45293
470_	45307	45317	45319	45329	45337	45341	45343	45361	45377	45389

	.0	.1	.2	.3	.4	.5	.6	.7	.8	.9
471_	45403	45413	45427	45433	45439	45481	45491	45497	45503	45523
472_	45533	45541	45553	45557	45569	45587	45589	45599	45613	45631
473_	45641	45659	45667	45673	45677	45691	45697	45707	45737	45751
474_	45757	45763	45767	45779	45817	45821	45823	45827	45833	45841
475_	45853	45863	45869	45887	45893	45943	45949	45953	45959	45971
476_	45979	45989	46021	46027	46049	46051	46061	46073	46091	46093
477_	46099	46103	46133	46141	46147	46153	46171	46181	46183	46187
478_	46199	46219	46229	46237	46261	46271	46273	46279	46301	46307
479_	46309	46327	46337	46349	46351	46381	46399	46411	46439	46441
480_	46447	46451	46457	46471	46477	46489	46499	46507	46511	46523
481_	46549	46559	46567	46573	46589	46591	46601	46619	46633	46639
482_	46643	46649	46663	46679	46681	46687	46691	46703	46723	46727
483_	46747	46751	46757	46769	46771	46807	46811	46817	46819	46829
484_	46831	46853	46861	46867	46877	46889	46901	46919	46933	46957
485_	46993	46997	47017	47041	47051	47057	47059	47087	47093	47111
486_	47119	47123	47129	47137	47143	47147	47149	47161	47189	47207
487_	47221	47237	47251	47269	47279	47287	47293	47297	47303	47309
488_	47317	47339	47351	47353	47363	47381	47387	47389	47407	47417
489_	47419	47431	47441	47459	47491	47497	47501	47507	47513	47521
490_	47527	47533	47543	47563	47569	47581	47591	47599	47609	47623
491_	47629	47639	47653	47657	47659	47681	47699	47701	47711	47713
492_	47717	47737	47741	47743	47777	47779	47791	47797	47807	47809
493_	47819	47837	47843	47857	47869	47881	47903	47911	47917	47933
494_	47939	47947	47951	47963	47969	47977	47981	48017	48023	48029
495_	48049	48073	48079	48091	48109	48119	48121	48131	48157	48163
496_	48179	48187	48193	48197	48221	48239	48247	48259	48271	48281
497_	48299	48311	48313	48337	48341	48353	48371	48383	48397	48407
498_	48409	48413	48437	48449	48463	48473	48479	48481	48487	48491
499_	48497	48523	48527	48533	48539	48541	48563	48571	48589	48593
500_	48611	48619	48623	48647	48649	48661	48673	48677	48679	48731
501_	48733	48751	48757	48761	48767	48779	48781	48787	48799	48809
502_	48817	48821	48823	48847	48857	48859	48869	48871	48883	48889
503_	48907	48947	48953	48973	48989	48991	49003	49009	49019	49031
504_	49033	49037	49043	49057	49069	49081	49103	49109	49117	49121
505_	49123	49139	49157	49169	49171	49177	49193	49199	49201	49207
506_	49211	49223	49253	49261	49277	49279	49297	49307	49331	49333
507_	49339	49363	49367	49369	49391	49393	49409	49411	49417	49429
508_	49433	49451	49459	49463	49477	49481	49499	49523	49529	49531
509_	49537	49547	49549	49559	49597	49603	49613	49627	49633	49639
510_	49663	49667	49669	49681	49697	49711	49727	49739	49741	49747
511_	49757	49783	49787	49789	49801	49807	49811	49823	49831	49843
512_	49853	49871	49877	49891	49919	49921	49927	49937	49939	49943
513_	49957	49991	49993	49999	50021	50023	50033	50047	50051	50053
514_	50069	50077	50087	50093	50101	50111	50119	50123	50129	50131
515_	50147	50153	50159	50177	50207	50221	50227	50231	50261	50263
516_	50273	50287	50291	50311	50321	50329	50333	50341	50359	50363
517_	50377	50383	50387	50411	50417	50423	50441	50459	50461	50497
518_	50503	50513	50527	50539	50543	50549	50551	50581	50587	50591
519_	50593	50599	50627	50647	50651	50671	50683	50707	50723	50741
520_	50753	50767	50773	50777	50789	50821	50833	50839	50849	50857
521_	50867	50873	50891	50893	50909	50923	50929	50951	50957	50969
522_	50971	50989	50993	51001	51031	51043	51047	51059	51061	51071
523_	51109	51131	51133	51137	51151	51157	51169	51193	51197	51199
524_	51203	51217	51229	51239	51241	51257	51263	51283	51287	51307
525_	51329	51341	51343	51347	51349	51361	51383	51407	51413	51419
526_	51421	51427	51431	51437	51439	51449	51461	51473	51479	51481

	_0	_1	_2	_3	_4	_5	_6	_7	_8	_9
527_	51487	51503	51511	51517	51521	51539	51551	51563	51577	51581
528_	51593	51599	51607	51613	51631	51637	51647	51659	51673	51679
529_	51683	51691	51713	51719	51721	51749	51767	51769	51787	51797
530_	51803	51817	51827	51829	51839	51853	51859	51869	51871	51893
531_	51899	51907	51913	51929	51941	51949	51971	51973	51977	51991
532_	52009	52021	52027	52051	52057	52067	52069	52081	52103	52121
533_	52127	52147	52153	52163	52177	52181	52183	52189	52201	52223
534_	52237	52249	52253	52259	52267	52289	52291	52301	52313	52321
535_	52361	52363	52369	52379	52387	52391	52433	52453	52457	52489
536_	52501	52511	52517	52529	52541	52543	52553	52561	52567	52571
537_	52579	52583	52609	52627	52631	52639	52667	52673	52691	52697
538_	52709	52711	52721	52727	52733	52747	52757	52769	52783	52807
539_	52813	52817	52837	52859	52861	52879	52883	52889	52901	52903
540_	52919	52937	52951	52957	52963	52967	52973	52981	52999	53003
541_	53017	53047	53051	53069	53077	53087	53089	53093	53101	53113
542_	53117	53129	53147	53149	53161	53171	53173	53189	53197	53201
543_	53231	53233	53239	53267	53269	53279	53281	53299	53309	53323
544_	53327	53353	53359	53377	53381	53401	53407	53411	53419	53437
545_	53441	53453	53479	53503	53507	53527	53549	53551	53569	53591
546_	53593	53597	53609	53611	53617	53623	53629	53633	53639	53653
547_	53657	53681	53693	53699	53719	53719	53731	53759	53773	53777
548_	53783	53791	53813	53819	53831	53849	53857	53861	53881	53887
549_	53891	53897	53899	53917	53923	53927	53939	53951	53959	53987
550_	53993	54001	54011	54013	54037	54049	54059	54083	54091	54101
551_	54121	54133	54139	54151	54163	54167	54181	54193	54217	54251
552_	54269	54277	54287	54293	54311	54319	54323	54331	54347	54361
553_	54367	54371	54377	54401	54403	54409	54413	54419	54421	54437
554_	54443	54449	54469	54493	54497	54499	54503	54517	54521	54539
555_	54541	54547	54559	54563	54577	54581	54583	54601	54617	54623
556_	54629	54631	54647	54667	54673	54679	54709	54713	54721	54727
557_	54751	54767	54773	54779	54787	54799	54829	54833	54851	54869
558_	54877	54881	54907	54917	54919	54941	54949	54959	54973	54979
559_	54983	55001	55009	55021	55049	55051	55057	55061	55073	55079
560_	55103	55109	55117	55127	55147	55163	55171	55201	55207	55213
561_	55217	55219	55229	55243	55249	55259	55291	55313	55331	55333
562_	55337	55339	55343	55351	55373	55381	55399	55411	55439	55441
563_	55457	55469	55487	55501	55511	55529	55541	55547	55579	55589
564_	55603	55609	55619	55621	55631	55633	55639	55661	55663	55667
565_	55673	55681	55691	55697	55711	55717	55721	55733	55763	55787
566_	55793	55799	55807	55813	55817	55819	55823	55829	55837	55843
567_	55849	55871	55889	55897	55901	55903	55921	55927	55931	55933
568_	55949	55967	55987	55997	56003	56009	56039	56041	56053	56081
569_	56087	56093	56099	56101	56113	56123	56131	56149	56167	56171
570_	56179	56197	56207	56209	56237	56239	56249	56263	56267	56269
571_	56299	56311	56333	56359	56369	56377	56383	56393	56401	56417
572_	56431	56437	56443	56453	56467	56473	56477	56479	56489	56501
573_	56503	56509	56519	56527	56531	56533	56543	56569	56591	56597
574_	56599	56611	56629	56633	56659	56663	56671	56681	56687	56701
575_	56711	56713	56731	56737	56747	56767	56773	56779	56783	56807
576_	56809	56813	56821	56827	56843	56857	56873	56891	56893	56897
577_	56909	56911	56921	56923	56929	56941	56951	56957	56963	56983
578_	56989	56993	56999	57037	57041	57047	57059	57073	57077	57089
579_	57097	57107	57119	57131	57139	57143	57149	57163	57173	57179
580_	57191	57193	57203	57221	57223	57241	57251	57259	57269	57271
581_	57283	57287	57301	57329	57331	57347	57349	57367	57373	57383
582_	57389	57397	57413	57427	57457	57467	57487	57493	57503	57527

	.0	.1	.2	.3	.4	.5	.6	.7	.8	.9
583_	57529	57557	57559	57571	57587	57593	57601	57637	57641	57649
584_	57653	57667	57679	57689	57697	57709	57713	57719	57727	57731
585_	57737	57751	57773	57781	57787	57791	57793	57803	57809	57829
586_	57839	57847	57853	57859	57881	57899	57901	57917	57923	57943
587_	57947	57973	57977	57991	58013	58027	58031	58043	58049	58057
588_	58061	58067	58073	58099	58109	58111	58129	58147	58151	58153
589_	58169	58171	58189	58193	58199	58207	58211	58217	58229	58231
590_	58237	58243	58271	58309	58313	58321	58337	58363	58367	58369
591_	58379	58391	58393	58403	58411	58417	58427	58439	58441	58451
592_	58453	58477	58481	58511	58537	58543	58549	58567	58573	58579
593_	58601	58603	58613	58631	58657	58661	58679	58687	58693	58699
594_	58711	58727	58733	58741	58757	58763	58771	58787	58789	58831
595_	58889	58897	58901	58907	58909	58913	58921	58937	58943	58963
596_	58967	58979	58991	58997	59009	59011	59021	59023	59029	59051
597_	59053	59063	59069	59077	59083	59093	59107	59113	59119	59123
598_	59141	59149	59159	59167	59183	59197	59207	59209	59219	59221
599_	59233	59239	59243	59263	59273	59281	59333	59341	59351	59357
600_	59359	59369	59377	59387	59393	59399	59407	59417	59419	59441
601_	59443	59447	59453	59467	59471	59473	59497	59509	59513	59539
602_	59557	59561	59567	59581	59611	59617	59621	59627	59629	59651
603_	59659	59663	59669	59671	59693	59699	59707	59723	59729	59743
604_	59747	59753	59771	59779	59791	59797	59809	59833	59863	59879
605_	59887	59921	59929	59951	59957	59971	59981	59999	60013	60017
606_	60029	60037	60041	60077	60083	60089	60091	60101	60103	60107
607_	60127	60133	60139	60149	60161	60167	60169	60209	60217	60223
608_	60251	60257	60259	60271	60289	60293	60317	60331	60337	60343
609_	60353	60373	60383	60397	60413	60427	60443	60449	60457	60493
610_	60497	60509	60521	60527	60539	60589	60601	60607	60611	60617
611_	60623	60631	60637	60647	60649	60659	60661	60679	60689	60703
612_	60719	60727	60733	60737	60757	60761	60763	60773	60779	60793
613_	60811	60821	60859	60869	60887	60889	60899	60901	60913	60917
614_	60919	60923	60937	60943	60953	60961	61001	61007	61027	61031
615_	61043	61051	61057	61091	61099	61121	61129	61141	61151	61153
616_	61169	61211	61231	61253	61261	61283	61291	61297	61331	
617_	61333	61339	61343	61357	61363	61379	61381	61403	61409	61417
618_	61441	61463	61469	61471	61483	61487	61493	61507	61511	61519
619_	61543	61547	61553	61559	61561	61583	61603	61609	61613	61627
620_	61631	61637	61643	61651	61657	61667	61673	61681	61687	61703
621_	61717	61723	61729	61751	61757	61781	61813	61819	61837	61843
622_	61861	61871	61879	61909	61927	61933	61949	61961	61967	61979
623_	61981	61987	61991	62003	62011	62017	62039	62047	62053	62057
624_	62071	62081	62099	62119	62129	62131	62137	62141	62143	62171
625_	62189	62191	62201	62207	62213	62219	62233	62273	62297	62299
626_	62303	62311	62323	62327	62347	62351	62383	62401	62417	62423
627_	62459	62467	62473	62477	62483	62497	62501	62507	62533	62539
628_	62549	62563	62581	62591	62597	62603	62617	62627	62633	62639
629_	62653	62659	62683	62687	62701	62723	62731	62743	62753	62761
630_	62773	62791	62801	62819	62827	62851	62861	62869	62873	62897
631_	62903	62921	62927	62929	62939	62969	62971	62981	62983	62987
632_	62989	63029	63031	63059	63067	63073	63079	63097	63103	63113
633_	63127	63131	63149	63179	63197	63199	63211	63241	63247	63277
634_	63281	63299	63311	63313	63317	63331	63337	63347	63353	63361
635_	63367	63377	63389	63391	63397	63409	63419	63421	63439	63443
636_	63463	63467	63473	63487	63493	63499	63521	63527	63533	63541
637_	63559	63577	63587	63589	63599	63601	63607	63611	63617	63629
638_	63647	63649	63659	63667	63671	63689	63691	63697	63703	63709

	_0	_1	_2	_3	_4	_5	_6	_7	_8	_9
639_	63719	63727	63737	63743	63761	63773	63781	63793	63799	63803
640_	63809	63823	63839	63841	63853	63857	63863	63901	63907	63913
641_	63929	63949	63977	63997	64007	64013	64019	64033	64037	64063
642_	64067	64081	64091	64109	64123	64151	64153	64157	64171	64187
643_	64189	64217	64223	64231	64237	64271	64279	64283	64301	64303
644_	64319	64327	64333	64373	64381	64399	64403	64433	64439	64451
645_	64453	64483	64489	64499	64513	64553	64567	64577	64579	64591
646_	64601	64609	64613	64621	64627	64633	64661	64663	64667	64679
647_	64693	64709	64717	64747	64763	64781	64783	64793	64811	64817
648_	64849	64853	64871	64877	64879	64891	64901	64919	64921	64927
649_	64937	64951	64969	64997	65003	65011	65027	65029	65033	65053
650_	65063	65071	65089	65099	65101	65111	65119	65123	65129	65141
651_	65147	65167	65171	65173	65179	65183	65203	65213	65239	65257
652_	65267	65269	65287	65293	65309	65323	65327	65353	65357	65371
653_	65381	65393	65407	65413	65419	65423	65437	65447	65449	65479
654_	65497	65519	65521	65537	65539	65543	65551	65557	65563	65579
655_	65581	65587	65599	65609	65617	65629	65633	65647	65651	65657
656_	65677	65687	65699	65701	65707	65713	65717	65719	65729	65731
657_	65761	65777	65789	65809	65827	65831	65837	65839	65843	65851
658_	65867	65881	65899	65921	65927	65929	65951	65957	65963	65981
659_	65983	65993	66029	66037	66041	66047	66067	66071	66083	66089
660_	66103	66107	66109	66137	66161	66169	66173	66179	66191	66221
661_	66239	66271	66293	66301	66337	66343	66347	66359	66361	66373
662_	66377	66383	66403	66413	66431	66449	66457	66463	66467	66491
663_	66499	66509	66523	66529	66533	66541	66553	66569	66571	66587
664_	66593	66601	66617	66629	66643	66653	66683	66697	66701	66713
665_	66721	66733	66739	66749	66751	66763	66791	66797	66809	66821
666_	66841	66851	66853	66863	66877	66883	66889	66919	66923	66931
667_	66943	66947	66949	66959	66973	66977	67003	67021	67033	67043
668_	67049	67057	67061	67073	67079	67103	67121	67129	67139	67141
669_	67153	67157	67169	67181	67187	67189	67211	67213	67217	67219
670_	67231	67247	67261	67271	67273	67289	67307	67339	67343	67349
671_	67369	67391	67399	67409	67411	67421	67427	67429	67433	67447
672_	67453	67477	67481	67489	67493	67499	67511	67523	67531	67537
673_	67547	67559	67567	67577	67579	67589	67601	67607	67619	67631
674_	67651	67679	67699	67709	67723	67733	67741	67751	67757	67759
675_	67763	67777	67783	67789	67801	67807	67819	67829	67843	67853
676_	67867	67883	67891	67901	67927	67931	67933	67939	67943	67957
677_	67961	67967	67979	67987	67993	68023	68041	68053	68059	68071
678_	68087	68099	68111	68113	68141	68147	68161	68171	68207	68209
679_	68213	68219	68227	68239	68261	68279	68281	68311	68329	68351
680_	68371	68389	68399	68437	68443	68447	68449	68473	68477	68483
681_	68489	68491	68501	68507	68521	68531	68539	68543	68567	68581
682_	68597	68611	68633	68639	68659	68669	68683	68687	68699	68711
683_	68713	68729	68737	68743	68749	68767	68771	68777	68791	68813
684_	68819	68821	68863	68879	68881	68891	68897	68899	68903	68909
685_	68917	68927	68947	68963	68993	69001	69011	69019	69029	69031
686_	69061	69067	69073	69109	69119	69127	69143	69149	69151	69163
687_	69191	69193	69197	69203	69221	69233	69239	69247	69257	69259
688_	69263	69313	69317	69337	69341	69371	69379	69383	69389	69401
689_	69403	69427	69431	69439	69457	69463	69467	69473	69481	69491
690_	69493	69497	69499	69539	69557	69593	69623	69653	69661	69677
691_	69691	69697	69709	69737	69739	69761	69763	69767	69779	69809
692_	69821	69827	69829	69833	69847	69857	69859	69877	69899	69911
693_	69929	69931	69941	69959	69991	69997	70001	70003	70009	70019
694_	70039	70051	70061	70067	70079	70099	70111	70117	70121	70123

	_0	_1	_2	_3	_4	_5	_6	_7	_8	_9
695_	70139	70141	70157	70163	70177	70181	70183	70199	70201	70207
696_	70223	70229	70237	70241	70249	70271	70289	70297	70309	70313
697_	70321	70327	70351	70373	70379	70381	70393	70423	70429	70439
698_	70451	70457	70459	70481	70487	70489	70501	70507	70529	70537
699_	70549	70571	70573	70583	70589	70607	70619	70621	70627	70639
700_	70657	70663	70667	70687	70709	70717	70729	70753	70769	70783
701_	70793	70823	70841	70843	70849	70853	70867	70877	70879	70891
702_	70901	70913	70919	70921	70937	70949	70951	70957	70969	70979
703_	70981	70991	70997	70999	71011	71023	71039	71059	71069	71081
704_	71089	71119	71129	71143	71147	71153	71161	71167	71171	71191
705_	71209	71233	71237	71249	71257	71261	71263	71287	71293	71317
706_	71327	71329	71333	71339	71341	71347	71353	71359	71363	71387
707_	71389	71399	71411	71413	71419	71429	71437	71443	71453	71471
708_	71473	71479	71483	71503	71527	71537	71549	71551	71563	71569
709_	71593	71597	71633	71647	71663	71671	71693	71699	71707	71711
710_	71713	71719	71741	71761	71777	71789	71807	71809	71821	71837
711_	71843	71849	71861	71867	71879	71881	71887	71899	71909	71917
712_	71933	71941	71947	71963	71971	71983	71987	71993	71999	72019
713_	72031	72043	72047	72053	72073	72077	72089	72091	72101	72103
714_	72109	72139	72161	72167	72169	72173	72211	72221	72223	72227
715_	72229	72251	72253	72269	72271	72277	72287	72307	72313	72337
716_	72341	72353	72367	72379	72383	72421	72431	72461	72467	72469
717_	72481	72493	72497	72503	72533	72547	72551	72559	72577	72613
718_	72617	72623	72643	72647	72649	72661	72671	72673	72679	72689
719_	72701	72707	72719	72727	72733	72739	72763	72767	72797	72817
720_	72823	72859	72869	72871	72883	72889	72893	72901	72907	72911
721_	72923	72931	72937	72949	72953	72959	72973	72977	72997	73009
722_	73013	73019	73037	73039	73043	73061	73063	73079	73091	73121
723_	73127	73133	73141	73181	73189	73237	73243	73259	73277	73291
724_	73303	73309	73327	73331	73351	73361	73363	73369	73379	73387
725_	73417	73421	73433	73453	73459	73471	73477	73483	73517	73523
726_	73529	73547	73553	73561	73571	73583	73589	73597	73607	73609
727_	73613	73637	73643	73651	73673	73679	73681	73693	73699	73709
728_	73721	73727	73751	73757	73771	73783	73819	73823	73847	73849
729_	73859	73867	73877	73883	73907	73907	73939	73943	73951	73961
730_	73973	73999	74017	74021	74027	74047	74051	74071	74077	74093
731_	74099	74101	74131	74143	74149	74159	74161	74167	74177	74189
732_	74197	74201	74203	74209	74219	74231	74257	74279	74287	74293
733_	74297	74311	74317	74323	74353	74357	74363	74377	74381	74383
734_	74411	74413	74419	74441	74449	74453	74471	74489	74507	74509
735_	74521	74527	74531	74551	74561	74567	74573	74587	74597	74609
736_	74611	74623	74653	74687	74699	74707	74713	74717	74719	74729
737_	74731	74747	74759	74761	74771	74779	74797	74821	74827	74831
738_	74843	74857	74861	74869	74873	74887	74891	74897	74903	74923
739_	74929	74933	74941	74959	75011	75013	75017	75029	75037	75041
740_	75079	75083	75109	75133	75149	75161	75167	75169	75181	75193
741_	75209	75211	75217	75223	75227	75239	75253	75269	75277	75289
742_	75307	75323	75329	75337	75347	75353	75367	75377	75389	75391
743_	75401	75403	75407	75431	75437	75479	75503	75511	75521	75527
744_	75533	75539	75541	75553	75557	75571	75577	75583	75611	75617
745_	75619	75629	75641	75653	75659	75679	75683	75689	75703	75707
746_	75709	75721	75731	75743	75767	75773	75781	75787	75793	75797
747_	75821	75833	75853	75869	75883	75913	75931	75937	75941	75967
748_	75979	75983	75989	75991	75997	76001	76003	76031	76039	76079
749_	76081	76091	76099	76103	76123	76129	76147	76157	76159	76163
750_	76207	76213	76231	76243	76249	76253	76259	76261	76283	76289

	.0	.1	.2	.3	.4	.5	.6	.7	.8	.9
751_	76303	76333	76343	76367	76369	76379	76387	76403	76421	76423
752_	76441	76463	76471	76481	76487	76493	76507	76511	76519	76537
753_	76541	76543	76561	76579	76597	76603	76607	76631	76649	76651
754_	76667	76673	76679	76697	76717	76733	76753	76771	76771	76777
755_	76781	76801	76819	76829	76831	76837	76847	76871	76873	76883
756_	76907	76913	76919	76943	76949	76961	76963	76991	77003	77017
757_	77023	77029	77041	77047	77069	77081	77093	77101	77137	77141
758_	77153	77167	77171	77191	77201	77213	77237	77239	77243	77249
759_	77261	77263	77267	77269	77279	77291	77317	77323	77339	77347
760_	77351	77359	77369	77377	77383	77417	77419	77431	77447	77471
761_	77477	77479	77489	77491	77509	77513	77521	77527	77543	77549
762_	77551	77557	77563	77569	77573	77587	77591	77611	77617	77621
763_	77641	77647	77659	77681	77687	77689	77699	77711	77713	77719
764_	77723	77731	77743	77747	77761	77773	77783	77797	77801	77813
765_	77839	77849	77863	77867	77893	77899	77929	77933	77951	77969
766_	77977	77983	77999	78007	78017	78031	78041	78049	78059	78079
767_	78101	78121	78137	78139	78157	78163	78167	78173	78179	78191
768_	78193	78203	78229	78233	78241	78259	78277	78283	78301	78307
769_	78311	78317	78341	78347	78367	78401	78427	78437	78439	78467
770_	78479	78487	78497	78509	78511	78517	78539	78541	78553	78569
771_	78571	78577	78583	78593	78607	78623	78643	78649	78653	78691
772_	78697	78707	78713	78721	78737	78779	78781	78787	78791	78797
773_	78803	78809	78823	78839	78853	78857	78877	78887	78889	78893
774_	78901	78919	78929	78941	78977	78979	78989	79031	79039	79043
775_	79063	79087	79103	79111	79133	79139	79147	79151	79153	79159
776_	79181	79187	79193	79201	79229	79231	79241	79259	79273	79279
777_	79283	79301	79309	79319	79333	79337	79349	79357	79367	79379
778_	79393	79397	79399	79411	79423	79427	79433	79451	79481	79493
779_	79531	79537	79549	79559	79561	79579	79589	79601	79609	79613
780_	79621	79627	79631	79633	79657	79669	79687	79691	79693	79697
781_	79699	79757	79769	79777	79801	79811	79813	79817	79823	79829
782_	79841	79843	79847	79861	79867	79873	79889	79901	79903	79907
783_	79939	79943	79967	79973	79979	79987	79997	79999	80021	80039
784_	80051	80071	80077	80107	80111	80141	80147	80149	80153	80167
785_	80173	80177	80191	80207	80209	80221	80231	80233	80239	80251
786_	80263	80273	80279	80287	80309	80317	80329	80341	80347	80363
787_	80369	80387	80407	80429	80447	80449	80471	80473	80489	80491
788_	80513	80527	80537	80557	80567	80599	80603	80611	80621	80627
789_	80629	80651	80657	80669	80671	80677	80681	80683	80687	80701
790_	80713	80737	80747	80749	80761	80777	80779	80783	80789	80803
791_	80809	80819	80831	80833	80849	80863	80897	80909	80911	80917
792_	80923	80929	80933	80953	80963	80989	81001	81013	81017	81019
793_	81023	81031	81041	81043	81047	81049	81071	81077	81083	81097
794_	81101	81119	81131	81157	81163	81173	81181	81197	81199	81203
795_	81223	81233	81239	81281	81283	81293	81299	81307	81331	81343
796_	81349	81353	81359	81371	81373	81401	81409	81421	81439	81457
797_	81463	81509	81517	81527	81533	81547	81551	81553	81559	81563
798_	81569	81611	81619	81629	81637	81647	81649	81667	81671	81677
799_	81689	81701	81703	81707	81727	81737	81749	81761	81769	81773
800_	81799	81817	81839	81847	81853	81869	81883	81899	81901	81919
801_	81929	81931	81937	81943	81953	81967	81971	81973	82003	82007
802_	82009	82013	82021	82031	82037	82039	82051	82067	82073	82129
803_	82139	82141	82153	82163	82171	82183	82189	82193	82207	82217
804_	82219	82223	82231	82237	82241	82261	82267	82279	82301	82307
805_	82339	82349	82351	82361	82373	82387	82393	82421	82457	82463
806_	82469	82471	82483	82487	82493	82499	82507	82529	82531	82549

	.0	.1	.2	.3	.4	.5	.6	.7	.8	.9
807_	82559	82561	82567	82571	82591	82601	82609	82613	82619	82633
808_	82651	82657	82699	82721	82723	82727	82729	82757	82759	82763
809_	82781	82787	82793	82799	82811	82813	82837	82847	82883	82889
810_	82891	82903	82913	82939	82963	82981	82997	83003	83009	83023
811_	83047	83059	83063	83071	83077	83089	83093	83101	83117	83137
812_	83177	83203	83207	83219	83221	83227	83231	83233	83243	83257
813_	83267	83269	83273	83299	83311	83339	83341	83357	83383	83389
814_	83399	83401	83407	83417	83423	83431	83437	83443	83449	83459
815_	83471	83477	83497	83537	83557	83561	83563	83579	83591	83597
816_	83609	83617	83621	83639	83641	83653	83663	83689	83701	83717
817_	83719	83737	83761	83773	83777	83791	83813	83833	83843	83857
818_	83869	83873	83891	83903	83911	83921	83933	83939	83969	83983
819_	83987	84011	84017	84047	84053	84059	84061	84067	84089	84121
820_	84127	84131	84137	84143	84163	84179	84181	84191	84199	84211
821_	84221	84223	84229	84239	84247	84263	84299	84307	84313	84317
822_	84319	84347	84349	84377	84389	84391	84401	84407	84421	84431
823_	84437	84443	84449	84457	84463	84467	84481	84499	84503	84509
824_	84521	84523	84533	84551	84559	84589	84629	84631	84649	84653
825_	84659	84673	84691	84697	84701	84713	84719	84731	84737	84751
826_	84761	84787	84793	84809	84811	84827	84857	84859	84869	84871
827_	84913	84919	84947	84961	84967	84977	84979	84991	85009	85021
828_	85027	85037	85049	85061	85081	85087	85091	85093	85103	85109
829_	85121	85133	85147	85159	85193	85199	85201	85213	85223	85229
830_	85237	85243	85247	85259	85297	85303	85313	85331	85333	85361
831_	85363	85369	85381	85411	85427	85429	85439	85447	85451	85453
832_	85469	85487	85513	85517	85523	85531	85549	85571	85577	85597
833_	85601	85607	85619	85621	85627	85639	85643	85661	85667	85669
834_	85691	85703	85711	85717	85733	85751	85781	85793	85817	85819
835_	85829	85831	85837	85843	85847	85853	85889	85903	85909	85931
836_	85933	85991	85999	86011	86017	86027	86029	86069	86077	86083
837_	86111	86113	86117	86131	86137	86143	86161	86171	86179	86183
838_	86197	86201	86209	86239	86243	86249	86257	86263	86269	86287
839_	86291	86293	86297	86311	86323	86341	86351	86353	86357	86369
840_	86371	86381	86389	86399	86413	86423	86441	86453	86461	86467
841_	86477	86491	86501	86509	86531	86533	86539	86561	86573	86579
842_	86587	86599	86627	86629	86677	86689	86693	86711	86719	86729
843_	86743	86753	86767	86771	86783	86813	86837	86843	86851	86857
844_	86861	86869	86923	86927	86929	86939	86951	86959	86969	86981
845_	86993	87011	87013	87037	87041	87049	87071	87083	87103	87107
846_	87119	87121	87133	87149	87151	87179	87181	87187	87211	87221
847_	87223	87251	87253	87257	87277	87281	87293	87299	87313	87317
848_	87323	87337	87359	87383	87403	87407	87421	87427	87433	87443
849_	87473	87481	87491	87509	87511	87517	87523	87539	87541	87547
850_	87553	87557	87559	87583	87587	87589	87613	87623	87629	87631
851_	87641	87643	87649	87671	87679	87683	87691	87697	87701	87719
852_	87721	87739	87743	87751	87767	87793	87797	87803	87811	87833
853_	87853	87869	87877	87881	87887	87911	87917	87931	87943	87959
854_	87961	87973	87977	87991	88001	88003	88007	88019	88037	88069
855_	88079	88093	88117	88129	88169	88177	88211	88223	88237	88241
856_	88259	88261	88289	88301	88321	88327	88337	88339	88379	88397
857_	88411	88423	88427	88463	88469	88471	88493	88499	88513	88523
858_	88547	88589	88591	88607	88609	88643	88651	88657	88661	88663
859_	88667	88681	88721	88729	88741	88747	88771	88789	88793	88799
860_	88801	88807	88811	88813	88817	88819	88843	88853	88861	88867
861_	88873	88883	88897	88903	88919	88937	88951	88969	88993	88997
862_	89003	89009	89017	89021	89041	89051	89057	89069	89071	89083

	_0	_1	_2	_3	_4	_5	_6	_7	_8	_9
863_	89087	89101	89107	89113	89119	89123	89137	89153	89189	89203
864_	89209	89213	89227	89231	89237	89261	89269	89273	89293	89303
865_	89317	89329	89363	89371	89381	89387	89393	89399	89413	89417
866_	89431	89443	89449	89459	89477	89491	89501	89513	89519	89521
867_	89527	89533	89561	89563	89567	89591	89597	89599	89603	89611
868_	89627	89633	89653	89657	89659	89669	89671	89681	89689	89753
869_	89759	89767	89779	89783	89797	89809	89819	89821	89833	89839
870_	89849	89867	89891	89897	89899	89909	89917	89923	89939	89959
871_	89963	89977	89983	89989	90001	90007	90011	90017	90019	90023
872_	90031	90053	90059	90067	90071	90073	90089	90107	90121	90127
873_	90149	90163	90173	90187	90191	90197	90199	90203	90217	90227
874_	90239	90247	90263	90271	90281	90289	90313	90353	90359	90371
875_	90373	90379	90397	90401	90403	90407	90437	90439	90469	90473
876_	90481	90499	90511	90523	90527	90529	90533	90547	90583	90599
877_	90617	90619	90631	90641	90647	90659	90677	90679	90697	90703
878_	90709	90731	90749	90787	90793	90803	90821	90823	90833	90841
879_	90847	90863	90887	90901	90907	90911	90917	90931	90947	90971
880_	90977	90989	90997	91009	91019	91033	91079	91081	91097	91099
881_	91121	91127	91129	91139	91141	91151	91153	91159	91163	91183
882_	91193	91199	91229	91237	91243	91249	91253	91283	91291	91297
883_	91303	91309	91331	91367	91369	91373	91381	91387	91393	91397
884_	91411	91423	91433	91453	91457	91459	91463	91493	91499	91513
885_	91529	91541	91571	91573	91577	91583	91591	91621	91631	91639
886_	91673	91691	91703	91711	91733	91753	91757	91771	91781	91801
887_	91807	91811	91813	91823	91837	91841	91867	91873	91909	91921
888_	91939	91943	91951	91957	91961	91967	91969	91997	92003	92009
889_	92033	92041	92051	92077	92083	92107	92111	92119	92143	92153
890_	92173	92177	92179	92189	92203	92219	92221	92227	92233	92237
891_	92243	92251	92269	92297	92311	92317	92333	92347	92353	92357
892_	92363	92369	92377	92381	92383	92387	92399	92401	92413	92419
893_	92431	92459	92461	92467	92479	92489	92503	92507	92551	92557
894_	92567	92569	92581	92593	92623	92627	92639	92641	92647	92657
895_	92669	92671	92681	92683	92693	92699	92707	92717	92723	92737
896_	92753	92761	92767	92779	92789	92791	92801	92809	92821	92831
897_	92849	92857	92861	92863	92867	92893	92899	92921	92927	92941
898_	92951	92957	92959	92987	92993	93001	93047	93053	93059	93077
899_	93083	93089	93097	93103	93113	93131	93133	93139	93151	93169
900_	93179	93187	93199	93229	93239	93241	93251	93253	93257	93263
901_	93281	93283	93287	93307	93319	93323	93329	93337	93371	93377
902_	93383	93407	93419	93427	93463	93479	93481	93487	93491	93493
903_	93497	93503	93523	93529	93553	93557	93559	93563	93581	93601
904_	93607	93629	93637	93683	93701	93703	93719	93739	93761	93763
905_	93787	93809	93811	93827	93851	93871	93887	93889	93893	93901
906_	93911	93913	93923	93937	93941	93949	93967	93971	93979	93983
907_	93997	94007	94009	94033	94049	94057	94063	94079	94099	94109
908_	94111	94117	94121	94151	94153	94169	94201	94207	94219	94229
909_	94253	94261	94273	94291	94307	94309	94321	94327	94331	94343
910_	94349	94351	94379	94397	94399	94421	94427	94433	94439	94441
911_	94447	94463	94477	94483	94513	94529	94531	94541	94543	94547
912_	94559	94561	94573	94583	94597	94603	94613	94621	94649	94651
913_	94687	94693	94709	94723	94727	94747	94771	94777	94781	94789
914_	94793	94811	94819	94823	94837	94841	94847	94849	94873	94889
915_	94903	94907	94933	94949	94951	94961	94993	94999	95003	95009
916_	95021	95027	95063	95071	95083	95087	95089	95093	95101	95107
917_	95111	95131	95143	95153	95177	95189	95191	95203	95213	95219
918_	95231	95233	95239	95257	95261	95267	95273	95279	95287	95311

	_0	_1	_2	_3	_4	_5	_6	_7	_8	_9
919_	95317	95327	95339	95369	95383	95393	95401	95413	95419	95429
920_	95441	95443	95461	95467	95471	95479	95483	95507	95527	95531
921_	95539	95549	95561	95569	95581	95597	95603	95617	95621	95629
922_	95633	95651	95701	95707	95713	95717	95723	95731	95737	95747
923_	95773	95783	95789	95791	95801	95803	95813	95819	95857	95869
924_	95873	95881	95891	95911	95917	95923	95929	95947	95957	95959
925_	95971	95987	95989	96001	96013	96017	96043	96053	96059	96079
926_	96097	96137	96149	96157	96167	96179	96181	96199	96211	96221
927_	96223	96233	96259	96263	96269	96281	96289	96293	96323	96329
928_	96331	96337	96353	96377	96401	96419	96431	96443	96451	96457
929_	96461	96469	96479	96487	96493	96497	96517	96527	96553	96557
930_	96581	96587	96589	96601	96643	96661	96667	96671	96697	96703
931_	96731	96737	96739	96749	96757	96763	96769	96779	96787	96797
932_	96799	96821	96823	96827	96847	96851	96857	96893	96907	96911
933_	96931	96953	96959	96973	96979	96989	96997	97001	97003	97007
934_	97021	97039	97073	97081	97103	97117	97127	97151	97157	97159
935_	97169	97171	97177	97187	97213	97231	97241	97259	97283	97301
936_	97303	97327	97367	97369	97373	97379	97381	97387	97397	97423
937_	97429	97441	97453	97459	97463	97499	97501	97511	97523	97547
938_	97549	97553	97561	97571	97577	97579	97583	97607	97609	97613
939_	97649	97651	97673	97687	97711	97729	97771	97777	97787	97789
940_	97813	97829	97841	97843	97847	97849	97859	97861	97871	97879
941_	97883	97919	97927	97931	97943	97961	97967	97973	97987	98009
942_	98011	98017	98041	98047	98057	98081	98101	98123	98129	98143
943_	98179	98207	98213	98221	98227	98251	98257	98269	98297	98299
944_	98317	98321	98323	98327	98347	98369	98377	98387	98389	98407
945_	98411	98419	98429	98443	98453	98459	98467	98473	98479	98491
946_	98507	98519	98533	98543	98561	98563	98573	98597	98621	98627
947_	98639	98641	98663	98669	98689	98711	98713	98717	98729	98731
948_	98737	98773	98779	98801	98807	98809	98837	98849	98867	98869
949_	98873	98887	98893	98897	98899	98909	98911	98927	98929	98939
950_	98947	98953	98963	98981	98993	98999	99013	99017	99023	99041
951_	99053	99079	99083	99089	99103	99109	99119	99131	99133	99137
952_	99139	99149	99173	99181	99191	99223	99233	99241	99251	99257
953_	99259	99277	99289	99317	99347	99349	99367	99371	99377	99391
954_	99397	99401	99409	99431	99439	99469	99487	99497	99523	99527
955_	99529	99551	99559	99563	99571	99577	99581	99607	99611	99623
956_	99643	99661	99667	99679	99689	99707	99709	99713	99719	99721
957_	99733	99761	99767	99787	99793	99809	99817	99823	99829	99833
958_	99839	99859	99871	99877	99881	99901	99907	99923	99929	99961
959_	99971	99989	99991	100003	100019	100043	100049	100057	100069	100103

2.4.13 FACTORIZATION TABLE

The following is a list of the factors of numbers up to and beyond 1,000. When a number is prime, it is shown in a bold face font.

	_0	_1	_2	_3	_4	_5	_6	_7	_8	_9
0_			**2**	**3**	2^2	**5**	$2 \cdot 3$	**7**	2^3	3^2
1_	$2 \cdot 5$	**11**	$2^2 \cdot 3$	**13**	$2 \cdot 7$	$3 \cdot 5$	2^4	**17**	$2 \cdot 3^2$	**19**
2_	$2^2 \cdot 5$	$3 \cdot 7$	$2 \cdot 11$	**23**	$2^3 \cdot 3$	5^2	$2 \cdot 13$	3^3	$2^2 \cdot 7$	**29**
3_	$2 \cdot 3 \cdot 5$	**31**	2^5	$3 \cdot 11$	$2 \cdot 17$	$5 \cdot 7$	$2^2 \cdot 3^2$	**37**	$2 \cdot 19$	$3 \cdot 13$
4_	$2^3 \cdot 5$	**41**	$2 \cdot 3 \cdot 7$	**43**	$2^2 \cdot 11$	$3^2 \cdot 5$	$2 \cdot 23$	**47**	$2^4 \cdot 3$	7^2
5_	$2 \cdot 5^2$	$3 \cdot 17$	$2^2 \cdot 13$	**53**	$2 \cdot 3^3$	$5 \cdot 11$	$2^3 \cdot 7$	$3 \cdot 19$	$2 \cdot 29$	**59**
6_	$2^2 \cdot 3 \cdot 5$	**61**	$2 \cdot 31$	$3^2 \cdot 7$	2^6		$5 \cdot 13$	$2 \cdot 3 \cdot 11$	**67**	$2^2 \cdot 17$	$3 \cdot 23$
7_	$2 \cdot 5 \cdot 7$	**71**	$2^3 \cdot 3^2$	**73**	$2 \cdot 37$	$3 \cdot 5^2$	$2^2 \cdot 19$	$7 \cdot 11$	$2 \cdot 3 \cdot 13$	**79**
8_	$2^4 \cdot 5$	3^4	$2 \cdot 41$	**83**	$2^2 \cdot 3 \cdot 7$	$5 \cdot 17$	$2 \cdot 43$	$3 \cdot 29$	$2^3 \cdot 11$	**89**
9_	$2 \cdot 3^2 \cdot 5$	$7 \cdot 13$	$2^2 \cdot 23$	$3 \cdot 31$	$2 \cdot 47$	$5 \cdot 19$	$2^5 \cdot 3$	**97**	$2 \cdot 7^2$	$3^2 \cdot 11$

	_0	_1	_2	_3	_4	_5	_6	_7	_8	_9
10_	$2^2 \cdot 5^2$	**101**	$2 \cdot 3 \cdot 17$	**103**	$2^3 \cdot 13$	$3 \cdot 5 \cdot 7$	$2 \cdot 53$	**107**	$2^2 \cdot 3^3$	**109**
11_	$2 \cdot 5 \cdot 11$	$3 \cdot 37$	$2^4 \cdot 7$	**113**	$2 \cdot 3 \cdot 19$	$5 \cdot 23$	$2^2 \cdot 29$	$3^2 \cdot 13$	$2 \cdot 59$	$7 \cdot 17$
12_	$2^3 \cdot 3 \cdot 5$	11^2	$2 \cdot 61$	$3 \cdot 41$	$2^2 \cdot 31$	5^3	$2 \cdot 3^2 \cdot 7$	**127**	2^7	$3 \cdot 43$
13_	$2 \cdot 5 \cdot 13$	**131**	$2^2 \cdot 3 \cdot 11$	$7 \cdot 19$	$2 \cdot 67$	$3^3 \cdot 5$	$2^3 \cdot 17$	**137**	$2 \cdot 3 \cdot 23$	**139**
14_	$2^2 \cdot 5 \cdot 7$	$3 \cdot 47$	$2 \cdot 71$	$11 \cdot 13$	$2^4 \cdot 3^2$	$5 \cdot 29$	$2 \cdot 73$	$3 \cdot 7^2$	$2^2 \cdot 37$	**149**
15_	$2 \cdot 3 \cdot 5^2$	**151**	$2^3 \cdot 19$	$3^2 \cdot 17$	$2 \cdot 7 \cdot 11$	$5 \cdot 31$	$2^2 \cdot 3 \cdot 13$	**157**	$2 \cdot 79$	$3 \cdot 53$
16_	$2^5 \cdot 5$	$7 \cdot 23$	$2 \cdot 3^4$	**163**	$2^2 \cdot 41$	$3 \cdot 5 \cdot 11$	$2 \cdot 83$	**167**	$2^3 \cdot 3 \cdot 7$	13^2
17_	$2 \cdot 5 \cdot 17$	$3^2 \cdot 19$	$2^2 \cdot 43$	**173**	$2 \cdot 3 \cdot 29$	$5^2 \cdot 7$	$2^4 \cdot 11$	$3 \cdot 59$	$2 \cdot 89$	**179**
18_	$2^2 \cdot 3^2 \cdot 5$	**181**	$2 \cdot 7 \cdot 13$	$3 \cdot 61$	$2^3 \cdot 23$	$5 \cdot 37$	$2 \cdot 3 \cdot 31$	$11 \cdot 17$	$2^2 \cdot 47$	$3^3 \cdot 7$
19_	$2 \cdot 5 \cdot 19$	**191**	$2^6 \cdot 3$	**193**	$2 \cdot 97$	$3 \cdot 5 \cdot 13$	$2^2 \cdot 7^2$	**197**	$2 \cdot 3^2 \cdot 11$	**199**
20_	$2^3 \cdot 5^2$	$3 \cdot 67$	$2 \cdot 101$	$7 \cdot 29$	$2^2 \cdot 3 \cdot 17$	$5 \cdot 41$	$2 \cdot 103$	$3^2 \cdot 23$	$2^4 \cdot 13$	$11 \cdot 19$
21_	$2 \cdot 3 \cdot 5 \cdot 7$	**211**	$2^2 \cdot 53$	$3 \cdot 71$	$2 \cdot 107$	$5 \cdot 43$	$2^3 \cdot 3^3$	$7 \cdot 31$	$2 \cdot 109$	$3 \cdot 73$
22_	$2^2 \cdot 5 \cdot 11$	$13 \cdot 17$	$2 \cdot 3 \cdot 37$	**223**	$2^5 \cdot 7$	$3^2 \cdot 5^2$	$2 \cdot 113$	**227**	$2^2 \cdot 3 \cdot 19$	**229**
23_	$2 \cdot 5 \cdot 23$	$3 \cdot 7 \cdot 11$	$2^3 \cdot 29$	**233**	$2 \cdot 3^2 \cdot 13$	$5 \cdot 47$	$2^2 \cdot 59$	$3 \cdot 79$	$2 \cdot 7 \cdot 17$	**239**
24_	$2^4 \cdot 3 \cdot 5$	**241**	$2 \cdot 11^2$	3^5	$2^2 \cdot 61$	$5 \cdot 7^2$	$2 \cdot 3 \cdot 41$	$13 \cdot 19$	$2^3 \cdot 31$	$3 \cdot 83$
25_	$2 \cdot 5^3$	**251**	$2^2 \cdot 3^2 \cdot 7$	$11 \cdot 23$	$2 \cdot 127$	$3 \cdot 5 \cdot 17$	2^8	**257**	$2 \cdot 3 \cdot 43$	$7 \cdot 37$
26_	$2^2 \cdot 5 \cdot 13$	$3^2 \cdot 29$	$2 \cdot 131$	**263**	$2^3 \cdot 3 \cdot 11$	$5 \cdot 53$	$2 \cdot 7 \cdot 19$	$3 \cdot 89$	$2^2 \cdot 67$	**269**
27_	$2 \cdot 3^3 \cdot 5$	**271**	$2^4 \cdot 17$	$3 \cdot 7 \cdot 13$	$2 \cdot 137$	$5^2 \cdot 11$	$2^2 \cdot 3 \cdot 23$	**277**	$2 \cdot 139$	$3^2 \cdot 31$
28_	$2^3 \cdot 5 \cdot 7$	**281**	$2 \cdot 3 \cdot 47$	**283**	$2^2 \cdot 71$	$3 \cdot 5 \cdot 19$	$2 \cdot 11 \cdot 13$	$7 \cdot 41$	$2^5 \cdot 3^2$	17^2
29_	$2 \cdot 5 \cdot 29$	$3 \cdot 97$	$2^2 \cdot 73$	**293**	$2 \cdot 3 \cdot 7^2$	$5 \cdot 59$	$2^3 \cdot 37$	$3^3 \cdot 11$	$2 \cdot 149$	$13 \cdot 23$
30_	$2^2 \cdot 3 \cdot 5^2$	$7 \cdot 43$	$2 \cdot 151$	$3 \cdot 101$	$2^4 \cdot 19$	$5 \cdot 61$	$2 \cdot 3^2 \cdot 17$	**307**	$2^2 \cdot 7 \cdot 11$	$3 \cdot 103$
31_	$2 \cdot 5 \cdot 31$	**311**	$2^3 \cdot 3 \cdot 13$	**313**	$2 \cdot 157$	$3^2 \cdot 5 \cdot 7$	$2^2 \cdot 79$	**317**	$2 \cdot 3 \cdot 53$	$11 \cdot 29$
32_	$2^6 \cdot 5$	$3 \cdot 107$	$2 \cdot 7 \cdot 23$	$17 \cdot 19$	$2^2 \cdot 3^4$	$5^2 \cdot 13$	$2 \cdot 163$	$3 \cdot 109$	$2^3 \cdot 41$	$7 \cdot 47$
33_	$2 \cdot 3 \cdot 5 \cdot 11$	**331**	$2^2 \cdot 83$	$3^2 \cdot 37$	$2 \cdot 167$	$5 \cdot 67$	$2^4 \cdot 3 \cdot 7$	**337**	$2 \cdot 13^2$	$3 \cdot 113$
34_	$2^2 \cdot 5 \cdot 17$	$11 \cdot 31$	$2 \cdot 3^2 \cdot 19$	7^3	$2^3 \cdot 43$	$3 \cdot 5 \cdot 23$	$2 \cdot 173$	**347**	$2^2 \cdot 3 \cdot 29$	**349**
35_	$2 \cdot 5^2 \cdot 7$	$3^3 \cdot 13$	$2^5 \cdot 11$	**353**	$2 \cdot 3 \cdot 59$	$5 \cdot 71$	$2^2 \cdot 89$	$3 \cdot 7 \cdot 17$	$2 \cdot 179$	**359**
36_	$2^3 \cdot 3^2 \cdot 5$	19^2	$2 \cdot 181$	$3 \cdot 11^2$	$2^2 \cdot 7 \cdot 13$	$5 \cdot 73$	$2 \cdot 3 \cdot 61$	**367**	$2^4 \cdot 23$	$3^2 \cdot 41$
37_	$2 \cdot 5 \cdot 37$	$7 \cdot 53$	$2^2 \cdot 3 \cdot 31$	**373**	$2 \cdot 11 \cdot 17$	$3 \cdot 5^3$	$2^3 \cdot 47$	$13 \cdot 29$	$2 \cdot 3^3 \cdot 7$	**379**
38_	$2^2 \cdot 5 \cdot 19$	$3 \cdot 127$	$2 \cdot 191$	**383**	$2^7 \cdot 3$	$5 \cdot 7 \cdot 11$	$2 \cdot 193$	$3^2 \cdot 43$	$2^2 \cdot 97$	**389**
39_	$2 \cdot 3 \cdot 5 \cdot 13$	$17 \cdot 23$	$2^3 \cdot 7^2$	$3 \cdot 131$	$2 \cdot 197$	$5 \cdot 79$	$2^2 \cdot 3^2 \cdot 11$	**397**	$2 \cdot 199$	$3 \cdot 7 \cdot 19$
40_	$2^4 \cdot 5^2$	**401**	$2 \cdot 3 \cdot 67$	$13 \cdot 31$	$2^2 \cdot 101$	$3^4 \cdot 5$	$2 \cdot 7 \cdot 29$	$11 \cdot 37$	$2^3 \cdot 3 \cdot 17$	**409**
41_	$2 \cdot 5 \cdot 41$	$3 \cdot 137$	$2^2 \cdot 103$	$7 \cdot 59$	$2 \cdot 3^2 \cdot 23$	$5 \cdot 83$	$2^5 \cdot 13$	$3 \cdot 139$	$2 \cdot 11 \cdot 19$	**419**
42_	$2^2 \cdot 3 \cdot 5 \cdot 7$	**421**	$2 \cdot 211$	$3^2 \cdot 47$	$2^3 \cdot 53$	$5^2 \cdot 17$	$2 \cdot 3 \cdot 71$	$7 \cdot 61$	$2^2 \cdot 107$	$3 \cdot 11 \cdot 13$
43_	$2 \cdot 5 \cdot 43$	**431**	$2^4 \cdot 3^3$	**433**	$2 \cdot 7 \cdot 31$	$3 \cdot 5 \cdot 29$	$2^2 \cdot 109$	$19 \cdot 23$	$2 \cdot 3 \cdot 73$	**439**
44_	$2^3 \cdot 5 \cdot 11$	$3^2 \cdot 7^2$	$2 \cdot 13 \cdot 17$	**443**	$2^2 \cdot 3 \cdot 37$	$5 \cdot 89$	$2 \cdot 223$	$3 \cdot 149$	$2^6 \cdot 7$	**449**
45_	$2 \cdot 3^2 \cdot 5^2$	$11 \cdot 41$	$2^2 \cdot 113$	$3 \cdot 151$	$2 \cdot 227$	$5 \cdot 7 \cdot 13$	$2^3 \cdot 3 \cdot 19$	**457**	$2 \cdot 229$	$3^3 \cdot 17$
46_	$2^2 \cdot 5 \cdot 23$	**461**	$2 \cdot 3 \cdot 7 \cdot 11$	**463**	$2^4 \cdot 29$	$3 \cdot 5 \cdot 31$	$2 \cdot 233$	**467**	$2^2 \cdot 3^2 \cdot 13$	$7 \cdot 67$
47_	$2 \cdot 5 \cdot 47$	$3 \cdot 157$	$2^3 \cdot 59$	$11 \cdot 43$	$2 \cdot 3 \cdot 79$	$5^2 \cdot 19$	$2^2 \cdot 7 \cdot 17$	$3^2 \cdot 53$	$2 \cdot 239$	**479**
48_	$2^5 \cdot 3 \cdot 5$	$13 \cdot 37$	$2 \cdot 241$	$3 \cdot 7 \cdot 23$	$2^2 \cdot 11^2$	$5 \cdot 97$	$2 \cdot 3^5$	**487**	$2^3 \cdot 61$	$3 \cdot 163$
49_	$2 \cdot 5 \cdot 7^2$	**491**	$2^2 \cdot 3 \cdot 41$	$17 \cdot 29$	$2 \cdot 13 \cdot 19$	$3^2 \cdot 5 \cdot 11$	$2^4 \cdot 31$	$7 \cdot 71$	$2 \cdot 3 \cdot 83$	**499**
50_	$2^2 \cdot 5^3$	$3 \cdot 167$	$2 \cdot 251$	**503**	$2^3 \cdot 3^2 \cdot 7$	$5 \cdot 101$	$2 \cdot 11 \cdot 23$	$3 \cdot 13^2$	$2^2 \cdot 127$	**509**
51_	$2 \cdot 3 \cdot 5 \cdot 17$	$7 \cdot 73$	2^9	$3^3 \cdot 19$	$2 \cdot 257$	$5 \cdot 103$	$2^2 \cdot 3 \cdot 43$	$11 \cdot 47$	$2 \cdot 7 \cdot 37$	$3 \cdot 173$
52_	$2^3 \cdot 5 \cdot 13$	**521**	$2 \cdot 3^2 \cdot 29$	**523**	$2^2 \cdot 131$	$3 \cdot 5^2 \cdot 7$	$2 \cdot 263$	$17 \cdot 31$	$2^4 \cdot 3 \cdot 11$	23^2
53_	$2 \cdot 5 \cdot 53$	$3^2 \cdot 59$	$2^2 \cdot 7 \cdot 19$	$13 \cdot 41$	$2 \cdot 3 \cdot 89$	$5 \cdot 107$	$2^3 \cdot 67$	$3 \cdot 179$	$2 \cdot 269$	$7^2 \cdot 11$
54_	$2^2 \cdot 3^3 \cdot 5$	**541**	$2 \cdot 271$	$3 \cdot 181$	$2^5 \cdot 17$	$5 \cdot 109$	$2 \cdot 3 \cdot 7 \cdot 13$	**547**	$2^2 \cdot 137$	$3^2 \cdot 61$
55_	$2 \cdot 5^2 \cdot 11$	$19 \cdot 29$	$2^3 \cdot 3 \cdot 23$	$7 \cdot 79$	$2 \cdot 277$	$3 \cdot 5 \cdot 37$	$2^2 \cdot 139$	**557**	$2 \cdot 3^2 \cdot 31$	$13 \cdot 43$
56_	$2^4 \cdot 5 \cdot 7$	$3 \cdot 11 \cdot 17$	$2 \cdot 281$	**563**	$2^2 \cdot 3 \cdot 47$	$5 \cdot 113$	$2 \cdot 283$	$3^4 \cdot 7$	$2^3 \cdot 71$	**569**
57_	$2 \cdot 3 \cdot 5 \cdot 19$	**571**	$2^2 \cdot 11 \cdot 13$	$3 \cdot 191$	$2 \cdot 7 \cdot 41$	$5^2 \cdot 23$	$2^6 \cdot 3^2$	**577**	$2 \cdot 17^2$	$3 \cdot 193$
58_	$2^2 \cdot 5 \cdot 29$	$7 \cdot 83$	$2 \cdot 3 \cdot 97$	$11 \cdot 53$	$2^3 \cdot 73$	$3^2 \cdot 5 \cdot 13$	$2 \cdot 293$	**587**	$2^2 \cdot 3 \cdot 7^2$	$19 \cdot 31$
59_	$2 \cdot 5 \cdot 59$	$3 \cdot 197$	$2^4 \cdot 37$	**593**	$2 \cdot 3^3 \cdot 11$	$5 \cdot 7 \cdot 17$	$2^2 \cdot 149$	$3 \cdot 199$	$2 \cdot 13 \cdot 23$	**599**
60_	$2^3 \cdot 3 \cdot 5^2$	**601**	$2 \cdot 7 \cdot 43$	$3^2 \cdot 67$	$2^2 \cdot 151$	$5 \cdot 11^2$	$2 \cdot 3 \cdot 101$	**607**	$2^5 \cdot 19$	$3 \cdot 7 \cdot 29$
61_	$2 \cdot 5 \cdot 61$	$13 \cdot 47$	$2^2 \cdot 3^2 \cdot 17$	**613**	$2 \cdot 307$	$3 \cdot 5 \cdot 41$	$2^3 \cdot 7 \cdot 11$	**617**	$2 \cdot 3 \cdot 103$	**619**

	_0	_1	_2	_3	_4	_5	_6	_7	_8	_9
62_	$2^2\cdot5\cdot31$	$3^3\cdot23$	$2\cdot311$	$7\cdot89$	$2^4\cdot3\cdot13$	5^4	$2\cdot313$	$3\cdot11\cdot19$	$2^2\cdot157$	$17\cdot37$
63_	$2\cdot3^2\cdot5\cdot7$	**631**	$2^3\cdot79$	$3\cdot211$	$2\cdot317$	$5\cdot127$	$2^2\cdot3\cdot53$	$7^2\cdot13$	$2\cdot11\cdot29$	$3^2\cdot71$
64_	$2^7\cdot5$	**641**	$2\cdot3\cdot107$	**643**	$2^2\cdot7\cdot23$	$3\cdot5\cdot43$	$2\cdot17\cdot19$	**647**	$2^3\cdot3^4$	$11\cdot59$
65_	$2\cdot5^2\cdot13$	$3\cdot7\cdot31$	$2^2\cdot163$	**653**	$2\cdot3\cdot109$	$5\cdot131$	$2^4\cdot41$	$3^2\cdot73$	$2\cdot7\cdot47$	**659**
66_	$2^2\cdot3\cdot5\cdot11$	**661**	$2\cdot331$	$3\cdot13\cdot17$	$2^3\cdot83$	$5\cdot7\cdot19$	$2\cdot3^2\cdot37$	$23\cdot29$	$2^2\cdot167$	$3\cdot223$
67_	$2\cdot5\cdot67$	$11\cdot61$	$2^5\cdot3\cdot7$	**673**	$2\cdot337$	$3^3\cdot5^2$	$2^2\cdot13^2$	**677**	$2\cdot3\cdot113$	$7\cdot97$
68_	$2^3\cdot5\cdot17$	$3\cdot227$	$2\cdot11\cdot31$	**683**	$2^2\cdot3^2\cdot19$	$5\cdot137$	$2\cdot7^3$	$3\cdot229$	$2^4\cdot43$	$13\cdot53$
69_	$2\cdot3\cdot5\cdot23$	**691**	$2^2\cdot173$	$3^2\cdot7\cdot11$	$2\cdot347$	$5\cdot139$	$2^3\cdot3\cdot29$	$17\cdot41$	$2\cdot349$	$3\cdot233$
70_	$2^2\cdot5^2\cdot7$	**701**	$2\cdot3^3\cdot13$	$19\cdot37$	$2^6\cdot11$	$3\cdot5\cdot47$	$2\cdot353$	$7\cdot101$	$2^2\cdot3\cdot59$	**709**
71_	$2\cdot5\cdot71$	$3^2\cdot79$	$2^3\cdot89$	$23\cdot31$	$2\cdot3\cdot7\cdot17$	$5\cdot11\cdot13$	$2^2\cdot179$	$3\cdot239$	$2\cdot359$	**719**
72_	$2^4\cdot3^2\cdot5$	$7\cdot103$	$2\cdot19^2$	$3\cdot241$	$2^2\cdot181$	$5^2\cdot29$	$2\cdot3\cdot11^2$	**727**	$2^3\cdot7\cdot13$	3^6
73_	$2\cdot5\cdot73$	$17\cdot43$	$2^2\cdot3\cdot61$	**733**	$2\cdot367$	$3\cdot5\cdot7^2$	$2^5\cdot23$	$11\cdot67$	$2\cdot3^2\cdot41$	**739**
74_	$2^2\cdot5\cdot37$	$3\cdot13\cdot19$	$2\cdot7\cdot53$	**743**	$2^3\cdot3\cdot31$	$5\cdot149$	$2\cdot373$	$3^2\cdot83$	$2^2\cdot11\cdot17$	$7\cdot107$
75_	$2\cdot3\cdot5^3$	**751**	$2^4\cdot47$	$3\cdot251$	$2\cdot13\cdot29$	$5\cdot151$	$2^2\cdot3^3\cdot7$	**757**	$2\cdot379$	$3\cdot11\cdot23$
76_	$2^3\cdot5\cdot19$	**761**	$2\cdot3\cdot127$	$7\cdot109$	$2^2\cdot191$	$3^2\cdot5\cdot17$	$2\cdot383$	$13\cdot59$	$2^8\cdot3$	**769**
77_	$2\cdot5\cdot7\cdot11$	$3\cdot257$	$2^2\cdot193$	**773**	$2\cdot3^2\cdot43$	$5^2\cdot31$	$2^3\cdot97$	$3\cdot7\cdot37$	$2\cdot389$	$19\cdot41$
78_	$2^2\cdot3\cdot5\cdot13$	$11\cdot71$	$2\cdot17\cdot23$	$3^3\cdot29$	$2^4\cdot7^2$	$5\cdot157$	$2\cdot3\cdot131$	**787**	$2^2\cdot197$	$3\cdot263$
79_	$2\cdot5\cdot79$	$7\cdot113$	$2^3\cdot3^2\cdot11$	$13\cdot61$	$2\cdot397$	$3\cdot5\cdot53$	$2^2\cdot199$	**797**	$2\cdot3\cdot7\cdot19$	$17\cdot47$
80_	$2^5\cdot5^2$	$3^2\cdot89$	$2\cdot401$	$11\cdot73$	$2^2\cdot3\cdot67$	$5\cdot7\cdot23$	$2\cdot13\cdot31$	$3\cdot269$	$2^3\cdot101$	**809**
81_	$2\cdot3^4\cdot5$	**811**	$2^2\cdot7\cdot29$	$3\cdot271$	$2\cdot11\cdot37$	$5\cdot163$	$2^4\cdot3\cdot17$	$19\cdot43$	$2\cdot409$	$3^2\cdot7\cdot13$
82_	$2^2\cdot5\cdot41$	**821**	$2\cdot3\cdot137$	**823**	$2^3\cdot103$	$3\cdot5^2\cdot11$	$2\cdot7\cdot59$	**827**	$2^2\cdot3^2\cdot23$	**829**
83_	$2\cdot5\cdot83$	$3\cdot277$	$2^6\cdot13$	$7^2\cdot17$	$2\cdot3\cdot139$	$5\cdot167$	$2^2\cdot11\cdot19$	$3^3\cdot31$	$2\cdot419$	**839**
84_	$2^3\cdot3\cdot5\cdot7$	29^2	$2\cdot421$	$3\cdot281$	$2^2\cdot211$	$5\cdot13^2$	$2\cdot3^2\cdot47$	$7\cdot11^2$	$2^4\cdot53$	$3\cdot283$
85_	$2\cdot5^2\cdot17$	$23\cdot37$	$2^2\cdot3\cdot71$	**853**	$2\cdot7\cdot61$	$3^2\cdot5\cdot19$	$2^3\cdot107$	**857**	$2\cdot3\cdot11\cdot13$	**859**
86_	$2^2\cdot5\cdot43$	$3\cdot7\cdot41$	$2\cdot431$	**863**	$2^5\cdot3^3$	$5\cdot173$	$2\cdot433$	$3\cdot17^2$	$2^2\cdot7\cdot31$	$11\cdot79$
87_	$2\cdot3\cdot5\cdot29$	$13\cdot67$	$2^3\cdot109$	$3^2\cdot97$	$2\cdot19\cdot23$	$5^3\cdot7$	$2^2\cdot3\cdot73$	**877**	$2\cdot439$	$3\cdot293$
88_	$2^4\cdot5\cdot11$	**881**	$2\cdot3^2\cdot7^2$	**883**	$2^2\cdot13\cdot17$	$3\cdot5\cdot59$	$2\cdot443$	**887**	$2^3\cdot3\cdot37$	$7\cdot127$
89_	$2\cdot5\cdot89$	$3^4\cdot11$	$2^2\cdot223$	$19\cdot47$	$2\cdot3\cdot149$	$5\cdot179$	$2^7\cdot7$	$3\cdot13\cdot23$	$2\cdot449$	$29\cdot31$
90_	$2^2\cdot3^2\cdot5^2$	$17\cdot53$	$2\cdot11\cdot41$	$3\cdot7\cdot43$	$2^3\cdot113$	$5\cdot181$	$2\cdot3\cdot151$	**907**	$2^2\cdot227$	$3^2\cdot101$
91_	$2\cdot5\cdot7\cdot13$	**911**	$2^4\cdot3\cdot19$	$11\cdot83$	$2\cdot457$	$3\cdot5\cdot61$	$2^2\cdot229$	$7\cdot131$	$2\cdot3^3\cdot17$	**919**
92_	$2^3\cdot5\cdot23$	$3\cdot307$	$2\cdot461$	$13\cdot71$	$2^2\cdot3\cdot7\cdot11$	$5^2\cdot37$	$2\cdot463$	$3^2\cdot103$	$2^5\cdot29$	**929**
93_	$2\cdot3\cdot5\cdot31$	$7^2\cdot19$	$2^2\cdot233$	$3\cdot311$	$2\cdot467$	$5\cdot11\cdot17$	$2^3\cdot3^2\cdot13$	**937**	$2\cdot7\cdot67$	$3\cdot313$
94_	$2^2\cdot5\cdot47$	**941**	$2\cdot3\cdot157$	$23\cdot41$	$2^4\cdot59$	$3^3\cdot5\cdot7$	$2\cdot11\cdot43$	**947**	$2^2\cdot3\cdot79$	$13\cdot73$
95_	$2\cdot5^2\cdot19$	$3\cdot317$	$2^3\cdot7\cdot17$	**953**	$2\cdot3^2\cdot53$	$5\cdot191$	$2^2\cdot239$	$3\cdot11\cdot29$	$2\cdot479$	$7\cdot137$
96_	$2^6\cdot3\cdot5$	31^2	$2\cdot13\cdot37$	$3^2\cdot107$	$2^2\cdot241$	$5\cdot193$	$2\cdot3\cdot7\cdot23$	**967**	$2^3\cdot11^2$	$3\cdot17\cdot19$
97_	$2\cdot5\cdot97$	**971**	$2^2\cdot3^5$	$7\cdot139$	$2\cdot487$	$3\cdot5^2\cdot13$	$2^4\cdot61$	**977**	$2\cdot3\cdot163$	$11\cdot89$
98_	$2^2\cdot5\cdot7^2$	$3^2\cdot109$	$2\cdot491$	**983**	$2^3\cdot3\cdot41$	$5\cdot197$	$2\cdot17\cdot29$	$3\cdot7\cdot47$	$2^2\cdot13\cdot19$	$23\cdot43$
99_	$2\cdot3^2\cdot5\cdot11$	**991**	$2^5\cdot31$	$3\cdot331$	$2\cdot7\cdot71$	$5\cdot199$	$2^2\cdot3\cdot83$	**997**	$2\cdot499$	$3^3\cdot37$
100_	$2^3\cdot5^3$	$7\cdot11\cdot13$	$2\cdot3\cdot167$	$17\cdot59$	$2^2\cdot251$	$3\cdot5\cdot67$	$2\cdot503$	$19\cdot53$	$2^4\cdot3^2\cdot7$	**1009**
101_	$2\cdot5\cdot101$	$3\cdot337$	$2^2\cdot11\cdot23$	**1013**	$2\cdot3\cdot13^2$	$5\cdot7\cdot29$	$2^3\cdot127$	$3^2\cdot113$	$2\cdot509$	**1019**
102_	$2^2\cdot3\cdot5\cdot17$	**1021**	$2\cdot7\cdot73$	$3\cdot11\cdot31$	2^{10}	$5^2\cdot41$	$2\cdot3^3\cdot19$	$13\cdot79$	$2^2\cdot257$	$3\cdot7^3$
103_	$2\cdot5\cdot103$	**1031**	$2^3\cdot3\cdot43$	**1033**	$2\cdot11\cdot47$	$3^2\cdot5\cdot23$	$2^2\cdot7\cdot37$	$17\cdot61$	$2\cdot3\cdot173$	**1039**
104_	$2^4\cdot5\cdot13$	$3\cdot347$	$2\cdot521$	$7\cdot149$	$2^2\cdot3^2\cdot29$	$5\cdot11\cdot19$	$2\cdot523$	$3\cdot349$	$2^3\cdot131$	**1049**
105_	$2\cdot3\cdot5^2\cdot7$	**1051**	$2^2\cdot263$	$3^4\cdot13$	$2\cdot17\cdot31$	$5\cdot211$	$2^5\cdot3\cdot11$	$7\cdot151$	$2\cdot23^2$	$3\cdot353$
106_	$2^2\cdot5\cdot53$	**1061**	$2\cdot3^2\cdot59$	**1063**	$2^3\cdot7\cdot19$	$3\cdot5\cdot71$	$2\cdot13\cdot41$	$11\cdot97$	$2^2\cdot3\cdot89$	**1069**
107_	$2\cdot5\cdot107$	$3^2\cdot7\cdot17$	$2^4\cdot67$	$29\cdot37$	$2\cdot3\cdot179$	$5^2\cdot43$	$2^2\cdot269$	$3\cdot359$	$2\cdot7^2\cdot11$	$13\cdot83$
108_	$2^3\cdot3^3\cdot5$	$23\cdot47$	$2\cdot541$	$3\cdot19^2$	$2^2\cdot271$	$5\cdot7\cdot31$	$2\cdot3\cdot181$	**1087**	$2^6\cdot17$	$3^2\cdot11^2$
109_	$2\cdot5\cdot109$	**1091**	$2^2\cdot3\cdot7\cdot13$	**1093**	$2\cdot547$	$3\cdot5\cdot73$	$2^3\cdot137$	**1097**	$2\cdot3^2\cdot61$	$7\cdot157$
110_	$2^2\cdot5^2\cdot11$	$3\cdot367$	$2\cdot19\cdot29$	**1103**	$2^4\cdot3\cdot23$	$5\cdot13\cdot17$	$2\cdot7\cdot79$	$3^3\cdot41$	$2^2\cdot277$	**1109**
111_	$2\cdot3\cdot5\cdot37$	$11\cdot101$	$2^3\cdot139$	$3\cdot7\cdot53$	$2\cdot557$	$5\cdot223$	$2^2\cdot3^2\cdot31$	**1117**	$2\cdot13\cdot43$	$3\cdot373$
112_	$2^5\cdot5\cdot7$	$19\cdot59$	$2\cdot3\cdot11\cdot17$	**1123**	$2^2\cdot281$	$3^2\cdot5^3$	$2\cdot563$	$7^2\cdot23$	$2^3\cdot3\cdot47$	**1129**

2.4.14   FACTORIZATION OF $2^m - 1$

$2^3 - 1 = 7$	$2^{19} - 1 = 524287$
$2^4 - 1 = 3 \times 5$	$2^{20} - 1 = 3 \times 5^2 \times 11 \times 31 \times 41$
$2^5 - 1 = 31$	$2^{21} - 1 = 7^2 \times 127 \times 337$
$2^6 - 1 = 3^2 \times 7$	$2^{22} - 1 = 3 \times 23 \times 89 \times 683$
$2^7 - 1 = 127$	$2^{23} - 1 = 47 \times 178481$
$2^8 - 1 = 3 \times 5 \times 17$	$2^{24} - 1 = 3^2 \times 5 \times 7 \times 13 \times 17 \times 241$
$2^9 - 1 = 7 \times 73$	$2^{25} - 1 = 31 \times 601 \times 1801$
$2^{10} - 1 = 3 \times 11 \times 31$	$2^{26} - 1 = 3 \times 2731 \times 8191$
$2^{11} - 1 = 23 \times 89$	$2^{27} - 1 = 7 \times 73 \times 262657$
$2^{12} - 1 = 3^2 \times 5 \times 7 \times 13$	$2^{28} - 1 = 3 \times 5 \times 29 \times 43 \times 113 \times 127$
$2^{13} - 1 = 8191$	$2^{29} - 1 = 233 \times 1103 \times 2089$
$2^{14} - 1 = 3 \times 43 \times 127$	$2^{30} - 1 = 3^2 \times 7 \times 11 \times 31 \times 151 \times 331$
$2^{15} - 1 = 7 \times 31 \times 151$	$2^{31} - 1 = 2147483647$
$2^{16} - 1 = 3 \times 5 \times 17 \times 257$	$2^{32} - 1 = 3 \times 5 \times 17 \times 257 \times 65537$
$2^{17} - 1 = 131071$	$2^{33} - 1 = 7 \times 23 \times 89 \times 599479$
$2^{18} - 1 = 3^3 \times 7 \times 19 \times 73$	$2^{34} - 1 = 3 \times 43691 \times 131071$

2.4.15   EULER TOTIENT FUNCTION

2.4.15.1   Definitions

1. $\phi(n)$ the *totient function* is the number of integers not exceeding and relatively prime to n.
2. $\sigma(n)$ is the *sum of the divisors* of n.
3. $\tau(n)$ is the *number of divisors* of n. (Also called the $d(n)$ function.)

Define $\sigma_k(n)$ to be the kth divisor function, the sum of the kth powers of the divisors of n. Then $\tau(n) = \sigma_0(n)$ and $\sigma(n) = \sigma_1(n)$.

EXAMPLE   The numbers less than 6 and relatively prime to 6 are $\{1, 5\}$. Hence $\phi(6) = 2$. The divisors of 6 are $\{1, 2, 3, 6\}$. There are $\tau(6) = 4$ divisors. The sum of these numbers is $\sigma(6) = 1 + 2 + 3 + 6 = 12$.

2.4.15.2   Properties of the totient function

1. ϕ is a multiplicative function: if $(n, m) = 1$, then $\phi(nm) = \phi(m)\phi(n)$.

2. If p is prime, then $\phi(p) = p - 1$.

3. Gauss's theorem states: $n = \sum_{d|n} \phi(d)$.

4. When $n = \prod_i p_i^{\alpha_i}$, and the $\{p_i\}$ are prime

$$\sigma_k(n) = \sum_{d|n} d^k = \prod_i \frac{p_i^{k(\alpha_i + 1)} - 1}{p_i^k - 1}. \tag{2.4.17}$$

5. Generating functions

$$\sum_{n=1}^{\infty} \frac{\sigma_k(n)}{n^s} = \zeta(s)\zeta(s-k)$$

$$\sum_{n=1}^{\infty} \frac{\phi(n)}{n^s} = \frac{\zeta(s-1)}{\zeta(s)}$$

(2.4.18)

6. A *perfect number* n satisfies $\sigma(n) = 2n$. The integer n is an even perfect number if, and only if, $n = 2^{m-1}(2^m - 1)$, where m is a positive integer such that $M_m = 2^m - 1$ is a Mersenne prime. The sequence of perfect numbers is $\{6, 28, 496, \ldots\}$ (see page 31), corresponding to $m = 2, 3, 5, \ldots$. It is not known whether there exists an odd perfect number.

2.4.15.3 Table of totient function values

n	$\phi(n)$	$\tau(n)$	$\sigma(n)$	n	$\phi(n)$	$\tau(n)$	$\sigma(n)$	n	$\phi(n)$	$\tau(n)$	$\sigma(n)$	n	$\phi(n)$	$\tau(n)$	$\sigma(n)$
1	0	1	1	2	1	2	3	3	2	2	4	4	2	3	7
5	4	2	6	6	2	4	12	7	6	2	8	8	4	4	15
9	6	3	13	10	4	4	18	11	10	2	12	12	4	6	28
13	12	2	14	14	6	4	24	15	8	4	24	16	8	5	31
17	16	2	18	18	6	6	39	19	18	2	20	20	8	6	42
21	12	4	32	22	10	4	36	23	22	2	24	24	8	8	60
25	20	3	31	26	12	4	42	27	18	4	40	28	12	6	56
29	28	2	30	30	8	8	72	31	30	2	32	32	16	6	63
33	20	4	48	34	16	4	54	35	24	4	48	36	12	9	91
37	36	2	38	38	18	4	60	39	24	4	56	40	16	8	90
41	40	2	42	42	12	8	96	43	42	2	44	44	20	6	84
45	24	6	78	46	22	4	72	47	46	2	48	48	16	10	124
49	42	3	57	50	20	6	93	51	32	4	72	52	24	6	98
53	52	2	54	54	18	8	120	55	40	4	72	56	24	8	120
57	36	4	80	58	28	4	90	59	58	2	60	60	16	12	168
61	60	2	62	62	30	4	96	63	36	6	104	64	32	7	127
65	48	4	84	66	20	8	144	67	66	2	68	68	32	6	126
69	44	4	96	70	24	8	144	71	70	2	72	72	24	12	195
73	72	2	74	74	36	4	114	75	40	6	124	76	36	6	140
77	60	4	96	78	24	8	168	79	78	2	80	80	32	10	186
81	54	5	121	82	40	4	126	83	82	2	84	84	24	12	224
85	64	4	108	86	42	4	132	87	56	4	120	88	40	8	180
89	88	2	90	90	24	12	234	91	72	4	112	92	44	6	168
93	60	4	128	94	46	4	144	95	72	4	120	96	32	12	252
97	96	2	98	98	42	6	171	99	60	6	156	100	40	9	217
101	100	2	102	102	32	8	216	103	102	2	104	104	48	8	210
105	48	8	192	106	52	4	162	107	106	2	108	108	36	12	280
109	108	2	110	110	40	8	216	111	72	4	152	112	48	10	248
113	112	2	114	114	36	8	240	115	88	4	144	116	56	6	210
117	72	6	182	118	58	4	180	119	96	4	144	120	32	16	360

n	$\phi(n)$	$\tau(n)$	$\sigma(n)$	n	$\phi(n)$	$\tau(n)$	$\sigma(n)$	n	$\phi(n)$	$\tau(n)$	$\sigma(n)$	n	$\phi(n)$	$\tau(n)$	$\sigma(n)$
121	110	3	133	122	60	4	186	123	80	4	168	124	60	6	224
125	100	4	156	126	36	12	312	127	126	2	128	128	64	8	255
129	84	4	176	130	48	8	252	131	130	2	132	132	40	12	336
133	108	4	160	134	66	4	204	135	72	8	240	136	64	8	270
137	136	2	138	138	44	8	288	139	138	2	140	140	48	12	336
141	92	4	192	142	70	4	216	143	120	4	168	144	48	15	403
145	112	4	180	146	72	4	222	147	84	6	228	148	72	6	266
149	148	2	150	150	40	12	372	151	150	2	152	152	72	8	300
153	96	6	234	154	60	8	288	155	120	4	192	156	48	12	392
157	156	2	158	158	78	4	240	159	104	4	216	160	64	12	378
161	132	4	192	162	54	10	363	163	162	2	164	164	80	6	294
165	80	8	288	166	82	4	252	167	166	2	168	168	48	16	480
169	156	3	183	170	64	8	324	171	108	6	260	172	84	6	308
173	172	2	174	174	56	8	360	175	120	6	248	176	80	10	372
177	116	4	240	178	88	4	270	179	178	2	180	180	48	18	546
181	180	2	182	182	72	8	336	183	120	4	248	184	88	8	360
185	144	4	228	186	60	8	384	187	160	4	216	188	92	6	336
189	108	8	320	190	72	8	360	191	190	2	192	192	64	14	508
193	192	2	194	194	96	4	294	195	96	8	336	196	84	9	399
197	196	2	198	198	60	12	468	199	198	2	200	200	80	12	465
201	132	4	272	202	100	4	306	203	168	4	240	204	64	12	504
205	160	4	252	206	102	4	312	207	132	6	312	208	96	10	434
209	180	4	240	210	48	16	576	211	210	2	212	212	104	6	378
213	140	4	288	214	106	4	324	215	168	4	264	216	72	16	600
217	180	4	256	218	108	4	330	219	144	4	296	220	80	12	504
221	192	4	252	222	72	8	456	223	222	2	224	224	96	12	504
225	120	9	403	226	112	4	342	227	226	2	228	228	72	12	560
229	228	2	230	230	88	8	432	231	120	8	384	232	112	8	450
233	232	2	234	234	72	12	546	235	184	4	288	236	116	6	420
237	156	4	320	238	96	8	432	239	238	2	240	240	64	20	744
241	240	2	242	242	110	6	399	243	162	6	364	244	120	6	434
245	168	6	342	246	80	8	504	247	216	4	280	248	120	8	480
249	164	4	336	250	100	8	468	251	250	2	252	252	72	18	728
253	220	4	288	254	126	4	384	255	128	8	432	256	128	9	511
257	256	2	258	258	84	8	528	259	216	4	304	260	96	12	588
261	168	6	390	262	130	4	396	263	262	2	264	264	80	16	720
265	208	4	324	266	108	8	480	267	176	4	360	268	132	6	476
269	268	2	270	270	72	16	720	271	270	2	272	272	128	10	558
273	144	8	448	274	136	4	414	275	200	6	372	276	88	12	672
277	276	2	278	278	138	4	420	279	180	6	416	280	96	16	720
281	280	2	282	282	92	8	576	283	282	2	284	284	140	6	504
285	144	8	480	286	120	8	504	287	240	4	336	288	96	18	819
289	272	3	307	290	112	8	540	291	192	4	392	292	144	6	518
293	292	2	294	294	84	12	684	295	232	4	360	296	144	8	570
297	180	8	480	298	148	4	450	299	264	4	336	300	80	18	868
301	252	4	352	302	150	4	456	303	200	4	408	304	144	10	620
305	240	4	372	306	96	12	702	307	306	2	308	308	120	12	672

2.5 VECTOR ALGEBRA

2.5.1 NOTATION FOR VECTORS AND SCALARS

A *vector* is an ordered n-tuple of values. A vector is usually represented by a lower-case, bold-faced letter, such as $\mathbf{v}$. The individual components of a vector $\mathbf{v}$ are typically denoted by a lower-case letter along with a subscript identifying the relative position of the component in the vector, such as $\mathbf{v} = [v_1, v_2, \ldots, v_n]$. In this case, the vector is said to be n-dimensional. If the n individual components of the vector are real numbers, then $\mathbf{v} \in \mathbb{R}^n$. Similarly, if the n components of $\mathbf{v}$ are complex, then $\mathbf{v} \in \mathbb{C}^n$.

Subscripts are also typically used to identify individual vectors within a set of vectors all belonging to the same type. For example, a set of n velocity vectors can be denoted by $\{\mathbf{v}_1, \ldots, \mathbf{v}_n\}$. In this case, a bold-face type is used on the individual members of the set to signify these elements of the set are vectors and not vector components.

Two vectors, $\mathbf{v}$ and $\mathbf{u}$, are said to be equal if all their components are equal. The negative of a vector, written as $-\mathbf{v}$, is one that acts in a direction opposite to $\mathbf{v}$, but is of equal magnitude.

2.5.2 PHYSICAL VECTORS

Any quantity that is completely determined by its magnitude is called a *scalar*. For example, mass, density, and temperature are scalars. Any quantity that is completely determined by its magnitude and direction is called, in physics, a vector. We often use a three-dimensional vector to represent a physical vector. Examples of physical vectors include velocity, acceleration, and force. A physical vector is represented by a directed line segment, the length of which represents the magnitude of the vector. Two vectors are said to be *parallel* if they have exactly the same direction, i.e., the angle between the two vectors equals zero.

2.5.3 FUNDAMENTAL DEFINITIONS

1. A *row vector* is a vector whose components are aligned horizontally. A *column vector* has its components aligned vertically. The *transpose* operator, denoted by the superscript ^{T}, switches the orientation of a vector between horizontal and vertical.

EXAMPLE

$$\mathbf{v} = [1, 2, 3, 4], \quad \mathbf{v}^{\mathrm{T}} = \begin{bmatrix} 1 \\ 2 \\ 3 \\ 4 \end{bmatrix}, \quad (\mathbf{v}^{\mathrm{T}})^{\mathrm{T}} = [1, 2, 3, 4].$$

row vector   column vector   row vector

Vectors are traditionally written with either parentheses or with square brackets.

2. Two vectors, $\mathbf{v}$ and $\mathbf{u}$, are said to be *orthogonal* if $\mathbf{v}^{\mathrm{T}}\mathbf{u} = 0$. (This is also written $\mathbf{v} \cdot \mathbf{u} = 0$, where the "·" denotes an inner product; see page 133.)

3. A set of vectors $\{\mathbf{v}_1, \ldots, \mathbf{v}_n\}$ is said to be *orthogonal* if $\mathbf{v}_i^{\mathrm{T}}\mathbf{v}_j = 0$ for all $i \neq j$.

4. A set of orthogonal vectors $\{\mathbf{v}_1, \ldots, \mathbf{v}_m\}$ is said to be *orthonormal* if, in addition to possessing the property of orthogonality, the set possesses the property that $\mathbf{v}_i^{\mathrm{T}}\mathbf{v}_i = 1$ for all $1 \leq i \leq m$.

2.5.4   LAWS OF VECTOR ALGEBRA

1. The vector sum of $\mathbf{v}$ and $\mathbf{u}$, represented by $\mathbf{v} + \mathbf{u}$, results in another vector of the same dimension, and is calculated by simply adding corresponding vector components, e.g., if $\mathbf{v}, \mathbf{u} \in \mathbb{R}^n$, then $\mathbf{v} + \mathbf{u} = [v_1 + u_1, \ldots, v_n + u_n]$.

2. The vector subtraction of $\mathbf{u}$ from $\mathbf{v}$, represented by $\mathbf{v} - \mathbf{u}$, is equivalent to the addition of $\mathbf{v}$ and $-\mathbf{u}$.

3. If $r > 0$ is a scalar, then the scalar multiplication $r\mathbf{v}$ (equal to $\mathbf{v}r$) represents a scaling by a factor r of the vector $\mathbf{v}$ in the same direction as $\mathbf{v}$. That is, the multiplicative scalar is distributed to each component of $\mathbf{v}$.

4. If $0 \leq r < 1$, then the scalar multiplication of r and $\mathbf{v}$ shrinks the length of $\mathbf{v}$, multiplication by $r = 1$ leaves $\mathbf{v}$ unchanged, and, if $r > 1$, then $r\mathbf{v}$ stretches the length of $\mathbf{v}$. When $r < 0$, scalar multiplication of r and $\mathbf{v}$ has the same effect on the magnitude (length) of $\mathbf{v}$ as when $r > 0$, but results in a vector oriented in the direction opposite to $\mathbf{v}$.

EXAMPLE

$$4 \begin{bmatrix} 1 \\ 0 \\ 3 \end{bmatrix} = \begin{bmatrix} 4 \\ 0 \\ 12 \end{bmatrix}, \quad -4 \begin{bmatrix} 1 \\ 0 \\ 3 \end{bmatrix} = \begin{bmatrix} -4 \\ 0 \\ -12 \end{bmatrix}.$$

5. If r and s are scalars, and $\mathbf{v}$, $\mathbf{u}$, and $\mathbf{w}$ are vectors, the following rules of algebra are valid:

$$\begin{aligned}
\mathbf{v} + \mathbf{u} &= \mathbf{u} + \mathbf{v}, \\
(r + s)\mathbf{v} &= r\mathbf{v} + s\mathbf{v} = \mathbf{v}r + \mathbf{v}s = \mathbf{v}(r + s), \\
r(\mathbf{v} + \mathbf{u}) &= r\mathbf{v} + r\mathbf{u}, \\
\mathbf{v} + (\mathbf{u} + \mathbf{w}) &= (\mathbf{v} + \mathbf{u}) + \mathbf{w} = \mathbf{v} + \mathbf{u} + \mathbf{w}.
\end{aligned} \tag{2.5.1}$$

2.5.5 VECTOR NORMS

1. A *norm* is the vector analog to the measure of absolute value for real scalars. Norms provide a distance measure for a vector space.

2. A vector norm applied to a vector $\mathbf{v}$ is denoted by a double bar notation $\|\mathbf{v}\|$. (Single bar notation, $|\mathbf{v}|$, is also sometimes used).

3. A norm on a vector space equips it with a *metric space* structure.

4. The properties of a vector norm are:

   (a) For any vector $\mathbf{v} \neq \mathbf{0}$, $\|\mathbf{v}\| > 0$,

   (b) $\|\gamma\mathbf{v}\| = |\gamma|\,\|\mathbf{v}\|$, and

   (c) $\|\mathbf{v} + \mathbf{u}\| \leq \|\mathbf{v}\| + \|\mathbf{u}\|$ (triangle inequality).

5. The three most commonly used vector norms on $\mathbb{R}^n$ or $\mathbb{C}^n$ are

   (a) The L_1 norm is defined as $\|\mathbf{v}\|_1 = |v_1| + \cdots + |v_n| = \sum_{i=1}^{n} |v_i|$.

   (b) The L_2 norm (Euclidean norm) is defined as

   $$\|\mathbf{v}\|_2 = (|v_1|^2 + |v_2|^2 + \cdots + |v_n|^2)^{1/2} = \left(\sum_{i=1}^{n} v_i^2 \right)^{1/2}. \qquad (2.5.2)$$

   (c) The L_∞ norm is defined as $\|\mathbf{v}\|_\infty = \max_{1 \leq i \leq n} |v_i|$.

6. In the absence of any subscript, the norm $\|\cdot\|$ is usually assumed to be the L_2 (Euclidean) norm.

7. A *unit vector* with respect to a particular norm $\|\cdot\|$ is a vector that satisfies the property that $\|\mathbf{v}\| = 1$, and is sometimes denoted by $\hat{\mathbf{v}}$.

2.5.6 DOT, SCALAR, OR INNER PRODUCT

1. The *dot* (or *scalar* or *inner product*) of two vectors of the same dimension, represented by $\mathbf{v} \cdot \mathbf{u}$ or $\mathbf{v}^T\mathbf{u}$, has two common definitions, depending upon the context in which this product is encountered.

   (a) In vector calculus and physics, the dot or scalar product is defined by

   $$\mathbf{v} \cdot \mathbf{u} = \|\mathbf{v}\|\,\|\mathbf{u}\| \cos\theta, \qquad (2.5.3)$$

   where θ represents the angle between the vectors $\mathbf{v}$ and $\mathbf{u}$.

(b) In optimization, linear algebra, and computer science, the inner product of two vectors, **u** and **v**, is equivalently defined as

$$\mathbf{u}^{\mathrm{T}}\mathbf{v} = \sum_{i=1}^{n} u_i v_i = u_1 v_1 + \cdots + u_n v_n. \tag{2.5.4}$$

From the first definition, it is apparent that the inner product of two perpendicular, or orthogonal, vectors is zero, since the cosine of $90°$ is zero.

2. The inner product of two parallel vectors (with $\mathbf{u} = r\mathbf{v}$) is given by $\mathbf{v} \cdot \mathbf{u} = r \left\| \mathbf{v} \right\|^2$. For example, when $r > 0$,

$$\mathbf{v} \cdot \mathbf{u} = \left\| \mathbf{v} \right\| \left\| \mathbf{u} \right\| \cos 0 = \left\| \mathbf{v} \right\| \left\| \mathbf{u} \right\| = \left\| \mathbf{v} \right\| \left\| r\mathbf{v} \right\| = r \left\| \mathbf{v} \right\|^2. \tag{2.5.5}$$

3. The dot product is distributive, e.g.,

$$(\mathbf{v} + \mathbf{u}) \cdot \mathbf{w} = \mathbf{v} \cdot \mathbf{w} + \mathbf{u} \cdot \mathbf{w}. \tag{2.5.6}$$

4. For $\mathbf{v}, \mathbf{u}, \mathbf{w} \in \mathbb{R}^n$ with $n > 1$,

$$\mathbf{v}^{\mathrm{T}}\mathbf{u} = \mathbf{v}^{\mathrm{T}}\mathbf{w} \qquad \not\Rightarrow \qquad \mathbf{u} = \mathbf{w}. \tag{2.5.7}$$

However, it *is* valid to conclude that

$$\mathbf{v}^{\mathrm{T}}\mathbf{u} = \mathbf{v}^{\mathrm{T}}\mathbf{w} \qquad \Rightarrow \qquad \mathbf{v}^{\mathrm{T}}(\mathbf{u} - \mathbf{w}) = 0, \tag{2.5.8}$$

i.e., the vector **v** is orthogonal to the vector $(\mathbf{u} - \mathbf{w})$.

FIGURE 2.1
Depiction of right-hand rule.

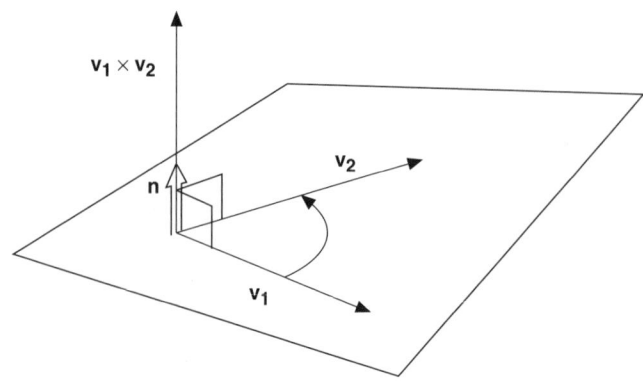

2.5.7 VECTOR OR CROSS PRODUCT

1. The *vector* (or *cross product*) of two non-zero three-dimensional vectors $\mathbf{v}$ and $\mathbf{u}$ is defined as

$$\mathbf{v} \times \mathbf{u} = \hat{\mathbf{n}} \, \|\mathbf{v}\| \, \|\mathbf{u}\| \sin \theta, \tag{2.5.9}$$

where $\hat{\mathbf{n}}$ is the unit *normal* vector (i.e., vector perpendicular to both $\mathbf{v}$ and $\mathbf{u}$) in the direction adhering to the *right-hand rule* (see Figure 2.1) and θ is the angle between $\mathbf{v}$ and $\mathbf{u}$.

2. If $\mathbf{v}$ and $\mathbf{u}$ are parallel, then $\mathbf{v} \times \mathbf{u} = \mathbf{0}$.

3. The quantity $\|\mathbf{v}\| \, \|\mathbf{u}\| \, |\sin \theta|$ represents the area of the parallelogram determined by $\mathbf{v}$ and $\mathbf{u}$.

4. The following rules apply for vector products:

$$
\begin{aligned}
(\gamma \mathbf{v}) \times (\alpha \mathbf{u}) &= (\gamma \alpha) \mathbf{v} \times \mathbf{u}, \\
\mathbf{v} \times \mathbf{u} &= -\mathbf{u} \times \mathbf{v}, \\
\mathbf{v} \times (\mathbf{u} + \mathbf{w}) &= \mathbf{v} \times \mathbf{u} + \mathbf{v} \times \mathbf{w}, \\
(\mathbf{v} + \mathbf{u}) \times \mathbf{w} &= \mathbf{v} \times \mathbf{w} + \mathbf{u} \times \mathbf{w}, \\
\mathbf{v} \times (\mathbf{u} \times \mathbf{w}) &= \mathbf{u}(\mathbf{w} \cdot \mathbf{v}) - \mathbf{w}(\mathbf{v} \cdot \mathbf{u}), \\
(\mathbf{v} \times \mathbf{u}) \cdot (\mathbf{w} \times \mathbf{z}) &= (\mathbf{v} \cdot \mathbf{w})(\mathbf{u} \cdot \mathbf{z}) - (\mathbf{v} \cdot \mathbf{z})(\mathbf{u} \cdot \mathbf{w}), \\
(\mathbf{v} \times \mathbf{u}) \times (\mathbf{w} \times \mathbf{z}) &= [\mathbf{v} \cdot (\mathbf{u} \times \mathbf{z})]\mathbf{w} - [\mathbf{v} \cdot (\mathbf{u} \times \mathbf{w})]\mathbf{z} \\
&= [\mathbf{v} \cdot (\mathbf{w} \times \mathbf{z})]\mathbf{u} - [\mathbf{u} \cdot (\mathbf{w} \times \mathbf{z})]\mathbf{v}.
\end{aligned}
\tag{2.5.10}
$$

5. The pairwise cross products of the unit vectors $\hat{\mathbf{i}}, \hat{\mathbf{j}}$, and $\hat{\mathbf{k}}$, corresponding to the directions of $\mathbf{v} = v_1 \hat{\mathbf{i}} + v_2 \hat{\mathbf{j}} + v_3 \hat{\mathbf{k}}$, are given by

$$
\begin{aligned}
\hat{\mathbf{i}} \times \hat{\mathbf{j}} &= -(\hat{\mathbf{j}} \times \hat{\mathbf{i}}) = \hat{\mathbf{k}}, \\
\hat{\mathbf{j}} \times \hat{\mathbf{k}} &= -(\hat{\mathbf{k}} \times \hat{\mathbf{j}}) = \hat{\mathbf{i}}, \\
\hat{\mathbf{k}} \times \hat{\mathbf{i}} &= -(\hat{\mathbf{i}} \times \hat{\mathbf{k}}) = \hat{\mathbf{j}}, \quad \text{and} \\
\hat{\mathbf{i}} \times \hat{\mathbf{i}} &= \hat{\mathbf{j}} \times \hat{\mathbf{j}} = \hat{\mathbf{k}} \times \hat{\mathbf{k}} = \mathbf{0}.
\end{aligned}
\tag{2.5.11}
$$

6. If $\mathbf{v} = v_1 \hat{\mathbf{i}} + v_2 \hat{\mathbf{j}} + v_3 \hat{\mathbf{k}}$ and $\mathbf{u} = u_1 \hat{\mathbf{i}} + u_2 \hat{\mathbf{j}} + u_3 \hat{\mathbf{k}}$, then

$$
\begin{aligned}
\mathbf{v} \times \mathbf{u} &= \begin{vmatrix} \hat{\mathbf{i}} & \hat{\mathbf{j}} & \hat{\mathbf{k}} \\ v_1 & v_2 & v_3 \\ u_1 & u_2 & u_3 \end{vmatrix} \\
&= (v_2 u_3 - u_2 v_3)\hat{\mathbf{i}} + (v_3 u_1 - u_3 v_1)\hat{\mathbf{j}} + (v_1 u_2 - u_1 v_2)\hat{\mathbf{k}}.
\end{aligned}
\tag{2.5.12}
$$

2.5.8   SCALAR AND VECTOR TRIPLE PRODUCTS

1. The *scalar triple product* involving three three-dimensional vectors $\mathbf{v}$, $\mathbf{u}$, and $\mathbf{w}$, sometimes denoted by $[\mathbf{vuw}]$ (not to be confused with a matrix containing three columns $\begin{bmatrix} \mathbf{v} & \mathbf{u} & \mathbf{w} \end{bmatrix}$), can be computed using the determinant

$$
[\mathbf{vuw}] = \mathbf{v} \cdot (\mathbf{u} \times \mathbf{w}) = \mathbf{v} \cdot \left[\begin{vmatrix} u_2 & u_3 \\ w_2 & w_3 \end{vmatrix} \hat{\mathbf{i}} - \begin{vmatrix} u_1 & u_3 \\ w_1 & w_3 \end{vmatrix} \hat{\mathbf{j}} + \begin{vmatrix} u_1 & u_2 \\ w_1 & w_2 \end{vmatrix} \hat{\mathbf{k}} \right]
$$

$$
= v_1 \begin{vmatrix} u_2 & u_3 \\ w_2 & w_3 \end{vmatrix} - v_2 \begin{vmatrix} u_1 & u_3 \\ w_1 & w_3 \end{vmatrix} + v_3 \begin{vmatrix} u_1 & u_2 \\ w_1 & w_2 \end{vmatrix}
$$

$$
= \begin{vmatrix} v_1 & v_2 & v_3 \\ u_1 & u_2 & u_3 \\ w_1 & w_2 & w_3 \end{vmatrix} \tag{2.5.13}
$$

$$
= \|v\| \, \|u\| \, \|w\| \cos \phi \sin \theta,
$$

where θ is the angle between $\mathbf{u}$ and $\mathbf{w}$, and ϕ is the angle between $\mathbf{v}$ and the normal to the plane defined by $\mathbf{u}$ and $\mathbf{w}$.

2. The absolute value of a triple scalar product calculates the volume of the parallelepiped determined by the three vectors. The result is independent of the order in which the triple product is taken.

3. $(\mathbf{v} \times \mathbf{u}) \times (\mathbf{w} \times \mathbf{z}) = [\mathbf{vwz}]\mathbf{u} - [\mathbf{uwz}]\mathbf{v} = [\mathbf{vuz}]\mathbf{w} - [\mathbf{vuw}]\mathbf{z}$.

4. The *vector triple product* involving three three-dimensional vectors $\mathbf{v}$, $\mathbf{u}$, and $\mathbf{w}$, given by $\mathbf{v} \times (\mathbf{u} \times \mathbf{w})$, results in a vector, perpendicular to $\mathbf{v}$, lying in the plane of $\mathbf{u}$ and $\mathbf{w}$, and is defined as

$$
\mathbf{v} \times (\mathbf{u} \times \mathbf{w}) = (\mathbf{v} \cdot \mathbf{w})\mathbf{u} - (\mathbf{v} \cdot \mathbf{u})\mathbf{w},
$$

$$
= \begin{vmatrix} \hat{\mathbf{i}} & \hat{\mathbf{j}} & \hat{\mathbf{k}} \\ v_1 & v_2 & v_3 \\ \begin{vmatrix} u_2 & u_3 \\ w_2 & w_3 \end{vmatrix} & \begin{vmatrix} u_3 & u_1 \\ w_3 & w_1 \end{vmatrix} & \begin{vmatrix} u_1 & u_2 \\ w_1 & w_2 \end{vmatrix} \end{vmatrix}. \tag{2.5.14}
$$

5. Given three non-coplanar reference vectors $\mathbf{v}$, $\mathbf{u}$, and $\mathbf{w}$, the *reciprocal system* is given by $\mathbf{v}^*$, $\mathbf{u}^*$, and $\mathbf{w}^*$, where

$$
\mathbf{v}^* = \frac{\mathbf{u} \times \mathbf{w}}{[\mathbf{vuw}]}, \qquad \mathbf{u}^* = \frac{\mathbf{w} \times \mathbf{v}}{[\mathbf{vuw}]}, \qquad \mathbf{w}^* = \frac{\mathbf{v} \times \mathbf{u}}{[\mathbf{vuw}]}. \tag{2.5.15}
$$

Note that

$$
1 = \mathbf{v} \cdot \mathbf{v}^* = \mathbf{u} \cdot \mathbf{u}^* = \mathbf{w} \cdot \mathbf{w}^*
$$

$$
\text{and } 0 = \mathbf{v} \cdot \mathbf{u}^* = \mathbf{v} \cdot \mathbf{w}^* = \mathbf{u} \cdot \mathbf{v}^*, \text{ etc.} \tag{2.5.16}
$$

The system $\hat{\mathbf{i}}, \hat{\mathbf{j}}, \hat{\mathbf{k}}$ is its own reciprocal.

2.6 LINEAR AND MATRIX ALGEBRA

2.6.1 DEFINITIONS

1. An $m \times n$ *matrix* is a two-dimensional array of numbers consisting of m rows and n columns. By convention, a matrix is denoted by a capital letter emphasized with italics, as in A, B, D, or boldface, **A**, **B**, **D**. Sometimes a matrix has a subscript denoting the dimensions of the matrix, e.g., $A_{2\times3}$. If A is a real $n \times m$ matrix, then we write $A \in \mathbb{R}^{n\times m}$. Higher dimensional matrices, although less frequently encountered, are accommodated in a similar fashion, e.g., a three-dimensional matrix $A_{m\times n\times p}$, and so on.

2. $A_{m\times n}$ is called *rectangular* if $m \neq n$.

3. $A_{m\times n}$ is called *square* if $m = n$.

4. A particular component (equivalently: element) of a matrix is denoted by the lower-case letter of that which names the matrix, along with two subscripts corresponding to the row i and column j location of the component in the array, e.g.,

$$A_{m\times n} \quad \text{has components } a_{ij};$$
$$B_{m\times n} \quad \text{has components } b_{ij}.$$

For example, a_{23} is the component in the second row and third column of matrix A.

5. Any component a_{ij} with $i = j$ is called a *diagonal* component.

6. The diagonal alignment of components in a matrix extending from the upper left to the lower right is called the *principal* or *main* diagonal.

7. Any component a_{ij} with $i \neq j$ is called an *off-diagonal* component.

8. Two matrices A and B are said to be equal if they have the same number of rows (m) and columns (n), and $a_{ij} = b_{ij}$ for all $1 \leq i \leq m$, $1 \leq j \leq n$.

9. An $m \times 1$ dimensional matrix is called a *column vector*. Similarly, a $1 \times n$ dimensional matrix is called a *row vector*.

10. A column (row) vector with all components equal to zero is called a *null* vector and is usually denoted by **0**.

11. A column vector with all components equal to one is often denoted by **e**. The analogous row vector is denoted by $\mathbf{e}^{\mathrm{T}}$.

12. The standard basis consists of the vectors $\{\mathbf{e}_1, \mathbf{e}_2, \ldots, \mathbf{e}_n\}$ where $\mathbf{e}_i$ is an $n \times 1$ vector of all zeros, except for the ith component, which is one.

13. The scalar $\mathbf{x}^T\mathbf{x} = \sum_{i=1}^{n} x_i^2$ is the sum of squares of all components of the vector $\mathbf{x}$.

14. The *weighted* sum of squares is defined by $\mathbf{x}^T D_w \mathbf{x} = \sum_{i=1}^{n} w_i x_i^2$, when $\mathbf{x}$ has n components and the diagonal matrix D_w is of dimension $(n \times n)$.

15. If Q is a square matrix, then $\mathbf{x}^T Q \mathbf{x}$ is called a *quadratic* form.

16. An $n \times n$ matrix A is called *non-singular*, or *invertible*, or *regular*, if there exists an $n \times n$ matrix B such that $AB = BA = I$. The unique matrix B satisfying this condition is called the *inverse* of A, and is denoted by A^{-1}.

17. The scalar $\mathbf{x}^T\mathbf{y} = \sum_{i=1}^{n} x_i y_i$, the inner product of $\mathbf{x}$ and $\mathbf{y}$, is the sum of products of the components of $\mathbf{x}$ by those of $\mathbf{y}$.

18. The *weighted* sum of products is $\mathbf{x}^T D_w \mathbf{y} = \sum_{i=1}^{n} w_i x_i y_i$, when $\mathbf{x}$ and $\mathbf{y}$ have n components, and the diagonal matrix D_w is $(n \times n)$.

19. The map $\mathbf{x} \mapsto \mathbf{x}^T Q \mathbf{y}$ is called a *bilinear* form, where Q is a matrix of appropriate dimension.

20. The *transpose* of an $m \times n$ matrix A, denoted by A^T, is an $n \times m$ matrix with rows and columns interchanged, so that the (i, j) component of A is the (j, i) component of A^T, and $(A^T)_{ji} = (A)_{ij} = a_{ij}$.

21. The *Hermitian conjugate* of a matrix A, denoted by A^H, is obtained by transposing A and replacing each element by its complex conjugate. Hence, if $a_{kl} = u_{kl} + iv_{kl}$, then $(A^H)_{kl} = u_{lk} - iv_{ik}$, with $i = \sqrt{-1}$.

22. If Q is a square matrix, then the map $\mathbf{x} \mapsto \mathbf{x}^H Q \mathbf{x}$ is called a *Hermitian* form.

2.6.2 TYPES OF MATRICES

1. A square matrix with all components off the principal diagonal equal to zero is called a *diagonal matrix*, typically denoted by the letter D with a subscript indicating the typical element in the principal diagonal.

EXAMPLE

$$D_a = \begin{bmatrix} a_{11} & 0 & 0 \\ 0 & a_{22} & 0 \\ 0 & 0 & a_{33} \end{bmatrix}, \qquad D_\lambda = \begin{bmatrix} \lambda_{11} & 0 & 0 \\ 0 & \lambda_{22} & 0 \\ 0 & 0 & \lambda_{33} \end{bmatrix}.$$

2. A *zero*, or *null, matrix* is one whose elements are all zero (notation is "0").

3. The *identity matrix*, denoted by I, is the diagonal matrix with $a_{ij} = 1$ for all $i = j$, and $a_{ij} = 0$ for $i \neq j$. The $n \times n$ identity matrix is denoted I_n.

4. The *elementary matrix*, E_{ij}, is defined differently in different contexts:

   (a) Elementary matrices have the form $E_{ij} = \mathbf{e}_i \mathbf{e}_j^T$. Hence, $A = \sum_i \sum_j a_{ij} E_{ij}$.

(b) Elementary matrices are also written as $E = I - \alpha uv^{\mathrm{T}}$, where I is the identity matrix, α is a scalar, and u and v are vectors of the same dimension. In this context, the elementary matrix is referred to as a *rank-one modification* of an identity matrix.

(c) In Gaussian elimination, the matrix that subtracts a multiple ℓ of row j from row i is called $E_{ij} = I - \ell \mathbf{e}_i \mathbf{e}_j^{\mathrm{T}}$, with 1's on the diagonal and the number $-\ell$ in row i column j.

EXAMPLE

$$E_{31} = I - 5\mathbf{e}_3 \mathbf{e}_1^{\mathrm{T}} = \begin{bmatrix} 1 & 0 & 0 & 0 \\ 0 & 1 & 0 & 0 \\ -5 & 0 & 1 & 0 \\ 0 & 0 & 0 & 1 \end{bmatrix}.$$

5. A matrix with all components above the principal diagonal equal to zero is called a *lower triangular matrix*.

EXAMPLE

$$A = \begin{bmatrix} a_{11} & 0 & 0 \\ a_{21} & a_{22} & 0 \\ a_{31} & a_{32} & a_{33} \end{bmatrix} \qquad \text{is lower triangular.}$$

6. A matrix with all components below the principal diagonal equal to zero is called an *upper triangular matrix*. (The transpose of a lower triangular matrix is an upper triangular matrix.)

7. A matrix whose components are arranged in m rows and a single column is called a *column matrix*, or *column vector*, and is typically denoted using boldface, lower-case letters, e.g., $\mathbf{a}$ and $\mathbf{b}$.

8. A matrix whose components are arranged in n columns and a single row is called a *row matrix*, or *row vector*, and is typically denoted as a transposed column vector, e.g., $\mathbf{a}^{\mathrm{T}}$ and $\mathbf{b}^{\mathrm{T}}$.

9. A square matrix is called *symmetric* if $A = A^{\mathrm{T}}$.

10. A square matrix is called *skew symmetric* if $A^{\mathrm{T}} = -A$.

11. A square matrix A is called *Hermitian* if $A = A^{\mathrm{H}}$. A square matrix A is called *skew-Hermitian* if $A^{\mathrm{H}} = -A$. All real symmetric matrices are Hermitian.

12. A square matrix Q with orthonormal columns is said to be *orthogonal*.[1] It follows directly that the rows of Q must also be orthonormal, so that $QQ^{\mathrm{T}} = Q^{\mathrm{T}}Q = I$, or $Q^{\mathrm{T}} = Q^{-1}$. The determinant of an orthogonal matrix is ± 1.

A *rotation matrix* is an orthogonal matrix whose determinant is equal to $+1$.

13. An $m \times n$ matrix A with orthonormal columns has the property $A^{\mathrm{T}}A = I$.

---

[1]Note the inconsistency in terminology that has persisted.

14. A square matrix is called *unitary* if $A^H A = I$. A real unitary matrix is orthogonal. The eigenvalues of a unitary matrix all have an absolute value of one.

15. A square matrix is called a *permutation matrix* if its columns are a permutation of the columns of I. A permutation matrix is orthogonal.

16. A square matrix is called *idempotent* if $AA = A^2 = A$.

17. A square matrix is called a *projection matrix* if it is both Hermitian and idempotent: $A^H = A^2 = A$.

18. A square matrix is called *normal* if $A^H A = AA^H$. The following matrices are normal: diagonal, Hermitian, unitary, skew-Hermitian.

19. A square matrix is called *nilpotent* to index k if $A^k = 0$ but $A^{k-1} \neq 0$. The eigenvalues of a nilpotent matrix are all zero.

20. A *principal sub-matrix* of a symmetric matrix A is formed by deleting rows and columns of A simultaneously, e.g., row 1 and column 1; row 9 and column 9, etc.

21. A square matrix whose elements are constant along each diagonal is called a *Toeplitz* matrix.

EXAMPLE

$$A = \begin{bmatrix} a & d & e \\ b & a & d \\ c & b & a \end{bmatrix} \qquad \text{and} \qquad M = \begin{bmatrix} 4 & 0 & 1 \\ -11 & 4 & 0 \\ 3 & -11 & 4 \end{bmatrix} \qquad (2.6.1)$$

are Toeplitz matrices. Notice that Toeplitz matrices are symmetric about a diagonal extending from the upper right-hand corner element to the lower left-hand corner element. This type of symmetry is called *persymmetry*.

22. A *Vandermonde* matrix is a square matrix $V \in \mathbb{R}^{(n+1) \times (n+1)}$ in which each column contains unit increasing powers of a single matrix value:

$$V = \begin{bmatrix} 1 & 1 & \cdots & 1 \\ v_1 & v_2 & \cdots & v_{(n+1)} \\ v_1^2 & v_2^2 & \cdots & v_{(n+1)}^2 \\ \vdots & \vdots & & \vdots \\ v_1^n & v_2^n & \cdots & v_{(n+1)}^n \end{bmatrix}. \qquad (2.6.2)$$

23. A square matrix U is said to be in *upper Hessenberg* form if $u_{ij} = 0$ whenever $i > j+1$. An upper Hessenberg matrix is essentially an upper triangular matrix with an extra non-zero element immediately below the main diagonal entry in each column of U. For example,

$$U = \begin{bmatrix} u_{11} & u_{12} & u_{13} & u_{14} \\ b_{21} & u_{22} & u_{23} & u_{24} \\ 0 & b_{32} & u_{33} & u_{34} \\ 0 & 0 & b_{43} & u_{44} \end{bmatrix} \qquad \text{is upper Hessenberg.}$$

24. If the sum of the components of each column of a matrix $A \in \mathbb{R}^{n \times n}$ equals one, then A is called a *Markov* matrix.

25. A *circulant* matrix is an $n \times n$ matrix of the form

$$
C = \begin{bmatrix}
c_0 & c_1 & c_2 & \cdots & c_{n-2} & c_{n-1} \\
c_{n-1} & c_0 & c_1 & \cdots & c_{n-3} & c_{n-2} \\
c_{n-2} & c_{n-1} & c_0 & \cdots & c_{n-4} & c_{n-3} \\
\vdots & & & & & \vdots \\
c_1 & c_2 & c_3 & \cdots & c_{n-1} & c_0
\end{bmatrix},
\tag{2.6.3}
$$

where the components c_{ij} are such that $(j - i) = k \mod n$ have the same value c_k. These components comprise the kth *stripe* of C.

26. A matrix $A = (a_{ij})$ has *lower bandwidth* p if $a_{ij} = 0$ whenever $i > j + p$ and *upper bandwidth* q if $a_{ij} = 0$ whenever $j > i + q$. When they are equal, they are the *bandwidth* of A. A diagonal matrix has bandwidth 0. A *tridiagonal matrix* has bandwidth 1. An upper (resp. lower) triangular matrix has upper (resp. lower) bandwidth of $n - 1$ (resp. $m - 1$).

27. If a rotation is defined by the matrix $A = \begin{bmatrix} a_{11} & a_{12} & a_{13} \\ a_{21} & a_{22} & a_{23} \\ a_{31} & a_{32} & a_{33} \end{bmatrix}$ then its fixed axis of rotation is given by $\mathbf{v} = \mathbf{i}(a_{23} - a_{32}) + \mathbf{j}(a_{31} - a_{13}) + \mathbf{k}(a_{12} - a_{21})$.

28. A *Householder transformation*, or *Householder reflection*, is an $n \times n$ matrix H of the form $H = I - (2\mathbf{u}\mathbf{u}^{\mathsf{T}})/(\mathbf{u}^{\mathsf{T}}\mathbf{u})$, where the *Householder vector* $\mathbf{u} \in \mathbb{R}^n$ is non-zero.

29. A *Givens rotation* is defined as a rank-two correction to the identity matrix given by

$$
G(i, k, \theta) = \begin{bmatrix}
1 & \cdots & 0 & \cdots & 0 & \cdots & 0 \\
\vdots & \ddots & \vdots & & \vdots & & \vdots \\
0 & \cdots & c & \cdots & s & \cdots & 0 \\
\vdots & & \vdots & \ddots & \vdots & & \vdots \\
0 & \cdots & -s & \cdots & c & \cdots & 0 \\
\vdots & & \vdots & & \vdots & \ddots & \vdots \\
0 & \cdots & 0 & \cdots & 0 & \cdots & 1
\end{bmatrix}
\begin{matrix} \\ \\ i \\ \\ k \\ \\ \\ \end{matrix},
\tag{2.6.4}
$$
$$
 \quad\quad i \quad\quad\quad k
$$

where $c = \cos\theta$ and $s = \sin\theta$ for some angle θ. Premultiplication by $G(i, k, \theta)^{\mathsf{T}}$ induces a counterclockwise rotation of θ radians in the (i, k) plane. For $\mathbf{x} \in \mathbb{R}^n$ and $\mathbf{y} = G(i, k, \theta)^{\mathsf{T}}\mathbf{x}$, the components of $\mathbf{y}$ are given by

$$
y_j = \begin{cases}
cx_i - sx_k, & \text{for } j = i \\
sx_i + cx_k, & \text{for } j = k \\
x_j, & \text{for } j \neq i, k.
\end{cases}
\tag{2.6.5}
$$

2.6.3 CONFORMABILITY FOR ADDITION AND MULTIPLICATION

1. Two matrices A and B can be added (subtracted) if they are of the same dimension. The result is a matrix of the same dimension. If $C = A + B$ then $c_{ij} = a_{ij} + b_{ij}$.

EXAMPLE

$$A_{2\times 3} + B_{2\times 3} = \begin{bmatrix} 3 & 2 & -1 \\ 4 & 0 & 9 \end{bmatrix} + \begin{bmatrix} 11 & -2 & 3 \\ 0 & 1 & 1 \end{bmatrix} = \begin{bmatrix} 14 & 0 & 2 \\ 4 & 1 & 10 \end{bmatrix}.$$

2. Multiplication of a matrix or a vector by a scalar is achieved by multiplying each component by that scalar. If $B = \alpha A$, then $b_{ij} = \alpha a_{ij}$ for all components.

3. The matrix multiplication AB is defined if the number of columns of A is equal to the number of rows of B.

4. The multiplication of two matrices $A_{m\times n}$ and $B_{n\times q}$ results in a matrix $C_{m\times q}$ whose components are defined as

$$c_{ij} = \sum_{k=1}^{n} a_{ik}b_{kj} \tag{2.6.6}$$

for $i = 1, 2, \ldots, m$ and $j = 1, 2, \ldots, q$. Each c_{ij} is the result of the inner (dot) product of the ith row of A with the jth column of B.

j^{th} column of B

i^{th} row of A

$(ij)^{th}$ element of C

This rule applies similarly for matrix multiplication involving more than two matrices. If $ABCD = E$ then

$$e_{ij} = \sum_{k}\sum_{l}\sum_{m} a_{ik}b_{kl}c_{lm}d_{mj}. \tag{2.6.7}$$

The second subscript for each matrix component must coincide with the first subscript of the next one.

EXAMPLE

$$\begin{bmatrix} 2 & -1 & 3 \\ -4 & 1 & 4 \end{bmatrix} \begin{bmatrix} 5 & -3 & -3 \\ 2 & 2 & -1 \\ -7 & 1 & 5 \end{bmatrix} = \begin{bmatrix} -13 & -5 & 10 \\ -46 & 18 & 31 \end{bmatrix}.$$

5. In general, matrix multiplication is not commutative: $AB \neq BA$.

6. Matrix multiplication is associative: $A(BC) = (AB)C$.

7. The distributive law of multiplication and addition holds: $C(A + B) = CA + CB$ and $(A + B)C = AC + BC$.

8. Both the transpose operator and the Hermitian operator reverse the order of matrix multiplication: $(ABC)^{\mathrm{T}} = C^{\mathrm{T}}B^{\mathrm{T}}A^{\mathrm{T}}$ and $(ABC)^{\mathrm{H}} = C^{\mathrm{H}}B^{\mathrm{H}}A^{\mathrm{H}}$.

9. *Strassen algorithm*: The matrix product $\begin{bmatrix} a_{11} & a_{12} \\ a_{21} & a_{22} \end{bmatrix} \begin{bmatrix} b_{11} & b_{12} \\ b_{21} & b_{22} \end{bmatrix} = \begin{bmatrix} c_{11} & c_{12} \\ c_{21} & c_{22} \end{bmatrix}$ can be computed in the following way:

$$m_1 = (a_{12} - a_{22})(b_{21} + b_{22}), \qquad c_{11} = m_1 + m_2 - m_4 + m_6,$$
$$m_2 = (a_{11} + a_{22})(b_{11} + b_{22}), \qquad c_{12} = m_4 + m_5,$$
$$m_3 = (a_{11} - a_{21})(b_{11} + b_{12}), \qquad c_{21} = m_6 + m_7, \text{ and}$$
$$m_4 = (a_{11} + a_{12})b_{22}, \qquad c_{22} = m_2 - m_3 + m_5 - m_7.$$
$$m_5 = a_{11}(b_{12} - b_{22}),$$
$$m_6 = a_{22}(b_{21} - b_{11}),$$
$$m_7 = (a_{21} + a_{22})b_{11},$$

This computation uses 7 multiplications and 18 additions and subtractions. Using this formula recursively allows multiplication of two $n \times n$ matrices using $O(n^{\log_2 7}) = O(n^{2.807\cdots})$ scalar multiplications.

10. The order in which matrices are grouped together for multiplication can change the number of scalar multiplications required. The straightforward number of scalar multiplications required to multiply matrix $X_{a \times b}$ by matrix $Y_{b \times c}$ is abc, without using clever algorithms such as Strassen's. For example, consider the matrix product $P = A_{10 \times 100}B_{100 \times 5}C_{5 \times 50}$. The parenthesization $P = ((AB)C)$ requires $(10 \times 100 \times 5) + (10 \times 5 \times 50) = 7{,}500$ scalar multiplications. The parenthesization $P = (A(BC))$ requires $(10 \times 100 \times 50) + (100 \times 5 \times 50) = 75{,}000$ scalar multiplications.

11. Pre-multiplication by a diagonal matrix scales the rows

$$\begin{bmatrix} d_{11} & 0 & \cdots & 0 \\ 0 & d_{22} & & 0 \\ \vdots & & \ddots & \\ 0 & 0 & & d_{nn} \end{bmatrix} \begin{bmatrix} a_{11} & \cdots & a_{1m} \\ a_{21} & \cdots & a_{2m} \\ \vdots & \ddots & \vdots \\ a_{n1} & \cdots & a_{nm} \end{bmatrix} = \begin{bmatrix} d_{11}a_{11} & \cdots & d_{11}a_{1m} \\ d_{22}a_{21} & \cdots & d_{22}a_{2m} \\ \vdots & \ddots & \vdots \\ d_{nn}a_{n1} & \cdots & d_{nn}a_{nm} \end{bmatrix}$$

12. Post-multiplication by a diagonal matrix scales the columns

$$\begin{bmatrix} a_{11} & \cdots & a_{1m} \\ a_{21} & \cdots & a_{2m} \\ \vdots & \ddots & \vdots \\ a_{n1} & \cdots & a_{nm} \end{bmatrix} \begin{bmatrix} d_{11} & 0 & \cdots & 0 \\ 0 & d_{22} & & 0 \\ \vdots & & \ddots & \\ 0 & 0 & & d_{mm} \end{bmatrix} = \begin{bmatrix} d_{11}a_{11} & \cdots & d_{mm}a_{1m} \\ d_{11}a_{21} & \cdots & d_{mm}a_{2m} \\ \vdots & \ddots & \vdots \\ d_{11}a_{n1} & \cdots & d_{mm}a_{nm} \end{bmatrix}$$

2.6.4 DETERMINANTS AND PERMANENTS

1. The *determinant* of a square matrix A, denoted by $|A|$ or $\det(A)$ or $\det A$, is a scalar function of A defined as

$$
\det(A) = \det \begin{bmatrix} a_{11} & a_{12} & \cdots & a_{1n} \\ a_{21} & a_{22} & \cdots & a_{2n} \\ \vdots & \vdots & \ddots & \vdots \\ a_{n1} & a_{n2} & \cdots & a_{nn} \end{bmatrix} = \begin{vmatrix} a_{11} & a_{12} & \cdots & a_{1n} \\ a_{21} & a_{22} & \cdots & a_{2n} \\ \vdots & \vdots & \ddots & \vdots \\ a_{n1} & a_{n2} & \cdots & a_{nn} \end{vmatrix} \quad (2.6.8)
$$

$$
= \sum_{\sigma} \operatorname{sgn}(\sigma) a_{1,\sigma(1)}\, a_{2,\sigma(2)} \cdots a_{n,\sigma(n)}
$$

$$
= \sum (-1)^{\delta} a_{1 i_1} a_{2 i_2} \cdots a_{n i_n}
$$

where the sum is taken either

(a) over all permutations σ of $\{1, 2, \ldots, n\}$ and the signum function, $\operatorname{sgn}(\sigma)$, is (-1) raised to the power of the number of successive transpositions required to change the permutation σ to the identity permutation; or

(b) over all permutations $i_1 \neq i_2 \neq \cdots \neq i_n$, and δ denotes the number of transpositions necessary to bring the sequence $(i_1, i_2, \ldots, i_n)$ back into the natural order $(1, 2, \ldots, n)$.

2. For a 2×2 matrix, $\begin{vmatrix} a_{11} & a_{12} \\ a_{21} & a_{22} \end{vmatrix} = a_{11} a_{22} - a_{12} a_{21}$.

For a 3×3 matrix,

$$
\begin{vmatrix} a_{11} & a_{12} & a_{13} \\ a_{21} & a_{22} & a_{23} \\ a_{31} & a_{32} & a_{33} \end{vmatrix} = a_{11} a_{22} a_{33} + a_{12} a_{23} a_{31} + a_{13} a_{21} a_{32}
$$

$$
- a_{13} a_{22} a_{31} - a_{11} a_{23} a_{32} - a_{12} a_{21} a_{33}. \quad (2.6.9)
$$

3. The determinant of the identity matrix is one.

4. Note that $|A|\,|B| = |AB|$ and $|A| = |A^{\mathrm{T}}|$.

5. Interchanging two rows (or columns) of a matrix changes the sign of its determinant.

6. A determinant does not change its value if a linear combination of other rows (or columns) is added to or subtracted from any given row (or column).

7. Multiplying an entire row (or column) of A by a scalar γ causes the determinant to be multiplied by the same scalar γ.

8. For an $n \times n$ matrix A, $|\gamma A| = \gamma^n |A|$.

9. If $\det(A) = 0$, then A is *singular*; if $\det(A) \neq 0$, then A is *non-singular* or *invertible*.

10. $\det(A^{-1}) = 1/\det(A)$.

11. When the edges of a parallelepiped P are defined by the rows (or columns) of A, the absolute value of the determinant of A measures the volume of P. Thus, if any row (or column) of A is dependent upon another row (or column) of A, the determinant of A equals zero.

12. The *cofactor* of a square matrix A, $\text{cof}_{ij}(A)$, is the determinant of a sub-matrix obtained by striking the ith row and the jth column of A and choosing a positive (negative) sign if $(i+j)$ is even (odd).

EXAMPLE

$$\text{cof}_{23}\begin{bmatrix} 2 & 4 & 3 \\ 6 & 1 & 5 \\ -2 & 1 & 3 \end{bmatrix} = (-1)^{2+3}\begin{vmatrix} 2 & 4 \\ -2 & 1 \end{vmatrix} = -(2+8) = -10. \qquad (2.6.10)$$

13. Let a_{ij} denote the components of A and a^{ij} those of A^{-1}. Then,

$$a^{ij} = \text{cof}_{ji}(A)/|A|. \qquad (2.6.11)$$

14. *Partitioning of determinants*: Let $A = \begin{bmatrix} B & C \\ D & E \end{bmatrix}$. Assuming all inverses exist, then

$$|A| = |E|\ \left|(B - CE^{-1}D)\right| = |B|\ \left|(E - DB^{-1}C)\right|. \qquad (2.6.12)$$

15. *Laplace development*: The determinant of A is a combination of row i (column j) and the cofactors of row i (column j), i.e.,

$$\begin{aligned} |A| &= a_{i1}\text{cof}_{i1}(A) + a_{i2}\text{cof}_{i2}(A) + \cdots + a_{in}\text{cof}_{in}(A), \\ &= a_{1j}\text{cof}_{1j}(A) + a_{2j}\text{cof}_{2j}(A) + \cdots + a_{nj}\text{cof}_{nj}(A), \end{aligned} \qquad (2.6.13)$$

for any row i or any column j.

16. Omitting the signum function in Equation (2.6.8) yields the definition of *permanent* of A, given by per $A = \sum_{\sigma} a_{1,\sigma(1)} \cdots a_{n,\sigma(n)}$. Properties of the permanent include:

  (a) If A is an $m \times n$ matrix and B is an $n \times m$ matrix, then

$$|\text{per}(AB)|^2 \le \text{per}(AA^{\text{H}})\,\text{per}(B^{\text{H}}B). \qquad (2.6.14)$$

  (b) If P and Q are permutation matrices, then per $PAQ = $ per A.

  (c) If D and G are diagonal matrices, then per $DAG = $ per D per A per G.

2.6.5 MATRIX NORMS

1. The mapping $g : \mathbb{R}^{m \times n} \Rightarrow \mathbb{R}$ is a *matrix norm* if g satisfies the same three properties as a vector norm:

   (a) $g(A) \geq 0$ for all A and $g(A) = 0$ if and only if $A \equiv 0$, so that (in norm notation) $\|A\| > 0$ for all non-zero A.

   (b) For two matrices $A, B \in \mathbb{R}^{m \times n}$, $g(A + B) \leq g(A) + g(B)$, so that $\|A + B\| \leq \|A\| + \|B\|$.

   (c) $g(rA) = |r| g(A)$, where $r \in \mathbb{R}$, so that $\|\gamma A\| = |\gamma| \ \|A\|$.

2. The most common matrix norms are the L_p *matrix* norm and the *Frobenius* norm.

3. The L_p *norm* of a matrix A is the number defined by

$$\|A\|_p = \sup_{\mathbf{x} \neq \mathbf{0}} \frac{\|A\mathbf{x}\|_p}{\|\mathbf{x}\|_p} \tag{2.6.15}$$

   where $\|\cdot\|_p$ represents one of the L_p (vector) norms with $p = 1, 2,$ or ∞.

   (a) The L_1 norm of $A_{m \times n}$ is defined as $\|A\|_1 = \max_{1 \leq j \leq n} \sum_{i=1}^{m} |a_{ij}|$.

   (b) The L_2 norm of A is the square root of the greatest eigenvalue of $A^T A$, (i.e., $\|A\|_2^2 = \lambda_{\max}(A^T A)$), which is the same as the largest singular value of A, $\|A\|_2 = \sigma_1(A)$.

   (c) The L_∞ norm is defined as $\|A\|_\infty = \max_{1 \leq i \leq m} \sum_{j=1}^{n} |a_{ij}|$.

4. The *Frobenius* or *Hilbert–Schmidt norm* of a matrix A is defined as

$$\|A\|_F = \sqrt{\sum_{i=1}^{m} \sum_{j=1}^{n} |a_{ij}|^2} \tag{2.6.16}$$

   which satisfies $\|A\|_F^2 = \text{tr}(A^T A)$. Since $\mathbb{R}^{m \times n}$ is isomorphic to $\mathbb{R}^{mn}$, the Frobenius norm can be interpreted as the L_2 norm of an $nm \times 1$ column vector in which each column of A is appended to the next in succession. (See the Vector operation ("Vec") in Section 2.6.17.)

5. When A is symmetric, then $\|A\|_2 = \max_j |\lambda_j|$, where $\{\lambda_j\}$ are the eigenvalues of A.

6. The following properties hold:

$$\frac{1}{\sqrt{m}} \|A\|_1 \leq \|A\|_2 \leq \sqrt{n} \|A\|_1 \,,$$

$$\max_{i,j} |a_{ij}| \leq \|A\|_2 \leq \sqrt{mn} \max_{i,j} |a_{ij}|, \quad \text{and} \tag{2.6.17}$$

$$\frac{1}{\sqrt{n}} \|A\|_\infty \leq \|A\|_2 \leq \sqrt{m} \|A\|_\infty \,.$$

7. The matrix p norms satisfy the additional property of *consistency*, defined as $\|AB\|_p \leq \|A\|_p \|B\|_p$.

8. The Frobenius norm is *compatible* with the vector 2 norm, i.e., $\|A\mathbf{x}\|_F \leq \|A\|_F \|\mathbf{x}\|_2$. The Frobenius norm also satisfies $\|A\|_2 \leq \|A\|_F \leq \sqrt{n} \|A\|_2$.

2.6.6 SINGULARITY, RANK, AND INVERSES

1. An $n \times n$ matrix A is called *singular* if there exists a vector $\mathbf{x} \neq \mathbf{0}$ such that $A\mathbf{x} = \mathbf{0}$ or $A^T\mathbf{x} = \mathbf{0}$. (Note that $\mathbf{x} = \mathbf{0}$ means that all components of $\mathbf{x}$ are zero.) If a matrix is not singular, it is called *non-singular*. (This is consistent with note 16 on page 138.)

2. $(AB)^{-1} = B^{-1}A^{-1}$, provided all inverses exist.

3. $(A^{-1})^T = (A^T)^{-1}$.

4. $(\gamma A)^{-1} = (1/\gamma)A^{-1}$.

5. If D_w is a diagonal matrix, then $D_w^{-1} = D_{1/w}$.

   EXAMPLE If $D_w = \begin{bmatrix} w_{11} & 0 \\ 0 & w_{22} \end{bmatrix}$ then $D_w^{-1} = D_{1/w} = \begin{bmatrix} \frac{1}{w_{11}} & 0 \\ 0 & \frac{1}{w_{22}} \end{bmatrix}$.

6. Partitioning: Let $A = \begin{bmatrix} B & C \\ D & E \end{bmatrix}$. Assuming that all inverses exist, then $A^{-1} = \begin{bmatrix} X & Y \\ Z & U \end{bmatrix}$, where

$$X = (B - CE^{-1}D)^{-1}, \qquad U = (E - DB^{-1}C)^{-1},$$
$$Y = -B^{-1}CU, \qquad Z = -E^{-1}DX.$$

7. The inverse of a 2×2 matrix is as follows (defined when $ad \neq bc$):

$$\begin{bmatrix} a & b \\ c & d \end{bmatrix}^{-1} = \frac{1}{ad - bc} \begin{bmatrix} d & -b \\ -c & a \end{bmatrix}$$

8. If A and B are both invertible, then

$$(A + B)^{-1} = B^{-1}\left(A^{-1} + B^{-1}\right)^{-1} A^{-1} = A^{-1}\left(A^{-1} + B^{-1}\right)^{-1} B^{-1}. \tag{2.6.18}$$

9. The *row rank* of a matrix A is defined as the number of linearly independent rows of A. Likewise, the *column rank* equals the number of linearly independent columns of A. For any matrix, the row rank equals the column rank.

10. If $A \in \mathbb{R}^{n \times n}$ has rank of n, then A is said to have *full rank*.

11. A square matrix is invertible if, and only if, it has full rank.

12. $\text{Rank}(AB) \leq \min[\text{rank}(A), \text{rank}(B)]$.

13. $\text{Rank}(A^T A) = \text{rank}(AA^T) = \text{rank}(A)$.

2.6.7 SYSTEMS OF LINEAR EQUATIONS

1. Suppose that A is a matrix. Then $A\mathbf{x} = \mathbf{b}$ is a *system of linear equations*. If A is square and non-singular, there exists a unique solution $\mathbf{x} = A^{-1}\mathbf{b}$.

2. For the linear system of equations involving n variables and m equations, written as $A\mathbf{x} = \mathbf{b}$ or

$$
\begin{aligned}
a_{11}x_1 + a_{12}x_2 + \cdots + a_{1m}x_m &= b_1, \\
a_{21}x_1 + a_{22}x_2 + \cdots + a_{2m}x_m &= b_2, \\
&\ \ \vdots \\
a_{n1}x_1 + a_{n2}x_2 + \cdots + a_{nm}x_m &= b_n,
\end{aligned}
\tag{2.6.19}
$$

the possible outcomes when searching for a solution are:

(a) A unique solution exists, and the system is called *consistent*.
(b) No such solution exists and the system is called *inconsistent*.
(c) Multiple solutions exist, the system has an infinite number of solutions, and the system is called *undetermined*.

3. The solvability cases of the linear systems $A\mathbf{x} = \mathbf{b}$ (when A is $m \times n$) are:

(a) If $\text{rank}(A) = m = n$, then there is a unique solution.
(b) If $\text{rank}(A) = m < n$, then there is an exact solution with free parameters.
(c) If $\text{rank}(A) = n < m$, then either there is a unique solution, or there is a unique least-squares solution.
(d) If $\text{rank}(A) < m < n$, or $\text{rank}(A) < n < m$, or $\text{rank}(A) < n = m$: then either there is an exact solution with free parameters, or there are non-unique least-squares solutions.

4. If the system of equations $A\mathbf{x} = \mathbf{b}$ is undetermined, then we may find the $\mathbf{x}$ that minimizes $\|A\mathbf{x} - \mathbf{b}\|_p$ for some p.

   EXAMPLE   If $A = \begin{bmatrix} 1 & 1 & 1 \end{bmatrix}^{\mathsf{T}}$ and $\mathbf{b} = \begin{bmatrix} b_1 & b_2 & b_3 \end{bmatrix}^{\mathsf{T}}$ with $b_1 \geq b_2 \geq b_3 \geq 0$ then the minimum of $\|A\mathbf{x} - \mathbf{b}\|_p$ in different norms is:

$$
\begin{aligned}
p = 1 &\longrightarrow \quad \mathbf{x}_{\text{optimal}} = b_2 \\
p = 2 &\longrightarrow \quad \mathbf{x}_{\text{optimal}} = (b_1 + b_2 + b_3)/3 \\
p = \infty &\longrightarrow \quad \mathbf{x}_{\text{optimal}} = (b_1 + b_3)/2
\end{aligned}
\tag{2.6.20}
$$

5. For the system of linear equations $A\mathbf{x} = \mathbf{b}$ (with A square and non-singular), the sensitivity of the solution $\mathbf{x}$ to pertubations in A and $\mathbf{b}$ is given in terms of the *condition number* of A defined by

$$
\text{cond}(A) = \|A^{-1}\| \, \|A\|,
\tag{2.6.21}
$$

where $\|\cdot\|$ is any of the L_p norms. In all cases, $\text{cond}(A) \geq 1$.

When cond(A) is equal to one, A is said to be *perfectly conditioned*. Matrices with small condition numbers are called *well-conditioned*. If cond(A) is large, then A is called *ill-conditioned*.

6. For the system of equations $(A + \epsilon F)\mathbf{x}(\epsilon) = (\mathbf{b} + \epsilon \mathbf{f})$, the solution satisfies

$$\frac{\|\mathbf{x}(\epsilon) - \mathbf{x}(0)\|}{\|\mathbf{x}(0)\|} \leq \text{cond}(A) \left(\epsilon \frac{\|F\|}{\|A\|} + \epsilon \frac{\|\mathbf{f}\|}{\|\mathbf{b}\|} \right) + O(\epsilon^2). \qquad (2.6.22)$$

7. When A is singular, the definition of condition number is modified slightly, incorporating the pseudo-inverse of A, and is defined by cond(A) $= \|A^+\| \, \|A\|$ (see page 151).

8. The size of the determinant of a square matrix A is *not* related to the condition number of A. For example, the $n \times n$ matrices below have $\kappa_\infty(B_n) = n2^{n-1}$ and $\det(B_n) = 1$; $\kappa_p(D_n) = 1$ and $\det(D_n) = 10^{-n}$.

$$B_n = \begin{bmatrix} 1 & -1 & \cdots & -1 \\ 0 & 1 & & -1 \\ \vdots & & \ddots & \\ 0 & 0 & & 1 \end{bmatrix}, \qquad D_n = \text{diag}(10^{-1}, \ldots, 10^{-1}).$$

9. Let $A = (a_{ij})$ be an $n \times n$ matrix. Using the L_2 condition number, cond $A = \max_j |\lambda_j(A)| / \min_i |\lambda_i(A)|$:

Matrix $A_{n \times n} = (a_{ij})$	Condition number
A is orthogonal	cond $(A) = 1$
$a_{ij} = \sqrt{2/(n+1)} \sin(ij\pi/(n+1))$	cond $(A) = 1$
$a_{ij} = n\delta_{ij} + 1$	cond $(A) = 2$
$a_{ij} = (i+j)/p, \, n = p-1, \, p$ a prime	cond $(A) = \sqrt{n+1}$
The circulant whose first row is $(1, 2, \ldots, n)$	cond $(A) \sim n$
$a_{ij} = \begin{cases} i/j & \text{if } i \leq j \\ j/i & \text{if } i > j \end{cases}$	cond $(A) \sim cn^{1+\epsilon}$, $0 \leq \epsilon \leq 1$
$a_{ij} = \begin{cases} -2 & \text{if } i = j \\ 1 & \text{if } |i-j| = 1 \\ 0 & \text{if } |i-j| \geq 2 \end{cases}$	cond $(A) \sim 4n^2/\pi^2$
$a_{ij} = 2\min(i,j) - 1$	cond $(A) \sim 16n^2/\pi^2$
$a_{ij} = (i+j-1)^{-1}$ (Hilbert matrix)	log cond $(A) \sim Kn$, $K \approx 3.5$

2.6.8   LINEAR SPACES AND LINEAR MAPPINGS

1. Let $R(A)$ and $N(A)$ denote, respectively, the range space and null space of an $m \times n$ matrix A. They are defined by:

$$\begin{aligned} R(A) &= \{\mathbf{y} \mid \mathbf{y} = A\mathbf{x}; \text{ for some } \mathbf{x} \in \mathbb{R}^n\}, \\ N(A) &= \{\mathbf{x} \in \mathbb{R}^n \mid A\mathbf{x} = \mathbf{0}\}. \end{aligned} \qquad (2.6.23)$$

2. The *projection matrix*, onto a subspace S, denoted P_S, is the unique matrix possessing the three properties:

   (a) $P_S = P_S^T$;
   (b) $P_S^2 = P_S$ (the projection matrix is *idempotent*);
   (c) The vector $\mathbf{b}_S$ lies in the subspace S if, and only if, $\mathbf{b}_S = P_S\mathbf{v}$ for some vector $\mathbf{v}$. In other words, $\mathbf{b}_S$ can be written as a linear combination of the columns of P_S.

3. When the $m \times n$ matrix A (with $n \leq m$) has rank n, the projection of A onto the subspaces of A is given by:

$$
\begin{aligned}
P_{R(A)} &= A(A^T A)^{-1} A^T, \\
P_{R(A^T)} &= I, \\
P_{N(A^T)} &= I - A(A^T A)^{-1} A^T.
\end{aligned}
\tag{2.6.24}
$$

When A is of rank m, the projection of A onto the subspaces of A is given by:

$$
\begin{aligned}
P_{R(A)} &= I, \\
P_{R(A^T)} &= A^T(AA^T)^{-1} A, \\
P_{N(A)} &= I - A^T(AA^T)^{-1} A.
\end{aligned}
\tag{2.6.25}
$$

4. When A is not of full rank, the matrix $A\tilde{A}$ satisfies the requirements for a projection matrix. The matrix $\tilde{A}$ is the coefficient matrix of the system of equations $\mathbf{x}_+ = \tilde{A}\mathbf{b}$, generated by the *least-squares* problem $\min \|\mathbf{b} - A\mathbf{x}\|_2^2$. Thus,

$$
\begin{aligned}
P_{R(A)} &= A\tilde{A}, \\
P_{R(A^T)} &= \tilde{A}A, \\
P_{N(A)} &= I - \tilde{A}A, \\
P_{N(A^T)} &= I - A\tilde{A}.
\end{aligned}
\tag{2.6.26}
$$

5. A matrix $B \in \mathbb{R}^{n \times n}$ is called *similar* to a matrix $A \in \mathbb{R}^{n \times n}$ if $B = T^{-1}AT$ for some non-singular matrix T.

6. If B is similar to A, then B has the same eigenvalues as A.

7. If B is similar to A and if $\mathbf{x}$ is an eigenvector of A, then $\mathbf{y} = T^{-1}\mathbf{x}$ is an eigenvector of B corresponding to the same eigenvalue.

2.6.9 TRACES

1. The *trace* of an $n \times n$ matrix A, usually denoted as $\operatorname{tr}(A)$, is defined as the sum of the n diagonal components of A.

2. The trace of an $n \times n$ matrix A equals the sum of the n eigenvalues of A, i.e.,
   $\operatorname{tr} A = a_{11} + a_{22} + \cdots + a_{nn} = \lambda_1 + \lambda_2 + \cdots + \lambda_n$.

3. The trace of a 1×1 matrix, a scalar, is itself.

4. If $A \in \mathbb{R}^{m \times k}$ and $B \in \mathbb{R}^{k \times m}$, then $\operatorname{tr}(AB) = \operatorname{tr}(BA)$.

5. If $A \in \mathbb{R}^{m \times k}$, $B \in \mathbb{R}^{k \times r}$, and $C \in \mathbb{R}^{r \times m}$, then $\operatorname{tr}(ABC) = \operatorname{tr}(BCA) = \operatorname{tr}(CAB)$. For example, if $B = \mathbf{b}$ is a column vector and $C = \mathbf{c}^{\mathrm{T}}$ is a row vector, then $\operatorname{tr}(A\mathbf{b}\mathbf{c}^{\mathrm{T}}) = \operatorname{tr}(\mathbf{b}\mathbf{c}^{\mathrm{T}}A) = \operatorname{tr}(\mathbf{c}^{\mathrm{T}}A\mathbf{b})$.

6. $\operatorname{tr}(A + \gamma B) = \operatorname{tr}(A) + \gamma \operatorname{tr}(B)$, where γ is a scalar.

7. $\operatorname{tr}(AB) = (\operatorname{Vec} A^{\mathrm{T}})^{\mathrm{T}} \operatorname{Vec} B$ (see Section 2.6.17 for the "Vec" operation)

2.6.10 GENERALIZED INVERSES

1. Every matrix A (singular or non-singular, rectangular or square) has a *generalized inverse*, or *pseudo-inverse*, A^+ defined by the *Moore–Penrose* conditions:

$$
\begin{aligned}
AA^+A &= A, \\
A^+AA^+ &= A^+, \\
(AA^+)^{\mathrm{T}} &= AA^+, \\
(A^+A)^{\mathrm{T}} &= A^+A.
\end{aligned}
\qquad (2.6.27)
$$

There is a unique pseudo-inverse satisfying the conditions in (2.6.27).

2. If, and only if, A is square and non-singular, then $A^+ = A^{-1}$.

3. For a square singular matrix A, $AA^+ \neq I$, and $A^+A \neq I$.

4. The pseudo-inverse is the unique solution to $\min_X \|AX - I_m\|_F$.

5. The least-squares problem is to find the $\mathbf{x}$ that minimizes $\|\mathbf{y} - A\mathbf{x}\|_2$. The $\mathbf{x}$ of least norm is $\mathbf{x} = A^+\mathbf{y}$.

6. If A is a rectangular $m \times n$ matrix of rank n, with $m > n$, then A^+ is of order $n \times m$ and $A^+A = I \in \mathbb{R}^{n \times n}$. In this case A^+ is called a *left inverse*, and $AA^+ \neq I$.

7. If A is a rectangular $m \times n$ matrix of rank m, with $m < n$, then A^+ is of order $n \times m$ and $AA^+ = I \in \mathbb{R}^{m \times m}$. In this case A^+ is called a *right inverse*, and $A^+A \neq I$.

8. The matrices AA^+ and A^+A are idempotent.

9. *Computing the pseudo-inverse*: The pseudo-inverse of $A_{m \times n}$ can be determined by the singular value decomposition $A = U\Sigma V^T$. If A has rank $r > 0$, then $\Sigma_{m \times n}$ has r positive singular values (σ_i) along the main diagonal extending from the upper left-hand corner and the remaining components of Σ are zero. Then $A^+ = (U\Sigma V^T)^+ = (V^T)^+\Sigma^+ U^+ = V\Sigma^+ U^T$ since $(V^T)^+ = V$ and $U^+ = U^T$ because of their orthogonality. The components σ_i^+ in Σ^+ are

$$\sigma_i^+ = \begin{cases} 1/\sigma_i, & \text{if } \sigma_i \neq 0 ; \\ 0, & \text{if } \sigma_i = 0. \end{cases} \tag{2.6.28}$$

10. The pseudo-inverse is ill-conditioned with respect to rank-changing perturbations. For example:

$$\left(\begin{bmatrix} 1 & -1 \\ 2 & -2 \end{bmatrix} + \epsilon \begin{bmatrix} 1 & 0 \\ 0 & 2 \end{bmatrix} \right)^+ = \begin{cases} \dfrac{1}{\epsilon^2} \begin{bmatrix} -1 & \frac{1}{2} \\ -1 & \frac{1}{2} \end{bmatrix} + \dfrac{1}{\epsilon} \begin{bmatrix} 1 & 0 \\ 0 & \frac{1}{2} \end{bmatrix}, & \epsilon \neq 0, \\[2ex] \dfrac{1}{10} \begin{bmatrix} 1 & 2 \\ -1 & -2 \end{bmatrix}, & \epsilon = 0. \end{cases}$$

(Note that the pseudo-inverse is the same as the inverse when $\epsilon \neq 0$.)

EXAMPLE   If $A = \begin{bmatrix} 1 & 1 & 1 & 1 \\ 1 & 1 & 1 & 1 \\ 1 & 1 & 1 & 1 \end{bmatrix}$ then $A^+ = \dfrac{1}{12} \begin{bmatrix} 1 & 1 & 1 \\ 1 & 1 & 1 \\ 1 & 1 & 1 \\ 1 & 1 & 1 \end{bmatrix}$.

2.6.11 EIGENSTRUCTURE

1. If A is a square $n \times n$ matrix, then the nth degree polynomial defined by $\det(A - \lambda I) = 0$ is called the *characteristic polynomial*, or *characteristic equation* of A.

2. The n roots (not necessarily distinct) of the characteristic polynomial are called the *eigenvalues* (or *characteristic roots*) of A. Therefore, the values, λ_i, $i = 1, \ldots, n$, are eigenvalues if, and only if, $|A - \lambda_i I| = 0$.

3. The characteristic polynomial $\det(A - \lambda I) = \sum_{i=0}^{n} r_i \lambda^i$ has the properties

$$r_n = (-1)^n,$$

$$r_{n-1} = -r_n \operatorname{tr}(A),$$

$$r_{n-2} = -\frac{1}{2} \left[r_{n-1} \operatorname{tr}(A) + r_n \operatorname{tr}(A^2) \right],$$

$$r_{n-3} = -\frac{1}{3} \left[r_{n-2} \operatorname{tr}(A) + r_{n-1} \operatorname{tr}(A^2) + r_n \operatorname{tr}(A^3) \right],$$

$$\vdots$$

$$r_0 = -\frac{1}{n} \left[\sum_{p=1}^{n} r_p \operatorname{tr}(A^p) \right].$$

4. Each eigenvalue λ has a corresponding *eigenvector* $\mathbf{x}$ (different from $\mathbf{0}$) that solves the system $A\mathbf{x} = \lambda\mathbf{x}$, or $(A - \lambda I)\mathbf{x} = \mathbf{0}$.

5. If $\mathbf{x}$ solves $A\mathbf{x} = \lambda\mathbf{x}$, then so does $\gamma\mathbf{x}$, where γ is an arbitrary scalar.

6. *Cayley–Hamilton theorem*: Any matrix A satisfies its own characteristic equation. That is $\sum_{i=0}^{n} r_i A^i = 0$.

7. The eigenvalues of a triangular (or diagonal) matrix are the diagonal components of the matrix.

8. The eigenvalues of idempotent matrices are zeros and ones.

9. If A is a real matrix with positive eigenvalues, then

$$\lambda_{\min}(AA^{\mathsf{T}}) \le [\lambda_{\min}(A)]^2 \le [\lambda_{\max}(A)]^2 \le \lambda_{\max}(AA^{\mathsf{T}}), \qquad (2.6.29)$$

where $\lambda_{\min}$ denotes the smallest and $\lambda_{\max}$ the largest eigenvalue.

10. If all the eigenvalues of a real symmetric matrix are distinct, then their associated eigenvectors are also distinct (linearly independent).

11. Real symmetric and Hermitian matrices have real eigenvalues.

12. The determinant of a matrix is equal to the product of the eigenvalues. That is, if A has the eigenvalues $\lambda_1, \lambda_2, \ldots, \lambda_n$, then $\det(A) = \lambda_1 \lambda_2 \cdots \lambda_n$.

13. The following table shows the eigenvalues of specific matrices

matrix	eigenvalues
diagonal matrix	diagonal elements
upper or lower triangular	diagonal elements
A is $n \times n$ and nilpotent	0 (n times)
A is $n \times n$ and idempotent of rank r	1 (r times); and 0 ($n - r$ times)
$(a - b)I_n + bJ_n$, where J_n is the $n \times n$ matrix of all 1's	$a + (n - 1)b$; and $a - b$ ($n - 1$ times)

14. Let A have the eigenvalues $\{\lambda_1, \lambda_2, \ldots, \lambda_n\}$. The eigenvalues of some functions of A are shown below:

matrix	eigenvalues
A^{T}	eigenvalues of A
A^{H}	complex conjugates of $\{\lambda_1^k, \lambda_2^k, \ldots, \lambda_n^k\}$
A^k, k an integer	$\{\lambda_1^k, \lambda_2^k, \ldots, \lambda_n^k\}$
A^{-k}, k an integer, A non-singular	$\{\lambda_1^{-k}, \lambda_2^{-k}, \ldots, \lambda_n^{-k}\}$
$q(A)$, q is a polynomial	$\{q(\lambda_1^k), q(\lambda_2^k), \ldots, q(\lambda_n^k)\}$
SAS^{-1}, S non-singular	eigenvalues of A
AB, where A is $m \times n$, B is $n \times m$, and $m \ge n$	eigenvalues of BA; and 0 ($m - n$ times)

2.6.12 MATRIX DIAGONALIZATION

1. If $A \in \mathbb{R}^{n \times n}$ possesses n linearly independent eigenvectors $\mathbf{x}_1, \ldots, \mathbf{x}_n$, then A can be diagonalized as $S^{-1}AS = \Lambda = \mathrm{diag}(\lambda_1, \ldots, \lambda_n)$, where the eigenvectors of A are chosen to comprise the columns of S.

2. If $A \in \mathbb{R}^{n \times n}$ can be diagonalized into $S^{-1}AS = \Lambda$, then $A^k = S\Lambda^k S^{-1}$, or $\Lambda^k = S^{-1}A^k S$.

3. *Spectral decomposition*: Any real symmetric matrix $A \in \mathbb{R}^{n \times n}$ can be diagonalized into the form $A = U\Lambda U^{\mathrm{T}}$, where Λ is the diagonal matrix of ordered eigenvalues of A such that $\lambda_1 \geq \lambda_2 \geq \cdots \geq \lambda_n$, and the columns of U are the corresponding n *orthonormal* eigenvectors of A.

   That is, if $A \in \mathbb{R}^{n \times n}$ is symmetric, then a real orthogonal matrix Q exists such that $Q^{\mathrm{T}}AQ = \mathrm{diag}(\lambda_1, \ldots, \lambda_n)$.

4. The *spectral radius* of a real symmetric matrix A, commonly denoted by $\rho(A)$, is defined as $\rho(A) = \max_{1 \leq i \leq n} |\lambda_i(A)|$.

5. If $A \in \mathbb{R}^{n \times n}$ and $B \in \mathbb{R}^{n \times n}$ are diagonalizable, then they share a common eigenvector matrix S if and only if $AB = BA$. (Not every eigenvector of A need be an eigenvector for B, e.g., the above equation is always true if $A = I$.)

6. *Schur decomposition*: If $A \in \mathbb{C}^{n \times n}$, then a unitary matrix $Q \in \mathbb{C}^{n \times n}$ exists such that $Q^{\mathrm{H}}AQ = D + N$, where $D = \mathrm{diag}(\lambda_1, \ldots, \lambda_n)$ and $N \in \mathbb{C}^{n \times n}$ is strictly upper triangular. The matrix Q can be chosen so that the eigenvalues λ_i appear in any order along the diagonal.

7. If $A \in \mathbb{R}^{n \times n}$ possesses $s \leq n$ linearly independent eigenvectors, it is similar to a matrix with s *Jordan blocks* (for some matrix M)

$$
J = M^{-1}AM = \begin{bmatrix} J_1 & & \mathbf{0} \\ & \ddots & \\ \mathbf{0} & & \ddots \\ & & J_s \end{bmatrix},
$$

where each Jordan block J_i is an upper triangular matrix with (a) the single eigenvalue λ_i repeated n_i times along the main diagonal; (b) $(n_i - 1)$ 1's appearing above the diagonal entries; and (c) all other components zero:

$$
J_i = \begin{bmatrix} \lambda_i & 1 & & \mathbf{0} \\ & \ddots & 1 & \\ & & \ddots & 1 \\ \mathbf{0} & & & \lambda_i \end{bmatrix}. \tag{2.6.30}
$$

2.6.13 MATRIX EXPONENTIALS

1. *Matrix exponentiation* is defined as (the series always converges):

$$e^{At} = I + At + \frac{(At)^2}{2!} + \frac{(At)^3}{3!} + \cdots. \tag{2.6.31}$$

2. Common properties of matrix exponentials are:

   (a) $\left(e^{As}\right)\left(e^{At}\right) = e^{A(s+t)}$,

   (b) $\left(e^{At}\right)\left(e^{-At}\right) = I$,

   (c) $\frac{d}{dt}e^{At} = Ae^{At}$,

   (d) When A and B are square matrices, the commutator of A and B is $C = [B, A] = BA - AB$. Then $e^{(A+B)} = e^A e^B e^{C/2}$ provided that $[C, A] = [C, B] = \mathbf{0}$ (i.e., each of A and B commute with their commutator). In particular, if A and B commute then $e^{A+B} = e^A e^B$.

3. For a matrix $A \in \mathbb{R}^{n \times n}$, the determinant of e^{At} is given by:

$$\det\left(e^{At}\right) = e^{\lambda_1 t} e^{\lambda_2 t} \cdots e^{\lambda_n t} = e^{\operatorname{tr}(At)}. \tag{2.6.32}$$

4. The diagonalization of e^{At}, when A is diagonalizable, is given by $e^{At} = Se^{Dt}S^{-1}$ where the columns of S consist of the eigenvectors of A, and the entries of the diagonal matrix D are the corresponding eigenvalues of A, that is, $A = SDS^{-1}$.

2.6.14 QUADRATIC FORMS

1. For a symmetric matrix A, the map $\mathbf{x} \mapsto \mathbf{x}^{\mathrm{T}} A \mathbf{x}$ is called a *pure quadratic form*. It has the form

$$\mathbf{x}^{\mathrm{T}} A \mathbf{x} = \sum_{i=1}^{n} \sum_{j=1}^{n} a_{ij} x_i x_j = a_{11} x_1^2 + a_{12} x_1 x_2 + a_{21} x_2 x_1 + \cdots + a_{nn} x_n^2. \tag{2.6.33}$$

2. For A symmetric, the gradient of $\mathbf{x}^{\mathrm{T}} A \mathbf{x} / \mathbf{x}^{\mathrm{T}} \mathbf{x}$ equals zero if, and only if, $\mathbf{x}$ is an eigenvector of A. Thus, the stationary values of this function (where the gradient vanishes) are the eigenvalues of A.

3. The ratio of two quadratic forms (B non-singular) $u(\mathbf{x}) = (\mathbf{x}^{\mathrm{T}} A \mathbf{x})/(\mathbf{x}^{\mathrm{T}} B \mathbf{x})$ attains stationary values at the eigenvalues of $B^{-1}A$. In particular,

$$u_{\max} = \lambda_{\max}(B^{-1}A), \qquad \text{and} \qquad u_{\min} = \lambda_{\min}(B^{-1}A).$$

4. A matrix A is *positive definite* if $\mathbf{x}^{\mathrm{T}} A \mathbf{x} > 0$ for all $\mathbf{x} \neq \mathbf{0}$.

5. A matrix A is *positive semi-definite* if $\mathbf{x}^T A \mathbf{x} \geq 0$ for all $\mathbf{x}$.

6. For a real, symmetric matrix $A \in \mathbb{R}^{n \times n}$, the following are necessary and sufficient conditions to establish the positive definiteness of A:

    (a) All eigenvalues of A have $\lambda_i > 0$, for $i = 1, \ldots, n$, and
    (b) The upper-left sub-matrices of A, called the *principal sub-matrices*, defined by $A_1 = \begin{bmatrix} a_{11} \end{bmatrix}$,

$$A_2 = \begin{bmatrix} a_{11} & a_{12} \\ a_{21} & a_{22} \end{bmatrix}, \qquad \ldots, \qquad A_n = \begin{bmatrix} a_{11} & a_{12} & \cdots & a_{1n} \\ a_{21} & a_{22} & \cdots & a_{2n} \\ \vdots & \vdots & & \vdots \\ a_{n1} & a_{n2} & \cdots & a_{nn} \end{bmatrix},$$

    have $\det(A_k) > 0$, for all $k = 1, \ldots, n$.

7. If A is positive definite, then all of the principal sub-matrices of A are also positive definite. Additionally, all diagonal entries of A are positive.

8. For a real, symmetric matrix $A \in \mathbb{R}^{n \times n}$, the following are necessary and sufficient conditions to establish the positive semi-definiteness of A:

    (a) All eigenvalues of A have $\lambda_i \geq 0$, for $i = 1, \ldots, n$,
    (b) The principal sub-matrices of A have $\det A_k \geq 0$, for all $k = 1, \ldots, n$.

9. If A is positive semi-definite, then all of the principal sub-matrices of A are also positive semi-definite. Additionally, all diagonal entries of A are non-negative.

10. If the matrix Q is positive definite, then $(\mathbf{x}^T - \mathbf{x}_0^T) Q^{-1} (\mathbf{x} - \mathbf{x}_0) = 1$ is the equation of an ellipsoid with its center at $\mathbf{x}_0^T$. The lengths of the semiaxes are equal to the square roots of the eigenvalues of Q; see page 330.

2.6.15 MATRIX FACTORIZATIONS

1. Singular value decomposition (SVD): Any $m \times n$ matrix A can be written as the product $A = U \Sigma V^T$, where U is an $m \times m$ orthogonal matrix, V is an $n \times n$ orthogonal matrix, and $\Sigma = \mathrm{diag}(\sigma_1, \sigma_2, \ldots, \sigma_p)$, with $p = \min(m, n)$ and $\sigma_1 \geq \sigma_2 \geq \cdots \geq \sigma_p \geq 0$. The values σ_i, $i = 1, \ldots, p$, are called the *singular values* of A.

    (a) When $\mathrm{rank}(A) = r > 0$, A has exactly r positive singular values, and $\sigma_{r+1} = \cdots = \sigma_p = 0$.
    (b) When A is a symmetric $n \times n$ matrix, then $\sigma_1 = |\lambda_1|, \ \ldots, \ \sigma_n = |\lambda_n|$, where $\lambda_1, \lambda_2, \ldots, \lambda_n$ are the eigenvalues of A.

(c) When A is an $m \times n$ matrix, if $m \geq n$ then the singular values of A are the square roots of the eigenvalues of $A^T A$. Otherwise, they are the square roots of the eigenvalues of AA^T.

2. Any non-singular $m \times n$ matrix A of maximal rank can be factored as $PA = LU$, where P is a permutation matrix, L is lower triangular, and U is upper triangular.

3. QR factorization: If all the columns of $A \in \mathbb{R}^{m \times n}$ are linearly independent, then A can be factored as $A = QR$, where $Q \in \mathbb{R}^{m \times n}$ has orthonormal columns and $R \in \mathbb{R}^{n \times n}$ is upper triangular and non-singular.

4. If $A \in \mathbb{R}^{n \times n}$ is symmetric positive definite, then

$$A = LDL^T = LD^{1/2}D^{1/2}L^T = \left(LD^{1/2}\right)\left(LD^{1/2}\right)^T = GG^T \quad (2.6.34)$$

where L is a lower triangular matrix and D is a diagonal matrix. The factorization $A = GG^T$ is called the *Cholesky factorization*, and the matrix G is commonly referred to as the *Cholesky triangle*.

2.6.16 THEOREMS

1. *Frobenius–Perron theorem:* If $A > 0$ (i.e., A is positive definite), then there exists a $\lambda_0 > 0$ and $\mathbf{x}_0 > \mathbf{0}$ such that

   (a) $A\mathbf{x}_0 = \lambda_0 \mathbf{x}_0$,
   (b) if λ is any other eigenvalue of A, $\lambda \neq \lambda_0$, then $|\lambda| < \lambda_0$, and
   (c) λ_0 is an eigenvalue with geometric and algebraic multiplicity equal to one.

2. If $A \geq 0$, and $A^k > 0$ for some positive integer k, then the results of the Frobenius–Perron theorem apply to A.

3. *Courant–Fischer minimax theorem:* If $\lambda_i(A)$ denotes the ith largest eigenvalue of a matrix $A = A^T \in \mathbb{R}^{n \times n}$, then

$$\lambda_j(A) = \max_{S_j} \min_{\mathbf{0} \neq \mathbf{x} \in S_j} \frac{\mathbf{x}^T A\mathbf{x}}{\mathbf{x}^T \mathbf{x}} \qquad j = 1, \ldots, n \qquad (2.6.35)$$

where $\mathbf{x} \in \mathbb{R}^n$ and S_j is a j-dimensional subspace.

From this follows *Raleigh's principle*: The quotient $R(\mathbf{x}) = \mathbf{x}^T A\mathbf{x}/\mathbf{x}^T \mathbf{x}$ is minimized by the eigenvector $\mathbf{x} = \mathbf{x}_1$ corresponding to the smallest eigenvalue λ_1 of A. The minimum of $R(\mathbf{x})$ is λ_1, that is,

$$\min R(\mathbf{x}) = \min \frac{\mathbf{x}^T A\mathbf{x}}{\mathbf{x}^T \mathbf{x}} = R(\mathbf{x}_1) = \frac{\mathbf{x}_1^T A\mathbf{x}_1}{\mathbf{x}_1^T \mathbf{x}_1} = \frac{\mathbf{x}_1^T \lambda_1 \mathbf{x}_1}{\mathbf{x}_1^T \mathbf{x}_1} = \lambda_1. \qquad (2.6.36)$$

4. *Cramer's rule:* The jth component of $\mathbf{x} = A^{-1}\mathbf{b}$ is given by

$$x_j = \frac{\det B_j}{\det A}, \qquad \text{where}$$

$$B_j = \begin{bmatrix} a_{11} & a_{12} & \cdots & a_{1,j-1} & b_1 & a_{1,j+1} & \cdots & a_{1n} \\ \vdots & \vdots & & \vdots & \vdots & \vdots & & \vdots \\ a_{n1} & a_{n2} & \cdots & a_{n,j-1} & b_n & a_{n,j+1} & \cdots & a_{nn} \end{bmatrix} \qquad (2.6.37)$$

The vector $\mathbf{b} = \begin{bmatrix} b_1 & \cdots & b_n \end{bmatrix}^{\mathrm{T}}$ replaces the jth column of the matrix A to form the matrix B_j.

5. *Sylvester's law of inertia:* For a symmetric matrix $A \in \mathbb{R}^{n \times n}$, the matrices A and $C^{\mathrm{T}} A C$, for C non-singular, have the same number of positive, negative, and zero eigenvalues.

6. *Gerschgorin circle theorem:* Each eigenvalue of an arbitrary $n \times n$ matrix $A = (a_{ij})$ lies in at least one of the circles $\{C_1, C_2, \ldots, C_n\}$ in the complex plane, where circle C_i has center a_{ii} and radius ρ_i given by $\rho_i = \sum_{\substack{j=1 \\ j \neq i}}^{n} |a_{ij}|$.

2.6.17 THE VECTOR OPERATION

The matrix $A_{m \times n}$ can be represented as a collection of $m \times 1$ column vectors: $A = \begin{bmatrix} \mathbf{a}_1 & \mathbf{a}_2 & \cdots & \mathbf{a}_n \end{bmatrix}$. Define Vec A as the matrix of size $nm \times 1$ (i.e., a vector) by

$$\text{Vec } A = \begin{bmatrix} \mathbf{a}_1 \\ \mathbf{a}_2 \\ \vdots \\ \mathbf{a}_n \end{bmatrix}. \qquad (2.6.38)$$

This operator has the following properties:

1. $\operatorname{tr} AB = \left(\text{Vec } A^{\mathrm{T}} \right)^{\mathrm{T}} \text{Vec } B$.

2. The permutation matrix U that associates Vec X and Vec X^{T} (that is, Vec $X^{\mathrm{T}} = U$ Vec X) is given by:

$$U = \begin{bmatrix} \text{Vec } E_{11}^{\mathrm{T}} & \text{Vec } E_{21}^{\mathrm{T}} & \cdots & \text{Vec } E_{n1}^{\mathrm{T}} \end{bmatrix} = \sum_{r,s} E_{rs} \otimes E_{rs}^{\mathrm{T}}. \qquad (2.6.39)$$

3. $\text{Vec}(AYB) = (B^{\mathrm{T}} \otimes A) \text{Vec } Y$.

4. If A and B are both of size $n \times n$, then

   (a) Vec $AB = (I_n \otimes A) \text{Vec } B$.
   (b) Vec $AB = (B^{\mathrm{T}} \otimes A) \text{Vec } I_n$.

2.6.18   KRONECKER PRODUCTS

If the matrix $A = (a_{ij})$ has size $m \times n$, and the matrix $B = (b_{ij})$ has size $r \times s$, then the *Kronecker product* (sometimes called the *tensor product*) of these matrices, denoted $A \otimes B$, is defined as the partitioned matrix

$$A \otimes B = \begin{bmatrix} a_{11}B & a_{12}B & \cdots & a_{1n}B \\ a_{21}B & a_{22}B & \cdots & a_{2n}B \\ \vdots & \vdots & & \vdots \\ a_{m1}B & a_{m2}B & \cdots & a_{mn}B \end{bmatrix}. \tag{2.6.40}$$

Hence, the matrix $A \otimes B$ has size $mr \times ns$.

EXAMPLE   If $A = \begin{bmatrix} a_{11} & a_{12} \\ a_{21} & a_{22} \end{bmatrix}$ and $B = \begin{bmatrix} b_{11} & b_{12} \\ b_{21} & b_{22} \end{bmatrix}$, then

$$A \otimes B = \begin{bmatrix} a_{11}B & a_{12}B \\ a_{21}B & a_{22}B \end{bmatrix} = \begin{bmatrix} a_{11}b_{11} & a_{11}b_{12} & a_{12}b_{11} & a_{12}b_{12} \\ a_{11}b_{21} & a_{11}b_{22} & a_{12}b_{21} & a_{12}b_{22} \\ a_{21}b_{11} & a_{21}b_{12} & a_{22}b_{11} & a_{22}b_{12} \\ a_{21}b_{21} & a_{21}b_{22} & a_{22}b_{21} & a_{22}b_{22} \end{bmatrix}. \tag{2.6.41}$$

The Kronecker product has the following properties:

1. If $\mathbf{z}$ and $\mathbf{w}$ are vectors of appropriate dimensions, then
   $A\mathbf{z} \otimes B\mathbf{w} = (A \otimes B)(\mathbf{z} \otimes \mathbf{w})$.

2. If α is a scalar, then $(\alpha A) \otimes B = A \otimes (\alpha B) = \alpha(A \otimes B)$.

3. The Kronecker product is distributive with respect to addition:

   (a) $(A + B) \otimes C = A \otimes C + B \otimes C$, and
   (b) $A \otimes (B + C) = A \otimes B + A \otimes C$.

4. The Kronecker product is associative: $A \otimes (B \otimes C) = (A \otimes B) \otimes C$.

5. $(A \otimes B)^{\mathrm{T}} = A^{\mathrm{T}} \otimes B^{\mathrm{T}}$.

6. The *mixed product rule*: If the dimensions of the matrices are such that the following expressions exist, then $(A \otimes B)(C \otimes D) = AC \otimes BD$.

7. If the inverses exist, then $(A \otimes B)^{-1} = A^{-1} \otimes B^{-1}$.

8. If $\{\lambda_i\}$ and $\{\mathbf{x}_i\}$ are the eigenvalues and the corresponding eigenvectors for A, and $\{\mu_j\}$ and $\{\mathbf{y}_j\}$ are the eigenvalues and the corresponding eigenvectors for B, then $A \otimes B$ has eigenvalues $\{\lambda_i \mu_j\}$ with corresponding eigenvectors $\{\mathbf{x}_i \otimes \mathbf{y}_j\}$.

9. If matrix A has size $n \times n$ and B has size $m \times m$, then $\det(A \otimes B) = (\det A)^m (\det B)^n$.

10. If f is an analytic matrix function and A has size $n \times n$, then

(a) $f(I_n \otimes A) = I_n \otimes f(A)$, and

(b) $f(A \otimes I_n) = f(A) \otimes I_n$.

11. $\text{tr}(A \otimes B) = (\text{tr } A)(\text{tr } B)$.

12. If A, B, C, and D are matrices with A similar to C and B similar to D, then $A \otimes B$ is similar to $C \otimes D$.

13. If $C(t) = A(t) \otimes B(t)$, then $\frac{dC}{dt} = \frac{dA}{dt} \otimes B + A \otimes \frac{dB}{dt}$.

2.6.19 KRONECKER SUMS

If the matrix $A = (a_{ij})$ has size $n \times n$ and matrix $B = (b_{ij})$ has size $m \times m$, then the *Kronecker sum* of these matrices, denoted $A \oplus B$, is defined[2] as

$$A \oplus B = A \otimes I_m + I_n \otimes B. \tag{2.6.42}$$

The Kronecker sum has the following properties:

1. If A has eigenvalues $\{\lambda_i\}$ and B has eigenvalues $\{\mu_j\}$, then $A \oplus B$ has eigenvalues $\{\lambda_i + \mu_j\}$.

2. The matrix equation $AX + XB = C$ may be equivalently written as $(B^{\text{T}} \oplus A) \text{Vec } X = \text{Vec } C$, where Vec is defined in Section 2.6.17.

3. $e^{A \oplus B} = e^A \otimes e^B$.

2.7 ABSTRACT ALGEBRA

2.7.1 DEFINITIONS

1. A *binary operation on a set S* is a function $\star : S \times S \to S$.

2. An *algebraic structure* $(S, \star_1, \ldots, \star_n)$ consists of a non-empty set S with one or more binary operations $\star_i$ defined on S. If the operations are understood, then the binary operations need not be mentioned explicitly.

3. The *order* of an algebraic structure S is the number of elements in S, written $|S|$.

---

[2] Note that $A \oplus B$ is also used to denote the $(m + n) \times (m + n)$ matrix $\begin{bmatrix} A & 0 \\ 0 & B \end{bmatrix}$.

4. A binary operation $\star$ on an algebraic structure $(S, \star)$ may have the following properties:

   (a) *Associative*: $a \star (b \star c) = (a \star b) \star c$ for all $a, b, c \in S$.
   (b) *Identity*: there exists an element $e \in S$ (*identity element* of S) such that $e \star a = a \star e = a$ for all $a \in S$.
   (c) *Inverse*: $a^{-1} \in S$ is an *inverse* of a if $a \star a^{-1} = a^{-1} \star a = e$.
   (d) *Commutative* (or *abelian*): if $a \star b = b \star a$ for all $a, b \in S$.

5. A *semigroup* $(S, \star)$ consists of a non-empty set S and an associative binary operation $\star$ on S.

6. A *monoid* $(S, \star)$ consists of a non-empty set S with an identity element and an associative binary operation $\star$.

2.7.1.1 Examples of semigroups and monoids

1. The sets $\mathbb{N} = \{0, 1, 2, 3, \ldots\}$ (natural numbers), $\mathbb{Z} = \{0, \pm 1, \pm 2, \ldots\}$ (integers), $\mathbb{Q}$ (rational numbers), $\mathbb{R}$ (real numbers), and $\mathbb{C}$ (complex numbers) where $\star$ is either addition or multiplication are semigroups and monoids.

2. The set of positive integers under addition is a semigroup but not a monoid.

3. If A is any non-empty set, then the set of all functions $f : A \to A$ where $\star$ is the composition of functions is a semigroup and a monoid.

4. Given a set S, the set of all strings of elements of S, where $\star$ is concatenation of strings, is a monoid (the identity is λ, the empty string).

2.7.2 GROUPS

1. A *group* $(G, \star)$ consists of a set G with a binary operation $\star$ defined on G such that $\star$ satisfies the associative, identity, and inverse laws. *Note:* The operation $\star$ is often written as $+$ (an *additive group*) or as $\cdot$ or $\times$ (a *multiplicative group*).

   (a) If $+$ is used, the identity is written 0 and the inverse of a is written $-a$. Usually, in this case, the group is commutative. The following notation is then used: $na = \underbrace{a + \ldots + a}_{n \text{ times}}$.

   (b) If multiplicative notation is used, $a \star b$ is often written ab, the identity is often written 1, and the inverse of a is written a^{-1}.

2. The *order* of $a \in G$ is the smallest positive integer n such that $a^n = 1$ where $a^n = a \cdot a \cdots a$ (n times) (or $a + a + \cdots + a = 0$ if G is written additively). If there is no such integer, the element has *infinite order*. In a finite group of order n, each element has some order k (depending on the particular element) and it must be that k divides n.

3. $(H, \star)$ is a subgroup of $(G, \star)$ if $H \subseteq G$ and $(H, \star)$ is a group (using the same binary operation as in $(G, \star)$).

4. The *cyclic subgroup* $\langle a \rangle$ generated by $a \in G$ is the subgroup $\{a^n \mid n \in \mathbb{Z}\} = \{\ldots, a^{-2} = (a^{-1})^2, a^{-1}, a^0 = e, a, a^2, \ldots\}$. The element a is a *generator* of $\langle a \rangle$. A group G is *cyclic* if there is a $a \in G$ such that $G = \langle a \rangle$.

5. If H is a subgroup of a group G, then a *left* [*right*] *coset* of H in G is the set $aH = \{ah \mid h \in H\}$ [$Ha = \{ha \mid h \in H\}$].

6. A *normal subgroup* of a group G is a subgroup H such that $aH = Ha$ for all $a \in G$.

7. A *simple group* is a group $G \neq \{e\}$ with only G and $\{e\}$ as normal subgroups.

8. If H is a normal subgroup of G, then the *quotient group* (or *factor group*) *of G modulo H* is the group $G/H = \{aH \mid a \in G\}$, with binary operation $aH \cdot bH = (ab)H$.

9. A finite group G is *solvable* if there is a sequence of subgroups $G_1, G_2, \ldots, G_k$, with $G_1 = G$ and $G_k = \{e\}$, such that each G_{i+1} is a normal subgroup of G_i and G_i/G_{i+1} is abelian.

2.7.2.1   Facts about groups

1. The identity element is unique.

2. Each element has exactly one inverse.

3. Each of the equations $a \star x = b$ and $x \star a = b$ has exactly one solution, $x = a^{-1} \star b$ and $x = b \star a^{-1}$.

4. $(a^{-1})^{-1} = a$.

5. $(a \star b)^{-1} = b^{-1} \star a^{-1}$.

6. The *left* (respectively *right*) *cancellation law* holds in all groups: If $a \star b = a \star c$ then $b = c$ (respectively, if $b \star a = c \star a$ then $b = c$).

7. *Lagrange's theorem*: If G is a finite group and H is a subgroup of G, then the order of H divides the order of G.

8. Every group of prime order is abelian and hence simple.

9. Every cyclic group is abelian.

10. Every abelian group is solvable.

11. *Feit–Thompson theorem*: All groups of odd order are solvable. Hence, all finite non-Abelian simple groups have even order.

12. Finite simple groups are of the following types:

    (a) $\mathbb{Z}_p$ (p prime)
    (b) A group of Lie type
    (c) A_n ($n \geq 5$)
    (d) Sporadic groups (see table on page 191)

2.7.2.2 Examples of groups

1. $\mathbb{Z}$, $\mathbb{Q}$, $\mathbb{R}$, and $\mathbb{C}$, with $\star$ the addition of numbers, are additive groups.

2. For n a positive integer, $n\mathbb{Z} = \{nz \mid z \in \mathbb{Z}\}$ is an additive group.

3. $\mathbb{Q} - \{0\} = \mathbb{Q}^*$, $\mathbb{R} - \{0\} = \mathbb{R}^*$, $\mathbb{C} - \{0\} = \mathbb{C}^*$, with $\star$ the multiplication of numbers, are multiplicative groups.

4. $\mathbb{Z}_n = \mathbb{Z}/n\mathbb{Z} = \{0, 1, 2, \ldots, n - 1\}$ is a group where $\star$ is addition modulo n.

5. $\mathbb{Z}_n^* = \{k \mid k \in \mathbb{Z}_n, k$ has a multiplicative inverse (under multiplication modulo n) in $\mathbb{Z}_n\}$ is a group under multiplication modulo n. If p is prime, $\mathbb{Z}_p^*$ is cyclic. If p is prime and $a \in \mathbb{Z}_p^*$ has order (index) $p - 1$, then a is a *primitive root modulo* p. See the tables on pages 192 and 193 for power residues and primitive roots.

6. If $(G_1, \star_1), (G_2, \star_2), \ldots, (G_n, \star_n)$ are groups, the (*direct*) *product group* is $(G_1 \times G_2 \times \cdots \times G_n, \star) = \{(a_1, a_2, \ldots, a_n) \mid a_i \in G_i, i = 1, 2, \ldots, n\}$ where $\star$ is defined by

$$(a_1, a_2, \ldots, a_n) \star (b_1, b_2, \ldots, b_n) = (a_1 \star_1 b_1, a_2 \star_2 b_2, \ldots, a_n \star_n b_n).$$

7. All $m \times n$ matrices with real entries form a group under addition of matrices.

8. All $n \times n$ matrices with real entries and non-zero determinants form a group under matrix multiplication.

9. All 1–1, onto functions $f : S \to S$ (*permutations* of S), where S is any non-empty set, form a group under composition of functions. See Section 2.7.9.

   In particular, if $S = \{1, 2, 3, \ldots, n\}$, the group of permutations of S is called the *symmetric group*, S_n. In S_n, each permutation can be written as a product of cycles. A *cycle* is a permutation $\sigma = (i_1\ i_2\ \cdots\ i_k)$, where $\sigma(i_1) = i_2, \sigma(i_2) = i_3, \ldots, \sigma(i_k) = i_1$. Each cycle of length greater than 1 can be written as a product of transpositions (cycles of length 2). A permutation is *even* (*odd*) if it can be written as the product of an even (odd) number of transpositions. (Every permutation is either even or odd.) The set of all even permutations in S_n is a normal subgroup, A_n, of S_n. The group A_n is called the *alternating group* on n elements.

10. Given a regular polygon, the *dihedral group* D_n is the group of all symmetries of the polygon, that is, the group composed of the set of all rotations around the center of the polygon through angles of $360k/n$ degrees (where $k = 0, 1, 2, \ldots, n - 1$), together with all reflections in lines passing through a vertex and the center of the polygongon, using composition of functions. Alternately, $D_n = \{a^i b^j \mid i = 0, 1; j = 0, 1, \ldots, n - 1; aba^{-1} = b^{-1}\}$.

2.7.3 RINGS

2.7.3.1 Definitions

1. A *ring* $(R, +, \cdot)$ consists of a non-empty set R and two binary operations, $+$ and $\cdot$, such that $(R, +)$ is an abelian group, the operation $\cdot$ is associative, and the *left distributive law* $a(b + c) = (ab) + (ac)$ and the *right distributive law* $(a + b)c = (ac) + (bc)$ hold for all $a, b, c \in R$.

2. A subset S of a ring R is a *subring* of R if S is a ring using the same operations used in R with the same unit.

3. A ring R is a *commutative* ring if the multiplication operation is commutative: $ab = ba$ for all $a, b \in R$.

4. A ring R (with $R \neq \{0\}$) is a *ring with unity* if there is an element 1 (called *unity*) such that $a1 = 1a = a$ for all $a \in R$.

5. A *unit* in a ring with unity is an element a with a multiplicative inverse a^{-1} (that is, $aa^{-1} = a^{-1}a = 1$).

6. If $a \neq 0$, $b \neq 0$, and $ab = 0$, then a is a *left divisor of zero* and b is a *right divisor of zero*.

7. A subset I of a ring $(R, +, \cdot)$ is a (two-sided) *ideal* of R if $(I, +)$ is a subgroup of $(R, +)$ and I is closed under left and right multiplication by elements of R (if $x \in I$ and $r \in R$, then $rx \in I$ and $xr \in I$).

8. An ideal $I \subseteq R$ is

   (a) *Proper*: if $I \neq \{0\}$ and $I \neq R$
   (b) *Maximal*: if I is proper and if there is no proper ideal properly containing I
   (c) *Prime*: if $ab \in I$ implies that a or $b \in I$
   (d) *Principal*: if there is $a \in R$ such that I is the intersection of all ideals containing a.

9. If I is an ideal in a ring R, then a *coset* is a set $r + I = \{r + a \mid a \in I\}$.

10. If I is an ideal in a ring R, then the *quotient ring* is the ring $R/I = \{r + I \mid r \in R\}$, where $(r + I) + (s + I) = (r + s) + I$ and $(r + I)(s + I) = (rs) + I$.

11. An *integral domain* $(R, +, \cdot)$ is a commutative ring with unity such that cancellations hold: if $ab = ac$ then $b = c$ (respectively, if $ba = ca$ then $b = c$) for all $a, b, c \in R$, where $a \neq 0$. (Equivalently, an integral domain is a commutative ring with unity that has no divisors of zero.)

12. If R is an integral domain, then a non-zero element $r \in R$ that is not a unit is *irreducible* if $r = ab$ implies that either a or b is a unit.

13. If R is an integral domain, a non-zero element $r \in R$ that is not a unit is a *prime* if, whenever $r|ab$, then either $r|a$ or $r|b$ ($x|y$ means that there is an element $z \in R$ such that $y = zx$.).

14. A *unique factorization domain* (UFD) is an integral domain such that every non-zero element that is not a unit can be written uniquely as the product of irreducible elements (except for factors that are units and except for the order in which the factor appears).

15. A *principal ideal domain* (PID) is an integral domain in which every ideal is a principal ideal.

16. A *division ring* is a ring in which every non-zero element has a multiplicative inverse (that is, every non-zero element is a unit). (Equivalently, a division ring is a ring in which the non-zero elements form a multiplicative group.) A non-commutative division ring is called a *skew field*.

2.7.3.2 Facts about rings

1. The set of all units of a ring is a group under the multiplication defined on the ring.

2. Every principal ideal domain is a unique factorization domain.

3. If R is a commutative ring with unity, then every maximal ideal is a prime ideal.

4. If R is a commutative ring with unity, then R is a field if and only if the only ideals of R are R and $\{0\}$.

5. If R is a commutative ring with unity and $I \neq R$ is an ideal, then R/I is an integral domain if and only if I is a prime ideal.

6. If R is a commutative ring with unity, then I is a maximal ideal if and only if R/I is a field.

7. If $f(x) \in F[x]$ (where F is a field) and the ideal generated by $f(x)$ is not $\{0\}$, then the ideal is maximal if and only if $f(x)$ is irreducible over F.

8. There are exactly four normed division rings; they have dimensions 1, 2, 4, and 8. They are the real numbers, the complex numbers, the quaternions, and the octonions. The quaternions are non-commutative and the octonions are non-associative.

2.7.3.3 Examples of rings

1. $\mathbb{Z}$ (integers), $\mathbb{Q}$ (rational numbers), $\mathbb{R}$ (real numbers), and $\mathbb{C}$ (complex numbers) are rings, with ordinary addition and multiplication of numbers.

2. $\mathbb{Z}_n$ is a ring, with addition and multiplication modulo n.

3. If $\sqrt{n}$ is not an integer, then $\mathbb{Z}[\sqrt{n}] = \{a + b\sqrt{n} \mid a, b \in \mathbb{Z}\}$, where $(a + b\sqrt{n}) + (c + d\sqrt{n}) = (a + c) + (b + d)\sqrt{n}$ and $(a + b\sqrt{n})(c + d\sqrt{n}) = (ac + nbd) + (ad + bc)\sqrt{n}$ is a ring.

4. The set of *Gaussian integers* $\mathbb{Z}[i] = \{a + bi \mid a, b \in \mathbb{Z}\}$ is a ring, with the usual definitions of addition and multiplication of complex numbers.

5. The *polynomial ring* in one variable over a ring R is the ring $R[x] = \{a_n x^n + \cdots + a_1 x + a_0 \mid a_i \in R;\ i = 0, 1, \ldots, n;\ n \in \mathcal{N}\}$. (Elements of $R[x]$ are added and multiplied using the usual rules for addition and multiplication of polynomials.) The *degree* of a polynomial $a_n x^n + \cdots + a_1 x + a_0$ with $a_n \neq 0$ is n. A polynomial is *monic* if $a_n = 1$. A polynomial $f(x)$ is *irreducible over R* if $f(x)$ cannot be factored as a product of polynomials in $R[x]$ of degree less than the degree of $f(x)$. A monic irreducible polynomial $f(x)$ of degree k in $\mathbb{Z}_p[x]$ (p prime) is *primitive* if the order of x in $\mathbb{Z}_p[x]/(f(x))$ is $p^k - 1$, where $(f(x)) = \{f(x)g(x) \mid g(x) \in \mathbb{Z}_p[x]\}$ (the ideal generated by $f(x)$).

   For example, the polynomial $x^2 + 1$ is

   (a) Irreducible in $\mathbb{R}[x]$ because $x^2 + 1$ has no real root
   (b) Reducible in $\mathbb{C}[x]$ because $x^2 + 1 = (x - i)(x + i)$
   (c) Reducible in $\mathbb{Z}_2[x]$ because $x^2 + 1 = (x + 1)^2$
   (d) Reducible in $\mathbb{Z}_5[x]$ because $x^2 + 1 = (x + 2)(x + 3)$

   See the table on page 194.

6. The *division ring of quaternions* is the ring $(\{a + bi + cj + dk \mid a, b, c, d \in \mathbb{R}\}, +, \cdot)$, where operations are carried out using the rules for polynomial addition and multiplication and the defining relations for the quaternion group Q (see page 182).

7. Every octonion is a real linear combination of the unit octonions $\{1, e_1, e_2, e_3, e_4, e_5, e_6, e_7\}$. Their properties include: (a) $e_i^2 = -1$; (b) $e_i e_j = -e_j e_i$ when $i \neq j$; (c) the index doubling identity: $e_i e_j = e_k \implies e_{2i} e_{2j} = e_{2k}$; and (d) the index cycling identity: $e_i e_j = e_k \implies e_{i+1} e_{j+1} = e_{k+1}$ where the indices are computed modulo 7. The full multiplication table is as follows:

	1	e_1	e_2	e_3	e_4	e_5	e_6	e_7
1	1	e_1	e_2	e_3	e_4	e_5	e_6	e_7
e_1	e_1	-1	e_4	e_7	$-e_2$	e_6	$-e_5$	$-e_3$
e_2	e_2	$-e_4$	-1	e_5	e_1	$-e_3$	e_7	$-e_6$
e_3	e_3	$-e_7$	$-e_5$	-1	e_6	e_2	$-e_4$	e_1
e_4	e_4	e_2	$-e_1$	$-e_6$	-1	e_7	e_3	$-e_5$
e_5	e_5	$-e_6$	e_3	$-e_2$	$-e_7$	-1	e_1	e_4
e_6	e_6	e_5	$-e_7$	e_4	$-e_3$	$-e_1$	-1	e_2
e_7	e_7	e_3	e_6	$-e_1$	e_5	$-e_4$	$-e_2$	-1

8. The following table gives examples of rings with additional properties:

ring	commuta-tive ring with unity	integral domain	principal ideal domain	Euclidean domain	division ring	field
$\mathbb{Q}, \mathbb{R}, \mathbb{C}$	yes	yes	yes	yes	yes	yes
$\mathbb{Z}$	yes	yes	yes	yes	no	no
$\mathbb{Z}_p$ (p prime)	yes	yes	yes	yes	yes	yes
$\mathbb{Z}_n$ (n composite)	yes	no	no	no	no	no
$\mathbb{Z}[x]$	yes	yes	no	no	no	no
$\mathcal{M}_{n \times n}$	no	no	no	no	no	no

2.7.4 FIELDS

2.7.4.1 Definitions

1. A *field* $(F, +, \cdot)$ is a commutative ring with unity such that each non-zero element of F has a multiplicative inverse (equivalently, a field is a commutative division ring).

2. The *characteristic* of a field (or a ring) is the smallest positive integer n such that $1 + 1 + \cdots + 1 = 0$ (n summands). If no such n exists, the field has characteristic 0 (or characteristic ∞).

3. Field K is an *extension field* of the field F if F is a subfield of K (i.e., $F \subseteq K$, and F is a field using the same operations used in K).

2.7.4.2 Examples of fields

1. $\mathbb{Q}, \mathbb{R}$, and $\mathbb{C}$ with ordinary addition and multiplication are fields.

2. $\mathbb{Z}_p$ (p a prime) is a field under addition and multiplication modulo p.

3. $F[x]/(f(x))$ is a field, provided that F is a field and $f(x)$ is a non-constant polynomial irreducible in $F[x]$.

2.7.5 QUADRATIC FIELDS

2.7.5.1 Definitions

1. A complex number is an *algebraic integer* if it is a root of a polynomial with integer coefficients that has a leading coefficient of 1.

2. If d is a square-free integer, then $\mathbb{Q}(\sqrt{d}) = \{a + b\sqrt{d}\}$, where a and b are rational numbers, is called a *quadratic field*. If $d > 0$ then $\mathbb{Q}(\sqrt{d})$ is a *real quadratic field*; if $d < 0$ then $\mathbb{Q}(\sqrt{d})$ is an *imaginary quadratic field*.

3. The *integers* of an algebraic number field are the algebraic integers that belong to this number field.

4. If $\{\alpha, \beta, \gamma\}$ are integers in $\mathbb{Q}(\sqrt{d})$ such that $\alpha\gamma = \beta$ then we say that α *divides* β; written $\alpha|\beta$.

5. An integer ϵ in $\mathbb{Q}(\sqrt{d})$ is a *unit* if it divides 1.

6. If $\alpha = a + b\sqrt{d}$ then

    (a) the *conjugate* of α is $\overline{\alpha} = a - b\sqrt{d}$.
    (b) the *norm* of α is $N(\alpha) = \alpha\overline{\alpha} = a^2 - db^2$.

7. If α is an integer of $\mathbb{Q}(\sqrt{d})$ and if ϵ is a unit of $\mathbb{Q}(\sqrt{d})$, then the number $\epsilon\alpha$ is an *associate* of α. A *prime* in $\mathbb{Q}(\sqrt{d})$ is an integer of $\mathbb{Q}(\sqrt{d})$ that is only divisible by the units and its associates.

8. A quadratic field $\mathbb{Q}(\sqrt{d})$ is a *Euclidean field* if, given integers α and β in $\mathbb{Q}(\sqrt{d})$ with $\beta \neq 0$, there are integers γ and δ in $\mathbb{Q}(\sqrt{d})$ such that $\alpha = \gamma\beta + \delta$ and $|N(\delta)| < |N(\beta)|$.

9. A quadratic field $\mathbb{Q}(\sqrt{d})$ has the *unique factorization property* if, whenever α is a non-zero, non-unit, integer in $\mathbb{Q}(\sqrt{d})$ with $\alpha = \epsilon\pi_1\pi_2\cdots\pi_r = \epsilon'\pi_1'\pi_2'\cdots\pi_s'$ where ϵ and ϵ' are units, then $r = s$ and the primes π_i and π_j' can be paired off into pairs of associates.

2.7.5.2 Facts about quadratic fields

1. The integers of $\mathbb{Q}(\sqrt{d})$ are of the form

    (a) $a + b\sqrt{d}$, with a and b integers, if $d \equiv 2 \pmod 4$ or $d \equiv 3 \pmod 4$.

    (b) $a + b\left(\frac{\sqrt{d}-1}{2}\right)$, with a and b integers, if $d \equiv 1 \pmod 4$.

2. Norms are positive in imaginary quadratic fields, but not necessarily positive in real quadratic fields. It is always true that $N(\alpha\beta) = N(\alpha)N(\beta)$.

3. If α is an integer in $\mathbb{Q}(\sqrt{d})$ and $N(\alpha)$ is an integer that is prime, then α is prime.

4. The number of units in $\mathbb{Q}(\sqrt{d})$ is as follows:

    (a) If $d = -3$, there are 6 units: ± 1, $\pm\frac{-1+\sqrt{-3}}{2}$, and $\pm\frac{-1-\sqrt{-3}}{2}$.

    (b) If $d = -1$, there are 4 units: ± 1 and $\pm i$.

    (c) If $d < 0$ and $d \neq -1$ and $d \neq -3$ there are 2 units: ± 1.

    (d) If $d > 0$ there are infinitely many units. There is a fundamental unit, ϵ_0, such that all other units have the form $\pm\epsilon_0^n$ where n is an integer.

5. The quadratic field $\mathbb{Q}(\sqrt{d})$ is Euclidean if and only if d is one of the following: $-11, -7, -3, -2, -1, 2, 3, 5, 6, 7, 11, 13, 17, 19, 21, 29, 33, 37, 41, 57, 73$.

6. If $d < 0$ then the imaginary quadratic field $\mathbb{Q}\left(\sqrt{d}\right)$ has the unique factorization property if and only if d is one of the following: $-1, -2, -3, -7, -11, -19, -43, -163$.

7. Of the 60 real quadratic fields $\mathbb{Q}(\sqrt{d})$ with $2 \leq d \leq 100$, exactly 38 of them have the unique factorization property: $d = 2, 3, 5, 6, 7, 11, 13, 14, 17, 19, 21, 22, 23, 29, 31, 33, 37, 38, 41, 43, 46, 47, 53, 57, 59, 61, 62, 67, 69, 71, 73, 77, 83, 86, 89, 93, 94,$ and 97.

2.7.5.3   Examples of quadratic fields

1. The algebraic integers of $\mathbb{Q}(\sqrt{-1})$ are of the form $a + bi$ where a and b are integers; they are called the *Gaussian integers*.

2. The number $1 + \sqrt{2}$ is a fundamental unit of $\mathbb{Q}(\sqrt{2})$. Hence, all units in $\mathbb{Q}(\sqrt{2})$ have the form $\pm(1 + \sqrt{2})^n$ for $n = 0, \pm1, \pm2, \ldots$.

3. The field $\mathbb{Q}\left(\sqrt{-5}\right)$ is not a unique factorization domain. This is illustrated by $6 = 2 \cdot 3 = (1 + \sqrt{-5}) \cdot (1 - \sqrt{-5})$, yet each of $\{2, 3, 1 + \sqrt{-5}, 1 - \sqrt{-5}\}$ is prime in this field.

4. The field $\mathbb{Q}\left(\sqrt{10}\right)$ is not a unique factorization domain. This is illustrated by $6 = 2 \cdot 3 = (4 + \sqrt{10}) \cdot (4 - \sqrt{10})$, yet each of $\{2, 3, 4 + \sqrt{10}, 4 - \sqrt{10}\}$ is prime in this field.

2.7.6   FINITE FIELDS

2.7.6.1   Facts about finite fields

1. If p is prime, then the ring $\mathbb{Z}_p$ is a finite field.

2. If p is prime and n is a positive integer, then there is exactly one field (up to isomorphism) with p^n elements. This field is denoted $GF(p^n)$ or F_{p^n} and is called a *Galois field*. (See the table on page 190.)

3. For F a finite field, there is a prime p and a positive integer n such that F has p^n elements. The prime number p is the characteristic of F. The field F is a *finite extension* of $\mathbb{Z}_p$, that is, F is a finite dimensional vector space over $\mathbb{Z}_p$.

4. If F is a finite field, then the set of non-zero elements of F under multiplication is a cyclic group. A generator of this group is a *primitive element*.

5. There are $\phi(p^n - 1)/n$ primitive polynomials of degree n ($n > 1$) over $GF(p)$, where ϕ is the Euler ϕ-function. (See table on page 128.)

6. There are $\dfrac{\sum_{j|k}\mu(k/j)p^{nj}}{k}$ irreducible polynomials of degree k over $GF(p^n)$, where μ is the Möbius function.

7. If F is a finite field where $|F| = k$ and $p(x)$ is a polynomial of degree n irreducible over F, then the field $F[x]/(p(x))$ has order k^n. If α is a root of $p(x) \in F[x]$ of degree $n \geq 1$, then $F[x]/(p(x)) = \{c_{n-1}\alpha^{n-1} + \cdots + c_1\alpha + c_0 \mid c_i \in F \text{ for all } i\}$.

8. When p is a prime, $GF(p^n)$ can be viewed as a vector space of dimension n over F_p. A basis of F_{p^n} of the form $\{\alpha, \alpha^p, \alpha^{p^2}, \ldots, \alpha^{p^{n-1}}\}$ is called a *normal basis*. If α is a primitive element of F_{p^n}, then the basis is said to be a *primitive normal basis*. Such an α satisfies a primitive normal polynomial of degree n over F_p.

Degree	Primitive normal polynomials		
n	$p = 2$	$p = 3$	$p = 5$
2	$x^2 + x + 1$	$x^2 + x + 2$	$x^2 + x + 2$
3	$x^3 + x^2 + 1$	$x^3 + 2x^2 + 1$	$x^3 + x^2 + 2$
4	$x^4 + x^3 + 1$	$x^4 + x^3 + 2$	$x^4 + x^3 + 4x + 2$
5	$x^5 + x^4 + x^2 + x + 1$	$x^5 + 2x^4 + 1$	$x^5 + 2x^4 + 3$
6	$x^6 + x^5 + 1$	$x^6 + x^5 + x^3 + 2$	$x^6 + x^5 + 2$
7	$x^7 + x^6 + 1$	$x^7 + x^6 + x^2 + 1$	$x^7 + x^6 + 2$

2.7.7 HOMOMORPHISMS AND ISOMORPHISMS

2.7.7.1 Definitions

1. A *group homomorphism* from group G_1 to group G_2 is a function $\varphi : G_1 \to G_2$ such that $\varphi(ab) = \varphi(a)\varphi(b)$ for all $a, b \in G_1$. *Note*: $a\varphi$ is often written instead of $\varphi(a)$.

2. A *character* of a group G is a group homomorphism $\chi : G \to \mathbb{C}^*$ (non-zero complex numbers under multiplication). (See table on page 191.)

3. A *ring homomorphism* from ring R_1 to ring R_2 is a function $\varphi : R_1 \to R_2$ such that $\varphi(a + b) = \varphi(a) + \varphi(b)$ and $\varphi(ab) = \varphi(a)\varphi(b)$ for all $a, b \in R_1$.

4. An *isomorphism* from group (ring) S_1 to group (ring) S_2 is a group (ring) homomorphism $\varphi : S_1 \to S_2$ that is 1-1 and onto S_2. If an isomorphism exists, then S_1 is said to be *isomorphic* to S_2. Write $S_1 \cong S_2$. (See the table on page 176 for numbers of non-isomorphic groups and the table on page 178 for examples of groups of orders less than 16.)

5. An *automorphism* of S is an isomorphism $\varphi : S \to S$.

6. The *kernel* of a group homomorphism $\varphi : G_1 \to G_2$ is $\varphi^{-1}(e) = \{g \in G_1 \mid \varphi(g) = e\}$. The *kernel* of a ring homomorphism $\varphi : R_1 \to R_2$ is $\varphi^{-1}(0) = \{r \in R_1 \mid \varphi(r) = 0\}$.

2.7.7.2   Facts about homomorphisms and isomorphisms

1. If $\varphi : G_1 \rightarrow G_2$ is a group homomorphism, then $\varphi(G_1)$ is a subgroup of G_2.

2. *Fundamental homomorphism theorem for groups*: If $\varphi : G_1 \rightarrow G_2$ is a group homomorphism with kernel K, then K is a normal subgroup of G_1 and $G_1/K \cong \varphi(G_1)$.

3. If G is a cyclic group of infinite order, then $G \cong (\mathbb{Z}, +)$.

4. If G is a cyclic group of order n, then $G \cong (\mathbb{Z}_n, +)$.

5. If p is prime, then there is only one group (up to isomorphism) of order p, the group $(\mathbb{Z}_p, +)$.

6. *Cayley's theorem*: If G is a finite group of order n, then G is isomorphic to some subgroup of the group of permutations on n objects.

7. $\mathbb{Z}_m \times \mathbb{Z}_n \cong \mathbb{Z}_{mn}$ if and only if m and n are relatively prime.

8. If $n = n_1 \cdot n_2 \cdot \ldots \cdot n_k$ where each n_i is a power of a different prime, then $\mathbb{Z}_n \cong \mathbb{Z}_{n_1} \times \mathbb{Z}_{n_2} \times \cdots \times \mathbb{Z}_{n_k}$.

9. *Fundamental theorem of finite abelian groups*: Every finite abelian group G (order ≥ 2) is isomorphic to a product of cyclic groups where each cyclic group has order a power of a prime. That is, there is a unique set $\{n_1, \ldots, n_k\}$ where each n_i is a power of some prime such that $G \cong \mathbb{Z}_{n_1} \times \mathbb{Z}_{n_2} \times \cdots \times \mathbb{Z}_{n_k}$.

10. *Fundamental theorem of finitely generated abelian groups*: If G is a finitely generated abelian group, then there is a unique integer $n \geq 0$ and a unique set $\{n_1, \ldots, n_k\}$ where each n_i is a power of some prime such that $G \cong \mathbb{Z}_{n_1} \times \mathbb{Z}_{n_2} \times \cdots \times \mathbb{Z}_{n_k} \times \mathbb{Z}^n$ (G is *finitely generated* if there are $a_1, a_2, \ldots, a_n \in G$ such that every element of G can be written as $a_{k_1}^{\epsilon_1} a_{k_2}^{\epsilon_2} \cdots a_{k_j}^{\epsilon_j}$ where $k_i \in \{1, \ldots, n\}$ (the k_i are not necessarily distinct) and $\epsilon_i \in \{1, -1\}$).

11. *Fundamental homomorphism theorem for rings*: If $\varphi : R_1 \rightarrow R_2$ is a ring homomorphism with kernel K, then K is an ideal in R_1 and $R_1/K \cong \varphi(R_1)$.

2.7.8   MATRIX CLASSES THAT ARE GROUPS

In the following examples, the group operation is ordinary matrix multiplication:

1. $GL(n, \mathbb{C})$ all complex non-singular $n \times n$ matrices
2. $GL(n, \mathbb{R})$ all real non-singular $n \times n$ matrices
3. $O(n)$ all $n \times n$ matrices A with $AA^T = I$, also called the *orthogonal group*
4. $SL(n, \mathbb{C})$ all complex $n \times n$ matrices of determinant 1, also called the *unimodular group* or the *special linear group*
5. $SL(n, \mathbb{R})$ all real $n \times n$ matrices of determinant 1

6. $SO(2)$ rotations of the plane: matrices of the form
$$A(\theta) = \begin{bmatrix} \cos\theta & -\sin\theta \\ \sin\theta & \cos\theta \end{bmatrix}$$

7. $SO(n)$ rotations of n-dimensional space
8. $SU(n)$ all $n \times n$ unitary matrices of determinant 1
9. $U(n)$ all $n \times n$ unitary matrices with $UU^H = I$

2.7.9   PERMUTATION GROUPS

Name	Symbol	Order	Definition
Symmetric group	S_n	$n!$	All permutations on $\{1, 2, \ldots, n\}$
Alternating group	A_n	$n!/2$	All even permutations on $\{1, 2, \ldots, n\}$
Cyclic group	C_n	n	Generated by $(12 \cdots n)$
Dihedral group	D_n	$2n$	Generated by $(12 \cdots n)$ and $(1n)(2\,n-1)(3\,n-2) \cdots$
Identity group	E_n	1	$(1)(2) \cdots (n)$ is the only permutation

EXAMPLE   With $p = 3$ elements, the identity permutation is (123), and:
$$A_3 = \{(123), (231), (312)\},$$
$$C_3 = \{(123), (231), (312)\},$$
$$D_3 = \{(231), (213), (132), (321), (312), (123)\},$$
$$E_3 = \{(123)\} \text{ and}$$
$$S_3 = \{(231), (213), (132), (321), (312), (123)\}.$$

2.7.9.1   Creating new permutation groups

Let A have permutations $\{X_i\}$, order n, degree d, let B have permutations $\{Y_j\}$, order m, degree e, and let C (a function of A and B) have permutations $\{W_k\}$, order p, degree f.

Name	Definition	Permutation	Order	Degree
Sum	$C = A + B$	$W = X \cup Y$	$p = mn$	$f = d + e$
Product	$C = A \times B$	$W = X \times Y$	$p = mn$	$f = de$
Composition	$C = A[B]$	$W = X \times Y$	$p = mn^d$	$f = de$
Power	$C = B^A$	$W = Y^X$	$p = mn$	$f = e^d$

2.7.9.2   Polya theory

Let π be a permutation. Define Inv (π) to be the number of invariant elements (i.e., mapped to themselves) in π. Define cyc (π) as the number of cycles in π. Suppose π has b_1 cycles of length 1, b_2 cycles of length 2, ..., b_k cycles of length k in its unique cycle decomposition. Then π can be encoded as the expression $x_1^{b_1} x_2^{b_2} \cdots x_k^{b_k}$. Summing these expressions for all permutations in the group G, and normalizing by the number of elements in G results in the *cycle index* of the group G:

$$P_G(x_1, x_2, \ldots, x_l) = \frac{1}{|G|} \sum_{\pi \in G} \left(x_1^{b_1} x_2^{b_2} \cdots x_k^{b_k} \right). \qquad (2.7.1)$$

1. *Burnside's Lemma*: Let G be a group of permutations of a set A, and let S be the equivalence relation on A induced by G. Then the number of equivalence classes in A is given by $\dfrac{1}{|G|} \sum_{\pi \in G} \text{Inv}\,(\pi)$.

2. *Special case of Polya's theorem*: Let R be an m element set of colors. Let G be a group of permutations $\{\pi_1, \pi_2, \dots\}$ of the set A. Let $C(A, R)$ be the set of colorings of the elements of A using colors in R. Then the number of distinct colorings in $C(A, R)$ is given by

$$\frac{1}{|G|}\left[m^{\text{cyc}\,(\pi_1)} + m^{\text{cyc}\,(\pi_2)} + \dots \right].$$

3. *Polya's theorem*: Let G be a group of permutations on a set A with cycle index $P_G(x_1, x_2, \dots, x_k)$. Let $C(A, R)$ be the collection of all colorings of A using colors in R. If w is a weight assignment on R, then the pattern inventory of colorings in $C(A, R)$ is given by

$$P_G\left(\sum_{r \in R} w(r), \sum_{r \in R} w^2(r), \cdots \sum_{r \in R} w^k(r) \right).$$

EXAMPLES

1. Consider necklaces constructed of $2k$ beads. Since a necklace can be flipped over, the appropriate permutation group is $G = \{\pi_1, \pi_2\}$ with $\pi_1 = (1)(2)\dots(2k)$ and $\pi_2 = \left(1 \quad 2k\right)\left(2 \quad 2k-1\right)\left(3 \quad 2k-2\right)\dots\left(k \quad k+1\right)$. Hence, $\text{cyc}\,(\pi_1) = 2k$, $\text{cyc}\,(\pi_2) = k$, and the cycle index is $P_G(x_1, x_2) = \left(x_1^{2k} + x_2^k\right)/2$. Using r colors, the number of distinct necklaces is $(r^{2k} + r^k)/2$.

   For a 4 bead necklace ($k = 2$) using $r = 2$ colors (say $w_1 = b$ for "black" and $w_2 = w$ for "white"), the $(2^4 + 2^2)/2 = 10$ different necklaces are $\{bbbb\}$, $\{bbbw\}$, $\{bbwb\}$, $\{bbww\}$, $\{bwbw\}$, $\{bwwb\}$, $\{wbbw\}$, $\{bwww\}$, $\{wbww\}$, and $\{wwww\}$. The pattern inventory of colorings, $P_G(\sum_i w_i, \sum_i w_i^2) = \left((b+w)^4 + (b^2+w^2)^2\right)/2 = b^4 + 2b^3w + 4b^2w^2 + 2bw^3 + w^4$, tells how many colorings of each type there are.

2. Consider coloring the corners of a square. If the squares can be rotated, but not reflected, then the number of distinct colorings, using k colors, is $\frac{1}{4}\left(k^4 + k^2 + 2k\right)$. If the squares can be rotated and reflected, then the number of distinct colorings, using k colors, is $\frac{1}{8}\left(k^4 + 2k^3 + 3k^2 + 2k\right)$.

   (a) If $k = 2$ colors are used, then there are 6 distinct classes of colorings whether reflections are allowed, or not. These classes are the same in both cases. The 16 colorings of a square with 2 colors form 6 distinct classes as shown:

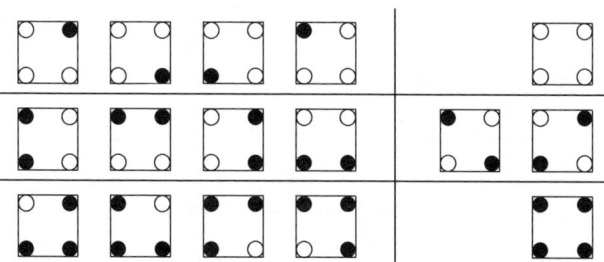

(b) If $k = 3$ colors are used, then there are 21 distinct classes of colorings if reflections are allowed, and 24 distinct classes of colorings if reflections are not allowed. Shown below are representative elements of each of these classes:

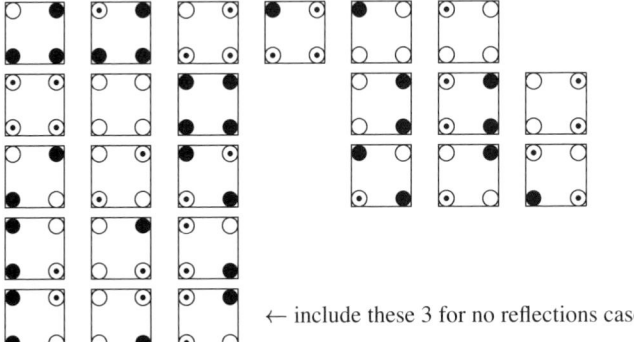

← include these 3 for no reflections case

2.7.9.3 Polya theory tables

1. Number of distinct corner colorings of regular polygons using rotations and reflections, or rotations only, with no more than k colors:

	rotations & reflections			rotations only		
object	$k = 2$	$k = 3$	$k = 4$	$k = 2$	$k = 3$	$k = 4$
triangle	4	10	20	4	11	24
square	6	21	55	6	24	70
pentagon	8	39	136	8	51	208
hexagon	13	92	430	14	130	700

2. Coloring regular 2- and 3- dimensional objects with no more than k colors:

tetrahedron	
corners of a tetrahedron	$\frac{1}{12}\left(k^6 + 3k^4 + 8k^2\right)$
edges of a tetrahedron	$\frac{1}{12}\left(k^6 + 3k^4 + 8k^2\right)$
faces of a tetrahedron	$\frac{1}{12}\left(k^4 + 11k^2\right)$
cube	
corners of a cube	$\frac{1}{24}\left(k^8 + 17k^4 + 6k^2\right)$
edges of a cube	$\frac{1}{24}\left(k^{12} + 6k^7 + 3k^6 + 8k^4 + 6k^3\right)$
faces of a cube	$\frac{1}{24}\left(k^6 + 3k^4 + 12k^3 + 8k^2\right)$

corners of a triangle	
with rotations	$\frac{1}{3}\left(k^3 + 2k\right)$
with rotations and reflections	$\frac{1}{6}\left(k^3 + 3k^2 + 2k\right)$
corners of a square	
with rotations	$\frac{1}{4}\left(k^4 + k^2 + 2k\right)$
with rotations and reflections	$\frac{1}{8}\left(k^4 + 2k^3 + 3k^2 + 2k\right)$
corners of a pentagon	
with rotations	$\frac{1}{5}\left(k^5 + 4k\right)$
with rotations and reflections	$\frac{1}{10}\left(k^5 + 5k^3 + 4k\right)$
corners of a hexagon	
with rotations	$\frac{1}{6}\left(k^6 + k^3 + 2k^2 + 2k\right)$
with rotations and reflections	$\frac{1}{12}\left(k^6 + 3k^4 + 4k^3 + 2k^2 + 2k\right)$
corners of a regular polygon	
with rotations	$\displaystyle\frac{1}{n}\sum_{d\mid n}\phi(d)k^{\frac{n}{d}}$
with rotations and reflections (n even)	$\displaystyle\frac{1}{2n}\sum_{d\mid n}\phi(d)k^{\frac{n}{d}} + \frac{1}{4}\left(k^{\frac{n}{2}} + k^{\frac{n+2}{2}}\right)$
with rotations and reflections (n odd)	$\displaystyle\frac{1}{2n}\sum_{d\mid n}\phi(d)k^{\frac{n}{d}} + \frac{1}{2}k^{\frac{n+1}{2}}$

3. The cycle index $P(x_1, x_2, \dots)$ and number of black-white colorings of regular objects under all permutations

corners of a triangle	
cycle index	$\frac{1}{6}\left(x_1^3 + 3x_1 x_2 + 2x_3\right)$
pattern inventory	$1b^3 + 1b^2 w + 1bw^2 + 1w^3$
corners of a square	
cycle index	$\frac{1}{8}\left(x_1^4 + 2x_2^2 x_2 + 3x_2^2 + 2x_4\right)$
pattern inventory	$1b^4 + 1b^3 w + 2b^2 w^2 + 1bw^3 + 1w^4$
corners of a pentagon	
cycle index	$\frac{1}{10}\left(x_1^5 + 4x_5 + 5x_1 x_2^2\right)$
pattern inventory	$1b^5 + 1b^4 w + 2b^3 w^2 + 2b^2 w^3 + 1bw^4 + 1w^5$

corners of a cube

cycle index $\quad \frac{1}{24}\left(x_1^8 + 6x_4^2 + 9x_2^4 + 8x_1^2 x_3^2\right)$

pattern inventory $\quad b^8 + b^7 w + 3b^6 w^2 + 3b^5 w^3 + 7b^4 w^4$

$\qquad\qquad\qquad +3b^3 w^5 + 3b^2 w^6 + bw^7 + w^8$

Note that the pattern inventory for the black-white colorings is given by $P((b+w),(b^2+w^2),(b^3+w^3),\dots)$.

2.7.10 TABLES

2.7.10.1 Number of non-isomorphic groups of different orders

The $10n + k$ entry is found by looking at row n and the column k. There are 10,494,213 non-isomorphic groups with 512 elements.

	_0	_1	_2	_3	_4	_5	_6	_7	_8	_9
0_		1	1	1	2	1	2	1	5	2
1_	2	1	5	1	2	1	14	1	5	1
2_	5	2	2	1	15	2	2	5	4	1
3_	4	1	51	1	2	1	14	1	2	2
4_	14	1	6	1	4	2	2	1	52	2
5_	5	1	5	1	15	2	13	2	2	1
6_	13	1	2	4	267	1	4	1	5	1
7_	4	1	50	1	2	3	4	1	6	1
8_	52	15	2	1	15	1	2	1	12	1
9_	10	1	4	2	2	1	231	1	5	2
10_	16	1	4	1	14	2	2	1	45	1
11_	6	2	43	1	6	1	5	4	2	1
12_	47	2	2	1	4	5	16	1	2328	2
13_	4	1	10	1	2	5	15	1	4	1
14_	11	1	2	1	197	1	2	6	5	1
15_	13	1	12	2	4	2	18	1	2	1
16_	238	1	55	1	5	2	2	1	57	2
17_	4	5	4	1	4	2	42	1	2	1
18_	37	1	4	2	12	1	6	1	4	13
29_	4	1	1543	1	2	2	12	1	10	1
20_	52	2	2	2	12	2	2	2	51	1
21_	12	1	5	1	2	1	177	1	2	2
22_	15	1	6	1	197	6	2	1	15	1

	_0	_1	_2	_3	_4	_5	_6	_7	_8	_9
23_	4	2	14	1	16	1	4	2	4	1
24_	208	1	5	67	5	2	4	1	12	1
25_	15	1	46	2	2	1	56092	1	6	1
26_	15	2	2	1	39	1	4	1	4	1
27_	30	1	54	5	2	4	10	1	2	4
28_	40	1	4	1	4	2	4	1	1045	2
29_	4	2	5	1	23	1	14	5	2	1
30_	49	2	2	1	42	2	10	1	9	2
31_	6	1	61	1	2	4	4	1	4	1
32_	1640	1	4	1	176	2	2	2	15	1
33_	12	1	4	5	2	1	228	1	5	1
34_	15	1	18	5	12	1	2	1	12	1
35_	10	14	195	1	4	2	5	2	2	1
36_	162	2	2	3	11	1	6	1	42	2
37_	4	1	15	1	4	7	12	1	60	1
38_	11	2	2	1	20169	2	2	4	5	1
39_	12	1	44	1	2	1	30	1	2	5
40_	221	1	6	1	5	16	6	1	46	1
41_	6	1	4	1	10	1	235	2	4	1
42_	41	1	2	2	14	2	4	1	4	2
43_	4	1	775	1	4	1	5	1	6	1
44_	51	13	4	1	18	1	2	1	1396	1
45_	34	1	5	2	2	1	54	1	2	5
46_	11	1	12	1	51	4	2	1	55	1
47_	4	2	12	1	6	2	11	2	2	1
48_	1213	1	2	2	12	1	261	1	14	2
49_	10	1	12	1	4	4	42	2	4	1
50_	56	1	2	1	202	2	6	6	4	1
51_	8	1	*	15	2	1	15	1	4	1

2.7.10.2 Number of non-isomorphic Abelian groups of different orders

The $10n + k$ entry is in row $n\_$ and column $\_k$.

	_0	_1	_2	_3	_4	_5	_6	_7	_8	_9
0_		1	1	1	2	1	1	1	3	2
1_	1	1	2	1	1	1	5	1	2	1
2_	2	1	1	1	3	2	1	3	2	1
3_	1	1	7	1	1	1	4	1	1	1
4_	3	1	1	1	2	2	1	1	5	2
5_	2	1	2	1	3	1	3	1	1	1

2.7.10.3 Names of groups of small order

Order n	Distinct groups of order n
1	$\{e\}$
2	C_2
3	C_3
4	$C_4, \quad V = C_2 \times C_2$
5	C_5
6	$C_6, \quad D_3$
7	C_7
8	$C_8, \quad C_2 \times C_4, \quad C_2 \times C_2 \times C_2, \quad D_4, \quad Q$
9	$C_9, \quad C_3 \times C_3$
10	$C_{10}, \quad D_5$
11	C_{11}
12	$C_{12}, \quad C_2 \times C_6, \quad T = C_3 \rtimes C_4, \quad D_6, \quad A_4$
13	C_{13}
14	$C_{14}, \quad D_7$
15	C_{15}

2.7.10.4 Representations of groups of small order

In all cases the identity element, $\{1\}$, forms a group of order 1.

1. **C_2, the cyclic group of order 2**
   Generator: a with relation $a^2 = 1$

	1	a
1	1	a
a	a	1

   Elements   *Subgroups*

   (a) order 2: a (a) order 2: $\{1, a\}$

2. **C_3, the cyclic group of order 3**
   Generator: a with relation $a^3 = 1$

	1	a	a^2
1	1	a	a^2
a	a	a^2	1
a^2	a^2	1	a

   Elements   *Subgroups*

   (a) order 3: a, a^2 (a) order 3: $\{1, a, a^2\}$

3. **C_4, the cyclic group of order 4**
   Generator: a with relation $a^4 = 1$

	1	a	a^2	a^3
1	1	a	a^2	a^3
a	a	a^2	a^3	1
a^2	a^2	a^3	1	a
a^3	a^3	1	a	a^2

Elements

  (a) order 4: a, a^3

  (b) order 2: a^2

Subgroups

  (a) order 4: $\{1, a, a^2, a^3\}$

  (b) order 2: $\{1, a^2\}$

4. V, the Klein four group

Generators: a, b with relations $a^2 = 1$, $b^2 = 1$, $ba = ab$:

	1	a	b	ab
1	1	a	b	ab
a	a	1	ab	b
b	b	ab	1	a
ab	ab	b	a	1

Elements

  (a) order 2: a, b, ab

Subgroups

  (a) order 4: $\{1, a, b, ab\}$

  (b) order 2: $\{1, a\}, \{1, b\}, \{1, ab\}$

5. C_5, the cyclic group of order 5

Generator: a with relation $a^5 = 1$

	1	a	a^2	a^3	a^4
1	1	a	a^2	a^3	a^4
a	a	a^2	a^3	a^4	1
a^2	a^2	a^3	a^4	1	a
a^3	a^3	a^4	1	a	a^2
a^4	a^4	1	a	a^2	a^3

Elements

  (a) order 5: a, a^2, a^3, a^4

Subgroups

  (a) order 5: $\{1, a, a^2, a^3, a^4\}$

6. C_6, the cyclic group of order 6

Generator: a with relation $a^6 = 1$

	1	a	a^2	a^3	a^4	a^5
1	1	a	a^2	a^3	a^4	a^5
a	a	a^2	a^3	a^4	a^5	1
a^2	a^2	a^3	a^4	a^5	1	a
a^3	a^3	a^4	a^5	1	a	a^2
a^4	a^4	a^5	1	a	a^2	a^3
a^5	a^5	1	a	a^2	a^3	a^4

Elements

  (a) order 6: a, a^5

  (b) order 3: a^2, a^4

  (c) order 2: a^3

Subgroups

  (a) order 6: $\{1, a, a^2, a^3, a^4, a^5\}$

  (b) order 3: $\{1, a^2, a^4\}$

  (c) order 2: $\{1, a^3\}$

7. S_3, the symmetric group on three elements

Generators: a, b with relations $a^3 = 1, b^2 = 1, ba = a^{-1}b$

	1	a	a^2	b	ab	a^2b
1	1	a	a^2	b	ab	a^2b
a	a	a^2	1	ab	a^2b	b
a^2	a^2	1	a	a^2b	b	ab
b	b	a^2b	ab	1	a^2	a
ab	ab	b	a^2b	a	1	a^2
a^2b	a^2b	ab	b	a^2	a	1

Elements

(a) order 3: a, a^2

(b) order 2: b, ab, a^2b

Subgroups

(a) order 6: $\{1, a, a^2, b, ab, a^2b\}$ (is a normal subgroup)

(b) order 3: $\{1, a, a^2\}$ (is a normal subgroup)

(c) order 2: $\{1, b\}, \{1, ab\}, \{1, a^2b\}$

8. C_7, the cyclic group of order 7

Generator: a with relation $a^7 = 1$

	1	a	a^2	a^3	a^4	a^5	a^6
1	1	a	a^2	a^3	a^4	a^5	a^6
a	a	a^2	a^3	a^4	a^5	a^6	1
a^2	a^2	a^3	a^4	a^5	a^6	1	a
a^3	a^3	a^4	a^5	a^6	1	a	a^2
a^4	a^4	a^5	a^6	1	a	a^2	a^3
a^5	a^5	a^6	1	a	a^2	a^3	a^4
a^6	a^6	1	a	a^2	a^3	a^4	a^5

Elements

(a) order 7: $a, a^2, a^3, a^4, a^5, a^6$

Subgroups

(a) order 7: $\{1, a, a^2, a^3, a^4, a^5, a^6\}$

9. C_8, the cyclic group of order 8

Generator: a with relation $a^8 = 1$

	1	a	a^2	a^3	a^4	a^5	a^6	a^7
1	1	a	a^2	a^3	a^4	a^5	a^6	a^7
a	a	a^2	a^3	a^4	a^5	a^6	a^7	1
a^2	a^2	a^3	a^4	a^5	a^6	a^7	1	a
a^3	a^3	a^4	a^5	a^6	a^7	1	a	a^2
a^4	a^4	a^5	a^6	a^7	1	a	a^2	a^3
a^5	a^5	a^6	a^7	1	a	a^2	a^3	a^4
a^6	a^6	a^7	1	a	a^2	a^3	a^4	a^5
a^7	a^7	1	a	a^2	a^3	a^4	a^5	a^6

Elements

  (a) order 8: a, a^3, a^5, a^7

  (b) order 4: a^2, a^6

  (c) order 2: a^4

Subgroups

  (a) order 8:

    $\left\{1, a, a^2, a^3, a^4, a^5, a^6, a^7\right\}$

  (b) order 4: $\left\{1, a^2, a^4, a^6\right\}$

  (c) order 2: $\left\{1, a^4\right\}$

10. $C_4 \times C_2$, **the direct product of a cyclic group of order 4 and a cyclic group of order 2**

    Generators: a, b with relations $a^4 = 1, b^2 = 1$, and $ba = ab$

	1	a	a^2	a^3	b	ab	a^2b	a^3b
1	1	a	a^2	a^3	b	ab	a^2b	a^3b
a	a	a^2	a^3	1	ab	a^2b	a^3b	b
a^2	a^2	a^3	1	a	a^2b	a^3b	b	ab
a^3	a^3	1	a	a^2	a^3b	b	ab	a^2b
b	b	ab	a^2b	a^3b	1	a	a^2	a^3
ab	ab	a^2b	a^3b	b	a	a^2	a^3	1
a^2b	a^2b	a^3b	b	ab	a^2	a^3	1	a
a^3b	a^3b	b	ab	a^2b	a^3	1	a	a^2

Elements

  (a) order 4: a, a^3, ab, a^3b

  (b) order 2: a^2, b, a^2b

Subgroups

  (a) order 8: $\left\{1, a, a^2, a^3, b, ab, a^2b, a^3b\right\}$

  (b) order 4: $\left\{1, a, a^2, a^3\right\}$ $\left\{1, ab, a^2, a^3b\right\}$ $\left\{1, a^2, b, a^2b\right\}$

  (c) order 2: $\left\{1, a^2\right\}$, $\{1, b\}$, $\left\{1, a^2b\right\}$

11. $C_2 \times C_2 \times C_2$, **the direct product of 3 cyclic groups of order 2**

    Generators: a, b, c with relations $a^2 = 1, b^2 = 1, c^2 = 1, ba = ab, ca = ac, cb = bc$

	1	a	b	ab	c	ac	bc	abc
1	1	a	b	ab	c	ac	bc	abc
a	a	1	ab	b	ac	c	abc	bc
b	b	ab	1	a	bc	abc	c	ac
ab	ab	b	a	1	abc	bc	ac	c
c	c	ac	bc	abc	1	a	b	ab
ac	ac	c	abc	bc	a	1	ab	b
bc	bc	abc	c	ac	b	ab	1	a
abc	abc	bc	ac	c	ab	b	a	1

Elements

  (a) order 2: a, b, ab, c, ac, bc, abc

Subgroups

(a) order 8: $\{1, a, b, ab, c, ac, bc, abc\}$
(b) order 4: $\{1, a, b, ab\}$, $\{1, a, c, ac\}$, $\{1, a, bc, abc\}$, $\{1, b, c, bc\}$,
   $\{1, b, ac, abc\}$, $\{1, ab, c, abc\}$, $\{1, ab, ac, bc\}$
(c) order 2: $\{1, a\}$, $\{1, b\}$, $\{1, ab\}$, $\{1, c\}$, $\{1, ac\}$, $\{1, bc\}$, $\{1, abc\}$

12. D_4, the dihedral group of order 8

Generators: a, b with relations $a^4 = 1$, $b^2 = 1$, $ba = a^{-1}b$

	1	a	a^2	a^3	b	ab	a^2b	a^3b
1	1	a	a^2	a^3	b	ab	a^2b	a^3b
a	a	a^2	a^3	1	ab	a^2b	a^3b	b
a^2	a^2	a^3	1	a	a^2b	a^3b	b	ab
a^3	a^3	1	a	a^2	a^3b	b	ab	a^2b
b	b	a^3b	a^2b	ab	1	a^3	a^2	a
ab	ab	b	a^3b	a^2b	a	1	a^3	a^2
a^2b	a^2b	ab	b	a^3b	a^2	a	1	a^3
a^3b	a^3b	a^2b	ab	b	a^3	a^2	a	1

Elements

(a) order 4: a, a^3
(b) order 2: a^2, b, ab, a^2b, a^3b

Subgroups

(a) order 8: $\{1, a, a^2, a^3, b, ab, a^2b, a^3b\}$
(b) order 4: $\{1, a^2, b, a^2b\}$, $\{1, a, a^2, a^3\}$, $\{1, a^2, ab, a^3b\}$
(c) order 2: $\{1, b\}$, $\{1, a^2b\}$, $\{1, a^2\}$, $\{1, ab\}$, $\{1, a^3b\}$

Normal subgroups

(a) order 8: $\{1, a, a^2, a^3, b, ab, a^2b, a^3b\}$
(b) order 4: $\{1, a^2, b, a^2b\}$, $\{1, a, a^2, a^3\}$, $\{1, a^2, ab, a^3b\}$
(c) order 2: $\{1, a^2\}$

13. Q, the quaternion group (of order 8)

Generators: a, b with relations $a^4 = 1$, $b^2 = a^2$, $ba = a^{-1}b$

	1	a	a^2	a^3	b	ab	a^2b	a^3b
1	1	a	a^2	a^3	b	ab	a^2b	a^3b
a	a	a^2	a^3	1	ab	a^2b	a^3b	b
a^2	a^2	a^3	1	a	a^2b	a^3b	b	ab
a^3	a^3	1	a	a^2	a^3b	b	ab	a^2b
b	b	a^3b	a^2b	ab	a^2	a	1	a^3
ab	ab	b	a^3b	a^2b	a^3	a^2	a	1
a^2b	a^2b	ab	b	a^3b	1	a^3	a^2	a
a^3b	a^3b	a^2b	ab	b	a	1	a^3	a^2

Elements

(a) order 4: $a, a^3, b, ab, a^2b, a^3b$
(b) order 2: a^2

Subgroups

(a) order 8: $\left\{1, a, a^2, a^3, b, ab, a^2b, a^3\right\}$
(b) order 4: $\left\{1, a, a^2, a^3\right\}, \left\{1, b, a^2, a^2b\right\}, \left\{1, ab, a^2, a^3b\right\}$
(c) order 2: $\left\{1, a^2\right\}$

Normal subgroups

(a) order 8: $\left\{1, a, a^2, a^3, b, ab, a^2b, a^3\right\}$
(b) order 4: $\left\{1, a, a^2, a^3\right\}, \left\{1, b, a^2, a^2b\right\}, \left\{1, ab, a^2, a^3b\right\}$
(c) order 2: $\left\{1, a^2\right\}$

Notes

- Q can be defined as the set $\{1, -1, i, -i, j, -j, k, -k\}$ where multiplication is defined by:

$$i^2 = j^2 = k^2 = -1, \; ij = -ji = k, \; jk = -kj = i, \; ki = -ik = j$$
$$(2.7.2)$$

$\times$	$+1$	-1	$+i$	$-i$	$+j$	$-j$	$+k$	$-k$
$+1$	$+1$	-1	$+i$	$-i$	$+j$	$-j$	$+k$	$-k$
-1	-1	$+1$	$-i$	$+i$	$-j$	$+j$	$-k$	$+k$
$+i$	$+i$	$-i$	-1	$+1$	$+k$	$-k$	$-j$	$+j$
$-i$	$-i$	$+i$	$+1$	-1	$-k$	$+k$	$+j$	$-j$
$+j$	$+j$	$-j$	$-k$	$+k$	-1	$+1$	$+i$	$-i$
$-j$	$-j$	$+j$	$+k$	$-k$	$+1$	-1	$-i$	$+i$
$+k$	$+k$	$-k$	$+j$	$-j$	$-i$	$+i$	-1	$+1$
$-k$	$-k$	$+k$	$-j$	$+j$	$+i$	$-i$	$+1$	-1

Elements

(a) order 4: $i, -i, j, -j, k, -k$
(b) order 2: -1

Subgroups   *(all of them are normal subgroups)*

(a) order 8: $\{1, -1, i, -i, j, -j, k, -k\}$
(b) order 4: $\{1, i, -1, -i\}, \{1, j, -1, -j\}, \{1, k, -1, -k\}$
(c) order 2: $\{1, -1\}$

- Q can be defined as the group composed of the 8 matrices:

$$+1_q = \begin{bmatrix} 1 & 0 \\ 0 & 1 \end{bmatrix}, -1_q = \begin{bmatrix} -1 & 0 \\ 0 & -1 \end{bmatrix}, +i_q = \begin{bmatrix} -i & 0 \\ 0 & i \end{bmatrix}, -i_q = \begin{bmatrix} i & 0 \\ 0 & -i \end{bmatrix},$$

$$+j_q = \begin{bmatrix} 0 & i \\ i & 0 \end{bmatrix}, -j_q = \begin{bmatrix} 0 & -i \\ -i & 0 \end{bmatrix}, +k_q = \begin{bmatrix} 0 & 1 \\ -1 & 0 \end{bmatrix}, -k_q = \begin{bmatrix} 0 & -1 \\ 1 & 0 \end{bmatrix}$$

where a subscript of q indicates a quaternion element, $i^2 = -1$, and matrix multiplication is the group operation.

14. **C_9, the cyclic group of order 9**

Generator: a with relation $a^9 = 1$

	1	a	a^2	a^3	a^4	a^5	a^6	a^7	a^8
1	1	a	a^2	a^3	a^4	a^5	a^6	a^7	a^8
a	a	a^2	a^3	a^4	a^5	a^6	a^7	a^8	1
a^2	a^2	a^3	a^4	a^5	a^6	a^7	a^8	1	a
a^3	a^3	a^4	a^5	a^6	a^7	a^8	1	a	a^2
a^4	a^4	a^5	a^6	a^7	a^8	1	a	a^2	a^3
a^5	a^5	a^6	a^7	a^8	1	a	a^2	a^3	a^4
a^6	a^6	a^7	a^8	1	a	a^2	a^3	a^4	a^5
a^7	a^7	a^8	1	a	a^2	a^3	a^4	a^5	a^6
a^8	a^8	1	a	a^2	a^3	a^4	a^5	a^6	a^7

Elements

(a) order 9: $a, a^2, a^4, a^5, a^6, a^7$

(b) order 3: a^3, a^6

Subgroups

(a) order 9: $\left\{1, a, a^2, a^3, a^4, a^5, a^6, a^7, a^8\right\}$

(b) order 3: $\left\{1, a^3, a^6\right\}$

15. **$C_3 \times C_3$, the direct product of two cyclic groups of order 3**

Generators: a, b with relations $a^3 = 1, b^3 = 1, ba = ab$

	1	a	a^2	b	ab	a^2b	b^2	ab^2	a^2b^2
1	1	a	a^2	b	ab	a^2b	b^2	ab^2	a^2b^2
a	a	a^2	1	ab	a^2b	b	ab^2	a^2b^2	b^2
a^2	a^2	1	a	a^2b	b	ab	a^2b^2	b^2	ab^2
b	b	ab	a^2b	b^2	ab^2	a^2b^2	1	a	a^2
ab	ab	a^2b	b	ab^2	a^2b^2	b^2	a	a^2	1
a^2b	a^2b	b	ab	a^2b^2	b^2	ab^2	a^2	1	a
b^2	b^2	ab^2	a^2b^2	1	a	a^2	b	ab	a^2b
ab^2	ab^2	a^2b^2	b^2	a	a^2	1	ab	a^2b	b
a^2b^2	a^2b^2	b^2	ab^2	a^2	1	a	a^2b	b	ab

Elements

(a) order 3: $a, a^2, b, ab, a^2b, b^2, ab^2, a^2b^2$

Subgroups

(a) order 3: $\left\{1, a, a^2\right\}, \left\{1, b, b^2\right\}, \left\{1, ab, a^2b^2\right\}, \left\{1, a^2b, ab^2\right\}$

16. **C_{10}, the cyclic group of order 10**

Generator: a with relation $a^{10} = 1$

	1	a	a^2	a^3	a^4	a^5	a^6	a^7	a^8	a^9
1	1	a	a^2	a^3	a^4	a^5	a^6	a^7	a^8	a^9
a	a	a^2	a^3	a^4	a^5	a^6	a^7	a^8	a^9	1
a^2	a^2	a^3	a^4	a^5	a^6	a^7	a^8	a^9	1	a
a^3	a^3	a^4	a^5	a^6	a^7	a^8	a^9	1	a	a^2
a^4	a^4	a^5	a^6	a^7	a^8	a^9	1	a	a^2	a^3
a^5	a^5	a^6	a^7	a^8	a^9	1	a	a^2	a^3	a^4
a^6	a^6	a^7	a^8	a^9	1	a	a^2	a^3	a^4	a^5
a^7	a^7	a^8	a^9	1	a	a^2	a^3	a^4	a^5	a^6
a^8	a^8	a^9	1	a	a^2	a^3	a^4	a^5	a^6	a^7
a^9	a^9	1	a	a^2	a^3	a^4	a^5	a^6	a^7	a^8

Elements

(a) order 10: a, a^3, a^7, a^9

(b) order 5: a^2, a^4, a^6, a^8

(c) order 2: a^5

Subgroups

(a) order 10: $\left\{1, a, a^2, a^3, a^4, a^5, a^6, a^7, a^8, a^9\right\}$

(b) order 5: $\left\{1, a^2, a^4, a^6, a^8\right\}$

(c) order 2: $\left\{1, a^5\right\}$

17. **D_5, the dihedral group of order 10**

Generators: a, b with relations $a^5 = 1, b^2 = 1, ba = a^{-1}b$

	1	a	a^2	a^3	a^4	b	ab	a^2b	a^3b	a^4b
1	1	a	a^2	a^3	a^4	b	ab	a^2b	a^3b	a^4b
a	a	a^2	a^3	a^4	1	ab	a^2b	a^3b	a^4b	b
a^2	a^2	a^3	a^4	1	a	a^2b	a^3b	a^4b	b	ab
a^3	a^3	a^4	1	a	a^2	a^3b	a^4b	b	ab	a^2b
a^4	a^4	1	a	a^2	a^3	a^4b	b	ab	a^2b	a^3b
b	b	a^4b	a^3b	a^2b	ab	1	a^4	a^3	a^2	a
ab	ab	b	a^4b	a^3b	a^2b	a	1	a^4	a^3	a^2
a^2b	a^2b	ab	b	a^4b	a^3b	a^2	a	1	a^4	a^3
a^3b	a^3b	a^2b	ab	b	a^4b	a^3	a^2	a	1	a^4
a^4b	a^4b	a^3b	a^2b	ab	b	a^4	a^3	a^2	a	1

Elements

(a) order 5: a, a^2, a^3, a^4

(b) order 2: b, ab, a^2b, a^3b, a^4b

Subgroups

(a) order 10: $\left\{1, a, a^2, a^3, a^4, b, ab, a^2b, a^3b, a^4b\right\}$

(b) order 5: $\left\{1, a, a^2, a^3, a^4\right\}$

(c) order 2: $\{1, b\}, \{1, ab\}\ \left\{1, a^2b\right\}, \left\{1, a^3b\right\}, \left\{1, a^4b\right\}$

Normal subgroups

  (a) order 10: $\left\{1, a, a^2, a^3, a^4, b, ab, a^2b, a^3b, a^4b\right\}$

  (b) order 5: $\left\{1, a, a^2, a^3, a^4\right\}$

18. D_6, the dihedral group of order 12

The multiplication table for D_6 (given below)

Generators: a, b with relations $a^6 = 1, b^2 = 1, ba = a^{-1}b$

	1	a	a^2	a^3	a^4	a^5	b	ab	a^2b	a^3b	a^4b	a^5b
1	1	a	a^2	a^3	a^4	a^5	b	ab	a^2b	a^3b	a^4b	a^5b
a	a	a^2	a^3	a^4	a^5	1	ab	a^2b	a^3b	a^4b	a^5b	b
a^2	a^2	a^3	a^4	a^5	1	a	a^2b	a^3b	a^4b	a^5b	b	ab
a^3	a^3	a^4	a^5	1	a	a^2	a^3b	a^4b	a^5b	b	ab	a^2b
a^4	a^4	a^5	1	a	a^2	a^3	a^4b	a^5b	b	ab	a^2b	a^3b
a^5	a^5	1	a	a^2	a^3	a^4	a^5b	b	ab	a^2b	a^3b	a^4b
b	b	a^5b	a^4b	a^3b	a^2b	ab	1	a^5	a^4	a^3	a^2	a
ab	ab	b	a^5b	a^4b	a^3b	a^2b	a	1	a^5	a^4	a^3	a^2
a^2b	a^2b	ab	b	a^5b	a^4b	a^3b	a^2	a	1	a^5	a^4	a^3
a^3b	a^3b	a^2b	ab	b	a^5b	a^4b	a^3	a^2	a	1	a^5	a^4
a^4b	a^4b	a^3b	a^2b	ab	b	a^5b	a^4	a^3	a^2	a	1	a^5
a^5b	a^5b	a^4b	a^3b	a^2b	ab	b	a^5	a^4	a^3	a^2	a	1

Elements

  (a) order 6: a, a^5

  (b) order 3: a^2, a^4

  (c) order 2: $a^3, b, ab, a^2b, a^3b, a^4b, a^5b$

Subgroups

  (a) order 12: $\left\{1, a, a^2, a^3, a^4, a^5, b, ab, a^2b, a^3b, a^4b, a^5b\right\}$

  (b) order 6: $\left\{1, a, a^2, a^3, a^4, a^5\right\}$ $\left\{1, a^2, a^4, b, a^2b, a^4b\right\}$,
      $\left\{1, a^2, a^4, ab, a^3b, a^5b\right\}$

  (c) order 4: $\left\{1, a^3, b, a^3b\right\}$, $\left\{1, a^3, ab, a^4b\right\}$, $\left\{1, a^3, a^2b, a^5b\right\}$

  (d) order 3: $\left\{1, a^2, a^4\right\}$

  (e) order 2: $\left\{1, a^3\right\}$, $\left\{1, b\right\}$, $\left\{1, ab\right\}$, $\left\{1, a^2b\right\}$, $\left\{1, a^3b\right\}$, $\left\{1, a^4b\right\}$,
      $\left\{1, a^5b\right\}$

Normal subgroups

  (a) order 12: $\left\{1, a, a^2, a^3, a^4, a^5, b, ab, a^2b, a^3b, a^4b, a^5b\right\}$

  (b) order 6: $\left\{1, a, a^2, a^3, a^4, a^5\right\}$, $\left\{1, a^2, a^4, b, a^2b, a^4b\right\}$,
      $\left\{1, a^2, a^4, ab, a^3b, a^5b\right\}$

  (c) order 3: $\left\{1, a^2, a^4\right\}$

  (d) order 2: $\left\{1, a^3\right\}$

Conjugacy classes: $\{1\}, \{a^3\}, \{a, a^5\}, \{a^2, a^4\}, \{b, a^2b, a^4b\}, \{ab, a^3b, a^5b\}$

19. $Z_3 \rtimes Z_4$, **the semidirect product of a cyclic group of order 4 acting on a cyclic group of order 3**   (Note: $\rtimes$ denotes the semidirect product.)

Generators: a, b with relations $a^6 = 1$, $b^2 = a^3$, $ba = a^{-1}b$

	1	a	a^2	a^3	a^4	a^5	b	ab	a^2b	a^3b	a^4b	a^5b
1	1	a	a^2	a^3	a^4	a^5	b	ab	a^2b	a^3b	a^4b	a^5b
a	a	a^2	a^3	a^4	a^5	1	ab	a^2b	a^3b	a^4b	a^5b	b
a^2	a^2	a^3	a^4	a^5	1	a	a^2b	a^3b	a^4b	a^5b	b	ab
a^3	a^3	a^4	a^5	1	a	a^2	a^3b	a^4b	a^5b	b	ab	a^2b
a^4	a^4	a^5	1	a	a^2	a^3	a^4b	a^5b	b	ab	a^2b	a^3b
a^5	a^5	1	a	a^2	a^3	a^4	a^5b	b	ab	a^2b	a^3b	a^4b
b	b	a^5b	a^4b	a^3b	a^2b	ab	a^3	a^2	a	1	a^5	a^4
ab	ab	b	a^5b	a^4b	a^3b	a^2b	a^4	a^3	a^2	a	1	a^5
a^2b	a^2b	ab	b	a^5b	a^4b	a^3b	a^5	a^4	a^3	a^2	a	1
a^3b	a^3b	a^2b	ab	b	a^5b	a^4b	1	a^5	a^4	a^3	a^2	a
a^4b	a^4b	a^3b	a^2b	ab	b	a^5b	a	1	a^5	a^4	a^3	a^2
a^5b	a^5b	a^4b	a^3b	a^2b	ab	b	a^2	a	1	a^5	a^4	a^3

Elements

(a) order 6: a, a^5
(b) order 4: $b, ab, a^2b, a^3b, a^4b, a^5b$
(c) order 3: a^2, a^4
(d) order 2: a^3

Subgroups

(a) order 12: $\{1, a, a^2, a^3, a^4, a^5, b, ab, a^2b, a^3b, a^4b, a^5b\}$
(b) order 6: $\{1, a, a^2, a^3, a^4, a^5\}$
(c) order 4: $\{1, b, a^3, a^3b\}, \{1, ab, a^3, a^4b\}, \{1, a^2b, a^3, a^5b\}$
(d) order 3: $\{1, a^2, a^4\}$
(e) order 2: $\{1, a^3\}$

Normal subgroups

(a) order 12: $\{1, a, a^2, a^3, a^4, a^5, b, ab, a^2b, a^3b, a^4b, a^5b\}$
(b) order 6: $\{1, a, a^2, a^3, a^4, a^5\}$
(c) order 3: $\{1, a^2, a^4\}$
(d) order 2: $\{1, a^3\}$

Conjugacy classes: $\{1\}, \{a^3\}, \{a, a^5\}, \{a^2, a^4\}, \{b, a^2b, a^4b\}, \{ab, a^3b, a^5b\}$

- Alternate representation, the group can be presented as follows,
  Generators: x, y with relations $x^3 = 1$, $y^4 = 1$, $yx = x^{-1}y$

	1	x	x^2	y	xy	x^2y	y^2	xy^2	x^2y^2	y^3	xy^3	x^2y^3
1	1	x	x^2	y	xy	x^2y	y^2	xy^2	x^2y^2	y^3	xy^3	x^2y^3
x	x	x^2	1	xy	x^2y	y	xy^2	x^2y^2	y^2	xy^3	x^2y^3	y^3
x^2	x^2	1	x	x^2y	y	xy	x^2y^2	y^2	xy^2	x^2y^3	y^3	xy^3
y	y	x^2y	xy	y^2	x^2y^2	xy^2	y^3	x^2y^3	xy^3	1	x^2	x
xy	xy	y	x^2y	xy^2	y^2	x^2y^2	xy^3	y^3	x^2y^3	x	1	x^2
x^2y	x^2y	xy	y	x^2y^2	xy^2	y^2	x^2y^3	xy^3	y^3	x^2	x	1
y^2	y^2	xy^2	x^2y^2	y^3	xy^3	x^2y^3	1	x	x^2	y	xy	x^2y
xy^2	xy^2	x^2y^2	y^2	xy^3	x^2y^3	y^3	x	x^2	1	xy	x^2y	y
x^2y^2	x^2y^2	y^2	xy^2	x^2y^3	y^3	xy^3	x^2	1	x	x^2y	y	xy
y^3	y^3	x^2y^3	xy^3	1	x^2	x	y	x^2y	xy	y^2	x^2y^2	xy^2
xy^3	xy^3	y^3	x^2y^3	x	1	x^2	xy	y	x^2y	xy^2	y^2	x^2y^2
x^2y^3	x^2y^3	xy^3	y^3	x^2	x	1	x^2y	xy	y	x^2y^2	xy^2	y^2

Elements

    (a) order 6: x^2y^2, xy^2

    (b) order 4: y, xy, x^2y, y^3, xy^3, x^2y^3

    (c) order 3: x, x^2

    (d) order 2: y^2

Subgroups

    (a) order 12: $\left\{1, x, x^2, y, xy, x^2y, y^2, xy^2, x^2y^2, y^3, xy^3, x^2y^3\right\}$

    (b) order 6: $\left\{1, x^2y^2, x, y^2, x^2, xy^2\right\}$

    (c) order 4: $\left\{1, y, y^2, y^3\right\}$, $\left\{1, xy, y^2, xy^3\right\}$, $\left\{1, x^2y, y^2, x^2y^3\right\}$

    (d) order 3: $\left\{1, x, x^2\right\}$

    (e) order 2: $\left\{1, y^2\right\}$

Normal subgroups

    (a) order 12: $\left\{1, x, x^2, y, xy, x^2y, y^2, xy^2, x^2y^2, y^3, xy^3, x^2y^3\right\}$

    (b) order 6: $\left\{1, x^2y^2, x, y^2, x^2, xy^2\right\}$

    (c) order 3: $\left\{1, x, x^2\right\}$

    (d) order 2: $\left\{1, y^2\right\}$

Conjugacy classes: $\{1\}$, $\{y^2\}$, $\{x, x^2\}$, $\{xy^2, x^2y^2\}$, $\{y, xy, x^2y\}$, $\{y^3, xy^3, x^2y^3\}$

20. **A_4, the alternating group on 4 elements**

    Generators: a, b, c with relations $a^2 = 1$, $b^2 = 1$, $c^3 = 1$, $ba = ab$, $ca = abc$, $cb = ac$

	1	a	b	ab	c	ac	bc	abc	c^2	ac^2	bc^2	abc^2
1	1	a	b	ab	c	ac	bc	abc	c^2	ac^2	bc^2	abc^2
a	a	1	ab	b	ac	c	abc	bc	ac^2	c^2	abc^2	bc^2
b	b	ab	1	a	bc	abc	c	ac	bc^2	abc^2	c^2	ac^2
ab	ab	b	a	1	abc	bc	ac	c	abc^2	bc^2	ac^2	c^2
c	c	abc	ac	bc	c^2	abc^2	ac^2	bc^2	1	ab	a	b
ac	ac	bc	c	abc	ac^2	bc^2	c^2	abc^2	a	b	1	ab
bc	bc	ac	abc	c	bc^2	ac^2	abc^2	c^2	b	a	ab	1
abc	abc	c	bc	ac	abc^2	c^2	bc^2	ac^2	ab	1	b	a
c^2	c^2	bc^2	abc^2	ac^2	1	b	ab	a	c	bc	abc	ac
ac^2	ac^2	abc^2	bc^2	c^2	a	ab	b	1	ac	abc	bc	c
bc^2	bc^2	c^2	ac^2	abc^2	b	1	a	ab	bc	c	ac	abc
abc^2	abc^2	ac^2	c^2	bc^2	ab	a	1	b	abc	ac	c	bc

Elements

1. order 3: $c, ac, bc, abc, c^2, ac^2, bc^2, abc^2$
2. order 2: a, b, ab

Subgroups

1. order 12: $\{1, a, b, ab, c, ac, bc, abc, c^2, ac^2, bc^2, abc^2\}$
2. order 4: $\{1, a, b, ab\}$
3. order 3: $\{1, c, c^2\}, \{1, ac, bc^2\}, \{1, bc, abc^2\}, \{1, abc, ac^2\}$
4. order 2: $\{1, a\}, \{1, b\}, \{1, ab\}$

Normal subgroups

1. order 12: $\{1, a, b, ab, c, ac, bc, abc, c^2, ac^2, bc^2, abc^2\}$
2. order 4: $\{1, a, b, ab\}$

Conjugacy classes: $\{1\}, \{a, b, ab\}, \{c, ac, bc, abc\}, \{c^2, ac^2, bc^2, abc^2\}$

2.7.10.5 Small finite fields

In the following, the entries under α^i denote the coefficient of powers of α. For example, the last entry of the $p(x) = x^3 + x^2 + 1$ table is 1 1 0. That is: $\alpha^6 \equiv 1\alpha^2 + 1\alpha^1 + 0\alpha^0$ modulo $p(\alpha)$, where the coefficients are taken modulo 2.

$q = 4$	$x^2 + x + 1$		$q = 8$	$x^3 + x + 1$		$q = 8$	$x^3 + x^2 + 1$
i	α^i		i	α^i		i	α^i
0	0 1		0	0 0 1		0	0 0 1
1	1 0		1	0 1 0		1	0 1 0
2	1 1		2	1 0 0		2	1 0 0
			3	0 1 1		3	1 0 1
			4	1 1 0		4	1 1 1
			5	1 1 1		5	0 1 1
			6	1 0 1		6	1 1 0

$q = 16$		$x^4 + x + 1$			$q = 16$		$x^4 + x^3 + 1$	
i	α^i	7	1 0 1 1		i	α^i	7	0 1 1 1
0	0 0 0 1	8	0 1 0 1		0	0 0 0 1	8	1 1 1 0
1	0 0 1 0	9	1 0 1 0		1	0 0 1 0	9	0 1 0 1
2	0 1 0 0	10	0 1 1 1		2	0 1 0 0	10	1 0 1 0
3	1 0 0 0	11	1 1 1 0		3	1 0 0 0	11	1 1 0 1
4	0 0 1 1	12	1 1 1 1		4	1 0 0 1	12	0 0 1 1
5	0 1 1 0	13	1 1 0 1		5	1 0 1 1	13	0 1 1 0
6	1 1 0 0	14	1 0 0 1		6	1 1 1 1	14	1 1 0 0

2.7.10.6   Addition and multiplication tables for F_2, F_3, F_4, and F_8

1. F_2 addition and multiplication:

+	0	1
0	0	1
1	1	0

·	0	1
0	0	0
1	0	1

2. F_3 addition and multiplication:

+	0	1	2
0	0	1	2
1	1	2	0
2	2	0	1

·	0	1	2
0	0	0	0
1	0	1	2
2	0	2	1

3. F_4 addition and multiplication

+	0	1	α	$\alpha + 1$
0	0	1	α	$\alpha + 1$
1	1	0	$\alpha + 1$	α
α	α	$\alpha + 1$	0	1
$\alpha + 1$	$\alpha + 1$	α	1	0

·	0	1	α	$\alpha + 1$
0	0	0	0	0
1	0	1	α	$\alpha + 1$
α	0	α	$\alpha + 1$	1
$\alpha + 1$	0	$\alpha + 1$	1	α

4. F_8 addition and multiplication (using strings of 0s and 1s to represent the polynomials: $0 = 000$, $1 = 001$, $\alpha = 010$, $\alpha + 1 = 011$, $\alpha^2 = 100$, $\alpha^2 + \alpha = 110$, $\alpha^2 + 1 = 101$, $\alpha^2 + \alpha + 1 = 111$):

+	000	001	010	011	100	101	110	111
000	000	001	010	011	100	101	110	111
001	001	000	011	010	101	100	111	110
010	010	011	000	001	110	111	100	101
011	011	010	001	000	111	110	101	100
100	100	101	110	111	000	001	010	011
101	101	100	111	110	001	000	011	010
110	110	111	100	101	010	011	000	001
111	111	110	101	100	011	010	001	000

·	000	001	010	011	100	101	110	111
000	000	000	000	000	000	000	000	000
001	000	001	010	011	100	101	110	111
010	000	010	100	110	011	001	111	101
011	000	011	110	101	111	100	001	010
100	000	100	011	111	110	010	101	001
101	000	101	001	100	010	111	011	110
110	000	110	111	001	101	011	010	100
111	000	111	101	010	001	110	100	011

2.7.10.7 Linear characters

A *linear character* of a finite group G is a homomorphism from G to the multiplicative group of the non-zero complex numbers. Let χ be a linear character of G, ι the identity of G and $g, h \in G$. Then

$$\chi(\iota) = 1$$
$$\chi(g \star h) = \chi(g)\chi(h)$$
$$\chi(g) = e^{2\pi ik/n} \qquad \text{for some integer } k, \text{ where } n \text{ is the order of } g$$
$$\chi(g^{-1}) = \overline{\chi(g)}$$

The trivial character of G maps every element to 1. If χ_1 and χ_2 are two linear characters, then so is $\chi = \chi_1\chi_2$ defined by $\chi(g) = \chi_1(g)\chi_2(g)$. In fact, the linear characters form a group.

Group	Characters
$\mathbb{Z}_n$	For $m = 0, 1, \ldots, n-1$, $\chi_m : k \mapsto e^{2\pi ikm/n}$
G Abelian	$G \cong \mathbb{Z}_{n_1} \times \mathbb{Z}_{n_2} \times \cdots \times \mathbb{Z}_{n_j}$ with each n_i a power of a prime. For $m_j = 0, 1, \ldots, n_j - 1$ and $g_j = (0, \ldots, 0, 1, 0, \ldots, 0)$ $\chi_{m_1, m_2, \ldots, m_n} : g_j \mapsto e^{2\pi im_j/n_j}$
D_n dihedral	For $x = \pm 1, y = \begin{cases} \pm 1 & \text{if } n \text{ even,} \\ 1 & \text{if } n \text{ odd,} \end{cases}$ $\chi_{x,y} : a \mapsto x, b \mapsto y$. (See definition of D_n.)
Quaternions	For $x \pm 1$ and $y = \pm 1$ or ∓ 1, $\chi_{x,y} : \begin{bmatrix} 0 & 1 \\ -1 & 0 \end{bmatrix} \mapsto x, \qquad \chi_{x,y} : \begin{bmatrix} -i & 0 \\ 0 & i \end{bmatrix} \mapsto y$
S_n symmetric	The trivial character and sgn, where sgn is the signum function on permutations. (See Section 1.8.2.)

2.7.10.8 List of all sporadic finite simple groups

These are the sporadic finite simple groups that are not in any of the standard classes (see page 162).

Group	Order
M_{11}	$2^4 \cdot 3^2 \cdot 5 \cdot 11$
M_{12}	$2^6 \cdot 3^3 \cdot 5 \cdot 11$
M_{22}	$2^7 \cdot 3^2 \cdot 5 \cdot 7 \cdot 11$
M_{23}	$2^7 \cdot 3^2 \cdot 5 \cdot 7 \cdot 11 \cdot 23$
M_{24}	$2^{10} \cdot 3^3 \cdot 5 \cdot 7 \cdot 11 \cdot 23$
J_1	$2^3 \cdot 3 \cdot 5 \cdot 7 \cdot 11 \cdot 19$
J_2	$2^7 \cdot 3^3 \cdot 5^2 \cdot 7$

Group	Order
J_3	$2^7 \cdot 3^5 \cdot 5 \cdot 17 \cdot 19$
J_4	$2^{21} \cdot 3^3 \cdot 5 \cdot 7 \cdot 11^3 \cdot 23 \cdot 29 \cdot 31 \cdot 37 \cdot 43$
HS	$2^9 \cdot 3^2 \cdot 5^3 \cdot 7 \cdot 11$
Mc	$2^7 \cdot 3^6 \cdot 5^3 \cdot 11$
Suz	$2^{13} \cdot 3^7 \cdot 5^2 \cdot 7 \cdot 11 \cdot 13$
Ru	$2^{14} \cdot 3^3 \cdot 5^3 \cdot 7 \cdot 13 \cdot 29$
He	$2^{10} \cdot 3^3 \cdot 5^2 \cdot 7^3 \cdot 17$
Ly	$2^8 \cdot 3^7 \cdot 5^6 \cdot 7 \cdot 11 \cdot 31 \cdot 37 \cdot 67$
ON	$2^9 \cdot 3^4 \cdot 5 \cdot 7^3 \cdot 11 \cdot 19 \cdot 31$
$.1$	$2^{21} \cdot 3^9 \cdot 5^4 \cdot 7^2 \cdot 11 \cdot 13 \cdot 23$
$.2$	$2^{18} \cdot 3^6 \cdot 5^3 \cdot 7 \cdot 11 \cdot 23$
$.3$	$2^{10} \cdot 3^7 \cdot 5^3 \cdot 7 \cdot 11 \cdot 23$
$M(22)$	$2^{17} \cdot 3^9 \cdot 5^2 \cdot 7 \cdot 11 \cdot 13$
$M(23)$	$2^{18} \cdot 3^{13} \cdot 5^2 \cdot 7 \cdot 11 \cdot 13 \cdot 17 \cdot 23$
$M(24)'$	$2^{21} \cdot 3^{16} \cdot 5^2 \cdot 7^3 \cdot 11 \cdot 13 \cdot 23 \cdot 29$
F_5	$2^{15} \cdot 3^{10} \cdot 5^3 \cdot 7^2 \cdot 13 \cdot 19 \cdot 31$
F_3	$2^{14} \cdot 3^6 \cdot 5^6 \cdot 7 \cdot 11 \cdot 19$
F_2	$2^{41} \cdot 3^{13} \cdot 5^6 \cdot 7^2 \cdot 11 \cdot 13 \cdot 17 \cdot 19 \cdot 23 \cdot 31 \cdot 47$
F_1	$2^{46} \cdot 3^{20} \cdot 5^9 \cdot 7^6 \cdot 11^2 \cdot 13^3 \cdot 17 \cdot 19 \cdot 23 \cdot 29 \cdot 31 \cdot 41 \cdot 47 \cdot 59 \cdot 71$

2.7.10.9   Indices and power residues

For $\mathbb{Z}_n^*$ the following table lists the index (order) of a and the *power residues* a, a^2, ..., $a^{\text{index}(a)} = 1$ for each element a, where $(a, n) = 1$.

Group	Element	Index	Power residues
$\mathbb{Z}_2^*$	1	1	1
$\mathbb{Z}_3^*$	1	1	1
	2	2	2,1
$\mathbb{Z}_4^*$	1	1	1
	3	2	3,1
$\mathbb{Z}_5^*$	1	1	1
	2	4	2,4,3,1
	3	4	3,4,2,1
	4	2	4,1
$\mathbb{Z}_6^*$	1	1	1
	5	2	5,1
$\mathbb{Z}_7^*$	1	1	1
	2	3	2,4,1
	3	6	3,2,6,4,5,1
	4	3	4,2,1
	5	6	5,4,6,2,3,1
	6	2	6,1
$\mathbb{Z}_8^*$	1	1	1
	3	2	3,1
	5	2	5,1
	7	2	7,1

Group	Element	Index	Power residues
$\mathbb{Z}_9^*$	1	1	1
	2	6	2,4,8,7,5,1
	4	3	4,7,1
	5	6	5,7,8,4,2,1
	7	3	7,4,1
	8	2	8,1
$\mathbb{Z}_{10}^*$	1	1	1
	3	4	3,9,7,1
	7	4	7,9,3,1
	9	2	9,1
$\mathbb{Z}_{11}^*$	1	1	1
	2	10	2,4,8,5,10, 9,7,3,6,1
	3	5	3,9,5,4,1
	4	5	4,5,9,3,1
	5	5	5,3,4,9,1
	6	10	6,3,7,9,10, 5,8,4,2,1
	7	10	7,5,2,3,10, 4,6,9,8,1
	8	10	8,9,6,4,10, 3,2,5,7,1
	9	5	9,4,3,5,1
	10	2	10,1

2.7.10.10   Power residues in $\mathbf{Z}_p$

For prime $p < 40$, the following table lists the minimal primitive root a and the power residues of a. These can be used to find $a^m \pmod{p}$ for any $(a, p) = 1$. For example, to find $3^7 \pmod{11}$ ($a = 3, m = 7$), look in row $p = 11$ until the power of a that is equal to 3 is found. In this case $2^8 \equiv 3 \pmod{11}$. This means that $3^7 \equiv (2^8)^7 \equiv 2^{56} \equiv (2^{10})^5 \cdot 2^6 \equiv 2^6 \equiv 9 \pmod{11}$.

p	a	Power residues									
3	2		.0	.1	.2	.3	.4	.5	.6	.7	.8	.9
		0.	1	2	1							
5	2		.0	.1	.2	.3	.4	.5	.6	.7	.8	.9
		0.	1	2	4	3	1					
7	3		.0	.1	.2	.3	.4	.5	.6	.7	.8	.9
		0.	1	3	2	6	4	5	1			
11	2		.0	.1	.2	.3	.4	.5	.6	.7	.8	.9
		0.	1	2	4	8	5	10	9	7	3	6
		1.	1									
13	2		.0	.1	.2	.3	.4	.5	.6	.7	.8	.9
		0.	1	2	4	8	3	6	12	11	9	5
		1.	10	7	1							
17	3		.0	.1	.2	.3	.4	.5	.6	.7	.8	.9
		0.	1	3	9	10	13	5	15	11	16	14
		1.	8	7	4	12	2	6	1			
19	2		.0	.1	.2	.3	.4	.5	.6	.7	.8	.9
		0.	1	2	4	8	16	13	7	14	9	18
		1.	17	15	11	3	6	12	5	10	1	
23	5		.0	.1	.2	.3	.4	.5	.6	.7	.8	.9
		0.	1	5	2	10	4	20	8	17	16	11
		1.	9	22	18	21	13	19	3	15	6	7
		2.	12	14	1							
29	2		.0	.1	.2	.3	.4	.5	.6	.7	.8	.9
		0.	1	2	4	8	16	3	6	12	24	19
		1.	9	18	7	14	28	27	25	21	13	26
		2.	23	17	5	10	20	11	22	15	1	
31	3		.0	.1	.2	.3	.4	.5	.6	.7	.8	.9
		0.	1	3	9	27	19	26	16	17	20	29
		1.	25	13	8	24	10	30	28	22	4	12
		2.	5	15	14	11	2	6	18	23	7	21
		3.	1									
37	2		.0	.1	.2	.3	.4	.5	.6	.7	.8	.9
		0.	1	2	4	8	16	32	27	17	34	31
		1.	25	13	26	15	30	23	9	18	36	35
		2.	33	29	21	5	10	20	3	6	12	24
		3.	11	22	7	14	28	19	1			

2.7.10.11 Table of primitive monic polynomials

In the table below, the elements in each string are the coefficients of the polynomial after the highest power of x. (For example, 564 represents $x^3 + 5x^2 + 6x + 4$.)

Field	Degree	Primitive polynomials					
F_2	1	0	1				
	2	11					
	3	011	101				
	4	0011	1001				
	5	00101	01001	01111	10111	11011	11101
	6	000101	011011	100001	100111	101101	110011
F_3	1	0	1				
	2	12	22				
	3	021	121	201	211		
	4	0012	0022	1002	1122	1222	2002
		2112	2212				
F_5	1	0	2	3			
	2	12	23	33	42		
	3	032	033	042	043	102	113
		143	203	213	222	223	242
		302	312	322	323	343	403
		412	442				
F_7	1	0	2	4			
	2	13	23	25	35	45	53
		55	63				
	3	032	052	062	112	124	152
		154	214	242	262	264	304
		314	322	334	352	354	362
		422	432	434	444	504	524
		532	534	542	552	564	604
		612	632	644	654	662	664

2.7.10.12 Table of irreducible polynomials in $\mathbb{Z}_2[x]$

Each polynomial is represented by its coefficients (which are either 0 or 1), beginning with the highest power. For example, $x^4 + x + 1$ is represented as 10011.

degree 1:	10	11				
degree 2:	111					
degree 3:	1011	1101				
degree 4:	10011	11001	11111			
degree 5:	100101	101001	101111	110111	111011	111101
degree 6:	1000011	1001001	1010111	1011011	1100001	1100111
	1101101	1110011	1110101			
degree 7:	10000011	10001001	10001111	10010001	10011101	10100111
	10101011	10111001	10111111	11000001	11001011	11010011
	11010101	11100101	11101111	11110001	11110111	11111101
degree 8:	100011011	100011101	100101011	100101101	100111001	100111111
	101001101	101011111	101100011	101100101	101101001	101110001
	101110111	101111011	110000111	110001011	110001101	110011111
	110100011	110101001	110110001	110111101	111000011	111001111
	111010111	111011101	111100111	111110011	111110101	111111001

2.7.10.13 Table of primitive roots

The number of integers not exceeding and relatively prime to the integer n is $\phi(n)$ (see page 128). These integers form a group under multiplication module n; the group is cyclic if, and only if, $n = 1, 2, 4$ or n is of the form p^k or $2p^k$, where p is an odd prime. The number g is a *primitive root* of n if it generates that group, i.e., if $\{g, g^2, \ldots, g^{\phi(n)}\}$ are distinct modulo n. There are $\phi(\phi(n))$ primitive roots of n.

1. If g is a primitive root of p and $g^{p-1} \not\equiv 1 \pmod{p^2}$, then g is a primitive root of p^k for all k.

2. If $g^{p-1} \equiv 1 \pmod{p^2}$ then $g + p$ is a primitive root of p^k for all k.

3. If g is a primitive root of p^k, then either g or $g + p^k$, whichever is odd, is a primitive root of $2p^k$.

4. If g is a primitive root of n, then g^k is a primitive root of n if, and only if, k and $\phi(n)$ are relatively prime, i.e., $(\phi(n), k) = 1$.

In the following table,

- g denotes the least primitive root of p
- G denotes the least negative primitive root of p
- ϵ denotes whether $10, -10$, or both, are primitive roots of p

p	$p-1$	g	G	ϵ	p	$p-1$	g	G	ϵ
3	2	2	-1	—	5	2^2	2	-2	—
7	$2 \cdot 3$	3	-2	10	11	$2 \cdot 5$	2	-3	—
13	$2^2 \cdot 3$	2	-2	—	17	2^4	3	-3	± 10
19	$2 \cdot 3^2$	2	-4	10	23	$2 \cdot 11$	5	-2	10
29	$2^2 \cdot 7$	2	-2	± 10	31	$2 \cdot 3 \cdot 5$	3	-7	-10
37	$2^2 \cdot 3^2$	2	-2	—	41	$2^3 \cdot 5$	6	-6	—
43	$2 \cdot 3 \cdot 7$	3	-9	-10	47	$2 \cdot 23$	5	-2	10
53	$2^2 \cdot 13$	2	-2	—	59	$2 \cdot 29$	2	-3	10
61	$2^2 \cdot 3 \cdot 5$	2	-2	± 10	67	$2 \cdot 3 \cdot 11$	2	-4	-10
71	$2 \cdot 5 \cdot 7$	7	-2	-10	73	$2^3 \cdot 3^2$	5	-5	—
79	$2 \cdot 3 \cdot 13$	3	-2	—	83	$2 \cdot 41$	2	-3	-10
89	$2^3 \cdot 11$	3	-3	—	97	$2^5 \cdot 3$	5	-5	± 10
101	$2^2 \cdot 5^2$	2	-2	—	103	$2 \cdot 3 \cdot 17$	5	-2	—
107	$2 \cdot 53$	2	-3	-10	109	$2^2 \cdot 3^3$	6	-6	± 10
113	$2^4 \cdot 7$	3	-3	± 10	127	$2 \cdot 3^2 \cdot 7$	3	-9	—
131	$2 \cdot 5 \cdot 13$	2	-3	10	137	$2^3 \cdot 17$	3	-3	—
139	$2 \cdot 3 \cdot 23$	2	-4	—	149	$2^2 \cdot 37$	2	-2	± 10
151	$2 \cdot 3 \cdot 5^2$	6	-5	-10	157	$2^2 \cdot 3 \cdot 13$	5	-5	—
163	$2 \cdot 3^4$	2	-4	-10	167	$2 \cdot 83$	5	-2	10

2.7.10.14   Table of factorizations of $x^n - 1$

n	Factorization of $x^n - 1 \bmod 2$
1	$-1 + x$
2	$(-1 + x)(1 + x)$
3	$(-1 + x)\left(1 + x + x^2\right)$
4	$(-1 + x)(1 + x)\left(1 + x^2\right)$
5	$(-1 + x)\left(1 + x + x^2 + x^3 + x^4\right)$
6	$(-1 + x)(1 + x)\left(1 - x + x^2\right)\left(1 + x + x^2\right)$
7	$(-1 + x)\left(1 + x + x^2 + x^3 + x^4 + x^5 + x^6\right)$
8	$(-1 + x)(1 + x)\left(1 + x^2\right)\left(1 + x^4\right)$
9	$(-1 + x)\left(1 + x + x^2\right)\left(1 + x^3 + x^6\right)$
10	$(-1 + x)(1 + x)\left(1 - x + x^2 - x^3 + x^4\right)\left(1 + x + x^2 + x^3 + x^4\right)$
11	$(-1 + x)\left(1 + x + x^2 + x^3 + x^4 + x^5 + x^6 + x^7 + x^8 + x^9 + x^{10}\right)$
12	$(-1 + x)(1 + x)\left(1 + x^2\right)\left(1 - x + x^2\right)\left(1 + x + x^2\right)\left(1 - x^2 + x^4\right)$
13	$(-1 + x)\left(\sum_{k=0}^{12} x^k\right)$
14	$(-1 + x)(1 + x)\left(1 - x + x^2 - x^3 + x^4 - x^5 + x^6\right)$ $\left(1 + x + x^2 + x^3 + x^4 + x^5 + x^6\right)$
15	$(-1 + x)\left(1 + x + x^2\right)\left(1 + x + x^2 + x^3 + x^4\right)$ $\left(1 - x + x^3 - x^4 + x^5 - x^7 + x^8\right)$
16	$(-1 + x)(1 + x)\left(1 + x^2\right)\left(1 + x^4\right)\left(1 + x^8\right)$
17	$(-1 + x)\left(\sum_{k=0}^{16} x^k\right)$
18	$(-1 + x)(1 + x)\left(1 - x + x^2\right)\left(1 + x + x^2\right)\left(1 - x^3 + x^6\right)\left(1 + x^3 + x^6\right)$
19	$(-1 + x)\left(\sum_{k=0}^{18} x^k\right)$
20	$(-1 + x)(1 + x)\left(1 + x^2\right)\left(1 - x + x^2 - x^3 + x^4\right)$ $\left(1 + x + x^2 + x^3 + x^4\right)\left(1 - x^2 + x^4 - x^6 + x^8\right)$
21	$(-1 + x)\left(1 + x + x^2\right)\left(1 + x + x^2 + x^3 + x^4 + x^5 + x^6\right)$ $\left(1 - x + x^3 - x^4 + x^6 - x^8 + x^9 - x^{11} + x^{12}\right)$
22	$(-1 + x)\left(1 - x + x^2 - x^3 + x^4 - x^5 + x^6 - x^7 + x^8 - x^9 + x^{10}\right)$ $(1 + x)\left(1 + x + x^2 + x^3 + x^4 + x^5 + x^6 + x^7 + x^8 + x^9 + x^{10}\right)$
23	$(-1 + x)\left(\sum_{k=0}^{22} x^k\right)$
24	$(-1 + x)(1 + x)\left(1 + x^2\right)\left(1 - x + x^2\right)\left(1 + x + x^2\right)\left(1 + x^4\right)$ $\left(1 - x^2 + x^4\right)\left(1 - x^4 + x^8\right)$
25	$(-1 + x)\left(1 + x + x^2 + x^3 + x^4\right)\left(1 + x^5 + x^{10} + x^{15} + x^{20}\right)$
26	$(-1 + x)(1 + x)$ $\left(1 - x + x^2 - x^3 + x^4 - x^5 + x^6 - x^7 + x^8 - x^9 + x^{10} - x^{11} + x^{12}\right)$ $\left(1 + x + x^2 + x^3 + x^4 + x^5 + x^6 + x^7 + x^8 + x^9 + x^{10} + x^{11} + x^{12}\right)$
27	$(-1 + x)\left(1 + x + x^2\right)\left(1 + x^3 + x^6\right)\left(1 + x^9 + x^{18}\right)$
28	$(-1 + x)(1 + x)\left(1 + x^2\right)\left(1 - x + x^2 - x^3 + x^4 - x^5 + x^6\right)$ $\left(1 + x + x^2 + x^3 + x^4 + x^5 + x^6\right)\left(1 - x^2 + x^4 - x^6 + x^8 - x^{10} + x^{12}\right)$
29	$(-1 + x)\left(\sum_{k=0}^{28} x^k\right)$
30	$(-1 + x)(1 + x)\left(1 - x + x^2\right)\left(1 + x + x^2\right)\left(1 - x + x^2 - x^3 + x^4\right)$ $\left(1 + x + x^2 + x^3 + x^4\right)\left(1 - x + x^3 - x^4 + x^5 - x^7 + x^8\right)$ $\left(1 + x - x^3 - x^4 - x^5 + x^7 + x^8\right)$
32	$(-1 + x)(1 + x)\left(1 + x^2\right)\left(1 + x^4\right)\left(1 + x^8\right)\left(1 + x^{16}\right)$

Chapter **3**

Discrete Mathematics

3.1   SYMBOLIC LOGIC **199**
    3.1.1   Propositional calculus *199*
    3.1.2   Tautologies *199*
    3.1.3   Truth tables as functions *200*
    3.1.4   Rules of inference *200*
    3.1.5   Deductions *201*
    3.1.6   Predicate calculus *201*

3.2   SET THEORY **202**
    3.2.1   Sets *202*
    3.2.2   Set operations and relations *202*
    3.2.3   Connection between sets and probability *203*
    3.2.4   Venn diagrams *203*
    3.2.5   Paradoxes and theorems of set theory *203*
    3.2.6   Inclusion/Exclusion *204*
    3.2.7   Partially ordered sets *204*

3.3   COMBINATORICS **206**
    3.3.1   Sample selection *206*
    3.3.2   Balls into cells *206*
    3.3.3   Binomial coefficients *208*
    3.3.4   Multinomial coefficients *209*
    3.3.5   Arrangements and derangements *210*
    3.3.6   Partitions *210*
    3.3.7   Bell numbers *211*
    3.3.8   Catalan numbers *212*
    3.3.9   Stirling cycle numbers *212*
    3.3.10   Stirling subset numbers *213*
    3.3.11   Tables *215*

3.4   GRAPHS **219**
    3.4.1   Notation *219*
    3.4.2   Basic definitions *219*
    3.4.3   Constructions *228*
    3.4.4   Fundamental results *231*
    3.4.5   Tree diagrams *241*

3.5   COMBINATORIAL DESIGN THEORY **241**
    3.5.1   t-Designs *241*
    3.5.2   Balanced incomplete block designs (BIBDs) *245*

1-58488-291-3/02/$0.00+$1.50
© 2003 CRC Press, Inc.

3.5.3   *Difference sets* 246
3.5.4   *Finite geometry* 247
3.5.5   *Steiner triple systems* 249
3.5.6   *Hadamard matrices* 249
3.5.7   *Latin squares* 251
3.5.8   *Room squares* 252
3.5.9   *Costas arrays* 252

3.6   **COMMUNICATION THEORY** **253**
    3.6.1   *Information theory* 253
    3.6.2   *Block coding* 256
    3.6.3   *Source coding for English text* 260
    3.6.4   *Morse code* 260
    3.6.5   *Gray code* 261
    3.6.6   *Finite fields* 261
    3.6.7   *Binary sequences* 263

3.7   **DIFFERENCE EQUATIONS** **265**
    3.7.1   *The calculus of finite differences* 265
    3.7.2   *Existence and uniqueness* 265
    3.7.3   *Linear independence: general solution* 266
    3.7.4   *Homogeneous equations with constant coefficients* 267
    3.7.5   *Non-homogeneous equations* 268
    3.7.6   *Generating functions and Z transforms* 268
    3.7.7   *Closed-form solutions for special equations* 269

3.8   **DISCRETE DYNAMICAL SYSTEMS AND CHAOS** **272**
    3.8.1   *Chaotic one-dimensional maps* 272
    3.8.2   *Logistic map* 272
    3.8.3   *Julia sets and the Mandelbrot set* 273

3.9   **GAME THEORY** **274**
    3.9.1   *Two person non-cooperative matrix games* 274
    3.9.2   *Voting power* 278

3.10   **OPERATIONS RESEARCH** **280**
    3.10.1   *Linear programming* 280
    3.10.2   *Duality and complementary slackness* 285
    3.10.3   *Linear integer programming* 287
    3.10.4   *Branch and bound* 287
    3.10.5   *Network flow methods* 288
    3.10.6   *Assignment problem* 288
    3.10.7   *Dynamic programming* 289
    3.10.8   *Shortest path problem* 290
    3.10.9   *Heuristic search techniques* 290

3.1 SYMBOLIC LOGIC

3.1.1 PROPOSITIONAL CALCULUS

Propositional calculus is the study of statements: how they are combined and how to determine their truth. Statements (or propositions) are combined by means of *connectives* such as *and* ($\wedge$), *or* ($\vee$), *not* ($\neg$, or sometimes $\sim$), *implies* ($\rightarrow$), and *if and only if* ($\longleftrightarrow$). Propositions are denoted by letters $\{p, q, r, \ldots\}$. For example, if p is the statement "$x = 3$", and q the statement "$y = 4$", then $p \vee \neg q$ would be interpreted as "$x = 3$ or $y \neq 4$". To determine the truth of a statement, *truth tables* are used. Using T (for true) and F (for false), the truth tables for these connectives are as follows:

p	q	$p \wedge q$	$p \vee q$	$p \rightarrow q$	$p \longleftrightarrow q$		p	$\neg p$
T	T	T	T	T	T		T	F
T	F	F	T	F	F		F	T
F	T	F	T	T	F			
F	F	F	F	T	T			

The proposition $p \rightarrow q$ can be read "*If p then q*" or, less often, "*q if p*". The table shows that "$p \vee q$" is an *inclusive or* because it is true even when p and q are both true. Thus, the statement "I'm watching TV or I'm doing homework" is a true statement if the narrator happens to be both watching TV and doing homework. Note that $p \rightarrow q$ is false only when p is true and q is false. Thus, *a false statement implies any statement* and *a true statement is implied by any statement*.

3.1.2 TAUTOLOGIES

A statement such as $(p \rightarrow (q \wedge r)) \vee \neg p$ is a *compound statement* composed of the *atomic propositions* p, q, and r. The letters P, Q, and R are used to designate compound statements. A *tautology* is a compound statement which always is true, regardless of the truth values of the atomic statements used to define it. For example, a simple tautology is $(\neg\neg p) \longleftrightarrow p$. Tautologies are logical truths. Some examples are as follows:

Law of the excluded middle	$p \vee \neg p$
De Morgan's laws	$\neg(p \vee q) \longleftrightarrow (\neg p \wedge \neg q)$
	$\neg(p \wedge q) \longleftrightarrow (\neg p \vee \neg q)$
Modus ponens	$(p \wedge (p \rightarrow q)) \rightarrow q$
Contrapositive law	$(p \rightarrow q) \longleftrightarrow (\neg q \rightarrow \neg p)$
Reductio ad absurdum	$(\neg p \rightarrow p) \rightarrow p$
Elimination of cases	$((p \vee q) \wedge \neg p) \rightarrow q$
Transitivity of implication	$((p \rightarrow q) \wedge (q \rightarrow r)) \rightarrow (p \rightarrow r)$
Proof by cases	$((p \rightarrow q) \wedge (\neg p \rightarrow q)) \rightarrow q$

Idempotent laws	$p \wedge p \longleftrightarrow p$; $p \vee p \longleftrightarrow p$
Commutative laws	$(p \wedge q) \longleftrightarrow (q \wedge p)$; $(p \vee q) \longleftrightarrow (q \vee p)$
Associative laws	$(p \wedge (q \wedge r)) \longleftrightarrow ((p \wedge q) \wedge r)$
	$(p \vee (q \vee r)) \longleftrightarrow ((p \vee q) \vee r)$

3.1.3 TRUTH TABLES AS FUNCTIONS

If we assign the value 1 to T, and 0 to F, then the truth table for $p \wedge q$ is simply the value pq. This can be done with all the connectives as follows:

Connective	Arithmetic function
$p \wedge q$	pq
$p \vee q$	$p + q - pq$
$p \rightarrow q$	$1 - p + pq$
$p \longleftrightarrow q$	$1 - p - q + 2pq$
$\neg p$	$1 - p$

These formulae may be used to verify tautologies, because, from this point of view, a tautology is a function whose value is identically 1. In using them, it is useful to remember that $pp = p^2 = p$, since $p = 0$ or $p = 1$.

3.1.4 RULES OF INFERENCE

A *rule of inference* in propositional calculus is a method of arriving at a valid (true) *conclusion*, given certain statements, assumed to be true, which are called the *hypotheses*. For example, suppose that P and Q are compound statements. Then if P and $P \rightarrow Q$ are true, then Q must necessarily be true. This follows from the *modus ponens* tautology in the above list of tautologies. We write this rule of inference $P, P \rightarrow Q \Rightarrow Q$. It is also classically written

$$\frac{\begin{array}{c} P \\ P \rightarrow Q \end{array}}{Q}$$

Some examples of rules of inferences follow, all derived from the above list of tautologies:

Modus ponens	$P, P \rightarrow Q \Rightarrow Q$
Contrapositive	$P \rightarrow Q \Rightarrow \neg Q \rightarrow \neg P$
Modus tollens	$P \rightarrow Q, \neg Q \Rightarrow \neg P$
Transitivity	$P \rightarrow Q, Q \rightarrow R \Rightarrow P \rightarrow R$
Elimination of cases	$P \vee Q, \neg P \Rightarrow Q$
"And" usage	$P \wedge Q \Rightarrow P, Q$

3.1.5 DEDUCTIONS

A *deduction* from hypotheses is a list of statements, each one of which is either one of the hypotheses, a tautology, or follows from previous statements in the list by a valid rule of inference. It follows that if the hypotheses are true, then the conclusion must be true. Suppose, for example, that we are given *hypotheses* $\neg q \to p$, $q \to \neg r$, r; it is required to deduce the *conclusion* p. A deduction showing this, with reasons for each step is as follows:

	Statement	Reason
1.	$q \to \neg r$	Hypothesis
2.	r	Hypothesis
3.	$\neg q$	Modus tollens (1,2)
4.	$\neg q \to p$	Hypothesis
5.	p	Modus ponens (3,4)

3.1.6 PREDICATE CALCULUS

Unlike propositional calculus, which may be considered the skeleton of logical discourse, *predicate calculus* is the language in which most mathematical reasoning takes place. It uses the symbols of propositional calculus, with the exception of the propositional variables p, q, ... Predicate calculus uses the *universal quantifier* $\forall$, the *existential quantifier* $\exists$, predicates $P(x), Q(x, y), \ldots$, variables $x, y, \ldots$, and assumes a *universe* U from which the variables are taken. The quantifiers are illustrated in the following table.

Symbol	Read as	Usage	Interpretation
$\exists$	There exists an	$\exists x(x > 10)$	There is an x such that $x > 10$
$\forall$	For all	$\forall x(x^2 + 1 \neq 0)$	For all x, $x^2 + 1 \neq 0$

Predicates are variable statements which may be true or false, depending on the values of its variable. In the above table, "$x > 10$" is a predicate in the one variable x as is "$x^2 + 1 \neq 0$". Without a given universe, we have no way of deciding whether a statement is true or false. Thus $\forall x(x^2 + 1 \neq 0)$ is true if the universe U is the real numbers, but false if U is the complex numbers. A useful rule for manipulating quantifiers is

$$\neg \forall x P(x) \longleftrightarrow \exists x \neg P(x).$$

For example, it is not true that all people are mortal if, and only if, there is a person who is immortal. Here the universe U is the set of people, and $P(x)$ is the predicate "x is mortal". This works with more than one quantifier. Thus,

$$\neg \forall x \exists y P(x, y) \longleftrightarrow \exists x \forall y \neg P(x, y).$$

For example, if it is not true that every person loves someone, then it follows that there is a person who loves no one (and vice versa).

Fermat's last theorem, stated in terms of the predicate calculus (U = the positive integers), is

$$\forall n \forall a \forall b \forall c[(n > 2) \rightarrow (a^n + b^n \neq c^n)].$$

It was proven in 1995; its proof is extremely complicated. One does not expect a simple deduction, as in the propositional calculus.

In 1931, Gödel proved the *Gödel Incompleteness Theorem*. This states that, in any logical system complex enough to contain arithmetic, it will always be possible to find a true result which is not formally provable using predicate logic. This result was especially startling because the notion of truth and provability had been often identified with each other.

3.2 SET THEORY

3.2.1 SETS

A *set* is a collection of objects. Some examples of sets are

1. The population of Cleveland on January 1, 1995
2. The real numbers between 0 and 1 inclusive
3. The prime numbers 2, 3, 5, 7, 11, . . .
4. The numbers 1, 2, 3, and 4
5. All of the formulae in this book

3.2.2 SET OPERATIONS AND RELATIONS

If x is an element in a set A, then we write $x \in A$ (read "x is in A"). If x is not in A, we write $x \notin A$. When considering sets, a set U, called the *universe*, is chosen, from which all elements are taken. The *null set* or *empty set* $\emptyset$ is the set containing no elements. Thus, $x \notin \emptyset$ for all $x \in U$. Some relations on sets are as follows:

Relation	Read as	Definition
$A \subseteq B$	A is contained in B	Any element of A is also an element of B
$A = B$	A equals B	$(A \subseteq B) \wedge (B \subseteq A)$

Some basic operations on sets are as follows:

Operation	Read as	Definition
$A \cup B$	A union B	The elements in A or in B
$A \cap B$	A intersection B	The elements in both A and B
$A - B$	A minus B	The elements in A which are not in B
A' or $\overline{A}$ or A^C	Complement of A	The elements in U which are not in A
$\mathbf{P}(A)$ or 2^A	Power set of A	The collection of all subsets of A
$A \oplus B$	Symmetric difference of A and B	The elements of A and B that are not in both A and B (i.e., the union minus the intersection)

3.2.3 CONNECTION BETWEEN SETS AND PROBABILITY

Set concept	Probability concept
Set	Event
Set containing a single element	Indecomposable, elementary, or atomic event
Set containing more than one element	Compound event
Universal set or space	Sample space
Complement of a set	Non-occurrence of an event
Function on the universal set	Random variable
Measure of a set	Probability of an event
Integral with respect to the measure	Expectation or expected value

3.2.4 VENN DIAGRAMS

The operations and relations on sets can be illustrated by *Venn diagrams*. The diagrams below show a few possibilities.

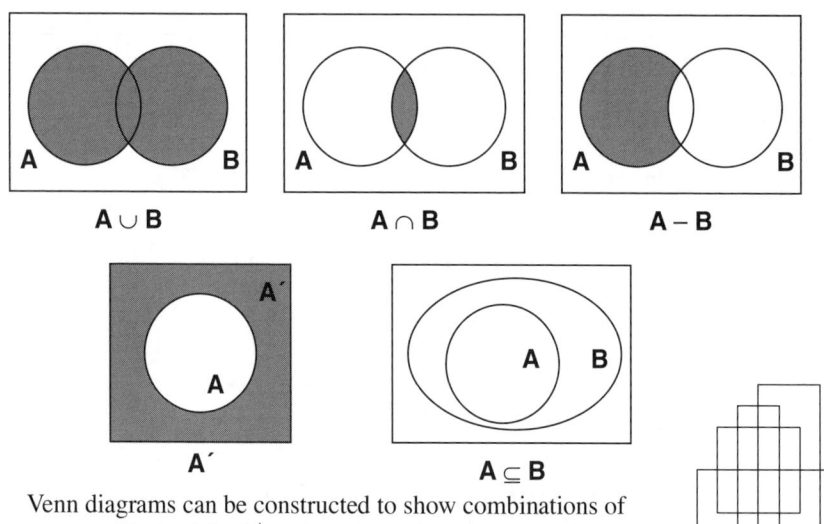

Venn diagrams can be constructed to show combinations of many events. Each of the 2^4 regions created by the rectangles in the diagram to the right represents a different combination.

3.2.5 PARADOXES AND THEOREMS OF SET THEORY

3.2.5.1 Russell's paradox

In about 1900, Bertrand Russell presented a paradox, paraphrased as follows: since the elements of sets can be arbitrary, sets can contain sets as elements. Therefore, a set can possibly be a member of itself. (For example, the set of all sets would be a member of itself. Another example is the collection of all sets that can be described in fewer than 50 words.) Now let A be the set of all sets which are *not* members of themselves. Then if A is a member of itself, it is not a member of itself. And if A is not a member of itself, then, by definition, A is a member of itself. This paradox leads to a much more careful evaluation of how sets should be defined.

3.2.5.2   Infinite sets and the continuum hypothesis

Georg Cantor showed how the number of elements of infinite sets can be counted, much as finite sets. He used the symbol $\aleph_0$ (read "aleph null") for the number of integers and introduced larger infinite numbers such as $\aleph_1$, $\aleph_2$, and so on. Cantor introduced a consistent arithmetic on infinite cardinals and a way of comparing infinite cardinals. A few of his results were as follows:

$$\aleph_0 + \aleph_0 = \aleph_0, \qquad (\aleph_0)^2 = \aleph_0, \qquad 2^{\aleph_0} = \aleph_0^{\aleph_0} > \aleph_0.$$

Cantor showed that $\mathbf{c} = 2^{\aleph_0} > \aleph_0$, where $\mathbf{c}$ is the cardinality of real numbers. The *continuum hypothesis* asks whether or not $\mathbf{c} = \aleph_1$, the first infinite cardinal greater than $\aleph_0$. In 1963, Paul J. Cohen showed that this result is independent of the other axioms of set theory. In his words, "... the truth or falsity of the continuum hypothesis ... cannot be determined by set theory as we know it today".

3.2.6   INCLUSION/EXCLUSION

Let $\{a_1, a_2, \ldots, a_r\}$ be properties that the elements of a set may or may not have. If the set has N objects, then the number of objects having exactly m properties (with $m \le r$), e_m, is given by

$$e_m = s_m - \binom{m+1}{1} s_{m+1} + \binom{m+2}{2} s_{m+2} - \binom{m+3}{3} s_{m+3} + \cdots \qquad (3.2.1)$$
$$\cdots + (-1)^p \binom{m+p}{p} s_{m+p} \cdots + (-1)^{r-m} \binom{m+(r-m)}{(r-m)} s_r.$$

Here $s_t = \sum N(a_{i_1} a_{i_2} \cdots a_{i_t})$ is the number of elements that have $a_{i_1}, \ldots$. When $m = 0$, this is the usual inclusion/exclusion rule:

$$e_0 = s_0 - s_1 + s_2 - \cdots + (-1)^r s_r,$$
$$= N - \sum_i N(a_i) + \sum_{i,j \text{ distinct}} N(a_i a_j) - \sum_{i,j,k \text{ distinct}} N(a_i a_j a_k) + \cdots \qquad (3.2.2)$$

3.2.7   PARTIALLY ORDERED SETS

Consider a set S and a relation on it. Given any two elements x and y in S, we can determine whether or not x is "related" to y; if it is, "$x \preceq y$". The relation "$\preceq$" will be a *partial order* on S if it satisfies the following three conditions:

reflexive	$s \preceq s$ for every $s \in S$,
antisymmetric	$s \preceq t$ and $t \preceq s$ imply $s = t$, and
transitive	$s \preceq t$ and $t \preceq u$ imply $s \preceq u$.

FIGURE 3.1

Left: Hasse diagram for integers up to 12 with $x \preceq y$ meaning "the number x divides the number y". Right: Hasse diagram for the power set of $\{a, b, c\}$ with $x \preceq y$ meaning "the set x is a subset of the set y".

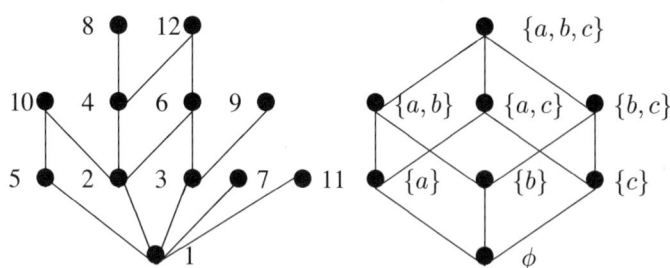

If $\preceq$ is a partial order on S, then the pair $(S, \preceq)$ is called a *partially ordered set* or a *poset*. Given the partial order $\preceq$ on the set S, define the relation $\prec$ by

$$x \prec y \qquad \text{if and only if} \qquad x \preceq y \text{ and } x \neq y.$$

We say that the element t *covers* the element s if $s \prec t$ and there is no element u with $s \prec u \prec t$. A *Hasse diagram* of the poset $(S, \preceq)$ is a figure consisting of the elements of S with a line segment directed generally upward from s to t whenever t covers s. (See Figure 3.1.)

Two elements x and y in a poset $(S, \preceq)$ are said to be *comparable* if either $x \preceq y$ or $y \preceq x$. If every pair of elements in a poset is comparable, then $(S, \preceq)$ is a *chain*. An *antichain* is a poset in which no two elements are comparable (i.e., $x \preceq y$ if and only if $x = y$ for all x and y in the antichain). A *maximal chain* is a chain that is not properly contained in another chain (and similarly for a *maximal antichain*).

EXAMPLES

1. Let S be the set of natural numbers up to 12 and let "$x \preceq y$" mean "the number x divides the number y". Then $(S, \preceq)$ is a poset with the Hasse diagram shown in Figure 3.1 (left). Observe that the elements 2 and 4 are comparable, but elements 2 and 5 are not comparable.

2. Let S be the set of all subsets of the set $\{a, b, c\}$ and let "$x \preceq y$" mean "the set x is contained in the set y". Then $(S, \preceq)$ is a poset with the Hasse diagram shown in Figure 3.1 (right).

3. There are 16 different isomorphism types of posets of size 4. There are 5 different isomorphism types of posets of size 3, shown below:

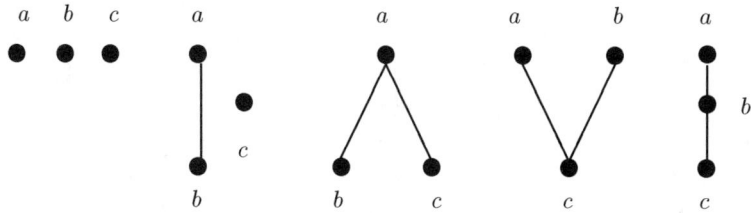

3.3 COMBINATORICS

3.3.1 SAMPLE SELECTION

There are four diffeent ways in which a sample of r elements can be obtained from a set of n distinguishable objects.

Order counts?	Repetitions allowed?	The sample is called an	Number of ways to choose the sample
No	No	r-combination	$C(n,r)$
Yes	No	r-permutation	$P(n,r)$
No	Yes	r-combination with replacement	$C^R(n,r)$
Yes	Yes	r-permutation with replacement	$P^R(n,r)$

where

$$C(n,r) = \binom{n}{r} = \frac{n!}{r!\,(n-r)!},$$

$$P(n,r) = (n)_r = n^{\underline{r}} = \frac{n!}{(n-r)!}, \tag{3.3.1}$$

$$C^R(n,r) = C(n+r-1,r) = \frac{(n+r-1)!}{r!(n-1)!}, \text{ and}$$

$$P^R(n,r) = n^r.$$

EXAMPLE   There are four ways to choose a 2-element sample from the set $\{a,b\}$:

r-combination	$C(2,2) = 1$	ab
r-permutation	$P(2,2) = 2$	ab and ba
r-combination with replacement	$C^R(2,2) = 3$	$aa, ab,$ and bb
r-permutation with replacement	$P^R(2,2) = 4$	$aa, ab, ba,$ and bb

3.3.2 BALLS INTO CELLS

There are eight different ways in which n balls can be placed into k cells:

Distinguish the balls?	Distinguish the cells?	Can cells be empty?	Number of ways to place n balls into k cells
Yes	Yes	Yes	k^n
Yes	Yes	No	$k!\left\{{n \atop k}\right\}$
Yes	No	Yes	$\left\{{n \atop 1}\right\} + \left\{{n \atop 2}\right\} + \ldots + \left\{{n \atop k}\right\}$
Yes	No	No	$\left\{{n \atop k}\right\}$
No	Yes	Yes	$C(k+n-1,n) = \binom{k+n-1}{n}$
No	Yes	No	$C(n-1,k-1) = \binom{n-1}{k-1}$
No	No	Yes	$p_1(n) + p_2(n) + \ldots + p_k(n)$
No	No	No	$p_k(n)$

where $\left\{ {n \atop k} \right\}$ is the Stirling subset number (see page 213) and $p_k(n)$ is the number of partitions of the number n into exactly k integer pieces (see page 210).

Given n distinguishable balls and k distinguishable cells, the number of ways in which we can place n_1 balls into cell 1, n_2 balls into cell 2, ..., n_k balls into cell k, is given by the multinomial coefficient $\binom{n}{n_1, n_2, \ldots, n_k}$ (see page 209).

EXAMPLE Consider placing $n = 3$ balls into $k = 2$ cells. Let $\{A, B, C\}$ denote the names of the balls (when needed) and $\{a, b, c\}$ denote the names of the cells (when needed). A cell will be denoted <u>like this</u>. Begin with: Are the balls distinguishable?

1. Yes, the balls are distinguishable. Are the cells distinguishable?

   (a) Yes, the cells are distinguishable. Can the cells be empty?

      i. Yes. Number of ways is $k^n = 2^3 = 8$:

      ii. No. Number of ways is $k! \left\{ {n \atop k} \right\} = 2! \left\{ {3 \atop 2} \right\} = 6$:

   (b) No, the cells are not distinguishable. Can the cells be empty?

      i. Yes. Number of ways is $\left\{ {n \atop 1} \right\} + \ldots + \left\{ {n \atop k} \right\} = \left\{ {3 \atop 1} \right\} + \left\{ {3 \atop 2} \right\} = 1 + 3 = 4$:

      ii. No. Number of ways is $\left\{ {n \atop k} \right\} = \left\{ {3 \atop 2} \right\} = 3$:

2. No, the balls are not distinguishable. Are the cells distinguishable?

   (a) Yes, the cells are distinguishable. Can the cells be empty?

      i. Yes. Number of ways is $\binom{k+n-1}{n} = \binom{4}{3} = 4$:

      ii. No. Number of ways is $\binom{n-1}{k-1} = \binom{2}{1} = 2$:

   (b) No, the cells are not distinguishable. Can the cells be empty?

      i. Yes. Number of ways is $p_1(n) + \ldots + p_k(n) = p_1(3) + p_2(3) = 1 + 1 = 2$:

      ii. No. Number of ways is $p_k(n) = p_2(3) = 1$:

3.3.3 BINOMIAL COEFFICIENTS

The binomial coefficient $C(n, m) = \binom{n}{m}$ is the number of ways of choosing m objects from a collection of n distinct objects without regard to order:

EXAMPLE   For the 5 element set $\{a, b, c, d, e\}$ there are $\binom{5}{3} = \frac{5!}{3!2!} = 10$ subsets containing exactly three elements. They are:

$$\{a, b, c\}, \quad \{a, b, d\}, \quad \{a, b, e\}, \quad \{a, c, d\}, \quad \{a, c, e\},$$
$$\{a, d, e\}, \quad \{b, c, d\}, \quad \{b, c, e\}, \quad \{b, d, e\}, \quad \{c, d, e\}.$$

Properties of binomial coefficients include:

1. $\dbinom{n}{m} = \dfrac{n!}{m!(n-m)!} = \dfrac{n(n-1)\cdots(n-m+1)}{m!} = \dbinom{n}{n-m}.$

2. $\dbinom{n}{0} = \dbinom{n}{n} = 1$ and $\dbinom{n}{1} = n.$

3. $\dbinom{2n}{n} = \dfrac{2^n(2n-1)!!}{n!} = \dfrac{2^n(2n-1)(2n-3)\cdots 3 \cdot 1}{n!}.$

4. If n and m are integers, and $m > n$, then $\binom{n}{m} = 0.$

5. The recurrence relation: $\binom{n+1}{m} = \binom{n}{m} + \binom{n}{m-1}.$

6. Two generating functions for binomial coefficients are $\sum_{m=0}^{n} \binom{n}{m} x^m = (1+x)^n$ for $n = 1, 2, \ldots$, and $\sum_{n=m}^{\infty} \binom{n}{m} x^{n-m} = (1-x)^{-m-1}.$

7. The *Vandermonde convolution* is $\dbinom{x+y}{n} = \displaystyle\sum_{k=0}^{n} \dbinom{x}{k}\dbinom{y}{n-k}.$

3.3.3.1 Pascal's triangle

The binomial coefficients $\binom{n}{k}$ can be arranged in a triangle in which each number is the sum of the two numbers above it. For example $\binom{3}{2} = \binom{2}{1} + \binom{2}{2}.$

```
          1
         1 1
        1 2 1
       1 3 3 1
      1 4 6 4 1
     1 5 10 10 5 1
    1 6 15 20 15 6 1
   1 7 21 35 35 21 7 1
  1 8 28 56 70 56 28 8 1
```

$$\binom{0}{0}$$
$$\binom{1}{0} \quad \binom{1}{1}$$
$$\binom{2}{0} \quad \binom{2}{1} \quad \binom{2}{2}$$
$$\binom{3}{0} \quad \binom{3}{1} \quad \binom{3}{2} \quad \binom{3}{3}$$
$$\binom{4}{0} \quad \binom{4}{1} \quad \binom{4}{2} \quad \binom{4}{3} \quad \binom{4}{4}$$

3.3.3.2 Binomial coefficient relationships

The binomial coefficients satisfy

$$\binom{n+1}{m+1} = \binom{n}{m} + \binom{n}{m+1},$$

$$\binom{n}{m} = \binom{n}{n-m},$$

$$\binom{n+m+1}{n+1} = \binom{n}{n} + \binom{n+1}{n} + \binom{n+2}{n} + \ldots + \binom{n+m}{n},$$

$$\binom{2n}{n} = \binom{n}{0}^2 + \binom{n}{1}^2 + \ldots + \binom{n}{n}^2,$$

$$\binom{m+n}{p} = \binom{m}{0}\binom{n}{p} + \binom{m}{1}\binom{n}{p-1} + \ldots + \binom{m}{p}\binom{n}{0},$$

$$2^n = \binom{n}{0} + \binom{n}{1} + \ldots + \binom{n}{n},$$

$$0 = \binom{n}{0} - \binom{n}{1} + \ldots + (-1)^n \binom{n}{n} \qquad \text{for } n \geq 1,$$

$$2^{n-1} = \binom{n}{0} + \binom{n}{2} + \binom{n}{4} + \ldots \qquad \text{for } n \geq 1,$$

$$2^{n-1} = \binom{n}{1} + \binom{n}{3} + \binom{n}{5} + \ldots \qquad \text{for } n \geq 1,$$

$$0 = 1\binom{n}{1} - 2\binom{n}{2} + \ldots + (-1)^{n+1} n \binom{n}{n} \qquad \text{for } n \geq 1.$$

3.3.4 MULTINOMIAL COEFFICIENTS

The multinomial coefficient $\binom{n}{n_1,n_2,\ldots,n_k}$ (also written $C(n; n_1, n_2, \ldots, n_k)$) is the number of ways of choosing n_1 objects, then n_2 objects, $\ldots$, then n_k objects from a collection of n distinct objects without regard to order. This requires that $\sum_{j=1}^{k} n_j = n$. The multinomial symbol is numerically evaluated as

$$\binom{n}{n_1, n_2, \ldots, n_k} = \frac{n!}{n_1! \, n_2! \cdots n_k!}. \qquad (3.3.2)$$

EXAMPLE   The number of ways to choose 2 objects, then 1 object, then 1 object from the set $\{a, b, c, d\}$ is $\binom{4}{2,1,1} = 12$; they are as follows (vertical bars show the ordered selections):

$\mid ab \mid c \mid d \mid$,	$\mid ab \mid d \mid c \mid$,	$\mid ac \mid b \mid d \mid$,	$\mid ac \mid d \mid b \mid$,
$\mid ad \mid b \mid c \mid$,	$\mid ad \mid c \mid b \mid$,	$\mid bc \mid a \mid d \mid$,	$\mid bc \mid d \mid a \mid$,
$\mid bd \mid a \mid c \mid$,	$\mid bd \mid c \mid a \mid$,	$\mid cd \mid a \mid b \mid$,	$\mid cd \mid b \mid a \mid$.

3.3.5 ARRANGEMENTS AND DERANGEMENTS

The number of ways to arrange n distinct objects in a row is $n!$; this is the number of permutations of n objects. For example, for the three objects $\{a, b, c\}$, the number of arrangements is $3! = 6$. These permutations are: abc, bac, cab, acb, bca, and cba.

The number of ways to arrange n objects (assuming that there are k types of objects and n_i copies of each object of type i) is the multinomial coefficient $\binom{n}{n_1, n_2, \ldots, n_k}$. For example, for the set $\{a, a, b, c\}$ the parameters are $n = 4$, $k = 3$, $n_1 = 2$, $n_2 = 1$, and $n_3 = 1$. Hence, there are $\binom{4}{2,1,1} = \frac{4!}{2!\,1!\,1!} = 12$ arrangements; they are

$$aabc, \quad aacb, \quad abac, \quad abca, \quad acab, \quad acba,$$
$$baac, \quad baca, \quad bcaa, \quad caab, \quad caba, \quad cbaa.$$

A *derangement* is a permutation of objects, in which object i is not in the ith location. For example, all of the derangements of $\{1, 2, 3, 4\}$ are

$$2143, \quad 2341, \quad 2413,$$
$$3142, \quad 3412, \quad 3421,$$
$$4123, \quad 4312, \quad 4321.$$

The number of derangements of n elements, D_n, satisfies the recursion relation, $D_n = (n - 1)(D_{n-1} + D_{n-2})$, with the initial values $D_1 = 0$ and $D_2 = 1$. Hence,

$$D_n = n! \left(1 - \frac{1}{1!} + \frac{1}{2!} - \frac{1}{3!} + \ldots + (-1)^n \frac{1}{n!} \right).$$

The numbers D_n are also called *sub-factorials* or *rencontres numbers*. For large values of n, $D_n/n! \sim e^{-1} \approx 0.37$. Hence more than one of every three permutations is a derangement.

n	1	2	3	4	5	6	7	8	9	10
D_n	0	1	2	9	44	265	1854	14833	133496	1334961

3.3.6 PARTITIONS

A partition of a number n is a representation of n as the sum of any number of positive integral parts. The number of partitions of n is denoted $p(n)$. For example:
$$5 = 4 + 1 = 3 + 2 = 3 + 1 + 1 = 2 + 2 + 1 = 2 + 1 + 1 + 1 = 1 + 1 + 1 + 1 + 1$$
so that $p(5) = 7$.

1. The number of partitions of n into exactly m parts is equal to the number of partitions of n into parts the largest of which is exactly m; this is denoted $p_m(n)$. For example, $p_2(5) = 2$ and $p_3(5) = 2$. Note that $\sum_m p_m(n) = p(n)$.

2. The number of partitions of n into at most m parts is equal to the number of partitions of n into parts which do not exceed m.

The generating functions for $p(n)$ is

$$1 + \sum_{n=1}^{\infty} p(n)x^n = \frac{1}{(1-x)(1-x^2)(1-x^3)\cdots} \tag{3.3.3}$$

n	1	2	3	4	5	6	7	8	9	10
$p(n)$	1	2	3	5	7	11	15	22	30	42
n	11	12	13	14	15	16	17	18	19	20
$p(n)$	56	77	101	135	176	231	297	385	490	627
n	21	22	23	24	25	26	27	28	29	30
$p(n)$	792	1002	1255	1575	1958	2436	3010	3718	4565	5604
n	31	32	33	34	35	40	45	50		
$p(n)$	6842	8349	10143	12310	14883	37338	89134	204226		

A table of $p_m(n)$ values. The columns sum to $p(n)$.

m	$n=1$	2	3	4	5	6	7	8	9	10	11
1	1	1	1	1	1	1	1	1	1	1	1
2		1	1	2	2	3	3	4	4	5	5
3			1	1	2	3	4	5	7	8	10
4				1	1	2	3	5	6	9	11
5					1	1	2	3	5	7	10
6						1	1	2	3	5	7
7							1	1	2	3	5
8								1	1	2	3
9									1	1	2
10										1	1

3.3.7 BELL NUMBERS

The nth Bell number, B_n, denotes the number of partitions of a set with n elements. Computationally, the Bell numbers may be written in terms of the Stirling subset numbers (page 213), $B_n = \sum_{m=1}^{n} \left\{ {n \atop m} \right\}$.

n	1	2	3	4	5	6	7	8	9	10
B_n	1	2	5	15	52	203	877	4140	21147	115975

1. A generating function for Bell numbers is $\sum_{n=0}^{\infty} B_n x^n = \exp\left(e^x - 1\right) - 1$. This gives *Dobinski's formula*: $B_n = e^{-1} \sum_{m=0}^{\infty} m^n / m!$.

2. For large values of n, $B_n \sim n^{-1/2} [\lambda(n)]^{n+1/2} e^{\lambda(n)-n-1}$ where $\lambda(n)$ is defined by the relation: $\lambda(n) \log \lambda(n) = n$.

EXAMPLE   There are $B_4 = 15$ different ways to partition the 4 element set $\{a, b, c, d\}$:

$\{a\}, \{c\}, \{b, d\},$ $\{a\}, \{d\}, \{b, c\},$ $\{b\}, \{c\}, \{a, d\},$ $\{b\}, \{d\}, \{a, c\},$
$\{c\}, \{d\}, \{a, b\},$ $\{a, b, c, d\},$ $\{a, b\}, \{c, d\},$ $\{a, c\}, \{b, d\},$
$\{a, d\}, \{b, c\},$ $\{a\}, \{b\}, \{c\}, \{d\},$ $\{a\}, \{b, c, d\},$ $\{b\}, \{a, c, d\},$
$\{c\}, \{a, b, d\},$ $\{d\}, \{a, b, c\},$ $\{a\}, \{b\}, \{c, d\}.$

3.3.8 CATALAN NUMBERS

The Catalan numbers are $C_n = \dfrac{1}{n-1}\dbinom{2n-2}{n-2}$. There is the recurrence relation:

$$C_n = C_0 C_{n-1} + C_1 C_{n-2} + \ldots + C_{n-1} C_0. \qquad (3.3.4)$$

n	0	1	2	3	4	5	6	7	8	9	10
C_n	$1/2$	1	1	2	5	14	42	132	429	1430	4862

EXAMPLE   Given the product $A_1 A_2 \ldots A_n$, the number of ways to pair terms keeping the original order is C_n. For example, with $n = 4$, there are $C_4 = 5$ ways to group the terms; they are $(A_1 A_2)(A_3 A_4)$, $((A_1 A_2) A_3) A_4$, $(A_1 (A_2 A_3)) A_4$, $A_1 ((A_2 A_3) A_4)$, and $A_1 (A_2 (A_3 A_4))$.

3.3.9 STIRLING CYCLE NUMBERS

The number $\begin{bmatrix} n \\ k \end{bmatrix}$, called a *Stirling cycle number*, is the number of permutations of n symbols which have exactly k non-empty cycles.

EXAMPLE   For the 4 element set $\{a, b, c, d\}$, there are $\begin{bmatrix} 4 \\ 2 \end{bmatrix} = 11$ permutations containing exactly 2 cycles. They are

$$\begin{pmatrix} 1\,2\,3\,4 \\ 2\,3\,1\,4 \end{pmatrix} = (123)(4), \qquad \begin{pmatrix} 1\,2\,3\,4 \\ 3\,1\,2\,4 \end{pmatrix} = (132)(4), \qquad \begin{pmatrix} 1\,2\,3\,4 \\ 3\,2\,4\,1 \end{pmatrix} = (134)(2),$$

$$\begin{pmatrix} 1\,2\,3\,4 \\ 4\,2\,1\,3 \end{pmatrix} = (143)(2), \qquad \begin{pmatrix} 1\,2\,3\,4 \\ 2\,4\,3\,1 \end{pmatrix} = (124)(3), \qquad \begin{pmatrix} 1\,2\,3\,4 \\ 4\,1\,3\,2 \end{pmatrix} = (142)(3),$$

$$\begin{pmatrix} 1\,2\,3\,4 \\ 1\,3\,4\,2 \end{pmatrix} = (234)(1), \qquad \begin{pmatrix} 1\,2\,3\,4 \\ 1\,4\,2\,3 \end{pmatrix} = (243)(1), \qquad \begin{pmatrix} 1\,2\,3\,4 \\ 2\,1\,4\,3 \end{pmatrix} = (12)(34),$$

$$\begin{pmatrix} 1\,2\,3\,4 \\ 3\,4\,1\,2 \end{pmatrix} = (13)(24), \qquad \begin{pmatrix} 1\,2\,3\,4 \\ 4\,3\,2\,1 \end{pmatrix} = (14)(23).$$

3.3.9.1 Properties of Stirling cycle numbers $\begin{bmatrix} n \\ k \end{bmatrix}$

1. $\begin{bmatrix} n \\ k \end{bmatrix} = (n-1) \begin{bmatrix} n-1 \\ k \end{bmatrix} + n \begin{bmatrix} n-1 \\ k-1 \end{bmatrix}$ for $k > 0$.

2. $\begin{bmatrix} n \\ 0 \end{bmatrix} = \begin{cases} 1 & \text{if } n = 0 \\ 0 & \text{if } n \neq 0 \end{cases}$ 3. $\displaystyle\sum_{k=0}^{n} \begin{bmatrix} n \\ k \end{bmatrix} = n!$

4. $\begin{bmatrix} n \\ k \end{bmatrix} = \displaystyle\sum_{m=0}^{n-k} (-1)^m \binom{n-1+m}{n-k+m} \binom{2n-k}{n-m-k} \left\{ \begin{matrix} n-m-k \\ m \end{matrix} \right\}$

    where $\left\{ \begin{matrix} n-m-k \\ k \end{matrix} \right\}$ is a Stirling subset number.

5. $\sum_{n=0}^{\infty} s(n,k)\frac{x^n}{n!} = \frac{(\log(1+x))^k}{k!}$ for $|x| < 1$. Here $s(n,k)$ is a *Stirling number of the first kind* and can be written as $s(n,k) = (-1)^{n-k}\left[{n \atop k}\right]$.

6. The factorial polynomial, defined as $x^{(n)} = x(x-1)\cdots(x-n+1)$ with $x^{(0)} = 1$, can be written as

$$x^{(n)} = \sum_{k=0}^{n} s(n,k)x^k = s(n,1)x + s(n,2)x^2 + \ldots + s(n,n)x^n \qquad (3.3.5)$$

For example: $x^{(3)} = x(x-1)(x-2) = 2x - 3x^2 + x^3 = \left[{3 \atop 1}\right]x - \left[{3 \atop 2}\right]x^2 + \left[{3 \atop 3}\right]x^3$.

3.3.9.2 Table of Stirling cycle numbers $\left[{n \atop k}\right]$

	k							
n	0	1	2	3	4	5	6	7
0	1							
1	0	1						
2	0	1	1					
3	0	2	3	1				
4	0	6	11	6	1			
5	0	24	50	35	10	1		
6	0	120	274	225	85	15	1	
7	0	720	1764	1624	735	175	21	1
8	0	5040	13068	13132	6769	1960	322	28
9	0	40320	109584	118124	67284	22449	4536	546
10	0	362880	1026576	1172700	723680	269325	63273	9450

3.3.10 STIRLING SUBSET NUMBERS

The *Stirling subset number*, $\left\{{n \atop k}\right\}$, is the number of ways to partition n into k blocks. Equivalently, it is the number of ways that n distinguishable balls can be placed into k indistinguishable cells, with no cell empty.

EXAMPLE Placing the 4 distinguishable balls $\{a, b, c, d\}$ into 2 indistinguishable cells, so that no cell is empty, can be done in $\left\{{4 \atop 2}\right\} = 7$ ways. These are (vertical bars delineate the cells)

$$|\, ab \,|\, cd \,|, \quad |\, ad \,|\, bc \,|, \quad |\, ac \,|\, bd \,|, \quad |\, a \,|\, bcd \,|,$$
$$|\, b \,|\, acd \,|, \quad |\, c \,|\, abd \,|, \quad |\, d \,|\, abc \,|\,.$$

3.3.10.1 Properties of Stirling subset numbers $\left\{{n \atop k}\right\}$

1. Stirling subset numbers are also called *Stirling numbers of the second kind*, and are denoted by $S(n,k)$.

2. $\left\{ {n \atop 0} \right\} = \begin{cases} 1 & \text{if } n = 0 \\ 0 & \text{if } n \neq 0. \end{cases}$

3. $\left\{ {n \atop 1} \right\} = \left\{ {n \atop n} \right\} = 1.$

4. $\left\{ {n \atop 2} \right\} = 2^{n-1} - 1.$

5. $\displaystyle\sum_{n=0}^{\infty} \left\{ {n \atop k} \right\} \frac{x^n}{n!} = \frac{(e^x - 1)^k}{k!}.$

6. $\left\{ {n \atop k} \right\} = k \left\{ {n-1 \atop k} \right\} + \left\{ {n-1 \atop k-1} \right\}$ for $k > 0$.

7. $\left\{ {n \atop k} \right\} = \displaystyle\frac{1}{k!} \sum_{m=0}^{k} (-1)^{k-m} \binom{k}{m} m^n.$

8. Ordinary powers can be expanded in terms of factorial polynomials. If $n > 0$, then

$$x^n = \sum_{k=0}^{n} \left\{ {n \atop k} \right\} x^{(k)} = \left\{ {n \atop 0} \right\} x^{(0)} + \left\{ {n \atop 1} \right\} x^{(1)} + \ldots + \left\{ {n \atop n} \right\} x^{(n)}. \tag{3.3.6}$$

For example,

$$\begin{aligned} x^3 &= \left\{ {3 \atop 0} \right\} x^{(0)} + \left\{ {3 \atop 1} \right\} x^{(1)} + \left\{ {3 \atop 2} \right\} x^{(2)} + \left\{ {3 \atop 3} \right\} x^{(3)} \\ &= \left\{ {3 \atop 0} \right\} + \left\{ {3 \atop 1} \right\} x + \left\{ {3 \atop 2} \right\} x(x-1) + \left\{ {3 \atop 3} \right\} x(x-1)(x-2) \\ &= 0 + x + 3(x^2 - x) + (x^3 - 3x^2 + 2x). \end{aligned} \tag{3.3.7}$$

3.3.10.2 Table of Stirling subset numbers $\left\{ {n \atop k} \right\}$

	k							
n	0	1	2	3	4	5	6	7
0	1							
1	0	1						
2	0	1	1					
3	0	1	3	1				
4	0	1	7	6	1			
5	0	1	15	25	10	1		
6	0	1	31	90	65	15	1	
7	0	1	63	301	350	140	21	1
8	0	1	127	966	1701	1050	266	28
9	0	1	255	3025	7770	6951	2646	462
10	0	1	511	9330	34105	42525	22827	5880
11	0	1	1023	28501	145750	246730	179487	63987
12	0	1	2047	86526	611501	1379400	1323652	627396
13	0	1	4095	261625	2532530	7508501	9321312	5715424
14	0	1	8191	788970	10391745	40075035	63436373	49329280
15	0	1	16383	2375101	42355950	210766920	420693273	408741333

3.3.11 TABLES

3.3.11.1 Permutations $P(n, m)$

These tables contain the number of permutations of n distinct things taken m at a time, given by $P(n, m) = \dfrac{n!}{(n - m)!} = n(n - 1) \cdots (n - m + 1)$.

					m				
n	0	1	2	3	4	5	6	7	8
0	1								
1	1	1							
2	1	2	2						
3	1	3	6	6					
4	1	4	12	24	24				
5	1	5	20	60	120	120			
6	1	6	30	120	360	720	720		
7	1	7	42	210	840	2520	5040	5040	
8	1	8	56	336	1680	6720	20160	40320	40320
9	1	9	72	504	3024	15120	60480	181440	362880
10	1	10	90	720	5040	30240	151200	604800	1814400
11	1	11	110	990	7920	55440	332640	1663200	6652800
12	1	12	132	1320	11880	95040	665280	3991680	19958400
13	1	13	156	1716	17160	154440	1235520	8648640	51891840
14	1	14	182	2184	24024	240240	2162160	17297280	121080960
15	1	15	210	2730	32760	360360	3603600	32432400	259459200

		m			
n	9	10	11	12	13
9	362880				
10	3628800	3628800			
11	19958400	39916800	39916800		
12	79833600	239500800	479001600	479001600	
13	259459200	1037836800	3113510400	6227020800	6227020800
14	726485760	3632428800	14529715200	43589145600	87178291200
15	1816214400	10897286400	54486432000	217945728000	653837184000

3.3.11.2 Combinations $C(n, m) = \binom{n}{m}$

These tables contains the number of combinations of n distinct things taken m at a time, given by $C(n, m) = \dbinom{n}{m} = \dfrac{n!}{m!(n - m)!}$.

				m				
n	0	1	2	3	4	5	6	7
1	1	1						
2	1	2	1					
3	1	3	3	1				
4	1	4	6	4	1			
5	1	5	10	10	5	1		

| | m | | | | | | | |
n	0	1	2	3	4	5	6	7
6	1	6	15	20	15	6	1	
7	1	7	21	35	35	21	7	1
8	1	8	28	56	70	56	28	8
9	1	9	36	84	126	126	84	36
10	1	10	45	120	210	252	210	120
11	1	11	55	165	330	462	462	330
12	1	12	66	220	495	792	924	792
13	1	13	78	286	715	1287	1716	1716
14	1	14	91	364	1001	2002	3003	3432
15	1	15	105	455	1365	3003	5005	6435
16	1	16	120	560	1820	4368	8008	11440
17	1	17	136	680	2380	6188	12376	19448
18	1	18	153	816	3060	8568	18564	31824
19	1	19	171	969	3876	11628	27132	50388
20	1	20	190	1140	4845	15504	38760	77520
21	1	21	210	1330	5985	20349	54264	116280
22	1	22	231	1540	7315	26334	74613	170544
23	1	23	253	1771	8855	33649	100947	245157
24	1	24	276	2024	10626	42504	134596	346104
25	1	25	300	2300	12650	53130	177100	480700
26	1	26	325	2600	14950	65780	230230	657800
27	1	27	351	2925	17550	80730	296010	888030
28	1	28	378	3276	20475	98280	376740	1184040
29	1	29	406	3654	23751	118755	475020	1560780
30	1	30	435	4060	27405	142506	593775	2035800
31	1	31	465	4495	31465	169911	736281	2629575
32	1	32	496	4960	35960	201376	906192	3365856
33	1	33	528	5456	40920	237336	1107568	4272048
34	1	34	561	5984	46376	278256	1344904	5379616
35	1	35	595	6545	52360	324632	1623160	6724520
36	1	36	630	7140	58905	376992	1947792	8347680
37	1	37	666	7770	66045	435897	2324784	10295472
38	1	38	703	8436	73815	501942	2760681	12620256
39	1	39	741	9139	82251	575757	3262623	15380937
40	1	40	780	9880	91390	658008	3838380	18643560

| | m | | | | |
n	8	9	10	11	12
8	1				
9	9	1			
10	45	10	1		
11	165	55	11	1	
12	495	220	66	12	1
13	1287	715	286	78	13
14	3003	2002	1001	364	91

			m		
n	8	9	10	11	12
15	6435	5005	3003	1365	455
16	12870	11440	8008	4368	1820
17	24310	24310	19448	12376	6188
18	43758	48620	43758	31824	18564
19	75582	92378	92378	75582	50388
20	125970	167960	184756	167960	125970
21	203490	293930	352716	352716	293930
22	319770	497420	646646	705432	646646
23	490314	817190	1144066	1352078	1352078
24	735471	1307504	1961256	2496144	2704156
25	1081575	2042975	3268760	4457400	5200300
26	1562275	3124550	5311735	7726160	9657700
27	2220075	4686825	8436285	13037895	17383860
28	3108105	6906900	13123110	21474180	30421755
29	4292145	10015005	20030010	34597290	51895935
30	5852925	14307150	30045015	54627300	86493225
31	7888725	20160075	44352165	84672315	141120525
32	10518300	28048800	64512240	129024480	225792840
33	13884156	38567100	92561040	193536720	354817320
34	18156204	52451256	131128140	286097760	548354040
35	23535820	70607460	183579396	417225900	834451800
36	30260340	94143280	254186856	600805296	1251677700
37	38608020	124403620	348330136	854992152	1852482996
38	48903492	163011640	472733756	1203322288	2707475148
39	61523748	211915132	635745396	1676056044	3910797436
40	76904685	273438880	847660528	2311801440	5586853480

			m		
n	13	14	15	16	17
13	1				
14	14	1			
15	105	15	1		
16	560	120	16	1	
17	2380	680	136	17	1
18	8568	3060	816	153	18
19	27132	11628	3876	969	171
20	77520	38760	15504	4845	1140
21	203490	116280	54264	20349	5985
22	497420	319770	170544	74613	26334
23	1144066	817190	490314	245157	100947
24	2496144	1961256	1307504	735471	346104
25	5200300	4457400	3268760	2042975	1081575
26	10400600	9657700	7726160	5311735	3124550
27	20058300	20058300	17383860	13037895	8436285

	m				
n	13	14	15	16	17
28	37442160	40116600	37442160	30421755	21474180
29	67863915	77558760	77558760	67863915	51895935
30	119759850	145422675	155117520	145422675	119759850
31	206253075	265182525	300540195	300540195	265182525
32	347373600	471435600	565722720	601080390	565722720
33	573166440	818809200	1037158320	1166803110	1166803110
34	927983760	1391975640	1855967520	2203961430	2333606220
35	1476337800	2319959400	3247943160	4059928950	4537567650
36	2310789600	3796297200	5567902560	7307872110	8597496600
37	3562467300	6107086800	9364199760	12875774670	15905368710
38	5414950296	9669554100	15471286560	22239974430	28781143380
39	8122425444	15084504396	25140840660	37711260990	51021117810
40	12033222880	23206929840	40225345056	62852101650	88732378800

3.3.11.3   Fractional binomial coefficients $\binom{a}{k}$

				k		
a	0	1	2	3	4	5
1/2	1	1/2	$-1/8$	1/16	$-5/128$	7/256
1/3	1	1/3	$-1/9$	5/81	$-10/243$	22/729
1/4	1	1/4	$-3/32$	7/128	$-77/2048$	231/8192
1/5	1	1/5	$-2/25$	6/125	$-21/625$	399/15625
1/6	1	1/6	$-5/72$	55/1296	$-935/31104$	4301/186624
1/7	1	1/7	$-3/49$	13/343	$-65/2401$	351/16807
1/8	1	1/8	$-7/128$	35/1024	$-805/32768$	4991/262144
1/9	1	1/9	$-4/81$	68/2187	$-442/19683$	3094/177147
2/3	1	2/3	$-1/9$	4/81	$-7/243$	14/729
2/5	1	2/5	$-3/25$	8/125	$-26/625$	468/15625
2/7	1	2/7	$-5/49$	20/343	$-95/2401$	494/16807
2/9	1	2/9	$-7/81$	112/2187	$-700/19683$	4760/177147
3/4	1	3/4	$-3/32$	5/128	$-45/2048$	117/8192
3/5	1	3/5	$-3/25$	7/125	$-21/625$	357/15625
3/7	1	3/7	$-6/49$	22/343	$-99/2401$	495/16807
3/8	1	3/8	$-15/128$	65/1024	$-1365/32768$	7917/262144
4/5	1	4/5	$-2/25$	4/125	$-11/625$	176/15625
4/7	1	4/7	$-6/49$	20/343	$-85/2401$	408/16807
4/9	1	4/9	$-10/81$	140/2187	$-805/19683$	5152/177147
5/6	1	5/6	$-5/72$	35/1296	$-455/31104$	1729/186624
5/7	1	5/7	$-5/49$	15/343	$-60/2401$	276/16807
5/8	1	5/8	$-15/128$	55/1024	$-1045/32768$	5643/262144
5/9	1	5/9	$-10/81$	130/2187	$-715/19683$	4433/177147
6/7	1	6/7	$-3/49$	8/343	$-30/2401$	132/16807
7/8	1	7/8	$-7/128$	21/1024	$-357/32768$	1785/262144
7/9	1	7/9	$-7/81$	77/2187	$-385/19683$	2233/177147
8/9	1	8/9	$-4/81$	40/2187	$-190/19683$	1064/177147

3.4 GRAPHS

3.4.1 NOTATION

3.4.1.1 Notation for graphs

E	edge set	V	vertex set
G	graph	ϕ	incidence mapping

3.4.1.2 Graph invariants

$\mathrm{Aut}(G)$	automorphism group	$c(G)$	circumference
$d(u,v)$	distance between two vertices	$\mathrm{diam}(G)$	diameter
$\deg x$	degree of a vertex	$e(G)$	size
$\mathrm{ecc}(x)$	eccentricity	$\mathrm{gir}(G)$	girth
$\mathrm{rad}(G)$	radius	$P_G(x)$	chromatic polynomial
$Z(G)$	center	$\alpha(G)$	independence number
$\chi(G)$	chromatic number	$\chi'(G)$	chromatic index
$\delta(G)$	minimum degree	$\Delta(G)$	maximum degree
$\gamma(G)$	genus	$\tilde{\gamma}(G)$	crosscap number
$\kappa(G)$	vertex connectivity	$\lambda(G)$	edge connectivity
$\nu(G)$	crossing number	$\overline{\nu}(G)$	rectilinear crossing number
$\theta(G)$	thickness	$\Upsilon(G)$	arboricity
$\omega(G)$	clique number	$|G|$	order

3.4.1.3 Examples of graphs

C_n	cycle	O_n	odd graph
$\overline{K}_n$	empty graph	P_n	path
K_n	complete graph	Q_n	cube
$K_{m,n}$	complete bipartite graph	S_n	star
$K_n^{(m)}$	Kneser graphs	$T_{n,k}$	Turán graph
M_n	Möbius ladder	W_n	wheel

3.4.2 BASIC DEFINITIONS

There are two standard definitions of graphs, a general definition and a more common simplification. Except where otherwise indicated, this book uses the simplified definition, according to which a *graph* is an ordered pair (V, E) consisting of an arbitrary set V and a set E of 2-element subsets of V. Each element of V is called a *vertex* (plural *vertices*). Each element of E is called an *edge*.

According to the general definition, a *graph* is an ordered triple $G = (V, E, \phi)$ consisting of arbitrary sets V and E and an *incidence mapping* ϕ that assigns to each element $e \in E$ a non-empty set $\phi(e) \subseteq V$ of cardinality at most two. Again, the elements of V are called *vertices* and the elements of E are called *edges*. A *loop* is an edge e for which $|\phi(e)| = 1$. A graph has *multiple edges* if edges $e \neq e'$ exist for which $\phi(e) = \phi(e')$.

A (general) graph is called *simple* if it has neither loops nor multiple edges. Because each edge in a simple graph can be identified with the two-element set $\phi(e) \subseteq V$, the simplified definition of graph given above is just an alternative definition of a simple graph.

The word *multigraph* is used to discuss general graphs with multiple edges but no loops. Occasionally the word *pseudograph* is used to emphasize that the graphs under discussion may have both loops and multiple edges. Every graph $G = (V, E)$ considered here is *finite*, i.e., both V and E are finite sets.

Specialized graph terms include the following:

acyclic: A graph is *acyclic* if it has no cycles.

adjacency: Two distinct vertices v and w in a graph are *adjacent* if the pair $\{v, w\}$ is an edge. Two distinct edges are *adjacent* if their intersection is non-empty, i.e., if there is a vertex incident with both of them.

adjacency matrix: For an ordering $v_1, v_2, \ldots, v_n$ of the vertices of a graph $G = (V, E)$ of order $|G| = n$, there is a corresponding $n \times n$ *adjacency matrix* $A = (a_{ij})$ defined as follows:

$$a_{ij} = \begin{cases} 1 & \text{if } \{v_i, v_j\} \in E; \\ 0 & \text{otherwise.} \end{cases} \tag{3.4.1}$$

arboricity: The *arboricity* $\Upsilon(G)$ of a graph G is the minimum number of edge-disjoint spanning forests into which G can be partitioned.

automorphism: An *automorphism* of a graph is a permutation of its vertices that is an isomorphism.

automorphism group: The composition of two automorphisms is again an automorphism; with this binary operation, the automorphisms of a graph G form a group $\text{Aut}(G)$ called the *automorphism group* of G.

ball: The *ball* of radius k about a vertex u in a graph is the set

$$B(u, k) = \{v \in V \mid d(u, v) \leq k\}. \tag{3.4.2}$$

See also *sphere* and *neighborhood*.

block: A *block* is a graph with no cut vertex. A *block* of a graph is a maximal subgraph that is a block.

boundary operator: The *boundary operator* for a graph is the linear mapping from 1-chains (elements of the edge space) to 0-chains (elements of the vertex space) that sends each edge to the indicator mapping the set of two vertices incident with it. See also *vertex space* and *edge space*.

bridge: A *bridge* is an edge in a connected graph whose removal would disconnect the graph.

cactus: A *cactus* is a connected graph, each of whose blocks is a cycle.

cage: An (r, n)-*cage* is a graph of minimal order among r-regular graphs with girth n. A $(3, n)$-cage is also called an n-cage.

center: The *center* $Z(G)$ of a graph $G = (V, E)$ consists of all vertices whose eccentricity equals the radius of G:

$$Z(G) = \{v \in V(G) \mid \operatorname{ecc}(v) = \operatorname{rad}(G)\}. \tag{3.4.3}$$

Each vertex in the center of G is called a *central vertex*.

characteristic polynomial: All adjacency matrices of a graph G have the same characteristic polynomial, which is called the *characteristic polynomial* of G.

chromatic index: The *chromatic index* $\chi'(G)$ is the least k for which there exists a proper k-coloring of the edges of G; in other words, it is the least number of matchings into which the edge set can be decomposed.

chromatic number: The *chromatic number* $\chi(G)$ of a graph G is the least k for which there exists a proper k-coloring of the vertices of G; in other words, it is the least k for which G is k-partite. See also *multipartite*.

chromatic polynomial: For a graph G of order $|G| = n$ with exactly k connected components, the *chromatic polynomial* of G is the unique polynomial $P_G(x)$ for which $P_G(m)$ is the number of proper colorings of G with m colors for each positive integer m.

circuit: A *circuit* in a graph is a trail whose first and last vertices are identical.

circulant graph: A graph G is a *circulant graph* if its adjacency matrix is a circulant matrix; that is, the rows are circular shifts of one another.

circumference: The circumference of a graph is the length of its longest cycle.

clique: A *clique* is a set S of vertices for which the induced subgraph $G[S]$ is complete.

clique number: The *clique number* $\omega(G)$ of a graph G is the largest cardinality of a clique in G.

coboundary operator: The *coboundary operator* for a graph is the linear mapping from 0-chains (elements of the vertex space) to 1-chains (elements of the edge space) that sends each vertex to the indicator mapping of the set of edges incident with it.

cocycle vector: A cut vector is sometimes called a *cocycle vector*.

coloring: A partition of the vertex set of a graph is called a *coloring*, and the blocks of the partition are called *color classes*. A coloring with k color classes is called a *k-coloring*. A coloring is *proper* if no two adjacent vertices belong to the same color class. See also *chromatic number* and *chromatic polynomial*.

complement: The *complement* $\overline{G}$ of a graph $G = (V, E)$ has vertex set V and edge set $\binom{V}{2} \setminus E$; that is, its edges are exactly the pairs of vertices that are not edges of G.

complete graph: A graph is *complete* if every pair of distinct vertices is an edge; K_n denotes a complete graph with n vertices.

component: A *component* of a graph is a maximal connected subgraph.

connectedness: A graph is said to be *connected* if each pair of vertices is joined by a walk; otherwise, the graph is *disconnected*. A graph is *k-connected* if it has order at least $k + 1$ and each pair of vertices is joined by k pairwise internally disjoint paths.

connectivity: The *connectivity* $\kappa(G)$ of G is the largest k for which G is k-connected.

contraction: To *contract* an edge $\{v, w\}$ of a graph G is to construct a new graph G' from G by removing the edge $\{v, w\}$ and identifying the vertices v and w. A graph G is *contractible* to a graph H if H can be obtained from G via the contraction of one or more edges of G.

cover: A set $S \subseteq V$ is a *vertex cover* if every edge of G is incident with some vertex in S. A set $T \subseteq E$ is an *edge cover* of a graph $G = (V, E)$ if each vertex of G is incident to at least one edge in T.

crosscap number: The *crosscap number* $\tilde{\gamma}(G)$ of a graph G is the least g for which G has an embedding in a non-orientable surface obtained from the sphere by adding g crosscaps. See also *genus*.

crossing: A *crossing* is a point lying in images of two edges of a drawing of a graph on a surface.

crossing number: The *crossing number* $\nu(G)$ of a graph G is the minimum number of crossings among all drawings of G in the plane. The *rectilinear crossing number* $\overline{\nu}(G)$ of a graph G is the minimum number of crossings among all drawings of G in the plane for which the image of each edge is a straight line segment.

cubic: A graph is a *cubic* graph if it is regular of degree 3.

cut: For each partition $V = V_1 \uplus V_2$ of the vertex set of a graph $G = (V, E)$ into two disjoint blocks, the set of all edges joining a vertex in V_1 to a vertex in V_2 is called a *cut*.

cut space: The *cut space* of a graph G is the subspace of the edge space of G spanned by the cut vectors.

cut vector: The *cut vector* corresponding to a cut C of a graph $G = (V, E)$ is the mapping $v \colon E \to GF(2)$ in the edge space of G

$$v(e) = \begin{cases} 1, & e \in C, \\ 0, & \text{otherwise.} \end{cases} \tag{3.4.4}$$

cut vertex: A *cut vertex* of a connected graph is a vertex whose removal, along with all edges incident with it, leaves a disconnected graph.

cycle: A *cycle* is a circuit, each pair of whose vertices other than the first and the last are distinct.

cycle space: The *cycle space* of a graph G is the subspace of the edge space of G consisting of all 1-chains with boundary 0. An indicator mapping of a set of edges with which each vertex is incident an even number of times is called a *cycle vector*. The cycle space is the span of the cycle vectors.

degree: The *degree* $\deg x$ of a vertex x in a graph is the number of vertices adjacent to it. The maximum and minimum degrees of vertices in a graph G are denoted $\Delta(G)$ and $\delta(G)$, respectively.

degree sequence: A sequence $(d_1, \ldots, d_n)$ is a *degree sequence* of a graph if there is some ordering $v_1, \ldots, v_n$ of the vertices for which d_i is the degree of v_i for each i.

diameter: The *diameter* of G is the maximum distance between two vertices of G; thus it is also the maximum eccentricity of a vertex in G.

digraph: A *digraph* is a directed graph; one in which each edge has a direction.

distance: The *distance* $d(u, v)$ between vertices u and v in a graph G is the minimum among the lengths of u, v-paths in G, or ∞ if there is no u, v-path.

drawing: A *drawing* of a graph G in a surface S consists of a one-to-one mapping from the vertices of G to points of S and a one-to-one mapping from the edges of G to open arcs in X so that (i) no image of an edge contains an image of some vertex, (ii) the image of each edge $\{v, w\}$ joins the images of v and w, (iii) the images of adjacent edges are disjoint, (iv) the images of two distinct edges never have more than one point in common, and (v) no point of the surface lies in the images of more than two edges.

eccentricity: The *eccentricity* $\text{ecc}(x)$ of a vertex x in a graph G is the maximum distance from x to a vertex of G.

edge connectivity: The *edge connectivity* of G, denoted $\lambda(G)$, is the minimum number of edges whose removal results in a disconnected graph.

edge space: The *edge space* of a graph $G = (V, E)$ is the vector space of all mappings from E to the two-element field $GF(2)$. Elements of the edge space are called 1-*chains*.

embedding: An *embedding* of a graph G in a topological space X consists of an assignment of the vertices of G to distinct points of X and an assignment of the edges of G to disjoint open arcs in X so that no arc representing an edge contains some point representing a vertex and so that each arc representing an edge joins the points representing the vertices incident with the edge. See also *drawing*.

end vertex: A vertex of degree 1 in a graph is called an *end vertex*.

Eulerian circuits and trails: A trail or circuit that includes every edge of a graph is said to be *Eulerian*, and a graph is *Eulerian* if it has an Eulerian circuit.

even: A graph is *even* if the degree of every vertex is even.

factor of a graph: A *factor* of a graph G is a spanning subgraph of G. A factor in which every vertex has the same degree k is called a k-factor. If $G_1, G_2, \ldots,$ G_k $(k \geq 2)$ are edge-disjoint factors of the graph G, and if $\bigcup_{i=1}^{k} E(G_i) = E(G)$, then G is said to be *factored* into $G_1, G_2, \ldots, G_k$ and we write $G = G_1 \oplus G_2 \oplus \cdots \oplus G_k$.

forest: A *forest* is an acyclic simple graph; see also *tree*.

genus: The *genus* $\gamma(G)$ (plural form *genera*) of a graph G is the least g for which G has an embedding in an orientable surface of genus g. See also *crosscap number*.

girth: The *girth* $\mathrm{gir}(G)$ of a graph G is the minimum length of a cycle in G, or ∞ if G is acyclic.

Hamiltonian cycles and paths: A path or cycle through all the vertices of a graph is said to be *Hamiltonian*. A graph is *Hamiltonian* if it has a Hamiltonian cycle.

homeomorphic graphs: Two graphs are *homeomorphic* to one another if there is a third graph of which each is a subdivision.

identification of vertices: To *identify* vertices v and w of a graph G is to construct a new graph G' from G by removing the vertices v and w and all the edges of G incident with them and introducing a new vertex u and new edges joining u to each vertex that was adjacent to v or to w in G. See also *contraction*.

incidence: A vertex v and an edge e are *incident* with one another if $v \in e$.

incidence matrix: For an ordering $v_1, v_2, \ldots, v_n$ of the vertices and an ordering $e_1, e_2, \ldots, e_m$ of the edges of a graph $G = (V, E)$ with order $|G| = n$ and

FIGURE 3.2

Three graphs that are isomorphic.

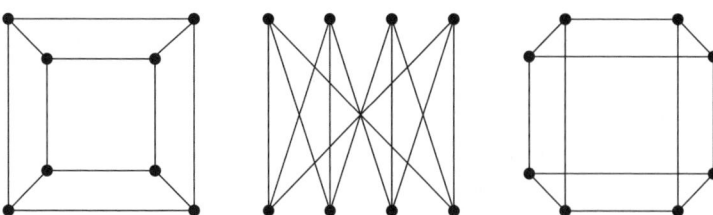

size $e(G) = m$, there is a corresponding $n \times m$ *incidence matrix* $B = (b_{ij})$ defined as follows:

$$b_{ij} = \begin{cases} 1, & \text{if } v_i \text{ and } e_j \text{ are incident,} \\ 0, & \text{otherwise.} \end{cases} \tag{3.4.5}$$

independence number: The *independence number* $\alpha(G)$ of a graph $G = (V, E)$ is the largest cardinality of an independent subset of V.

independent set: A set $S \subseteq V$ is said to be *independent* if the induced subgraph $G[S]$ is empty. See also *matching*.

internally disjoint paths: Two paths in a graph with the same initial vertex v and terminal vertex w are *internally disjoint* if they have no internal vertex in common.

isolated vertex: A vertex is *isolated* if it is adjacent to no other vertex.

isomorphism: An *isomorphism* between the two graphs $G = (V_G, E_G)$ and $H = (V_H, E_H)$ is a bijective mapping $\psi \colon V_G \to V_H$ for which $\{x, y\} \in E_G$ if and only if $\{\psi(x), \psi(y)\} \in E_H$. If there is an isomorphism between G and H, then G and H are said to be *isomorphic* to one another; this is denoted as $G \cong H$. Figure 3.2 contains three graphs that are isomorphic.

labeled graph: Graph theorists sometimes speak loosely of *labeled graphs* of order n and *unlabeled graphs* of order n to distinguish between graphs with a fixed vertex set of cardinality n and the family of isomorphism classes of such graphs. Thus, one may refer to *labeled graphs* to indicate an intention to distinguish between any two graphs that are *distinct* (i.e., have different vertex sets and/or different edge sets). One may refer to *unlabeled graphs* to indicate the intention to view any two distinct but isomorphic graphs as the 'same' graph, and to distinguish only between non-isomorphic graphs.

matching: A *matching* in a graph is a set of edges, no two having a vertex in common. A *maximal matching* is a matching that is not a proper subset of any other matching. A *maximum matching* is a matching of greatest cardinality. For a matching M, an *M-alternating path* is a path whose every other edge belongs to M, and an M-*augmenting* path is an M-alternating path whose first and last edges do not belong to M. A matching *saturates* a vertex if the vertex belongs to some edge of the matching.

monotone graph property: A property $\mathcal{P}$ that a graph may or may not enjoy is said to be *monotone* if, whenever H is a graph enjoying $\mathcal{P}$, every supergraph G of H with $|G| = |H|$ also enjoys $\mathcal{P}$.

multipartite graph: A graph is k-*partite* if its vertex set can be partitioned into k disjoint sets called *color classes* in such a way that every edge joins vertices in two different color classes (see also *coloring*). A two-partite graph is called *bipartite*.

neighbor: Adjacent vertices v and w in a graph are said to be *neighbors* of one another.

neighborhood: The sphere $S(x, 1)$ is called the *neighborhood* of x, and the ball $B(x, 1)$ is called the *closed neighborhood* of x.

order: The *order* $|G|$ of a graph $G = (V, E)$ is the number of vertices in G; in other words, $|G| = |V|$.

path: A *path* is a walk whose vertices are distinct.

perfect graph: A graph is *perfect* if $\chi(H) = \omega(H)$ for all induced subgraphs H of G.

planarity: A graph is *planar* if it has a proper embedding in the plane.

radius: The *radius* $\mathrm{rad}(G)$ of a graph G is the minimum vertex eccentricity in G.

regularity: A graph is k-*regular* if each of its vertices has degree k. A graph is *strongly regular* with parameters (k, λ, μ) if (i) it is k-regular, (ii) every pair of adjacent vertices has exactly λ common neighbors, and (iii) every pair of non-adjacent vertices has exactly μ common neighbors. A graph $G = (V, E)$ of order $|G| \geq 3$ is called *highly regular* if there exists an $n \times n$ matrix $C = (c_{ij})$, where $2 \leq n < |G|$, called a *collapsed adjacency matrix*, so that, for each vertex v of G, there is a partition of V into n subsets $V_1 = \{v\}, V_2, \ldots, V_n$ so that every vertex $y \in V_j$ is adjacent to exactly c_{ij} vertices in V_i. Every highly regular graph is regular.

rooted graph: A *rooted graph* is an ordered pair (G, v) consisting of a graph G and a distinguished vertex v of G called the *root*.

self-complementary: A graph is *self-complementary* if it is isomorphic to its complement.

similarity: Two vertices u and v of a graph G are *similar* (in symbols $u \sim v$) if there is an automorphism α of G for which $\alpha(u) = v$. Similarly, two edges (u, v) and (a, b) in the graph G are *similar* if an automorphism α of G exists for which $\{\alpha(u), \alpha(v)\} = \{a, b\}$.

size: The *size* $e(G)$ of a graph $G = (V, E)$ is the number of edges of G, that is, $e(G) = |E|$.

spectrum: The *spectrum* of a graph G is the spectrum of its characteristic polynomial, i.e., the non-decreasing sequence of $|G|$ eigenvalues of the characteristic polynomial of G. Since adjacency matrices are real symmetric, their spectrum is real.

sphere: The *sphere* of radius k about a vertex u is the set

$$S(u, k) = \{v \in V \mid d(u, v) = k\}. \tag{3.4.6}$$

See also *ball* and *neighborhood*.

subdivision: To *subdivide* an edge $\{v, w\}$ of a graph G is to construct a new graph G' from G by removing the edge $\{v, w\}$ and introducing new vertices x_i and new edges $\{v, x_1\}$, $\{x_k, w\}$ and $\{x_i, x_{i+1}\}$ for $1 \le i < k$. A *subdivision* of a graph is a graph obtained by subdividing one or more edges of the graph.

subgraph: A graph $H = (V_H, E_H)$ is a *subgraph* of a graph $G = (V_G, E_G)$ (in symbols, $H \preceq G$), if $V_H \subseteq V_G$ and $E_H \subseteq E_G$. In that case, G is a *supergraph* of H (in symbols, $G \succeq H$). If $V_H = V_G$, then H is called a *spanning subgraph* of G. For each set $S \subseteq V_G$, the subgraph $G[S]$ of G *induced* by S is the unique subgraph of G with vertex set S for which every edge of G incident with two vertices in S is also an edge of $G[S]$.

symmetry: A graph is *vertex symmetric* if every pair of vertices is similar. A graph is *edge symmetric* if every pair of edges is similar. A graph is *symmetric* if it is both vertex and edge symmetric.

2-switch: For vertices v, w, x, y in a graph G for which $\{v, w\}$ and $\{x, y\}$ are edges, but $\{v, y\}$ and $\{x, w\}$ are not edges, the construction of a new graph G' from G via the removal of edges $\{v, w\}$ and $\{x, y\}$ together with the insertion of the edges $\{v, y\}$ and $\{x, w\}$ is called a *2-switch*.

thickness: The *thickness* $\theta(G)$ of a graph G is the least k for which G is a union of k planar graphs.

trail: A *trail* in a graph is a walk whose edges are distinct.

tree: A *tree* is a connected forest, i.e., a connected acyclic graph. A spanning subgraph of a graph G that is a tree is called a *spanning tree* of G.

triangle: A 3-cycle is called a triangle.

trivial graph: A *trivial* graph is a graph with exactly one vertex and no edges.

unicyclic graph: A unicyclic graph is a connected graph that contains exactly one cycle.

vertex space: The *vertex space* of a graph G is the vector space of all mappings from V to the two-element field $GF(2)$. The elements of the vertex space are called 0-*chains*.

walk: A *walk* in a graph is an alternating sequence $v_0, e_1, v_1, \ldots, e_k, v_k$ of vertices v_i and edges e_i for which e_i is incident with v_{i-1} and with v_i for each i. Such a walk is said to have *length* k and to *join* v_0 and v_k. The vertices v_0 and v_k are called the *initial vertex* and *terminal vertex* of the walk; the remaining vertices are called *internal vertices* of the walk.

3.4.3 CONSTRUCTIONS

3.4.3.1 Operations on graphs

For graphs $G_1 = (V_1, E_1)$ and $G_2 = (V_2, E_2)$, there are several binary operations that yield a new graph from G_1 and G_2. The following table gives the names of some of those operations and the orders and sizes of the resulting graphs.

Operation producing G		Order $\|G\|$	Size $e(G)$
Composition	$G_1[G_2]$	$\|G_1\| \cdot \|G_2\|$	$\|G_1\|e(G_2) + \|G_2\|^2 e(G_1)$
Conjunction	$G_1 \wedge G_2$	$\|G_1\| \cdot \|G_2\|$	
Edge sum[a]	$G_1 \oplus G_2$	$\|G_1\| = \|G_2\|$	$\leq (e(G_1) + e(G_2))$
Join	$G_1 + G_2$	$\|G_1\| + \|G_2\|$	$e(G_1) + e(G_2) + \|G_1\| \cdot \|G_2\|$
Product	$G_1 \times G_2$	$\|G_1\| \cdot \|G_2\|$	$\|G_1\|e(G_2) + \|G_2\|e(G_1)$
Union	$G_1 \cup G_2$	$\|G_1\| + \|G_2\|$	$e(G_1) + e(G_2)$

[a]When applicable.

composition: For graphs $G_1 = (V_1, E_1)$ and $G_2 = (V_2, E_2)$, the *composition* $G = G_1[G_2]$ is the graph with vertex set $V_1 \times V_2$ whose edges are (1) the pairs $\{(u, v), (u, w)\}$ with $u \in V_1$ and $\{v, w\} \in E_2$ and (2) the pairs $\{(t, u), (v, w)\}$ for which $\{t, v\} \in E_1$.

conjunction: The conjunction $G_1 \wedge G_2$ of two graphs $G_1 = (V_1, E_1)$ and $G_2 = (V_2, E_2)$ is the graph $G_3 = (V_3, E_3)$ for which $V_3 = V_1 \times V_2$ and for which vertices $\mathbf{e}_1 = (u_1, u_2)$ and $\mathbf{e}_2 = (v_1, v_2)$ in V_3 are adjacent in G_3 if, and only if, u_1 is adjacent to v_1 in G_1 and u_2 is adjacent to v_2 in G_2.

edge difference: For graphs $G_1 = (V, E_1)$ and $G_2 = (V, E_2)$ with the same vertex set V, the *edge difference* $G_1 - G_2$ is the graph with vertex set V and edge set $E_1 \setminus E_2$.

edge sum: For graphs $G_1 = (V, E_1)$ and $G_2 = (V, E_2)$ with the same vertex set V, the *edge sum* of G_1 and G_2 is the graph $G_1 \oplus G_2$ with vertex set V and edge set $E_1 \cup E_2$. Sometimes the edge sum is denoted $G_1 \cup G_2$.

join: For graphs $G_1 = (V_1, E_1)$ and $G_2 = (V_2, E_2)$ with $V_1 \cap V_2 = \emptyset$, the *join* $G_1 + G_2 = G_2 + G_1$ is the graph obtained from the union of G_1 and G_2 by adding edges joining each vertex in V_1 to each vertex in V_2.

power: For a graph $G = (V, E)$, the kth *power* G^k is the graph with the same vertex set V whose edges are the pairs $\{u, v\}$ for which $d(u, v) \leq k$ in G. The *square* of G is G^2.

product: For graphs $G_1 = (V_1, E_1)$ and $G_2 = (V_2, E_2)$, the *product* $G_1 \times G_2$ has vertex set $V_1 \times V_2$; its edges are all of the pairs $\{(u, v), (u, w)\}$ for which $u \in V_1$ and $\{v, w\} \in E_2$ and all of the pairs $\{(t, v), (u, v)\}$ for which $\{t, u\} \in E_1$ and $v \in V_2$.

union: For graphs $G_1 = (V_1, E_1)$ and $G_2 = (V_2, E_2)$ with $V_1 \cap V_2 = \emptyset$, the *union* of G_1 and G_2 is the graph $G_1 \cup G_2 = (V_1 \cup V_2, E_1 \cup E_2)$. The union is sometimes called the *disjoint union* to distinguish it from the *edge sum*.

3.4.3.2 Graphs described by one parameter

complete graph, K_n: A complete graph of order n is a graph isomorphic to the graph K_n with vertex set $\{1, 2, \ldots, n\}$ whose every pair of vertices is an edge. The graph K_n has size $\binom{n}{2}$ and is Hamiltonian. If G is a graph g of order n, then $K_n = G \oplus \overline{G}$.

cube, Q_n: An *n-cube* is a graph isomorphic to the graph Q_n whose vertices are the 2^n binary n-vectors and whose edges are the pairs of vectors that differ in exactly one place. It is an n-regular bipartite graph of order 2^n and size $n2^{n-1}$. An equivalent recursive definition, $Q_1 = K_2$ and $Q_n = Q_{n-1} \times K_2$.

cycle, C_n: A *cycle* of order n is a graph isomorphic to the graph C_n with vertex set $\{0, 1, \ldots, n-1\}$ whose edges are the pairs $\{v_i, v_{i+1}\}$ with $0 \le i < n$ and arithmetic modulo n. The cycle C_n has size n and is Hamiltonian.

The graph C_n is a special case of a circulant graph. The graph C_3 is called a *triangle*; the graph C_4 is called a *square*.

empty graph: A graph is *empty* if it has no edges; $\overline{K}_n$ denotes an empty graph of order n.

Kneser graphs, $K_n^{(m)}$: For $n \ge 2m$, the *Kneser graph* $K_n^{(m)}$ is the complement of the intersection graph of the m-subsets of an n-set. The *odd graph* O_m is the Kneser graph $K_{2m+1}^{(m)}$. The *Petersen graph* is the odd graph $O_2 = K_5^{(2)}$.

ladder: A *ladder* is a graph of the form $P_n \times P_2$. The *Möbius ladder* M_n is the graph obtained from the ladder $P_n \times P_2$ by joining the opposite end vertices of the two copies of P_n.

path, P_n: A *path* of order n is a graph isomorphic to the graph P_n whose vertex set is $\{1, \ldots, n\}$ and whose edges are the pairs $\{v_i, v_{i+1}\}$ with $1 \le i < n$. A path of order n has size $n - 1$ and is a tree.

star, S_n: A *star* of order n is a graph isomorphic to the graph $S_n = K_{1,n}$. It has a vertex cover consisting of a single vertex, its size is n, and it is a complete bipartite graph and a tree.

wheel, W_n: The wheel W_n of order $n \ge 4$ consists of a cycle of order $n - 1$ and an additional vertex adjacent to every vertex in the cycle. Equivalently, $W_n = C_{n-1} + K_1$. This graph has size $2(n - 1)$.

3.4.3.3 Graphs described by two parameters

complete bipartite graph, $K_{n,m}$:   The complete bipartite graph $K_{n,m}$ is the graph $\overline{K}_n + \overline{K}_m$. Its vertex set can be partitioned into two color classes of cardinalities n and m, respectively, so that each vertex in one color class is adjacent to every vertex in the other color class. The graph $K_{n,m}$ has order $n + m$ and size nm.

planar mesh: A graph of the form $P_n \times P_m$ is called a *planar mesh*.

prism: A graph of the form $C_m \times P_n$ is called a *prism*.

Toeplitz graph, TN(w, s):   The Toeplitz graph TN(w, s) is defined in terms of its adjacency matrix $A = (a_{ij})$, for which

$$a_{ij} = \begin{cases} 1, & \text{if } |i - j| = 1 \pmod{w}, \\ 0, & \text{otherwise.} \end{cases} \tag{3.4.7}$$

The Toeplitz graph is of order $ws + 2$, size $(s + 1)(w - s + 2)/2$, and girth 3 or 4; it is $(s + 1)$-regular and Hamiltonian. Moreover, TN$(1, s) = K_{s+2}$ and TN$(w, 1) = C_{w+2}$.

toroidal mesh: A graph of the form $C_m \times C_n$ with $m \geq 2$ and $n \geq 2$ is called a *toroidal mesh*.

Turán graph, $T_{n,k}$:   The *Turán graph* $T_{n,k}$ is the complete k-partite graph in which the cardinalities of any two color classes differ by, at most, one. It has $n - k \lfloor n/k \rfloor$ color classes of cardinality $\lfloor n/k \rfloor + 1$ and $k - n + k \lfloor n/k \rfloor$ color classes of cardinality $\lfloor n/k \rfloor$. Note that $\omega(T_{n,k}) = k$.

3.4.3.4 Graphs described by three or more parameters

Cayley graph: For a group Γ and a set X of generators of Γ, the *Cayley graph* of the pair (Γ, X) is the graph with vertex set Γ in which $\{\alpha, \beta\}$ is an edge if either $\alpha^{-1}\beta \in X$ or $\beta^{-1}\alpha \in X$.

complete multipartite graph, $K_{n_1, n_2, \ldots, n_k}$:   The *complete k-partite graph* $K_{n_1, n_2, \ldots, n_k}$ is the graph $\overline{K}_{n_1} + \cdots + \overline{K}_{n_k}$. It is a a k-partite graph with color classes V_i of cardinalities $|V_i| = n_i$ for which every pair of vertices in two distinct color classes is an edge. The graph $K_{n_1, n_2, \ldots, n_k}$ has order $\sum_{i=1}^{k} n_k$ and size $\sum_{1 \leq i < j \leq k} n_i n_j$.

double loop graph, DLG$(n; a, b)$:   The double loop graph DLG$(n; a, b)$ (with a and b between 1 and $(n - 1)/2$) consists of n vertices with every vertex i connected by an edge to the vertices $i \pm a$ and $i \pm b$ (modulo n). The name comes from the following fact: If GCD$(a, b, n) = 1$, then DLG$(n; a, b)$ is Hamiltonian and, additionally, DLG$(n; a, b)$ can be decomposed into two Hamiltonian cycles. These graphs are also known as circulant graphs.

FIGURE 3.3
Examples of graphs with 6 or 7 vertices.

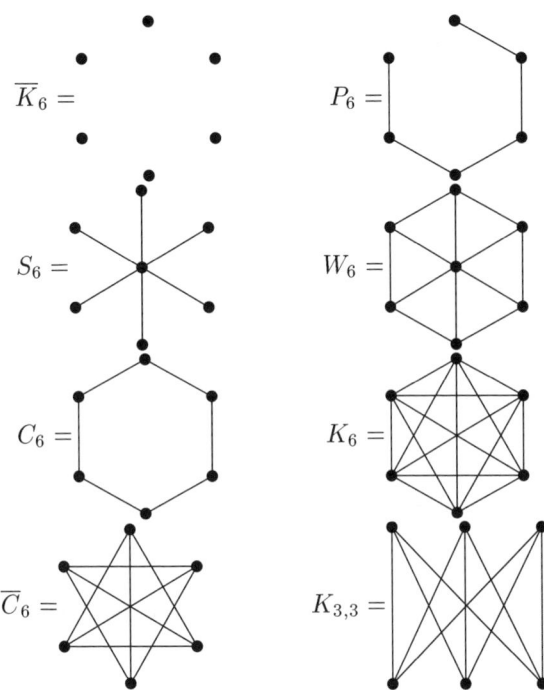

intersection graph: For a family $F = \{S_1, \ldots, S_n\}$ of subsets of a set S, the *intersection graph* of F is the graph with vertex set F in which $\{S_i, S_j\}$ is an edge if and only if $S_i \cap S_j \neq \emptyset$. Each graph G is an intersection graph of some family of subsets of a set of cardinality at most $\lfloor |G|^2/4 \rfloor$.

interval graph: An *interval graph* is an intersection graph of a family of intervals on the real line.

3.4.4 FUNDAMENTAL RESULTS

3.4.4.1 Walks and connectivity

1. Every x, y walk includes all the edges of some x, y path.

2. Some path in G has length $\delta(G)$.

3. Connectivity is a monotone graph property. If more edges are added to a connected graph, the new graph is itself connected.

4. A graph is disconnected if, and only if, it is the union of two graphs.

5. The sets S for which $G[S]$ is a component partition of the vertex set V.

6. Every vertex of a graph lies in at least one block.

7. For every graph G, $0 \leq \kappa(G) \leq |G| - 1$.

8. For all integers a, b, c with $0 < a \leq b \leq c$, a graph G exists with $\kappa(G) = a$, $\lambda(G) = b$, and $\delta(G) = c$.

9. For any graph G, $\kappa(G) \leq \lambda(G) \leq \delta(G)$.

10. *Menger's theorem*: Suppose that G is a connected graph of order greater than k. Then G is k-connected if, and only if, it is impossible to disconnect G by removing fewer than k vertices, and G is k-edge connected if, and only if, it is impossible to disconnect G by removing fewer than k edges.

11. If G is a connected graph with a bridge, then $\lambda(G) = 1$. If G has order n and is r-regular with $r \geq n/2$, then $\lambda(G) = r$.

3.4.4.2 Trees

1. A graph is a tree if, and only if, it is acyclic and has size $|G| - 1$.

2. A graph is a tree if, and only if, it is connected and has size $|G| - 1$.

3. A graph is a tree if, and only if, each of its edges is a bridge.

4. A graph is a tree if, and only if, each vertex of degree greater than 1 is a cut vertex.

5. A graph is a tree if, and only if, each pair of its vertices is joined by exactly one path.

6. Every tree of order greater than 1 has at least two end vertices.

7. The center of a tree consists of one vertex or two adjacent vertices.

8. For each graph G, every tree with at most $\delta(G)$ edges is a subgraph of G.

9. Every connected graph has a spanning tree.

10. *Kirchhoff matrix-tree theorem*: Let G be a connected graph and let A be an adjacency matrix for G. Obtain a matrix M from $-A$ by replacing each term a_{ii} on the main diagonal with $\deg v_i$. Then all cofactors of M have the same value, which is the number of spanning trees of G.

11. *Nash–Williams arboricity theorem*: For a graph G and for each $n \leq |G|$, define $e_n(G) = \max\{e(H) : H \preceq G, \text{ and } |H| = n\}$. Then

$$\Upsilon(G) = \max_n \left\lceil \frac{e_n}{n-1} \right\rceil. \tag{3.4.8}$$

3.4.4.3 Circuits and cycles

1. *Euler's theorem*: A multigraph is Eulerian if and only if it is connected and even.

2. If G is Hamiltonian, and if G' is obtained from G by removing a non-empty set S of vertices, then the number of components of G' is at most $|S|$.

3. *Ore's theorem*: If G is a graph for which $\deg v + \deg w \geq |G|$ whenever v and w are non-adjacent vertices, then G is Hamiltonian.

4. *Dirac's theorem*: If G is a graph of order $|G| \geq 3$ and $\deg v \geq |G|/2$ for each vertex v, then G is Hamiltonian.

5. (Erdös–Chvátal) If $\alpha(G) \leq \kappa(G)$, then G is Hamiltonian.

6. Every 4-connected planar graph is Hamiltonian.

7. The following table tells which members of several families of graphs are Eulerian or Hamiltonian:

graph	Is it Eulerian?	Is it Hamiltonian?
B_n	yes	yes, for $n \geq 1$
C_n	yes	yes, for $n \geq 1$
D_n	yes, for n even	yes, for $n \geq 2$
K_n	yes, for odd n	yes, for $n \geq 3$
$K_{m,n}$	yes, for m and n both even	yes, for $m = n$
Q_n	yes, for n even	yes, for $n \geq 2$
W_n	no	yes, for $n \geq 2$

3.4.4.4 Cliques and independent sets

1. A set $S \subseteq V$ is a vertex cover if, and only if, $V \setminus S$ is an independent set.

2. *Turán's theorem*: If $|G| = n$ and $\omega(G) \leq k$, then $e(G) \leq e(T_{n,k})$.

3. *Ramsey's theorem*: For all positive integers k and l, there is a least integer $R(k, l)$ for which every graph of order at least $R(k, l)$ has either a clique of cardinality k or an independent set of cardinality l. For $k \geq 2$ and $l \geq 2$, $R(k, l) \leq R(k, l - 1) + R(k - 1, l)$.

$R(k, l)$	$l = 1$	2	3	4	5	6	7
$k = 1$	1	1	1	1	1	1	1
2	1	2	3	4	5	6	7
3	1	3	6	9	14	18	23

3.4.4.5 Colorings and partitions

1. Every graph G is k-partite for some k; in particular, G is $|G|$-partite.

2. Every graph G has a bipartite subgraph H for which $e(H) \geq e(G)/2$.

3. $P_G(x) = P_{G-e}(x) - P_{G\backslash e}(x)$.

4. *Brooks' theorem*: If G is a connected graph that is neither a complete graph nor a cycle of odd length, then $\chi(G) \leq \Delta(G)$.

5. For all positive integers g and c, a graph G exists with $\chi(G) \geq c$ and $\mathrm{gir}(G) \geq g$.

6. *Nordhaus–Gaddum bounds*: For every graph G,

$$2\sqrt{|G|} \leq \chi(G) + \chi(\bar{G}) \leq |G| + 1, \quad \text{and}$$

$$|G| \leq \chi(G)\chi(\bar{G}) \quad \leq \left(\frac{|G| + 1}{2}\right)^2.$$

7. *Szekeres–Wilf theorem*: For every graph $G = (V, E)$,

$$\chi(G) \leq 1 + \max_{S \subseteq V} \delta(G[S]).$$

8. (König) If G is bipartite, then $\chi'(G) = \Delta(G)$.

9. *Vizing's theorem*: For every graph G, $\Delta(G) \leq \chi'(G) \leq \Delta(G) + 1$.

10. The following table gives the chromatic numbers and chromatic polynomials of various graphs:

G	$\chi(G)$	$P_G(x)$
K_n	n	$x(x-1)\cdots(x-n+1)$
$\overline{K}_n$	1	x^n
T_n	2	$x(x-1)^{n-1}$
P_n	2	$x(x-1)^{n-1}$
C_4	2	$x(x-1)(x^2-3x+3)$

11. The following table gives the chromatic numbers and edge-chromatic numbers of various graphs:

G	$\chi(G)$	$\chi_1(G)$
C_n with n even, $n \geq 2$	2	2
C_n with n odd, $n \geq 3$	3	3
K_n with n even, $n \geq 2$	n	$n-1$
K_n with n odd, $n \geq 3$	n	n
$K_{m,n}$ with $m, n \geq 1$	2	$\max(m, n)$
$K_{m_1,\dots,m_k}$ with $m_i \geq 1$	k	$\max(m_1, \dots, m_k)$
P_n	2	2
Petersen graph	3	4
W_n with n even, $n \geq 2$	3	n
W_n with n odd, $n \geq 3$	4	n

12. (Appel–Haken) *Four-color theorem*: $\chi(G) \leq 4$ for every planar graph G.

13. For each graph G of order $|G| = n$ and size $e(G) = m$ with exactly k components, the chromatic polynomial is of the form

$$P_G(x) = \sum_{i=0}^{n-k} (-1)^i a_i x^{n-i}, \qquad (3.4.9)$$

with $a_0 = 1$, $a_1 = m$ and every a_i positive.

14. Not every polynomial is a chromatic polynomial. For example $P(x) = x^4 - 4x^3 + 3x^2$ is not a chromatic polynomial.

15. Sometimes a class of chromatic polynomials can only come from a specific class of graphs. For example:

   (a) If $P_G(x) = x^n$, then $G = \overline{K}_n$.
   (b) If $P_G(x) = (x)_n$, then $G = K_n$.
   (c) If $P_G(x) = x(x-1) \cdots (x-r+2)(x-r+1)^2(x-r)^{n-r-1}$ for a graph of order $n \geq r+1$, then G can be obtained from an r-tree T of order n by deleting an edge contained in exactly $r - 1$ triangles of T.

3.4.4.6 Distance

1. A metric space (X, d) is the metric space associated with a connected graph with vertex set X if, and only if, it satisfies two conditions: (i) $d(u, v)$ is a non-negative integer for all $u, v \in X$, and (ii) whenever $d(u, v) \geq 2$, some element of X lies between u and v. The edges of the graph are the pairs $\{u, v\} \subseteq X$ for which $d(u, v) = 1$. (In an arbitrary metric space (X, d), a point $v \in X$ is said to lie *between* distinct points $u \in X$ and $w \in X$ if it satisfies the *triangle equality* $d(u, w) = d(u, v) + d(v, w)$.)

2. If $G = (V, E)$ is connected, then distance is always finite, and d is a metric on V. Note that $\deg(x) = |S(x, 1)|$.

3. *Moore bound*: For every connected graph G,

$$|G| \leq 1 + \Delta(G) \sum_{i=1}^{\operatorname{diam}(G)} (\Delta(G) - 1)^i. \qquad (3.4.10)$$

A graph for which the Moore bound holds exactly is called a *Moore graph* with parameters $(|G|, \Delta(G), \operatorname{diam}(G))$. Every Moore graph is regular. If G is a Moore graph with parameters (n, r, d), then $(n, r, d) = (n, n-1, 1)$ (in which case G is complete), $(n, r, d) = (2m + 1, 2, m)$ (in which case G is a $(2m + 1)$-cycle), or $(n, r, d) \in \{(10, 3, 2), (50, 7, 2), (3250, 57, 2)\}$.

3.4.4.7 Drawings, embeddings, planarity, and thickness

1. Every graph has an embedding in $\mathbb{R}^3$ for which the arcs representing edges are all straight line segments. Such an embedding can be constructed by using distinct points on the curve $\{(t, t^2, t^3) : 0 \le t \le 1\}$ as representatives for the vertices.

2. For $n \ge 2$,

$$\gamma(Q_n) = (n - 4)2^{n-3} + 1, \text{ and}$$
$$\tilde{\gamma}(Q_n) = (n - 4)2^{n-2} + 2.$$

(3.4.11)

3. For $r, s \ge 2$,

$$\gamma(K_{r,s}) = \left\lceil \frac{(r - 2)(s - 2)}{4} \right\rceil, \text{ and}$$
$$\tilde{\gamma}(K_{r,s}) = \left\lceil \frac{(r - 2)(s - 2)}{2} \right\rceil.$$

(3.4.12)

4. For $n \ge 3$,

$$\gamma(K_n) = \left\lceil \frac{(n - 3)(n - 4)}{12} \right\rceil, \text{ and}$$
$$\tilde{\gamma}(K_n) = \left\lceil \frac{(n - 3)(n - 4)}{6} \right\rceil.$$

(3.4.13)

5. *Heawood map coloring theorem*: The greatest chromatic number among graphs of genus n is

$$\max\{\chi(G) \mid \gamma(G) = n\} = \left\lceil \frac{7 + \sqrt{1 + 48n}}{2} \right\rceil.$$

6. *Kuratowski's theorem*: A graph is planar if and only if it has no subgraph homeomorphic to K_5 or $K_{3,3}$.

7. A graph is planar if and only if it does not have a subgraph contractible to K_5 or $K_{3,3}$.

8. The graph K_n is non-planar if and only if $n \ge 5$.

9. Every planar graph can be embedded in the plane so that every edge is a straight line segment; this is a *Fary embedding*.

10. The four-color theorem states that any planar graph is four colorable.

11. For every graph G of order $|G| \ge 3$, $\theta(G) \ge \left\lceil \dfrac{e(G)}{3|G| - 6} \right\rceil$.

12. The complete graphs K_9 and K_{10} have thickness 3; for $n \notin \{9, 10\}$,

$$\theta(K_n) = \left\lfloor \frac{n + 7}{6} \right\rfloor.$$

(3.4.14)

13. The n-cube has thickness $\theta(Q_n) = \lfloor n/4 \rfloor + 1$.

14. For every planar graph G, $\nu(G) = \bar{\nu}(G)$. That equality does not hold for all graphs: $\nu(K_8) = 18$, and $\bar{\nu}(K_8) = 19$.

3.4.4.8 Vertex degrees

1. *Handshaking lemma*: For every graph G, $\sum_{v \in V} \deg v = 2e(G)$.

2. Every 2-switch preserves the degree sequence.

3. If G and H have the same degree sequence, then H can be obtained from G via a sequence of 2-switches.

4. *Havel theorem*: The values $\{d_1, d_2, \ldots, d_n\}$ with $d_1 \geq d_2 \geq \cdots \geq d_n > 0$ are a degree sequence if and only if the sequence obtained by deleting d_1 and subtracting 1 from each of the next d_1 largest values (i.e., $\{d_2 - 1, d_3 - 1, \ldots, d_{d_1+1} - 1, d_{d_1+2}, \ldots, d_n\}$) is a degree sequence.

3.4.4.9 Algebraic methods

1. The bipartite graphs $K_{n,n}$ are circulant graphs.

2. For a graph G with exactly k connected components, the cycle space has dimension $e(G) - |G| + k$, and the cut space has dimension $|G| - k$.

3. In the kth power $A^k = (a^k_{ij})$ of the adjacency matrix, each entry a^k_{ij} is the number of v_i, v_j walks of length k.

4. The incidence matrix of a graph G is totally unimodular if, and only if, G is bipartite.

5. Every odd graph is vertex-transitive.

6. The smallest graph that is vertex symmetric, but is not edge symmetric, is the prism $K_3 \times K_2$. The smallest graph that is edge symmetric, but is not vertex symmetric, is $S_2 = P_3 = K_{1,2}$.

7. The spectrum of a disconnected graph is the union of the spectra of its components.

8. The sum of the eigenvalues in the spectrum of a graph is zero.

9. The number of distinct eigenvalues in the spectrum of a connected graph is greater than the diameter of the graph.

10. The largest eigenvalue in the spectrum of a graph G is, at most, $\Delta(G)$, with equality if, and only if, G is regular.

11. (Wilf) If G is a connected graph and its largest eigenvalue is λ, then $\chi(G) \leq 1 + \lambda$. Moreover, equality holds if, and only if, G is a complete graph or a cycle of odd length.

12. (Hoffman) If G is a connected graph of order n with spectrum $\lambda_1 \geq \cdots \geq \lambda_n$, then $\chi(G) \geq 1 - \lambda_1/\lambda_n$.

13. *Integrality condition*: If G is a strongly regular graph with parameters (k, λ, μ), then the quantities

$$\frac{1}{2}\left(|G| - 1 \pm \frac{(|G| - 1)(\mu - \lambda) - 2k}{\sqrt{(\mu - \lambda)^2 + 4(k - \mu)}}\right) \qquad (3.4.15)$$

are non-negative integers.

14. The following table gives the automorphism groups of various graphs:

G	$\mathrm{Aut}(G)$
C_n	D_n
$\overline{K}_n$	S_n
K_n	S_n
$K_{1,n}$	$E_1 + S_n$
P_n	C_n

15. A graph and its complement have the same group, $\mathrm{Aut}(G) = \mathrm{Aut}(\overline{G})$.

16. *Frucht's theorem*: Every finite group is the automorphism group of some graph.

17. If G and G' are edge isomorphic, then G and G' are not necessarily required to be isomorphic. For example, the graphs C_3 and S_3 are edge isomorphic, but not isomorphic.

18. If the graph G has order n, then the order of its automorphism group $|\mathrm{Aut}(G)|$ is a divisor of $n!$. The order of the automorphism group equals $n!$ if and only if $G \simeq K_n$ or $G \simeq \overline{K}_n$.

3.4.4.10 Matchings

1. A matching M is a maximum matching if, and only if, there is no M-augmenting path.

2. *Hall's theorem*: A bipartite graph G with bipartition (B_1, B_2) has a matching that saturates every vertex in B_1 if, and only if, $|S(A, 1)| \geq |A|$ for every $A \subseteq B_1$.

3. *König's theorem*: In a bipartite graph, the cardinality of a maximum matching equals the cardinality of a minimum vertex cover.

3.4.4.11   Enumeration

1. The number of labeled graphs of order n is $2^{\binom{n}{2}}$.

2. The number of labeled graphs of order n and size m is $\binom{\binom{n}{2}}{m}$.

3. The number of different ways in which a graph G of order n can be labeled is $n!/|\mathrm{Aut}(G)|$.

4. *Cayley's formula*: The number of labeled trees of order n is n^{n-2}.

5. The number of labeled trees of order n with exactly t end vertices is $\frac{n!}{t!} S(n-2, n-t)$ for $2 \le t \le n-1$ (where $S(,)$ is a Stirling number of the second kind).

6. The following table lists the number of graphs, number of digraphs, number of trees (t_n), number of rooted trees (T_n), of different orders.

Order	Graphs	Digraphs	Trees (t_n)	Rooted trees (T_n)
1	1	1	1	1
2	2	3	1	1
3	4	16	1	2
4	11	218	2	4
5	34	9608	3	9
6	156	1540944	6	20
7	1044	882033440	11	48
8	12346	1793359192848	23	115
9	274668		47	286
10	12005168		106	719

7. Define the generating functions $T(x) = \sum_{n=0}^{\infty} T_n x^n$ and $t(x) = \sum_{n=0}^{\infty} t_n x^n$. Then $T(x) = x \exp\left(\sum_{r=1}^{\infty} \frac{1}{r} T(x^r)\right)$, and $t(x) = T(x) - \frac{1}{2}\left[T^2(x) - T(x^2)\right]$.

8. The following table lists the number of isomorphism classes of graphs of order n and size m.

m	$n = 1$	2	3	4	5	6	7	8	9
0	1	1	1	1	1	1	1	1	1
1		1	1	1	1	1	1	1	1
2			1	2	2	2	2	2	1
3			1	3	4	5	5	5	5
4				2	6	9	10	11	11
5				1	6	15	21	24	25
6					1	6	21	41	56	63
7						4	24	65	115	148
8						2	24	97	221	345
9						1	21	131	402	771
10						1	15	148	663	1637

m	$n=1$	2	3	4	5	6	7	8	9
11						9	148	980	3252
12						5	131	1312	5995
13						2	97	1557	10120
14						1	65	1646	15615
15						1	41	1557	21933
16							21	1312	27987
17							10	980	32403
18							5	663	34040

9. The following table lists the number of isomorphism classes of digraphs with n vertices and m arcs.

m	$n=1$	2	3	4	5
0	1	1	1	1	1
1		2	6	12	20
2		1	15	66	190
3			20	220	1,140
4			15	495	4,845
5			6	792	15,504
6			1	924	38,760
7				792	77,520

10. The following table gives the number of labeled graphs of order n having various properties:

n	1	2	3	4	5	6	7	8
All	1	2	8	64	1 024	2^{15}	2^{21}	2^{28}
Connected	1	1	4	38	728	26 704	1 866 256	251 548 592
Even	1	1	2	8	64	1 024	2^{15}	2^{21}
Trees	1	1	3	16	125	1 296	16 807	262 144

11. The following table gives the numbers of isomorphism classes of graphs of order n exhibiting various properties:

n	1	2	3	4	5	6	7	8
All	1	2	4	11	34	156	1 044	12 346
Connected	1	1	2	6	21	112	853	11 117
Even	1	1	2	3	7	16	54	243
Eulerian	1	0	1	1	4	8	37	184
Blocks	0	1	1	3	10	56	468	7 123
Trees	1	1	1	2	3	6	11	23
Rooted trees	1	1	2	4	9	20	48	115

3.4.4.12 Descriptions of graphs with few vertices

The small graphs can be described in terms of the operations on page 228. Let $\mathcal{G}_{n,m}$ denote the family of isomorphism classes of graphs of order n and size m. Then

$$\mathcal{G}_{1,0} = \{K_1\}, \qquad\qquad \mathcal{G}_{4,0} = \{\overline{K}_4\},$$
$$\mathcal{G}_{2,0} = \{\overline{K}_2\}, \qquad\qquad \mathcal{G}_{4,1} = \{P_2 \cup \overline{K}_2\},$$
$$\mathcal{G}_{2,1} = \{K_2\}, \qquad\qquad \mathcal{G}_{4,2} = \{P_3 \cup \overline{K}_1, P_2 \cup P_2\},$$
$$\mathcal{G}_{3,0} = \{\overline{K}_3\}, \qquad\qquad \mathcal{G}_{4,3} = \{P_4, K_3 \cup \overline{K}_1, K_{1,3}\},$$
$$\mathcal{G}_{3,1} = \{K_2 \cup K_1\}, \qquad\quad \mathcal{G}_{4,4} = \{C_4, (K_2 \cup K_1) + K_1\},$$
$$\mathcal{G}_{3,2} = \{P_3\}, \qquad\qquad \mathcal{G}_{4,5} = \{K_4 - e\},$$
$$\mathcal{G}_{3,3} = \{K_3\}, \qquad\qquad \mathcal{G}_{4,6} = \{K_4\}$$

3.4.5 TREE DIAGRAMS

Let $T_{n,m}$ denote the mth isomorphism class of trees of order n. Figure 3.4 (page 242) depicts trees of order at most 7. Figure 3.5 (page 243) depicts trees of order 8.

3.5 COMBINATORIAL DESIGN THEORY

Combinatorial design theory is the study of families of subsets with various prescribed regularity properties. An *incidence structure* is an ordered pair $(X, \mathcal{B})$:

1. $X = \{x_1, \ldots, x_v\}$ is a set of *points*.
2. $\mathcal{B} = \{B_1, \ldots, B_b\}$ is a set of *blocks* or *lines*; each $B_j \subseteq X$.
3. The *replication number* r_i of x_i is the number of blocks that contain x_i.
4. The size of B_j is k_j.

Counting the number of pairs (x, B) with $x \in B$ yields $\sum_{i=1}^{v} r_i = \sum_{j=1}^{b} k_j$. The *incidence matrix* of an incidence structure is the $v \times b$ matrix $A = (a_{ij})$ with $a_{ij} = 1$ if $x_i \in B_j$ and 0 otherwise.

3.5.1 t-DESIGNS

The incidence structure $(X, \mathcal{B})$ is called a t-(v, k, λ) *design* if

1. For all j, $k_j = k$ and $1 < k < v$, and
2. Any subset of t points is contained in exactly λ blocks.

FIGURE 3.4
Trees with 7 or fewer vertices.

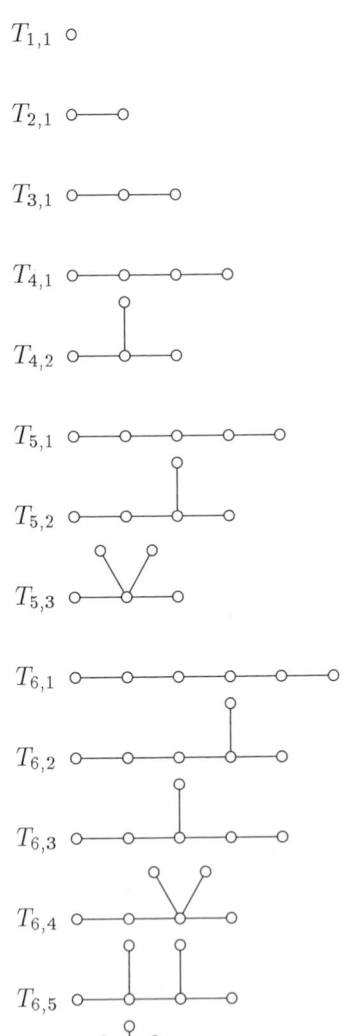

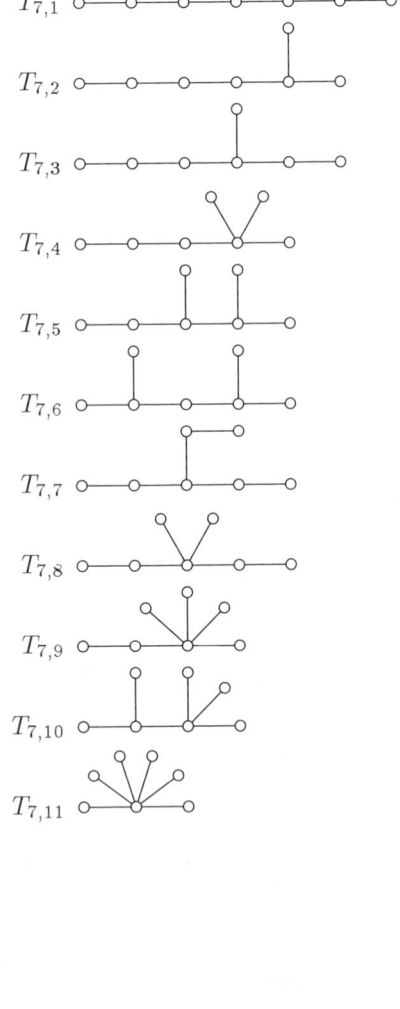

FIGURE 3.5
Trees with 8 vertices.

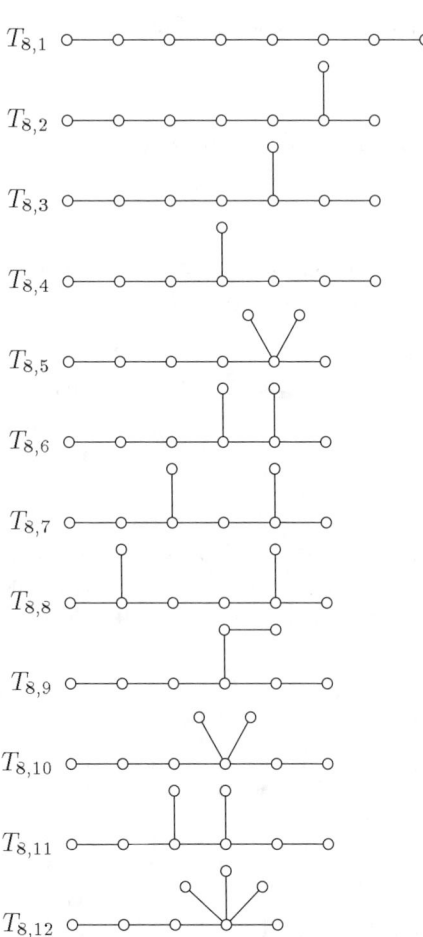

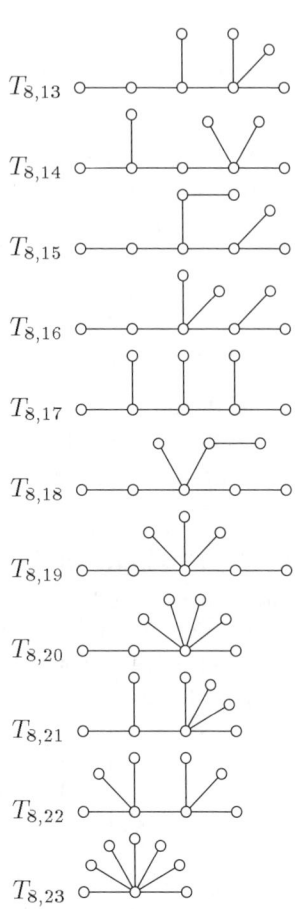

A 1-design is equivalent to a $v \times b$ 0-1 matrix with constant row and column sums. Every t-(v, k, λ) design is also a ℓ-(v, k, λ_ℓ) design ($1 \leq \ell \leq t$), where

$$\lambda_\ell = \lambda \binom{v-\ell}{t-\ell} \Big/ \binom{k-\ell}{t-\ell}. \tag{3.5.1}$$

A necessary condition for the existence of a t-(v, k, λ) design is that λ_ℓ must be an integer for all ℓ, $1 \leq \ell \leq t$. Another necessary condition is the generalized Fisher's inequality: if $t = 2s$, then $b \geq \binom{v}{s}$.

3.5.1.1 Related designs

The existence of a t-(v, k, λ) design also implies the existence of the following designs:

Complementary design
    Let $\mathcal{B}_C = \{X \backslash B \mid B \in \mathcal{B}\}$. Then the incidence structure $(X, \mathcal{B}_C)$ is a t-$(v, v-k, \lambda\binom{v-t}{k} \big/ \binom{v-t}{k-t})$ design (provided $v \geq k+t$).

Derived design
    Fix $x \in X$ and let $\mathcal{B}_D = \{B \backslash \{x\} \mid B \in \mathcal{B} \text{ with } x \in B\}$. Then the incidence structure $(X \backslash \{x\}, \mathcal{B}_D)$ is a $(t-1)$-$(v-1, k-1, \lambda)$ design.

Residual design
    Fix $x \in X$ and let $\mathcal{B}_R = \{B \mid B \in \mathcal{B} \text{ with } x \notin B\}$. Then the incidence structure $(X \backslash \{x\}, \mathcal{B}_R)$ is a $(t-1)$-$(v-1, k-1, \lambda\binom{v-t}{k-t+1} \big/ \binom{v-t}{k-t})$ design.

3.5.1.2 The Mathieu 5-design

The following are the 132 blocks of a 5-(12,6,1) design. The blocks are the supports of the weight-6 codewords in the ternary Golay code (page 258). Similarly, the supports of the 759 weight-8 codewords in the binary Golay code form the blocks of a 5-(24,8,1) design.

0	1	2	3	4	11	0	1	2	3	5	10	0	1	2	3	6	8	0	1	2	3	7	9
0	1	2	4	5	9	0	1	2	4	6	10	0	1	2	4	7	8	0	1	2	5	6	7
0	1	2	5	8	11	0	1	2	6	9	11	0	1	2	7	10	11	0	1	2	8	9	10
0	1	3	4	5	8	0	1	3	4	6	7	0	1	3	4	9	10	0	1	3	5	6	9
0	1	3	5	7	11	0	1	3	6	10	11	0	1	3	7	8	10	0	1	3	8	9	11
0	1	4	5	6	11	0	1	4	5	7	10	0	1	4	6	8	9	0	1	4	7	9	11
0	1	4	8	10	11	0	1	5	6	8	10	0	1	5	7	8	9	0	1	5	9	10	11
0	1	6	7	8	11	0	1	6	7	9	10	0	2	3	4	5	7	0	2	3	4	6	9
0	2	3	4	8	10	0	2	3	5	6	11	0	2	3	5	8	9	0	2	3	6	7	10
0	2	3	7	8	11	0	2	3	9	10	11	0	2	4	5	6	8	0	2	4	5	10	11
0	2	4	6	7	11	0	2	4	7	9	10	0	2	4	8	9	11	0	2	5	6	9	10
0	2	5	7	8	10	0	2	5	7	9	11	0	2	6	7	8	9	0	2	6	8	10	11
0	3	4	5	6	10	0	3	4	5	9	11	0	3	4	6	8	11	0	3	4	7	8	9
0	3	4	7	10	11	0	3	5	6	7	8	0	3	5	7	9	10	0	3	5	8	10	11
0	3	6	7	9	11	0	3	6	8	9	10	0	4	5	6	7	9	0	4	5	7	8	11
0	4	5	8	9	10	0	4	6	7	8	10	0	4	6	9	10	11	0	5	6	7	10	11
0	5	6	8	9	11	0	7	8	9	10	11	1	2	3	4	5	6	1	2	3	4	7	10
1	2	3	4	8	9	1	2	3	5	7	8	1	2	3	5	9	11	1	2	3	6	7	11

1	2	3	6	9	10	1	2	3	8	10	11	1	2	4	5	7	11	1	2	4	5	8	10
1	2	4	6	7	9	1	2	4	6	8	11	1	2	4	9	10	11	1	2	5	6	8	9
1	2	5	6	10	11	1	2	5	7	9	10	1	2	6	7	8	10	1	2	7	8	9	11
1	3	4	5	7	9	1	3	4	5	10	11	1	3	4	6	8	10	1	3	4	6	9	11
1	3	4	7	8	11	1	3	5	6	7	10	1	3	5	6	8	11	1	3	5	8	9	10
1	3	6	7	8	9	1	3	7	9	10	11	1	4	5	6	7	8	1	4	5	6	9	10
1	4	5	8	9	11	1	4	6	7	10	11	1	4	7	8	9	10	1	5	6	7	9	11
1	5	7	8	10	11	1	6	8	9	10	11	2	3	4	5	8	11	2	3	4	5	9	10
2	3	4	6	7	8	2	3	4	6	10	11	2	3	4	7	9	11	2	3	5	6	7	9
2	3	5	6	8	10	2	3	5	7	10	11	2	3	6	8	9	11	2	3	7	8	9	10
2	4	5	6	7	10	2	4	5	6	9	11	2	4	5	7	8	9	2	4	6	8	9	10
2	4	7	8	10	11	2	5	6	7	8	11	2	5	8	9	10	11	2	6	7	9	10	11
3	4	5	6	7	11	3	4	5	6	8	9	3	4	5	7	8	10	3	4	6	7	9	10
3	4	8	9	10	11	3	5	6	9	10	11	3	5	7	8	9	11	3	6	7	8	10	11
4	5	6	8	10	11	4	5	7	9	10	11	4	6	7	8	9	11	5	6	7	8	9	10

3.5.2 BALANCED INCOMPLETE BLOCK DESIGNS (BIBDS)

Balanced incomplete block designs (BIBDs) are t-designs with $t = 2$, so that every pair of points is on the same number of blocks. The relevant parameters are v, b, r, k, and λ with

$$vr = bk \qquad \text{and} \qquad v(v - 1)\lambda = bk(k - 1). \tag{3.5.2}$$

If A is the $v \times b$ incidence matrix, then $AA^{\mathrm{T}} = (r - \lambda)I_v + \lambda J_v$, where I_n is the $n \times n$ identity matrix and J_n is the $n \times n$ matrix of all ones.

3.5.2.1 Symmetric designs

Fisher's inequality states that $b \geq v$. If $b = v$ (equivalently, $r = k$), then the BIBD is called a *symmetric design*, denoted as a (v, k, λ)-design. The incidence matrix for a symmetric design satisfies

$$J_v A = k J_v = A J_v \quad \text{and} \quad A^{\mathrm{T}} A = (k - \lambda)I_v + \lambda J_v, \tag{3.5.3}$$

that is, any two blocks intersect in λ points. The dualness of symmetric designs can be summarized by the following:

$$
\begin{array}{lcl}
v \text{ points} & \longleftrightarrow & v \text{ blocks,} \\
k \text{ blocks on a point} & \longleftrightarrow & k \text{ points in a block, and} \\
\text{Any two points on } \lambda \text{ blocks} & \longleftrightarrow & \text{Any two blocks share } \lambda \text{ points.}
\end{array}
$$

Some necessary conditions for symmetric designs are

1. If v is even, then $k - \lambda$ is a square integer.

2. *Bruck–Ryser–Chowla theorem*: If v is odd, then the following equation has integer solutions (not all zero):

$$x^2 = (k - \lambda)y^2 + (-1)^{(v-1)/2}\lambda z^2.$$

3.5.2.2 Existence table for BIBDs

Some of the most fruitful construction methods for BIBD are dealt with in separate sections, difference sets (page 246), finite geometry (page 247), Steiner triple systems (page 249), and Hadamard matrices (page 249). The table below gives all parameters for which BIBDs exist with $k \leq v/2$ and $b \leq 30$.

v	b	r	k	λ	v	b	r	k	λ	v	b	r	k	λ
6	10	5	3	2	10	18	9	5	4	15	30	14	7	6
6	20	10	3	4	10	30	9	3	2	16	16	6	6	2
6	30	15	3	6	10	30	12	4	4	16	20	5	4	1
7	7	3	3	1	11	11	5	5	2	16	24	9	6	3
7	14	6	3	2	11	22	10	5	4	16	30	15	8	7
7	21	9	3	3	12	22	11	6	5	19	19	9	9	4
7	28	12	3	4	13	13	4	4	1	21	21	5	5	1
8	14	7	4	3	13	26	6	3	1	21	30	10	7	3
8	28	14	4	6	13	26	8	4	2	23	23	11	11	5
9	12	4	3	1	13	26	12	6	5	25	25	9	9	3
9	18	8	4	3	14	26	13	7	6	25	30	6	5	1
9	24	8	3	2	15	15	7	7	3	27	27	13	13	6
10	15	6	4	2										

3.5.3 DIFFERENCE SETS

Let G be a finite group of order v (see page 161). A subset D of size k is a (v, k, λ)-*difference set* in G if every non-identity element of G can be written λ times as a "difference" $d_1 d_2^{-1}$ with d_1 and d_2 in D. If G is the cyclic group $\mathbb{Z}_v$, then the difference set is a *cyclic difference set*. The *order* of a difference set is $n = k - \lambda$. For example, $\{1, 2, 4\}$ is a $(7, 3, 1)$ cyclic difference set of order 2.

The existence of a (v, k, λ)-difference set implies the existence of a (v, k, λ)-design. The points are the elements of G and the blocks are the translates of D: all sets $Dg = \{dg : d \in D\}$ for $g \in G$. Note that each translate Dg is itself a difference set.

EXAMPLES

1. Here are the 7 blocks for a $(7, 3, 1)$-design based on $D = \{1, 2, 4\}$:

$$1\,2\,4 \quad 2\,3\,5 \quad 3\,4\,6 \quad 4\,5\,0 \quad 5\,6\,1 \quad 6\,0\,2 \quad 0\,1\,3$$

2. A $(16, 6, 2)$-difference set in $G = \mathbb{Z}_2 \oplus \mathbb{Z}_2 \oplus \mathbb{Z}_2 \oplus \mathbb{Z}_2$ is

$$0000 \quad 0001 \quad 0010 \quad 0100 \quad 1000 \quad 1111$$

3. A $(21,5,1)$-difference set in $G = \langle a, b \,|\, a^3 = b^7 = 1, a^{-1}ba = a^4 \rangle$ is $\{a, a^2, b, b^2, b^4\}$.

3.5.3.1 Some families of cyclic difference sets

Paley: Let v be a prime congruent to 3 modulo 4. Then the non-zero squares in $\mathbb{Z}_v$ form a $(v, (v-1)/2, (v-3)/4)$-difference set. Example: $(v, k, \lambda) = (11, 5, 2)$.

Stanton–Sprott: Let $v = p(p + 2)$, where p and $p + 2$ are both primes. Then there is a $(v, (v - 1)/2, (v - 3)/4)$-difference set. Example: $(v, k, \lambda) = (35, 17, 8)$.

Biquadratic residues (I): If $v = 4a^2 + 1$ is a prime with a odd, then the non-zero fourth powers modulo v form a $(v, (v - 1)/4, (v - 5)/16)$-difference set. Example: $(v, k, \lambda) = (37, 9, 2)$.

Biquadratic residues (II): If $v = 4a^2 + 9$ is a prime with a odd, then zero and the fourth powers modulo v form a $(v, (v + 3)/4, (v + 3)/16)$-difference set. Example: $(v, k, \lambda) = (13, 4, 1)$.

Singer: If q is a prime power, then there exists a $\left(\dfrac{q^m - 1}{q - 1}, \dfrac{q^{m-1} - 1}{q - 1}, \dfrac{q^{m-2} - 1}{q - 1} \right)$-difference set for all $m \geq 3$.

3.5.3.2 Existence table of cyclic difference sets

This table gives all cyclic difference sets for $k \leq v/2$ and $v \leq 50$ up to equivalence by translation and multiplication by a number relatively prime to v.

v	k	λ	n	Difference set
7	3	1	2	1 2 4
11	5	2	3	1 3 4 5 9
13	4	1	3	0 1 3 9
15	7	3	4	0 1 2 4 5 8 10
19	9	4	5	1 4 5 6 7 9 11 16 17
21	5	1	4	3 6 7 12 14
23	11	5	6	1 2 3 4 6 8 9 12 13 16 18
31	6	1	5	1 5 11 24 25 27
31	15	7	8	1 2 3 4 6 8 12 15 16 17 23 24 27 29 30
				1 2 4 5 7 8 9 10 14 16 18 19 20 25 28
35	17	8	9	0 1 3 4 7 9 11 12 13 14 16 17 21 27 28 29 33
37	9	2	7	1 7 9 10 12 16 26 33 34
40	13	4	9	1 2 3 5 6 9 14 15 18 20 25 27 35
43	21	10	11	1 2 3 4 5 8 11 12 16 19 20 21 22 27 32 33 35 37 39 41 42
				1 4 6 9 10 11 13 14 15 16 17 21 23 24 25 31 35 36 38 40 41
47	23	11	12	1 2 3 4 6 7 8 9 12 14 16 17 18 21 24 25 27 28 32 34 36 37
				42

3.5.4 FINITE GEOMETRY

3.5.4.1 Affine planes

A finite *affine plane* is a finite set of *points* together with subsets of points called *lines* that satisfy the axioms:

1. Any two points are on exactly one line.

2. *(Parallel postulate)* Given a point P and a line L not containing P, there is exactly one line through P that does not intersect L.
3. There are four points, no three of which are collinear.

These axioms are sufficient to show that a finite affine plane is a BIBD (see page 245) with

$$v = n^2 \qquad b = n^2 + n \qquad r = n + 1 \qquad k = n \qquad \lambda = 1$$

(n is the *order* of the plane). The lines of a projective plane can be divided into $n + 1$ parallel classes each containing n lines. A sufficient condition for affine planes to exist is for n to be a prime power.

Below are two views of the affine plane of order 2 showing the parallel classes.

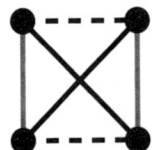 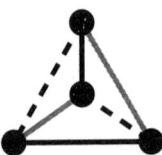

Below is the affine plane of order 3 showing the parallel classes.

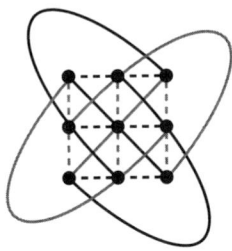

3.5.4.2   Projective planes

A finite *projective plane* is a finite set of points together with subsets of points called *lines* that satisfy the axioms:

1. Any two points are on exactly one line.
2. Any two lines intersect in exactly one point.
3. There are four points, no three of which are collinear.

These axioms are sufficient to show that a finite projective plane is a symmetric design (see page 245) with

$$v = n^2 + n + 1 \qquad k = n + 1 \qquad \lambda = 1 \qquad\qquad (3.5.4)$$

(n is the *order* of the plane). A sufficient condition for projective planes to exist is for n to be a prime power.

A projective plane of order n can be constructed from an affine plane of order n by adding a *line at infinity*. A line of $n+1$ new points is added to the affine plane. For each parallel class, one distinct new point is added to each line. The construction works in reverse: removing any one line from a projective plane of order n and its points leaves an affine plane of order n. To the right is the projective plane of order 2. The center circle functions as a line at infinity; removing it produces the affine plane of order 2.

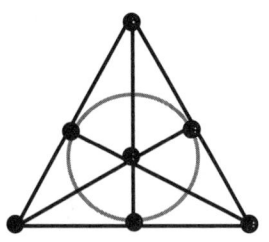

3.5.5 STEINER TRIPLE SYSTEMS

A *Steiner triple system* (STS) is a 2-(v,3,1) design. In particular, STSs are BIBDs (see page 245). STSs exist if, and only if, $v \equiv 1$ or 3 (mod 6). The number of blocks in an STS is $b = v(v-1)/6$.

3.5.5.1 Some families of Steiner triple systems

$v = 2^m - 1$: Take as points all non-zero vectors over $\mathbb{Z}_2$ of length m. A block consists of any set of three distinct vectors $\{x, y, z\}$ such that $x + y + z = 0$.

$v = 3^m$: Take as points all vectors over $\mathbb{Z}_3$ of length m. A block consists of any set of three distinct vectors $\{x, y, z\}$ such that $x + y + z = 0$.

3.5.5.2 Resolvable Steiner triple systems

An STS is *resolvable* if the blocks can be divided into parallel classes such that each point occurs in exactly one block per class. A resolvable STS exists if and only if $v \equiv 3$ (mod 6). For example, the affine plane of order 3 is a resolvable STS with $v = 9$ (see page 247).

A resolvable STS with $v = 15$ ($b = 35$) is known as the *Kirkman schoolgirl problem* and dates from 1850. Here is an example. Each column of 5 triples is a parallel class:

a b i	a c j	a d k	a e l	a f m	a g n	a h o
c d f	d e g	e f h	f g b	g h c	h b d	b c e
g j o	h k i	b l j	c m k	d n l	e o m	f i n
e k n	f l o	g m i	h n j	b o k	c i l	d j m
h l m	b m n	c n o	d o i	e i j	f j k	g k l

3.5.6 HADAMARD MATRICES

A *Hadamard matrix* of order n is an $n \times n$ matrix H with entries ± 1 such that $HH^T = nI_n$. In order for a Hadamard matrix to exist, n must be 1, 2, or a multiple of 4. It is conjectured that this condition is also sufficient. If H_1 and H_2 are Hadamard matrices, then so is the Kronecker product $H_1 \otimes H_2$.

3.5.6.1 Some Hadamard matrices

We use "−" to denote -1.

$n = 2$

$$\begin{bmatrix} 1 & 1 \\ 1 & - \end{bmatrix}$$

$n = 4$

$$\begin{bmatrix} 1 & 1 & 1 & 1 \\ 1 & - & 1 & - \\ 1 & 1 & - & - \\ 1 & - & - & 1 \end{bmatrix}$$

$n = 8$

$$\begin{bmatrix}
1 & 1 & 1 & 1 & 1 & 1 & 1 & 1 \\
1 & - & 1 & - & 1 & - & 1 & - \\
1 & 1 & - & - & 1 & 1 & - & - \\
1 & - & - & 1 & 1 & - & - & 1 \\
1 & 1 & 1 & 1 & - & - & - & - \\
1 & - & 1 & - & - & 1 & - & 1 \\
1 & 1 & - & - & - & - & 1 & 1 \\
1 & - & - & 1 & - & 1 & 1 & -
\end{bmatrix}$$

$n = 12$

$$\begin{bmatrix}
1 & 1 & 1 & 1 & 1 & 1 & 1 & 1 & 1 & 1 & 1 & 1 \\
1 & 1 & 1 & - & - & 1 & - & - & 1 & - & - & 1 \\
1 & 1 & 1 & - & 1 & - & - & 1 & - & - & 1 & - \\
1 & - & - & - & 1 & 1 & 1 & 1 & 1 & - & - & - \\
1 & - & 1 & 1 & - & - & 1 & 1 & - & - & - & 1 \\
1 & 1 & - & 1 & - & - & 1 & - & 1 & - & 1 & - \\
1 & - & - & - & - & - & 1 & 1 & 1 & 1 & 1 & 1 \\
1 & - & 1 & - & - & 1 & 1 & - & - & 1 & 1 & - \\
1 & 1 & - & - & 1 & - & 1 & - & - & 1 & - & 1 \\
1 & - & - & 1 & 1 & 1 & - & - & - & - & 1 & 1 \\
1 & - & 1 & 1 & 1 & - & - & - & - & 1 & 1 & - \\
1 & 1 & - & 1 & - & 1 & - & 1 & - & 1 & - & -
\end{bmatrix}$$

$n = 16$

$$\begin{bmatrix}
1 & 1 & 1 & 1 & 1 & 1 & 1 & 1 & 1 & 1 & 1 & 1 & 1 & 1 & 1 & 1 \\
1 & - & 1 & - & 1 & - & 1 & - & 1 & - & 1 & - & 1 & - & 1 & - \\
1 & 1 & - & - & 1 & 1 & - & - & 1 & 1 & - & - & 1 & 1 & - & - \\
1 & - & - & 1 & 1 & - & - & 1 & 1 & - & - & 1 & 1 & - & - & 1 \\
1 & 1 & 1 & 1 & - & - & - & - & 1 & 1 & 1 & 1 & - & - & - & - \\
1 & - & 1 & - & - & 1 & 1 & - & 1 & - & 1 & - & - & 1 & - & 1 \\
1 & 1 & - & - & - & - & 1 & 1 & 1 & 1 & - & - & - & - & 1 & 1 \\
1 & - & - & 1 & - & 1 & 1 & - & 1 & - & - & 1 & - & 1 & 1 & - \\
1 & 1 & 1 & 1 & 1 & 1 & 1 & 1 & - & - & - & - & - & - & - & - \\
1 & - & 1 & - & 1 & - & 1 & - & - & 1 & - & 1 & - & 1 & - & 1 \\
1 & 1 & - & - & 1 & 1 & - & - & - & - & 1 & 1 & - & - & 1 & 1 \\
1 & - & - & 1 & 1 & - & - & 1 & - & 1 & 1 & - & - & 1 & 1 & - \\
1 & 1 & 1 & 1 & - & - & - & - & - & - & - & - & 1 & 1 & 1 & 1 \\
1 & - & 1 & - & - & 1 & - & 1 & - & 1 & - & 1 & 1 & - & 1 & - \\
1 & 1 & - & - & - & - & 1 & 1 & - & - & 1 & 1 & 1 & 1 & - & - \\
1 & - & - & 1 & - & 1 & 1 & - & - & 1 & 1 & - & 1 & - & - & 1
\end{bmatrix}$$

3.5.6.2 Designs and Hadamard matrices

Without loss of generality, a Hadamard matrix can be assumed to have a first row and column consisting of all $+1$s.

BIBDs: Delete the first row and column. The points of the design are the remaining column indices. Each row produces a block of the design, namely those indices where the entry is $+1$. The resulting design is an $(n-1, (n-2)/2, (n-4)/4)$ symmetric design (see page 249).

3-Designs: The points are the indices of the columns. Each row, except the first row, yields two blocks, one block for those indices where the entries are $+1$ and one block for those indices where the entries are -1. The resulting design is a 3-$(n, n/2, (n-2)/2)$ design.

3.5.7 LATIN SQUARES

A *Latin square* of size n is an $n \times n$ array $S = [s_{ij}]$ of n symbols such that every symbol appears exactly once in each row and column. Two Latin squares S and T are *orthogonal* if every pair of symbols occurs exactly once as a pair (s_{ij}, t_{ij}). Let $M(n)$ be the maximum size of a set of mutually orthogonal Latin squares (MOLS).

1. $M(n) \leq n - 1$.
2. $M(n) = n - 1$ if n is a prime power.
3. $M(n_1 n_2) \geq \min(M(n_1), M(n_2))$.
4. $M(6) = 1$ (i.e., there are no two MOLS of size 6).
5. $M(n) \geq 2$ for all $n \geq 3$ except $n = 6$. (Latin squares of all sizes exist.)

The existence of $n - 1$ MOLS of size n is equivalent to the existence of an affine plane of order n (see page 247).

3.5.7.1 Examples of mutually orthogonal Latin squares

These are complete sets of MOLS for $n = 3, 4$, and 5.

$n = 3$

0	1	2
1	2	0
2	0	1

0	1	2
2	0	1
1	2	0

$n = 4$

0	1	2	3
1	0	3	2
2	3	0	1
3	2	1	0

0	1	2	3
2	3	0	1
3	2	1	0
1	0	3	2

0	1	2	3
3	2	1	0
1	0	3	2
2	3	0	1

$n = 5$

0	1	2	3	4
1	2	3	4	0
2	3	4	0	1
3	4	0	1	2
4	0	1	2	3

0	1	2	3	4
2	3	4	0	1
4	0	1	2	3
1	2	3	4	0
3	4	0	1	2

0	1	2	3	4
3	4	0	1	2
1	2	3	4	0
4	0	1	2	3
2	3	4	0	1

0	1	2	3	4
4	0	1	2	3
3	4	0	1	2
2	3	4	0	1
1	2	3	4	0

These are two superimposed MOLS for $n = 7, 8, 9$, and 10.

$n = 7$

00	11	22	33	44	55	66
16	20	31	42	53	64	05
25	36	40	51	62	03	14
34	45	56	60	01	12	23
43	54	65	06	10	21	32
52	63	04	15	26	30	41
61	02	13	24	35	46	50

$n = 8$

00	11	22	33	44	55	66	77
12	03	30	21	56	47	74	65
24	35	06	17	60	71	42	53
33	22	11	00	77	66	55	44
46	57	64	75	02	13	20	31
57	46	75	64	13	02	31	20
65	74	47	56	21	30	03	12
71	60	53	42	35	24	17	06

$n = 9$

00	11	22	33	44	55	66	77	88
12	20	01	45	53	34	78	86	67
21	02	10	54	35	43	87	68	76
36	47	58	60	71	82	03	14	25
48	56	37	72	80	61	15	23	04
57	38	46	81	62	70	24	05	13
63	74	85	06	17	28	30	41	52
75	83	64	18	26	07	42	50	31
84	65	73	27	08	16	51	32	40

$n = 10$

00	67	58	49	91	83	75	12	24	36
76	11	07	68	59	92	84	23	35	40
85	70	22	17	08	69	93	34	46	51
94	86	71	33	27	18	09	45	50	62
19	95	80	72	44	37	28	56	61	03
38	29	96	81	73	55	47	60	02	14
57	48	39	90	82	74	66	01	13	25
21	32	43	54	65	06	10	77	88	99
42	53	64	05	16	20	31	89	97	78
63	04	15	26	30	41	52	98	79	87

3.5.8 ROOM SQUARES

A *Room square* of side n is an $n \times n$ array with entries either empty or consisting of an unordered pair of symbols from a symbol set of size $n + 1$ with the requirements:

1. Each symbol appears exactly once in each row and column.
2. Every unordered pair occurs exactly once in the array.

Room squares exist if and only if n is odd and $n \geq 7$.

A Room square yields a construction of a round-robin tournament between $n+1$ opponents over n rounds and played at n locales:

1. Rows of the square represent rounds in the tournament.
2. Columns in the square represent locales.
3. Each pair represents one competition.

Then each team plays exactly once in each round, at each locale, and against each opponent.

EXAMPLE   This is a Room square of side 7:

01			26		57	34
45	02			37		61
72	56	03			41	
	13	67	04			52
63		24	71	05		
	74		35	12	06	
		15		46	23	07

3.5.9 COSTAS ARRAYS

An $n \times n$ *Costas array* is an array of zeros and ones whose two-dimensional autocorrelation function is n at the origin and no more than 1 anywhere else. There are B_n basic Costas arrays; there are C_n arrays when rotations and flips are allowed. Each array can be interpreted as a permutation.

n	1	2	3	4	5	6	7	8	9	10	11	12
B_n	1	1	1	2	6	17	13	17	30	60	555	990
C_n	1	2	4	12	40	116	200	444	760	2160	4368	7852

n	13	14	15	16	17	18	19	20
B_n	1616	2168	2467	2648	2294	1892	1283	810
C_n	12828	17252	19612	21104	18276	15096	10240	6464

B_n: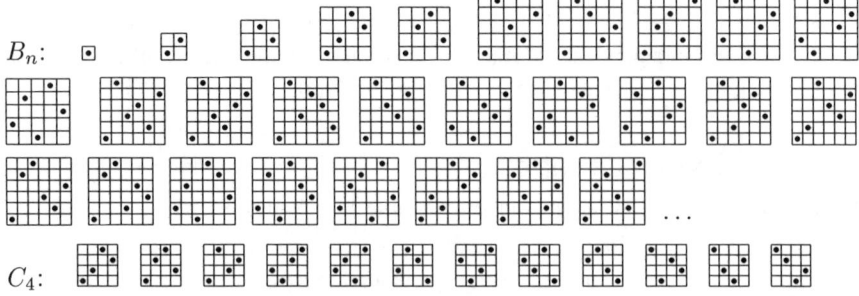

C_4: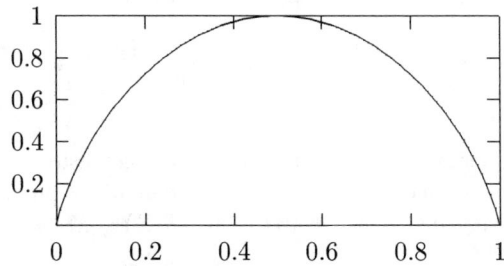

3.6 COMMUNICATION THEORY

3.6.1 INFORMATION THEORY

3.6.1.1 Definitions

Let $\mathbf{p}_X = (p_{x_1}, p_{x_2}, \ldots, p_{x_n})$ be the probability distribution of the discrete random variable X with $\text{Prob}\,(X = x_i) = p_{x_i}$. The *entropy* of the distribution is

$$H(\mathbf{p}_X) = -\sum_{x_i} p_{x_i} \log_2 p_{x_i}. \qquad (3.6.1)$$

The units for entropy are *bits*. Entropy measures how much information is gained from learning the value of X. When X takes only two values, $\mathbf{p} = (p, 1 - p)$, then

$$H(\mathbf{p}_X) = H(p, 1 - p) = -p \log_2 p - (1 - p) \log_2 (1 - p). \qquad (3.6.2)$$

This is also denoted $H(p)$. The range of $H(p)$ is from 0 to 1 with a maximum at $p = 0.5$. Below is a plot of p versus $H(p)$. The maximum of $H(\mathbf{p}_X)$ is $\log_2 n$ and is obtained when X is uniformly distributed, taking n values.

Given two discrete random variables X and Y, $\mathbf{p}_{X \times Y}$ is the joint distribution of X and Y. The *mutual information* of X and Y is defined by

$$I(X,Y) = H(\mathbf{p}_X) + H(\mathbf{p}_Y) - H(\mathbf{p}_{X \times Y}). \tag{3.6.3}$$

Note that (a) $I(X,Y) = I(Y,X)$; (b) $I(X,Y) \geq 0$; and (c) $I(X,Y) = 0$ if, and only if, X and Y are independent. Mutual information gives the amount of information that learning a value of X says about the value of Y (and vice versa).

3.6.1.2   Continuous entropy

For a d-dimensional continuous random variable $\mathbf{X}$, the entropy is

$$h(\mathbf{X}) = - \int_{\mathbb{R}^d} p(\mathbf{x}) \log p(\mathbf{x}) \, d\mathbf{x}. \tag{3.6.4}$$

Continuous entropy is not the limiting case of the entropy of a discrete random variable. In fact, if X is the limit of the one-dimensional discrete random variable $\{X_n\}$, and the entropy of X is finite, then

$$\lim_{n \to \infty} (H(X_n) - n \log 2) = h(X). \tag{3.6.5}$$

If $\mathbf{X}$ and $\mathbf{Y}$ are continuous d-dimensional random variables with density functions $p(\mathbf{x})$ and $q(\mathbf{y})$, then the *relative entropy* is

$$H(\mathbf{X}, \mathbf{Y}) = \int_{\mathbb{R}^d} p(\mathbf{x}) \log \frac{p(\mathbf{x})}{q(\mathbf{x})} \, d\mathbf{x}. \tag{3.6.6}$$

A d-dimensional normal (or Gaussian) random variable $N(\mathbf{a}, \Gamma)$ has the density function

$$g(\mathbf{x}) = \frac{1}{(2\pi)^{d/2} \sqrt{|\Gamma|}} \exp\left(-\frac{1}{2} (\mathbf{x} - \mathbf{a})^{\mathrm{T}} \Gamma^{-1} (\mathbf{x} - \mathbf{a})\right) \tag{3.6.7}$$

where $\mathbf{a}$ is the vector of means and Γ is the positive definite covariance matrix.

1. If $\mathbf{X} = (X_1, X_2, \ldots, X_d)$ is a d-dimensional normal random vector with distribution $N(\mathbf{a}, \Gamma)$ then $h(\mathbf{X}) = \frac{1}{2} \log\left((2\pi e)^d |\Gamma|\right)$.

2. If $\mathbf{X}$ and $\mathbf{Y}$ are d-dimensional normal random vectors with distributions $N(\mathbf{a}, \Gamma)$ and $N(\mathbf{b}, \Delta)$ then

$$H(\mathbf{X}, \mathbf{Y}) = \frac{1}{2} \left(\log \frac{|\Delta|}{|\Gamma|} + \mathrm{tr}\left(\Gamma \left(\Delta^{-1} - \Gamma^{-1}\right)\right) + (\mathbf{a} - \mathbf{b})^{\mathrm{T}} \Delta^{-1} (\mathbf{a} - \mathbf{b})\right) \tag{3.6.8}$$

3. If $\mathbf{X}$ is a d-dimensional normal random vector with distribution $N(\mathbf{a}, \Gamma)$, and if $\mathbf{Y}$ is a d-dimensional random vector with a continuous probability distribution having the same covariance matrix Γ, then $h(\mathbf{X}) \leq h(\mathbf{Y})$.

3.6.1.3 Channel capacity

The *transition probabilities* are defined by $t_{x,y} = \text{Prob}\,(Y = y \mid X = x)$. The distribution $\mathbf{p}_X$ determines $\mathbf{p}_Y$ by $p_y = \sum t_{x,y} p_x$. The matrix $T = (t_{x,y})$ is the *transition matrix*. The matrix T defines a *channel* given by a transition diagram (input is X, output is Y). For example (here X and Y only take two values),

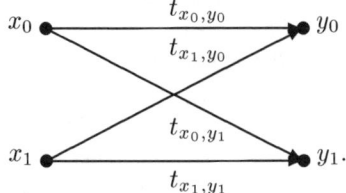

The *capacity* of the channel is defined as

$$C = \max_{\mathbf{p}_X} I(X, Y). \tag{3.6.9}$$

A channel is *symmetric* if each row is a permutation of the first row and each column is a permutation of the first column. The capacity of a symmetric channel is $C = \log_2 n - H(\mathbf{p})$, where $\mathbf{p}$ is the first row; the capacity is achieved when $\mathbf{p}_X$ represents equally likely inputs. The channel shown on the left is symmetric; both channels achieve capacity with equally likely inputs.

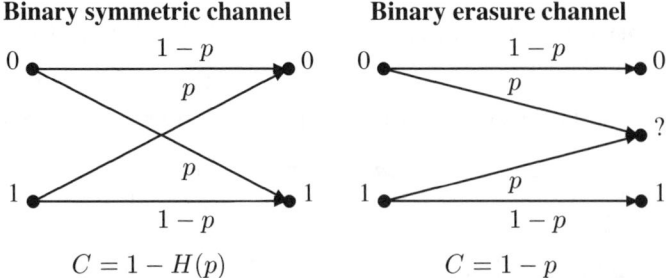

3.6.1.4 Shannon's theorem

Let both X and Y be discrete random variables with values in an alphabet A. A *code* is a set of *codewords* (n-tuples with entries from A) that is in one-to-one correspondence with a set of M messages. The *rate* R of the code is defined as $\frac{1}{n} \log_2 M$. Assume that the codeword is sent via a channel with transition matrix T by sending each vector element independently. Define

$$e = \max_{\text{all codewords}} \text{Prob}\,((\text{codeword incorrectly decoded})). \tag{3.6.10}$$

Shannon's coding theorem states:

1. If $R < C$, then there is a sequence of codes with rate R and $n \to \infty$ such that $e \to 0$.

2. If $R \geq C$, then e is always bounded away from 0.

3.6.2   BLOCK CODING

3.6.2.1   Definitions

A *code* C over an alphabet A is a set of vectors of a fixed length n with entries from A. Let A be the finite field GF(q) (see Section 2.7.6.1). If C is a vector space over A, then C is a *linear code*; the *dimension* k of a linear code is its dimension as a vector space. The *Hamming distance* $d_H(\mathbf{u}, \mathbf{v})$ between two vectors, $\mathbf{u}$ and $\mathbf{v}$, is the number of places in which they differ. For a vector $\mathbf{u}$ over GF(q), define the *weight*, wt($\mathbf{u}$), as the number of non-zero components. Then $d_H(\mathbf{u}, \mathbf{v}) = \text{wt}(\mathbf{u} - \mathbf{v})$. The minimum Hamming distance between two distinct vectors in a code C is called the *minimum distance d*. A code can detect e errors if $e < d$. A code can correct t errors if $2t + 1 \le d$.

3.6.2.2   Coding diagram for linear codes

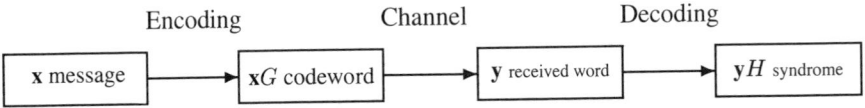

Encoding   Channel   Decoding

| x message | xG codeword | y received word | yH syndrome |

1. A *message* $\mathbf{x}$ consists of k information symbols.

2. The message is encoded as $\mathbf{x}G \in C$, where G is a $k \times n$ matrix called the generating matrix.

3. After transmission over a channel, a (possibly corrupted) vector $\mathbf{y}$ is received.

4. There exists a *parity check matrix* H such that $\mathbf{c} \in C$ if and only if $\mathbf{c}H = \mathbf{0}$. Thus the *syndrome* $\mathbf{z} = \mathbf{y}H$ can be used to try to decode $\mathbf{y}$.

5. If G has the form $[I\ A]$, where I is the $k \times k$ identity matrix, then $H = \begin{bmatrix} -A \\ I \end{bmatrix}$.

Note: in coding theory unspecified vectors are usually row vectors.

3.6.2.3   Cyclic codes

A linear code C of length n is *cyclic* if $(a_0, \ldots, a_{n-1}) \in C$ implies $(a_{n-1}, a_0, \ldots, a_{n-2}) \in C$. To each codeword $(a_0, a_1, \ldots, a_{n-1}) \in C$ is associated the polynomial $a(x) = \sum_{i=0}^{n-1} a_i x^i$. Every cyclic code has a *generating polynomial* $g(x)$ such that $a(x)$ corresponds to a codeword if, and only if, $a(x) \equiv d(x)g(x) \pmod{x^n - 1}$ for some $d(x)$. The *roots* of a cyclic code are roots of $g(x)$ in some extension field GF(q') with primitive element α.

1. *BCH Bound:* If a cyclic code C has roots $\alpha^i, \alpha^{i+1}, \ldots, \alpha^{i+d-2}$, then the minimum distance of C is at least d.

2. *Binary BCH codes* (BCH stands for Bose, Ray-Chaudhuri, and Hocquenghem): Fix m, define $n = 2^m - 1$, and let α be a primitive element in GF(2^m). Define $f_i(x)$ as the minimum binary polynomial of α^i. Then

$$g(x) = \text{LCM}\,(f_1(x), \ldots, f_{2e}(x)) \qquad (3.6.11)$$

defines a generating polynomial for a binary BCH code of length n and minimum distance at least $\delta = 2e + 1$ (δ is called the *designed distance*). The code dimension is at least $n - me$.

3. *Dual code:* Given a code C, the dual code is $C^{\perp} = \{\mathbf{a} \mid \mathbf{a} \cdot \mathbf{x} = 0 \text{ for all } \mathbf{x} \in C\}$. The code $C^{\perp}$ is an $(n, n - k)$ linear code over the same field. A code is *self-dual* if $C = C^{\perp}$.

4. *MDS codes:* A linear code that meets the Singleton bound, $n + 1 = k + d$, is called *MDS* (for *maximum distance separable*). Any k columns of a generating matrix of an MDS code are linearly independent.

5. *Reed–Solomon codes:* Let α be a primitive element for GF(q) and $n = q - 1$. The generating polynomial $g(x) = (x - \alpha)(x - \alpha^2) \cdots (x - \alpha^{d-1})$ defines a cyclic MDS code with distance d and dimension $k = n - d + 1$.

6. *Hexacode:* The *hexacode* is a $(6, 3, 4)$ self-dual MDS code over GF(4). Let the finite field of four elements be $\{0, 1, a, b\}$ with $b = a^2 = a + 1$. The code is generated by the vectors $(1, 0, 0, 1, b, a)$, $(0, 1, 0, 1, a, b)$, and $(0, 0, 1, 1, 1, 1)$. The 64 codewords are:

```
000000  1001ba  a00a1b  b00ba1  0101ab  110011  a10bb0  b10a0a
0a0ab1  1a0b0b  aa00aa  ba0110  0b0b1a  1b0aa0  ab0101  bb00bb
001111  1010ab  a01b0a  b01ab0  0110ba  111100  a11aa1  b11b1b
0a1ba0  1a1a1a  aa11bb  ba1001  0b1a0b  1b1bb1  ab1010  bb11aa
00aaaa  10ab10  a0a0b1  b0a10b  01ab01  11aabb  a1a11a  b1a0a0
0aa01b  1aa1a1  aaaa00  baabba  0ba1b0  1ba00a  ababab  bbaa11
00bbbb  10ba01  a0b1a0  b0b01a  01ba10  11bbaa  a1b00b  b1b1b1
0ab10a  1ab0b0  aabb11  babaab  0bb0a1  1bb11b  abbaba  bbbb00.
```

7. *Perfect codes:* A linear code is *perfect* if it satisfies the Hamming bound, $q^{n-k} = \sum_{i=0}^{e} \binom{n}{i}(q - 1)^i$. The binary Hamming codes and Golay codes are perfect.

8. *Binary Hamming codes:* These codes have parameters $\{n = 2^m - 1, k = 2^m - 1 - m, d = 3\}$. The parity check matrix is the $2^m - 1 \times m$ matrix whose rows are all of the binary m-tuples in a fixed order. The generating and parity check matrices for the $(7, 4)$ Hamming code are

$$G = [I \quad A] = \begin{bmatrix} 1 & 0 & 0 & 0 & 1 & 1 & 0 \\ 0 & 1 & 0 & 0 & 1 & 0 & 1 \\ 0 & 0 & 1 & 0 & 0 & 1 & 1 \\ 0 & 0 & 0 & 1 & 1 & 1 & 1 \end{bmatrix}, \qquad H = \begin{bmatrix} -A \\ I \end{bmatrix} = \begin{bmatrix} 1 & 1 & 0 \\ 1 & 0 & 1 \\ 0 & 1 & 1 \\ 1 & 1 & 1 \\ 1 & 0 & 0 \\ 0 & 1 & 0 \\ 0 & 0 & 1 \end{bmatrix}.$$

$$(3.6.12)$$

9. *Binary Golay code:* This has the parameters $\{n = 24, k = 12, d = 8\}$. The generating matrix is

$$
G = \begin{bmatrix} I & A \end{bmatrix} = \begin{bmatrix}
1\,0\,0\,0\,0\,0\,0\,0\,0\,0\,0\,0 & 0\,1\,1\,1\,1\,1\,1\,1\,1\,1\,1\,1 \\
0\,1\,0\,0\,0\,0\,0\,0\,0\,0\,0\,0 & 1\,1\,1\,0\,1\,1\,1\,0\,0\,0\,1\,0 \\
0\,0\,1\,0\,0\,0\,0\,0\,0\,0\,0\,0 & 1\,1\,0\,1\,1\,1\,0\,0\,0\,1\,0\,1 \\
0\,0\,0\,1\,0\,0\,0\,0\,0\,0\,0\,0 & 1\,0\,1\,1\,1\,0\,0\,0\,1\,0\,1\,1 \\
0\,0\,0\,0\,1\,0\,0\,0\,0\,0\,0\,0 & 1\,1\,1\,1\,0\,0\,0\,1\,0\,1\,1\,0 \\
0\,0\,0\,0\,0\,1\,0\,0\,0\,0\,0\,0 & 1\,1\,1\,0\,0\,0\,1\,0\,1\,1\,0\,1 \\
0\,0\,0\,0\,0\,0\,1\,0\,0\,0\,0\,0 & 1\,1\,0\,0\,0\,1\,0\,1\,1\,0\,1\,1 \\
0\,0\,0\,0\,0\,0\,0\,1\,0\,0\,0\,0 & 1\,0\,0\,0\,1\,0\,1\,1\,0\,1\,1\,1 \\
0\,0\,0\,0\,0\,0\,0\,0\,1\,0\,0\,0 & 1\,0\,0\,1\,0\,1\,1\,0\,1\,1\,1\,0 \\
0\,0\,0\,0\,0\,0\,0\,0\,0\,1\,0\,0 & 1\,0\,1\,0\,1\,1\,0\,1\,1\,1\,0\,0 \\
0\,0\,0\,0\,0\,0\,0\,0\,0\,0\,1\,0 & 1\,1\,0\,1\,1\,0\,1\,1\,1\,0\,0\,0 \\
0\,0\,0\,0\,0\,0\,0\,0\,0\,0\,0\,1 & 1\,0\,1\,1\,0\,1\,1\,1\,0\,0\,0\,1
\end{bmatrix}. \tag{3.6.13}
$$

10. *Ternary Golay code:* This has the parameters $\{n = 12, k = 6, d = 6\}$. The generating matrix is

$$
G = \begin{bmatrix} I & A \end{bmatrix} = \begin{bmatrix}
1\,0\,0\,0\,0\,0 & 0\,1\,1\,1\,1\,1 \\
0\,1\,0\,0\,0\,0 & 1\,0\,1\,2\,2\,1 \\
0\,0\,1\,0\,0\,0 & 1\,1\,0\,1\,2\,2 \\
0\,0\,0\,1\,0\,0 & 1\,2\,1\,0\,1\,2 \\
0\,0\,0\,0\,1\,0 & 1\,2\,2\,1\,0\,1 \\
0\,0\,0\,0\,0\,1 & 1\,1\,2\,2\,1\,0
\end{bmatrix}. \tag{3.6.14}
$$

3.6.2.4   Bounds

Bounds for block codes investigate the trade-offs between the length n, the number of codewords M, the minimum distance d, and the alphabet size q. The number of errors that can be corrected is e with $2e + 1 \le d$. If the code is linear, then the bounds concern the dimension k with $M = q^k$.

1. *Hamming* or *sphere-packing bound:* $M \le q^n / \sum_{i=0}^{e} \binom{n}{i}(q-1)^i$.

2. *Plotkin bound:* Suppose that $d > n(q-1)/q$. Then $M \le \frac{qd}{qd - n(q-1)}$.

3. *Singleton bound:* For any code, $M \le q^{n-d+1}$; if the code is linear, then $k + d \le n + 1$.

4. *Varsharmov–Gilbert bound:* There is a block code with minimum distance at least d and $M \ge q^n / \sum_{i=0}^{d-1} \binom{n}{i}(q-1)^i$.

3.6.2.5 Table of binary BCH codes

n	k	d
7	4	3
15	11	3
	7	5
	5	7
31	26	3
	21	5
	16	7
	11	11
	6	15
63	57	3
	51	5

n	k	d
63	45	7
	39	9
	36	11
	30	13
	24	15
	18	21
	16	23
	10	27
	7	31
127	120	3
	113	5
	106	7
	99	9

n	k	d
127	92	11
	85	13
	78	15
	71	19
	64	21
	57	23
	50	27
	43	31
	36	31
	29	≥ 43
	22	47
	15	55
	8	63

3.6.2.6 Table of best binary codes

Let $A(n, d)$ be the number of codewords[1] in the largest binary code of length n and minimum distance d. Note that $A(n - 1, d - 1) = A(n, d)$ if d is odd and $A(n, 2) = 2^{n-1}$ (given, e.g., by even weight words).

n	$d = 4$	$d = 6$	$d = 8$	$d = 10$
6	4	2	1	1
7	8	2	1	1
8	16	2	2	1
9	20	4	2	1
10	40	6	2	2
11	72	12	2	2
12	144	24	4	2
13	256	32	4	2
14	512	64	8	2
15	1024	128	16	4
16	2048	256	32	4
17	2720–3276	256–340	36–37	6
18	5312–6552	512–680	64–74	10
19	10496–13104	1024–1288	128–144	20
20	20480–26208	2048–2372	256–279	40
21	36864–43690	2560–4096	512	42–48
22	73728–87380	4096–6942	1024	50–88
23	147456–173784	8192–13774	2048	76–150
24	294912–344636	16384–24106	4096	128-280

---

[1] Data from *Sphere Packing, Lattices and Groups* by J. H. Conway and N. J. A. Sloane, 3rd ed., Springer-Verlag, New York, 1998.

3.6.3 SOURCE CODING FOR ENGLISH TEXT

English text has, on average, 4.08 bits/character.

Letter	Probability	Huffman code	Alphabetical code
Space	0.1859	000	00
A	0.0642	0100	0100
B	0.0127	011111	010100
C	0.0218	11111	010101
D	0.0317	01011	01011
E	0.1031	101	0110
F	0.0208	001100	011100
G	0.0152	011101	011101
H	0.0467	1110	01111
I	0.0575	1000	1000
J	0.0008	0111001110	1001000
K	0.0049	01110010	1001001
L	0.0321	01010	100101
M	0.0198	001101	10011
N	0.0574	1001	1010
O	0.0632	0110	1011
P	0.0152	011110	110000
Q	0.0008	0111001101	110001
R	0.0484	1101	11001
S	0.0514	0010	1101
T	0.0796	0010	1110
U	0.0228	11110	111100
V	0.0083	0111000	111101
W	0.0175	001110	111110
X	0.0013	0111001100	1111110
Y	0.0164	001111	11111110
Z	0.0005	0111001111	11111111
Cost	4.0799	4.1195	4.1978

3.6.4 MORSE CODE

The international version of Morse code is

A	• —		K	— • —		U			• • —		1			• — — — —
B	— • • •		L	• — • •		V			• • • —		2			• • — — —
C	— • — •		M	— —		W			• — —		3			• • • — —
D	— • •		N	— •		X			— • • —		4			• • • • —
E	•		O	— — —		Y			— • — —		5			• • • • •
F	• • — •		P	• — — •		Z			— — • •		6			— • • • •
G	— — •		Q	— — • —							7			— — • • •
H	• • • •		R	• — •		Period			• — • — • —		8			— — — • •
I	• •		S	• • •		Comma			— — • • — —		9			— — — — •
J	• — — —		T	—		Question			• • — — • •		0			— — — — —

3.6.5 GRAY CODE

A *Gray code* is a sequence ordering such that a small change in the sequence number results in a small change in the sequence.

EXAMPLES

1. The sixteen 4-bit strings $\{0000, 0001, \ldots, 1111\}$ can be ordered so that adjacent bit strings differ in only 1 bit:

Sequence number	Bit string	Sequence number	Bit string
0	0000	8	1100
1	0001	9	1101
2	0011	10	1111
3	0010	11	1110
4	0110	12	1010
5	0111	13	1011
6	0101	14	1001
7	0100	15	1000

2. The subsets of $\{a, b, c\}$ can be ordered so that adjacent subsets differ by only the insertion or deletion of a single element:

$$\phi, \quad \{a\}, \quad \{a, b\}, \quad \{b\}, \quad \{b, c\}, \quad \{a, b, c\}, \quad \{a, c\}, \quad \{c\}.$$

3.6.6 FINITE FIELDS

Pertinent definitions for finite fields may be found in Section 2.7.6 on page 169.

3.6.6.1 Irreducible polynomials

Let $N_q(n)$ be the number of monic irreducible polynomials of degree n over GF(q). Then

$$q^n = \sum_{d \mid n} d N_q(d) \quad \text{and} \quad N_q(n) = \frac{1}{n} \sum_{d \mid n} \mu\left(\frac{n}{d}\right) q^d, \tag{3.6.15}$$

where $\mu(\cdot)$ is the number theoretic Möbius function (see page 102).

3.6.6.2 Table of binary irreducible polynomials

The table lists the non-zero coefficients of binary irreducible polynomials, e.g., 2 1 0 corresponds to $x^2 + x^1 + x^0 = x^2 + x + 1$. The *exponent* of an irreducible polynomial is the smallest L such that $f(x)$ divides $x^L - 1$. A "P" after the exponent indicates that the polynomial is primitive.

$f(x)$	Exponent	$f(x)$	Exponent	$f(x)$	Exponent
2 1 0	3 P	7 3 2 1 0	127 P	8 6 4 3 2 1 0	255 P
3 1 0	7 P	7 4 0	127 P	8 6 5 1 0	255 P
3 2 0	7 P	7 4 3 2 0	127 P	8 6 5 2 0	255 P
4 1 0	15 P	7 5 2 1 0	127 P	8 6 5 3 0	255 P
4 2 0	15 P	7 5 3 1 0	127 P	8 6 5 4 0	255 P
4 3 2 1 0	5	7 5 4 3 0	127 P	8 6 5 4 2 1 0	85
5 2 0	31 P	7 5 4 3 2 1 0	127 P	8 6 5 4 3 1 0	85
5 3 0	31 P	7 6 0	127 P	8 7 2 1 0	255 P
5 3 2 1 0	31 P	7 6 3 1 0	127 P	8 7 3 1 0	85
5 4 2 1 0	31 P	7 6 4 1 0	127 P	8 7 3 2 0	255 P
5 4 3 1 0	31 P	7 6 4 2 0	127 P	8 7 4 3 2 1 0	51
5 4 3 2 0	31 P	7 6 5 2 0	127 P	8 7 5 1 0	85
6 1 0	63 P	7 6 5 3 2 1 0	127 P	8 7 5 3 0	255 P
6 3 0	9	7 6 5 4 0	127 P	8 7 5 4 0	51
6 4 2 1 0	21	7 6 5 4 2 1 0	127 P	8 7 5 4 3 2 0	85
6 4 3 1 0	63 P	7 6 5 4 3 2 0	127 P	8 7 6 1 0	255 P
6 5 0	63 P	8 4 3 1 0	51	8 7 6 3 2 1 0	255 P
6 5 2 1 0	63 P	8 4 3 2 0	255 P	8 7 6 4 2 1 0	17
6 5 3 2 0	63 P	8 5 3 1 0	255 P	8 7 6 4 3 2 0	85
6 5 4 1 0	63	8 5 3 2 0	255 P	8 7 6 5 2 1 0	255 P
6 5 4 2 0	21 P	8 5 4 3 0	17	8 7 6 5 4 1 0	51
7 1 0	127 P	8 5 4 3 2 1 0	85	8 7 6 5 4 2 0	255 P
7 3 0	127 P	8 6 3 2 0	255 P	8 7 6 5 4 3 0	85

3.6.6.3   Table of binary primitive polynomials

Listed below[2] are primitive polynomials, with the least number of non-zero terms, of degree from 1 to 64. Only the exponents of the non-zero terms are listed, e.g., 2 1 0 corresponds to $x^2 + x + 1$.

$f(x)$	$f(x)$	$f(x)$	$f(x)$	$f(x)$
1 0	14 5 3 1 0	27 5 2 1 0	40 5 4 3 0	53 6 2 1 0
2 1 0	15 1 0	28 3 0	41 3 0	54 6 5 4 3 2 0
3 1 0	16 5 3 2 0	29 2 0	42 5 4 3 2 1 0	55 6 2 1 0
4 1 0	17 3 0	30 6 4 1 0	43 6 4 3 0	56 7 4 2 0
5 2 0	18 5 2 1 0	31 3 0	44 6 5 2 0	57 5 3 2 0
6 1 0	19 5 2 1 0	32 7 5 3 2 1 0	45 4 3 1 0	58 6 5 1 0
7 1 0	20 3 0	33 6 4 1 0	46 8 5 3 2 1 0	59 6 5 4 3 1 0
8 4 3 2 0	21 2 0	34 7 6 5 2 1 0	47 5 0	60 1 0
9 4 0	22 1 0	35 2 0	48 7 5 4 2 1 0	61 5 2 1 0
10 3 0	23 5 0	36 6 5 4 2 1 0	49 6 5 4 0	62 6 5 3 0
11 2 0	24 4 3 1 0	37 5 4 3 2 1 0	50 4 3 2 0	63 1 0
12 6 4 1 0	25 3 0	38 6 5 1 0	51 6 3 1 0	64 4 3 1 0
13 4 3 1 0	26 6 2 1 0	39 4 0	52 3 0	

---

[2]Taken in part from "Primitive Polynomials (Mod 2)", E. J. Watson, *Math. Comp.*, **16**, 368–369, 1962.

3.6.7 BINARY SEQUENCES

3.6.7.1 Barker sequences

A *Barker sequence* is a sequence $(s_1, s_2, \ldots, s_N)$ with $s_j = \pm 1$ such that $\sum_{j=1}^{N-i} s_j s_{j+i} = \pm 1$ or 0, for $i = 1, \ldots, N - 1$. The following table lists all known Barker sequences (up to reversal, multiplication by -1, and multiplying alternate values by -1).

Length	Barker sequence
2	+1 +1
3	+1 +1 −1
4	+1 +1 +1 −1
4	+1 +1 −1 +1
5	+1 +1 +1 −1 +1
7	+1 +1 +1 −1 −1 +1 −1
11	+1 +1 +1 −1 −1 −1 +1 −1 −1 +1 −1
13	+1 +1 +1 +1 +1 −1 −1 +1 +1 −1 +1 −1 +1

3.6.7.2 Periodic sequences

Let $\mathbf{s} = (s_0, s_1, \ldots, s_{N-1})$ be a periodic sequence with period N. A (left) *shift* of $\mathbf{s}$ is the sequence $(s_1, \ldots, s_{N-1}, s_0)$. For τ relatively prime to N, the *decimation* of $\mathbf{s}$ is the sequence $(s_0, s_\tau, s_{2\tau}, \ldots)$, which also has period N. The *periodic autocorrelation* is defined as the vector $(a_0, \ldots, a_{N-1})$, with

$$a_i = \sum_{j=0}^{N-1} s_j s_{j+i}, \qquad \text{(subscripts taken modulo } N\text{).} \tag{3.6.16}$$

An autocorrelation is *two-valued* if all values are equal except possibly for the 0th term.

3.6.7.3 The m-sequences

A binary *m-sequence* of length $N = 2^r - 1$ is the sequence of period N defined by

$$(s_0, s_1, \ldots, s_{N-1}), \qquad s_i = \text{Tr}(\alpha^i), \tag{3.6.17}$$

where α is a primitive element of $\text{GF}(2^r)$ and Tr is the trace function from $\text{GF}(2^r)$ to $\text{GF}(2)$ defined by $\text{Tr}(\beta) = \beta + \beta^2 + \beta^4 + \cdots + \beta^{2^{r-1}}$.

1. All m-sequences of a given length are equivalent under decimation.
2. Binary m-sequences have a two-valued autocorrelation (with the identification that $0 \leftrightarrow +1$ and $1 \leftrightarrow -1$).
3. All m-sequences possess the *span property*: all binary r-tuples occur in the sequence except the all-zeros r-tuple.

The existence of a binary sequence of length $2^n - 1$ with a two-valued autocorrelation is equivalent to the existence of a cyclic difference set with parameters $(2^n - 1, 2^{n-1} - 1, 2^{n-2} - 1)$.

3.6.7.4 Shift registers

Below are examples of the two types of shift registers used to generate binary m-sequences. The generating polynomial in each case is $x^4 + x + 1$, the initial register loading is 1 0 0 0, and the generated sequence is $\{0, 0, 0, 1, 0, 0, 1, 1, 0, 1, 0, 1, 1, 1, 1, \ldots\}$.

Additive shift register	Multiplicative shift register

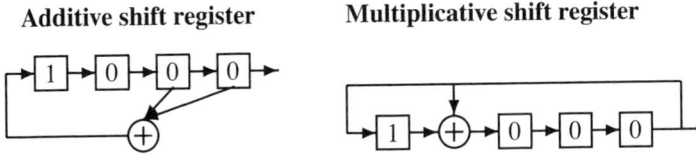

3.6.7.5 Binary sequences with two-valued autocorrelation

The following table lists all binary sequences with two-valued periodic autocorrelation of length $2^n - 1$ for $n = 3$ to 8 (up to shifts, decimations, and complementation). The table indicates the positions of 0; the remaining values are 1. An S indicates that that sequence has the span property.

n	Positions of 0
3 S	1 2 4
4 S	0 1 2 4 5 8 10
5 S	1 2 3 4 6 8 12 15 16 17 23 24 27 29 30
5	1 2 4 5 7 8 9 10 14 16 18 19 20 25 28
6 S	0 1 2 3 4 6 7 8 9 12 13 14 16 18 19 24 26 27 28 32 33 35 36 38 41 45 48 49 52 54 56
6	0 1 2 3 4 5 6 8 9 10 12 16 17 18 20 23 24 27 29 32 33 34 36 40 43 45 46 48 53 54 58
7	1 2 4 8 9 11 13 15 16 17 18 19 21 22 25 26 30 31 32 34 35 36 37 38 41 42 44 47 49 50 52 60 61 62 64 68 69 70 71 72 73 74 76 79 81 82 84 87 88 94 98 99 100 103 104 107 113 115 117 120 121 122 124
7	1 2 3 4 5 6 7 8 10 12 14 16 19 20 23 24 25 27 28 32 33 38 40 46 47 48 50 51 54 56 57 61 63 64 65 66 67 73 75 76 77 80 87 89 92 94 95 96 97 100 101 102 107 108 111 112 114 117 119 122 123 125 126
7 S	1 2 3 4 6 7 8 9 12 14 15 16 17 18 24 27 28 29 30 31 32 34 36 39 47 48 51 54 56 58 60 61 62 64 65 67 68 71 72 77 78 79 83 87 89 94 96 97 99 102 103 105 107 108 112 113 115 116 117 120 121 122 124
7	1 2 3 4 6 7 8 9 12 13 14 16 17 18 19 24 25 26 27 28 31 32 34 35 36 38 47 48 50 51 52 54 56 61 62 64 65 67 68 70 72 73 76 77 79 81 87 89 94 96 97 100 102 103 104 107 108 112 115 117 121 122 124
7	1 2 3 4 5 6 8 9 10 12 15 16 17 18 19 20 24 25 27 29 30 32 33 34 36 38 39 40 48 50 51 54 55 58 59 60 64 65 66 68 71 72 73 76 77 78 80 83 89 91 93 96 99 100 102 105 108 109 110 113 116 118 120
7	1 2 3 4 5 6 8 10 11 12 16 19 20 21 22 24 25 27 29 32 33 37 38 39 40 41 42 44 48 49 50 51 54 58 63 64 65 66 69 73 74 76 77 78 80 82 83 84 88 89 95 96 98 100 102 105 108 111 116 119 123 125 126
8 S	0 1 2 3 4 6 7 8 12 13 14 16 17 19 23 24 25 26 27 28 31 32 34 35 37 38 41 45 46 48 49 50 51 52 54 56 59 62 64 67 68 70 73 74 75 76 82 85 90 92 96 98 99 100 102 103 104 105 108 111 112 113 118 119 123 124 127 128 129 131 134 136 137 139 140 141 143 145 146 148 150 152 153 157 161 164 165 170 177 179 180 183 184 187 189 191 192 193 196 197 198 199 200 204 206 208 210 216 217 219 221 222 223 224 226 227 236 237 238 239 241 246 247 248 251 253 254
8	0 1 2 4 7 8 9 11 14 16 17 18 19 21 22 23 25 27 28 29 32 33 34 35 36 38 42 43 44 46 49 50 51 54 56 58 61 64 66 68 69 70 71 72 76 79 81 84 85 86 87 88 89 92 93 95 97 98 99 100 101 102 108 112 113 116 117 119 122 125 128 131 132 133 136 137 138 139 140 141 142 144 145 149 152 153 158 162 163 167 168 170 171 172 174 175 176 177 178 184 186 187 190 193 194 196 197 198 200 202 204 209 211 213 215 216 221 224 226 232 233 234 235 238 244 245 250
8	0 1 2 3 4 6 8 12 13 15 16 17 24 25 26 27 29 30 31 32 34 35 39 47 48 50 51 52 54 57 58 59 60 61 62 64 67 68 70 71 78 79 85 91 94 96 99 100 102 103 104 107 108 109 114 116 118 119 120 121 122 124 127 128 129 134 135 136 140 141 142 143 145 147 151 153 156 157 158 161 163 167 170 173 177 179 181 182 187 188 191 192 195 198 199 200 201 203 204 206 208 209 211 214 216 217 218 221 223 225 227 228 229 232 233 236 238 239 240 241 242 244 247 248 251 253 254
8	0 1 2 3 4 6 7 8 11 12 14 15 16 17 21 22 23 24 25 28 29 30 32 34 35 37 41 42 44 46 47 48 50 51 56 58 60 64 68 69 70 71 73 74 81 82 84 85 88 91 92 94 96 97 100 102 107 109 111 112 113 116 119 120 121 123 127 128 129 131 133 135 136 138 139 140 142 145 146 148 151 153 162 163 164 168 170 173 176 181 182 183 184 187 188 189 191 192 193 194 195 197 200 203 204 209 214 218 219 221 222 223 224 225 226 229 232 237 238 239 240 242 246 247 251 253 254

3.7   DIFFERENCE EQUATIONS

3.7.1   THE CALCULUS OF FINITE DIFFERENCES

1. $\Delta(f(x)) = f(x + h) - f(x)$ (forward difference).
2. $\Delta^2(f(x)) = f(x + 2h) - 2f(x + h) + f(x)$.
3. $\Delta^n(f(x)) = \Delta\left(\Delta^{n-1}(f(x))\right) = \sum_{k=0}^{n}(-1)^k \binom{n}{k} f(x + (n - k)h)$.
4. $\Delta(cf(x)) = c\Delta(f(x))$.
5. $\Delta(f(x) + g(x)) = \Delta(f(x)) + \Delta(g(x))$.
6. $\Delta(f(x)g(x)) = g(x)\Delta(f(x)) + f(x)\Delta(g(x))$.
7. $\Delta\left(\frac{f(x)}{g(x)}\right) = \dfrac{g(x)\Delta(f(x)) - f(x)\Delta(g(x))}{g(x)g(x + h)}$, provided that $g(x)g(x + h) \neq 0$.
8. $\Delta^n(x^n) = n!\,h^n, \quad n = 0, 1, \ldots$.

3.7.2   EXISTENCE AND UNIQUENESS

A *difference equation of order* k has the form

$$x_{n+k} = f(x_n, x_{n+1}, \ldots, x_{n+(k-1)}, n) \tag{3.7.1}$$

where f is a given function and k is a positive integer. A *solution* to Equation (3.7.1) is a sequence of numbers $\{x_n\}_{n=0}^{\infty}$ which satisfies the equation. Any constant solution of Equation (3.7.1) is called an *equilibrium solution*.

A *linear* difference equation of order k has the form

$$a_n^{(k)} x_{n+k} + a_n^{(k-1)} x_{n+(k-1)} + \cdots + a_n^{(1)} x_{n-1} + a_n^{(0)} x_n = g_n, \tag{3.7.2}$$

where k is a positive integer and the coefficients $a_n^{(0)}, \ldots, a_n^{(k)}$ along with $\{g_n\}$ are known. If the sequence g_n is identically zero, then Equation (3.7.2) is called *homogeneous*; otherwise, it is called *non-homogeneous*. If the coefficients $a_n^{(0)}, \ldots, a_n^{(k)}$ are constants (i.e., do not depend on n), Equation (3.7.2) is a *difference equation with constant coefficients*; otherwise it is a *difference equation with variable coefficients*.

THEOREM 3.7.1   *(Existence and uniqueness)*

Consider the initial-value problem (IVP)

$$
\begin{aligned}
x_{n+k} + b_n^{(k-1)} x_{n+(k-1)} + \cdots + b_n^{(1)} x_{n+1} + b_n^{(0)} x_n &= f_n, \\
x_i = \alpha_i, \qquad i = 0, 1, \ldots, k - 1,
\end{aligned}
\tag{3.7.3}
$$

for $n = 0, 1, \ldots$, where $b_n^{(i)}$ and f_n are given sequences with $b_n^{(0)} \neq 0$ for all n and the $\{\alpha_i\}$ are given initial conditions. Then the above equations have exactly one solution.

3.7.3   LINEAR INDEPENDENCE: GENERAL SOLUTION

The sequences $x^{(1)}, x^{(2)}, \ldots, x^{(k)}$ (sequence $x^{(i)}$ has the terms $\{x_1^{(i)}, x_2^{(i)}, x_2^{(i)}, \ldots \}$) are *linearly dependent* if constants $c_1, c_2, \ldots, c_k$ (not all of them zero) exist such that

$$\sum_{i=1}^{k} c_i x_n^{(i)} = 0 \quad \text{for } n = 0, 1, \ldots. \tag{3.7.4}$$

Otherwise the sequences $x^{(1)}, x^{(2)}, \ldots, x^{(k)}$ are *linearly independent*.

The *Casoratian* of the k sequences $x^{(1)}, x^{(2)}, \ldots, x^{(k)}$ is the $k \times k$ determinant

$$C\left(x_n^{(1)}, x_n^{(2)}, \ldots, x_n^{(k)}\right) = \begin{vmatrix} x_n^{(1)} & x_n^{(2)} & \cdots & x_n^{(k)} \\ x_{n+1}^{(1)} & x_{n+1}^{(2)} & \cdots & x_{n+1}^{(k)} \\ \cdots\cdots\cdots\cdots\cdots\cdots\cdots\cdots \\ x_{n+k-1}^{(1)} & x_{n+k-1}^{(2)} & \cdots & x_{n+k-1}^{(k)} \end{vmatrix}. \tag{3.7.5}$$

THEOREM 3.7.2

The solutions $x^{(1)}, x^{(2)}, \ldots, x^{(k)}$ of the linear homogeneous difference equation,

$$x_{n+k} + b_n^{(k-1)} x_{n+(k-1)} + \cdots + b_n^{(1)} x_{n+1} + b_n^{(0)} x_n = 0, \qquad n = 0, 1, \ldots, \tag{3.7.6}$$

are linearly independent if, and only if, their Casoratian is different from zero for $n = 0$.

Note that the solutions to Equation (3.7.6) form a k-dimensional vector space. The set $\{x^{(1)}, x^{(2)}, \ldots, x^{(k)}\}$ is a *fundamental system* of solutions for Equation (3.7.6) if, and only if, the sequences $x^{(1)}, x^{(2)}, \ldots, x^{(k)}$ are linearly independent solutions of the homogeneous difference Equation (3.7.6).

THEOREM 3.7.3

Consider the non-homogeneous linear difference equation

$$x_{n+k} + b_n^{(k-1)} x_{n+(k-1)} + \cdots + b_n^{(1)} x_{n+1} + b_n^{(0)} x_n = d_n, \qquad n = 0, 1, \ldots \tag{3.7.7}$$

where $b_n^{(i)}$ and d_n are given sequences. Let $x_n^{(h)}$ be the general solution of the corresponding homogeneous equation

$$x_{n+k} + b_n^{(k-1)} x_{n+(k-1)} + \cdots + b_n^{(1)} x_{n+1} + b_n^{(0)} x_n = 0, \qquad n = 0, 1, \ldots,$$

and let $x_n^{(p)}$ be a particular solution of Equation (3.7.7). Then $x_n^{(p)} + x_n^{(h)}$ is the general solution of Equation (3.7.7).

THEOREM 3.7.4 (*Superposition principle*)

Let $x^{(1)}$ and $x^{(2)}$ be solutions of the non-homogeneous linear difference equations

$$x_{n+k} + b_n^{(k-1)} x_{n+(k-1)} + \cdots + b_n^{(1)} x_{n+1} + b_n^{(0)} x_n = \alpha_n, \quad n = 0, 1, \ldots,$$

and

$$x_{n+k} + b_n^{(k-1)} x_{n+(k-1)} + \cdots + b_n^{(1)} x_{n+1} + b_n^{(0)} x_n = \beta_n, \quad n = 0, 1, \ldots,$$

respectively, where $b^{(i)}$ and $\{\alpha_n\}$ and $\{\beta_n\}$ are given sequences. Then $x^{(1)} + x^{(2)}$ is a solution of the equation

$$x_{n+k} + b_n^{(k-1)} x_{n+(k-1)} + \cdots + b_n^{(1)} x_{n+1} + b_n^{(0)} x_n = \alpha_n + \beta_n, \quad n = 0, 1, \ldots.$$

3.7.4 HOMOGENEOUS EQUATIONS WITH CONSTANT COEFFICIENTS

The results given below for second-order linear difference equations extend naturally to higher order equations.

Consider the second-order linear homogeneous difference equation,

$$\alpha_2 x_{n+2} + \alpha_1 x_{n+1} + \alpha_0 x_n = 0, \tag{3.7.8}$$

where the $\{\alpha_i\}$ are real constant coefficients with $\alpha_2 \alpha_0 \neq 0$. The *characteristic equation* corresponding to Equation (3.7.8) is defined as the quadratic equation

$$\alpha_2 \lambda^2 + \alpha_1 \lambda + \alpha_0 = 0. \tag{3.7.9}$$

The solutions λ_1, λ_2 of the characteristic equation are the *eigenvalues* or the *characteristic roots* of Equation (3.7.8).

THEOREM 3.7.5

Let λ_1 and λ_2 be the eigenvalues of Equation (3.7.8). Then the general solution of Equation (3.7.8) is given as described below with arbitrary constants c_1 and c_2.

Case 1: $\lambda_1 \neq \lambda_2$ with $\lambda_1, \lambda_2 \in \mathbb{R}$ *(real and distinct roots).*
      The general solution is given by $x_n = c_1 \lambda_1^n + c_2 \lambda_2^n$.

Case 2: $\lambda_1 = \lambda_2 \in \mathbb{R}$ *(real and equal roots) .*
      The general solution is given by $x_n = c_1 \lambda_1^n + c_2 n \lambda_1^n$.

Case 3: $\lambda_1 = \overline{\lambda_2}$ *(complex conjugate roots).*
      Suppose that $\lambda_1 = re^{i\phi}$. The general solution is given by

$$x_n = c_1 r^n \cos(n\phi) + c_2 r^n \sin(n\phi).$$

The constants $\{c_1, c_2\}$ are determined from the initial conditions.

EXAMPLE   The unique solution of the initial-value problem

$$F_{n+2} = F_{n+1} + F_n, \qquad n = 0, 1, \ldots,$$
$$F_0 = 0, \qquad F_1 = 1,$$

(3.7.10)

is the Fibonacci sequence. The equation $\lambda^2 = \lambda + 1$ has the real and distinct roots $\lambda_{1,2} = \frac{1 \pm \sqrt{5}}{2}$. Using Theorem 3.7.5 the solution is

$$F_n = \frac{1}{\sqrt{5}} \left[\left(\frac{1 + \sqrt{5}}{2} \right)^n - \left(\frac{1 - \sqrt{5}}{2} \right)^n \right], \qquad n = 0, 1, \ldots.$$

(3.7.11)

3.7.5   NON-HOMOGENEOUS EQUATIONS

THEOREM 3.7.6   *(Variation of parameters)*

Consider the difference equation, $x_{n+2} + \alpha_n x_{n+1} + \beta_n x_n = \gamma_n$, where $\{\alpha_n\}$, $\{\beta_n\}$, and $\{\gamma_n\}$ are given sequences with $\beta_n \neq 0$. Let $x^{(1)}$ and $x^{(2)}$ be two linearly independent solutions of the homogeneous equation corresponding to this equation. A particular solution $x^{(p)}$ has the component values $x_n^{(p)} = x_n^{(1)} v_n^{(1)} + x_n^{(2)} v_n^{(2)}$ where the sequences $v^{(1)}$ and $v^{(2)}$ satisfy the following system of equations:

$$x_{n+1}^{(1)} \left(v_{n+1}^{(1)} - v_n^{(1)} \right) + x_{n+1}^{(2)} \left(v_{n+1}^{(2)} - v_n^{(2)} \right) = 0, \quad and$$
$$x_{n+2}^{(1)} \left(v_{n+1}^{(1)} - v_n^{(1)} \right) + x_{n+2}^{(2)} \left(v_{n+1}^{(2)} - v_n^{(2)} \right) = \gamma_n.$$

(3.7.12)

3.7.6   GENERATING FUNCTIONS AND Z TRANSFORMS

Generating functions can be used to solve initial-value problems of difference equations in the same way that Laplace transforms are used to solve initial-value problems of differential equations.

The *generating function* of the sequence $\{x_n\}$, denoted by $G[x_n]$, is defined by the infinite series

$$G[x_n] = \sum_{n=0}^{\infty} x_n s^n$$

(3.7.13)

provided that the series converges for $|s| < r$, for some positive number r. The following are useful properties of the generating function:

1. *Linearity:* $G\left[c_1 x_n + c_2 y_n\right] = c_1 G\left[x_n\right] + c_2 G\left[y_n\right].$

2. *Translation invariance:* $G\left[x_{n+k}\right] = \dfrac{1}{s^k} \left(G\left[x_n\right] - \displaystyle\sum_{n=0}^{k-1} x_n s^n \right).$

3. *Uniqueness:* $G\left[x_n\right] = G\left[y_n\right] \Longleftrightarrow x_n = y_n \quad \text{for} \quad n = 0, 1, \ldots.$

The *Z-transform* of a sequence $\{x_n\}$ is denoted by $\mathcal{Z}[x_n]$ and is defined by the infinite series,

$$\mathcal{Z}[x_n] = \sum_{n=0}^{\infty} \frac{x_n}{z^n}, \tag{3.7.14}$$

provided that the series converges for $|z| > r$, for some positive number r.

Comparing the definitions for the generating function and the Z-transform one can see that they are connected because Equation (3.7.14) can be obtained from Equation (3.7.13) by setting $s = z^{-1}$.

Generating functions for some common sequences

$\{x_n\}$	$G[x_n]$	$\{x_n\}$	$G[x_n]$
1	$\dfrac{1}{1-s}$	$\sin(\beta n)$	$\dfrac{s\sin\beta}{1-2s\cos\beta+s^2}$
a^n	$\dfrac{1}{1-as}$	$\cos(\beta n)$	$\dfrac{1-s\cos\beta}{1-2s\cos\beta+s^2}$
na^n	$\dfrac{as}{(1-as)^2}$	$a^n\sin(\beta n)$	$\dfrac{as\sin\beta}{1-2as\cos\beta+a^2s^2}$
$n^p a^n$	$\left(s\dfrac{d}{ds}\right)^p \dfrac{1}{1-as}$	$a^n\cos(\beta n)$	$\dfrac{1-as\cos\beta}{1-2as\cos\beta+a^2s^2}$
n	$\dfrac{s}{(1-s)^2}$	x_{n+1}	$\dfrac{1}{s}\left(G[x_n]-x_0\right)$
$n+1$	$\dfrac{1}{(1-s)^2}$	x_{n+2}	$\dfrac{1}{s^2}\left(G[x_n]-x_0-sx_1\right)$
n^p	$\left(s\dfrac{d}{ds}\right)^p \dfrac{1}{1-s}$	x_{n+k}	$\dfrac{1}{s^k}\left(G[x_n]-\sum_{n=0}^{k-1}x_n s^n\right)$

3.7.7 CLOSED-FORM SOLUTIONS FOR SPECIAL EQUATIONS

In general, it is difficult to find a closed-form solution for a difference equation which is not linear of order one or linear of any order with constant coefficients. A few special difference equations which possess closed-form solutions are presented below.

3.7.7.1 First order equation

The general solution of the first-order linear difference equation with variable coefficients,

$$x_{n+1} - \alpha_n x_n = \beta_n, \qquad n = 0, 1, \ldots, \tag{3.7.15}$$

is given by

$$x_n = \left(\prod_{k=0}^{n-1} \alpha_k \right) x_0 + \sum_{m=0}^{n-2} \left(\prod_{k=m+1}^{n-1} \alpha_k \right) \beta_m + \beta_{n-1}, \qquad n = 0, 1, \ldots, \quad (3.7.16)$$

where x_0 is an arbitrary constant.

3.7.7.2 Riccati equation

Consider the non-linear first-order equation,

$$x_{n+1} = \frac{\alpha_n x_n + \beta_n}{\gamma_n x_n + \delta_n}, \qquad n = 0, 1, \ldots, \qquad (3.7.17)$$

where $\alpha_n, \beta_n, \gamma_n, \delta_n$ are given sequences of real numbers with

$$\gamma_n \neq 0 \quad \text{and} \quad \begin{vmatrix} \alpha_n & \beta_n \\ \gamma_n & \delta_n \end{vmatrix} \neq 0, \qquad n = 0, 1, \ldots. \qquad (3.7.18)$$

The following statements are true:

1. The change of variables,

$$\frac{u_{n+1}}{u_n} = \gamma_n x_n + \delta_n, \qquad n = 0, 1, \ldots,$$

$$u_0 = 1, \qquad\qquad\qquad\qquad\qquad (3.7.19)$$

reduces Equation (3.7.17) to the linear second-order equation,

$$u_{n+2} = A_n u_n + B_n, \qquad n = 0, 1, \ldots,$$

$$u_0 = 1, \qquad\qquad\qquad\qquad\qquad (3.7.20)$$

$$u_1 = \gamma_0 x_0 + \delta_0,$$

where $A_n = \delta_{n+1} + \alpha_n \dfrac{\gamma_{n+1}}{\gamma_n}$, and $B_n = (\beta_n \gamma_n - \alpha_n \delta_n) \dfrac{\gamma_{n+1}}{\gamma_n}$.

2. Let $x^{(p)}$ be a particular solution of Equation (3.7.17). The change of variables,

$$v_n = \frac{1}{x_n - x_n^{(p)}}, \qquad n = 0, 1, \ldots, \qquad (3.7.21)$$

reduces Equation (3.7.17) to the linear first-order equation,

$$v_{n+1} + C_n v_n + D_n = 0, \qquad n = 0, 1, \ldots, \qquad (3.7.22)$$

where $C_n = \dfrac{\left(\gamma_n x_n^{(p)} + \delta_n \right)^2}{\beta_n \gamma_n - \alpha_n \delta_n}$, and $D_n = \dfrac{\gamma_n \left(\gamma_n x_n^{(p)} + \delta_n \right)}{\beta_n \gamma_n - \alpha_n \delta_n}$.

3. Let $x^{(1)}$ and $x^{(2)}$ be two particular solutions of Equation (3.7.17) with $x_n^{(1)} \neq x_n^{(2)}$ for $n = 0, 1, \dots$. Then the change of variables,

$$w_n = \frac{1}{x_n - x_n^{(1)}} + \frac{1}{x_n^{(1)} - x_n^{(2)}}, \qquad n = 0, 1, \dots, \qquad (3.7.23)$$

reduces Equation (3.7.17) to the linear homogeneous first-order equation,

$$w_{n+1} + E_n w_n = 0, \qquad n = 0, 1, \dots, \qquad (3.7.24)$$

where $E_n = \dfrac{\left(\gamma_n x_n^{(1)} + \delta_n\right)^2}{\beta_n \gamma_n - \alpha_n \delta_n}$.

3.7.7.3 Logistic equation

Consider the initial-value problem

$$x_{n+1} = r x_n \left(1 - \frac{x_n}{k}\right), \qquad n = 0, 1, \dots, \qquad (3.7.25)$$

$$x_0 = \alpha, \qquad\qquad \text{with } \alpha \in [0, k],$$

where r and k are positive numbers with $r \leq 4$. The following are true:

1. When $r = k = 4$, Equation (3.7.25) reduces to

$$x_{n+1} = 4x_n - x_n^2. \qquad (3.7.26)$$

If $\alpha = 4\sin^2(\theta)$ with $\theta \in [0, \frac{\pi}{2}]$, then Equation (3.7.26) has the closed-form solution

$$x_{n+1} = 4\sin^2(2^{n+1}\theta), \qquad n = 0, 1, \dots, \qquad (3.7.27)$$

$$x_0 = 4\sin^2(\theta), \qquad\qquad \text{with } \theta \in \left[0, \frac{\pi}{2}\right].$$

2. When $r = 4$ and $k = 1$, Equation (3.7.25) reduces to

$$x_{n+1} = 4x_n - 4x_n^2. \qquad (3.7.28)$$

If $\alpha = \sin^2(\theta)$ with $\theta \in [0, \frac{\pi}{2}]$, then Equation (3.7.28) has the closed-form solution

$$x_{n+1} = \sin^2(2^{n+1}\theta), \qquad n = 0, 1, \dots, \qquad (3.7.29)$$

$$x_0 = \sin^2(\theta), \qquad\qquad \text{with } \theta \in \left[0, \frac{\pi}{2}\right].$$

3.8   DISCRETE DYNAMICAL SYSTEMS AND CHAOS

A dynamical system described by a function $f : M \to M$ is *chaotic* if

1. f is *transitive*—that is, for any pair of non-empty open sets U and V in M there exists a positive constant k such that $f^k(U) \cap V$ is not empty (here $f^k = \underbrace{f \circ f \circ \cdots \circ f}_{k \text{ times}}$); and

2. The *periodic points* of f are dense in M; and

3. f has a sensitive dependence on initial conditions—that is, there is a positive number δ (depending only on f and M) such that in every non-empty open subset of M there is a pair of points whose eventual iterates under f are separated by a distance of at least δ.

Some systems depend on a parameter and become chaotic for some values of that parameter. There are various routes to chaos, one of them is via period doubling bifurcations.

Let the distance between successive bifurcations of a process be d_k. The limiting ratio $\delta = \lim_{k \to \infty} d_k/d_{k+1}$ is constant in many situations and is equal to Feigenbaum's constant $\delta \approx 4.6692016091029$.

3.8.1   CHAOTIC ONE-DIMENSIONAL MAPS

1. Logistic map: $x_{n+1} = 4x_n(1 - x_n)$ with $x_0 \in [0, 1]$.

   Solution is $x_n = \dfrac{1}{2} - \dfrac{1}{2}\cos[2^n \cos^{-1}(1 - 2x_0)]$.

2. Tent map: $x_{n+1} = 1 - 2\left|x_n - \frac{1}{2}\right|$ with $x_0 \in [0, 1]$.

   Solution is $x_n = \dfrac{1}{\pi}\cos^{-1}[\cos(2^n \pi x_0)]$.

3. Baker transformation: $x_{n+1} = 2x_n \pmod 1$ with $x_0 \in [0, 1]$.

   Solution is $x_n = \dfrac{1}{\pi}\cot^{-1}[\cot(2^n \pi x_0)]$.

3.8.2   LOGISTIC MAP

Consider $u_{n+1} = f(u_n) = au_n(1 - u_n)$ with $a \in [0, 4]$. Note that if $0 \le u_0 \le 1$ then $0 \le u_n \le 1$. The fixed points satisfy $u = f(u) = au(1 - u)$; they are $u = 0$ and $u = (a - 1)/a$.

1. If $a = 0$ then $u_n = 0$.
2. If $0 < a \le 1$ then $u_n \to 0$.
3. If $1 < a < 3$ then $u_n \to (a - 1)/a$.
4. If $3 < a < 3.449\ldots$ then u_n oscillates between the two roots of $u = f(f(u))$ which are not roots of $u = f(u)$, that is, $u_{\pm} = (a + 1 \pm \sqrt{a^2 - 2a - 3})/2a$.

The location of the final state is summarized by the following diagram (the horizontal axis is the a value).

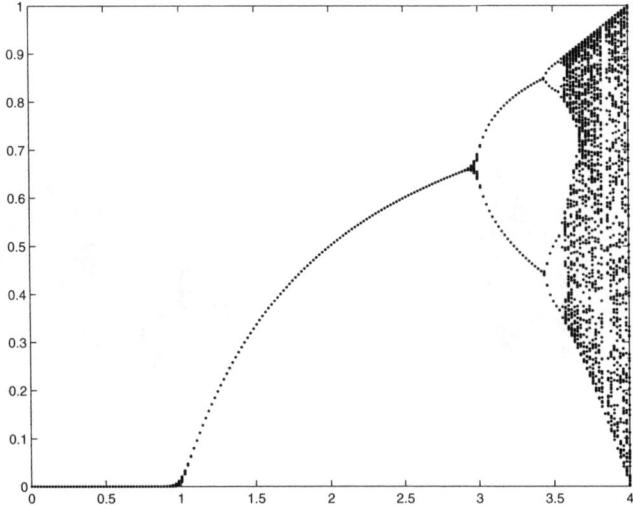

Minimal values of a at which a cycle with a given number of points appears:

n	2^n points in a cycle	minimal a
1	2	3
2	4	3.449490...
3	8	3.544090...
4	16	3.564407...
5	32	3.568750...
6	64	3.56969...
7	128	3.56989...
8	256	3.569934...
9	512	3.569943...
10	1024	3.5699451...
11	2048	3.569945557...

3.8.3   JULIA SETS AND THE MANDELBROT SET

For the function $f(x) = x^2 + c$ consider the iterates of all complex points z, $z_{n+1} = f(z_n)$ with $z_0 = z$. For each z, either the iterates remain bounded (z is in the *prisoner set*) or they escape to infinity (z is in the *escape set*). The *Julia set* J_c is the boundary between these two sets. Using lighter colors to indicate a "faster" escape to infinity, Figure 3.6 shows two Julia sets. One of these Julia sets is connected, the other is disconnected. The Mandelbrot set, M, is the set of those complex values c for which J_c is a connected set (see Figure 3.7). Alternately, the *Mandelbrot* set consists of all points c for which the discrete dynamical system, $z_{n+1} = z_n^2 + c$ with $z_0 = 0$, converges.

The boundary of the Mandelbrot set is a *fractal*. There is not universal agreement on the definition of "fractal". One definition is that it is a set whose fractal dimension differs from its topological dimension.

FIGURE 3.6
Connected Julia set for $c = -0.5i$ (left). Disconnected Julia set for $c = -\frac{3}{4}(1 + i)$ (right). (Julia sets are the black objects.)

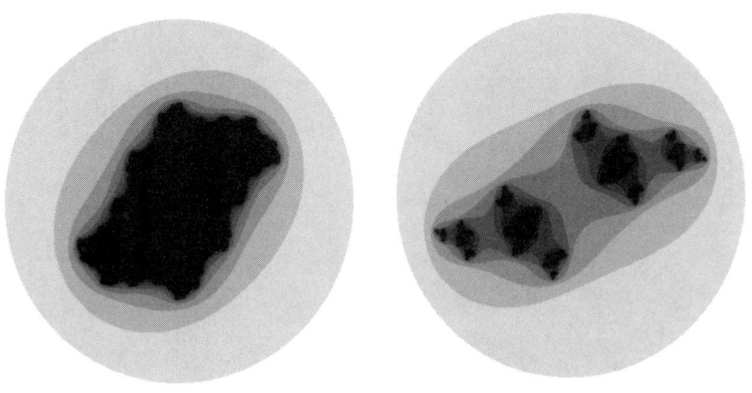

FIGURE 3.7
The Mandelbrot set. The leftmost point has the coordinates $(-2, 0)$.

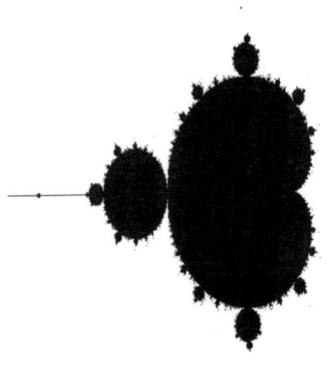

3.9 GAME THEORY

3.9.1 TWO PERSON NON-COOPERATIVE MATRIX GAMES

Matrix games idealize situations with participants having different goals. Given matrices $A = (a_{i,j})$ and $B = (b_{i,j})$ consider a game played as follows. After Alice chooses action i (one choice out of n possible actions) and simultaneously Bob chooses action j (one choice out of m possible actions) Alice and Bob receive a payoff of $a_{i,j}$ and $b_{i,j}$, respectively. Define (i, j) as the *outcome* of the game. If the players have a *mixed* or *random strategy* then Alice selects $\mathbf{x}^T = (x_1, x_2, \ldots, x_n)$ where x_i corresponds to the probability that she chooses action i and Bob selects $\mathbf{y}^T = (y_1, y_2, \ldots, y_m)$ where y_j corresponds to the probability that he chooses action j. Here, the players seek to maximize their average payoff; that is, Alice wants to maximize $\mathbf{x}^T A \mathbf{y} = \sum_i \sum_j x_i a_{i,j} y_j$ and Bob wants to maximize $\mathbf{x}^T B \mathbf{y} = \sum_i \sum_j x_i b_{i,j} y_j$.

If $a_{i,j} \leq a_{i^*,j}$ for all i and j then i^* is a *dominant* strategy for Alice. Similarly, if $b_{i,j} \leq b_{i,j^*}$ for all i and j then j^* is a *dominant* strategy for Bob. An outcome (i^*, j^*) is *Pareto optimal* if, for all i and j, the relation $a_{i^*,j^*} < a_{i,j}$ implies $b_{i^*,j^*} > b_{i,j}$.

If $A + B = 0$ then the game is a *zero sum game* and Bob equivalently is trying to minimize Alice's payoff. If $A + B \neq 0$ then the game is a *non-zero sum* game and there is potential for mutual gain or loss.

3.9.1.1   The pure zero sum game

For a zero sum game, $A + B = 0$. The outcome $(\hat{i}, \hat{j})$ is an *equilibrium* if $a_{i,\hat{j}} \leq a_{\hat{i},\hat{j}} \leq a_{\hat{i},j}$ for all i and j. (Note that this implies $b_{\hat{i},j} \leq b_{\hat{i},\hat{j}} \leq b_{i,\hat{j}}$.)

1. If $(\hat{i}, \hat{j})$ is an equilibrium then $\max_i \min_j a_{i,j} = \min_j \max_i a_{i,j} = a_{\hat{i},\hat{j}}$.

2. For all A, $\max_i \min_j a_{i,j} \leq \min_j \max_i a_{i,j}$.

3. For some A, there is no equilibrium. For example, if $A = \left[\begin{smallmatrix} 1 & 2 \\ 2 & 1 \end{smallmatrix}\right]$ then $1 = \max_i \min_j a_{i,j} < \min_j \max_i a_{i,j} = 2$.

4. For some A, there might be an equilibrium which is not a dominant strategy for a player. For example, if $A = \left[\begin{smallmatrix} 3 & 4 \\ 2 & 1 \end{smallmatrix}\right]$, then the outcome (1,1) is an equilibrium with the property that Alice has the dominant strategy $i = 1$ as $a_{1,1} \geq a_{2,1}$ and $a_{1,2} \geq a_{2,2}$, but there is no dominant strategy for Bob as $a_{1,1} < a_{1,2}$ but $a_{2,1} > a_{2,2}$.

5. For all A, all outcomes are Pareto optimal as $a_{\hat{i},\hat{j}} < a_{i,j}$ implies
$$b_{\hat{i},\hat{j}} = -a_{\hat{i},\hat{j}} > -a_{i,j} = b_{i,j}.$$

6. For a given A, there can be multiple equilibria. If (i_1, j_1) and (i_2, j_2) are each an equilibrium then $a_{i_1,j_1} = a_{i_2,j_2} = a_{i_1,j_2} = a_{i_2,j_1}$ and (i_1, j_2) and (i_2, j_1) are also equilibria. For example, if $A = \left[\begin{smallmatrix} 3 & 4 & 3 \\ 2 & 1 & 2 \\ 3 & 4 & 3 \end{smallmatrix}\right]$ then the outcomes (1,1), (1,3), (3,1) and (3,3) are each an equilibrium.

7. Order of actions: If Alice chooses her action before Bob (or commits to an action first) and she chooses i then Bob will choose action[3] $\beta(i) = \arg\min_j a_{i,j}$. Hence Alice will choose $\hat{i} = \arg\max_i a_{i,\beta(i)} = \arg\max_i \min_j a_{i,j}$ which is said to be the *maximin* strategy. Alternatively, if Bob chooses his action before Alice he will choose the *minimax strategy* $\check{j} = \arg\min_j \max_i a_{i,j}$. A player should never prefer to choose an action first, but if there is an equilibrium, the advantage of taking the second action can be eliminated.

---

[3]The "arg" function refers to the index.

3.9.1.2 The mixed zero sum game

The outcome $(\widehat{\mathbf{x}}, \widehat{\mathbf{y}})$ is an *equilibrium* for the mixed zero sum game if

$$\mathbf{x}^{\mathrm{T}} A \widehat{\mathbf{y}} \leq \widehat{\mathbf{x}}^{\mathrm{T}} A \widehat{\mathbf{y}} \leq \widehat{\mathbf{x}}^{\mathrm{T}} A \mathbf{y} \qquad \text{for all } \mathbf{x} \in S_x \text{ and all } \mathbf{y} \in S_y$$

where $S_x = \{\mathbf{x} \mid 0 \leq x_i \text{ and } \sum_i x_i = 1\}$ and $S_y = \{\mathbf{y} \mid 0 \leq y_j \text{ and } \sum_j y_j = 1\}$.

1. If $(\widehat{\mathbf{x}}, \widehat{\mathbf{y}})$ is an equilibrium then

$$\max_{\mathbf{x} \in S_x} \min_{\mathbf{y} \in S_y} \mathbf{x}^{\mathrm{T}} A \mathbf{y} = \min_{\mathbf{y} \in S_y} \max_{\mathbf{x} \in S_x} \mathbf{x}^{\mathrm{T}} A \mathbf{y} = \widehat{\mathbf{x}}^{\mathrm{T}} A \widehat{\mathbf{y}}.$$

2. For all A, there exists at least one mixed strategy equilibrium. If there is more than one equilibrium, the average payoff to each player is independent of which equilibrium is used.

3. If $\min_j \sum_i x_i a_{i,j} \leq \min_j \sum_i \widehat{x}_i a_{i,j}$ for all $\mathbf{x} \in S_x$ and $\max_i \sum_j a_{i,j} y_j \geq \max_i \sum_j a_{i,j} \widehat{y}_j$ for all $\mathbf{y} \in S_y$ then $(\widehat{\mathbf{x}}, \widehat{\mathbf{y}})$ is an equilibrium.

4. If $A \geq 0$ then an equilibrium strategy for Alice solves $\max_{\mathbf{x}} v$ subject to $v \leq \sum_i x_i a_{i,j}$ for all j, with $\sum_i x_i = 1$, and $x_i \geq 0$. Here, the optimal value of v will correspond to the average payoff to Alice in equilibrium. If $x'_i = x_i / v$ this is equivalent to solving

$$\text{minimize } \sum_i x'_i \quad \text{subject to} \quad \begin{cases} \sum_i x'_i a_{i,j} \geq 1, & \text{for all } j, \\ x'_i \geq 0, & \text{for all } i. \end{cases}$$

Similarly, if $\mathbf{y}' = (y'_1, \ldots, y'_n)$ is a solution to

$$\text{maximize } \sum_j y'_j \quad \text{subject to} \quad \begin{cases} \sum_j a_{i,j} y'_j \leq 1, & \text{for all } i, \\ y'_j \geq 0, & \text{for all } j, \end{cases}$$

then $\mathbf{y} = \mathbf{y}' / \sum_j y'_j$ is the equilibrium strategy for Bob. The payoff to Alice in equilibrium is $v = 1/\sum_i x'_i = 1/\sum_j y'_j$. These optimization problems are an example of *linear programming* (see page 280).

For example, if $A = \left[\begin{smallmatrix} 1 & 2 \\ 2 & 1 \end{smallmatrix}\right]$ then the solution to

$$\text{maximize } x'_1 + x'_2 \quad \text{subject to} \quad \begin{cases} x'_1 + 2x'_2 \geq 1, & x'_1 \geq 0, \\ 2x'_1 + x'_2 \geq 1, & x'_2 \geq 0, \end{cases}$$

is $x'_1 = x'_2 = \frac{1}{3}$ so the payoff to Alice in equilibrium is $\frac{3}{2}$ and the equilibrium is $(x_1, x_2) = \left(\frac{1}{2}, \frac{1}{2}\right)$.

5. If A is 2×2 and $a_{1,1} < a_{1,2}$, $a_{1,1} < a_{2,1}$, $a_{2,2} < a_{2,1}$, and $a_{2,2} < a_{1,2}$ (so neither player has a dominant strategy) then the mixed strategy equilibrium is

$$\widehat{\mathbf{x}} = \left(\frac{a_{2,1} - a_{2,2}}{a_{1,2} + a_{2,1} - a_{1,1} - a_{2,2}}, \frac{a_{1,2} - a_{1,1}}{a_{1,2} + a_{2,1} - a_{1,1} - a_{2,2}} \right),$$

$$\widehat{\mathbf{y}} = \left(\frac{a_{1,2} - a_{2,2}}{a_{1,2} + a_{2,1} - a_{1,1} - a_{2,2}}, \frac{a_{2,1} - a_{1,1}}{a_{1,2} + a_{2,1} - a_{1,1} - a_{2,2}} \right)$$

$$(3.9.1)$$

and the payoff to Alice in equilibrium is

$$\max_{\widehat{\mathbf{x}}\in S_x} \min_{\widehat{\mathbf{y}}\in S_y} \sum_i \sum_j \widehat{x}_i a_{i,j} \widehat{y}_j = \min_{\widehat{\mathbf{y}}\in S_y} \max_{\widehat{\mathbf{x}}\in S_x} \sum_i \sum_j \widehat{x}_i a_{i,j} \widehat{y}_j$$

$$= \frac{a_{1,2}a_{2,1} - a_{1,1}a_{2,2}}{a_{1,2} + a_{2,1} - a_{1,1} - a_{2,2}}. \tag{3.9.2}$$

For example, if $A = \left[\begin{smallmatrix} 1 & 2 \\ 2 & 1 \end{smallmatrix}\right]$ then the solution is $\widehat{\mathbf{x}} = \widehat{\mathbf{y}} = (\frac{1}{2}, \frac{1}{2})$ and the payoff to Alice in equilibrium is $\frac{3}{2}$.

6. Order of actions: It is never an advantage to take the first action. However, if the first player uses a mixed strategy equilibrium then the advantages of taking the second action can always be eliminated.

3.9.1.3 The non-zero sum game

The outcome $(\widehat{i}, \widehat{j})$ is a *Nash equilibrium* if $a_{i,\widehat{j}} \leq a_{\widehat{i},\widehat{j}}$ for all i and $b_{\widehat{i},j} \leq b_{\widehat{i},\widehat{j}}$ for all j. For the mixed strategy game the outcome $(\widehat{\mathbf{x}}, \widehat{\mathbf{y}})$ is a Nash equilibrium if $\mathbf{x}^T A \widehat{\mathbf{y}} \leq \widehat{\mathbf{x}}^T A \widehat{\mathbf{y}}$ for all $\mathbf{x} \in S_x$ and $\widehat{\mathbf{x}}^T A \mathbf{y} \leq \widehat{\mathbf{x}}^T A \widehat{\mathbf{y}}$ for all $\mathbf{y} \in S_y$.

1. For all A and B, there exists at least one mixed strategy Nash equilibrium.

2. For all A and B, $\max_i \sum_j a_{i,j} \widehat{y}_j \leq \sum_i \sum_j \widehat{x}_i a_{i,j} \widehat{y}_j$ and $\max_j \sum_i \widehat{x}_i b_{i,j} \leq \sum_i \sum_j \widehat{x}_i b_{i,j} \widehat{y}_j$ is a necessary and sufficient condition for $(\widehat{\mathbf{x}}, \widehat{\mathbf{y}})$ to be a Nash equilibrium. For example, if $A = \left[\begin{smallmatrix} 4 & 1 \\ 1 & 3 \end{smallmatrix}\right]$ and $B = \left[\begin{smallmatrix} 3 & 1 \\ 1 & 4 \end{smallmatrix}\right]$ then there are three Nash equilibra: (i) $\{\widehat{\mathbf{x}} = (0, 1), \widehat{\mathbf{y}} = (0, 1)\}$, (ii) $\{\widehat{\mathbf{x}} = (\frac{3}{5}, \frac{2}{5}), \widehat{\mathbf{y}} = (\frac{2}{5}, \frac{3}{5})\}$, and (iii) $\{\widehat{\mathbf{x}} = (1, 0), \widehat{\mathbf{y}} = (1, 0)\}$.

3. For all A and B, if $\sum_j a_{i,j} \widehat{y}_j$ is a constant for all i and $\sum_i \widehat{x}_i b_{i,j}$ is a constant for all j (i.e., each player chooses an action to make the other indifferent to their action) then $(\widehat{\mathbf{x}}, \widehat{\mathbf{y}})$ is a Nash equilibrium.

4. For some A and B if (i_1, j_1) and (i_2, j_2) are each a (pure strategy) Nash equilibrium then, unlike the case for a zero sum game, a_{i_1,j_1} need not equal a_{i_2,j_2}, b_{i_1,j_1} need not equal b_{i_2,j_2}, and neither (i_1, j_2) nor (i_2, j_1) need be a Nash equilibrium. For example, if $A = \left[\begin{smallmatrix} 4 & 1 \\ 1 & 3 \end{smallmatrix}\right]$ and $B = \left[\begin{smallmatrix} 3 & 1 \\ 1 & 4 \end{smallmatrix}\right]$ then both of the outcomes (1,1) and (2,2) are Nash equilibria yet $a_{1,1} \neq a_{2,2}$, $b_{1,1} \neq b_{2,2}$ and neither (1,2) nor (2,1) is a Nash equilibrium.

5. *Prisoners' Dilemma*: A game in which there is a dominant strategy for both players but it is not Pareto optimal. For example, if $A = \left[\begin{smallmatrix} 10 & 2 \\ 15 & 5 \end{smallmatrix}\right]$ and $B = \left[\begin{smallmatrix} 10 & 15 \\ 2 & 5 \end{smallmatrix}\right]$ then the dominant (and equilibrium) outcome is (2,2) since $a_{2,j} > a_{1,j}$ for all j and $b_{i,2} > b_{i,1}$ for all i. Here, Alice and Bob receive a payoff of 5 although the outcome (1,1) would be preferred by both because each would receive a payoff of 10. The name is derived from the possibility that the potential jail sentence for two people accused of a joint crime could be constructed to create a dominant strategy for both to choose to confess to the crime, yet both would serve less jail time if neither confesses.

6. *Braess Paradox*: A game in which the Nash equilibrium has a worse payoff for all players than the Nash equilibrium which would result if there were fewer possible actions. For example, if $A = \begin{bmatrix} 10 & 2 & 1 \\ 15 & 5 & 2 \\ 20 & 6 & 4 \end{bmatrix}$ and $B = \begin{bmatrix} 10 & 15 & 20 \\ 2 & 5 & 6 \\ 1 & 2 & 4 \end{bmatrix}$ then the Nash equilibrium is (3,3) which is worse for both players than the Nash equilibrium (2,2) which would occur if the third option for each player was unavailable. This paradox was originally constructed to demonstrate that the equilibrium distribution of flows in a traffic network would be preferred by everyone if a traffic link, e.g., a bridge, weren't there.

7. Order of actions: If Alice chooses action i before Bob chooses an action then Bob will choose $\beta(i) = \arg\max_j b_{i,j}$ and hence Alice will choose $\hat{i} = \arg\max_i a_{i,\beta(i)}$. Alternatively, if Bob chooses an action before Alice, he will choose $\hat{j} = \arg\max_j b_{\alpha(j),j}$ where $\alpha(j) = \arg\max_i a_{i,j}$. Unlike the zero sum game there might be an advantage to choosing the action first, e.g., for $A = \begin{bmatrix} 4 & 1 \\ 1 & 3 \end{bmatrix}$ and $B = \begin{bmatrix} 3 & 1 \\ 1 & 4 \end{bmatrix}$ if Alice is first the outcome will be (1,1) with a payoff of 4 to Alice and 3 to Bob. If Bob is first the outcome will be (2,2) with a payoff of 4 to Bob and 3 to Alice.

3.9.2 VOTING POWER

Matrix games implicitly exclude the possibility of binding agreements among the players to resolve their differing goals. When such agreement exist, it is sometimes possible to quantify the relative power of the players.

3.9.2.1 Voting power definitions

A *weighted voting* game is represented by the vector $[q; w_1, w_2, \ldots, w_n]$; this means:

1. There are n players.
2. Player i has w_i votes (with $w_i > 0$).
3. A *coalition* is a subset of players.
4. A coalition S is *winning* if $\sum_{i \in S} w_i \geq q$, where q is the *quota*.
5. A game is *proper* if $\frac{1}{2} \sum w_i < q$.

A player

1. can have *veto power*: no coalition can win without this player
2. is a *dictator*: has more votes than the quota
3. is a *dummy*: cannot affect any coalitions

3.9.2.2 Shapley–Shubik power index

Consider all permutations of players. Scan each permutation from beginning to end; add together the votes that each player contributes. Eventually a total of at least q will be arrived at, this occurs at the *pivotal* player. The *Shapley–Shubik power index* (ϕ) of player i is the number of permutations for which player i is pivotal, divided by the total number of permutations.

EXAMPLE

1. Consider the $[5; 4, 2, 1, 1]$ game; the players are {A,B,C,D}.

2. For the $4! = 24$ permutations of four players the pivotal player is underlined:

$$
\begin{array}{cccc}
A\underline{B}CD & B\underline{A}CD & C\underline{A}BD & D\underline{A}BC \\
A\underline{B}DC & B\underline{A}DC & C\underline{A}DB & D\underline{A}CB \\
A\underline{C}BD & BC\underline{A}D & CB\underline{A}D & DB\underline{A}C \\
A\underline{C}DB & BCD\underline{A} & CBD\underline{A} & DBC\underline{A} \\
A\underline{D}BC & BD\underline{A}C & CD\underline{A}B & DC\underline{A}B \\
A\underline{D}CB & BDC\underline{A} & CDBA & DCB\underline{A}
\end{array}
$$

3. Hence player A has power $\phi(A) = \frac{18}{24} = 0.75$.

4. The other three players have equal power of $\frac{2}{24} \approx 0.083$.

3.9.2.3 Banzhaf power index

Consider all 2^N possible coalitions of players. For each coalition, if player i can change the winningness of the coalition, by either entering it or leaving it, then i is *marginal* or *swing*. The *Banzhaf power index* (β) of player i is proportional to the number of times he is marginal; the total power of all players is 1.

EXAMPLE

1. Consider the $[5; 4, 2, 1, 1]$ game; the players are {A,B,C,D}.

2. There are 16 subsets of four players; each player is "in" (I) or is "out" (O) of a coalition. For each coalition the marginal players are listed (in total there are 20 marginal players).

O O O O	$\Rightarrow$	{}	I O O O $\Rightarrow$	{B,C,D}
O O O I	$\Rightarrow$	{A}	I O O I $\Rightarrow$	{A,D}
O O I O	$\Rightarrow$	{A}	I O I O $\Rightarrow$	{A,C}
O O I I	$\Rightarrow$	{A}	I O I I $\Rightarrow$	{A}
O I O O	$\Rightarrow$	{A}	I I O O $\Rightarrow$	{A,B}
O I O I	$\Rightarrow$	{A}	I I O I $\Rightarrow$	{A}
O I I O	$\Rightarrow$	{A}	I I I O $\Rightarrow$	{A}
O I I I	$\Rightarrow$	{A}	I I I I $\Rightarrow$	{A}

3. Player A is marginal 14 times, and has power $\beta(A) = \frac{14}{20} = 0.7$.

4. The players B, C, and D are each marginal 2 times and have equal power of $\frac{2}{20} = 0.1$.

3.9.2.4 Voting power examples

1. For the game $[51; 49, 48, 3]$ the winning coalitions are: $\{1, 2, 3\}, \{1, 2\}, \{1, 3\}$, and $\{2, 3\}$. These are the same winning coalitions as the game $[3; 1, 1, 1]$. Hence, all players have equal power by either index, even though the number of votes each player has is different.

2. The original EEC (1958) had France, Germany, Italy, Belgium, The Netherlands, and Luxembourg. They voted as $[12; 4, 4, 4, 2, 2, 1]$. Therefore:

$$
\phi = \frac{1}{60}(14, 14, 14, 9, 9, 0) \quad \text{and} \quad \beta = \frac{1}{21}(5, 5, 5, 3, 3, 0). \tag{3.9.3}
$$

3. The *UN security council* has 15 members. The five permanent members have veto power. For a motion to pass, it must be supported by at least 9 members of the council and it must not be vetoed. A game representation is: $[39; \underbrace{7,7,7,7,7}_{\text{5 members}}, \underbrace{1,1,1,1,1,1,1,1,1,1}_{\text{10 members}}]$.

   (a) Shapley–Shubik powers: of each permanent member $\phi_{\text{major}} = \frac{421}{2145} \approx 0.196$, of each minor member $\phi_{\text{minor}} = \frac{4}{2145} \approx 0.002$.

   (b) Banzhaf powers: $\beta_{\text{major}} = \frac{106}{635} \approx 0.167$ and $\beta_{\text{minor}} = \frac{21}{1270} \approx 0.017$.

4. Changing the quota in a game may change the powers of the players:

   (a) For $[6; 4, 3, 2]$ have $\phi = \frac{1}{6}(4, 1, 1)$.
   (b) For $[7; 4, 3, 2]$ have $\phi = \frac{1}{2}(1, 1, 0)$.
   (c) For $[8; 4, 3, 2]$ have $\phi = \frac{1}{3}(1, 1, 1)$.

5. For the n-player game $[q; a, 1, 1, 1, 1, \ldots, 1]$ with $1 < a \leq q \leq n$ we find $\phi_{\text{major}} = a/n$ and $\phi_{\text{minor}} = (n - a)/n(n - 1)$.

6. Four-person committee, one member is chair. Use majority rule until deadlock, then chair decides. This is a $[3; 2, 1, 1, 1]$ game so that $\phi = \frac{1}{6}(3, 1, 1, 1)$.

7. Five-person committee, with two co-chairs. Need a majority, and at least one co-chair. This is a $[7; 3, 3, 2, 2, 2]$ game so that $\phi = \frac{1}{12}(3, 3, 2, 2, 2)$ and $\beta = \frac{1}{29}(7, 7, 5, 5, 5)$.

8. For the game $[7; 5, 3, 3, 1]$ the powers are $\phi = \beta = \frac{1}{12}(5, 3, 3, 1)$.

3.10 OPERATIONS RESEARCH

Operations research integrates mathematical modeling and analysis with engineering in an effort to design and control systems.

3.10.1 LINEAR PROGRAMMING

Linear programming (LP) is a technique for modeling problems with linear objective functions and linear constraints. The *standard form* for an LP model with n decision variables and m resource constraints is

$$\text{Minimize} \quad \sum_{j=1}^{n} c_j x_j \qquad \textit{(objective function)},$$

$$\text{Subject to} \quad \begin{cases} x_j \geq 0 \quad \text{for } j = 1, \cdots, n, & \text{(non-negativity requirement)}, \\ \sum_{j=1}^{n} a_{ij} x_j = b_i \quad \text{for } i = 1, \cdots, m, & \textit{(constraint functions)}, \end{cases}$$

$$(3.10.1)$$

where x_j is the amount of decision variable j used, c_j is decision $j's$ per unit contribution to the objective, a_{ij} is decision $j's$ per unit usage of resource i, and b_i is the total amount of resource i to be used.

Let $\mathbf{x}$ represent the $(n \times 1)$ vector $(x_1, x_2, \cdots, x_n)^{\mathrm{T}}$, $\mathbf{c}$ the $(n \times 1)$ vector $(c_1, c_2, \cdots, c_n)^{\mathrm{T}}$, $\mathbf{b}$ the $(m \times 1)$ vector $(b_1, b_2, \cdots, b_m)^{\mathrm{T}}$, A the $(m \times n)$ matrix (a_{ij}), and $\mathbf{A}_{.j}$ the $(n \times 1)$ column of A associated with x_j. Then the standard model, written in matrix notation, is "minimize $\mathbf{c}^{\mathrm{T}}\mathbf{x}$ subject to $A\mathbf{x} = \mathbf{b}$ and $\mathbf{x} \geq \mathbf{0}$". A vector $\mathbf{x}$ is called *feasible* if, and only if, $A\mathbf{x} = \mathbf{b}$ and $\mathbf{x} \geq \mathbf{0}$.

3.10.1.1 Modeling in LP

LP is an appropriate modeling technique if the following four assumptions are satisfied by the situation:

1. All data coefficients are known with certainty.
2. There is a single objective.
3. The problem relationships are linear functions of the decisions.
4. The decisions can take on continuous values.

Branches of optimization such as stochastic programming, multi-objective programming, non-linear programming, and integer programming have developed in operations research to allow a richer variety of models and solution techniques for situations where the assumptions required for LP are inappropriate.

1. *Product mix problem* — Consider a company that has three products to sell. Each product requires four operations and the per unit data are given in the following table:

Product	Drilling	Assembly	Finishing	Packing	Profit
A	2	3	1	2	45
B	3	6	2	4	90
C	2	1	4	1	55
Hours available	480	960	540	320	

Let x_A, x_B, and x_C represent the number of units of A, B, and C manufactured daily. A model to maximize profit subject to the labor restrictions is

$$\text{Maximize} \quad 45x_A + 90x_B + 55x_C \qquad \text{(total profit)},$$

Subject to:
$$\begin{cases} 2x_A + 3x_B + 2x_C \leq 480 & \text{(drilling hours)}, \\ 3x_A + 6x_B + 1x_C \leq 960 & \text{(assembly hours)}, \\ 1x_A + 2x_B + 4x_C \leq 540 & \text{(finishing hours)}, \\ 2x_A + 4x_B + 1x_C \leq 320 & \text{(packing hours)}, \\ x_A \geq 0, \quad x_B \geq 0, \quad x_C \geq 0. \end{cases}$$

$$(3.10.2)$$

2. *Maximum flow through a network* — Consider the directed network in Figure 3.8. Node S is the source node and node T is the terminal node. On each arc shipping material up to the arc capacity C_{ij} is permitted. Material is neither created nor destroyed at nodes other than S and T. The goal is to maximize the amount of material that can be shipped through the network from S to T. Letting x_{ij} represent the amount of material shipped from node i to node j, a model that determines the maximum flow is shown below.

FIGURE 3.8
Directed network modeling a flow problem.

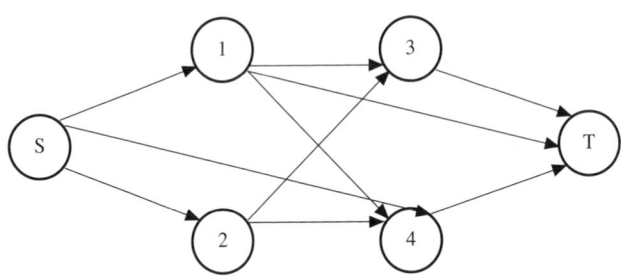

$$\text{Maximize} \quad x_{1T} + x_{3T} + x_{4T},$$

Subject to
$$\begin{cases} x_{S1} = x_{13} + x_{14} + x_{1T} & \text{(node 1 conservation)}, \\ x_{S2} = x_{23} + x_{24} & \text{(node 2 conservation)}, \\ x_{13} + x_{23} = x_{3T} & \text{(node 3 conservation)}, \\ x_{S4} + x_{24} + x_{14} = x_{4T} & \text{(node 4 conservation)}, \\ 0 \leq x_{ij} \leq C_{ij} & \text{for all pairs } (i,j) \quad \text{(arc capacity)}. \end{cases}$$

$$(3.10.3)$$

3.10.1.2   Transformation to standard form

Any LP model can be transformed to standard form as follows:

Original model	Standard form
change "maximize objective"	to "minimize objective"

multiply c_j by -1

change "$\leq$" constraint	to "$=$" constraint

add slack variable(s) to $\displaystyle\sum_{j=1}^{n} a_{ij}x_j \leq b_i$

(for example: $\displaystyle\sum_{j=1}^{n} a_{ij}x_j + S_i = b_i.$)

change "$\geq$" constraint	to "$=$" constraint

subtract surplus variable(s) from $\displaystyle\sum_{j=1}^{n} a_{ij}x_j \geq b_i$

(for example: $\displaystyle\sum_{j=1}^{n} a_{ij}x_j - U_i = b_i$)

LP requires that the slack and surplus variables are non-negative; therefore, decisions are feasible to the new constraint if, and only if, they are feasible to the original constraint. All slack and surplus variables have $c_j = 0$.

3.10.1.3 Solving LP models: simplex method

Assume that there is at least one feasible $\mathbf{x}$ vector, and that A has rank m. Geometrically, because all constraints are linear, the set of feasible $\mathbf{x}$ forms a convex polyhedral set (bounded or unbounded) that must have at least one extreme point. The motivation for the simplex method for solving LP models is the following:

> For any LP model with a bounded optimal solution, an optimal solution exists at an extreme point of the feasible set.

Given a feasible solution $\mathbf{x}$, let $\mathbf{x}^B$ be the components of $\mathbf{x}$ with $x_j > 0$ and $\mathbf{x}^N$ be the components with $x_j = 0$. Associated with $\mathbf{x}^B$, define B as the columns of A associated with each x_j in $\mathbf{x}^B$. For example, if x_2, x_4, and S_1 are positive in $\mathbf{x}$, then $x^B = (x_2, x_4, S_1)^{\mathrm{T}}$, and B is the matrix with columns $\mathbf{A}_{.2}, \mathbf{A}_{.4}, \mathbf{A}_{.S_1}$. Define N as the remaining columns of A, i.e., those associated with $\mathbf{x}^N$. A *basic feasible solution* (BFS) is a feasible solution where the columns of B are linearly independent. The following theorem relates a BFS with extreme points:

> A feasible solution $\mathbf{x}$ is at an extreme point of the feasible region if, and only if, $\mathbf{x}$ is a BFS.

The following simplex method finds an optimal solution to the LP by finding the optimal partition of $\mathbf{x}$ into $\mathbf{x}^B$ and $\mathbf{x}^N$:

Step (1) Find an initial basic feasible solution. Define $\mathbf{x}^B, \mathbf{x}^N, B, N, \mathbf{c}^B$, and $\mathbf{c}^N$ as above.

Step (2) Compute the vector $\mathbf{c}' = (\mathbf{c}^N - \mathbf{c}^B B^{-1} N)$. If $\mathbf{c}'_j \geq \mathbf{0}$ for all j, then stop; the solution $\mathbf{x}^B = B^{-1}\mathbf{b}$ is optimal with objective value $\mathbf{c}^B B^{-1}\mathbf{b}$. Otherwise, select the variable x_j in $\mathbf{x}^N$ with the most negative c'_j value, and go to Step (3).

Step (3) Compute $\mathbf{A}'_{\cdot j} = B^{-1}\mathbf{A}_{\cdot j}$. If $A'_{\cdot j} \leq \mathbf{0}$ for all j, then stop; the problem is unbounded and the objective can decrease to $-\infty$. Otherwise, compute $\mathbf{b}' = B^{-1}\mathbf{b}$ and find $\min\limits_{i|a'_{ij}>0} \dfrac{b'_i}{a'_{ij}}$. Assume the minimum ratio occurs in row r. Insert x_j into the rth position of $\mathbf{x}^B$, take the variable that was in this position, and move it to $\mathbf{x}^N$. Update B, N, $\mathbf{c}^B$, and $\mathbf{c}^N$ accordingly. Return to Step (2).

Ties in the selections in Steps (2) and (3) can be broken arbitrarily. The unboundedness signal suggests that the model is missing constraints or that there has been an incorrect data entry or computational error, because, in real problems, the profit or cost cannot be unbounded. For maximization problems, only Step (2) changes. The solution is optimal when $c'_j \leq 0$ for all j, then choose the variable with the maximum c'_j value to move into $\mathbf{x}^B$. Effective methods for updating B^{-1} in each iteration of Step (3) exist to ease the computational burden.

To find an initial basic feasible solution define a variable A_i and add this variable to the left hand side of constraint i transforming to $\sum_{j=1}^{n} a_{ij} + A_i = b_i$. The new constraint is equivalent to the original constraint if, and only if, $A_i = 0$. Also, because A_i appears only in constraint i and there are m A_i variables, the columns corresponding to the A_i variables are of rank m. We now solve a "new" LP model with the adjusted constraints and the new objective "minimize $\sum_{i=1}^{m} A_i$". If the optimal solution to this new model is 0, the solution is a basic feasible solution to the original problem and we can use it in step (1). Otherwise, no basic feasible solution exists for the original problem.

3.10.1.4   Solving LP models: interior point method

An alternative method to investigating extreme points is to cut through the middle of the polyhedron and go directly towards the optimal solution. Extreme points, however, provide an efficient method of determining movement directions (Step (1) of simplex method) and movement distances (Step (2) of simplex method). There were no effective methods on the interior of the feasible set until Karmarkar's method was developed in 1984. The method assumes that the model has the following form:

Minimize $\sum_{j=1}^{n} c_j x_j,$

Subject to
$$\begin{cases} \sum_{j=1}^{n} a_{ij} x_j = 0 & \text{for } i = 1, \cdots, m, \\ \sum_{j=1}^{n} x_j = 1, \\ x_j \geq 0, & \text{for } j = 1, \cdots, n. \end{cases} \qquad (3.10.4)$$

Also, assume that the optimal objective value is 0 and that $x_j = 1/n$ for $j = 1, \cdots, n$ is feasible. Any model can be transformed so that these assumptions hold.

The following *centering transformation*, relative to the kth estimate of solution vector $\mathbf{x}^k$, takes any feasible solution vector $\mathbf{x}$ and transforms it to $\mathbf{y}$ such that the $\mathbf{x}^k$ is transformed to the center of the feasible simplex: $\mathbf{y}_j = \dfrac{x_j/x_j^k}{\sum_{r=1}^{n}(x_r/x_r^k)}$. Let Diag$(x^k)$ represent an $n \times n$ matrix with off-diagonal entries equal to 0 and the diagonal entry in row j equal to x_j^k. The formal algorithm is as follows:

Step (1) Initialize $x_j^0 = 1/n$ and set the iteration count $k = 0$.

Step (2) If $\sum_{j=1}^{n} x_j^k c_j$ is sufficiently close to 0, then stop; $\mathbf{x}^k$ is optimal. Otherwise go to Step (3).

Step (3) Move from the center of the transformed space in an improving direction using

$$\mathbf{y}^{k+1} = \left[\frac{1}{n}, \frac{1}{n}, \cdots, \frac{1}{n} \right]^{\text{T}} - \frac{\theta(I - P^T(PP^T)^{-1}P)[\text{Diag}(\mathbf{x}^k)]\mathbf{c}^T}{||C_p||\sqrt{n(n-1)}},$$

where $||C_p||$ is the length of the vector $(I - P^T(PP^T)^{-1}P)[\text{Diag}(\mathbf{x}^k)]\mathbf{c}^T$, P is an $(m+1) \times n$ matrix whose first m rows are $A[\text{Diag}(\mathbf{x}^k)]$ and whose last row is a vector of $1's$, and θ is a parameter that must be between 0 and 1. Go to Step (4).

Step (4) Find the new point $\mathbf{x}^{k+1}$ in the original space by applying the inverse transformation of the centering transformation to $\mathbf{y}^{k+1}$. Set $k = k+1$ and return to Step (2).

The method is guaranteed to converge to the optimal solution when $\theta = \frac{1}{4}$ is used.

3.10.2 DUALITY AND COMPLEMENTARY SLACKNESS

Define y_i as the *dual variable (shadow price)* representing the purchase price for a unit of resource i. The *dual problem* to the primal model (maximize objective, all constraints of the form "$\leq$") is

$$\text{Minimize} \quad \sum_{i=1}^{m} b_i y_i,$$

$$\text{Subject to} \quad \begin{cases} \displaystyle\sum_{i=1}^{m} a_{ij} y_i \geq c_j & \text{for } j = 1, \ldots, n, \\ y_i \geq 0, & \text{for } i = 1, \ldots, m. \end{cases} \tag{3.10.5}$$

The objective minimizes the amount of money spent to obtain the resources. The constraints ensure that the marginal cost of the resources is greater than or equal to the marginal profit for each product.

The following results link the dual model (minimization) with its primal model (maximization).

1. *Weak duality theorem*: Assume that $\mathbf{x}$ and $\mathbf{y}$ are feasible solutions to the respective primal and dual problems. Then

$$\sum_{j=1}^{n} c_j x_j \leq \sum_{i=1}^{m} b_i y_i.$$

2. *Strong duality theorem*: Assume that the primal model has a finite optimal solution $\mathbf{x}^*$. Then the dual has a finite optimal solution $\mathbf{y}^*$, and

$$\sum_{j=1}^{n} c_j x_j^* = \sum_{i=1}^{m} b_i y_i^*.$$

3. *Complementary slackness theorem*: Assume that x and y are feasible solutions to the respective primal and dual problems. Then, x is optimal for the primal and y is optimal for the dual if and only if:

$$y_i \cdot \left(b_i - \sum_{j=1}^{n} a_{ij} x_j \right) = 0, \qquad \text{for } i = 1, \cdots, m, \quad \text{and}$$

$$x_j \cdot \left(\sum_{i=1}^{m} a_{ij} y_i - c_j \right) = 0, \qquad \text{for } j = 1, \cdots, n. \tag{3.10.6}$$

3.10.3   LINEAR INTEGER PROGRAMMING

Linear integer programming models result from restricting the decisions in linear programming models to be integer valued. The standard form is

$$\text{Minimize} \quad \sum_{j=1}^{n} c_j x_j \quad \text{(objective function)},$$

$$\text{Subject to} \quad \begin{cases} x_j \geq 0, & \text{and integer for } j = 1, \ldots, n, \\ \sum_{j=1}^{n} a_{ij} x_j = b_i & \text{for } i = 1, \ldots, m \quad \text{(constraint functions).} \end{cases} \quad (3.10.7)$$

As long as the variable values are bounded, then the general model can be transformed into a model where all variable values are restricted to either 0 or 1. Therefore, algorithms that can solve 0–1 integer programming models are sufficient for most applications.

3.10.4   BRANCH AND BOUND

Branch and bound implicitly enumerates all feasible integer solutions to find the optimal solution. The main idea is to break the feasible set into subsets (branching) and then evaluate the best solution in each subset or determine that the subset cannot contain the optimal solution (bounding). When a subset is evaluated, it is said to be *fathomed*. The following algorithm performs the branching by partitioning on variables with fractional values and uses a linear programming relaxation to generate a bound on the best solution in a subset:

Step (1) Assume that a feasible integer solution, called the *incumbent*, is known whose objective function value is z (initially, z may be set to infinity if no feasible solution is known). Set p, the subset counter, equal to 1. Set the original model as the first problem in the subset list.

Step (2) If $p = 0$, then stop. The incumbent solution is the optimal solution. Otherwise go to Step (3).

Step (3) Solve the LP relaxation of the pth problem in the subset list (allow all integer valued variables to take on continuous values). Denote the LP objective value by v. If $v \geq z$ or the LP is infeasible, then set $p = p - 1$ (fathom by bound or infeasibility), and return to Step (2). If the LP solution is integer valued, then update the incumbent to the LP solution, set $z = \min(z, v)$ and $p = p - 1$, and return to Step (2). Otherwise, go to Step (4).

Step (4) Take any variable x_j with fractional value in the LP solution. Replace problem p with two problems created by individually adding the constraints $x_j \leq \lfloor x_j \rfloor$ and $x_j \geq \lceil x_j \rceil$ to problem p. Add these two problems to the bottom of the subset list replacing the pth problem, set $p = p + 1$, and go to Step (2).

3.10.5   NETWORK FLOW METHODS

A *network* consists of N, the set of nodes, and A, the set of arcs. Each arc (i, j) defines a connection from node i to node j. Depending on the application, arc (i, j) may have an associated cost and upper and lower capacity on flow.

Decision problems on networks can often be modeled using linear programming models, and these models usually have the property that solutions from the simplex method are integer valued (the *total unimodularity property*). Because of the underlying graphical structure, more efficient algorithms are also available. We present the augmenting path algorithm for the maximum flow problem and the Hungarian method for the assignment problem.

3.10.5.1   Maximum flow

Let x_{ij} represent the flow on arc (i, j), c_{ij} the flow capacity of (i, j), S be the source node, and T be the terminal node. The maximum flow problem is to ship as much flow from S to T without violating the capacity on any arc, and all flow sent into node i must leave i (for $i \neq S, T$). The following algorithm solves the problem by continually adding flow-carrying paths until no path can be found:

Step (1) Initialize $x_{ij} = 0$ for all (i, j).

Step (2) Find a flow-augmenting path from S to T using the following labeling method. Start by labeling S with a^*. From any labeled node i, label node j with the label i if j is unlabeled and $x_{ij} < c_{ij}$ (*forward labeling arc*). From any labeled node i, label node j with the label i if j is unlabeled and $x_{ji} > 0$ (*backward labeling arc*). Perform labeling until no additional nodes can be labeled. If T cannot be labeled, then stop. The current x_{ij} values are optimal. Otherwise, go to Step (3).

Step (3) There is a path from S to T where flow is increased on the forward labeling arcs, decreased on the backward labeling arcs, and gets more flow from S to T. Let F be the minimum of $c_{ij} - x_{ij}$ over all forward labeling arcs and of x_{ij} over all backward labeling arcs. Set $x_{ij} = x_{ij} + F$ for the forward arcs and $x_{ij} = x_{ij} - F$ for the backward arcs. Return to Step (2).

The algorithm terminates with a set of arcs with $x_{ij} = c_{ij}$ and if these are deleted, then S and T are in two disconnected pieces of the network. The algorithm finds the maximum flow by finding the minimum capacity set of arcs that disconnects S and T (minimum capacity cutset).

3.10.6   ASSIGNMENT PROBLEM

Consider a set J of jobs and a set I of employees. Each employee can do 1 job, and each job must be done by 1 employee. If job j is assigned to employee i, then the cost to the company is c_{ij}. The problem is to assign employees to jobs to minimize the overall cost.

This problem can be formulated as an optimization problem on a bipartite graph where the jobs are one part and the employees are the other. Let m be the cardinality of J and I (they must be equal cardinality sets, otherwise there is no feasible solution), and let C be the $m \times m$ matrix of costs c_{ij}. The following algorithm solves for the optimal assignment.

Step (1) Find $l_i = \min_j c_{ij}$ for each row i. Let $c_{ij} = c_{ij} - l_i$. Find $n_j = \min_i c_{ij}$ for each column j. Let $c_{ij} = c_{ij} - n_j$.

Step (2) Construct a graph with nodes for S, T, and each element of the sets J and I. Construct an arc from S to each node in I, and set its capacity to 1. Construct an arc from each node in J to T, and set its capacity to 1. If $c_{ij} = 0$, then construct an arc from $i \in I$ to $j \in J$, and set its capacity to 2. Solve a maximum flow problem on the constructed graph. If m units of flow can go through the network, then stop. The maximum flow solution on the arcs between I and J represents the optimal assignment. Otherwise, go to Step (3).

Step (3) Update C using the following rules based on the labels in the solution to the maximum flow problem: Let L_I and L_J be the set of elements of I and J respectively with labels when the maximum flow algorithm terminates. Let $\delta = \min_{i \in L_I, j \in (J - L_J)} c_{ij}$; note that $\delta > 0$. For $i \in L_I$ and $j \in (J - L_J)$, set $c_{ij} = c_{ij} - \delta$. For $i \in (I - L_I)$ and $j \in L_J$, set $c_{ij} = c_{ij} + \delta$. Leave all other c_{ij} values unchanged. Return to Step (2).

In Step (3), the algorithm creates new arcs, eliminates some unused arcs, and leaves unchanged arcs with $x_{ij} = 1$. When returning to Step (2), you can solve the next maximum flow problem by adding and deleting the appropriate arcs and starting with the flows and labels of the preceding execution of the maximum flow algorithm.

3.10.7 DYNAMIC PROGRAMMING

Dynamic programming is a technique for determining a sequence of optimal decisions for a system or process that operates over time and requires successive dependent decisions. The following five properties are required for using dynamic programming:

1. The system can be characterized by a set of parameters called *state variables*.
2. At each decision point or *stage* of the process, there is a choice of actions.
3. Given the current state and the decision, it is possible to specify how the state will evolve before the next decision.
4. Only the current state matters, not the path by which the system arrived at the state (termed *time separability*).
5. An objective function depends on the state and the decisions made.

The time separability requirement is necessary to formulate a functional form for the decision problem. Let $f_t^*(i)$ denote the optimal objective value to take the system from stage t to the end of the process, given that the process is now in state i, $A_t(i)$ is the set of decisions possible at stage t and state i, a_j is a particular action in $A_t(i)$, $\Delta_{it}(a_j)$ is the change from state i between stage t and stage $t + 1$ based on the action taken, and $c_{it}(a_j)$ is the immediate impact on the objective of taking action a_j at stage t and state i.

The *principle of optimality* states:

> An optimal sequence of decisions has the property that, whatever the initial state and initial decision are, the remaining decisions must be an optimal policy based on the state resulting from the initial information.

Using the principle, Bellman's equations are

$$f_t^*(i) = \min_{a_j \in A_t(i)} \left(f_t^*(i) = \min_{a_j \in A_t(i)} \left\{ c_{it}(a_j) + f_{t+1}^*[\Delta_{it}(a_j)] \right\} \right), \qquad \text{for all } i.$$

3.10.8 SHORTEST PATH PROBLEM

Consider a network (N, A) where N is the set of nodes, A the set of arcs, and d_{ij} represents the "distance" of traveling on arc (i, j) (if no arc exists between i and j, $d_{ij} = \infty$). For any two nodes R and S, the shortest path problem is to find the shortest distance route through the network from R to S. Let the state space be N and a stage representing travel along one arc. $f^*(i)$ is the optimal distance from node i to S. The resultant recursive equations to solve are

$$f^*(i) = \min_{j \in N} [d_{ij} + f^*(j)], \qquad \text{for all } i.$$

Dijkstra's algorithm can be used successively to approximate the solution to the equations when $d_{ij} > 0$ for all (i, j).

Step (1) Set $f^*(S) = 0$ and $f^*(i) = d_{iS}$ for all $i \in N$. Let P be the set of permanently labeled nodes; $P = \{S\}$. Let $T = N - P$ be the set of temporarily labeled nodes.

Step (2) Find $i \in T$ with $f^*(i) = \min_{j \in T} f^*(j)$. Set $T = T - \{i\}$ and $P = P \cup \{i\}$. If $T = \emptyset$ (the empty set), then stop; $f^*(R)$ is the optimal path length. Otherwise, go to Step (3).

Step (3) Set $f^*(j) = \min[f^*(j), f^*(i) + d_{ij}]$ for all $j \in T$. Return to Step (2).

3.10.9 HEURISTIC SEARCH TECHNIQUES

Heuristic search techniques are commonly used in algorithms for combinatorial optimization problems. The method starts with an initial vector x_0 and attempts to find improved solutions. Define $\mathbf{x}$ to be the current solution vector, $f(\mathbf{x})$ to be its objective value, and $N(\mathbf{x})$ to be its neighborhood. For the remainder of this section, assume that we seek to find the minimum value of $f(\mathbf{x})$. In each iteration of a neighborhood search algorithm, we generate a number of solutions $\mathbf{x}' \in N(\mathbf{x})$, and compare $f(\mathbf{x})$ with each $f(\mathbf{x}')$. If $f(\mathbf{x}')$ is a better value than $f(\mathbf{x})$, then we update the current solution to $\mathbf{x}'$ (termed *accepting* $\mathbf{x}'$) and we iterate again searching from $\mathbf{x}'$. The process continues until none of the generated neighbors of the current solution yield lower solutions. Common improvement procedures such as *pairwise interchange* and "k-opt" are specific instances of neighborhood search methods. Note that at each iteration, we move to a better solution or we terminate with the best solution seen so far.

The key issues involved in designing a neighborhood search method are the definition of the neighborhood of $\mathbf{x}$, and the number of neighbors to generate in each iteration. When the neighborhood of $\mathbf{x}$ is easily computed and evaluated, then one can generate the entire neighborhood to ensure finding a better solution if one exists in the neighborhood. It is also possible to generate only a portion of the neighborhood (however, this could lead to premature termination with a poorer solution). In general, deterministic neighborhood search can only guarantee finding a local minimum solution to the optimization problem. One can make multiple runs, each with different initial solution vectors, to increase the chances of finding the global minimum solution.

3.10.9.1 Simulated annealing (SA)

Simulated annealing is neighborhood search method that uses randomization to avoid terminating at a locally optimal point. In each iteration of SA, we generate a single neighbor $\mathbf{x}'$ of $\mathbf{x}$. If $f(\mathbf{x}') \leq f(\mathbf{x})$, then we accept $\mathbf{x}'$. Otherwise, we accept $\mathbf{x}'$ with a probability that depends upon $f(\mathbf{x}') - f(\mathbf{x})$, and a non-stationary control parameter.

Define the following notation:

1. C_k is the kth control parameter,
2. L_k is the maximum number of neighbors evaluated while the kth control parameter is in use, and
3. I_k is the counter for the number of solutions currently evaluated at the kth control parameter.

To initialize the algorithm, we assume that we are given the sequences $\{C_k\}$ (termed the *cooling schedule*) and $\{L_k\}$ such that $C_k \to 0$ as $k \to \infty$, and an initial value $\mathbf{x}_0$. The SA algorithm is (assuming that we are trying to minimize $f(\mathbf{x})$):

Step (1) Set $\mathbf{x} = \mathbf{x}_0$, $k = 1$ and $I_k = 0$.

Step (2) Generate a neighbor $\mathbf{x}'$ of $\mathbf{x}$, compute $f(\mathbf{x}')$ and increment I_k by 1.

Step (3) If $f(\mathbf{x}') \leq f(\mathbf{x})$, then replace $\mathbf{x}$ with $\mathbf{x}'$ and go to Step (5).

Step (4) If $f(\mathbf{x}') > f(\mathbf{x})$, then with probability $\exp\left(\frac{f(\mathbf{x}) - f(\mathbf{x}')}{C_k}\right)$, replace $\mathbf{x}$ with $\mathbf{x}'$.

Step (5) If $I_k = L_k$, then increment k by 1, reset $I_k = 0$, and check for the termination criterion. If the termination criterion is met, then stop, $\mathbf{x}$ is the solution; if not, then go to Step (2).

This SA description uses an exponential form for the acceptance probability for inferior solutions; this was proposed by Metropolis *et al.* The algorithm terminates when C_k approaches 0, or there has been no improvement in the solution over a number of C_k values. In the above SA description we only store the current solution, not the best solution encountered. In practice, however, in Step (2), we also compare $f(\mathbf{x}')$ with an incumbent solution. If the incumbent solution is worse, then we replace the incumbent with $\mathbf{x}'$; otherwise we retain the current incumbent. Note that SA only generates one neighbor of $\mathbf{x}$ in each iteration. However, we may not accept $\mathbf{x}'$ and therefore may generate another neighbor of $\mathbf{x}$.

Several issues related to parameter setting, neighborhood definition, and solution updating must be resolved when implementing SA. For example, the sequences of C_k and L_k values are critical for successful application. If C_k goes to zero too quickly, then the algorithm can easily get stuck in a local minimum solution. If C_1 is large and C_k tends to zero too slowly, then the algorithm requires more computation to achieve convergence. To date, the majority of the analytical results and empirical studies in the SA literature consider the effects of the C_k schedule, the L_k schedule, and the initial solution $\mathbf{x}_0$. The method can be shown to converge with probability one to the set of optimal solutions as long as $C_k \to 0$ as $k \to \infty$ and $\sum_{k=0}^{\infty} C_k < \infty$.

3.10.9.2   Tabu search

Tabu search (TS) is similar to simulated annealing. A problem in SA is that one can start to climb out of a local minimum solution, only to return via a sequence of "better" solutions that leads directly back to where the algorithm has already searched. Also, the neighborhood structure may permit moving to extremely poor solutions where one knows that no optimal solutions exist. Finally, the modeler may have insight on where to look for good solutions and SA has no easy mechanism for enabling the search of specific areas. Tabu search tries to remedy each of these deficiencies.

The key terminology in TS is $S(\mathbf{x})$, the set of moves from $\mathbf{x}$. This is similar to the neighborhood of $\mathbf{x}$, and is all the solutions that you can get to from $\mathbf{x}$ in one *move*. The terminology "move" is similar to the "step" in simulated annealing, but can be more general because it can be applied to both continuous and discrete variable problems. Let s be a move in $S(\mathbf{x})$. For example, consider a 5-city traveling salesman problem where $\mathbf{x}$ is a vector containing the sequence of cities visited, including the return to the initial city. Let $\mathbf{x} = (2, 1, 5, 3, 4, 2)$. If we define a move to be an "adjacent pairwise interchange", then the set of moves $S(\mathbf{x})$ is:

$$\{(1, 2, 5, 3, 4, 1), (2, 5, 1, 3, 4, 2), (2, 1, 3, 5, 4, 2), (2, 1, 5, 4, 3, 2), (4, 1, 5, 3, 2, 4)\}$$

For another example, if we use a standard non-linear programming direction–step size search algorithm, then one can construct a family of moves of the form $S(\mathbf{x}) = \mathbf{x} + u\mathbf{d}$. Here, u is a step size scalar and $\mathbf{d}$ is the direction of movement and the family depends on the values of u and $\mathbf{d}$ selected.

To run the method, define T as the *tabu set*. This is a set of moves that one does not want the method to use. Also, define OPT to be the function that selects a particular $s \in S(\mathbf{x})$ that creates an eventual improvement in the objective. Then

Step (1) Find initial incumbent $\mathbf{x}_0$. Set $\mathbf{x} = \mathbf{x}^*$, $k = 0$, $T = \phi$.

Step (2) If $S(\mathbf{x}) - T = \phi$, then go to Step (4). Otherwise set $k = k + 1$ and select $s_k \in S(\mathbf{x}) - T$ such that $s_k(\mathbf{x}) = OPT(s \mid s \in S(\mathbf{x}) - T)$.

Step (3) Let $\mathbf{x} = s_k(\mathbf{x})$. If $f(\mathbf{x}) < f(\mathbf{x}_0)$, then $\mathbf{x}_0 = \mathbf{x}$.

Step (4) If a chosen number of iterations has elapsed either in total or since $\mathbf{x}_0$ was last improved, or if $S(\mathbf{x}) - T = \phi$ from Step (2), then stop. Otherwise, update T (if necessary) and return to Step (2).

The method is more effective if the user understands the solution space and can guide the search somewhat. Often the tabu list contains solutions that were previously visited or solutions that would reverse properties of good solutions. Early in the method, it is important that the search space is evaluated in a coarse manner so that one does not skip an area where the optimal solution is located. Management of the tabu list is also critical; the list will be expanding and contracting as you obtain more information about the solution space. The ability to solve the OPT problem quickly in Step (2) also helps the method by enabling the evaluation of more candidate solutions. Tabu search can move to inferior solutions temporarily when OPT returns a solution that has a worse objective value than $f(\mathbf{x})$ and, in fact, this happens every time when $\mathbf{x}$ is a local minimum solution. For more details, see Glover.

3.10.9.3 Genetic algorithms

A different approach to heuristically solving difficult combinatorial optimization problems mimics evolutionary theory within an algorithmic process. A population of individuals is represented by K various feasible solutions $\mathbf{x}_k$ for $k = 1, \ldots, K$. The collection of such solutions at any iteration of a genetic algorithm is referred to as a *generation*, and the individual elements of each solution $\mathbf{x}_k$ are called *chromosomes*. For instance, in the context of the traveling salesman problem, a generation would consist of a set of traveling salesman tours, and the chromosomes of an individual solution would each represent a city.

To continue drawing parallels with the evolutionary process, each iteration of a genetic algorithm creates a new generation by computing new solutions based on the previous population. More specifically, an individual of the new generation is created from (usually two) parent solutions by means of a *crossover operator*. The crossover operator prescribes methodology by which a solution is created by combining characteristics of the parent solutions. The selection of the crossover operator is one of the most important aspects of designing an algorithm.

The rules for composing a new generation differ among various implementations, but often consist of selecting some of the best solutions from the previous generation along with some new solutions created by crossovers from the previous generation. Additionally, these solutions may mutate from generation to generation in order to introduce new elements and chromosomal patterns into the population. The objective value of each new solution in the new generation is computed, and the best solution found thus far in the algorithm is updated if applicable. The creation of the new generation of solutions concludes a genetic algorithm iteration. The algorithm stops once some termination criteria are reached (e.g., after a specified number of generations are evolved, or perhaps if no new best solution was recorded in the last q generations). A typical genetic algorithm for minimization is:

Step (1) Choose a population size K, a maximum number of generations Q, and a number of survivors $S < K$, where K, Q, and S are all integers. Also, choose some mutation probability p (typically, p is set to be a small value, often close to 0.05). Create some initial set of solutions $\mathbf{x}_k$, for $k = 1, \ldots, K$, and let Generation 0 consist of these solutions. Calculate the objective function of each solution in Generation 0, and let $\mathbf{x}^*$ with objective function $f(\mathbf{x}^*)$ denote the best such solution. Initialize the generation counter $i = 0$.

Step (2) Copy the best (according to objective function value) S solutions from Generation i into Generation $i + 1$.

Step (3) Create the remaining $K - S$ solutions for Generation $i + 1$ by executing a crossover operation on randomly selected parents from Generation i. For each new solution $\mathbf{x}_k$ created, calculate its objective function value $f(\mathbf{x}_k)$. If $f(\mathbf{x}_k) < f(\mathbf{x}^*)$, then set $\mathbf{x}^* = \mathbf{x}_k$ and $f(\mathbf{x}^*) = f(\mathbf{x}_k)$.

Step (4) For $k = 1, \ldots, K$, mutate solution $\mathbf{x}_k$ in Generation i with probability p. Calculate the new objective function value $f(\mathbf{x}_k)$, and if $f(\mathbf{x}_k) < f(\mathbf{x}^*)$, then set $\mathbf{x}^* = \mathbf{x}_k$ and $f(\mathbf{x}^*) = f(\mathbf{x}_k)$.

Step (5) Set $i = i + 1$. If $i = Q$, then terminate with solution $\mathbf{x}^*$. Otherwise, return to Step 2.

Genetic algorithm implementations differ because of the application area and the processes required by the algorithm. Three specific processes required by genetic algorithms are discussed below. For more details, see Goldberg.

1. *Creating the initial population* — Although often overlooked as an important part of executing genetic algorithms, careful consideration should be given to the creation of the initial set of solutions for Generation 0. For instance, we may employ a rudimentary constructive heuristic to generate a set of good initial solutions rather than using some blindly random approach. However, care must also be taken to ensure that the set of heuristic solutions generated is sufficiently diverse. That is, if all initial solutions are nearly identical, then the solutions created in the next generation may closely resemble those of the previous generation, thus limiting the scope of the genetic algorithm search space. Hence, one may penalize solutions having too close a resemblance to previously generated solutions in the initial step.

2. *The crossover process* — The crossover operator is the most important consideration in designing a genetic algorithm. While the selection of the parents for the crossover operation is done randomly, preference should be given to parent solutions having better quality objective function values (imitating mating of the most fit individuals, as in evolution theory). However, feasibility restrictions are sometimes implied by the structure of a solution, and must therefore apply to the output of the crossover operator.

EXAMPLE   For example, in the traveling salesman problem, one such feasibility restriction requires that a solution must be a permutation of integers. Consider the following two parent solutions, where the return to the first city is implied:

$$(1, 3, 5, 2, 4, 6) \quad \text{and} \quad (1, 2, 3, 6, 5, 4).$$

A crossover operator that simply takes the first three chromosomes from the first parent and the last three chromosomes from the last parent would result in the solution (1, 3, 5, 6, 5, 4), which is infeasible to the permutation restriction. An alternative operator might modify the foregoing operator with the following modification: post-process the solution by replacing repeated chromosomes with omitted chromosomes. In the prior example, city 5 is repeated while city 2 is omitted, and thus one of the following two solutions would be generated:

$$(1, 3, 2, 6, 5, 4) \quad \text{or} \quad (1, 3, 5, 6, 2, 4).$$

3. *The mutation operator* — A common behavior of genetic algorithms executed without the mutation operator is that the solutions within the same generation converge to a small set of distinct solutions, from which radically different (and perhaps optimal) solutions cannot be created via the given crossover operator. The purpose of the mutation operator is to inject diverse elements into future iterations. Generally speaking, the mutation operator is often a small change within the solution, such as a pairwise interchange on a solution to the traveling salesman problem. However, the mutation operator must generate significant enough change in the solution to ensure that there exists a chance of having this modification propagate into solutions in future generations.

Chapter 4

Geometry

4.1 COORDINATE SYSTEMS IN THE PLANE *299*
    4.1.1 *Convention* *299*
    4.1.2 *Substitutions and transformations* *299*
    4.1.3 *Cartesian coordinates in the plane* *301*
    4.1.4 *Polar coordinates in the plane* *302*
    4.1.5 *Homogeneous coordinates in the plane* *303*
    4.1.6 *Oblique coordinates in the plane* *303*

4.2 PLANE SYMMETRIES OR ISOMETRIES *305*
    4.2.1 *Formulae for symmetries: Cartesian coordinates* *305*
    4.2.2 *Formulae for symmetries: homogeneous coordinates* *306*
    4.2.3 *Formulae for symmetries: polar coordinates* *307*
    4.2.4 *Crystallographic groups* *307*
    4.2.5 *Classifying the crystallographic groups* *312*

4.3 OTHER TRANSFORMATIONS OF THE PLANE *312*
    4.3.1 *Similarities* *312*
    4.3.2 *Affine transformations* *313*
    4.3.3 *Projective transformations* *314*

4.4 LINES *314*
    4.4.1 *Lines with prescribed properties* *315*
    4.4.2 *Distances* *316*
    4.4.3 *Angles* *316*
    4.4.4 *Concurrence and collinearity* *317*

4.5 POLYGONS *317*
    4.5.1 *Triangles* *318*
    4.5.2 *Quadrilaterals* *322*
    4.5.3 *Regular polygons* *323*

4.6 CONICS *325*
    4.6.1 *Alternative characterization* *325*
    4.6.2 *The general quadratic equation* *328*
    4.6.3 *Additional properties of ellipses* *330*
    4.6.4 *Additional properties of hyperbolas* *331*
    4.6.5 *Additional properties of parabolas* *333*
    4.6.6 *Circles* *333*

1-58488-291-3/02/$0.00+$1.50
© 2003 CRC Press, Inc.

4.7   SPECIAL PLANE CURVES **336**
    4.7.1   Algebraic curves *336*
    4.7.2   Roulettes (spirograph curves) *340*
    4.7.3   Curves in polar coordinates *342*
    4.7.4   Spirals *342*
    4.7.5   The Peano curve and fractal curves *343*
    4.7.6   Fractal objects *344*
    4.7.7   Classical constructions *345*

4.8   COORDINATE SYSTEMS IN SPACE **345**
    4.8.1   Cartesian coordinates in space *345*
    4.8.2   Cylindrical coordinates in space *346*
    4.8.3   Spherical coordinates in space *346*
    4.8.4   Relations between Cartesian, cylindrical, and spherical coordinates *348*
    4.8.5   Homogeneous coordinates in space *348*

4.9   SPACE SYMMETRIES OR ISOMETRIES **348**
    4.9.1   Formulae for symmetries: Cartesian coordinates *349*
    4.9.2   Formulae for symmetries: homogeneous coordinates *351*

4.10   OTHER TRANSFORMATIONS OF SPACE **352**
    4.10.1   Similarities *352*
    4.10.2   Affine transformations *352*
    4.10.3   Projective transformations *353*

4.11   DIRECTION ANGLES AND DIRECTION COSINES **353**

4.12   PLANES **354**
    4.12.1   Planes with prescribed properties *354*
    4.12.2   Concurrence and coplanarity *355*

4.13   LINES IN SPACE **355**
    4.13.1   Distances *356*
    4.13.2   Angles *356*
    4.13.3   Concurrence, coplanarity, parallelism *357*

4.14   POLYHEDRA **357**
    4.14.1   Convex regular polyhedra *358*
    4.14.2   Polyhedra nets *360*

4.15   CYLINDERS **361**

4.16   CONES **361**

4.17   SURFACES OF REVOLUTION: THE TORUS **363**

4.18   QUADRICS **364**
    4.18.1   Spheres *365*

4.19   SPHERICAL GEOMETRY & TRIGONOMETRY **368**
    4.19.1   Right spherical triangles *368*

4.19.2   *Oblique spherical triangles* *369*

4.20 DIFFERENTIAL GEOMETRY **373**
    4.20.1   *Curves* *373*
    4.20.2   *Surfaces* *376*

4.21 ANGLE CONVERSION **381**

4.22 KNOTS UP TO EIGHT CROSSINGS **382**

4.1 COORDINATE SYSTEMS IN THE PLANE

4.1.1 CONVENTION

When we talk about "the point with coordinates (x, y)" or "the curve with equation $y = f(x)$", we always mean Cartesian coordinates. If a formula involves other coordinates, this fact will be stated explicitly.

4.1.2 SUBSTITUTIONS AND TRANSFORMATIONS

Formulae for changes in coordinate systems can lead to confusion because (for example) moving the coordinate axes *up* has the same effect on equations as moving objects *down* while the axes stay fixed. (To read the next paragraph, you can move your eyes down or slide the page up.)

    To avoid confusion, we will carefully distinguish between transformations of the plane and substitutions, as explained below. Similar considerations will apply to transformations and substitutions in three dimensions (Section 4.8).

4.1.2.1 Substitutions

A *substitution*, or *change of coordinates*, relates the coordinates of a point in one coordinate system to those of *the same point in a different coordinate system*. Usually one coordinate system has the superscript $'$ and the other does not, and we write

$$\begin{cases} x = F_x(x', y'), \\ y = F_y(x', y'), \end{cases} \qquad \text{or} \qquad (x, y) = F(x', y') \qquad (4.1.1)$$

(where subscripts and primes are not derivatives, they are coordinates). This means: given the equation of an object in the unprimed coordinate system, one obtains the equation of the *same* object in the primed coordinate system by substituting $F_x(x', y')$ for x and $F_y(x', y')$ for y in the equation. For instance, suppose the primed coordinate system is obtained from the unprimed system by moving the x axis up a distance d. Then $x = x'$ and $y = y' + d$. The circle with equations $x^2 + y^2 = 1$ in the unprimed system has equations $x'^2 + (y' + d)^2 = 1$ in the primed system. Thus, transforming an implicit equation in (x, y) into one in (x', y') is immediate.

The point $P = (a, b)$ in the unprimed system, with equation $x = a$, $y = b$, has equation $F_x(x', y') = a$, $F_y(x', y') = b$ in the new system. To get the primed coordinates explicitly, one must solve for x' and y' (in the example just given we have $x' = a$, $y' + d = b$, which yields $x' = a$, $y' = b - d$). Therefore, if possible, we give the *inverse equations*

$$\begin{cases} x' = G_{x'}(x, y), \\ y' = G_{y'}(x, y) \end{cases} \quad \text{or} \quad (x', y') = G(x, y),$$

which are equivalent to Equation (4.1.1) if $G(F(x', y')) = (x', y')$ and $F(G(x, y)) = (x, y)$. Then to go from the unprimed to the primed system, one merely inserts the known values of x and y into these equations. This is also the best strategy when dealing with a curve expressed parametrically, that is, $x = x(t)$, $y = y(t)$.

4.1.2.2 Transformations

A *transformation* associates with each point (x, y) a *different point in the same co-ordinate system*; we denote this by

$$(x, y) \mapsto F(x, y), \tag{4.1.2}$$

where F is a map from the plane to itself (a two-component function of two variables). For example, translating down by a distance d is accomplished by $(x, y) \mapsto (x, y - d)$ (see Section 4.2). Thus, the action of the transformation on a point whose coordinates are known (or on a curve expressed parametrically) can be immediately computed.

If, on the other hand, we have an object (say a curve) defined *implicitly* by the equation $C(x, y) = 0$, finding the equation of the transformed object requires using the *inverse transformation*

$$(x, y) \mapsto G(x, y) \tag{4.1.3}$$

defined by $G(F(x, y)) = (x, y)$ and $F(G(x, y)) = (x, y)$. The equation of the transformed object is $C(G(x, y)) = 0$. For instance, if C is the circle with equation $x^2 + y^2 = 1$ and we are translating down by a distance d, the inverse transformation is $(x, y) \mapsto (x, y + d)$ (translating up), and the equation of the translated circle is $x^2 + (y + d)^2 = 1$. Compare to the example following Equation (4.1.1).

4.1.2.3 Using transformations to change coordinates

Usually, we will not give formulae of the form (4.1.1) for changes between two coordinate systems of the same type, because they can be immediately derived from the corresponding formulae (4.1.2) for transformations, which are given in Section 4.2. We give two examples for clarity.

FIGURE 4.1

Change of coordinates by a rotation.

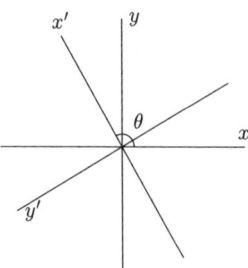

Let the two Cartesian coordinate systems (x, y) and (x', y') be related as follows: They have the same origin, and the positive x'-axis is obtained from the positive x-axis by a (counterclockwise) rotation through an angle θ (Figure 4.1). If a point has coordinates (x, y) in the unprimed system, its coordinates (x', y') in the primed system are the same as the coordinates in the unprimed system of a point that undergoes the *inverse rotation*, that is, a rotation by an angle $\alpha = -\theta$. According to Equation (4.2.2) (page 305), this transformation acts as follows:

$$(x, y) \mapsto \begin{bmatrix} \cos\theta & \sin\theta \\ -\sin\theta & \cos\theta \end{bmatrix} (x, y) = (x\cos\theta + y\sin\theta,\ -x\sin\theta + y\cos\theta). \quad (4.1.4)$$

Therefore the right-hand side of Equation (4.1.4) is (x', y'), and the desired substitution is

$$\begin{aligned} x' &= x\cos\theta + y\sin\theta, \\ y' &= -x\sin\theta + y\cos\theta. \end{aligned} \quad (4.1.5)$$

Switching the roles of the primed and unprimed systems we get the equivalent substitution

$$\begin{aligned} x &= x'\cos\theta - y'\sin\theta, \\ y &= x'\sin\theta + y'\cos\theta \end{aligned} \quad (4.1.6)$$

(because the x-axis is obtained from the x'-axis by a rotation through an angle $-\theta$).

Similarly, let the two Cartesian coordinate systems (x, y) and (x', y') differ by a translation: x is parallel to x' and y to y', and the origin of the second system coincides with the point (x_0, y_0) of the first system. The coordinates (x, y) and (x', y') of a point are related by

$$\begin{aligned} x &= x' + x_0, & x' &= x - x_0, \\ y &= y' + y_0, & y' &= y - y_0. \end{aligned} \quad (4.1.7)$$

4.1.3 CARTESIAN COORDINATES IN THE PLANE

In *Cartesian coordinates* (or *rectangular coordinates*), the "address" of a point P is given by two real numbers indicating the positions of the perpendicular projections from the point to two fixed, perpendicular, graduated lines, called the *axes*. If one coordinate is denoted x and the other y, the axes are called the *x-axis* and the *y-axis*, and we write $P = (x, y)$. Usually the x-axis is horizontal, with x increasing to the right, and the y-axis is vertical, with y increasing vertically up. The point $x = 0$, $y = 0$, where the axes intersect, is the *origin*. See Figure 4.2.

FIGURE 4.2
In Cartesian coordinates, $P_1 = (4, 3)$, $P_2 = (-1.3, 2.5)$, $P_3 = (-1.5, -1.5)$, $P_4 = (3.5, -1)$, and $P_5 = (4.5, 0)$. The axes divide the plane into four quadrants. P_1 is in the first quadrant, P_2 in the second, P_3 in the third, and P_4 in the fourth. P_5 is on the positive x-axis.

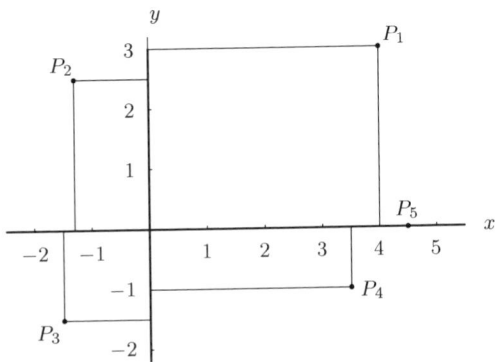

4.1.4 POLAR COORDINATES IN THE PLANE

In *polar coordinates* a point P is also characterized by two numbers: the distance $r \geq 0$ to a fixed *pole* or *origin* O, and the angle θ that the ray OP makes with a fixed ray originating at O, which is generally drawn pointing to the right (this is called the *initial ray*). The angle θ is defined only up to a multiple of $360°$ or 2π radians. In addition, it is sometimes convenient to relax the condition $r > 0$ and allow r to be a signed distance; so (r, θ) and $(-r, \theta + 180°)$ represent the same point (Figure 4.3).

FIGURE 4.3
Among the possible sets of polar coordinates for P are $(10, 30°)$, $(10, 390°)$ and $(10, -330°)$. Among the sets of polar coordinates for Q are $(2.5, 210°)$ and $(-2.5, 30°)$.

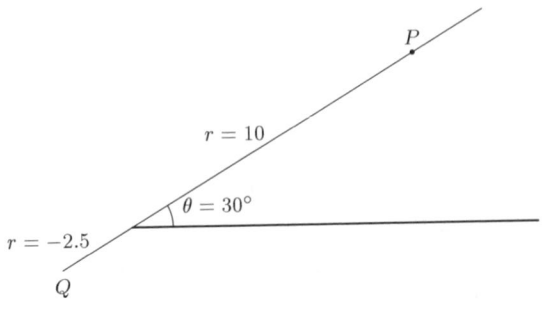

4.1.4.1 Relations between Cartesian and polar coordinates

Consider a system of polar coordinates and a system of Cartesian coordinates with the same origin. Assume that the initial ray of the polar coordinate system coincides with the positive x-axis, and that the ray $\theta = 90°$ coincides with the positive y-axis. Then the polar coordinates (r, θ) with $r > 0$ and the Cartesian coordinates (x, y) of the same point are related as follows (x and y are assumed positive when using the $\tan^{-1}$ formula):

$$\begin{cases} x = r\cos\theta, \\ y = r\sin\theta, \end{cases} \qquad \begin{cases} r = \sqrt{x^2 + y^2}, \\ \theta = \tan^{-1}\dfrac{y}{x}, \end{cases} \qquad \begin{cases} \sin\theta = \dfrac{y}{\sqrt{x^2 + y^2}}, \\ \cos\theta = \dfrac{x}{\sqrt{x^2 + y^2}}. \end{cases}$$

4.1.5 HOMOGENEOUS COORDINATES IN THE PLANE

A triple of real numbers $(x : y : t)$, with $t \neq 0$, is a set of *homogeneous coordinates* for the point P with Cartesian coordinates $(x/t, \, y/t)$. Thus the same point has many sets of homogeneous coordinates: $(x : y : t)$ and $(x' : y' : t')$ represent the same point if and only if there is some real number α such that $x' = \alpha x, y' = \alpha y, z' = \alpha z$.

When we think of the same triple of numbers as the Cartesian coordinates of a point in three-dimensional space (page 345), we write it as (x, y, t) instead of $(x : y : t)$. The connection between the point in space with Cartesian coordinates (x, y, t) and the point in the plane with homogeneous coordinates $(x : y : t)$ becomes apparent when we consider the plane $t = 1$ in space, with Cartesian coordinates given by the first two coordinates x, y (Figure 4.4). The point (x, y, t) in space can be connected to the origin by a line L that intersects the plane $t = 1$ in the point with Cartesian coordinates $(x/t, \, y/t)$ or homogeneous coordinates $(x : y : t)$.

Homogeneous coordinates are useful for several reasons. One of the most important is that they allow one to unify all symmetries of the plane (as well as other transformations) under a single umbrella. All of these transformations can be regarded as linear maps in the space of triples $(x : y : t)$, and so can be expressed in terms of matrix multiplications (see page 306).

If we consider triples $(x : y : t)$ such that at least one of x, y, t is non-zero, we can name not only the points in the plane but also points "at infinity". Thus, $(x : y : 0)$ represents the point at infinity in the direction of the ray emanating from the origin going through the point (x, y).

4.1.6 OBLIQUE COORDINATES IN THE PLANE

The following generalization of Cartesian coordinates is sometimes useful. Consider two *axes* (graduated lines), intersecting at the *origin* but not necessarily perpendicularly. Let the angle between them be ω. In this system of *oblique coordinates*, a point P is given by two real numbers indicating the positions of the projections from the point to each axis, in the direction of the other axis (see Figure 4.5). The first axis (x-axis) is generally drawn horizontally. The case $\omega = 90°$ yields a Cartesian coordinate system.

FIGURE 4.4

The point P with spatial coordinates (x, y, t) projects to the point Q with spatial coordinates $(x/t, y/t, 1)$. The plane Cartesian coordinates of Q are $(x/t, y/t)$, and $(x : y : t)$ is one set of homogeneous coordinates for Q. Any point on the line L (except for the origin O) would also project to Q.

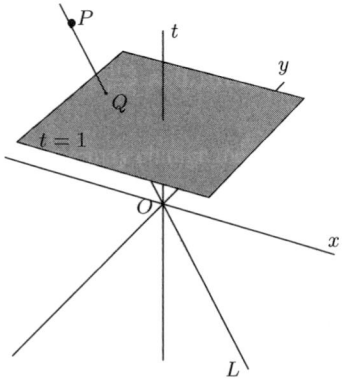

FIGURE 4.5

In oblique coordinates, $P_1 = (4, 3)$, $P_2 = (-1.3, 2.5)$, $P_3 = (-1.5, -1.5)$, $P_4 = (3.5, -1)$, and $P_5 = (4.5, 0)$. Compare to Figure 4.2.

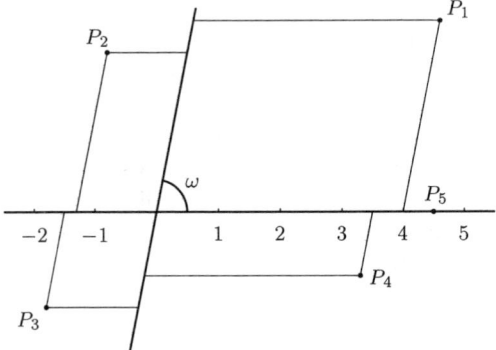

4.1.6.1   Relations between two oblique coordinate systems

Let the two oblique coordinate systems (x, y) and (x', y'), with angles ω and ω', share the same origin, and suppose the positive x'-axis makes an angle θ with the positive x-axis. The coordinates (x, y) and (x', y') of a point in the two systems are related by

$$x = \frac{x' \sin(\omega - \theta) + y' \sin(\omega - \omega' - \theta)}{\sin \omega},$$

$$y = \frac{x' \sin \theta + y' \sin(\omega' + \theta)}{\sin \omega}.$$

(4.1.8)

This formula also covers passing from a Cartesian system to an oblique system and vice versa, by taking $\omega = 90°$ or $\omega' = 90°$.

The relation between two oblique coordinate systems that differ by a translation is the same as for Cartesian systems. See Equation (4.1.7).

4.2 PLANE SYMMETRIES OR ISOMETRIES

A transformation of the plane (invertible map of the plane to itself) that preserves distances is called an *isometry* of the plane. Every isometry of the plane is of one of the following types:

1. The *identity* (which leaves every point fixed)
2. A *translation* by a vector **v**
3. A *rotation* through an angle α around a point P
4. A *reflection* in a line L
5. A *glide-reflection* in a line L with displacement d

Although the identity is a particular case of a translation and a rotation, and reflections are particular cases of glide-reflections, it is more intuitive to consider each case separately.

4.2.1 FORMULAE FOR SYMMETRIES: CARTESIAN COORDINATES

In the formulae below, a multiplication between a matrix and a pair of coordinates should be carried out regarding the pair as a column vector (or a matrix with two rows and one column). Thus $\begin{bmatrix} a & b \\ c & d \end{bmatrix} (x, y) = (ax + by,\ cx + dy)$.

1. *Translation* by (x_0, y_0):

$$(x, y) \mapsto (x + x_0,\ y + y_0). \tag{4.2.1}$$

2. *Rotation* through α (counterclockwise) around the origin:

$$(x, y) \mapsto \begin{bmatrix} \cos\alpha & -\sin\alpha \\ \sin\alpha & \cos\alpha \end{bmatrix} (x, y). \tag{4.2.2}$$

3. *Rotation* through α (counterclockwise) around an arbitrary point (x_0, y_0):

$$(x, y) \mapsto (x_0, y_0) + \begin{bmatrix} \cos\alpha & -\sin\alpha \\ \sin\alpha & \cos\alpha \end{bmatrix} (x - x_0,\ y - y_0). \tag{4.2.3}$$

4. *Reflection:*

$$
\begin{aligned}
\text{in the } x\text{-axis:} &\quad (x, y) \mapsto (x, -y), \\
\text{in the } y\text{-axis:} &\quad (x, y) \mapsto (-x, y), \\
\text{in the diagonal } x = y: &\quad (x, y) \mapsto (y, x).
\end{aligned}
\tag{4.2.4}
$$

5. *Reflection* in a line with equation $ax + by + c = 0$:

$$(x, y) \mapsto \frac{1}{a^2 + b^2} \left(\begin{bmatrix} b^2 - a^2 & -2ab \\ -2ab & a^2 - b^2 \end{bmatrix} (x, y) - (2ac, 2bc) \right). \quad (4.2.5)$$

6. *Reflection* in a line going through (x_0, y_0) and making an angle α with the x-axis:

$$(x, y) \mapsto (x_0, y_0) + \begin{bmatrix} \cos 2\alpha & \sin 2\alpha \\ \sin 2\alpha & -\cos 2\alpha \end{bmatrix} (x - x_0, y - y_0). \quad (4.2.6)$$

7. *Glide-reflection* in a line L with displacement d: Apply first a reflection in L, then a translation by a vector of length d in the direction of L, that is, by the vector

$$\frac{1}{a^2 + b^2} (\pm ad, \mp bd) \quad (4.2.7)$$

if L has equation $ax + by + c = 0$.

4.2.2 FORMULAE FOR SYMMETRIES: HOMOGENEOUS COORDINATES

All isometries of the plane can be expressed in homogeneous coordinates in terms of multiplication by a matrix. This fact is useful in implementing these transformations on a computer. It also means that the successive application of transformations reduces to matrix multiplication. The corresponding matrices are as follows:

1. *Translation* by (x_0, y_0):

$$T_{(x_0, y_0)} = \begin{bmatrix} 1 & 0 & x_0 \\ 0 & 1 & y_0 \\ 0 & 0 & 1 \end{bmatrix}. \quad (4.2.8)$$

2. *Rotation* through α around the origin:

$$R_\alpha = \begin{bmatrix} \cos \alpha & -\sin \alpha & 0 \\ \sin \alpha & \cos \alpha & 0 \\ 0 & 0 & 1 \end{bmatrix}. \quad (4.2.9)$$

3. *Reflection* in a line going through the origin and making an angle α with the x-axis:

$$M_\alpha = \begin{bmatrix} \cos 2\alpha & \sin 2\alpha & 0 \\ \sin 2\alpha & -\cos 2\alpha & 0 \\ 0 & 0 & 1 \end{bmatrix}. \quad (4.2.10)$$

From this one can deduce all other transformations.

EXAMPLE   To find the matrix for a rotation through α around an arbitrary point $P = (x_0, y_0)$, we apply a translation by $-(x_0, y_0)$ to move P to the origin, a rotation through α around the origin, and then a translation by (x_0, y_0):

$$T_{(x_0, y_0)} R_\alpha T_{-(x_0, y_0)} = \begin{bmatrix} \cos\alpha & -\sin\alpha & x_0 - x_0\cos\alpha + y_0\sin\alpha \\ \sin\alpha & \cos\alpha & y_0 - y_0\cos\alpha - x_0\sin\alpha \\ 0 & 0 & 1 \end{bmatrix} \qquad (4.2.11)$$

(notice the order of the multiplication).

4.2.3   FORMULAE FOR SYMMETRIES: POLAR COORDINATES

1. *Rotation around the origin* through an angle α:

$$(r, \theta) \mapsto (r, \ \theta + \alpha). \qquad (4.2.12)$$

2. *Reflection in a line through the origin* and making an angle α with the positive x-axis:

$$(r, \theta) \mapsto (r, \ 2\alpha - \theta). \qquad (4.2.13)$$

4.2.4   CRYSTALLOGRAPHIC GROUPS

A group of symmetries of the plane that is doubly infinite is a *wallpaper group*, or *crystallographic group*. There are 17 types of such groups, corresponding to 17 essentially distinct ways to tile the plane in a doubly periodic pattern. (There are also 230 three-dimensional crystallographic groups.)

The simplest crystallographic group involves translations only (page 309, top left). The others involve, in addition to translations, one or more of the other types of symmetries (rotations, reflections, glide-reflections). The *Conway notation* for crystallographic groups is based on the types of non-translational symmetries occurring in the "simplest description" of the group:

1. $^\circ$ indicates a translations only,
2. * indicates a reflection (mirror symmetry),
3. $^\times$ a glide-reflection,
4. a number n indicates a rotational symmetry of order n (rotation by $360\,^\circ/n$).

In addition, if a number n comes after the *, the center of the corresponding rotation lies on mirror lines, so that the symmetry there is actually dihedral of order $2n$.

Thus the group ** in the table below (page 309, middle left) has two inequivalent lines of mirror symmetry; the group 333 (page 311, top right) has three inequivalent centers of order-3 rotation; the group $22\,^*$ (page 309, bottom right) has two inequivalent centers of order-2 rotation as well as mirror lines; and ^{*632} (page 311, bottom right) has points of dihedral symmetry of order $12 (= 2 \times 6)$, 6, and 4.

The following table gives the groups in the Conway notation and in the notation traditional in crystallography. It also gives the quotient space of the plane by the action of the group. The entry "4,4,2 turnover" means the surface of a triangular puff pastry with corner angles $45°(= 180°/4)$, $45°$ and $90°$. The entry "4,4,2 turnover slit along 2,4" means the same surface, slit along the edge joining a $45°$ vertex to the $90°$ vertex. Open edges are silvered (mirror lines); such edges occur exactly for those groups whose Conway notation includes a *.

The last column of the table gives the dimension of the space of inequivalent groups of the given type (equivalent groups are those that can be obtained from one another by proportional scaling or rigid motion). For instance, there is a group of type ° for every shape parallelogram, and there are two degrees of freedom for the choice of such a shape (say the ratio and angle between sides). Thus, the ° group of page 309 (top left) is based on a square fundamental domain, while for the ° group of page 311 (top left) a fundamental parallelogram would have the shape of two juxtaposed equilateral triangles. These two groups are inequivalent, although they are of the same type.

Look on page	Conway	Cryst	Quotient space	Dim
309 top left; 311 top left	°	p1	Torus	2
309 top right	××	pg	Klein bottle	1
309 middle left	**	pm	Cylinder	1
309 middle right	×*	cm	Möbius strip	1
309 bottom left	22×	pgg	Nonorientable football	1
309 bottom right	22*	pmg	Open pillowcase	1
310 top left	2222	p2	Closed pillowcase	2
310 top right	2*22	cmm	4,4,2 turnover, slit along 4,4	1
310 middle left	*2222	pmm	Square	1
310 middle right	442	p4	4,4,2 turnover	0
310 bottom left	4*2	p4g	4,4,2 turnover, slit along 4,2	0
310 bottom right	*442	p4m	4,4,2 triangle	0
311 top right	333	p3	3,3,3 turnover	0
311 middle left	*333	p3m1	3,3,3 triangle	0
311 middle right	3*3	p31m	6,3,2 turnover, slit along 3,2	0
311 bottom left	632	p6	6,3,2 turnover	0
311 bottom right	*632	p6m	6,3,2 triangle	0

The figures on pages 309–311 show wallpaper patterns based on each of the 17 types of crystallographic groups (two patterns are shown for the °, or translations-only, type). Thin lines bound *unit cells*, or *fundamental domains*. When solid, they represent lines of mirror symmetry, and are fully determined. When dashed, they represent arbitrary boundaries, which can be shifted so as to give different fundamental domains. One can even make these lines into curves, provided the symmetry is respected. Dots at the intersections of thin lines represent centers of rotational symmetry.

Some of the relationships between the types are made obvious by the patterns. For instance, on the first row of page 309, we see that the group on the right, of type ××, contains the one on the left, of type °, with index two. However, there are more relationships than can be indicated in a single set of pictures. For instance, there is a group of type ×× hiding in any group of type 3*3.

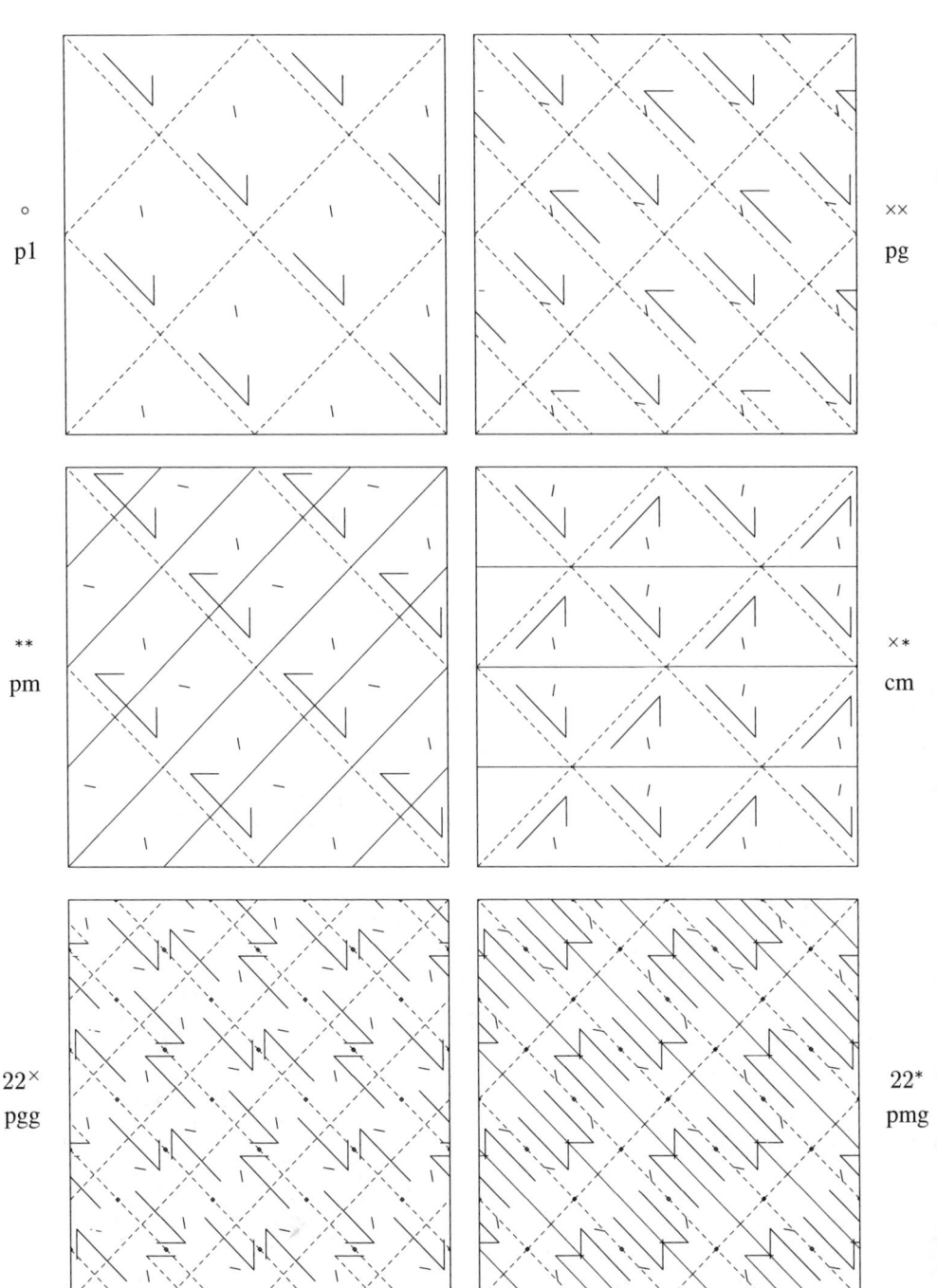

$\circ$
p1

$\times\times$
pg

$**$
pm

$\times*$
cm

$22\times$
pgg

22^*
pmg

2222
p2

2*22
cmm

*2222
pmm

442
p4

4*2
p4g

*442
p4m

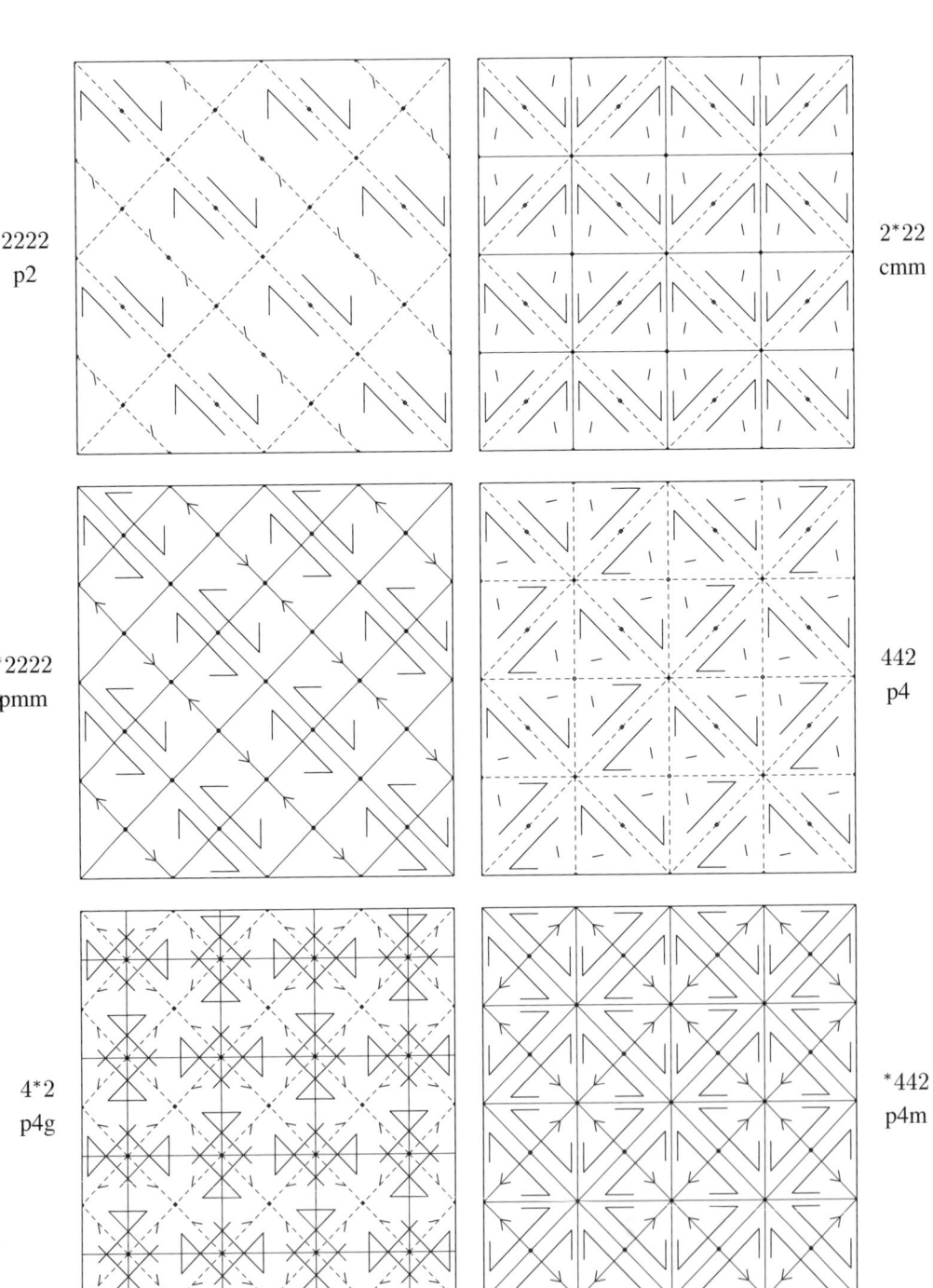

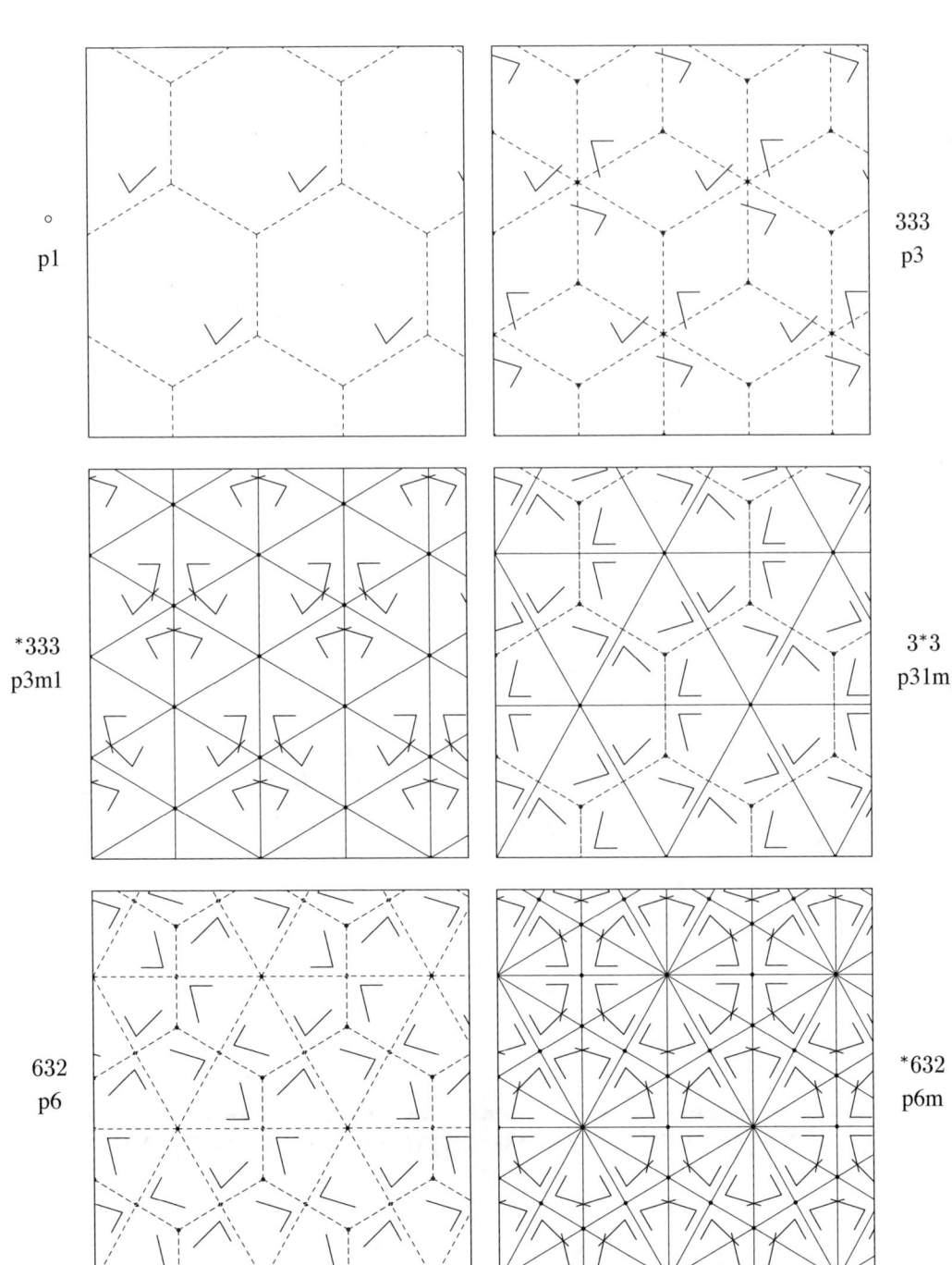

p1

333
p3

*333
p3m1

3*3
p31m

632
p6

*632
p6m

4.2.5 CLASSIFYING THE CRYSTALLOGRAPHIC GROUPS

To classify an image representing a crystallographic group, answer the following sequence of questions starting with: "What is the minimal rotational invariance?".

- None
  Is there a reflection?

  - No.
    Is there a glide-reflection?
    * No: p1 (page 309)
    * Yes: pg (page 309)

  - Yes.
    Is there a glide-reflection in an axis that is not a reflection axis?
    * No: pm (page 309)
    * Yes: cm (page 309)

- 2-fold (180° rotation)
  Is there a reflection?

  - No.
    Is there a glide-reflection?
    * No: p2 (page 310)
    * Yes: pgg (page 309)

  - Yes.
    Are there reflections in two directions?
    * No: pmg (page 309)
    * Yes: Are all rotation centers on reflection axes?
      · No: cmm (page 310)

      · Yes: pmm (page 310)

- 3-fold (120° rotation)
  Is there a reflection?

  - No: p3 (page 311)
  - Yes.
    Are all centers of threefold rotations on reflection axes?
    * No: p31m (page 311)
    * Yes: p3m1 (page 311)

- 4-fold (90° rotation)
  Is there a reflection?

  - No: p4 (page 310)
  - Yes.
    Are there four reflection axes?
    * No: p4g (page 310)
    * Yes: p4m (page 310)

- 6-fold (60° rotation)
  Is there a reflection?

  - No: p6 (page 311)
  - Yes: p6m (page 311)

4.3 OTHER TRANSFORMATIONS OF THE PLANE

4.3.1 SIMILARITIES

A transformation of the plane that preserves shapes is called a *similarity*. Every similarity of the plane is obtained by composing a *proportional scaling transformation* (also known as a *homothety*) with an isometry. A proportional scaling transformation centered at the origin has the form

$$(x, y) \mapsto (ax, ay), \tag{4.3.1}$$

where $a \neq 0$ is the *scaling factor* (a real number). The corresponding matrix in *homogeneous coordinates* is

$$H_a = \begin{bmatrix} a & 0 & 0 \\ 0 & a & 0 \\ 0 & 0 & 1 \end{bmatrix}. \tag{4.3.2}$$

In *polar coordinates*, the transformation is $(r, \theta) \mapsto (ar, \theta)$.

4.3.2 AFFINE TRANSFORMATIONS

A transformation that preserves lines and parallelism (maps parallel lines to parallel lines) is an *affine transformation*. There are two important particular cases of such transformations:

A *non-proportional scaling transformation* centered at the origin has the form $(x, y) \mapsto (ax, by)$, where $a, b \neq 0$ are the *scaling factors* (real numbers). The corresponding matrix in *homogeneous coordinates* is

$$H_{a,b} = \begin{bmatrix} a & 0 & 0 \\ 0 & b & 0 \\ 0 & 0 & 1 \end{bmatrix}. \tag{4.3.3}$$

A *shear* preserving horizontal lines has the form $(x, y) \mapsto (x + ry, y)$, where r is the *shearing factor* (see Figure 4.6). The corresponding matrix in *homogeneous coordinates* is

$$S_r = \begin{bmatrix} 1 & r & 0 \\ 0 & 1 & 0 \\ 0 & 0 & 1 \end{bmatrix}. \tag{4.3.4}$$

Every affine transformation is obtained by composing a scaling transformation with an isometry, or a shear with a homothety and an isometry.

FIGURE 4.6
A shear with factor $r = \frac{1}{2}$.

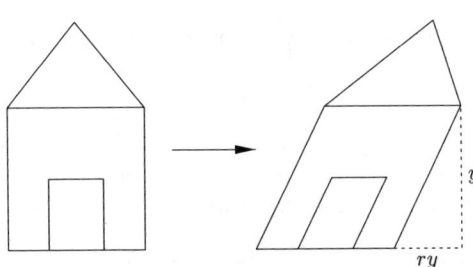

FIGURE 4.7

A perspective transformation with center O, mapping the plane P to the plane Q. The transformation is not defined on the line L, where P intersects the plane parallel to Q and going through O.

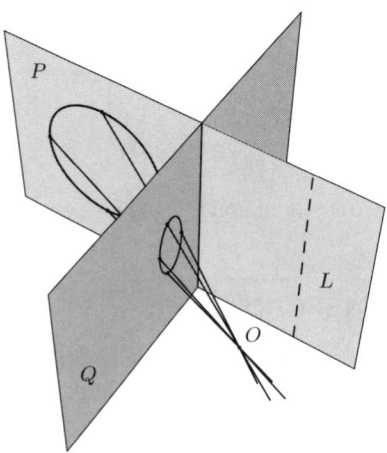

4.3.3 PROJECTIVE TRANSFORMATIONS

A transformation that maps lines to lines (but does not necessarily preserve parallelism) is a *projective transformation*. Any plane projective transformation can be expressed by an invertible 3×3 matrix in homogeneous coordinates; conversely, any invertible 3×3 matrix defines a projective transformation of the plane. Projective transformations (if not affine) are not defined on all of the plane but only on the complement of a line (the missing line is "mapped to infinity").

A common example of a projective transformation is given by a *perspective transformation* (Figure 4.7). Strictly speaking this gives a transformation from one plane to another, but, if we identify the two planes by (for example) fixing a Cartesian system in each, we get a projective transformation from the plane to itself.

4.4 LINES

The (Cartesian) equation of a *straight line* is linear in the coordinates x and y:

$$ax + by + c = 0. \tag{4.4.1}$$

The *slope* of this line is $-a/b$, the *intersection with the x-axis* (or *x-intercept*) is $x = -c/a$, and the *intersection with the y-axis* (or *y*-intercept) is $y = -c/b$. If $a = 0$, the line is parallel to the x-axis, and if $b = 0$, then the line is parallel to the y-axis.

(In an *oblique coordinate system*, everything in the preceding paragraph remains true, except for the value of the slope.)

When $a^2 + b^2 = 1$ and $c \leq 0$ in the equation $ax + by + c = 0$, the equation is said to be in *normal form*. In this case $-c$ is the *distance of the line to the origin*, and ω (with $\cos \omega = a$ and $\sin \omega = b$) is the angle that the perpendicular dropped to the line from the origin makes with the positive x-axis (Figure 4.8, with $p = -c$).

To reduce an arbitrary equation $ax + by + c = 0$ to normal form, divide by $\pm\sqrt{a^2 + b^2}$, where the sign of the radical is chosen opposite the sign of c when $c \neq 0$ and the same as the sign of b when $c = 0$.

FIGURE 4.8
The normal form of the line L is $x \cos \omega + y \sin \omega = p$.

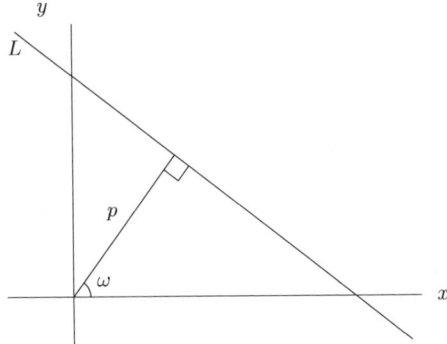

4.4.1   LINES WITH PRESCRIBED PROPERTIES

1. Line of slope m intersecting the x-axis at $x = x_0$: $y = m(x - x_0)$.

2. Line of slope m intersecting the y-axis at $y = y_0$: $y = mx + y_0$.

3. Line intersecting the x-axis at $x = x_0$ and the y-axis at $y = y_0$:

$$\frac{x}{x_0} + \frac{y}{y_0} = 1. \tag{4.4.2}$$

(This formula remains true in *oblique coordinates*.)

4. Line of slope m passing though (x_0, y_0): $y - y_0 = m(x - x_0)$.

5. Line passing through points (x_0, y_0) and (x_1, y_1):

$$\frac{y - y_1}{x - x_1} = \frac{y_0 - y_1}{x_0 - x_1} \quad \text{or} \quad \begin{vmatrix} x & y & 1 \\ x_0 & y_0 & 1 \\ x_1 & y_1 & 1 \end{vmatrix} = 0. \tag{4.4.3}$$

(These formulae remain true in *oblique coordinates*.)

6. *Slope* of line going through points (x_0, y_0) and (x_1, y_1): $\dfrac{y_1 - y_0}{x_1 - x_0}$.

7. Line passing through points with *polar coordinates* (r_0, θ_0) and (r_1, θ_1):

$$r(r_0 \sin(\theta - \theta_0) - r_1 \sin(\theta - \theta_1)) = r_0 r_1 \sin(\theta_1 - \theta_0). \quad (4.4.4)$$

4.4.2 DISTANCES

The *distance* between two points in the plane is the *length of the line segment* joining the two points. If the points have *Cartesian coordinates* (x_0, y_0) and (x_1, y_1), this distance is

$$\sqrt{(x_1 - x_0)^2 + (y_1 - y_0)^2}. \quad (4.4.5)$$

If the points have *polar coordinates* (r_0, θ_0) and (r_1, θ_1), this distance is

$$\sqrt{r_0^2 + r_1^2 - 2r_0 r_1 \cos(\theta_0 - \theta_1)}. \quad (4.4.6)$$

If the points have *oblique coordinates* (x_0, y_0) and (x_1, y_1), this distance is

$$\sqrt{(x_1 - x_0)^2 + (y_1 - y_0)^2 + 2(x_1 - x_0)(y_1 - y_0) \cos \omega}, \quad (4.4.7)$$

where ω is the angle between the axes (Figure 4.5).

The point $k\%$ of the way from $P_0 = (x_0, y_0)$ to $P_1 = (x_1, y_1)$ is

$$\left(\frac{kx_1 + (100 - k)x_0}{100}, \frac{ky_1 + (100 - k)y_0}{100} \right). \quad (4.4.8)$$

(The same formula also works in oblique coordinates.) This point divides the segment $P_0 P_1$ in the ratio $k : (100 - k)$. As a particular case, the *midpoint* of $P_0 P_1$ is given by $\left(\frac{1}{2}(x_0 + x_1), \frac{1}{2}(y_0 + y_1) \right)$.

The *distance* from the point (x_0, y_0) to the line $ax + by + c = 0$ is

$$\left| \frac{ax_0 + by_0 + c}{\sqrt{a^2 + b^2}} \right|. \quad (4.4.9)$$

4.4.3 ANGLES

The *angle* between two lines $a_0 x + b_0 y + c_0 = 0$ and $a_1 x + b_1 y + c_1 = 0$ is

$$\tan^{-1} \frac{b_1}{a_1} - \tan^{-1} \frac{b_0}{a_0} = \tan^{-1} \frac{a_0 b_1 - a_1 b_0}{a_0 a_1 + b_0 b_1}. \quad (4.4.10)$$

In particular, the two lines are *parallel* when $a_0 b_1 = a_1 b_0$, and *perpendicular* when $a_0 a_1 = -b_0 b_1$.

The *angle* between two lines of slopes m_0 and m_1 is $\tan^{-1}(m_1) - \tan^{-1}(m_0)$ (or $\tan^{-1}((m_1 - m_0)/(1 + m_0 m_1))$). In particular, the two lines are *parallel* when $m_0 = m_1$ and *perpendicular* when $m_0 m_1 = -1$.

4.4.4   CONCURRENCE AND COLLINEARITY

Three lines $a_0 x + b_0 y + c_0 = 0$, $a_1 x + b_1 y + c_1 = 0$, and $a_2 x + b_2 y + c_2 = 0$ are *concurrent* (i.e., intersect at a single point) if and only if

$$\begin{vmatrix} a_0 & b_0 & c_0 \\ a_1 & b_1 & c_1 \\ a_2 & b_2 & c_2 \end{vmatrix} = 0. \qquad (4.4.11)$$

(This remains true in *oblique coordinates*.)

Three points (x_0, y_0), (x_1, y_1), and (x_2, y_2) are *collinear* (i.e., all three points are on a straight line) if and only if

$$\begin{vmatrix} x_0 & y_0 & 1 \\ x_1 & y_1 & 1 \\ x_2 & y_2 & 1 \end{vmatrix} = 0. \qquad (4.4.12)$$

(This remains true in *oblique coordinates*.)

Three points with polar coordinates (r_0, θ_0), (r_1, θ_1), and (r_2, θ_2) are collinear if and only if

$$r_1 r_2 \sin(\theta_2 - \theta_1) + r_0 r_1 \sin(\theta_1 - \theta_0) + r_2 r_0 \sin(\theta_0 - \theta_2) = 0. \qquad (4.4.13)$$

4.5   POLYGONS

Given $k \geq 3$ points $A_1, \ldots, A_k$ in the plane, in a certain order, we obtain a *k-sided polygon* or *k-gon* by connecting each point to the next, and the last to the first, with a line segment. The points A_i are the *vertices* and the segments $A_i A_{i+1}$ are the *sides* or *edges* of the polygon. When $k = 3$ we have a *triangle*, when $k = 4$ we have a *quadrangle* or *quadrilateral*, and so on (see page 324 for names of regular polygons). Here we will assume that all polygons are *simple*: this means that no consecutive edges are on the same line and no two edges intersect (except that consecutive edges intersect at the common vertex) (see Figure 4.9).

FIGURE 4.9
Two simple quadrilaterals (left and middle) and one that is not simple (right). We will treat only simple polygons.

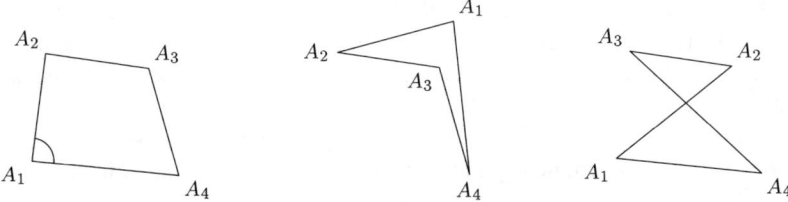

When we refer to the *angle* at a vertex A_k we have in mind the interior angle (as marked in the leftmost polygon in Figure 4.9). We denote this angle by the same symbol as the vertex. The complement of A_k is the *exterior angle* at that vertex; geometrically, it is the angle between one side and the extension of the adjacent side. *In any k-gon, the sum of the angles equals* $2(k-2)$ *right angles, or* $2(k-2) \times 90°$; for example, the sum of the angles of a triangle is $180°$.

The *area* of a polygon whose vertices A_i have coordinates (x_i, y_i), for $1 \leq i \leq k$, is the absolute value of

$$
\begin{aligned}
\text{area} &= \tfrac{1}{2}(x_1 y_2 - x_2 y_1) + \cdots + \tfrac{1}{2}(x_{k-1} y_k - x_k y_{k-1}) + \tfrac{1}{2}(x_k y_1 - x_1 y_k), \\
&= \frac{1}{2} \sum_{i=1}^{k} (x_i y_{i+1} - x_{i+1} y_i),
\end{aligned} \tag{4.5.1}
$$

where in the summation we take $x_{k+1} = x_1$ and $y_{k+1} = y_1$. In particular, for a triangle we have

$$
\text{area} = \tfrac{1}{2}(x_1 y_2 - x_2 y_1 + x_2 y_3 - x_3 y_2 + x_3 y_1 - x_1 y_3) = \frac{1}{2} \begin{vmatrix} x_1 & y_1 & 1 \\ x_2 & y_2 & 1 \\ x_3 & y_3 & 1 \end{vmatrix}. \tag{4.5.2}
$$

In *oblique coordinates* with angle ω between the axes, the area is as given above, multiplied by $\sin \omega$.

If the vertices have *polar coordinates* (r_i, θ_i), for $1 \leq i \leq k$, the area is the absolute value of

$$
\text{area} = \frac{1}{2} \sum_{i=1}^{k} r_i r_{i+1} \sin(\theta_{i+1} - \theta_i), \tag{4.5.3}
$$

where we take $r_{k+1} = r_1$ and $\theta_{k+1} = \theta_1$.

Formulae for specific polygons in terms of side lengths, angles, etc., are given on the following pages.

4.5.1 TRIANGLES

Because the angles of a triangle add up to $180°$, at least two of them must be acute (less than $90°$). In an *acute triangle* all angles are acute. A *right triangle* has one right angle, and an *obtuse triangle* has one obtuse angle.

The *altitude* corresponding to a side is the perpendicular dropped to the line containing that side from the opposite vertex. The *bisector* of a vertex is the line that divides the angle at that vertex into two equal parts. The *median* is the segment joining a vertex to the midpoint of the opposite side. See Figure 4.10.

Every triangle also has an *inscribed circle* tangent to its sides and interior to the triangle (in other words, any three non-concurrent lines determine a circle). The center of this circle is the point of intersection of the bisectors. We denote the radius of the inscribed circle by r.

Every triangle has a *circumscribed circle* going through its vertices; in other words, any three non-collinear points determine a circle. The point of intersection of

FIGURE 4.10

Notations for an arbitrary triangle of sides a, b, c and vertices A, B, C. The altitude corresponding to C is h_c, the median is m_c, the bisector is t_c. The radius of the circumscribed circle is R, that of the inscribed circle is r.

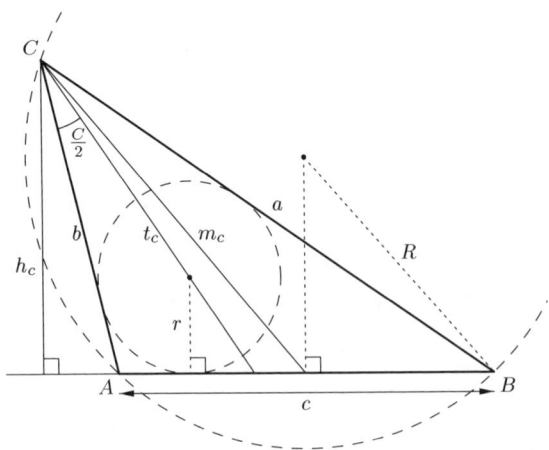

the medians is the center of mass of the triangle (considered as an area in the plane). We denote the radius of the circumscribed circle by R.

Introduce the following notations for an *arbitrary triangle* of vertices A, B, C and sides a, b, c (see Figure 4.10). Let h_c, t_c, and m_c be the lengths of the altitude, bisector, and median originating in vertex C, let r and R be as usual the radii of the inscribed and circumscribed circles, and let s be the semi-perimeter: $s = \frac{1}{2}(a+b+c)$. Then

$$A + B + C = 180°$$

$$c^2 = a^2 + b^2 - 2ab\cos C \quad \text{(law of cosines)},$$

$$a = b\cos C + c\cos B,$$

$$\frac{a}{\sin A} = \frac{b}{\sin B} = \frac{c}{\sin C} \quad \text{(law of sines)},$$

$$\text{area} = \tfrac{1}{2}h_c c = \tfrac{1}{2}ab\sin C = \frac{c^2 \sin A \sin B}{2\sin C} = rs = \frac{abc}{4R},$$

$$= \sqrt{s(s-a)(s-b)(s-c)} \quad \text{(Heron)},$$

$$r = c\sin(\tfrac{1}{2}A)\sin(\tfrac{1}{2}B)\sec(\tfrac{1}{2}C) = \frac{ab\sin C}{2s} = (s-c)\tan(\tfrac{1}{2}C),$$

$$= \left(\frac{1}{h_a} + \frac{1}{h_b} + \frac{1}{h_c}\right)^{-1},$$

$$R = \frac{c}{2\sin C} = \frac{abc}{4\,\text{area}},$$

$$h_c = a\sin B = b\sin A = \frac{2\,\text{area}}{c},$$

$$t_c = \frac{2ab}{a+b} \cos \tfrac{1}{2}C = \sqrt{ab\left(1 - \frac{c^2}{(a+b)^2}\right)}, \quad \text{and}$$

$$m_c = \sqrt{\tfrac{1}{2}a^2 + \tfrac{1}{2}b^2 - \tfrac{1}{4}c^2}.$$

A triangle is *equilateral* if all of its sides have the same length, or, equivalently, if all of its angles are the same (and equal to 60°). It is *isosceles* if two sides are the same, or, equivalently, if two angles are the same. Otherwise it is *scalene*.

For an *equilateral triangle* of side a we have

$$\text{area} = \tfrac{1}{4}a^2\sqrt{3}, \quad r = \tfrac{1}{6}a\sqrt{3}, \quad R = \tfrac{1}{3}a\sqrt{3}, \quad h = \tfrac{1}{2}a\sqrt{3}, \qquad (4.5.4)$$

where h is any altitude. The altitude, the bisector, and the median for each vertex coincide.

For an *isosceles triangle*, the altitude for the uonequal side is also the corresponding bisector and median, but this is not true for the other two altitudes. Many formulae for an isosceles triangle of sides a, a, c can be immediately derived from those for a right triangle of legs $a, \tfrac{1}{2}c$ (see Figure 4.11, left).

For a *right triangle*, the *hypotenuse* is the longest side and is opposite the right angle; the *legs* are the two shorter sides adjacent to the right angle. The altitude for each leg equals the other leg. In Figure 4.11 (right), h denotes the altitude for the hypotenuse, while m and n denote the segments into which this altitude divides the hypotenuse.

The following formulae apply for a right triangle:

$$A + B = 90°, \qquad\qquad c^2 = a^2 + b^2 \quad \textit{(Pythagoras)},$$
$$r = \frac{ab}{a+b+c}, \qquad\qquad R = \tfrac{1}{2}c,$$
$$a = c\sin A = c\cos B, \qquad b = c\sin B = c\cos A,$$
$$mc = b^2, \qquad\qquad nc = a^2,$$
$$\text{area} = \tfrac{1}{2}ab, \qquad\qquad hc = ab.$$

FIGURE 4.11

Left: an isosceles triangle can be divided into two congruent right triangles. Right: notations for a right triangle.

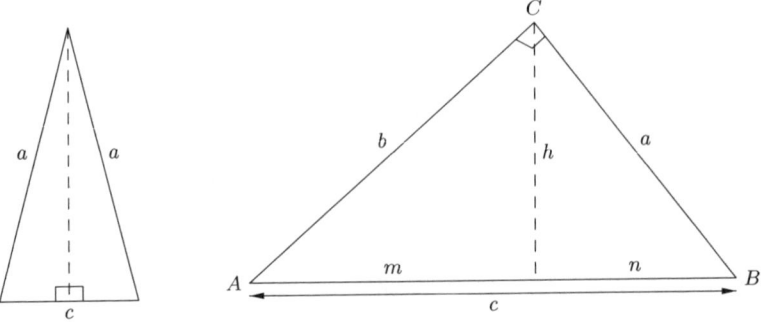

FIGURE 4.12

Left: Ceva's theorem. Right: Menelaus's theorem.

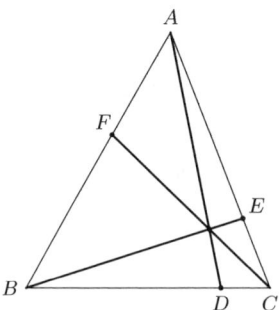 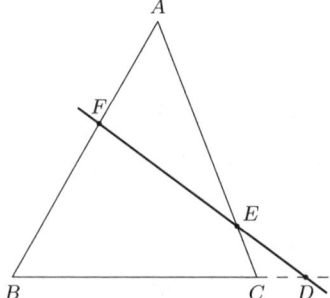

The hypotenuse is a diameter of the circumscribed circle. The median joining the midpoint of the hypotenuse (the center of the circumscribed circle) to the right angle makes angles $2A$ and $2B$ with the hypotenuse.

Additional facts about triangles:

1. In any triangle, the longest side is opposite the largest angle, and the shortest side is opposite the smallest angle. This follows from the law of sines.

2. *Ceva's theorem* (see Figure 4.12, left): In a triangle ABC, let D, E, and F be points on the lines BC, CA, and AB, respectively. Then the lines AD, BE, and CF are concurrent if, and only if, the signed distances $BD, CE, \ldots$ satisfy

$$BD \cdot CE \cdot AF = DC \cdot EA \cdot FB. \qquad (4.5.5)$$

This is so in three important particular cases: when the three lines are the medians, when they are the bisectors, and when they are the altitudes.

3. *Menelaus's theorem* (see Figure 4.12, right): In a triangle ABC, let D, E, and F be points on the lines BC, CA, and AB, respectively. Then D, E, and F are collinear if, and only if, the signed distances $BD, CE, \ldots$ satisfy

$$BD \cdot CE \cdot AF = -DC \cdot EA \cdot FB. \qquad (4.5.6)$$

4. Each side of a triangle is less than the sum of the other two. For any three lengths such that each is less than the sum of the other two, there is a triangle with these side lengths.

5. Determining if a point is inside a triangle

Given a triangle's vertices $\{P_0, P_1, P_2\}$ and the test point P_3, place P_0 at the origin by subtracting its coordinates from each of the others. Then compute (here $P_i = (x_i, y_i)$)

$$
\begin{aligned}
a &= x_1 y_2 - x_2 y_1, \\
b &= x_1 y_3 - x_3 y_1, \\
c &= x_2 y_3 - x_3 y_2
\end{aligned}
\qquad (4.5.7)
$$

The point P_3 is inside the triangle $\{P_0, P_1, P_2\}$ if and only if

$$ab > 0 \quad \text{and} \quad ac < 0 \quad \text{and} \quad a(a - b + c) > 0 \tag{4.5.8}$$

4.5.2 QUADRILATERALS

The following formulae give the area of a *general quadrilateral* (see Figure 4.13, left, for the notation).

$$\begin{aligned}
\text{area} &= \tfrac{1}{2}pq \sin \theta = \tfrac{1}{4}(b^2 + d^2 - a^2 - c^2) \tan \theta \\
&= \tfrac{1}{4}\sqrt{4p^2q^2 - (b^2 + d^2 - a^2 - c^2)^2} \\
&= \sqrt{(s-a)(s-b)(s-c)(s-d) - abcd \cos \tfrac{1}{2}(A + C)}.
\end{aligned} \tag{4.5.9}$$

FIGURE 4.13

Left: notation for a general quadrilateral; in addition $s = \tfrac{1}{2}(a + b + c + d)$. *Right: a parallelogram.*

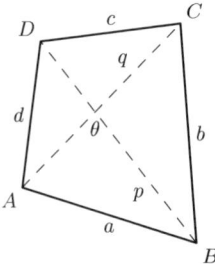
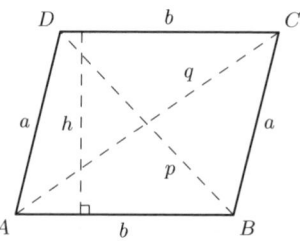

Often, however, it is easiest to compute the area by dividing the quadrilateral into triangles. One can also divide the quadrilateral into triangles to compute one side given the other sides and angles, etc.

More formulae can be given for *special cases* of quadrilaterals. In a *parallelogram*, opposite sides are parallel and the diagonals intersect in the middle (Figure 4.13, right). It follows that opposite sides have the same length and that two consecutive angles add up to $180°$. In the notation of the figure, we have

$$A = C, \quad B = D, \qquad\qquad A + B = 180°,$$
$$h = a \sin A = a \sin B, \qquad\qquad \text{area} = bh,$$
$$p = \sqrt{a^2 + b^2 - 2ab \cos A}, \qquad q = \sqrt{a^2 + b^2 - 2ab \cos B}.$$

(All this follows from the triangle formulae applied to the triangles ABD and ABC.)

Two particular cases of parallelograms are

1. The *rectangle* □, where all angles equal $90°$. The diagonals of a rectangle have the same length. The general formulae for parallelograms reduce to

$$h = a, \quad \text{area} = ab, \quad \text{and} \quad p = q = \sqrt{a^2 + b^2}, \tag{4.5.10}$$

2. The *rhombus* or *diamond* $\Diamond$, where adjacent sides have the same length ($a = b$). The diagonals of a rhombus are perpendicular. In addition to the general formulae for parallelograms, we have area $= \frac{1}{2}pq$ and $p^2 + q^2 = 4a^2$.

The *square* or regular quadrilateral is both a rectangle and a rhombus.

A quadrilateral is a *trapezoid* if two sides are parallel. In the notation of the figure on the right we have

$$A + D = B + C = 180°, \quad \text{area} = \tfrac{1}{2}(AB + CD)h.$$

The diagonals of a quadrilateral with consecutive sides a, b, c, d are perpendicular if and only if $a^2 + c^2 = b^2 + d^2$.

A quadrilateral is *cyclic* if it can be inscribed in a circle, that is, if its four vertices belong to a single, circumscribed, circle. This is possible if and only if the sum of opposite angles is $180°$. If R is the radius of the circumscribed circle, we have (in the notation of Figure 4.13, left)

$$\text{area} = \sqrt{(s-a)(s-b)(s-c)(s-d)} = \tfrac{1}{2}(ac + bd)\sin\theta,$$

$$= \frac{\sqrt{(ac+bd)(ad+bc)(ab+cd)}}{4R} \quad (\textit{Brahmagupta}),$$

$$p = \sqrt{\frac{(ac+bd)(ab+cd)}{(ad+bc)}},$$

$$R = \frac{1}{4}\sqrt{\frac{(ac+bd)(ad+bc)(ab+cd)}{(s-a)(s-b)(s-c)(s-d)}},$$

$$\sin\theta = \frac{2\,\text{area}}{ac+bd},$$

$$pq = ac + bd \quad (\textit{Ptolemy}).$$

A quadrilateral is *circumscribable* if it has an inscribed circle (that is, a circle tangent to all four sides). Its area is rs, where r is the radius of the inscribed circle and s is as above.

For a quadrilateral that is both cyclic and circumscribable, we have the following additional equalities, where m is the distance between the centers of the inscribed and circumscribed circles:

$$a + c = b + d, \qquad\qquad\qquad \text{area} = \sqrt{abcd} = rs,$$

$$R = \frac{1}{4}\sqrt{\frac{(ac+bd)(ad+bc)(ab+cd)}{abcd}}, \quad \frac{1}{r^2} = \frac{1}{(R-m)^2} + \frac{1}{(R+m)^2}.$$

4.5.3   REGULAR POLYGONS

A polygon is *regular* if all its sides are equal and all its angles are equal. Either condition implies the other in the case of a triangle, but not in general. (A rhombus has equal sides but not necessarily equal angles, and a rectangle has equal angles but not necessarily equal sides.)

For a k-sided regular polygon of side a, let θ be the angle at any vertex, and r and R the radii of the inscribed and circumscribed circles (r is called the *apothem*). As usual, let $s = \frac{1}{2}ka$ be the half-perimeter. Then

$$\theta = \left(\frac{k-2}{k}\right) 180°,$$

$$a = 2r \tan \frac{180°}{k} = 2R \sin \frac{180°}{k},$$

$$\text{area} = \frac{1}{4}ka^2 \cot \frac{180°}{k} = kr^2 \tan \frac{180°}{k}$$

$$= \frac{1}{2}kR^2 \sin \frac{360°}{k},$$

$$r = \frac{1}{2}a \cot \frac{180°}{k},$$

$$R = \frac{1}{2}a \csc \frac{180°}{k}.$$

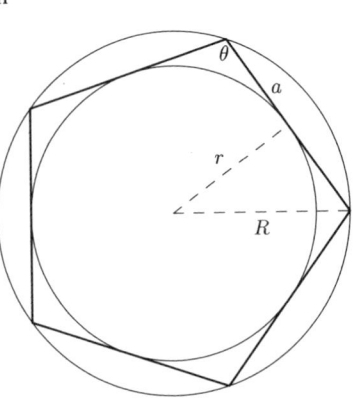

Name	k	area	r	R
Equilateral triangle	3	$0.43301\,a^2$	$0.28868\,a$	$0.57735\,a$
Square	4	a^2	$0.50000\,a$	$0.70711\,a$
Regular pentagon	5	$1.72048\,a^2$	$0.68819\,a$	$0.85065\,a$
Regular hexagon	6	$2.59808\,a^2$	$0.86603\,a$	a
Regular heptagon	7	$3.63391\,a^2$	$1.03826\,a$	$1.15238\,a$
Regular octagon	8	$4.82843\,a^2$	$1.20711\,a$	$1.30656\,a$
Regular nonagon	9	$6.18182\,a^2$	$1.37374\,a$	$1.46190\,a$
Regular decagon	10	$7.69421\,a^2$	$1.53884\,a$	$1.61803\,a$
Regular undecagon	11	$9.36564\,a^2$	$1.70284\,a$	$1.77473\,a$
Regular dodecagon	12	$11.19625\,a^2$	$1.86603\,a$	$1.93185\,a$

If a_k denotes the side of a k-sided regular polygon inscribed in a circle of radius R, we have

$$a_{2k} = \sqrt{2R^2 - R\sqrt{4R^2 - a_k^2}}. \tag{4.5.11}$$

If A_k denotes the side of a k-sided regular polygon circumscribed about the same circle,

$$A_{2k} = \frac{2RA_k}{2R + \sqrt{4R^2 + A_k^2}}. \tag{4.5.12}$$

In particular,

$$A_{2k} = \frac{a_k A_k}{a_k + A_k}, \quad a_{2k} = \sqrt{\frac{a_k A_{2k}}{2}}. \tag{4.5.13}$$

The areas s_k, s_{2k}, S_k, and S_{2k} of the same polygons satisfy

$$s_{2k} = \sqrt{s_k S_k}, \quad S_{2k} = \frac{2s_{2k}S_k}{s_{2k} + S_k}. \tag{4.5.14}$$

4.6 CONICS

A *conic* (or *conic section*) is a plane curve that can be obtained by intersecting a right circular cone (page 361) with a plane that does not go through the vertex of the cone. There are three possibilities, depending on the relative positions of the cone and the plane (Figure 4.14). If no line of the cone is parallel to the plane, then the intersection is a closed curve, called an *ellipse*. If one line of the cone is parallel to the plane, the intersection is an open curve whose two ends are asymptotically parallel; this is called a *parabola*. Finally, there may be two lines in the cone parallel to the plane; the curve in this case has two open segments, and is called a *hyperbola*.

FIGURE 4.14
A section of a cone by a plane can yield an ellipse (left), a parabola (middle) or a hyperbola (right).

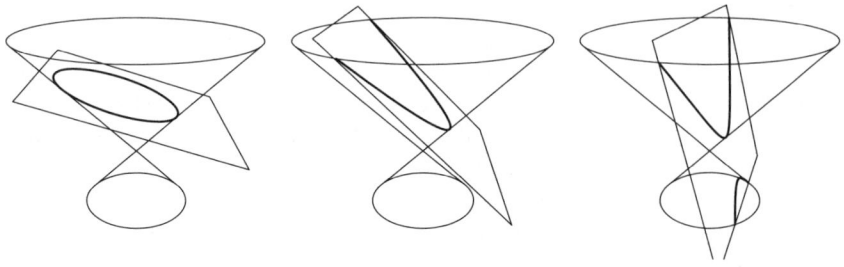

4.6.1 ALTERNATIVE CHARACTERIZATION

Assume given a point F in the plane, a line d not going through F, and a positive real number e. The set of points P such that the distance PF is e times the distance from P to d (measured along a perpendicular) is a conic. We call F the *focus*, d the *directrix*, and e the *eccentricity* of the conic. If $e < 1$ we have an ellipse, if $e = 1$ a parabola, and if $e > 1$ a hyperbola (Figure 4.15). This construction gives all conics except the circle, which is a particular case of the ellipse according to the earlier definition (we can recover it by taking the limit $e \to 0$).

For any conic, a line perpendicular to d and passing through F is an axis of symmetry. The ellipse and the hyperbola have an additional axis of symmetry, perpendicular to the first, so that there is an alternate focus and directrix, F' and d', obtained as the reflection of F and d with respect to this axis. (By contrast, the focus and directrix are uniquely defined for a parabola.)

The simplest analytic form for the ellipse and hyperbola is obtained when the two symmetry axes coincide with the coordinate axes. The ellipse in Figure 4.16 has equation

$$\frac{x^2}{a^2} + \frac{y^2}{b^2} = 1,$$

(4.6.1)

FIGURE 4.15

Definition of conics by means of the ratio (eccentricity) between the distance to a point and the distance to a line. On the left, e = .7 (ellipse); in the middle, e = 1 (parabola); on the right, e = 2 (hyperbola).

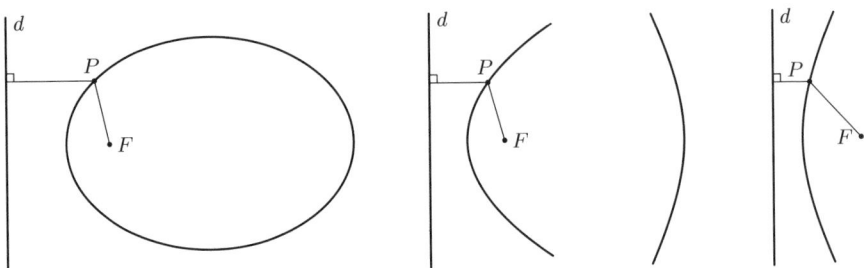

with $b < a$. The x-axis is the *major axis*, and the y-axis is the *minor axis*. These names are also applied to the segments, determined on the axes by the ellipse, and to the lengths of these segments: $2a$ for the major axis and $2b$ for the minor. The *vertices* are the intersections of the major axis with the ellipse and have coordinates $(a, 0)$ and $(-a, 0)$. The distance from the center to either *focus* is $\sqrt{a^2 - b^2}$, and the sum of the distances from a point in the ellipse to the foci is $2a$. The *latera recta* (in the singular, *latus rectum*) are the chords perpendicular to the major axis and going through the foci; their length is $2b^2/a$. The *eccentricity* is $\sqrt{a^2 - b^2}/a$. All ellipses of the same eccentricity are similar; in other words, the shape of an ellipse depends only on the ratio b/a. The distance from the center to either *directrix* is $a^2/\sqrt{a^2 - b^2}$.

The hyperbola in Figure 4.17 has equation

$$\frac{x^2}{a^2} - \frac{y^2}{b^2} = 1. \tag{4.6.2}$$

FIGURE 4.16

Ellipse with major semiaxis a and minor semiaxis b. Here $b/a = 0.6$.

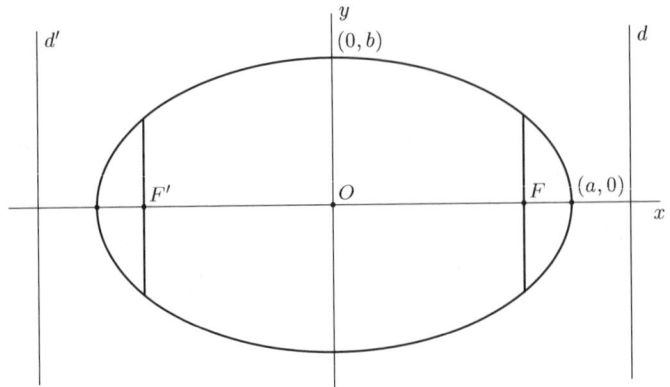

FIGURE 4.17
Hyperbola with transverse semiaxis a and conjugate semiaxis b. Here $b/a = 0.4$.

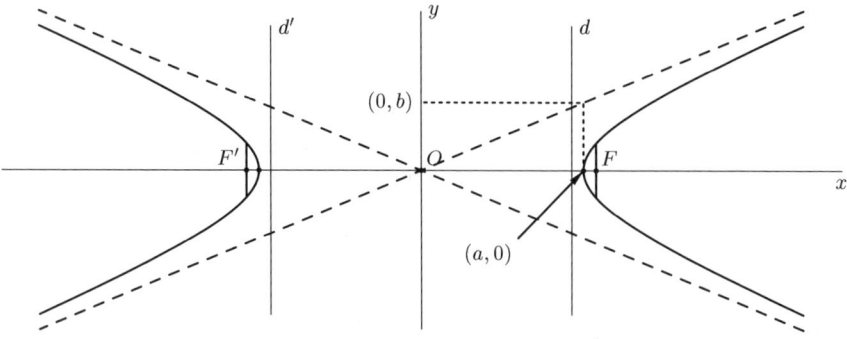

The x-axis is the *transverse axis*, and the y-axis is the *conjugate axis*. The *vertices* are the intersections of the transverse axis with the hyperbola and have coordinates $(a, 0)$ and $(-a, 0)$. The segment thus determined, or its length $2a$, is also called the transverse axis, while the length $2b$ is also called the conjugate axis. The distance from the center to either *focus* is $\sqrt{a^2 + b^2}$, and the difference between the distances from a point in the hyperbola to the foci is $2a$. The *latera recta* are the chords perpendicular to the transverse axis and going through the foci; their length is $2b^2/a$. The *eccentricity* is $\sqrt{a^2 + b^2}/a$. The distance from the center to either *directrix* is $a^2/\sqrt{a^2 + b^2}$. The legs of the hyperbola approach the *asymptotes*, lines of slope $\pm b/a$ that cross at the center.

All hyperbolas of the same eccentricity are similar; in other words, the shape of a hyperbola depends only on the ratio b/a. Unlike the case of the ellipse (where the major axis, containing the foci, is always longer than the minor axis), the two axes of a hyperbola can have arbitrary lengths. When they have the same length, so that $a = b$, the asymptotes are perpendicular, and $e = \sqrt{2}$, the hyperbola is called *rectangular*.

The simplest analytic form for the parabola is obtained when the axis of symmetry coincides with one coordinate axis, and the *vertex* (the intersection of the axis with the curve) is at the origin. The equation of the parabola on the right is

$$y^2 = 4ax, \qquad (4.6.3)$$

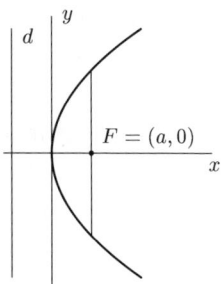

where a is the distance from the vertex to the focus, or, which is the same, from the vertex to the directrix. The *latus rectum* is the chord perpendicular to the axis and going through the focus; its length is $4a$. All parabolas are similar: they can be made identical by scaling, translation, and rotation.

4.6.2 THE GENERAL QUADRATIC EQUATION

The analytic equation for a conic in arbitrary position is the following:

$$Ax^2 + By^2 + Cxy + Dx + Ey + F = 0, \qquad (4.6.4)$$

where at least one of A, B, C is nonzero. To reduce this to one of the forms given previously, perform the following steps (note that the decisions are based on the most recent values of the coefficients, taken after all the transformations so far):

1. If $C \neq 0$, simultaneously perform the substitutions $x \mapsto qx + y$ and $y \mapsto qy - x$, where

$$q = \sqrt{\left(\frac{B - A}{C}\right)^2 + 1} \; + \; \frac{B - A}{C}. \qquad (4.6.5)$$

   Now $C = 0$. (This step corresponds to rotating and scaling about the origin.)

2. If $B = 0$, interchange x and y. Now $B \neq 0$.

3. If $E \neq 0$, perform the substitution $y \mapsto y - \frac{1}{2}(E/B)$. (This corresponds to translating in the y direction.) Now $E = 0$.

4. If $A = 0$:

   (a) If $D \neq 0$, perform the substitution $x \mapsto x - (F/D)$ (translation in the x direction), and divide the equation by B to get Equation (4.6.3). The conic is a *parabola*.

   (b) If $D = 0$, the equation gives a *degenerate conic*. If $F = 0$, we have the line $y = 0$ with multiplicity two. If $F < 0$, we have two parallel lines $y = \pm\sqrt{F/B}$. If $F > 0$ we have two imaginary lines; the equation has no solution within the real numbers.

5. If $A \neq 0$:

   (a) If $D \neq 0$, perform the substitution $x \mapsto x - \frac{1}{2}(D/A)$. Now $D = 0$. (This corresponds to translating in the x direction.)

   (b) If $F \neq 0$, divide the equation by F to get a form with $F = 1$.

      i. If A and B have opposite signs, the conic is a *hyperbola*; to get to Equation (4.6.2), interchange x and y, if necessary, so that A is positive; then make $a = 1/\sqrt{A}$ and $b = 1/\sqrt{B}$.

      ii. If A and B are both positive, the conic is an *ellipse*; to get to Equation (4.6.1), interchange x and y, if necessary, so that $A \leq B$, then make $a = 1/\sqrt{A}$ and $b = 1/\sqrt{B}$. The *circle* is the particular case $a = b$.

      iii. If A and B are both negative, we have an *imaginary ellipse*; the equation has no solution in real numbers.

(c) If $F = 0$, the equation again represents a *degenerate conic*: when A and B have different signs, we have a pair of lines $y = \pm\sqrt{-B/A}\,x$, and, when they have the same sign, we get a point (the origin).

EXAMPLE   We work out an example for clarity. Suppose the original equation is

$$4x^2 + y^2 - 4xy + 3x - 4y + 1 = 0. \tag{4.6.6}$$

In step 1 we apply the substitutions $x \mapsto 2x + y$ and $y \mapsto 2y - x$. This gives $25x^2 + 10x - 5y + 1 = 0$. Next we interchange x and y (step 2) and get $25y^2 + 10y - 5x + 1 = 0$. Replacing y by $y - \frac{1}{5}$ in step 3, we get $25y^2 - 5x = 0$. Finally, in step 4a we divide the equation by 25, thus giving it the form of Equation (4.6.3) with $a = \frac{1}{20}$. We have reduced the conic to a parabola with vertex at the origin and focus at $(\frac{1}{20}, 0)$. To locate the features of the original curve, we work our way back along the chain of substitutions (recall the convention about substitutions and transformations from Section 4.1.2):

Substitution	$y \mapsto y - \frac{1}{5}$	$\begin{array}{c} x \mapsto y \\ y \mapsto x \end{array}$	$\begin{array}{c} x \mapsto 2x + y \\ y \mapsto 2y - x \end{array}$
Vertex	$(0, 0)$	$(0, -\frac{1}{5})$	$(-\frac{1}{5}, 0)$	$(-\frac{2}{5}, -\frac{1}{5})$
Focus	$(\frac{1}{20}, 0)$	$(\frac{1}{20}, -\frac{1}{5})$	$(-\frac{1}{5}, \frac{1}{20})$	$(-\frac{7}{20}, \frac{6}{20})$

We conclude that the original curve, Equation (4.6.6), is a parabola with vertex $(-\frac{2}{5}, -\frac{1}{5})$ and focus $(-\frac{7}{20}, \frac{6}{20})$.

An alternative analysis of Equation (4.6.4) consists in forming the quantities

$$\Delta = \begin{vmatrix} A & \frac{1}{2}C & \frac{1}{2}D \\ \frac{1}{2}C & B & \frac{1}{2}E \\ \frac{1}{2}D & \frac{1}{2}E & F \end{vmatrix}, \quad J = \begin{vmatrix} A & \frac{1}{2}C \\ \frac{1}{2}C & B \end{vmatrix}, \quad I = A + B,$$

$$K = \begin{vmatrix} A & \frac{1}{2}D \\ \frac{1}{2}D & F \end{vmatrix} + \begin{vmatrix} B & \frac{1}{2}E \\ \frac{1}{2}E & F \end{vmatrix}, \tag{4.6.7}$$

and finding the appropriate case in the following table, where an entry in parentheses indicates that the equation has no solution in real numbers:

Δ	J	Δ/I	K	Type of conic
$\neq 0$	< 0			Hyperbola
$\neq 0$	0			Parabola
$\neq 0$	> 0	< 0		Ellipse
$\neq 0$	> 0	> 0		(Imaginary ellipse)
0	< 0			Intersecting lines
0	> 0			Point
0	0		< 0	Distinct parallel lines
0	0		> 0	(Imaginary parallel lines)
0	0		0	Coincident lines

For the central conics (the ellipse, the hyperbola, intersecting lines, and the point), the center (x_0, y_0) is the solution of the system of equations

$$2Ax + Cy + D = 0,$$
$$Cx + 2By + E = 0,$$

namely

$$(x_0, y_0) = \left(\frac{2BD - CE}{C^2 - 4AB}, \frac{2AE - CD}{C^2 - 4AB} \right),$$ (4.6.8)

and the axes have slopes q and $-1/q$, where q is given by Equation (4.6.5). (The value $-1/q$ can be obtained from Equation (4.6.5) by simply placing a minus sign before the radical.) The length of the semiaxis with slope q is

$$\sqrt{\frac{|\Delta|}{|Jr|}}, \quad \text{where } r = \frac{1}{2}(A + B + \sqrt{(B - A)^2 + C^2});$$ (4.6.9)

note that r is one of the eigenvalues of the matrix of which J is the determinant. To obtain the other semiaxis, take the other eigenvalue (change the sign of the radical in the expression of r just given).

EXAMPLE   Consider the equation $3x^2 + 4xy - 2y^2 + 3x - 2y + 7 = 0$. We have

$$\Delta = \begin{vmatrix} 6 & 4 & 3 \\ 4 & -4 & -2 \\ 3 & -2 & 14 \end{vmatrix} = -596 \neq 0$$

$$J = \begin{vmatrix} 6 & 4 \\ 4 & -4 \end{vmatrix} = -40 < 0$$

We conclude that this is a hyperbola.

4.6.3   ADDITIONAL PROPERTIES OF ELLIPSES

Let C be the ellipse with equation $x^2/a^2 + y^2/b^2 = 1$, with $a > b$, and let $F, F' = (\pm\sqrt{a^2 - b^2}, 0)$ be its foci (see Figure 4.16).

1. A *parametric representation* for C is given by $(a \cos \theta, b \sin \theta)$. The *area* of the shaded sector on the right is $\frac{1}{2}ab\theta = \frac{1}{2}ab \cos^{-1}(x/a)$. The *length* of the arc from $(a, 0)$ to the point $(a \cos \theta, b \sin \theta)$ is given by the elliptic integral

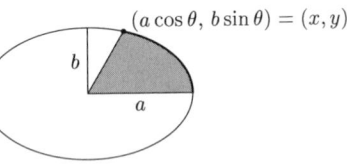

$(a \cos \theta, b \sin \theta) = (x, y)$

$$a \int_0^\theta \sqrt{1 - e^2 \cos^2 \phi} \, d\phi = a \left(E\left(\frac{\pi}{2}, e\right) - E\left(\frac{\pi}{2} - \theta, e\right) \right),$$ (4.6.10)

where e is the eccentricity. (See page 569 for elliptic integrals.) Setting $\theta = 2\pi$ results in

$$\text{area of } C = \pi ab, \quad \text{perimeter of } C = 4a\, E(\pi/2, e).$$ (4.6.11)

Note the approximation: perimeter of $C \approx 2a \left[2 + (\pi - 2) \left(\frac{b}{a}\right)^{1.456} \right]$.

2. Given an ellipse in the form $Ax^2 + By^2 + Cxy = 1$, form the matrix $D = \begin{bmatrix} A & C/2 \\ C/2 & B \end{bmatrix}$. Let the eigenvalues of D be $\{\lambda_1, \lambda_2\}$ and let $\{\mathbf{v}_1, \mathbf{v}_2\}$ be the corresponding unit eigenvectors (choose them orthogonal if $\lambda_1 = \lambda_2$). Then the major and minor semi-axes are given by $\mathbf{v}_i = \frac{\mathbf{u}_i}{\sqrt{\lambda_i}}$ and

(a) The area of the ellipse is $\dfrac{\pi}{\sqrt{\lambda_1 \lambda_2}} = \dfrac{2\pi}{\sqrt{4AB - C^2}}$.

(b) The ellipse has the parametric representation $\mathbf{x}(t) = \cos(t)\mathbf{v}_1 + \sin(t)\mathbf{v}_2$.

(c) The rectangle with vertices $(\pm\mathbf{v}_1, \pm\mathbf{v}_2)$ is tangent to the ellipse.

3. A *rational parametric representation* for C is given by $\left(a\,\dfrac{1 - t^2}{1 + t^2}, \dfrac{2bt}{1 + t^2} \right)$.

4. The *polar equation* for C in the usual polar coordinate system is

$$r = \frac{ab}{\sqrt{a^2 \sin^2 \theta + b^2 \cos^2 \theta}}. \tag{4.6.12}$$

With respect to a coordinate system with origin at a focus, the equation is

$$r = \frac{l}{1 \pm e \cos \theta}, \tag{4.6.13}$$

where $l = b^2/a$ is half the latus rectum. (Use the $+$ sign for the focus with positive x-coordinate and the $-$ sign for the focus with negative x-coordinate.)

5. Let P be any point of C. The *sum of the distances* PF and PF' is constant and equal to $2a$.

6. Let P be any point of C. Then the rays PF and PF' make the same angle with the tangent to C at P. Thus any light ray originating at F and reflected in the ellipse will go through F'.

7. Let T be any line tangent to C. The product of the distances from F and F' to T is constant and equals b^2.

8. *Lahire's theorem*: Let D and D' be fixed lines in the plane, and consider a third moving line on which three points P, P', and P'' are marked. If we constrain P to lie in D and P' to lie in D', then P'' describes an ellipse.

4.6.4   ADDITIONAL PROPERTIES OF HYPERBOLAS

Let C be the hyperbola with equation $x^2/a^2 - y^2/b^2 = 1$, and let

$$F, F' = (\pm\sqrt{a^2 + b^2}, 0) \tag{4.6.14}$$

be its foci (see Figure 4.17). The *conjugate hyperbola* of C is the hyperbola C' with equation $-x^2/a^2 + y^2/b^2 = 1$. It has the same asymptotes as C, the same axes (transverse and conjugate axes being interchanged), and its eccentricity e' is related to that of C by $e'^{-2} + e^{-2} = 1$.

1. A *parametric representation* for C is given by $(a \sec\theta, b\tan\theta)$. A different parametric representation, which gives one branch only, is $(a\cosh\theta, b\sinh\theta)$. The *area* of the shaded sector on the right is

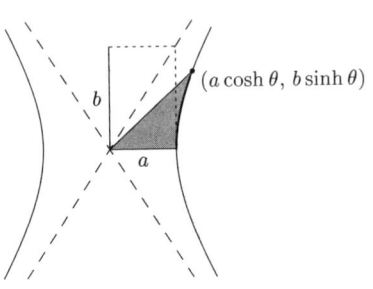

$$\tfrac{1}{2}ab\theta = \tfrac{1}{2}ab\cosh^{-1}(x/a)$$

$$= \tfrac{1}{2}ab\log\frac{x + \sqrt{x^2 - a^2}}{a}$$

where $x = a\cosh\theta$. The *length* of the arc from the point $(a,0)$ to the point $(a\cosh\theta, b\sinh\theta)$ is given by the elliptic integral

$$a\int_0^\theta \sqrt{e^2\cosh^2\phi - 1}\,d\phi = -bi\,E\!\left(\theta i, \frac{ea}{b}\right) = a\int_1^x \sqrt{\frac{e^2\xi^2 - a^2}{\xi^2 - a^2}}\,d\xi,$$

$$(4.6.15)$$

where e is the eccentricity and $i = \sqrt{-1}$.

2. A *rational parametric representation* for C is given by

$$\left(a\,\frac{1+t^2}{1-t^2}, \frac{2bt}{1-t^2}\right). \qquad (4.6.16)$$

3. The *polar equation* for C in the usual polar coordinate system is

$$r = \frac{ab}{\sqrt{a^2\sin^2\theta - b^2\cos^2\theta}}. \qquad (4.6.17)$$

With respect to a system with origin at a focus, the equation is

$$r = \frac{l}{1 \pm e\cos\theta}, \qquad (4.6.18)$$

where $l = b^2/a$ is half the latus rectum. (Use the $-$ sign for the focus with positive x-coordinate and the $+$ sign for the focus with negative x-coordinate.)

4. Let P be any point of C. The unsigned *difference between the distances PF and PF'* is constant and equal to $2a$.

5. Let P be any point of C. Then the rays PF and PF' make the same angle with the tangent to C at P. Thus any light ray originating at F and reflected in the hyperbola will appear to emanate from F'.

6. Let T be any line tangent to C. The product of the distances from F and F' to T is constant and equals b^2.

7. Let P be any point of C. The area of the parallelogram formed by the asymptotes and the parallels to the asymptotes going through P is constant and equals $\tfrac{1}{2}ab$.

8. Let L be any line in the plane. If L intersects C at P and P' and intersects the asymptotes at Q and Q', the distances PQ and $P'Q'$ are the same. If L is tangent to C we have $P = P'$, so that the point of tangency bisects the segment QQ'.

4.6.5 ADDITIONAL PROPERTIES OF PARABOLAS

Let C be the parabola with equation $y^2 = 4ax$, and let $F = (a, 0)$ be its focus.

1. Let $P = (x, y)$ and $P' = (x', y')$ be points on C. The area bounded by the chord PP' and the corresponding arc of the parabola is

$$\frac{|y' - y|^3}{24a}. \tag{4.6.19}$$

It equals four-thirds of the area of the triangle PQP', where Q is the point on C whose tangent is parallel to the chord PP' (formula due to *Archimedes*).

2. The *length* of the arc from $(0, 0)$ to the point (x, y) is

$$\frac{y}{4}\sqrt{4 + \frac{y^2}{a^2}} + a\sinh^{-1}\left(\frac{y}{2a}\right) = \frac{y}{4}\sqrt{4 + \frac{y^2}{a^2}} + a\log\frac{y + \sqrt{y^2 + 4a^2}}{2a}. \tag{4.6.20}$$

3. The *polar equation* for C in the usual polar coordinate system is

$$r = \frac{4a\cos\theta}{\sin^2\theta}. \tag{4.6.21}$$

With respect to a coordinate system with origin at F, the equation is

$$r = \frac{l}{1 - \cos\theta}, \tag{4.6.22}$$

where $l = 2a$ is half the latus rectum.

4. Let P be any point of C. Then the ray PF and the horizontal line through P make the same angle with the tangent to C at P. Thus light rays parallel to the axis and reflected in the parabola converge onto F (principle of the *parabolic reflector*).

4.6.6 CIRCLES

The set of points in a plane whose distance to a fixed point (the *center*) is a fixed positive number (the *radius*) is a *circle*. A circle of radius r and center (x_0, y_0) is described by the equation

$$(x - x_0)^2 + (y - y_0)^2 = r^2, \tag{4.6.23}$$

or

$$x^2 + y^2 - 2xx_0 - 2yy_0 + x_0^2 + y_0^2 - r^2 = 0. \tag{4.6.24}$$

Conversely, an equation of the form

$$x^2 + y^2 + 2dx + 2ey + f = 0 \tag{4.6.25}$$

defines a circle if $d^2 + e^2 > f$; the center is $(-d, -e)$ and the radius is $\sqrt{d^2 + e^2 - f}$.

Three points not on the same line determine a unique circle. If the points have coordinates (x_1, y_1), (x_2, y_2), and (x_3, y_3), then the equation of the circle is

$$\begin{vmatrix} x^2 + y^2 & x & y & 1 \\ x_1^2 + y_1^2 & x_1 & y_1 & 1 \\ x_2^2 + y_2^2 & x_2 & y_2 & 1 \\ x_3^2 + y_3^2 & x_3 & y_3 & 1 \end{vmatrix} = 0. \tag{4.6.26}$$

A *chord* of a circle is a line segment between two of its points (Figure 4.18). A *diameter* is a chord that goes through the center, or the length of such a chord (therefore the diameter is twice the radius). Given two points $P_1 = (x_1, y_1)$ and $P_2 = (x_2, y_2)$, there is a unique circle whose diameter is $P_1 P_2$; its equation is

$$(x - x_1)(x - x_2) + (y - y_1)(y - y_2) = 0. \tag{4.6.27}$$

The *length* or *circumference* of a circle of radius r is $2\pi r$, and the *area* is πr^2. The length of the *arc of circle* subtended by an angle θ, shown as s in Figure 4.18, is $r\theta$. (All angles are measured in radians.) Other relations between the radius, the arc length, the chord, and the areas of the corresponding *sector* and *segment* are, in the notation of Figure 4.18,

$$d = R\cos\tfrac{1}{2}\theta = \tfrac{1}{2}c\cot\tfrac{1}{2}\theta = \tfrac{1}{2}\sqrt{4R^2 - c^2},$$

$$c = 2R\sin\tfrac{1}{2}\theta = 2d\tan\tfrac{1}{2}\theta = 2\sqrt{R^2 - d^2} = \sqrt{4h(2R - h)},$$

$$\theta = \frac{s}{R} = 2\cos^{-1}\frac{d}{R} = 2\tan^{-1}\frac{c}{2d} = 2\sin^{-1}\frac{c}{2R},$$

$$\text{area of sector} = \tfrac{1}{2}Rs = \tfrac{1}{2}R^2\theta,$$

$$\text{area of segment} = \tfrac{1}{2}R^2(\theta - \sin\theta) = \tfrac{1}{2}(Rs - cd) = R^2\cos^{-1}\frac{d}{R} - d\sqrt{R^2 - d^2}$$

$$= R^2\cos^{-1}\frac{R - h}{R} - (R - h)\sqrt{2Rh - h^2}.$$

Other properties of circles:

1. If the central angle AOB equals θ, the angle ACB, where C is any point on the circle, equals $\tfrac{1}{2}\theta$ or $180° - \tfrac{1}{2}\theta$ (Figure 4.19, left). Conversely, given a segment AB, the set of points that "see" AB under a fixed angle is an arc of a circle (Figure 4.19, right). In particular, the set of points that see AB under a right angle is a circle with diameter AB.

2. Let P_1, P_2, P_3, P_4 be points in the plane, and let d_{ij}, for $1 \le i, j \le 4$, be the distance between P_i and P_j. A necessary and sufficient condition for all of the

FIGURE 4.18

The arc of a circle subtended by the angle θ is s; the chord is c; the sector is the whole slice of the pie; the segment is the cap bounded by the arc and the chord (that is, the slice minus the triangle).

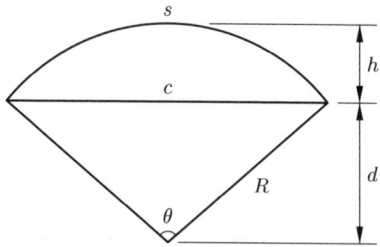

FIGURE 4.19

Left: the angle ACB equals $\frac{1}{2}\theta$ for any C in the long arc AB; ADB equals $180° - \frac{1}{2}\theta$ for any D in the short arc AB. Right: the locus of points, from which the segment AB subtends a fixed angle θ, is an arc of the circle.

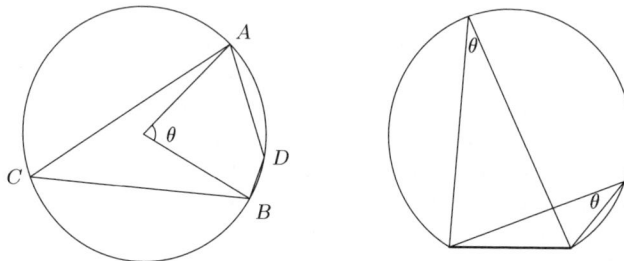

points to lie on the same circle (or line) is that one of the following equalities be satisfied:

$$\pm d_{12}d_{34} \pm d_{13}d_{24} \pm d_{14}d_{23} = 0. \tag{4.6.28}$$

This is equivalent to Ptolemy's formula for cyclic quadrilaterals (page 323).

3. In *oblique coordinates* with angle ω, a circle of center (x_0, y_0) and radius r is described by the equation

$$(x - x_0)^2 + (y - y_0)^2 + 2(x - x_0)(y - y_0)\cos\omega = r^2. \tag{4.6.29}$$

4. In *polar coordinates*, the equation for a circle centered at the pole and having radius a is $r = a$. The equation for a circle of radius a passing through the pole and with center at the point $(r, \theta) = (a, \theta_0)$ is $r = 2a\cos(\theta - \theta_0)$. The equation for a circle of radius a and with center at the point $(r, \theta) = (r_0, \theta_0)$ is

$$r^2 - 2r_0 r\cos(\theta - \theta_0) + r_0^2 - a^2 = 0. \tag{4.6.30}$$

5. If a line intersects a circle of center O at points A and B, the segments OA and OB make equal angles with the line. In particular, a tangent line is perpendicular to the radius that goes through the point of tangency.

6. Fix a circle and a point P in the plane, and consider a line through P that intersects the circle at A and B (with $A = B$ for a tangent). Then the product of the distances $PA \cdot PB$ is the same for all such lines. It is called the *power* of P with respect to the circle.

4.7 SPECIAL PLANE CURVES

4.7.1 ALGEBRAIC CURVES

Curves that can be given in implicit form as $f(x, y) = 0$, where f is a polynomial, are called *algebraic*. The degree of f is called the degree or *order* of the curve. Thus, conics (page 325) are algebraic curves of degree two. Curves of degree three already have a great variety of shapes, and only a few common ones will be given here.

The simplest case is the curve which is a graph of a polynomial of degree three: $y = ax^3 + bx^2 + cx + d$, with $a \neq 0$. This curve is a (general) *cubic parabola* (Figure 4.20), symmetric with respect to the point B where $x = -b/3a$.

The equation of a *semi-cubic parabola* (Figure 4.21, left) is $y^2 = kx^3$; by proportional scaling one can take $k = 1$. This curve should not be confused with the *cissoid of Diocles* (Figure 4.21, middle), whose equation is $(a - x)y^2 = x^3$ with $a \neq 0$.

The latter is asymptotic to the line $x = a$, whereas the semi-cubic parabola has no asymptotes. The cissoid's points are characterized by the equality $OP = AB$ in Figure 4.21, middle. One can take $a = 1$ by proportional scaling.

FIGURE 4.20
The general cubic parabola for $a > 0$. For $a < 0$, reflect in a horizontal line.

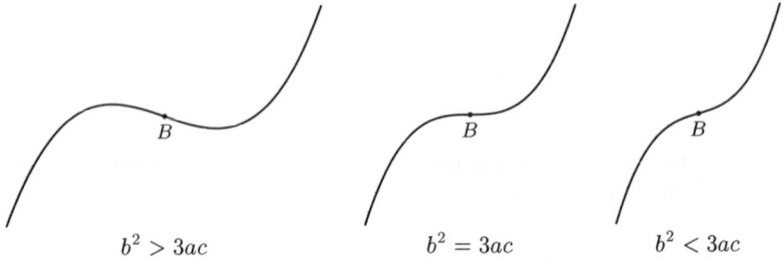

$b^2 > 3ac$ $b^2 = 3ac$ $b^2 < 3ac$

FIGURE 4.21

The semi-cubic parabola, the cissoid of Diocles, and the witch of Agnesi.

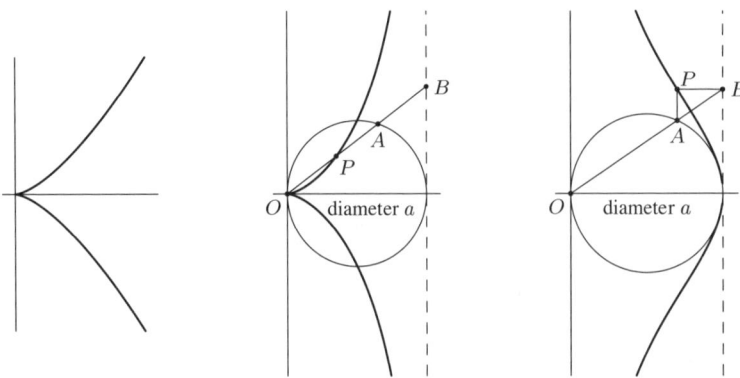

More generally, any curve of degree three with equation $(x - x_0)y^2 = f(x)$, where f is a polynomial, is symmetric with respect to the x-axis and asymptotic to the line $x = x_0$. In addition to the cissoid, the following particular cases are important:

1. The *witch of Agnesi* has equation $xy^2 = a^2(a - x)$, with $a \neq 0$, and is characterized by the geometric property shown in Figure 4.21, right. The same property provides the parametric representation $x = a\cos^2\theta$, $y = a\tan\theta$. Once more, proportional scaling reduces to the case $a = 1$.

2. The *folium of Descartes* (Figure 4.22, left) is described by equation $(x - a)y^2 = -x^2(\frac{1}{3}x + a)$, with $a \neq 0$ (reducible to $a = 1$ by proportional scaling). By rotating $135°$ (right) we get the alternative and more familiar equation $x^3 + y^3 = cxy$, where $c = \frac{1}{3}\sqrt{2}a$. The folium of Descartes is a *rational curve*, that is, it is parametrically represented by rational functions. In the tilted position, the equation is $x = ct/(1 + t^3)$, $y = ct^2/(1 + t^3)$ (so that $t = y/x$).

3. The *strophoid's* equation is $(x - a)y^2 = -x^2(x + a)$, with $a \neq 0$ (reducible to $a = 1$ by proportional scaling). It satisfies the property $AP = AP' = OA$ in Figure 4.22, right; this means that POP' is a right angle. The strophoid's polar representation is $r = -a\cos 2\theta \sec\theta$, and the rational parametric representation is $x = a(t^2 - 1)/(t^2 + 1)$, $y = at(t^2 - 1)/(t^2 + 1)$ (so that $t = y/x$).

Among the important curves of degree four are the following:

1. A *Cassini's oval* is characterized by the following condition: Given two *foci* F and F', a distance $2a$ apart, a point P belongs to the curve if the product of the distances PF and PF' is a constant k^2. If the foci are on the x-axis and equidistant from the origin, the curve's equation is $(x^2 + y^2 + a^2)^2 - 4a^2x^2 = k^4$. Changes in a correspond to rescaling, while the value of k/a controls

FIGURE 4.22

The folium of Descartes in two positions, and the strophoid.

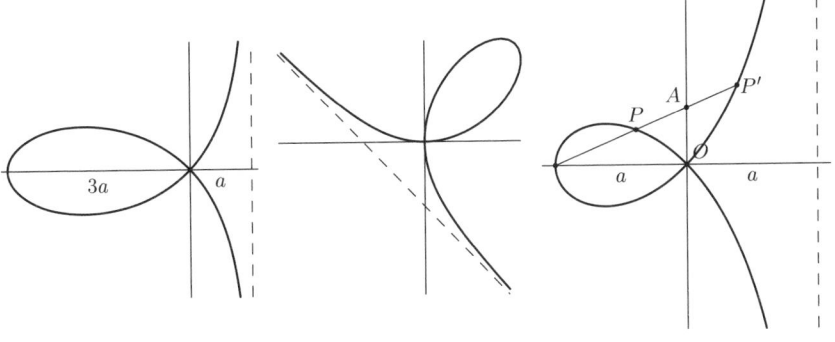

FIGURE 4.23

Cassini's ovals for $k = 0.5a$, $0.9a$, a, $1.1a$ and $1.5a$ (from the inside to the outside). The foci (dots) are at $x = a$ and $x = -a$. The black curve, $k = a$, is also called Bernoulli's lemniscate.

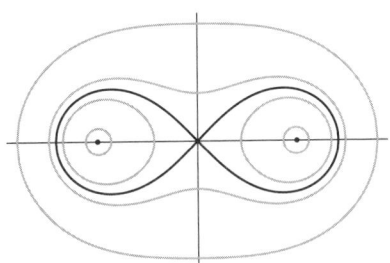

the shape: the curve has one smooth segment and one with a self-intersection, or two segments depending on whether k is greater than, equal to, or smaller than a (Figure 4.23). The case $k = a$ is also known as the *lemniscate* (of Jakob Bernoulli); the equation reduces to $(x^2 + y^2)^2 = a^2(x^2 - y^2)$, and upon a $45°$ rotation to $(x^2 + y^2)^2 = 2a^2xy$. Each Cassini oval is the section of a torus of revolution by a plane parallel to the axis of revolution.

2. A *conchoid of Nichomedes* is the set of points such that the signed distance AP in Figure 4.24, left, equals a fixed real number k (the line L and the origin O being fixed). If L is the line $x = a$, the conchoid's polar equation is $r = a \sec \theta + k$. Once more, a is a scaling parameter, and the value of k/a controls the shape: when $k > -a$ the curve is smooth, when $k = -a$ there is a cusp, and when $k < -a$ there is a self-intersection. The curves for k and $-k$ can also be considered two leaves of the same conchoid, with Cartesian equation $(x - a)^2(x^2 + y^2) = k^2x^2$.

3. A *limaçon of Pascal* is the set of points such that the distance AP in Figure 4.25, left, equals a fixed positive number k measured on either side (the

FIGURE 4.24

Defining property of the conchoid of Nichomedes (left), and curves for $k = \pm 0.5a$, $k = \pm a$, and $k = \pm 1.5a$ (right).

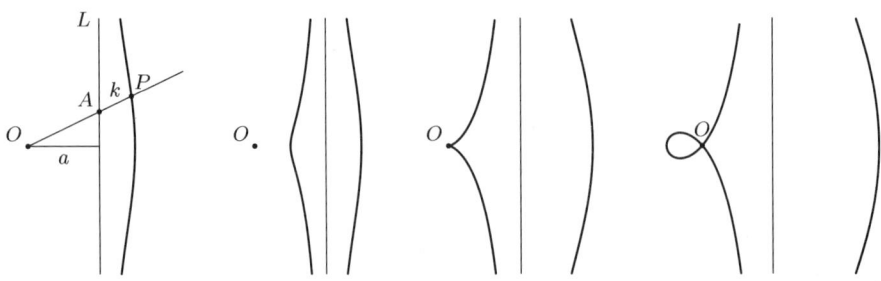

FIGURE 4.25

Defining property of the limaçon of Pascal (left), and curves for $k = 1.5a$, $k = a$, and $k = 0.5a$ (right). The middle curve is the cardioid; the one on the right a trisectrix.

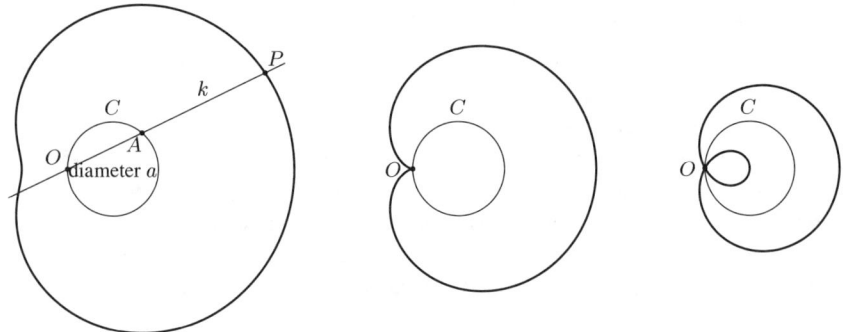

circle C and the origin O being fixed). If C has diameter a and center at $(0, \frac{1}{2}a)$, the limaçon's polar equation is $r = a\cos\theta + k$, and its Cartesian equation is

$$(x^2 + y^2 - ax)^2 = k^2(x^2 + y^2). \tag{4.7.1}$$

The value of k/a controls the shape, and there are two particularly interesting cases. For $k = a$, we get a *cardioid* (see also page 341). For $a = \frac{1}{2}k$, we get a curve that can be used to *trisect* an arbitrary angle α. If we draw a line L through the center of the circle C making an angle α with the positive x-axis, and if we call P the intersection of L with the limaçon $a = \frac{1}{2}k$, the line from O to P makes an angle with L equal to $\frac{1}{3}\alpha$.

Hypocycloids and epicycloids with rational ratios (see next section) are also algebraic curves, generally of higher degree.

FIGURE 4.26

Cycloid (top) and trochoids with $k = 0.5a$ and $k = 1.5a$, where k is the distance PQ from the center of the rolling circle to the pole.

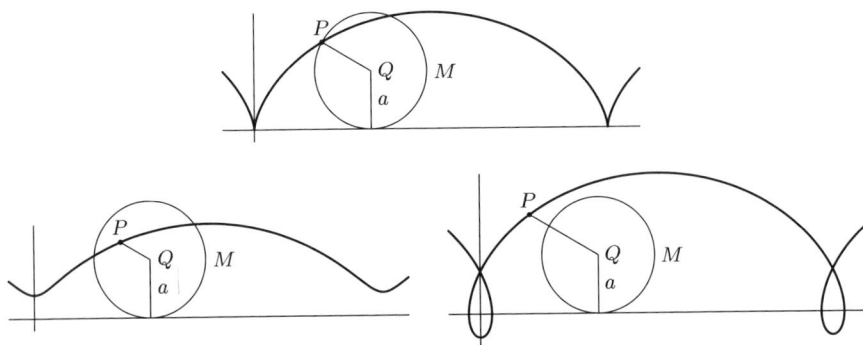

4.7.2 ROULETTES (SPIROGRAPH CURVES)

Suppose given a fixed curve C and a moving curve M, which rolls on C without slipping. The curve drawn by a point P kept fixed with respect to M is called a *roulette*, of which P is the *pole*.

The most important examples of roulettes arise when M is a circle and C is a straight line or a circle, but an interesting additional example is provided by the *catenary* $y = a \cosh(x/a)$, which arises by rolling the parabola $y = x^2/(4a)$ on the x-axis with pole the focus of the parabola (that is, $P = (0, a)$ in the initial position). The catenary is the shape taken under the action of gravity by a chain or string of uniform density whose ends are held in the air.

A circle rolling on a straight line gives a *trochoid*, with the *cycloid* as a special case when the pole P lies on the circle (Figure 4.26). If the moving circle M has radius a and the distance from the pole P to the center of M is k, the trochoid's parametric equation is

$$x = a\phi - k \sin \phi, \qquad y = a - k \cos \phi. \tag{4.7.2}$$

The cycloid, therefore, has the parametric equation

$$x = a(\phi - \sin \phi), \qquad y = a(1 - \cos \phi). \tag{4.7.3}$$

One can eliminate ϕ to get x as a (multivalued) function of y, which takes the following form for the cycloid:

$$x = \pm \left(a \cos^{-1} \left(\frac{a - y}{a} \right) - \sqrt{2ay - y^2} \right) \tag{4.7.4}$$

The length of one arch of the cycloid is $8a$, and the area under the arch is $3\pi a^2$.

A trochoid is also called a *curtate cycloid* when $k < a$ (that is, when P is inside the circle) and a *prolate cycloid* when $k > a$.

FIGURE 4.27

Left: initial configuration for epitrochoid (black) and configuration at parameter value θ (gray). Middle: epicycloid with $b = \frac{1}{2}a$ (nephroid). Right: epicycloid with $b = a$ (cardioid).

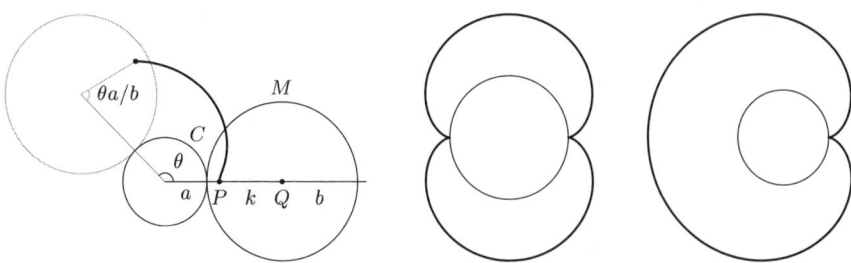

A circle rolling on another circle and exterior to it gives an *epitrochoid*. If a is the radius of the fixed circle, b that of the rolling circle, and k is the distance from P to the center of the rolling circle, the parametric equation of the epitrochoid is

$$x = (a + b)\cos\theta - k\cos((1 + b/a)\theta), \qquad y = (a + b)\sin\theta - k\sin((1 + b/a)\theta). \tag{4.7.5}$$

These equations assume that, at the start, everything is aligned along the positive x-axis, as in Figure 4.27, left. Usually one considers the case when a/b is a rational number, say $a/b = p/q$ where p and q are relatively prime. Then the rolling circle returns to its original position after rotating q times around the fixed circle, and the epitrochoid is a closed curve—in fact, an algebraic curve. One also usually takes $k = b$, so that P lies on the rolling circle; the curve in this case is called an *epicycloid*. The middle diagram in Figure 4.27 shows the case $b = k = \frac{1}{2}a$, called the *nephroid*; this curve is the cross section of the caustic of a spherical mirror. The diagram on the right shows the case $b = k = a$, which gives the cardioid (compare to Figure 4.25, middle).

Hypotrochoids and *hypocycloids* are defined in the same way as epitrochoids and epicycloids, but the rolling circle is inside the fixed one. The parametric equation of the hypotrochoid is

$$x = (a - b)\cos\theta + k\cos((a/b - 1)\theta), \qquad y = (a - b)\sin\theta - k\sin((a/b - 1)\theta), \tag{4.7.6}$$

where the letters have the same meaning as for the epitrochoid. Usually one takes a/b rational and $k = b$. There are several interesting particular cases:

- $b = k = a$ gives a point.
- $b = k = \frac{1}{2}a$ gives a diameter of the circle C.
- $b = k = \frac{1}{3}a$ gives the *deltoid* (Figure 4.28, left), whose algebraic equation is

$$(x^2 + y^2)^2 - 8ax^3 + 24axy^2 + 18a^2(x^2 + y^2) - 27a^4 = 0. \tag{4.7.7}$$

FIGURE 4.28
The hypocycloids with $a = 3b$ (deltoid) and $a = 4b$ (astroid).

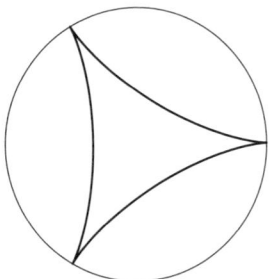
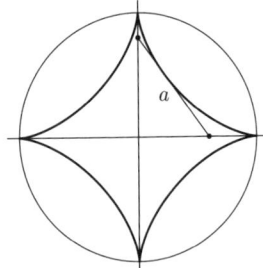

- $b = k = \frac{1}{4}a$ gives the *astroid* (Figure 4.28, right), an algebraic curve of degree six whose equation can be reduced to $x^{2/3} + y^{2/3} = a^{2/3}$. The figure illustrates another property of the astroid: its tangent intersects the coordinate axes at points that are always the same distance a apart. Otherwise said, the astroid is the envelope of a moving segment of fixed length whose endpoints are constrained to lie on the two coordinate axes.

4.7.3   CURVES IN POLAR COORDINATES

polar equation	type of curve	
$r = a$	circle	
$r = a\cos\theta$	circle	
$r = a\sin\theta$	circle	
$r^2 - 2br\cos(\theta - \beta) + (b^2 - a^2) = 0$	circle at (b, β) of radius a	
$r = \dfrac{k}{1 - e\cos\theta}$	$\begin{cases} e = 1 & \text{parabola} \\ 0 < e < 1 & \text{ellipse} \\ e > 1 & \text{hyperbola} \end{cases}$	

4.7.4   SPIRALS

A number of interesting curves have polar equation $r = f(\theta)$, where f is a monotonic function (always increasing or decreasing). This property leads to a spiral shape. The *logarithmic spiral* or *Bernoulli spiral* (Figure 4.29, left) is self-similar: by rotation the curve can be made to match any scaled copy of itself. Its equation is $r = ke^{a\theta}$; the angle between the radius from the origin and the tangent to the curve is constant and equal to $\phi = \cot^{-1} a$. A curve parameterized by arc length and such that the radius of curvature is proportional to the parameter at each point is a Bernoulli spiral.

In the *Archimedean spiral* or *linear spiral* (Figure 4.29, middle), the spacing between intersections along a ray from the origin is constant. The equation of this spiral is $r = a\theta$; by scaling one can take $a = 1$. It has an inner endpoint, in contrast

FIGURE 4.29

The Bernoulli or logarithmic spiral (left), the Archimedes or linear spiral (middle), and the Cornu spiral (right).

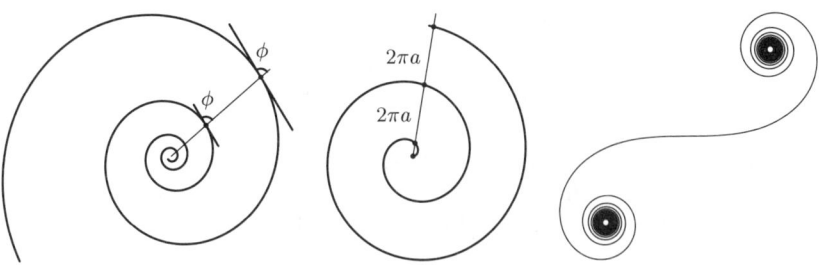

with the logarithmic spiral, which spirals down to the origin without reaching it. The *Cornu spiral* or *clothoid* (Figure 4.29, right), important in optics and engineering, has the following parametric representation in Cartesian coordinates:

$$X = aC(t) = a \int_0^t \cos(\tfrac{1}{2}\pi s^2)\, ds, \qquad y = aS(t) = a \int_0^t \sin(\tfrac{1}{2}\pi s^2)\, ds. \quad (4.7.8)$$

(C and S are the so-called Fresnel integrals; see page 547.) A curve parameterized by arc length and such that the radius of curvature is inversely proportional to the parameter at each point is a Cornu spiral (compare to the Bernoulli spiral).

4.7.5 THE PEANO CURVE AND FRACTAL CURVES

There are curves (in the sense of continuous maps from the real line to the plane) that completely cover a two-dimensional region of the plane. We give a construction of such a *Peano curve*, adapted from David Hilbert's example. The construction is inductive and is based on replacement rules. We consider building blocks of six

shapes: ⌣ ⌣ ⌣ ⌣ ◖— , the length of the straight segments being twice the radius of the curved ones. A sequence of these patterns, end-to-end, represents a curve, if we disregard the gray and black half-disks. The replacement rules are the following:

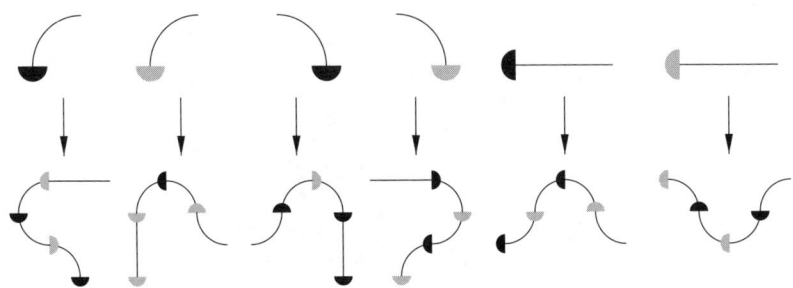

The rules are applied taking into account the way each piece is turned. Here we apply the replacement rules to a particular initial pattern:

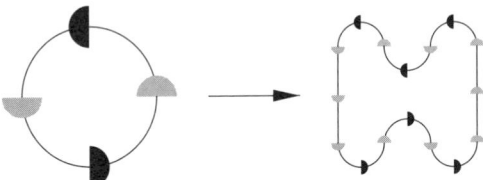

(We scale the result so it has the same size as the original.) Applying the process repeatedly gives, in the limit, the Peano curve. Note that the sequence converges uniformly and thus the limit function is continuous. Here are the first five steps:

The same idea of replacement rules leads to many interesting fractal, and often self-similar, curves. For example, the substitution ⎯⎯ → ⎯⋀⎯ leads to the *Koch snowflake* when applied to an initial equilateral triangle, like this (the first three stages and the sixth are shown):

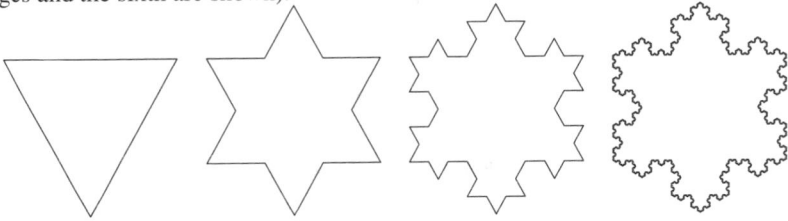

4.7.6 FRACTAL OBJECTS

Given an object X, if $n(\epsilon)$ open sets of diameter of ϵ are required to cover X, then the capacity dimension of X is

$$d_{\text{capacity}} = \lim_{\epsilon \to 0} \frac{\ln n(\epsilon)}{\ln \epsilon} \qquad (4.7.9)$$

indicating that $n(\epsilon)$ scales as $\epsilon^{d_{\text{capacity}}}$. Note that

$$d_{\text{correlation}} \leq d_{\text{information}} \leq d_{\text{capacity}}. \qquad (4.7.10)$$

The capacity dimension of various objects:

Object	Dimension	Object	Dimension
Logistic equation	0.538	Sierpiński sieve	$\frac{\ln 3}{\ln 2} \approx 1.5850$
Cantor set	$\frac{\ln 2}{\ln 3} \approx 0.6309$	Pentaflake	$\frac{\ln 2 + \ln 3}{\ln(1+\phi)3} \approx 1.8617$
Koch snowflake	$\frac{2\ln 2}{\ln 3} \approx 1.2619$	Sierpiński carpet	$\frac{3\ln 2}{\ln 3} \approx 1.8928$
Cantor dust	$\frac{\ln 5}{\ln 3} \approx 1.4650$	Tetrix	2
Minkowski sausage	$\frac{3}{2} = 1.5$	Menger sponge	$\frac{2\ln 2 + \ln 5}{\ln 3} \approx 2.7268$

4.7.7 CLASSICAL CONSTRUCTIONS

The ancient Greeks used straightedges and compasses to find the solutions to numerical problems. For example, they found square roots by constructing the geometric mean of two segments. Three famous problems that have been proved intractable by this method are:

1. The trisection of an arbitrary angle.
2. The squaring of the circle (the construction of a square whose area is equal to that of a given circle).
3. The doubling of the cube (the construction of a cube with double the volume of a given cube).

A regular polygon inscribed in the unit circle can be constructed by straightedge and compass alone if, and only if, n has the form $n = 2^\ell p_1 p_2 \dots p_k$, where ℓ is a nonnegative integer and the $\{p_i\}$ are zero or more distinct Fermat primes (primes of the form $2^{2^m} + 1$). The only known Fermat primes are for 3, 5, 17, 257, and 65537, corresponding to $m = 0, 1, 2, 3, 4$. Thus, regular polygons can be constructed for $n = 3, 4, 5, 6, 8, 10, 12, 15, 16, 17, 20, 24, \dots, 257, \dots$.

4.8 COORDINATE SYSTEMS IN SPACE

4.8.0.1 Conventions

When we talk about "the point with coordinates (x, y, z)" or "the surface with equation $f(x, y, z)$", we always mean Cartesian coordinates. If a formula involves another type of coordinates, this fact will be stated explicitly. Note that Section 4.1.2 has information on substitutions and transformations relevant to the three-dimensional case.

4.8.1 CARTESIAN COORDINATES IN SPACE

In *Cartesian coordinates* (or *rectangular coordinates*), a point P is referred to by three real numbers, indicating the positions of the perpendicular projections from the point to three fixed, perpendicular, graduated lines, called the *axes*. If the coordinates are denoted x, y, z, in that order, the axes are called the *x-axis*, etc., and we write $P = (x, y, z)$. Often the x-axis is imagined to be horizontal and pointing roughly toward the viewer (out of the page), the y-axis also horizontal and pointing more or less to the right, and the z-axis vertical, pointing up. The system is called *right-handed* if it can be rotated so the three axes are in this position. Figure 4.30 shows a right-handed system. The point $x = 0$, $y = 0$, $z = 0$ is the *origin*, where the three axes intersect.

FIGURE 4.30

In Cartesian coordinates, $P = (4.2, 3.4, 2.2)$.

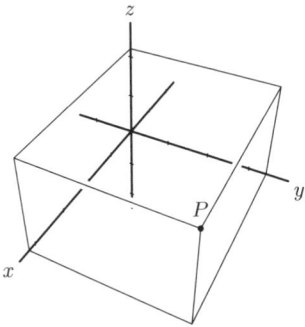

FIGURE 4.31

Among the possible sets (r, θ, z) of cylindrical coordinates for P are $(10, 30°, 5)$ and $(10, 390°, 5)$.

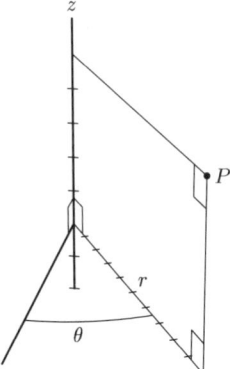

4.8.2   CYLINDRICAL COORDINATES IN SPACE

To define *cylindrical coordinates*, we take an axis (usually called the *z-axis*) and a perpendicular plane, on which we choose a ray (the *initial ray*) originating at the intersection of the plane and the axis (the *origin*). The coordinates of a point P are the polar coordinates (r, θ) of the projection of P on the plane, and the coordinate z of the projection of P on the axis (Figure 4.31). See Section 4.1.4 for remarks on the values of r and θ.

4.8.3   SPHERICAL COORDINATES IN SPACE

To define *spherical coordinates*, we take an axis (the *polar axis*) and a perpendicular plane (the *equatorial plane*), on which we choose a ray (the *initial ray*) originating at the intersection of the plane and the axis (the *origin O*). The coordinates of a point P are the distance ρ from P to the origin, the angle ϕ (*zenith*) between the line OP

and the positive polar axis, and the angle θ (*azimuth*) between the initial ray and the projection of OP to the equatorial plane. See Figure 4.32. As in the case of polar and cylindrical coordinates, θ is only defined up to multiples of $360\,°$, and likewise ϕ. Usually ϕ is assigned a value between 0 and $180\,°$, but values of ϕ between $180\,°$ and $360\,°$ can also be used; the triples (ρ, ϕ, θ) and $(\rho,\ 360\,° - \phi,\ 180\,° + \theta)$ represent the same point. Similarly, one can extend ρ to negative values; the triples (ρ, ϕ, θ) and $(-\rho,\ 180\,° - \phi,\ 180\,° + \theta)$ represent the same point.

FIGURE 4.32
A set of spherical coordinates for P is $(\rho, \theta, \phi) = (10, 60°, 30°)$.

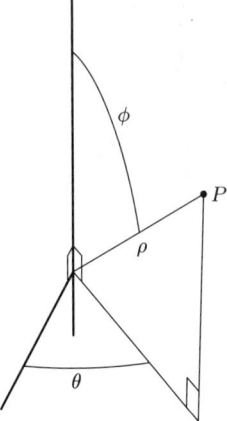

FIGURE 4.33
Standard relations between Cartesian, cylindrical, and spherical coordinate systems. The origin is the same for all three. The positive z-axes of the Cartesian and cylindrical systems coincide with the positive polar axis of the spherical system. The initial rays of the cylindrical and spherical systems coincide with the positive x-axis of the Cartesian system, and the rays $\theta = 90°$ coincide with the positive y-axis.

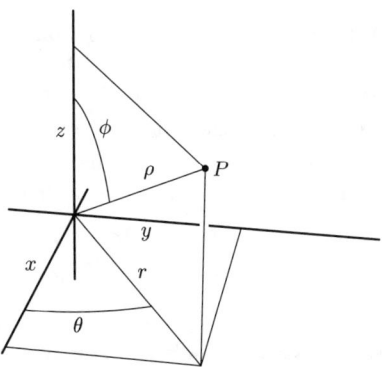

4.8.4 RELATIONS BETWEEN CARTESIAN, CYLINDRICAL, AND SPHERICAL COORDINATES

Consider a Cartesian, a cylindrical, and a spherical coordinate system, related as shown in Figure 4.33. The Cartesian coordinates (x, y, z), the cylindrical coordinates (r, θ, z), and the spherical coordinates (ρ, ϕ, θ) of a point are related as follows (where the $\tan^{-1}$ function must be interpreted correctly in all quadrants):

$$
\text{cart} \leftrightarrow \text{cyl} \quad
\begin{cases}
x = r \cos \theta, \\
y = r \sin \theta, \\
z = z,
\end{cases}
\quad
\begin{cases}
r = \sqrt{x^2 + y^2}, \\
\theta = \tan^{-1} \dfrac{y}{x}, \\
z = z,
\end{cases}
\quad
\begin{cases}
\sin \theta = \dfrac{y}{\sqrt{x^2 + y^2}}, \\
\cos \theta = \dfrac{x}{\sqrt{x^2 + y^2}}, \\
z = z.
\end{cases}
$$

$$
\text{cyl} \leftrightarrow \text{sph} \quad
\begin{cases}
r = \rho \sin \phi, \\
z = \rho \cos \phi, \\
\theta = \theta,
\end{cases}
\quad
\begin{cases}
\rho = \sqrt{r^2 + z^2}, \\
\phi = \tan^{-1} \dfrac{r}{z}, \\
\theta = \theta,
\end{cases}
\quad
\begin{cases}
\sin \phi = \dfrac{r}{\sqrt{r^2 + z^2}}, \\
\cos \phi = \dfrac{z}{\sqrt{r^2 + z^2}}, \\
\theta = \theta.
\end{cases}
$$

$$
\text{cart} \leftrightarrow \text{sph} \quad
\begin{cases}
x = \rho \cos \theta \sin \phi, \\
y = \rho \sin \theta \sin \phi, \\
z = \rho \cos \phi,
\end{cases}
\quad
\begin{cases}
\rho = \sqrt{x^2 + y^2 + z^2}, \\
\theta = \tan^{-1} \dfrac{y}{x}, \\
\phi = \tan^{-1} \dfrac{\sqrt{x^2 + y^2}}{z} \\
\quad = \cos^{-1} \dfrac{z}{\sqrt{x^2 + y^2 + z^2}}.
\end{cases}
$$

4.8.5 HOMOGENEOUS COORDINATES IN SPACE

A quadruple of real numbers $(x : y : z : t)$, with $t \neq 0$, is a set of *homogeneous coordinates* for the point P with Cartesian coordinates $(x/t, y/t, z/t)$. Thus the same point has many sets of homogeneous coordinates: $(x : y : z : t)$ and $(x' : y' : z' : t')$ represent the same point if, and only if, there is some real number α such that $x' = \alpha x$, $y' = \alpha y$, $z' = \alpha z$, $t' = \alpha t$. If P has Cartesian coordinates (x_0, y_0, z_0), one set of homogeneous coordinates for P is $(x_0, y_0, z_0, 1)$.

Section 4.1.5 has more information on the relationship between Cartesian and homogeneous coordinates. Section 4.9.2 has formulae for space transformations in homogeneous coordinates.

4.9   SPACE SYMMETRIES OR ISOMETRIES

A transformation of space (invertible map of space to itself) that preserves distances is called an *isometry* of space. Every isometry of space is a composition of transformations of the following types:

1. The *identity* (which leaves every point fixed)
2. A *translation* by a vector **v**
3. A *rotation* through an angle α around a line L
4. A *screw motion* through an angle α around a line L, with displacement d
5. A *reflection* in a plane P
6. A *glide-reflection* in a plane P with displacement vector **v**
7. A *rotation-reflection* (rotation through an angle α around a line L composed with reflection in a plane perpendicular to L).

   The identity is a particular case of a translation and of a rotation; rotations are particular cases of screw motions; reflections are particular cases of glide-reflections. However, as in the plane case, it is more intuitive to consider each case separately.

4.9.1   FORMULAE FOR SYMMETRIES: CARTESIAN COORDINATES

In the formulae below, multiplication between a matrix and a triple of coordinates should be carried out regarding the triple as a column vector (or a matrix with three rows and one column).

1. *Translation* by (x_0, y_0, z_0):
$$(x, y, z) \mapsto (x + x_0, \, y + y_0, \, z + z_0). \qquad (4.9.1)$$

2. *Rotation* through α (counterclockwise) around the line through the origin with direction cosines a, b, c (see page 353): $(x, y, z) \mapsto M(x, y, z)$, where M is the matrix
$$\begin{bmatrix} a^2(1 - \cos\alpha) + \cos\alpha & ab(1 - \cos\alpha) - c\sin\alpha & ac(1 - \cos\alpha) + b\sin\alpha \\ ab(1 - \cos\alpha) + c\sin\alpha & b^2(1 - \cos\alpha) + \cos\alpha & bc(1 - \cos\alpha) - a\sin\alpha \\ ac(1 - \cos\alpha) - b\sin\alpha & bc(1 - \cos\alpha) + a\sin\alpha & c^2(1 - \cos\alpha) + \cos\alpha \end{bmatrix}.$$
$$(4.9.2)$$

3. *Rotation* through α (counterclockwise) around the line with direction cosines a, b, c through an arbitrary point (x_0, y_0, z_0):
$$(x, y, z) \mapsto (x_0, y_0, z_0) + M(x - x_0, \, y - y_0, \, z - z_0), \qquad (4.9.3)$$
where M is given by Equation (4.9.2).

4. *Arbitrary rotations and Euler angles*: Any rotation of space fixing the origin can be decomposed as a rotation by ϕ about the z-axis, followed by a rotation by θ about the y-axis, followed by a rotation by ψ about the z-axis. The numbers ϕ, θ, and ψ are called the *Euler angles* of the composite rotation, which acts as: $(x, y, z) \mapsto M(x, y, z)$, where M is the matrix given by
$$\begin{bmatrix} \cos\phi\cos\theta\cos\psi - \sin\phi\sin\psi & -\sin\phi\cos\theta\cos\psi - \cos\phi\sin\psi & \sin\theta\cos\psi \\ \cos\phi\cos\theta\sin\psi + \sin\phi\cos\psi & -\sin\phi\cos\theta\sin\psi + \cos\phi\cos\psi & \sin\theta\sin\psi \\ -\cos\phi\sin\theta & \sin\phi\sin\theta & \cos\theta \end{bmatrix}.$$
$$(4.9.4)$$

FIGURE 4.34
The coordinate rays Ox, Oy, Oz, together with their images $O\xi$, $O\eta$, $O\zeta$ under a rotation, fix the Euler angles associated with that rotation, as follows: $\theta = zO\zeta$, $\psi = xOr = yOs$, and $\phi = sO\eta$. (Here the ray Or is the projection of $O\zeta$ to the xy-plane. The ray Os is determined by the intersection of the xy- and $\xi\eta$-planes.)

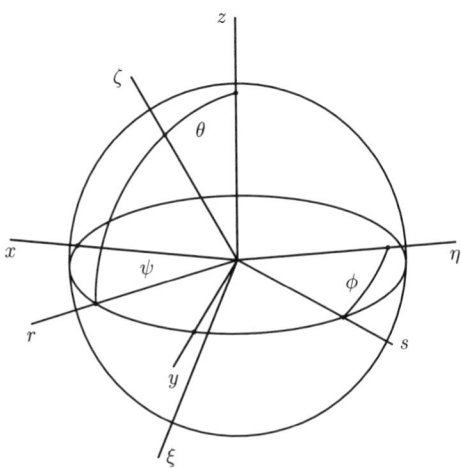

(An alternative decomposition, more natural if we think of the coordinate system as a rigid trihedron that rotates in space, is the following: a rotation by ψ about the z-axis, followed by a rotation by θ about the *rotated* y-axis, followed by a rotation by ϕ about the *rotated* z-axis. Note that the order is reversed.)

Provided that θ is not a multiple of $180°$, the decomposition of a rotation in this form is unique (apart from the ambiguity arising from the possibility of adding a multiple of $360°$ to any angle). Figure 4.34 shows how the Euler angles can be read off geometrically.

Warning: Some references define Euler angles differently; the most common variation is that the second rotation is taken about the x-axis instead of about the y-axis.

5. *Screw motion* with angle α and displacement d around the line with direction cosines a, b, c through an arbitrary point (x_0, y_0, z_0):

$$(x, y, z) \mapsto (x_0 + ad, y_0 + bd, z_0 + cd) + M(x - x_0, y - y_0, z - z_0), \quad (4.9.5)$$

where M is given by (4.9.2).

6. *Reflection*

$$\begin{aligned}
\text{in the } xy\text{-plane:} \quad & (x, y, z) \mapsto (x, y, -z). \\
\text{in the } xz\text{-plane:} \quad & (x, y, z) \mapsto (x, -y, z). \\
\text{in the } yz\text{-plane:} \quad & (x, y, z) \mapsto (-x, y, z).
\end{aligned} \qquad (4.9.6)$$

7. *Reflection* in a plane with equation $ax + by + cz + d = 0$:

$$(x, y, z) \mapsto \frac{1}{a^2 + b^2 + c^2} \left(M(x_0, y_0, z_0) - (2ad, 2bd, 2cd) \right), \qquad (4.9.7)$$

where M is the matrix

$$M = \begin{bmatrix} -a^2 + b^2 + c^2 & -2ab & -2ac \\ -2ab & a^2 - b^2 + c^2 & -2bc \\ -2ac & -2bc & a^2 + b^2 - c^2 \end{bmatrix}. \qquad (4.9.8)$$

8. *Reflection* in a plane going through (x_0, y_0, z_0) and whose normal has direction cosines a, b, c:

$$(x, y, z) \mapsto (x_0 + y_0 + z_0) + M(x - x_0, y - y_0, z - z_0), \qquad (4.9.9)$$

where M is as in (4.9.8).

9. *Glide-reflection* in a plane P with displacement vector $\mathbf{v}$: Apply first a reflection in P, then a translation by the vector $\mathbf{v}$.

4.9.2 FORMULAE FOR SYMMETRIES: HOMOGENEOUS COORDINATES

All isometries of space can be expressed in homogeneous coordinates in terms of multiplication by a matrix. As in the case of plane isometries (Section 4.2.2), this means that the successive application of transformations reduces to matrix multiplication. (In the formulae below, $\left[\begin{smallmatrix} M & 0 \\ 0 & 1 \end{smallmatrix} \right]$ is the 4×4 projective matrix obtained from the 3×3 matrix M by adding a row and a column as stated.)

1. *Translation* by (x_0, y_0, z_0):
$$\begin{bmatrix} 1 & 0 & 0 & x_0 \\ 0 & 1 & 0 & y_0 \\ 0 & 0 & 1 & z_0 \\ 0 & 0 & 0 & 1 \end{bmatrix}.$$

2. *Rotation* through the origin:
$$\begin{bmatrix} M & 0 \\ 0 & 1 \end{bmatrix},$$
where M is given in (4.9.2) or (4.9.4), as the case may be.

3. *Reflection* in a plane through the origin:
$$\begin{bmatrix} M & 0 \\ 0 & 1 \end{bmatrix},$$
where M is given in (4.9.8).

From this one can deduce all other transformations, as in the case of plane transformations (see page 305).

4.10   OTHER TRANSFORMATIONS OF SPACE

4.10.1   SIMILARITIES

A transformation of space that preserves shapes is called a *similarity*. Every similarity of space is obtained by composing a *proportional scaling transformation* (also known as a *homothety*) with an isometry. A proportional scaling transformation centered at the origin has the form

$$(x, y, z) \mapsto (ax, ay, az), \tag{4.10.1}$$

where $a \neq 0$ is the scaling factor (a real number). The corresponding matrix in *homogeneous coordinates* is

$$H_a = \begin{bmatrix} a & 0 & 0 & 0 \\ 0 & a & 0 & 0 \\ 0 & 0 & a & 0 \\ 0 & 0 & 0 & 1 \end{bmatrix}. \tag{4.10.2}$$

In *cylindrical coordinates*, the transformation is $(r, \theta, z) \mapsto (ar, \theta, az)$. In *spherical coordinates*, it is $(r, \phi, \theta) \mapsto (ar, \phi, \theta)$.

4.10.2   AFFINE TRANSFORMATIONS

A transformation that preserves lines and parallelism (maps parallel lines to parallel lines) is an *affine transformation*. There are two important particular cases of such transformations:

1. A *non-proportional scaling transformation* centered at the origin has the form $(x, y, z) \mapsto (ax, by, cz)$, where $a, b, c \neq 0$ are the scaling factors (real numbers). The corresponding matrix in *homogeneous coordinates* is

$$H_{a,b,c} = \begin{bmatrix} a & 0 & 0 & 0 \\ 0 & b & 0 & 0 \\ 0 & 0 & c & 0 \\ 0 & 0 & 0 & 1 \end{bmatrix}. \tag{4.10.3}$$

2. A *shear* in the x-direction and preserving horizontal planes has the form $(x, y, z) \mapsto (x + rz, y, z)$, where r is the shearing factor. The corresponding matrix in *homogeneous coordinates* is

$$S_r = \begin{bmatrix} 1 & 0 & r & 0 \\ 0 & 1 & 0 & 0 \\ 0 & 0 & 1 & 0 \\ 0 & 0 & 0 & 1 \end{bmatrix}. \tag{4.10.4}$$

Every affine transformation is obtained by composing a scaling transformation with an isometry, or one or two shears with a homothety and an isometry.

4.10.3 PROJECTIVE TRANSFORMATIONS

A transformation that maps lines to lines (but does not necessarily preserve parallelism) is a *projective transformation.* Any spatial projective transformation can be expressed by an invertible 4×4 matrix in homogeneous coordinates; conversely, any invertible 4×4 matrix defines a projective transformation of space. Projective transformations (if not affine) are not defined on all of space, but only on the complement of a plane (the missing plane is "mapped to infinity").

The following particular case is often useful, especially in computer graphics, in *projecting a scene* from space to the plane. Suppose an observer is at the point $E = (x_0, y_0, z_0)$ of space, looking toward the origin $O = (0, 0, 0)$. Let P, the *screen*, be the plane through O and perpendicular to the ray EO. Place a rectangular coordinate system $\xi\eta$ on P with origin at O so that the positive η-axis lies in the half-plane determined by E and the positive z-axis of space (that is, the z-axis is pointing "up" as seen from E). Then consider the transformation that associates with a point $X = (x, y, z)$ the triple (ξ, η, ζ), where (ξ, η) are the coordinates of the point, where the line EX intersects P (the *screen coordinates* of X as seen from E), and ζ is the inverse of the signed distance from X to E along the line EO (this distance is the *depth* of X as seen from E). This is a projective transformation, given by the matrix

$$\begin{bmatrix} -r^2 y_0 & r^2 x_0 & 0 & 0 \\ -r x_0 z_0 & -r y_0 z_0 & r\rho^2 & 0 \\ 0 & 0 & 0 & r\rho \\ -\rho x_0 & -\rho y_0 & -\rho z_0 & r^2 \rho \end{bmatrix} \qquad (4.10.5)$$

with $\rho = \sqrt{x_0^2 + y_0^2}$ and $r = \sqrt{x_0^2 + y_0^2 + z_0^2}$.

4.11 DIRECTION ANGLES AND DIRECTION COSINES

Given a vector (a, b, c) in three-dimensional space, the *direction cosines* of this vector are

$$\cos\alpha = \frac{a}{\sqrt{a^2 + b^2 + c^2}},$$
$$\cos\beta = \frac{b}{\sqrt{a^2 + b^2 + c^2}}, \qquad (4.11.1)$$
$$\cos\gamma = \frac{c}{\sqrt{a^2 + b^2 + c^2}}.$$

Here the *direction angles* α, β, γ are the angles that the vector makes with the positive x-, y- and z-axes, respectively. In formulae, usually the direction cosines appear, rather than the direction angles. We have

$$\cos^2\alpha + \cos^2\beta + \cos^2\gamma = 1. \qquad (4.11.2)$$

4.12 PLANES

The (Cartesian) equation of a *plane* is linear in the coordinates x, y, and z:

$$ax + by + cz + d = 0. \qquad (4.12.1)$$

The *normal direction* to this plane is (a, b, c). The *intersection* of this plane with the x-axis, or *x-intercept*, is $x = -d/a$, the *y-intercept* is $y = -d/b$, and the *z-intercept* is $z = -d/c$. The plane is vertical (perpendicular to the xy-plane) if $c = 0$. It is perpendicular to the x-axis if $b = c = 0$, and likewise for the other coordinates.

When $a^2 + b^2 + c^2 = 1$ and $d \leq 0$ in the equation $ax + by + cz + d = 0$, the equation is said to be in *normal form*. In this case $-d$ is the *distance of the plane to the origin*, and (a, b, c) are the *direction cosines* of the normal.

To reduce an arbitrary equation $ax + by + cz + d = 0$ to normal form, divide by $\pm\sqrt{a^2 + b^2 + c^2}$, where the sign of the radical is chosen opposite the sign of d when $d \neq 0$, the same as the sign of c when $d = 0$ and $c \neq 0$, and the same as the sign of b otherwise.

4.12.1 PLANES WITH PRESCRIBED PROPERTIES

1. Plane through (x_0, y_0, z_0) and perpendicular to the direction (a, b, c):

$$a(x - x_0) + b(y - y_0) + c(z - z_0) = 0. \qquad (4.12.2)$$

2. Plane through (x_0, y_0, z_0) and parallel to the two directions (a_1, b_1, c_1) and (a_2, b_2, c_2):

$$\begin{vmatrix} x - x_0 & y - y_0 & z - z_0 \\ a_1 & b_1 & c_1 \\ a_2 & b_2 & c_2 \end{vmatrix} = 0. \qquad (4.12.3)$$

3. Plane through (x_0, y_0, z_0) and (x_1, y_1, z_1) and parallel to the direction (a, b, c):

$$\begin{vmatrix} x - x_0 & y - y_0 & z - z_0 \\ x_1 - x_0 & y_1 - y_0 & z_1 - z_0 \\ a & b & c \end{vmatrix} = 0. \qquad (4.12.4)$$

4. Plane going through (x_0, y_0, z_0), (x_1, y_1, z_1), and (x_2, y_2, z_2):

$$\begin{vmatrix} x & y & z & 1 \\ x_0 & y_0 & z_0 & 1 \\ x_1 & y_1 & z_1 & 1 \\ x_2 & y_2 & z_2 & 1 \end{vmatrix} = 0 \quad \text{or} \quad \begin{vmatrix} x - x_0 & y - y_0 & z - z_0 \\ x_1 - x_0 & y_1 - y_0 & z_1 - z_0 \\ x_2 - x_0 & y_2 - y_0 & z_2 - z_0 \end{vmatrix} = 0.$$

$$(4.12.5)$$

(The last three formulae remain true in *oblique coordinates*.)

5. The *distance* from the point (x_0, y_0, z_0) to the plane $ax + by + cz + d = 0$ is

$$\left| \frac{ax_0 + by_0 + cz_0 + d}{\sqrt{a^2 + b^2 + c^2}} \right|. \qquad (4.12.6)$$

6. The *angle* between two planes $a_0 x + b_0 y + c_0 z + d_0 = 0$ and $a_1 x + b_1 y + c_1 z + d_1 = 0$ is

$$\cos^{-1} \frac{a_0 a_1 + b_0 b_1 + c_0 c_1}{\sqrt{a_0^2 + b_0^2 + c_0^2} \sqrt{a_1^2 + b_1^2 + c_1^2}}. \tag{4.12.7}$$

In particular, the two planes are *parallel* when $a_0 : b_0 : c_0 = a_1 : b_1 : c_1$, and *perpendicular* when $a_0 a_1 + b_0 b_1 + c_0 c_1 = 0$.

4.12.2 CONCURRENCE AND COPLANARITY

Four planes $a_0 x + b_0 y + c_0 z + d_0 = 0$, $a_1 x + b_1 y + c_1 z + d_1 = 0$, $a_2 x + b_2 y + c_2 z + d_2 = 0$, and $a_3 x + b_3 y + c_3 z + d_3 = 0$ are *concurrent* (share a point) if and only if

$$\begin{vmatrix} a_0 & b_0 & c_0 & d_0 \\ a_1 & b_1 & c_1 & d_1 \\ a_2 & b_2 & c_2 & d_2 \\ a_3 & b_3 & c_3 & d_3 \end{vmatrix} = 0. \tag{4.12.8}$$

Four points (x_0, y_0, z_0), (x_1, y_1, z_1), (x_2, y_2, z_2), and (x_3, y_3, z_3) are *coplanar* (lie on the same plane) if and only if

$$\begin{vmatrix} x_0 & y_0 & z_0 & 1 \\ x_1 & y_1 & z_1 & 1 \\ x_2 & y_2 & z_2 & 1 \\ x_3 & y_3 & z_3 & 1 \end{vmatrix} = 0. \tag{4.12.9}$$

(Both of these assertions remain true in *oblique coordinates*.)

4.13 LINES IN SPACE

Two planes that are not parallel or coincident intersect in a *straight line*, such that one can express a line by a pair of linear equations

$$\left. \begin{array}{l} ax + by + cz + d = 0 \\ a'x + b'y + c'z + d' = 0 \end{array} \right\} \tag{4.13.1}$$

such that $bc' - cb'$, $ca' - ac'$, and $ab' - ba'$ are not all zero. The line thus defined is parallel to the vector $(bc' - cb', ca' - ac', ab' - ba')$. The *direction cosines* of the line are those of this vector. See Equation (4.11.1). (The direction cosines of a line are only defined up to a simultaneous change in sign, because the opposite vector still gives the same line.)

The following particular cases are important:

1. Line through (x_0, y_0, z_0) parallel to the vector (a, b, c):

$$\frac{x - x_0}{a} = \frac{y - y_0}{b} = \frac{z - z_0}{c}. \tag{4.13.2}$$

2. Line through (x_0, y_0, z_0) and (x_1, y_1, z_1):
$$\frac{x - x_0}{x_1 - x_0} = \frac{y - y_0}{y_1 - y_0} = \frac{z - z_0}{z_1 - z_0}. \qquad (4.13.3)$$
This line is parallel to the vector $(x_1 - x_0,\ y_1 - y_0,\ z_1 - z_0)$.

4.13.1 DISTANCES

1. The *distance* between two points in space is the *length of the line segment* joining them. The distance between the points (x_0, y_0, z_0) and (x_1, y_1, z_1) is
$$\sqrt{(x_1 - x_0)^2 + (y_1 - y_0)^2 + (z_1 - z_0)^2}. \qquad (4.13.4)$$

2. The point $k\%$ of the way from $P_0 = (x_0, y_0, z_0)$ to $P_1 = (x_1, y_1, z_1)$ is
$$\left(\frac{kx_1 + (100 - k)x_0}{100}, \frac{ky_1 + (100 - k)y_0}{100}, \frac{kz_1 + (100 - k)z_0}{100} \right). \qquad (4.13.5)$$
(The same formula also applies in oblique coordinates.) This point divides the segment $P_0 P_1$ in the ratio $k : (100 - k)$. As a particular case, the *midpoint* of $P_0 P_1$ is given by
$$\left(\frac{x_1 + x_0}{2}, \frac{y_1 + y_0}{2}, \frac{z_1 + z_0}{2} \right). \qquad (4.13.6)$$

3. The *distance* between the point (x_0, y_0, z_0) and the line through (x_1, y_1, z_1) in direction (a, b, c):
$$\sqrt{\frac{\left| \begin{matrix} y_0 - y_1 & z_0 - z_1 \\ b & c \end{matrix} \right|^2 + \left| \begin{matrix} z_0 - z_1 & x_0 - x_1 \\ c & a \end{matrix} \right|^2 + \left| \begin{matrix} x_0 - x_1 & y_0 - y_1 \\ a & b \end{matrix} \right|^2}{a^2 + b^2 + c^2}} \qquad (4.13.7)$$

4. The *distance* between the line through (x_0, y_0, z_0) in direction (a_0, b_0, c_0) and the line through (x_1, y_1, z_1) in direction (a_1, b_1, c_1):
$$\frac{\left| \begin{matrix} x_1 - x_0 & y_1 - y_0 & z_1 - z_0 \\ a_0 & b_0 & c_0 \\ a_1 & b_1 & c_1 \end{matrix} \right|}{\sqrt{\left| \begin{matrix} b_0 & c_0 \\ b_1 & c_1 \end{matrix} \right|^2 + \left| \begin{matrix} c_0 & a_0 \\ c_1 & a_1 \end{matrix} \right|^2 + \left| \begin{matrix} a_0 & b_0 \\ a_1 & b_1 \end{matrix} \right|^2}}. \qquad (4.13.8)$$

4.13.2 ANGLES

Angle between lines with directions (x_0, y_0, z_0) and (x_1, y_1, z_1):
$$\cos^{-1} \frac{a_0 a_1 + b_0 b_1 + c_0 c_1}{\sqrt{a_0^2 + b_0^2 + c_0^2} \sqrt{a_1^2 + b_1^2 + c_1^2}}. \qquad (4.13.9)$$
In particular, the two lines are *parallel* when $a_0 : b_0 : c_0 = a_1 : b_1 : c_1$, and *perpendicular* when $a_0 a_1 + b_0 b_1 + c_0 c_1 = 0$.

Angle between lines with direction angles $\alpha_0, \beta_0, \gamma_0$ and $\alpha_1, \beta_1, \gamma_1$:

$$\cos^{-1}(\cos \alpha_0 \cos \alpha_1 + \cos \beta_0 \cos \beta_1 + \cos \gamma_0 \cos \gamma_1). \tag{4.13.10}$$

4.13.3 CONCURRENCE, COPLANARITY, PARALLELISM

Two lines, each specified by point and direction, are *coplanar* if, and only if, the determinant in the numerator of Equation (4.13.8) is zero. In this case they are *concurrent* (if the denominator is non-zero) or *parallel* (if the denominator is zero).

Three lines with directions (a_0, b_0, c_0), (a_1, b_1, c_1), and (a_2, b_2, c_2) are *parallel to a common plane* if and only if

$$\begin{vmatrix} a_0 & b_0 & c_0 \\ a_1 & b_1 & c_1 \\ a_2 & b_2 & c_2 \end{vmatrix} = 0. \tag{4.13.11}$$

4.14 POLYHEDRA

For any polyhedron topologically equivalent to a sphere—in particular, for any *convex polyhedron*—the *Euler formula* holds:

$$v - e + f = 2, \tag{4.14.1}$$

where v is the number of vertices, e is the number of edges, and f is the number of faces.

Many common polyhedra are particular cases of cylinders (Section 4.15) or cones (Section 4.16). A cylinder with a polygonal base (the base is also called a directrix) is called a *prism*. A cone with a polygonal base is called a *pyramid*. A frustum of a cone with a polygonal base is called a *truncated pyramid*. Formulae (4.15.1), (4.16.1), and (4.16.2) give the volumes of a general prism, pyramid, and truncated pyramid.

A prism whose base is a parallelogram is a *parallelepiped*. The *volume* of a parallelepiped with one vertex at the origin and adjacent vertices at (x_1, y_1, z_1), (x_2, y_2, z_2), and (x_3, y_3, z_3) is given by

$$\text{volume} = \begin{vmatrix} x_1 & y_1 & z_1 \\ x_2 & y_2 & z_2 \\ x_3 & y_3 & z_3 \end{vmatrix}. \tag{4.14.2}$$

The *rectangular parallelepiped* is a particular case: all of its faces are rectangles. If the side lengths are a, b, c, the *volume* is abc, the *total surface area* is $2(ab + ac + bc)$, and each *diagonal* has length $\sqrt{a^2 + b^2 + c^2}$. When $a = b = c$ we get the *cube*. See Section 4.14.1. A pyramid whose base is a triangle is a *tetrahedron*. The *volume* of a tetrahedon with one vertex at the origin and the other vertices at (x_1, y_1, z_1),

(x_2, y_2, z_2), and (x_3, y_3, z_3) is given by

$$\text{volume} = \frac{1}{6} \begin{vmatrix} x_1 & y_1 & z_1 \\ x_2 & y_2 & z_2 \\ x_3 & y_3 & z_3 \end{vmatrix}. \tag{4.14.3}$$

In a tetrahedron with vertices P_0, P_1, P_2, P_3, let d_{ij} be the distance (edge length) from P_i to P_j. Form the determinants

$$\Delta = \begin{vmatrix} 0 & 1 & 1 & 1 & 1 \\ 1 & 0 & d_{01}^2 & d_{02}^2 & d_{03}^2 \\ 1 & d_{01}^2 & 0 & d_{12}^2 & d_{13}^2 \\ 1 & d_{02}^2 & d_{12}^2 & 0 & d_{23}^2 \\ 1 & d_{03}^2 & d_{13}^2 & d_{23}^2 & 0 \end{vmatrix} \quad \text{and} \quad \Gamma = \begin{vmatrix} 0 & d_{01}^2 & d_{02}^2 & d_{03}^2 \\ d_{01}^2 & 0 & d_{12}^2 & d_{13}^2 \\ d_{02}^2 & d_{12}^2 & 0 & d_{23}^2 \\ d_{03}^2 & d_{13}^2 & d_{23}^2 & 0 \end{vmatrix}. \tag{4.14.4}$$

Then the *volume* of the tetrahedron is $\sqrt{|\Delta|/288}$, and the radius of the *circumscribed sphere* is $\frac{1}{2}\sqrt{|\Gamma/2\Delta|}$.

Expanding the determinant we find that the volume V satisfies the formula:

$$\begin{aligned} 144V^2 = & - d_{01}^2 d_{12}^2 d_{02}^2 - d_{01}^2 d_{13}^2 d_{03}^2 - d_{12}^2 d_{13}^2 d_{23}^2 - d_{02}^2 d_{03}^2 d_{23}^2 \\ & + d_{01}^2 d_{02}^2 d_{13}^2 + d_{12}^2 d_{02}^2 d_{13}^2 + d_{01}^2 d_{12}^2 d_{03}^2 + d_{12}^2 d_{02}^2 d_{03}^2 \\ & + d_{12}^2 d_{13}^2 d_{03}^2 + d_{02}^2 d_{13}^2 d_{03}^2 + d_{01}^2 d_{12}^2 d_{23}^2 + d_{01}^2 d_{02}^2 d_{23}^2 \\ & + d_{01}^2 d_{13}^2 d_{23}^2 + d_{02}^2 d_{13}^2 d_{23}^2 + d_{01}^2 d_{03}^2 d_{23}^2 + d_{12}^2 d_{03}^2 d_{23}^2 \\ & - d_{02}^2 d_{02}^2 d_{13}^2 - d_{02}^2 d_{13}^2 d_{13}^2 \\ & - d_{12}^2 d_{12}^2 d_{03}^2 - d_{12}^2 d_{03}^2 d_{03}^2 \\ & - d_{01}^2 d_{01}^2 d_{23}^2 - d_{01}^2 d_{23}^2 d_{23}^2. \end{aligned} \tag{4.14.5}$$

(Mnemonic: Each of the first four negative terms corresponds to a closed path around a face; each positive term to an open path along three consecutive edges; each remaining negative term to a pair of opposite edges with weights 2 and 1. All such edge combinations are represented.)

For an arbitrary tetrahedron, let P be a vertex and let a, b, c be the lengths of the edges converging on P. If A, B, C are the angles between the same three edges, the volume of the tetrahedron is

$$V = \frac{1}{6}abc\sqrt{1 - \cos^2 A - \cos^2 B - \cos^2 C + 2\cos A \cos B \cos C}. \tag{4.14.6}$$

4.14.1   CONVEX REGULAR POLYHEDRA

Figure 4.35 shows the five regular polyhedra, or *Platonic solids*. In the following tables and formulae, a is the length of an edge, θ the dihedral angle at each edge, R the radius of the circumscribed sphere, r the radius of the inscribed sphere, V the volume, S the total surface area, v the total number of vertices, e the total number of edges, f the total number of faces, p the number of edges in a face (3 for equilateral triangles, 4 for squares, 5 for regular pentagons), and q the number of edges meeting at a vertex.

FIGURE 4.35
The Platonic solids. Top: the tetrahedron (self-dual). Middle: the cube and the octahedron (dual to one another). Bottom: the dodecahedron and the icosahedron (dual to one another).

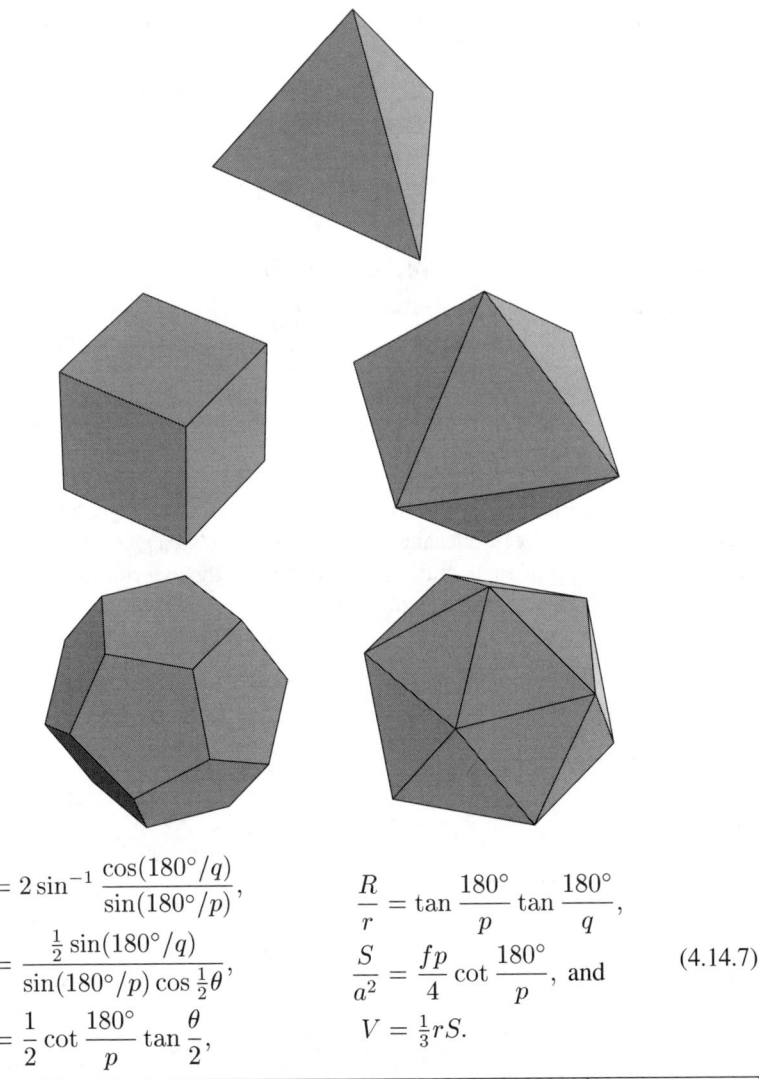

$$\theta = 2\sin^{-1}\frac{\cos(180°/q)}{\sin(180°/p)},$$

$$\frac{R}{a} = \frac{\frac{1}{2}\sin(180°/q)}{\sin(180°/p)\cos\frac{1}{2}\theta},$$

$$\frac{r}{a} = \frac{1}{2}\cot\frac{180°}{p}\tan\frac{\theta}{2},$$

$$\frac{R}{r} = \tan\frac{180°}{p}\tan\frac{180°}{q},$$

$$\frac{S}{a^2} = \frac{fp}{4}\cot\frac{180°}{p}, \text{ and}$$

$$V = \tfrac{1}{3}rS.$$

(4.14.7)

Name	v	e	f	p	q	$\sin\theta$	θ
Regular tetrahedron	4	6	4	3	3	$2\sqrt{2}/3$	$70°31'44''$
Cube	8	12	6	4	3	1	$90°$
Regular octahedron	6	12	8	3	4	$2\sqrt{2}/3$	$109°28'16''$
Regular dodecahedron	20	30	12	5	3	$2/\sqrt{5}$	$116°33'54''$
Regular icosahedron	12	30	20	3	5	$2/3$	$138°11'23''$

Name	R/a		r/a	
Tetrahedron	$\sqrt{6}/4$	0.612372	$\sqrt{6}/12$	0.204124
Cube	$\sqrt{3}/2$	0.866025	$\frac{1}{2}$	0.5
Octahedron	$\sqrt{2}/2$	0.707107	$\sqrt{6}/6$	0.408248
Dodecahedron	$\frac{1}{4}(\sqrt{15}+\sqrt{3})$	1.401259	$\frac{1}{20}\sqrt{250+110\sqrt{5}}$	1.113516
Icosahedron	$\frac{1}{4}\sqrt{10+2\sqrt{5}}$	0.951057	$\frac{1}{12}\sqrt{42+18\sqrt{5}}$	0.755761

Name	S/a^2		V/a^3	
Tetrahedron	$\sqrt{3}$	1.73205	$\sqrt{2}/12$	0.117851
Cube	6	6.	1	1.
Octahedron	$2\sqrt{3}$	3.46410	$\sqrt{2}/3$	0.471405
Dodecahedron	$3\sqrt{25+10\sqrt{5}}$	20.64573	$\frac{1}{4}(15+7\sqrt{5})$	7.663119
Icosahedron	$5\sqrt{3}$	8.66025	$\frac{5}{12}(3+\sqrt{5})$	2.181695

4.14.2 POLYHEDRA NETS

Nets for the five Platonic solids are shown: a) tetrahedron, b) octahedron, c) icosahedron, d) cube, and e) dodecahedron. Paper models can be made by making an enlarged photocopy of each, cutting them out along the exterior lines, folding on the interior lines, and using tape to join the edges.

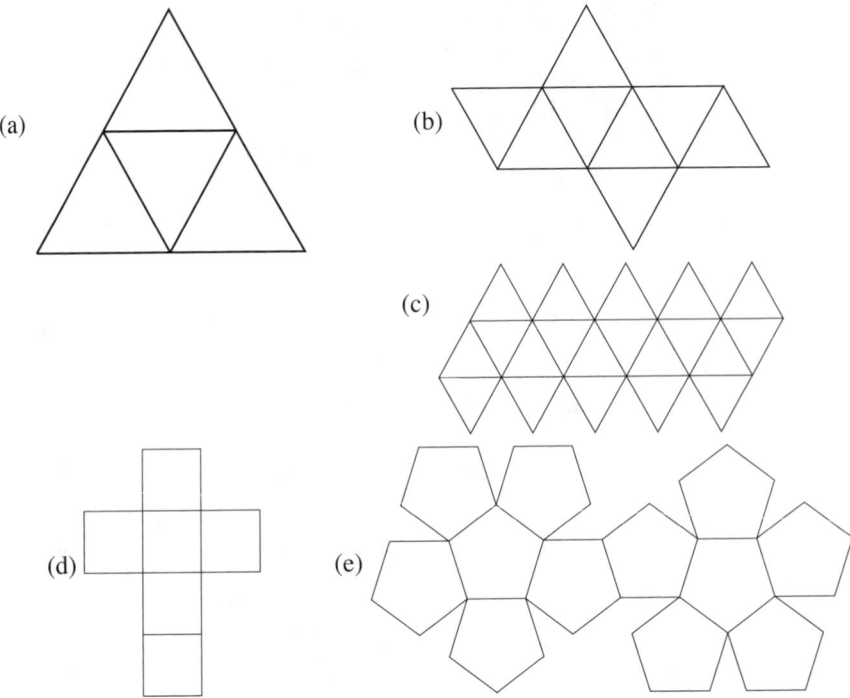

FIGURE 4.36

Left: an oblique cylinder with generator L and directrix C. Right: a right circular cylinder.

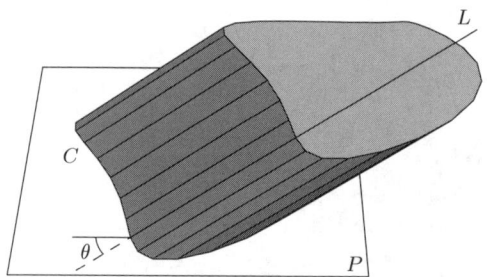
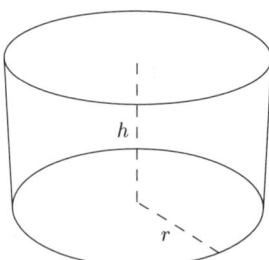

4.15 CYLINDERS

Given a line L and a curve C in a plane P, the *cylinder* with *generator* L and *directrix* C is the surface obtained by moving L parallel to itself, so that a point of L is always on C. If L is parallel to the z-axis, the surface's implicit equation does not involve the variable z. Conversely, any implicit equation that does not involve one of the variables (or that can be brought to that form by a change of coordinates) represents a cylinder.

If C is a simple closed curve, we also apply the word *cylinder* to the solid enclosed by the surface generated in this way (Figure 4.36, left). The *volume* contained between P and a plane P' parallel to P is

$$V = Ah = Al\sin\theta, \tag{4.15.1}$$

where A is the area in the plane P enclosed by C, h is the distance between P and P' (measured perpendicularly), l is the length of the segment of L contained between P and P', and θ is the angle that L makes with P. When $\theta = 90°$ we have a *right cylinder*, and $h = l$. For a right cylinder, the *lateral area* between P and P' is hs, where s is the length (circumference) of C.

The most important particular case is the *right circular cylinder* (often simply called a *cylinder*). If r is the radius of the base and h is the altitude (Figure 4.36, right), the *lateral area* is $2\pi rh$, the *total area* is $2\pi r(r+h)$, and the *volume* is $\pi r^2 h$. The *implicit equation* of this surface can be written $x^2 + y^2 = r^2$; see also page 364.

4.16 CONES

Given a curve C in a plane P and a point O not in P, the *cone* with *vertex* O and *directrix* C is the surface obtained as the union of all rays that join O with points of C. If O is the origin and the surface is given implicitly by an algebraic equation, that equation is homogeneous (all terms have the same total degree in the variables). Conversely, any homogeneous implicit equation (or one that can be made homogeneous by a change of coordinates) represents a cone.

FIGURE 4.37

Top: a cone with vertex O and directrix C. Bottom left: a right circular cone. Bottom right: A frustum of the latter.

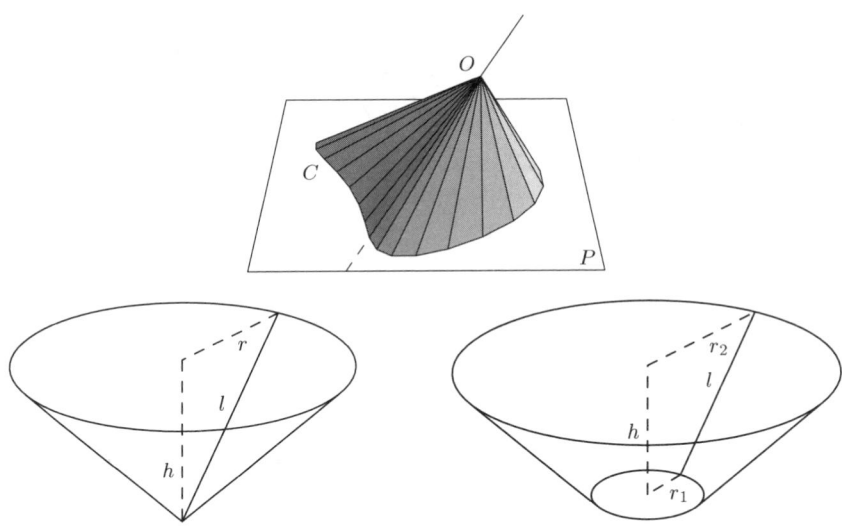

If C is a simple closed curve, we also apply the word *cone* to the solid enclosed by the surface generated in this way (Figure 4.37, top). The *volume* contained between P and the vertex O is

$$V = \tfrac{1}{3}Ah, \tag{4.16.1}$$

where A is the area in the plane P enclosed by C and h is the distance from O and P (measured perpendicularly).

The solid contained between P and a plane P' parallel to P (on the same side of the vertex) is called a *frustum*. It's volume is

$$V = \tfrac{1}{3}h(A + A' + \sqrt{AA'}), \tag{4.16.2}$$

where A and A' are the areas enclosed by the sections of the cone by P and P' (often called the *bases* of the frustum), and h is the distance between P and P'.

The most important particular case of a cone is the *right circular cone* (often simply called a *cone*). If r is the radius of the base, h is the altitude, and l is the length between the vertex and a point on the base circle (Figure 4.37, bottom left), the following relationships apply:

$$l = \sqrt{r^2 + h^2},$$

$$\text{Lateral area} = \pi r l = \pi r \sqrt{r^2 + h^2},$$

$$\text{Total area} = \pi r(l + r) = \pi r(r + \sqrt{r^2 + h^2}), \text{ and}$$

$$\text{Volume} = \tfrac{1}{3}\pi r^2 h.$$

The *implicit equation* of this surface can be written $x^2 + y^2 = z^2$; see also Section 4.18.

For a *frustum* of a right circular cone (Figure 4.37, bottom right),

$$l = \sqrt{(r_1 - r_2)^2 + h^2},$$
$$\text{Lateral area} = \pi(r_1 + r_2)l,$$
$$\text{Total area} = \pi(r_1^2 + r_2^2 + (r_1 + r_2)l), \quad \text{and}$$
$$\text{Volume} = \tfrac{1}{3}\pi h(r_1^2 + r_2^2 + r_1 r_2).$$

4.17 SURFACES OF REVOLUTION: THE TORUS

A *surface of revolution* is formed by the rotation of a planar curve C about an axis in the plane of the curve and not cutting the curve. The *Pappus–Guldinus theorem* says that:

1. The *area of the surface of revolution* on a curve C is equal to the product of the length of C and the length of the path traced by the centroid of C (which is 2π times the distance from this centroid to the axis of revolution).

2. The *volume bounded by the surface of revolution* on a simple closed curve C is equal to the product of the area bounded by C and the length of the path traced by the centroid of the area bounded by C.

When C is a circle, the surface obtained is a *circular torus* or *torus of revolution* (Figure 4.38). Let r be the radius of the revolving circle and let R be the distance from its center to the axis of rotation. The *area* of the torus is $4\pi^2 Rr$, and its *volume* is $2\pi^2 Rr^2$.

FIGURE 4.38
A torus of revolution.

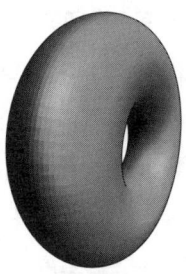
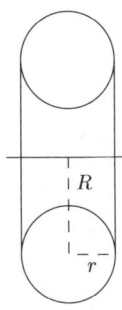

4.18 QUADRICS

A surface defined by an algebraic equation of degree two is called a *quadric*. Spheres, circular cylinders, and circular cones are quadrics. By means of a rigid motion, any quadric can be transformed into a quadric having one of the following equations (where $a, b, c \neq 0$):

1. Real ellipsoid: $x^2/a^2 + y^2/b^2 + z^2/c^2 = 1$
2. Imaginary ellipsoid: $x^2/a^2 + y^2/b^2 + z^2/c^2 = -1$
3. Hyperboloid of one sheet: $x^2/a^2 + y^2/b^2 - z^2/c^2 = 1$
4. Hyperboloid of two sheets: $x^2/a^2 + y^2/b^2 - z^2/c^2 = -1$
5. Real quadric cone: $x^2/a^2 + y^2/b^2 - z^2/c^2 = 0$
6. Imaginary quadric cone: $x^2/a^2 + y^2/b^2 + z^2/c^2 = 0$
7. Elliptic paraboloid: $x^2/a^2 + y^2/b^2 + 2z = 0$
8. Hyperbolic paraboloid: $x^2/a^2 - y^2/b^2 + 2z = 0$
9. Real elliptic cylinder: $x^2/a^2 + y^2/b^2 = 1$
10. Imaginary elliptic cylinder: $x^2/a^2 + y^2/b^2 = -1$
11. Hyperbolic cylinder: $x^2/a^2 - y^2/b^2 = 1$
12. Real intersecting planes: $x^2/a^2 - y^2/b^2 = 0$
13. Imaginary intersecting planes: $x^2/a^2 + y^2/b^2 = 0$
14. Parabolic cylinder: $x^2 + 2y = 0$
15. Real parallel planes: $x^2 = 1$
16. Imaginary parallel planes: $x^2 = -1$
17. Coincident planes: $x^2 = 0$

Surfaces with Equations 9–17 are cylinders over the plane curves of the same equation (Section 4.15). Equations 2, 6, 10, and 16 have no real solutions, so that they do not describe surfaces in real three-dimensional space. A surface with Equation 5 can be regarded as a cone (Section 4.16) over a conic C (any ellipse, parabola, or hyperbola can be taken as the directrix; there is a two-parameter family of essentially distinct cones over it, determined by the position of the vertex with respect to C). The surfaces with Equations 1, 3, 4, 7, and 8 are shown in Figure 4.39.

The surfaces with Equations 1–6 are *central quadrics*; in the form given, the center is at the origin. The quantities a, b, c are the *semi-axes*.

The *volume of the ellipsoid* with semi-axes a, b, c is $\frac{4}{3}\pi abc$. When two of the semi-axes are the same, we can also write the *area of the ellipsoid* in closed-form. Suppose $b = c$, so the ellipsoid $x^2/a^2 + (y^2 + z^2)/b^2 = 1$ is the surface of revolution obtained by rotating the ellipse $x^2/a^2 + y^2/b^2 = 1$ around the x-axis. Its area is

$$2\pi b^2 + \frac{2\pi a^2 b}{\sqrt{a^2 - b^2}}\sin^{-1}\frac{\sqrt{a^2 - b^2}}{a} = 2\pi b^2 + \frac{\pi a^2 b}{\sqrt{b^2 - a^2}}\log\frac{b + \sqrt{b^2 - a^2}}{b - \sqrt{b^2 - a^2}}.$$

$$(4.18.1)$$

The two quantities are equal, but only one avoids complex numbers, depending on whether $a > b$ or $a < b$. When $a > b$, we have a *prolate spheroid*, that is, an ellipse rotated around its major axis; when $a < b$ we have an *oblate spheroid*, which is an ellipse rotated around its minor axis.

Given a general quadratic equation in three variables,

$$ax^2 + by^2 + cz^2 + 2fyz + 2gzx + 2hxy + 2px + 2qy + 2rz + d = 0, \quad (4.18.2)$$

one can determine the type of conic by consulting the table:

ρ_3	ρ_4	Δ	k signs	K signs	Type of quadric
3	4	< 0			Real ellipsoid
3	4	> 0	Same		Imaginary ellipsoid
3	4	> 0	Opp		Hyperboloid of one sheet
3	4	< 0	Opp		Hyperboloid of two sheets
3	3		Opp		Real quadric cone
3	3		Same		Imaginary quadric cone
2	4	< 0	Same		Elliptic paraboloid
2	4	> 0	Opp		Hyperbolic paraboloid
2	3		Same	Opp	Real elliptic cylinder
2	3		Same	Same	Imaginary elliptic cylinder
2	3		Opp		Hyperbolic cylinder
2	2		Opp		Real intersecting planes
2	2		Same		Imaginary intersecting planes
1	3				Parabolic cylinder
1	2			Opp	Real parallel planes
1	2			Same	Imaginary parallel planes
1	1				Coincident planes

The columns have the following meaning. Let

$$e = \begin{bmatrix} a & h & g \\ h & b & f \\ g & f & c \end{bmatrix} \quad \text{and} \quad E = \begin{bmatrix} a & h & g & p \\ h & b & f & q \\ g & f & c & r \\ p & q & r & d \end{bmatrix}. \quad (4.18.3)$$

Let ρ_3 and ρ_4 be the ranks of e and E, and let Δ be the determinant of E. The column "k signs" refers to the non-zero eigenvalues of e, that is, the roots of

$$\begin{vmatrix} a - x & h & g \\ h & b - x & f \\ g & f & c - x \end{vmatrix} = 0; \quad (4.18.4)$$

if all non-zero eigenvalues have the same sign, choose "same", otherwise "opposite". Similarly, "K signs" refers to the sign of the non-zero eigenvalues of E.

4.18.1 SPHERES

The set of points in space whose distance to a fixed point (the *center*) is a fixed number (the *radius*) is a *sphere*. A circle of radius r and center (x_0, y_0, z_0) is defined by the equation

$$(x - x_0)^2 + (y - y_0)^2 + (z - z_0)^2 = r^2, \quad (4.18.5)$$

or

$$x^2 + y^2 + z^2 - 2xx_0 - 2yy_0 - 2zz_0 + x_0^2 + y_0^2 + z_0^2 - r^2 = 0. \quad (4.18.6)$$

FIGURE 4.39

The five non-degenerate real quadrics. Top left: ellipsoid. Top right: hyperboloid of two sheets (one facing up and one facing down). Bottom left: elliptic paraboloid. Bottom middle: hyperboloid of one sheet. Bottom right: hyperbolic paraboloid.

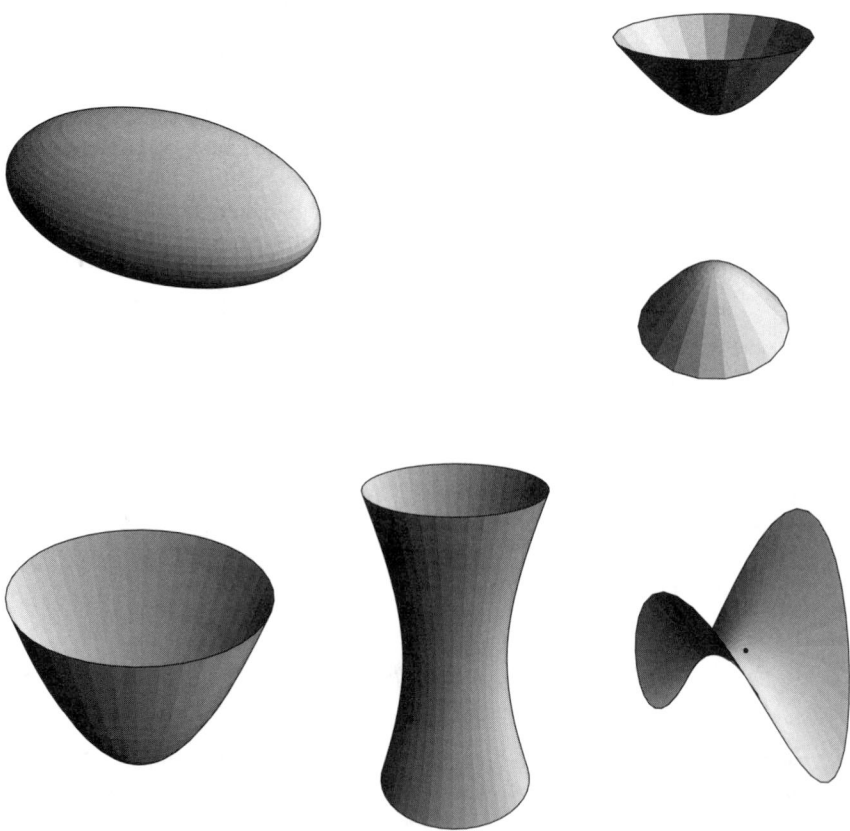

Conversely, an equation of the form

$$x^2 + y^2 + z^2 + 2dx + 2ey + 2fz + g = 0 \qquad (4.18.7)$$

defines a sphere if $d^2 + e^2 + f^2 > g$; the center is $(-d, -e, -f)$ and the radius is $\sqrt{d^2 + e^2 + f^2 - g}$.

1. Four points not in the same plane determine a unique sphere. If the points have coordinates (x_1, y_1, z_1), (x_2, y_2, z_2), (x_3, y_3, z_3), and (x_4, x_4, z_4), the

FIGURE 4.40
Left: a spherical cap. Middle: a spherical zone (of two bases). Right: a spherical segment.

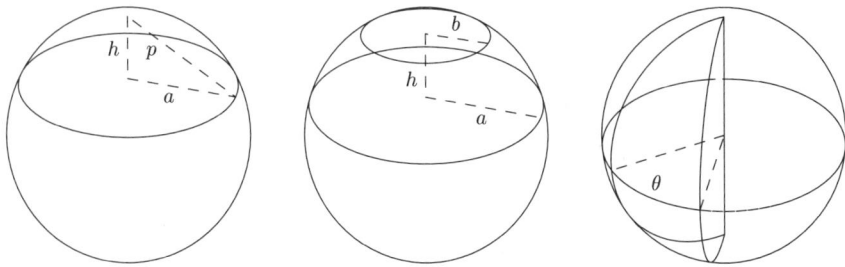

equation of the sphere is

$$
\begin{vmatrix}
x^2 + y^2 + z^2 & x & y & z & 1 \\
x_1^2 + y_1^2 + z_1^2 & x_1 & y_1 & z_1 & 1 \\
x_2^2 + y_2^2 + z_2^2 & x_2 & y_2 & z_2 & 1 \\
x_3^2 + y_3^2 + z_3^2 & x_3 & y_3 & z_3 & 1 \\
x_4^2 + y_4^2 + z_4^2 & x_4 & y_4 & z_4 & 1
\end{vmatrix} = 0.
\qquad (4.18.8)
$$

2. Given two points $P_1 = (x_1, y_1, z_1)$ and $P_2 = (x_2, y_2, z_2)$, there is a unique sphere whose diameter is $P_1 P_2$; its equation is

$$
(x - x_1)(x - x_2) + (y - y_1)(y - y_2) + (z - z_1)(z - z_2) = 0. \quad (4.18.9)
$$

3. The *area* of a sphere of radius r is $4\pi r^2$, and the *volume* is $\frac{4}{3}\pi r^3$.

4. The *area of a spherical polygon* (that is, of a polygon on the sphere whose sides are arcs of great circles) is

$$
S = \left(\sum_{i=1}^{n} \theta_i - (n - 2)\pi \right) r^2,
\qquad (4.18.10)
$$

where r is the radius of the sphere, n is the number of vertices, and θ_i are the internal angles of the polygons in radians. In particular, the sum of the angles of a spherical triangle is always greater than $\pi = 180°$, and the excess is proportional to the area.

4.18.1.1 Spherical cap

Let the radius be r (Figure 4.40, left). The *area* of the curved region is $2\pi r h = \pi p^2$. The *volume* of the cap is $\frac{1}{3}\pi h^2(3r - h) = \frac{1}{6}\pi h(3a^2 + h^2)$.

4.18.1.2 Spherical zone (of two bases)

Let the radius be r (Figure 4.40, middle). The *area* of the curved region (called a *spherical zone*) is $2\pi r h$. The *volume* of the zone is $\frac{1}{6}\pi h(3a^2 + 3b^2 + h^2)$.

4.18.1.3 Spherical segment and lune

Let the radius be r (Figure 4.40, right). The *area* of the curved region (called a *spherical segment* or *lune*) is $2r^2\theta$, the angle being measured in radians. The *volume* of the segment is $\frac{2}{3}r^3\theta$.

4.18.1.4 Volume and area of spheres

If the volume of an n-dimensional sphere of radius r is $V_n(r)$ and its surface area is $S_n(r)$, then

$$V_n(r) = \frac{2\pi r^2}{n}V_{n-2}(r) = \frac{2\pi^{n/2}r^n}{n\Gamma\left(\frac{n}{2}\right)} = \frac{\pi^{n/2}r^n}{\left(\frac{n}{2}\right)!},$$

(4.18.11)

$$S_n(r) = \frac{n}{r}V_n(r) = \frac{d}{dr}[V_n(r)].$$

Hence, the area of a circle is $V_2 = \pi r^2 \approx 3.1416r^2$, the volume of a 3-dimensional sphere is $V_3 = \frac{4}{3}\pi r^3 \approx 4.1888r^3$, the volume of a 4-dimensional sphere is $V_4 = \frac{1}{2}\pi^2 r^4 \approx 4.9348r^4$, the circumference of a circle is $S_2 = 2\pi r$, and the surface area of a sphere is $S_3 = 4\pi r^2$.

For large values of n,

$$V_n(r) \approx \frac{n^{-(n+1)/2}}{\sqrt{\pi}}(2\pi e)^{n/2}r^n.$$

(4.18.12)

4.19 SPHERICAL GEOMETRY & TRIGONOMETRY

The angles in a spherical triangle do not have to add up to 180 degrees. It is possible for a spherical triangle to have 3 right angles.

4.19.1 RIGHT SPHERICAL TRIANGLES

Let a, b, and c be the sides of a right spherical triangle with opposite angles A, B, and C, respectively, where each side is measured by the angle subtended at the center of the sphere. Assume that $C = \pi/2 = 90°$ (see Figure 4.41, left). Then,

$$\sin a = \tan b \cot B = \sin A \sin c, \qquad \cos A = \tan b \cot c = \cos a \sin B,$$
$$\sin b = \tan a \cot A = \sin B \sin c, \qquad \cos B = \tan a \cot c = \cos b \sin A,$$
$$\cos c = \cos A \cot B = \cos a \cos b.$$

4.19.1.1 Napier's rules of circular parts

Arrange the five quantities a, b, co-A (this is the complement of A), co-c, co-B of a right spherical triangle with right angle at C, in cyclic order as pictured in

FIGURE 4.41
Right spherical triangle (left) and diagram for Napier's rule (right).

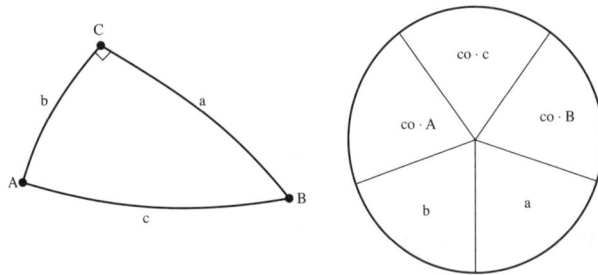

Figure 4.41, right. If any one of these quantities is designated a *middle part*, then two of the other parts are *adjacent* to it, and the remaining two parts are *opposite* to it. The formulae above for a right spherical triangle may be recalled by the following two rules:

1. The sine of any middle part is equal to the product of the *tangents* of the two *adjacent* parts.
2. The sine of any middle part is equal to the product of the *cosines* of the two *opposite* parts.

4.19.1.2   Rules for determining quadrant

1. A leg and the angle opposite to it are always of the same quadrant.
2. If the hypotenuse is less than $90°$, the legs are of the same quadrant.
3. If the hypotenuse is greater than $90°$, the legs are of unlike quadrants.

4.19.2   OBLIQUE SPHERICAL TRIANGLES

In the following:

- a, b, c represent the sides of any spherical triangle.
- A, B, C represent the corresponding opposite angles.
- a', b', c', A', B', C' are the corresponding parts of the polar triangle.
- $s = (a + b + c)/2$.
- $S = (A + B + C)/2$.
- Δ is the area of spherical triangle.
- E is the spherical excess of the triangle.
- R is the radius of the sphere upon which the triangle lies.

$$0° < a + b + c < 360°, \qquad 180° < A + B + C < 540°,$$
$$E = A + B + C - 180°, \qquad \Delta = \pi R^2 E / 180.$$

$$\tan \frac{1}{4} E = \sqrt{\tan \frac{s}{2} \tan \frac{1}{2}(s-a) \tan \frac{1}{2}(s-b) \tan \frac{1}{2}(s-c)}.$$

$$A = 180° - a', \qquad B = 180° - b', \qquad C = 180° - c',$$
$$a = 180° - A', \qquad b = 180° - B', \qquad c = 180° - C'.$$

4.19.2.1   Spherical law of sines

$$\frac{\sin a}{\sin A} = \frac{\sin b}{\sin B} = \frac{\sin c}{\sin C}.$$

4.19.2.2   Spherical law of cosines for sides

$$\cos a = \cos b \cos c + \sin b \sin c \cos A,$$
$$\cos b = \cos c \cos a + \sin c \sin a \cos B,$$
$$\cos c = \cos a \cos b + \sin a \sin b \cos C.$$

4.19.2.3   Spherical law of cosines for angles

$$\cos A = -\cos B \cos C + \sin B \sin C \cos a,$$
$$\cos B = -\cos C \cos A + \sin C \sin A \cos b,$$
$$\cos C = -\cos A \cos B + \sin A \sin B \cos c.$$

4.19.2.4   Spherical law of tangents

$$\frac{\tan \frac{1}{2}(B-C)}{\tan \frac{1}{2}(B+C)} = \frac{\tan \frac{1}{2}(b-c)}{\tan \frac{1}{2}(b+c)},$$
$$\frac{\tan \frac{1}{2}(C-A)}{\tan \frac{1}{2}(C+A)} = \frac{\tan \frac{1}{2}(c-a)}{\tan \frac{1}{2}(c+a)},$$
$$\frac{\tan \frac{1}{2}(A-B)}{\tan \frac{1}{2}(A+B)} = \frac{\tan \frac{1}{2}(a-b)}{\tan \frac{1}{2}(a+b)}.$$

4.19.2.5   Spherical half angle formulae

Define $k^2 = (\tan r)^2 = \dfrac{\sin(s-a) \sin(s-b) \sin(s-c)}{\sin s}$. Then

$$\tan\left(\frac{A}{2}\right) = \frac{k}{\sin(s-a)},$$

$$\tan\left(\frac{B}{2}\right) = \frac{k}{\sin(s-b)}, \qquad (4.19.1)$$

$$\tan\left(\frac{C}{2}\right) = \frac{k}{\sin(s-c)}.$$

4.19.2.6 Spherical half side formulae

Define $K^2 = (\tan R)^2 = \dfrac{-\cos S}{\cos(S-A)\cos(S-B)\cos(S-C)}$. Then

$$\tan(a/2) = K\cos(S-A),$$
$$\tan(b/2) = K\cos(S-B), \qquad (4.19.2)$$
$$\tan(c/2) = K\cos(S-C).$$

4.19.2.7 Gauss's formulae

$$\frac{\sin\frac{1}{2}(a-b)}{\sin\frac{1}{2}c} = \frac{\sin\frac{1}{2}(A-B)}{\cos\frac{1}{2}C}, \qquad \frac{\cos\frac{1}{2}(a-b)}{\cos\frac{1}{2}c} = \frac{\sin\frac{1}{2}(A+B)}{\cos\frac{1}{2}C},$$

$$\frac{\sin\frac{1}{2}(a+b)}{\sin\frac{1}{2}c} = \frac{\cos\frac{1}{2}(A-B)}{\sin\frac{1}{2}C}, \qquad \frac{\cos\frac{1}{2}(a+b)}{\cos\frac{1}{2}c} = \frac{\cos\frac{1}{2}(A+B)}{\sin\frac{1}{2}C}.$$

4.19.2.8 Napier's analogs

$$\frac{\sin\frac{1}{2}(A-B)}{\sin\frac{1}{2}(A+B)} = \frac{\tan\frac{1}{2}(a-b)}{\tan\frac{1}{2}c}, \qquad \frac{\sin\frac{1}{2}(a-b)}{\sin\frac{1}{2}(a+b)} = \frac{\tan\frac{1}{2}(A-B)}{\cot\frac{1}{2}C},$$

$$\frac{\cos\frac{1}{2}(A-B)}{\cos\frac{1}{2}(A+B)} = \frac{\tan\frac{1}{2}(a+b)}{\tan\frac{1}{2}c}, \qquad \frac{\cos\frac{1}{2}(a-b)}{\cos\frac{1}{2}(a+b)} = \frac{\tan\frac{1}{2}(A+B)}{\cot\frac{1}{2}C}.$$

4.19.2.9 Rules for determining quadrant

1. If $A > B > C$, then $a > b > c$.
2. A side (angle) which differs by more than $90°$ from another side (angle) is in the same quadrant as its opposite angle (side).
3. Half the sum of any two sides and half the sum of the opposite angles are in the same quadrant.

4.19.2.10   Summary of solution of oblique spherical triangles

Given	Solution	Check
Three sides	Half-angle formulae	Law of sines
Three angles	Half-side formulae	Law of sines
Two sides and included angle	Napier's analogies (to find sum and difference of unknown angles); then law of sines (to find remaining side).	Gauss's formulae
Two angles and included side	Napier's analogies (to find sum and difference of unknown sides); then law of sines (to find remaining angle).	Gauss's formulae
Two sides and an opposite angle	Law of sines (to find an angle); then Napier's analogies (to find remaining angle and side). Note the number of solutions.	Gauss's formulae
Two angles and an opposite side	Law of sines (to find a side); then Napier's analogies (to find remaining side and angle). Note the number of solutions.	Gauss's formulae

4.19.2.11   Haversine formulae

$$\text{hav } a = \frac{1 - \cos a}{2} = \sin^2 \frac{a}{2}$$
$$= \text{hav}(b - c) + \sin b \sin c \, \text{hav } A.$$
$$\text{hav } A = \frac{\sin(s - b) \sin(s - c)}{\sin b \sin c},$$
$$= \frac{\text{hav } a - \text{hav}(b - c)}{\sin b \sin c},$$
$$= \text{hav}[180° - (B + C)] + \sin B \sin C \, \text{hav } a.$$

4.19.2.12   Finding the distance between two points on the earth

To find the distance between two points on the surface of a spherical earth, let point P_1 have a (latitude, longitude) of (ϕ_1, θ_1) and point P_2 have a (latitude, longitude) of (ϕ_2, θ_2). Two different computational methods are as follows:

1. Let A be the North pole and let B and C be the points P_1 and P_2. Then the spherical law of cosines for sides gives the central angle, a, subtended by the desired distance:

$$\cos(a) = \cos(b) \cos(c) + \sin(b) \sin(c) \cos(A)$$

where the angle A is the difference in longitudes, and b and c are the angles of the points from the pole (i.e., $90° -$ latitude). Scale by $R_\oplus$ (the radius of the earth) to get the desired distance.

2. In (x, y, z) space (with $+z$ being the North pole) points P_1 and P_2 are represented as vectors from the center of the earth in spherical coordinates:

$$\mathbf{v}_1 = \begin{bmatrix} R_\oplus \cos(\phi_1) \cos(\theta_1) & R_\oplus \cos(\phi_1) \sin(\theta_1) & R_\oplus \sin(\phi_1) \end{bmatrix},$$
$$\mathbf{v}_2 = \begin{bmatrix} R_\oplus \cos(\phi_2) \cos(\theta_2) & R_\oplus \cos(\phi_2) \sin(\theta_2) & R_\oplus \sin(\phi_2) \end{bmatrix}.$$

(4.19.3)

The angle between these vectors, α, is given by

$$\cos \alpha = \frac{\mathbf{v}_1 \cdot \mathbf{v}_2}{|\mathbf{v}_1| |\mathbf{v}_2|} = \frac{\mathbf{v}_1 \cdot \mathbf{v}_2}{R_\oplus^2} = \cos(\phi_1) \cos(\phi_2) \cos(\theta_1 - \theta_2) + \sin(\phi_1) \sin(\phi_2)$$

$$= 2 \tan^{-1} \sqrt{\frac{b}{1-b}}$$

where $b = \sin^2 \left(\frac{\phi_1 - \phi_2}{2} \right) + \cos(\phi_1) \cos(\phi_2) \sin^2 \left(\frac{\theta_1 - \theta_2}{2} \right)$. (The formula using b is more accurate numerically when P_1 and P_2 are close.) The great circle distance between P_1 and P_2 is then $R_\oplus \alpha$.

EXAMPLE   The angle between $P_{\text{New York}}$ with ($\phi_1 = 40.78°, \theta_1 = 73.97°$) and P_{Beijing} with ($\phi_2 = 39.93°, \theta_2 = 243.58°$) is $\alpha = 98.8°$. Using $R_\oplus = 6367$ the great circle distance between New York and Beijing is about 11,000 km.

4.20 DIFFERENTIAL GEOMETRY

4.20.1 CURVES

4.20.1.1 Definitions

1. A *regular parametric representation of class* C^k, $k \geq 1$, is a vector valued function $\mathbf{f} : I \to \mathbb{R}^3$, where $I \subset \mathbb{R}$ is an interval that satisfies (i) $\mathbf{f}$ is of class C^k (i.e., has continuous k^{th} order derivatives), and (ii) $\mathbf{f}'(t) \neq 0$, for all $t \in I$. In terms of a standard basis of $\mathbb{R}^3$, we write $\mathbf{x} = \mathbf{f}(t) = (f_1(t), f_2(t), f_3(t))$, where the real valued functions f_i, $i = 1, 2, 3$ are the *component functions* of $\mathbf{f}$.

2. An *allowable change of parameter of class* C^k is any C^k function $\phi : J \to I$, where J is an interval and $\phi(J) \subset I$, that satisfies $\phi'(\tau) \neq 0$, for all $\tau \in J$.

3. A C^k regular parametric representation $\mathbf{f}$ is *equivalent* to a C^k regular parametric representation $\mathbf{g}$ if and only if an allowable change of parameter ϕ exists so that $\phi(I_g) = I_f$, and $\mathbf{g}(\tau) = \mathbf{f}(\phi(\tau))$, for all $\tau \in I_g$.

4. A *regular curve* C *of class* C^k is an equivalence class of C^k regular parametric representation under the equivalence relation on the set of regular parametric representations defined above.

5. The *arc length* of any regular curve C defined by the regular parametric representation $\mathbf{f}$, with $I_f = [a, b]$, is defined by

$$L = \int_a^b \left| \mathbf{f}'(u) \right| \, du. \tag{4.20.1}$$

An *arc length parameter* along C is defined by

$$s = \alpha(t) = \pm \int_c^t \left| \mathbf{f}'(u) \right| \, du. \tag{4.20.2}$$

The choice of sign is arbitrary and c is any number in I_f.

6. A *natural representation of class* C^k of the regular curve defined by the regular parametric representation $\mathbf{f}$ is defined by $\mathbf{g}(s) = \mathbf{f}(\alpha^{-1}(s))$, for all $s \in [0, L]$.

7. A *property* of a regular curve C is any property of a regular parametric representation representing C which is invariant under any allowable change of parameter.

8. Let $\mathbf{g}$ be a natural representation of a regular curve C. The following quantities may be defined at each point $\mathbf{x} = \mathbf{g}(s)$ of C:

Binormal line	$\mathbf{y} = \lambda\mathbf{b}(s) + \mathbf{x}$
Curvature	$\kappa(s) = \mathbf{n}(s) \cdot \mathbf{k}(s)$
Curvature vector	$\mathbf{k}(s) = \dot{\mathbf{t}}(s)$
Moving trihedron	$\{\mathbf{t}(s), \mathbf{n}(s), \mathbf{b}(s)\}$
Normal plane	$(\mathbf{y} - \mathbf{x}) \cdot \mathbf{t}(s) = 0$
Osculating plane	$(\mathbf{y} - \mathbf{x}) \cdot \mathbf{b}(s) = 0$
Osculating sphere	$(\mathbf{y} - \mathbf{c}) \cdot (\mathbf{y} - \mathbf{c}) = r^2$ where
	$\mathbf{c} = \mathbf{x} + \rho(s)\mathbf{n}(s) - (\dot{\kappa}(s)/(\kappa^2(s)\tau(s)))\mathbf{b}(s)$
	and $r^2 = \rho^2(s) + \kappa^2(s)/(\kappa^4(s)\tau^2(s))$
Principal normal line	$\mathbf{y} = \lambda\mathbf{n}(s) + \mathbf{x}$
Principal normal unit vector	$\mathbf{n}(s) = \pm\mathbf{k}(s)/|\mathbf{k}(s)|$, for $\mathbf{k}(s) \neq 0$ defined to be continuous along C
Radius of curvature	$\rho(s) = 1/|\kappa(s)|$, when $\kappa(s) \neq 0$
Rectifying plane	$(\mathbf{y} - \mathbf{x}) \cdot \mathbf{n}(s) = 0$
Tangent line	$\mathbf{y} = \lambda\mathbf{t}(s) + \mathbf{x}$
Torsion	$\tau(s) = -\mathbf{n}(s) \cdot \dot{\mathbf{b}}(s)$
Unit binormal vector	$\mathbf{b}(s) = \mathbf{t}(s) \times \mathbf{n}(s)$
Unit tangent vector	$\mathbf{t}(s) = \dot{\mathbf{g}}(s)$ with $\left(\dot{\mathbf{g}}(s) = \frac{d\mathbf{g}}{ds}\right)$

4.20.1.2 Results

The arc length L and the arc length parameter s of any regular parametric representation $\mathbf{f}$ are *invariant* under any allowable change of parameter. Thus, L is a property of the regular curve C defined by $\mathbf{f}$.

The arc length parameter satisfies $\frac{ds}{dt} = \alpha'(t) = \pm|\mathbf{f}'(t)|$, which implies that $|\mathbf{f}'(s)| = 1$, if and only if t is an arc length parameter. Thus, arc length parameters are uniquely determined up to the transformation $s \mapsto \tilde{s} = \pm s + s_0$, where s_0 is any constant.

The curvature, torsion, tangent line, normal plane, principal normal line, rectifying plane, binormal line, and osculating plane are properties of the regular curve C defined by any regular parametric representation $\mathbf{f}$.

If $\mathbf{x} = \mathbf{f}(t)$ is any regular representation of a regular curve C, the following results hold at point $\mathbf{f}(t)$ of C:

$$|\kappa| = \frac{|\mathbf{x}'' \times \mathbf{x}'|}{|\mathbf{x}'|^3}, \qquad \tau = \frac{\mathbf{x}' \cdot (\mathbf{x}'' \times \mathbf{x}''')}{|\mathbf{x}' \times \mathbf{x}''|^2}. \qquad (4.20.3)$$

The vectors of the moving trihedron satisfy the *Serret–Frenet equations*

$$\dot{\mathbf{t}} = \kappa \mathbf{n}, \quad \dot{\mathbf{n}} = -\kappa \mathbf{t} + \tau \mathbf{b}, \quad \dot{\mathbf{b}} = -\tau \mathbf{n}. \qquad (4.20.4)$$

For any plane curve represented parametrically by $\mathbf{x} = \mathbf{f}(t) = (t, f(t), 0)$,

$$|\kappa| = \frac{\left| \frac{d^2 x}{dt^2} \right|}{\left(1 + \left(\frac{dx}{dt} \right)^2 \right)^{3/2}}. \qquad (4.20.5)$$

Expressions for the curvature vector and curvature of a *plane curve* corresponding to different representations are given in the following table:

Representation	Curvature vector $\mathbf{k}$	Curvature, $|\kappa| = \rho^{-1}$
$x = f(t)$, $y = g(t)$	$\dfrac{(\dot{x}\ddot{y} - \dot{y}\ddot{x})}{(\dot{x}^2 + \dot{y}^2)^2}(-\dot{y}, \dot{x})$	$\dfrac{|\dot{x}\ddot{y} - \dot{y}\ddot{x}|}{(\dot{x}^2 + \dot{y}^2)^{3/2}}$
$y = f(x)$	$\dfrac{y''}{(1 + y'^2)^2}(-y', 1)$	$\dfrac{|y''|}{(1 + y'^2)^{3/2}}$
$r = f(\theta)$	$\dfrac{(r^2 + 2r'^2 - rr'')(-\dot{r}\sin\theta - r\cos\theta,}{(r^2 + r'^2)^2} \quad \dot{r}\cos\theta - r\sin\theta)$	$\dfrac{r^2 + 2r'^2 - rr''}{(r^2 + r'^2)^{3/2}}$

The equation of the *osculating circle* of a plane curve is given by

$$(\mathbf{y} - \mathbf{c}) \cdot (\mathbf{y} - \mathbf{c}) = \rho^2, \qquad (4.20.6)$$

where $\mathbf{c} = \mathbf{x} + \rho^2 \mathbf{k}$ is the *center of curvature*.

THEOREM 4.20.1 *(Fundamental existence and uniqueness theorem)*

Let $\kappa(s)$ and $\tau(s)$ be any continuous functions defined for all $s \in [a, b]$. Then there exists, up to a congruence, a unique space curve C for which κ is the curvature function, τ is the torsion function, and s an arc length parameter along C.

4.20.1.3 Example

A regular parametric representation of the *circular helix* is given by $\mathbf{x} = \mathbf{f}(t) = (a \cos t, a \sin t, bt)$, for all $t \in \mathbb{R}$, where $a > 0$ and $b \neq 0$ are constant. By successive differentiation,

$$\mathbf{x}' = (-a \sin t, \quad a \cos t, b),$$
$$\mathbf{x}'' = (-a \cos t, -a \sin t, 0), \qquad (4.20.7)$$
$$\mathbf{x}''' = (\ a \sin t, -a \cos t, 0),$$

so that $\frac{ds}{dt} = |\mathbf{x}'| = \sqrt{a^2 + b^2}$. Hence,

1. *Arc length parameter:* $s = \alpha(t) = t(a^2 + b^2)^{\frac{1}{2}}$
2. *Curvature vector:* $\mathbf{k} = \frac{d\mathbf{t}}{ds} = \frac{d\mathbf{t}}{dt}\frac{dt}{ds} = (a^2 + b^2)^{-1}(-a \cos t, -a \sin t, 0)$

3. *Curvature*: $\kappa = |\mathbf{k}| = a(a^2 + b^2)^{-1}$
4. *Principal normal unit vector*: $\mathbf{n} = \mathbf{k}/|\mathbf{k}| = (-\cos t, -\sin t, 0)$
5. *Unit tangent vector*: $\mathbf{t} = \frac{\mathbf{x}'}{|\mathbf{x}'|} = (a^2 + b^2)^{-\frac{1}{2}}(-a\sin t, a\cos t, b)$
6. *Unit binormal vector*:

$$\mathbf{b} = \mathbf{t} \times \mathbf{n} = (a^2 + b^2)^{-\frac{1}{2}}(b\sin t, b\cos t, a)$$

$$\dot{\mathbf{b}} = \frac{dt}{ds}\frac{d\mathbf{b}}{dt} = b(a^2 + b^2)^{-1}(\cos t, \sin t, 0)$$

7. *Torsion*: $\tau = -\mathbf{n} \cdot \dot{\mathbf{b}} = b(a^2 + b^2)^{-1}$

The values of $|\kappa|$ and τ can be verified using the formulae in (4.20.3). The sign of (the invariant) τ determines whether the helix is right handed, $\tau > 0$, or left handed, $\tau < 0$.

4.20.2   SURFACES

4.20.2.1   Definitions

1. A *coordinate patch of class* C^k, $k \geq 1$ on a surface $S \subset \mathbb{R}^3$ is a vector valued function $f : U \to S$, where $U \subset \mathbb{R}^2$ is an open set, that satisfies (i) $\mathbf{f}$ is class C^k on U, (ii) $\frac{\partial \mathbf{f}}{\partial u}(u, v) \times \frac{\partial \mathbf{f}}{\partial v}(u, v) \neq 0$, for all $(u, v) \in U$, and (iii) $\mathbf{f}$ is one-to-one and bi-continuous on U.

2. In terms of a standard basis of $\mathbb{R}^3$ we write $\mathbf{x} = \mathbf{f}(u, v) = (f_1(u, v), f_2(u, v), f_3(u, v))$, where the real valued functions $\{f_1, f_2, f_3\}$ are the *component functions* of $\mathbf{f}$. The notation $\mathbf{x}_1 = \mathbf{x}_u = \frac{\partial \mathbf{f}}{\partial u}$, $\mathbf{x}_2 = \mathbf{x}_v = \frac{\partial \mathbf{f}}{\partial v}$, $u^1 = u, u^2 = v$, is frequently used.

3. A *Monge patch* is a coordinate patch where $\mathbf{f}$ has the form $\mathbf{f}(u, v) = (u, v, f(u, v))$, where f is a real valued function of class C^k.

4. The *u-parameter curves* $v = v_0$ on S are the images of the lines $v = v_0$ in U. They are parametrically represented by $\mathbf{x} = \mathbf{f}(u, v_0)$. The *v-parameter curves* $u = u_0$ are defined similarly.

5. An *allowable parameter transformation* of class C^k is a one-to-one function $\phi : U \to V$, where $U, V \subset \mathbb{R}^2$ are open, that satisfies

$$\det \begin{bmatrix} \frac{\partial \phi^1}{\partial u}(u, v) & \frac{\partial \phi^1}{\partial v}(u, v) \\ \frac{\partial \phi^2}{\partial u}(u, v) & \frac{\partial \phi^2}{\partial v}(u, v) \end{bmatrix} \neq 0, \tag{4.20.8}$$

for all $(u, v) \in U$, where the real valued functions, ϕ^1 and ϕ^2, defined by $\phi(u, v) = (\phi^1(u, v), \phi^2(u, v))$ are the component functions of ϕ. One may also write the parameter transformation as $\tilde{u}^1 = \phi^1(u^1, u^2)$, $\tilde{u}^2 = \phi^2(u^1, u^2)$.

6. A *local property* of surface S is any property of a coordinate patch that is invariant under any allowable parameter transformation.

7. Let $\mathbf{f}$ define a coordinate patch on a surface S. The following quantities may be defined at each point $\mathbf{x} = \mathbf{f}(u, v)$ on the patch:

Asymptotic direction	A direction $du : dv$ for which $\kappa_n = 0$
Asymptotic line	A curve on S whose tangent line at each point coincides with an asymptotic direction
Dupin's indicatrix	$ex_1^2 + 2fx_1 x_2 + gx_2^2 = \pm 1$
Elliptic point	$eg - f^2 > 0$
First fundamental form	$\begin{aligned} I = d\mathbf{x} \cdot d\mathbf{x} &= g_{\alpha\beta}(u, v)\, du^\alpha\, du^\beta \\ &= E(u, v)\, du^2 + 2F(u, v)\, du\, dv + G(u, v)\, dv^2 \end{aligned}$
First fundamental metric coefficients	$\begin{cases} E(u, v) = g_{11}(u, v) = \mathbf{x}_1 \cdot \mathbf{x}_1 \\ F(u, v) = g_{12}(u, v) = \mathbf{x}_1 \cdot \mathbf{x}_2 \\ G(u, v) = g_{22}(u, v) = \mathbf{x}_2 \cdot \mathbf{x}_2 \end{cases}$
Fundamental differential	$d\mathbf{x} = \mathbf{x}_\alpha\, du^\alpha = \mathbf{x}_u\, du + \mathbf{x}_v\, dv$ (a repeated upper and lower index signifies a summation over the range $\alpha = 1, 2$)
Gaussian curvature	$K = \kappa_1 \kappa_2 = \dfrac{eg - f^2}{EG - F^2}$
Geodesic curvature vector of curve C on S through x	$\mathbf{k}_g = \mathbf{k} - (\mathbf{k} \cdot \mathbf{n})\mathbf{n} = [\ddot{u}^\alpha + \Gamma^\alpha_{\beta\gamma} \dot{u}^\beta \dot{u}^\gamma]\mathbf{x}_\alpha$ where $\Gamma^\alpha_{\beta\gamma}$ denote the Christoffel symbols of the second kind for the metric $g_{\alpha\beta}$, defined in Section 5.10
Geodesic on S	A curve on S which satisfies $\mathbf{k}_g = 0$ at each point
Hyperbolic point	$eg - f^2 < 0$
Line of curvature	A curve on S whose tangent line at each point coincides with a principal direction
Mean curvature	$H = \dfrac{\kappa_1 + \kappa_2}{2} = \dfrac{gE + eG - 2fF}{2(EG - F^2)}$
Normal curvature in the du : dv direction	$\kappa_n = \mathbf{k} \cdot \mathbf{n} = \dfrac{II}{I}$
Normal curvature vector of curve C on S through x	$\mathbf{k}_n = (\mathbf{k} \cdot \mathbf{n})\mathbf{n}$
Normal line	$\mathbf{y} = \lambda \mathcal{N} + \mathbf{x}$
Normal vector	$\mathcal{N} = \mathbf{x}_u \times \mathbf{x}_v$
Parabolic point	$eg - f^2 = 0$ not all of $e, f, g = 0$
Planar point	$e = f = g = 0$
Principal curvatures	The extreme values κ_1 and κ_2 of κ_n
Principal directions	The perpendicular directions $du : dv$ in which κ_n attains its extreme values
Second fundamental form	$\begin{aligned} II = -d\mathbf{x} \cdot d\mathbf{n} &= b_{\alpha\beta}(u, v)\, du^\alpha\, du^\beta \\ &= e(u, v)\, du^2 + 2f(u, v)\, du\, dv + g(u, v)\, dv^2 \end{aligned}$
Second fundamental metric coefficients	$\begin{cases} e(u, v) = b_{11}(u, v) = \mathbf{x}_{11} \cdot \mathbf{n} \\ f(u, v) = b_{12}(u, v) = \mathbf{x}_{12} \cdot \mathbf{n} \\ g(u, v) = b_{22}(u, v) = \mathbf{x}_{22} \cdot \mathbf{n} \end{cases}$

Tangent plane	$(\mathbf{y} - \mathbf{x}) \cdot \mathcal{N} = 0$, or $\mathbf{y} = \mathbf{x} + \lambda \mathbf{x}_u + \mu \mathbf{x}_v$
Unit normal vector	$\mathbf{N} = \dfrac{\mathbf{x}_u \times \mathbf{x}_v}{|\mathbf{x}_u \times \mathbf{x}_v|}$
Umbilical point	$\kappa_n = $ constant for all directions $du : dv$

4.20.2.2   Results

1. The tangent plane, normal line, first fundamental form, second fundamental form, and all derived quantities thereof are local properties of any surface S.

2. The transformation laws for the first and second fundamental metric coefficients under any allowable parameter transformation are given respectively by

$$\tilde{g}_{\alpha\beta} = g_{\gamma\delta} \frac{\partial u^\gamma}{\partial \tilde{u}^\alpha} \frac{\partial u^\delta}{\partial \tilde{u}^\beta}, \quad \text{and} \quad \tilde{b}_{\alpha\beta} = b_{\gamma\delta} \frac{\partial u^\gamma}{\partial \tilde{u}^\alpha} \frac{\partial u^\delta}{\partial \tilde{u}^\beta}. \tag{4.20.9}$$

Thus $g_{\alpha\beta}$ and $b_{\alpha\beta}$ are the components of type $(0, 2)$ tensors.

3. $I \geq 0$ for all directions $du : dv$; $I = 0$ if and only if $du = dv = 0$.

4. The angle θ between two tangent lines to S at $\mathbf{x} = \mathbf{f}(u, v)$ defined by the directions $du : dv$ and $\delta u : \delta v$ is given by

$$\cos \theta = \frac{g_{\alpha\beta} \, du^\alpha \delta u^\beta}{(g_{\alpha\beta} \, du^\alpha \, du^\beta)^{\frac{1}{2}} (g_{\alpha\beta} \delta u^\alpha \delta u^\beta)^{\frac{1}{2}}}. \tag{4.20.10}$$

The angle between the u-parameter curves and the v-parameter curves is given by $\cos \theta = F(u, v)/(E(u, v)G(u, v))^{\frac{1}{2}}$. The u-parameter and v-parameter curves are orthogonal if and only if $F(u, v) = 0$.

5. The arc length of a curve C on S, defined by $\mathbf{x} = \mathbf{f}(u^1(t), u^2(t))$, with $a \leq t \leq b$, is given by

$$L = \int_a^b \sqrt{g_{\alpha\beta}(u^1(t), u^2(t)) \dot{u}^\alpha \dot{u}^\beta} \, dt \tag{4.20.11}$$

$$= \int_a^b \sqrt{E(u(t), v(t)) \dot{u}^2 + 2F(u(t), v(t)) \dot{u}\dot{v} + G(u(t), v(t)) \dot{v}^2} \, dt.$$

6. The area of $S = \mathbf{f}(U)$ is given by

$$A = \iint_U \sqrt{\det(g_{\alpha\beta}(u^1, u^2))} \, du^1 \, du^2$$

$$= \iint_U \sqrt{E(u, v)G(u, v) - F^2(u, v)} \, du \, dv. \tag{4.20.12}$$

7. The principal curvatures are the roots of the *characteristic equation*, $\det(b_{\alpha\beta} - \lambda g_{\alpha\beta}) = 0$, which may be written as $\lambda^2 - b_{\alpha\beta} g^{\alpha\beta} \lambda + b/g = 0$, where $g^{\alpha\beta}$ is the inverse of $g_{\alpha\beta}$, $b = \det(b_{\alpha\beta})$, and $g = \det(g_{\alpha\beta})$. The expanded form of the characteristic equation is

$$(EG - F^2)\lambda^2 - (eG - 2fF + gE)\lambda + eg - f^2 = 0. \tag{4.20.13}$$

8. The principal directions $du : dv$ are obtained by solving the homogeneous equation,

$$b_{1\alpha}g_{2\beta}\, du^\alpha\, du^\beta - b_{2\alpha}g_{1\beta}\, du^\alpha\, du^\beta = 0, \qquad (4.20.14)$$

or

$$(eF - fE)\, du^2 + (eG - gE)\, du\, dv + (fG - gF)\, dv^2 = 0. \qquad (4.20.15)$$

9. *Rodrigues formula*: $du : dv$ is a principal direction with principal curvature κ if, and only if, $d\mathbf{N} + \kappa\, d\mathbf{x} = 0$.

10. A point $\mathbf{x} = \mathbf{f}(u, v)$ on S is an umbilical point if and only if there exists a constant k such that $b_{\alpha\beta}(u, v) = kg_{\alpha\beta}(u, v)$.

11. The principal directions at $\mathbf{x}$ are orthogonal if $\mathbf{x}$ is not an umbilical point.

12. The u- and v-parameter curves at any non-umbilical point $\mathbf{x}$ are tangent to the principal directions if and only if $f(u, v) = F(u, v) = 0$. If $\mathbf{f}$ defines a coordinate patch without umbilical points, the u- and v-parameter curves are lines of curvature if and only if $f = F = 0$.

13. If $f = F = 0$ on a coordinate patch, the principal curvatures are given by $\kappa_1 = e/E$, $\kappa_2 = g/G$. It follows that the Gaussian and mean curvatures have the forms

$$K = \frac{eg}{EG}, \quad \text{and} \quad H = \frac{1}{2}\left(\frac{e}{E} + \frac{g}{G}\right). \qquad (4.20.16)$$

14. The *Gauss equation*: $\mathbf{x}_{\alpha\beta} = \Gamma_{\alpha\beta}^\gamma \mathbf{x}_\gamma + b_{\alpha\beta}\mathbf{n}$.

15. The *Weingarten equation*: $\mathbf{n}_\alpha = -b_{\alpha\beta}g^{\beta\gamma}\mathbf{x}_\gamma$.

16. The *Gauss–Mainardi–Codazzi equations*: $b_{\alpha\beta}b_{\gamma\delta} - b_{\alpha\gamma}b_{\beta\delta} = R_{\delta\alpha\beta\gamma}$, $b_{\alpha\beta,\gamma} - b_{\alpha\gamma,\beta} + \Gamma_{\alpha\beta}^\delta b_{\delta\gamma} - \Gamma_{\alpha\gamma}^\delta b_{\delta\beta} = 0$, where $R_{\delta\alpha\beta\gamma}$ denotes the Riemann curvature tensor defined in Section 5.10.3.

THEOREM 4.20.2 *(Gauss's theorema egregium)*

The Gaussian curvature K depends only on the components of the first fundamental metric $g_{\alpha\beta}$ and their derivatives.

THEOREM 4.20.3 *(Fundamental theorem of surface theory)*

If $g_{\alpha\beta}$ and $b_{\alpha\beta}$ are sufficiently differentiable functions of u and v which satisfy the Gauss–Mainardi–Codazzi equations, $\det(g_{\alpha\beta}) > 0$, $g_{11} > 0$, and $g_{22} > 0$, then a surface exists with $I = g_{\alpha\beta}\, du^\alpha\, du^\beta$ and $II = b_{\alpha\beta}\, du^\alpha\, du^\beta$ as its first and second fundamental forms. This surface is unique up to a congruence.

4.20.2.3 Example: paraboloid of revolution

A Monge patch for a paraboloid of revolution is given by $\mathbf{x} = \mathbf{f}(u, v) = (u, v, u^2 + v^2)$, for all $(u, v) \in U = \mathbb{R}^2$. By successive differentiation one obtains $\mathbf{x}_u = (1, 0, 2u)$, $\mathbf{x}_v = (0, 1, 2v)$, $\mathbf{x}_{uu} = (0, 0, 2)$, $\mathbf{x}_{uv} = (0, 0, 0)$, and $\mathbf{x}_{vv} = (0, 0, 2)$.

1. *Unit normal vector*: $\mathbf{n} = (1 + 4u^2 + 4v^2)^{-\frac{1}{2}}(-2u, -2v, 1)$.

2. *First fundamental coefficients*: $E(u, v) = g_{11}(u, v) = 1 + 4u^2$, $F(u, v) = g_{12}(u, v) = 4uv$, $G(u, v) = g_{22}(u, v) = 1 + 4v^2$.

3. *First fundamental form*: $I = (1 + 4u^2)\, du^2 + 8uv\, du\, dv + (1 + 4v^2)\, dv^2$. Since $F(u, v) = 0 \Rightarrow u = 0$ or $v = 0$, it follows that the u-parameter curve $v = 0$ is orthogonal to any v-parameter curve, and the v-parameter curve $u = 0$ is orthogonal to any u-parameter curve. Otherwise the u- and v-parameter curves are *not* orthogonal.

4. *Second fundamental coefficients*: $e(u, v) = b_{11}(u, v) = 2(1 + 4u^2 + 4v^2)^{-\frac{1}{2}}$, $f(u, v) = b_{12}(u, v) = 0$, $g(u, v) = b_{22}(u, v) = 2(1 + 4u^2 + 4v^2)^{-\frac{1}{2}}$.

5. *Second fundamental form*: $II = 2(1 + 4u^2 + 4v^2)^{-\frac{1}{2}}(du^2 + dv^2)$.

6. *Classification of points*: $e(u, v)g(u, v) = 4(1 + 4u^2 + 4v^2) > 0$ implies that all points on S are elliptic points. The point $(0, 0, 0)$ is the only umbilical point.

7. *Equation for the principal directions*: $uv\, du^2 + (v^2 - u^2)\, du\, dv + uv\, dv^2 = 0$ factors to read $(u\, du + v\, dv)(v\, du - u\, dv) = 0$.

8. *Lines of curvature*: Integrate the differential equations, $u\, dv + v\, dv = 0$, and $v\, du - v\, du = 0$, to obtain, respectively, the equations of the lines of curvature, $u^2 + v^2 = r^2$, and $u/v = \cot \theta$, where r and θ are constant.

9. *Characteristic equation*: $(1 + 4u^2 + 4v^2)\lambda^2 - 4(1 + 2u^2 + 2v^2)(1 + 4u^2 + 4v^2)^{-\frac{1}{2}}\lambda + 4(1 + 4u^2 + 4v^2)^{-1} = 0$.

10. *Principal curvatures*: $\kappa_1 = 2(1 + 4u^2 + 4v^2)^{-\frac{1}{2}}$, $\kappa_2 = 2(1 + 4u^2 + 4v^2)^{-\frac{3}{2}}$. The paraboloid of revolution may also be represented by $\mathbf{x} = \widehat{\mathbf{f}}(r, \theta) = (r \cos \theta, r \sin \theta, r^2)$. In this representation the r- and θ-parameter curves are lines of curvature.

11. *Gaussian curvature*: $K = 4(1 + 4u^2 + 4v^2)^{-2}$.

12. *Mean curvature*: $H = 2(1 + 2u^2 + 2v^2)(1 + 4u^2 + 4v^2)^{-\frac{3}{2}}$.

4.21 ANGLE CONVERSION

Degrees	Radians	Minutes	Radians	Seconds	Radians
1°	0.01745 33	1′	0.00029 089	1″	0.00000 48481
2°	0.03490 66	2′	0.00058 178	2″	0.00000 96963
3°	0.05235 99	3′	0.00087 266	3″	0.00001 45444
4°	0.06981 32	4′	0.00116 355	4″	0.00001 93925
5°	0.08726 65	5′	0.00145 444	5″	0.00002 42407
6°	0.10471 98	6′	0.00174 533	6″	0.00002 90888
7°	0.12217 30	7′	0.00203 622	7″	0.00003 39370
8°	0.13962 63	8′	0.00232 711	8″	0.00003 87851
9°	0.15707 96	9′	0.00261 799	9″	0.00004 36332
10°	0.17453 29	10′	0.00290 888	10″	0.00004 84814

Rad.	Deg.	Min.	Sec.	Deg.	Rad.	Deg.	Min.	Sec.	Deg.
1	57°	17′	44.8″	57.2958	0.1	5°	43′	46.5″	5.7296
2	114°	35′	29.6″	114.5916	0.2	11°	27′	33.0″	11.4592
3	171°	53′	14.4″	171.8873	0.3	17°	11′	19.4″	17.1887
4	229°	10′	59.2″	229.1831	0.4	22°	55′	5.9″	22.9183
5	286°	28′	44.0″	286.4789	0.5	28°	38′	52.4″	28.6479
6	343°	46′	28.8″	343.7747	0.6	34°	22′	38.9″	34.3775
7	401°	4′	13.6″	401.0705	0.7	40°	6′	25.4″	40.1070
8	458°	21′	58.4″	458.3662	0.8	45°	50′	11.8″	45.8366
9	515°	39′	43.3″	515.6620	0.9	51°	33′	58.3″	51.5662

Rad.	Deg.	Min.	Sec.	Deg.	Rad.	Deg.	Min.	Sec.	Deg.
0.01	0°	34′	22.6″	0.5730	0.001	0°	3′	26.3″	0.0573
0.02	1°	8′	45.3″	1.1459	0.002	0°	6′	52.5″	0.1146
0.03	1°	43′	7.9″	1.7189	0.003	0°	10′	18.8″	0.1719
0.04	2°	17′	30.6″	2.2918	0.004	0°	13′	45.1″	0.2292
0.05	2°	51′	53.2″	2.8648	0.005	0°	17′	11.3″	0.2865
0.06	3°	26′	15.9″	3.4377	0.006	0°	20′	37.6″	0.3438
0.07	4°	0′	38.5″	4.0107	0.007	0°	24′	3.9″	0.4011
0.08	4°	35′	1.2″	4.5837	0.008	0°	27′	30.1″	0.4584
0.09	5°	9′	23.8″	5.1566	0.009	0°	30′	56.4″	0.5157

4.22   KNOTS UP TO EIGHT CROSSINGS

n	3	4	5	6	7	8	9	10	11
Number of knots with n crossings	1	1	2	3	7	21	49	165	552

Image by Charlie Gunn and David Broman. Copyright The Geometry Center, University of Minnesota. With permission.

Chapter 5

Continuous
Mathematics

5.1 DIFFERENTIAL CALCULUS **385**
   5.1.1 *Limits* *385*
   5.1.2 *Derivatives* *386*
   5.1.3 *Derivatives of common functions* *386*
   5.1.4 *Derivative formulae* *387*
   5.1.5 *Derivative theorems* *388*
   5.1.6 *The two-dimensional chain rule* *388*
   5.1.7 *l'Hôspital's rule* *389*
   5.1.8 *Maxima and minima of functions* *389*
   5.1.9 *Vector calculus* *390*
   5.1.10 *Matrix and vector derivatives* *391*

5.2 DIFFERENTIAL FORMS **395**
   5.2.1 *Products of 1-forms* *395*
   5.2.2 *Differential 2-forms* *396*
   5.2.3 *The 2-forms in $\mathbb{R}^n$* *396*
   5.2.4 *Higher dimensional forms* *397*
   5.2.5 *The exterior derivative* *397*
   5.2.6 *Properties of the exterior derivative* *398*

5.3 INTEGRATION **398**
   5.3.1 *Definitions* *398*
   5.3.2 *Properties of integrals* *399*
   5.3.3 *Methods of evaluating integrals* *400*
   5.3.4 *Types of integrals* *404*
   5.3.5 *Integral inequalities* *406*
   5.3.6 *Convergence tests* *407*
   5.3.7 *Variational principles* *407*
   5.3.8 *Continuity of integral antiderivatives* *407*
   5.3.9 *Asymptotic integral evaluation* *408*
   5.3.10 *Special functions defined by integrals* *408*
   5.3.11 *Applications of integration* *409*
   5.3.12 *Moments of inertia for various bodies* *410*
   5.3.13 *Tables of integrals* *411*

5.4 TABLE OF INDEFINITE INTEGRALS **412**
   5.4.1 *Elementary forms* *412*

1-58488-291-3/02/$0.00+$1.50
© 2003 CRC Press, Inc.

5.4.2	Forms containing $a + bx$	413
5.4.3	Forms containing $c^2 \pm x^2$ and $x^2 - c^2$	415
5.4.4	Forms containing $a + bx$ and $c + dx$	415
5.4.5	Forms containing $a + bx^n$	416
5.4.6	Forms containing $c^3 \pm x^3$	418
5.4.7	Forms containing $c^4 \pm x^4$	418
5.4.8	Forms containing $a + bx + cx^2$	419
5.4.9	Forms containing $\sqrt{a + bx}$	420
5.4.10	Forms containing $\sqrt{a + bx}$ and $\sqrt{c + dx}$	421
5.4.11	Forms containing $\sqrt{x^2 \pm a^2}$	422
5.4.12	Forms containing $\sqrt{a^2 - x^2}$	424
5.4.13	Forms containing $\sqrt{a + bx + cx^2}$	426
5.4.14	Forms containing $\sqrt{2ax - x^2}$	428
5.4.15	Miscellaneous algebraic forms	428
5.4.16	Forms involving trigonometric functions	430
5.4.17	Forms involving inverse trigonometric functions	439
5.4.18	Logarithmic forms	441
5.4.19	Exponential forms	443
5.4.20	Hyperbolic forms	445
5.4.21	Bessel functions	448
5.5	**TABLE OF DEFINITE INTEGRALS**	**448**
5.5.1	Table of semi-integrals	455
5.6	**ORDINARY DIFFERENTIAL EQUATIONS**	**456**
5.6.1	Linear differential equations	456
5.6.2	Solution techniques overview	461
5.6.3	Integrating factors	462
5.6.4	Variation of parameters	462
5.6.5	Green's functions	463
5.6.6	Table of Green's functions	463
5.6.7	Transform techniques	464
5.6.8	Named ordinary differential equations	465
5.6.9	Liapunov's direct method	466
5.6.10	Lie groups	466
5.6.11	Stochastic differential equations	467
5.6.12	Types of critical points	468
5.7	**PARTIAL DIFFERENTIAL EQUATIONS**	**468**
5.7.1	Classifications of PDEs	468
5.7.2	Named partial differential equations	469
5.7.3	Transforming partial differential equations	469
5.7.4	Well-posedness of PDEs	470
5.7.5	Green's functions	471
5.7.6	Quasi-linear equations	472
5.7.7	Separation of variables	472
5.7.8	Solutions of Laplace's equation	474
5.7.9	Solutions to the wave equation	475
5.7.10	Particular solutions to some PDEs	477
5.8	**EIGENVALUES**	**477**
5.9	**INTEGRAL EQUATIONS**	**478**
5.9.1	Definitions	478

5.9.2   *Connection to differential equations* 479
5.9.3   *Fredholm alternative* 479
5.9.4   *Special equations with solutions* 480

5.10 **TENSOR ANALYSIS** **482**
    5.10.1   *Definitions* 482
    5.10.2   *Algebraic tensor operations* 483
    5.10.3   *Differentiation of tensors* 484
    5.10.4   *Metric tensor* 486
    5.10.5   *Results* 487
    5.10.6   *Examples of tensors* 489

5.11 **ORTHOGONAL COORDINATE SYSTEMS** **492**
    5.11.1   *Table of orthogonal coordinate systems* 494

5.12 **CONTROL THEORY** **497**

5.1   DIFFERENTIAL CALCULUS

5.1.1   LIMITS

If $\lim\limits_{x \to a} f(x) = A < \infty$ and $\lim\limits_{x \to a} g(x) = B < \infty$ then

1. $\lim\limits_{x \to a} (f(x) \pm g(x)) = A \pm B$

2. $\lim\limits_{x \to a} f(x)g(x) = AB$

3. $\lim\limits_{x \to a} \dfrac{f(x)}{g(x)} = \dfrac{A}{B}$ (if $B \neq 0$)

4. $\lim\limits_{x \to a} [f(x)]^{g(x)} = A^B$ (if $A > 0$)

5. $\lim\limits_{x \to a} h(f(x)) = h(A)$ (if h continuous)

6. If $f(x) \leq g(x)$, then $A \leq B$

7. If $A = B$ and $f(x) \leq h(x) \leq g(x)$, then $\lim\limits_{x \to a} h(x) = A$

EXAMPLES

1. $\lim\limits_{x \to \infty} \left(1 + \dfrac{t}{x}\right)^x = e^t$

2. $\lim\limits_{x \to \infty} x^{1/x} = 1$

3. $\lim\limits_{x \to \infty} \dfrac{(\log x)^p}{x^q} = 0 \text{ (if } q > 0)$

4. $\lim\limits_{x \to 0+} x^p |\log x|^q = 0 \text{ (if } p > 0)$

5. $\lim\limits_{x \to 0} \dfrac{\sin ax}{x} = a$

6. $\lim\limits_{x \to 0} \dfrac{a^x - 1}{x} = \log a$

7. $\lim\limits_{x \to 0} \dfrac{\log(1 + x)}{x} = 1$

5.1.2 DERIVATIVES

The *derivative* of the function $f(x)$, written $f'(x)$, is defined as

$$f'(x) = \lim_{\Delta x \to 0} \frac{f(x + \Delta x) - f(x)}{\Delta x} \tag{5.1.1}$$

if the limit exists. If $y = f(x)$, then $\frac{dy}{dx} = f'(x)$. The nth derivative is

$$y^{(n)} = \frac{dy^{(n-1)}}{dx} = \frac{d}{dx}\left(\frac{d^{n-1}y}{dx^{n-1}}\right) = \frac{d^n y}{dx^n} = f^{(n)}(x).$$

The second and third derivatives are usually written as y'' and y'''. Sometimes the fourth and fifth derivatives are written as $y^{(iv)}$ and $y^{(v)}$.

The partial derivative of $f(x, y)$ with respect to x, written $f_x(x, y)$ or $\frac{\partial f}{\partial x}$, is defined as

$$f_x(x, y) = \lim_{\Delta x \to 0} \frac{f(x + \Delta x, y) - f(x, y)}{\Delta x}. \tag{5.1.2}$$

5.1.3 DERIVATIVES OF COMMON FUNCTIONS

Let a be a constant.

$f(x)$	$f'(x)$	$f(x)$	$f'(x)$	$f(x)$	$f'(x)$
$\sin x$	$\cos x$	$\sinh x$	$\cosh x$	x^a	ax^{a-1}
$\cos x$	$-\sin x$	$\cosh x$	$\sinh x$	$\frac{1}{x^a}$	$-\frac{a}{x^{a+1}}$
$\tan x$	$\sec^2 x$	$\tanh x$	$\operatorname{sech}^2 x$	$\sqrt{x}$	$\frac{1}{2\sqrt{x}}$
$\csc x$	$-\csc x \cot x$	$\operatorname{csch} x$	$-\operatorname{csch} x \coth x$	$\ln |x|$	$1/x$
$\sec x$	$\sec x \tan x$	$\operatorname{sech} x$	$-\operatorname{sech} x \tanh x$	e^x	e^x
$\cot x$	$-\csc^2 x$	$\coth x$	$-\operatorname{csch}^2 x$	$a^x \ (a > 0)$	$a^x \ln a$
$\sin^{-1} x$	$\frac{1}{\sqrt{1-x^2}}$	$\sinh^{-1} x$	$\frac{1}{\sqrt{x^2+1}}$		
$\cos^{-1} x$	$-\frac{1}{\sqrt{1-x^2}}$	$\cosh^{-1} x$	$\frac{1}{\sqrt{x^2-1}}$		
$\tan^{-1} x$	$\frac{1}{1+x^2}$	$\tanh^{-1} x$	$\frac{1}{1-x^2}$		
$\csc^{-1} x$	$-\frac{1}{x\sqrt{x^2-1}}$	$\operatorname{csch}^{-1} x$	$\frac{-1}{|x|\sqrt{1+x^2}}$		
$\sec^{-1} x$	$\frac{1}{x\sqrt{x^2-1}}$	$\operatorname{sech}^{-1} x$	$\frac{-1}{|x|\sqrt{1-x^2}}$		
$\cot^{-1} x$	$-\frac{1}{1+x^2}$	$\coth^{-1} x$	$\frac{1}{1-x^2}$		

5.1.4 DERIVATIVE FORMULAE

Let u, v, w be functions of x, and let a, c, and n be constants. Appropriate non-zero values, differentiability, and invertability are assumed.

(a) $\dfrac{d}{dx}(a) = 0$

(b) $\dfrac{d}{dx}(x) = 1$

(c) $\dfrac{d}{dx}(au) = a\dfrac{du}{dx}$

(d) $\dfrac{d}{dx}(u+v) = \dfrac{du}{dx} + \dfrac{dv}{dx}$

(e) $\dfrac{d}{dx}(uv) = v\dfrac{du}{dx} + u\dfrac{dv}{dx}$

(f) $\dfrac{d}{dx}(uvw) = uv\dfrac{dw}{dx} + uw\dfrac{dv}{dx} + vw\dfrac{du}{dx}$

(g) $\dfrac{d}{dx}\left(\dfrac{u}{v}\right) = \dfrac{1}{v}\dfrac{du}{dx} - \dfrac{u}{v^2}\dfrac{dv}{dx} = \dfrac{v(du/dx) - u(dv/dx)}{v^2}$

(h) $\dfrac{d}{dx}(u^n) = nu^{n-1}\dfrac{du}{dx}$

(i) $\dfrac{d}{dx}(u^v) = vu^{v-1}\dfrac{du}{dx} + (\log_e u)u^v\dfrac{dv}{dx}$

(j) $\dfrac{d}{dx}(\sqrt{u}) = \dfrac{1}{2\sqrt{u}}\dfrac{du}{dx}$

(k) $\dfrac{d}{dx}(\log_e u) = \dfrac{1}{u}\dfrac{du}{dx}$

(l) $\dfrac{d}{dx}(\log_a u) = (\log_a e)\dfrac{1}{u}\dfrac{du}{dx}$

(m) $\dfrac{d}{dx}\left(\dfrac{1}{u}\right) = -\dfrac{1}{u^2}\dfrac{du}{dx}$

(n) $\dfrac{d}{dx}\left(\dfrac{1}{u^n}\right) = -\dfrac{n}{u^{n+1}}\dfrac{du}{dx}$

(o) $\dfrac{d}{dx}\left(\dfrac{u^n}{v^m}\right) = \dfrac{u^{n-1}}{v^{m+1}}\left(nv\dfrac{du}{dx} - mu\dfrac{dv}{dx}\right)$

(p) $\dfrac{d}{dx}(u^n v^m) = u^{n-1}v^{m-1}\left(nv\dfrac{du}{dx} + mu\dfrac{dv}{dx}\right)$

(q) $\dfrac{d}{dx}(f(u)) = \dfrac{df}{du}(u) \cdot \dfrac{du}{dx}$

(r) $\dfrac{d^2}{dx^2}(f(u)) = \dfrac{df}{du}(u) \cdot \dfrac{d^2u}{dx^2} + \dfrac{d^2f}{du^2}(u) \cdot \left(\dfrac{du}{dx}\right)^2$

(s) $\dfrac{d^n}{dx^n}(uv) = \binom{n}{0}v\dfrac{d^n u}{dx^n} + \binom{n}{1}\dfrac{dv}{dx}\dfrac{d^{n-1}u}{dx^{n-1}} + \cdots + \binom{n}{n}\dfrac{d^n v}{dx^n}u$

(t) $\dfrac{d}{dx}\displaystyle\int_c^x f(t)\,dt = f(x)$

(u) $\dfrac{d}{dx}\displaystyle\int_x^c f(t)\,dt = -f(x)$

(v) $\dfrac{dx}{dy} = \left(\dfrac{dy}{dx}\right)^{-1}$ and $\dfrac{d^2x}{dy^2} = -\dfrac{d^2y}{dx^2}\bigg/\left(\dfrac{dy}{dx}\right)^3$

(w) If $F(x, y) = 0$, then $\dfrac{dy}{dx} = -\dfrac{F_x}{F_y}$ and

$$\frac{d^2y}{dx^2} = -\frac{\left(F_{xx}F_y^2 - 2F_{xy}F_xF_y + F_{yy}F_x^2\right)}{F_y^3}$$

(x) Leibniz's rule gives the derivative of an integral:

$$\frac{d}{dx}\left(\int_{f(x)}^{g(x)} h(x, t)\, dt\right) = g'(x)h(x, g(x)) - f'(x)h(x, f(x)) + \int_{f(x)}^{g(x)} \frac{\partial h}{\partial x}(x, t)\, dt.$$

(y) If $x = x(t)$ and $y = y(t)$ then (the dots denote differentiation with respect to t):

$$\frac{dy}{dx} = \frac{\dot{y}(t)}{\dot{x}(t)}, \qquad\qquad \frac{d^2y}{dx^2} = \frac{\dot{x}\ddot{y} - \ddot{x}\dot{y}}{(\dot{x})^3}.$$

5.1.5   DERIVATIVE THEOREMS

1. *Fundamental theorem of calculus*:   Suppose f is continuous on $[a, b]$.

   (a) If G is defined as $G(x) = \displaystyle\int_a^x f(t)\, dt$ for all x in $[a, b]$, then G is an antiderivative of f on $[a, b]$.

   (b) If F is any antiderivative of f, then $\displaystyle\int_a^b f(t)\, dt = F(b) - F(a)$. (Recall that $\int_a^b f(x)\, dx$ is defined as a limit of Riemann sums.)

2. *Intermediate value theorem*:   If $f(x)$ is continuous on $[a, b]$ and if $f(a) \neq f(b)$, then f takes on every value between $f(a)$ and $f(b)$ in the interval (a, b).

3. *Rolle's theorem*:   If $f(x)$ is continuous on $[a, b]$ and differentiable on (a, b), and if $f(a) = f(b)$, then $f'(c) = 0$ for at least one number c in (a, b).

4. *Mean value theorem*:   If $f(x)$ is continuous on $[a, b]$ and differentiable on (a, b), then a number c exists in (a, b) such that $f(b) - f(a) = (b - a)f'(c)$.

5.1.6   THE TWO-DIMENSIONAL CHAIN RULE

If $x = x(t)$, $y = y(t)$, and $z = z(x, y)$, then

$$\frac{dz}{dt} = \frac{\partial z}{\partial x}\frac{dx}{dt} + \frac{\partial z}{\partial y}\frac{dy}{dt}, \text{ and}$$

$$\frac{d^2z}{dt^2} = \frac{\partial z}{\partial x}\frac{d^2x}{dt^2} + \frac{dx}{dt}\left(\frac{\partial^2 z}{\partial x^2}\frac{dx}{dt} + \frac{\partial^2 z}{\partial x \partial y}\frac{dy}{dt}\right) \tag{5.1.3}$$

$$+ \frac{\partial z}{\partial y}\frac{d^2y}{dt^2} + \frac{dy}{dt}\left(\frac{\partial^2 z}{\partial y^2}\frac{dy}{dt} + \frac{\partial^2 z}{\partial x \partial y}\frac{dx}{dt}\right).$$

If $x = x(u, v)$, $y = y(u, v)$, and $z = z(x, y)$, then

$$\frac{\partial z}{\partial u} = \frac{\partial z}{\partial x}\frac{\partial x}{\partial u} + \frac{\partial z}{\partial y}\frac{\partial y}{\partial u} \quad \text{and} \quad \frac{\partial z}{\partial v} = \frac{\partial z}{\partial x}\frac{\partial x}{\partial v} + \frac{\partial z}{\partial y}\frac{\partial y}{\partial v}. \tag{5.1.4}$$

If $u = u(x, y)$, $v = v(x, y)$, and $f = f(x, y)$, then the partial derivative of f with respect to u, holding v constant, written $\left(\frac{\partial f}{\partial u}\right)_v$, can be expressed as

$$\left(\frac{\partial f}{\partial u}\right)_v = \left(\frac{\partial f}{\partial x}\right)_y \left(\frac{\partial x}{\partial u}\right)_v + \left(\frac{\partial f}{\partial y}\right)_x \left(\frac{\partial y}{\partial u}\right)_v. \tag{5.1.5}$$

5.1.7 L'HÔSPITAL'S RULE

If $f(x)$ and $g(x)$ are differentiable in the neighborhood of point a, and if $f(x)$ and $g(x)$ both tend to 0 or ∞ as $x \to a$, then

$$\lim_{x \to a} \frac{f(x)}{g(x)} = \lim_{x \to a} \frac{f'(x)}{g'(x)} \tag{5.1.6}$$

if the right hand side exists.

EXAMPLES

$$\lim_{x \to 0} \frac{x - \sin x}{x^3} = \lim_{x \to 0} \frac{1 - \cos x}{3x^2} = \lim_{x \to 0} \frac{\sin x}{6x} = \lim_{x \to 0} \frac{\cos x}{6} = \frac{1}{6},$$

$$\lim_{x \to \infty} \frac{x^n}{e^x} = \lim_{x \to \infty} \frac{nx^{n-1}}{e^x} = \lim_{x \to \infty} \frac{n(n-1)x^{n-2}}{e^x} = \cdots = \lim_{x \to \infty} \frac{n!}{e^x} = 0.$$

5.1.8 MAXIMA AND MINIMA OF FUNCTIONS

1. If a function $f(x)$ has a local extremum at a number c, then either $f'(c) = 0$ or $f'(c)$ does not exist.

2. If $f'(c) = 0$, $f(x)$ is differentiable on an open interval containing c, and

   (a) if $f''(c) < 0$, then f has a local maximum at c;
   (b) if $f''(c) > 0$, then f has a local minimum at c.

5.1.8.1 Lagrange multipliers

To find the extreme values of the function $f(x_1, x_2, \ldots, x_n) = f(\mathbf{x})$ subject to the m side constraints $\mathbf{g}(\mathbf{x}) = 0$, introduce an m-dimensional vector of Lagrange multipliers λ and define $F(\mathbf{x}, \lambda) = f(\mathbf{x}) + \lambda^{\mathrm{T}}\mathbf{g}(\mathbf{x})$. Then the extreme values of F with respect to all of its arguments are found by solving:

$$\frac{\partial F}{\partial x_i} = \frac{\partial f}{\partial x_i} + \lambda^{\mathrm{T}}\frac{\partial \mathbf{g}}{\partial x_i} = 0, \quad \text{and} \quad \frac{\partial F}{\partial \lambda_j} = g_j = 0. \tag{5.1.7}$$

For the extreme values of $f(x, y)$ subject to $g(x, y) = 0$ (i.e., $n = 2$ and $m = 1$):

$$f_x + \lambda g_x = 0, \quad f_y + \lambda g_y = 0, \quad \text{and} \quad g = 0. \tag{5.1.8}$$

EXAMPLE   To find the points on the unit circle (given by $g(x,y) = (x-1)^2 + (y - 2)^2 - 1 = 0$) that are closest and furthest from the origin (the distance squared from the origin is $f(x,y) = x^2 + y^2$) can be determined by solving the three (non-linear) algebraic equations:

$$2x + 2\lambda(x-1) = 0, \qquad 2y + 2\lambda(y-2) = 0, \qquad (x-1)^2 + (y-2)^2 = 1.$$

The solutions are $x = 1 + 1/\sqrt{5}, y = 2 + 2/\sqrt{5}, \lambda = -1 - \sqrt{5}$ (furthest), and $x = 1 - 1/\sqrt{5}, y = 2 - 2/\sqrt{5}, \lambda = \sqrt{5} - 1$ (closest).

5.1.9   VECTOR CALCULUS

1. Definitions of "div", "grad", and "curl" are on page 493.

2. A vector field $\mathbf{F}$ is *irrotational* if $\nabla \times \mathbf{F} = 0$. A vector field $\mathbf{F}$ is *solenoidal* if $\nabla \cdot \mathbf{F} = 0$.

3. In Cartesian coordinates, $\nabla = \left(\frac{\partial}{\partial x}, \frac{\partial}{\partial y}, \frac{\partial}{\partial z} \right) = \mathbf{i}\frac{\partial}{\partial x} + \mathbf{j}\frac{\partial}{\partial y} + \mathbf{k}\frac{\partial}{\partial z}$. If u and v are scalars and $\mathbf{F}$ and $\mathbf{G}$ are vectors in $\mathbb{R}^3$, then

$$\nabla(u + v) = \nabla u + \nabla v,$$
$$\nabla(uv) = u\nabla v + v\nabla u,$$
$$\nabla(\mathbf{F} + \mathbf{G}) = \nabla\mathbf{F} + \nabla\mathbf{G},$$
$$\nabla(\mathbf{F} \cdot \mathbf{G}) = (\mathbf{F} \cdot \nabla)\mathbf{G} + (\mathbf{G} \cdot \nabla)\mathbf{F} + \mathbf{F} \times (\nabla \times \mathbf{G}) + \mathbf{G} \times (\nabla \times \mathbf{F}),$$
$$\nabla \cdot (u\mathbf{F}) = u(\nabla \cdot \mathbf{F}) + \mathbf{F} \cdot \nabla u,$$
$$\nabla \cdot (\mathbf{F} \times \mathbf{G}) = \mathbf{G} \cdot (\nabla \times \mathbf{F}) - \mathbf{F} \cdot (\nabla \times \mathbf{G}),$$
$$\nabla \times (u\mathbf{F}) = u(\nabla \times \mathbf{F}) + (\nabla u) \times \mathbf{F},$$
$$\nabla \times (\mathbf{F} + \mathbf{G}) = \nabla \times \mathbf{F} + \nabla \times \mathbf{G}$$
$$\nabla \times (\mathbf{F} \times \mathbf{G}) = \mathbf{F}(\nabla \cdot \mathbf{G}) - \mathbf{G}(\nabla \cdot \mathbf{F}) + (\mathbf{G} \cdot \nabla)\mathbf{F} - (\mathbf{F} \cdot \nabla)\mathbf{G},$$
$$\mathbf{F} \cdot \frac{d\mathbf{F}}{dt} = |\mathbf{F}| \frac{d|\mathbf{F}|}{dt},$$
$$\nabla \times (\nabla \times \mathbf{F}) = \nabla(\nabla \cdot \mathbf{F}) - \nabla^2 \mathbf{F},$$
$$\nabla \times (\nabla u) = \mathbf{0},$$
$$\nabla \cdot (\nabla \times \mathbf{F}) = 0, \quad \text{and}$$
$$\nabla^2(uv) = u\nabla^2 v + 2(\nabla u) \cdot (\nabla v) + v\nabla^2 u.$$

Note that $\nabla \cdot u = \sum_{i=1}^{3} \frac{\partial u}{\partial x_i}$ is a scalar while $\mathbf{F} \cdot \nabla = \sum_{i=1}^{3} F_i \frac{\partial}{\partial x_i}$ is an operator.

4. If $r = |\mathbf{r}|$, $\mathbf{a}$ is a constant vector, and n is an integer, then

Φ	$\nabla\Phi$	$\nabla^2\Phi$
$\mathbf{a} \cdot \mathbf{r}$	$\mathbf{a}$	0
r^n	$nr^{n-2}\mathbf{r}$	$n(n+1)r^{n-2}$
$\log r$	$\mathbf{r}/r^2$	$1/r^2$

$\mathbf{F}$	$\nabla \cdot \mathbf{F}$	$\nabla \times \mathbf{F}$	$(\mathbf{G} \cdot \nabla)\mathbf{F}$
$\mathbf{r}$	3	0	$\mathbf{G}$
$\mathbf{a} \times \mathbf{r}$	0	$2\mathbf{a}$	$\mathbf{a} \times \mathbf{G}$
$\mathbf{a}r^n$	$nr^{n-2}(\mathbf{r} \cdot \mathbf{a})$	$nr^{n-2}(\mathbf{r} \times \mathbf{a})$	$nr^{n-2}(\mathbf{r} \cdot \mathbf{G})\mathbf{a}$
$\mathbf{r}r^n$	$(n+3)r^n$	0	$r^n\mathbf{G} + nr^{n-2}(\mathbf{r} \cdot \mathbf{G})\mathbf{r}$
$\mathbf{a} \log r$	$\mathbf{r} \cdot \mathbf{a}/r^2$	$\mathbf{r} \times \mathbf{a}/r^2$	$(\mathbf{G} \cdot \mathbf{r})\mathbf{a}/r^2$

$\mathbf{F}$	$\nabla^2 \mathbf{F}$	$\nabla\nabla \cdot \mathbf{F}$
$\mathbf{r}$	0	0
$\mathbf{a} \times \mathbf{r}$	0	0
$\mathbf{a}r^n$	$n(n+1)r^{n-2}\mathbf{a}$	$nr^{n-2}\mathbf{a} + n(n-2)r^{n-4}(\mathbf{r} \cdot \mathbf{a})\mathbf{r}$
$\mathbf{r}r^n$	$n(n+3)r^{n-2}\mathbf{r}$	$n(n+3)r^{n-2}\mathbf{r}$
$\mathbf{a} \log r$	$\mathbf{a}/r^2$	$[r^2\mathbf{a} - 2(\mathbf{r} \cdot \mathbf{a})\mathbf{r}]/r^4$

$$\frac{d}{dt}(\mathbf{F} + \mathbf{G}) = \frac{d\mathbf{F}}{dt} + \frac{d\mathbf{G}}{dt},$$

$$\frac{d}{dt}(\mathbf{F} \cdot \mathbf{G}) = \mathbf{F} \cdot \frac{d\mathbf{G}}{dt} + \frac{d\mathbf{F}}{dt} \cdot \mathbf{G},$$

$$\frac{d}{dt}(\mathbf{F} \times \mathbf{G}) = \mathbf{F} \times \frac{d\mathbf{G}}{dt} + \frac{d\mathbf{F}}{dt} \times \mathbf{G},$$

$$\frac{d}{dt}(\mathbf{V}_1 \times \mathbf{V}_2 \times \mathbf{V}_3) = \left(\frac{d\mathbf{V}_1}{dt}\right) \times (\mathbf{V}_2 \times \mathbf{V}_3) + \mathbf{V}_1 \times \left(\left(\frac{d\mathbf{V}_2}{dt}\right) \times \mathbf{V}_3\right),$$

$$+ \mathbf{V}_1 \times \left(\mathbf{V}_2 \times \left(\frac{d\mathbf{V}_3}{dt}\right)\right),$$

$$\frac{d}{dt}[\mathbf{V}_1\mathbf{V}_2\mathbf{V}_3] = \left[\left(\frac{d\mathbf{V}_1}{dt}\right)\mathbf{V}_2\mathbf{V}_3\right] + \left[\mathbf{V}_1\left(\frac{d\mathbf{V}_2}{dt}\right)\mathbf{V}_3\right] + \left[\mathbf{V}_1\mathbf{V}_2\left(\frac{d\mathbf{V}_3}{dt}\right)\right],$$

where $[\mathbf{V}_1\mathbf{V}_2\mathbf{V}_3] = \mathbf{V}_1 \cdot (\mathbf{V}_2 \times \mathbf{V}_3)$ is the scalar triple product (see page 136).

5.1.10 MATRIX AND VECTOR DERIVATIVES

5.1.10.1 Definitions

1. The derivative of the row vector $\mathbf{y} = \begin{bmatrix} y_1 & y_2 & \cdots & y_m \end{bmatrix}$ with respect to the scalar x is

$$\frac{\partial \mathbf{y}}{\partial x} = \begin{bmatrix} \dfrac{\partial y_1}{\partial x} & \dfrac{\partial y_2}{\partial x} & \cdots & \dfrac{\partial y_m}{\partial x} \end{bmatrix}. \tag{5.1.9}$$

2. The derivative of a scalar y with respect to the vector $\mathbf{x}$ is

$$\frac{\partial y}{\partial \mathbf{x}} = \begin{bmatrix} \dfrac{\partial y}{\partial x_1} \\ \dfrac{\partial y}{\partial x_2} \\ \vdots \\ \dfrac{\partial y}{\partial x_n} \end{bmatrix}. \tag{5.1.10}$$

3. Let $\mathbf{x}$ be a $n \times 1$ vector and let $\mathbf{y}$ be a $m \times 1$ vector. The derivative of $\mathbf{y}$ with respect to $\mathbf{x}$ is the matrix

$$\frac{\partial \mathbf{y}}{\partial \mathbf{x}} = \begin{bmatrix} \frac{\partial y_1}{\partial x_1} & \frac{\partial y_2}{\partial x_1} & \cdots & \frac{\partial y_m}{\partial x_1} \\ \frac{\partial y_1}{\partial x_2} & \frac{\partial y_2}{\partial x_2} & \cdots & \frac{\partial y_m}{\partial x_2} \\ \vdots & \vdots & \ddots & \vdots \\ \frac{\partial y_1}{\partial x_n} & \frac{\partial y_2}{\partial x_n} & \cdots & \frac{\partial y_m}{\partial x_n} \end{bmatrix}. \tag{5.1.11}$$

In multivariate analysis, if $\mathbf{x}$ and $\mathbf{y}$ have the same dimension, then the absolute value of the determinant of $\frac{\partial \mathbf{y}}{\partial \mathbf{x}}$ is called the *Jacobian* of the transformation determined by $\mathbf{y} = \mathbf{y}(\mathbf{x})$, written $\frac{\partial(y_1, y_2, \ldots, y_n)}{\partial(x_1, x_2, \ldots, x_n)}$.

4. The Jacobian of the derivatives $\frac{\partial \phi}{\partial x_1}, \frac{\partial \phi}{\partial x_2}, \ldots \frac{\partial \phi}{\partial x_n}$ of the function $\phi(x_1, \ldots, x_n)$ with respect to $x_1, \ldots, x_n$ is called the *Hessian H* of ϕ:

$$H = \begin{vmatrix} \frac{\partial^2 \phi}{\partial x_1^2} & \frac{\partial^2 \phi}{\partial x_1 \partial x_2} & \frac{\partial^2 \phi}{\partial x_1 \partial x_3} & \cdots & \frac{\partial^2 \phi}{\partial x_1 \partial x_n} \\ \frac{\partial^2 \phi}{\partial x_2 \partial x_1} & \frac{\partial^2 \phi}{\partial x_2^2} & \frac{\partial^2 \phi}{\partial x_2 \partial x_3} & \cdots & \frac{\partial^2 \phi}{\partial x_2 \partial x_n} \\ \vdots & \vdots & \vdots & & \vdots \\ \frac{\partial^2 \phi}{\partial x_n \partial x_1} & \frac{\partial^2 \phi}{\partial x_n \partial x_2} & \frac{\partial^2 \phi}{\partial x_n \partial x_3} & \cdots & \frac{\partial^2 \phi}{\partial x_n^2} \end{vmatrix}$$

5. The derivative of the matrix $A(t) = (a_{ij}(t))$, with respect to the scalar t, is the matrix $\frac{dA(t)}{dt} = \left(\frac{da_{ij}(t)}{dt} \right)$.

6. If $X = (x_{ij})$ is a $m \times n$ matrix and if y is a scalar function of X, then the derivative of y with respect to X is (here, $E_{ij} = \mathbf{e}_i \mathbf{e}_j^\mathsf{T}$):

$$\frac{\partial y}{\partial X} = \begin{bmatrix} \frac{\partial y}{\partial x_{11}} & \frac{\partial y}{\partial x_{12}} & \cdots & \frac{\partial y}{\partial x_{1n}} \\ \frac{\partial y}{\partial x_{21}} & \frac{\partial y}{\partial x_{22}} & \cdots & \frac{\partial y}{\partial x_{2n}} \\ \vdots & \vdots & \ddots & \vdots \\ \frac{\partial y}{\partial x_{m1}} & \frac{\partial y}{\partial x_{m2}} & \cdots & \frac{\partial y}{\partial x_{mn}} \end{bmatrix} = \sum_{\substack{1 \leq i \leq m \\ 1 \leq j \leq n}} E_{ij} \frac{\partial y}{\partial x_{ij}}. \tag{5.1.12}$$

7. If $Y = (y_{ij})$ is a $p \times q$ matrix and X is a $m \times n$ matrix, then the derivative of Y with respect to X is

$$\frac{\partial Y}{\partial X} = \begin{bmatrix} \frac{\partial Y}{\partial x_{11}} & \frac{\partial Y}{\partial x_{12}} & \cdots & \frac{\partial Y}{\partial x_{1n}} \\ \frac{\partial Y}{\partial x_{21}} & \frac{\partial Y}{\partial x_{22}} & \cdots & \frac{\partial Y}{\partial x_{2n}} \\ \vdots & \vdots & \ddots & \vdots \\ \frac{\partial Y}{\partial x_{m1}} & \frac{\partial Y}{\partial x_{m2}} & \cdots & \frac{\partial Y}{\partial x_{mn}} \end{bmatrix} = \sum_{\substack{1 \leq i \leq m \\ 1 \leq j \leq n}} E_{ij} \otimes \frac{\partial Y}{\partial x_{ij}}. \tag{5.1.13}$$

5.1.10.2 Properties

y (a scalar or a vector)	$\frac{\partial y}{\partial \mathbf{x}}$ (recall, $\mathbf{x}$ is a vector)
$\mathbf{x}^{\mathrm{T}}$	I
$A\mathbf{x}$	A^{T}
$\mathbf{x}^{\mathrm{T}} A$	A
$\mathbf{x}^{\mathrm{T}}\mathbf{x}$	$2\mathbf{x}$
$\mathbf{x}^{\mathrm{T}} A\mathbf{x}$	$A\mathbf{x} + A^{\mathrm{T}}\mathbf{x}$

1.

(with A constant)

2. $\frac{\partial A\mathbf{x}}{\partial \mathbf{x}^{\mathrm{T}}} = A$

y (a scalar, vector, or matrix)	$\frac{\partial y}{\partial X}$ (recall, X is a matrix)
YZ	$Y\frac{dZ}{dX} + \frac{dY}{dX}Z$
AXB	$A^{\mathrm{T}}B^{\mathrm{T}}$
$\mathbf{a}^{\mathrm{T}}X^{\mathrm{T}}X\mathbf{b}$	$X(\mathbf{ab}^{\mathrm{T}} + \mathbf{ba}^{\mathrm{T}})$
$\mathbf{a}^{\mathrm{T}}X^{\mathrm{T}}X\mathbf{a}$	$2X\mathbf{aa}^{\mathrm{T}}$
$\mathbf{a}^{\mathrm{T}}X^{\mathrm{T}}CX\mathbf{b}$	$C^{\mathrm{T}}X\mathbf{ab}^{\mathrm{T}} + CX\mathbf{ba}^{\mathrm{T}}$
$\mathbf{a}^{\mathrm{T}}X^{\mathrm{T}}CX\mathbf{a}$	$(C + C^{\mathrm{T}})X\mathbf{aa}^{\mathrm{T}}$
$(X\mathbf{a} + \mathbf{b})^{\mathrm{T}}C(X\mathbf{a} + \mathbf{b})$	$(C + C^{\mathrm{T}})(X\mathbf{a} + \mathbf{b})\mathbf{a}^{\mathrm{T}}$
$\mathrm{tr}(X)$	I
$\mathrm{tr}(A^{\mathrm{T}}X), \mathrm{tr}(XA^{\mathrm{T}}), \mathrm{tr}(AX^{\mathrm{T}}),$ or $\mathrm{tr}(X^{\mathrm{T}}A)$	A
$\mathrm{tr}(XAX^{\mathrm{T}})$	$X^{\mathrm{T}}(A + A^{\mathrm{T}})$
$\mathrm{tr}(X^{\mathrm{T}}AX)$	$(A + A^{\mathrm{T}})X$
$\mathrm{tr}(X^{\mathrm{T}}AXB)$	$AXB + A^{\mathrm{T}}XB^{\mathrm{T}}$
$\det(X)$ or $\det(X^{\mathrm{T}})$	$\det(X)(X^{-1})^{\mathrm{T}}$
$\log|X|$	$(X^{-1})^{\mathrm{T}}$

3.

(with $\{\mathbf{a}, \mathbf{b}, A, B\}$ constants, $Y = Y(X)$, and $Z = Z(X)$)

y (a scalar, vector, or matrix)	$\frac{dy}{dt}$ (here t is a scalar)
AB	$\frac{dA}{dt}B + A\frac{dB}{dt}$
$A \otimes B$	$\frac{dA}{dt} \otimes B + A \otimes \frac{dB}{dt}$
A^{-1}	$-A\frac{dA}{dt}A^{-1}$

4.

(with $A = A(t)$ and $B = B(t)$)

5. If $Y = AX^{-1}B$, then

    (a) $\frac{\partial Y}{\partial x_{rs}} = -AX^{-1}E_{rs}X^{-1}B$.

    (b) $\frac{\partial y_{ij}}{\partial X} = -(X^{-1})^{\mathrm{T}}A^{\mathrm{T}}E_{ij}B^{\mathrm{T}}(X^{-1})^{\mathrm{T}}$.

6. If $Y = AXB$ then $\frac{\partial Y}{\partial x_{ij}} = A E_{ij} B$ where $E_{ij} = \mathbf{e}_i \mathbf{e}_j^\mathsf{T}$ has the same size as X.

7. If $Y = AX^\mathsf{T}B$, then $\frac{\partial y_{ij}}{\partial X} = B E_{ij}^\mathsf{T} A$.

8. If $Y = X^\mathsf{T}AX$, then

   (a) $\frac{\partial Y}{\partial x_{rs}} = E_{rs}^\mathsf{T} AX + X^\mathsf{T} A E_{rs}$.

   (b) $\frac{\partial y_{ij}}{\partial X} = AX E_{ij}^\mathsf{T} + A^\mathsf{T} X E_{ij}$.

9. If $\mathbf{y} = \text{Vec}\, Y$ and $\mathbf{x} = \text{Vec}\, X$ (see page 158), then

   (a) If $Y = AX$, then $\frac{\partial \mathbf{y}}{\partial \mathbf{x}} = I \otimes A^\mathsf{T}$.

   (b) If $Y = XA$, then $\frac{\partial \mathbf{y}}{\partial \mathbf{x}} = A \otimes I$.

   (c) If $Y = AX^{-1}B$, then $\frac{\partial \mathbf{y}}{\partial \mathbf{x}} = -(X^{-1}B) \otimes (X^{-1})^\mathsf{T} A^\mathsf{T}$.

10. The derivative of the determinant of a matrix can be written:

    (a) If Y_{ij} is the cofactor of element y_{ij} in $|Y|$, then $\frac{\partial |Y|}{\partial x_{rs}} = \sum_i \sum_j Y_{ij} \frac{\partial y_{ij}}{\partial x_{rs}}$.

    (b) If all the components (x_{ij}) of X are independent, then $\frac{\partial |X|}{\partial X} = |X|\,(X^{-1})^\mathsf{T}$.

11. Derivatives of powers of matrices are obtained as follows:

    (a) If $Y = X^r$, then $\frac{\partial Y}{\partial x_{rs}} = \sum_{k=0}^{r-1} X^k E_{rs} X^{n-k-1}$.

    (b) If $Y = X^{-r}$, then $\frac{\partial Y}{\partial x_{rs}} = -X^{-r} \left(\sum_{k=0}^{r-1} X^k E_{rs} X^{n-k-1} \right) X^{-r}$.

    (c) The nth derivative of the rth power of the matrix A^{-1}, in terms of derivatives of the matrix A, is

    $$\frac{d^n A^{-r}}{dx^n} = n! \left(\sum_{k=1}^{n} (-1)^k \frac{\mathcal{P}_{i_1}}{i_1!} \frac{\mathcal{P}_{i_2}}{i_2!} \cdots \frac{\mathcal{P}_{i_k}}{i_k!} \right) A^{-r} \tag{5.1.14}$$

    where $\mathcal{P}_i = A^{-r} \frac{d^i A^r}{dx^i}$ and the summation is taken over all positive integers $(i_1, i_2, \ldots, i_k)$, distinct or otherwise, such that $\sum_{m=1}^{k} i_m = n$. Setting $n = r = 1$ results in

    $$\frac{dA^{-1}}{dx} = -A^{-1} \frac{dA}{dx} A^{-1}. \tag{5.1.15}$$

12. Derivative formulae:

    (a) If $\mathbf{z} = \mathbf{z}(\mathbf{y}(\mathbf{x}))$, then $\dfrac{\partial \mathbf{z}}{\partial \mathbf{y}} = \dfrac{\partial \mathbf{z}}{\partial \mathbf{x}} \dfrac{\partial \mathbf{x}}{\partial \mathbf{y}}$.

    (b) If X and Y are matrices, then $\left(\frac{\partial Y}{\partial X} \right)^\mathsf{T} = \frac{\partial Y^\mathsf{T}}{\partial X^\mathsf{T}}$.

    (c) If X, Y, and Z are matrices of size $m \times n$, $n \times v$, and $p \times q$, then $\frac{\partial (XY)}{\partial Z} = \frac{\partial X}{\partial Z}(I_q \otimes Y) + (I_p \otimes X)\frac{\partial Y}{\partial Z}$.

5.2 DIFFERENTIAL FORMS

Define $dx_k(\cdot)$ to be the function that assigns a vector its kth coordinate; that is, for the vector $\mathbf{a} = (a_1, \ldots, a_k, \ldots, a_n)$, we have $dx_k(\mathbf{a}) = a_k$. Geometrically, $dx_k(\mathbf{a})$ is the length, with appropriate sign, of the projection of $\mathbf{a}$ on the kth coordinate axis. When the $\{F_i\}$ are functions, the following linear combination of the functions $\{dx_k\}$

$$\omega_{\mathbf{x}} = F_1(\mathbf{x})\, dx_1 + F_2(\mathbf{x})\, dx_2 + \cdots + F_n(\mathbf{x})\, dx_n \qquad (5.2.1)$$

produces a new function $\omega_{\mathbf{x}}$. This function acts on vectors $\mathbf{a}$ as

$$\omega_{\mathbf{x}}(\mathbf{a}) = F_1(\mathbf{x})\, dx_1(\mathbf{a}) + F_2(\mathbf{x})\, dx_2(\mathbf{a}) + \cdots + F_n(\mathbf{x})\, dx_n(\mathbf{a}). \qquad (5.2.2)$$

Such a function is a *differential 1-form* or a *1-form*. For example:

1. If $\mathbf{a} = (-2, 0, 4)$ then $dx_1(\mathbf{a}) = -2$, $dx_2(\mathbf{a}) = 0$, and $dx_3(\mathbf{a}) = 4$.

2. If in $\mathbb{R}^2$, $\omega_{\mathbf{x}} = \omega_{(x,y)} = x^2\, dx + y^2\, dy$, then $\omega_{(x,y)}(a, b) = ax^2 + by^2$ and $\omega_{(1,-3)}(a, b) = a + 9b$.

3. If $f(\mathbf{x})$ is a differentiable function, then $\nabla_{\mathbf{x}} f$, the differential of f at $\mathbf{x}$, is a 1-form. Note that $\nabla_{\mathbf{x}} f$ acting on $\mathbf{a} = (a_1, a_2, a_3)$ is

$$\nabla_{\mathbf{x}} f(\mathbf{a}) = \frac{\partial f}{\partial x_1}(\mathbf{x})\, dx_1(\mathbf{a}) + \frac{\partial f}{\partial x_2}(\mathbf{x})\, dx_2(\mathbf{a}) + \frac{\partial f}{\partial x_3}(\mathbf{x})\, dx_3(\mathbf{a})$$

$$= \frac{\partial f}{\partial x_1}(\mathbf{x})a_1 + \frac{\partial f}{\partial x_2}(\mathbf{x})a_2 + \frac{\partial f}{\partial x_3}(\mathbf{x})a_3.$$

5.2.1 PRODUCTS OF 1-FORMS

The basic 1-forms in $\mathbb{R}^3$ are dx_1, dx_2, and dx_3. The *wedge product* (or *exterior product*) $dx_1 \wedge dx_2$ is defined so that it is a function of ordered pairs of vectors in $\mathbb{R}^2$. Geometrically, $dx_1 \wedge dx_2(\mathbf{a}, \mathbf{b})$ will be the area of the parallelogram spanned by the projections of $\mathbf{a}$ and $\mathbf{b}$ into the (x_1, x_2)-plane. The sign of the area is determined so that if the projections of $\mathbf{a}$ and $\mathbf{b}$ have the same orientation as the positive x_1 and x_2 axes, then the area is positive; it is negative when these orientations are opposite. Thus, if $\mathbf{a} = (a_1, a_2, a_3)$ and $\mathbf{b} = (b_1, b_2, b_3)$, then

$$dx_1 \wedge dx_2(\mathbf{a}, \mathbf{b}) = \det \begin{bmatrix} a_1 & b_1 \\ a_2 & b_2 \end{bmatrix} = a_1 b_2 - a_2 b_1, \qquad (5.2.3)$$

and the determinant automatically gives the correct sign. This generalizes to

$$dx_i \wedge dx_j(\mathbf{a}, \mathbf{b}) = \det \begin{bmatrix} dx_i(\mathbf{a}) & dx_i(\mathbf{b}) \\ dx_j(\mathbf{a}) & dx_j(\mathbf{b}) \end{bmatrix} = \det \begin{bmatrix} a_i & b_i \\ a_j & b_j \end{bmatrix}. \qquad (5.2.4)$$

1. If ω and μ are 1-forms, and f and g are real-valued functions, then $f\omega + g\mu$ is a 1-form.

2. If ω, ν, and μ are 1-forms, then $(f\omega + g\nu) \wedge \mu = f\,\omega \wedge \mu + g\,\nu \wedge \mu$.

3. $dx_i \wedge dx_j = -dx_j \wedge dx_i$

4. $dx_i \wedge dx_i = 0$

5. $dx_i \wedge dx_j(\mathbf{b}, \mathbf{a}) = -dx_i \wedge dx_j(\mathbf{a}, \mathbf{b})$

5.2.2 DIFFERENTIAL 2-FORMS

In $\mathbb{R}^3$, the most general linear combination of the functions $dx_i \wedge dx_j$ has the form $c_1\,dx_2 \wedge dx_3 + c_2\,dx_3 \wedge dx_1 + c_3\,dx_1 \wedge dx_2$. If $\mathbf{F} = (F_1, F_2, F_3)$ is a vector field, then the function of ordered pairs,

$$\tau_{\mathbf{x}}(\mathbf{a}, \mathbf{b}) = F_1(\mathbf{x})\,dx_2 \wedge dx_3 + F_2(\mathbf{x})\,dx_3 \wedge dx_1 + F_3(\mathbf{x})\,dx_1 \wedge dx_2, \quad (5.2.5)$$

is a *differential 2-form* or *2-form*.

EXAMPLES

1. For the specific 2-form $\tau_{\mathbf{x}} = 2\,dx_2 \wedge dx_3 + dx_3 \wedge dx_1 + 5\,dx_1 \wedge dx_2$, if $\mathbf{a} = (1, 2, 3)$ and $\mathbf{b} = (0, 1, 1)$, then

$$\tau_{\mathbf{x}}(\mathbf{a}, \mathbf{b}) = 2 \det \begin{bmatrix} 2 & 1 \\ 3 & 1 \end{bmatrix} + \det \begin{bmatrix} 3 & 1 \\ 1 & 0 \end{bmatrix} + 5 \det \begin{bmatrix} 1 & 0 \\ 2 & 1 \end{bmatrix}$$

$$= 2 \cdot (-1) + 1 \cdot (-1) + 5 \cdot (1) = 2$$

independent of $\mathbf{x}$. Note that $\mathbf{a} \times \mathbf{b} = \det \begin{bmatrix} \mathbf{i} & \mathbf{j} & \mathbf{k} \\ 1 & 2 & 3 \\ 0 & 1 & 1 \end{bmatrix} = (-1, -1, 1)$, and so $\tau_{\mathbf{x}}(\mathbf{a}, \mathbf{b}) = (2, 1, 5) \cdot (\mathbf{a} \times \mathbf{b})$.

2. When changing from Cartesian coordinates to polar coordinates, the element of area dA can be written

$$\begin{aligned} dA &= dx \wedge dy \\ &= (-r \sin \theta\, d\theta + \cos \theta\, dr) \wedge (r \cos \theta\, d\theta + \sin \theta\, dr) \\ &= -r^2 \sin \theta \cos \theta\, (d\theta \wedge d\theta) + \sin \theta \cos \theta\, (dr \wedge dr) \quad (5.2.6) \\ &\quad - r \sin^2 \theta\, (d\theta \wedge dr) + r \cos^2 \theta\, (dr \wedge d\theta) \\ &= r\, dr \wedge d\theta. \end{aligned}$$

5.2.3 THE 2-FORMS IN $\mathbb{R}^n$

Every 2-form can be written as a linear combination of "basic 2-forms". For example, in $\mathbb{R}^2$ there is only one basic 2-form (which may be taken to be $dx_1 \wedge dx_2$) and in $\mathbb{R}^3$ there are 3 basic 2-forms (possibly the set $\{dx_1 \wedge dx_2, dx_2 \wedge dx_3, dx_3 \wedge dx_1\}$). The exterior product of any two 1-forms (in, say, $\mathbb{R}^n$) is found by multiplying the 1-forms as if there were ordinary polynomials in the variables $dx_1, \ldots, dx_n$, and then simplifying using the rules for $dx_i \wedge dx_j$.

EXAMPLE   Denoting the basic 1-forms in $\mathbb{R}^3$ as dx, dy, and dz then

$$
\begin{aligned}
(x\, dx + y^2\, dy) \wedge (dx + x\, dy) &= x\, (dx \wedge dx) + y^2\, (dy \wedge dx), \\
&\quad + x^2\, (dx \wedge dy) + xy^2\, (dy \wedge dy), \\
&= 0 - y^2\, (dx \wedge dy) + x^2\, (dx \wedge dy) + 0, \\
&= (x^2 - y^2)\, dx \wedge dy.
\end{aligned}
\tag{5.2.7}
$$

5.2.4   HIGHER DIMENSIONAL FORMS

The meaning of the basic 3-form $dx_1 \wedge dx_2 \wedge dx_3$ is that of a signed volume function. Thus, if $\mathbf{a} = (a_1, a_2, a_3)$, $\mathbf{b} = (b_1, b_2, b_3)$, and $\mathbf{c} = (c_1, c_2, c_3)$ then

$$
dx_1 \wedge dx_2 \wedge dx_3(\mathbf{a}, \mathbf{b}, \mathbf{c}) = \det \begin{bmatrix} a_1 & b_1 & c_1 \\ a_2 & b_2 & c_2 \\ a_3 & b_3 & c_3 \end{bmatrix}
\tag{5.2.8}
$$

which is a 3-dimensional oriented volume of the parallelepiped defined by the vectors $\mathbf{a}$, $\mathbf{b}$, and $\mathbf{c}$.

For an ordered p-tuple of vectors in $\mathbb{R}^n$, $(\mathbf{a}_1, \mathbf{a}_2, \ldots, \mathbf{a}_p)$, where $p \geq 1$

$$
dx_{k_1} \wedge dx_{k_2} \wedge \cdots \wedge dx_{k_p}(\mathbf{a}_1, \ldots, \mathbf{a}_p) = \det(dx_{k_i}(\mathbf{a}_j))_{\substack{i=1,\ldots,p \\ j=1,\ldots,p}}.
\tag{5.2.9}
$$

This equation defines the basic p-forms in $\mathbb{R}^n$, of which the general p-forms are linear combinations. Properties include:

1. The interchange of adjacent factors in a basic p-form changes the sign of the form. If ω^p is a p-form in $\mathbb{R}^n$ and ω^q is a q-form in $\mathbb{R}^n$, then $\omega^p \wedge \omega^q = (-1)^{pq}\, \omega^q \wedge \omega^p$.

2. A basic p-form with a repeated factor is zero.

3. If $p > n$, then any p-form is identically zero.

4. The general p-form can be written $\omega^p = \sum_{i_1 < \cdots < i_p} f_{i_1,\ldots,i_p} dx_{i_1} \wedge \cdots \wedge dx_{i_k}$, where $1 \leq i_k \leq n$ for $k = 1, \ldots, p$. This sum has $\binom{n}{p}$ distinct non-zero terms.

5.2.5   THE EXTERIOR DERIVATIVE

The exterior differentiation operator is denoted by d. When d is applied to a scalar function $f(\mathbf{x})$, the result is the 1-form that is equivalent to the usual "total differential" $df = \nabla_{\mathbf{x}} f = \frac{\partial f}{\partial x_1} dx_1 + \cdots + \frac{\partial f}{\partial x_n} dx_n$. For the 1-form $\omega^1 = f_1\, dx + \cdots + f_n\, dx_n$ the *exterior derivative* is $d\omega^1 = (df_1) \wedge dx_1 + \cdots + (df_n) \wedge dx_n$. This generalizes to higher dimensional forms.

EXAMPLES

1. If $f(x_1, x_2) = x_1^2 + x_2^3$, then $df = d(x_1^2 + x_2^3) = 2x_1\, dx_1 + 3x_2^2\, dx_2$.

2. If $\omega^1_{(x_1,x_2)} = x_1 x_2\, dx_1 + (x_1^2 + x_2^2)\, dx_2$, then $d\omega^1$ is given by

$$dw^1 = d(x_1 x_2\, dx_1 + (x_1^2 + x_2^2)\, dx_2)$$
$$= (x_2\, dx_1 + x_1\, dx_2) \wedge dx_1 + (2x_1\, dx_1 + 2x_2\, dx_2) \wedge dx_2$$
$$= x_1\, dx_1 \wedge dx_2.$$

5.2.6   PROPERTIES OF THE EXTERIOR DERIVATIVE

1. If $f_1(x_1, x_2)$ and $f_2(x_1, x_2)$ are differentiable functions, then

$$df_1 \wedge df_2 = \det\left(\frac{\partial(f_1, f_2)}{\partial(x_1, x_2)} \right) dx_1 \wedge dx_2.$$

2. If ω^p and ω^q represent a p-form and a q-form, then

$$d(\omega^p \wedge \omega^q) = (d\omega^p) \wedge \omega^q + (-1)^{pq}\omega^p \wedge (d\omega^q).$$

3. If ω^p is a p-form with at least two derivatives, then $d(d\omega^p) = 0$.

   (a) The relation $d(d\omega^0) = 0$ is equivalent to

$$\mathrm{curl}(\mathrm{grad}\, f) = \nabla \times (\nabla f) = 0. \qquad (5.2.10)$$

   (b) The relation $d(d\omega^1) = 0$ is equivalent to

$$\mathrm{div}(\mathrm{curl}\,\mathbf{F}) = \nabla \cdot (\nabla \mathbf{F}) = 0. \qquad (5.2.11)$$

5.3   INTEGRATION

5.3.1   DEFINITIONS

The following definitions apply to the expression $I = \int_a^b f(x)\, dx$:

1. The *integrand*   is $f(x)$.
2. The *upper limit*   is b.
3. The *lower limit*   is a.
4. I is *the integral of $f(x)$ from a to b.*

It is conventional to indicate the indefinite integral of a function represented by a lowercase letter by the corresponding uppercase letter. For example, $F(x) = \int_a^x f(t)\, dt$ and $G(x) = \int_a^x g(t)\, dt$. Note that all functions that differ from $F(x)$ by a constant are also indefinite integrals of $f(x)$. We will use the following notation:

1. $\int f(x)\, dx$ indefinite integral of $f(x)$ (also written $\int^x f(t)\, dt$)

2. $\int_a^b f(x)\, dx$ definite integral of $f(x)$, defined as (for a continuous function)

$$\lim_{n\to\infty}\left(\frac{b-a}{n}\sum_{k=1}^{n} f\left[a+\frac{k}{n}(b-a)\right]\right)$$

3. $\oint_C f(x)\, dx$ definite integral of $f(x)$, taken along the contour C

4. $\int_a^\infty f(x)\, dx$ defined as $\displaystyle\lim_{R\to\infty}\int_a^R f(x)\, dx$

5. $\int_{-\infty}^\infty f(x)\, dx$ defined as the limit of $\int_{-S}^R f(x)\, dx$ as R and S independently go to ∞

6. *Improper integral*   integral for which the region of integration is not bounded, or the integrand is not bounded

7. *Cauchy principal value*

(a) The Cauchy principal value of $\int_a^b f(x)\, dx$, denoted $⨍_a^b f(x)\, dx$, is defined as $\displaystyle\lim_{\epsilon\to 0+}\left(\int_a^{c-\epsilon} f(x)\, dx + \int_{c+\epsilon}^b f(x)\, dx\right)$, assuming that f is singular only at c.

(b) The Cauchy principal value of the integral $\int_{-\infty}^\infty f(x)\, dx$ is defined as the limit of $\displaystyle\int_{-R}^R f(x)\, dx$ as $R\to\infty$.

8. If, at the complex point $z=a$, $f(z)$ is either analytic or has an isolated singularity, then the residue of $f(z)$ at $z=a$ is given by the contour integral $\mathrm{Res}_f(a) = \frac{1}{2\pi i}\oint_C f(\xi)\, d\xi$, where C is a closed contour around a in a positive direction.

5.3.2 PROPERTIES OF INTEGRALS

Indefinite integrals have the properties (here F is the antiderivative of f):

1. $\int [af(x)+bg(x)]\, dx = a\int f(x)\, dx + b\int g(x)\, dx$ (linearity).

2. $\int f(x)g(x)\, dx = F(x)g(x) - \int F(x)g'(x)\, dx$ (integration by parts).

3. $\int f(g(x))g'(x)\, dx = F(g(x))$ (substitution).

4. $\int f(ax+b)\, dx = \frac{1}{a}F(ax+b)$.

5. If $f(x)$ is an odd function and $F(0)=0$, then $F(x)$ is an even function.

6. If $f(x)$ is an even function and $F(0) = 0$, then $F(x)$ is an odd function.

7. If $f(x)$ has a finite number of discontinuities, then the integral $\int f(x)\,dx$ is the sum of the integrals over those subintervals where $f(x)$ is continuous (provided they exist).

8. *Fundamental theorem of calculus* If $f(x)$ is bounded, and integrable on $[a, b]$, and there exists a function $F(x)$ such that $F'(x) = f(x)$ for $a \leq x \leq b$ then

$$\int_a^x f(x)\,dx = F(x) \,\big|_a^x = F(x) - F(a) \qquad (5.3.1)$$

for $a \leq x \leq b$.

Definite integrals have the properties:

1. $\int_a^a f(x)\,dx = 0$.
2. $\int_a^b f(x)\,dx = -\int_b^a f(x)\,dx$.
3. $\int_a^b f(x)\,dx + \int_b^c f(x)\,dx = \int_a^c f(x)\,dx$ \hfill (additivity).
4. $\int_a^b [cf(x) + dg(x)]\,dx = c\int_a^b f(x)\,dx + d\int_a^b g(x)\,dx$ \hfill (linearity).

5.3.3   METHODS OF EVALUATING INTEGRALS

5.3.3.1   Substitution

Substitution can be used to change integrals to simpler forms. When the transform $t = g(x)$ is chosen, the integral $I = \int f(t)\,dt$ becomes $I = \int f(g(x))\,dt = \int f(g(x))g'(x)\,dx$. Several precautions must be taken when using substitutions:

1. Be sure to make the substitution in the dx term, as well as everywhere else in the integral.

2. Be sure that the function substituted is one-to-one and differentiable. If this is not the case, then the integral must be restricted in such a way as to make it true.

3. With definite integrals, the limits should also be expressed in terms of the new dependent variables. With indefinite integrals, it is necessary to perform the reverse substitution to obtain the answer in terms of the original independent variable. This may also be done for definite integrals, but it is usually easier to change the limits.

EXAMPLE   Consider the integral

$$I = \int \frac{x^4}{\sqrt{a^2 - x^2}}\,dx \qquad (5.3.2)$$

for $a \neq 0$. Here we choose to make the substitution $x = |a| \sin\theta$. From this we find $dx = |a| \cos\theta\,d\theta$ and

$$\sqrt{a^2 - x^2} = \sqrt{a^2 - a^2 \sin^2\theta} = |a|\sqrt{1 - \sin^2\theta} = |a|\,|\cos\theta|. \qquad (5.3.3)$$

Note the absolute value signs. It is very important to interpret the square root radical consistently as the positive square root. Thus $\sqrt{x^2} = |x|$. Failure to observe this is a common cause of errors.

Note that the substitution used above is not a one-to-one function, that is, it does not have a unique inverse. Thus the range of θ must be restricted in such a way as to make the function one-to-one. In this case we can solve for θ to obtain

$$\theta = \sin^{-1} \frac{x}{|a|}. \tag{5.3.4}$$

This will be unique if we restrict the inverse sine to the principal values $-\frac{\pi}{2} \le \theta \le \frac{\pi}{2}$.

Thus, the integral becomes (with $dx = |a| \cos \theta \, d\theta$)

$$I = \int \frac{a^4 \sin^4 \theta}{|a| \, |\cos \theta|} |a| \cos \theta \, d\theta. \tag{5.3.5}$$

Now, however, in the range of values chosen for θ, we find that $\cos \theta$ is always non-negative. Thus, we may remove the absolute value signs from $\cos \theta$ in the denominator. Then the $\cos \theta$ terms cancel and the integral becomes

$$I = a^4 \int \sin^4 \theta \, d\theta. \tag{5.3.6}$$

By application of integration formula #283 on page 430, and simplifications, this is integrated to obtain

$$I = -\frac{a^4}{4} \sin^3 \theta \cos \theta - \frac{3a^4}{8} \sin \theta \cos \theta + \frac{3a^4}{8} \theta + C. \tag{5.3.7}$$

To obtain an evaluation of I as a function of x, we must transform variables from θ to x. We have

$$\cos \theta = \pm\sqrt{1 - \sin^2 \theta} = \pm\sqrt{1 - \frac{x^2}{a^2}} = \pm\frac{\sqrt{a^2 - x^2}}{|a|}. \tag{5.3.8}$$

Because of the previously recorded fact that $\cos \theta$ is non-negative for our range of θ, we may omit the $\pm$ sign. Using $\sin \theta = x/|a|$ and $\cos \theta = \sqrt{a^2 - x^2}/|a|$ we can evaluate Equation (5.3.7) to obtain the final result,

$$I = \int \frac{x^4}{\sqrt{a^2 - x^2}} = -\frac{x^3}{4}\sqrt{a^2 - x^2} - \frac{3a^2 x}{8}\sqrt{a^2 - x^2} + \frac{3a^4}{8} \sin^{-1} \frac{x}{|a|} + C. \tag{5.3.9}$$

5.3.3.2 Partial fraction decomposition

Every integral of the form $\int R(x) \, dx$, where R is a rational function, can be evaluated (in principle) in terms of elementary functions. The technique is to factor the denominator of R and create a partial fraction decomposition. Then each resulting sub-integral is elementary.

EXAMPLE   Consider $I = \int \dfrac{2x^3 - 10x^2 + 13x - 4}{x^2 - 5x + 6} \, dx$. This can be written as

$$I = \int \left(2x + \frac{x - 4}{x^2 - 5x + 6} \right) dx = \int \left(2x + \frac{2}{x - 2} - \frac{1}{x - 3} \right) dx$$

which can be readily integrated $I = x^2 + 2\ln(x - 2) - \ln(x - 3)$.

5.3.3.3 Useful transformations

The following transformations may make evaluation of an integral easier:

1. $\int f\left(x, \sqrt{x^2 + a^2}\right) dx = a \int f(a\tan u, a\sec u) \sec^2 u\, du$
   $$\text{when } u = \tan^{-1}\tfrac{x}{a} \text{ and } a > 0.$$

2. $\int f\left(x, \sqrt{x^2 - a^2}\right) dx = a \int f(a\sec u, a\tan u) \sec u \tan u\, du$
   $$\text{when } u = \sec^{-1}\tfrac{x}{a} \text{ and } a > 0.$$

3. $\int f\left(x, \sqrt{a^2 - x^2}\right) dx = a \int f(a\sin u, a\cos u) \cos u\, du$
   $$\text{when } u = \sin^{-1}\tfrac{x}{a} \text{ and } a > 0.$$

4. $\int f(\sin x)\, dx = 2 \int f\left(\frac{2z}{1+z^2}\right) \frac{dz}{1+z^2}$ when $z = \tan\frac{x}{2}$.

5. $\int f(\cos x)\, dx = 2 \int f\left(\frac{1-z^2}{1+z^2}\right) \frac{dz}{1+z^2}$ when $z = \tan\frac{x}{2}$.

6. $\int f(\cos x)\, dx = -\int f(v)\frac{dv}{\sqrt{1-v^2}}$ when $v = \cos x$.

7. $\int f(\sin x)\, dx = \int f(u)\frac{du}{\sqrt{1-u^2}}$ when $u = \sin x$.

8. $\int f(\sin x, \cos x)\, dx = \int f\left(u, \sqrt{1-u^2}\right) \frac{du}{\sqrt{1-u^2}}$ when $u = \sin x$.

9. $\int f(\sin x, \cos x)\, dx = 2 \int f\left(\frac{2z}{1+z^2}, \frac{1-z^2}{1+z^2}\right) \frac{dz}{1+z^2}$ when $z = \tan\frac{x}{2}$.

10. $\int_{-\infty}^{\infty} f(u)\, dx = \int_{-\infty}^{\infty} f(x)\, dx$ when $u = x - \sum_{j=1}^{n} \frac{a_j}{x - c_j}$ where $\{a_i\}$ is any
    sequence of positive constants and the $\{c_j\}$ are any real constants whatsoever.

Several transformations of the integral $\int_0^{\infty} f(x)\, dx$, with an infinite integration range, to an integral with a finite integration range, are shown:

$t(x)$	$x(t)$	$\frac{dx}{dt}$	Finite interval integral
e^{-x}	$-\log t$	$-\frac{1}{t}$	$\int_0^1 \frac{f(-\log t)}{t}\, dt$
$\frac{x}{1+x}$	$\frac{t}{1-t}$	$\frac{1}{(1-t)^2}$	$\int_0^1 f\left(\frac{t}{1-t}\right) \frac{dt}{(1-t)^2}$
$\tanh x$	$\frac{1}{2}\log\frac{1+t}{1-t}$	$\frac{1}{1-t^2}$	$\int_0^1 f\left(\frac{1}{2}\log\frac{1+t}{1-t}\right) \frac{dt}{1-t^2}$

5.3.3.4 Integration by parts

In one dimension, the integration by parts formula is

$$\int u\, dv = uv - \int v\, du \qquad \text{for indefinite integrals,} \qquad (5.3.10)$$

$$\int_a^b u\, dv = uv \Big|_a^b - \int_a^b v\, du \qquad \text{for definite integrals.} \qquad (5.3.11)$$

When evaluating a given integral by this method, u and v must be chosen so that the form $\int u\, dv$ becomes identical to the given integral. This is usually accomplished by specifying u and dv and deriving du and v. Then the integration by parts formula will produce a boundary term and another integral to be evaluated. If u and v were well-chosen, then this second integral may be easier to evaluate.

EXAMPLE   Consider the integral

$$I = \int x \sin x \, dx.$$

Two obvious choices for the integration by parts formula are $\{u = x, \, dv = \sin x \, dx\}$ and $\{u = \sin x, \, dv = x \, dx\}$. We try each of them in turn.

1. Using $\{u = x, \, dv = \sin x \, dx\}$, we compute $du = dx$ and $v = \int dv = \int \sin x \, dx = -\cos x$. Hence, we can represent I in the alternative form as

$$I = \int x \sin x \, dx = \int u \, dv = uv - \int v \, du = -x \cos x + \int \cos x \, dx.$$

In this representation of I, we must evaluate the last integral. Because we know $\int \cos x \, dx = \sin x$ the final result is $I = \sin x - x \cos x$.

2. Using $\{u = \sin x, \, dv = x \, dx\}$ we compute $du = \cos x \, dx$ and $v = \int dv = \int x \, dx = x^2/2$. Hence, we can represent I in the alternative form as

$$I = \int x \sin x \, dx = \int u \, dv = uv - \int v \, du = \frac{x^2}{2} \sin x - \int \frac{x^2}{2} \cos x \, dx.$$

In this case, we have actually made the problem "worse" since the remaining integral appearing in I is "harder" than the one we started with.

EXAMPLE   Consider the integral

$$I = \int e^x \sin x \, dx.$$

We choose to use the integration by parts formula with $u = e^x$ and $dv = \sin x \, dx$. From these we compute $du = e^x \, dx$ and $v = \int dv = \int \sin x \, dx = -\cos x$. Hence, we can represent I in the alternative form as

$$I = \int e^x \sin x \, dx = \int u \, dv = uv - \int v \, du = -e^x \cos x + \int e^x \cos x \, dx.$$

If we write this as

$$I = -e^x \cos x + J \qquad \text{with} \qquad J = \int e^x \cos x \, dx, \tag{5.3.12}$$

then we can apply integration by parts to J using $\{u = e^x, \, dv = \cos x \, dx\}$. From these we compute $du = e^x \, dx$ and $v = \int dv = \int \cos x \, dx = \sin x$. Hence, we can represent J in the alternative form as

$$J = \int e^x \cos x \, dx = \int u \, dv = uv - \int v \, du = e^x \sin x - \int e^x \sin x \, dx.$$

If we write this as

$$J = e^x \sin x - I, \tag{5.3.13}$$

then we can solve the linear equations (5.3.12) and (5.3.13) simultaneously to determine both I and J. We find

$$I = \int e^x \sin x \, dx = \frac{1}{2} \left(e^x \sin x - e^x \cos x \right), \quad \text{and}$$

$$J = \int e^x \cos x \, dx = \frac{1}{2} \left(e^x \sin x + e^x \cos x \right). \tag{5.3.14}$$

5.3.3.5   Extended integration by parts rule

The following rule is obtained by $n + 1$ successive applications of integration by parts. Let

$$g_1(x) = \int g(x)\, dx, \qquad g_2(x) = \int g_1(x)\, dx,$$

$$g_3(x) = \int g_2(x)\, dx, \qquad \dots, \qquad g_m(x) = \int g_{m-1}(x)\, dx. \tag{5.3.15}$$

Then

$$\int f(x)g(x)\, dx = f(x)g_1(x) - f'(x)g_2(x) + f''(x)g_3(x) - \dots$$

$$+ (-1)^n f^{(n)}(x)g_{n+1}(x) + (-1)^{n+1} \int f^{(n+1)}(x)g_{n+1}(x)\, dx. \tag{5.3.16}$$

5.3.4   TYPES OF INTEGRALS

5.3.4.1   Line and surface integrals

A *line integral* is a definite integral whose path of integration is along a specified curve; it can be evaluated by reducing it to ordinary integrals. If $f(x, y)$ is continuous on C, and the integration contour C is parameterized by $(\phi(t), \psi(t))$ as t varies from a to b, then

$$\int_C f(x, y)\, dx = \int_a^b f\Big(\phi(t), \psi(t)\Big)\, \phi'(t)\, dt,$$

$$\int_C f(x, y)\, dy = \int_a^b f\Big(\phi(t), \psi(t)\Big)\, \psi'(t)\, dt. \tag{5.3.17}$$

In a simply-connected domain, the line integral $I = \int_C X\, dx + Y\, dy + Z\, dz$ is independent of the closed curve C if, and only if, $\mathbf{u} = (X, Y, Z)$ is a gradient vector, $\mathbf{u} = \operatorname{grad} F$ (that is, $F_x = X$, $F_y = Y$, and $F_z = Z$).

Green's theorem: Let D be a domain of the xy plane, and let C be a piecewise-smooth, simple closed curve in D whose interior R is also in D. Let $P(x, y)$ and $Q(x, y)$ be functions defined in D with continuous first partial derivatives in D. Then

$$\oint_C (P\, dx + Q\, dy) = \iint_R \left(\frac{\partial Q}{\partial x} - \frac{\partial P}{\partial y} \right) dx\, dy. \tag{5.3.18}$$

The above theorem may be written in the two alternative forms (using $\mathbf{u} = P(x, y)\mathbf{i} + Q(x, y)\mathbf{j}$ and $\mathbf{v} = Q(x, y)\mathbf{i} - P(x, y)\mathbf{j}$),

$$\oint_C \mathbf{u}_T\, ds = \iint_R \operatorname{curl} \mathbf{u}\, dx\, dy \quad \text{and} \quad \oint_C \mathbf{v}_n\, ds = \iint_R \operatorname{div} \mathbf{v}\, dx\, dy. \tag{5.3.19}$$

The first equation above is a simplification of Stokes's theorem; the second equation is the divergence theorem.

Stokes's theorem: Let S be a piecewise-smooth oriented surface in space, whose boundary C is a piecewise-smooth simple closed curve, directed in accordance with the given orientation of S. Let $\mathbf{u} = L\mathbf{i} + M\mathbf{j} + N\mathbf{k}$ be a vector field with continuous and differentiable components in a domain D of space including S. Then, $\int_C u_T \, ds = \iint_S (\mathrm{curl}\,\mathbf{u}) \cdot \mathbf{n} \, d\sigma$, where $\mathbf{n}$ is the chosen unit normal vector on S, that is

$$\int_C L \, dx + M \, dy + N \, dz = \iint_S \left(\frac{\partial N}{\partial y} - \frac{\partial M}{\partial z} \right) dy \, dz$$

$$+ \left(\frac{\partial L}{\partial z} - \frac{\partial N}{\partial x} \right) dz \, dx + \left(\frac{\partial M}{\partial x} - \frac{\partial L}{\partial y} \right) dx \, dy. \quad (5.3.20)$$

Divergence theorem: Let $\mathbf{v} = L\mathbf{i} + M\mathbf{j} + N\mathbf{k}$ be a vector field in a domain D of space. Let L, M, and N be continuous with continuous derivatives in D. Let S be a piecewise-smooth surface in D that forms the complete boundary of a bounded closed region R in D. Let $\mathbf{n}$ be the outer normal of S with respect to R. Then $\iint_S v_n \, d\sigma = \iiint_R \mathrm{div}\,\mathbf{v} \, dx \, dy \, dz$, that is

$$\iint_S L \, dy \, dz + M \, dz \, dx + N \, dx \, dy$$

$$= \iiint_R \left(\frac{\partial L}{\partial x} + \frac{\partial M}{\partial y} + \frac{\partial N}{\partial z} \right) dx \, dy \, dz. \quad (5.3.21)$$

If D is a three-dimensional domain with boundary B, let dV represent the volume element of D, let dS represent the surface element of B, and let $d\mathbf{S} = \mathbf{n}\,dS$, where $\mathbf{n}$ is the outer normal vector of the surface B. Then Gauss's formulae are

$$\iiint_D \nabla \cdot \mathbf{A} \, dV = \iint_B d\mathbf{S} \cdot \mathbf{A} = \iint_B (\mathbf{n} \cdot \mathbf{A}) \, dS,$$

$$\iiint_D \nabla \times \mathbf{A} \, dV = \iint_B d\mathbf{S} \times \mathbf{A} = \iint_B (\mathbf{n} \times \mathbf{A}) \, dS, \quad \text{and} \quad (5.3.22)$$

$$\iiint_D \nabla\phi \, dV = \iint_B \phi \, d\mathbf{S},$$

where ϕ is an arbitrary scalar and $\mathbf{A}$ is an arbitrary vector.

Green's theorems also relate a volume integral to a surface integral: Let V be a volume with surface S, which we assume is simple and closed. Define n as the outward normal to S. Let ϕ and ψ be scalar functions which, together with $\nabla^2 \phi$ and $\nabla^2 \psi$, are defined in V and on S. Then

1. Green's first theorem states that

$$\int_S \phi \frac{\partial \psi}{\partial n} \, dS = \int_V \left(\phi\nabla^2\psi + \nabla\phi \cdot \nabla\psi \right) dV. \quad (5.3.23)$$

2. Green's second theorem states that

$$\int_S \left(\phi \frac{\partial \psi}{\partial n} - \psi \frac{\partial \phi}{\partial n} \right) dS = \int_V \left(\phi \nabla^2 \psi - \psi \nabla^2 \phi \right) dV. \qquad (5.3.24)$$

5.3.4.2   Contour integrals

If $f(z)$ is analytic in the region inside of the simple closed curve C (with proper orientation), then

1. The Cauchy–Goursat integral theorem is $\oint_C f(\xi)\, d\xi = 0$.

2. Cauchy's integral formula is

$$f(z) = \frac{1}{2\pi i} \oint_C \frac{f(\xi)}{\xi - z}\, d\xi \qquad \text{and} \qquad f'(z) = \frac{1}{2\pi i} \oint_C \frac{f(\xi)}{(\xi - z)^2}\, d\xi. \qquad (5.3.25)$$

In general, $f^{(n)}(z) = \dfrac{n!}{2\pi i} \oint_C \dfrac{f(\xi)}{(\xi - z)^{n+1}}\, d\xi.$

The *residue theorem*: For every simple closed contour $\mathcal{C}$ enclosing at most a finite number of (necessarily isolated) singularities $\{z_1, z_2, \ldots, z_n\}$ of a (single-valued) analytic function $f(z)$ continuous on $\mathcal{C}$,

$$\frac{1}{2\pi i} \oint_C f(\xi)\, d\xi = \sum_{k=1}^{n} \operatorname{Res}_f(z_k). \qquad (5.3.26)$$

5.3.5   INTEGRAL INEQUALITIES

1. *Schwartz inequality:* $\int_a^b |fg| \leq \sqrt{\left(\int_a^b |f|^2 \right) \left(\int_a^b |g|^2 \right)}.$

2. *Minkowski's inequality:*

$$\left(\int_a^b |f + g|^p \right)^{1/p} \leq \left(\int_a^b |f|^p \right)^{1/p} + \left(\int_a^b |g|^p \right)^{1/p} \quad \text{when } p \geq 1.$$

3. *Hölder's inequality:*

$$\int_a^b |fg| \leq \left[\int_a^b |f|^p \right]^{1/p} \left[\int_a^b |g|^q \right]^{1/q} \quad \text{when } \tfrac{1}{p} + \tfrac{1}{q} = 1, p > 1, \text{ and } q > 1.$$

4. $\left| \int_a^b f(x)\, dx \right| \leq \int_a^b |f(x)|\, dx \leq \left(\max\limits_{x \in [a,b]} |f(x)| \right) (b - a)$ assuming $a \leq b$.

5. If $f(x) \leq g(x)$ on the interval $[a, b]$, then $\int_a^b f(x)\, dx \leq \int_a^b g(x)\, dx$.

5.3.6 CONVERGENCE TESTS

1. If $\int_a^b |f(x)|\, dx$ is convergent, and f is integrable, then $\int_a^b f(x)\, dx$ is convergent.

2. If $0 \le f(x) \le g(x)$, $\int_a^b g(x)\, dx$ is convergent, and f is integrable, then $\int_a^b f(x)\, dx$ is convergent.

3. If $0 \le g(x) \le f(x)$ and $\int_a^b g(x)\, dx$ is divergent, then $\int_a^b f(x)\, dx$ is divergent.

The following integrals may be used, for example, with the above tests:

(a) $\displaystyle\int_2^\infty \frac{dx}{x(\log x)^p}$ and $\displaystyle\int_1^\infty \frac{dx}{x^p}$ are convergent when $p > 1$, and divergent when $p \le 1$.

(b) $\displaystyle\int_0^1 \frac{dx}{x^p}$ is convergent when $p < 1$, and divergent when $p \ge 1$.

5.3.7 VARIATIONAL PRINCIPLES

If J depends on a function $g(\mathbf{x})$ and its derivatives through an integral of the form $J(g) = \int F(g, \nabla g, \dots)\, dx$, then J will be stationary to small perturbations of g if F satisfies the corresponding Euler–Lagrange equation.

Function	Euler–Lagrange equation
$\int_R F\,(x,y,y')\, dx$	$\dfrac{\partial F}{\partial y} - \dfrac{d}{dx}\left(\dfrac{\partial F}{\partial y'}\right) = 0$
$\int_R F\,(x,y,y',\dots,y^{(n)})\, dx$	$\dfrac{\partial F}{\partial y} - \dfrac{d}{dx}\left(\dfrac{\partial F}{\partial y'}\right) + \dfrac{d^2}{dx^2}\left(\dfrac{\partial F}{\partial y''}\right) -$
	$\cdots + (-1)^n \dfrac{d^n}{dx^n}\left(\dfrac{\partial F}{\partial y^{(n)}}\right) = 0$
$\iint_R \left[a\left(\dfrac{\partial u}{\partial x}\right)^2 + b\left(\dfrac{\partial u}{\partial x}\right)^2 + cu^2 + 2fu \right] dx\, dy$	$\dfrac{\partial}{\partial x}\left(a\dfrac{\partial u}{\partial x}\right) + \dfrac{\partial}{\partial y}\left(b\dfrac{\partial u}{\partial y}\right) - cu = f$
$\iint_R F(x,y,u,u_x,u_y,u_{xx},u_{xy},u_{yy})\, dx\, dy$	$\dfrac{\partial F}{\partial u} - \dfrac{\partial}{\partial x}\left(\dfrac{\partial F}{\partial u_x}\right) - \dfrac{\partial}{\partial y}\left(\dfrac{\partial F}{\partial u_y}\right) +$
	$\dfrac{\partial^2}{\partial x^2}\left(\dfrac{\partial F}{\partial u_{xx}}\right) + \dfrac{\partial^2}{\partial x \partial y}\left(\dfrac{\partial F}{\partial u_{xy}}\right) +$
	$\dfrac{\partial^2}{\partial y^2}\left(\dfrac{\partial F}{\partial u_{yy}}\right) = 0$

5.3.8 CONTINUITY OF INTEGRAL ANTIDERIVATIVES

Consider the following different antiderivatives of an integral

$$F(x) = \int f(x)\, dx = \int \frac{3}{5 - 4\cos x}\, dx = \begin{cases} 2\tan^{-1}\left(3\tan\frac{x}{2}\right) \\ 2\tan^{-1}(3\sin x/(\cos x + 1)) \\ -\tan^{-1}(-3\sin x/(5\cos x - 4)) \\ 2\tan^{-1}\left(3\tan\frac{x}{2}\right) + 2\pi\left\lfloor \frac{x}{2\pi} + \frac{1}{2}\right\rfloor \end{cases}$$

where $\lfloor \cdot \rfloor$ denotes the floor function. These antiderivatives are all "correct" because differentiating any of them results in the original integrand (except at isolated points). However, if we desire $\int_0^{4\pi} f(x)\,dx = F(4\pi) - F(0)$ to hold, then only the last antiderivative is correct. This is true because the other antiderivatives of $F(x)$ are discontinuous when x is a multiple of π.

In general, if $\hat{F}(x) = \int^x f(t)\,dt$ is a discontinuous evaluation (with $\hat{F}(x)$ discontinuous at the single point $x = b$), then a continuous evaluation on a finite interval is given by $\int_a^c f(x)\,dx = F(c) - F(a)$, where

$$F(x) = \hat{F}(x) - \hat{F}(a) + H(x-b)\left[\lim_{x\to b-} \hat{F}(x) - \lim_{x\to b+} \hat{F}(x) \right] \qquad (5.3.27)$$

and where $H(\cdot)$ is the Heaviside function. For functions with an infinite number of discontinuities, note that $\displaystyle\sum_{n=1}^{\infty} H(x - pn - q) = \left\lfloor \frac{x-q}{p} \right\rfloor$.

5.3.9   ASYMPTOTIC INTEGRAL EVALUATION

1. *Laplace's method*: If $f'(x_0) = 0$, $f''(x_0) < 0$, and $\lambda \to \infty$, then

$$I_{x_0}(\lambda) \equiv \int_{x_0-\epsilon}^{x_0+\epsilon} g(x)e^{\lambda f(x)}\,dx \sim g(x_0)e^{\lambda f(x_0)}\sqrt{\frac{2\pi}{\lambda|f''(x_0)|}} + \dots$$

$$(5.3.28)$$

Hence, if points of local maximum $\{x_i\}$ satisfy $f'(x_i) = 0$ and $f''(x_i) < 0$, then $\int_{-\infty}^{\infty} g(x)e^{\lambda f(x)}\,dx \sim \sum_i I_{x_i}(\lambda)$.

2. *Method of stationary phase*: If $f(x_0) \neq 0$, $f'(x_0) = 0$, $f''(x_0) \neq 0$, $g(x_0) \neq 0$, and $\lambda \to \infty$, then

$$J_{x_0,\epsilon}(\lambda) = \int_{x_0-\epsilon}^{x_0+\epsilon} g(x)e^{i\lambda f(x)}\,dx$$

$$\sim g(x_0)\sqrt{\frac{2\pi}{\lambda|f''(x_0)|}}\exp\left[i\lambda f(x_0) + \frac{i\pi}{4}\operatorname{sgn} f''(x_0) \right] + \dots \qquad (5.3.29)$$

5.3.10   SPECIAL FUNCTIONS DEFINED BY INTEGRALS

Not all integrals of elementary functions (sines, cosines, rational functions, and others) can be evaluated in terms of elementary functions. For example, the integral $\int e^{-x^2}\,dx$ is represented by the special function "erf(x)" (see page 545). Other useful functions include dilogarithms (see page 551) and elliptic integrals (see page 568).

The dilogarithm function is defined by $\mathrm{Li}_2(x) = -\int_0^x \ln(1-t)/t\,dt$. All integrals of the form $\int^x P(x, \sqrt{R})\log Q(x, \sqrt{R})\,dx$, where P and Q are rational functions and $R = A + Bx + Cx^2$, can be evaluated in terms of elementary functions and dilogarithms.

All integrals of the form $\int_x R(x, \sqrt{T(x)})\, dx$, where R is a rational function of its arguments and $T(x)$ is a third- or fourth-order polynomial, can be integrated in terms of elementary functions and elliptic functions.

5.3.11 APPLICATIONS OF INTEGRATION

1. Using Green's theorems, the area bounded by the simple, closed, positively-oriented contour $\mathcal{C}$ is

$$\text{area} = \oint_{\mathcal{C}} x\, dy = -\oint_{\mathcal{C}} y\, dx. \tag{5.3.30}$$

2. Arc length:

   (a) $s = \displaystyle\int_{x_1}^{x_2} \sqrt{1 + y'^2}\, dx$ for $y = f(x)$.

   (b) $s = \displaystyle\int_{t_1}^{t_2} \sqrt{\dot{\phi}^2 + \dot{\psi}^2}\, dt$ for $x = \phi(t),\, y = \psi(t)$.

   (c) $s = \displaystyle\int_{\theta_1}^{\theta_2} \sqrt{r^2 + \left(\frac{dr}{d\phi}\right)^2}\, d\theta = \int_{r_1}^{r_2} \sqrt{1 + r^2\left(\frac{dr}{d\phi}\right)^2}\, dr$ for $r = f(\theta)$.

3. Surface area for surfaces of revolution:

   (a) $A = 2\pi \displaystyle\int_{x_1}^{x_2} f(x)\sqrt{1 + (f'(x))^2}\, dx$ when $y = f(x)$ is rotated about the x-axis.

   (b) $A = 2\pi \displaystyle\int_{y_1}^{y_2} x\sqrt{1 + (f'(x))^2}\, dy$ when $y = f(x)$ is rotated about the y-axis and f is one-to-one.

   (c) $A = 2\pi \displaystyle\int_{t_1}^{t_2} \psi\sqrt{\dot{\phi}^2 + \dot{\psi}^2}\, dt$ for $x = \phi(t),\, y = \psi(t)$ rotated about the x-axis.

   (d) $A = 2\pi \displaystyle\int_{t_1}^{t_2} \phi\sqrt{\dot{\phi}^2 + \dot{\psi}^2}\, dt$ for $x = \phi(t),\, y = \psi(t)$ rotated about the y-axis.

   (e) $A = 2\pi \displaystyle\int_{\phi_1}^{\phi_2} r\sin\phi\sqrt{r^2 + \left(\frac{dr}{d\phi}\right)^2}\, d\phi$ for $r = r(\phi)$ rotated about the x-axis.

   (f) $A = 2\pi \displaystyle\int_{\phi_1}^{\phi_2} r\cos\phi\sqrt{r^2 + \left(\frac{dr}{d\phi}\right)^2}\, d\phi$ for $r = r(\phi)$ rotated about the y-axis.

4. Volumes of revolution:

   (a) $V = \pi \displaystyle\int_{x_1}^{x_2} f^2(x)\, dx$ for $y = f(x)$ rotated about the x-axis.

(b) $V = \pi \displaystyle\int_{x_1}^{x_2} x^2 f'(x)\, dx$ for $y = f(x)$ rotated about the y-axis.

(c) $V = \pi \displaystyle\int_{y_1}^{y_2} g^2(y)\, dy$ for $x = g(y)$ rotated about the y-axis.

(d) $V = \pi \displaystyle\int_{t_1}^{t_2} \psi^2 \dot\phi\, dt$ for $x = \phi(t)$, $y = \psi(t)$ rotated about the x-axis.

(e) $V = \pi \displaystyle\int_{t_1}^{t_2} \phi^2 \dot\psi\, dt$ for $x = \phi(t)$, $y = \psi(t)$ rotated about the y-axis.

(f) $V = \pi \displaystyle\int_{\phi_1}^{\phi_2} \sin^2\phi \left(\frac{dr}{d\phi}\cos\phi - r\sin\phi\right) d\phi$ for $r = f(\phi)$ rotated about the x-axis.

(g) $V = \pi \displaystyle\int_{\phi_1}^{\phi_2} \cos^2\phi \left(\frac{dr}{d\phi}\sin\phi - r\cos\phi\right) d\phi$ for $r = f(\phi)$ rotated about the y-axis.

5. The area enclosed by the curve $x^{b/c} + y^{b/c} = a^{b/c}$, where $a > 0$, c is an odd integer, and b is an even integer, is $A = \dfrac{2ca^2}{b} \dfrac{\left[\Gamma\left(\frac{c}{b}\right)\right]^2}{\Gamma\left(\frac{2c}{b}\right)}$.

6. The integral $I = \iiint\limits_{R} x^{h-1} y^{m-1} z^{n-1}\, dV$, where R is the region of space bounded by the coordinate planes and that portion of the surface $\left(\dfrac{x}{a}\right)^p + \left(\dfrac{y}{b}\right)^q + \left(\dfrac{z}{c}\right)^k = 1$, in the first octant, and where $\{h, m, n, p, q, k, a, b, c\}$ are all positive real numbers, is given by

$$\int_0^a x^{h-1}\, dx \int_0^{b\left[1-\left(\frac{x}{a}\right)^p\right]^{\frac{1}{q}}} y^{m-1}\, dy \int_0^{c\left[1-\left(\frac{x}{a}\right)^p-\left(\frac{y}{b}\right)^q\right]^{\frac{1}{k}}} z^{n-1}\, dz$$

$$= \frac{a^h b^m c^n}{pqk} \frac{\Gamma\left(\frac{h}{p}\right)\Gamma\left(\frac{m}{q}\right)\Gamma\left(\frac{n}{k}\right)}{\Gamma\left(\frac{h}{p}+\frac{m}{q}+\frac{n}{k}+1\right)}.$$

5.3.12 MOMENTS OF INERTIA FOR VARIOUS BODIES

	Body	Axis	Moment of inertia
(1)	Uniform thin rod	Normal to the length, at one end	$m\frac{l^2}{3}$
(2)	Uniform thin rod	Normal to the length, at the center	$m\frac{l^2}{12}$
(3)	Thin rectangular sheet, sides a and b	Through the center parallel to b	$m\frac{a^2}{12}$
(4)	Thin rectangular sheet, sides a and b	Through the center perpendicular to the sheet	$m\frac{a^2+b^2}{12}$
			continued on next page

	Body	Axis	Moment of inertia
(5)	Thin circular sheet of radius r	Normal to the plate through the center	$m\dfrac{r^2}{2}$
(6)	Thin circular sheet of radius r	Along any diameter	$m\dfrac{r^2}{4}$
(7)	Thin circular ring, radii r_1 and r_2	Through center normal to plane of ring	$m\dfrac{r_1^2+r_2^2}{2}$
(8)	Thin circular ring, radii r_1 and r_2	Along any diameter	$m\dfrac{r_1^2+r_2^2}{4}$
(9)	Rectangular parallelepiped, edges a, b, and c	Through center perpendicular to face ab (parallel to edge c)	$m\dfrac{a^2+b^2}{12}$
(10)	Sphere, radius r	Any diameter	$m\dfrac{2}{5}r^2$
(11)	Spherical shell, external radius r_1, internal radius r_2	Any diameter	$m\dfrac{2}{5}\dfrac{r_1^5-r_2^5}{r_1^3-r_2^3}$
(12)	Spherical shell, very thin, mean radius r	Any diameter	$m\dfrac{2}{3}r^2$
(13)	Right circular cylinder of radius r, length l	Longitudinal axis of the slide	$m\dfrac{r^2}{2}$
(14)	Right circular cylinder of radius r, length l	Transverse diameter	$m\left(\dfrac{r^2}{4}+\dfrac{l^2}{12}\right)$
(15)	Hollow circular cylinder, radii r_1 and r_2, length l	Longitudinal axis of the figure	$m\dfrac{r_1^2+r_2^2}{2}$
(16)	Thin cylindrical shell, length l, mean radius r	Longitudinal axis of the figure	mr^2
(17)	Hollow circular cylinder, radii r_1 and r_2, length l	Transverse diameter	$m\left(\dfrac{r_1^2+r_2^2}{4}+\dfrac{l^2}{12}\right)$
(18)	Hollow circular cylinder, very thin, length l, mean radius r	Transverse diameter	$m\left(\dfrac{r^2}{2}+\dfrac{l^2}{12}\right)$
(19)	Elliptic cylinder, length l, transverse semiaxes a and b	Longitudinal axis	$m\left(\dfrac{a^2+b^2}{4}\right)$
(20)	Right cone, altitude h, radius of base r	Axis of the figure	$m\dfrac{3}{10}r^2$
(21)	Spheroid of revolution, equatorial radius r	Polar axis	$m\dfrac{2}{5}r^2$
(22)	Ellipsoid, axes $2a$, $2b$, $2c$	Axis $2a$	$m\dfrac{b^2+c^2}{5}$

5.3.13   TABLES OF INTEGRALS

Many extensive compilations of integrals tables exist. No matter how extensive the integral table, it is fairly uncommon to find the exact integral desired. Usually some form of transformation will have to be made. The simplest type of transformation is substitution. Simple forms of substitutions, such as $y = ax$, are employed, almost unconsciously, by experienced users of integral tables. Finding the right substitution is largely a matter of intuition and experience.

We adopt the following conventions in the integral tables:

1. A constant of integration must be included with all indefinite integrals.

2. All angles are measured in radians; inverse trigonometric and hyperbolic functions represent principal values.

3. Logarithmic expressions are to base $e = 2.71828\ldots$, unless otherwise specified, and are to be evaluated for the absolute value of the arguments involved therein.

4. The natural logarithm function is denoted as $\log x$.

5. The variables n and m usually denote integers. The denominator of the expressions shown is not allowed to be zero; this may require that $a \neq 0$ or $m \neq n$ or some other similar statement.

6. When inverse trigonometric functions occur in the integrals, be sure that any replacements made for them are strictly in accordance with the rules for such functions. This causes little difficulty when the argument of the inverse trigonometric function is positive, because all angles involved are in the first quadrant. However, if the argument is negative, special care must be used. Thus, if $u > 0$ then

$$\sin^{-1} u = \cos^{-1} \sqrt{1 - u^2} = \csc^{-1} \frac{1}{u} = \ldots.$$

However, if $u < 0$, then

$$\sin^{-1} u = -\cos^{-1} \sqrt{1 - u^2} = -\pi - \csc^{-1} \frac{1}{u} = \ldots.$$

5.4 TABLE OF INDEFINITE INTEGRALS

5.4.1 ELEMENTARY FORMS

1. $\displaystyle\int a\, dx = ax.$

2. $\displaystyle\int a\, f(x)\, dx = a \int f(x)\, dx.$

3. $\displaystyle\int \phi(y(x))\, dx = \int \frac{\phi(y)}{y'}\, dy,$ where $y' = \frac{dy}{dx}.$

4. $\displaystyle\int (u + v)\, dx = \int u\, dx + \int v\, dx,$ where u and v are any functions of x.

5. $\displaystyle\int u\, dv = u \int dv - \int v\, du = uv - \int v\, du.$

6. $\displaystyle\int u \frac{dv}{dx}\, dx = uv - \int v \frac{du}{dx}\, dx.$

7. $\int x^n \, dx = \dfrac{x^{n+1}}{n+1}$, except when $n = -1$.

8. $\int \dfrac{dx}{x} = \log x$.

9. $\int \dfrac{f'(x)}{f(x)} \, dx = \log f(x), \ (df(x) = f'(x) \, dx)$.

10. $\int \dfrac{f'(x)}{2\sqrt{f(x)}} \, dx = \sqrt{f(x)}, \ (df(x) = f'(x) \, dx)$.

11. $\int e^x \, dx = e^x$.

12. $\int e^{ax} \, dx = \dfrac{e^{ax}}{a}$.

13. $\int b^{ax} \, dx = \dfrac{b^{ax}}{a \log b}, \ b > 0$.

14. $\int \log x \, dx = x \log x - x$.

15. $\int a^x \, dx = \dfrac{a^x}{\log a}, \ a > 0$.

16. $\int \dfrac{dx}{a^2 + x^2} = \dfrac{1}{a} \tan^{-1} \dfrac{x}{a}$.

17. $\int \dfrac{dx}{a^2 - x^2} = \begin{cases} \dfrac{1}{a} \tanh^{-1} \dfrac{x}{a}, \\ \text{or} \\ \dfrac{1}{2a} \log \dfrac{a+x}{a-x}, \quad a^2 > x^2. \end{cases}$

18. $\int \dfrac{dx}{x^2 - a^2} = \begin{cases} -\dfrac{1}{a} \coth^{-1} \dfrac{x}{a}, \\ \text{or} \\ \dfrac{1}{2a} \log \dfrac{x-a}{x+a}, \quad x^2 > a^2. \end{cases}$

19. $\int \dfrac{dx}{\sqrt{a^2 - x^2}} = \begin{cases} \sin^{-1} \dfrac{x}{|a|}, \\ \text{or} \\ -\cos^{-1} \dfrac{x}{|a|}, \quad a^2 > x^2. \end{cases}$

20. $\int \dfrac{dx}{\sqrt{x^2 \pm a^2}} = \log\left(x + \sqrt{x^2 \pm a^2}\right)$.

21. $\int \dfrac{dx}{x\sqrt{x^2 - a^2}} = \dfrac{1}{|a|} \sec^{-1} \dfrac{x}{a}$.

22. $\int \dfrac{dx}{x\sqrt{a^2 \pm x^2}} = -\dfrac{1}{a} \log\left(\dfrac{a + \sqrt{a^2 \pm x^2}}{x}\right)$.

5.4.2   FORMS CONTAINING $a + bx$

23. $\int (a + bx)^n \, dx = \dfrac{(a + bx)^{n+1}}{(n+1)b}, \ n \neq -1$.

24. $\int x(a + bx)^n \, dx = \dfrac{1}{b^2(n+2)} (a + bx)^{n+2} - \dfrac{a}{b^2(n+1)} (a + bx)^{n+1}$,

$$n \neq -1, \ n \neq -2.$$

25. $\displaystyle\int x^2(a+bx)^n\,dx = \frac{1}{b^3}\left[\frac{(a+bx)^{n+3}}{n+3} - 2a\frac{(a+bx)^{n+2}}{n+2} + a^2\frac{(a+bx)^{n+1}}{n+1}\right],$
$$n \neq -1, \qquad n \neq -2,\ n \neq -3.$$

26. $\displaystyle\int x^m(a+bx)^n\,dx =$
$$\begin{cases} \dfrac{x^{m+1}(a+bx)^n}{m+n+1} + \dfrac{an}{m+n+1}\displaystyle\int x^m(a+bx)^{n-1}\,dx, \\[2mm] \text{or} \\[2mm] \dfrac{1}{a(n+1)}\left[-x^{m+1}(a+bx)^{n+1} + (m+n+2)\displaystyle\int x^m(a+bx)^{n+1}\,dx\right], \\[2mm] \text{or} \\[2mm] \dfrac{1}{b(m+n+1)}\left[x^m(a+bx)^{n+1} - ma\displaystyle\int x^{m-1}(a+bx)^n\,dx\right]. \end{cases}$$

27. $\displaystyle\int \frac{dx}{a+bx} = \frac{1}{b}\log|a+bx|.$

28. $\displaystyle\int \frac{dx}{(a+bx)^2} = -\frac{1}{b(a+bx)}.$

29. $\displaystyle\int \frac{dx}{(a+bx)^3} = -\frac{1}{2b(a+bx)^2}.$

30. $\displaystyle\int \frac{x}{a+bx}\,dx = \begin{cases} \dfrac{1}{b^2}\left[a+bx - a\log(a+bx)\right], \\[2mm] \text{or} \\[2mm] \dfrac{x}{b} - \dfrac{a}{b^2}\log(a+bx). \end{cases}$

31. $\displaystyle\int \frac{x}{(a+bx)^2}\,dx = \frac{1}{b^2}\left[\log(a+bx) + \frac{a}{a+bx}\right].$

32. $\displaystyle\int \frac{x}{(a+bx)^n}\,dx = \frac{1}{b^2}\left[\frac{-1}{(n-2)(a+bx)^{n-2}} + \frac{a}{(n-1)(a+bx)^{n-1}}\right],$
$$n \neq 1,\ n \neq 2.$$

33. $\displaystyle\int \frac{x^2}{a+bx}\,dx = \frac{1}{b^3}\left(\frac{1}{2}(a+bx)^2 - 2a(a+bx) + a^2\log(a+bx)\right).$

34. $\displaystyle\int \frac{x^2}{(a+bx)^2}\,dx = \frac{1}{b^3}\left(a+bx - 2a\log(a+bx) - \frac{a^2}{a+bx}\right).$

35. $\displaystyle\int \frac{x^2}{(a+bx)^3}\,dx = \frac{1}{b^3}\left(\log(a+bx) + \frac{2a}{a+bx} - \frac{a^2}{2(a+bx)^2}\right).$

36. $\displaystyle\int \frac{x^2}{(a+bx)^n}\,dx = \frac{1}{b^3}\left[\frac{-1}{(n-3)(a+bx)^{n-3}} + \frac{2a}{(n-2)(a+bx)^{n-2}}\right.$
$$\left. - \frac{a^2}{(n-1)(a+bx)^{n-1}}\right],\quad n\neq 1,\ n\neq 2,\ n\neq 3.$$

37. $\displaystyle\int \frac{dx}{x(a+bx)} = -\frac{1}{a}\log\frac{a+bx}{x}.$

38. $\displaystyle\int \frac{dx}{x(a+bx)^2} = \frac{1}{a(a+bx)} - \frac{1}{a^2}\log\frac{a+bx}{x}.$

39. $\displaystyle\int \frac{dx}{x(a+bx)^3} = \frac{1}{a^3}\left[\frac{1}{2}\left(\frac{2a+bx}{a+bx}\right)^2 - \log\frac{a+bx}{x}\right].$

40. $\displaystyle\int \frac{dx}{x^2(a+bx)} = -\frac{1}{ax} + \frac{b}{a^2}\log\frac{a+bx}{x}.$

41. $\displaystyle\int \frac{dx}{x^3(a+bx)} = \frac{2bx-a}{2a^2x^2} + \frac{b^2}{a^3}\log\frac{x}{a+bx}.$

42. $\displaystyle\int \frac{dx}{x^2(a+bx)^2} = -\frac{a+2bx}{a^2x(a+bx)} + \frac{2b}{a^3}\log\frac{a+bx}{x}.$

5.4.3 FORMS CONTAINING $c^2 \pm x^2$ AND $x^2 - c^2$

43. $\displaystyle\int \frac{dx}{c^2+x^2} = \frac{1}{c}\tan^{-1}\frac{x}{c}.$

44. $\displaystyle\int \frac{dx}{c^2-x^2} = \frac{1}{2c}\log\frac{c+x}{c-x}, \quad c^2 > x^2.$

45. $\displaystyle\int \frac{dx}{x^2-c^2} = \frac{1}{2c}\log\frac{x-c}{x+c}, \quad x^2 > c^2.$

46. $\displaystyle\int \frac{x}{c^2\pm x^2}\,dx = \pm\frac{1}{2}\log(c^2\pm x^2).$

47. $\displaystyle\int \frac{x}{(c^2\pm x^2)^{n+1}}\,dx = \mp\frac{1}{2n(c^2\pm x^2)^n}, \quad n \neq 0.$

48. $\displaystyle\int \frac{dx}{(c^2\pm x^2)^n} = \frac{1}{2c^2(n-1)}\left[\frac{x}{(c^2\pm x^2)^{n-1}} + (2n-3)\int\frac{dx}{(c^2\pm x^2)^{n-1}}\right].$

49. $\displaystyle\int \frac{dx}{(x^2-c^2)^n} = \frac{1}{2c^2(n-1)}\left[-\frac{x}{(x^2-c^2)^{n-1}} - (2n-3)\int\frac{dx}{(x^2-c^2)^{n-1}}\right].$

50. $\displaystyle\int \frac{x}{x^2-c^2}\,dx = \frac{1}{2}\log(x^2-c^2).$

51. $\displaystyle\int \frac{x}{(x^2-c^2)^{n+1}}\,dx = -\frac{1}{2n(x^2-c^2)^n}.$

5.4.4 FORMS CONTAINING $a + bx$ AND $c + dx$

$u = a+bx, \quad v = c+dx, \quad \text{and } k = ad - bc. \quad \text{(If } k = 0, \text{ then } v = (c/a)u.)$

52. $\displaystyle\int \frac{dx}{uv} = \frac{1}{k}\log\left(\frac{v}{u}\right).$

53. $\displaystyle\int \frac{x}{uv}\,dx = \frac{1}{k}\left(\frac{a}{b}\log u - \frac{c}{d}\log v\right).$

54. $\displaystyle\int \frac{dx}{u^2v} = \frac{1}{k}\left(\frac{1}{u} + \frac{d}{k}\log\frac{v}{u}\right).$

55. $\displaystyle\int \frac{x}{u^2v}\,dx = -\frac{a}{bku} - \frac{c}{k^2}\log\frac{v}{u}.$

56. $\displaystyle\int \frac{x^2}{u^2v}\,dx = \frac{a^2}{b^2ku} + \frac{1}{k^2}\left(\frac{c^2}{d}\log v + \frac{a(k-bc)}{b^2}\log u\right).$

57. $\displaystyle\int \frac{dx}{u^nv^m} = \frac{1}{k(m-1)}\left[\frac{-1}{u^{n-1}v^{m-1}} - b(m+n-2)\int\frac{dx}{u^nv^{m-1}}\right].$

58. $\displaystyle\int \frac{u}{v}\,dx = \frac{bx}{d} + \frac{k}{d^2}\log v.$

59. $\displaystyle \int \frac{u^m}{v^n}\, dx = \begin{cases} -\dfrac{1}{k(n-1)}\left[\dfrac{u^{m+1}}{v^{n-1}} + b(n-m-2)\displaystyle\int \dfrac{u^m}{v^{n-1}}\, dx\right], \\[4pt] \text{or} \\[4pt] -\dfrac{1}{d(n-m-1)}\left[\dfrac{u^m}{v^{n-1}} + mk\displaystyle\int \dfrac{u^{m-1}}{v^n}\, dx\right], \\[4pt] \text{or} \\[4pt] -\dfrac{1}{d(n-1)}\left[\dfrac{u^m}{v^{n-1}} - mb\displaystyle\int \dfrac{u^{m-1}}{v^{n-1}}\, dx\right]. \end{cases}$

5.4.5   FORMS CONTAINING $a + bx^n$

60. $\displaystyle \int \frac{dx}{a + bx^2} = \frac{1}{\sqrt{ab}}\tan^{-1}\frac{x\sqrt{ab}}{a}, \quad ab > 0.$

61. $\displaystyle \int \frac{dx}{a + bx^2} = \begin{cases} \dfrac{1}{2\sqrt{-ab}}\log\dfrac{a + x\sqrt{-ab}}{a - x\sqrt{-ab}}, & ab < 0, \\[4pt] \text{or} \\[4pt] \dfrac{1}{\sqrt{-ab}}\tanh^{-1}\dfrac{x\sqrt{-ab}}{a}, & ab < 0. \end{cases}$

62. $\displaystyle \int \frac{dx}{a^2 + b^2 x^2}\, dx = \frac{1}{ab}\tan^{-1}\frac{bx}{a}.$

63. $\displaystyle \int \frac{x}{a + bx^2}\, dx = \frac{1}{2b}\log(a + bx^2).$

64. $\displaystyle \int \frac{x^2}{a + bx^2}\, dx = \frac{x}{b} - \frac{a}{b}\int \frac{dx}{a + bx^2}.$

65. $\displaystyle \int \frac{dx}{(a + bx^2)^2} = \frac{x}{2a(a + bx^2)} + \frac{1}{2a}\int \frac{dx}{a + bx^2}.$

66. $\displaystyle \int \frac{dx}{a^2 - b^2 x^2} = \frac{1}{2ab}\log\frac{a + bx}{a - bx}.$

67. $\displaystyle \int \frac{dx}{(a + bx^2)^{m+1}} =$

$\begin{cases} \dfrac{1}{2ma}\dfrac{x}{(a + bx^2)^m} + \dfrac{2m-1}{2ma}\displaystyle\int \dfrac{dx}{(a + bx^2)^m}, \\[6pt] \text{or} \\[6pt] \dfrac{(2m)!}{(m!)^2}\left[\dfrac{x}{2a}\displaystyle\sum_{r=1}^{m}\dfrac{r!(r-1)!}{(4a)^{m-r}(2r)!(a + bx^2)^r} + \dfrac{1}{(4a)^m}\displaystyle\int \dfrac{dx}{a + bx^2}\right]. \end{cases}$

68. $\displaystyle \int \frac{x\, dx}{(a + bx^2)^{m+1}} = -\frac{1}{2bm(a + bx^2)^m}, \quad m \neq 0.$

69. $\displaystyle \int \frac{x^2\, dx}{(a + bx^2)^{m+1}} = -\frac{x}{2mb(a + bx^2)^m} + \frac{1}{2mb}\int \frac{dx}{(a + bx^2)^m}, \quad m \neq 0.$

70. $\displaystyle \int \frac{dx}{x(a + bx^2)} = \frac{1}{2a}\log\frac{x^2}{a + bx^2}.$

71. $\displaystyle \int \frac{dx}{x^2(a + bx^2)} = -\frac{1}{ax} - \frac{b}{a}\int \frac{dx}{a + bx^2}.$

72. $\displaystyle \int \frac{dx}{x(a + bx^2)^{m+1}} = \begin{cases} \dfrac{1}{2am(a + bx^2)^m} + \dfrac{1}{a}\displaystyle\int \dfrac{dx}{x(a + bx^2)^m}, \\[6pt] \text{or} \\[6pt] \dfrac{1}{2a^{m+1}}\left[\displaystyle\sum_{r=1}^{m}\dfrac{a^r}{r(a + bx^2)^r} + \log\dfrac{x^2}{a + bx^2}\right]. \end{cases}$

73. $\displaystyle\int \frac{dx}{x^2\,(a+bx^2)^{m+1}} = \frac{1}{a}\int \frac{dx}{x^2(a+bx^2)^m} - \frac{b}{a}\int \frac{dx}{(a+bx^2)^{m+1}}.$

74. $\displaystyle\int \frac{dx}{a+bx^3} = \frac{k}{3a}\left[\frac{1}{2}\log\frac{(k+x)^3}{a+bx^3} + \sqrt{3}\tan^{-1}\frac{2x-k}{k\sqrt{3}}\right], \quad k=\sqrt[3]{\frac{a}{b}}.$

75. $\displaystyle\int \frac{x\,dx}{a+bx^3} = \frac{1}{3bk}\left[\frac{1}{2}\log\frac{a+bx^3}{(k+x)^3} + \sqrt{3}\tan^{-1}\frac{2x-k}{k\sqrt{3}}\right], \quad k=\sqrt[3]{\frac{a}{b}}.$

76. $\displaystyle\int \frac{x^2\,dx}{a+bx^3} = \frac{1}{3b}\log a+bx^3.$

77. $\displaystyle\int \frac{dx}{a+bx^4} =$

$$\begin{cases} \dfrac{k}{2a}\left[\dfrac{1}{2}\log\dfrac{x^2+2kx+2k^2}{x^2-2kx+2k^2} + \tan^{-1}\dfrac{2kx}{2k^2-x^2}\right], & ab>0,\ k=\left(\dfrac{a}{4b}\right)^{1/4}, \\ \text{or} \\ \dfrac{k}{2a}\left[\dfrac{1}{2}\log\dfrac{x+k}{x-k} + \tan^{-1}\dfrac{x}{k}\right], & ab<0,\ k=\left(-\dfrac{a}{b}\right)^{1/4}. \end{cases}$$

78. $\displaystyle\int \frac{x}{a+bx^4}\,dx = \frac{1}{2bk}\tan^{-1}\frac{x^2}{k}, \quad ab>0,\ k=\sqrt{\frac{a}{b}}.$

79. $\displaystyle\int \frac{x}{a+bx^4}\,dx = \frac{1}{4bk}\log\frac{x^2-k}{x^2+k}, \quad ab<0,\ k=\sqrt{-\frac{a}{b}}.$

80. $\displaystyle\int \frac{x^2}{a+bx^4}\,dx = \frac{1}{4bk}\left[\frac{1}{2}\log\frac{x^2-2kx+2k^2}{x^2+2kx+2k^2} + \tan^{-1}\frac{2kx}{2k^2-x^2}\right],$

$ab>0,\ k=\left(\frac{a}{4b}\right)^{1/4}.$

81. $\displaystyle\int \frac{x^2\,dx}{a+bx^4} = \frac{1}{4bk}\left[\log\frac{x-k}{x+k} + 2\tan^{-1}\frac{x}{k}\right], \quad ab<0,\ k=\sqrt[4]{-\frac{a}{b}}.$

82. $\displaystyle\int \frac{x^3\,dx}{a+bx^4} = \frac{1}{4b}\log\left(a+bx^4\right).$

83. $\displaystyle\int \frac{dx}{x(a+bx^n)} = \frac{1}{an}\log\frac{x^n}{a+bx^n}, \quad n\neq 0.$

84. $\displaystyle\int \frac{dx}{(a+bx^n)^{m+1}} = \frac{1}{a}\int \frac{dx}{(a+bx^n)^m} - \frac{b}{a}\int \frac{x^n\,dx}{(a+bx^n)^{m+1}}.$

85. $\displaystyle\int \frac{x^m\,dx}{(a+bx^n)^{p+1}} = \frac{1}{b}\int \frac{x^{m-n}\,dx}{(a+bx^n)^p} - \frac{a}{b}\int \frac{x^{m-n}\,dx}{(a+bx^n)^{p+1}}.$

86. $\displaystyle\int \frac{dx}{x^m(a+bx^n)^{p+1}} = \frac{1}{a}\int \frac{dx}{x^m(a+bx^n)^p} - \frac{b}{a}\int \frac{dx}{x^{m-n}(a+bx^n)^{p+1}}.$

87. $\displaystyle\int x^m(a+bx^n)^p\,dx =$

$$\begin{cases} \dfrac{1}{b(np+m+1)}\left[x^{m-n+1}(a+bx^n)^{p+1} - a(m-n+1)\displaystyle\int x^{m-n}(a+bx^n)^p\,dx\right], \\ \text{or} \\ \dfrac{1}{np+m+1}\left[x^{m+1}(a+bx^n)^p + anp\displaystyle\int x^m(a+bx^n)^{p-1}\,dx\right], \\ \text{or} \\ \dfrac{1}{a(m+1)}\left[x^{m+1}(a+bx^n)^{p+1} - b(m+1+np+n)\displaystyle\int x^{m+n}(a+bx^n)^p\,dx\right], \\ \text{or} \\ \dfrac{1}{an(p+1)}\left[-x^{m+1}(a+bx^n)^{p+1} + (m+1+np+n)\displaystyle\int x^m(a+bx^n)^{p+1}\,dx\right]. \end{cases}$$

5.4.6 FORMS CONTAINING $c^3 \pm x^3$

88. $\displaystyle \int \frac{dx}{c^3 \pm x^3} = \pm \frac{1}{6c^2} \log \left(\frac{(c \pm x)^3}{c^3 \pm x^3} \right) + \frac{1}{c^2\sqrt{3}} \tan^{-1} \frac{2x \mp c}{c\sqrt{3}}.$

89. $\displaystyle \int \frac{dx}{(c^3 \pm x^3)^2} = \frac{x}{3c^3(c^3 \pm x^3)} + \frac{2}{3c^3} \int \frac{dx}{c^3 \pm x^3}.$

90. $\displaystyle \int \frac{dx}{(c^3 \pm x^3)^{n+1}} = \frac{1}{3nc^3} \left[\frac{x}{(c^3 \pm x^3)^n} + (3n - 1) \int \frac{dx}{(c^3 \pm x^3)^n} \right], \quad n \neq 0.$

91. $\displaystyle \int \frac{x \, dx}{c^3 \pm x^3} = \frac{1}{6c} \log \frac{c^3 \pm x^3}{(c \pm x)^3} \pm \frac{1}{c\sqrt{3}} \tan^{-1} \frac{2x \mp c}{c\sqrt{3}}.$

92. $\displaystyle \int \frac{x \, dx}{(c^3 \pm x^3)^2} = \frac{x^2}{3c^3(c^3 \pm x^3)} + \frac{1}{3c^3} \int \frac{x \, dx}{c^3 \pm x^3}.$

93. $\displaystyle \int \frac{x \, dx}{(c^3 \pm x^3)^{n+1}} = \frac{1}{3nc^3} \left[\frac{x^2}{(c^3 \pm x^3)^n} + (3n - 2) \int \frac{x \, dx}{(c^3 \pm x^3)^n} \right], \quad n \neq 0.$

94. $\displaystyle \int \frac{x^2 \, dx}{c^3 \pm x^3} = \pm \frac{1}{3} \log (c^3 \pm x^3).$

95. $\displaystyle \int \frac{x^2 \, dx}{(c^3 \pm x^3)^{n+1}} = \mp \frac{1}{3n(c^3 \pm x^3)^n}, \quad n \neq 0.$

96. $\displaystyle \int \frac{dx}{x(c^3 \pm x^3)} = \frac{1}{3c^3} \log \frac{x^3}{c^3 \pm x^3}.$

97. $\displaystyle \int \frac{dx}{x(c^3 \pm x^3)^2} = \frac{1}{3c^3(c^3 \pm x^3)} + \frac{1}{3c^6} \log \frac{x^3}{c^3 \pm x^3}.$

98. $\displaystyle \int \frac{dx}{x(c^3 \pm x^3)^{n+1}} = \frac{1}{3nc^3(c^3 \pm x^3)^n} + \frac{1}{c^3} \int \frac{dx}{x(c^3 \pm x^3)^n}, \quad n \neq 0.$

99. $\displaystyle \int \frac{dx}{x^2(c^3 \pm x^3)} = -\frac{1}{c^3 x} \mp \frac{1}{c^3} \int \frac{x \, dx}{(c^3 \pm x^3)}.$

100. $\displaystyle \int \frac{dx}{x^2(c^3 \pm x^3)^{n+1}} = \frac{1}{c^3} \int \frac{dx}{x^2(c^3 \pm x^3)^n} \mp \frac{1}{c^3} \int \frac{x \, dx}{(c^3 \pm x^3)^{n+1}}.$

5.4.7 FORMS CONTAINING $c^4 \pm x^4$

101. $\displaystyle \int \frac{dx}{c^4 + x^4} = \frac{1}{2c^3\sqrt{2}} \left[\frac{1}{2} \log \left(\frac{x^2 + cx\sqrt{2} + c^2}{x^2 - cx\sqrt{2} + c^2} \right) + \tan^{-1} \frac{cx\sqrt{2}}{c^2 - x^2} \right].$

102. $\displaystyle \int \frac{dx}{c^4 - x^4} = \frac{1}{2c^3} \left[\frac{1}{2} \log \frac{c + x}{c - x} + \tan^{-1} \frac{x}{c} \right].$

103. $\displaystyle \int \frac{x \, dx}{c^4 + x^4} = \frac{1}{2c^2} \tan^{-1} \frac{x^2}{c^2}.$

104. $\displaystyle \int \frac{x \, dx}{c^4 - x^4} = \frac{1}{4c^2} \log \frac{c^2 + x^2}{c^2 - x^2}.$

105. $\displaystyle \int \frac{x^2 \, dx}{c^4 + x^4} = \frac{1}{2c\sqrt{2}} \left[\frac{1}{2} \log \left(\frac{x^2 - cx\sqrt{2} + c^2}{x^2 + cx\sqrt{2} + c^2} \right) + \tan^{-1} \frac{cx\sqrt{2}}{c^2 - x^2} \right].$

106. $\displaystyle \int \frac{x^2 \, dx}{c^4 - x^4} = \frac{1}{2c} \left[\frac{1}{2} \log \frac{c + x}{c - x} - \tan^{-1} \frac{x}{c} \right].$

107. $\int \dfrac{x^3\,dx}{c^4 \pm x^4} = \pm\dfrac{1}{4}\log\left(c^4 \pm x^4\right).$

5.4.8   FORMS CONTAINING $a + bx + cx^2$

$$X = a + bx + cx^2 \quad \text{and} \quad q = 4ac - b^2.$$

If $q = 0$, then $X = c\left(x + \frac{b}{2c}\right)^2$ and other formulae should be used.

108. $\displaystyle\int \dfrac{dx}{X} = \begin{cases} \dfrac{2}{\sqrt{q}}\tan^{-1}\dfrac{2cx+b}{\sqrt{q}}, & q > 0, \\[2mm] \text{or} \\[1mm] \dfrac{-2}{\sqrt{-q}}\tanh^{-1}\dfrac{2cx+b}{\sqrt{-q}}, & q < 0, \\[2mm] \text{or} \\[1mm] \dfrac{1}{\sqrt{-q}}\log\dfrac{2cx+b-\sqrt{-q}}{2cx+b+\sqrt{-q}}, & q < 0. \end{cases}$

109. $\displaystyle\int \dfrac{dx}{X^2} = \dfrac{2cx+b}{qX} + \dfrac{2c}{q}\int \dfrac{dx}{X}.$

110. $\displaystyle\int \dfrac{dx}{X^3} = \dfrac{2cx+b}{q}\left(\dfrac{1}{2X^2} + \dfrac{3c}{qX}\right) + \dfrac{6c^2}{q^2}\int \dfrac{dx}{X}.$

111. $\displaystyle\int \dfrac{dx}{X^{n+1}} = \begin{cases} \dfrac{2cx+b}{nqX^n} + \dfrac{2(2n-1)c}{qn}\int \dfrac{dx}{X^n}, \\[2mm] \text{or} \\[1mm] \dfrac{(2n)!}{(n!)^2}\left(\dfrac{c}{q}\right)^n\left[\dfrac{2cx+b}{q}\sum\limits_{r=1}^{n}\left(\dfrac{q}{cX}\right)^r\left(\dfrac{(r-1)!r!}{(2r)!}\right) + \int \dfrac{dx}{X}\right]. \end{cases}$

112. $\displaystyle\int \dfrac{x\,dx}{X} = \dfrac{1}{2c}\log X - \dfrac{b}{2c}\int \dfrac{dx}{X}.$

113. $\displaystyle\int \dfrac{x\,dx}{X^2} = -\dfrac{bx+2a}{qX} - \dfrac{b}{q}\int \dfrac{dx}{X}.$

114. $\displaystyle\int \dfrac{x\,dx}{X^{n+1}} = -\dfrac{2a+bx}{nqX^n} - \dfrac{b(2n-1)}{nq}\int \dfrac{dx}{X^n}, \quad n \neq 0.$

115. $\displaystyle\int \dfrac{x^2\,dx}{X} = \dfrac{x}{c} - \dfrac{b}{2c^2}\log X + \dfrac{b^2-2ac}{2c^2}\int \dfrac{dx}{X}.$

116. $\displaystyle\int \dfrac{x^2\,dx}{X^2} = \dfrac{(b^2-2ac)x+ab}{cqX} + \dfrac{2a}{q}\int \dfrac{dx}{X}.$

117. $\displaystyle\int \dfrac{x^m\,dx}{X^{n+1}} = -\dfrac{x^{m-1}}{(2n-m+1)cX^n} - \dfrac{n-m+1}{2n-m+1}\dfrac{b}{c}\int \dfrac{x^{m-1}}{X^{n+1}}\,dx$

$$+ \dfrac{m-1}{2n-m+1}\dfrac{a}{c}\int \dfrac{x^{m-2}}{X^{n+1}}\,dx.$$

118. $\displaystyle\int \dfrac{dx}{xX} = \dfrac{1}{2a}\log\dfrac{x^2}{X} - \dfrac{b}{2a}\int \dfrac{dx}{X}.$

119. $\displaystyle\int \dfrac{dx}{x^2X} = \dfrac{b}{2a^2}\log\dfrac{X}{x^2} - \dfrac{1}{ax} + \left(\dfrac{b^2}{2a^2} - \dfrac{c}{a}\right)\int \dfrac{dx}{X}.$

120. $\displaystyle\int \dfrac{dx}{xX^n} = \dfrac{1}{2a(n-1)X^{n-1}} - \dfrac{b}{2a}\int \dfrac{dx}{X^n} + \dfrac{1}{a}\int \dfrac{dx}{xX^{n-1}}, \quad n \neq 1.$

121. $\displaystyle\int \frac{dx}{x^m X^{n+1}} = -\frac{1}{(m-1)ax^{m-1}X^n} - \frac{n+m-1}{m-1}\frac{b}{a}\int \frac{dx}{x^{m-1}X^{n+1}}$
$$-\frac{2n+m-1}{m-1}\frac{c}{a}\int \frac{dx}{x^{m-2}X^{n+1}}.$$

5.4.9 FORMS CONTAINING $\sqrt{a+bx}$

122. $\displaystyle\int \sqrt{a+bx}\,dx = \frac{2}{3b}\sqrt{(a+bx)^3}.$

123. $\displaystyle\int x\sqrt{a+bx}\,dx = -\frac{2(2a-3bx)}{15b^2}\sqrt{(a+bx)^3}.$

124. $\displaystyle\int x^2\sqrt{a+bx}\,dx = \frac{2(8a^2-12abx+15b^2x^2)}{105b^3}\sqrt{(a+bx)^3}.$

125. $\displaystyle\int x^m\sqrt{a+bx}\,dx =$

$$\begin{cases} \dfrac{2}{b(2m+3)}\left[x^m\sqrt{(a+bx)^3} - ma\int x^{m-1}\sqrt{a+bx}\,dx\right], \\ \text{or} \\ \dfrac{2}{b^{m+1}}\sqrt{a+bx}\displaystyle\sum_{r=0}^{m}\dfrac{m!(-a)^{m-r}}{r!(m-r)!(2r+3)}(a+bx)^{r+1}. \end{cases}$$

126. $\displaystyle\int \frac{\sqrt{a+bx}}{x}\,dx = 2\sqrt{a+bx} + a\int \frac{dx}{x\sqrt{a+bx}}.$

127. $\displaystyle\int \frac{\sqrt{a+bx}}{x^m}\,dx = -\frac{1}{(m-1)a}\left[\frac{\sqrt{(a+bx)^3}}{x^{m-1}} + \frac{(2m-5)b}{2}\int \frac{\sqrt{a+bx}}{x^{m-1}}\,dx\right].$

128. $\displaystyle\int \frac{dx}{\sqrt{a+bx}} = \frac{2\sqrt{a+bx}}{b}.$

129. $\displaystyle\int \frac{x\,dx}{\sqrt{a+bx}} = -\frac{2(2a-bx)}{3b^2}\sqrt{a+bx}.$

130. $\displaystyle\int \frac{x^2\,dx}{\sqrt{a+bx}} = \frac{2(8a^2-4abx+3b^2x^2)}{15b^3}\sqrt{a+bx}.$

131. $\displaystyle\int \frac{x^m\,dx}{\sqrt{a+bx}} = \begin{cases} \dfrac{2}{(2m+1)b}\left[x^m\sqrt{a+bx} - ma\int \dfrac{x^{m-1}}{\sqrt{a+bx}}\,dx\right], \\ \text{or} \\ \dfrac{2(-a)^m\sqrt{a+bx}}{b^{m+1}}\displaystyle\sum_{r=0}^{m}\dfrac{(-1)^r m!(a+bx)^r}{(2r+1)r!(m-r)!a^r}. \end{cases}$

132. $\displaystyle\int \frac{dx}{x\sqrt{a+bx}} = \begin{cases} \dfrac{2}{\sqrt{-a}}\tan^{-1}\sqrt{\dfrac{a+bx}{-a}}, & a<0, \\ \text{or} \\ \dfrac{1}{\sqrt{a}}\log\left(\dfrac{\sqrt{a+bx}-\sqrt{a}}{\sqrt{a+bx}+\sqrt{a}}\right), & a>0. \end{cases}$

133. $\displaystyle\int \frac{dx}{x^2\sqrt{a+bx}} = -\frac{\sqrt{a+bx}}{ax} - \frac{b}{2a}\int \frac{dx}{x\sqrt{a+bx}}.$

134. $\displaystyle\int \frac{dx}{x^n\sqrt{a+bx}} =$

$$\begin{cases} -\dfrac{\sqrt{a+bx}}{(n-1)ax^{n-1}} - \dfrac{(2n-3)b}{(2n-2)a}\displaystyle\int \dfrac{dx}{x^{n-1}\sqrt{a+bx}}, \\ \text{or} \\ \dfrac{(2n-2)!}{[(n-1)!]^2}\left[-\dfrac{\sqrt{a+bx}}{a}\displaystyle\sum_{r=1}^{n-1} \dfrac{r!(r-1)!}{x^r(2r)!}\left(-\dfrac{b}{4a}\right)^{n-r-1} + \left(-\dfrac{b}{4a}\right)^{n-1}\displaystyle\int \dfrac{dx}{x\sqrt{a+bx}} \right]. \end{cases}$$

135. $\displaystyle\int (a+bx)^{\pm n/2}\, dx = \frac{2(a+bx)^{(2\pm n)/2}}{b(2\pm n)}.$

136. $\displaystyle\int x(a+bx)^{\pm n/2}\, dx = \frac{2}{b^2}\left[\frac{(a+bx)^{(4\pm n)/2}}{4\pm n} - \frac{a(a+bx)^{(2\pm n)/2}}{2\pm n} \right].$

137. $\displaystyle\int \frac{dx}{x(a+bx)^{n/2}} = \frac{1}{a}\int \frac{dx}{x(a+bx)^{(n-2)/2}} - \frac{b}{a}\int \frac{dx}{(a+bx)^{n/2}}.$

138. $\displaystyle\int \frac{(a+bx)^{n/2}}{x}\, dx = b\int (a+bx)^{(n-2)/2}\, dx + a\int \frac{(a+bx)^{(n-2)/2}}{x}\, dx.$

5.4.10 FORMS CONTAINING $\sqrt{a+bx}$ AND $\sqrt{c+dx}$

and $\sqrt{c+dx}$

$$u = a + bx, \qquad v = c + dx, \qquad k = ad - bc.$$

If $k = 0$, then $v = \frac{c}{a}u$, and other formulae should be used.

139. $\displaystyle\int \frac{dx}{\sqrt{uv}} = \begin{cases} \dfrac{2}{\sqrt{bd}}\tanh^{-1}\dfrac{\sqrt{bduv}}{bv}, & bd > 0,\ k < 0, \\[2mm] \dfrac{2}{\sqrt{bd}}\tanh^{-1}\dfrac{\sqrt{bduv}}{du}, & bd > 0,\ k > 0, \\[2mm] \dfrac{1}{\sqrt{bd}}\log\dfrac{(bv+\sqrt{bduv})^2}{v}, & bd > 0, \\[2mm] \dfrac{2}{\sqrt{-bd}}\tan^{-1}\dfrac{\sqrt{-bduv}}{bv}, & bd < 0, \\[2mm] -\dfrac{1}{\sqrt{-bd}}\sin^{-1}\left(\dfrac{2bdx+ad+bc}{|k|}\right), & bd < 0. \end{cases}$

140. $\displaystyle\int \sqrt{uv}\, dx = \frac{k+2bv}{4bd}\sqrt{uv} - \frac{k^2}{8bd}\int \frac{dx}{\sqrt{uv}}.$

141. $\displaystyle\int \frac{dx}{v\sqrt{u}} = \begin{cases} \dfrac{1}{\sqrt{kd}}\log\dfrac{d\sqrt{u}-\sqrt{kd}}{d\sqrt{u}+\sqrt{kd}}, & kd > 0, \\[2mm] \text{or} \\[2mm] \dfrac{1}{\sqrt{kd}}\log\dfrac{\left(d\sqrt{u}-\sqrt{kd}\right)^2}{v}, & kd > 0, \\[2mm] \text{or} \\[2mm] \dfrac{2}{\sqrt{-kd}}\tan^{-1}\dfrac{d\sqrt{u}}{\sqrt{-kd}}, & kd < 0. \end{cases}$

142. $\displaystyle\int \frac{x\, dx}{\sqrt{uv}} = \frac{\sqrt{uv}}{bd} - \frac{ad+bc}{2bd}\int \frac{dx}{\sqrt{uv}}.$

143. $\int \dfrac{dx}{v\sqrt{uv}} = -\dfrac{2\sqrt{uv}}{kv}.$

144. $\int \dfrac{v\,dx}{\sqrt{uv}} = \dfrac{\sqrt{uv}}{b} - \dfrac{k}{2b}\int \dfrac{dx}{\sqrt{uv}}.$

145. $\int \sqrt{\dfrac{v}{u}}\,dx = \dfrac{v}{|v|}\int \dfrac{v\,dx}{\sqrt{uv}}.$

146. $\int v^m\sqrt{u}\,dx = \dfrac{1}{(2m+3)d}\left(2v^{m+1}\sqrt{u} + k\int \dfrac{v^m\,dx}{\sqrt{u}}\right).$

147. $\int \dfrac{dx}{v^m\sqrt{u}} = -\dfrac{1}{(m-1)k}\left(\dfrac{\sqrt{u}}{v^{m-1}} + \left(m - \dfrac{3}{2}\right)b\int \dfrac{dx}{v^{m-1}\sqrt{u}}\right), \quad m \neq 1.$

148. $\displaystyle\int \dfrac{v^m}{\sqrt{u}}\,dx = \begin{cases} \dfrac{2}{b(2m+1)}\left(v^m\sqrt{u} - mk\int \dfrac{v^{m-1}}{\sqrt{u}}\,dx\right), \\ \text{or} \\ \dfrac{2(m!)^2\sqrt{u}}{b(2m+1)!}\displaystyle\sum_{r=0}^{m}\left(-\dfrac{4k}{b}\right)^{m-r}\dfrac{(2r)!}{(r!)^2}v^r. \end{cases}$

5.4.11   FORMS CONTAINING $\sqrt{x^2 \pm a^2}$

149. $\int \sqrt{x^2 \pm a^2}\,dx = \dfrac{1}{2}\left[x\sqrt{x^2 \pm a^2} \pm a^2 \log\left(x + \sqrt{x^2 \pm a^2}\right)\right].$

150. $\int \dfrac{dx}{\sqrt{x^2 \pm a^2}} = \log\left(x + \sqrt{x^2 \pm a^2}\right).$

151. $\int \dfrac{dx}{x\sqrt{x^2 - a^2}} = \dfrac{1}{|a|}\sec^{-1}\dfrac{x}{a}.$

152. $\int \dfrac{dx}{x\sqrt{x^2 + a^2}} = -\dfrac{1}{a}\log\left(\dfrac{a + \sqrt{x^2 + a^2}}{x}\right).$

153. $\int \dfrac{\sqrt{x^2 + a^2}}{x}\,dx = \sqrt{x^2 + a^2} - a\log\left(\dfrac{a + \sqrt{x^2 + a^2}}{x}\right).$

154. $\int \dfrac{\sqrt{x^2 - a^2}}{x}\,dx = \sqrt{x^2 - a^2} - |a|\sec^{-1}\dfrac{x}{a}.$

155. $\int \dfrac{x}{\sqrt{x^2 \pm a^2}}\,dx = \sqrt{x^2 \pm a^2}.$

156. $\int x\sqrt{x^2 \pm a^2}\,dx = \dfrac{1}{3}\sqrt{(x^2 \pm a^2)^3}.$

157. $\int \sqrt{(x^2 \pm a^2)^3}\,dx = \dfrac{1}{4}\left[x\sqrt{(x^2 \pm a^2)^3} \pm \dfrac{3a^2 x}{2}\sqrt{x^2 \pm a^2} \right.$
$$\left. + \dfrac{3a^4}{2}\log\left(x + \sqrt{x^2 \pm a^2}\right)\right].$$

158. $\int \dfrac{dx}{\sqrt{(x^2 \pm a^2)^3}} = \dfrac{\pm x}{a^2\sqrt{x^2 \pm a^2}}.$

159. $\int \dfrac{x}{\sqrt{(x^2 \pm a^2)^3}}\,dx = \dfrac{-1}{\sqrt{x^2 \pm a^2}}.$

160. $\int x\sqrt{(x^2 \pm a^2)^3}\,dx = \dfrac{1}{5}\sqrt{(x^2 \pm a^2)^5}.$

161. $\int x^2\sqrt{x^2 \pm a^2}\,dx = \dfrac{x}{4}\sqrt{(x^2 \pm a^2)^3} \mp \dfrac{a^2}{8}x\sqrt{x^2 \pm a^2} - \dfrac{a^4}{8}\log\left(x + \sqrt{x^2 \pm a^2}\right).$

162. $\int x^3 \sqrt{x^2 + a^2}\, dx = \dfrac{1}{15}(3x^2 - 2a^2)\sqrt{(x^2 + a^2)^3}.$

163. $\int x^3 \sqrt{x^2 - a^2}\, dx = \dfrac{1}{5}\sqrt{(x^2 - a^2)^5} + \dfrac{a^2}{3}\sqrt{(x^2 - a^2)^3}.$

164. $\int \dfrac{x^2}{\sqrt{x^2 \pm a^2}}\, dx = \dfrac{x}{2}\sqrt{x^2 \pm a^2} \mp \dfrac{a^2}{2}\log\left(x + \sqrt{x^2 \pm a^2}\right).$

165. $\int \dfrac{x^3}{\sqrt{x^2 \pm a^2}}\, dx = \dfrac{1}{3}\sqrt{(x^2 \pm a^2)^3} \mp a^2\sqrt{x^2 \pm a^2}.$

166. $\int \dfrac{dx}{x^2\sqrt{x^2 \pm a^2}}\, dx = \mp\dfrac{\sqrt{x^2 \pm a^2}}{a^2 x}.$

167. $\int \dfrac{dx}{x^3\sqrt{x^2 + a^2}}\, dx = -\dfrac{\sqrt{x^2 + a^2}}{2a^2 x^2} + \dfrac{1}{2a^3}\log\left(\dfrac{a + \sqrt{x^2 + a^2}}{x}\right).$

168. $\int \dfrac{dx}{x^3\sqrt{x^2 - a^2}}\, dx = \dfrac{\sqrt{x^2 - a^2}}{2a^2 x^2} + \dfrac{1}{2|a|^3}\sec^{-1}\dfrac{x}{a}.$

169. $\int x^2\sqrt{(x^2 \pm a^2)^3}\, dx = \dfrac{x}{6}\sqrt{(x^2 \pm a^2)^5} \mp \dfrac{a^2 x}{24}\sqrt{(x^2 \pm a^2)^3} - \dfrac{a^4 x}{16}\sqrt{x^2 \pm a^2}$
$$\mp \dfrac{a^6}{16}\log\left(x + \sqrt{x^2 \pm a^2}\right).$$

170. $\int x^3\sqrt{(x^2 \pm a^2)^3}\, dx = \dfrac{1}{7}\sqrt{(x^2 \pm a^2)^7} \mp \dfrac{a^2}{5}\sqrt{(x^2 \pm a^2)^5}.$

171. $\int \dfrac{\sqrt{x^2 \pm a^2}}{x^2}\, dx = -\dfrac{\sqrt{x^2 \pm a^2}}{x} + \log\left(x + \sqrt{x^2 \pm a^2}\right).$

172. $\int \dfrac{\sqrt{x^2 + a^2}}{x^3}\, dx = -\dfrac{\sqrt{x^2 + a^2}}{2x^2} - \dfrac{1}{2a}\log\left(\dfrac{a + \sqrt{x^2 + a^2}}{x}\right).$

173. $\int \dfrac{\sqrt{x^2 - a^2}}{x^3}\, dx = -\dfrac{\sqrt{x^2 - a^2}}{2x^2} + \dfrac{1}{2|a|}\sec^{-1}\dfrac{x}{a}.$

174. $\int \dfrac{\sqrt{x^2 \pm a^2}}{x^4}\, dx = \mp\dfrac{\sqrt{(x^2 \pm a^2)^3}}{3a^2 x^3}.$

175. $\int \dfrac{x^2\, dx}{\sqrt{(x^2 \pm a^2)^3}} = -\dfrac{x}{\sqrt{x^2 \pm a^2}} + \log\left(x + \sqrt{x^2 \pm a^2}\right).$

176. $\int \dfrac{x^3\, dx}{\sqrt{(x^2 \pm a^2)^3}} = \sqrt{x^2 \pm a^2} \pm \dfrac{a^2}{\sqrt{x^2 \pm a^2}}.$

177. $\int \dfrac{dx}{x\sqrt{(x^2 + a^2)^3}} = \dfrac{1}{a^2\sqrt{x^2 + a^2}} - \dfrac{1}{a^3}\log\left(\dfrac{a + \sqrt{x^2 + a^2}}{x}\right).$

178. $\int \dfrac{dx}{x\sqrt{(x^2 - a^2)^3}} = -\dfrac{1}{a^2\sqrt{x^2 - a^2}} - \dfrac{1}{|a^3|}\sec^{-1}\dfrac{x}{a}.$

179. $\int \dfrac{dx}{x^2\sqrt{(x^2 \pm a^2)^3}} = -\dfrac{1}{a^4}\left[\dfrac{\sqrt{x^2 \pm a^2}}{x} + \dfrac{x}{\sqrt{x^2 \pm a^2}}\right].$

180. $\int \dfrac{dx}{x^3\sqrt{(x^2 + a^2)^3}} = -\dfrac{3 + a^2}{2a^4\sqrt{x^2 + a^2}} + \dfrac{3}{2a^5}\log\left(\dfrac{a + \sqrt{x^2 + a^2}}{x}\right).$

181. $\int \dfrac{dx}{x^3\sqrt{(x^2 - a^2)^3}} = \dfrac{1}{2a^2 x^2\sqrt{x^2 - a^2}} - \dfrac{3}{2a^4\sqrt{x^2 - a^2}} - \dfrac{3}{2|a^5|}\sec^{-1}\dfrac{x}{a}.$

182. $\int \dfrac{x^m\, dx}{\sqrt{x^2 \pm a^2}} = \dfrac{1}{m}x^{m-1}\sqrt{x^2 \pm a^2} \mp \dfrac{m-1}{m}a^2\int \dfrac{x^{m-2}}{\sqrt{x^2 \pm a^2}}\, dx.$

183. $\displaystyle\int \frac{x^{2m}\,dx}{\sqrt{x^2 \pm a^2}} = \frac{(2m)!}{2^{2m}(m!)^2}\left[\sqrt{x^2 \pm a^2}\sum_{r=1}^{m}\frac{r!(r-1)!}{(2r)!}(\mp a^2)^{m-r}(2x)^{2r-1}\right.$
$$\left.+(\mp a^2)^m\log\left(x+\sqrt{x^2 \pm a^2}\right)\right].$$

184. $\displaystyle\int \frac{x^{2m+1}\,dx}{\sqrt{x^2 \pm a^2}} = \sqrt{x^2 \pm a^2}\sum_{r=0}^{m}\frac{(2r)!(m!)^2}{(2m+1)!(r!)^2}(\mp 4a^2)^{m-r}x^{2r}.$

185. $\displaystyle\int \frac{dx}{x^m\sqrt{x^2 \pm a^2}} = \mp\frac{\sqrt{x^2 \pm a^2}}{(m-1)a^2 x^{m-1}} \mp \frac{(m-2)}{(m-1)a^2}\int\frac{dx}{x^{m-2}\sqrt{x^2 \pm a^2}}.$

186. $\displaystyle\int \frac{dx}{x^{2m}\sqrt{x^2 \pm a^2}} = \sqrt{x^2 \pm a^2}\sum_{r=0}^{m-1}\frac{(m-1)!m!(2r)!2^{2m-2r-1}}{(r!)^2(2m)!(\mp a^2)^{m-r}x^{2r+1}}.$

187. $\displaystyle\int \frac{dx}{x^{2m+1}\sqrt{x^2 + a^2}} = \frac{(2m)!}{(m!)^2}\left[\frac{\sqrt{x^2 + a^2}}{a^2}\sum_{r=1}^{m}(-1)^{m-r+1}\frac{r!(r-1)!}{2(2r)!(4a^2)^{m-r}x^{2r}}\right.$
$$\left.+\frac{(-1)^{m+1}}{2^{2m}a^{2m+1}}\log\left(\frac{\sqrt{x^2+a^2}+a}{x}\right)\right].$$

188. $\displaystyle\int \frac{dx}{x^{2m+1}\sqrt{x^2 - a^2}} = \frac{(2m)!}{(m!)^2}\left[\frac{\sqrt{x^2 - a^2}}{a^2}\sum_{r=1}^{m}\frac{r!(r-1)!}{2(2r)!(4a^2)^{m-r}x^{2r}}\right.$
$$\left.+\frac{1}{2^{2m}|a|^{2m+1}}\sec^{-1}\frac{x}{a}\right].$$

189. $\displaystyle\int \frac{dx}{(x-a)\sqrt{x^2-a^2}} = -\frac{\sqrt{x^2-a^2}}{a(x-a)}.$

190. $\displaystyle\int \frac{dx}{(x+a)\sqrt{x^2-a^2}} = \frac{\sqrt{x^2-a^2}}{a(x+a)}.$

5.4.12 FORMS CONTAINING $\sqrt{a^2 - x^2}$

191. $\displaystyle\int \sqrt{a^2 - x^2}\,dx = \frac{1}{2}\left(x\sqrt{a^2 - x^2} + a^2\sin^{-1}\frac{x}{|a|}\right).$

192. $\displaystyle\int \frac{dx}{\sqrt{a^2 - x^2}} = \sin^{-1}\frac{x}{|a|} = -\cos^{-1}\frac{x}{|a|}.$

193. $\displaystyle\int \frac{dx}{x\sqrt{a^2 - x^2}} = -\frac{1}{a}\log\left(\frac{a+\sqrt{a^2-x^2}}{x}\right).$

194. $\displaystyle\int \frac{\sqrt{a^2 - x^2}}{x}\,dx = \sqrt{a^2 - x^2} - a\log\left(\frac{a+\sqrt{a^2-x^2}}{x}\right).$

195. $\displaystyle\int \frac{x}{\sqrt{a^2 - x^2}}\,dx = -\sqrt{a^2 - x^2}.$

196. $\displaystyle\int x\sqrt{a^2 - x^2}\,dx = -\frac{1}{3}\sqrt{(a^2-x^2)^3}.$

197. $\displaystyle\int \sqrt{(a^2 - x^2)^3}\,dx = \frac{1}{4}\left(x\sqrt{(a^2-x^2)^3} + \frac{3a^2 x}{2}\sqrt{a^2-x^2} + \frac{3a^4}{2}\sin^{-1}\frac{x}{|a|}\right).$

198. $\displaystyle\int \frac{dx}{\sqrt{(a^2 - x^2)^3}} = \frac{x}{a^2\sqrt{a^2-x^2}}.$

199. $\displaystyle\int \frac{x}{\sqrt{(a^2 - x^2)^3}}\,dx = \frac{1}{\sqrt{a^2-x^2}}.$

200. $\int x\sqrt{(a^2-x^2)^3}\,dx = -\frac{1}{5}\sqrt{(a^2-x^2)^5}.$

201. $\int x^2\sqrt{a^2-x^2}\,dx = -\frac{x}{4}\sqrt{(a^2-x^2)^3} + \frac{a^2}{8}\left(x\sqrt{a^2-x^2} + a^2\sin^{-1}\frac{x}{|a|}\right).$

202. $\int x^3\sqrt{a^2-x^2}\,dx = \left(-\frac{1}{5}x^2 - \frac{2}{15}a^2\right)\sqrt{(a^2-x^2)^3}.$

203. $\int x^2\sqrt{(a^2-x^2)^3}\,dx = -\frac{1}{6}x\sqrt{(a^2-x^2)^5} + \frac{a^2x}{24}\sqrt{(a^2-x^2)^3}$
$$+ \frac{a^4x}{16}\sqrt{a^2-x^2} + \frac{a^6}{16}\sin^{-1}\frac{x}{|a|}.$$

204. $\int x^3\sqrt{(a^2-x^2)^3}\,dx = \frac{1}{7}\sqrt{(a^2-x^2)^7} - \frac{a^2}{5}\sqrt{(a^2-x^2)^5}.$

205. $\int \frac{x^2}{\sqrt{a^2-x^2}}\,dx = -\frac{x}{2}\sqrt{a^2-x^2} + \frac{a^2}{2}\sin^{-1}\frac{x}{|a|}.$

206. $\int \frac{dx}{x^2\sqrt{a^2-x^2}} = -\frac{\sqrt{a^2-x^2}}{a^2x}.$

207. $\int \frac{\sqrt{a^2-x^2}}{x^2}\,dx = -\frac{\sqrt{a^2-x^2}}{x} - \sin^{-1}\frac{x}{|a|}.$

208. $\int \frac{\sqrt{a^2-x^2}}{x^3}\,dx = -\frac{\sqrt{a^2-x^2}}{2x^2} + \frac{1}{2a}\log\left(\frac{a+\sqrt{a^2-x^2}}{x}\right).$

209. $\int \frac{\sqrt{a^2-x^2}}{x^4}\,dx = -\frac{\sqrt{(a^2-x^2)^3}}{3a^2x^3}.$

210. $\int \frac{x^2\,dx}{\sqrt{(a^2-x^2)^3}} = \frac{x}{\sqrt{a^2-x^2}} - \sin^{-1}\frac{x}{|a|}.$

211. $\int \frac{x^3\,dx}{\sqrt{a^2-x^2}} = -\frac{2}{3}\sqrt{(a^2-x^2)^3} - x^2\sqrt{a^2-x^2}.$

212. $\int \frac{x^3\,dx}{\sqrt{(a^2-x^2)^3}} = 2\sqrt{a^2-x^2} + \frac{x^2}{\sqrt{a^2-x^2}} = \frac{a^2}{\sqrt{a^2-x^2}} + \sqrt{a^2-x^2}.$

213. $\int \frac{dx}{x^3\sqrt{a^2-x^2}} = -\frac{\sqrt{a^2-x^2}}{2a^2x^2} - \frac{1}{2a^3}\log\left(\frac{a+\sqrt{a^2-x^2}}{x}\right).$

214. $\int \frac{dx}{x\sqrt{(a^2-x^2)^3}} = \frac{1}{a^2\sqrt{a^2-x^2}} - \frac{1}{a^3}\log\left(\frac{a+\sqrt{a^2-x^2}}{x}\right).$

215. $\int \frac{dx}{x^2\sqrt{(a^2-x^2)^3}} = \frac{1}{a^4}\left(-\frac{\sqrt{a^2-x^2}}{x} + \frac{x}{\sqrt{a^2-x^2}}\right).$

216. $\int \frac{dx}{x^3\sqrt{(a^2-x^2)^3}} = \frac{3-a^2}{2a^4\sqrt{a^2-x^2}} - \frac{3}{2a^5}\log\left(\frac{a+\sqrt{a^2-x^2}}{x}\right).$

217. $\int \frac{x^m}{\sqrt{a^2-x^2}}\,dx = -\frac{x^{m-1}\sqrt{a^2-x^2}}{m} + \frac{(m-1)a^2}{m}\int \frac{x^{m-2}}{\sqrt{a^2-x^2}}\,dx.$

218. $\int \frac{x^{2m}}{\sqrt{a^2-x^2}}\,dx = \frac{(2m)!}{(m!)^2}\left[-\sqrt{a^2-x^2}\sum_{r=1}^m \frac{r!(r-1)!}{2^{2m-2r+1}(2r)!}a^{2m-2r}x^{2r-1}\right.$
$$\left.+ \frac{a^{2m}}{2^{2m}}\sin^{-1}\frac{x}{|a|}\right].$$

219. $\int \dfrac{x^{2m+1}}{\sqrt{a^2 - x^2}}\, dx = -\sqrt{a^2 - x^2} \sum\limits_{r=0}^{m} \dfrac{(2r)!(m!)^2}{(2m+1)!(r!)^2}(4a^2)^{m-r} x^{2r}.$

220. $\int \dfrac{dx}{x^m \sqrt{a^2 - x^2}} = -\dfrac{\sqrt{a^2 - x^2}}{(m-1)a^2 x^{m-1}} + \dfrac{(m-2)}{(m-1)a^2} \int \dfrac{dx}{x^{m-2}\sqrt{a^2 - x^2}}.$

221. $\int \dfrac{dx}{x^{2m}\sqrt{a^2 - x^2}} = -\sqrt{a^2 - x^2} \sum\limits_{r=0}^{m-1} \dfrac{(m-1)!m!(2r)!2^{2m-2r-1}}{(r!)^2(2m)!a^{2m-2r}x^{2r+1}}.$

222. $\int \dfrac{dx}{x^{2m+1}\sqrt{a^2 - x^2}} = \dfrac{(2m)!}{(m!)^2}\left[-\dfrac{\sqrt{a^2 - x^2}}{a^2} \sum\limits_{r=1}^{m} \dfrac{r!(r-1)!}{2(2r)!(4a^2)^{m-r}x^{2r}} \right.$

$$\left. + \dfrac{1}{2^{2m}a^{2m+1}} \log\left(\dfrac{a - \sqrt{a^2 - x^2}}{x}\right) \right].$$

223. $\int \dfrac{dx}{(b^2 - x^2)\sqrt{a^2 - x^2}} =$

$$\begin{cases} \dfrac{1}{2b\sqrt{a^2 - b^2}} \log\left(\dfrac{(b\sqrt{a^2 - x^2} + x\sqrt{a^2 - b^2})^2}{b^2 - x^2}\right), & a^2 > b^2, \\ \text{or} \\ \dfrac{1}{b\sqrt{b^2 - a^2}} \tan^{-1} \dfrac{x\sqrt{b^2 - a^2}}{b\sqrt{a^2 - x^2}}, & b^2 > a^2. \end{cases}$$

224. $\int \dfrac{dx}{(b^2 + x^2)\sqrt{a^2 - x^2}} = \dfrac{1}{b\sqrt{a^2 + b^2}} \tan^{-1} \dfrac{x\sqrt{a^2 + b^2}}{b\sqrt{a^2 - x^2}}.$

225. $\int \dfrac{\sqrt{a^2 - x^2}}{b^2 + x^2}\, dx = \dfrac{\sqrt{a^2 + b^2}}{|b|} \sin^{-1} \dfrac{x\sqrt{a^2 + b^2}}{|a|\sqrt{x^2 + b^2}} - \sin^{-1} \dfrac{x}{|a|}, \quad b^2 > a^2.$

5.4.13 FORMS CONTAINING $\sqrt{a + bx + cx^2}$

$$X = a + bx + cx^2, \quad q = 4ac - b^2, \quad \text{and} \quad k = 4c/q.$$

$$\text{If } q = 0, \text{ then } \sqrt{X} = \sqrt{c}\left|x + \tfrac{b}{2c}\right|.$$

226. $\int \dfrac{dx}{\sqrt{X}} = \begin{cases} \dfrac{1}{\sqrt{c}} \log\left(\dfrac{2\sqrt{cX} + 2cx + b}{\sqrt{q}}\right), & c > 0, \\ \text{or} \\ \dfrac{1}{\sqrt{c}} \sinh^{-1} \dfrac{2cx + b}{\sqrt{q}}, & c > 0, \\ \text{or} \\ -\dfrac{1}{\sqrt{-c}} \sin^{-1} \dfrac{2cx + b}{\sqrt{-q}}, & c < 0. \end{cases}$

227. $\int \dfrac{dx}{X\sqrt{X}} = \dfrac{2(2cx + b)}{q\sqrt{X}}.$

228. $\int \dfrac{dx}{X^2\sqrt{X}} = \dfrac{2(2cx + b)}{3q\sqrt{X}}\left(\dfrac{1}{X} + 2k\right).$

229. $\int \dfrac{dx}{X^n\sqrt{X}} = \begin{cases} \dfrac{2(2cx + b)\sqrt{X}}{(2n-1)qX^n} + \dfrac{2k(n-1)}{2n-1} \int \dfrac{dx}{X^{n-1}\sqrt{X}}, \\ \text{or} \\ \dfrac{(2cx + b)(n!)(n-1)!4^n k^{n-1}}{q(2n)!\sqrt{X}} \sum\limits_{r=0}^{n-1} \dfrac{(2r)!}{(4kX)^r(r!)^2}. \end{cases}$

230. $\int \sqrt{X}\, dx = \dfrac{(2cx+b)\sqrt{X}}{4c} + \dfrac{1}{2k}\int \dfrac{dx}{\sqrt{X}}.$

231. $\int X\sqrt{X}\, dx = \dfrac{(2cx+b)\sqrt{X}}{8c}\left(X + \dfrac{3}{2k}\right) + \dfrac{3}{8k^2}\int \dfrac{dx}{\sqrt{X}}.$

232. $\int X^2\sqrt{X}\, dx = \dfrac{(2cx+b)\sqrt{X}}{12c}\left(X^2 + \dfrac{5X}{4k} + \dfrac{15}{8k^2}\right) + \dfrac{5}{16k^3}\int \dfrac{dx}{\sqrt{X}}.$

233. $\int X^n\sqrt{X}\, dx =$

$$\begin{cases} \dfrac{(2cx+b)X^n\sqrt{X}}{4(n+1)c} + \dfrac{2n+1}{2(n+1)k}\int X^{n-1}\sqrt{X}\, dx, \\ \quad\text{or} \\ \dfrac{(2n+2)!}{[(n+1)!]^2\,(4k)^{n+1}}\left[\dfrac{k(2cx+b)\sqrt{X}}{c}\sum_{r=0}^{n}\dfrac{r!(r+1)!(4kX)^r}{(2r+2)!} + \int \dfrac{dx}{\sqrt{X}}\right]. \end{cases}$$

234. $\int \dfrac{x\, dx}{\sqrt{X}} = \dfrac{\sqrt{X}}{c} - \dfrac{b}{2c}\int \dfrac{dx}{\sqrt{X}}.$

235. $\int \dfrac{x\, dx}{X\sqrt{X}} = -\dfrac{2(bx+2a)}{q\sqrt{X}}.$

236. $\int \dfrac{x\, dx}{X^n\sqrt{X}} = -\dfrac{\sqrt{X}}{(2n-1)cX^n} - \dfrac{b}{2c}\int \dfrac{dx}{X^n\sqrt{X}}.$

237. $\int \dfrac{x^2\, dx}{\sqrt{X}} = \left(\dfrac{x}{2c} - \dfrac{3b}{4c^2}\right)\sqrt{X} + \dfrac{3b^2-4ac}{8c^2}\int \dfrac{dx}{\sqrt{X}}.$

238. $\int \dfrac{x^2\, dx}{X\sqrt{X}} = \dfrac{(2b^2-4ac)x+2ab}{cq\sqrt{X}} + \dfrac{1}{c}\int \dfrac{dx}{\sqrt{X}}.$

239. $\int \dfrac{x^2\, dx}{X^n\sqrt{X}} = \dfrac{(2b^2-4ac)x+2ab}{(2n-1)cqX^{n-1}\sqrt{X}} + \dfrac{4ac+(2n-3)b^2}{(2n-1)cq}\int \dfrac{dx}{X^{n-1}\sqrt{X}}.$

240. $\int \dfrac{x^3\, dx}{\sqrt{X}} = \left(\dfrac{x^2}{3c} - \dfrac{5bx}{12c^2} + \dfrac{5b^2}{8c^3} - \dfrac{2a}{3c^2}\right)\sqrt{X} + \left(\dfrac{3ab}{4c^2} - \dfrac{5b^3}{16c^3}\right)\int \dfrac{dx}{\sqrt{X}}.$

241. $\int \dfrac{x^n\, dx}{\sqrt{X}} = \dfrac{1}{nc}x^{n-1}\sqrt{X} - \dfrac{(2n-1)b}{2nc}\int \dfrac{x^{n-1}\, dx}{\sqrt{X}} - \dfrac{(n-1)a}{nc}\int \dfrac{x^{n-2}\, dx}{\sqrt{X}}.$

242. $\int x\sqrt{X}\, dx = \dfrac{X\sqrt{X}}{3c} - \dfrac{b(2cx+b)}{8c^2}\sqrt{X} - \dfrac{b}{4ck}\int \dfrac{dx}{\sqrt{X}}.$

243. $\int xX\sqrt{X}\, dx = \dfrac{X^2\sqrt{X}}{5c} - \dfrac{b}{2c}\int X\sqrt{X}\, dx.$

244. $\int xX^n\sqrt{X}\, dx = \dfrac{X^{n+1}\sqrt{X}}{(2n+3)c} - \dfrac{b}{2c}\int X^n\sqrt{X}\, dx.$

245. $\int x^2\sqrt{X}\, dx = \left(x - \dfrac{5b}{6c}\right)\dfrac{X\sqrt{X}}{4c} + \dfrac{5b^2-4ac}{16c^2}\int \sqrt{X}\, dx.$

246. $\int \dfrac{dx}{x\sqrt{X}} = \begin{cases} \dfrac{1}{\sqrt{-a}}\sin^{-1}\left(\dfrac{bx+2a}{|x|\sqrt{-q}}\right), & a < 0, \\ \quad\text{or} \\ -\dfrac{2\sqrt{X}}{bx}, & a = 0, \\ \quad\text{or} \\ -\dfrac{1}{\sqrt{a}}\log\left(\dfrac{2\sqrt{aX}+bx+2a}{x}\right), & a > 0. \end{cases}$

247. $\int \dfrac{dx}{x^2\sqrt{X}} = -\dfrac{\sqrt{X}}{ax} - \dfrac{b}{2a}\int \dfrac{dx}{x\sqrt{X}}.$

248. $\int \dfrac{\sqrt{X}}{x}\,dx = \sqrt{X} + \dfrac{b}{2}\int \dfrac{dx}{\sqrt{X}} + a\int \dfrac{dx}{x\sqrt{X}}.$

249. $\int \dfrac{\sqrt{X}}{x^2}\,dx = -\dfrac{\sqrt{X}}{x} + \dfrac{b}{2}\int \dfrac{dx}{x\sqrt{X}} + c\int \dfrac{dx}{\sqrt{X}}.$

5.4.14   FORMS CONTAINING $\sqrt{2ax - x^2}$

250. $\int \sqrt{2ax - x^2}\,dx = \dfrac{1}{2}\left[(x-a)\sqrt{2ax-x^2} + a^2\sin^{-1}\dfrac{x-a}{|a|}\right].$

251. $\int \dfrac{dx}{\sqrt{2ax - x^2}} = \begin{cases} \cos^{-1}\left(\dfrac{a-x}{|a|}\right), \\ \text{or} \\ \sin^{-1}\left(\dfrac{x-a}{|a|}\right). \end{cases}$

252. $\int x^n\sqrt{2ax - x^2}\,dx =$

$\begin{cases} -\dfrac{x^{n-1}\sqrt{(2ax-x^2)^3}}{n+2} + \dfrac{(2n+1)a}{n+2}\int x^{n-1}\sqrt{2ax-x^2}\,dx, \\ \text{or} \\ \sqrt{2ax-x^2}\left[\dfrac{x^{n+1}}{n+2} - \displaystyle\sum_{r=0}^{n}\dfrac{(2n+1)!(r!)^2 a^{n-r+1}}{2^{n-r}(2r+1)!(n+2)!n!}x^r\right] + \dfrac{(2n+1)!a^{n+2}}{2^n n!(n+2)!}\sin^{-1}\left(\dfrac{x-a}{|a|}\right). \end{cases}$

253. $\int \dfrac{\sqrt{2ax - x^2}}{x^n}\,dx = \dfrac{\sqrt{(2ax-x^2)^3}}{(3-2n)ax^n} + \dfrac{n-3}{(2n-3)a}\int \dfrac{\sqrt{2ax-x^2}}{x^{n-1}}\,dx.$

254. $\int \dfrac{x^n\,dx}{\sqrt{2ax - x^2}} =$

$\begin{cases} -\dfrac{x^{n-1}\sqrt{2ax-x^2}}{n} + \dfrac{a(2n-1)}{n}\int \dfrac{x^{n-1}}{\sqrt{2ax-x^2}}\,dx, \\ \text{or} \\ -\sqrt{2ax-x^2}\displaystyle\sum_{r=1}^{n}\dfrac{(2n)!r!(r-1)!a^{n-r}}{2^{n-r}(2r)!(n!)^2}x^{r-1} + \dfrac{(2n)!a^n}{2^n(n!)^2}\sin^{-1}\left(\dfrac{x-a}{|a|}\right). \end{cases}$

255. $\int \dfrac{dx}{x^n\sqrt{2ax - x^2}} = \begin{cases} \dfrac{\sqrt{2ax-x^2}}{a(1-2n)x^n} + \dfrac{n-1}{(2n-1)a}\int \dfrac{dx}{x^{n-1}\sqrt{2ax-x^2}}, \\ \text{or} \\ -\sqrt{2ax-x^2}\displaystyle\sum_{r=0}^{n-1}\dfrac{2^{n-r}(n-1)!n!(2r)!}{(2n)!(r!)^2 a^{n-r}x^{r+1}}. \end{cases}$

256. $\int \dfrac{dx}{\sqrt{(2ax - x^2)^3}} = \dfrac{x-a}{a^2\sqrt{2ax-x^2}}.$

257. $\int \dfrac{x\,dx}{\sqrt{(2ax - x^2)^3}} = \dfrac{x}{a\sqrt{2ax-x^2}}.$

5.4.15   MISCELLANEOUS ALGEBRAIC FORMS

258. $\int \dfrac{dx}{\sqrt{2ax + x^2}} = \log\left(x + a + \sqrt{2ax + x^2}\right).$

259. $\displaystyle\int \sqrt{ax^2 + c}\, dx = \begin{cases} \dfrac{x}{2}\sqrt{ax^2 + c} + \dfrac{c}{2\sqrt{-a}}\sin^{-1}\left(x\sqrt{-\dfrac{a}{c}}\right), & a < 0, \\[2ex] \text{or} \\[1ex] \dfrac{x}{2}\sqrt{ax^2 + c} + \dfrac{c}{2\sqrt{a}}\log\left(x\sqrt{a} + \sqrt{ax^2 + c}\right), & a > 0. \end{cases}$

260. $\displaystyle\int \sqrt{\dfrac{1 + x}{1 - x}}\, dx = \sin^{-1} x - \sqrt{1 - x^2}.$

261. $\displaystyle\int \dfrac{dx}{x\sqrt{ax^n + c}} = \begin{cases} \dfrac{1}{n\sqrt{c}}\log\dfrac{\sqrt{ax^n + c} - \sqrt{c}}{\sqrt{ax^n + c} + \sqrt{c}}, \\[2ex] \text{or} \\[1ex] \dfrac{2}{n\sqrt{c}}\log\dfrac{\sqrt{ax^n + c} - \sqrt{c}}{\sqrt{x^n}}, & c > 0, \\[2ex] \text{or} \\[1ex] \dfrac{2}{n\sqrt{-c}}\sec^{-1}\sqrt{-\dfrac{ax^n}{c}}, & c < 0. \end{cases}$

262. $\displaystyle\int \dfrac{dx}{\sqrt{ax^2 + c}} = \begin{cases} \dfrac{1}{\sqrt{-a}}\sin^{-1}\left(x\sqrt{-\dfrac{a}{c}}\right), & a < 0, \\[2ex] \text{or} \\[1ex] \dfrac{1}{\sqrt{a}}\log\left(x\sqrt{a} + \sqrt{ax^2 + c}\right), & a > 0. \end{cases}$

263. $\displaystyle\int (ax^2 + c)^{m+1/2}\, dx =$

$\begin{cases} \dfrac{x(ax^2 + c)^{m+1/2}}{2(m + 1)} + \dfrac{(2m + 1)c}{2(m + 1)}\displaystyle\int (ax^2 + c)^{m-1/2}\, dx, \\[2ex] \text{or} \\[1ex] x\sqrt{ax^2 + c}\displaystyle\sum_{r=0}^{m}\dfrac{(2m + 1)!(r!)^2 c^{m-r}}{2^{2m-2r+1}m!(m + 1)!(2r + 1)!}(ax^2 + c)^r \\[3ex] \qquad + \dfrac{(2m + 1)!c^{m+1}}{2^{2m+1}m!(m + 1)!}\displaystyle\int \dfrac{dx}{\sqrt{ax^2 + c}}. \end{cases}$

264. $\displaystyle\int x(ax^2 + c)^{m+1/2}\, dx = \dfrac{(ax^2 + c)^{m+3/2}}{(2m + 3)a}.$

265. $\displaystyle\int \dfrac{(ax^2 + c)^{m+1/2}}{x}\, dx =$

$\begin{cases} \dfrac{(ax^2 + c)^{m+1/2}}{2m + 1} + c\displaystyle\int \dfrac{(ax^2 + c)^{m-1/2}}{x}\, dx, \\[2ex] \text{or} \\[1ex] \sqrt{ax^2 + c}\displaystyle\sum_{r=0}^{m}\dfrac{c^{m-r}(ax^2 + c)^r}{2r + 1} + c^{m+1}\displaystyle\int \dfrac{dx}{x\sqrt{ax^2 + c}}. \end{cases}$

266. $\displaystyle\int \dfrac{dx}{(ax^2 + c)^{m+1/2}} =$

$\begin{cases} \dfrac{x}{(2m - 1)c(ax^2 + c)^{m-1/2}} + \dfrac{2m - 2}{(2m - 1)c}\displaystyle\int \dfrac{dx}{(ax^2 + c)^{m-1/2}}, \\[2ex] \text{or} \\[1ex] \dfrac{x}{\sqrt{ax^2 + c}}\displaystyle\sum_{r=0}^{m-1}\dfrac{2^{2m-2r-1}(m - 1)!m!(2r)!}{(2m)!(r!)^2 c^{m-r}(ax^2 + c)^r}. \end{cases}$

267. $\displaystyle\int \dfrac{dx}{x^m\sqrt{ax^2 + c}} = -\dfrac{\sqrt{ax^2 + c}}{(m - 1)cx^{m-1}} - \dfrac{(m - 2)a}{(m - 1)c}\displaystyle\int \dfrac{dx}{x^{m-2}\sqrt{ax^2 + c}}, \quad m \neq 1.$

268. $\int \dfrac{1+x^2}{(1-x^2)\sqrt{1+x^4}}\,dx = \dfrac{1}{\sqrt{2}}\log\left(\dfrac{x\sqrt{2}+\sqrt{1+x^4}}{1-x^2}\right).$

269. $\int \dfrac{1-x^2}{(1+x^2)\sqrt{1+x^4}}\,dx = \dfrac{1}{\sqrt{2}}\tan^{-1}\dfrac{x\sqrt{2}}{\sqrt{1+x^4}}.$

270. $\int \dfrac{dx}{x\sqrt{x^n+a^2}} = -\dfrac{2}{na}\log\left(\dfrac{a+\sqrt{x^n+a^2}}{\sqrt{x^n}}\right).$

271. $\int \dfrac{dx}{x\sqrt{x^n-a^2}} = -\dfrac{2}{na}\sin^{-1}\dfrac{a}{\sqrt{x^n}}.$

272. $\int \sqrt{\dfrac{x}{a^3-x^3}}\,dx = \dfrac{2}{3}\sin^{-1}\left(\dfrac{x}{a}\right)^{3/2}.$

5.4.16 FORMS INVOLVING TRIGONOMETRIC FUNCTIONS

273. $\int \sin ax\,dx = -\dfrac{1}{a}\cos ax.$

274. $\int \cos ax\,dx = \dfrac{1}{a}\sin ax.$

275. $\int \tan ax\,dx = -\dfrac{1}{a}\log\cos ax = \dfrac{1}{a}\log\sec ax.$

276. $\int \cot ax\,dx = \dfrac{1}{a}\log\sin ax = -\dfrac{1}{a}\log\csc ax.$

277. $\int \sec ax\,dx = \dfrac{1}{a}\log\left(\sec ax + \tan ax\right) = \dfrac{1}{a}\log\tan\left(\dfrac{\pi}{4}+\dfrac{ax}{2}\right).$

278. $\int \csc ax\,dx = \dfrac{1}{a}\log\left(\csc ax - \cot ax\right) = \dfrac{1}{a}\log\tan\dfrac{ax}{2}.$

279. $\int \sin^2 ax\,dx = \dfrac{x}{2} - \dfrac{1}{2a}\cos ax\sin ax = \dfrac{x}{2} - \dfrac{1}{4a}\sin 2ax.$

280. $\int \sin^3 ax\,dx = -\dfrac{1}{3a}(\cos ax)(\sin^2 ax + 2).$

281. $\int \sin^4 ax\,dx = \dfrac{3x}{8} - \dfrac{\sin 2ax}{4a} + \dfrac{\sin 4ax}{32a}.$

282. $\int \sin^n ax\,dx = -\dfrac{\sin^{n-1} ax\cos ax}{na} + \dfrac{n-1}{n}\int \sin^{n-2} ax\,dx.$

283. $\int \sin^{2m} ax\,dx = -\dfrac{\cos ax}{a}\sum_{r=0}^{m-1}\dfrac{(2m)!(r!)^2}{2^{2m-2r}(2r+1)!(m!)^2}\sin^{2r+1} ax + \dfrac{(2m)!}{2^{2m}(m!)^2}x.$

284. $\int \sin^{2m+1} ax\,dx = -\dfrac{\cos ax}{a}\sum_{r=0}^{m-1}\dfrac{2^{2m-2r}(m!)^2(2r)!}{(2m+1)!(r!)^2}\sin^{2r} ax.$

285. $\int \cos^2 ax\,dx = \dfrac{1}{2}x + \dfrac{1}{2a}\sin ax\cos ax = \dfrac{1}{2}x + \dfrac{1}{4a}\sin 2ax.$

286. $\int \cos^3 ax\,dx = \dfrac{1}{3a}\sin ax(\cos^2 ax + 2).$

287. $\int \cos^4 ax\,dx = \dfrac{3}{8}x + \dfrac{\sin 2ax}{4a} + \dfrac{\sin 4ax}{32a}.$

288. $\int \cos^n ax\,dx = \dfrac{1}{na}\cos^{n-1} ax\sin ax + \dfrac{n-1}{n}\int \cos^{n-2} ax\,dx.$

289. $\displaystyle\int \cos^{2m} ax\, dx = \frac{\sin ax}{a} \sum_{r=0}^{m-1} \frac{(2m)!(r!)^2}{2^{2m-2r}(2r+1)!(m!)^2} \cos^{2r+1} ax + \frac{(2m)!}{2^{2m}(m!)^2}x.$

290. $\displaystyle\int \cos^{2m+1} ax\, dx = \frac{\sin ax}{a} \sum_{r=0}^{m} \frac{2^{2m-2r}(m!)^2(2r)!}{(2m+1)!(r!)^2} \cos^{2r} ax.$

291. $\displaystyle\int \frac{dx}{\sin^2 ax} = \int \operatorname{cosec}^2 ax\, dx = -\frac{1}{a}\cot ax.$

292. $\displaystyle\int \frac{dx}{\sin^m ax} = \int \operatorname{cosec}^m ax\, dx = -\frac{1}{a(m-1)}\frac{\cos ax}{\sin^{m-1} ax} + \frac{m-2}{m-1}\int \frac{dx}{\sin^{m-2} ax}.$

293. $\displaystyle\int \frac{dx}{\sin^{2m} ax} = \int \operatorname{cosec}^{2m} ax\, dx = -\frac{1}{a}\cos ax \sum_{r=0}^{m-1} \frac{2^{2m-2r-1}(m-1)!m!(2r)!}{(2m)!(r!)^2 \sin^{2r+1} ax}.$

294. $\displaystyle\int \frac{dx}{\sin^{2m+1} ax} = \int \operatorname{cosec}^{2m+1} ax\, dx =$

$$-\frac{1}{a}\cos ax \sum_{r=0}^{m-1} \frac{(2m)!(r!)^2}{2^{2m-2r}(2r+1)!(m!)^2 \sin^{2r+2} ax} + \frac{1}{a}\frac{(2m)!}{2^{2m}(m!)^2}\log\tan\frac{ax}{2}.$$

295. $\displaystyle\int \frac{dx}{\cos^2 ax} = \int \sec^2 ax\, dx = \frac{1}{a}\tan ax.$

296. $\displaystyle\int \frac{dx}{\cos^m ax} = \int \sec^m ax\, dx = \frac{1}{a(m-1)}\frac{\sin ax}{\cos^{m-1} ax} + \frac{m-2}{m-1}\int \frac{dx}{\cos^{m-2} ax}.$

297. $\displaystyle\int \frac{dx}{\cos^{2m} ax} = \int \sec^{2m} ax\, dx = \frac{1}{a}\sin ax \sum_{r=0}^{m-1} \frac{2^{2m-2r-1}(m-1)!m!(2r)!}{(2m)!(r!)^2 \cos^{2r+1} ax}.$

298. $\displaystyle\int \frac{dx}{\cos^{2m+1} ax} = \int \sec^{2m+1} ax\, dx = \frac{1}{a}\frac{(2m)!}{2^{2m}(m!)^2}\log(\sec ax + \tan ax)$

$$+\frac{1}{a}\sin ax \sum_{r=0}^{m-1} \frac{(2m)!(r!)^2}{2^{2m-2r}(m!)^2(2r+1)!\cos^{2r+2} ax}.$$

299. $\displaystyle\int (\sin mx)(\sin nx)\, dx = \frac{\sin(m-n)x}{2(m-n)} - \frac{\sin(m+n)x}{2(m+n)},\quad m^2 \neq n^2.$

300. $\displaystyle\int (\cos mx)(\cos nx)\, dx = \frac{\sin(m-n)x}{2(m-n)} + \frac{\sin(m+n)x}{2(m+n)},\quad m^2 \neq n^2.$

301. $\displaystyle\int (\sin ax)(\cos ax)\, dx = \frac{1}{2a}\sin^2 ax.$

302. $\displaystyle\int (\sin mx)(\cos nx)\, dx = -\frac{\cos(m-n)x}{2(m-n)} - \frac{\cos(m+n)x}{2(m+n)},\quad m^2 \neq n^2.$

303. $\displaystyle\int (\sin^2 ax)(\cos^2 ax)\, dx = -\frac{1}{32a}\sin 4ax + \frac{x}{8}.$

304. $\displaystyle\int (\sin ax)(\cos^m ax)\, dx = -\frac{\cos^{m+1} ax}{(m+1)a}.$

305. $\displaystyle\int (\sin^m ax)(\cos ax)\, dx = \frac{\sin^{m+1} ax}{(m+1)a}.$

306. $\displaystyle\int (\cos^m ax)(\sin^n ax)\, dx =$

$$\begin{cases} \dfrac{\cos^{m-1} ax \sin^{n+1} ax}{(m+n)a} + \dfrac{m-1}{m+n}\displaystyle\int (\cos^{m-2} ax)(\sin^n ax)\, dx, \\[2ex] \text{or} \\[1ex] -\dfrac{\cos^{m+1} ax \sin^{n-1} ax}{(m+n)a} + \dfrac{n-1}{m+n}\displaystyle\int (\cos^m ax)(\sin^{n-2} ax)\, dx. \end{cases}$$

307. $\displaystyle\int \frac{\cos^m ax}{\sin^n ax}\,dx = \begin{cases} -\dfrac{\cos^{m+1} ax}{a(n-1)\sin^{n-1} ax} - \dfrac{m-n+2}{n-1}\displaystyle\int \dfrac{\cos^m ax}{\sin^{n-2} ax}\,dx, \\ \text{or} \\ \dfrac{\cos^{m-1} ax}{a(m-n)\sin^{n-1} ax} + \dfrac{m-1}{m-n}\displaystyle\int \dfrac{\cos^{m-2} ax}{\sin^n ax}\,dx. \end{cases}$

308. $\displaystyle\int \frac{\sin^m ax}{\cos^n ax}\,dx = \begin{cases} \dfrac{\sin^{m+1} ax}{a(n-1)\cos^{n-1} ax} - \dfrac{m-n+2}{n-1}\displaystyle\int \dfrac{\sin^m ax}{\cos^{n-2} ax}\,dx, \\ \text{or} \\ -\dfrac{\sin^{m-1} ax}{a(m-n)\cos^{n-1} ax} + \dfrac{m-1}{m-n}\displaystyle\int \dfrac{\sin^{m-2} ax}{\cos^n ax}\,dx. \end{cases}$

309. $\displaystyle\int \frac{\sin ax}{\cos^2 ax}\,dx = \frac{1}{a\cos ax} = \frac{\sec ax}{a}.$

310. $\displaystyle\int \frac{\sin^2 ax}{\cos ax}\,dx = -\frac{1}{a}\sin ax + \frac{1}{a}\log\tan\left(\frac{\pi}{4}+\frac{ax}{2}\right).$

311. $\displaystyle\int \frac{\cos ax}{\sin^2 ax}\,dx = -\frac{\csc ax}{a} = -\frac{1}{a\sin ax}.$

312. $\displaystyle\int \frac{dx}{(\sin ax)(\cos ax)} = \frac{1}{a}\log\tan ax.$

313. $\displaystyle\int \frac{dx}{(\sin ax)(\cos^2 ax)} = \frac{1}{a}\left(\sec ax + \log\tan\frac{ax}{2}\right).$

314. $\displaystyle\int \frac{dx}{(\sin ax)(\cos^n ax)} = \frac{1}{a(n-1)\cos^{n-1} ax} + \int \frac{dx}{(\sin ax)(\cos^{n-2} ax)}.$

315. $\displaystyle\int \frac{dx}{(\sin^2 ax)(\cos ax)} = -\frac{1}{a}\csc ax + \frac{1}{a}\log\tan\left(\frac{\pi}{4}+\frac{ax}{2}\right).$

316. $\displaystyle\int \frac{dx}{(\sin^2 ax)(\cos^2 ax)} = -\frac{2}{a}\cot 2ax.$

317. $\displaystyle\int \frac{dx}{\sin^m ax \cos^n ax} =$
$\begin{cases} -\dfrac{1}{a(m-1)\sin^{m-1} ax \cos^{n-1} ax} + \dfrac{m+n-2}{m-1}\displaystyle\int \dfrac{dx}{\sin^{m-2} ax \cos^n ax}, \\ \text{or} \\ \dfrac{1}{a(n-1)\sin^{m-1} ax \cos^{n-1} ax} + \dfrac{m+n-2}{n-1}\displaystyle\int \dfrac{dx}{\sin^m ax \cos^{n-2} ax}. \end{cases}$

318. $\displaystyle\int \sin(a+bx)\,dx = -\frac{1}{b}\cos(a+bx).$

319. $\displaystyle\int \cos(a+bx)\,dx = \frac{1}{b}\sin(a+bx).$

320. $\displaystyle\int \frac{dx}{1\pm\sin ax} = \mp\frac{1}{a}\tan\left(\frac{\pi}{4}\mp\frac{ax}{2}\right).$

321. $\displaystyle\int \frac{dx}{1+\cos ax} = \frac{1}{a}\tan\frac{ax}{2}.$

322. $\displaystyle\int \frac{dx}{1-\cos ax} = -\frac{1}{a}\cot\frac{ax}{2}.$

323. $\displaystyle\int \frac{dx}{a+b\sin x} = \begin{cases} \dfrac{2}{\sqrt{a^2-b^2}}\tan^{-1}\left(\dfrac{a\tan\frac{x}{2}+b}{\sqrt{a^2-b^2}}\right), \\ \text{or} \\ \dfrac{1}{\sqrt{b^2-a^2}}\log\left(\dfrac{a\tan\frac{x}{2}+b-\sqrt{b^2-a^2}}{a\tan\frac{x}{2}+b+\sqrt{b^2-a^2}}\right). \end{cases}$

324. $\int \dfrac{dx}{a+b\cos x} = $
$$\begin{cases} \dfrac{2}{\sqrt{a^2-b^2}}\tan^{-1}\dfrac{\sqrt{a^2-b^2}\,\tan\frac{x}{2}}{a+b}, \\[2mm] \text{or} \\[2mm] \dfrac{1}{\sqrt{b^2-a^2}}\log\left(\dfrac{\sqrt{b^2-a^2}\,\tan\frac{x}{2}+a+b}{\sqrt{b^2-a^2}\,\tan\frac{x}{2}-a-b}\right). \end{cases}$$

325. $\int \dfrac{dx}{a+b\sin x+c\cos x} = $
$$\begin{cases} \dfrac{1}{\sqrt{b^2+c^2-a^2}}\log\left(\dfrac{b-\sqrt{b^2+c^2-a^2}+(a-c)\tan\frac{x}{2}}{b+\sqrt{b^2+c^2-a^2}+(a-c)\tan\frac{x}{2}}\right), & a\neq c,\ a^2<b^2+c^2, \\[2mm] \text{or} \\[2mm] \dfrac{2}{\sqrt{a^2-b^2-c^2}}\tan^{-1}\dfrac{b+(a-c)\tan\frac{x}{2}}{\sqrt{a^2-b^2-c^2}}, & a^2>b^2+c^2, \\[2mm] \text{or} \\[2mm] \dfrac{1}{a}\left[\dfrac{a-(b+c)\sin x-(b-c)\sin x}{a-(b+c)\sin x+(b-c)\sin x}\right]. & a^2=b^2+c^2. \end{cases}$$

326. $\int \dfrac{\sin^2 x}{a+b\cos^2 x}\,dx = \dfrac{1}{b}\sqrt{\dfrac{a+b}{a}}\tan^{-1}\left(\sqrt{\dfrac{a}{a+b}}\tan x\right)-\dfrac{x}{b},\quad ab>0,\ |a|>|b|.$

327. $\int \dfrac{dx}{a^2\cos^2 x+b^2\sin^2 x} = \dfrac{1}{ab}\tan^{-1}\left(\dfrac{b\tan x}{a}\right).$

328. $\int \dfrac{\cos^2 cx}{a^2+b^2\sin^2 cx}\,dx = \dfrac{\sqrt{a^2+b^2}}{ab^2c}\tan^{-1}\dfrac{\sqrt{a^2+b^2}\,\tan cx}{a}-\dfrac{x}{b^2}.$

329. $\int \dfrac{\sin cx\cos cx}{a\cos^2 cx+b\sin^2 cx}\,dx = \dfrac{1}{2c(b-a)}\log\left(a\cos^2 cx+b\sin^2 cx\right),\ a\neq b.$

330. $\int \dfrac{\cos cx}{a\cos cx+b\sin cx}\,dx = $
$$\int \dfrac{dx}{a+b\tan cx} = \dfrac{1}{c(a^2+b^2)}\left[acx+b\log\left(a\cos cx+b\sin cx\right)\right].$$

331. $\int \dfrac{\sin cx}{a\cos cx+b\sin cx}\,dx = $
$$\int \dfrac{dx}{b+a\cot cx} = \dfrac{1}{c(a^2+b^2)}\left[bcx-a\log\left(a\cos cx+b\sin cx\right)\right].$$

332. $\int \dfrac{dx}{a\cos^2 x+2b\cos x\sin x+c\sin^2 x} = $
$$\begin{cases} \dfrac{1}{2\sqrt{b^2-ac}}\log\left(\dfrac{c\tan x+b-\sqrt{b^2-ac}}{c\tan x+b+\sqrt{b^2-ac}}\right), & b^2>ac, \\[2mm] \text{or} \\[2mm] \dfrac{1}{\sqrt{ac-b^2}}\tan^{-1}\left(\dfrac{c\tan x+b}{\sqrt{ac-b^2}}\right), & b^2<ac, \\[2mm] \text{or} \\[2mm] -\dfrac{1}{c\tan x+b}, & b^2=ac. \end{cases}$$

333. $\int \dfrac{\sin ax}{1\pm\sin ax}\,dx = \pm x+\dfrac{1}{a}\tan\left(\dfrac{\pi}{4}\mp\dfrac{ax}{2}\right).$

334. $\int \dfrac{dx}{(\sin ax)(1\pm\sin ax)} = \dfrac{1}{a}\tan\left(\dfrac{\pi}{4}\mp\dfrac{ax}{2}\right)+\dfrac{1}{a}\log\tan\dfrac{ax}{2}.$

335. $\int \dfrac{dx}{(1+\sin ax)^2} = -\dfrac{1}{2a}\tan\left(\dfrac{\pi}{4}-\dfrac{ax}{2}\right)-\dfrac{1}{6a}\tan^3\left(\dfrac{\pi}{4}-\dfrac{ax}{2}\right).$

336. $\displaystyle\int \frac{dx}{(1-\sin ax)^2} = \frac{1}{2a}\cot\left(\frac{\pi}{4}-\frac{ax}{2}\right) + \frac{1}{6a}\cot^3\left(\frac{\pi}{4}-\frac{ax}{2}\right).$

337. $\displaystyle\int \frac{\sin ax}{(1+\sin ax)^2}\,dx = -\frac{1}{2a}\tan\left(\frac{\pi}{4}-\frac{ax}{2}\right) + \frac{1}{6a}\tan^3\left(\frac{\pi}{4}-\frac{ax}{2}\right).$

338. $\displaystyle\int \frac{\sin ax}{(1-\sin ax)^2}\,dx = -\frac{1}{2a}\cot\left(\frac{\pi}{4}-\frac{ax}{2}\right) + \frac{1}{6a}\cot^3\left(\frac{\pi}{4}-\frac{ax}{2}\right).$

339. $\displaystyle\int \frac{\sin x}{a+b\sin x}\,dx = \frac{x}{b} - \frac{a}{b}\int \frac{dx}{a+b\sin x}.$

340. $\displaystyle\int \frac{dx}{(\sin x)(a+b\sin x)} = \frac{1}{a}\log\tan\frac{x}{2} - \frac{b}{a}\int \frac{dx}{a+b\sin x}.$

341. $\displaystyle\int \frac{dx}{(a+b\sin x)^2} = \begin{cases} \dfrac{b\cos x}{(a^2-b^2)(a+b\sin x)} + \dfrac{a}{a^2-b^2}\displaystyle\int \dfrac{dx}{a+b\sin x}, \\[1ex] \text{or} \\[1ex] \dfrac{a\cos x}{(b^2-a^2)(a+b\sin x)} + \dfrac{b}{b^2-a^2}\displaystyle\int \dfrac{dx}{a+b\sin x}. \end{cases}$

342. $\displaystyle\int \frac{dx}{a^2+b^2\sin^2 cx} = \frac{1}{ac\sqrt{a^2+b^2}}\tan^{-1}\left(\frac{\sqrt{a^2+b^2}\,\tan cx}{a}\right).$

343. $\displaystyle\int \frac{dx}{a^2-b^2\sin^2 cx} = \begin{cases} \dfrac{1}{ac\sqrt{a^2-b^2}}\tan^{-1}\left(\dfrac{\sqrt{a^2-b^2}\,\tan cx}{a}\right), & a^2>b^2, \\[1ex] \text{or} \\[1ex] \dfrac{1}{2ac\sqrt{b^2-a^2}}\log\left(\dfrac{\sqrt{b^2-a^2}\,\tan cx + a}{\sqrt{b^2-a^2}\,\tan cx - a}\right), & a^2<b^2. \end{cases}$

344. $\displaystyle\int \frac{\cos ax}{1+\cos ax}\,dx = x - \frac{1}{a}\tan\frac{ax}{2}.$

345. $\displaystyle\int \frac{\cos ax}{1-\cos ax}\,dx = -x - \frac{1}{a}\cot\frac{ax}{2}.$

346. $\displaystyle\int \frac{dx}{(\cos ax)(1+\cos ax)} = \frac{1}{a}\log\tan\left(\frac{\pi}{4}+\frac{ax}{2}\right) - \frac{1}{a}\tan\frac{ax}{2}.$

347. $\displaystyle\int \frac{dx}{(\cos ax)(1-\cos ax)} = \frac{1}{a}\log\tan\left(\frac{\pi}{4}+\frac{ax}{2}\right) - \frac{1}{a}\cot\frac{ax}{2}.$

348. $\displaystyle\int \frac{dx}{(1+\cos ax)^2} = \frac{1}{2a}\tan\frac{ax}{2} + \frac{1}{6a}\tan^3\frac{ax}{2}.$

349. $\displaystyle\int \frac{dx}{(1-\cos ax)^2} = -\frac{1}{2a}\cot\frac{ax}{2} - \frac{1}{6a}\cot^3\frac{ax}{2}.$

350. $\displaystyle\int \frac{\cos ax}{(1+\cos ax)^2}\,dx = \frac{1}{2a}\tan\frac{ax}{2} - \frac{1}{6a}\tan^3\frac{ax}{2}.$

351. $\displaystyle\int \frac{\cos ax}{(1-\cos ax)^2}\,dx = \frac{1}{2a}\cot\frac{ax}{2} - \frac{1}{6a}\cot^3\frac{ax}{2}.$

352. $\displaystyle\int \frac{\cos x}{a+b\cos x}\,dx = \frac{x}{b} - \frac{a}{b}\int \frac{dx}{a+b\cos x}.$

353. $\displaystyle\int \frac{dx}{(\cos x)(a+b\cos x)} = \frac{1}{a}\log\tan\left(\frac{x}{2}+\frac{\pi}{4}\right) - \frac{b}{a}\int \frac{dx}{a+b\cos x}.$

354. $\displaystyle\int \frac{dx}{(a+b\cos x)^2} = \frac{b\sin x}{(b^2-a^2)(a+b\cos x)} - \frac{a}{b^2-a^2}\int \frac{dx}{a+b\cos x}.$

355. $\displaystyle\int \frac{\cos x}{(a+b\cos x)^2}\,dx = \frac{a\sin x}{(a^2-b^2)(a+b\cos x)} - \frac{b}{a^2-b^2}\int \frac{dx}{a+b\cos x}.$

356. $\int \dfrac{dx}{a^2 + b^2 - 2ab\cos cx} = \dfrac{2}{c(a^2 - b^2)} \tan^{-1}\left(\dfrac{a+b}{a-b}\tan\dfrac{cx}{2}\right).$

357. $\int \dfrac{dx}{a^2 + b^2\cos^2 cx} = \dfrac{1}{ac\sqrt{a^2+b^2}} \tan^{-1}\dfrac{a\tan cx}{\sqrt{a^2+b^2}}.$

358. $\int \dfrac{dx}{a^2 - b^2\cos^2 cx} = \begin{cases} \dfrac{1}{ac\sqrt{a^2-b^2}} \tan^{-1}\left(\dfrac{a\tan cx}{\sqrt{a^2-b^2}}\right), & a^2 > b^2, \\ \text{or} \\ \dfrac{1}{2ac\sqrt{b^2-a^2}} \log\left(\dfrac{a\tan cx - \sqrt{b^2-a^2}}{a\tan cx + \sqrt{b^2-a^2}}\right), & b^2 > a^2. \end{cases}$

359. $\int \dfrac{\sin ax}{1 \pm \cos ax}\,dx = \mp\dfrac{1}{a} \log\left(1 \pm \cos ax\right).$

360. $\int \dfrac{\cos ax}{1 \pm \sin ax}\,dx = \pm\dfrac{1}{a} \log\left(1 \pm \sin ax\right).$

361. $\int \dfrac{dx}{(\sin ax)(1 \pm \cos ax)} = \pm\dfrac{1}{2a(1 \pm \cos ax)} + \dfrac{1}{2a} \log\tan\dfrac{ax}{2}.$

362. $\int \dfrac{dx}{(\cos ax)(1 \pm \sin ax)} = \mp\dfrac{1}{2a(1 \pm \sin ax)} + \dfrac{1}{2a} \log\tan\left(\dfrac{ax}{2} + \dfrac{\pi}{4}\right).$

363. $\int \dfrac{\sin ax}{(\cos ax)(1 \pm \cos ax)}\,dx = \dfrac{1}{a} \log\left(\sec ax \pm 1\right).$

364. $\int \dfrac{\cos ax}{(\sin ax)(1 \pm \sin ax)}\,dx = -\dfrac{1}{a} \log\left(\csc ax \pm 1\right).$

365. $\int \dfrac{\sin ax}{(\cos ax)(1 \pm \sin ax)}\,dx = \dfrac{1}{2a(1 \pm \sin ax)} \pm \dfrac{1}{2a} \log\tan\left(\dfrac{ax}{2} + \dfrac{\pi}{4}\right).$

366. $\int \dfrac{\cos ax}{(\sin ax)(1 \pm \cos ax)}\,dx = -\dfrac{1}{2a(1 \pm \cos ax)} \pm \dfrac{1}{2a} \log\tan\dfrac{ax}{2}.$

367. $\int \dfrac{dx}{\sin ax \pm \cos ax} = \dfrac{1}{a\sqrt{2}} \log\tan\left(\dfrac{ax}{2} \pm \dfrac{\pi}{8}\right).$

368. $\int \dfrac{dx}{(\sin ax \pm \cos ax)^2} = \dfrac{1}{2a} \tan\left(ax \mp \dfrac{\pi}{4}\right).$

369. $\int \dfrac{dx}{1 + \cos ax \pm \sin ax} = \pm\dfrac{1}{a} \log\left(1 \pm \tan\dfrac{ax}{2}\right).$

370. $\int \dfrac{dx}{a^2\cos^2 cx - b^2\sin^2 cx} = \dfrac{1}{2abc} \log\left(\dfrac{b\tan cx + a}{b\tan cx - a}\right).$

371. $\int x\sin ax\,dx = \dfrac{1}{a^2} \sin ax - \dfrac{x}{a} \cos ax.$

372. $\int x^2\sin ax\,dx = \dfrac{2x}{a^2} \sin ax + \dfrac{2 - a^2x^2}{a^3} \cos ax.$

373. $\int x^3\sin ax\,dx = \dfrac{3a^2x^2 - 6}{a^4} \sin ax + \dfrac{6x - a^2x^3}{a^3} \cos ax.$

374. $\int x^m \sin ax\,dx =$

$\begin{cases} -\dfrac{1}{a}x^m \cos ax + \dfrac{m}{a} \displaystyle\int x^{m-1} \cos ax\,dx, \\ \text{or} \\ \cos ax \displaystyle\sum_{r=0}^{\left\lfloor\frac{m}{2}\right\rfloor} \dfrac{(-1)^{r+1} m!}{(m-2r)!} \dfrac{x^{m-2r}}{a^{2r+1}} + \sin ax \displaystyle\sum_{r=0}^{\left\lfloor\frac{m-1}{2}\right\rfloor} \dfrac{(-1)^r m!}{(m-2r-1)!} \dfrac{x^{m-2r-1}}{a^{2r+2}}. \end{cases}$

375. $\int x \cos ax \, dx = \dfrac{1}{a^2} \cos ax + \dfrac{x}{a} \sin ax.$

376. $\int x^2 \cos ax \, dx = \dfrac{2x}{a^2} \cos ax + \dfrac{a^2 x^2 - 2}{a^3} \sin ax.$

377. $\int x^3 \cos ax \, dx = \dfrac{3a^2 x^2 - 6}{a^4} \cos ax + \dfrac{a^2 x^3 - 6x}{a^3} \sin ax.$

378. $\int x^m \cos ax \, dx =$

$$\begin{cases} \dfrac{x^m}{a} \sin ax - \dfrac{m}{a} \displaystyle\int x^{m-1} \sin ax \, dx, \\[2mm] \quad \text{or} \\[2mm] \sin ax \displaystyle\sum_{r=0}^{\lfloor \frac{m}{2} \rfloor} \dfrac{(-1)^r m!}{(m-2r)!} \dfrac{x^{m-2r}}{a^{2r+1}} + \cos ax \displaystyle\sum_{r=0}^{\lfloor \frac{m-1}{2} \rfloor} \dfrac{(-1)^r m!}{(m-2r-1)!} \dfrac{x^{m-2r-1}}{a^{2r+2}}. \end{cases}$$

379. $\int \dfrac{\sin ax}{x} \, dx = \displaystyle\sum_{n=0}^{\infty} (-1)^n \dfrac{(ax)^{2n+1}}{(2n+1)(2n+1)!}.$

380. $\int \dfrac{\cos ax}{x} \, dx = \displaystyle\sum_{n=0}^{\infty} (-1)^n \dfrac{(ax)^{2n}}{(2n)(2n)!}.$

381. $\int x \sin^2 ax \, dx = \dfrac{x^2}{4} - \dfrac{x}{4a} \sin 2ax - \dfrac{1}{8a^2} \cos 2ax.$

382. $\int x^2 \sin^2 ax \, dx = \dfrac{x^3}{6} - \left(\dfrac{x^2}{4a} - \dfrac{1}{8a^3} \right) \sin 2ax - \dfrac{x}{4a^2} \cos 2ax.$

383. $\int x \sin^3 ax \, dx = \dfrac{x}{12a} \cos 3ax - \dfrac{1}{36a^2} \sin 3ax - \dfrac{3x}{4a} \cos ax + \dfrac{3}{4a^2} \sin ax.$

384. $\int x \cos^2 ax \, dx = \dfrac{x^2}{4} + \dfrac{x}{4a} \sin 2ax + \dfrac{1}{8a^2} \cos 2ax.$

385. $\int x^2 \cos^2 ax \, dx = \dfrac{x^3}{6} + \left(\dfrac{x^2}{4a} - \dfrac{1}{8a^3} \right) \sin 2ax + \dfrac{x}{4a^2} \cos 2ax.$

386. $\int x \cos^3 ax \, dx = \dfrac{x}{12a} \sin 3ax + \dfrac{1}{36a^2} \cos 3ax + \dfrac{3x}{4a} \sin ax + \dfrac{3}{4a^2} \cos ax.$

387. $\int \dfrac{\sin ax}{x^m} \, dx = \dfrac{\sin ax}{(1-m)x^{m-1}} + \dfrac{a}{m-1} \int \dfrac{\cos ax}{x^{m-1}} \, dx.$

388. $\int \dfrac{\cos ax}{x^m} \, dx = \dfrac{\cos ax}{(1-m)x^{m-1}} + \dfrac{a}{1-m} \int \dfrac{\sin ax}{x^{m-1}} \, dx.$

389. $\int \dfrac{x}{1 \pm \sin ax} \, dx = \mp \dfrac{x \cos ax}{a(1 \pm \sin ax)} + \dfrac{1}{a^2} \log (1 \pm \sin ax).$

390. $\int \dfrac{x}{1 + \cos ax} \, dx = \dfrac{x}{a} \tan \dfrac{ax}{2} + \dfrac{2}{a^2} \log \cos \dfrac{ax}{2}.$

391. $\int \dfrac{x}{1 - \cos ax} \, dx = -\dfrac{x}{a} \cot \dfrac{ax}{2} + \dfrac{2}{a^2} \log \sin \dfrac{ax}{2}.$

392. $\int \dfrac{x + \sin x}{1 + \cos x} \, dx = x \tan \dfrac{x}{2}.$

393. $\int \dfrac{x - \sin x}{1 - \cos x} \, dx = -x \cot \dfrac{x}{2}.$

394. $\int \sqrt{1 - \cos ax} \, dx = -\dfrac{2 \sin ax}{a\sqrt{1 - \cos ax}} = -\dfrac{2\sqrt{2}}{a} \cos \dfrac{ax}{2}.$

395. $\displaystyle\int \sqrt{1+\cos ax}\,dx = \frac{2\sin ax}{a\sqrt{1+\cos ax}} = \frac{2\sqrt{2}}{a}\sin\frac{ax}{2}.$

For the following six integrals, each k represents an integer.

396. $\displaystyle\int \sqrt{1+\sin x}\,dx = \begin{cases} 2\left(\sin\dfrac{x}{2} - \cos\dfrac{x}{2}\right), & (8k-1)\dfrac{\pi}{2} < x \le (8k+3)\dfrac{\pi}{2}, \\ \text{or} \\ -2\left(\sin\dfrac{x}{2} - \cos\dfrac{x}{2}\right), & (8k+3)\dfrac{\pi}{2} < x \le (8k+7)\dfrac{\pi}{2}. \end{cases}$

397. $\displaystyle\int \sqrt{1-\sin x}\,dx = \begin{cases} 2\left(\sin\dfrac{x}{2} + \cos\dfrac{x}{2}\right), & (8k-3)\dfrac{\pi}{2} < x \le (8k+1)\dfrac{\pi}{2}, \\ \text{or} \\ -2\left(\sin\dfrac{x}{2} + \cos\dfrac{x}{2}\right), & (8k+1)\dfrac{\pi}{2} < x \le (8k+5)\dfrac{\pi}{2}. \end{cases}$

398. $\displaystyle\int \frac{dx}{\sqrt{1-\cos x}} = \begin{cases} \sqrt{2}\log\tan\dfrac{x}{4}, & 4k\pi < x \le (4k+2)\pi, \\ \text{or} \\ -\sqrt{2}\log\tan\dfrac{x}{4}, & (4k+2)\pi < x \le (4k+4)\pi. \end{cases}$

399. $\displaystyle\int \frac{dx}{\sqrt{1+\cos x}} = \begin{cases} \sqrt{2}\log\tan\left(\dfrac{x+\pi}{4}\right), & (4k-1)\pi < x \le (4k+1)\pi, \\ \text{or} \\ -\sqrt{2}\log\tan\left(\dfrac{x+\pi}{4}\right), & (4k+1)\pi < x \le (4k+3)\pi. \end{cases}$

400. $\displaystyle\int \frac{dx}{\sqrt{1-\sin x}} = \begin{cases} \sqrt{2}\log\tan\left(\dfrac{x}{4} - \dfrac{\pi}{8}\right), & (8k+1)\dfrac{\pi}{2} < x \le (8k+5)\dfrac{\pi}{2}, \\ \text{or} \\ -\sqrt{2}\log\tan\left(\dfrac{x}{4} - \dfrac{\pi}{8}\right), & (8k+5)\dfrac{\pi}{2} < x \le (8k+9)\dfrac{\pi}{2}. \end{cases}$

401. $\displaystyle\int \frac{dx}{\sqrt{1+\sin x}} = \begin{cases} \sqrt{2}\log\tan\left(\dfrac{x}{4} + \dfrac{\pi}{8}\right), & (8k-1)\dfrac{\pi}{2} < x \le (8k+3)\dfrac{\pi}{2}, \\ \text{or} \\ -\sqrt{2}\log\tan\left(\dfrac{x}{4} + \dfrac{\pi}{8}\right), & (8k+3)\dfrac{\pi}{2} < x \le (8k+7)\dfrac{\pi}{2}. \end{cases}$

402. $\displaystyle\int \tan^2 ax\,dx = \frac{1}{a}\tan ax - x.$

403. $\displaystyle\int \tan^3 ax\,dx = \frac{1}{2a}\tan^2 ax + \frac{1}{a}\log\cos ax.$

404. $\displaystyle\int \tan^4 ax\,dx = \frac{1}{3a}\tan^3 ax - \frac{1}{a}\tan ax + x.$

405. $\displaystyle\int \tan^n ax\,dx = \frac{1}{a(n-1)}\tan^{n-1} ax - \int \tan^{n-2} ax\,dx.$

406. $\displaystyle\int \cot^2 ax\,dx = -\frac{1}{a}\cot ax - x.$

407. $\displaystyle\int \cot^3 ax\,dx = -\frac{1}{2a}\cot^2 ax - \frac{1}{a}\log\sin ax.$

408. $\displaystyle\int \cot^4 ax\,dx = -\frac{1}{3a}\cot^3 ax + \frac{1}{a}\cot ax + x.$

409. $\displaystyle\int \cot^n ax\,dx = -\frac{1}{a(n-1)}\cot^{n-1} ax - \int \cot^{n-2} ax\,dx.$

410. $\displaystyle\int \frac{x}{\sin^2 ax}\,dx = \int x\csc^2 ax\,dx = -\frac{x\cot ax}{a} + \frac{1}{a^2}\log\sin ax.$

411. $\displaystyle\int \frac{x}{\sin^n ax}\, dx = \int x\csc^n ax\, dx = -\frac{x\cos ax}{a(n-1)\sin^{n-1} ax}$

$$-\frac{1}{a^2(n-1)(n-2)\sin^{n-2} ax} + \frac{n-2}{n-1}\int \frac{x}{\sin^{n-2} ax}\, dx.$$

412. $\displaystyle\int \frac{x}{\cos^2 ax}\, dx = \int x\sec^2 ax\, dx = \frac{x}{a}\tan ax + \frac{1}{a^2}\log\cos ax.$

413. $\displaystyle\int \frac{x}{\cos^n ax}\, dx = \int x\sec^n ax\, dx = \frac{x\sin ax}{a(n-1)\cos^{n-1} ax}$

$$-\frac{1}{a^2(n-1)(n-2)\cos^{n-2} ax} + \frac{n-2}{n-1}\int \frac{x}{\cos^{n-2} ax}\, dx.$$

414. $\displaystyle\int \frac{\sin ax}{\sqrt{1+b^2\sin^2 ax}}\, dx = -\frac{1}{ab}\sin^{-1}\frac{b\cos ax}{\sqrt{1+b^2}}.$

415. $\displaystyle\int \frac{\sin ax}{\sqrt{1-b^2\sin^2 ax}}\, dx = -\frac{1}{ab}\log\left(b\cos ax + \sqrt{1-b^2\sin^2 ax}\right).$

416. $\displaystyle\int (\sin ax)\sqrt{1+b^2\sin^2 ax}\, dx = -\frac{\cos ax}{2a}\sqrt{1+b^2\sin^2 ax} - \frac{1+b^2}{2ab}\sin^{-1}\frac{b\cos ax}{\sqrt{1+b^2}}.$

417. $\displaystyle\int (\sin ax)\sqrt{1-b^2\sin^2 ax}\, dx = -\frac{\cos ax}{2a}\sqrt{1-b^2\sin^2 ax}$

$$-\frac{1-b^2}{2ab}\log\left(b\cos ax + \sqrt{1-b^2\sin^2 ax}\right).$$

418. $\displaystyle\int \frac{\cos ax}{\sqrt{1+b^2\sin^2 ax}}\, dx = \frac{1}{ab}\log\left(b\sin ax + \sqrt{1+b^2\sin^2 ax}\right).$

419. $\displaystyle\int \frac{\cos ax}{\sqrt{1-b^2\sin^2 ax}}\, dx = \frac{1}{ab}\sin^{-1}(b\sin ax).$

420. $\displaystyle\int (\cos ax)\sqrt{1+b^2\sin^2 ax}\, dx = \frac{\sin ax}{2a}\sqrt{1+b^2\sin^2 ax}$

$$+\frac{1}{2ab}\log\left(b\sin ax + \sqrt{1+b^2\sin^2 ax}\right).$$

421. $\displaystyle\int (\cos ax)\sqrt{1-b^2\sin^2 ax}\, dx = \frac{\sin ax}{2a}\sqrt{1-b^2\sin^2 ax} + \frac{1}{2ab}\sin^{-1}(b\sin ax).$

For the following integral, k represents an integer and $a > |b|$

422. $\displaystyle\int \frac{dx}{\sqrt{a+b\tan^2 cx}} =$

$$\begin{cases} \dfrac{1}{c\sqrt{a-b}}\sin^{-1}\left(\sqrt{\dfrac{a-b}{a}}\sin cx\right), & (4k-1)\dfrac{\pi}{2} < x \le (4k+1)\dfrac{\pi}{2}, \\[2mm] \text{or} \\[2mm] \dfrac{-1}{c\sqrt{a-b}}\sin^{-1}\left(\sqrt{\dfrac{a-b}{a}}\sin cx\right), & (4k+1)\dfrac{\pi}{2} < x \le (4k+3)\dfrac{\pi}{2}. \end{cases}$$

423. $\displaystyle\int \cos^n x\, dx = \frac{1}{2^{n-1}}\sum_{k=0}^{\frac{n}{2}-1}\binom{n}{k}\frac{\sin[(n-2k)x]}{(n-2k)} + \frac{1}{2^n}\binom{n}{\frac{n}{2}}x,$ n is an even integer.

424. $\displaystyle\int \cos^n x\, dx = \frac{1}{2^{n-1}}\sum_{k=0}^{\frac{n-1}{2}}\binom{n}{k}\frac{\sin[(n-2k)x]}{(n-2k)},$ n is an odd integer.

425. $\displaystyle\int \sin^n x\, dx = \frac{1}{2^{n-1}}\sum_{k=0}^{\frac{n}{2}-1}\binom{n}{k}\frac{\sin\left([(n-2k)(\frac{\pi}{2}-x)]\right)}{(2k-n)} + \frac{1}{2^n}\binom{n}{\frac{n}{2}}x,$

n is an even integer.

426. $\displaystyle\int \sin^n x\, dx = \frac{1}{2^{n-1}} \sum_{k=0}^{\frac{n-1}{2}} \binom{n}{k} \frac{\sin\left(\left[(n-2k)(\frac{\pi}{2}-x)\right]\right)}{(2k-n)}$, n is an odd integer.

5.4.17 FORMS INVOLVING INVERSE TRIGONOMETRIC FUNCTIONS

427. $\displaystyle\int \sin^{-1} ax\, dx = x \sin^{-1} ax + \frac{\sqrt{1-a^2x^2}}{a}$.

428. $\displaystyle\int \cos^{-1} ax\, dx = x \cos^{-1} ax - \frac{\sqrt{1-a^2x^2}}{a}$.

429. $\displaystyle\int \tan^{-1} ax\, dx = x \tan^{-1} ax - \frac{1}{2a} \log\left(1+a^2x^2\right)$.

430. $\displaystyle\int \cot^{-1} ax\, dx = x \cot^{-1} ax + \frac{1}{2a} \log\left(1+a^2x^2\right)$.

431. $\displaystyle\int \sec^{-1} ax\, dx = x \sec^{-1} ax - \frac{1}{a} \log\left(ax + \sqrt{a^2x^2-1}\right)$.

432. $\displaystyle\int \csc^{-1} ax\, dx = x \csc^{-1} ax + \frac{1}{a} \log\left(ax + \sqrt{a^2x^2-1}\right)$.

433. $\displaystyle\int \left(\sin^{-1} \frac{x}{a}\right) dx = x \sin^{-1} \frac{x}{a} + \sqrt{a^2-x^2}$, $a > 0$.

434. $\displaystyle\int \left(\cos^{-1} \frac{x}{a}\right) dx = x \cos^{-1} \frac{x}{a} - \sqrt{a^2-x^2}$, $a > 0$.

435. $\displaystyle\int \left(\tan^{-1} \frac{x}{a}\right) dx = x \tan^{-1} \frac{x}{a} - \frac{a}{2} \log\left(a^2+x^2\right)$.

436. $\displaystyle\int \left(\cot^{-1} \frac{x}{a}\right) dx = x \cot^{-1} \frac{x}{a} + \frac{a}{2} \log\left(a^2+x^2\right)$.

437. $\displaystyle\int x \sin^{-1}(ax)\, dx = \frac{1}{4a^2}\left((2a^2x^2-1)\sin^{-1}(ax) + ax\sqrt{1-a^2x^2}\right)$.

438. $\displaystyle\int x \cos^{-1}(ax)\, dx = \frac{1}{4a^2}\left((2a^2x^2-1)\cos^{-1}(ax) - ax\sqrt{1-a^2x^2}\right)$.

439. $\displaystyle\int x^n \sin^{-1}(ax)\, dx = \frac{x^{n+1}}{n+1} \sin^{-1}(ax) - \frac{a}{n+1}\int \frac{x^{n+1}}{\sqrt{1-a^2x^2}}\, dx$, $n \neq -1$.

440. $\displaystyle\int x^n \cos^{-1} ax\, dx = \frac{x^{n+1}}{n+1} \cos^{-1}(ax) + \frac{a}{n+1}\int \frac{x^{n+1}}{\sqrt{1-a^2x^2}}\, dx$, $n \neq -1$.

441. $\displaystyle\int x \tan^{-1}(ax)\, dx = \frac{1+a^2x^2}{2a^2} \tan^{-1}(ax) - \frac{x}{2a}$.

442. $\displaystyle\int x^n \tan^{-1}(ax)\, dx = \frac{x^{n+1}}{n+1} \tan^{-1}(ax) - \frac{a}{n+1}\int \frac{x^{n+1}}{1+a^2x^2}\, dx$.

443. $\displaystyle\int x \cot^{-1}(ax)\, dx = \frac{1+a^2x^2}{2a^2} \cot^{-1}(ax) + \frac{x}{2a}$.

444. $\displaystyle\int x^n \cot^{-1}(ax)\, dx = \frac{x^{n+1}}{n+1} \cot^{-1}(ax) + \frac{a}{n+1}\int \frac{x^{n+1}}{1+a^2x^2}\, dx$.

445. $\displaystyle\int \frac{\sin^{-1}(ax)}{x^2}\, dx = a \log\left(\frac{1-\sqrt{1-a^2x^2}}{x}\right) - \frac{\sin^{-1}(ax)}{x}$.

446. $\displaystyle\int \frac{\cos^{-1}(ax)}{x^2}\, dx = -\frac{1}{x} \cos^{-1}(ax) + a \log\left(\frac{1+\sqrt{1-a^2x^2}}{x}\right)$.

447. $\int \dfrac{\tan^{-1}(ax)}{x^2}\, dx = -\dfrac{1}{x}\tan^{-1}(ax) - \dfrac{a}{2}\log\left(\dfrac{1+a^2x^2}{x^2}\right).$

448. $\int \dfrac{\cot^{-1}(ax)}{x^2}\, dx = -\dfrac{1}{x}\cot^{-1}(ax) - \dfrac{a}{2}\log\left(\dfrac{x^2}{1+a^2x^2}\right).$

449. $\int (\sin^{-1}(ax))^2\, dx = x(\sin^{-1}(ax))^2 - 2x + \dfrac{2\sqrt{1-a^2x^2}}{a}\sin^{-1}(ax).$

450. $\int (\cos^{-1}(ax))^2\, dx = x(\cos^{-1}(ax))^2 - 2x - \dfrac{2\sqrt{1-a^2x^2}}{a}\cos^{-1}(ax).$

451. $\int (\sin^{-1}(ax))^n\, dx =$

$$\begin{cases} x(\sin^{-1}(ax))^n + \dfrac{n\sqrt{1-a^2x^2}}{a}(\sin^{-1}(ax))^{n-1} - n(n-1)\displaystyle\int (\sin^{-1}(ax))^{n-2}\, dx, \\[2mm] \text{or} \\[2mm] \displaystyle\sum_{r=0}^{\lfloor\frac{n}{2}\rfloor} \dfrac{(-1)^r n!}{(n-2r)!}x(\sin^{-1}ax)^{n-2r} + \sum_{r=0}^{\lfloor\frac{n-1}{2}\rfloor}(-1)^r\dfrac{n!\sqrt{1-a^2x^2}}{(n-2r-1)!a}(\sin^{-1}ax)^{n-2r-1}. \end{cases}$$

452. $\int (\cos^{-1}(ax))^n\, dx =$

$$\begin{cases} x(\cos^{-1}(ax))^n - \dfrac{n\sqrt{1-a^2x^2}}{a}(\cos^{-1}(ax))^{n-1} - n(n-1)\displaystyle\int (\cos^{-1}(ax))^{n-2}\, dx, \\[2mm] \text{or} \\[2mm] \displaystyle\sum_{r=0}^{\lfloor\frac{n}{2}\rfloor} \dfrac{(-1)^r n!}{(n-2r)!}x(\cos^{-1}ax)^{n-2r} - \sum_{r=0}^{\lfloor\frac{n-1}{2}\rfloor}(-1)^r\dfrac{n!\sqrt{1-a^2x^2}}{(n-2r-1)!a}(\cos^{-1}ax)^{n-2r-1}. \end{cases}$$

453. $\int \dfrac{\sin^{-1}ax}{\sqrt{1-a^2x^2}}\, dx = \dfrac{1}{2a}\left(\sin^{-1}ax\right)^2.$

454. $\int \dfrac{x^n \sin^{-1}ax}{\sqrt{1-a^2x^2}}\, dx = -\dfrac{x^{n-1}}{na^2}\sqrt{1-a^2x^2}\,\sin^{-1}ax + \dfrac{x^n}{n^2a}$
$$+\dfrac{n-1}{na^2}\int \dfrac{x^{n-2}\sin^{-1}ax}{\sqrt{1-a^2x^2}}\, dx.$$

455. $\int \dfrac{\cos^{-1}ax}{\sqrt{1-a^2x^2}}\, dx = -\dfrac{1}{2a}\left(\cos^{-1}ax\right)^2.$

456. $\int \dfrac{x^n \cos^{-1}ax}{\sqrt{1-a^2x^2}}\, dx = -\dfrac{x^{n-1}}{na^2}\sqrt{1-a^2x^2}\,\cos^{-1}ax - \dfrac{x^n}{n^2a}$
$$+\dfrac{n-1}{na^2}\int \dfrac{x^{n-2}\cos^{-1}ax}{\sqrt{1-a^2x^2}}\, dx.$$

457. $\int \dfrac{\tan^{-1}ax}{1+a^2x^2}\, dx = \dfrac{1}{2a}\left(\tan^{-1}ax\right)^2.$

458. $\int \dfrac{\cot^{-1}ax}{1+a^2x^2}\, dx = -\dfrac{1}{2a}\left(\cot^{-1}ax\right)^2.$

459. $\int x\sec^{-1}ax\, dx = \dfrac{x^2}{2}\sec^{-1}ax - \dfrac{1}{2a^2}\sqrt{a^2x^2-1}.$

460. $\int x^n\sec^{-1}ax\, dx = \dfrac{x^{n+1}}{n+1}\sec^{-1}ax - \dfrac{1}{n+1}\int \dfrac{x^n}{\sqrt{a^2x^2-1}}\, dx.$

461. $\int \dfrac{\sec^{-1}ax}{x^2}\, dx = -\dfrac{\sec^{-1}ax}{x} + \dfrac{\sqrt{a^2x^2-1}}{x}.$

462. $\int x \csc^{-1} ax \, dx = \dfrac{x^2}{2} \csc^{-1} ax + \dfrac{1}{2a^2} \sqrt{a^2 x^2 - 1}.$

463. $\int x^n \csc^{-1} ax \, dx = \dfrac{x^{n+1}}{n+1} \csc^{-1} ax + \dfrac{1}{n+1} \int \dfrac{x^n}{\sqrt{a^2 x^2 - 1}} \, dx.$

464. $\int \dfrac{\csc^{-1} ax}{x^2} \, dx = -\dfrac{\csc^{-1} ax}{x} - \dfrac{\sqrt{a^2 x^2 - 1}}{x}.$

5.4.18 LOGARITHMIC FORMS

465. $\int \log x \, dx = x \log x - x.$

466. $\int x \log x \, dx = \dfrac{x^2}{2} \log x - \dfrac{x^2}{4}.$

467. $\int x^2 \log x \, dx = \dfrac{x^3}{3} \log x - \dfrac{x^3}{9}.$

468. $\int x^n \log x \, dx = \dfrac{x^{n+1}}{n+1} \log x - \dfrac{x^{n+1}}{(n+1)^2}.$

469. $\int (\log x)^2 \, dx = x(\log x)^2 - 2x \log x + 2x.$

470. $\int (\log x)^n \, dx = \begin{cases} x(\log x)^n - n \int (\log x)^{n-1} \, dx, & n \neq -1, \\ \text{or} \\ (-1)^n n! x \sum\limits_{r=0}^{n} \dfrac{(-\log x)^r}{r!}, & n \neq -1. \end{cases}$

471. $\int \dfrac{(\log x)^n}{x} \, dx = \dfrac{1}{n+1} (\log x)^{n+1}, \quad n \neq -1.$

472. $\int \dfrac{dx}{\log x} = \log (\log x) + \log x + \dfrac{(\log x)^2}{2 \cdot 2!} + \dfrac{(\log x)^3}{3 \cdot 3!} + \cdots.$

473. $\int \dfrac{dx}{x \log x} = \log (\log x).$

474. $\int \dfrac{dx}{x(\log x)^n} = \dfrac{1}{(1-n)(\log x)^{n-1}}, \quad n \neq 1.$

475. $\int \dfrac{x^m \, dx}{(\log x)^n} = \dfrac{x^{m+1}}{(1-n)(\log x)^{n-1}} + \dfrac{m+1}{n-1} \int \dfrac{x^m \, dx}{(\log x)^{n-1}}, \quad n \neq 1.$

476. $\int x^m (\log x)^n \, dx = \begin{cases} \dfrac{x^{m+1}(\log x)^n}{m+1} - \dfrac{n}{m+1} \int x^m (\log x)^{n-1} \, dx, \\ \text{or} \\ (-1)^n \dfrac{n!}{m+1} x^{m+1} \sum\limits_{r=0}^{n} \dfrac{(-\log x)^r}{r!(m+1)^{n-r}}. \end{cases}$

477. $\int x^p \cos (b \log x) \, dx = \dfrac{x^{p+1}}{(p+1)^2 + b^2} \left[b \sin (b \log x) + (p+1) \cos (b \log x) \right].$

478. $\int x^p \sin (b \log x) \, dx = \dfrac{x^{p+1}}{(p+1)^2 + b^2} \left[(p+1) \sin (b \log x) - b \cos (b \log x) \right].$

479. $\int \log (ax + b) \, dx = \dfrac{ax+b}{a} \log (ax+b) - x.$

480. $\int \dfrac{\log(ax+b)}{x^2}\,dx = \dfrac{a}{b}\log x - \dfrac{ax+b}{bx}\log(ax+b).$

481. $\int x^m \log(ax+b)\,dx = \dfrac{1}{m+1}\left[x^{m+1} - \left(-\dfrac{b}{a}\right)^{m+1}\right]\log(ax+b)$

$$-\dfrac{1}{m+1}\left(-\dfrac{b}{a}\right)^{m+1}\sum_{r=1}^{m+1}\dfrac{1}{r}\left(-\dfrac{ax}{b}\right)^r.$$

482. $\int \dfrac{\log(ax+b)}{x^m}\,dx = -\dfrac{1}{m-1}\dfrac{\log(ax+b)}{x^{m-1}} + \dfrac{1}{m-1}\left(-\dfrac{a}{b}\right)^{m-1}\log\dfrac{ax+b}{x}$

$$+\dfrac{1}{m-1}\left(-\dfrac{a}{b}\right)^{m-1}\sum_{r=1}^{m-2}\dfrac{1}{r}\left(-\dfrac{b}{ax}\right)^r, \quad m > 2.$$

483. $\int \log\dfrac{x+a}{x-a}\,dx = (x+a)\log(x+a) - (x-a)\log(x-a).$

484. $\int x^m \log\dfrac{x+a}{x-a}\,dx = \dfrac{x^{m+1}-(-a)^{m+1}}{m+1}\log(x+a) - \dfrac{x^{m+1}-a^{m+1}}{m+1}\log(x-a)$

$$+\dfrac{2a^{m+1}}{m+1}\sum_{r=1}^{\lfloor\frac{m+1}{2}\rfloor}\dfrac{1}{m-2r+2}\left(\dfrac{x}{a}\right)^{m-2r+2}.$$

485. $\int \dfrac{1}{x^2}\log\dfrac{x+a}{x-a}\,dx = \dfrac{1}{x}\log\dfrac{x-a}{x+a} - \dfrac{1}{a}\log\dfrac{x^2-a^2}{x^2}.$

For the following two integrals, $X = a + bx + cx^2$.

486. $\int \log X\,dx =$

$$\begin{cases} \left(x+\dfrac{b}{2c}\right)\log X - 2x + \dfrac{\sqrt{4ac-b^2}}{c}\tan^{-1}\dfrac{2cx+b}{\sqrt{4ac-b^2}}, & b^2-4ac < 0, \\[2ex] \text{or} \\[1ex] \left(x+\dfrac{b}{2c}\right)\log X - 2x + \dfrac{\sqrt{b^2-4ac}}{c}\tanh^{-1}\dfrac{2cx+b}{\sqrt{b^2-4ac}}, & b^2-4ac > 0. \end{cases}$$

487. $\int x^n \log X\,dx = \dfrac{x^{n+1}}{n+1}\log X - \dfrac{2c}{n+1}\int\dfrac{x^{n+2}}{X}\,dx - \dfrac{b}{n+1}\int\dfrac{x^{n+1}}{X}\,dx, \quad n \neq -1.$

488. $\int \log(x^2+a^2)\,dx = x\log(x^2+a^2) - 2x + 2a\tan^{-1}\dfrac{x}{a}.$

489. $\int \log(x^2-a^2)\,dx = x\log(x^2-a^2) - 2x + a\log\dfrac{x+a}{x-a}.$

490. $\int x\log(x^2+a^2)\,dx = \dfrac{1}{2}\left(x^2+a^2\right)\log\left(x^2+a^2\right) - \dfrac{1}{2}x^2.$

491. $\int \log\left(x+\sqrt{x^2\pm a^2}\right)\,dx = x\log\left(x+\sqrt{x^2\pm a^2}\right) - \sqrt{x^2\pm a^2}.$

492. $\int x\log\left(x+\sqrt{x^2\pm a^2}\right)\,dx = \left(\dfrac{x^2}{2}\pm\dfrac{a^2}{4}\right)\log\left(x+\sqrt{x^2\pm a^2}\right) - \dfrac{x\sqrt{x^2\pm a^2}}{4}.$

493. $\int x^m \log\left(x+\sqrt{x^2\pm a^2}\right)\,dx = \dfrac{x^{m+1}}{m+1}\log\left(x+\sqrt{x^2\pm a^2}\right)$

$$-\dfrac{1}{m+1}\int\dfrac{x^{m+1}}{\sqrt{x^2\pm a^2}}\,dx.$$

494. $\int \dfrac{\log\left(x+\sqrt{x^2+a^2}\right)}{x^2}\,dx = -\dfrac{\log\left(x+\sqrt{x^2+a^2}\right)}{x} - \dfrac{1}{a}\log\dfrac{a+\sqrt{x^2+a^2}}{x}.$

495. $\displaystyle \int \frac{\log\left(x+\sqrt{x^2-a^2}\right)}{x^2}\,dx = -\frac{\log\left(x+\sqrt{x^2-a^2}\right)}{x} + \frac{1}{|a|}\sec^{-1}\frac{x}{a}.$

496. $\displaystyle \int x^n \log\left(x^2-a^2\right) dx = \frac{1}{n+1}\left[x^{n+1}\log\left(x^2-a^2\right) - a^{n+1}\log\left(x-a\right)\right.$

$$\left. -(-a)^{n+1}\log\left(x+a\right) - 2\sum_{r=0}^{\left\lfloor \frac{n}{2}\right\rfloor} \frac{a^{2r}x^{n-2r+1}}{n-2r+1}\right].$$

5.4.19 EXPONENTIAL FORMS

497. $\displaystyle \int e^x\,dx = e^x.$

498. $\displaystyle \int e^{-x}\,dx = -e^{-x}.$

499. $\displaystyle \int e^{ax}\,dx = \frac{e^{ax}}{a}.$

500. $\displaystyle \int xe^{ax}\,dx = \frac{e^{ax}}{a^2}(ax-1).$

501. $\displaystyle \int x^m e^{ax}\,dx = \begin{cases} \dfrac{x^m e^{ax}}{a} - \dfrac{m}{a}\displaystyle\int x^{m-1}e^{ax}\,dx, \\[2mm] \text{or} \\[2mm] e^{ax}\displaystyle\sum_{r=0}^{m}(-1)^r\,\dfrac{m!\,x^{m-r}}{(m-r)!\,a^{r+1}}. \end{cases}$

502. $\displaystyle \int \frac{e^{ax}}{x}\,dx = \log x + \frac{ax}{1!} + \frac{a^2 x^2}{2\cdot 2!} + \frac{a^3 x^3}{3\cdot 3!} + \dots.$

503. $\displaystyle \int \frac{e^{ax}}{x^m}\,dx = \frac{1}{1-m}\frac{e^{ax}}{x^{m-1}} + \frac{a}{m-1}\int \frac{e^{ax}}{x^{m-1}}\,dx, \quad m \neq 1.$

504. $\displaystyle \int e^{ax}\log x\,dx = \frac{e^{ax}\log x}{a} - \frac{1}{a}\int \frac{e^{ax}}{x}\,dx.$

505. $\displaystyle \int \frac{dx}{1+e^x} = x - \log\left(1+e^x\right) = \log\frac{e^x}{1+e^x}.$

506. $\displaystyle \int \frac{dx}{a+be^{px}} = \frac{x}{a} - \frac{1}{ap}\log\left(a+be^{px}\right).$

507. $\displaystyle \int \frac{dx}{ae^{mx}+be^{-mx}} = \frac{1}{m\sqrt{ab}}\tan^{-1}\left(e^{mx}\sqrt{\frac{a}{b}}\right), \quad a>0,\ b>0.$

508. $\displaystyle \int \frac{dx}{ae^{mx}-be^{-mx}} = \begin{cases} \dfrac{1}{2m\sqrt{ab}}\log\left(\dfrac{\sqrt{a}e^{mx}-\sqrt{b}}{\sqrt{a}e^{mx}+\sqrt{b}}\right), & a>0,\ b>0, \\[3mm] \text{or} \\[2mm] \dfrac{-1}{m\sqrt{ab}}\tanh^{-1}\left(\sqrt{\dfrac{a}{b}}e^{mx}\right), & a>0,\ b>0. \end{cases}$

509. $\displaystyle \int \left(a^x - a^{-x}\right) dx = \frac{a^x + a^{-x}}{\log a}.$

510. $\displaystyle \int \frac{e^{ax}}{b+ce^{ax}}\,dx = \frac{1}{ac}\log\left(b+ce^{ax}\right).$

511. $\displaystyle \int \frac{xe^{ax}}{(1+ax)^2}\,dx = \frac{e^{ax}}{a^2(1+ax)}.$

512. $\int x e^{-x^2}\, dx = -\dfrac{1}{2} e^{-x^2}.$

513. $\int e^{ax} \sin(bx)\, dx = \dfrac{e^{ax}\left[a\sin(bx) - b\cos(bx)\right]}{a^2 + b^2}.$

514. $\int e^{ax} \sin(bx)\sin(cx)\, dx = \dfrac{e^{ax}\left[(b-c)\sin(b-c)x + a\cos(b-c)x\right]}{2\left[a^2 + (b-c)^2\right]}$
$$-\dfrac{e^{ax}\left[(b+c)\sin(b+c)x + a\cos(b+c)x\right]}{2\left[a^2 + (b+c)^2\right]}.$$

515. $\int e^{ax} \sin(bx)\cos(cx)\, dx = \dfrac{e^{ax}\left[a\sin(b-c)x - (b-c)\cos(b-c)x\right]}{2\left[a^2 + (b-c)^2\right]}$
$$+\dfrac{e^{ax}\left[a\sin(b+c)x - (b+c)\cos(b+c)x\right]}{2\left[a^2 + (b+c)^2\right]}.$$

516. $\int e^{ax} \sin(bx)\sin(bx+c)\, dx = \dfrac{e^{ax}\cos c}{2a} - \dfrac{e^{ax}\left[a\cos 2bx + c + 2b\sin 2bx + c\right]}{2\left[a^2 + 4b^2\right]}.$

517. $\int e^{ax} \sin(bx)\cos(bx+c)\, dx = -\dfrac{e^{ax}\sin c}{2a} + \dfrac{e^{ax}\left[a\sin 2bx + c - 2b\cos 2bx + c\right]}{2\left[a^2 + 4b^2\right]}.$

518. $\int e^{ax} \cos(bx)\, dx = \dfrac{e^{ax}}{a^2 + b^2}\left[a\cos(bx) + b\sin(bx)\right].$

519. $\int e^{ax} \cos(bx)\cos(cx)\, dx = \dfrac{e^{ax}\left[(b-c)\sin(b-c)x + a\cos(b-c)x\right]}{2\left[a^2 + (b-c)^2\right]}$
$$+\dfrac{e^{ax}\left[(b+c)\sin(b+c)x + a\cos(b+c)x\right]}{2\left[a^2 + (b+c)^2\right]}.$$

520. $\int e^{ax} \cos(bx)\cos(bx+c)\, dx = \dfrac{e^{ax}\cos c}{2a} + \dfrac{e^{ax}\left[a\cos 2bx + c + 2b\sin 2bx + c\right]}{2\left[a^2 + 4b^2\right]}.$

521. $\int e^{ax} \cos(bx)\sin(bx+c)\, dx = \dfrac{e^{ax}\sin c}{2a} + \dfrac{e^{ax}\left[a\sin 2bx + c - 2b\cos 2bx + c\right]}{2\left[a^2 + 4b^2\right]}.$

522. $\int e^{ax} \sin^n(bx)\, dx = \dfrac{1}{a^2 + n^2 b^2}\left[(a\sin(bx) - nb\cos(bx))e^{ax}\sin^{n-1}(bx)\right.$
$$\left.+n(n-1)b^2 \int e^{ax}\sin^{n-2}(bx)\, dx\right].$$

523. $\int e^{ax} \cos^n(bx)\, dx = \dfrac{1}{a^2 + n^2 b^2}\left[(a\cos(bx) + nb\sin(bx))e^{ax}\cos^{n-1}(bx)\right.$
$$\left.+n(n-1)b^2 \int e^{ax}\cos^{n-2}(bx)\, dx\right].$$

524. $\int x^m e^x \sin x\, dx = \dfrac{1}{2} x^m e^x (\sin x - \cos x) - \dfrac{m}{2}\int x^{m-1} e^x \sin x\, dx$
$$+\dfrac{m}{2}\int x^{m-1} e^x \cos x\, dx.$$

525. $\int x^m e^{ax} \sin bx\, dx = x^m e^{ax}\dfrac{a\sin(bx) - b\cos(bx)}{a^2 + b^2}$
$$-\dfrac{m}{a^2 + b^2}\int x^{m-1} e^{ax}(a\sin(bx) - b\cos(bx))\, dx.$$

526. $\int x^m e^x \cos x\, dx = \dfrac{1}{2} x^m e^x (\sin x + \cos x) - \dfrac{m}{2}\int x^{m-1} e^x \sin x\, dx$
$$-\dfrac{m}{2}\int x^{m-1} e^x \cos x\, dx.$$

527. $\int x^m e^{ax} \cos bx\, dx = x^m e^{ax}\dfrac{a\cos(bx) + b\sin(bx)}{a^2 + b^2}$
$$-\dfrac{m}{a^2 + b^2}\int x^{m-1} e^{ax}(a\cos(bx) + b\sin(bx))\, dx.$$

528. $\displaystyle\int e^{ax}\cos^m x\sin^n x\,dx =$

$$
\begin{cases}
\dfrac{e^{ax}(\cos^{m-1} x)(\sin^n x)\,[a\cos x + (m+n)\sin x]}{(m+n)^2 + a^2} \\[2mm]
\quad -\dfrac{na}{(m+n)^2 + a^2}\int e^{ax}(\cos^{m-1} x)(\sin^{n-1} x)\,dx \\[2mm]
\quad +\dfrac{(m-1)(m+n)}{(m+n)^2 + a^2}\int e^{ax}(\cos^{m-2} x)(\sin^n x)\,dx, \\[3mm]
\qquad\text{or} \\[2mm]
\dfrac{e^{ax}(\cos^m x)(\sin^{n-1} x)\,[a\sin x - (m+n)\cos x]}{(m+n)^2 + a^2} \\[2mm]
\quad +\dfrac{ma}{(m+n)^2 + a^2}\int e^{ax}(\cos^{m-1} x)(\sin^{n-1} x)\,dx \\[2mm]
\quad +\dfrac{(n-1)(m+n)}{(m+n)^2 + a^2}\int e^{ax}(\cos^m x)(\sin^{n-2} x)\,dx, \\[3mm]
\qquad\text{or} \\[2mm]
\dfrac{e^{ax}(\cos^{m-1} x)(\sin^{n-1} x)\,[a\sin x\cos x + m\sin^2 x - n\cos^2 x]}{(m+n)^2 + a^2} \\[2mm]
\quad +\dfrac{m(m-1)}{(m+n)^2 + a^2}\int e^{ax}(\cos^{m-2} x)(\sin^n x)\,dx \\[2mm]
\quad +\dfrac{n(n-1)}{(m+n)^2 + a^2}\int e^{ax}(\cos^m x)(\sin^{n-2} x)\,dx, \\[3mm]
\qquad\text{or} \\[2mm]
\dfrac{e^{ax}(\cos^{m-1} x)(\sin^{n-1} x)\,[a\sin x\cos x + m\sin^2 x - n\cos^2 x]}{(m+n)^2 + a^2} \\[2mm]
\quad +\dfrac{m(m-1)}{(m+n)^2 + a^2}\int e^{ax}(\cos^{m-2} x)(\sin^{n-2} x)\,dx \\[2mm]
\quad +\dfrac{(n-m)(n+m-1)}{(m+n)^2 + a^2}\int e^{ax}(\cos^m x)(\sin^{n-2} x)\,dx.
\end{cases}
$$

529. $\displaystyle\int xe^{ax}\sin(bx)\,dx = \frac{xe^{ax}}{a^2 + b^2}\,[a\sin(bx) - b\cos(bx)]$
$$-\frac{e^{ax}}{(a^2 + b^2)^2}\,\left[(a^2 - b^2)\sin bx - 2ab\cos(bx)\right].$$

530. $\displaystyle\int xe^{ax}\cos(bx)\,dx = \frac{xe^{ax}}{a^2 + b^2}\,[a\cos(bx) + b\sin(bx)]$
$$-\frac{e^{ax}}{(a^2 + b^2)^2}\,\left[(a^2 - b^2)\cos bx + 2ab\sin(bx)\right].$$

531. $\displaystyle\int \frac{e^{ax}}{\sin^n x}\,dx = -\frac{e^{ax}\,[a\sin x + (n-2)\cos x]}{(n-1)(n-2)\sin^{n-1} x} + \frac{a^2 + (n-2)^2}{(n-1)(n-2)}\int \frac{e^{ax}}{\sin^{n-2} x}\,dx.$

532. $\displaystyle\int \frac{e^{ax}}{\cos^n x}\,dx = -\frac{e^{ax}\,[a\cos x - (n-2)\sin x]}{(n-1)(n-2)\cos^{n-1} x} + \frac{a^2 + (n-2)^2}{(n-1)(n-2)}\int \frac{e^{ax}}{\cos^{n-2} x}\,dx.$

533. $\displaystyle\int e^{ax}\tan^n x\,dx = e^{ax}\frac{\tan^{n-1} x}{n-1} - \frac{a}{n-1}\int e^{ax}\tan^{n-1} x\,dx - \int e^{ax}\tan^{n-2} x\,dx.$

5.4.20 HYPERBOLIC FORMS

534. $\displaystyle\int \sinh x\,dx = \cosh x.$

535. $\displaystyle\int \cosh x\,dx = \sinh x.$

536. $\int \tanh x \, dx = \log \cosh x.$

537. $\int \coth x \, dx = \log \sinh x.$

538. $\int \operatorname{sech} x \, dx = \tan^{-1}(\sinh x).$

539. $\int \operatorname{csch} x \, dx = \log \tanh\left(\dfrac{x}{2}\right).$

540. $\int x \sinh x \, dx = x \cosh x - \sinh x.$

541. $\int x^n \sinh x \, dx = x^n \cosh x - n \int x^{n-1}(\cosh x) \, dx.$

542. $\int x \cosh x \, dx = x \sinh x - \cosh x.$

543. $\int x^n \cosh x \, dx = x^n \sinh x - n \int x^{n-1}(\sinh x) \, dx.$

544. $\int \operatorname{sech} x \tanh x \, dx = -\operatorname{sech} x.$

545. $\int \operatorname{csch} x \coth x \, dx = -\operatorname{csch} x.$

546. $\int \sinh^2 x \, dx = \dfrac{\sinh 2x}{4} - \dfrac{x}{2}.$

547. $\int \sinh^m x \cosh^n x \, dx =$

$$\begin{cases} \dfrac{1}{m+n} \sinh^{m+1} x \cosh^{n-1} x + \dfrac{n-1}{m+n} \int \sinh^m x \cosh^{n-2} x \, dx, & m+n \neq 0, \\ \text{or} \\ \dfrac{1}{m+n} \sinh^{m-1} x \cosh^{n+1} x - \dfrac{m-1}{m+n} \int \sinh^{m-2} x \cosh^n x \, dx, & m+n \neq 0. \end{cases}$$

548. $\int \dfrac{dx}{(\sinh^m x)(\cosh^n x)} =$

$$\begin{cases} -\dfrac{1}{(m-1)(\sinh^{m-1} x)(\cosh^{n-1} x)} - \dfrac{m+n-2}{m-1} \int \dfrac{dx}{(\sinh^{m-2} x)(\cosh^n x)} \, dx, & m \neq 1, \\ \text{or} \\ \dfrac{1}{(n-1)(\sinh^{m-1} x)(\cosh^{n-1} x)} + \dfrac{m+n-2}{n-1} \int \dfrac{dx}{(\sinh^m x)(\cosh^{n-2} x)} \, dx, & n \neq 1. \end{cases}$$

549. $\int \tanh^2 x \, dx = x - \tanh x.$

550. $\int \tanh^n x \, dx = -\dfrac{\tanh^{n-1} x}{n-1} + \int (\tanh^{n-2} x) \, dx, \quad n \neq 1.$

551. $\int \operatorname{sech}^2 x \, dx = \tanh x.$

552. $\int \cosh^2 x \, dx = \dfrac{\sinh 2x}{4} + \dfrac{x}{2}.$

553. $\int \coth^2 x \, dx = x - \coth x.$

554. $\int \coth^n x \, dx = -\dfrac{\coth^{n-1} x}{n-1} + \int \coth^{n-2} x \, dx, \quad n \neq 1.$

555. $\int \operatorname{csch}^2 x \, dx = -\coth x.$

556. $\int (\sinh mx)(\sinh nx) \, dx = \dfrac{\sinh{(m+n)x}}{2(m+n)} - \dfrac{\sinh{(m-n)x}}{2(m-n)}, \quad m^2 \neq n^2.$

557. $\int (\cosh mx)(\cosh nx) \, dx = \dfrac{\sinh{(m+n)x}}{2(m+n)} + \dfrac{\sinh{(m-n)x}}{2(m-n)}, \quad m^2 \neq n^2.$

558. $\int (\sinh mx)(\cosh nx) \, dx = \dfrac{\cosh{(m+n)x}}{2(m+n)} + \dfrac{\cosh{(m-n)x}}{2(m-n)}, \quad m^2 \neq n^2.$

559. $\int \left(\sinh^{-1} \dfrac{x}{a} \right) dx = x \sinh^{-1} \dfrac{x}{a} - \sqrt{x^2 + a^2}, \quad a > 0.$

560. $\int x \left(\sinh^{-1} \dfrac{x}{a} \right) dx = \left(\dfrac{x^2}{2} + \dfrac{a^2}{4} \right) \sinh^{-1} \dfrac{x}{a} - \dfrac{x}{4} \sqrt{x^2 + a^2}, \quad a > 0.$

561. $\int x^n \sinh^{-1} x \, dx = \dfrac{x^{n+1}}{n+1} \sinh^{-1} x - \dfrac{1}{n+1} \int \dfrac{x^{n+1}}{\sqrt{1+x^2}} \, dx, \quad n \neq -1.$

562. $\int^z \cosh^{-1} \dfrac{x}{a} \, dx = \begin{cases} z \cosh^{-1} \dfrac{z}{a} - \sqrt{z^2 - a^2}, & \cosh^{-1} \dfrac{z}{a} > 0, \\ \text{or} \\ z \cosh^{-1} \dfrac{z}{a} + \sqrt{z^2 - a^2}, & \cosh^{-1} \dfrac{z}{a} < 0, \ a > 0. \end{cases}$

563. $\int x \left(\cosh^{-1} \dfrac{x}{a} \right) dx = \left(\dfrac{x^2}{2} - \dfrac{a^2}{4} \right) \cosh^{-1} \dfrac{x}{a} - \dfrac{x}{4} \sqrt{x^2 - a^2}.$

564. $\int x^n \cosh^{-1} x \, dx = \dfrac{x^{n+1}}{n+1} \cosh^{-1} x - \dfrac{1}{n+1} \int \dfrac{x^{n+1}}{\sqrt{x^2 - 1}} \, dx, \quad n \neq -1.$

565. $\int \left(\tanh^{-1} \dfrac{x}{a} \right) dx = x \tanh^{-1} \dfrac{x}{a} + \dfrac{a}{2} \log{(a^2 - x^2)}, \quad \left| \dfrac{x}{a} \right| < 1.$

566. $\int \left(\coth^{-1} \dfrac{x}{a} \right) dx = x \coth^{-1} \dfrac{x}{a} + \dfrac{a}{2} \log{(x^2 - a^2)}, \quad \left| \dfrac{x}{a} \right| > 1.$

567. $\int x \left(\tanh^{-1} \dfrac{x}{a} \right) dx = \dfrac{x^2 - a^2}{2} \tanh^{-1} \dfrac{x}{a} + \dfrac{ax}{2}, \quad \left| \dfrac{x}{a} \right| < 1.$

568. $\int x^n \tanh^{-1} x \, dx = \dfrac{x^{n+1}}{n+1} \tanh^{-1} x - \dfrac{1}{n+1} \int \dfrac{x^{n+1}}{1 - x^2} \, dx, \quad n \neq -1.$

569. $\int x \left(\coth^{-1} \dfrac{x}{a} \right) dx = \dfrac{x^2 - a^2}{2} \coth^{-1} \dfrac{x}{a} + \dfrac{ax}{2}, \quad \left| \dfrac{x}{a} \right| > 1.$

570. $\int x^n \coth^{-1} x \, dx = \dfrac{x^{n+1}}{n+1} \coth^{-1} x + \dfrac{1}{n+1} \int \dfrac{x^{n+1}}{x^2 - 1} \, dx, \quad n \neq -1.$

571. $\int \operatorname{sech}^{-1} x \, dx = x \operatorname{sech}^{-1} x + \sin^{-1} x.$

572. $\int x \operatorname{sech}^{-1} x \, dx = \dfrac{x^2}{2} \operatorname{sech}^{-1} x - \dfrac{1}{2} \sqrt{1 - x^2}.$

573. $\int x^n \operatorname{sech}^{-1} x \, dx = \dfrac{x^{n+1}}{n+1} \operatorname{sech}^{-1} x + \dfrac{1}{n+1} \int \dfrac{x^n}{\sqrt{1 - x^2}} \, dx, \quad n \neq -1.$

574. $\int \operatorname{csch}^{-1} x \, dx = x \operatorname{csch}^{-1} x + \dfrac{x}{|x|} \sinh^{-1} x.$

575. $\int x \operatorname{csch}^{-1} x \, dx = \dfrac{x^2}{2} \operatorname{csch}^{-1} x + \dfrac{1}{2} \dfrac{x}{|x|} \sqrt{1 + x^2}.$

576. $\displaystyle\int x^n \operatorname{csch}^{-1} x\, dx = \frac{x^{n+1}}{n+1} \operatorname{csch}^{-1} x + \frac{1}{n+1}\frac{x}{|x|}\int \frac{x^n}{\sqrt{1+x^2}}\, dx,\ \ n \neq -1.$

5.4.21 BESSEL FUNCTIONS

$Z_p(x)$ represents any of the Bessel functions $\{J_p(x),\, Y_p(x),\, K_p(x),\, I_p(x)\}$.

577. $\displaystyle\int x^{p+1} Z_p(x)\, dx = x^{p+1} Z_{p+1}(x).$

578. $\displaystyle\int x^{-p+1} Z_p(x)\, dx = -x^{-p+1} Z_{p-1}(x).$

579. $\displaystyle\int x\, [Z_p(ax)]^2\, dx = \frac{x^2}{2}\left[[Z_p(ax)]^2 - Z_{p-1}(ax)Z_{p+1}(ax)\right].$

580. $\displaystyle\int Z_1(x)\, dx = -Z_0(x).$

581. $\displaystyle\int x Z_0(x)\, dx = x Z_1(x).$

5.5 TABLE OF DEFINITE INTEGRALS

582. $\displaystyle\int_0^\infty x^{n-1} e^{-x}\, dx = \Gamma(n), \quad \operatorname{Re} n > 0.$

583. $\displaystyle\int_0^\infty x^n p^{-x}\, dx = \frac{n!}{(\log p)^{n+1}}, \quad p > 0,\ n \text{ is a non-negative integer}.$

584. $\displaystyle\int_0^\infty x^{n-1} e^{-(a+1)x}\, dx = \frac{\Gamma(n)}{(a+1)^n}, \quad n > 0,\ a > -1.$

585. $\displaystyle\int_0^1 x^m \left(\log \frac{1}{x}\right)^n dx = \frac{\Gamma(n+1)}{(m+1)^{n+1}}, \quad m > -1,\ n > -1.$

586. $\displaystyle\int_0^1 x^{m-1}(1-x)^{n-1}\, dx = \int_0^\infty \frac{x^{m-1}}{(1+x)^{m+n}} = \frac{\Gamma(m)\Gamma(n)}{\Gamma(m+n)}, \quad n > 0,\ m > 0.$

587. $\displaystyle\int_a^b (x-a)^m(b-x)^n\, dx = (b-a)^{m+n+1}\frac{\Gamma(m+1)\Gamma(n+1)}{\Gamma(m+n+2)},$
$$m > -1,\ n > -1,\ b > a.$$

588. $\displaystyle\int_1^\infty \frac{dx}{x^m} = \frac{1}{m-1}, \quad m > 1.$

589. $\displaystyle\int_0^\infty \frac{dx}{(1+x)x^p} = \pi \csc p\pi, \quad 0 < p < 1.$

590. $\displaystyle\int_0^\infty \frac{dx}{(1-x)x^p} = -\pi \cot p\pi, \quad 0 < p < 1.$

591. $\displaystyle\int_0^1 \frac{x^p}{(1-x)^p}\, dx = p\pi \csc p\pi, \quad |p| < 1.$

592. $\displaystyle\int_0^1 \frac{x^p}{(1-x)^{p+1}}\, dx = \int_0^1 \frac{(1-x)^p}{x^{p+1}}\, dx = -\pi \operatorname{cosec} p\pi, \quad -1 < p < 0.$

593. $\displaystyle\int_0^\infty \frac{x^{p-1}}{1+x}\, dx = \frac{\pi}{\sin p\pi}, \quad 0 < p < 1.$

594. $\displaystyle\int_0^\infty \frac{x^{m-1}}{1+x^n}\,dx = \frac{\pi}{n\sin\frac{m\pi}{n}},\quad 0 < m < n.$

595. $\displaystyle\int_0^\infty \frac{x^a}{(m+x^b)^c}\,dx = \frac{m^{(a+1-bc)/b}}{b}\cdot\frac{\Gamma\left(\frac{a+1}{b}\right)\Gamma\left(c-\frac{a+1}{b}\right)}{\Gamma(c)},$
$$a > -1,\ b > 0,\ m > 0,\ c > \tfrac{a+1}{b}.$$

596. $\displaystyle\int_0^\infty \frac{dx}{(1+x)\sqrt{x}} = \pi.$

597. $\displaystyle\int_0^\infty \frac{a}{a^2+x^2}\,dx = \begin{cases} \dfrac{\pi}{2}, & a > 0, \\ \text{or} \\ 0, & a = 0, \\ \text{or} \\ -\dfrac{\pi}{2}, & a < 0. \end{cases}$

598. $\displaystyle\int_0^a \left(a^2-x^2\right)^{n/2}\,dx = \int_{-a}^a \frac{1}{2}\left(a^2-x^2\right)^{n/2}\,dx = \frac{n!!}{(n+1)!!}\frac{\pi}{2}a^{n+1},$
$$a > 0,\ n \text{ is an odd integer}.$$

599. $\displaystyle\int_0^a x^m\left(a^2-x^2\right)^{n/2}\,dx = \frac{1}{2}a^{m+n+1}\frac{\Gamma\left(\frac{m+1}{2}\right)\Gamma\left(\frac{n+2}{2}\right)}{\Gamma\left(\frac{m+n+3}{2}\right)},\quad a > 0,\ m > -1,\ n > -2.$

600. $\displaystyle\int_0^{\pi/2}\sin^n x\,dx = \int_0^{\pi/2}\cos^n x\,dx = \begin{cases} \dfrac{\sqrt{\pi}\,\Gamma\left(\frac{n+1}{2}\right)}{2\,\Gamma\left(\frac{n+2}{2}\right)}, & n > -1, \\ \text{or} \\ \dfrac{(n-1)!!}{n!!}\dfrac{\pi}{2}, & n \neq 0,\ n \text{ is an even integer}, \\ \text{or} \\ \dfrac{(n-1)!!}{n!!}, & n \neq 1,\ n \text{ is an odd integer}. \end{cases}$

601. $\displaystyle\int_0^\infty \frac{\sin ax}{x}\,dx = \begin{cases} \dfrac{\pi}{2}, & a > 0, \\ \text{or} \\ 0, & a = 0, \\ \text{or} \\ -\dfrac{\pi}{2}, & a < 0. \end{cases}$

602. $\displaystyle\int_0^\infty \frac{\cos x}{x}\,dx = \infty.$

603. $\displaystyle\int_0^\infty \frac{\tan x}{x}\,dx = \frac{\pi}{2}.$

604. $\displaystyle\int_0^\infty \frac{\tan ax}{x}\,dx = \frac{\pi}{2},\quad a > 0.$

605. $\displaystyle\int_0^\pi \sin(nx)\sin(mx)\,dx = \int_0^\pi \cos(nx)\cos(mx)\,dx = 0,$
$$n \neq m,\ n \text{ is an integer},\ m \text{ is an integer}.$$

606. $\displaystyle\int_0^{\pi/n}\sin(nx)\cos(nx)\,dx = \int_0^\pi \sin(nx)\cos(nx)\,dx = 0,\quad n \text{ is an integer}.$

607. $\displaystyle\int_0^\pi \sin ax\cos bx\,dx = \begin{cases} \dfrac{2a}{a^2-b^2}, & a-b \text{ is an odd integer}. \\ \text{or} \\ 0, & a-b \text{ is an even integer}. \end{cases}$

608. $\displaystyle\int_0^\infty \frac{\sin x \cos ax}{x}\, dx = \begin{cases} 0, & |a| > 1, \\ & \text{or} \\ \dfrac{\pi}{4}, & |a| = 1, \\ & \text{or} \\ \dfrac{\pi}{2}, & |a| < 1. \end{cases}$

609. $\displaystyle\int_0^\infty \frac{\sin ax \sin bx}{x^2}\, dx = \begin{cases} \dfrac{\pi a}{2}, & 0 < a \le b, \\ & \text{or} \\ \dfrac{\pi b}{2}, & 0 < b \le a. \end{cases}$

610. $\displaystyle\int_0^\pi \sin^2 mx\, dx = \int_0^\pi \cos^2 mx\, dx = \frac{\pi}{2}, \quad m \text{ is an integer.}$

611. $\displaystyle\int_0^\infty \frac{\sin^2 px}{x^2}\, dx = \frac{\pi\, |p|}{2}.$

612. $\displaystyle\int_0^\infty \frac{\sin x}{x^p}\, dx = \frac{\pi}{2\Gamma(p)\sin(p\pi/2)}, \quad 0 < p < 1.$

613. $\displaystyle\int_0^\infty \frac{\cos x}{x^p}\, dx = \frac{\pi}{2\Gamma(p)\cos(p\pi/2)}, \quad 0 < p < 1.$

614. $\displaystyle\int_0^\infty \frac{1 - \cos px}{x^2}\, dx = \frac{\pi\, |p|}{2}.$

615. $\displaystyle\int_0^\infty \frac{\sin px \cos qx}{x}\, dx = \begin{cases} 0, & q > p > 0, \\ & \text{or} \\ \dfrac{\pi}{2}, & p > q > 0, \\ & \text{or} \\ \dfrac{\pi}{4}, & p = q > 0. \end{cases}$

616. $\displaystyle\int_0^\infty \frac{\cos mx}{x^2 + a^2}\, dx = \frac{\pi}{2\,|a|}e^{-|ma|}.$

617. $\displaystyle\int_0^\infty \cos x^2\, dx = \int_0^\infty \sin x^2\, dx = \frac{1}{2}\sqrt{\frac{\pi}{2}}.$

618. $\displaystyle\int_0^\infty \sin(ax^n)\, dx = \frac{1}{na^{1/n}}\Gamma\left(\frac{1}{n}\right)\sin\frac{\pi}{2n}, \quad n > 1.$

619. $\displaystyle\int_0^\infty \cos(ax^n)\, dx = \frac{1}{na^{1/n}}\Gamma\left(\frac{1}{n}\right)\cos\frac{\pi}{2n}, \quad n > 1.$

620. $\displaystyle\int_0^\infty \frac{\sin x}{\sqrt{x}}\, dx = \int_0^\infty \frac{\cos x}{\sqrt{x}}\, dx = \sqrt{\frac{\pi}{2}}.$

621. $\displaystyle\int_0^\infty \frac{\sin^3 x}{x}\, dx = \frac{\pi}{4}.$

622. $\displaystyle\int_0^\infty \frac{\sin^3 x}{x^2}\, dx = \frac{3}{4}\log 3.$

623. $\displaystyle\int_0^\infty \frac{\sin^3 x}{x^3}\, dx = \frac{3\pi}{8}.$

624. $\displaystyle\int_0^\infty \frac{\sin^4 x}{x^4}\, dx = \frac{\pi}{3}.$

625. $\displaystyle\int_0^{\pi/2} \frac{dx}{1 + a\cos x}\, dx = \frac{\cos^{-1} a}{\sqrt{1 - a^2}}, \quad |a| < 1.$

626. $\displaystyle\int_0^\pi \frac{dx}{a + b\cos x}\, dx = \frac{\pi}{\sqrt{a^2 - b^2}}, \quad a > b \geq 0.$

627. $\displaystyle\int_0^{2\pi} \frac{dx}{1 + a\cos x}\, dx = \frac{2\pi}{\sqrt{1 - a^2}}, \quad |a| < 1.$

628. $\displaystyle\int_0^\infty \frac{\cos ax - \cos bx}{x}\, dx = \log\left|\frac{b}{a}\right|.$

629. $\displaystyle\int_0^{\pi/2} \frac{dx}{a^2 \sin^2 x + b^2 \cos^2 x}\, dx = \frac{\pi}{2\,|ab|}.$

630. $\displaystyle\int_0^{\pi/2} \frac{dx}{(a^2 \sin^2 x + b^2 \cos^2 x)^2}\, dx = \frac{\pi(a^2 + b^2)}{4a^3 b^3}, \quad a > 0,\ b > 0.$

631. $\displaystyle\int_0^{\pi/2} \sin^{n-1} x \cos^{m-1} x\, dx = \frac{1}{2} B\left(\frac{n}{2}\right)\frac{m}{2},$

$\qquad\qquad\qquad\qquad\qquad$ m is a positive integer, n is a positive integer.

632. $\displaystyle\int_0^{\pi/2} \sin^{2n+1} x\, dx = \frac{(2n)!!}{(2n+1)!!}, \quad n$ is a positive integer.

633. $\displaystyle\int_0^{\pi/2} \sin^{2n} x\, dx = \frac{(2n-1)!!}{(2n)!!}\frac{\pi}{2}, \quad n$ is a positive integer.

634. $\displaystyle\int_0^{\pi/2} \frac{x}{\sin x}\, dx = 2\left(\frac{1}{1^2} - \frac{1}{3^2} + \frac{1}{5^2} - \frac{1}{7^2} + \cdots\right).$

635. $\displaystyle\int_0^{\pi/2} \frac{dx}{1 + \tan^m x}\, dx = \frac{\pi}{4}, \quad m$ is a non-negative integer.

636. $\displaystyle\int_0^{\pi/2} \sqrt{\cos x}\, dx = \frac{(2\pi)^{3/2}}{(\Gamma(1/4))^2}.$

637. $\displaystyle\int_0^{\pi/2} \tan^h x\, dx = \frac{\pi}{2\cos\left(\frac{h\pi}{2}\right)}, \quad 0 < h < 1.$

638. $\displaystyle\int_0^{\pi/2} \frac{\tan^{-1} ax - \tan^{-1} bx}{x}\, dx = \frac{\pi}{2}\log\frac{a}{b}, \quad a > 0,\ b > 0.$

639. $\displaystyle\int_0^\infty e^{-ax}\, dx = \frac{1}{a}, \quad a > 0.$

640. $\displaystyle\int_0^\infty \frac{e^{-ax} - e^{-bx}}{x}\, dx = \log\frac{b}{a}, \quad a > 0,\ b > 0.$

641. $\displaystyle\int_0^\infty x^n e^{-ax}\, dx = \begin{cases} \dfrac{\Gamma(n+1)}{a^{n+1}}, & a > 0,\ n > -1, \\[2mm] \text{or} \\[2mm] \dfrac{n!}{a^{n+1}}, & a > 0,\ n \text{ is a positive integer.} \end{cases}$

642. $\displaystyle\int_0^\infty x^n e^{-ax^p}\, dx = \frac{\Gamma((n+1)/p)}{pa^{(n+1)/p}}, \quad a > 0,\ p > 0,\ n > -1.$

643. $\displaystyle\int_0^\infty e^{-a^2 x^2}\, dx = \frac{1}{2a}\sqrt{\pi}, \quad a > 0.$

644. $\displaystyle\int_0^b e^{-ax^2}\, dx = \frac{1}{2}\sqrt{\frac{\pi}{a}}\, \mathrm{erf}\left(b\sqrt{a}\right), \quad a > 0.$

645. $\displaystyle\int_b^\infty e^{-ax^2}\, dx = \frac{1}{2}\sqrt{\frac{\pi}{a}}\, \mathrm{erfc}\left(b\sqrt{a}\right), \quad a > 0.$

646. $\displaystyle\int_0^\infty xe^{-x^2}\,dx = \frac{1}{2}.$

647. $\displaystyle\int_0^\infty x^2 e^{-x^2}\,dx = \frac{\sqrt{\pi}}{4}.$

648. $\displaystyle\int_0^\infty x^{2n} e^{-ax^2}\,dx = \frac{(2n-1)!!}{2(2a)^n}\sqrt{\frac{\pi}{a}}, \quad a>0,\ n>0.$

649. $\displaystyle\int_0^\infty x^{2n+1} e^{-ax^2}\,dx = \frac{n!}{2a^{n+1}}, \quad a>0,\ n>-1.$

650. $\displaystyle\int_0^1 x^m e^{-ax}\,dx = \frac{m!}{a^{m+1}}\left[1 - e^{-a}\sum_{r=0}^m \frac{a^r}{r!}\right].$

651. $\displaystyle\int_0^\infty e^{\left(-x^2 - a^2/x^2\right)}\,dx = \frac{e^{-2|a|}\sqrt{\pi}}{2}.$

652. $\displaystyle\int_0^\infty e^{\left(-ax^2 - b/x^2\right)}\,dx = \frac{1}{2}\sqrt{\frac{\pi}{a}}\,e^{-2\sqrt{ab}}, \quad a>0,\ b>0.$

653. $\displaystyle\int_0^\infty \sqrt{x}\,e^{-ax}\,dx = \frac{1}{2a}\sqrt{\frac{\pi}{a}}, \quad a>0.$

654. $\displaystyle\int_0^\infty \frac{e^{-ax}}{\sqrt{x}}\,dx = \sqrt{\frac{\pi}{a}}, \quad a>0.$

655. $\displaystyle\int_0^\infty e^{-ax}\cos mx\,dx = \frac{a}{a^2+m^2}, \quad a>0.$

656. $\displaystyle\int_0^\infty e^{-ax}\cos(bx+c)\,dx = \frac{a\cos c - b\sin c}{a^2+b^2}, \quad a>0.$

657. $\displaystyle\int_0^\infty e^{-ax}\sin mx\,dx = \frac{m}{a^2+m^2}, \quad a>0.$

658. $\displaystyle\int_0^\infty e^{-ax}\sin(bx+c)\,dx = \frac{b\cos c + a\sin c}{a^2+b^2}, \quad a>0.$

659. $\displaystyle\int_0^\infty xe^{-ax}\sin bx\,dx = \frac{2ab}{(a^2+b^2)^2}, \quad a>0.$

660. $\displaystyle\int_0^\infty xe^{-ax}\cos bx\,dx = \frac{a^2-b^2}{(a^2+b^2)^2}, \quad a>0.$

661. $\displaystyle\int_0^\infty x^n e^{-ax}\sin bx\,dx = \frac{n!\left[(a+ib)^{n+1} - (a-ib)^{n+1}\right]}{2i(a^2+b^2)^{n+1}}, \quad a>0.$

662. $\displaystyle\int_0^\infty x^n e^{-ax}\cos bx\,dx = \frac{n!\left[(a-ib)^{n+1} + (a+ib)^{n+1}\right]}{2(a^2+b^2)^{n+1}}, \quad a>0,\ n>-1.$

663. $\displaystyle\int_0^\infty \frac{e^{-ax}\sin x}{x}\,dx = \cot^{-1} a, \quad a>0.$

664. $\displaystyle\int_0^\infty e^{-a^2 x^2}\cos bx\,dx = \frac{\sqrt{\pi}}{2\,|a|}\exp^{-b^2/(4a^2)}, \quad ab>0.$

665. $\displaystyle\int_0^\infty e^{-x\cos\phi}x^{b-1}\sin(x\sin\phi)\,dx = \Gamma(b)\sin(b\phi), \quad b>0,\ -\frac{\pi}{2}<\phi<\frac{\pi}{2}.$

666. $\displaystyle\int_0^\infty e^{-x\cos\phi}x^{b-1}\cos(x\sin\phi)\,dx = \Gamma(b)\cos(b\phi), \quad b>0,\ -\frac{\pi}{2}<\phi<\frac{\pi}{2}.$

667. $\displaystyle\int_0^\infty x^{b-1}\cos x\,dx = \Gamma(b)\cos\left(\frac{b\pi}{2}\right), \quad 0<b<1.$

668. $\displaystyle\int_0^\infty x^{b-1}\sin x\,dx = \Gamma(b)\sin\left(\frac{b\pi}{2}\right),\quad 0 < b < 1.$

669. $\displaystyle\int_0^1 (\log x)^n\,dx = (-1)^n n!,\quad n > -1.$

670. $\displaystyle\int_0^1 \sqrt{\log\frac{1}{x}}\,dx = \frac{\sqrt{\pi}}{2}.$

671. $\displaystyle\int_0^1 \left(\log\frac{1}{x}\right)^n\,dx = n!.$

672. $\displaystyle\int_0^1 x\log(1-x)\,dx = -\frac{3}{4}.$

673. $\displaystyle\int_0^1 x\log(1+x)\,dx = \frac{1}{4}.$

674. $\displaystyle\int_0^1 x^m(\log x)^n\,dx = \frac{(-1)^n\Gamma(n+1)}{(m+1)^{m+1}},\quad m > -1,\ n \text{ is a non-negative integer.}$

675. $\displaystyle\int_0^1 \frac{\log x}{1+x}\,dx = -\frac{\pi^2}{12}.$

676. $\displaystyle\int_0^1 \frac{\log x}{1-x}\,dx = -\frac{\pi^2}{6}.$

677. $\displaystyle\int_0^1 \frac{\log(1+x)}{x}\,dx = \frac{\pi^2}{12}.$

678. $\displaystyle\int_0^1 \frac{\log(1-x)}{x}\,dx = -\frac{\pi^2}{6}.$

679. $\displaystyle\int_0^1 (\log x)\log(1+x)\,dx = 2 - 2\log 2 - \frac{\pi^2}{12}.$

680. $\displaystyle\int_0^1 (\log x)\log(1-x)\,dx = 2 - \frac{\pi^2}{6}.$

681. $\displaystyle\int_0^1 \frac{\log x}{1-x^2}\,dx = -\frac{\pi^2}{8}.$

682. $\displaystyle\int_0^1 \log\left(\frac{1+x}{1-x}\right)\frac{dx}{x} = \frac{\pi^2}{4}.$

683. $\displaystyle\int_0^1 \frac{\log x}{\sqrt{1-x^2}}\,dx = -\frac{\pi}{2}\log 2.$

684. $\displaystyle\int_0^1 x^m\left[\log\left(\frac{1}{x}\right)\right]^n\,dx = \frac{\Gamma(n+1)}{(m+1)^{n+1}},\quad m > -1,\ n > -1.$

685. $\displaystyle\int_0^1 \frac{x^p - x^q}{\log x}\,dx = \log\left(\frac{p+1}{q+1}\right),\quad p > -1,\ q > -1.$

686. $\displaystyle\int_0^1 \frac{dx}{\sqrt{\log(-\log x)}} = \sqrt{\pi}.$

687. $\displaystyle\int_0^\infty \log\left(\frac{e^x + 1}{e^x - 1}\right)\,dx = \frac{\pi^2}{4}.$

688. $\displaystyle\int_0^{\pi/2} \log\sin x\,dx = \int_0^{\pi/2} \log\cos x\,dx = -\frac{\pi}{2}\log 2.$

689. $\displaystyle\int_0^{\pi/2} \log\sec x\,dx = \int_0^{\pi/2} \log\csc x\,dx = \frac{\pi}{2}\log 2.$

690. $\displaystyle\int_0^\pi x \log \sin x \, dx = -\frac{\pi^2}{2} \log 2.$

691. $\displaystyle\int_0^{\pi/2} (\sin x) \log \sin x \, dx = \log 2 - 1.$

692. $\displaystyle\int_0^{\pi/2} \log \tan x \, dx = 0.$

693. $\displaystyle\int_0^\pi \log (a \pm b \cos x) \, dx = \pi \log \left(\frac{a + \sqrt{a^2 - b^2}}{2} \right), \quad a \geq b.$

694. $\displaystyle\int_0^\pi \log (a^2 - 2ab \cos x + b^2) \, dx = \begin{cases} 2\pi \log a, & a \geq b > 0, \\ \text{or} \\ 2\pi \log b, & b \geq a > 0. \end{cases}$

695. $\displaystyle\int_0^\infty \frac{\sin ax}{\sinh bx} \, dx = \frac{\pi}{2b} \tanh \frac{a\pi}{2 |b|}.$

696. $\displaystyle\int_0^\infty \frac{\cos ax}{\cosh bx} \, dx = \frac{\pi}{2b} \operatorname{sech} \frac{a\pi}{2b}.$

697. $\displaystyle\int_0^\infty \frac{dx}{\cosh ax} = \frac{\pi}{2 |a|}.$

698. $\displaystyle\int_0^\infty \frac{x}{\sinh ax} \, dx = \frac{\pi^2}{4a^2}, \quad a \geq 0.$

699. $\displaystyle\int_0^\infty e^{-ax} \cosh (bx) \, dx = \frac{a}{a^2 - b^2}, \quad |b| < a.$

700. $\displaystyle\int_0^\infty e^{-ax} \sinh (bx) \, dx = \frac{b}{a^2 - b^2}, \quad |b| < a.$

701. $\displaystyle\int_0^\infty \frac{\sinh ax}{e^{bx} + 1} \, dx = \frac{\pi}{2b} \csc \frac{a\pi}{b} - \frac{1}{2a}, \quad b \geq 0.$

702. $\displaystyle\int_0^\infty \frac{\sinh ax}{e^{bx} - 1} \, dx = \frac{1}{2a} - \frac{\pi}{2b} \cot \frac{a\pi}{b}, \quad b \geq 0.$

703. $\displaystyle\int_0^{\pi/2} \frac{dx}{\sqrt{1 - k^2 \sin^2 x}} = \frac{\pi}{2} \left[1 + \left(\frac{1}{2} \right)^2 k^2 + \left(\frac{1 \cdot 3}{2 \cdot 4} \right)^2 k^4 + \left(\frac{1 \cdot 3 \cdot 5}{2 \cdot 4 \cdot 6} \right)^2 k^6 + \ldots \right],$
$$k^2 < 1.$$

704. $\displaystyle\int_0^{\pi/2} \frac{dx}{(1 - k^2 \sin^2 x)^{3/2}} = \frac{\pi}{2} \left[1 + \left(\frac{1}{2} \right)^2 3k^2 + \left(\frac{1 \cdot 3}{2 \cdot 4} \right)^2 5k^4 \right.$
$$\left. + \left(\frac{1 \cdot 3 \cdot 5}{2 \cdot 4 \cdot 6} \right)^2 7k^6 + \ldots \right], \quad k^2 < 1.$$

705. $\displaystyle\int_0^{\pi/2} \sqrt{1 - k^2 \sin^2 x} \, dx = \frac{\pi}{2} \left[1 - \left(\frac{1}{2} \right)^2 k^2 - \left(\frac{1 \cdot 3}{2 \cdot 4} \right)^2 \frac{k^4}{3} \right.$
$$\left. - \left(\frac{1 \cdot 3 \cdot 5}{2 \cdot 4 \cdot 6} \right)^2 \frac{k^6}{5} - \ldots \right], \quad k^2 < 1.$$

706. $\displaystyle\int_0^\infty e^{-x} \log x \, dx = -\gamma.$

707. $\displaystyle\int_0^\infty e^{-x^2} \log x \, dx = -\frac{\sqrt{\pi}}{4} (\gamma + 2 \log 2).$

708. $\displaystyle\int_0^\infty \left(\frac{1}{1 - e^{-x}} - \frac{1}{x} \right) e^{-x} \, dx = \gamma.$

709. $\int_0^\infty \frac{1}{x}\left(\frac{1}{1-e^{-x}}-\frac{1}{x}\right)\,dx=\gamma.$

5.5.1 TABLE OF SEMI-INTEGRALS

f	$\dfrac{d^{-1/2}f}{dx^{-1/2}}$
(1) 0	0
(2) 1	$2\sqrt{\frac{x}{\pi}}$
(3) $x^{-1/2}$	$\sqrt{\pi}$
(4) x	$\frac{4x^{2/3}}{3\sqrt{\pi}}$
(5) $x^n,\ n=0,1,2,\dots$	$\frac{(n!)^2(4x)^{n+1/2}}{(2n+1)!\sqrt{\pi}}$
(6) $x^p,\ p>-1$	$\frac{\Gamma(p+1)}{\Gamma(p+\frac{3}{2})}x^{p+1/2}$
(7) $\sqrt{1+x}$	$\sqrt{\frac{x}{\pi}}+\frac{(1+x)\tan^{-1}\left(\sqrt{x}\right)}{\sqrt{\pi}}$
(8) $\frac{1}{\sqrt{1+x}}$	$\frac{2}{\sqrt{\pi}}\tan^{-1}\left(\sqrt{x}\right)$
(9) $\frac{1}{1+x}$	$\frac{2\sinh^{-1}\left(\sqrt{x}\right)}{\sqrt{\pi(1+x)}}$
(10) e^x	$e^x\,\mathrm{erf}\left(\sqrt{x}\right)$
(11) $e^x\,\mathrm{erf}\left(\sqrt{x}\right)$	e^x-1
(12) $\sin\left(\sqrt{x}\right)$	$\sqrt{\pi x}\,J_1\left(\sqrt{x}\right)$
(13) $\cos\left(\sqrt{x}\right)$	$\sqrt{\pi x}\,H_{-1}\left(\sqrt{x}\right)$
(14) $\sinh\left(\sqrt{x}\right)$	$\sqrt{\pi x}\,I_1\left(\sqrt{x}\right)$
(15) $\cosh\left(\sqrt{x}\right)$	$\sqrt{\pi x}\,L_{-1}\left(\sqrt{x}\right)$
(16) $\frac{\sin\left(\sqrt{x}\right)}{\sqrt{x}}$	$\sqrt{\pi}\,H_0\left(\sqrt{x}\right)$
(17) $\frac{\cos\left(\sqrt{x}\right)}{\sqrt{x}}$	$\sqrt{\pi}\,J_0\left(\sqrt{x}\right)$
(18) $\log x$	$2\sqrt{\frac{x}{\pi}}\left[\log(4x)-2\right]$
(19) $\frac{\log x}{\sqrt{x}}$	$\sqrt{\pi}\log\left(\frac{x}{4}\right)$

5.6 ORDINARY DIFFERENTIAL EQUATIONS

5.6.1 LINEAR DIFFERENTIAL EQUATIONS

A linear differential equation is one that can be written in the form

$$b_n(x)y^{(n)} + b_{n-1}(x)y^{(n-1)} + \cdots + b_1(x)y' + b_0(x)y = R(x) \qquad (5.6.1)$$

or $p(D)y = R(x)$, where D is the differentiation operator ($Dy = dy/dx$), $p(D)$ is a polynomial in D with coefficients $\{b_i\}$ depending on x, and $R(x)$ is an arbitrary function. In this notation, a power of D denotes repeated differentiation, that is, $D^n y = d^n y/dx^n$. For such an equation, the general solution has the form

$$y(x) = y_h(x) + y_p(x) \qquad (5.6.2)$$

where $y_h(x)$ is a homogeneous solution and $y_p(x)$ is the particular solution. These functions satisfy $p(D)y_h = 0$ and $p(D)y_p = R(x)$.

5.6.1.1 Vector representation

Equation (5.6.1) can be written in the form $\dfrac{d\mathbf{y}}{dx} = A(x)\mathbf{y} + \mathbf{r}(x)$ where

$$\mathbf{y} = \begin{bmatrix} y \\ y' \\ y'' \\ \vdots \\ y^{(n-1)} \end{bmatrix}, \quad A(x) = \begin{bmatrix} 0 & 1 & 0 & \cdots & 0 \\ 0 & 0 & 1 & & 0 \\ \vdots & \vdots & & \ddots & \\ 0 & 0 & 0 & & 1 \\ -\frac{b_0}{b_n} & -\frac{b_1}{b_n} & -\frac{b_2}{b_n} & \cdots & -\frac{b_{n-1}}{b_n} \end{bmatrix}, \quad \mathbf{r}(x) = \begin{bmatrix} 0 \\ 0 \\ \vdots \\ 0 \\ \frac{R}{b_n} \end{bmatrix}.$$

5.6.1.2 Homogeneous solution

For the special case of a linear differential equation with constant coefficients (i.e., the $\{b_i\}$ in Equation (5.6.1) are constants), the procedure for finding the homogeneous solution is as follows:

1. Factor the polynomial $p(D)$ into real and complex linear factors, just as if D were a variable instead of an operator.

2. For each non-repeated linear factor of the form $(D - a)$, where a is real, write a term of the form ce^{ax}, where c is an arbitrary constant.

3. For each repeated real linear factor of the form $(D - a)^m$, write the following sum of m terms

$$c_1 e^{ax} + c_2 x e^{ax} + c_3 x^2 e^{ax} + \cdots + c_m x^{m-1} e^{ax} \qquad (5.6.3)$$

where the c_i's are arbitrary constants.

4. For each non-repeated complex conjugate pair of factors of the form $(D - a + ib)(D - a - ib)$, write the following two terms

$$c_1 e^{ax} \cos bx + c_2 e^{ax} \sin bx. \tag{5.6.4}$$

5. For each repeated complex conjugate pair of factors of the form $(D - a + ib)^m (D - a - ib)^m$, write the following $2m$ terms

$$c_1 e^{ax} \cos bx + c_2 e^{ax} \sin bx + c_3 x e^{ax} \cos bx + c_4 x e^{ax} \sin bx + \ldots$$
$$+ c_{2m-1} x^{m-1} e^{ax} \cos bx + c_{2m} x^{m-1} e^{ax} \sin bx. \tag{5.6.5}$$

6. The sum of all the terms thus written is the homogeneous solution.

EXAMPLE   For the linear equation

$$y^{(7)} - 14 y^{(6)} + 81 y^{(5)} - 252 y^{(4)} + 455 y^{(3)} - 474 y'' + 263 y' - 60 y = 0,$$

$p(D)$ factors as $p(D) = (D - 1)^3 (D - (2 + i))(D - (2 - i))(D - 3)(D - 4)$. The roots are thus $\{1, 1, 1, 2 + i, 2 - i, 3, 4\}$. Hence, the homogeneous solution has the form

$$y_h(x) = \left(c_0 + c_1 x + c_2 x^2 \right) e^x + \left(c_3 \sin x + c_4 \cos x \right) e^{2x} + c_5 e^{3x} + c_6 e^{4x}$$

where $\{c_0, \ldots, c_6\}$ are arbitrary constants.

5.6.1.3   Particular solutions

The following are solutions for some specific ordinary differential equations. In these tables we assume that $P(x)$ is a polynomial of degree n and $\{a, b, p, q, r, s\}$ are constants. In all of these tables, when using "cos" instead of "sin" in $R(x)$, use the given result, but replace "sin" by "cos", and replace "cos" by "$-\sin$".
The numbers on the left hand side are for reference.

	If $R(x)$ is	A particular solution to $y' - ay = R(x)$ is
(1)	e^{rx}	$e^{rx}/(r - a)$.
(2)	$\sin sx$	$-\frac{a \sin sx + s \cos sx}{a^2 + s^2} = -\left(a^2 + s^2 \right)^{-1/2} \sin \left(sx + \tan^{-1} \frac{s}{a} \right)$.
(3)	$P(x)$	$-\frac{1}{a} \left[P(x) + \frac{P'(x)}{a} + \frac{P''(x)}{a^2} + \cdots + \frac{P^{(n)}(x)}{a^n} \right]$.
(4)	$e^{rx} \sin sx$	Replace a by $a - r$ in formula (2) and multiply by e^{rx}.
(5)	$P(x) e^{rx}$	Replace a by $a - r$ in formula (3) and multiply by e^{rx}.
(6)	$P(x) \sin sx$	$-\sin sx \left[\frac{a}{a^2 + s^2} P(x) + \frac{a^2 - s^2}{(a^2 + s^2)^2} P'(x) \right.$ $\left. + \cdots + \frac{a^k - \binom{k}{2} a^{k-2} s^2 + \binom{k}{4} a^{k-4} s^4 - \cdots}{(a^2 + s^2)^k} P^{(k-1)}(x) + \cdots \right]$ $-\cos sx \left[\frac{s}{a^2 + s^2} P(x) + \frac{2as}{(a^2 + s^2)^2} P'(x) \right.$ $\left. + \cdots + \frac{\binom{k}{1} a^{k-1} s - \binom{k}{3} a^{k-3} s^3 + \cdots}{(a^2 + s^2)^k} P^{(k-1)}(x) + \cdots \right]$.
(7)	$P(x) e^{rx} \sin sx$	Replace a by $a - r$ in formula (6) and multiply by e^{rx}.

continued on next page

		continued from last page
	If $R(x)$ is	A particular solution to $y' - ay = R(x)$ is
(8)	e^{ax}	xe^{ax}.
(9)	$e^{ax} \sin sx$	$-\dfrac{e^{ax} \cos sx}{s}$.
(10)	$P(x)e^{ax}$	$e^{ax} \int^x P(z)\, dz$.
(11)	$P(x)e^{ax} \sin sx$	$\dfrac{e^{ax} \sin sx}{s}\left[\dfrac{P'(x)}{s} - \dfrac{P'''(x)}{s^3} + \dfrac{P^{(5)}(x)}{s^5} + \cdots\right]$ $-\dfrac{e^{ax} \cos sx}{s}\left[P(x) - \dfrac{P''(x)}{s^2} + \dfrac{P^{(4)}(x)}{s^4} + \cdots\right]$.

	If $R(x)$ is	A particular solution to $y'' - 2ay' + a^2y = R(x)$ is
(12)	e^{rx}	$\dfrac{e^{rx}}{(r-a)^2}$.
(13)	$\sin sx$	$\dfrac{(a^2 - s^2)\sin sx + 2as \cos sx}{(a^2 + s^2)^2}$ $= \dfrac{1}{a^2 + s^2}\sin\left(sx + \tan^{-1}\dfrac{2as}{a^2 - s^2}\right)$.
(14)	$P(x)$	$\dfrac{1}{a^2}\left[P(x) + \dfrac{2P'(x)}{a} + \dfrac{3P''(x)}{a^2} + \cdots + \dfrac{(n+1)P^{(n)}(x)}{a^n}\right]$.
(15)	$e^{rx} \sin sx$	Multiply formula (13) by e^{rx} and replace a by $a - r$.
(16)	$P(x)e^{rx}$	Multiply formula (14) by e^{rx} and replace a by $a - r$.
(17)	$P(x) \sin sx$	$\sin sx\left[\dfrac{a^2-s^2}{(a^2+s^2)^2}P(x) + 2\dfrac{a^2-3as^2}{(a^2+s^2)^3}P'(x)\right.$ $\left.+\cdots+(k-1)\dfrac{a^k - \binom{k}{2}a^{k-2}s^2 + \binom{k}{4}a^{k-4}s^4 - \cdots}{(a^2+s^2)^k}P^{(k-2)}(x) + \cdots\right]$ $+ \cos sx\left[\dfrac{2as}{(a^2+s^2)^2}P(x) + 2\dfrac{3a^2s - s^3}{(a^2+s^2)^3}P'(x)\right.$ $\left.+\cdots+(k-1)\dfrac{\binom{k}{1}a^{k-1}s - \binom{k}{3}a^{k-3}s^3 + \cdots}{(a^2+s^2)^k}P^{(k-2)}(x) + \cdots\right]$.
(18)	$P(x)e^{rx} \sin sx$	Multiply formula (17) by e^{rx} and replace a by $a - r$.
(19)	e^{ax}	$\dfrac{x^2 e^{ax}}{2}$.
(20)	$e^{ax} \sin sx$	$-\dfrac{e^{ax} \sin sx}{s^2}$.
(21)	$P(x)e^{ax}$	$e^{ax} \int^x \int^y P(z)\, dz\, dy$.
(22)	$P(x)e^{ax} \sin sx$	$-\dfrac{e^{ax} \sin sx}{s^2}\left[P(x) - \dfrac{3P''(x)}{s^2} + \dfrac{5P^{(4)}(x)}{s^4} + \cdots\right]$ $-\dfrac{e^{ax} \cos sx}{s^2}\left[\dfrac{2P(x)}{s} - \dfrac{4P'''(x)}{s^3} + \dfrac{6P^{(5)}(x)}{s^5} + \cdots\right]$.

	If $R(x)$ is	A particular solution to $y'' + qy = R(x)$ is
(23)	e^{rx}	$\dfrac{e^{rx}}{r^2 + q}$.
(24)	$\sin sx$	$\dfrac{\sin sx}{q - s^2}$.
(25)	$P(x)$	$\dfrac{1}{q}\left[P(x) - \dfrac{P''(x)}{q} + \dfrac{P^{(4)}(x)}{q^2} + \cdots + (-1)^k \dfrac{P^{(2k)}(x)}{q^k} + \cdots\right]$.
(26)	$e^{rx}\sin sx$	$\dfrac{(r^2 - s^2 + q)e^{rx}\sin sx - 2rse^{rx}\cos sx}{(r^2 - s^2 + q)^2 + (2rs)^2}$
		$= \dfrac{e^{rx}}{\sqrt{(r^2 - s^2 + q)^2 + (2rs)^2}}\sin\left[sx - \tan^{-1}\dfrac{2rs}{r^2 - s^2 + q}\right]$.
(27)	$P(x)e^{rx}$	$\dfrac{e^{rx}}{q+r^2}\left[P(x) - \dfrac{2r}{q+r^2}P'(x) + \dfrac{3r^2-q}{(q+r^2)^2}P''(x)\right.$
		$\left. + \cdots + (-1)^{k-1}\dfrac{\binom{k}{1}r^{k-1} - \binom{k}{3}r^{k-3}q + \cdots}{(q+r^2)^{k-1}}P^{(k-1)}(x) + \cdots\right]$.
(28)	$P(x)\sin sx$	$\dfrac{\sin sx}{q-s^2}\left[P(x) - \dfrac{3s^2+q}{(q-s^2)^2}P''(x)\right.$
		$\left. + \cdots + (-1)^k\dfrac{\binom{2k+1}{1}s^{2k} + \binom{2k+1}{3}s^{2k-2}q + \cdots}{(q-s^2)^{2k}}P^{(2k)}(x) + \cdots\right]$
		$- \dfrac{s\cos sx}{q-s^2}\left[\dfrac{2P'(x)}{(q-s^2)} - \dfrac{4s^2+4q}{(q-s^2)^3}P'''(x)\right.$
		$\left. + \cdots + (-1)^{k+1}\dfrac{\binom{2k}{1}s^{2k-2} + \binom{2k}{3}s^{2k-4}q + \cdots}{(q-s^2)^{2k-1}}P^{(2k-1)}(x) + \cdots\right]$.

	If $R(x)$ is	A particular solution to $y'' + b^2y = R(x)$ is
(29)	$\sin bx$	$-\dfrac{x\cos bx}{2b}$.
(30)	$P(x)\sin bx$	$\dfrac{\sin bx}{(2b)^2}\left[P(x) - \dfrac{P''(x)}{(2b)^2} + \dfrac{P^{(4)}(x)}{(2b)^4} + \cdots\right]$
		$- \dfrac{\cos bx}{2b}\displaystyle\int\left[P(x) - \dfrac{P''(x)}{(2b)^2} + \cdots\right]dx.$

	If $R(x)$ is	A particular solution to $y'' + py' + qy = R(x)$ is
(31)	e^{rx}	$\dfrac{e^{rx}}{r^2 + pr + q}$.
(32)	$\sin sx$	$\dfrac{(q-s^2)\sin sx - ps\cos sx}{(q-s^2)^2 + (ps)^2} = \dfrac{1}{\sqrt{(q-s^2)^2 + (ps)^2}}\sin\left(sx - \tan^{-1}\dfrac{ps}{q-s^2}\right)$.
(33)	$P(x)$	$\dfrac{1}{q}\left[P(x) - \dfrac{p}{q}P'(x) + \dfrac{p^2-q}{q^2}P''(x) - \dfrac{p^2-2pq}{q^2}P'''(x)\right.$
		$\left. + \cdots + (-1)^n\dfrac{p^n - \binom{n-1}{1}p^{n-2}q + \binom{n-2}{2}p^{n-4}q^2 - \cdots}{q^n}P^{(n)}(x)\right]$.
(34)	$e^{rx}\sin sx$	Multiply formula (32) by e^{rx}, replace p by $p + 2r$, and replace q by $q + pr + r^2$.
(35)	$P(x)e^{rx}$	Multiply formula (33) by e^{rx}, replace p by $p + 2r$, and replace q by $q + pr + r^2$.

If $R(x)$ is	A particular solution to $(D-a)^n y = R(x)$ is
(36) e^{rx}	$e^{rx}/(r-a)^n$.
(37) $\sin sx$	$\frac{(-1)^n}{(a^2+s^2)^n}\left[\left(a^n - \binom{n}{2}a^{n-2}s^2 + \binom{n}{4}a^{n-4}s^4 - \dots\right)\sin sx + \left(\binom{n}{1}a^{n-1}s + \binom{n}{3}a^{n-3}s^3 + \dots\right)\cos sx\right].$
(38) $P(x)$	$\frac{(-1)^n}{a^n}\left[P(x) + \binom{n}{1}\frac{P'(x)}{a} + \binom{n+1}{2}\frac{P''(x)}{a^2} + \binom{n+2}{3}\frac{P'''(x)}{a^3} + \dots\right].$
(39) $e^{rx}\sin sx$	Multiply formula (37) by e^{rx} and replace a by $a - r$.
(40) $P(x)e^{rx}$	Multiply formula (38) by e^{rx} and replace a by $a - r$.

5.6.1.4 Second-order linear constant coefficient equation

Consider $ay'' + by' + cy = 0$, where a, b, and c are real constants. Let m_1 and m_2 be the roots of $am^2 + bm + c = 0$. There are three forms of the solution:

1. If m_1 and m_2 are real and distinct, then $y(x) = c_1 e^{m_1 x} + c_2 e^{m_2 x}$
2. If m_1 and m_2 are real and equal, then $y(x) = c_1 e^{m_1 x} + c_2 x e^{m_1 x}$
3. If $m_1 = p + iq$ and $m_2 = p - iq$ (with $p = -b/2a$ and $q = \sqrt{4ac - b^2}/2a$), then $y(x) = e^{px}(c_1 \cos qx + c_2 \sin qx)$

Consider $ay'' + by' + cy = R(x)$, where a, b, and c are real constants. Let m_1 and m_2 be as above.

1. If m_1 and m_2 are real and distinct, then $y(x) = C_1 e^{m_1 x} + C_2 e^{m_2 x}$
   $+e^{m_1 x}/(m_1 - m_2)\int^x e^{-m_1 z}R(z)\,dz + e^{m_2 x}/(m_2 - m_1)\int^x e^{-m_2 z}R(z)\,dz.$
2. If m_1 and m_2 are real and equal, then $y(x) = C_1 e^{m_1 x} + C_2 x e^{m_1 x}$
   $+x e^{m_1 x}\int^x e^{-m_1 z}R(z)\,dz - e^{m_1 x}\int^x z e^{-m_1 z}R(z)\,dz.$
3. If $m_1 = p + iq$ and $m_2 = p - iq$, then $y(x) = e^{px}(c_1 \cos qx + c_2 \sin qx) + e^{px}\sin qx/q\int^x e^{-pz}R(z)\cos qz\,dz - e^{px}\cos qx/q\int^x e^{-pz}R(z)\sin qz\,dz.$

5.6.1.5 Damping: none, under, over, and critical

Consider the linear ordinary differential equation $x'' + \mu x' + x = 0$. If the damping coefficient μ is positive, then all solutions decay to $x = 0$. If $\mu = 0$, the system is undamped and the solution oscillates without decaying. The value of μ such that the roots of the characteristic equation $\lambda^2 + \mu\lambda + 1 = 0$ are real and equal is the critical damping coefficient. If μ is less than (greater than) the critical damping coefficient, then the system is under (over) damped.

Consider four cases with the same initial values: $y(0) = 1$ and $y'(0) = 0$.

1. $y'' + 3y' + y = 0$ Overdamped
2. $y'' + 2y' + y$ Critically damped
3. $y'' + 0.2y' + y = 0$ Underdamped
4. $y'' + y = 0$ Undamped

Illustration of different types of damping.

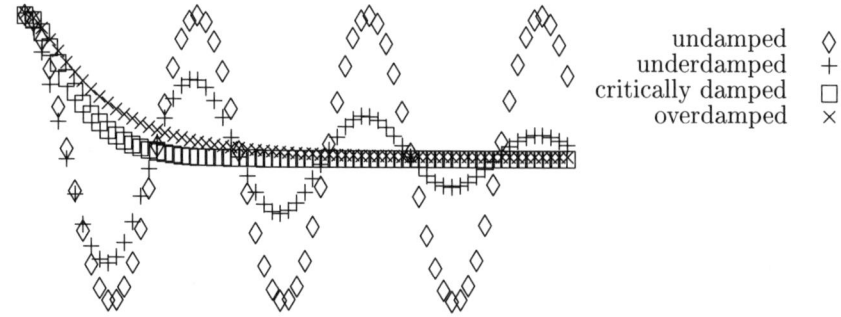

undamped	◇
underdamped	+
critically damped	☐
overdamped	×

5.6.2 SOLUTION TECHNIQUES OVERVIEW

Differential equation	Solution or solution technique
Autonomous equation $f(y^{(n)}, y^{(n-1)}, \ldots, y'', y', y) = 0$	Change dependent variable to $u(y) = y'(x)$
Bernoulli's equation $y' + f(x)y = g(x)y^n$	Change dependent variable to $v(x) = (y(x))^{1-n}$
Clairaut's equation $f(xy' - y) = g(y')$	One solution is $f(xC - y) = g(C)$
Constant coefficient equation $a_0 y^{(n)} + a_1 y^{(n-1)} + \ldots$ $+ a_{n-1}y' + a_n y = 0$	There are solutions of the form $y = x^k e^{\lambda x}$. See Section 5.6.1.2.
Dependent variable missing $f(y^{(n)}, y^{(n-1)}, \ldots, y'', y', x) = 0$	Change dependent variable to $u(x) = y'(x)$
Euler's equation $a_0 x^n y^{(n)} + a_1 x^{n-1} y^{(n-1)} + \ldots$ $+ a_{n-1}xy' + a_n y = 0$	Change independent variable to $x = e^t$
Exact equation $M(x, y)\,dx + N(x, y)\,dy = 0$ with $\dfrac{\partial M}{\partial y} = \dfrac{\partial N}{\partial x}$	Integrate $M(x, y)$ with respect to x holding y constant, call this $m(x, y)$. Then $m(x, y) + \displaystyle\int \left(N - \frac{\partial m}{\partial y}\right) dy = C$
Homogeneous equation $y' = f\left(\dfrac{y}{x}\right)$	$\ln x = \displaystyle\int \frac{dv}{f(v) - v} + C$ unless $f(v) = v$, in which case $y = Cx$.
Linear first-order equation $y' + f(x)y = g(x)$	$y(x) =$ $e^{-\int^x f(t)\,dt} \left[\displaystyle\int^x e^{\int^z f(t)\,dt} g(z)\,dz + C \right]$

Differential equation	Solution or solution technique
Reducible to homogeneous $(a_1x + b_1y + c_1)\,dx$ $\qquad + (a_2x + b_2y + c_2)\,dy = 0$ **with** $a_1/a_2 \neq b_1/b_2$	Change variables to $u = a_1x + b_1y + c$ and $v = a_2x + b_2y + c$
Reducible to separable $(a_1x + b_1y + c_1)\,dx$ $\qquad + (a_2x + b_2y + c_2)\,dy = 0$ **with** $a_1/a_2 = b_1/b_2$	Change dependent variable to $u(x) = a_1x + b_1y$
Separation of variables $y' = f(x)g(y)$	$\displaystyle \int \frac{dy}{g(y)} = \int f(x)\,dx + C$

5.6.3 INTEGRATING FACTORS

An integrating factor is a multiplicative term that makes a differential equation have the form of an exact equation. If the equation $M(x,y)\,dx + N(x,y)\,dy = 0$ is not in the form of an exact differential equation (i.e., $M_y \neq N_x$) then it may be put into this form by multiplying by an integrating factor.

1. If $\frac{1}{N}\left(\frac{\partial M}{\partial y} - \frac{\partial N}{\partial x}\right) = f(x)$, a function of x alone, then $u(x) = \exp\left(\int^x f(z)\,dz\right)$ is an integrating factor.

2. If $\frac{1}{M}\left(\frac{\partial M}{\partial y} - \frac{\partial N}{\partial x}\right) = g(y)$, a function of y alone, then $u(y) = \exp\left(\int^y g(z)\,dz\right)$ is an integrating factor.

EXAMPLE The equation $\frac{y}{x}\,dx + dy = 0$ has $\{M = y/x, N = 1\}$ and $f(x) = 1/x$. Hence $u(x) = \exp\left(\int^x \frac{1}{z}\,dz\right) = \exp(\log x) = x$ is an integrating factor. Multiplying the original equation by $u(x)$ results in $y\,dx + x\,dy = 0$ or $d(xy) = 0$.

5.6.4 VARIATION OF PARAMETERS

If the linear second-order equation $L[y] = y'' + P(x)y' + Q(x)y = R(x)$ has the independent homogeneous solutions $u(x)$ and $v(x)$ (i.e., $L[u] = 0 = L[v]$), then the solution to the original equation is given by

$$y(x) = -u(x)\int \frac{v(x)R(x)}{W(u,v)}\,dx + v(x)\int \frac{u(x)R(x)}{W(u,v)}\,dx, \qquad (5.6.6)$$

where $W(u,v) = uv' - u'v = \left|\begin{smallmatrix} u & v \\ u' & v' \end{smallmatrix}\right|$ is the Wronskian.

EXAMPLE The homogeneous solutions to $y'' + y = \csc x$ are $u(x) = \sin x$ and $v(x) = \cos x$. Here, $W(u,v) = -1$. Hence, $y(x) = \sin x \log(\sin x) - x \cos x$.

If the linear third order equation $L[y] = y''' + P_2(x)y'' + P(x)y' + Q(x)y = R(x)$ has the homogeneous solutions $y_1(x)$, $y_2(x)$, and $y_3(x)$ (i.e., $L[y_i] = 0$), then the solution to the original equation is given by

$$
y(x) = y_1(x) \int \frac{\begin{vmatrix} 0 & y_2 & y_3 \\ 0 & y_2' & y_3' \\ R & y_2'' & y_3'' \end{vmatrix}}{W(y_1, y_2, y_3)} \, dx + y_2(x) \int \frac{\begin{vmatrix} y_1 & 0 & y_3 \\ y_1' & 0 & y_3' \\ y_1'' & R & y_3'' \end{vmatrix}}{W(y_1, y_2, y_3)} \, dx
$$

$$
+ y_3(x) \int \frac{\begin{vmatrix} y_1 & y_2 & 0 \\ y_1' & y_2' & 0 \\ y_1'' & y_2'' & R \end{vmatrix}}{W(y_1, y_2, y_3)} \, dx
$$

(5.6.7)

where $W(y_1, y_2, y_3) = \begin{vmatrix} y_1 & y_2 & y_3 \\ y_1' & y_2' & y_3' \\ y_1'' & y_2'' & y_3'' \end{vmatrix}$ is the Wronskian.

5.6.5 GREEN'S FUNCTIONS

Let $L[y] = f(x)$ be a linear differential equation of order n, on (x_0, x_1), for $y(x)$ with the linear homogeneous boundary conditions $\{B_i[y] = \sum_{j=0}^{n-1}(a_{ij}y^{(j)}(x_0) + b_{ij}y^{(j)}(x_1)) = 0\}$, for $i = 1, 2, \ldots, n$. If there is a Green's function $G(x; z)$ that satisfies

$$
\begin{aligned}
L[G(x; z)] &= \delta(x - z), \\
B_i[G(x; z)] &= 0,
\end{aligned}
$$

(5.6.8)

where δ is Dirac's delta function, then the solution of the original system can be written as $y(x) = \int G(x; z)f(z)\,dz$, integrated over an appropriate region.

EXAMPLE To solve $y'' = f(x)$ with $y(0) = 0$ and $y(L) = 0$, the appropriate Green's function is

$$
G(x; z) = \begin{cases} \dfrac{x(z - L)}{L} & \text{for } 0 \le x \le z, \\[2mm] \dfrac{z(x - L)}{L} & \text{for } z \le x \le L. \end{cases}
$$

(5.6.9)

Hence, the solution is

$$
y(x) = \int_0^L G(x; z)\, f(z)\, dz = \int_0^x \frac{z(x - L)}{L}\, f(z)\, dz + \int_x^L \frac{x(z - L)}{L}\, f(z)\, dz. \quad (5.6.10)
$$

5.6.6 TABLE OF GREEN'S FUNCTIONS

For the following, the Green's function is $G(x, \xi)$ when $x \le \xi$ and $G(\xi, x)$ when $x \ge \xi$.

1. For the equation $\dfrac{d^2y}{dx^2} = f(x)$ with

    (a) $y(0) = y(1) = 0$, $G(x, \xi) = -(1 - \xi)x$,

    (b) $y(0) = 0, y'(1) = 0$, $G(x, \xi) = -x$,

    (c) $y(0) = -y(1), y'(0) = -y'(1)$, $G(x, \xi) = -\frac{1}{2}(x - \xi) - \frac{1}{4}$, and

    (d) $y(-1) = y(1) = 0$, $G(x, \xi) = -\frac{1}{2}(x - \xi - x\xi + 1)$.

2. For the equation $\dfrac{d^2y}{dx^2} - y = f(x)$ with y finite in $(-\infty, \infty)$,

$G(x, \xi) = -\frac{1}{2}e^{x-\xi}$.

3. For the equation $\dfrac{d^2y}{dx^2} + k^2y = f(x)$ with

   (a) $y(0) = y(1) = 0$, $G(x, \xi) = -\dfrac{\sin kx \sin k(1 - \xi)}{k \sin k}$,

   (b) $y(-1) = y(1), y'(-1) = y'(1)$, and $G(x, \xi) = \dfrac{\cos k(x - \xi + 1)}{2k \sin k}$.

4. For the equation $\dfrac{d^2y}{dx^2} - k^2y = f(x)$ with

   (a) $y(0) = y(1) = 0$, $G(x, \xi) = -\dfrac{\sinh kx \sinh k(1 - \xi)}{k \sinh k}$,

   (b) $y(-1) = y(1), y'(-1) = y'(1)$, and $G(x, \xi) = -\dfrac{\cosh k(x - \xi + 1)}{2k \sinh k}$.

5. For the equation $\dfrac{d}{dx}\left(x\dfrac{dy}{dx}\right) = f(x)$, with $y(0)$ finite and $y(1) = 0$,
   $G(x, \xi) = \ln \xi$.

6. For the equation $\dfrac{d}{dx}\left(x\dfrac{dy}{dx}\right) - \dfrac{m^2}{x}y = f(x)$, with $y(0)$ finite and $y(1) = 0$,
   $G(x, \xi) = -\frac{1}{2m}\left[\left(\frac{x}{\xi}\right)^m - (x\xi)^m\right]$, $(m = 1, 2, \dots)$.

7. For the equation $\dfrac{d}{dx}\left((1 - x^2)\dfrac{dy}{dx}\right) - \dfrac{m^2}{1 - x^2}y = f(x)$, with $y(-1)$ and $y(1)$
   finite, $G(x, \xi) = -\frac{1}{2m}\left(\frac{1+x}{1-x}\frac{1-\xi}{1+\xi}\right)^{m/2}$, $(m = 1, 2, \dots)$.

8. For the equation $\dfrac{d^4y}{dx^4} = f(x)$, with $y(0) = y'(0) = y(1) = y'(1) = 0$,
   $G(x, \xi) = -\frac{x^2(\xi-1)^2}{6}(2x\xi + x - 3\xi)$.

5.6.7 TRANSFORM TECHNIQUES

Transforms can sometimes be used to solve linear differential equations. Laplace transforms (page 585) are appropriate for initial-value problems, while Fourier transforms (page 576) are appropriate for boundary-value problems.

EXAMPLE   Consider the linear second-order equation $y'' + y = p(x)$, with the initial conditions $y(0) = 0$ and $y'(0) = 0$. Multiplying this equation by e^{-sx}, and integrating with respect to x from 0 to ∞, results in

$$\int_0^\infty e^{-sx} y''(x)\, dx + \int_0^\infty e^{-sx} y(x)\, dx = \int_0^\infty e^{-sx} p(x)\, dx.$$

Integrating by parts, and recognizing that $Y(s) = \mathcal{L}[y(x)] = \int_0^\infty e^{-sx} y(x)\, dx$ is the Laplace transform of y, this simplifies to

$$(s^2 + 1)Y(s) = \int_0^\infty e^{-sx} p(x)\, dx = \mathcal{L}[p(x)].$$

If $p(x) \equiv 1$, then $\mathcal{L}[p(x)] = s^{-1}$. The table of Laplace transforms (entry 20 in the table on page 606) shows that the $y(x)$ corresponding to $Y(s) = 1/[s(1 + s^2)]$ is $y(x) = \mathcal{L}^{-1}[Y(s)] = 1 - \cos x$.

5.6.8   NAMED ORDINARY DIFFERENTIAL EQUATIONS

1. Airy equation: $y'' = xy$
   Solution: $y = c_1 \operatorname{Ai}(x) + c_2 \operatorname{Bi}(x)$

2. Bernoulli equation: $y' = a(x)y^n + b(x)y$

3. Bessel equation: $x^2 y'' + xy' + (\lambda^2 x^2 - n^2)y = 0$
   Solution: $y = c_1 J_n(\lambda x) + c_2 Y_n(\lambda x)$

4. Bessel equation (transformed): $x^2 y'' + (2p + 1)xy' + (\lambda^2 x^{2r} + \beta^2)y = 0$
   Solution: $y = x^{-p} \left[c_1 J_{q/r}\left(\dfrac{\lambda}{r} x^r\right) + c_2 Y_{q/r}\left(\dfrac{\lambda}{r} x^r\right) \right] \quad q \equiv \sqrt{p^2 - \beta^2}$

5. Bôcher equation: $y'' + \dfrac{1}{2}\left[\dfrac{m_1}{x - a_1} + \cdots + \dfrac{m_{n-1}}{x - a_{n-1}} \right] y'$
   $\qquad + \dfrac{1}{4}\left[\dfrac{A_0 + A_1 x + \cdots + A_l x^l}{(x - a_1)^{m_1}(x - a_2)^{m_2} \cdots (x - a_{n-1})^{m_{n-1}}} \right] y = 0$

6. Duffing's equation: $y'' + y + \epsilon y^3 = 0$

7. Emden–Fowler equation: $(x^p y')' \pm x^\sigma y^n = 0$

8. Hypergeometric equation: $y'' + \left(\dfrac{1 - \alpha - \alpha'}{x - a} + \dfrac{1 - \beta - \beta'}{x - b} + \dfrac{1 - \gamma - \gamma'}{x - c} \right) y'$
   $\qquad - \left(\dfrac{\alpha\alpha'}{(x-a)(b-c)} + \dfrac{\beta\beta'}{(x-b)(c-a)} + \dfrac{\gamma\gamma'}{(x-c)(a-b)} \right) \dfrac{(a-b)(b-c)(c-a)}{(x-a)(x-b)(x-c)} u = 0$
   Solution: $y = P \left\{ \begin{array}{ccc} a & b & c \\ \alpha & \beta & \gamma \\ \alpha' & \beta' & \gamma' \end{array} \; x \right\}$ (Riemann's P function)

9. Legendre equation: $(1 - x^2)y'' - 2xy' + n(n + 1)y = 0$
   Solution: $y = c_1 P_n(x) + c_2 Q_n(x)$

10. Mathieu equation: $y'' + (a - 2q \cos 2x)y = 0$

11. Painlevé transcendent (first equation): $y'' = 6y^2 + x$

12. Parabolic cylinder equation: $y'' + (ax^2 + bx + c)y = 0$

13. Riccati equation: $y' = a(x)y^2 + b(x)y + c(x)$

5.6.9 LIAPUNOV'S DIRECT METHOD

If, as $\mathbf{x}(t)$ evolves, a function $V = V(\mathbf{x})$ can be found so that $V(\mathbf{x}(t)) > 0$ and $\frac{dV(\mathbf{x}(t))}{dt} < 0$ for $\mathbf{x} \neq \mathbf{0}$, then the system is asymptotically stable: $V[\mathbf{x}(t)] \to 0$ as $t \to \infty$.

EXAMPLE   For the non-linear system of differential equations with $a > 0$

$$\dot{x}_1 = -ax_1 - x_1x_2^2, \qquad \dot{x}_2 = -ax_2 + x_1^2x_2,$$

define $V = x_1^2 + x_2^2$. Since $\dot{V} = -2a(x_1^2 + x_2^2) = -2aV$ we find $V = V(\mathbf{x}(t)) = V_0e^{-2at}$. Hence $x_1(t)$ and $x_2(t)$ both decay to 0.

5.6.10 LIE GROUPS

The invertible transformation $\{\overline{x} = \phi(x, y, a), \overline{y} = \psi(x, y, a)\}$ forms a one-parameter group if $\phi(\overline{x}, \overline{y}, b) = \phi(x, y, a + b)$ and $\psi(\overline{x}, \overline{y}, b) = \psi(x, y, a + b)$. For small a, these transformations become

$$\overline{x} = x + a\xi(x, y) + O(a^2) \qquad \text{and} \qquad \overline{y} = y + a\eta(x, y) + O(a^2) \qquad (5.6.11)$$

where the infinitesimal generator is $X = \xi(x, y)\frac{\partial}{\partial x} + \eta(x, y)\frac{\partial}{\partial y}$. If $D = \frac{\partial}{\partial x} + y'\frac{\partial}{\partial y} + y''\frac{\partial}{\partial y'} + \ldots$, then the derivatives of the new variables are

$$\overline{y}' = \frac{d\overline{y}}{d\overline{x}} = \frac{D\psi}{D\phi} = \frac{\psi_x + y'\psi_y}{\phi_x + y'\phi_y} = P(x, y, y', a) = y' + a\zeta_1 + O(a^2), \quad \text{and}$$

$$\overline{y}'' = \frac{d\overline{y}'}{d\overline{x}} = \frac{DP}{D\phi} = \frac{P_x + y'P_y + y''P_{y'}}{\phi_x + y'\phi_y} = y'' + a\zeta_2 + O(a^2),$$

$(5.6.12)$

where

$$\zeta_1 = D(\eta) - y'D(\xi) = \eta_x + (\eta_y - \xi_x)y' - y'^2\xi_y, \quad \text{and}$$

$$\zeta_2 = D(\zeta_1) - y''D(\xi) = \eta_{xx} + (2\eta_{xy} - \xi_{xx})y' + (\eta_{yy} - 2\xi_{xy})y'^2 \qquad (5.6.13)$$

$$- y'^3\xi_{yy} + (\eta_y - 2\xi_x - 3y'\xi_y)y''.$$

The prolongations of X are $X^{(1)} = X + \zeta_1\frac{\partial}{\partial y'}$ and $X^{(2)} = X^{(1)} + \zeta_2\frac{\partial}{\partial y''}$. For a given differential equation, the different infinitesimal generators will generate an r-dimensional Lie group (L_r).

For the equation $F(x, y, y', y'') = 0$ to be invariant under the action of the above group, $X^{(2)}F\ |_{F=0} = 0$. When $F = y'' - f(x, y, y')$, this determining equation becomes

$$\eta_{xx} + (2\eta_{xy} - \xi_{xx})y' + (\eta_{yy} - 2\xi_{xy})y'^2 - y'^3\xi_y y$$
$$+ (\eta_y - 2\xi_x - 3y'\xi_y)f - \left[\eta_x + (\eta_y - \xi_x)y' - y'^2\right]f_{y'}$$
$$- \xi f_x - \eta f_y = 0. \quad (5.6.14)$$

Given the two generators $X_1 = \xi_1 \frac{\partial}{\partial x} + \eta_1 \frac{\partial}{\partial y}$ and $X_2 = \xi_2 \frac{\partial}{\partial x} + \eta_2 \frac{\partial}{\partial y}$, the pseudoscalar product is $X_1 \vee X_2 = \xi_1\eta_2 - \xi_2\eta_1$, and the commutator is $[X_1, X_2] = X_1 X_2 - X_2 X_1$. By a suitable choice of basis, any two-dimensional Lie algebra can be reduced to one of four types:

No.	Commutator	Pseudoscalar	Typified by
I	$[X_1, X_2] = 0$	$X_1 \vee X_2 \neq 0$	$\{X_1 = \frac{\partial}{\partial x},\ X_2 = \frac{\partial}{\partial y}\}$
II	$[X_1, X_2] = 0$	$X_1 \vee X_2 = 0$	$\{X_1 = \frac{\partial}{\partial y},\ X_2 = x\frac{\partial}{\partial y}\}$
III	$[X_1, X_2] = X_1$	$X_1 \vee X_2 \neq 0$	$\{X_1 = \frac{\partial}{\partial y},\ X_2 = x\frac{\partial}{\partial x} + y\frac{\partial}{\partial y}\}$
IV	$[X_1, X_2] = X_1$	$X_1 \vee X_2 = 0$	$\{X_1 = \frac{\partial}{\partial y},\ X_2 = y\frac{\partial}{\partial y}\}$

5.6.10.1 Integrating second-order ordinary differential equations

An algorithm for integrating second-order ordinary differential equations is given by:

1. Determine the admitted Lie algebra L_r, where r is the dimension.
2. If $r < 2$, then Lie groups are not useful for the given equation. If $r > 2$, determine a subalgebra $L_2 \subset L_r$.
3. From the commutator and pseudoscalar product, change the basis to obtain one of the four cases in the above table.
4. Introduce canonical variables specified by the change of basis into the original differential equation. Integrate this new equation.
5. Rewrite the solution in terms of the original variables.

5.6.11 STOCHASTIC DIFFERENTIAL EQUATIONS

A stochastic differential equation for the unknown $X(t)$ has the form (here, a and b are given):

$$dX(t) = a(X(t))\,dt + b(X(t))\,dB(t) \quad (5.6.15)$$

where $B(t)$ is a Brownian motion. Brownian motion has a Gaussian probability distribution and independent increments. The probability density function $f_{X(t)}$ for $X(t)$ satisfies the forward Kolmogorov equation

$$\frac{\partial}{\partial t} f_{X(t)}(x) = \frac{1}{2}\frac{\partial^2}{\partial x^2}\left[b^2(x)f_{X(t)}(x)\right] - \frac{\partial}{\partial x}\left[a(x)f_{X(t)}(x)\right]. \quad (5.6.16)$$

The conditional expectation of the function $\phi(X(t))$, $u(t, x) = \mathrm{E}\left[\phi(X(t)) \mid X(0) = x\right]$, satisfies

$$\frac{\partial}{\partial t}u(t, x) = \frac{1}{2}b^2(x)\frac{\partial^2}{\partial x^2}u(t, x) + a(x)\frac{\partial}{\partial x}u(t, x) \quad \text{with} \quad u(0, x) = \phi(x).$$

(5.6.17)

5.6.12 TYPES OF CRITICAL POINTS

An ODE may have several types of critical points; these include improper node, deficient improper node, proper node, saddle, center, and focus. See Figure 5.1.

FIGURE 5.1

Types of critical points. Clockwise from upper left: center, improper node, deficient improper node, spiral, star, saddle.

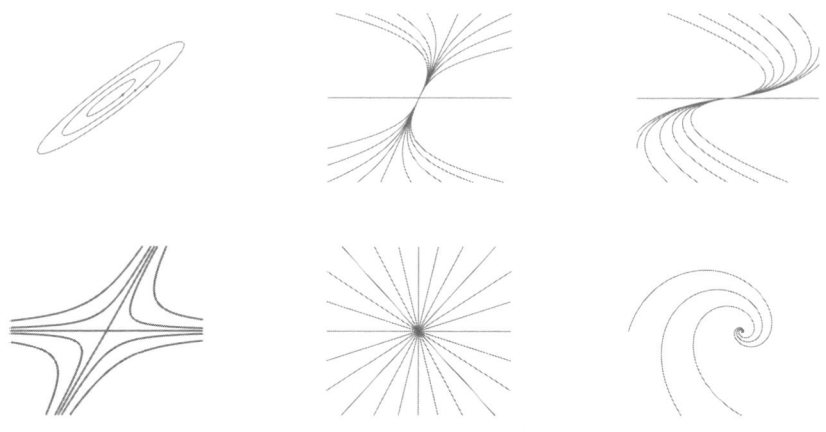

5.7 PARTIAL DIFFERENTIAL EQUATIONS

5.7.1 CLASSIFICATIONS OF PDES

Consider second-order partial differential equations, with two independent variables, of the form

$$A(x, y)\frac{\partial^2 u}{\partial x^2} + B(x, y)\frac{\partial^2 u}{\partial x \partial y} + C(x, y)\frac{\partial^2 u}{\partial y^2} = \Psi\left(u, \frac{\partial u}{\partial x}, \frac{\partial u}{\partial y}, x, y\right). \quad (5.7.1)$$

If $\begin{bmatrix} B^2 - 4AC > 0 \\ B^2 - 4AC = 0 \\ B^2 - 4AC < 0 \end{bmatrix}$ at some point (x, y), then Equation (5.7.1) is $\begin{bmatrix} \text{hyperbolic} \\ \text{parabolic} \\ \text{elliptic} \end{bmatrix}$ at that point. If an equation is of the same type at all points, then the equation is simply of that type.

5.7.2 NAMED PARTIAL DIFFERENTIAL EQUATIONS

1. Biharmonic equation: $\quad\quad\quad\quad \nabla^4 u = 0$
2. Burgers' equation: $\quad\quad\quad\quad\quad u_t + u u_x = \nu u_{xx}$
3. Diffusion (or heat) equation: $\quad \nabla(c(\mathbf{x}, t) \nabla u) = u_t$
4. Hamilton–Jacobi equation: $\quad\quad V_t + H(t, \mathbf{x}, V_{x_1}, \ldots, V_{x_n}) = 0$
5. Helmholtz equation: $\quad\quad\quad \nabla^2 u + k^2 u = 0$
6. Korteweg de Vries equation: $\quad u_t + u_{xxx} - 6 u u_x = 0$
7. Laplace's equation: $\quad\quad\quad\quad \nabla^2 u = 0$
8. Navier–Stokes equations: $\quad\quad \mathbf{u}_t + (\mathbf{u} \cdot \nabla) \mathbf{u} = -\frac{\nabla P}{\rho} + \nu \nabla^2 \mathbf{u}$
9. Poisson equation: $\quad\quad\quad\quad \nabla^2 u = -4\pi\rho(\mathbf{x})$
10. Schrödinger equation: $\quad\quad -\frac{\hbar^2}{2m} \nabla^2 u + V(\mathbf{x}) u = i\hbar u_t$
11. Sine–Gordon equation: $\quad\quad u_{xx} - u_{yy} \pm \sin u = 0$
12. Tricomi equation: $\quad\quad\quad\quad u_{yy} = y u_{xx}$
13. Wave equation: $\quad\quad\quad\quad\quad c^2 \nabla^2 u = u_{tt}$
14. Telegraph equation: $\quad\quad\quad\quad u_{xx} = a u_{tt} + b u_t + c u$

5.7.3 TRANSFORMING PARTIAL DIFFERENTIAL EQUATIONS

To transform a partial differential equation, construct a new function which depends upon new variables, and then differentiate with respect to the old variables to see how the derivatives transform.

EXAMPLE Consider transforming

$$f_{xx} + f_{yy} + x f_y = 0, \quad\quad\quad (5.7.2)$$

from the $\{x, y\}$ variables to the $\{u, v\}$ variables, where $\{u = x, v = x/y\}$. Note that the inverse transformation is given by $\{x = u, y = u/v\}$.

First, define $g(u, v)$ as the function $f(x, y)$ when written in the new variables, that is

$$f(x, y) = g(u, v) = g\left(x, \frac{x}{y}\right). \quad\quad\quad (5.7.3)$$

Now create the needed derivative terms, carefully applying the chain rule. For example, differentiating Equation (5.7.3) with respect to x results in

$$f_x(x, y) = g_u \frac{\partial}{\partial x}(u) + g_v \frac{\partial}{\partial x}(v) = g_1 \frac{\partial}{\partial x}(x) + g_2 \frac{\partial}{\partial x}\left(\frac{x}{y}\right)$$

$$= g_1 + g_2 \frac{1}{y} = g_1 + \frac{v}{u} g_2,$$

where a subscript of "1" ("2") indicates a derivative with respect to the first (second) argument of the function $g(u, v)$, that is, $g_1(u, v) = g_u(u, v)$. Use of this "slot notation" tends to minimize errors. In like manner

$$f_y(x, y) = g_u \frac{\partial}{\partial y}(u) + g_v \frac{\partial}{\partial y}(v) = g_1 \frac{\partial}{\partial y}(x) + g_2 \frac{\partial}{\partial y}\left(\frac{x}{y}\right)$$

$$= -\frac{x}{y^2}g_2 = -\frac{v^2}{u}g_2.$$

The second-order derivatives can be calculated similarly:

$$f_{xx}(x, y) = \frac{\partial}{\partial x}(f_x(x, y)) = \frac{\partial}{\partial x}\left(g_1 + \frac{1}{y}g_2\right)$$

$$= g_{11} + \frac{2v}{u}g_{12} + \frac{v^2}{u^2}g_{22},$$

$$f_{xy}(x, y) = \frac{\partial}{\partial x}\left(-\frac{x}{y^2}g_2\right) = -\frac{u^2}{v^2}g_2 - \frac{u^3}{v^3}g_{12} - \frac{u^2}{v^2}g_{22},$$

$$f_{yy}(x, y) = \frac{\partial}{\partial y}\left(-\frac{x}{y^2}g_2\right) = \frac{2v^3}{u^2}g_2 + \frac{v^4}{u^2}g_{22}.$$

Finally, Equation (5.7.2) in the new variables has the form,

$$0 = f_{xx} + f_{yy} + x f_y$$

$$= \left(g_{11} + \frac{2v}{u}g_{12} + \frac{v^2}{u^2}g_{22}\right) + \left(\frac{2v^3}{u^2}g_2 + \frac{v^4}{u^2}g_{22}\right) + (u)\left(-\frac{v^2}{u}g_2\right)$$

$$= \frac{v^2(2v - u^2)}{u^2}g_v + g_{uu} + \frac{2v}{u}g_{uv} + \frac{v^2(1 + v^2)}{u^2}g_{vv}.$$

5.7.4 WELL-POSEDNESS OF PDES

Partial differential equations involving $u(\mathbf{x})$ usually have the following types of boundary conditions:

1. Dirichlet conditions: $u = 0$ on the boundary
2. Neumann conditions: $\frac{\partial u}{\partial n} = 0$ on the boundary
3. Cauchy conditions: u and $\frac{\partial u}{\partial n}$ specified on the boundary

A well-posed differential equation meets these conditions:

1. The solution exists.
2. The solution is unique.
3. The solution is stable (i.e., the solution depends continuously on the boundary conditions and initial conditions).

Type of boundary conditions	Type of equation		
	Elliptic	Hyperbolic	Parabolic
Dirichlet			
Open surface	Undetermined	Undetermined	Unique, stable solution in one direction
Closed surface	Unique, stable solution	Undetermined	Undetermined
Neumann			
Open surface	Undetermined	Undetermined	Unique, stable solution in one direction
Closed surface	Overdetermined	Overdetermined	Overdetermined
Cauchy			
Open surface	Not physical results	Unique, stable solution	Overdetermined
Closed surface	Overdetermined	Overdetermined	Overdetermined

5.7.5 GREEN'S FUNCTIONS

In the following, $\mathbf{r} = (x, y, z)$, $\mathbf{r}_0 = (x_0, y_0, z_0)$, $R^2 = (x - x_0)^2 + (y - y_0)^2 + (z - z_0)^2$, $P^2 = (x - x_0)^2 + (y - y_0)^2$.

1. For the potential equation $\nabla^2 G + k^2 G = -4\pi\delta(\mathbf{r} - \mathbf{r}_0)$, with the radiation condition (outgoing waves only), the solution is

$$G = \begin{cases} \frac{2\pi i}{k} e^{ik|x - x_0|} & \text{in one dimension,} \\ i\pi H_0^{(1)}(kP) & \text{in two dimensions, and} \\ \frac{e^{ikR}}{R} & \text{in three dimensions,} \end{cases} \qquad (5.7.4)$$

where $H_0^{(1)}(\cdot)$ is a Hankel function (see page 559).

2. For the n-dimensional diffusion equation

$$\nabla^2 G - a^2 \frac{\partial G}{\partial t} = -4\pi\delta(\mathbf{r} - \mathbf{r}_0)\delta(t - t_0), \qquad (5.7.5)$$

with the initial condition $G = 0$ for $t < t_0$, and the boundary condition $G = 0$ at $r = \infty$, the solution is

$$G = \frac{4\pi}{a^2} \left(\frac{a}{2\sqrt{\pi(t - t_0)}} \right)^n \exp\left(-\frac{a^2|\mathbf{r} - \mathbf{r}_0|^2}{4(t - t_0)} \right). \qquad (5.7.6)$$

3. For the wave equation

$$\nabla^2 G - \frac{1}{c^2} \frac{\partial^2 G}{\partial t^2} = -4\pi\delta(\mathbf{r} - \mathbf{r}_0)\delta(t - t_0), \qquad (5.7.7)$$

with the initial conditions $G = G_t = 0$ for $t < t_0$, and the boundary condition $G = 0$ at $r = \infty$, the solution is

$$G = \begin{cases} 2c\pi H\left[(t - t_0) - \frac{|x - x_0|}{c}\right] & \text{in one space dimension,} \\ \frac{2c}{\sqrt{c^2(t-t_0)^2 - P^2}} H\left[(t - t_0) - \frac{P}{c}\right] & \text{in two space dimensions, and} \\ \frac{1}{R}\delta\left[\frac{R}{c} - (t - t_0)\right] & \text{in three space dimensions.} \end{cases}$$

(5.7.8)

where $H(\cdot)$ is the Heaviside function.

5.7.6 QUASI-LINEAR EQUATIONS

Consider the first-order quasi-linear differential equation for $u(\mathbf{x}) = u(x_1, \ldots, x_N)$,

$$a_1(\mathbf{x}, u)u_{x_1} + a_2(\mathbf{x}, u)u_{x_2} + \cdots + a_N(\mathbf{x}, u)u_{x_N} = b(\mathbf{x}, u). \tag{5.7.9}$$

Defining $\frac{\partial x_k}{\partial s} = a_k(\mathbf{x}, u)$, for $k = 1, 2, \ldots, N$, the original equation becomes $\frac{du}{ds} = b(\mathbf{x}, u)$. To solve the original system, the ordinary differential equations for $u(s, \mathbf{t})$ and the $\{x_k(s, \mathbf{t})\}$ must be solved. Their initial conditions can often be parameterized as (with $\mathbf{t} = (t_1, \ldots, t_{N-1})$)

$$u(s = 0, \mathbf{t}) = v(\mathbf{t}),$$
$$x_1(s = 0, \mathbf{t}) = h_1(\mathbf{t}),$$
$$x_2(s = 0, \mathbf{t}) = h_2(\mathbf{t}),$$
$$\vdots$$
$$x_N(s = 0, \mathbf{t}) = h_N(\mathbf{t}),$$

(5.7.10)

from which the solution follows. This results in an implicit solution.

EXAMPLE For the equation $u_x + x^2 u_y = -yu$ with $u = f(y)$ when $x = 0$, the corresponding equations are

$$\frac{\partial x}{\partial s} = 1, \qquad \frac{\partial y}{\partial s} = x^2, \qquad \frac{du}{ds} = -yu.$$

The original initial data can be written parametrically as $x(s = 0, t) = 0$, $y(s = 0, t_1) = t_1$, and $u(s = 0, t_1) = f(t_1)$. Solving for x results in $x(s, t_1) = s$. The equation for y can then be integrated to yield $y(s, t_1) = \frac{s^3}{3} + t_1$. Finally, the equation for u is integrated to obtain $u(s, t_1) = f(t_1)\exp\left(-\frac{s^4}{12} - st_1\right)$. These solutions constitute an implicit solution of the original system.

In this case, it is possible to eliminate the s and t_1 variables analytically to obtain the explicit solution: $u(x, y) = f\left(y - \frac{x^3}{3}\right)\exp\left(\frac{x^4}{4} - xy\right)$.

5.7.7 SEPARATION OF VARIABLES

A solution of a linear PDE is attempted in the form $u(\mathbf{x}) = u(x_1, x_2, \cdots, x_n) = X_1(x_1)X_2(x_2)\ldots X_n(x_n)$. Logical reasoning may determine the $\{X_i\}$.

1. For example, the diffusion or heat equation in a circle is

$$\frac{\partial u}{\partial t} = \nabla^2 u = \frac{1}{r}\frac{\partial}{\partial r}\left(r\frac{\partial u}{\partial r}\right) + \frac{1}{r^2}\frac{\partial^2 u}{\partial \theta^2} \qquad (5.7.11)$$

for the unknown $u(t, r, \theta)$, where t is time and (r, θ) are polar coordinates. If $u(t, r, \theta) = T(t)R(r)\Theta(\theta)$, then

$$\frac{1}{rR}\frac{d}{dr}\left(r\frac{dR}{dr}\right) + \frac{1}{r^2\Theta}\frac{d^2\Theta}{d\theta^2} - \frac{1}{T}\frac{dT}{dt} = 0. \qquad (5.7.12)$$

Logical reasoning leads to

$$\frac{1}{T}\frac{dT}{dt} = -\lambda, \qquad \frac{1}{\Theta}\frac{d^2\Theta}{d\theta^2} = -\rho, \qquad r\frac{d}{dr}\left(r\frac{dR}{dr}\right) + (-\rho + r^2\lambda)R = 0,$$

where λ and ρ are unknown constants. Solving these ordinary differential equations yields the general solution,

$$u(t, r, \theta) = \int_{-\infty}^{\infty} d\lambda \int_{-\infty}^{\infty} d\rho\, e^{-\lambda t}\left[B(\lambda, \rho)\sin(\sqrt{\rho}\theta) + C(\lambda, \rho)\cos(\sqrt{\rho}\theta)\right]$$
$$\times \left[D(\lambda, \rho)J_{\sqrt{\rho}}(\sqrt{\lambda}r) + E(\lambda, \rho)Y_{\sqrt{\rho}}(\sqrt{\lambda}r)\right].$$
$$(5.7.13)$$

Boundary conditions are required to determine the $\{B, C, D, E\}$.

2. A necessary and sufficient condition for a system with Hamiltonian $H = \frac{1}{2}(p_x^2 + p_y^2) + V(x, y)$, to be separable in elliptic, polar, parabolic, or Cartesian coordinates, is that the expression,

$$(V_{yy} - V_{xx})(-2axy - b'y - bx + d) + 2V_{xy}(ay^2 - ax^2 + by - b'x + c - c')$$
$$+ V_x(6ay + 3b) + V_y(-6ax - 3b'), \qquad (5.7.14)$$

vanishes for some constants $(a, b, b', c, c', d) \neq (0, 0, 0, c, c, 0)$.

3. Consider the orthogonal coordinate system $\{u^1, u^2, u^3\}$, with the metric $\{g_{ii}\}$, and $g = g_{11}g_{22}g_{33}$. The Stäckel matrix is defined as

$$S = \begin{bmatrix} \Phi_{11}(u^1) & \Phi_{12}(u^1) & \Phi_{13}(u^1) \\ \Phi_{21}(u^2) & \Phi_{22}(u^2) & \Phi_{23}(u^2) \\ \Phi_{31}(u^3) & \Phi_{32}(u^3) & \Phi_{33}(u^3) \end{bmatrix}, \qquad (5.7.15)$$

where the $\{\Phi_{ij}\}$ are tabulated in different cases. The determinant of S can be written as $s = \Phi_{11}M_{11} + \Phi_{21}M_{21} + \Phi_{31}M_{33}$, where

$$M_{11} = \begin{vmatrix} \Phi_{22} & \Phi_{23} \\ \Phi_{32} & \Phi_{33} \end{vmatrix}, \qquad M_{21} = -\begin{vmatrix} \Phi_{12} & \Phi_{13} \\ \Phi_{32} & \Phi_{33} \end{vmatrix}, \qquad M_{31} = \begin{vmatrix} \Phi_{12} & \Phi_{13} \\ \Phi_{22} & \Phi_{23} \end{vmatrix}.$$
$$(5.7.16)$$

If the separability conditions, $g_{ii} = s/M_{i1}$ and $\sqrt{g}/s = f_1(u^1)f_2(u^2)f_3(u^3)$, are met then the Helmholtz equation $\nabla^2 W + \lambda^2 W = 0$ separates with $W = X_1(u^1)X_2(u^2)X_3(u^3)$. Here the $\{X_i\}$ are defined by

$$\frac{1}{f_i}\frac{d}{du^i}\left(f_i\frac{dX_i}{du^i}\right) + X_i\sum_{j=1}^{3}\alpha_j\Phi_{ij} = 0, \tag{5.7.17}$$

with $\alpha_1 = \lambda^2$, and α_2 and α_3 arbitrary.

(a) Necessary and sufficient conditions for the separation of the Laplace equation ($\nabla^2 W = 0$) are

$$\frac{g_{ii}}{g_{jj}} = \frac{M_{j1}}{M_{i1}} \quad \text{and} \quad \frac{\sqrt{g}}{g_{ii}} = f_1(u^1)f_2(u^2)f_3(u^3)M_{i1}. \tag{5.7.18}$$

(b) Necessary and sufficient conditions for the separation of the scalar Hemlholtz equation are

$$g_{ii} = \frac{S}{M_{i1}} \quad \text{and} \quad \frac{\sqrt{g}}{S} = f_1(u^1)f_2(u^2)f_3(u^3). \tag{5.7.19}$$

EXAMPLE   In parabolic coordinates $\{\mu, \nu, \psi\}$ the metric coefficients are $g_{11} = g_{22} = \mu^2 + \nu^2$ and $g_{33} = \mu^2\nu^2$. Hence, $\sqrt{g} = \mu\nu(\mu^2 + \nu^2)$. For the Stäckel matrix

$$S = \begin{bmatrix} \mu^2 & -1 & -1/\mu^2 \\ \nu^2 & 1 & -1/\nu^2 \\ 0 & 0 & 1 \end{bmatrix} \tag{5.7.20}$$

(for which $s = \mu^2 + \nu^2$, $M_{11} = M_{21} = 1$, and $M_{31} = \mu^{-2} + \nu^{-2}$), the separability condition holds with $f_1 = \mu$, $f_2 = \nu$, and $f_3 = 1$. Hence, the Helmholtz equation separates in parabolic coordinates. The separated equations are

$$\frac{1}{\mu}\frac{d}{d\mu}\left(\mu\frac{dX_1}{d\mu}\right) + X_1\left(\alpha_1\mu^2 - \alpha_2 - \frac{\alpha_3}{\mu^2}\right) = 0,$$

$$\frac{1}{\nu}\frac{d}{d\nu}\left(\nu\frac{dX_2}{d\nu}\right) + X_2\left(\alpha_1\nu^2 + \alpha_2 - \frac{\alpha_3}{\nu^2}\right) = 0, \text{ and} \tag{5.7.21}$$

$$\frac{d^2 X_3}{d\psi^2} + \alpha_3 X_3 = 0,$$

where $W = X_1(\mu)X_2(\nu)X_3(\psi)$.

5.7.8   SOLUTIONS OF LAPLACE'S EQUATION

1. If $\nabla^2 u = 0$ in a circle of radius R and $u(R,\theta) = f(\theta)$, for $0 \leq \theta < 2\pi$, then $u(r,\theta)$ is

$$u(r,\theta) = \frac{1}{2\pi}\int_0^{2\pi}\frac{R^2 - r^2}{R^2 - 2Rr\cos(\theta - \phi) + r^2}f(\phi)\,d\phi.$$

2. If $\nabla^2 u = 0$ in a sphere of radius one and $u(1, \theta, \phi) = f(\theta, \phi)$, then

$$u(r, \theta, \phi) = \frac{1}{4\pi} \int_0^\pi \int_0^{2\pi} f(\Theta, \Phi) \frac{1 - r^2}{(1 - 2r \cos \gamma + r^2)^{3/2}} \sin \Theta \, d\Theta \, d\Phi,$$

where $\cos \gamma = \cos \theta \cos \Theta + \sin \theta \sin \Theta \cos(\phi - \Phi)$.

3. If $\nabla^2 u = 0$ in the half plane $y \geq 0$, and $u(x, 0) = f(x)$, then

$$u(x, y) = \frac{1}{\pi} \int_{-\infty}^\infty \frac{f(t)y}{(x - t)^2 + y^2} \, dt.$$

4. If $\nabla^2 u = 0$ in the half space $z \geq 0$, and $u(x, y, 0) = f(x, y)$, then

$$u(x, y, z) = \frac{z}{2\pi} \int_{-\infty}^\infty \int_{-\infty}^\infty \frac{f(\zeta, \eta)}{[(x - \zeta)^2 + (y - \eta)^2 + z^2]^{3/2}} \, d\zeta \, d\eta.$$

5.7.9 SOLUTIONS TO THE WAVE EQUATION

1. Consider the wave equation $\frac{\partial^2 u}{\partial t^2} = \nabla^2 u = \frac{\partial^2 u}{\partial x_1^2} + \cdots + \frac{\partial^2 u}{\partial x_n^2}$, with $\mathbf{x} = (x_1, \ldots, x_n)$ and the initial data $u(0, \mathbf{x}) = f(\mathbf{x})$ and $u_t(0, \mathbf{x}) = g(\mathbf{x})$. When n is odd (and $n \geq 3$), the solution is

$$u(t, \mathbf{x}) = \frac{1}{1 \cdot 3 \cdots (n - 2)} \left\{ \frac{\partial}{\partial t} \left(\frac{\partial}{t \, \partial t} \right)^{(n-3)/2} t^{n-2} \omega[f; \mathbf{x}, t] \right.$$
$$\left. + \left(\frac{\partial}{t \, \partial t} \right)^{(n-3)/2} t^{n-2} \omega[g; \mathbf{x}, t] \right\}, \tag{5.7.22}$$

where $\omega[h; \mathbf{x}, t]$ is the average of $h(\mathbf{x})$ over the surface of an n-dimensional sphere of radius t centered at $\mathbf{x}$; that is, $\omega[h; \mathbf{x}, t] = \frac{1}{\sigma_{n-1}(t)} \int h(\zeta) \, d\Omega$, where $|\zeta - \mathbf{x}|^2 = t^2$, $\sigma_{n-1}(t)$ is the surface area of the n-dimensional sphere of radius t, and $d\Omega$ is an element of area.

When n is even, the solution is given by

$$u(t, \mathbf{x}) = \frac{1}{2 \cdot 4 \cdots (n - 2)} \left\{ \frac{\partial}{\partial t} \left(\frac{\partial}{t \, \partial t} \right)^{(n-2)/2} \int_0^t \omega[f; \mathbf{x}, \rho] \frac{\rho^{n-1} \, d\rho}{\sqrt{t^2 - \rho^2}} \right.$$
$$\left. + \left(\frac{\partial}{t \, \partial t} \right)^{(n-2)/2} \int_0^t \omega[g; \mathbf{x}, \rho] \frac{\rho^{n-1} \, d\rho}{\sqrt{t^2 - \rho^2}} \right\}, \tag{5.7.23}$$

where $\omega[h; \mathbf{x}, t]$ is defined as above. Since this expression is integrated over ρ, the values of f and g must be known everywhere in the *interior* of the n-dimensional sphere.

Using u_n for the solution in n dimensions, the above simplify to

$$u_1(x,t) = \frac{1}{2}[f(x-t) + f(x+t)] + \frac{1}{2}\int_{x-t}^{x+t} g(\zeta)\, d\zeta, \qquad (5.7.24)$$

$$u_2(\mathbf{x},t) = \frac{1}{2\pi}\frac{\partial}{\partial t}\iint_{R(t)} \frac{f(x_1+\zeta_1, x_2+\zeta_2)}{\sqrt{t^2 - \zeta_1^2 - \zeta_2^2}}\, d\zeta_1\, d\zeta_2$$

$$+ \frac{1}{2\pi}\iint_{R(t)} \frac{g(x_1+\zeta_1, x_2+\zeta_2)}{\sqrt{t^2 - \zeta_1^2 - \zeta_2^2}}\, d\zeta_1\, d\zeta_2, \quad \text{and} \qquad (5.7.25)$$

$$u_3(\mathbf{x},t) = \frac{\partial}{\partial t}\left(t\omega[f;\mathbf{x},t]\right) + t\omega[g;\mathbf{x},t], \qquad (5.7.26)$$

where $R(t)$ is the region $\{(\zeta_1, \zeta_2) \mid \zeta_1^2 + \zeta_2^2 \le t^2\}$ and

$$\omega[h;\mathbf{x},t] = \frac{1}{4\pi}\int_0^{2\pi}\int_0^{\pi} h(x_1 + t\sin\theta\cos\phi, x_2 + t\sin\theta\sin\phi, \qquad (5.7.27)$$

$$x_3 + t\cos\theta) \times \sin\theta\, d\theta\, d\phi.$$

2. The solution of the one-dimensional wave equation

$$\begin{aligned}
v_{tt} &= c^2 v_{xx} \\
v(0,t) &= 0, & \text{for } 0 < t < \infty, \qquad (5.7.28) \\
v(x,0) &= f(x), & \text{for } 0 \le x < \infty, \\
v_t(x,0) &= g(x), & \text{for } 0 \le x < \infty,
\end{aligned}$$

is

$$v(x,t) = \begin{cases} \frac{1}{2}[f(x+ct) + f(x-ct)] + \frac{1}{2c}\int_{x-ct}^{x+ct} g(\zeta)\, d\zeta, & \text{for } x \ge ct, \\ \frac{1}{2}[f(x+ct) - f(ct-x)] + \frac{1}{2c}\int_{ct-x}^{x+ct} g(\zeta)\, d\zeta, & \text{for } x < ct. \end{cases}$$
$$(5.7.29)$$

3. The solution of the inhomogeneous wave equation

$$\frac{\partial^2 u}{\partial t^2} - \frac{\partial^2 u}{\partial x^2} - \frac{\partial^2 u}{\partial y^2} - \frac{\partial^2 u}{\partial z^2} = F(t,x,y,z), \qquad (5.7.30)$$

with the initial conditions $u(0,x,y,z) = 0$ and $u_t(0,x,y,z) = 0$, is

$$u(t,x,y,z) = \frac{1}{4\pi}\iiint_{\rho \le t} \frac{F(t-\rho, \zeta, \eta, \xi)}{\rho}\, d\zeta\, d\eta\, d\xi, \qquad (5.7.31)$$

with $\rho = \sqrt{(x-\zeta)^2 + (y-\eta)^2 + (z-\xi)^2}$.

5.7.10 PARTICULAR SOLUTIONS TO SOME PDES

In these tables, we assume that $P(x)$ is a polynomial of degree n.

If $R(x)$ is	A particular solution to $z_x + mz_y = R(x,y)$ is
(1) e^{ax+by}	$e^{ax+by}/(a+mb)$.
(2) $f(ax+by)$	$\int f(y)\,du/(a+mb), \quad u = ax+by.$
(3) $f(y-mx)$	$xf(y-mx)$.
(4) $\phi(x,y)f(y-mx)$	evaluate $f(y-mx)\int \phi(x,a+mx)\,dx;$ then substitute $a = y-mx$.

If $R(x)$ is	A particular solution to $z_x + mz_y - kz = R(x,y)$ is
(5) e^{ax+by}	$e^{ax+by}/(a+mb-k)$.
(6) $\sin(ax+by)$	$-\dfrac{(a+bm)\cos(ax+by)+k\sin(ax+by)}{(a+bm)^2+k^2}$.
(7) $e^{\alpha x+\beta y}\sin(ax+by)$	Replace k by $k-\alpha-m\beta$ in formula (6) and multiply by $e^{\alpha x+\beta y}$.
(8) $e^{xk}f(ax+by)$	$e^{kx}\int f(y)\,du/(a+mb), \quad u = ax+by.$
(9) $f(y-mx)$	$-f(y-mx)/k$.
(10) $P(x)f(y-mx)$	$-\frac{1}{k}f(y-mx)\Big[P(x)+\frac{P'(x)}{k}+\frac{P''(x)}{k^2}+\cdots+\frac{P^{(n)}(x)}{k^n}\Big]$.
(11) $e^{kx}f(y-mx)$	$xe^{kx}f(y-mx)$.

5.8 EIGENVALUES

The *eigenvalues* $\{\lambda_i\}$ of the differential operator L are the solutions of $L[u_i] = \lambda_i u_i$. Given a geometric shape, the eigenvalues of the Dirichlet problem are called eigenfrequencies (that is, $\nabla^2 u = \lambda u$ with $u = 0$ on the boundary).

(a) You cannot hear the shape of a drum. The following figures have the same eigenfrequencies.

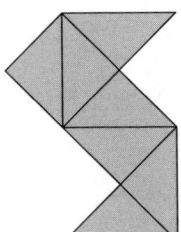
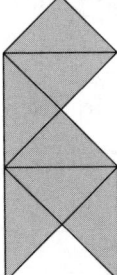

(b) You cannot hear the shape of a two-piece band. The following pairs of figures have the same eigenfrequencies.

5.9 INTEGRAL EQUATIONS

5.9.1 DEFINITIONS

$$h(x)u(x) = f(x) + \lambda \int_a^{b(x)} k(x,t)G[u(t);t]\,dt. \qquad (5.9.1)$$

- $k(x,t)$ kernel
- $u(x)$ function to be determined
- $h(x)$, $f(x)$, $b(x)$, $G[z,t]$ given functions
- λ eigenvalue

5.9.1.1 Classification of integral equations

1. Linear $G[u(x);x] = u(x)$
2. Volterra $b(x) = x$
3. Fredholm $b(x) = b$
4. First kind $h(x) = 0$
5. Second kind $h(x) = 1$
6. Third kind $h(x) \neq \text{constant}$
7. Homogeneous $f(x) = 0$
8. Singular $a = -\infty, \ b = \infty$

5.9.1.2 Classification of kernels

1. Symmetric $k(x,t) = k(t,x)$
2. Hermitian $k(x,t) = \overline{k(t,x)}$
3. Separable/degenerate $k(x,t) = \sum_{i=1}^{n} a_i(x)b_i(t), \ n < \infty$
4. Difference $k(x,t) = k(x-t)$
5. Cauchy $k(x,t) = \frac{1}{x-t}$
6. Singular $k(x,t) \to \infty$ as $t \to x$
7. Hilbert–Schmidt $\int_a^b \int_a^b |k(x,t)|^2 \, dx\,dt < \infty$

5.9.2 CONNECTION TO DIFFERENTIAL EQUATIONS

The initial-value problem

$$u''(x) + A(x)u'(x) + B(x)u(x) = g(x), \quad x > a,$$
$$u(a) = c_1, \quad u'(a) = c_2, \tag{5.9.2}$$

is equivalent to the Volterra integral equation,

$$u(x) = f(x) + \int_a^x k(x,t)u(t)\, dt, \quad x \geq a,$$
$$f(x) = \int_a^x (x - t)g(t)\, dt + (x - a)[A(a)c_1 + c_2] + c_1, \tag{5.9.3}$$
$$k(x,t) = (t - x)[B(t) - A'(t)] - A(t).$$

The boundary-value problem

$$u''(x) + A(x)u'(x) + B(x)u(x) = g(x), \quad a < x < b,$$
$$u(a) = c_1, \quad u(b) = c_2, \tag{5.9.4}$$

is equivalent to the Fredholm integral equation

$$u(x) = f(x) + \int_a^b k(x,t)u(t)\, dt, \quad a \leq x \leq b,$$

$$f(x) = c_1 + \int_a^x (x - t)g(t)\, dt + \frac{x - a}{b - a}\left[c_2 - c_1 - \int_a^b (b - t)g(t)\, dt\right],$$

$$k(x,t) = \begin{cases} \frac{x-b}{b-a}\{A(t) - (a - t)[A'(t) - B(t)]\}, & x > t, \\ \frac{x-a}{b-a}\{A(t) - (b - t)[A'(t) - B(t)]\}, & x < t. \end{cases}$$
$$\tag{5.9.5}$$

5.9.3 FREDHOLM ALTERNATIVE

For $u(x) = f(x) + \lambda \int_a^b k(x,t)u(t)\, dt$ with $\lambda \neq 0$, consider the solutions to $u_H(x) = \lambda \int_a^b k(x,t)u_H(t)\, dt$.

1. If the only solution is $u_H(x) = 0$, then there is a unique solution $u(x)$.

2. If $u_H(x) \neq 0$, then there is no solution unless $\int_a^b u_H^*(t)f(t)\, dt = 0$ for all $u_H^*(x)$ such that $u_H^*(x) = \lambda \int_a^b k(t,x)u_H^*(t)\, dt$. In this case, there are infinitely many solutions.

5.9.4 SPECIAL EQUATIONS WITH SOLUTIONS

1. Generalized Abel equation: $\displaystyle\int_0^x \frac{u(t)\,dt}{[h(x)-h(t)]^\alpha} = f(x)$

   Solution is

   $$u(x) = \frac{\sin(\alpha\pi)}{\pi}\frac{d}{dx}\int_0^x \frac{h'(t)f(t)\,dt}{[h(x)-h(t)]^{1-\alpha}} \qquad (5.9.6)$$

   where $0 \le x \le 1$, $0 \le \alpha < 1$, $0 \le h(x) \le 1$, $h'(x) > 0$, and $h'(x)$ is continuous.

2. Cauchy equation: $\displaystyle \mu u(x) = f(x) + \int_0^1 \frac{u(t)}{t-x}\,dt$

   Solution is

   $$u(x) = \begin{cases} \dfrac{x^\gamma \sin^2(\pi\gamma)}{\pi^2}\dfrac{d}{dx}\displaystyle\int_x^1 \dfrac{ds}{(s-x)^\gamma}\int_0^s \dfrac{t^{-\gamma}f(t)}{(s-t)^{1-\gamma}}\,dt, & \mu < 0, \\[3mm] \dfrac{(1-x)^\gamma \sin^2(\pi\gamma)}{\pi^2}\dfrac{d}{dx}\displaystyle\int_0^x \dfrac{ds}{(x-s)^\gamma}\int_s^1 \dfrac{(1-t)^{-\gamma}f(t)}{(t-s)^{1-\gamma}}\,dt, & \mu > 0, \end{cases} \qquad (5.9.7)$$

   where $0 < x < 1$, μ is real, $\mu \ne 0$, $|\mu| = \pi\cot(\pi\gamma)$, $0 < \gamma < 1/2$, and the integral is a Cauchy principal value integral.

3. Volterra equation with difference kernel: $\displaystyle u(x) = f(x) + \lambda\int_0^x k(x-t)u(t)\,dt$

   Solution is

   $$u(x) = \mathcal{L}^{-1}\left[\frac{F(s)}{1-\lambda K(s)}\right], \qquad (5.9.8)$$

   for $x \ge 0$, $\mathcal{L}[f(x)] = F(s)$ and $\mathcal{L}[k(x)] = K(s)$.

4. Singular equation with difference kernel: $\displaystyle u(x) = f(x) + \lambda\int_{-\infty}^\infty k(x-t)u(t)\,dt$

   Solution is

   $$u(x) = \frac{1}{\sqrt{2\pi}}\int_{-\infty}^\infty e^{-i\alpha x}\frac{F(\alpha)}{1-\lambda K(\alpha)}\,d\alpha \qquad (5.9.9)$$

   where $-\infty < x < \infty$, $F(\alpha) = \mathcal{F}[f(x)]$, and $K(\alpha) = \mathcal{F}[k(x)]$.

5. Fredholm equation with separable kernel:

   $$u(x) = f(x) + \lambda\int_a^b \sum_{k=1}^n a_k(x)b_k(t)u(t)\,dt$$

   Solution is

   $$u(x) = f(x) + \lambda\sum_{k=1}^n c_k a_k(x), \qquad (5.9.10)$$

   with $c_m = \displaystyle\int_a^b b_m(t)f(t)\,dt + \lambda\sum_{k=1}^n c_k\int_a^b b_m(t)a_k(t)\,dt$ where $a \le x \le b$, $n < \infty$, and $m = 1, 2, \ldots, n$.

6. Fredholm equation with symmetric kernel: $u(x) = f(x) + \lambda \int_a^b k(x,t)u(t)\, dt$

   Solve $u_n(x) = \lambda_n \int_a^b k(x,t)u_n(t)\, dt$ for $\{u_n, \lambda_n\}_{n=1,2,\ldots}$. Then

   (a) For $\lambda \neq \lambda_n$, solution is $u(x) = f(x) + \lambda \sum_{n=1}^{\infty} \dfrac{u_n(x) \int_a^b f(t)u_n(t)\, dt}{(\lambda_n - \lambda) \int_a^b u_n^2(t)\, dt}$.

   (b) For $\lambda = \lambda_n$ and $\int_a^b f(t)u_m(t)\, dt = 0$ for all m, solutions are

   $$u(x) = f(x) + cu_m(x) + \lambda_m \sum_{\substack{n=1 \\ n\neq m}}^{\infty} \frac{u_n(x) \int_a^b f(t)u_n(t)\, dt}{(\lambda_n - \lambda_m) \int_a^b u_n^2(t)\, dt},$$

   for $m = 1, 2, \ldots$

   where $a \leq x \leq b$, and $k(x,t) = k(t,x)$ (see Section 5.9.3).

7. Volterra equation of second kind: $u(x) = f(x) + \lambda \int_a^x k(x,t)u(t)\, dt$

   Solution is

   $$u(x) = f(x) + \lambda \int_a^x \sum_{n=0}^{\infty} \lambda^n k_{n+1}(x,t) f(t)\, dt, \tag{5.9.11}$$

   where $k_1(x,t) = k(x,t)$, $k_{n+1}(x,t) = \int_t^x k(x,s)k_n(s,t)\, ds$, when $k(x,t)$ and $f(x)$ are continuous, $\lambda \neq 0$, and $x \geq a$.

8. Fredholm equation of second kind: resolvent kernel

   $$u(x) = f(x) + \lambda \int_a^b k(x,t)u(t)\, dt \tag{5.9.12}$$

   Solution is

   $$u(x) = f(x) + \lambda \int_a^b \frac{D(x,t;\lambda)}{D(\lambda)} f(t)\, dt, \tag{5.9.13}$$

   where

   $$D(\lambda) = \sum_{n=0}^{\infty} \frac{(-\lambda)^n c_n}{n!},$$

   $$c_0 = 1, \qquad c_n = \int_a^b A_{n-1}(t,t)\, dt, \quad n = 1, 2, \ldots, \tag{5.9.14}$$

   $$D(x,t;\lambda) = k(x,t) + \sum_{n=1}^{\infty} \frac{(-\lambda)^n}{n!} A_n(x,t),$$

   $$A_0(x,t) = k(x,t)$$

   $$A_n(x,t) = c_n k(x,t) - n \int_a^b k(x,s)A_{n-1}(s,t)\, ds$$

   where $k(x,t)$ and $f(x)$ are continuous, $\lambda \neq 0$, $a \leq x \leq b$, and $D(\lambda) \neq 0$.

9. Fredholm equation of second kind: Neumann series

$$u(x) = f(x) + \lambda \int_a^b k(x,t)u(t)\, dt \qquad (5.9.15)$$

Solution is

$$u(x) = f(x) + \sum_{n=1}^{\infty} \lambda^n \phi_n(x), \qquad (5.9.16)$$

where

$$\phi_n(x) = \int_a^b k_n(x,s)f(s)\, ds,$$

$$k_1(x,s) = k(x,s), \qquad (5.9.17)$$

$$\text{and } k_n(x,s) = \int_a^b k(x,t)k_{n-1}(t,s)\, dt, \quad n = 2,3,\dots,$$

where $|\lambda| < \left(\int_a^b \int_a^b k^2(x,t)\, dx\, dt \right)^{-1/2}$, $\lambda \neq 0$, and $a \leq x \leq b$.

5.10   TENSOR ANALYSIS

5.10.1   DEFINITIONS

1. An *n-dimensional coordinate manifold* of class C^k, $k \geq 1$, is a point set M together with the totality of allowable coordinate systems on M. An *allowable coordinate system* (ϕ, U) on M is a one-to-one mapping $\phi : U \to M$, where U is an open subset of $\mathbb{R}^n$. The n-tuple $(x^1, \dots, x^n) \in U$ give the *coordinates* of the corresponding point $\phi(x^1, \dots, x^n) \in M$. If $(\tilde{\phi}, \tilde{U})$ is a second coordinate system on M, then the one-to-one correspondence $\tilde{\phi}^{-1} \circ \phi : U \to \tilde{U}$, called a *coordinate transformation* on M, is assumed to be of class C^k. It may be written as

$$\tilde{x}^i = \tilde{f}^i(x^1, \dots, x^n), \quad i = 1, \dots, n, \qquad (5.10.1)$$

where the $\tilde{f}$ are defined by $(\tilde{\phi}^{-1} \circ \phi)(x^1, \dots, x^n) = (\tilde{f}^1(x^1, \dots, x^n), \dots, \tilde{f}^n(x^1, \dots, x^n))$. The coordinate transformation $\tilde{\phi}^{-1} \circ \phi$ has inverse $\phi^{-1} \circ \tilde{\phi}$, expressible in terms of the coordinates as

$$x^i = f^i(\tilde{x}^1, \dots, \tilde{x}^n), \quad i = 1, \dots, n. \qquad (5.10.2)$$

2. In the *Einstein summation convention* a repeated upper and lower index signifies summation over the range $k = 1, \dots, n$.

3. The Jacobian matrix of the transformation, $\frac{\partial \tilde{x}^i}{\partial x^j}$, satisfies $\frac{\partial \tilde{x}^i}{\partial x^k} \frac{\partial x^k}{\partial \tilde{x}^j} = \delta^i_j$ and $\frac{\partial x^i}{\partial \tilde{x}^k} \frac{\partial \tilde{x}^k}{\partial x^j} = \delta^i_j$, where $\delta^i_j = \begin{cases} 1, & i = j \\ 0, & i \neq j \end{cases}$ denotes the Kronecker delta. Note also that $\det\left(\frac{\partial \tilde{x}^i}{\partial x^j}\right) \neq 0$.

4. A function $F : M \to \mathbb{R}$ is called a *scalar invariant* on M. The *coordinate representation of F* in any coordinate system (ϕ, U) is defined by $f := F \circ \phi$. The coordinate representations $\tilde{f}$ of F with respect to a second coordinate system $(\tilde{\phi}, \tilde{U})$ is related to f by $\tilde{f}(\tilde{x}^1, \ldots, \tilde{x}^n) = f(f^1(\tilde{x}^1, \ldots, \tilde{x}^n), \ldots, f^n(\tilde{x}^1, \ldots, \tilde{x}^n))$.

5. A *parameterized curve* on M is a mapping $\gamma : I \to M$, where $I \subset \mathbb{R}$ is some interval. The *coordinate representation of γ* in any coordinate system (ϕ, U) is a mapping $g : I \to \mathbb{R}^n$ defined by $g = \phi^{-1} \circ \gamma$. The mapping g defines a parameterized curve in $\mathbb{R}^n$. The component functions of g denoted by g^i (for $i = 1, \ldots, n$) are defined by $g(t) = (g^1(t), \cdots, g^n(t))$. The curve γ is C^k if, and only if, the functions g^i are C^k for every coordinate system on M. The coordinate representation $\tilde{g}$ of γ with respect to a second coordinate system $(\tilde{\phi}, \tilde{u})$ is related to g by $\tilde{g}^i(t) = \tilde{f}^i(g^1(t), \cdots, g^n(t))$.

6. A *mixed tensor T of contravariant valence r, covariant valence s, and weight w at $p \in M$*, called a *tensor of type (r, s, w)*, is an object which, with respect to each coordinate system on M, is represented by n^{r+s} real numbers whose values in any two coordinate systems, ϕ and $\tilde{\phi}$, are related by

$$\widetilde{T}^{i_1 \cdots i_r}_{ j_1 \cdots j_s} = \left[\det\left(\frac{\partial x^i}{\partial \tilde{x}^j}\right)\right]^w T^{k_1 \cdots k_r}_{ \ell_1 \cdots \ell_s} \underbrace{\frac{\partial \tilde{x}^{i_1}}{\partial x^{k_1}} \cdots \frac{\partial \tilde{x}^{i_r}}{\partial x^{k_r}}}_{r \text{ factors}} \underbrace{\frac{\partial x^{\ell_1}}{\partial \tilde{x}^{j_1}} \cdots \frac{\partial x^{\ell_s}}{\partial \tilde{x}^{j_s}}}_{s \text{ factors}}.$$

(5.10.3)

The superscripts are called *contravariant indices* and the subscripts *covariant indices*. If $w \neq 0$, then T is said to be a *relative tensor*. If $w = 0$, then T is said to be an *absolute tensor* or a *tensor of type (r, s)*. In the sequel only absolute tensors, which will be called tensors, will be considered unless otherwise indicated. A *tensor field T of type (r, s)* is an assignment of a tensor of type (r, s) to each point of M. A tensor field T is C^k if its component functions are C^k for every coordinate system on M.

5.10.2 ALGEBRAIC TENSOR OPERATIONS

1. *Addition:* The components of the *sum* of the tensors T_1 and T_2 of type (r, s) are given by

$$T_3{}^{i_1 \cdots i_r}_{ j_1 \cdots j_s} = T_1{}^{i_1 \cdots i_r}_{ j_1 \cdots j_s} + T_2{}^{i_1 \cdots i_r}_{ j_1 \cdots j_s}.$$

(5.10.4)

2. *Multiplication:* The components of the *tensor* or *outer product* of a tensor T_1 of type (r, s) and a tensor T_2 of type (t, u) are given by

$$T_3{}^{i_1 \cdots i_r k_1 \cdots k_t}_{ j_1 \cdots j_s \ell_1 \cdots \ell_u} = T_1{}^{i_1 \cdots i_r}_{ j_1 \cdots j_s} T_2{}^{k_1 \cdots k_t}_{ \ell_1 \cdots \ell_u}.$$

(5.10.5)

3. *Contraction:* The components of the *contraction* of the tth contravariant index with the uth covariant index of a tensor T of type (r, s), with $rs \neq 0$, are given by $T^{i_1 \cdots i_{t-1} k i_{t+1} \cdots i_r}_{\quad\quad\quad\quad\quad j_1 \cdots j_{u-1} k j_{u+1} \cdots j_s}$.

4. *Permutation of indices:* Let T be any tensor of type $(0, r)$ and S_r the group of permutations of the set $\{1, \cdots, r\}$. The components of the tensor obtained by permuting the indices of T with any $\sigma \in S_r$, are given by $(\sigma T)_{i_1 \cdots i_r} = T_{i_{\sigma(1)} \cdots i_{\sigma(r)}}$. The *symmetric part* of T, denoted by $\mathcal{S}(T)$, is the tensor whose components are given by

$$\mathcal{S}(T)_{i_1 \cdots i_r} = T_{(i_1 \cdots i_r)} = \frac{1}{r!} \sum_{\sigma \in S_r} T_{i_{\sigma(1)} \cdots i_{\sigma(r)}}. \qquad (5.10.6)$$

The tensor T is said to be *symmetric* if, and only if, $T_{i_1 \cdots i_r} = T_{(i_1 \cdots i_r)}$. The *skew symmetric* part of T, denoted by $\mathcal{A}(T)$, is the tensor whose components are given by

$$\mathcal{A}(T)_{i_1 \cdots i_r} = T_{[i_1 \cdots i_r]} = \frac{1}{r!} \sum_{\sigma \in S_r} \text{sgn}(\sigma) T_{i_{\sigma(1)} \cdots i_{\sigma(r)}}, \qquad (5.10.7)$$

where $\text{sgn}(\sigma) = \pm 1$ according to whether σ is an even or odd permutation. The tensor T is said to be *skew symmetric* if, and only if, $T_{i_1 \cdots i_r} = T_{[i_1 \cdots i_r]}$. If $r = 2$, $\mathcal{S}(T)_{i_1 i_2} = \frac{1}{2}(T_{i_1 i_2} + T_{i_2 i_1})$ and $\mathcal{A}(T)_{i_1 i_2} = \frac{1}{2}(T_{i_1 i_2} - T_{i_2 i_1})$.

5.10.3 DIFFERENTIATION OF TENSORS

1. In tensor analysis a comma is used to denote partial differentiation and a semicolon to denote covariant differentiation. For example, $A_{ij\ldots l,n}$ denotes $\partial A_{ij\ldots l}/\partial x^n$.

2. A *linear connection* ∇ at $p \in M$ is an object which, with respect to each coordinate system on M, is represented by n^3 real numbers $\Gamma^i_{\ jk}$, called the *connection coefficients*, whose values in any two coordinate systems ϕ and $\tilde{\phi}$ are related by

$$\tilde{\Gamma}^i_{\ jk} = \Gamma^\ell_{\ mn} \frac{\partial \tilde{x}^i}{\partial x^\ell} \frac{\partial x^m}{\partial \tilde{x}^j} \frac{\partial x^n}{\partial \tilde{x}^k} + \frac{\partial^2 x^\ell}{\partial \tilde{x}^j \partial \tilde{x}^k} \frac{\partial \tilde{x}^i}{\partial x^\ell}. \qquad (5.10.8)$$

The quantities $\Gamma^i_{\ jk}$ are *not* the components of a tensor of type $(1, 2)$. A linear connection ∇ on M is an assignment of a linear connection to each point of M. A connection ∇ is C^k if its connection coefficients $\Gamma^i_{\ jk}$ are C^k in every coordinate system on M.

3. The components of the *covariant derivative* of a tensor field T of type (r, s), with respect to a connection ∇, are given by

$$\nabla_k T^{i_1 \cdots i_r}_{\quad j_1 \cdots j_s} = \partial_k T^{i_1 \cdots i_r}_{\quad j_1 \cdots j_s} + \Gamma^{i_1}_{\ \ell k} T^{\ell i_2 \cdots i_r}_{\quad j_1 \cdots j_s} + \cdots \qquad (5.10.9)$$
$$\cdots + \Gamma^{i_r}_{\ \ell k} T^{i_1 \cdots i_{r-1} \ell}_{\quad j_1 \cdots j_s} - \Gamma^\ell_{\ j_1 k} T^{i_1 \cdots i_r}_{\quad \ell j_2 \cdots j_s} \cdots - \Gamma^\ell_{\ j_s k} T^{i_1 \cdots i_r}_{\quad j_1 \cdots j_{s-1} \ell},$$

where

$$\partial_k T^{i_1 \cdots i_r}{}_{j_i \cdots j_s} = T^{i_1 \cdots i_r}{}_{j_1 \cdots j_s, k} = \frac{\partial T^{i_1 \cdots i_r}{}_{j_1 \cdots j_s}}{\partial x^k}. \tag{5.10.10}$$

This formula has this structure:

(a) Apart from the partial derivative term, there is a negative affine term for each covariant index and a positive affine term for each contravariant index.

(b) The second subscript in the Γ-symbols is always the differentiated index (k in this case).

4. Let $Y^i(t)$ be a contravariant vector field and $Z_i(t)$ a covariant vector field defined along a parameterized curve γ. The *absolute covariant derivatives* of Y^i and Z_i are defined as follows:

$$\begin{aligned} \frac{\delta Y^i}{\delta t} &= \frac{dY^i}{dt} + \Gamma^i{}_{jk} Y^j \frac{dx^k}{dt} \\ \frac{\delta Z_i}{\delta t} &= \frac{dZ_i}{dt} - \Gamma^j{}_{ik} Z_j \frac{dx^k}{dt} \end{aligned} \tag{5.10.11}$$

where x^i denotes the components of γ in the coordinate system ϕ. This derivative may also be defined for tensor fields of type (r, s) defined along γ. The derivative has the same structure as Equation (5.10.9).

5. A vector field $Y^i(t)$ is said to be *parallel along* a parameterized curve γ if $\frac{\delta Y^i}{\delta t} = 0$.

6. A parameterized curve γ in M is said to be an *affinely parameterized geodesic* if the component functions of γ satisfy

$$\frac{\delta}{\delta t}\left(\frac{dx^i}{dt}\right) = \frac{d^2 x^i}{dt^2} + \Gamma^i{}_{jk} \frac{dx^j}{dt} \frac{dx^k}{dt} = 0$$

which is equivalent to the statement that the tangent vector $\frac{dx^i}{dt}$ to γ is parallel along γ.

7. The components of the *torsion tensor* S of ∇ on M are defined by

$$S^i{}_{jk} = \Gamma^i{}_{jk} - \Gamma^i{}_{kj}. \tag{5.10.12}$$

8. The components of the *curvature tensor* R of ∇ on M are defined by

$$R^i{}_{jk\ell} = \partial_k \Gamma^i{}_{j\ell} - \partial_\ell \Gamma^i{}_{jk} + \Gamma^m{}_{j\ell}\Gamma^i{}_{mk} - \Gamma^m{}_{jk}\Gamma^i{}_{m\ell}. \tag{5.10.13}$$

In some references R is defined with the opposite sign.

9. The *Ricci tensor* of ∇ is defined by $R_{jk} = R^\ell{}_{jk\ell}$.

5.10.4   METRIC TENSOR

1. A *covariant metric tensor field* on M is a tensor field g_{ij} which satisfies $g_{ij} = g_{ji}$ and $g = \det(g_{ij}) \neq 0$ on M. The *contravariant metric* g^{ij} satisfies $g^{ik}g_{kj} = \delta^i_j$. The *line element* is expressible in terms of the metric tensor as $ds^2 = g_{ij}dx^i dx^j$.

2. *Signature of the metric:* For each $p \in M$, a coordinate system exists such that $g_{ij}(p) = \text{diag}\,(\underbrace{1, \cdots, 1}_{r}, \underbrace{-1, \cdots, -1}_{n-r})$. The *signature* of g_{ij} is defined to be $s = 2r - n$. It is independent of the coordinate system in which $g_{ij}(p)$ has the above diagonal form and is the same at every $p \in M$. A metric is said to be *positive definite* if $s = n$.

   A manifold admitting a positive definite metric is called a *Riemannian manifold*. A metric is said to be *indefinite* if $s \neq \pm n$. A manifold, admitting an indefinite metric is called a *pseudo-Riemannian manifold*. If $s = \pm(n-2)$ the metric is said to be *Lorentzian* and the corresponding manifold is called a *Lorentzian manifold*.

3. The *inner product* of a pair of vectors X^i and Y^j is given by $g_{ij}X^iY^j$. If $X^i = Y^i$, then $g_{ij}X^iX^j$ defines the "square" of the length of X^i. If g_{ij} is positive definite, then $g_{ij}X^iX^j \geq 0$ for all X^i, and $g_{ij}X^iX^j = 0$ if, and only if, $X^i = 0$. In the positive definite case, the *angle* θ between two tangent vectors X^i and Y^j is defined by $\cos\theta = g_{ij}X^iY^j/(g_{k\ell}X^kX^\ell g_{mn}Y^mY^n)^{\frac{1}{2}}$. If g is indefinite, $g_{ij}X^iX^j$ may have a positive, negative, or zero value. A non-zero vector X^i, satisfying $g_{ij}X^iX^j = 0$, is called a *null vector*. If g_{ij} is indefinite, it is not possible in general to define the angle between two tangent vectors.

4. *Operation of lowering indices:* The components of the tensor resulting from lowering the tth contravariant index of a tensor T of type (r, s), with $r \geq 1$, are given by

$$T^{i_1 \cdots i_{t-1} \cdot i_{t+1} \cdots i_r}_{\phantom{i_1 \cdots i_{t-1}} i_t \phantom{\cdot i_{t+1} \cdots i_r} j_1 \cdots j_s} = g_{i_t k} T^{i_1 \cdots i_{t-1} k i_{t+1} \cdots i_r}_{\phantom{i_1 \cdots i_{t-1} k i_{t+1} \cdots i_r} j_1 \cdots j_s}. \tag{5.10.14}$$

5. *Operation of raising indices:* The components of the tensor resulting from raising the tth covariant index of a tensor T of type (r, s), with $s \geq 1$, are given by

$$T^{ i_1 \cdots i_r}_{j_1 \cdots j_{t-1} \cdot j_{t+1} \cdots j_s} {}^{j_t} = g^{j_t k} T^{i_1 \cdots i_r}_{j_1 \cdots j_{t-1} k j_{t+1} \cdots j_s}. \tag{5.10.15}$$

6. The *arc length* of a parameterized curve $\gamma : I \to M$, where $I = [a, b]$, and ϕ is any coordinate system, is defined by

$$L = \int_a^b \sqrt{\epsilon g_{ij}(x^1(t), \cdots, x^n(t))\dot{x}^i \dot{x}^j}\, dt, \tag{5.10.16}$$

where $\epsilon = \text{sgn}(g_{ij}\dot{x}^i\dot{x}^j) = \pm 1$ and $\dot{x}^i = \frac{dx^i}{dt}$.

5.10.5 RESULTS

The following results hold on any manifold M admitting any connection ∇:

1. The covariant derivative operator ∇_k is linear with respect to tensor addition, satisfies the product rule with respect to tensor multiplication, and commutes with contractions of tensors.

2. If T is any tensor of type $(0, r)$, then

$$\nabla_{[k}T_{i_1\cdots i_r]} = T_{[i_1\cdots i_r, k]} - \frac{1}{2}\left(S^\ell_{[i_1 k}T_{|\ell|i_2\cdots i_r]} + \cdots + S^\ell_{[i_r k}T_{i_1\cdots i_{r-1}|\ell]}\right),$$
(5.10.17)

where $||$ indicates that the enclosed indices are excluded from the symmetrization. Thus $T_{[i_1\cdots i_r, k]}$ defines a tensor of type $(0, r + 1)$, and $\nabla_{[k}T_{i_1\cdots i_r]} = T_{[i_1\cdots i_r, k]}$ in the torsion free case. If $T_j = \nabla_j f = f_{,j}$, where f is any scalar invariant, then $\nabla_{[i}\nabla_{j]}f = \frac{1}{2}f_{,k}S^k{}_{ij}$. In the torsion free case, $\nabla_i\nabla_j f = \nabla_j\nabla_i f$.

3. If X^i is any contravariant vector field on M, then the identity $2\nabla_{[j}\nabla_{k]}X^i + \nabla_\ell X^i S^\ell_{jk} = X^\ell R^i{}_{\ell jk}$, called the *Ricci identity*, reduces to $2\nabla_{[j}\nabla_{k]}X^i = R^i{}_{\ell jk}X^\ell$, in the torsion free case. If Y_i is any covariant vector field, the Ricci identity has the form $2\nabla_{[i}\nabla_{j]}Y_k - \nabla_\ell Y_k S^\ell_{ij} = -Y_\ell R^\ell{}_{kij}$. The Ricci identity may be extended to tensor fields of type (r, s). For the tensor field $T^i{}_{jk}$, it has the form

$$2\nabla_{[i}\nabla_{j]}T^k{}_{\ell m} - \nabla_n T^k{}_{\ell m}S^n{}_{ij} = T^n{}_{\ell m}R^k{}_{nij} - T^k{}_{nm}R^n{}_{\ell ij} - T^k{}_{\ell n}R^n{}_{mij}.$$
(5.10.18)

If g is any metric tensor field, the above identity implies that $R_{(ij)k\ell} = \nabla_{[k}\nabla_{\ell]}g_{ij} - \frac{1}{2}\nabla_m g_{ij}S^m{}_{k\ell}$.

4. The torsion tensor S and curvature tensor R satisfy the following identities:

$$S^i{}_{(jk)} = 0, \qquad\qquad 0 = R^i{}_{j[kl;m]} - R^i{}_{jn[k}S^n{}_{\ell m]}, \qquad (5.10.19)$$

$$R^i{}_{j(k\ell)} = 0, \qquad\qquad R^i{}_{[jk\ell]} = -S^i{}_{[jk;\ell]} + S^i{}_{m[j}S^m{}_{k\ell]}. \qquad (5.10.20)$$

In the torsion free case, these identities reduce to the *cyclical identity* $R^i{}_{[jk\ell]} = 0$ and *Bianchi's identity* $R^i{}_{j[k\ell;m]} = 0$.

The following results hold for any pseudo-Riemannian manifold M with a metric tensor field g_{ij}:

1. A unique connection ∇ called the *Levi–Civita* or *pseudo-Riemannian connection* with vanishing torsion $(S^i{}_{jk} = 0)$ exists that satisfies $\nabla_i g_{jk} = 0$. It follows that $\nabla_i g^{jk} = 0$. *Christoffel symbols of the first kind.* The connection coefficients of ∇, called the *Christoffel symbols of the second kind*, are given by $\Gamma^i{}_{jk} = g^{i\ell}[jk, \ell]$, where $[jk, \ell] = \frac{1}{2}(g_{j\ell,k} + g_{k\ell,j} - g_{jk,\ell})$ are the $\Gamma^k{}_{jk} = \frac{1}{2}\partial_j(\log g) = |g|^{-\frac{1}{2}}\partial_j|g|^{\frac{1}{2}}$ and $g_{ij,k} = [ki, j] + [kj, i]$.

2. The operations of raising and lowering indices commute with the covariant derivative. For example if $X_i = g_{ij}X^j$, then $\nabla_k X_i = g_{ij}\nabla_k X^j$.

3. The *divergence* of a vector X^i is given by $\nabla_i X^i = |g|^{-\frac{1}{2}}\partial_i(|g|^{\frac{1}{2}}X^i)$.

4. The *Laplacian* of a scalar invariant f is given by
$$\Delta f = g^{ij}\nabla_i\nabla_j f = \nabla_i(g^{ij}\nabla_j f) = |g|^{-\frac{1}{2}}\partial_i(|g|^{\frac{1}{2}}g^{ij}\partial_j f).$$

5. The equations of an affinely parameterized geodesic may be written as
$$\frac{d}{dt}(g_{ij}\dot{x}^j) - \frac{1}{2}g_{jk,i}\dot{x}^j\dot{x}^k = 0.$$

6. Let X^i and Y^i be the components of any vector fields which are propagated in parallel along any parameterized curve γ. Then $\frac{d}{dt}(g_{ij}X^iY^j) = 0$, which implies that the inner product $g_{ij}X^iY^j$ is constant along γ. In particular, if $\dot{x}^i$ are the components of the tangent vector to γ, then $g_{ij}\dot{x}^i\dot{x}^j$ is constant along γ.

7. The *Riemann tensor*, defined by $R_{ijk\ell} = g_{im}R^m_{jk\ell}$, is given by
$$R_{ijk\ell} = [j\ell, i]_{,k} - [jk, i]_{,\ell} + [i\ell, m]\Gamma^m_{jk} - [ik, m]\Gamma^m_{j\ell}$$
$$= \frac{1}{2}(g_{i\ell,jk} + g_{jk,i\ell} - g_{j\ell,ik} - g_{ik,j\ell}) \qquad (5.10.21)$$
$$+ g^{mn}([i\ell, m][jk, n] - [ik, m][j\ell, n]).$$

It has the following symmetries:
$$R_{ij(k\ell)} = R_{(ij)k\ell} = 0, \quad R_{ijk\ell} = R_{k\ell ij}, \text{ and } R_{i[jk\ell]} = 0. \quad (5.10.22)$$

Consequently it has a maximum of $n^2(n^2 - 1)/12$ independent components.

8. The equations $R_{ijk\ell} = 0$ are necessary and sufficient conditions for M to be a *flat* pseudo-Riemannian manifold, that is, a manifold for which a coordinate system exists so that the components g_{ij} are *constant* on M.

9. The Ricci tensor is given by
$$R_{ij} = \partial_j\Gamma^k_{ik} - \partial_k\Gamma^k_{ij} + \Gamma^k_{i\ell}\Gamma^\ell_{kj} - \Gamma^k_{ij}\Gamma^\ell_{k\ell}$$
$$= \frac{1}{2}\partial_i\partial_j(\log|g|) - \frac{1}{2}\Gamma^k_{ij}\partial_k(\log|g|) - \partial_k\Gamma^k_{ij} + \Gamma^k_{im}\Gamma^m_{kj}.$$

It possesses the symmetry $R_{ij} = R_{ji}$, and thus has a maximum of $n(n+1)/2$ independent components.

10. The *scalar curvature* or *curvature invariant* is defined by $R = g^{ij}R_{ij}$.

11. The *Einstein tensor* is defined by $G_{ij} = R_{ij} - \frac{1}{2}Rg_{ij}$. In view of the Bianchi identity, it satisfies: $g^{jk}\nabla_j G_{ki} = 0$.

12. A *normal coordinate system* with origin $x_0 \epsilon M$ is defined by $\overset{0}{g}_{ij}\,x^j = g_{ij}x^j$, where a "0" affixed over a quantity indicates that the quantity is evaluated at x_0. The connection coefficients satisfy $\overset{0}{\Gamma}{}^i_{(j_1 j_2, j_3 \cdots j_r)} = 0$ (for $r = 2, 3, \ldots$) in any normal coordinate system. The equations of the geodesics through x_0 are given by $x^i = sk^i$, where s is an affine parameter and k^i is any constant vector.

5.10.6 EXAMPLES OF TENSORS

1. The components of the *gradient* of a scalar invariant $\frac{\partial f}{\partial x^i}$ define a tensor of type (0,1), since they transform as $\frac{\partial \tilde{f}}{\partial \tilde{x}^i} = \frac{\partial f}{\partial x^j} \frac{\partial x^j}{\partial \tilde{x}^i}$.

2. The components of the *tangent vector* to a parameterized curve $\frac{dx^i}{dt}$ define a tensor of type (1,0), because they transform as $\frac{d\tilde{x}^i}{dt} = \frac{dx^j}{dt} \frac{\partial \tilde{x}^i}{\partial x^j}$.

3. The determinant of the metric tensor g defines a relative scalar invariant of weight of $w = 2$, because it transforms as $\tilde{g} = \left| \frac{\partial x^i}{\partial \tilde{x}^j} \right|^2 g$.

4. The Kronecker deltas δ_i^j are the components of a constant absolute tensor of type $(1, 1)$, because $\delta_j^i = \delta_\ell^k \frac{\partial \tilde{x}^i}{\partial x^k} \frac{\partial x^\ell}{\partial \tilde{x}^j}$.

5. The permutation symbol defined by

$$e_{i_1 \cdots i_n} = \begin{cases} 1, & \text{if } i_1 \cdots i_n \text{ is an even permutation of } 1 \cdots n, \\ -1, & \text{if } i_1 \cdots i_n \text{ is an odd permutation of } 1 \cdots n, \text{ and} \\ 0 & \text{otherwise,} \end{cases}$$

(5.10.23)

satisfies $\left| \frac{\partial x^i}{\partial \tilde{x}^i} \right| e_{j_1 \cdots j_n} = e_{i_1 \cdots i_n} \frac{\partial x^{i_1}}{\partial \tilde{x}^{j_1}} \cdots \frac{\partial x^{i_n}}{\partial \tilde{x}^{j_n}}$. Hence it defines a tensor of type $(0, n, -1)$, that is, it is a relative tensor of weight $w = -1$. The contravariant permutation symbol $e^{i_1 \cdots i_n}$, defined in a similar way, is a relative tensor of weight $w = 1$.

6. The Levi–Civita symbol, $\epsilon_{i_1 \cdots i_n} = |g|^{\frac{1}{2}} e_{i_1 \cdots i_n}$, defines a covariant absolute tensor of valence n. The contravariant Levi–Civita tensor satisfies

$$\epsilon^{i_1 \cdots i_n} = g^{i_1 j_1} \cdots g^{i_n j_n} \epsilon_{j_1 \cdots j_n} = (-1)^{\frac{n-s}{2}} |g|^{-\frac{1}{2}} e^{i_1 \cdots i_n}. \quad (5.10.24)$$

Using this symbol, the *dual* of a covariant skew-symmetric tensor of valence r is defined by $*T_{i_1 \cdots i_{n-r}} = \frac{1}{r!} \epsilon^{j_1 \cdots j_r}{}_{i_1 \cdots i_{n-r}} T_{j_1 \cdots j_r}$.

7. *Cartesian tensors:* Let $M = E^3$ (i.e., Euclidean three-space) with metric tensor $g_{ij} = \delta_{ij}$ with respect to Cartesian coordinates. The components of a *Cartesian tensor* of valence r transform as

$$\tilde{T}_{i_1 \cdots i_r} = T_{j_1 \cdots j_r} O_{i_1 j_1} \cdots O_{i_r j_r}, \quad (5.10.25)$$

where O_{ij} are the components of a constant orthogonal matrix which satisfies $(O^{-1})_{ij} = (O^T)_{ij} = O_{ji}$. For Cartesian tensors, all indices are written as covariant, because *no* distinction is required between covariant and contravariant indices.

An *oriented Cartesian tensor* is a Cartesian tensor where the orthogonal matrix in the transformation law is restricted by $\det(O_{ij}) = 1$. The Levi–Civita symbol ϵ_{ijk} is an example of an oriented Cartesian tensor as is the cross product, $(X \times Y)_i = \epsilon_{ijk} X_j Y_k$, of two vectors. The connection coefficients satisfy

$\Gamma^i{}_{jk} = 0$ in every Cartesian coordinate system on E^3. Thus the partial derivatives of Cartesian tensors are themselves Cartesian tensors, that is, if $T_{i_1 \cdots i_r}$ is a Cartesian tensor, then so is $\partial_k T_{i_1 \cdots i_r}$. A particular example is the curl of a vector field X_i given by $(\operatorname{curl} X)_i = \epsilon_{ijk} \partial_j X_k$ which defines an oriented Cartesian tensor.

8. Note the useful relations: $\epsilon_{ijk}\epsilon_{klm} = \delta_{il}\delta_{jm} - \delta_{im}\delta_{jl}$, $\epsilon_{ikl}\epsilon_{klm} = 2\delta_{im}$, and $\epsilon_{ijk}\epsilon_{lmn} = \delta_{il}\delta_{jm}\delta_{kn} + \delta_{im}\delta_{jn}\delta_{kl} + \delta_{in}\delta_{jl}\delta_{km} - \delta_{in}\delta_{jm}\delta_{kl} - \delta_{im}\delta_{jl}\delta_{kn} - \delta_{il}\delta_{jn}\delta_{km}$.

9. The stress tensor E_{ij} and the strain tensor e_{ij} are examples of Cartesian tensors.

10. *Orthogonal curvilinear coordinates*: Let M be a 3-dimensional Riemannian manifold admitting a coordinate system $[x^1, x^2, x^3]$ such that the metric tensor has the form $g_{ii} = h_i^2(x^1, x^2, x^3)$ for $i = 1, \ldots, 3$ with $g_{ij} = g^{ij} = 0$ for $i \neq j$. The metric tensor on E^3 has this form with respect to orthogonal curvilinear coordinates. The non-zero components of various corresponding quantities corresponding to this metric are as follows:

(a) Covariant metric tensor

$$g_{11} = h_1{}^2, \qquad g_{22} = h_2{}^2, \qquad g_{33} = h_3{}^2 \qquad (5.10.26)$$

(b) Contravariant metric tensor

$$g^{11} = h_1{}^{-2}, \qquad g^{22} = h_2{}^{-2}, \qquad g^{33} = h_3{}^{-2} \qquad (5.10.27)$$

(c) Christoffel symbols of the first kind (note that $[ij, k] = 0$ if i, j, and k are all different),

$$
\begin{array}{lll}
[11, 1] = h_1 h_{1,1} & [11, 2] = -h_1 h_{1,2} & [11, 3] = -h_1 h_{1,3} \\
[12, 1] = h_1 h_{1,2} & [12, 2] = h_2 h_{2,1} & [13, 1] = h_1 h_{1,3} \\
[13, 3] = h_3 h_{3,1} & [22, 1] = -h_2 h_{2,1} & [22, 2] = h_2 h_{2,2} \\
[22, 3] = -h_2 h_{2,3} & [23, 2] = h_2 h_{2,3} & [23, 3] = h_3 h_{3,2} \\
[33, 1] = -h_3 h_{3,1} & [33, 2] = -h_3 h_{3,2} & [33, 3] = h_3 h_{3,3}.
\end{array}
$$

(d) Christoffel symbols of the second kind (note that $\Gamma^k{}_{ij} = 0$ if i, j, and k are all different),

$$
\begin{array}{lll}
\Gamma^1{}_{11} = h_1{}^{-1} h_{1,1} & \Gamma^1{}_{12} = h_1{}^{-1} h_{1,2} & \Gamma^1{}_{13} = h_1{}^{-1} h_{1,3} \\
\Gamma^1{}_{22} = -h_1{}^{-2} h_2 h_{2,1} & \Gamma^1{}_{33} = -h_1{}^{-2} h_3 h_{3,1} & \Gamma^2{}_{11} = -h_1 h_2{}^{-2} h_{1,2} \\
\Gamma^2{}_{12} = h_2{}^{-1} h_{2,1} & \Gamma^2{}_{22} = h_2{}^{-1} h_{2,2} & \Gamma^2{}_{23} = h_2{}^{-1} h_{2,3} \\
\Gamma^2{}_{33} = -h_2{}^{-2} h_3 h_{3,2} & \Gamma^3{}_{11} = -h_1 h_3{}^{-2} h_{1,3} & \Gamma^3{}_{13} = h_3{}^{-1} h_{3,1} \\
\Gamma^3{}_{22} = -h_2 h_3{}^{-2} h_{2,3} & \Gamma^3{}_{23} = h_3{}^{-1} h_{3,2} & \Gamma^3{}_{33} = h_3{}^{-1} h_{3,3}.
\end{array}
$$

(e) Vanishing Riemann tensor conditions (Lamé equations),

$$h_{1,2,3} - h_2{}^{-1}h_{1,2}h_{2,3} - h_3{}^{-1}h_{1,3}h_{3,2} = 0,$$

$$h_{2,1,3} - h_1{}^{-1}h_{1,3}h_{2,1} - h_3{}^{-1}h_{3,1}h_{2,3} = 0,$$

$$h_{3,1,2} - h_1{}^{-1}h_{1,2}h_{3,1} - h_2{}^{-1}h_{2,1}h_{3,2} = 0,$$

$$h_2 h_{2,3,3} + h_3 h_{3,2,2} + h_1{}^{-2}h_2 h_3 h_{2,1} h_{3,1} - h_2{}^{-1}h_3 h_{2,2} h_{3,2}$$
$$- h_2 h_3{}^{-1}h_{2,3} h_{3,3} = 0,$$

$$h_1 h_{1,3,3} + h_3 h_{3,1,1} + h_1 h_2{}^{-2}h_3 h_{1,2} h_{3,2} - h_1{}^{-1}h_3 h_{1,1} h_{3,1}$$
$$- h_1 h_3{}^{-1}h_{1,3} h_{3,3} = 0,$$

$$h_1 h_{1,2,2} + h_2 h_{2,1,1} + h_1 h_2 h_3{}^{-2}h_{1,3} h_{2,3} - h_1{}^{-1}h_2 h_{1,1} h_{2,1}$$
$$- h_1 h_2{}^{-1}h_{1,2} h_{2,2} = 0.$$

11. *The 2-sphere*: A coordinate system $[\theta, \phi]$ for the 2-sphere $x^2 + y^2 + z^2 = r^2$ is given by $x = r \sin\theta \cos\phi$, $y = r \sin\theta \sin\phi$, $z = r \cos\theta$, where $[\theta, \phi] \in U = (0, \pi) \times (0, 2\pi)$. This is a non-Euclidean space. The non-zero independent components of various quantities defined on the sphere are given below:

(a) Covariant metric tensor components are $g_{11} = r^2$, $g_{22} = r^2 \sin^2\theta$.

(b) Contravariant metric tensor components are $g^{11} = r^{-2}, g^{22} = r^{-2} \csc^2\theta$.

(c) Christoffel symbols of the first kind are $[12, 2] = r^2 \sin\theta \cos\theta$,
$[22, 1] = -r^2 \sin\theta \cos\theta$.

(d) Christoffel symbols of the second kind are $\Gamma^1{}_{22} = -\sin\theta \cos\theta$,
$\Gamma^2{}_{12} = -\cos\theta \csc\theta$.

(e) Covariant Riemann tensor components are $R_{1212} = r^2 \sin^2\theta$.

(f) Covariant Ricci tensor components are $R_{11} = -1$, $R_{22} = -\sin^2\theta$.

(g) Ricci scalar is $R = -2r^{-2}$.

12. *The 3-sphere*: A coordinate system $[\psi, \theta, \phi]$ for the 3-sphere $x^2 + y^2 + z^2 + w^2 = r^2$ is given by $x = r \sin\psi \sin\theta \cos\phi$, $y = r \sin\psi \sin\theta \sin\phi$, $z = r \sin\psi \cos\theta$, and $w = r \cos\psi$, where $[\psi, \theta, \phi] \in U = (0, \pi) \times (0, \pi) \times (0, 2\pi)$. The non-zero components of various quantities defined on the sphere are given below:

(a) Covariant metric tensor components

$$g_{11} = r^2, \qquad g_{22} = r^2 \sin^2\psi, \qquad g_{33} = r^2 \sin^2\psi \sin^2\theta \qquad (5.10.28)$$

(b) Contravariant metric tensor components

$$g^{11} = r^{-2}, \qquad g^{22} = r^{-2} \csc^2\psi, \qquad g^{33} = r^{-2} \csc^2\psi \csc^2\theta$$
$$(5.10.29)$$

(c) Christoffel symbols of the first kind,

$$[22,1] = -r^2 \sin\psi \cos\psi \qquad\qquad [33,1] = -r^2 \sin\psi \cos\psi \sin^2\theta$$
$$[12,2] = r^2 \sin\psi \cos\psi \qquad\qquad [33,2] = -r^2 \sin^2\psi \sin\theta \cos\theta$$
$$[13,3] = r^2 \sin\psi \cos\psi \sin^2\theta \qquad [23,3] = r^2 \sin^2\psi \sin\theta \cos\theta.$$

(d) Christoffel symbols of the second kind,

$$\Gamma^1_{22} = -\sin\psi \cos\psi \qquad\qquad \Gamma^1_{33} = -\sin\psi \cos\psi \sin^2\theta$$
$$\Gamma^2_{12} = \cot\psi \qquad\qquad\qquad \Gamma^2_{33} = -\sin\theta \cos\theta$$
$$\Gamma^3_{13} = \cot\psi \qquad\qquad\qquad \Gamma^3_{23} = \cot\theta.$$

(e) Covariant Riemann tensor components,

$$R_{1212} = r^2 \sin^2\psi, \quad R_{1313} = r^2 \sin^2\psi \sin^2\theta, \quad R_{2323} = r^2 \sin^4\psi \sin^2\theta.$$

(f) Covariant Ricci tensor components,

$$R_{11} = -2, \quad R_{22} = -2\sin^2\psi, \quad R_{33} = -2\sin^2\psi \sin^2\theta \,.$$

(g) The Ricci scalar is $R = -6r^{-2}$.

(h) Covariant Einstein tensor components

$$G_{11} = 1, \qquad G_{22} = \sin^2\psi, \qquad G_{33} = \sin^2\psi \sin^2\theta \qquad (5.10.30)$$

13. *Polar coordinates:* The line element is given by $ds^2 = dr^2 + r^2 d\theta^2$. Thus the metric tensor is $g_{ij} = \begin{pmatrix} 1 & 0 \\ 0 & r^2 \end{pmatrix}$, and the non-zero Christoffel symbols are $[21,2] = [12,2] = -[22,1] = r$.

5.11   ORTHOGONAL COORDINATE SYSTEMS

In an orthogonal coordinate system, let $\{\mathbf{a}_i\}$ denote the unit vectors in each of the three coordinate directions, and let $\{u_i\}$ denote distance along each of these axes. The coordinate system may be designated by the *metric coefficients* $\{g_{11}, g_{22}, g_{33}\}$, defined by

$$g_{ii} = \left(\frac{\partial x_1}{\partial u_i}\right)^2 + \left(\frac{\partial x_2}{\partial u_i}\right)^2 + \left(\frac{\partial x_3}{\partial u_i}\right)^2, \qquad (5.11.1)$$

where $\{x_1, x_2, x_3\}$ represent rectangular coordinates. Then define $g = g_{11}g_{22}g_{33}$.

Operations for orthogonal coordinate systems are sometimes written in terms of the $\{h_i\}$ functions, instead of the $\{g_{ii}\}$ functions. Here, $h_i = \sqrt{g_{ii}}$, so that $\sqrt{g} = h_1 h_2 h_3$. For example, in cylindrical coordinates, $\{x_1 = r\cos\theta, x_2 = r\sin\theta, x_3 = z\}$ with $\{u_1 = r, u_2 = \theta, u_3 = z\}$, we find $\{g_{11} = 1, g_{22} = r^2, g_{33} = 1\}$ so that $\{h_1 = 1, h_2 = r, h_3 = 1\}$.

In the following, ϕ represents a scalar, and $\mathbf{E} = E_1\mathbf{a}_1 + E_2\mathbf{a}_2 + E_3\mathbf{a}_3$ and $\mathbf{F} = F_1\mathbf{a}_1 + F_2\mathbf{a}_2 + F_3\mathbf{a}_3$ represent vectors.

$\operatorname{grad}\phi = \nabla\phi = $ the *gradient* of ϕ

$$= \frac{\mathbf{a}_1}{\sqrt{g_{11}}}\frac{\partial\phi}{\partial u_1} + \frac{\mathbf{a}_2}{\sqrt{g_{22}}}\frac{\partial\phi}{\partial u_2} + \frac{\mathbf{a}_3}{\sqrt{g_{33}}}\frac{\partial\phi}{\partial u_3}, \tag{5.11.2}$$

$\operatorname{div}\mathbf{E} = \nabla\cdot\mathbf{E} = $ the *divergence* of $\mathbf{E}$

$$= \frac{1}{\sqrt{g}}\left\{\frac{\partial}{\partial u_1}\left(\frac{gE_1}{g_{11}}\right) + \frac{\partial}{\partial u_2}\left(\frac{gE_2}{g_{22}}\right) + \frac{\partial}{\partial u_3}\left(\frac{gE_3}{g_{33}}\right)\right\}, \tag{5.11.3}$$

$\operatorname{curl}\mathbf{E} = \nabla\times\mathbf{E} = $ the *curl* of $\mathbf{E}$

$$= \mathbf{a}_1\frac{\Gamma_1}{\sqrt{g_{11}}} + \mathbf{a}_2\frac{\Gamma_2}{\sqrt{g_{22}}} + \mathbf{a}_3\frac{\Gamma_3}{\sqrt{g_{33}}} \tag{5.11.4}$$

$$= \begin{vmatrix} \frac{\mathbf{a}_1}{h_2 h_3} & \frac{\mathbf{a}_2}{h_1 h_3} & \frac{\mathbf{a}_3}{h_1 h_2} \\ \frac{\partial}{\partial u_1} & \frac{\partial}{\partial u_2} & \frac{\partial}{\partial u_3} \\ h_1 E_1 & h_2 E_2 & h_3 E_3 \end{vmatrix}, \tag{5.11.5}$$

$$[(\mathbf{F}\cdot\nabla)\,\mathbf{E}]_j = \sum_{i=1}^{3}\left[\frac{F_i}{h_i}\frac{\partial E_j}{\partial u_i} + \frac{E_i}{h_i h_j}\left(F_j\frac{\partial h_j}{\partial u_i} - F_i\frac{\partial h_i}{\partial u_j}\right)\right], \tag{5.11.6}$$

$\nabla^2\phi = $ the *Laplacian* of ϕ (sometimes written as $\Delta\phi$)

$$= \frac{1}{h_1 h_2 h_3}\left\{\frac{\partial}{\partial u_1}\left[\frac{h_2 h_3}{h_1}\frac{\partial\phi}{\partial u_1}\right] + \frac{\partial}{\partial u_2}\left[\frac{h_3 h_1}{h_2}\frac{\partial\phi}{\partial u_2}\right] + \frac{\partial}{\partial u_3}\left[\frac{h_1 h_2}{h_3}\frac{\partial\phi}{\partial u_3}\right]\right\}$$

$$= \frac{1}{\sqrt{g}}\left\{\frac{\partial}{\partial u_1}\left[\frac{\sqrt{g}}{g_{11}}\frac{\partial\phi}{\partial u_1}\right] + \frac{\partial}{\partial u_2}\left[\frac{\sqrt{g}}{g_{22}}\frac{\partial\phi}{\partial u_2}\right] + \frac{\partial}{\partial u_3}\left[\frac{\sqrt{g}}{g_{33}}\frac{\partial\phi}{\partial u_3}\right]\right\}, \tag{5.11.7}$$

$$\operatorname{grad}\operatorname{div}\mathbf{E} = \nabla(\nabla\cdot\mathbf{E}) = \frac{\mathbf{a}_1}{\sqrt{g_{11}}}\frac{\partial\Upsilon}{\partial x_1} + \frac{\mathbf{a}_2}{\sqrt{g_{22}}}\frac{\partial\Upsilon}{\partial x_2} + \frac{\mathbf{a}_3}{\sqrt{g_{33}}}\frac{\partial\Upsilon}{\partial x_3}, \tag{5.11.8}$$

$\operatorname{curl}\operatorname{curl}\mathbf{E} = \nabla\times(\nabla\times\mathbf{E})$

$$= \mathbf{a}_1\sqrt{\frac{g_{11}}{g}}\left[\frac{\partial\Gamma_3}{\partial x_2} - \frac{\partial\Gamma_2}{\partial x_3}\right] + \mathbf{a}_2\sqrt{\frac{g_{22}}{g}}\left[\frac{\partial\Gamma_1}{\partial x_3} - \frac{\partial\Gamma_3}{\partial x_1}\right] \tag{5.11.9}$$

$$+ \mathbf{a}_3\sqrt{\frac{g_{33}}{g}}\left[\frac{\partial\Gamma_2}{\partial x_1} - \frac{\partial\Gamma_1}{\partial x_2}\right],$$

$\text{✡}\mathbf{E} = \operatorname{grad}\operatorname{div}\mathbf{E} - \operatorname{curl}\operatorname{curl}\mathbf{E} = $ the *vector Laplacian* of $\mathbf{E}$

$$= \nabla(\nabla\cdot\mathbf{E}) - \nabla\times(\nabla\times\mathbf{E})$$

$$= \mathbf{a}_1\left\{\frac{1}{\sqrt{g_{11}}}\frac{\partial\Upsilon}{\partial x_1} + \sqrt{\frac{g_{11}}{g}}\left[\frac{\partial\Gamma_2}{\partial x_3} - \frac{\partial\Gamma_3}{\partial x_2}\right]\right\}$$

$$+ \mathbf{a}_2\left\{\frac{1}{\sqrt{g_{22}}}\frac{\partial\Upsilon}{\partial x_2} + \sqrt{\frac{g_{22}}{g}}\left[\frac{\partial\Gamma_3}{\partial x_1} - \frac{\partial\Gamma_1}{\partial x_3}\right]\right\} \tag{5.11.10}$$

$$+ \mathbf{a}_3\left\{\frac{1}{\sqrt{g_{33}}}\frac{\partial\Upsilon}{\partial x_3} + \sqrt{\frac{g_{33}}{g}}\left[\frac{\partial\Gamma_1}{\partial x_2} - \frac{\partial\Gamma_2}{\partial x_1}\right]\right\},$$

where Υ and $\mathbf{\Gamma} = (\Gamma_1, \Gamma_2, \Gamma_3)$ are defined by

$$
\Upsilon = \frac{1}{\sqrt{g}} \left\{ \frac{\partial}{\partial x_1} \left[E_1 \sqrt{\frac{g}{g_{11}}} \right] + \frac{\partial}{\partial x_2} \left[E_2 \sqrt{\frac{g}{g_{22}}} \right] + \frac{\partial}{\partial x_3} \left[E_3 \sqrt{\frac{g}{g_{33}}} \right] \right\},
$$

$$
\Gamma_1 = \frac{g_{11}}{\sqrt{g}} \left\{ \frac{\partial}{\partial x_2} \left(\sqrt{g_{33}} E_3 \right) - \frac{\partial}{\partial x_3} \left(\sqrt{g_{22}} E_2 \right) \right\},
$$

$$
\Gamma_2 = \frac{g_{22}}{\sqrt{g}} \left\{ \frac{\partial}{\partial x_3} \left(\sqrt{g_{11}} E_1 \right) - \frac{\partial}{\partial x_1} \left(\sqrt{g_{33}} E_3 \right) \right\}, \qquad (5.11.11)
$$

$$
\Gamma_3 = \frac{g_{33}}{\sqrt{g}} \left\{ \frac{\partial}{\partial x_1} \left(\sqrt{g_{22}} E_2 \right) - \frac{\partial}{\partial x_2} \left(\sqrt{g_{11}} E_1 \right) \right\}.
$$

5.11.1   TABLE OF ORTHOGONAL COORDINATE SYSTEMS

The $\{f_i\}$ listed below are the separated components of the Laplace or Helmholtz equations (see (5.7.18) and (5.7.19)). See Moon and Spencer for details.

The corresponding equations for the four separable coordinate systems which exist in two dimensions may be obtained from the coordinate systems 1–4 listed below by suppressing the z coordinate and assuming that all functions depend only on the x and y coordinates.

1. **Rectangular coordinates** $\{x, y, z\}$

   Ranges: $-\infty < x < \infty, -\infty < y < \infty, -\infty < z < \infty$.

   $g_{11} = g_{22} = g_{33} = \sqrt{g} = 1$,

   $f_1 = f_2 = f_3 = 1$.

   In this coordinate system the following notation is sometimes used:
   $\mathbf{i} = \mathbf{a}_x, \mathbf{j} = \mathbf{a}_y, \mathbf{k} = \mathbf{a}_z$.

   $$
   \text{grad } f = \mathbf{a}_x \frac{\partial f}{\partial x} + \mathbf{a}_y \frac{\partial f}{\partial y} + \mathbf{a}_z \frac{\partial f}{\partial z}
   $$

   $$
   \text{div } \mathbf{E} = \frac{\partial}{\partial x} (E_x) + \frac{\partial}{\partial y} (E_y) + \frac{\partial}{\partial z} (E_z)
   $$

   $$
   \text{curl } \mathbf{E} = \left(\frac{\partial E_z}{\partial y} - \frac{\partial E_y}{\partial z} \right) \mathbf{a}_x + \left(\frac{\partial E_x}{\partial z} - \frac{\partial E_z}{\partial x} \right) \mathbf{a}_y + \left(\frac{\partial E_y}{\partial x} - \frac{\partial E_x}{\partial y} \right) \mathbf{a}_z
   $$

   $$
   \nabla^2 f = \frac{\partial^2 f}{\partial x^2} + \frac{\partial^2 f}{\partial y^2} + \frac{\partial^2 f}{\partial z^2} \qquad (5.11.12)
   $$

   $$
   [(\mathbf{F} \cdot \nabla) \mathbf{E}]_x = F_x \frac{\partial E_x}{\partial x} + F_y \frac{\partial E_x}{\partial y} + F_z \frac{\partial E_x}{\partial z}
   $$

2. **Circular cylindrical coordinates** $\{r, \theta, z\}$

   Relations: $x = r \cos \theta, y = r \sin \theta, z = z$.

   Ranges: $0 \leq r < \infty, 0 \leq \theta < 2\pi, -\infty < z < \infty$.

$g_{11} = 1, g_{22} = r^2, g_{33} = 1, \sqrt{g} = r,$

$f_1 = r, f_2 = f_3 = 1.$

$$\operatorname{grad} f = \mathbf{a}_r \frac{\partial f}{\partial r} + \frac{\mathbf{a}_\theta}{r} \frac{\partial f}{\partial \theta} + \mathbf{a}_z \frac{\partial f}{\partial z}$$

$$\operatorname{div} \mathbf{E} = \frac{1}{r} \frac{\partial}{\partial r}(r E_r) + \frac{1}{r} \frac{\partial E_\theta}{\partial \theta} + \frac{\partial E_z}{\partial z}$$

$$(\operatorname{curl} \mathbf{E})_r = \frac{1}{r} \frac{\partial E_z}{\partial \theta} - \frac{\partial E_\theta}{\partial z}$$

$$(\operatorname{curl} \mathbf{E})_\theta = \frac{\partial E_r}{\partial z} - \frac{\partial E_z}{\partial r} \qquad (5.11.13)$$

$$(\operatorname{curl} \mathbf{E})_z = \frac{1}{r} \frac{\partial (r E_\theta)}{\partial r} - \frac{1}{r} \frac{\partial E_r}{\partial \theta}$$

$$\nabla^2 f = \frac{1}{r} \frac{\partial}{\partial r} \left(r \frac{\partial f}{\partial r} \right) + \frac{1}{r^2} \frac{\partial^2 f}{\partial \theta^2} + \frac{\partial^2 f}{\partial z^2}$$

3. **Elliptic cylindrical coordinates** $\{\eta, \psi, z\}$

   Relations: $x = a \cosh \eta \cos \psi, y = a \sinh \eta \sin \psi, z = z.$

   Ranges: $0 \le \eta < \infty, 0 \le \psi < 2\pi, -\infty < z < \infty.$

   $g_{11} = g_{22} = a^2 (\cosh^2 \eta - \cos^2 \psi), g_{33} = 1, \sqrt{g} = a^2 (\cosh^2 \eta - \cos^2 \psi),$

   $f_1 = f_2 = f_3 = 1.$

4. **Parabolic cylindrical coordinates** $\{\mu, \nu, z\}$

   Relations: $x = \frac{1}{2}(\mu^2 - \nu^2), y = \mu\nu, z = z.$

   Ranges: $0 \le \mu < \infty, -\infty < \nu < 2\pi, -\infty < z < \infty.$

   $g_{11} = g_{22} = \mu^2 + \nu^2, g_{33} = 1, \sqrt{g} = \mu^2 + \nu^2,$

   $f_1 = f_2 = f_3 = 1.$

5. **Spherical coordinates** $\{r, \theta, \psi\}$

   Relations: $x = r \sin \theta \cos \phi, y = r \sin \theta \sin \phi, z = r \cos \theta.$

   Ranges: $0 \le r < \infty, 0 \le \theta \le \pi, 0 \le \psi < 2\pi.$

   $g_{11} = 1, g_{22} = r^2, g_{33} = r^2 \sin^2 \theta, \sqrt{g} = r^2 \sin \theta,$

   $f_1 = r^2, f_2 = \sin \theta, f_3 = 1.$

$$\operatorname{grad} f = \mathbf{e}_r \frac{\partial f}{\partial r} + \frac{\mathbf{e}_\theta}{r} \frac{\partial f}{\partial \theta} + \frac{\mathbf{e}_\phi}{r \sin \theta} \frac{\partial f}{\partial \phi}$$

$$\operatorname{div} \mathbf{E} = \frac{1}{r^2} \frac{\partial}{\partial r}(r^2 E_r) + \frac{1}{r \sin \theta} \frac{\partial}{\partial \theta}(E_\theta \sin \theta) + \frac{1}{r \sin \theta} \frac{\partial E_\phi}{\partial \phi}$$

$$(\operatorname{curl} \mathbf{E})_r = \frac{1}{r \sin \theta} \left[\frac{\partial}{\partial \theta}(E_\phi \sin \theta) - \frac{\partial A_\theta}{\partial \phi} \right] \qquad (5.11.14)$$

$$(\operatorname{curl} \mathbf{E})_\theta = \frac{1}{r \sin \theta} \frac{\partial E_r}{\partial \phi} - \frac{1}{r} \frac{\partial (r E_\phi)}{\partial r}$$

$$(\operatorname{curl} \mathbf{E})_\phi = \frac{1}{r} \frac{\partial (r E_\theta)}{\partial r} - \frac{1}{r} \frac{\partial E_r}{\partial \theta}$$

$$\nabla^2 f = \frac{1}{r^2} \frac{\partial}{\partial r} \left(r^2 \frac{\partial f}{\partial r} \right) + \frac{1}{r^2 \sin \theta} \frac{\partial}{\partial \theta} \left(\sin \theta \frac{\partial f}{\partial \theta} \right) + \frac{1}{r^2 \sin^2 \theta} \frac{\partial^2 f}{\partial \phi^2}$$

6. **Prolate spheroidal coordinates** $\{\eta, \theta, \psi\}$

   Relations: $x = a \sinh \eta \sin \theta \cos \psi, y = a \sinh \eta \sin \theta \sin \psi, z = a \cosh \eta \cos \theta$.

   Ranges: $0 \le \eta < \infty, 0 \le \theta \le \pi, 0 \le \psi < 2\pi$.

   $g_{11} = g_{22} = a^2(\sinh^2 \eta + \sin^2 \theta)\ g_{33} = a^2 \sinh^2 \eta \sin^2 \theta$,
   $\sqrt{g} = a^3(\sinh^2 \eta + \sin^2 \theta) \sinh \eta \sin \theta$.

   $f_1 = \sinh \eta, f_2 = \sin \theta, f_3 = a$.

7. **Oblate spheroidal coordinates** $\{\eta, \theta, \psi\}$

   Relations: $x = a \cosh \eta \sin \theta \cos \psi, y = a \cosh \eta \sin \theta \sin \psi, z = a \sinh \eta \cos \theta$.

   Ranges: $0 \le \eta < \infty, 0 \le \theta \le \pi, 0 \le \psi < 2\pi$.

   $g_{11} = g_{22} = a^2(\cosh^2 \eta - \sin^2 \theta)\ g_{33} = a^2 \cosh^2 \eta \sin^2 \theta$,
   $\sqrt{g} = a^3(\cosh^2 \eta - \sin^2 \theta) \cosh \eta \sin \theta$.

   $f_1 = \cosh \eta, f_2 = \sin \theta, f_3 = a$.

8. **Parabolic coordinates** $\{\mu, \nu, \psi\}$

   Relations: $x = \mu\nu \cos \psi, y = \mu\nu \sin \psi, z = \frac{1}{2}(\mu^2 - \nu^2)$.

   Ranges: $0 \le \mu < \infty, 0 \le \nu \le \infty, 0 \le \psi < 2\pi$.

   $g_{11} = g_{22} = \mu^2 + \nu^2\ g_{33} = \mu^2 \nu^2, \sqrt{g} = \mu\nu(\mu^2 + \nu^2)$.

   $f_1 = \mu, f_2 = \nu, f_3 = 1$.

9. **Conical coordinates** $\{r, \theta, \lambda\}$

   Relations: $x^2 = (r\theta\lambda/bc)^2, y^2 = r^2(\theta^2 - b^2)(b^2 - \lambda^2)/[b^2(c^2 - b^2)], z^2 = r^2(c^2 - \theta^2)(c^2 - \lambda^2)/[c^2(c^2 - b^2)]$.

   Ranges: $0 \le r < \infty, b^2 < \theta^2 < c^2, 0 < \lambda^2 < b^2$.

$g_{11} = 1, g_{22} = r^2(\theta^2 - \lambda^2)/((\theta^2 - b^2)(c^2 - \theta^2)),$
$g_{33} = r^2(\theta^2 - \lambda^2)/((b^2 - \lambda^2)(c^2 - \lambda^2)),$
$\sqrt{g} = r^2(\theta^2 - \lambda^2)/\sqrt{(\theta^2 - b^2)(c^2 - \theta^2)(b^2 - \lambda^2)(c^2 - \lambda^2)}.$
$f_1 = r^2, f_2 = \sqrt{(\theta^2 - b^2)(c^2 - \theta^2)}, f_3 = \sqrt{(b^2 - \lambda^2)(c^2 - \lambda^2)}.$

10. **Ellipsoidal coordinates** $\{\eta, \theta, \lambda\}$

Relations: $x^2 = (\eta\theta\lambda/bc)^2, y^2 = (\eta^2 - b^2)(\theta^2 - b^2)(b^2 - \lambda^2)/[b^2(c^2 - b^2)],$
$z^2 = (\eta^2 - c^2)(c^2 - \theta^2)(c^2 - \lambda^2)/[c^2(c^2 - b^2)].$

Ranges: $c^2 \le \eta^2 < \infty, b^2 < \theta^2 < c^2, 0 < \lambda^2 < b^2.$

$g_{11} = (\eta^2 - \theta^2)(\eta^2 - \lambda^2)/((\eta^2 - b^2)(\eta^2 - c^2)),$
$g_{22} = (\theta^2 - \lambda^2)(\eta^2 - \theta^2)/((\theta^2 - b^2)(c^2 - \theta^2)),$
$g_{33} = (\eta^2 - \lambda^2)(\theta^2 - \lambda^2)/((b^2 - \lambda^2)(c^2 - \lambda^2)),$
$\sqrt{g} = \frac{(\eta^2 - \theta^2)(\eta^2 - \lambda^2)(\theta^2 - \lambda^2)}{\sqrt{(\eta^2 - b^2)(\eta^2 - c^2)(\theta^2 - b^2)(c^2 - \theta^2)(b^2 - \lambda^2)(c^2 - \lambda^2)}}.$
$f_1 = \sqrt{(\eta^2 - b^2)(\eta^2 - c^2)}, f_2 = \sqrt{(\theta^2 - b^2)(c^2 - \theta^2)},$
$f_3 = \sqrt{(b^2 - \lambda^2)(c^2 - \lambda^2)}.$

11. **Paraboloidal coordinates** $\{\mu, \nu, \lambda\}$

Relations: $x^2 = 4(\mu - b)(b - \nu)(b - \lambda)/(b - c),$
$y^2 = 4(\mu - c)(c - \nu)(\lambda - c)/(b - c), z^2 = \mu + \nu + \lambda - b - c.$

Ranges: $b < \mu < \infty, 0 < \nu < c, c < \lambda < b.$

$g_{11} = (\mu - \nu)(\mu - \lambda)/((\mu - b)(\mu - c)), g_{22} = (\mu - \nu)(\lambda - \nu)/((b - \nu)(c - \nu)),$
$g_{33} = (\lambda - \nu)(\mu - \lambda)/((b - \lambda)(\lambda - c)), \sqrt{g} = \frac{(\mu - \nu)(\mu - \lambda)(\lambda - \nu)}{\sqrt{(\mu - b)(\mu - c)(b - \nu)(c - \nu)(b - \lambda)(\lambda - c)}}.$
$f_1 = \sqrt{(\mu - b)(\mu - c)}, f_2 = \sqrt{(b - \nu)(c - \nu)}, f_3 = \sqrt{(b - \lambda)(\lambda - c)}.$

5.12 CONTROL THEORY

Let **x** be a state vector, let **y** be an observation vector, and let **u** be the control. The vectors **x**, **y**, and **u** have n, m, and p components, respectively. If a system evolves as:

$$\dot{\mathbf{x}} = A\mathbf{x} + B\mathbf{u}$$
$$\mathbf{y} = C\mathbf{x} + D\mathbf{u}$$

(5.12.1)

then, taking Laplace transforms, $\tilde{\mathbf{y}} = G(s)\tilde{\mathbf{u}}$ where $G(s)$ is the transfer function given by $G(s) = C(sI - A)^{-1}B + D$.

A system is said to be *controllable* if and only if for any times $\{t_0, t_1\}$ and any states $\{\mathbf{x}_0, \mathbf{x}_1\}$ there exists a control $\mathbf{u}(t)$ such that $\mathbf{x}(t_0) = \mathbf{x}_0$ and $\mathbf{x}(t_1) = \mathbf{x}_1$. The system is controllable if and only if

$$\text{rank}\begin{bmatrix} B & AB & A^2B & \dots & A^{n-1}B \end{bmatrix} = n.$$

(5.12.2)

If, given $\mathbf{u}(t)$ and $\mathbf{y}(t)$ on some interval $t_0 < t < t_1$, the value of $\mathbf{x}(t)$ can be deduced on that interval, then the system is said to be *observable*. Observability is equivalent to the condition

$$\text{rank} \begin{bmatrix} C^{\mathrm{T}} & A^{\mathrm{T}}C^{\mathrm{T}} & \cdots & \left(A^{(n-1)}\right)^{\mathrm{T}} C^{\mathrm{T}} \end{bmatrix} = n. \qquad (5.12.3)$$

If the control is bounded above and below (say $u_i^- < u_i < u_i^+$), then a *bang–bang* control is one for which $u_i = u_i^-$ or $u_i = u_i^+$. (That is, for every t, either $u_i(t) = u_i^-(t)$ or $u_i(t) = u_i^+(t)$; switches are possible.) A *bang–off–bang* control is one for which $u_i = 0$, $u_i = u_i^-$, or $u_i = u_i^+$.

A second frequently studied control problem is $\dot{\mathbf{x}} = \mathbf{f}(\mathbf{x}, \mathbf{u}, t)$, where $\mathbf{x}(t_0)$ and $\mathbf{x}(t_f)$ are specified, and there is a cost function, $J = \int_{t_0}^{t_f} \phi(\mathbf{x}, \mathbf{u}, t)$. The goal is to minimize the cost function. Defining the Hamiltonian $H(\mathbf{x}, \mathbf{z}, \mathbf{u}, t) = \phi + \mathbf{z} \cdot \mathbf{f}$, the optimal control (the one that minimizes the cost function) satisfies

$$\dot{\mathbf{x}} = \frac{\partial H}{\partial \mathbf{z}}, \quad \dot{\mathbf{z}} = -\frac{\partial H}{\partial \mathbf{x}}, \quad \mathbf{0} = \frac{\partial H}{\partial \mathbf{u}}. \qquad (5.12.4)$$

EXAMPLE   In the one-dimensional case, with $\dot{x} = -ax + u$, $x(0) = x_0$, $x(\infty) = 0$, and $J = \int_0^\infty (x^2 + u^2)\, dt$; the optimal control is given by $u = (a - \sqrt{1 + a^2})x^*(t)$ where $x^*(t) = x_0 e^{-\sqrt{1+a^2}\, t}$.

Chapter 6

Special Functions

6.1   TRIGONOMETRIC OR CIRCULAR FUNCTIONS 503
    6.1.1   *Definition of angles* 503
    6.1.2   *Characterization of angles* 503
    6.1.3   *Circular functions* 504
    6.1.4   *Circular functions of special angles* 506
    6.1.5   *Evaluating sines and cosines at multiples of* π 507
    6.1.6   *Symmetry and periodicity relationships* 507
    6.1.7   *Functions in terms of angles in the first quadrant* 507
    6.1.8   *One circular function in terms of another* 508
    6.1.9   *Circular functions in terms of exponentials* 508
    6.1.10   *Fundamental identities* 509
    6.1.11   *Angle sum and difference relationships* 509
    6.1.12   *Double-angle formulae* 509
    6.1.13   *Multiple-angle formulae* 510
    6.1.14   *Half-angle formulae* 510
    6.1.15   *Powers of circular functions* 511
    6.1.16   *Products of sine and cosine* 511
    6.1.17   *Sums of circular functions* 511

6.2   CIRCULAR FUNCTIONS AND PLANAR TRIANGLES 512
    6.2.1   *Right triangles* 512
    6.2.2   *General plane triangles* 512
    6.2.3   *Half-angle formulae* 514
    6.2.4   *Solution of triangles* 514
    6.2.5   *Tables of trigonometric functions* 516

6.3   INVERSE CIRCULAR FUNCTIONS 518
    6.3.1   *Definition in terms of an integral* 518
    6.3.2   *Principal values of the inverse circular functions* 518
    6.3.3   *Fundamental identities* 519
    6.3.4   *Functions of negative arguments* 519
    6.3.5   *Relationship to inverse hyperbolic functions* 519
    6.3.6   *Sum and difference of two inverse circular functions* 520

6.4   CEILING AND FLOOR FUNCTIONS 520

6.5   EXPONENTIAL FUNCTION 520
    6.5.1   *Exponentiation* 520
    6.5.2   *Definition of* e^z 521
    6.5.3   *Derivative and integral of* e^x 521

1-58488-291-3/02/$0.00+$1.50
© 2003 CRC Press, Inc.

6.6   LOGARITHMIC FUNCTIONS *522*
  6.6.1   *Definition of the natural logarithm* *522*
  6.6.2   *Logarithm of special values* *522*
  6.6.3   *Relating the logarithm to the exponential* *523*
  6.6.4   *Identities* *523*
  6.6.5   *Series expansions for the natural logarithm* *523*
  6.6.6   *Derivative and integration formulae* *523*

6.7   HYPERBOLIC FUNCTIONS *523*
  6.7.1   *Definitions of the hyperbolic functions* *524*
  6.7.2   *Range of values* *524*
  6.7.3   *Hyperbolic functions in terms of one another* *524*
  6.7.4   *Relations among hyperbolic functions* *525*
  6.7.5   *Relationship to circular functions* *525*
  6.7.6   *Series expansions* *525*
  6.7.7   *Symmetry relationships* *525*
  6.7.8   *Sum and difference formulae* *526*
  6.7.9   *Multiple argument relations* *526*
  6.7.10   *Sums of functions* *526*
  6.7.11   *Products of functions* *527*
  6.7.12   *Half–argument formulae* *527*
  6.7.13   *Differentiation formulae* *527*

6.8   INVERSE HYPERBOLIC FUNCTIONS *527*
  6.8.1   *Range of values* *527*
  6.8.2   *Relationships among inverse hyperbolic functions* *528*
  6.8.3   *Relationships with logarithmic functions* *529*
  6.8.4   *Relationships with circular functions* *529*
  6.8.5   *Sum and difference of functions* *529*

6.9   GUDERMANNIAN FUNCTION *530*
  6.9.1   *Fundamental identities* *530*
  6.9.2   *Derivatives of Gudermannian* *530*
  6.9.3   *Relationship to hyperbolic and circular functions* *531*
  6.9.4   *Numerical values of hyperbolic functions* *531*

6.10   ORTHOGONAL POLYNOMIALS *532*
  6.10.1   *Hermite polynomials* *532*
  6.10.2   *Jacobi polynomials* *533*
  6.10.3   *Laguerre polynomials* *533*
  6.10.4   *Generalized Laguerre polynomials* *533*
  6.10.5   *Legendre polynomials* *534*
  6.10.6   *Chebyshev polynomials, first kind* *534*
  6.10.7   *Chebyshev polynomials, second kind* *535*
  6.10.8   *Tables of orthogonal polynomials* *535*
  6.10.9   *Zernike polynomials* *537*
  6.10.10   *Spherical harmonics* *538*

6.11   GAMMA FUNCTION *540*
  6.11.1   *Recursion formula* *540*
  6.11.2   *Gamma function of special values* *541*
  6.11.3   *Properties* *541*
  6.11.4   *Asymptotic expansion* *542*
  6.11.5   *Logarithmic derivative of the gamma function* *543*

    6.11.6   Numerical values *543*

6.12 BETA FUNCTION **544**
    6.12.1   Numerical values of the beta function *545*

6.13 ERROR FUNCTIONS **545**
    6.13.1   Properties *545*
    6.13.2   Error function of special values *546*
    6.13.3   Expansions *546*
    6.13.4   Special cases *546*

6.14 FRESNEL INTEGRALS **547**
    6.14.1   Properties *547*
    6.14.2   Asymptotic expansion *548*
    6.14.3   Numerical values of error functions and Fresnel integrals *548*

6.15 SINE, COSINE, AND EXPONENTIAL INTEGRALS **549**
    6.15.1   Sine and cosine integrals *549*
    6.15.2   Exponential integrals *550*
    6.15.3   Logarithmic integral *550*
    6.15.4   Numerical values *551*

6.16 POLYLOGARITHMS **551**
    6.16.1   Polylogarithms of special values *552*
    6.16.2   Polylogarithm properties *552*

6.17 HYPERGEOMETRIC FUNCTIONS **552**
    6.17.1   Special cases *553*
    6.17.2   Properties *553*
    6.17.3   Recursion formulae *554*

6.18 LEGENDRE FUNCTIONS **554**
    6.18.1   Differential equation: Legendre function *554*
    6.18.2   Definition *555*
    6.18.3   Singular points *555*
    6.18.4   Relationships *555*
    6.18.5   Recursion relationships *556*
    6.18.6   Integrals *556*
    6.18.7   Polynomial case *556*
    6.18.8   Differential equation: associated Legendre function *557*
    6.18.9   Relationships between the associated and ordinary Legendre functions *558*
    6.18.10 Orthogonality relationship *558*
    6.18.11 Recursion relationships *558*

6.19 BESSEL FUNCTIONS **559**
    6.19.1   Differential equation *559*
    6.19.2   Singular points *559*
    6.19.3   Relationships *560*
    6.19.4   Series expansions *560*
    6.19.5   Recurrence relationships *560*
    6.19.6   Behavior as $z \to 0$ *561*
    6.19.7   Integrals *561*
    6.19.8   Fourier expansion *561*

6.19.9   *Auxiliary functions* *561*
6.19.10 *Inverse relationships* *562*
6.19.11 *Asymptotic expansions* *562*
6.19.12 *Zeros of Bessel functions* *563*
6.19.13 *Half order Bessel functions* *563*
6.19.14 *Modified Bessel functions* *564*
6.19.15 *Airy functions* *565*
6.19.16 *Numerical values for the Bessel functions* *567*

6.20 ELLIPTIC INTEGRALS **568**
6.20.1   *Definitions* *568*
6.20.2   *Properties* *569*
6.20.3   *Numerical values of the elliptic integrals* *570*

6.21 JACOBIAN ELLIPTIC FUNCTIONS **572**
6.21.1   *Properties* *572*
6.21.2   *Derivatives and integrals* *573*
6.21.3   *Series expansions* *573*

6.22 CLEBSCH–GORDAN COEFFICIENTS **574**

6.23 INTEGRAL TRANSFORMS: PRELIMINARIES **576**

6.24 FOURIER TRANSFORM **576**
6.24.1   *Existence* *577*
6.24.2   *Properties* *578*
6.24.3   *Inversion formula* *580*
6.24.4   *Poisson summation formula* *580*
6.24.5   *Shannon's sampling theorem* *581*
6.24.6   *Uncertainty principle* *581*
6.24.7   *Fourier sine and cosine transforms* *582*

6.25 DISCRETE FOURIER TRANSFORM (DFT) **582**
6.25.1   *Properties* *583*

6.26 FAST FOURIER TRANSFORM (FFT) **584**

6.27 MULTIDIMENSIONAL FOURIER TRANSFORM **585**

6.28 LAPLACE TRANSFORM **585**
6.28.1   *Existence and domain of convergence* *585*
6.28.2   *Properties* *586*
6.28.3   *Inversion formulae* *588*
6.28.4   *Convolution* *589*

6.29 HANKEL TRANSFORM **589**
6.29.1   *Properties* *590*

6.30 HARTLEY TRANSFORM **591**

6.31 HILBERT TRANSFORM **591**

6.31.1 *Existence* *592*
6.31.2 *Properties* *592*
6.31.3 *Relationship with the Fourier transform* *593*

6.32 Z-TRANSFORM **594**
6.32.1 *Examples* *595*
6.32.2 *Properties* *595*
6.32.3 *Inversion formula* *596*
6.32.4 *Convolution and product* *598*

6.33 TABLES OF TRANSFORMS **599**

6.1 TRIGONOMETRIC OR CIRCULAR FUNCTIONS

6.1.1 DEFINITION OF ANGLES

If two lines intersect and one line is rotated about the point of intersection, the angle of rotation is designated positive if the rotation is counterclockwise. Angles are commonly measured in units of radians or degrees. Degrees are a historical unit related to the calendar defined by a complete revolution equalling 360 degrees (the approximate number of days in a year), written $360°$. Radians are the angular unit usually used for mathematics and science. The radian measure of an angle is defined as the arc length traced by the tip of a rotating line divided by the length of that line. Thus a complete rotation of a line about the origin corresponds to 2π radians of rotation. It is a convenient convention that a full rotation of 2π radians is divided into four angular segments of $\pi/2$ each and that these are referred to as the four quadrants designated by Roman numerals I, II, III, and IV (see Figure 6.1).

6.1.2 CHARACTERIZATION OF ANGLES

A *right* angle is the angle between two perpendicular lines. It is equal to $\pi/2$ radians or 90 degrees. An *acute* angle is a positive angle less than $\pi/2$ radians. An *obtuse* angle is one between $\pi/2$ and π radians. A *convex* angle is one between 0 and π radians.

FIGURE 6.1

The four quadrants (left) and notation for trigonometric functions (right).

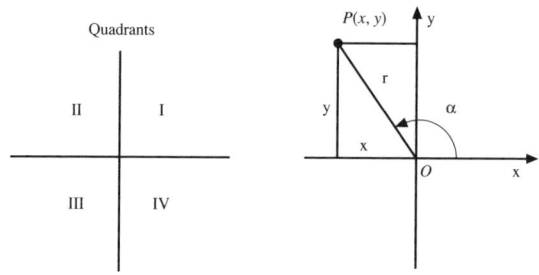

FIGURE 6.2

Definitions of angles.

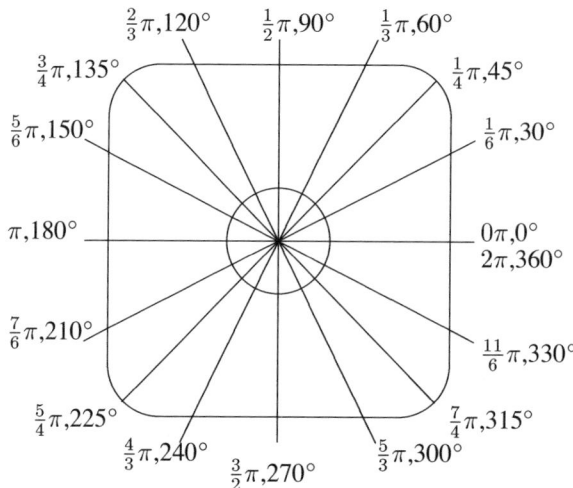

6.1.2.1   Relation between radians and degrees

The angle π radians corresponds to 180 degrees. Therefore,

$$\text{one radian} = \frac{180}{\pi} \text{ degrees} = 57.30 \text{ degrees},$$
$$\text{one degree} = \frac{\pi}{180} \text{ radians} = 0.01745 \text{ radians}.$$

$$(6.1.1)$$

6.1.3   CIRCULAR FUNCTIONS

Consider the rectangular coordinate system shown in Figure 6.1. The coordinate x is positive to the right of the origin and the coordinate y is positive above the origin. The radius vector $\mathbf{r}$ shown terminating on the point $P(x, y)$ is shown rotated by the angle α up from the x axis. The radius vector $\mathbf{r}$ has component values x and y.

The trigonometric or circular functions of the angle α are defined in terms of the signed coordinates x and y and the length r, which is always positive. Note that the coordinate x is negative in quadrants II and III and the coordinate y is negative in quadrants III and IV. The definitions of the trigonometric functions in terms of the Cartesian coordinates x and y of the point $P(x, y)$ are shown below. In formulae, the angle α is usually specified in radian measure.

$$\text{sine } \alpha = \sin \alpha = y/r, \qquad\qquad \text{cosine } \alpha = \cos \alpha = x/r,$$
$$\text{tangent } \alpha = \tan \alpha = y/x, \qquad\qquad \text{cotangent } \alpha = \cot \alpha = x/y, \qquad (6.1.2)$$
$$\text{cosecant } \alpha = \csc \alpha = r/y, \qquad\qquad \text{secant } \alpha = \sec \alpha = r/x.$$

There are also the following seldom used functions:

versed sine of α = versine of α = vers $\alpha = 1 - \cos \alpha$,
coversed sine of α = versed cosine of α = covers $\alpha = 1 - \sin \alpha$,
exsecant of α = exsec $\alpha = \sec \alpha - 1$,
haversine of α = hav $\alpha = \frac{1}{2}$ vers $\alpha = \frac{1}{2}(1 - \cos \alpha)$.

6.1.3.1   Signs in the four quadrants

Quadrant	sin	cos	tan	csc	sec	cot
I	+	+	+	+	+	+
II	+	−	−	+	−	−
III	−	−	+	−	−	+
IV	−	+	−	−	+	−

FIGURE 6.3

Sine and cosine; angles are in radians.

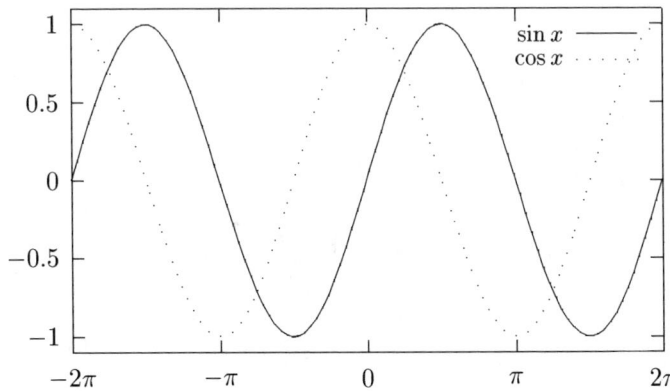

FIGURE 6.4

Tangent and cotangent; angles are in radians.

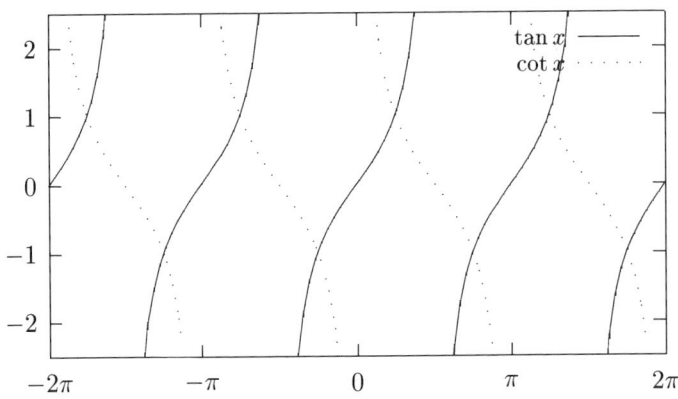

6.1.4   CIRCULAR FUNCTIONS OF SPECIAL ANGLES

Angle	$0 = 0°$	$\pi/12 = 15°$	$\pi/6 = 30°$	$\pi/4 = 45°$	$\pi/3 = 60°$
sin	0	$\frac{\sqrt{2}}{4}(\sqrt{3}-1)$	$1/2$	$\sqrt{2}/2$	$\sqrt{3}/2$
cos	1	$\frac{\sqrt{2}}{4}(\sqrt{3}+1)$	$\sqrt{3}/2$	$\sqrt{2}/2$	$1/2$
tan	0	$2-\sqrt{3}$	$\sqrt{3}/3$	1	$\sqrt{3}$
csc	∞	$\sqrt{2}(\sqrt{3}+1)$	2	$\sqrt{2}$	$2\sqrt{3}/3$
sec	1	$\sqrt{2}(\sqrt{3}-1)$	$2\sqrt{3}/3$	$\sqrt{2}$	2
cot	∞	$2+\sqrt{3}$	$\sqrt{3}$	1	$\sqrt{3}/3$

Angle	$5\pi/12 = 75°$	$\pi/2 = 90°$	$7\pi/12 = 105°$	$2\pi/3 = 120°$
sin	$\frac{\sqrt{2}}{4}(\sqrt{3}+1)$	1	$\frac{\sqrt{2}}{4}(\sqrt{3}+1)$	$\sqrt{3}/2$
cos	$\frac{\sqrt{2}}{4}(\sqrt{3}-1)$	0	$-\frac{\sqrt{2}}{4}(\sqrt{3}-1)$	$-1/2$
tan	$2+\sqrt{3}$	∞	$-(2+\sqrt{3})$	$-\sqrt{3}$
csc	$\sqrt{2}(\sqrt{3}-1)$	1	$\sqrt{2}(\sqrt{3}-1)$	$2\sqrt{3}/3$
sec	$\sqrt{2}(\sqrt{3}+1)$	∞	$-\sqrt{2}(\sqrt{3}+1)$	-2
cot	$2-\sqrt{3}$	0	$-(2-\sqrt{3})$	$-\sqrt{3}/3$

Angle	$3\pi/4 = 135°$	$5\pi/6 = 150°$	$11\pi/12 = 165°$	$\pi = 180°$
sin	$\sqrt{2}/2$	$1/2$	$\frac{\sqrt{2}}{4}(\sqrt{3}-1)$	0
cos	$-\sqrt{2}/2$	$-\sqrt{3}/2$	$-\frac{\sqrt{2}}{4}(\sqrt{3}+1)$	-1
tan	-1	$-\sqrt{3}/3$	$-(2-\sqrt{3})$	0
csc	$\sqrt{2}$	2	$\sqrt{2}(\sqrt{3}+1)$	∞
sec	$-\sqrt{2}$	$-2\sqrt{3}/3$	$-\sqrt{2}(\sqrt{3}-1)$	-1
cot	-1	$-\sqrt{3}$	$-(2+\sqrt{3})$	∞

6.1.5 EVALUATING SINES AND COSINES AT MULTIPLES OF π

The following table is useful for evaluating sines and cosines in multiples of π:

	n an integer	n even	n odd	$n/2$ odd	$n/2$ even
$\sin n\pi$	0	0	0	0	0
$\cos n\pi$	$(-1)^n$	$+1$	-1	$+1$	$+1$
$\sin n\pi/2$		0	$(-1)^{(n-1)/2}$	0	0
$\cos n\pi/2$		$(-1)^{n/2}$	0	-1	$+1$

	n odd	$n/2$ odd	$n/2$ even
$\sin n\pi/4$	$(-1)^{(n^2+4n+11)/8}/\sqrt{2}$	$(-1)^{(n-2)/4}$	0

Note the useful formulae (where $i^2 = -1$)

$$\sin \frac{n\pi}{2} = \frac{i^{n+1}}{2}\left[(-1)^n - 1\right], \qquad \cos \frac{n\pi}{2} = \frac{i^n}{2}\left[(-1)^n + 1\right]. \qquad (6.1.3)$$

6.1.6 SYMMETRY AND PERIODICITY RELATIONSHIPS

$$\sin(-\alpha) = -\sin\alpha, \qquad \cos(-\alpha) = +\cos\alpha, \qquad \tan(-\alpha) = -\tan\alpha. \quad (6.1.4)$$

When n is any integer,

$$\begin{aligned}
\sin(\alpha + n2\pi) &= \sin\alpha, \\
\cos(\alpha + n2\pi) &= \cos\alpha, \\
\tan(\alpha + n\pi) &= \tan\alpha.
\end{aligned} \qquad (6.1.5)$$

6.1.7 FUNCTIONS IN TERMS OF ANGLES IN THE FIRST QUADRANT

When n is any integer:

	$-\alpha$	$\dfrac{\pi}{2} \pm \alpha$	$\pi \pm \alpha$	$\dfrac{3\pi}{2} \pm \alpha$	$2n\pi \pm \alpha$
sin	$-\sin\alpha$	$\cos\alpha$	$\mp\sin\alpha$	$-\cos\alpha$	$\pm\sin\alpha$
cos	$\cos\alpha$	$\mp\sin\alpha$	$-\cos\alpha$	$\pm\sin\alpha$	$+\cos\alpha$
tan	$-\tan\alpha$	$\mp\cot\alpha$	$\pm\tan\alpha$	$\mp\cot\alpha$	$\pm\tan\alpha$
csc	$-\csc\alpha$	$\sec\alpha$	$\mp\csc\alpha$	$-\sec\alpha$	$\pm\csc\alpha$
sec	$\sec\alpha$	$\mp\csc\alpha$	$-\sec\alpha$	$\pm\csc\alpha$	$\sec\alpha$
cot	$-\cot\alpha$	$\mp\tan\alpha$	$\pm\cot\alpha$	$\mp\tan\alpha$	$\pm\cot\alpha$

6.1.8   ONE CIRCULAR FUNCTION IN TERMS OF ANOTHER

For $0 \leq x \leq \pi/2$,

	$\sin x$	$\cos x$	$\tan x$
$\sin x =$	$\sin x$	$\sqrt{1 - \cos^2 x}$	$\dfrac{\tan x}{\sqrt{1 + \tan^2 x}}$
$\cos x =$	$\sqrt{1 - \sin^2 x}$	$\cos x$	$\dfrac{1}{\sqrt{1 + \tan^2 x}}$
$\tan x =$	$\dfrac{\sin x}{\sqrt{1 - \sin^2 x}}$	$\dfrac{\sqrt{1 - \cos^2 x}}{\cos x}$	$\tan x$
$\csc x =$	$\dfrac{1}{\sin x}$	$\dfrac{1}{\sqrt{1 - \cos^2 x}}$	$\dfrac{\sqrt{1 + \tan^2 x}}{\tan x}$
$\sec x =$	$\dfrac{1}{\sqrt{1 - \sin^2 x}}$	$\dfrac{1}{\cos x}$	$\sqrt{1 + \tan^2 x}$
$\cot x =$	$\dfrac{\sqrt{1 - \sin^2 x}}{\sin x}$	$\dfrac{\cos x}{\sqrt{1 - \cos^2 x}}$	$\dfrac{1}{\tan x}$

	$\csc x$	$\sec x$	$\cot x$
$\sin x =$	$\dfrac{1}{\csc x}$	$\dfrac{\sqrt{\sec^2 x - 1}}{\sec x}$	$\dfrac{1}{\sqrt{1 + \cot^2 x}}$
$\cos x =$	$\dfrac{\sqrt{\csc^2 x - 1}}{\csc x}$	$\dfrac{1}{\sec x}$	$\dfrac{\cot x}{\sqrt{1 + \cot^2 x}}$
$\tan x =$	$\dfrac{1}{\sqrt{\csc^2 x - 1}}$	$\sqrt{\sec^2 x - 1}$	$\dfrac{1}{\cot x}$
$\csc x =$	$\csc x$	$\dfrac{\sec x}{\sqrt{\sec^2 x - 1}}$	$\sqrt{1 + \cot^2 x}$
$\sec x =$	$\dfrac{\csc x}{\sqrt{\csc^2 x - 1}}$	$\sec x$	$\dfrac{\sqrt{1 + \cot^2 x}}{\cot x}$
$\cot x =$	$\sqrt{\csc^2 x - 1}$	$\dfrac{1}{\sqrt{\sec^2 x - 1}}$	$\cot x$

6.1.9   CIRCULAR FUNCTIONS IN TERMS OF EXPONENTIALS

$$\cos z = \frac{e^{iz} + e^{-iz}}{2} \qquad\qquad e^{iz} = \cos z + i\sin z$$

$$\sin z = \frac{e^{iz} - e^{-iz}}{2i} \qquad\qquad e^{-iz} = \cos z - i\sin z$$

$$\tan z = \frac{\sin z}{\cos z} = \frac{e^{iz} - e^{-iz}}{i(e^{iz} + e^{-iz})} \qquad\qquad \text{where } i^2 = -1 \text{ and } z \text{ may be complex}$$

6.1.10 FUNDAMENTAL IDENTITIES

1. Reciprocal relations

$$\sin\alpha = \frac{1}{\csc\alpha}, \qquad \cos\alpha = \frac{1}{\sec\alpha}, \qquad \tan\alpha = \frac{\sin\alpha}{\cos\alpha} = \frac{1}{\cot\alpha},$$

$$\csc\alpha = \frac{1}{\sin\alpha}, \qquad \sec\alpha = \frac{1}{\cos\alpha}, \qquad \cot\alpha = \frac{\cos\alpha}{\sin\alpha} = \frac{1}{\tan\alpha}.$$

2. Pythagorean theorem

$$\sin^2 z + \cos^2 z = 1$$

$$\sec^2 z - \tan^2 z = 1$$

$$\csc^2 z - \cot^2 z = 1$$

3. Product relations

$$\sin\alpha = \tan\alpha\cos\alpha \qquad\qquad \cos\alpha = \cot\alpha\sin\alpha$$

$$\tan\alpha = \sin\alpha\sec\alpha \qquad\qquad \cot\alpha = \cos\alpha\csc\alpha$$

$$\sec\alpha = \csc\alpha\tan\alpha \qquad\qquad \csc\alpha = \sec\alpha\cot\alpha$$

4. Quotient relations

$$\sin\alpha = \frac{\tan\alpha}{\sec\alpha} \qquad \cos\alpha = \frac{\cot\alpha}{\csc\alpha} \qquad \tan\alpha = \frac{\sin\alpha}{\cos\alpha}$$

$$\csc\alpha = \frac{\sec\alpha}{\tan\alpha} \qquad \sec\alpha = \frac{\csc\alpha}{\cot\alpha} \qquad \cot\alpha = \frac{\cos\alpha}{\sin\alpha}$$

6.1.11 ANGLE SUM AND DIFFERENCE RELATIONSHIPS

$$\sin(\alpha \pm \beta) = \sin\alpha\cos\beta \pm \cos\alpha\sin\beta$$

$$\cos(\alpha \pm \beta) = \cos\alpha\cos\beta \mp \sin\alpha\sin\beta$$

$$\tan(\alpha \pm \beta) = \frac{\tan\alpha \pm \tan\beta}{1 \mp \tan\alpha\tan\beta}$$

$$\cot(\alpha \pm \beta) = \frac{\cot\alpha\cot\beta \mp 1}{\cot\beta \pm \cot\alpha}$$

6.1.12 DOUBLE-ANGLE FORMULAE

$$\sin 2\alpha = 2\sin\alpha\cos\alpha = \frac{2\tan\alpha}{1 + \tan^2\alpha}$$

$$\cos 2\alpha = 2\cos^2\alpha - 1 = 1 - 2\sin^2\alpha = \cos^2\alpha - \sin^2\alpha = \frac{1 - \tan^2\alpha}{1 + \tan^2\alpha}$$

$$\tan 2\alpha = \frac{2\tan\alpha}{1 - \tan^2\alpha}$$

$$\cot 2\alpha = \frac{\cot^2\alpha - 1}{2\cot\alpha}$$

6.1.13   MULTIPLE-ANGLE FORMULAE

$$\sin 3\alpha = -4\sin^3\alpha + 3\sin\alpha.$$

$$\sin 4\alpha = -8\sin^3\alpha\cos\alpha + 4\sin\alpha\cos\alpha.$$

$$\sin 5\alpha = 16\sin^5\alpha - 20\sin^3\alpha + 5\sin\alpha.$$

$$\sin 6\alpha = 32\sin\alpha\cos^5\alpha - 32\sin\alpha\cos^3\alpha + 6\sin\alpha\cos\alpha.$$

$$\sin n\alpha = 2\sin(n-1)\alpha\cos\alpha - \sin(n-2)\alpha.$$

$$\cos 3\alpha = 4\cos^3\alpha - 3\cos\alpha.$$

$$\cos 4\alpha = 8\cos^4\alpha - 8\cos^2\alpha + 1.$$

$$\cos 5\alpha = 16\cos^5\alpha - 20\cos^3\alpha + 5\cos\alpha.$$

$$\cos 6\alpha = 32\cos^6\alpha - 48\cos^4\alpha + 18\cos^2\alpha - 1.$$

$$\cos n\alpha = 2\cos(n-1)\alpha\cos\alpha - \cos(n-2)\alpha.$$

$$\tan 3\alpha = \frac{-\tan^3\alpha + 3\tan\alpha}{-3\tan^2\alpha + 1}.$$

$$\tan 4\alpha = \frac{-4\tan^3\alpha + 4\tan\alpha}{\tan^4\alpha - 6\tan^2\alpha + 1}.$$

$$\tan n\alpha = \frac{\tan(n-1)\alpha + \tan\alpha}{-\tan(n-1)\alpha\tan\alpha + 1}.$$

6.1.14   HALF-ANGLE FORMULAE

$$\cos\frac{\alpha}{2} = \pm\sqrt{\frac{1+\cos\alpha}{2}}$$

(positive if $\alpha/2$ is in quadrant I or IV, negative if in II or III).

$$\sin\frac{\alpha}{2} = \pm\sqrt{\frac{1-\cos\alpha}{2}}$$

(positive if $\alpha/2$ is in quadrant I or II, negative if in III or IV).

$$\tan\frac{\alpha}{2} = \frac{1-\cos\alpha}{\sin\alpha} = \frac{\sin\alpha}{1+\cos\alpha} = \pm\sqrt{\frac{1-\cos\alpha}{1+\cos\alpha}}$$

(positive if $\alpha/2$ is in quadrant I or III, negative if in II or IV).

$$\cot\frac{\alpha}{2} = \frac{1+\cos\alpha}{\sin\alpha} = \frac{\sin\alpha}{1-\cos\alpha} = \pm\sqrt{\frac{1+\cos\alpha}{1-\cos\alpha}}$$

(positive if $\alpha/2$ is in quadrant I or III, negative if in II or IV).

6.1.15 POWERS OF CIRCULAR FUNCTIONS

$$\sin^2 \alpha = \frac{1}{2}(1 - \cos 2\alpha).$$
$$\cos^2 \alpha = \frac{1}{2}(1 + \cos 2\alpha).$$

$$\sin^3 \alpha = \frac{1}{4}(-\sin 3\alpha + 3\sin \alpha).$$
$$\cos^3 \alpha = \frac{1}{4}(\cos 3\alpha + 3\cos \alpha).$$

$$\sin^4 \alpha = \frac{1}{8}(3 - 4\cos 2\alpha + \cos 4\alpha).$$
$$\cos^4 \alpha = \frac{1}{8}(3 + 4\cos 2\alpha + \cos 4\alpha).$$

$$\tan^2 \alpha = \frac{1 - \cos 2\alpha}{1 + \cos 2\alpha}.$$
$$\cot^2 \alpha = \frac{1 + \cos 2\alpha}{1 - \cos 2\alpha}.$$

6.1.16 PRODUCTS OF SINE AND COSINE

$$\cos \alpha \cos \beta = \frac{1}{2}\cos(\alpha - \beta) + \frac{1}{2}\cos(\alpha + \beta).$$

$$\sin \alpha \sin \beta = \frac{1}{2}\cos(\alpha - \beta) - \frac{1}{2}\cos(\alpha + \beta).$$

$$\sin \alpha \cos \beta = \frac{1}{2}\sin(\alpha - \beta) + \frac{1}{2}\sin(\alpha + \beta).$$

6.1.17 SUMS OF CIRCULAR FUNCTIONS

$$\sin \alpha \pm \sin \beta = 2\sin \frac{\alpha \pm \beta}{2} \cos \frac{\alpha \mp \beta}{2}.$$

$$\cos \alpha + \cos \beta = 2\cos \frac{\alpha + \beta}{2} \cos \frac{\alpha - \beta}{2}.$$

$$\cos \alpha - \cos \beta = -2\sin \frac{\alpha + \beta}{2} \sin \frac{\alpha - \beta}{2}.$$

$$\tan \alpha \pm \tan \beta = \frac{\sin \alpha \pm \beta}{\cos \alpha \cos \beta}.$$

$$\cot \alpha \pm \cot \beta = \frac{\sin \beta \pm \alpha}{\sin \alpha \sin \beta}.$$

$$\frac{\sin \alpha + \sin \beta}{\sin \alpha - \sin \beta} = \frac{\tan \frac{\alpha + \beta}{2}}{\tan \frac{\alpha - \beta}{2}}.$$

$$\frac{\sin \alpha + \sin \beta}{\cos \alpha - \cos \beta} = \cot \frac{-\alpha + \beta}{2}.$$

$$\frac{\sin \alpha + \sin \beta}{\cos \alpha + \cos \beta} = \tan \frac{\alpha + \beta}{2}.$$

$$\frac{\sin \alpha - \sin \beta}{\cos \alpha + \cos \beta} = \tan \frac{\alpha - \beta}{2}.$$

6.2   CIRCULAR FUNCTIONS AND PLANAR TRIANGLES

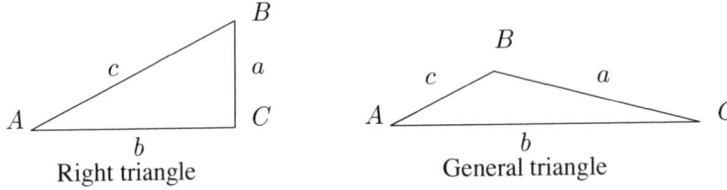

Right triangle   General triangle

6.2.1   RIGHT TRIANGLES

Let A, B, and C designate the vertices of a right triangle with C the right angle and a, b, and c the lengths of the sides opposite the corresponding vertices.

1. Trigonometric functions in terms of angle sides

$$\sin A = \frac{a}{c} = \frac{1}{\csc A},$$

$$\cos A = \frac{b}{c} = \frac{1}{\sec A}, \tag{6.2.1}$$

$$\tan A = \frac{a}{b} = \frac{1}{\cot A}.$$

2. The Pythagorean theorem states that $a^2 + b^2 = c^2$.

3. The sum of the interior angles equals π, i.e., $A + B + C = \pi$.

6.2.2   GENERAL PLANE TRIANGLES

Let A, B, and C designate the interior angles of a general triangle and let a, b, and c be the length of the sides opposite those angles.

1. Radius of the inscribed circle:

$$r = \sqrt{\frac{(s-a)(s-b)(s-c)}{s}}, \tag{6.2.2}$$

where the semi-perimeter is

$$s = \frac{1}{2}(a+b+c). \tag{6.2.3}$$

2. Radius of the circumscribed circle:

$$R = \frac{a}{2\sin A} = \frac{b}{2\sin B} = \frac{c}{2\sin C} = \frac{abc}{4(\text{Area})}.$$

3. Law of sines:

$$\frac{a}{\sin A} = \frac{b}{\sin B} = \frac{c}{\sin C}.$$

4. Law of cosines:

$$a^2 = c^2 + b^2 - 2bc\cos A, \qquad \cos A = \frac{c^2 + b^2 - a^2}{2bc}.$$

$$b^2 = a^2 + c^2 - 2ca\cos B, \qquad \cos B = \frac{a^2 + c^2 - b^2}{2ca}.$$

$$c^2 = b^2 + a^2 - 2ab\cos C, \qquad \cos C = \frac{b^2 + a^2 - c^2}{2ab}.$$

5. Triangle sides in terms of other components:

$$a = b\cos C + c\cos B,$$
$$c = b\cos A + a\cos B,$$
$$b = a\cos C + c\cos A.$$

6. Law of tangents:

$$\frac{a+b}{a-b} = \frac{\tan\frac{A+B}{2}}{\tan\frac{A-B}{2}},$$

$$\frac{b+c}{b-c} = \frac{\tan\frac{B+C}{2}}{\tan\frac{B-C}{2}},$$

$$\frac{a+c}{a-c} = \frac{\tan\frac{A+C}{2}}{\tan\frac{A-C}{2}}.$$

7. Area of general triangle:

$$\text{Area} = \frac{bc\sin A}{2} = \frac{ac\sin B}{2} = \frac{ab\sin C}{2},$$

$$= \frac{c^2\sin A\sin B}{2\sin C} = \frac{b^2\sin A\sin C}{2\sin B} = \frac{a^2\sin B\sin C}{2\sin A},$$

$$= rs = \frac{abc}{4R} = \sqrt{s(s-a)(s-b)(s-c)} \qquad \text{(Heron's formula)}.$$

8. Mollweide's formulae:

$$\frac{b-c}{a} = \frac{\sin\frac{1}{2}(B-C)}{\cos\frac{1}{2}A},$$

$$\frac{c-a}{b} = \frac{\sin\frac{1}{2}(C-A)}{\cos\frac{1}{2}B},$$

$$\frac{a-b}{c} = \frac{\sin\frac{1}{2}(A-B)}{\cos\frac{1}{2}C}.$$

9. Newton's formulae:

$$\frac{b+c}{a} = \frac{\cos\frac{1}{2}(B-C)}{\sin\frac{1}{2}A},$$

$$\frac{c+a}{b} = \frac{\cos\frac{1}{2}(C-A)}{\sin\frac{1}{2}B},$$

$$\frac{a+b}{c} = \frac{\cos\frac{1}{2}(A-B)}{\sin\frac{1}{2}C}.$$

6.2.3 HALF-ANGLE FORMULAE

$$\tan\frac{A}{2} = \frac{r}{s-a} \qquad \tan\frac{B}{2} = \frac{r}{s-b} \qquad \tan\frac{C}{2} = \frac{r}{s-c}$$

$$\sin\frac{A}{2} = \sqrt{\frac{(s-b)(s-c)}{bc}} \qquad\qquad \cos\frac{A}{2} = \sqrt{\frac{s(s-a)}{bc}}$$

$$\sin\frac{B}{2} = \sqrt{\frac{(s-c)(s-a)}{ca}} \qquad\qquad \cos\frac{B}{2} = \sqrt{\frac{s(s-b)}{ca}}$$

$$\sin\frac{C}{2} = \sqrt{\frac{(s-a)(s-b)}{ab}} \qquad\qquad \cos\frac{C}{2} = \sqrt{\frac{s(s-c)}{ab}}$$

6.2.4 SOLUTION OF TRIANGLES

A triangle is totally described by specifying any side and two additional parameters: either the remaining two sides (if they satisfy the triangle inequality), another side and the included angle, or two specified angles. If two sides are given and an angle that is not the included angle, then there might be 0, 1, or 2 such triangles. Two angles alone specify the shape of a triangle, but not its size, which requires specification of a side.

6.2.4.1 Three sides given

Formulae for any one of the angles:

$$\cos A = \frac{c^2 + b^2 - a^2}{2bc}, \qquad\qquad \sin A = \frac{2}{bc}\sqrt{s(s-a)(s-b)(s-c)},$$

$$\sin\frac{A}{2} = \sqrt{\frac{(s-b)(s-c)}{bc}}, \qquad\qquad \cos\frac{A}{2} = \sqrt{\frac{s(s-a)}{bc}},$$

$$\tan\frac{A}{2} = \sqrt{\frac{(s-b)(s-c)}{s(s-a)}} = \frac{r}{s-a}.$$

FIGURE 6.5

Different triangles requiring solution.

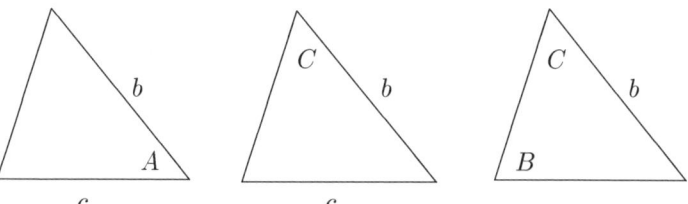

6.2.4.2 Given two sides (b, c) and the included angle (A)

See Figure 6.5, left. The remaining side and angles can be determined by repeated use of the law of cosines. For example,

1. "Non-logarithmic solution"; perform these steps sequentially:

    (a) $a^2 = b^2 + c^2 - 2bc \cos A$
    (b) $\cos B = (a^2 + c^2 - b^2)/2ca$
    (c) $\cos C = (a^2 + b^2 - c^2)/2ba$

2. "Logarithmic solution"; perform these steps sequentially:

    (a) $B + C = \pi - A$
    (b) $\tan \dfrac{(B - C)}{2} = \dfrac{b - c}{b + c} \tan \dfrac{(B + C)}{2}$
    (c) $B = \dfrac{B + C}{2} + \dfrac{B - C}{2}$
    (d) $C = \dfrac{B + C}{2} - \dfrac{B - C}{2}$
    (e) $a = \dfrac{b \sin A}{\sin B}$

6.2.4.3 Given two sides (b, c) and an angle (C), not the included angle

See Figure 6.5, middle. The remaining angles and side are determined by use of the law of sines and the fact that the sum of the angles is π ($A + B + C = \pi$).

$$\sin B = \frac{b \sin C}{c}, \qquad A = \pi - B - C, \qquad a = \frac{b \sin A}{\sin B}. \tag{6.2.4}$$

6.2.4.4 Given one side (b) and two angles (B, C)

See Figure 6.5, right. The third angle is specified by $A = \pi - B - C$. The remaining sides are found by

$$a = \frac{b \sin A}{\sin B}, \qquad c = \frac{b \sin C}{\sin B}. \tag{6.2.5}$$

6.2.5 TABLES OF TRIGONOMETRIC FUNCTIONS

(degrees) x	sin x	cos x	tan x	cot x	sec x	csc x
0	0	1	0	$\pm\infty$	1	$\pm\infty$
5	0.0872	0.9962	0.0875	11.4300	1.0038	11.4737
10	0.1736	0.9848	0.1763	5.6713	1.0154	5.7588
15	0.2588	0.9659	0.2680	3.7321	1.0353	3.8637
20	0.3420	0.9397	0.3640	2.7475	1.0642	2.9238
25	0.4226	0.9063	0.4663	2.1445	1.1034	2.3662
30	0.5000	0.8660	0.5774	1.7321	1.1547	2.0000
35	0.5736	0.8192	0.7002	1.4282	1.2208	1.7434
40	0.6428	0.7660	0.8391	1.1918	1.3054	1.5557
45	0.7071	0.7071	1.0000	1.0000	1.4142	1.4142
50	0.7660	0.6428	1.1918	0.8391	1.5557	1.3054
55	0.8192	0.5736	1.4282	0.7002	1.7434	1.2208
60	0.8660	0.5000	1.7321	0.5774	2.0000	1.1547
65	0.9063	0.4226	2.1445	0.4663	2.3662	1.1034
70	0.9397	0.3420	2.7475	0.3640	2.9238	1.0642
75	0.9659	0.2588	3.7321	0.2680	3.8637	1.0353
80	0.9848	0.1736	5.6713	0.1763	5.7588	1.0154
85	0.9962	0.0872	11.4300	0.0875	11.4737	1.0038
90	1	0	$\pm\infty$	0	$\pm\infty$	1
105	0.9659	-0.2588	-3.7321	-0.2680	-3.8637	1.0353
120	0.8660	-0.5000	-1.7321	-0.5774	-2.0000	1.1547
135	0.7071	-0.7071	-1.0000	-1.0000	-1.4142	1.4142
150	0.5000	-0.8660	-0.5774	-1.7321	-1.1547	2.0000
165	0.2588	-0.9659	-0.2680	-3.7321	-1.0353	3.8637
180	0	-1	0	$\pm\infty$	-1	$\pm\infty$
195	-0.2588	-0.9659	0.2680	3.7321	-1.0353	-3.8637
210	-0.5000	-0.8660	0.5774	1.7321	-1.1547	-2.0000
225	-0.7071	-0.7071	1.0000	1.0000	-1.4142	-1.4142
240	-0.8660	-0.5000	1.7321	0.5774	-2.0000	-1.1547
255	-0.9659	-0.2588	3.7321	0.2680	-3.8637	-1.0353
270	-1	0	$\pm\infty$	0	$\pm\infty$	-1
285	-0.9659	0.2588	-3.7321	-0.2680	3.8637	-1.0353
300	-0.8660	0.5000	-1.7321	-0.5774	2.0000	-1.1547
315	-0.7071	0.7071	-1.0000	-1.0000	1.4142	-1.4142
330	-0.5000	0.8660	-0.5774	-1.7321	1.1547	-2.0000
345	-0.2588	0.9659	-0.2680	-3.7321	1.0353	-3.8637
360	0	1	0	$\pm\infty$	1	$\pm\infty$

(radians) x	$\sin x$	$\cos x$	$\tan x$	$\cot x$	$\sec x$	$\csc x$
0	0	1	0	$\pm\infty$	1	$\pm\infty$
0.1	0.0998	0.9950	0.1003	9.9666	1.0050	10.0167
0.2	0.1987	0.9801	0.2027	4.9332	1.0203	5.0335
0.3	0.2955	0.9553	0.3093	3.2327	1.0468	3.3839
0.4	0.3894	0.9211	0.4228	2.3652	1.0857	2.5679
0.5	0.4794	0.8776	0.5463	1.8305	1.1395	2.0858
0.6	0.5646	0.8253	0.6841	1.4617	1.2116	1.7710
0.7	0.6442	0.7648	0.8423	1.1872	1.3075	1.5523
0.8	0.7174	0.6967	1.0296	0.9712	1.4353	1.3940
0.9	0.7833	0.6216	1.2602	0.7936	1.6087	1.2766
1.0	0.8415	0.5403	1.5574	0.6421	1.8508	1.1884
1.1	0.8912	0.4536	1.9648	0.5090	2.2046	1.1221
1.2	0.9320	0.3624	2.5722	0.3888	2.7597	1.0729
1.3	0.9636	0.2675	3.6021	0.2776	3.7383	1.0378
1.4	0.9854	0.1700	5.7979	0.1725	5.8835	1.0148
1.5	0.9975	0.0707	14.1014	0.0709	14.1368	1.0025
$\pi/2$	1	0	$\pm\infty$	0	$\pm\infty$	1
1.6	0.9996	-0.0292	-34.2325	-0.0292	-34.2471	1.0004
1.7	0.9917	-0.1288	-7.6966	-0.1299	-7.7613	1.0084
1.8	0.9738	-0.2272	-4.2863	-0.2333	-4.4014	1.0269
1.9	0.9463	-0.3233	-2.9271	-0.3416	-3.0932	1.0567
2.0	0.9093	-0.4161	-2.1850	-0.4577	-2.4030	1.0998
2.1	0.8632	-0.5048	-1.7098	-0.5848	-1.9808	1.1585
2.2	0.8085	-0.5885	-1.3738	-0.7279	-1.6992	1.2369
2.3	0.7457	-0.6663	-1.1192	-0.8935	-1.5009	1.3410
2.4	0.6755	-0.7374	-0.9160	-1.0917	-1.3561	1.4805
2.5	0.5985	-0.8011	-0.7470	-1.3386	-1.2482	1.6709
2.6	0.5155	-0.8569	-0.6016	-1.6622	-1.1670	1.9399
2.7	0.4274	-0.9041	-0.4727	-2.1154	-1.1061	2.3398
2.8	0.3350	-0.9422	-0.3555	-2.8127	-1.0613	2.9852
2.9	0.2392	-0.9710	-0.2464	-4.0584	-1.0299	4.1797
3.0	0.1411	-0.9900	-0.1425	-7.0153	-1.0101	7.0862
3.1	0.0416	-0.9991	-0.0416	-24.0288	-1.0009	24.0496
π	0	-1	0	$\pm\infty$	-1	$\pm\infty$

6.3   INVERSE CIRCULAR FUNCTIONS

6.3.1   DEFINITION IN TERMS OF AN INTEGRAL

$$\text{arc sin}(z) = \sin^{-1} z = \int_0^z \frac{dt}{\sqrt{1 - t^2}},$$

$$\text{arc cos}(z) = \cos^{-1} z = \int_z^1 \frac{dt}{\sqrt{1 - t^2}} = \frac{\pi}{2} - \sin^{-1} z, \tag{6.3.1}$$

$$\text{arc tan}(z) = \tan^{-1} z = \int_0^z \frac{dt}{1 + t^2} = \frac{\pi}{2} - \cot^{-1} z,$$

where z can be complex. The path of integration must not cross the real axis in the first two cases. In the third case, it must not cross the imaginary axis except possibly inside the unit circle. If $-1 \leq x \leq 1$, then $\sin^{-1} x$ and $\cos^{-1} x$ are real, $-\frac{\pi}{2} \leq \sin^{-1} x \leq \frac{\pi}{2}$, and $0 \leq \cos^{-1} x \leq \pi$.

$$\csc^{-1} z = \sin^{-1}(1/z),$$
$$\sec^{-1} z = \cos^{-1}(1/z),$$
$$\cot^{-1} z = \tan^{-1}(1/z), \tag{6.3.2}$$
$$\sec^{-1} z + \csc^{-1} z = \pi/2.$$

6.3.2   PRINCIPAL VALUES OF THE INVERSE CIRCULAR FUNCTIONS

The general solutions of $\{\sin t = z, \cos t = z, \tan t = z\}$ are, respectively:

$$t = \sin^{-1} z = (-1)^k t_0 + k\pi, \qquad \text{with } \sin t_0 = z,$$
$$t = \cos^{-1} z = \pm t_1 + 2k\pi, \qquad \text{with } \sin t_1 = z,$$
$$t = \tan^{-1} z = t_2 + k\pi, \qquad \text{with } \sin t_2 = z,$$

where k is an arbitrary integer. While "$\sin^{-1} x$" can denote, as above, any angle whose sin is x, the function $\sin^{-1} x$ usually denotes the *principal value*. The principal values of the inverse trigonometric functions are defined as follows:

1. When $\quad -1 \leq x \leq 1, \quad$ then $-\pi/2 \leq \sin^{-1} x \leq \pi/2$.
2. When $\quad -1 \leq x \leq 1, \quad$ then $\quad\; 0 \leq \cos^{-1} x \leq \pi$.
3. When $-\infty \leq x \leq \infty, \quad$ then $-\pi/2 \leq \tan^{-1} x \leq \pi/2$.
4. When $\quad\;\; 1 \leq x, \quad\quad\;\;$ then $\quad\; 0 \leq \csc^{-1} x \leq \pi/2$.
   When $\quad\quad\; x \leq -1, \quad$ then $-\pi/2 \leq \csc^{-1} x \leq 0$.
5. When $\quad\;\; 1 \leq x, \quad\quad\;\;$ then $\quad\; 0 \leq \sec^{-1} x \leq \pi/2$.
   When $\quad\quad\; x \leq -1, \quad$ then $\quad \pi/2 \leq \sec^{-1} x \leq \pi$.
6. When $-\infty \leq x \leq \infty, \quad$ then $\quad\; 0 \leq \cot^{-1} x \leq \pi$.

6.3.3 FUNDAMENTAL IDENTITIES

$$\sin^{-1} x + \cos^{-1} x = \pi/2.$$

$$\tan^{-1} x + \cot^{-1} x = \pi/2.$$

If $\alpha = \sin^{-1} x$, then

$$\sin\alpha = x, \qquad \cos\alpha = \sqrt{1-x^2}, \qquad \tan\alpha = \frac{x}{\sqrt{1-x^2}},$$

$$\csc\alpha = \frac{1}{x}, \qquad \sec\alpha = \frac{1}{\sqrt{1-x^2}}, \qquad \cot\alpha = \frac{\sqrt{1-x^2}}{x}.$$

If $\alpha = \cos^{-1} x$, then

$$\sin\alpha = \sqrt{1-x^2}, \qquad \cos\alpha = x, \qquad \tan\alpha = \frac{\sqrt{1-x^2}}{x},$$

$$\csc\alpha = \frac{1}{\sqrt{1-x^2}}, \qquad \sec\alpha = \frac{1}{x}, \qquad \cot\alpha = \frac{x}{\sqrt{1-x^2}}.$$

If $\alpha = \tan^{-1} x$, then

$$\sin\alpha = \frac{x}{\sqrt{1+x^2}}, \qquad \cos\alpha = \frac{1}{\sqrt{1+x^2}}, \qquad \tan\alpha = x,$$

$$\csc\alpha = \frac{\sqrt{1+x^2}}{x}, \qquad \sec\alpha = \sqrt{1+x^2}, \qquad \cot\alpha = \frac{1}{x}.$$

6.3.4 FUNCTIONS OF NEGATIVE ARGUMENTS

$$\sin^{-1}(-z) = -\sin^{-1} z, \qquad \sec^{-1}(-z) = \pi - \sec^{-1} z,$$
$$\cos^{-1}(-z) = \pi - \cos^{-1} z, \qquad \csc^{-1}(-z) = -\csc^{-1} z,$$
$$\tan^{-1}(-z) = -\tan^{-1} z, \qquad \cot^{-1}(-z) = \pi - \cot^{-1} z.$$

6.3.5 RELATIONSHIP TO INVERSE HYPERBOLIC FUNCTIONS

$$\sin^{-1} z = -i \sinh^{-1}(iz), \qquad \csc^{-1} z = i \operatorname{csch}^{-1}(iz),$$
$$\tan^{-1} z = -i \tanh^{-1}(iz), \qquad \cot^{-1} z = i \coth^{-1}(iz).$$

6.3.6 SUM AND DIFFERENCE OF TWO INVERSE CIRCULAR FUNCTIONS

$$\sin^{-1} z_1 \pm \sin^{-1} z_2 = \sin^{-1}\left(z_1\sqrt{1 - z_2^2} \pm z_2\sqrt{1 - z_1^2} \right).$$

$$\cos^{-1} z_1 \pm \cos^{-1} z_2 = \cos^{-1}\left(z_1 z_2 \mp \sqrt{(1 - z_2^2)(1 - z_1^2)} \right).$$

$$\tan^{-1} z_1 \pm \tan^{-1} z_2 = \tan^{-1}\left(\frac{z_1 \pm z_2}{1 \mp z_1 z_2} \right).$$

$$\sin^{-1} z_1 \pm \cos^{-1} z_2 = \sin^{-1}\left(z_1 z_2 \pm \sqrt{(1 - z_1^2)(1 - z_2^2)} \right),$$

$$= \cos^{-1}\left(z_2\sqrt{1 - z_1^2} \mp z_1\sqrt{1 - z_2^2} \right).$$

$$\tan^{-1} z_1 \pm \cot^{-1} z_2 = \tan^{-1}\left(\frac{z_1 z_2 \pm 1}{z_2 \mp z_1} \right) = \cot^{-1}\left(\frac{z_2 \mp z_1}{z_1 z_2 \pm 1} \right).$$

6.4 CEILING AND FLOOR FUNCTIONS

The *ceiling function* of x, denoted $\lceil x \rceil$, is the least integer that is not smaller than x. For example, $\lceil \pi \rceil = 4$, $\lceil 5 \rceil = 5$, and $\lceil -1.5 \rceil = -1$.

The *floor function* of x, denoted $\lfloor x \rfloor$, is the largest integer that is not larger than x. For example, $\lfloor \pi \rfloor = 3$, $\lfloor 5 \rfloor = 5$, and $\lfloor -1.5 \rfloor = -2$.

6.5 EXPONENTIAL FUNCTION

6.5.1 EXPONENTIATION

For a any real number and m a positive integer, the *exponential* a^m is defined as

$$a^m = \underbrace{a \cdot a \cdot a \cdots a}_{m \text{ terms}}. \tag{6.5.1}$$

The following three laws of exponents follow for $a \neq 0$:

1. $a^n \cdot a^m = a^{n+m}$.

2. $\dfrac{a^m}{a^n} = \begin{cases} a^{m-n}, & \text{if } m > n, \\ 1, & \text{if } m = n, \\ \frac{1}{a^{n-m}}, & \text{if } m < n. \end{cases}$

3. $(a^m)^n = a^{(mn)}$.

The nth *root function* is defined as the inverse of the nth power function:

$$\text{If } b^n = a, \text{ then } b = \sqrt[n]{a} = a^{(1/n)}. \tag{6.5.2}$$

If n is odd, there will be a unique real number satisfying the above definition of $\sqrt[n]{a}$, for any real value of a. If n is even, for positive values of a there will be two real values for $\sqrt[n]{a}$, one positive and one negative. By convention, the symbol $\sqrt[n]{a}$ means the positive value. If n is even and a is negative, then there are no real values for $\sqrt[n]{a}$.

To extend the definition to include a^t (for t not necessarily a positive integer) so as to maintain the laws of exponents, the following definitions are required (where we now restrict a to be positive; p and q are integers):

$$a^0 = 1 \qquad a^{p/q} = \sqrt[q]{a^p} \qquad a^{-t} = \frac{1}{a^t}. \tag{6.5.3}$$

With these restrictions, the second law of exponents can be written as $\dfrac{a^m}{a^n} = a^{m-n}$.

If $a > 1$, then the function a^x is monotonically increasing while, if $0 < a < 1$ then the function a^x is monotonically decreasing.

6.5.2 DEFINITION OF e^z

$$\exp(z) = e^z = \lim_{m \to \infty} \left(1 + \frac{z}{m}\right)^m$$

$$= 1 + z + \frac{z^2}{2!} + \frac{z^3}{3!} + \frac{z^4}{4!} + \dots. \tag{6.5.4}$$

If $z = x + iy$, then $e^z = e^x e^{iy} = e^x(\cos y + i \sin y)$.

The numerical value of e is given on page 15.

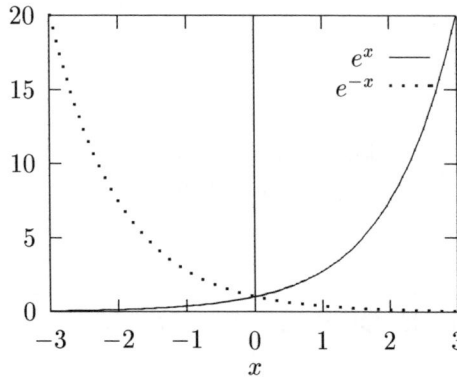

6.5.3 DERIVATIVE AND INTEGRAL OF e^x

The derivative of e^x is e^x. The integral of e^x is e^x.

6.6 LOGARITHMIC FUNCTIONS

6.6.1 DEFINITION OF THE NATURAL LOGARITHM

The *natural logarithm* (also known as the *Napierian logarithm*) of z is written as $\ln z$ or as $\log_e z$. It is sometimes written $\log z$ (this is also used to represent a "generic" logarithm, a logarithm to any base). One definition is

$$\ln z = \int_1^z \frac{dt}{t}, \tag{6.6.1}$$

where the integration path from 1 to z does not cross the origin or the negative real axis.

For complex values of z the natural logarithm, as defined above, can be represented in terms of its magnitude and phase. If $z = x + iy = re^{i\theta}$, then

$$\ln z = \ln r + i(\theta + 2k\pi), \tag{6.6.2}$$

for some $k = 0, \pm 1, \ldots$, where $r = \sqrt{x^2 + y^2}$, $x = r\cos\theta$, and $y = r\sin\theta$. Usually, the value of k is chosen so that $0 \le (\theta + 2k\pi) < 2\pi$.

6.6.1.1 Logarithms to a base other than e

The logarithmic function to the base a, written $\log_a$, is defined as

$$\log_a z = \frac{\log_b z}{\log_b a} = \frac{\ln z}{\ln a}. \tag{6.6.3}$$

Note the properties:

1. $\log_a a^p = p.$
2. $\log_a b = \dfrac{1}{\log_b a}.$
3. $\log_{10} z = \dfrac{\ln z}{\ln 10} = (\log_{10} e)\ln z \approx (0.4342944819)\ln z.$
4. $\ln z = (\ln 10)\log_{10} z \approx (2.3025850929)\log_{10} z.$

6.6.2 LOGARITHM OF SPECIAL VALUES

$$\ln 0 = -\infty, \qquad \ln 1 = 0, \qquad \ln e = 1,$$

$$\ln(-1) = i\pi + 2\pi i k, \qquad \ln(\pm i) = \pm\frac{i\pi}{2} + 2\pi i k.$$

6.6.3 RELATING THE LOGARITHM TO THE EXPONENTIAL

For real values of z the logarithm is a monotonic function, as is the exponential. Any monotonic function has a single-valued inverse function; the natural logarithm is the inverse of the exponential. If $x = e^y$, then $y = \ln x$, and $x = e^{\ln x}$. The same inverse relations hold for bases other than e. That is, if $u = a^w$, then $w = \log_a u$, and $u = a^{\log_a u}$.

6.6.4 IDENTITIES

$$\log_a z_1 z_2 = \log_a z_1 + \log_a z_2, \qquad \text{for } (-\pi < \arg z_1 + \arg z_2 < \pi).$$

$$\log_a \frac{z_1}{z_2} = \log_a z_1 - \log_a z_2, \qquad \text{for } (-\pi < \arg z_1 - \arg z_2 < \pi).$$

$$\log_a z^n = n \log_a z, \qquad \text{for } (-\pi < n \arg z < \pi), \text{ when } n \text{ is an integer.}$$

6.6.5 SERIES EXPANSIONS FOR THE NATURAL LOGARITHM

$$\ln (1 + z) = z - \frac{1}{2}z^2 + \frac{1}{3}z^3 - \dots, \qquad \text{for } |z| < 1.$$

$$\ln z = \left(\frac{z-1}{z}\right) + \frac{1}{2}\left(\frac{z-1}{z}\right)^2 + \frac{1}{3}\left(\frac{z-1}{z}\right)^3 + \dots, \qquad \text{for Re } z \geq \frac{1}{2}.$$

6.6.6 DERIVATIVE AND INTEGRATION FORMULAE

$$\frac{d \ln z}{dz} = \frac{1}{z}, \qquad \int \frac{dz}{z} = \ln z, \qquad \int \ln z \, dz = z \ln z - z.$$

6.7 HYPERBOLIC FUNCTIONS

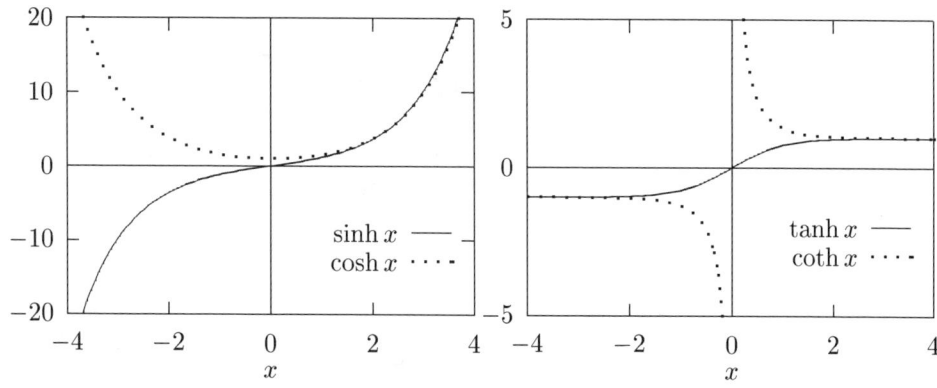

6.7.1 DEFINITIONS OF THE HYPERBOLIC FUNCTIONS

$$\sinh z = \frac{e^z - e^{-z}}{2}, \qquad\qquad \operatorname{csch} z = \frac{1}{\sinh z},$$

$$\cosh z = \frac{e^z + e^{-z}}{2}, \qquad\qquad \operatorname{sech} z = \frac{1}{\cosh z},$$

$$\tanh z = \frac{e^z - e^{-z}}{e^z + e^{-z}} = \frac{\sinh z}{\cosh z}, \qquad\qquad \coth z = \frac{1}{\tanh z}.$$

When $z = x + iy$,

$$\sinh z = \sinh x \cos y + i \cosh x \sin y, \qquad \tanh z = \frac{\sinh 2x + i \sin 2y}{\cosh 2x + \cos 2y},$$

$$\cosh z = \cosh x \cos y + i \sinh x \sin y, \qquad \coth z = \frac{\sinh 2x - i \sin 2y}{\cosh 2x - \cos 2y}.$$

6.7.2 RANGE OF VALUES

Function	Domain (interval of u)	Range (interval of function)	Remarks
$\sinh u$	$(-\infty, +\infty)$	$(-\infty, +\infty)$	
$\cosh u$	$(-\infty, +\infty)$	$[1, +\infty)$	
$\tanh u$	$(-\infty, +\infty)$	$(-1, +1)$	
$\operatorname{csch} u$	$(-\infty, 0)$	$(0, -\infty)$	Two branches,
	$(0, +\infty)$	$(+\infty, 0)$	pole at $u = 0$.
$\operatorname{sech} u$	$(-\infty, +\infty)$	$(0, 1]$	
$\coth u$	$(-\infty, 0)$	$(-1, -\infty)$	Two branches,
	$(0, +\infty)$	$(+\infty, 1)$	pole at $u = 0$.

6.7.3 HYPERBOLIC FUNCTIONS IN TERMS OF ONE ANOTHER

Function	$\sinh x$	$\cosh x$	$\tanh x$
$\sinh x =$	$\sinh x$	$\pm\sqrt{(\cosh x)^2 - 1}$	$\dfrac{\tanh x}{\sqrt{1 - (\tanh x)^2}}$
$\cosh x =$	$\sqrt{1 + (\sinh x)^2}$	$\cosh x$	$\dfrac{1}{\sqrt{1 - (\tanh x)^2}}$
$\tanh x =$	$\dfrac{\sinh x}{\sqrt{1 + (\sinh x)^2}}$	$\pm\dfrac{\sqrt{(\cosh x)^2 - 1}}{\cosh x}$	$\tanh x$
$\operatorname{csch} x =$	$\dfrac{1}{\sinh x}$	$\pm\dfrac{1}{\sqrt{(\cosh x)^2 - 1}}$	$\dfrac{\sqrt{1 - (\tanh x)^2}}{\tanh x}$
$\operatorname{sech} x =$	$\dfrac{1}{\sqrt{1 + (\sinh x)^2}}$	$\dfrac{1}{\cosh x}$	$\sqrt{1 - (\tanh x)^2}$
$\coth x =$	$\dfrac{\sqrt{1 + (\sinh x)^2}}{\sinh x}$	$\pm\dfrac{\cosh x}{\sqrt{(\cosh x)^2 - 1}}$	$\dfrac{1}{\tanh x}$

Function	$\operatorname{csch} x$	$\operatorname{sech} x$	$\coth x$
$\sinh x =$	$\dfrac{1}{\operatorname{csch} x}$	$\pm\dfrac{\sqrt{1-(\operatorname{sech} x)^2}}{\operatorname{sech} x}$	$\pm\dfrac{1}{\sqrt{(\coth x)^2-1}}$
$\cosh x =$	$\pm\dfrac{\sqrt{(\operatorname{csch} x)^2+1}}{\operatorname{csch} x}$	$\dfrac{1}{\operatorname{sech} x}$	$\pm\dfrac{\coth x}{\sqrt{(\coth x)^2-1}}$
$\tanh x =$	$\dfrac{1}{\sqrt{(\operatorname{csch} x)^2+1}}$	$\pm\sqrt{1-(\operatorname{sech} x)^2}$	$\dfrac{1}{\coth x}$
$\operatorname{csch} x =$	$\operatorname{csch} x$	$\pm\dfrac{\operatorname{sech} x}{\sqrt{1-(\operatorname{sech} x)^2}}$	$\pm\sqrt{(\coth x)^2-1}$
$\operatorname{sech} x =$	$\pm\dfrac{\operatorname{csch} x}{\sqrt{(\operatorname{csch} x)^2+1}}$	$\operatorname{sech} x$	$\pm\dfrac{\sqrt{(\coth x)^2-1}}{\coth x}$
$\coth x =$	$\sqrt{(\operatorname{csch} x)^2+1}$	$\pm\dfrac{1}{\sqrt{1-(\operatorname{sech} x)^2}}$	$\coth x$

6.7.4 RELATIONS AMONG HYPERBOLIC FUNCTIONS

$$e^{z}=\cosh z+\sinh z, \qquad e^{-z}=\cosh z-\sinh z,$$

$$(\cosh z)^2-(\sinh z)^2=(\tanh z)^2+(\operatorname{sech} z)^2=(\coth z)^2-(\operatorname{csch} z)^2=1.$$

6.7.5 RELATIONSHIP TO CIRCULAR FUNCTIONS

$$\cosh z=\cos iz, \qquad \sinh z=-i\sin iz, \qquad \tanh z=-i\tan iz.$$

6.7.6 SERIES EXPANSIONS

$$\cosh z=1+\frac{z^2}{2!}+\frac{z^4}{4!}+\frac{z^6}{6!}+\dots, \qquad |z|<\infty.$$

$$\sinh z=z+\frac{z^3}{3!}+\frac{z^5}{5!}+\frac{z^7}{7!}+\dots, \qquad |z|<\infty.$$

$$\tanh z=z-\frac{z^3}{3}+\frac{2\,z^5}{15}-\frac{17\,z^7}{315}+\dots, \qquad |z|<\frac{\pi}{2}.$$

6.7.7 SYMMETRY RELATIONSHIPS

$$\cosh(-z)=+\cosh z, \qquad \sinh(-z)=-\sinh z, \qquad \tanh(-z)=-\tanh z.$$

6.7.8 SUM AND DIFFERENCE FORMULAE

$$\cosh\left(z_1 \pm z_2\right) = \cosh z_1 \cosh z_2 \pm \sinh z_1 \sinh z_2,$$

$$\sinh\left(z_1 \pm z_2\right) = \sinh z_1 \cosh z_2 \pm \cosh z_1 \sinh z_2,$$

$$\tanh\left(z_1 \pm z_2\right) = \frac{\tanh z_1 \pm \tanh z_2}{1 \pm \tanh z_1 \tanh z_2} = \frac{\sinh 2z_1 \pm \sinh 2z_2}{\cosh 2z_1 \pm \cosh 2z_2},$$

$$\coth\left(z_1 \pm z_2\right) = \frac{1 \pm \coth z_1 \coth z_2}{\coth z_1 \pm \coth z_2} = \frac{\sinh 2z_1 \mp \sinh 2z_2}{\cosh 2z_1 - \cosh 2z_2}.$$

6.7.9 MULTIPLE ARGUMENT RELATIONS

$$\sinh 2\alpha = 2 \sinh \alpha \cosh \alpha = \frac{2 \tanh \alpha}{1 - \tanh^2 \alpha}.$$

$$\sinh 3\alpha = 3 \sinh \alpha + 4 \sinh^3 \alpha = \sinh \alpha (4 \cosh^2 \alpha - 1).$$

$$\sinh 4\alpha = 4 \sinh^3 \alpha \cosh \alpha + 4 \cosh^3 \alpha \sinh \alpha.$$

$$\cosh 2\alpha = \cosh^2 \alpha + \sinh^2 \alpha = 2 \cosh^2 \alpha - 1,$$

$$= 1 + 2 \sinh^2 \alpha = \frac{1 + \tanh^2 \alpha}{1 - \tanh^2 \alpha}.$$

$$\cosh 3\alpha = -3 \cosh \alpha + 4 \cosh^3 \alpha = \cosh \alpha (4 \sinh^2 \alpha + 1).$$

$$\cosh 4\alpha = \cosh^4 \alpha + 6 \sinh^2 \alpha \cosh^2 \alpha + 6 \sinh^4 \alpha.$$

$$\tanh 2\alpha = \frac{2 \tanh \alpha}{1 + \tanh^2 \alpha}.$$

$$\tanh 3\alpha = \frac{3 \tanh \alpha + \tanh^3 \alpha}{1 + 3 \tanh^2 \alpha}.$$

$$\coth 2\alpha = \frac{1 + \coth^2 \alpha}{2 \coth \alpha}.$$

$$\coth 3\alpha = \frac{3 \coth \alpha + \coth^3 \alpha}{1 + 3 \coth^2 \alpha}.$$

6.7.10 SUMS OF FUNCTIONS

$$\sinh u \pm \sinh w = 2 \sinh \frac{u \pm w}{2} \cosh \frac{u \mp w}{2},$$

$$\cosh u + \cosh w = 2 \cosh \frac{u + w}{2} \cosh \frac{u - w}{2},$$

$$\cosh u - \cosh w = 2 \sinh \frac{u + w}{2} \sinh \frac{u - w}{2},$$

$$\tanh u \pm \tanh w = \frac{\sinh u \pm w}{\cosh u \cosh w},$$

$$\coth u \pm \coth w = \frac{\sinh u \pm w}{\sinh u \sinh w}.$$

6.7.11 PRODUCTS OF FUNCTIONS

$$\sinh u \sinh w = \frac{1}{2}\left(\cosh(u+w) - \cosh(u-w)\right),$$

$$\sinh u \cosh w = \frac{1}{2}\left(\sinh(u+w) + \sinh(u-w)\right),$$

$$\cosh u \cosh w = \frac{1}{2}\left(\cosh(u+w) + \cosh(u-w)\right).$$

6.7.12 HALF–ARGUMENT FORMULAE

$$\sinh\frac{z}{2} = \pm\sqrt{\frac{\cosh z - 1}{2}}, \qquad\qquad \cosh\frac{z}{2} = +\sqrt{\frac{\cosh z + 1}{2}},$$

$$\tanh\frac{z}{2} = \pm\sqrt{\frac{\cosh z - 1}{\cosh z + 1}} = \frac{\sinh z}{\cosh z + 1}, \qquad \coth\frac{z}{2} = \pm\sqrt{\frac{\cosh z + 1}{\cosh z - 1}} = \frac{\sinh z}{\cosh z - 1}.$$

6.7.13 DIFFERENTIATION FORMULAE

$$\frac{d\sinh z}{dz} = \cosh z, \qquad\qquad \frac{d\cosh z}{dz} = \sinh z,$$

$$\frac{d\tanh z}{dz} = (\operatorname{sech} z)^2, \qquad\qquad \frac{d\operatorname{csch} z}{dz} = -\operatorname{csch} z \coth z,$$

$$\frac{d\operatorname{sech} z}{dz} = -\operatorname{sech} z \tanh z, \qquad\qquad \frac{d\coth z}{dz} = -(\operatorname{csch} z)^2.$$

6.8 INVERSE HYPERBOLIC FUNCTIONS

$$\cosh^{-1} z = \int_0^z \frac{dt}{\sqrt{t^2 - 1}}, \qquad \sinh^{-1} z = \int_0^z \frac{dt}{\sqrt{1 + t^2}}, \qquad \tanh^{-1} z = \int_0^z \frac{dt}{1 - t^2}.$$

6.8.1 RANGE OF VALUES

Function	Domain	Range	Remarks
$\sinh^{-1} u$	$(-\infty, +\infty)$	$(-\infty, +\infty)$	Odd function
$\cosh^{-1} u$	$[1, +\infty)$	$(-\infty, +\infty)$	Even function, double valued
$\tanh^{-1} u$	$(-1, +1)$	$(-\infty, +\infty)$	Odd function
$\operatorname{csch}^{-1} u$	$(-\infty, 0), (0, \infty)$	$(0, -\infty), (\infty, 0)$	Odd function, two branches, Pole at $u = 0$
$\operatorname{sech}^{-1} u$	$(0, 1]$	$(-\infty, +\infty)$	Double valued
$\coth^{-1} u$	$(-\infty, -1), (1, \infty)$	$(-\infty, 0), (\infty, 0)$	Odd function, two branches

6.8.2 RELATIONSHIPS AMONG INVERSE HYPERBOLIC FUNCTIONS

Function	$\sinh^{-1} x$	$\cosh^{-1} x$	$\tanh^{-1} x$
$\sinh^{-1} x =$	$\sinh^{-1} x$	$\pm\cosh^{-1}\sqrt{x^2+1}$	$\tanh^{-1}\dfrac{x}{\sqrt{1+x^2}}$
$\cosh^{-1} x =$	$\pm\sinh^{-1}\sqrt{x^2-1}$	$\cosh^{-1} x$	$\pm\tanh^{-1}\dfrac{\sqrt{x^2-1}}{x}$
$\tanh^{-1} x =$	$\sinh^{-1}\dfrac{x}{\sqrt{1-x^2}}$	$\pm\cosh^{-1}\dfrac{1}{\sqrt{1-x^2}}$	$\tanh^{-1} x$
$\operatorname{csch}^{-1} x =$	$\sinh^{-1}\dfrac{1}{x}$	$\pm\cosh^{-1}\dfrac{\sqrt{1+x^2}}{x}$	$\tanh^{-1}\dfrac{1}{\sqrt{1+x^2}}$
$\operatorname{sech}^{-1} x =$	$\pm\sinh^{-1}\dfrac{\sqrt{1-x^2}}{x}$	$\cosh^{-1}\dfrac{1}{x}$	$\pm\tanh^{-1}\sqrt{1-x^2}$
$\coth^{-1} x =$	$\sinh^{-1}\dfrac{1}{\sqrt{x^2-1}}$	$\pm\cosh^{-1}\dfrac{x}{\sqrt{x^2-1}}$	$\tanh^{-1}\dfrac{1}{x}$

Function	$\operatorname{csch}^{-1} x$	$\operatorname{sech}^{-1} x$	$\coth^{-1} x$
$\sinh^{-1} x =$	$\operatorname{csch}^{-1}\dfrac{1}{x}$	$\pm\operatorname{sech}^{-1}\dfrac{1}{\sqrt{1+x^2}}$	$\coth^{-1}\dfrac{\sqrt{1+x^2}}{x}$
$\cosh^{-1} x =$	$\pm\operatorname{csch}^{-1}\dfrac{1}{\sqrt{x^2-1}}$	$\operatorname{sech}^{-1}\dfrac{1}{x}$	$\pm\coth^{-1}\dfrac{x}{\sqrt{x^2-1}}$
$\tanh^{-1} x =$	$\operatorname{csch}^{-1}\dfrac{\sqrt{1-x^2}}{x}$	$\pm\operatorname{sech}^{-1}\sqrt{1-x^2}$	$\coth^{-1}\dfrac{1}{x}$
$\operatorname{csch}^{-1} x =$	$\operatorname{csch}^{-1} x$	$\pm\operatorname{sech}^{-1}\dfrac{x}{\sqrt{1+x^2}}$	$\coth^{-1}\sqrt{1+x^2}$
$\operatorname{sech}^{-1} x =$	$\pm\operatorname{csch}^{-1}\dfrac{x}{\sqrt{1-x^2}}$	$\operatorname{sech}^{-1} x$	$\pm\coth^{-1}\dfrac{1}{\sqrt{1-x^2}}$
$\coth^{-1} x =$	$\operatorname{csch}^{-1}\sqrt{x^2-1}$	$\operatorname{sech}^{-1}\dfrac{\sqrt{x^2-1}}{x}$	$\coth^{-1} x$

6.8.3 RELATIONSHIPS WITH LOGARITHMIC FUNCTIONS

$$\sinh^{-1} x = \log\left(x + \sqrt{x^2 + 1}\right), \qquad \operatorname{csch}^{-1} x = \log\left(\frac{1 \pm \sqrt{1 + x^2}}{x}\right),$$

$$\cosh^{-1} x = \log\left(x \pm \sqrt{x^2 - 1}\right), \qquad \operatorname{sech}^{-1} x = \log\left(\frac{1 \pm \sqrt{1 - x^2}}{x}\right),$$

$$\tanh^{-1} x = \frac{1}{2}\log\left(\frac{1 + x}{1 - x}\right), \qquad \operatorname{coth}^{-1} x = \frac{1}{2}\log\left(\frac{x + 1}{x - 1}\right).$$

6.8.4 RELATIONSHIPS WITH CIRCULAR FUNCTIONS

$$\sinh^{-1} x = -i\sin^{-1} ix \qquad\qquad \sinh^{-1} ix = +i\sin^{-1} x$$

$$\tanh^{-1} x = -i\tan^{-1} ix \qquad\qquad \tanh^{-1} ix = +i\tan^{-1} x$$

$$\operatorname{csch}^{-1} x = +i\csc^{-1} ix \qquad\qquad \operatorname{csch}^{-1} ix = -i\csc^{-1} x$$

$$\operatorname{coth}^{-1} x = +i\cot^{-1} ix \qquad\qquad \operatorname{coth}^{-1} ix = -i\cot^{-1} x$$

6.8.5 SUM AND DIFFERENCE OF FUNCTIONS

$$\sinh^{-1} x \pm \sinh^{-1} y = \sinh^{-1}\left(x\sqrt{1 + y^2} \pm y\sqrt{1 + x^2}\right),$$

$$\cosh^{-1} x \pm \cosh^{-1} y = \cosh^{-1}\left(xy \pm \sqrt{(y^2 - 1)(x^2 - 1)}\right),$$

$$\tanh^{-1} x \pm \tanh^{-1} y = \tanh^{-1}\left(\frac{x \pm y}{xy \pm 1}\right),$$

$$\sinh^{-1} x \pm \cosh^{-1} y = \sinh^{-1}\left(xy \pm \sqrt{(1 + x^2)(y^2 - 1)}\right),$$

$$= \cosh^{-1}\left(y\sqrt{1 + x^2} \pm x\sqrt{y^2 - 1}\right),$$

$$\tanh^{-1} x \pm \coth^{-1} y = \tanh^{-1}\left(\frac{xy \pm 1}{y \pm x}\right),$$

$$= \coth^{-1}\left(\frac{y \pm x}{xy \pm 1}\right).$$

6.9   GUDERMANNIAN FUNCTION

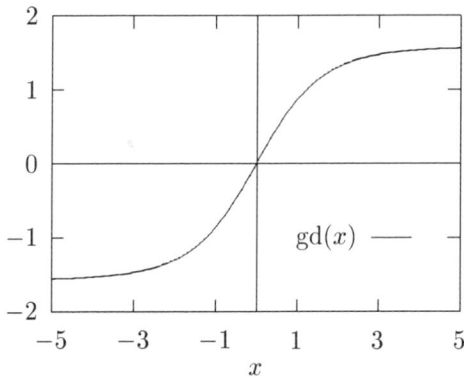

This function relates circular and hyperbolic functions without the use of functions of imaginary argument. The Gudermannian is a monotonic odd function which is asymptotic to $\pm\frac{\pi}{2}$ as $x \to \pm\infty$. It is zero at the origin.

$$\operatorname{gd} x = \text{ the Gudermannian of } x$$
$$= \int_0^x \frac{dt}{\cosh t} = 2\tan^{-1}\left(\tanh\frac{x}{2}\right) = 2\tan^{-1}e^x - \frac{\pi}{2}.$$

(6.9.1)

$$\operatorname{gd}^{-1} x = \text{ the inverse Gudermannian of } x$$
$$= \int_0^x \frac{dt}{\cos t} = \log\left[\tan\left(\frac{\pi}{4} + \frac{x}{2}\right)\right] = \log\left(\sec x + \tan x\right).$$

If $\operatorname{gd}(x + iy) = \alpha + i\beta$, then

$$\tan\alpha = \frac{\sinh x}{\cos y}, \qquad\qquad \tanh\beta = \frac{\sin y}{\cosh x},$$

$$\tanh x = \frac{\sin\alpha}{\cosh\beta}, \qquad\qquad \tan y = \frac{\sin\beta}{\cosh\alpha}.$$

6.9.1   FUNDAMENTAL IDENTITIES

$$\tanh\left(\frac{x}{2}\right) = \tan\left(\frac{\operatorname{gd} x}{2}\right),$$

$$e^x = \cosh x + \sinh x = \sec\operatorname{gd} x + \tan\operatorname{gd} x,$$

$$= \tan\left(\frac{\pi}{4} + \frac{\operatorname{gd} x}{2}\right) = \frac{1 + \sin\left(\operatorname{gd} x\right)}{\cos(\operatorname{gd} x)},$$

$$i\operatorname{gd}^{-1} x = \operatorname{gd}^{-1}(ix), \qquad \text{where } i = \sqrt{-1}.$$

6.9.2   DERIVATIVES OF GUDERMANNIAN

$$\frac{d(\operatorname{gd} x)}{dx} = \operatorname{sech} x \qquad \frac{d(\operatorname{gd}^{-1} x)}{dx} = \sec x.$$

6.9.3 RELATIONSHIP TO HYPERBOLIC AND CIRCULAR FUNCTIONS

$$\sinh x = \tan{(\operatorname{gd} x)}, \qquad \operatorname{csch} x = \cot{(\operatorname{gd} x)},$$
$$\cosh x = \sec{(\operatorname{gd} x)}, \qquad \operatorname{sech} x = \cos{(\operatorname{gd} x)},$$
$$\tanh x = \sin{(\operatorname{gd} x)}, \qquad \coth x = \csc{(\operatorname{gd} x)}.$$

6.9.4 NUMERICAL VALUES OF HYPERBOLIC FUNCTIONS

x	e^x	$\ln x$	$\operatorname{gd} x$	$\sinh x$	$\cosh x$	$\tanh x$
0	1	$-\infty$	0	0	1	0
0.1	1.1052	-2.3026	0.0998	0.1002	1.0050	0.0997
0.2	1.2214	-1.6094	0.1987	0.2013	1.0201	0.1974
0.3	1.3499	-1.2040	0.2956	0.3045	1.0453	0.2913
0.4	1.4918	-0.9163	0.3897	0.4108	1.0811	0.3799
0.5	1.6487	-0.6931	0.4804	0.5211	1.1276	0.4621
0.6	1.8221	-0.5108	0.5669	0.6367	1.1855	0.5370
0.7	2.0138	-0.3567	0.6490	0.7586	1.2552	0.6044
0.8	2.2255	-0.2231	0.7262	0.8881	1.3374	0.6640
0.9	2.4596	-0.1054	0.7985	1.0265	1.4331	0.7163
1.0	2.7183	-0.0000	0.8658	1.1752	1.5431	0.7616
1.1	3.0042	0.0953	0.9281	1.3356	1.6685	0.8005
1.2	3.3201	0.1823	0.9857	1.5095	1.8107	0.8337
1.3	3.6693	0.2624	1.0387	1.6984	1.9709	0.8617
1.4	4.0552	0.3365	1.0872	1.9043	2.1509	0.8854
1.5	4.4817	0.4055	1.1317	2.1293	2.3524	0.9051
1.6	4.9530	0.4700	1.1724	2.3756	2.5775	0.9217
1.7	5.4739	0.5306	1.2094	2.6456	2.8283	0.9354
1.8	6.0496	0.5878	1.2432	2.9422	3.1075	0.9468
1.9	6.6859	0.6419	1.2739	3.2682	3.4177	0.9562
2.0	7.3891	0.6931	1.3018	3.6269	3.7622	0.9640
2.1	8.1662	0.7419	1.3271	4.0219	4.1443	0.9705
2.2	9.0250	0.7885	1.3501	4.4571	4.5679	0.9757
2.3	9.9742	0.8329	1.3709	4.9370	5.0372	0.9801
2.4	11.0232	0.8755	1.3899	5.4662	5.5569	0.9837
2.5	12.1825	0.9163	1.4070	6.0502	6.1323	0.9866
2.6	13.4637	0.9555	1.4225	6.6947	6.7690	0.9890
2.7	14.8797	0.9933	1.4366	7.4063	7.4735	0.9910

continued on next page

continued from previous page						
x	e^x	$\ln x$	$\text{gd}\, x$	$\sinh x$	$\cosh x$	$\tanh x$
2.8	16.4446	1.0296	1.4493	8.1919	8.2527	0.9926
2.9	18.1741	1.0647	1.4609	9.0596	9.1146	0.9940
3.0	20.0855	1.0986	1.4713	10.0179	10.0677	0.9951
3.1	22.1980	1.1314	1.4808	11.0765	11.1215	0.9959
3.2	24.5325	1.1632	1.4893	12.2459	12.2866	0.9967
3.3	27.1126	1.1939	1.4971	13.5379	13.5748	0.9973
3.4	29.9641	1.2238	1.5041	14.9654	14.9987	0.9978
3.5	33.1155	1.2528	1.5104	16.5426	16.5728	0.9982
3.6	36.5982	1.2809	1.5162	18.2855	18.3128	0.9985
3.7	40.4473	1.3083	1.5214	20.2113	20.2360	0.9988
3.8	44.7012	1.3350	1.5261	22.3394	22.3618	0.9990
3.9	49.4024	1.3610	1.5303	24.6911	24.7113	0.9992
4.0	54.5982	1.3863	1.5342	27.2899	27.3082	0.9993

6.10   ORTHOGONAL POLYNOMIALS

Orthogonal polynomials are classes of polynomials, $\{p_n(x)\}$, which obey an orthogonality relationship of the form

$$\int_I w(x)\, p_n(x)\, p_m(x)\, dx = c_n \delta_{nm} \tag{6.10.1}$$

for a given *weight function* $w(x)$ and interval I.

6.10.1   HERMITE POLYNOMIALS

Symbol: $H_n(x)$.
Interval: $(-\infty, \infty)$.
Differential equation: $y'' - 2xy' + 2ny = 0$.
Explicit expression: $H_n(x) = \displaystyle\sum_{m=0}^{\lfloor n/2 \rfloor} \frac{(-1)^m n! (2x)^{n-2m}}{m!(n-2m)!}$.
Recurrence relation: $H_{n+1}(x) = 2x H_n(x) - 2n H_{n-1}(x)$.
Weight: e^{-x^2}.
Standardization: $H_n(x) = 2^n x^n + \dots$.
Norm: $\displaystyle\int_{-\infty}^{\infty} e^{-x^2} [H_n(x)]^2 \, dx = 2^n n! \sqrt{\pi}$.

Rodrigues' formula: $H_n(x) = (-1)^n e^{x^2} \dfrac{d^n}{dx^n}(e^{-x^2})$.

Generating function: $\displaystyle\sum_{n=0}^{\infty} H_n(x) \frac{z^n}{n!} = e^{-z^2 + 2zx}$.

Inequality: $|H_n(x)| < \sqrt{2^n e^{x^2} n!}$.

6.10.2 JACOBI POLYNOMIALS

Symbol: $P_n^{(\alpha,\beta)}(x)$.
Interval: $[-1, 1]$.
Differential equation: $(1-x^2)y'' + [\beta - \alpha - (\alpha+\beta+2)x]y' + n(n+\alpha+\beta+1)y = 0$.
Explicit expression:

$$P_n^{(\alpha,\beta)}(x) = \frac{1}{2^n} \sum_{m=0}^{n} \binom{n+\alpha}{m}\binom{n+\beta}{n-m}(x-1)^{n-m}(x+1)^m.$$

Recurrence relation: $2(n+1)(n+\alpha+\beta+1)(2n+\alpha+\beta)P_{n+1}^{(\alpha,\beta)}(x)$
$= (2n+\alpha+\beta+1)[(\alpha^2-\beta^2) + (2n+\alpha+\beta+2)(2n+\alpha+\beta)x]P_n^{(\alpha,\beta)}(x)$
$- 2(n+\alpha)(n+\beta)(2n+\alpha+\beta+2)P_{n-1}^{(\alpha,\beta)}(x)$.

Weight: $(1-x)^\alpha(1+x)^\beta$.

Standardization: $P_n^{(\alpha,\beta)}(1) = \binom{n+\alpha}{n}$.

Norm: $\displaystyle\int_{-1}^{1}(1-x)^\alpha(1+x)^\beta\left[P_n^{(\alpha,\beta)}(x)\right]^2 dx = \frac{2^{\alpha+\beta+1}\Gamma(n+\alpha+1)\Gamma(n+\beta+1)}{(2n+\alpha+\beta+1)n!\Gamma(n+\alpha+\beta+1)}$.

Rodrigues' formula: $P_n^{(\alpha,\beta)}(x) = \frac{(-1)^n}{2^n n!(1-x)^\alpha(1+x)^\beta}\frac{d^n}{dx^n}\left[(1-x)^{n+\alpha}(1+x)^{n+\beta}\right]$.

Generating function: $\displaystyle\sum_{n=0}^{\infty} P_n^{(\alpha,\beta)}(x)z^n = 2^{\alpha+\beta}R^{-1}(1-z+R)^{-\alpha}(1+z+R)^{-\beta}$,

where $R = \sqrt{1-2xz+z^2}$ and $|z| < 1$.

Inequality: $\displaystyle\max_{-1\leq x\leq 1}\left|P_n^{(\alpha,\beta)}(x)\right| = \begin{cases} \binom{n+q}{n} \sim n^q, & \text{if } q = \max(\alpha,\beta) \geq -\frac{1}{2}, \\ \left|P_n^{(\alpha,\beta)}(x')\right| \sim n^{-1/2}, & \text{if } q = \max(\alpha,\beta) < -\frac{1}{2}, \end{cases}$

where $\alpha, \beta > 1$ and x' (in the second result) is one of the two maximum points
nearest $(\beta - \alpha)/(\alpha + \beta + 1)$.

6.10.3 LAGUERRE POLYNOMIALS

Symbol: $L_n(x)$.
Interval: $[0, \infty)$.
$L_n(x)$ is the same as $L_n^{(0)}(x)$ (see the generalized Laguerre polynomials).

6.10.4 GENERALIZED LAGUERRE POLYNOMIALS

Symbol: $L_n^{(\alpha)}(x)$.
Interval: $[0, \infty)$.
Differential equation: $xy'' + (\alpha + 1 - x)y' + ny = 0$.

Explicit expression: $L_n^{(\alpha)}(x) = \displaystyle\sum_{m=0}^{n}\frac{(-1)^m}{m!}\binom{n+\alpha}{n-m}x^m$.

Recurrence relation:
$$(n + 1)L_{n+1}^{(\alpha)}(x) = [(2n + \alpha + 1) - x]L_n^{(\alpha)}(x) - (n + \alpha)L_{n-1}^{(\alpha)}(x) .$$

Weight: $x^\alpha e^{-x}$.

Standardization: $L_n^{(\alpha)}(x) = \dfrac{(-1)^n}{n!}x^n + \dots .$

Norm: $\displaystyle\int_0^\infty x^\alpha e^{-x} \left[L_n^{(\alpha)}(x)\right]^2 dx = \dfrac{\Gamma(n + \alpha + 1)}{n!}.$

Rodrigues' formula: $L_n^{(\alpha)}(x) = \dfrac{1}{n!x^\alpha e^{-x}} \dfrac{d^n}{dx^n}[x^{n+\alpha}e^{-x}].$

Generating function: $\displaystyle\sum_{n=0}^\infty L_n^{(\alpha)}(x)z^n = (1 - z)^{-\alpha-1} \exp\left(\dfrac{xz}{z - 1}\right).$

Inequality: $\left|L_n^{(\alpha)}(x)\right| \leq \begin{cases} \dfrac{\Gamma(n+\alpha+1)}{n!\Gamma(\alpha+1)}e^{x/2}, & \text{if } x \geq 0 \text{ and } \alpha > 0, \\ \left[2 - \dfrac{\Gamma(n+\alpha+1)}{n!\Gamma(\alpha+1)}\right]e^{x/2}, & \text{if } x \geq 0 \text{ and } -1 < \alpha < 0. \end{cases}$

Note that $\alpha > -1$ and $L_n^{(m)}(x) = (-1)^m \dfrac{d^m}{dx^m}\left[L_{n+m}(x)\right].$

6.10.5   LEGENDRE POLYNOMIALS

Symbol: $P_n(x)$.

Interval: $[-1, 1]$.

Differential equation: $(1 - x^2)y'' - 2xy' + n(n + 1)y = 0$.

Explicit expression: $P_n(x) = \dfrac{1}{2^n} \displaystyle\sum_{m=0}^{\lfloor n/2 \rfloor} (-1)^m \binom{n}{m}\binom{2n - 2m}{n} x^{n-2m}.$

Recurrence relation: $(n + 1)P_{n+1}(x) = (2n + 1)xP_n(x) - nP_{n-1}(x)$.

Weight: 1.

Standardization: $P_n(1) = 1$.

Norm: $\displaystyle\int_{-1}^1 [P_n(x)]^2 dx = \dfrac{2}{2n + 1}.$

Rodrigues' formula: $P_n(x) = \dfrac{(-1)^n}{2^n n!} \dfrac{d^n}{dx^n}[(1 - x^2)^n].$

Generating function: $\displaystyle\sum_{n=0}^\infty P_n(x)z^n = (1 - 2xz + z^2)^{-1/2},$

$$\text{for } -1 < x < 1 \text{ and } |z| < 1.$$

Inequality: $|P_n(x)| \leq 1$ for $-1 \leq x \leq 1$.

See Section 6.18.7 on page 556.

6.10.6   CHEBYSHEV POLYNOMIALS, FIRST KIND

Symbol: $T_n(x)$.

Interval: $[-1, 1]$.

Differential equation: $(1 - x^2)y'' - xy' + n^2 y = 0$.

Explicit expression: $T_n(x) = \cos\left(n\cos^{-1}x\right) = \dfrac{n}{2} \displaystyle\sum_{m=0}^{\lfloor n/2 \rfloor} (-1)^m \dfrac{(n - m - 1)!}{m!(n - 2m)!}(2x)^{n-2m}.$

Recurrence relation: $T_{n+1}(x) = 2xT_n(x) - T_{n-1}(x)$.
Weight: $(1 - x^2)^{-1/2}$.
Standardization: $T_n(1) = 1$.

Norm: $\displaystyle\int_{-1}^{1} (1 - x^2)^{-1/2} \left[T_n(x)\right]^2 \, dx = \begin{cases} \pi, & n = 0, \\ \pi/2, & n \neq 0. \end{cases}$

Rodrigues' formula: $\displaystyle T_n(x) = \frac{\sqrt{\pi(1 - x^2)}}{(-2)^n \Gamma(n + \frac{1}{2})} \frac{d^n}{dx^n}\left[(1 - x^2)^{n - 1/2}\right]$.

Generating function: $\displaystyle\sum_{n=0}^{\infty} T_n(x) z^n = \frac{1 - xz}{1 - 2xz + z^2}$, for $-1 < x < 1$ and $|z| < 1$.

Inequality: $|T_n(x)| \leq 1$, for $-1 \leq x \leq 1$.
Note that $T_n(x) = \frac{n! \sqrt{\pi}}{\Gamma(n + \frac{1}{2})} P_n^{(-1/2, -1/2)}(x)$.

6.10.7 CHEBYSHEV POLYNOMIALS, SECOND KIND

Symbol: $U_n(x)$.
Interval: $[-1, 1]$.
Differential equation: $(1 - x^2)y'' - 3xy' + n(n + 2)y = 0$.

Explicit expression: $\displaystyle U_n(x) = \sum_{m=0}^{\lfloor n/2 \rfloor} \frac{(-1)^m (m - n)!}{m!(n - 2m)!}(2x)^{n - 2m}$

$$U_n(\cos\theta) = \frac{\sin[(n + 1)\theta]}{\sin\theta}.$$

Recurrence relation: $U_{n+1}(x) = 2xU_n(x) - U_{n-1}(x)$.
Weight: $(1 - x^2)^{1/2}$.
Standardization: $U_n(1) = n + 1$.

Norm: $\displaystyle\int_{-1}^{1} (1 - x^2)^{1/2} \left[U_n(x)\right]^2 \, dx = \frac{\pi}{2}$.

Rodrigues' formula: $\displaystyle U_n(x) = \frac{(-1)^n (n + 1)\sqrt{\pi}}{(1 - x^2)^{1/2} 2^{n+1} \Gamma(n + \frac{3}{2})} \frac{d^n}{dx^n}[(1 - x^2)^{n + (1/2)}]$.

Generating function: $\displaystyle\sum_{n=0}^{\infty} U_n(x) z^n = \frac{1}{1 - 2xz + z^2}$, for $-1 < x < 1$ and $|z| < 1$.

Inequality: $|U_n(x)| \leq n + 1$, for $-1 \leq x \leq 1$.
Note that $U_n(x) = \frac{(n+1)! \sqrt{\pi}}{2\Gamma(n + \frac{3}{2})} P_n^{(1/2, 1/2)}(x)$.

6.10.8 TABLES OF ORTHOGONAL POLYNOMIALS

$H_0 = 1$ $x^{10} = (30240H_0 + 75600H_2 + 25200H_4 + 2520H_6 + 90H_8 + H_{10})/1024$
$H_1 = 2x$ $x^9 = (15120H_1 + 10080H_3 + 1512H_5 + 72H_7 + H_9)/512$
$H_2 = 4x^2 - 2$ $x^8 = (1680H_0 + 3360H_2 + 840H_4 + 56H_6 + H_8)/256$
$H_3 = 8x^3 - 12x$ $x^7 = (840H_1 + 420H_3 + 42H_5 + H_7)/128$

$$H_4 = 16x^4 - 48x^2 + 12$$
$$H_5 = 32x^5 - 160x^3 + 120x$$
$$H_6 = 64x^6 - 480x^4 + 720x^2 - 120$$
$$H_7 = 128x^7 - 1344x^5 + 3360x^3 - 1680x$$
$$H_8 = 256x^8 - 3584x^6 + 13440x^4 - 13440x^2 + 1680$$
$$H_9 = 512x^9 - 9216x^7 + 48384x^5 - 80640x^3 + 30240x$$
$$H_{10} = 1024x^{10} - 23040x^8 + 161280x^6 - 403200x^4 + 302400x^2 - 30240$$

$$x^6 = (120H_0 + 180H_2 + 30H_4 + H_6)/64$$
$$x^5 = (60H_1 + 20H_3 + H_5)/32$$
$$x^4 = (12H_0 + 12H_2 + H_4)/16$$
$$x^3 = (6H_1 + H_3)/8$$
$$x^2 = (2H_0 + H_2)/4$$
$$x = (H_1)/2$$
$$1 = H_0$$

$$L_0 = 1$$
$$L_1 = -x + 1$$
$$L_2 = (x^2 - 4x + 2)/2$$
$$L_3 = (-x^3 + 9x^2 - 18x + 6)/6$$
$$L_4 = (x^4 - 16x^3 + 72x^2 - 96x + 24)/24$$
$$L_5 = (-x^5 + 25x^4 - 200x^3 + 600x^2 - 600x + 120)/120$$
$$L_6 = (x^6 - 36x^5 + 450x^4 - 2400x^3 + 5400x^2 - 4320x + 720)/720$$

$$x^6 = 720L_0 - 4320L_1 + 10800L_2 - 14400L_3 + 10800L_4 - 4320L_5 + 720L_6$$
$$x^5 = 120L_0 - 600L_1 + 1200L_2 - 1200L_3 + 600L_4 - 120L_5$$
$$x^4 = 24L_0 - 96L_1 + 144L_2 - 96L_3 + 24L_4$$
$$x^3 = 6L_0 - 18L_1 + 18L_2 - 6L_3$$
$$x^2 = 2L_0 - 4L_1 + 2L_2$$
$$x = L_0 - L_1$$
$$1 = L_0$$

$$P_0 = 1$$
$$P_1 = x$$
$$P_2 = (3x^2 - 1)/2$$
$$P_3 = (5x^3 - 3x)/2$$
$$P_4 = (35x^4 - 30x^2 + 3)/8$$
$$P_5 = (63x^5 - 70x^3 + 15x)/8$$
$$P_6 = (231x^6 - 315x^4 + 105x^2 - 5)/16$$
$$P_7 = (429x^7 - 693x^5 + 315x^3 - 35x)/16$$
$$P_8 = (6435x^8 - 12012x^6 + 6930x^4 - 1260x^2 + 35)/128$$
$$P_9 = (12155x^9 - 25740x^7 + 18018x^5 - 4620x^3 + 315x)/128$$
$$P_{10} = (46189x^{10} - 109395x^8 + 90090x^6 - 30030x^4 + 3465x^2 - 63)/256$$

$$x^{10} = (4199P_0 + 16150P_2 + 15504P_4 + 7904P_6 + 2176P_8 + 256P_{10})/46189$$
$$x^9 = (3315P_1 + 4760P_3 + 2992P_5 + 960P_7 + 128P_9)/12155$$
$$x^8 = (715P_0 + 2600P_2 + 2160P_4 + 832P_6 + 128P_8)/6435$$
$$x^7 = (143P_1 + 182P_3 + 88P_5 + 16P_7)/429$$
$$x^6 = (33P_0 + 110P_2 + 72P_4 + 16P_6)/231$$
$$x^5 = (27P_1 + 28P_3 + 8P_5)/63$$
$$x^4 = (7P_0 + 20P_2 + 8P_4)/35$$
$$x^3 = (3P_1 + 2P_3)/5$$
$$x^2 = (P_0 + 2P_2)/3$$
$$x = P_1$$
$$1 = P_0$$

$$T_0 = 1$$
$$T_1 = x$$
$$T_2 = 2x^2 - 1$$
$$T_3 = 4x^3 - 3x$$
$$T_4 = 8x^4 - 8x^2 + 1$$
$$T_5 = 16x^5 - 20x^3 + 5x$$
$$T_6 = 32x^6 - 48x^4 + 18x^2 - 1$$
$$T_7 = 64x^7 - 112x^5 + 56x^3 - 7x$$
$$T_8 = 128x^8 - 256x^6 + 160x^4 - 32x^2 + 1$$
$$T_9 = 256x^9 - 576x^7 + 432x^5 - 120x^3 + 9x$$
$$T_{10} = 512x^{10} - 1280x^8 + 1120x^6 - 400x^4 + 50x^2 - 1$$

$$x^{10} = (126T_0 + 210T_2 + 120T_4 + 45T_6 + 10T_8 + T_{10})/512$$
$$x^9 = (126T_1 + 84T_3 + 36T_5 + 9T_7 + T_9)/256$$
$$x^8 = (35T_0 + 56T_2 + 28T_4 + 8T_6 + T_8)/128$$
$$x^7 = (35T_1 + 21T_3 + 7T_5 + T_7)/64$$
$$x^6 = (10T_0 + 15T_2 + 6T_4 + T_6)/32$$
$$x^5 = (10T_1 + 5T_3 + T_5)/16$$
$$x^4 = (3T_0 + 4T_2 + T_4)/8$$
$$x^3 = (3T_1 + T_3)/4$$
$$x^2 = (T_0 + T_2)/2$$
$$x = T_1$$
$$1 = T_0$$

$$U_0 = 1$$
$$U_1 = 2x$$
$$U_2 = 4x^2 - 1$$
$$U_3 = 8x^3 - 4x$$
$$U_4 = 16x^4 - 12x^2 + 1$$
$$U_5 = 32x^5 - 32x^3 + 6x$$
$$U_6 = 64x^6 - 80x^4 + 24x^2 - 1$$
$$U_7 = 128x^7 - 192x^5 + 80x^3 - 8x$$
$$U_8 = 256x^8 - 448x^6 + 240x^4 - 40x^2 + 1$$
$$U_9 = 512x^9 - 1024x^7 + 672x^5 - 160x^3 + 10x$$
$$U_{10} = 1024x^{10} - 2304x^8 + 1792x^6 - 560x^4 + 60x^2 - 1$$

$$x^{10} = (42U_0 + 90U_2 + 75U_4 + 35U_6 + 9U_8 + U_{10})/1024$$
$$x^9 = (42U_1 + 48U_3 + 27U_5 + 8U_7 + U_9)/512$$
$$x^8 = (14U_0 + 28U_2 + 20U_4 + 7U_6 + U_8)/256$$
$$x^7 = (14U_1 + 14U_3 + 6U_5 + U_7)/128$$
$$x^6 = (5U_0 + 9U_2 + 5U_4 + U_6)/64$$
$$x^5 = (5U_1 + 4U_3 + U_5)/32$$
$$x^4 = (2U_0 + 3U_2 + U_4)/16$$
$$x^3 = (2U_1 + U_3)/8$$
$$x^2 = (U_0 + U_2)/4$$
$$x = (U_1)/2$$
$$1 = U_0$$

6.10.8.1 Table of Jacobi polynomials

Notation: $(m)_n = m(m+1)\ldots(m+n-1)$.

$P_0^{(\alpha,\beta)}(x) = 1.$

$P_1^{(\alpha,\beta)}(x) = \dfrac{1}{2}\Big(2(\alpha+1) + (\alpha+\beta+2)(x-1)\Big).$

$P_2^{(\alpha,\beta)}(x) = \dfrac{1}{8}\Big(4(\alpha+1)_2 + 4(\alpha+\beta+3)(\alpha+2)(x-1) + (\alpha+\beta+3)_2(x-1)^2\Big).$

$P_3^{(\alpha,\beta)}(x) = \dfrac{1}{48}\Big(8(\alpha+1)_3 + 12(\alpha+\beta+4)(\alpha+2)_2(x-1)$

$\qquad + 6(\alpha+\beta+4)_2(\alpha+3)(x-1)^2 + (\alpha+\beta+4)_3(x-1)^3\Big).$

$P_4^{(\alpha,\beta)}(x) = \dfrac{1}{384}\Big(16(\alpha+1)_4 + 32(\alpha+\beta+5)(\alpha+2)_3(x-1)$

$\qquad + 24(\alpha+\beta+5)_2(\alpha+3)_2(x-1)^2 + 8(\alpha+\beta+5)_3(\alpha+4)(x-1)^3$

$\qquad + (\alpha+\beta+5)_4(x-1)^4\Big).$

6.10.9 ZERNIKE POLYNOMIALS

The *circle polynomials* or *Zernike polynomials* form a complete orthogonal set over the interior of the unit circle. They are given by

$$U_n^m(r,\theta) = R_n^m(r)e^{im\theta} \tag{6.10.2}$$

where $R_n^m(r)$ are *radial polynomials* (see below), n and m are integers with $n-|m|$ even and $0 \le |m| \le n$.

6.10.9.1 Properties

1. Orthogonality

$$\int_0^{2\pi}\int_0^1 \overline{U_n^m(r,\theta)}U_{n'}^{m'}(r,\theta)\,r\,dr\,d\theta = \frac{\pi}{n+1}\delta_{nn'}\delta_{mm'}$$

$$\int_0^1 R_n^m(r)R_{n'}^m(r)\,r\,dr = \frac{1}{2(n+1)}\delta_{nn'}. \tag{6.10.3}$$

2. Explicit formula for the radial polynomials

$$R_n^{\pm m}(r) = \frac{1}{\left(\frac{n-|m|}{2}\right)!\,r^{|m|}}\left\{\left(\frac{\partial}{\partial(r^2)}\right)^{\frac{n-|m|}{2}}\left[(r^2)^{\frac{n+|m|}{2}}(r^2-1)^{\frac{n-|m|}{2}}\right]\right\}$$

$$= \sum_{s=0}^{\frac{n-|m|}{2}} \frac{(-1)^s(n-s)!}{s!\left(\frac{n+|m|}{2}-s\right)!\left(\frac{n-|m|}{2}-s\right)!}r^{n-2s}. \tag{6.10.4}$$

3. Expansions in Zernike polynomials

(a) If $f(r, \theta)$ is a piecewise continuous function then

$$f(r, \theta) = \sum_{n=0}^{\infty} \sum_{m=-n}^{n} A_n^m U_n^m(r, \theta) \qquad \text{where } n - |m| \text{ is even.}$$

(6.10.5)

$$A_n^m = \overline{A_n^{-m}} = \frac{n+1}{\pi} \int_0^{2\pi} \int_0^1 \overline{U_n^m(r, \theta)} f(r, \theta) \, r \, dr \, d\theta.$$

(b) If $f(r, \theta)$ is a real piecewise continuous function then

$$f(r, \theta) = \sum_{n=0}^{\infty} \sum_{m=0}^{n} \left[C_n^m \cos(m\theta) + S_n^m \sin(m\theta) \right] R_n^m(r)$$

where $n - |m|$ is even.

(6.10.6)

$$\begin{bmatrix} C_n^m \\ S_n^m \end{bmatrix} = \frac{\epsilon_m (n+1)}{\pi} \int_0^{2\pi} \int_0^1 f(r, \theta) R_n^m(r) \begin{bmatrix} \cos(m\theta) \\ \sin(m\theta) \end{bmatrix} r \, dr \, d\theta$$

$$\text{where } \epsilon_m = \begin{cases} 1 & \text{if } m = 0 \\ 2 & \text{otherwise.} \end{cases}$$

6.10.9.2   Tables of Zernike polynomials

n	$m = 0$	2	4
0	1		
2	$2r^2 - 1$	r^2	
4	$6r^4 - 6r^2 + 1$	$4r^4 - 3r^2$	r^4
6	$20r^6 - 30r^4 + 12r^2 - 1$	$15r^6 - 20r^4 + 6r^2$	$6r^6 - 5r^4$
8	$70r^8 - 140r^6 + 90r^4 - 20r^2 + 1$	$56r^8 - 105r^6 + 60r^4 - 10r^2$	$28r^8 - 42r^6 + 15r^4$

n	$m = 1$	3	5
1	r		
3	$3r^3 - 2r$	r^3	
5	$10r^5 - 12r^3 + 3r$	$5r^5 - 4r^3$	r^5
7	$35r^7 - 60r^5 + 30r^3 - 4r$	$21r^7 - 30r^5 + 10r^3$	$7r^7 - 6r^5$
9	$126r^9 - 280r^7 + 210r^5 - 60r^3 + 5r$	$84r^9 - 168r^7 + 105r^5 - 20r^3$	$36r^9 - 56r^7 + 21r^5$

6.10.10   SPHERICAL HARMONICS

The *spherical harmonics* are defined by

$$Y_{lm}(\theta, \phi) = \sqrt{\frac{2l+1}{4\pi} \frac{(l-m)!}{(l+m)!}} P_l^m(\cos \theta) e^{im\phi}$$

(6.10.7)

for l an integer and $m = -l, -l + 1, \ldots, l - 1, l$. They satisfy

$$Y_{l,-m}(\theta, \phi) = (-1)^m \overline{Y_{l,m}(\theta, \phi)},$$

$$Y_{l0}(\theta, \phi) = \sqrt{\frac{2l+1}{4\pi}} P_l(\cos\theta), \qquad (6.10.8)$$

$$Y_{lm}\left(\frac{\pi}{2}, \phi\right) = \begin{cases} \sqrt{\frac{(2l+1)(l-m)!(l+m)!}{4\pi}} \frac{(-1)^{(l+m)/2} e^{im\phi}}{2^l \left(\frac{l-m}{2}\right)! \left(\frac{l+m}{2}\right)!}, & \frac{l+m}{2} \text{ integral,} \\ 0, & \frac{l+m}{2} \text{ not integral.} \end{cases}$$

The normalization and orthogonality conditions are

$$\int_0^{2\pi} d\phi \int_0^{\pi} \sin\theta \, d\theta \, \overline{Y_{l'm'}(\theta, \phi)} Y_{lm}(\theta, \phi) = \delta_{ll'} \delta_{mm'}, \qquad (6.10.9)$$

and

$$\int_0^{2\pi} d\phi \int_0^{\pi} \sin\theta \, d\theta \, \overline{Y_{l_1 m_2}(\theta, \phi)} Y_{l_2 m_2}(\theta, \phi) Y_{l_3 m_3}(\theta, \phi),$$

$$= \sqrt{\frac{(2l_2 + 1)(2l_3 + 1)}{4\pi(2l_1 + 1)}} \begin{pmatrix} l_1 & l_3 & l_1 \\ m_2 & m_3 & m_1 \end{pmatrix} \begin{pmatrix} l_1 & l_3 & l_1 \\ 0 & 0 & 0 \end{pmatrix}, \qquad (6.10.10)$$

where the terms on the right hand side are Clebsch–Gordan coefficients (see page 574). Because of the (distributional) completeness relation,

$$\sum_{l=0}^{\infty} \sum_{m=-l}^{l} Y_{lm}(\theta, \phi) \overline{Y_{lm}(\theta', \phi')} = \delta(\phi - \phi') \delta(\cos\theta - \cos\theta'), \qquad (6.10.11)$$

an arbitrary function $g(\theta, \phi)$ can be expanded in spherical harmonics as

$$g(\theta, \phi) = \sum_{l=0}^{\infty} \sum_{m=-l}^{l} A_{lm} Y_{lm}(\theta, \phi), \qquad A_{lm} = \int \overline{Y_{lm}(\theta, \phi)} g(\theta, \phi) \, d\Omega. \qquad (6.10.12)$$

In spherical coordinates,

$$\nabla^2 \left[f(r) Y_{lm}(\theta, \phi) \right] = \left[\frac{1}{r^2} \frac{d}{dr} \left(r^2 \frac{df}{dr} \right) - l(l+1) \frac{f(r)}{r^2} \right] Y_{lm}(\theta, \phi). \qquad (6.10.13)$$

6.10.10.1 Table of spherical harmonics

$$l = 0 \qquad Y_{00} = \frac{1}{\sqrt{4\pi}}.$$

$$l = 1 \qquad \begin{cases} Y_{11} = -\sqrt{\dfrac{3}{8\pi}} \sin\theta\, e^{i\phi}, \\[2ex] Y_{10} = \sqrt{\dfrac{3}{4\pi}} \cos\theta. \end{cases}$$

$$l = 2 \qquad \begin{cases} Y_{22} = \dfrac{1}{4}\sqrt{\dfrac{15}{2\pi}} \sin^2\theta\, e^{2i\phi}, \\[2ex] Y_{21} = -\sqrt{\dfrac{15}{8\pi}} \sin\theta \cos\theta\, e^{i\phi}, \\[2ex] Y_{20} = \dfrac{1}{2}\sqrt{\dfrac{5}{4\pi}} (3\cos^2\theta - 1). \end{cases}$$

$$l = 3 \qquad \begin{cases} Y_{33} = -\dfrac{1}{4}\sqrt{\dfrac{35}{4\pi}} \sin^3\theta\, e^{3i\phi}, \\[2ex] Y_{32} = \dfrac{1}{4}\sqrt{\dfrac{105}{2\pi}} \sin^2\theta \cos\theta\, e^{2i\phi}, \\[2ex] Y_{31} = -\dfrac{1}{4}\sqrt{\dfrac{21}{4\pi}} \sin\theta(5\cos^2\theta - 1)\, e^{i\phi}, \\[2ex] Y_{30} = \dfrac{1}{2}\sqrt{\dfrac{7}{4\pi}} (5\cos^3\theta - 3\cos\theta). \end{cases}$$

6.11   GAMMA FUNCTION

$$\Gamma(z) = \int_0^\infty t^{z-1} e^{-t}\, dt, \qquad z = x + iy, \qquad x > 0.$$

6.11.1   RECURSION FORMULA

$$\Gamma(z+1) = z\,\Gamma(z).$$

The relation $\Gamma(z) = \Gamma(z+1)/z$ can be used to extend the gamma function to the left half plane for all z except when z is a non-positive integer (i.e., $z \neq 0, -1, -2, \ldots$).

FIGURE 6.6

Graphs of $\Gamma(x)$ *and* $1/\Gamma(x)$ *for* x *real.* (From Temme, N. M., *Special Functions: An Introduction to the Classical Functions of Mathematical Physics*, John Wiley & Sons, New York, 1996. With permission.)

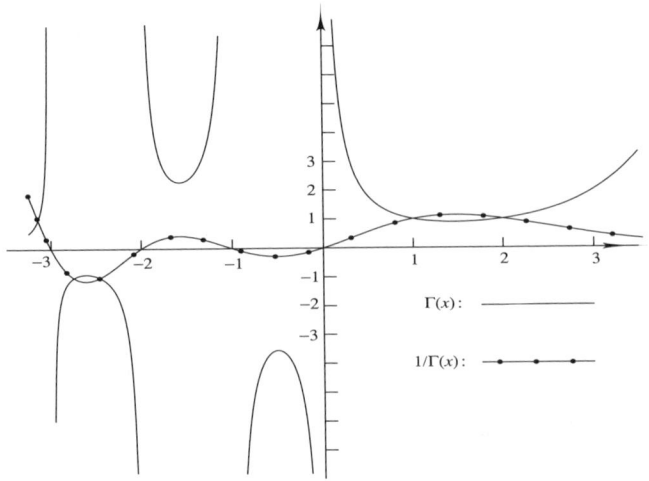

6.11.2 GAMMA FUNCTION OF SPECIAL VALUES

$$\Gamma(n+1) = n! \quad \text{if } n = 0, 1, 2, \ldots, \text{where } 0! = 1,$$

$$\Gamma(1) = 1, \quad \Gamma(2) = 1, \quad \Gamma(3) = 2, \quad \Gamma(\tfrac{1}{2}) = \sqrt{\pi},$$

$$\Gamma\left(m + \frac{1}{2}\right) = \frac{1 \cdot 3 \cdot 5 \cdots (2m-1)}{2^m} \sqrt{\pi}, \quad m = 1, 2, 3, \ldots,$$

$$\Gamma\left(-m + \frac{1}{2}\right) = \frac{(-1)^m 2^m}{1 \cdot 3 \cdot 5 \cdots (2m-1)} \sqrt{\pi}, \quad m = 1, 2, 3, \ldots.$$

$$\Gamma\left(\tfrac{1}{4}\right) = 3.62560\,99082, \qquad \Gamma\left(\tfrac{1}{3}\right) = 2.67893\,85347,$$

$$\Gamma\left(\tfrac{1}{2}\right) = \sqrt{\pi} = 1.77245\,38509, \qquad \Gamma\left(\tfrac{2}{3}\right) = 1.35411\,79394,$$

$$\Gamma\left(\tfrac{3}{4}\right) = 1.22541\,67024, \qquad \Gamma\left(\tfrac{3}{2}\right) = \sqrt{\pi}/2 = 0.88622\,69254.$$

6.11.3 PROPERTIES

1. Singular points:

    The gamma function has simple poles at $z = -n$ (for $n = 0, 1, 2, \ldots$), with the respective residues $(-1)^n/n!$; that is,

$$\lim_{z \to -n} (z+n)\Gamma(z) = \frac{(-1)^n}{n!}.$$

2. Definition by products:

$$\Gamma(z) = \lim_{n \to \infty} \frac{n!\, n^z}{z(z+1)\cdots(z+n)},$$

$$\frac{1}{\Gamma(z)} = z\, e^{\gamma z} \prod_{n=1}^{\infty} \left[(1 + z/n)\, e^{-z/n}\right], \qquad \gamma \text{ is Euler's constant.}$$

3. Other integrals:

$$\Gamma(z)\, \cos \frac{\pi z}{2} = \int_0^\infty t^{z-1}\, \cos t\, dt, \qquad 0 < \mathrm{Re}\, z < 1,$$

$$\Gamma(z)\, \sin \frac{\pi z}{2} = \int_0^\infty t^{z-1}\, \sin t\, dt, \qquad -1 < \mathrm{Re}\, z < 1.$$

4. Derivative at $x = 1$:

$$\Gamma'(1) = \int_0^\infty \ln t\, e^{-t}\, dt = -\gamma.$$

5. Multiplication formula:

$$\Gamma(2z) = \pi^{-1/2}\, 2^{2z-1}\, \Gamma(z)\, \Gamma\left(z + \frac{1}{2}\right).$$

6. Reflection formulae:

$$\Gamma(z)\, \Gamma(1-z) = \frac{\pi}{\sin \pi z},$$

$$\Gamma\left(\tfrac{1}{2} + z\right) \Gamma\left(\tfrac{1}{2} - z\right) = \frac{\pi}{\cos \pi z},$$

$$\Gamma(z-n) = (-1)^n \Gamma(z) \frac{\Gamma(1-z)}{\Gamma(n+1-z)} = \frac{(-1)^n\, \pi}{\sin \pi z\, \Gamma(n+1-z)}.$$

6.11.4   ASYMPTOTIC EXPANSION

For $z \to \infty$, $|\arg z| < \pi$:

$$\Gamma(z) \sim \sqrt{\frac{2\pi}{z}}\, z^z\, e^{-z} \left[1 + \frac{1}{12\, z} + \frac{1}{288\, z^2} - \frac{139}{51\,840\, z^3} + \cdots\right].$$

$$\ln \Gamma(z) \sim \ln\left(\sqrt{\frac{2\pi}{z}}\, z^z e^{-z}\right) + \sum_{n=1}^{\infty} \frac{B_{2n}}{2n\,(2n-1)} \frac{1}{z^{2n-1}} \tag{6.11.1}$$

$$\sim \ln\left(\sqrt{\frac{2\pi}{z}}\, z^z e^{-z}\right) + \frac{1}{12z} - \frac{1}{360z^3} + \frac{1}{1\,260z^5} - \frac{1}{1\,680z^7} + \cdots,$$

where B_n are the Bernoulli numbers. If we let $z = n$ a large positive integer, then a useful approximation for $n!$ is given by *Stirling's formula*,

$$\Gamma(n+1) = n! \sim \sqrt{2\pi n}\, n^n\, e^{-n}, \qquad n \to \infty. \tag{6.11.2}$$

6.11.5 LOGARITHMIC DERIVATIVE OF THE GAMMA FUNCTION

1. Definition:

$$\psi(z) = \frac{d}{dz} \ln \Gamma(z) = -\gamma + \sum_{n=0}^{\infty} \left(\frac{1}{n+1} - \frac{1}{z+n} \right), \quad z \neq 0, -1, -2, \ldots.$$

2. Special values:

$$\psi(1) = -\gamma, \quad \psi\left(\tfrac{1}{2}\right) = -\gamma - 2\ln 2.$$

3. Asymptotic expansion:

For $z \to \infty$, $|\arg z| < \pi$:

$$\psi(z) \sim \ln z - \frac{1}{2z} - \sum_{n=1}^{\infty} \frac{B_{2n}}{2nz^{2n}}$$

$$\sim \ln z - \frac{1}{2\,z} - \frac{1}{12\,z^2} + \frac{1}{120\,z^4} - \frac{1}{252\,z^6} + \ldots.$$

6.11.6 NUMERICAL VALUES

x	$\Gamma(x)$	$\ln \Gamma(x)$	$\psi(x)$	$\psi'(x)$
1.00	1.00000000	0.00000000	−0.57721566	1.64493407
1.04	0.97843820	−0.02179765	−0.51327488	1.55371164
1.08	0.95972531	−0.04110817	−0.45279934	1.47145216
1.12	0.94359019	−0.05806333	−0.39545533	1.39695222
1.16	0.92980307	−0.07278247	−0.34095315	1.32920818
1.20	0.91816874	−0.08537409	−0.28903990	1.26737721
1.24	0.90852106	−0.09593721	−0.23949368	1.21074707
1.28	0.90071848	−0.10456253	−0.19211890	1.15871230
1.32	0.89464046	−0.11133336	−0.14674236	1.11075532
1.36	0.89018453	−0.11632650	−0.10321006	1.06643142
1.40	0.88726382	−0.11961291	−0.06138454	1.02535659
1.44	0.88580506	−0.12125837	−0.02114267	0.98719773
1.48	0.88574696	−0.12132396	0.01762627	0.95166466
1.52	0.88703878	−0.11986657	0.05502211	0.91850353
1.56	0.88963920	−0.11693929	0.09113519	0.88749142
1.60	0.89351535	−0.11259177	0.12604745	0.85843189
1.64	0.89864203	−0.10687051	0.15983345	0.83115118
1.68	0.90500103	−0.09981920	0.19256120	0.80549511
1.72	0.91258058	−0.09147889	0.22429289	0.78132645

x	$\Gamma(x)$	$\ln \Gamma(x)$	$\psi(x)$	$\psi'(x)$
1.76	0.92137488	−0.08188828	0.25508551	0.75852269
1.80	0.93138377	−0.07108387	0.28499143	0.73697414
1.84	0.94261236	−0.05910015	0.31405886	0.71658233
1.88	0.95507085	−0.04596975	0.34233226	0.69725865
1.92	0.96877431	−0.03172361	0.36985272	0.67892313
1.96	0.98374254	−0.01639106	0.39665832	0.66150345
2.00	1.00000000	0.00000000	0.42278434	0.64493407

6.12 BETA FUNCTION

$$B(p,q) = \int_0^1 t^{p-1}(1-t)^{q-1}\,dt, \quad \text{Re } p > 0, \quad \text{Re } q > 0. \tag{6.12.1}$$

1. Relations:

$$B(p,q) = B(q,p),$$

$$B(p,q+1) = \frac{q}{p}\,B(p+1,q) = \frac{q}{p+q}\,B(p,q),$$

$$B(p,q)\,B(p+q,r) = \frac{\Gamma(p)\,\Gamma(q)\,\Gamma(r)}{\Gamma(p+q+r)}.$$

2. Relation with the gamma function:

$$B(p,q) = \frac{\Gamma(p)\,\Gamma(q)}{\Gamma(p+q)}.$$

3. Other integrals (in all cases Re $p > 0$ and Re $q > 0$):

$$B(p,q) = 2\int_0^{\pi/2} \sin^{2p-1}\theta\,\cos^{2q-1}\theta\,d\theta$$

$$= \int_0^{\infty} \frac{t^{p-1}}{(t+1)^{p+q}}\,dt$$

$$= \int_0^{\infty} e^{-pt}\left(1-e^{-t}\right)^{q-1}\,dt$$

$$= r^q(r+1)^p \int_0^1 \frac{t^{p-1}(1-t)^{q-1}}{(r+t)^{p+q}}\,dt, \quad r > 0.$$

6.12.1 NUMERICAL VALUES OF THE BETA FUNCTION

p	$q = 0.100$	0.200	0.300	0.400	0.500	0.600	0.700	0.800	0.900	1.000
0.1	19.715	14.599	12.831	11.906	11.323	10.914	10.607	10.365	10.166	10.000
0.2	14.599	9.502	7.748	6.838	6.269	5.872	5.576	5.345	5.157	5.000
0.3	12.831	7.748	6.010	5.112	4.554	4.169	3.883	3.661	3.482	3.333
0.4	11.906	6.838	5.112	4.226	3.679	3.303	3.027	2.813	2.641	2.500
0.5	11.323	6.269	4.554	3.679	3.142	2.775	2.506	2.299	2.135	2.000
0.6	10.914	5.872	4.169	3.303	2.775	2.415	2.154	1.954	1.796	1.667
0.7	10.607	5.576	3.883	3.027	2.506	2.154	1.899	1.705	1.552	1.429
0.8	10.365	5.345	3.661	2.813	2.299	1.954	1.705	1.517	1.369	1.250
0.9	10.166	5.157	3.482	2.641	2.135	1.796	1.552	1.369	1.226	1.111
1.0	10.000	5.000	3.333	2.500	2.000	1.667	1.429	1.250	1.111	1.000
1.2	9.733	4.751	3.099	2.279	1.791	1.468	1.239	1.069	0.938	0.833
1.4	9.525	4.559	2.921	2.113	1.635	1.321	1.101	0.938	0.813	0.714
1.6	9.355	4.404	2.779	1.982	1.513	1.208	0.994	0.837	0.718	0.625
1.8	9.213	4.276	2.663	1.875	1.415	1.117	0.909	0.758	0.644	0.556
2.0	9.091	4.167	2.564	1.786	1.333	1.042	0.840	0.694	0.585	0.500
2.2	8.984	4.072	2.480	1.710	1.264	0.979	0.783	0.641	0.536	0.455
2.4	8.890	3.989	2.406	1.644	1.205	0.925	0.734	0.597	0.495	0.417
2.6	8.805	3.915	2.340	1.586	1.153	0.878	0.692	0.558	0.460	0.385
2.8	8.728	3.848	2.282	1.534	1.107	0.837	0.655	0.525	0.430	0.357
3.0	8.658	3.788	2.230	1.488	1.067	0.801	0.622	0.496	0.403	0.333

6.13 ERROR FUNCTIONS

$$\operatorname{erf} x = \frac{2}{\sqrt{\pi}} \int_0^x e^{-t^2} \, dt,$$

$$\operatorname{erfc} x = \frac{2}{\sqrt{\pi}} \int_x^\infty e^{-t^2} \, dt.$$

The function $\operatorname{erf} x$ is known as the *error function*. The function $\operatorname{erfc} x$ is known as the *complementary error function*.

6.13.1 PROPERTIES

1. Relationships:

$$\operatorname{erf} x + \operatorname{erfc} x = 1, \qquad \operatorname{erf}(-x) = -\operatorname{erf} x, \qquad \operatorname{erfc}(-x) = 2 - \operatorname{erfc} x.$$

2. Relationship with normal probability function:

$$\frac{1}{\sqrt{2\pi}} \int_0^x e^{-\frac{1}{2}t^2} \, dt = \frac{1}{2} \operatorname{erf}\left(\frac{x}{\sqrt{2}}\right).$$

6.13.2 ERROR FUNCTION OF SPECIAL VALUES

$$\operatorname{erf}(\pm\infty) = \pm 1, \quad \operatorname{erfc}(-\infty) = 2, \quad \operatorname{erfc}\infty = 0,$$

$$\operatorname{erf} x_0 = \operatorname{erfc} x_0 = \frac{1}{2} \quad \text{if } x_0 \approx 0.476936.$$

6.13.3 EXPANSIONS

1. Series expansions:

$$\operatorname{erf} x = \frac{2}{\sqrt{\pi}} \sum_{n=0}^{\infty} \frac{(-1)^n \, x^{2n+1}}{(2n+1)\, n!} = \frac{2}{\sqrt{\pi}} \left(x - \frac{x^3}{3} + \frac{1}{2!} \frac{x^5}{5} - \frac{1}{3!} \frac{x^7}{7} + \cdots \right)$$

$$= \frac{2}{\sqrt{\pi}} \sum_{n=0}^{\infty} \frac{\Gamma\left(\frac{3}{2}\right) e^{-x^2}}{\Gamma\left(n + \frac{3}{2}\right)} x^{2n+1} = \frac{2}{\sqrt{\pi}} e^{-x^2} \left(x + \frac{2}{3} x^3 + \frac{4}{15} x^5 \cdots \right).$$

2. Asymptotic expansion:

   For $z \to \infty$, $|\arg z| < \frac{3}{4}\pi$,

$$\operatorname{erfc} z \sim \frac{2}{\sqrt{\pi}} \frac{e^{-z^2}}{2z} \sum_{n=0}^{\infty} \frac{(-1)^n \, (2n)!}{n!(2z)^{2n}}$$

$$\sim \frac{2}{\sqrt{\pi}} \frac{e^{-z^2}}{2z} \left(1 - \frac{1}{2\,z^2} + \frac{3}{4\,z^4} - \frac{15}{8\,z^6} + \cdots \right).$$

6.13.4 SPECIAL CASES

1. Dawson's integral

$$F(x) = e^{-x^2} \int_0^x e^{t^2} \, dt = -\frac{1}{2} i \sqrt{\pi}\, e^{-x^2} \, \operatorname{erf}(ix).$$

2. Plasma dispersion function

$$w(z) = e^{-z^2} \operatorname{erfc}(-iz)$$

$$= \frac{1}{\pi i} \int_{-\infty}^{\infty} \frac{e^{-t^2}}{t - z} \, dt, \quad \operatorname{Im} z > 0$$

$$= 2\, e^{-z^2} - w(-z)$$

$$= \sum_{n=0}^{\infty} \frac{(iz)^n}{\Gamma(n/2 + 1)}.$$

FIGURE 6.7

Cornu's spiral, formed from Fresnel functions, is the set $\{x, y, t\}$ where $x = C(t)$, $y = S(t)$, $t \geq 0$. (From Temme, N. M., *Special Functions: An Introduction to the Classical Functions of Mathematical Physics,* John Wiley & Sons, New York, 1996. With permission.)

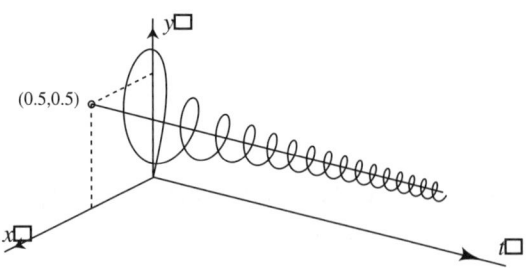

6.14 FRESNEL INTEGRALS

$$C(z) = \sqrt{\frac{2}{\pi}} \int_0^z \cos t^2 \, dt, \quad S(z) = \sqrt{\frac{2}{\pi}} \int_0^z \sin t^2 \, dt.$$

6.14.1 PROPERTIES

1. Relations:

$$C(z) + iS(z) = \frac{1+i}{2} \, \text{erf} \, \frac{(1-i)z}{\sqrt{2}}.$$

2. Limits:

$$\lim_{z \to \infty} C(z) = \frac{1}{2}, \quad \lim_{z \to \infty} S(z) = \frac{1}{2}.$$

3. Representations:

$$C(z) = \frac{1}{2} + f(z) \sin(z^2) - g(z) \cos(z^2),$$

$$S(z) = \frac{1}{2} - f(z) \cos(z^2) - g(z) \sin(z^2),$$

where

$$f(z) = \frac{1}{\pi\sqrt{2}} \int_0^\infty \frac{e^{-z^2 t}}{\sqrt{t}(t^2 + 1)} \, dt, \quad g(z) = \frac{1}{\pi\sqrt{2}} \int_0^\infty \frac{\sqrt{t}\, e^{-z^2 t}}{(t^2 + 1)} \, dt.$$

4. Cornu's spiral:

$$\sqrt{\frac{2}{\pi}} \int_z^\infty e^{it^2} \, dt = [g(z) + if(z)] \, e^{iz^2}.$$

6.14.2 ASYMPTOTIC EXPANSION

And for $z \to \infty$, $|\arg z| < \frac{1}{2}\pi$,

$$f(z) \sim \frac{1}{\pi\sqrt{2}} \sum_{n=0}^{\infty} (-1)^n \frac{\Gamma(2n+1/2)}{z^{2n+1/2}} = \frac{1}{\sqrt{2\pi z}} \left[1 - \frac{3}{4 z^2} + \frac{105}{16 z^4} - \cdots \right],$$

$$g(z) \sim \frac{1}{\pi\sqrt{2}} \sum_{n=0}^{\infty} (-1)^n \frac{\Gamma(2n+3/2)}{z^{2n+3/2}} = \frac{1}{2z\sqrt{2\pi z}} \left[1 - \frac{15}{4 z^2} + \frac{945}{16 z^4} - \cdots \right].$$

6.14.3 NUMERICAL VALUES OF ERROR FUNCTIONS AND FRESNEL INTEGRALS

x	$\mathrm{erf}(x)$	$e^{x^2}\mathrm{erfc}(x)$	$C(x)$	$S(x)$
0.0	0.00000000	1.00000000	0.00000000	0.00000000
0.2	0.22270259	0.80901952	0.15955138	0.00212745
0.4	0.42839236	0.67078779	0.31833776	0.01699044
0.6	0.60385609	0.56780472	0.47256350	0.05691807
0.8	0.74210096	0.48910059	0.61265370	0.13223984
1.0	0.84270079	0.42758358	0.72170592	0.24755829
1.2	0.91031398	0.37853742	0.77709532	0.39584313
1.4	0.95228512	0.33874354	0.75781398	0.55244498
1.6	0.97634838	0.30595299	0.65866707	0.67442706
1.8	0.98909050	0.27856010	0.50694827	0.71289443
2.0	0.99532227	0.25539568	0.36819298	0.64211874
2.2	0.99813715	0.23559296	0.32253723	0.49407286
2.4	0.99931149	0.21849873	0.40704642	0.36532279
2.6	0.99976397	0.20361325	0.55998756	0.36073841
2.8	0.99992499	0.19054888	0.64079292	0.48940140
3.0	0.99997791	0.17900115	0.56080398	0.61721360
3.2	0.99999397	0.16872810	0.41390216	0.58920847
3.4	0.99999848	0.15953536	0.39874249	0.44174492
3.6	0.99999964	0.15126530	0.53845493	0.39648758
3.8	0.99999992	0.14378884	0.60092662	0.52778933
4.0	0.99999998	0.13699946	0.47431072	0.59612656
4.2	1.00000000	0.13080849	0.41041217	0.46899697
4.4	1.00000000	0.12514166	0.54218734	0.41991084
4.6	1.00000000	0.11993626	0.56533023	0.55685845
4.8	1.00000000	0.11513908	0.42894668	0.54293254
5.0	1.00000000	0.11070464	0.48787989	0.42121705

FIGURE 6.8

Sine and cosine integrals Si(x) *and* Ci(x), *for* $0 \leq x \leq 8$. (From Temme, N. M., *Special Functions: An Introduction to the Classical Functions of Mathematical Physics*, John Wiley & Sons, New York, 1996. With permission.)

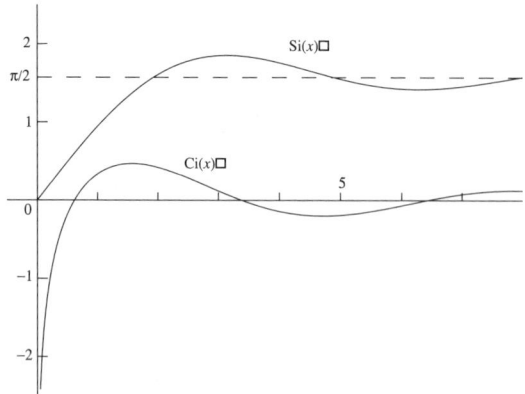

6.15 SINE, COSINE, AND EXPONENTIAL INTEGRALS

6.15.1 SINE AND COSINE INTEGRALS

$$\text{Si}(z) = \int_0^z \frac{\sin t}{t}\, dt, \quad \text{Ci}(z) = \gamma + \ln z + \int_0^z \frac{\cos t - 1}{t}\, dt,$$

where γ is Euler's constant.

1. Alternative definitions:

$$\text{Si}(z) = \frac{1}{2}\pi - \int_z^\infty \frac{\sin t}{t}\, dt, \quad \text{Ci}(z) = -\int_z^\infty \frac{\cos t}{t}\, dt.$$

2. Limits:

$$\lim_{z \to \infty} \text{Si}(z) = \frac{1}{2}\pi, \quad \lim_{z \to \infty} \text{Ci}(z) = 0.$$

3. Representations:

$$\text{Si}(z) = -f(z)\cos z - g(z)\sin z + \frac{1}{2}\pi,$$
$$\text{Ci}(z) = +f(z)\sin z - g(z)\cos z,$$

where

$$f(z) = \int_0^\infty \frac{e^{-zt}}{t^2 + 1}\, dt, \quad g(z) = \int_0^\infty \frac{te^{-zt}}{t^2 + 1}\, dt.$$

4. Asymptotic expansion:

   For $z \to \infty$, $|\arg z| < \pi$,

$$f(z) \sim \frac{1}{z} \sum_{n=0}^{\infty} (-1)^n \frac{(2n)!}{z^{2n}}, \qquad g(z) \sim \frac{1}{z^2} \sum_{n=0}^{\infty} (-1)^n \frac{(2n+1)!}{z^{2n}}.$$

6.15.2 EXPONENTIAL INTEGRALS

$$E_n(z) = \int_1^{\infty} \frac{e^{-zt}}{t^n} \, dt, \quad \operatorname{Re} z > 0, \quad n = 1, 2, \ldots$$

$$= \frac{z^{n-1} e^{-z}}{\Gamma(n)} \int_0^{\infty} \frac{e^{-zt} t^{n-1}}{t+1} \, dt, \quad \operatorname{Re} z > 0.$$

1. Special case:

$$E_1(z) = \int_z^{\infty} \frac{e^{-t}}{t} \, dt, \quad |\arg z| < \pi.$$

   For real values of $z = x$,

$$\operatorname{Ei}(x) = \int_{-\infty}^{x} \frac{e^t}{t} \, dt,$$

   where for $x > 0$ the integral should be interpreted as a Cauchy principal value integral.

2. Representations:

$$E_1(z) = -\gamma - \ln z + \int_0^z \frac{1 - e^{-t}}{t} \, dt,$$

$$E_1\left(z e^{\frac{1}{2}\pi i}\right) = -\gamma - \ln z - \operatorname{Ci}(z) + i \left[-\frac{1}{2}\pi + \operatorname{Si}(z)\right].$$

$$E_1(x) = -\operatorname{Ei}(-x), \qquad x > 0$$

6.15.3 LOGARITHMIC INTEGRAL

$$\operatorname{li}(x) = \int_0^x \frac{dt}{\ln t} = \operatorname{Ei}(\ln x),$$

where for $x > 1$ the integral should be interpreted as a Cauchy principal value integral.

6.15.4 NUMERICAL VALUES

x	$\mathrm{Si}(x)$	$\mathrm{Ci}(x)$	$e^x E_1(x)$	$e^{-x}\mathrm{Ei}(x)$	$\mathrm{li}(x)$
0.0	0.000000	$-\infty$	∞	$-\infty$	0.000000
0.2	0.199556	-1.042206	1.493349	-0.672801	-0.085126
0.4	0.396461	-0.378809	1.047828	0.070226	-0.252949
0.6	0.588129	-0.022271	0.827933	0.422520	-0.546851
0.8	0.772096	0.198279	0.691245	0.605424	-1.134012
1.0	0.946083	0.337404	0.596347	0.697175	$-\infty$
1.2	1.108047	0.420459	0.525935	0.735544	-0.933787
1.4	1.256227	0.462007	0.471293	0.741568	-0.144991
1.6	1.389180	0.471733	0.427488	0.727902	0.353748
1.8	1.505817	0.456811	0.391492	0.702498	0.732637
2.0	1.605413	0.422981	0.361329	0.670483	1.045164
2.2	1.687625	0.375075	0.335651	0.635192	1.315238
2.4	1.752486	0.317292	0.313502	0.598799	1.555671
2.6	1.800394	0.253337	0.294186	0.562705	1.774145
2.8	1.832097	0.186488	0.277179	0.527789	1.975643
3.0	1.848653	0.119630	0.262084	0.494576	2.163589
3.2	1.851401	0.055257	0.248588	0.463356	2.340436
3.4	1.841914	-0.004518	0.236446	0.434256	2.508008
3.6	1.821948	-0.057974	0.225460	0.407294	2.667700
3.8	1.793390	-0.103778	0.215471	0.382424	2.820603
4.0	1.758203	-0.140982	0.206346	0.359552	2.967585
4.2	1.718369	-0.169013	0.197976	0.338561	3.109354
4.4	1.675834	-0.187660	0.190270	0.319321	3.246490
4.6	1.632460	-0.197047	0.183151	0.301697	3.379479
4.8	1.589975	-0.197604	0.176554	0.285555	3.508729
5.0	1.549931	-0.190030	0.170422	0.270766	3.634588

6.16 POLYLOGARITHMS

$$\mathrm{Li}_1(z) = \int_0^z \frac{dt}{1-t} = -\ln(1-z), \qquad \text{logarithm,}$$

$$\mathrm{Li}_2(z) = \int_0^z \frac{\mathrm{Li}_1(t)}{t}\, dt = -\int_0^z \frac{\ln(1-t)}{t}\, dt, \qquad \text{dilogarithm,}$$

$$\mathrm{Li}_n(z) = \int_0^z \frac{\mathrm{Li}_{n-1}(t)}{t}\, dt, \quad n \geq 2, \qquad \text{polylogarithm,}$$

$$\mathrm{Li}_\nu(z) = \frac{z}{\Gamma(\nu)} \int_0^\infty \frac{t^{\nu-1}}{e^t - z}\, dt, \quad \mathrm{Re}\,\nu > 0,\ z \notin \{\mathrm{Re}\,z \in [1,\infty], \mathrm{Im}\,z = 0\}.$$

6.16.1 POLYLOGARITHMS OF SPECIAL VALUES

$$\mathrm{Li}_2(1) = \frac{\pi^2}{6}, \quad \mathrm{Li}_2(-1) = -\frac{\pi^2}{12}, \quad \mathrm{Li}_2(\tfrac{1}{2}) = \frac{\pi^2}{12} - \frac{(\ln 2)^2}{2},$$

$$\mathrm{Li}_\nu(1) = \zeta(\nu), \quad \mathrm{Re}\,\nu > 1 \quad \text{(Riemann zeta function)}.$$

6.16.2 POLYLOGARITHM PROPERTIES

1. Definition: For any complex ν

$$\mathrm{Li}_\nu(z) = \sum_{k=1}^{\infty} \frac{z^k}{k^\nu}, \quad |z| < 1.$$

2. Singular points:

   $z = 1$ is a singular point of $\mathrm{Li}_\nu(z)$.

3. Generating function:

$$\sum_{n=2}^{\infty} w^{n-1}\,\mathrm{Li}_n(z) = z \int_0^\infty \frac{e^{wt} - 1}{e^t - z}\,dt, \quad z \notin [1, \infty).$$

   The series converges for $|w| < 1$; the integral is defined for $\mathrm{Re}\,w < 1$.

4. Functional equations for dilogarithms:

$$\mathrm{Li}_2(z) + \mathrm{Li}_2(1 - z) = \frac{1}{6}\pi^2 - \ln z \,\ln(1 - z),$$

$$\frac{1}{2}\mathrm{Li}_2(x^2) = \mathrm{Li}_2(x) + \mathrm{Li}_2(-x),$$

$$\mathrm{Li}_2(-1/x) + \mathrm{Li}_2(-x) = -\frac{1}{6}\pi^2 - \frac{1}{2}(\ln x)^2,$$

$$2\mathrm{Li}_2(x) + 2\mathrm{Li}_2(y) + 2\mathrm{Li}_2(z) =$$
$$\mathrm{Li}_2(-xy/z) + \mathrm{Li}_2(-yz/x) + \mathrm{Li}_2(-zx/y),$$

   where $1/x + 1/y + 1/z = 1$.

6.17 HYPERGEOMETRIC FUNCTIONS

Recall the geometric series and binomial expansion ($|z| < 1$),

$$(1 - z)^{-1} = \sum_{n=0}^{\infty} z^n, \qquad (1 - z)^{-a} = \sum_{n=0}^{\infty} \binom{-a}{n}(-z)^n = \sum_{n=0}^{\infty} \frac{(a)_n}{n!} z^n,$$

where the shifted factorial, $(a)_n$, is defined in Section 1.2.6.

The *Gauss hypergeometric function*, F, is defined by:

$$
\begin{aligned}
F(a, b; c; z) &= \sum_{n=0}^{\infty} \frac{(a)_n (b)_n}{(c)_n\, n!} z^n \\
&= 1 + \frac{ab}{c} z + \frac{a(a+1)\, b(b+1)}{c(c+1)\, 2!} z^2 + \dots, \quad |z| < 1, \\
&= F(b, a; c; z)
\end{aligned}
\tag{6.17.1}
$$

where a, b and c may all assume complex values, $c \neq 0, -1, -2, \dots$.

6.17.1 SPECIAL CASES

1. $F(a, b; b; z) = (1 - z)^{-a}$
2. $F(1, 1; 2; z) = -\dfrac{\ln(1 - z)}{z}$
3. $F\left(\dfrac{1}{2}, 1; \dfrac{3}{2}; z^2\right) = \dfrac{1}{2z} \ln\left(\dfrac{1 + z}{1 - z}\right)$
4. $F\left(\dfrac{1}{2}, 1; \dfrac{3}{2}; -z^2\right) = \dfrac{\tan^{-1} z}{z}$
5. $F\left(\dfrac{1}{2}, \dfrac{1}{2}; \dfrac{3}{2}; z^2\right) = \dfrac{\sin^{-1} z}{z}$
6. $F\left(\dfrac{1}{2}, \dfrac{1}{2}; \dfrac{3}{2}; -z^2\right) = \dfrac{\ln(z + \sqrt{1 + z^2})}{z}$
7. Polynomial case; for $m = 0, 1, 2, \dots$

$$
F(-m, b; c; z) = \sum_{n=0}^{m} \frac{(-m)_n (b)_n}{(c)_n\, n!} z^n = \sum_{n=0}^{m} (-1)^n \binom{m}{n} \frac{(b)_n}{(c)_n} z^n.
\tag{6.17.2}
$$

6.17.2 PROPERTIES

1. Derivatives:

$$
\frac{d}{dz} F(a, b; c; z) = \frac{ab}{c} F(a+1, b+1; c+1; z),
$$

$$
\frac{d^n}{dz^n} F(a, b; c; z) = \frac{(a)_n (b)_n}{(c)_n} F(a+n, b+n; c+n; z).
$$

2. Special values; when $\operatorname{Re}(c - a - b) > 0$:

$$
F(a, b; c; 1) = \frac{\Gamma(c)\Gamma(c - a - b)}{\Gamma(c - a)\Gamma(c - b)}.
$$

3. Integral; when $\operatorname{Re} c > \operatorname{Re} b > 0$:

$$
F(a, b; c; z) = \frac{\Gamma(c)}{\Gamma(b)\Gamma(c - b)} \int_0^1 t^{b-1}(1 - t)^{c-b-1}(1 - tz)^{-a}\, dt.
$$

4. Functional relationships:

$$F(a,b;c;z) = (1-z)^{-a}F\left(a,c-b;c;\frac{z}{z-1}\right)$$
$$= (1-z)^{-b}F\left(c-a,b;c;\frac{z}{z-1}\right)$$
$$= (1-z)^{c-a-b}F(c-a,c-b;c;z).$$

5. Differential equation:

$$z(1-z)F'' + [c-(a+b+1)z]F' - abF = 0,$$

with (regular) singular points $z = 0, 1, \infty$.

6.17.3   RECURSION FORMULAE

Notation: F is $F(a,b;c;z)$; $F(a+), F(a-)$ are $F(a+1,b;c;z)$, $F(a-1,b;c;z)$, respectively, etc.

1. $(c-a)F(a-) + (2a-c-az+bz)F + a(z-1)F(a+) = 0$
2. $c(c-1)(z-1)F(c-)+c[c-1-(2c-a-b-1)z]F+(c-a)(c-b)zF(c+) = 0$
3. $c[a+(b-c)z]F - ac(1-z)F(a+) + (c-a)(c-b)zF(c+) = 0$
4. $c(1-z)F - cF(a-) + (c-b)zF(c+) = 0$
5. $(b-a)F + aF(a+) - bF(b+) = 0$
6. $(c-a-b)F + a(1-z)F(a+) - (c-b)F(b-) = 0$
7. $(c-a-1)F + aF(a+) - (c-1)F(c-) = 0$
8. $(b-a)(1-z)F - (c-a)F(a-) + (c-b)F(b-) = 0$
9. $[a-1+(b+1-c)z]F + (c-a)F(a-) - (c-1)(1-z)F(c-) = 0$

6.18   LEGENDRE FUNCTIONS

6.18.1   DIFFERENTIAL EQUATION: LEGENDRE FUNCTION

The *Legendre differential equation* is,

$$(1-z^2)w'' - 2zw' + \nu(\nu+1)w = 0.$$

The solutions $P_\nu(z), Q_\nu(z)$ can be given in terms of Gaussian hypergeometric functions.

FIGURE 6.9

Legendre functions $P_n(x)$, $n = 1, 2, 3$ (left) and $Q_n(x)$, $n = 0, 1, 2, 3$ (right) on the interval $[-1, 1]$. *(From Temme, N. M.,* Special Functions: An Introduction to the Classical Functions of Mathematical Physics, *John Wiley & Sons, New York, 1996. With permission.)*

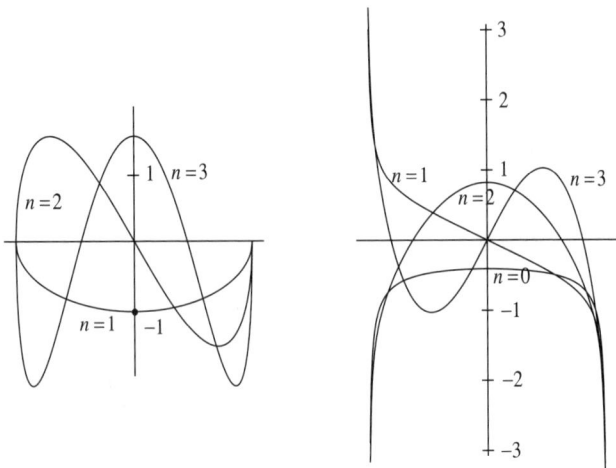

6.18.2   DEFINITION

$$P_\nu(z) = F\left(-\nu, \nu + 1; 1; \frac{1}{2} - \frac{1}{2}z\right),$$

$$Q_\nu(z) = \frac{\sqrt{\pi}\,\Gamma(\nu + 1)}{\Gamma\left(\nu + \frac{3}{2}\right)(2z)^{\nu+1}} F\left(\frac{1}{2}\nu + 1, \frac{1}{2}\nu + \frac{1}{2}; \nu + \frac{3}{2}; z^{-2}\right).$$

The Q_ν function is not defined if $\nu = -1, -2, \ldots$.

6.18.3   SINGULAR POINTS

$P_\nu(z)$ has a singular point at $z = -1$ and is analytic in the remaining part of the complex plane, with a branch cut along $(-\infty, -1]$. $Q_\nu(z)$ has singular points at $z = \pm 1$ and is analytic in the remaining part of the complex z-plane, with a branch cut along $(-\infty, +1]$.

6.18.4   RELATIONSHIPS

$$P_{-\nu-1}(z) = P_\nu(z),$$
$$Q_{-\nu-1}(z) = Q_\nu(z) - \pi \cot \nu\pi P_\nu(z).$$

6.18.5 RECURSION RELATIONSHIPS

$$(\nu + 1)P_{\nu+1}(z) = (2\nu + 1)zP_\nu(z) - \nu P_{\nu-1}(z),$$
$$(2\nu + 1)P_\nu(z) = P'_{\nu+1}(z) - P'_{\nu-1}(z),$$
$$(\nu + 1)P_\nu(z) = P'_{\nu+1}(z) - zP'_\nu(z),$$
$$\nu P_\nu(z) = zP'_\nu(z) - P'_{\nu-1}(z),$$
$$(1 - z^2)P'_\nu(z) = \nu P_{\nu-1}(z) - \nu z P_\nu(z).$$

The functions $Q_\nu(z)$ satisfy the same relations.

6.18.6 INTEGRALS

$$
\begin{aligned}
P_\nu(\cosh \alpha) &= \frac{2}{\pi} \int_0^\alpha \frac{\cosh(\nu + \frac{1}{2})\theta}{\sqrt{2\cosh\alpha - 2\cosh\theta}} \, d\theta \\
&= \frac{1}{\pi} \int_{-\alpha}^\alpha \frac{e^{-(\nu+1/2)\theta}}{\sqrt{2\cosh\alpha - 2\cosh\theta}} \, d\theta \\
&= \frac{1}{\pi} \int_0^\pi \frac{d\psi}{(\cosh\alpha + \sinh\alpha \, \cos\psi)^{\nu+1}} \\
&= \frac{1}{\pi} \int_0^\pi (\cosh\alpha + \sinh\alpha \, \cos\psi)^\nu \, d\psi.
\end{aligned}
\tag{6.18.1}
$$

$$
\begin{aligned}
P_\nu(\cos\beta) &= \frac{2}{\pi} \int_0^\beta \frac{\cos(\nu + \frac{1}{2})\theta}{\sqrt{2\cos\theta - 2\cos\beta}} \, d\theta \\
&= \frac{1}{\pi} \int_0^\pi \frac{d\psi}{(\cos\beta + i\sin\beta \, \cos\psi)^{\nu+1}} \\
&= \frac{1}{\pi} \int_0^\pi (\cos\beta + i\sin\beta \, \cos\psi)^\nu \, d\psi.
\end{aligned}
\tag{6.18.2}
$$

$$
\begin{aligned}
Q_\nu(z) &= 2^{-\nu-1} \int_{-1}^1 \frac{(1 - t^2)^\nu}{(z - t)^{\nu+1}} \, dt \qquad \text{Re } \nu > -1, \; |\arg z| < \pi, \; z \notin [-1, 1] \\
&= \int_0^\infty \left[z + \sqrt{z^2 - 1} \cosh\phi \right]^{-\nu-1} d\phi \\
&= \int_\alpha^\infty \frac{e^{-(\nu+1/2)\theta}}{\sqrt{2\cosh\theta - 2\cosh\alpha}} \, d\theta, \quad z = \cosh\alpha.
\end{aligned}
\tag{6.18.3}
$$

6.18.7 POLYNOMIAL CASE

Legendre polynomials are special cases (see Section 6.10.5) when $\nu = n = 0, 1, \ldots,$

$$
P_n(x) = F\left(-n, n + 1; 1; \frac{1}{2} - \frac{x}{2}\right)
\tag{6.18.4}
$$

$$
= \sum_{k=0}^m \frac{(-1)^k (2n - 2k)!}{2^n \, k! \, (n-k)! \, (n-2k)!} x^{n-2k}, \quad m = \begin{cases} \frac{1}{2}n, & \text{if } n \text{ even,} \\ \frac{1}{2}(n-1), & \text{if } n \text{ odd.} \end{cases}
$$

The Legendre polynomials satisfy $\displaystyle\int_{-1}^{1} P_n(x)P_m(x)\,dx = \frac{2}{2m+1}\delta_{nm}$.

The Legendre series representation is

$$f(x) = \sum_{n=0}^{\infty} A_n P_n(x), \qquad A_n = \frac{2n+1}{2}\int_{-1}^{1} f(x)P_n(x)\,dx. \tag{6.18.5}$$

For integer order, we distinguish two cases: $Q_n(x)$ (defined for $x \in (-1,1)$) and $Q_n(z)$ (defined for $\text{Re } z \notin [-1,1]$):

$$Q_0(x) = \frac{1}{2}\ln\frac{1+x}{1-x}, \qquad Q_1(x) = \frac{1}{2}x\ln\frac{1+x}{1-x} - 1, \tag{6.18.6}$$

and

$$Q_0(z) = \frac{1}{2}\ln\frac{z+1}{z-1}, \qquad Q_1(z) = \frac{1}{2}z\ln\frac{z+1}{z-1} - 1. \tag{6.18.7}$$

In both cases

$$Q_n(y) = P_n(y)Q_0(y) - \sum_{k=0}^{n-1} \frac{(2k+1)[1-(-1)^{n+k}]}{(n+k+1)(n-k)}P_k(y). \tag{6.18.8}$$

Legendre polynomials $P_n(x)$ and functions $Q_n(x)$, $x \in (-1,1)$.

n	$P_n(x)$	$Q_n(x)$
0	1	$\frac{1}{2}\ln[(1+x)/(1-x)]$
1	x	$P_1(x)Q_0(x) - 1$
2	$\frac{1}{2}(3x^2 - 1)$	$P_2(x)Q_0(x) - \frac{3}{2}x$
3	$\frac{1}{2}x(5x^2 - 3)$	$P_3(x)Q_0(x) - \frac{5}{2}x^2 + \frac{2}{3}$
4	$\frac{1}{8}(35x^4 - 30x^2 + 3)$	$P_4(x)Q_0(x) - \frac{35}{8}x^3 + \frac{55}{24}x$
5	$\frac{1}{8}x(63x^4 - 70x^2 + 15)$	$P_5(x)Q_0(x) - \frac{63}{8}x^4 + \frac{49}{8}x^2 - \frac{8}{15}$

6.18.8 DIFFERENTIAL EQUATION: ASSOCIATED LEGENDRE FUNCTION

The *associated Legendre differential equation* is

$$(1-z^2)y'' - 2zy' + \left[\nu(\nu+1) - \frac{\mu^2}{1-z^2}\right]y = 0.$$

The solutions $P_\nu^\mu(z)$, $Q_\nu^\mu(z)$, the *associated Legendre functions*, can be given in terms of Gauss hypergeometric functions. We only consider integer values of μ, ν, and replace them with m, n, respectively. Then the associated differential equation follows from the Legendre differential equation after it has been differentiated m times.

6.18.9   RELATIONSHIPS BETWEEN THE ASSOCIATED AND ORDINARY LEGENDRE FUNCTIONS

The following relationships are for $z \notin [-1, 1]$

$$P_n^m(z) = (z^2 - 1)^{\frac{1}{2}m} \frac{d^m}{dz^m} P_n(z),$$

$$= \frac{(-1)^m}{2^n n!} (z^2 - 1)^{\frac{1}{2}m} \frac{d^{n+m}}{dx^{n+m}} (z^2 - 1)^n,$$

$$P_n^{-m}(z) = \frac{(n-m)!}{(n+m)!} P_n^m(z),$$

$$Q_n^m(z) = (z^2 - 1)^{\frac{1}{2}m} \frac{d^m}{dz^m} Q_n(z),$$

$$Q_n^{-m}(z) = \frac{(n-m)!}{(n+m)!} Q_n^m(z),$$

$$P_n^{-m}(z) = (z^2 - 1)^{-\frac{1}{2}m} \underbrace{\int_1^z \cdots \int_1^z}_{m} P_n(z) \, (dz)^m,$$

$$Q_n^{-m}(z) = (-1)^m (z^2 - 1)^{-\frac{1}{2}m} \underbrace{\int_z^\infty \cdots \int_z^\infty}_{m} Q_n(z) \, (dz)^m,$$

$$P_{-n-1}^m(z) = P_n^m(z).$$

6.18.10   ORTHOGONALITY RELATIONSHIP

Let $n \geq m$, then

$$\int_{-1}^1 P_n^m(x) P_k^m(x) \, dx = \begin{cases} 0, & \text{if } k \neq n, \\[2mm] \dfrac{2}{2n+1} \dfrac{(n+m)!}{(n-m)!}, & \text{if } k = n. \end{cases}$$

6.18.11   RECURSION RELATIONSHIPS

$$P_n^{m+1}(z) + \frac{2mz}{\sqrt{z^2 - 1}} P_n^m(z) = (n - m + 1)(n + m) P_n^{m-1}(z),$$

$$(z^2 - 1) \frac{dP_n^m(z)}{dz} = mz P_n^m(z) + \sqrt{z^2 - 1} P_n^{m+1}(z),$$

$$(2n + 1) z P_n^m(z) = (n - m + 1) P_{n+1}^m(z) + (n + m) P_{n-1}^m(z),$$

$$(z^2 - 1) \frac{dP_n^m(z)}{dz} = (n - m + 1) P_{n+1}^m(z) - (n + 1) z P_n^m(z),$$

$$P_{n-1}^m(z) - P_{n+1}^m(z) = -(2n + 1) \sqrt{z^2 - 1} P_n^{m-1}(z).$$

The functions $Q_n^m(z)$ satisfy the same relations.

6.19 BESSEL FUNCTIONS

6.19.1 DIFFERENTIAL EQUATION

The *Bessel differential equation* is

$$z^2 y'' + z y' + (z^2 - \nu^2) y = 0.$$

The solutions are denoted with

$$J_\nu(z), \quad Y_\nu(z) \quad \text{(the ordinary \textit{Bessel functions})}$$

and

$$H_\nu^{(1)}(z), \quad H_\nu^{(2)}(z) \quad \text{(the \textit{Hankel functions})}.$$

Further solutions are

$$J_{-\nu}(z), \quad Y_{-\nu}(z), \quad H_{-\nu}^{(1)}(z), \quad H_{-\nu}^{(2)}(z).$$

When ν is an integer,

$$J_{-n}(z) = (-1)^n J_n(z), \quad n = 0, 1, 2, \ldots.$$

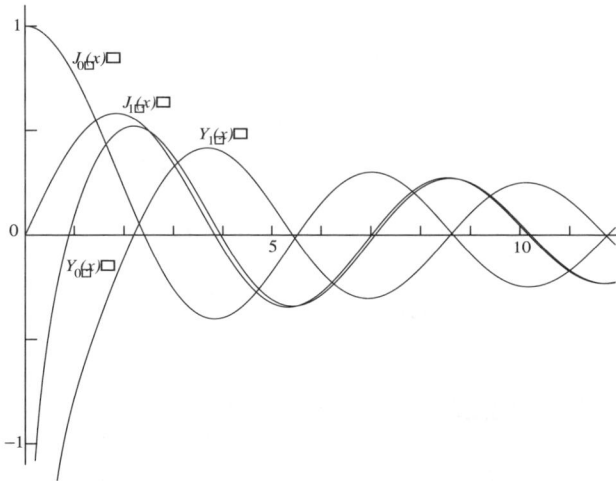

Bessel functions $J_0(x)$, $J_1(x)$, $Y_0(x)$, $Y_1(x)$, $0 \le x \le 12$.

6.19.2 SINGULAR POINTS

The Bessel differential equation has a regular singularity at $z = 0$ and an irregular singularity at $z = \infty$.

6.19.3 RELATIONSHIPS

$$H_\nu^{(1)}(z) = J_\nu(z) + iY_\nu(z), \quad H_\nu^{(2)}(z) = J_\nu(z) - iY_\nu(z).$$

Neumann function: If $\nu \neq 0, \pm 1, \pm 2, \ldots$

$$Y_\nu(z) = \frac{\cos \nu\pi \, J_\nu(z) - J_{-\nu}(z)}{\sin \nu\pi}.$$

When $\nu = n$ (integer) then the limit $\nu \to n$ should be taken in the right-hand side of this equation. Complete solutions to Bessel's equation may be written as

$$
\begin{array}{ll}
c_1 J_\nu(z) + c_2 J_{-\nu}(z), & \text{if } \nu \text{ is not an integer,} \\
c_1 J_\nu(z) + c_2 Y_\nu(z), & \text{for any value of } \nu, \\
c_1 H_\nu^{(1)}(z) + c_2 H_\nu^{(2)}(z), & \text{for any value of } \nu.
\end{array}
$$

6.19.4 SERIES EXPANSIONS

For any complex z,

$$J_\nu(z) = (\tfrac{1}{2} z)^\nu \sum_{n=0}^{\infty} \frac{(-1)^n (\tfrac{1}{2} z)^{2n}}{\Gamma(n + \nu + 1) \, n!},$$

$$J_0(z) = 1 - (\tfrac{1}{2} z)^2 + \frac{1}{2! \, 2!}(\tfrac{1}{2} z)^4 - \frac{1}{3! \, 3!}(\tfrac{1}{2} z)^6 + \cdots,$$

$$J_1(z) = \frac{1}{2} z \left[1 - \frac{1}{1! \, 2!}(\tfrac{1}{2} z)^2 + \frac{1}{2! \, 3!}(\tfrac{1}{2} z)^4 - \frac{1}{3! \, 4!}(\tfrac{1}{2} z)^6 + \cdots \right],$$

$$Y_n(z) = \frac{2}{\pi} J_n(z) \ln(\tfrac{1}{2} z) - \frac{(\tfrac{1}{2} z)^{-n}}{\pi} \sum_{k=0}^{n-1} \frac{(n - k - 1)!}{k!}(\tfrac{1}{2} z)^{2k} -$$

$$\frac{(\tfrac{1}{2} z)^n}{\pi} \sum_{k=0}^{\infty} [\psi(k+1) + \psi(n + k + 1)] \frac{(-1)^k (\tfrac{1}{2} z)^{2k}}{k! \, (n+k)!},$$

where ψ is the logarithmic derivative of the gamma function.

6.19.5 RECURRENCE RELATIONSHIPS

$$C_{\nu-1}(z) + C_{\nu+1}(z) = \frac{2\nu}{z} C_\nu(z),$$

$$C_{\nu-1}(z) - C_{\nu+1}(z) = 2C_\nu'(z),$$

$$C_\nu'(z) = C_{\nu-1}(z) - \frac{\nu}{z} C_\nu(z),$$

$$C_\nu'(z) = -C_{\nu+1}(z) + \frac{\nu}{z} C_\nu(z),$$

where $C_\nu(z)$ denotes one of the functions $J_\nu(z)$, $Y_\nu(z)$, $H_\nu^{(1)}(z)$, $H_\nu^{(2)}(z)$.

6.19.6   BEHAVIOR AS $z \to 0$

Let $\mathrm{Re}\, \nu > 0$, then

$$J_\nu(z) \sim \frac{(\frac{1}{2}z)^\nu}{\Gamma(\nu+1)}, \qquad\qquad Y_\nu(z) \sim -\frac{1}{\pi}\Gamma(\nu)\left(\frac{2}{z}\right)^\nu,$$

$$H_\nu^{(1)}(z) \sim \frac{1}{\pi i}\Gamma(\nu)\left(\frac{2}{z}\right)^\nu, \qquad H_\nu^{(2)}(z) \sim -\frac{1}{\pi i}\Gamma(\nu)\left(\frac{2}{z}\right)^\nu.$$

The same relations hold as $\mathrm{Re}\, \nu \to \infty$, with z fixed.

6.19.7   INTEGRALS

Let $\mathrm{Re}\, z > 0$ and ν be any complex number.

$$J_\nu(z) = \frac{1}{\pi}\int_0^\pi \cos(\nu\theta - z\sin\theta)\, d\theta - \frac{\sin\nu\pi}{\pi}\int_0^\infty e^{-\nu t - z\sinh t}\, dt$$

$$= \frac{(z/2)^\nu}{\sqrt{\pi}\,\Gamma(\nu+\frac{1}{2})}\int_{-1}^1 \left(1-t^2\right)^{\nu-\frac{1}{2}}\cos zt\, dt, \quad \mathrm{Re}\,\nu > -\tfrac{1}{2},\ z \text{ complex}, z \neq 0,$$

$$= \frac{2\left(\frac{1}{2}x\right)^{-\nu}}{\sqrt{\pi}\,\Gamma\left(\frac{1}{2}-\nu\right)}\int_1^\infty \frac{\sin xt}{(t^2-1)^{\nu+\frac{1}{2}}}\, dt, \qquad x > 0,\ |\mathrm{Re}\,\nu| < -\tfrac{1}{2},$$

$$Y_\nu(z) = \frac{1}{\pi}\int_0^\pi \sin(z\sin\theta - \nu\theta)\, d\theta - \int_0^\infty \left(e^{\nu t} + e^{-\nu t}\cos\nu\pi\right)e^{-z\sinh t}\, dt$$

$$= -\frac{2\left(\frac{1}{2}x\right)^{-\nu}}{\sqrt{\pi}\,\Gamma\left(\frac{1}{2}-\nu\right)}\int_1^\infty \frac{\cos xt}{(t^2-1)^{\nu+\frac{1}{2}}}\, dt, \qquad x > 0,\ |\mathrm{Re}\,\nu| < -\tfrac{1}{2}.$$

When $\nu = n$ (an integer), the second integral in the first relation disappears.

6.19.8   FOURIER EXPANSION

For any complex z,

$$e^{-iz\sin t} = \sum_{n=-\infty}^\infty e^{-int} J_n(z),$$

with Parseval relation

$$\sum_{n=-\infty}^\infty J_n^2(z) = 1.$$

6.19.9   AUXILIARY FUNCTIONS

Let $\chi = z - \left(\frac{1}{2}\nu + \frac{1}{4}\right)\pi$ and define

$$P(\nu, z) = \sqrt{\pi z/2}[\ J_\nu(z)\cos\chi + Y_\nu(z)\sin\chi],$$

$$Q(\nu, z) = \sqrt{\pi z/2}[-J_\nu(z)\sin\chi + Y_\nu(z)\cos\chi].$$

6.19.10 INVERSE RELATIONSHIPS

$$J_\nu(z) = \sqrt{2/(\pi z)}\,[P(\nu, z)\cos\chi - Q(\nu, z)\sin\chi],$$
$$Y_\nu(z) = \sqrt{2/(\pi z)}\,[P(\nu, z)\sin\chi + Q(\nu, z)\cos\chi].$$

For the Hankel functions,

$$H_\nu^{(1)}(z) = \sqrt{2/(\pi z)}\,[P(\nu, z) + iQ(\nu, z)]e^{i\chi},$$
$$H_\nu^{(2)}(z) = \sqrt{2/(\pi z)}\,[P(\nu, z) - iQ(\nu, z)]e^{-i\chi}.$$

The functions $P(\nu, z), Q(\nu, z)$ are the slowly varying components in the asymptotic expansions of the oscillatory Bessel and Hankel functions.

6.19.11 ASYMPTOTIC EXPANSIONS

Let (α, n) be defined by

$$
\begin{aligned}
(\alpha, n) &= \frac{2^{-2n}}{n!}\{(4\alpha^2 - 1)(4\alpha^2 - 3^2)\cdots(4\alpha^2 - (2n-1)^2)\} \\
&= \frac{\Gamma(\tfrac{1}{2} + \alpha + n)}{n!\,\Gamma(\tfrac{1}{2} + \alpha - n)}, \qquad n = 0, 1, 2, \ldots, \\
&= \frac{(-1)^n \cos(\pi\alpha)}{\pi n!}\Gamma(\tfrac{1}{2} + \alpha + n)\Gamma(\tfrac{1}{2} - \alpha + n),
\end{aligned}
$$

with recursion

$$(\alpha, n+1) = -\frac{(n+\tfrac{1}{2})^2 - \alpha^2}{n+1}(\alpha, n), \quad n = 1, 2, 3, \ldots, \quad (\alpha, 0) = 1.$$

Then, for $z \to \infty$,

$$P(\nu, z) \sim \sum_{n=0}^{\infty}(-1)^n \frac{(\nu, 2n)}{(2z)^{2n}}, \quad Q(\nu, z) \sim \sum_{n=0}^{\infty}(-1)^n \frac{(\nu, 2n+1)}{(2z)^{2n+1}}.$$

With $\mu = 4\nu^2$,

$$P(\nu, z) \sim 1 - \frac{(\mu - 1)(\mu - 9)}{2!\,(8z)^2} + \frac{(\mu - 1)(\mu - 9)(\mu - 25)(\mu - 49)}{4!\,(8z)^4} - \cdots,$$
$$Q(\nu, z) \sim \frac{\mu - 1}{8z} - \frac{(\mu - 1)(\mu - 9)(\mu - 25)}{3!\,(8z)^3} + \cdots.$$

For large positive values of x,

$$J_\nu(x) = \sqrt{2/(\pi x)}\,\left[\cos(x - \tfrac{1}{2}\nu\pi - \tfrac{1}{4}\pi) + O(x^{-1})\right],$$
$$Y_\nu(x) = \sqrt{2/(\pi x)}\,\left[\sin(x - \tfrac{1}{2}\nu\pi - \tfrac{1}{4}\pi) + O(x^{-1})\right].$$

6.19.12 ZEROS OF BESSEL FUNCTIONS

For $\nu \geq 0$, the zeros $j_{\nu,k}$ (and $y_{\nu,k}$) of $J_\nu(x)$ (and $Y_\nu(x)$) can be arranged as sequences

$$0 < j_{\nu,1} < j_{\nu,2} < \cdots < j_{\nu,n} < \cdots, \quad \lim_{n \to \infty} j_{\nu,n} = \infty,$$

$$0 < y_{\nu,1} < y_{\nu,2} < \cdots < y_{\nu,n} < \cdots, \quad \lim_{n \to \infty} y_{\nu,n} = \infty.$$

Between two consecutive positive zeros of $J_\nu(x)$, there is exactly one zero of $J_{\nu+1}(x)$. Conversely, between two consecutive positive zeros of $J_{\nu+1}(x)$, there is exactly one zero of $J_\nu(x)$. The same holds for the zeros of $Y_\nu(z)$. Moreover, between each pair of consecutive positive zeros of $J_\nu(x)$, there is exactly one zero of $Y_\nu(x)$, and conversely.

6.19.12.1 Asymptotic expansions of the zeros

When ν is fixed, $s \gg \nu$, and $\mu = 4\nu^2$,

$$j_{\nu,s} \sim \alpha - \frac{\mu - 1}{8\alpha} \left[1 - \frac{4(7\mu^2 - 31)}{3(8\alpha)^2} - \frac{32(83\mu^2 - 982\mu + 3779)}{15(8\alpha)^4} + \cdots \right]$$

where $\alpha = (s + \frac{1}{2}\nu - \frac{1}{4})\pi$; $y_{\nu,s}$ has the same asymptotic expansion with $\alpha = (s + \frac{1}{2}\nu - \frac{3}{4})\pi$.

n	$j_{0,n}$	$j_{1,n}$	$y_{0,n}$	$y_{1,n}$
1	2.40483	3.83171	0.89358	2.19714
2	5.52008	7.01559	3.95768	5.42968
3	8.65373	10.17347	7.08605	8.59601
4	11.79153	13.32369	10.22235	11.74915
5	14.93092	16.47063	13.36110	14.89744
6	18.07106	19.61586	16.50092	18.04340
7	21.21164	22.76008	19.64131	21.18807

Positive zeros $j_{\nu,n}, y_{\nu,n}$ of Bessel functions $J_\nu(x), Y_\nu(x)$, $\nu = 0, 1$.

6.19.13 HALF ORDER BESSEL FUNCTIONS

For integer values of n, let

$$j_n(z) = \sqrt{\pi/(2z)}\, J_{n+\frac{1}{2}}(z), \quad y_n(z) = \sqrt{\pi/(2z)}\, Y_{n+\frac{1}{2}}(z).$$

Then

$$j_0(z) = y_{-1}(z) = \frac{\sin z}{z}, \quad y_0(z) = -j_{-1}(z) = -\frac{\cos z}{z},$$

and, for $n = 0, 1, 2, \ldots$,

$$j_n(z) = (-z)^n \left[\frac{1}{z}\frac{d}{dz} \right]^n \frac{\sin z}{z}, \quad y_n(z) = -(-z)^n \left[\frac{1}{z}\frac{d}{dz} \right]^n \frac{\cos z}{z}.$$

6.19.13.1   Recursion relationships

The functions $j_n(z), y_n(z)$ both satisfy

$$z[f_{n-1}(z) + f_{n+1}(z)] = (2n + 1)f_n(z),$$
$$nf_{n-1}(z) - (n + 1)f_{n+1}(z) = (2n + 1)f_n'(z).$$

6.19.13.2   Differential equation

$$z^2 f'' + 2zf' + [z^2 - n(n + 1)]f = 0.$$

6.19.14   MODIFIED BESSEL FUNCTIONS

1. Differential equation

$$z^2 y'' + zy' - (z^2 + \nu^2)y = 0.$$

2. Solutions $I_\nu(z), K_\nu(z)$,

$$I_\nu(z) = \left(\frac{z}{2}\right)^\nu \sum_{n=0}^\infty \frac{(z/2)^{2n}}{\Gamma(n + \nu + 1)\, n!},$$

$$K_\nu(z) = \frac{\pi}{2} \frac{I_{-\nu}(z) - I_\nu(z)}{\sin \nu\pi},$$

where the right-hand side should be determined by l'Hôpital's rule when ν assumes integer values. When $n = 0, 1, 2, \ldots$,

$$K_n(z) = (-1)^{n+1} I_n(z) \ln \frac{z}{2} + \frac{1}{2} \left(\frac{2}{z}\right)^n \sum_{k=0}^{n-1} \frac{(n - k - 1)!}{k!} \left(-\frac{z^2}{4}\right)^k$$

$$+ \frac{(-1)^n}{2} \left(\frac{z}{2}\right)^n \sum_{k=0}^\infty [\psi(k + 1) + \psi(n + k + 1)] \frac{(z/2)^{2k}}{k!\,(n + k)!}.$$

3. Relations with the ordinary Bessel functions

$$I_\nu(z) = e^{-\frac{1}{2}\nu\pi i} J_\nu\left(ze^{\frac{1}{2}\pi i}\right), \qquad -\pi < \arg z \le \frac{\pi}{2},$$

$$I_\nu(z) = e^{\frac{3}{2}\nu\pi i} J_\nu\left(ze^{-\frac{3}{2}\pi i}\right), \qquad \frac{\pi}{2} < \arg z \le \pi,$$

$$K_\nu(z) = \frac{\pi}{2} i e^{\frac{1}{2}\nu\pi i} H_\nu^{(1)}\left(ze^{\frac{1}{2}\pi i}\right), \qquad -\pi < \arg z \le \frac{\pi}{2},$$

$$K_\nu(z) = -\frac{\pi}{2} i e^{-\frac{1}{2}\nu\pi i} H_\nu^{(2)}\left(ze^{-\frac{1}{2}\pi i}\right), \qquad -\frac{\pi}{2} < \arg z \le \pi,$$

$$Y_\nu\left(ze^{\frac{1}{2}\pi i}\right) = e^{\frac{1}{2}(\nu+1)\pi i} I_\nu(z) - \frac{2}{\pi} e^{-\frac{1}{2}\nu\pi i} K_\nu(z), \qquad -\pi < \arg z \le \frac{\pi}{2}.$$

For $n = 0, 1, 2, \ldots$,

$$I_n(z) = i^{-n} J_n(iz), \quad Y_n(iz) = i^{n+1} I_n(z) - \frac{2}{\pi} i^{-n} K_n(z),$$

$$I_{-n}(z) = I_n(z), \quad K_{-\nu}(z) = K_\nu(z), \quad \text{for any } \nu.$$

4. Recursion relationships

$$I_{\nu-1}(z) - I_{\nu+1}(z) = \frac{2\nu}{z}I_\nu(z), \qquad K_{\nu+1}(z) - K_{\nu-1}(z) = \frac{2\nu}{z}K_\nu(z),$$

$$I_{\nu-1}(z) + I_{\nu+1}(z) = 2I_\nu'(z), \qquad K_{\nu-1}(z) + K_{\nu+1}(z) = -2K_\nu'(z).$$

5. Integrals

$$I_\nu(z) = \frac{1}{\pi}\int_0^\pi e^{z\cos\theta}\cos(\nu\theta)\,d\theta - \frac{\sin\nu\pi}{\pi}\int_0^\infty e^{-\nu t - z\cosh t}\,dt$$

$$= \frac{(2z)^\nu e^z}{\sqrt{\pi}\,\Gamma\left(\nu+\frac{1}{2}\right)}\int_0^1 e^{-2zt}[t(1-t)]^{\nu-\frac{1}{2}}\,dt,$$

$$\operatorname{Re}\nu > -\tfrac{1}{2},\ z\ \text{complex},\ z\neq 0.$$

$$K_\nu(z) = \int_0^\infty e^{-z\cosh t}\cosh(\nu t)\,dt$$

$$= \frac{\sqrt{\pi}\left(\frac{2}{z}\right)^\nu e^{-z}}{\Gamma\left(\nu+\frac{1}{2}\right)}\int_0^\infty e^{-2zt}t^{\nu-\frac{1}{2}}(t+1)^{\nu-\frac{1}{2}}\,dt,$$

$$\operatorname{Re}\nu > -\tfrac{1}{2},\ \operatorname{Re}z > 0.$$

$$K_\nu(xz) = \frac{\Gamma\left(\nu+\frac{1}{2}\right)(2z)^\nu}{\sqrt{\pi}\,x^\nu}\int_0^\infty \frac{\cos xt\,dt}{(t^2+z^2)^{\nu+\frac{1}{2}}}\,dt,$$

$$\operatorname{Re}\nu > -\tfrac{1}{2},\ x > 0,\ |\arg z| < \tfrac{1}{2}\pi.$$

When $\nu = n$ (an integer), the second integral in the first relation disappears.

6.19.15 AIRY FUNCTIONS

1. Differential equation

$$y'' - zy = 0.$$

2. Solutions are $\mathrm{Ai}(z)$ and $\mathrm{Bi}(z)$:

$$\mathrm{Ai}(z) = c_1 f(z) - c_2 g(z),$$

$$\mathrm{Bi}(z) = \sqrt{3}\,[c_1 f(z) + c_2 g(z)]$$

where

$$f(z) = 1 + \frac{1}{3!}z^3 + \frac{1\cdot 4}{6!}z^6 + \frac{1\cdot 4\cdot 7}{9!}z^9 + \cdots,$$

$$g(z) = z + \frac{2}{4!}z^4 + \frac{2\cdot 5}{7!}z^7 + \frac{2\cdot 5\cdot 8}{10!}z^{10} + \cdots,$$

$$c_1 = \quad \mathrm{Ai}(0) \ = \mathrm{Bi}(0)/\sqrt{3} = 3^{-2/3}\Gamma(\tfrac{2}{3}) = 0.35502\,80538\,87817,$$

$$c_2 = -\,\mathrm{Ai}'(0) = \mathrm{Bi}'(0)/\sqrt{3} = 3^{-1/3}\Gamma(\tfrac{1}{3}) = 0.25881\,94037\,92807.$$

FIGURE 6.10

Graphs of the Airy functions $\mathrm{Ai}(x)$ *and* $\mathrm{Bi}(x)$, x *real.*

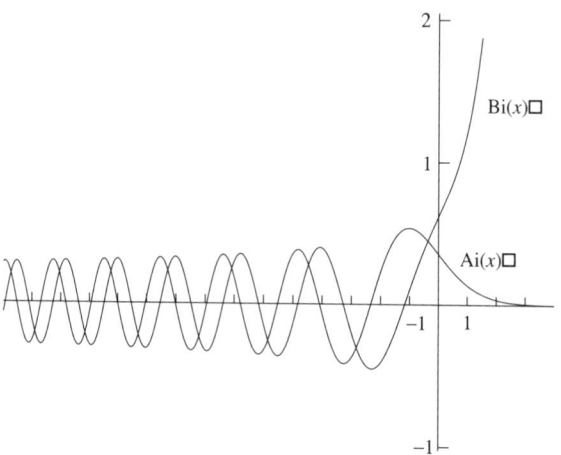

3. Wronskian relation

$$\mathrm{Ai}(z)\,\mathrm{Bi}'(z) - \mathrm{Ai}'(z)\,\mathrm{Bi}(z) = \frac{1}{\pi}.$$

4. Relations with the Bessel functions

   Let $\zeta = \frac{2}{3}z^{\frac{3}{2}}$, then

$$\mathrm{Ai}(z) = \frac{1}{3}\sqrt{z}\left[I_{-\frac{1}{3}}(\zeta) - I_{\frac{1}{3}}(\zeta)\right] = \frac{1}{\pi}\sqrt{\frac{z}{3}}K_{\frac{1}{3}}(\zeta).$$

$$\mathrm{Ai}(-z) = \frac{1}{3}\sqrt{z}\left[J_{\frac{1}{3}}(\zeta) + J_{-\frac{1}{3}}(\zeta)\right].$$

$$\mathrm{Bi}(z) = \sqrt{\frac{z}{3}}\left[I_{-\frac{1}{3}}(\zeta) + I_{\frac{1}{3}}(\zeta)\right].$$

$$\mathrm{Bi}(-z) = \sqrt{\frac{z}{3}}\left[J_{-\frac{1}{3}}(\zeta) - J_{\frac{1}{3}}(\zeta)\right].$$

5. Integrals for real x

$$\mathrm{Ai}(x) = \frac{1}{\pi}\int_0^\infty \cos\left(\tfrac{1}{3}t^3 + xt\right)\,dt,$$

$$\mathrm{Bi}(x) = \frac{1}{\pi}\int_0^\infty e^{-\frac{1}{3}t^3 + xt}\,dt + \frac{1}{\pi}\int_0^\infty \sin\left(\tfrac{1}{3}t^3 + xt\right)\,dt.$$

6. Asymptotic behavior

Let $\zeta = \frac{2}{3}z^{\frac{3}{2}}$. Then, for $z \to \infty$,

$$\mathrm{Ai}(z) = \frac{1}{2\sqrt{\pi}}z^{-\frac{1}{4}}e^{-\zeta}\left[1 + O\left(\zeta^{-1}\right)\right], \qquad |\arg z| < \pi,$$

$$\mathrm{Bi}(z) = \frac{1}{\sqrt{\pi}}z^{-\frac{1}{4}}e^{\zeta}\left[1 + O\left(\zeta^{-1}\right)\right], \qquad |\arg z| < \tfrac{1}{3}\pi,$$

$$\mathrm{Ai}(-z) = \frac{1}{\sqrt{\pi}}z^{-\frac{1}{4}}\left[\sin\left(\zeta + \tfrac{1}{4}\pi\right) + O\left(\zeta^{-1}\right)\right], \qquad |\arg z| < \frac{2}{3}\pi,$$

$$\mathrm{Bi}(-z) = -\frac{1}{\sqrt{\pi}}z^{-\frac{1}{4}}\left[\cos\left(\zeta + \tfrac{1}{4}\pi\right) + O\left(\zeta^{-1}\right)\right], \qquad |\arg z| < \frac{2}{3}\pi.$$

6.19.16 NUMERICAL VALUES FOR THE BESSEL FUNCTIONS

x	$J_0(x)$	$J_1(x)$	$Y_0(x)$	$Y_1(x)$
0.0	1.00000000	0.00000000	$-\infty$	$-\infty$
0.2	0.99002497	0.09950083	-1.08110532	-3.32382499
0.4	0.96039823	0.19602658	-0.60602457	-1.78087204
0.6	0.91200486	0.28670099	-0.30850987	-1.26039135
0.8	0.84628735	0.36884205	-0.08680228	-0.97814418
1.0	0.76519769	0.44005059	0.08825696	-0.78121282
1.2	0.67113274	0.49828906	0.22808350	-0.62113638
1.4	0.56685512	0.54194771	0.33789513	-0.47914697
1.6	0.45540217	0.56989594	0.42042690	-0.34757801
1.8	0.33998641	0.58151695	0.47743171	-0.22366487
2.0	0.22389078	0.57672481	0.51037567	-0.10703243
2.2	0.11036227	0.55596305	0.52078429	0.00148779
2.4	0.00250768	0.52018527	0.51041475	0.10048894
2.6	-0.09680495	0.47081827	0.48133059	0.18836354
2.8	-0.18503603	0.40970925	0.43591599	0.26354539
3.0	-0.26005195	0.33905896	0.37685001	0.32467442
3.2	-0.32018817	0.26134325	0.30705325	0.37071134
3.4	-0.36429560	0.17922585	0.22961534	0.40101529
3.6	-0.39176898	0.09546555	0.14771001	0.41539176
3.8	-0.40255641	0.01282100	0.06450325	0.41411469
4.0	-0.39714981	-0.06604333	-0.01694074	0.39792571
4.2	-0.37655705	-0.13864694	-0.09375120	0.36801281
4.4	-0.34225679	-0.20277552	-0.16333646	0.32597067
4.6	-0.29613782	-0.25655284	-0.22345995	0.27374524
4.8	-0.24042533	-0.29849986	-0.27230379	0.21356517
5.0	-0.17759677	-0.32757914	-0.30851763	0.14786314

x	$e^{-x}I_0(x)$	$e^{-x}I_1(x)$	$e^x K_0(x)$	$e^x K_1(x)$
0.0	1.00000000	0.00000000	∞	∞
0.2	0.82693855	0.08228312	2.14075732	5.83338603
0.4	0.69740217	0.13676322	1.66268209	3.25867388
0.6	0.59932720	0.17216442	1.41673762	2.37392004
0.8	0.52414894	0.19449869	1.25820312	1.91793030
1.0	0.46575961	0.20791042	1.14446308	1.63615349
1.2	0.41978208	0.21525686	1.05748453	1.44289755
1.4	0.38306252	0.21850759	0.98807000	1.30105374
1.6	0.35331500	0.21901949	0.93094598	1.19186757
1.8	0.32887195	0.21772628	0.88283353	1.10480537
2.0	0.30850832	0.21526929	0.84156822	1.03347685
2.2	0.29131733	0.21208773	0.80565398	0.97377017
2.4	0.27662232	0.20848109	0.77401814	0.92291367
2.6	0.26391400	0.20465225	0.74586824	0.87896728
2.8	0.25280553	0.20073741	0.72060413	0.84053006
3.0	0.24300035	0.19682671	0.69776160	0.80656348
3.2	0.23426883	0.19297862	0.67697511	0.77628028
3.4	0.22643140	0.18922985	0.65795227	0.74907206
3.6	0.21934622	0.18560225	0.64045596	0.72446066
3.8	0.21290013	0.18210758	0.62429158	0.70206469
4.0	0.20700192	0.17875084	0.60929767	0.68157595
4.2	0.20157738	0.17553253	0.59533899	0.66274241
4.4	0.19656556	0.17245023	0.58230127	0.64535587
4.6	0.19191592	0.16949973	0.57008720	0.62924264
4.8	0.18758620	0.16667571	0.55861332	0.61425660
5.0	0.18354081	0.16397227	0.54780756	0.60027386

6.20 ELLIPTIC INTEGRALS

6.20.1 DEFINITIONS

Any integral of the type $\int R(x, y)\, dx$, where $R(x, y)$ is a rational function of x and y, with y^2 being a polynomial of the third or fourth degree in x (that is $y^2 = a_0 x^4 + a_1 x^3 + a_2 x^2 + a_3 x + a_4$ with $|a_0| + |a_1| > 0$) is called an *elliptic integral*. All elliptic integrals can be reduced to three basic types.

1. *Elliptic integral of the first kind*

$$F(\phi, k) = \int_0^\phi \frac{d\theta}{\sqrt{1 - k^2 \sin^2 \theta}}$$

$$= \int_0^x \frac{dt}{\sqrt{(1 - t^2)(1 - k^2 t^2)}}, \quad x = \sin \phi, \quad k^2 < 1.$$

2. *Elliptic integral of the second kind*

$$E(\phi, k) = \int_0^\phi \sqrt{1 - k^2 \sin^2 \theta} \, d\theta$$

$$= \int_0^x \frac{\sqrt{1 - k^2 t^2}}{\sqrt{1 - t^2}} \, dt, \quad x = \sin \phi, \quad k^2 < 1.$$

3. *Elliptic integral of the third kind*

$$\Pi(n; \phi, k) = \int_0^\phi \frac{1}{1 + n \sin^2 \theta} \frac{d\theta}{\sqrt{1 - k^2 \sin^2 \theta}}$$

$$= \int_0^x \frac{1}{1 + nt^2} \frac{dt}{\sqrt{(1 - t^2)(1 - k^2 t^2)}}, \quad x = \sin \phi, \quad k^2 < 1,$$

where for $n < -1$ the integral should be interpreted as a Cauchy principal value integral.

6.20.2 PROPERTIES

1. The *complete elliptic integrals* of the first and second kinds are

$$K = K(k) = F\left(\frac{\pi}{2}, k\right) = \int_0^{\pi/2} \left(1 - k^2 \sin^2 t\right)^{-1/2} dt = \frac{\pi}{2} F\left(\frac{1}{2}, \frac{1}{2}; 1; k^2\right),$$

$$K(\alpha) = \int_0^{\pi/2} \left(1 - \sin^2 \alpha \sin^2 t\right)^{-1/2} dt,$$

$$E = E(k) = E\left(\frac{\pi}{2}, k\right) = \int_0^{\pi/2} \left(1 - k^2 \sin^2 t\right)^{1/2} dt = \frac{\pi}{2} F\left(-\frac{1}{2}, \frac{1}{2}; 1; k^2\right),$$

$$E(\alpha) = \int_0^{\pi/2} \left(1 - \sin^2 \alpha \sin^2 t\right)^{1/2} dt,$$

where $F\left(\pm\frac{1}{2}, \frac{1}{2}; 1; k^2\right)$ is the Gauss hypergeometric function.

2. *Complementary integrals*

   In these expressions, primes do not mean derivatives.

$$K' = K'(k) = K(k') = \int_0^{\pi/2} \left(1 - (1 - k^2) \sin^2 t\right)^{-1/2} dt = F\left(\frac{\pi}{2}, k'\right),$$

$$E' = E'(k) = E(k') = \int_0^{\pi/2} \left(1 - (1 - k^2) \sin^2 t\right)^{1/2} dt = E\left(\frac{\pi}{2}, k'\right),$$

k is called the *modulus*; $k' = \sqrt{1 - k^2}$ is called the *complementary modulus*.

3. The *Legendre relation* is

$$K E' + E K' - K K' = \frac{\pi}{2}.$$

4. Extension of the range of ϕ

$$F(\pi, k) = 2K, \qquad E(\pi, k) = 2E,$$

and, for $m = 0, 1, 2, \ldots,$

$$F(\phi + m\pi, k) = mF(\pi, k) + F(\phi, k) = 2mK + F(\phi, k),$$
$$E(\phi + m\pi, k) = mE(\pi, k) + E(\phi, k) = 2mE + E(\phi, k).$$

6.20.3 NUMERICAL VALUES OF THE ELLIPTIC INTEGRALS

$$F(\phi, \alpha) = \int_0^\phi \left(1 - \sin^2 \alpha \sin^2 t\right)^{-1/2} dt \qquad \text{(note that } k = \sin \alpha \text{)}$$

ϕ	$0°$	$10°$	$20°$	$30°$	$40°$	$50°$	$60°$	$70°$	$80°$	$90°$
$0°$	0.000	0.000	0.000	0.000	0.000	0.000	0.000	0.000	0.000	0.000
$10°$	0.175	0.175	0.175	0.175	0.175	0.175	0.175	0.175	0.175	0.175
$20°$	0.349	0.349	0.350	0.351	0.352	0.353	0.354	0.355	0.356	0.356
$30°$	0.524	0.524	0.526	0.529	0.533	0.538	0.542	0.546	0.548	0.549
$40°$	0.698	0.700	0.704	0.712	0.721	0.732	0.744	0.754	0.760	0.763
$50°$	0.873	0.876	0.884	0.898	0.917	0.940	0.965	0.988	1.004	1.011
$60°$	1.047	1.052	1.066	1.090	1.123	1.164	1.213	1.262	1.301	1.317
$70°$	1.222	1.229	1.250	1.285	1.337	1.407	1.494	1.596	1.692	1.735
$80°$	1.396	1.406	1.434	1.485	1.560	1.666	1.813	2.012	2.265	2.436
$90°$	1.571	1.583	1.620	1.686	1.787	1.936	2.157	2.505	3.153	∞

$$E(\phi, \alpha) = \int_0^\phi \left(1 - \sin^2 \alpha \sin^2 t\right)^{1/2} dt \qquad \text{(note that } k = \sin \alpha \text{)}$$

ϕ	$0°$	$10°$	$20°$	$30°$	$40°$	$50°$	$60°$	$70°$	$80°$	$90°$
$0°$	0.000	0.000	0.000	0.000	0.000	0.000	0.000	0.000	0.000	0.000
$10°$	0.175	0.175	0.174	0.174	0.174	0.174	0.174	0.174	0.174	0.174
$20°$	0.349	0.349	0.348	0.347	0.346	0.345	0.344	0.343	0.342	0.342
$30°$	0.524	0.523	0.521	0.518	0.514	0.510	0.506	0.503	0.501	0.500
$40°$	0.698	0.697	0.692	0.685	0.676	0.667	0.657	0.650	0.645	0.643
$50°$	0.873	0.870	0.861	0.848	0.832	0.813	0.795	0.780	0.770	0.766
$60°$	1.047	1.043	1.029	1.008	0.980	0.949	0.918	0.891	0.873	0.866
$70°$	1.222	1.215	1.195	1.163	1.122	1.075	1.027	0.983	0.951	0.940
$80°$	1.396	1.387	1.360	1.316	1.259	1.193	1.122	1.056	1.005	0.985
$90°$	1.571	1.559	1.524	1.467	1.393	1.306	1.211	1.118	1.040	1.000

α	$K(\alpha)$	$K'(\alpha)$	$E(\alpha)$	$E'(\alpha)$	k^2	$K(k)$	$K'(k)$	$E(k)$	$E'(k)$
$0°$	$\pi/2$	∞	$\pi/2$	1	0	$\pi/2$	∞	$\pi/2$	1
$5°$	1.574	3.832	1.568	1.013	0.05	1.591	2.908	1.551	1.060
$10°$	1.583	3.153	1.559	1.040	0.10	1.612	2.578	1.531	1.105
$15°$	1.598	2.768	1.544	1.076	0.15	1.635	2.389	1.510	1.143
$20°$	1.620	2.505	1.524	1.118	0.20	1.660	2.257	1.489	1.178
$25°$	1.649	2.309	1.498	1.164	0.25	1.686	2.157	1.467	1.211
$30°$	1.686	2.157	1.467	1.211	0.30	1.714	2.075	1.445	1.242
$35°$	1.731	2.035	1.432	1.259	0.35	1.744	2.008	1.423	1.271
$40°$	1.787	1.936	1.393	1.306	0.40	1.778	1.950	1.399	1.298
$45°$	1.854	1.854	1.351	1.351	0.45	1.814	1.899	1.375	1.325
$50°$	1.936	1.787	1.306	1.393	0.50	1.854	1.854	1.351	1.351
$55°$	2.035	1.731	1.259	1.432	0.55	1.899	1.814	1.325	1.375
$60°$	2.157	1.686	1.211	1.467	0.60	1.950	1.778	1.298	1.399
$65°$	2.309	1.649	1.164	1.498	0.65	2.008	1.744	1.271	1.423
$70°$	2.505	1.620	1.118	1.524	0.70	2.075	1.714	1.242	1.445
$75°$	2.768	1.598	1.076	1.544	0.75	2.157	1.686	1.211	1.467
$80°$	3.153	1.583	1.040	1.559	0.80	2.257	1.660	1.178	1.489
$85°$	3.832	1.574	1.013	1.568	0.85	2.389	1.635	1.143	1.510
$90°$	∞	$\pi/2$	1	$\pi/2$	0.90	2.578	1.612	1.105	1.531
					0.95	2.908	1.591	1.060	1.551
					1	∞	$\pi/2$	1	$\pi/2$

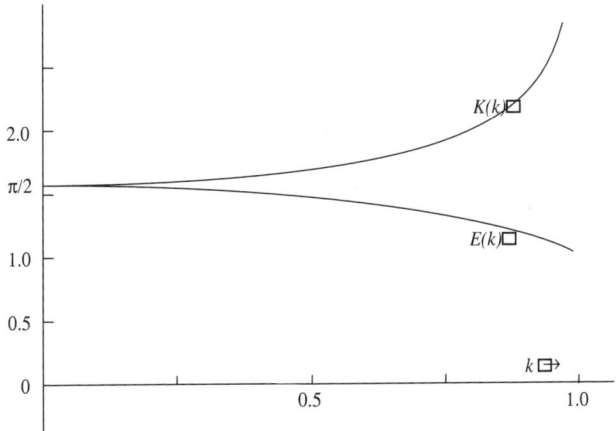

The complete elliptic integrals $E(k)$ and $K(k)$, $0 \le k \le 1$.[1]

[1] From Temme, N.M., *Special Functions: An Introduction to the Classical Functions of Mathematical Physics,* John Wiley & Sons, New York, 1996. With permission.

6.21   JACOBIAN ELLIPTIC FUNCTIONS

The *Jacobian Elliptic functions* are the inverses of elliptic integrals. If $u = F(\phi, k)$ (the elliptic integral of the first kind)

$$u = \int_0^\phi \frac{d\theta}{\sqrt{1 - k^2 \sin^2 \theta}} = \int_0^{\sin \phi} \frac{dt}{\sqrt{(1 - t^2)(1 - k^2 t^2)}}, \qquad (6.21.1)$$

with $k^2 < 1$, then the inverse function is

$$\phi = \text{am } u \qquad \text{(the } \textit{amplitude} \text{ of } u\text{)}. \qquad (6.21.2)$$

(Note that the parameter k is not always explicitly written.) The Jacobian elliptic functions are then defined as

1. $\text{sn } u = \text{sn}(u, k) = \sin \phi = \sin(\text{am } u)$

2. $\text{cn } u = \text{cn}(u, k) = \cos \phi = \cos(\text{am } u) = \sqrt{1 - \text{sn}^2 u}$

3. $\text{dn } u = \text{dn}(u, k) = \sqrt{1 - k^2 \sin^2 \phi} = \sqrt{1 - k^2 \text{sn}^2 u}$

Note that

$$u = \int_1^{\text{cn}(u,k)} \frac{dt}{\sqrt{(1 - t^2)(k'^2 + k^2 t^2)}},$$

$$u = \int_1^{\text{dn}(u,k)} \frac{dt}{\sqrt{(1 - t^2)(t^2 - k'^2)}}. \qquad (6.21.3)$$

6.21.1   PROPERTIES

1. Relationships

$$\text{sn}^2(u, k) + \text{cn}^2(u, k) = 1,$$
$$\text{dn}^2(u, k) + k^2 \text{sn}^2(u, k) = 1,$$
$$\text{dn}^2(u, k) - k^2 \text{cn}^2(u, k) = 1 - k^2 = k'^2.$$

2. Special values

(a) $\text{sn}(0, k) = 0,$ (e) $\text{sn}(u, 0) = \sin u,$ (h) $\text{sn}(u, 1) = \tanh u,$

(b) $\text{cn}(0, k) = 1,$ (f) $\text{cn}(u, 0) = \cos u,$ (i) $\text{cn}(u, 1) = \text{sech } u,$

(c) $\text{dn}(0, k) = 1,$ (g) $\text{dn}(u, 0) = 1,$ (j) $\text{dn}(u, 1) = \text{sech } u.$

(d) $\text{am}(0, k) = 0,$

3. Symmetry properties

    (a) $\text{sn}(-u) = -\text{sn}(u),$ (c) $\text{dn}(-u) = \text{dn}(u),$

    (b) $\text{cn}(-u) = \text{cn}(u),$ (d) $\text{am}(-u) = -\text{am}(u).$

4. Addition formulae

    (a) $\text{sn}(u \pm v) = \dfrac{\text{sn}\,u\,\text{cn}\,v\,\text{dn}\,v \pm \text{cn}\,u\,\text{sn}\,v\,\text{dn}\,u}{1 - k^2\,\text{sn}^2\,u\,\text{sn}^2\,v},$

    (b) $\text{cn}(u \pm v) = \dfrac{\text{cn}\,u\,\text{cn}\,v \mp \text{sn}\,u\,\text{dn}\,u\,\text{sn}\,v\,\text{dn}\,v}{1 - k^2\,\text{sn}^2\,u\,\text{sn}^2\,v},$

    (c) $\text{dn}(u \pm v) = \dfrac{\text{dn}\,u\,\text{dn}\,v \mp k^2\,\text{sn}\,u\,\text{cn}\,u\,\text{sn}\,v\,\text{cn}\,v}{1 - k^2\,\text{sn}^2\,u\,\text{sn}^2\,v}.$

5. The elliptic functions are *doubly periodic functions* with respect to the variable u. The periods of

$$\begin{aligned}
\text{sn}(u, k) \quad &\text{are} \quad 4K \quad \text{and} \quad 2iK', \\
\text{cn}(u, k) \quad &\text{are} \quad 4K \quad \text{and} \quad 2K + 2iK', \\
\text{dn}(u, k) \quad &\text{are} \quad 2K \quad \text{and} \quad 4iK'.
\end{aligned}$$

6.21.2   DERIVATIVES AND INTEGRALS

1. $\dfrac{d}{du}\,\text{sn}\,u = \text{cn}\,u\,\text{dn}\,u.$ 4. $\displaystyle\int \text{sn}\,u\,du = \frac{1}{k}\,(\text{dn}\,u - k\,\text{cn}\,u).$

2. $\dfrac{d}{du}\,\text{cn}\,u = -\text{sn}\,u\,\text{dn}\,u.$ 5. $\displaystyle\int \text{cn}\,u\,du = \frac{1}{k}\,\cos^{-1}(\text{dn}\,u).$

3. $\dfrac{d}{du}\,\text{dn}\,u = -k^2\,\text{cn}\,u\,\text{dn}\,u.$ 6. $\displaystyle\int \text{dn}\,u\,du = \text{am}\,u = \sin^{-1}(\text{sn}\,u).$

6.21.3   SERIES EXPANSIONS

$$\begin{aligned}
\text{sn}(u, k) &= u - (1 + k^2)\frac{u^3}{3!} + (1 + 14k^2 + k^4)\frac{u^5}{5!} \\
&\quad - (1 + 135k^2 + 135k^4 + k^6)\frac{u^7}{7!} + \cdots, \\
\text{cn}(u, k) &= 1 - \frac{u^2}{2!} + (1 + 4k^2)\frac{u^4}{4!} - (1 + 44k^2 + 16k^4)\frac{u^6}{6!} + \cdots, \\
\text{dn}(u, k) &= 1 - k^2\frac{u^2}{2!} + k^2(4 + k^2)\frac{u^4}{4!} - k^2(16 + 44k^2 + k^4)\frac{u^6}{6!} + \cdots.
\end{aligned}$$

$$(6.21.4)$$

Let the *nome* q be defined by $q = e^{-\pi K/K'}$ and $v = \pi u/(2K)$. Then

$$\text{sn}(u, k) = \frac{2\pi}{kK} \sum_{n=0}^{\infty} \frac{q^{n+\frac{1}{2}}}{1 - q^{2n+1}} \sin[(2n + 1)v],$$

$$\text{cn}(u, k) = \frac{2\pi}{kK} \sum_{n=0}^{\infty} \frac{q^{n+\frac{1}{2}}}{1 + q^{2n+1}} \cos[(2n + 1)v], \qquad (6.21.5)$$

$$\text{dn}(u, k) = \frac{\pi}{2K} + \frac{2\pi}{K} \sum_{n=1}^{\infty} \frac{q^n}{1 + q^{2n}} \cos(2nv).$$

6.22   CLEBSCH–GORDAN COEFFICIENTS

These coefficients arise in the integration of three spherical harmonic functions (see equation 6.10.10 on page 539).

$$\begin{pmatrix} j_1 & j_2 & j \\ m_1 & m_2 & m \end{pmatrix} = \delta_{m,m_1+m_2} \sqrt{\frac{(j_1 + j_2 - j)!(j + j_1 - j_2)!(j + j_2 - j_1)!(2j + 1)}{(j + j_1 + j_2 + 1)!}}$$

$$\times \sum_k \frac{(-1)^k \sqrt{(j_1 + m_1)!(j_1 - m_1)!(j_2 + m_2)!(j_2 - m_2)!(j + m)!(j - m)!}}{k!(j_1 + j_2 - j - k)!(j_1 - m_1 - k)!(j_2 + m_2 - k)!(j - j_2 + m_1 + k)!(j - j_1 - m_2 + k)!}.$$

1. Conditions:

   (a) Each of $\{j_1, j_2, j, m_1, m_2, m\}$ may be an integer, or half an integer. Additionally: $j > 0$, $j_1 > 0$, $j_2 > 0$ and $j + j_1 + j_2$ is an integer.
   (b) $j_1 + j_2 - j \geq 0$.
   (c) $j_1 - j_2 + j \geq 0$.
   (d) $-j_1 + j_2 + j \geq 0$.
   (e) $|m_1| \leq j_1$, $|m_2| \leq j_2$, $|m| \leq j$.

2. Special values:

   (a) $\begin{pmatrix} j_1 & j_2 & j \\ m_1 & m_2 & m \end{pmatrix} = 0$ if $m_1 + m_2 \neq m$.

   (b) $\begin{pmatrix} j_1 & 0 & j \\ m_1 & 0 & m \end{pmatrix} = \delta_{j_1,j}\delta_{m_1,m}$.

   (c) $\begin{pmatrix} j_1 & j_2 & j \\ 0 & 0 & 0 \end{pmatrix} = 0$ when $j_1 + j_2 + j$ is an odd integer.

   (d) $\begin{pmatrix} j_1 & j_1 & j \\ m_1 & m_1 & m \end{pmatrix} = 0$ when $2j_1 + j$ is an odd integer.

3. Symmetry relations: all of the following are equal to $\begin{pmatrix} j_1 & j_2 & j \\ m_1 & m_2 & m \end{pmatrix}$:

   (a) $\begin{pmatrix} j_2 & j_1 & j \\ -m_2 & -m_1 & -m \end{pmatrix}$,

(b) $(-1)^{j_1+j_2-j}\begin{pmatrix} j_2 & j_1 & j \\ m_1 & m_2 & m \end{pmatrix}$,

(c) $(-1)^{j_1+j_2-j}\begin{pmatrix} j_1 & j_2 & j \\ -m_1 & -m_2 & -m \end{pmatrix}$,

(d) $\sqrt{\frac{2j+1}{2j_1+1}}(-1)^{j_2+m_2}\begin{pmatrix} j & j_2 & j_1 \\ -m & m_2 & -m_1 \end{pmatrix}$,

(e) $\sqrt{\frac{2j+1}{2j_1+1}}(-1)^{j_1-m_1+j-m}\begin{pmatrix} j & j_2 & j_1 \\ m & -m_2 & m_1 \end{pmatrix}$,

(f) $\sqrt{\frac{2j+1}{2j_1+1}}(-1)^{j-m+j_1-m_1}\begin{pmatrix} j_2 & j & j_1 \\ m_2 & -m & -m_1 \end{pmatrix}$,

(g) $\sqrt{\frac{2j+1}{2j_2+1}}(-1)^{j_1-m_1}\begin{pmatrix} j_1 & j & j_2 \\ m_1 & -m & -m_2 \end{pmatrix}$,

(h) $\sqrt{\frac{2j+1}{2j_2+1}}(-1)^{j_1-m_1}\begin{pmatrix} j & j_1 & j_2 \\ m & -m_1 & m_2 \end{pmatrix}$.

By use of the symmetry relations, Clebsch–Gordan coefficients may be put in the standard form $j_1 \le j_2 \le j$ and $m \ge 0$.

m_2	m	j_1	j	$\begin{pmatrix} j_1 & \frac{1}{2} & j \\ m_1 & m_2 & m \end{pmatrix}$	
$-\frac{1}{2}$	0	$\frac{1}{2}$	1	$\frac{\sqrt{2}}{2}$	≈ 0.707107
0	$\frac{1}{2}$	$\frac{1}{2}$	1	$\frac{\sqrt{3}}{2}$	≈ 0.866025
$\frac{1}{2}$	0	$\frac{1}{2}$	1	$\frac{\sqrt{2}}{2}$	≈ 0.707107
$\frac{1}{2}$	$\frac{1}{2}$	$\frac{1}{2}$	1	$\frac{\sqrt{3}}{2}$	≈ 0.866025
$\frac{1}{2}$	1	$\frac{1}{2}$	1	1	≈ 1.000000

m_2	m	j_1	j	$\begin{pmatrix} j_1 & 1 & j \\ m_1 & m_2 & m \end{pmatrix}$	
-1	0	1	1	$\frac{\sqrt{2}}{2}$	≈ 0.707107
-1	0	1	2	$\frac{\sqrt{6}}{6}$	≈ 0.408248
$-\frac{1}{2}$	0	$\frac{1}{2}$	$\frac{3}{2}$	$\frac{\sqrt{2}}{2}$	≈ 0.707107
$-\frac{1}{2}$	$\frac{1}{2}$	1	1	$\frac{3}{4}$	≈ 0.750000
$-\frac{1}{2}$	$\frac{1}{2}$	1	2	$\frac{\sqrt{5}}{4}$	≈ 0.559017
0	0	1	2	$\frac{\sqrt{6}}{3}$	≈ 0.816496
0	0	$\frac{1}{2}$	$\frac{3}{2}$	$\frac{\sqrt{3}}{2}$	≈ 0.866025
0	$\frac{1}{2}$	$\frac{1}{2}$	$\frac{3}{2}$	$\frac{\sqrt{6}}{3}$	≈ 0.8164967
0	$\frac{1}{2}$	1	1	$\frac{\sqrt{2}}{4}$	≈ 0.353553
0	$\frac{1}{2}$	1	2	$\frac{\sqrt{10}}{4}$	≈ 0.790569
0	1	1	1	$\frac{\sqrt{2}}{2}$	≈ 0.707107

m_2	m	j_1	j	$\begin{pmatrix} j_1 & 1 & j \\ m_1 & m_2 & m \end{pmatrix}$	
0	1	1	2	$\frac{\sqrt{2}}{2}$	≈ 0.707107
$\frac{1}{2}$	0	$\frac{1}{2}$	$\frac{3}{2}$	$\frac{\sqrt{2}}{2}$	≈ 0.707107
$\frac{1}{2}$	$\frac{1}{2}$	1	1	$-\frac{\sqrt{2}}{4}$	≈ -0.353553
$\frac{1}{2}$	$\frac{1}{2}$	1	2	$\frac{\sqrt{10}}{4}$	≈ 0.790569
$\frac{1}{2}$	1	$\frac{1}{2}$	$\frac{3}{2}$	$\frac{\sqrt{30}}{6}$	≈ 0.912871
$\frac{1}{2}$	$\frac{3}{2}$	1	2	$\frac{\sqrt{105}}{12}$	≈ 0.853913
1	0	1	1	$-\frac{\sqrt{2}}{2}$	≈ -0.707107
1	0	1	2	$\frac{\sqrt{6}}{6}$	≈ 0.408248
1	$\frac{1}{2}$	$\frac{1}{2}$	$\frac{3}{2}$	$\frac{\sqrt{3}}{3}$	≈ 0.577350
1	$\frac{1}{2}$	1	1	$-\frac{3}{4}$	≈ -0.750000
1	$\frac{1}{2}$	1	2	$\frac{\sqrt{5}}{4}$	≈ 0.559017
1	1	$\frac{1}{2}$	$\frac{3}{2}$	$\frac{\sqrt{10}}{4}$	≈ 0.790569
1	1	1	1	$-\frac{\sqrt{2}}{2}$	≈ -0.707107
1	1	1	2	$\frac{\sqrt{2}}{2}$	≈ 0.707107
1	$\frac{3}{2}$	$\frac{1}{2}$	$\frac{3}{2}$	1	≈ 1.000000
1	$\frac{3}{2}$	1	2	$\frac{\sqrt{105}}{12}$	≈ 0.853913
1	2	1	2	1	≈ 1.000000

6.23   INTEGRAL TRANSFORMS: PRELIMINARIES

1. $I = (a, b)$ is an interval, where $-\infty \leq a < b \leq \infty$.

2. $L^1(I)$ is the set of all absolutely integrable functions on I. In particular, $L^1(\mathbb{R})$ is the set of all absolutely integrable functions on the real line $\mathbb{R}$.

3. $L^2(I)$ is the set of all square integrable functions on I (i.e., $\int_I |f(x)|^2 \, dx < \infty$).

4. If f is integrable over every finite closed subinterval of I, but not necessarily on I itself, we say that f is *locally integrable* on I. For example, the function $f(x) = 1/x$ is not integrable on the interval $I = (0, 1)$, yet it is locally integrable on it.

5. A function $f(x)$, defined on a closed interval $[a, b]$, is said to be of *bounded variation* if there is an $M > 0$ such that, for any partition $a = x_0 < x_1 < \cdots < x_n = b$, the following relation holds: $\sum_{i=1}^{n} |f(x_i) - f(x_{i-1})| \leq M$.

6. If f has a derivative f' at every point of $[a, b]$, then by the mean value theorem, for any $a \leq x < y \leq b$, we have $f(x) - f(y) = f'(z)(x - y)$, for some $x < z < y$. If f' is bounded, then f is of bounded variation.

7. The left limit of a function $f(x)$ at a point t (if it exists) will be denoted by $\lim_{x \to t^-} f(x) = f(t-)$, and likewise the right limit at t will be denoted by $\lim_{x \to t^+} f(x) = f(t+)$.

6.24   FOURIER TRANSFORM

The origin of the Fourier integral transformation can be traced to Fourier's celebrated work on the *Analytical Theory of Heat*, which appeared in 1822. Fourier's major finding was to show that an "arbitrary" function defined on a finite interval could be expanded in a trigonometric series (series of sinusoidal functions, see Section 1.4). In an attempt to extend his results to functions defined on the infinite interval $(-\infty, \infty)$, Fourier introduced what is now known as the Fourier integral transform.

The *Fourier integral transform* of a function $f(t)$ is defined by

$$\mathcal{F}(f)(\omega) = \hat{f}(\omega) = F(\omega) = \frac{1}{\sqrt{2\pi}} \int_{-\infty}^{\infty} f(t) e^{it\omega} \, dt, \qquad (6.24.1)$$

whenever the integral exists.

There is no universal agreement on the definition of the Fourier integral transform. Some authors take the kernel of the transformation as $e^{-it\omega}$, so that the kernel of the inverse transformation is $e^{it\omega}$. In either case, if we define the Fourier transform as

$$\hat{f}(\omega) = a \int_{-\infty}^{\infty} f(t) e^{\pm it\omega}\, dt,$$

then its inverse is $f(t) = b \int_{-\infty}^{\infty} \hat{f}(\omega) e^{\mp it\omega}\, d\omega,$ (6.24.2)

for some constants a and b, with $ab = 1/2\pi$. Again there is no agreement on the choice of the constants; sometimes one of them is taken as 1 so that the other is $1/(2\pi)$. For the sake of symmetry, we choose $a = b = 1/\sqrt{2\pi}$. The functions f and $\hat{f}$ are called a *Fourier transform pair*.

Another definition that is popular in the engineering literature is the one in which the kernel of the transform is taken as $e^{2\pi it\omega}$ (or $e^{-2\pi it\omega}$) so that the kernel of the inverse transform is $e^{-2\pi it\omega}$ (or $e^{2\pi it\omega}$). The main advantage of this definition is that the constants a and b disappear and the Fourier transform pair becomes

$$\hat{f}(\omega) = \int_{-\infty}^{\infty} f(t) e^{\pm 2\pi it\omega}\, dt \quad \text{and} \quad f(t) = \int_{-\infty}^{\infty} \hat{f}(\omega) e^{\mp 2\pi it\omega}\, d\omega. \quad (6.24.3)$$

The Fourier cosine and sine coefficients of $f(t)$ are defined by

$$a(\omega) = \frac{1}{\pi} \int_{-\infty}^{\infty} f(t) \cos\omega t\, dt \quad \text{and} \quad b(\omega) = \frac{1}{\pi} \int_{-\infty}^{\infty} f(t) \sin\omega t\, dt. \quad (6.24.4)$$

The Fourier cosine and sine coefficients are related to the Fourier cosine and sine integral transforms. For example, if f is even, then $a(\omega) = \sqrt{2/\pi} F_c(\omega)$ and, if f is odd, $b(\omega) = \sqrt{2/\pi} F_s(\omega)$ (see Section 6.24.7).

Two other integrals related to the Fourier integral transform are Fourier's repeated integral and the allied integral. *Fourier's repeated integral*, $S(f,t)$, of $f(t)$ is defined by

$$\begin{aligned} S(f,t) &= \int_0^{\infty} (a(\omega) \cos t\omega + b(\omega) \sin t\omega)\, d\omega, \\ &= \frac{1}{\pi} \int_0^{\infty} d\omega \int_{-\infty}^{\infty} f(x) \cos\omega(t-x)\, dx. \end{aligned} \quad (6.24.5)$$

The *allied Fourier integral*, $\tilde{S}(f,t)$, of f is defined by

$$\begin{aligned} \tilde{S}(f,t) &= \int_0^{\infty} (b(\omega) \cos t\omega - a(\omega) \sin t\omega)\, d\omega, \\ &= \frac{1}{\pi} \int_0^{\infty} d\omega \int_{-\infty}^{\infty} f(x) \sin\omega(x-t)\, dx. \end{aligned} \quad (6.24.6)$$

6.24.1   EXISTENCE

For the Fourier integral transform to exist, it is sufficient that f be absolutely integrable on $(-\infty, \infty)$, i.e., $f \in L^1(\mathbb{R})$.

THEOREM 6.24.1 *(Riemann–Lebesgue lemma)*

If $f \in L^1(\mathbb{R})$, then its Fourier transform $\hat{f}(\omega)$ is defined everywhere, uniformly continuous, and tends to zero as $\omega \to \pm\infty$.

The uniform continuity follows from the relationship

$$\left| \hat{f}(\omega + h) - \hat{f}(\omega) \right| \leq \frac{1}{\sqrt{2\pi}} \int_{-\infty}^{\infty} |f(t)| \left| e^{iht} - 1 \right| dt,$$

and the tendency toward zero as $\omega \to \pm\infty$ is a consequence of the Riemann–Lebesgue lemma.

THEOREM 6.24.2 *(Generalized Riemann–Lebesgue lemma)*

Let $f \in L^1(I)$, where $I = (a, b)$ is finite or infinite and let ω be a real variable. Let $a \leq a' < b' \leq b$ and $\hat{f}_\omega(\lambda, a', b') = \int_{a'}^{b'} f(t) e^{i\lambda\omega t} dt$. Then $\lim\limits_{\omega \to \pm\infty} \hat{f}_\omega(\lambda, a', b') = 0$, and the convergence is uniform in a' and b'. In particular, $\lim\limits_{\omega \to \pm\infty} \int_{-\infty}^{\infty} f(t) e^{i\omega t} dt = 0$.

6.24.2 PROPERTIES

1. *Linearity*: The Fourier transform is linear,

$$\mathcal{F}\left[af(t) + bg(t)\right](\omega) = a\mathcal{F}\left[f(t)\right](\omega) + b\mathcal{F}\left[g(t)\right](\omega) = a\hat{f}(\omega) + b\hat{g}(\omega),$$

   where a and b are complex numbers.

2. *Translation*: $\mathcal{F}\left[f(t - b)\right](\omega) = e^{ib\omega}\hat{f}(\omega)$.

3. *Dilation (scaling)*: $\mathcal{F}\left[f(at)\right](\omega) = \frac{1}{a}\hat{f}(\frac{\omega}{a}), \quad a > 0$.

4. *Translation and dilation*:

$$\mathcal{F}\left[f(at - b)\right](\omega) = \frac{1}{a}e^{ib\omega/a}\hat{f}\left(\frac{\omega}{a}\right), \quad a > 0.$$

5. *Complex conjugation*: $\mathcal{F}\left[\overline{f(t)}\right](\omega) = \overline{\hat{f}(-\omega)}$.

6. *Modulation*: $\mathcal{F}\left[e^{iat}f(t)\right](\omega) = \hat{f}(\omega + a)$, and

$$\mathcal{F}\left[e^{iat}f(bt)\right](\omega) = \frac{1}{b}\hat{f}\left(\frac{\omega + a}{b}\right), \quad b > 0. \qquad (6.24.7)$$

7. *Differentiation*: If $f^{(k)} \in L^1(\mathbb{R})$, for $k = 0, 1, 2, \cdots, n$ and $\lim_{|t| \to \infty} f^{(k)}(t) = 0$ for $k = 0, 1, 2, \cdots, n - 1$, then

$$\mathcal{F}\left[f^{(n)}(t)\right](\omega) = (-i\omega)^n \hat{f}(\omega). \qquad (6.24.8)$$

8. *Integration*: Let $f \in L^1(\mathbb{R})$, and define $g(x) = \int_{-\infty}^{x} f(t)\, dt$. If $g \in L^1(\mathbb{R})$, then $\hat{g}(\omega) = -\hat{f}(\omega)/(i\omega)$.

9. *Multiplication by polynomials*: If $t^k f(t) \in L^1(\mathbb{R})$ for $k = 0, 1, \ldots, n$, then

$$\mathcal{F}\left[t^k f(t)\right](\omega) = \frac{1}{(i)^k} \hat{f}^{(k)}(\omega), \qquad (6.24.9)$$

and hence,

$$\mathcal{F}\left[\left(\sum_{k=0}^{n} a_k t^k\right) f(t)\right](\omega) = \sum_{k=0}^{n} \frac{a_k}{(i)^k} \hat{f}^{(k)}(\omega). \qquad (6.24.10)$$

10. *Convolution*: The convolution operation, $\star$, associated with the Fourier transform is defined as

$$h(t) = (f \star g)(t) = \frac{1}{\sqrt{2\pi}} \int_{-\infty}^{\infty} f(x)g(t-x)\, dx,$$

where f and g are defined over the whole real line.

THEOREM 6.24.3

If f and g belong to $L^1(\mathbb{R})$, then so does h. Moreover, $\hat{h}(\omega) = \hat{f}(\omega)\hat{g}(\omega)$. If $\hat{f}$ and $\hat{g}$ belong to $L^1(\mathbb{R})$, then $(\hat{f} \star \hat{g})(\omega) = \widehat{(fg)}(\omega)$.

11. *Parseval's relation*: If $f, g \in L^2(\mathbb{R})$, and if F and G are the Fourier transforms of f and g respectively, then Parseval's relation is

$$\int_{-\infty}^{\infty} F(\omega)G(\omega)\, d\omega = \int_{-\infty}^{\infty} f(t)g(-t)\, dt. \qquad (6.24.11)$$

Replacing G by $\overline{G}$ (so that $g(-t)$ is replaced by $\overline{g}(t)$) results in a more convenient form of Parseval's relation

$$\int_{-\infty}^{\infty} F(\omega)\overline{G}(\omega)\, d\omega = \int_{-\infty}^{\infty} f(t)\overline{g}(t)\, dt. \qquad (6.24.12)$$

In particular, for $f = g$,

$$\int_{-\infty}^{\infty} |F(\omega)|^2\, d\omega = \int_{-\infty}^{\infty} |f(t)|^2\, dt. \qquad (6.24.13)$$

6.24.3   INVERSION FORMULA

Many of the theorems on the inversion of the Fourier transform are based on *Dini's condition* which can be stated as follows:

If $f \in L^1(\mathbb{R})$, then a necessary and sufficient condition for

$$S(f, x) = \lim_{\lambda \to \infty} \frac{1}{\pi} \int_{-\infty}^{\infty} f(t) \frac{\sin \lambda(x - t)}{(x - t)} \, dt = a \qquad (6.24.14)$$

is that

$$\lim_{\lambda \to \infty} \int_0^\delta \left(f(x + y) + f(x - y) - 2a \right) \frac{\sin \lambda y}{y} \, dy = 0, \qquad (6.24.15)$$

for any fixed $\delta > 0$.

By the Riemann–Lebesgue lemma, this condition is satisfied if

$$\int_0^\delta \left| \frac{f(x + y) + f(x - y) - 2a}{y} \right| \, dy < \infty, \qquad (6.24.16)$$

for some $\delta > 0$. In particular, condition (6.24.16) holds for $a = f(x)$, if f is differentiable at x, and for $a = [f(x+) + f(x-)]/2$, if f is of bounded variation in a neighborhood of x.

THEOREM 6.24.4   *(Inversion theorem)*

Let f be a locally integrable function, of bounded variation in a neighborhood of the point x. If f satisfies either one of the following conditions:

1. $f(t) \in L^1(\mathbb{R})$, *or*

2. $f(t)/(1 + |t|) \in L^1(\mathbb{R})$, *and the integral $\int_{-\infty}^{\infty} f(t) e^{i\omega t} \, dt$ converges uniformly on every finite interval of ω,*

then

$$\frac{1}{\sqrt{2\pi}} \int_{-\infty}^{\infty} \hat{f}(\omega) e^{-ix\omega} \, d\omega = \lim_{\lambda \to \infty} \frac{1}{\sqrt{2\pi}} \int_{-\lambda}^{\lambda} \hat{f}(\omega) e^{-ix\omega} \, d\omega$$

is equal to $[f(x+) + f(x-)]/2$ whenever the expression has meaning, to $f(x)$ whenever $f(x)$ is continuous at x, and to $f(x)$ almost everywhere. If f is continuous and of bounded variation in the interval (a, b), then the convergence is uniform in any interval interior to (a, b).

6.24.4   POISSON SUMMATION FORMULA

The Poisson summation formula may be written in the form

$$\frac{1}{\sqrt{2\pi}} \sum_{k=-\infty}^{\infty} f\left(t + \frac{k\pi}{\sigma}\right) = \frac{\sigma}{\pi} \sum_{k=-\infty}^{\infty} \hat{f}(2k\sigma) e^{-2ikt\sigma}, \quad \sigma > 0, \qquad (6.24.17)$$

provided that the two series converge. A sufficient condition for the validity of Equation (6.24.17) is that $f = O(1 + |t|)^{-\alpha}$ as $|t| \to \infty$, and $\hat{f} = O\left((1 + |\omega|)^{-\alpha}\right)$ as $|\omega| \to \infty$ for some $\alpha > 1$.

Another version of the Poisson summation formula is

$$\sum_{k=-\infty}^{\infty} \hat{f}(\omega + k\sigma)\overline{\hat{g}}(\omega + k\sigma) = \frac{1}{\sigma}\sum_{k=-\infty}^{\infty}\left(\int_{-\infty}^{\infty} f(t)\overline{g}\left(t - \frac{2\pi k}{\sigma}\right) dt\right) e^{2\pi i k \omega/\sigma}.$$

(6.24.18)

6.24.5 SHANNON'S SAMPLING THEOREM

If f is a function *band-limited* to $[-\sigma, \sigma]$, i.e.,

$$f(t) = \frac{1}{\sqrt{2\pi}}\int_{-\sigma}^{\sigma} F(\omega)e^{it\omega} \, d\omega,$$

with $F \in L^2(-\sigma, \sigma)$, then it can be reconstructed from its sample values at the points $t_k = (k\pi)/\sigma, k = 0, \pm 1, \pm 2, \cdots$, via the formula

$$f(t) = \sum_{k=-\infty}^{\infty} f(t_k)\frac{\sin \sigma(t - t_k)}{\sigma(t - t_k)},$$

(6.24.19)

with the series absolutely and uniformly convergent on compact sets.

The series on the right-hand side of Equation (6.24.19) can be written as $\sin \sigma t \sum_{k=-\infty}^{\infty} f(t_k)\frac{(-1)^k}{(\sigma t - k\pi)}$, which is a special case of a Cardinal series (these series have the form $\sin \sigma t \sum_{k=-\infty}^{\infty} C_k \frac{(-1)^k}{(\sigma t - k\pi)}$).

6.24.6 UNCERTAINTY PRINCIPLE

Let T and W be two real numbers defined by

$$T^2 = \frac{1}{E}\int_{-\infty}^{\infty} t^2|f(t)|^2 \, dt \quad \text{and} \quad W^2 = \frac{1}{E}\int_{-\infty}^{\infty} \omega^2|\hat{f}(\omega)|^2 \, d\omega,$$

(6.24.20)

where

$$E = \int_{-\infty}^{\infty} |f(t)|^2 \, dt = \int_{-\infty}^{\infty} |\hat{f}(\omega)|^2 \, d\omega.$$

(6.24.21)

Assuming that f is differentiable and $\lim_{|t|\to\infty} tf^2(t) = 0$, then $2TW \geq 1$, or

$$\left(\int_{-\infty}^{\infty} t^2|f(t)|^2 \, dt\right)^{1/2}\left(\int_{-\infty}^{\infty} \omega^2|\hat{f}(\omega)|^2 \, d\omega\right)^{1/2} \geq \frac{1}{2}\int_{-\infty}^{\infty} |f(t)|^2 \, dt. \quad (6.24.22)$$

This means that f and $\hat{f}$ cannot both be very small. Another related property of the Fourier transform is that, if either one of the functions f or $\hat{f}$ vanishes outside some finite interval, then the other one must trail on to infinity. In other words, they can not both vanish outside any finite interval.

6.24.7 FOURIER SINE AND COSINE TRANSFORMS

The *Fourier cosine transform, $F_c(\omega)$,* and the *Fourier sine transform, $F_s(\omega)$,* of $f(t)$ are defined for $\omega > 0$ as

$$
F_c(\omega) = \sqrt{\frac{2}{\pi}} \int_0^\infty f(t) \cos \omega t \, dt \quad \text{and} \quad F_s(\omega) = \sqrt{\frac{2}{\pi}} \int_0^\infty f(t) \sin \omega t \, dt.
$$
(6.24.23)

The inverse transforms have the same functional form:

$$
f(t) = \sqrt{\frac{2}{\pi}} \int_0^\infty F_c(\omega) \cos \omega t \, d\omega = \sqrt{\frac{2}{\pi}} \int_0^\infty F_s(\omega) \sin \omega t \, d\omega.
$$
(6.24.24)

If f is even, i.e., $f(t) = f(-t)$, then $F(\omega) = F_c(\omega)$, and if f is odd, i.e., $f(t) = -f(-t)$, then $F(\omega) = iF_s(\omega)$.

6.25   DISCRETE FOURIER TRANSFORM (DFT)

The *discrete Fourier transform* of the sequence $\{a_n\}_{n=0}^{N-1}$, where $N \geq 1$, is a sequence $\{A_m\}_{m=0}^{N-1}$, defined by

$$
A_m = \sum_{n=0}^{N-1} a_n (W_N)^{mn}, \qquad \text{for} \quad m = 0, 1, \cdots, N-1,
$$
(6.25.1)

where $W_N = e^{2\pi i/N}$. Note that $\sum_{m=0}^{N-1} W_N^{m(k-n)} = N\delta_{kn}$. For example, the DFT of the sequence $\{1, 0, 1, 1\}$ is $\{3, -i, 1, i\}$.

The inversion formula is

$$
a_n = \frac{1}{N} \sum_{m=0}^{N-1} A_m W_N^{-mn}, \quad n = 0, 1, \cdots, N-1.
$$
(6.25.2)

Equations (6.25.1) and (6.25.2) are called a discrete Fourier transform (DFT) pair of order N. The factor $1/N$ and the negative sign in the exponent of W_N that appear in Equation (6.25.2) are sometimes introduced in Equation (6.25.1) instead. We use the notation

$$
\mathcal{F}_N[(a_n)] = A_m \qquad \mathcal{F}_N^{-1}[(A_m)] = a_n,
$$
(6.25.3)

to indicate that the discrete Fourier transform of order N of the sequence $\{a_n\}$ is $\{A_m\}$ and that the inverse transform of $\{A_m\}$ is $\{a_n\}$.

Because $W_N^{\pm(m+N)n} = W_N^{\pm mn}$, Equations (6.25.1) and (6.25.2) can be used to extend the sequences $\{a_n\}_{n=0}^{N-1}$ and $\{A_m\}_{m=0}^{N-1}$, as periodic sequences with period N. This means that $A_{m+N} = A_m$, and $a_{n+N} = a_n$. This will be used in what follows without explicit note. Using this, the summation limits, 0 and $N-1$, can be replaced with n_1 and $n_1 + N - 1$, respectively, where n_1 is any integer. In the special case where $n_1 = -M$ and $N = 2M + 1$, Equations (6.25.1) and (6.25.2) become

$$A_m = \sum_{n=-M}^{M} a_n W_N^{mn}, \qquad \text{for} \quad m = -M, -M+1, \cdots, M-1, M, \quad (6.25.4)$$

and

$$a_n = \frac{1}{2M+1} \sum_{m=-M}^{M} A_m W_N^{-mn}, \qquad \text{for} \quad n = -M, -M+1, \cdots, M-1, M.$$
$$(6.25.5)$$

6.25.1 PROPERTIES

1. *Linearity:* The discrete Fourier transform is linear, that is

$$\mathcal{F}_N[\alpha(a_n) + \beta(b_n)] = \alpha A_m + \beta B_m,$$

for any complex numbers α and β, where the sum of two sequences is defined as $(a_n) + (b_n) = (a_n + b_n)$.

2. *Translation:* $\mathcal{F}_N[(a_{n-k})] = W_N^{mk} A_m$, or $e^{2\pi imk/N} A_m = \sum_{n=0}^{N-1} a_{n-k} W_N^{mn}$.

3. *Modulation:* $\mathcal{F}_N[(W_N^{nk} a_n)] = A_{m+k}$, or $A_{m+k} = \sum_{n=0}^{N-1} e^{2\pi ink/N} a_n W_N^{mn}$.

4. *Complex Conjugation:* $\mathcal{F}_N[(\bar{a}_{-n})] = \bar{A}_m$, or $\bar{A}_m = \sum_{n=0}^{N-1} \bar{a}_{-n} W_N^{mn}$.

5. *Symmetry:* $\mathcal{F}_N[(a_{-n})] = A_{-m}$, or $A_{-m} = \sum_{n=0}^{N-1} a_{-n} W_N^{mn}$.

6. *Convolution*: The convolution of the sequences $\{a_n\}_{n=0}^{N-1}$ and $\{b_n\}_{n=0}^{N-1}$ is the sequence $\{c_n\}_{n=0}^{N-1}$ given by

$$c_n = \sum_{k=0}^{N-1} a_k b_{n-k}. \qquad (6.25.6)$$

The *convolution relation of the DFT* is $\mathcal{F}_N[(c_n)] = \mathcal{F}_N[(a_n)]\mathcal{F}_N[(b_n)]$, or $C_m = A_m B_m$. A consequence of this and Equation (6.25.2), is the relation

$$\sum_{k=0}^{N-1} a_k b_{n-k} = \frac{1}{N} \sum_{m=0}^{N-1} A_m B_m W_N^{-mn}. \qquad (6.25.7)$$

7. *Parseval's relation*:

$$\sum_{n=0}^{N-1} a_n \bar{d}_n = \frac{1}{N} \sum_{m=0}^{N-1} A_m \bar{D}_m. \qquad (6.25.8)$$

In particular,

$$\sum_{n=0}^{N-1} |a_n|^2 = \frac{1}{N} \sum_{m=0}^{N-1} |A_m|^2. \qquad (6.25.9)$$

In (4) and (5), the fact that $\overline{W}_N = W_N^{-1}$ has been used. A sequence $\{a_n\}$ is said to be even if $\{a_{-n}\} = \{a_n\}$ and is said to be odd if $\{a_{-n}\} = \{-a_n\}$. The following are consequences of (4) and (5):

1. If $\{a_n\}$ is a sequence of real numbers (i.e., $\overline{a_n} = a_n$), then $\overline{A_m} = A_{-m}$.

2. $\{a_n\}$ is real and even if and only if $\{A_m\}$ is real and even.

3. $\{a_n\}$ is real and odd if and only if $\{A_m\}$ is pure imaginary and odd.

6.26 FAST FOURIER TRANSFORM (FFT)

To determine A_m for each $m = 0, 1, \cdots, M - 1$ (using Equation (6.25.1)), $M - 1$ multiplications are required. Hence the total number of multiplications required to determine all the A_m's is $(M - 1)^2$. This number can be reduced by using decimation.

Assuming M is even, we define $M = 2N$ and write

$$\mathcal{F}_{2N}[(a_n)] = A_m. \tag{6.26.1}$$

Now split $\{a_n\}$ into two sequences, one consisting of terms with even subscripts ($b_n = a_{2n}$) and one with odd subscripts ($c_n = a_{2n+1}$). Then

$$A_m = B_m + W_{2N}^m C_m. \tag{6.26.2}$$

For the evaluation of B_m and C_m, the total number of multiplications required is $2(N - 1)^2$. To determine A_m from Equation (6.26.2), we must calculate the product $W_{2N}^m C_m$, for each fixed m. Therefore, the total number of multiplications required to determine A_m from Equation (6.26.2) is $2(N - 1)^2 + 2N - 1 = 2N^2 - 2N + 1$.

But if we had determined A_m from (6.26.1), we would have performed $(2N - 1)^2 = 4N^2 - 4N + 1$ multiplications. Thus, splitting the sequence $\{a_n\}$ into two sequences and then applying the discrete Fourier transform reduces the number of multiplications required to evaluate A_m approximately by a factor of 2.

If N is even, this process can be repeated. Split $\{b_n\}$ and $\{c_n\}$ into two sequences, each of length $N/2$. Then B_m and C_m are determined in terms of four discrete Fourier transforms, each of order $N/2$. This process can be repeated $k - 1$ times if $M = 2^k$ for some positive integer k.

If we denote the required number of multiplications for the discrete Fourier transform of order $N = 2^k$ by $F(N)$, then $F(2N) = 2F(N) + N$ and $F(2) = 1$, which leads to $F(N) = \frac{N}{2} \log_2 N$.

6.27   MULTIDIMENSIONAL FOURIER TRANSFORM

If $\mathbf{x} = (x_1, x_2, \ldots, x_n)$ and $\mathbf{u} = (u_1, u_2, \ldots, u_n)$, then (see table on page 604):

1. Fourier transform

$$F(\mathbf{u}) = (2\pi)^{-n/2} \int \cdots \int_{\mathbb{R}^n} f(\mathbf{x}) e^{i(\mathbf{x} \cdot \mathbf{u})} \, d\mathbf{x}.$$

2. Inverse Fourier transform

$$f(\mathbf{x}) = (2\pi)^{-n/2} \int \cdots \int_{\mathbb{R}^n} F(\mathbf{u}) e^{-i(\mathbf{x} \cdot \mathbf{u})} \, d\mathbf{u}.$$

3. Parseval's relation

$$\int \cdots \int_{\mathbb{R}^n} f(\mathbf{x}) \overline{g(\mathbf{x})} \, d\mathbf{x} = \int \cdots \int_{\mathbb{R}^n} F(\mathbf{u}) \overline{G(\mathbf{u})} \, d\mathbf{u}.$$

6.28   LAPLACE TRANSFORM

The Laplace transformation dates back to the work of the French mathematician, Pierre Simon Marquis de Laplace (1749–1827), who used it in his work on probability theory in the 1780's.

The *Laplace transform* of a function $f(t)$ is defined as

$$F(s) = [\mathcal{L}f](s) = \int_0^{\infty} f(t) e^{-st} \, dt \tag{6.28.1}$$

(also written as $[\mathcal{L}f(t)]$ and $[\mathcal{L}(f(t))]$), whenever the integral exists for at least one value of s. The transform variable, s, can be taken as a complex number. We say that f is Laplace transformable or the Laplace transformation is applicable to f if $[\mathcal{L}f]$ exists for at least one value of s. The integral on the right-hand side of Equation (6.28.1) is called the *Laplace integral* of f.

6.28.1   EXISTENCE AND DOMAIN OF CONVERGENCE

Sufficient conditions for the existence of the Laplace transform are

1. f is a locally integrable function on $[0, \infty)$, i.e., $\int_0^a |f(t)| \, dt < \infty$, for any $a > 0$.

2. f is of (real) *exponential type*, i.e., for some constants $M, t_0 > 0$ and real γ, f satisfies

$$|f(t)| \leq M e^{\gamma t}, \qquad \text{for all } t \geq t_0. \tag{6.28.2}$$

If f is a locally integrable function on $[0, \infty)$ and of (real) exponential type γ, then the Laplace integral of f, $\int_0^{\infty} f(t) e^{-st} \, dt$, converges absolutely for $\operatorname{Re} s > \gamma$ and uniformly for $\operatorname{Re} s \geq \gamma_1$, for any $\gamma_1 > \gamma$. Consequently, $F(s)$ is analytic in the half-plane $\Omega = \{s \in \mathbb{C} : \operatorname{Re} s > \gamma\}$. It can be shown that if $F(s)$ exists for some s_0, then it also exists for any s for which $\operatorname{Re} s > \operatorname{Re} s_0$. The actual domain of existence of the Laplace transform may be larger than the one given above. For example, the function $f(t) = \cos e^t$ is of real exponential type zero, but $F(s)$ exists for $\operatorname{Re} s > -1$.

If $f(t)$ is a locally integrable function on $[0, \infty)$, not of exponential type, and

$$\int_0^\infty f(t)e^{-s_0 t}\, dt \tag{6.28.3}$$

converges for some complex number s_0, then the Laplace integral

$$\int_0^\infty f(t)e^{-st}\, dt \tag{6.28.4}$$

converges in the region Re $s >$ Re s_0 and also converges uniformly in the region $|\arg(s - s_0)| \le \theta < \frac{\pi}{2}$. Moreover, if Equation (6.28.3) diverges, then so does Equation (6.28.4) for Re $s <$ Re s_0.

6.28.2 PROPERTIES

1. *Linearity:* $\mathcal{L}(\alpha f + \beta g) = \alpha\mathcal{L}(f) + \beta\mathcal{L}(g) = \alpha F + \beta G$, for any constants α and β.

2. *Dilation:* $[\mathcal{L}\,(f(at))]\,(s) = \dfrac{1}{a}F\left(\dfrac{s}{a}\right)$, for $a > 0$.

3. *Multiplication by exponential functions:*

$$\left[\mathcal{L}\left(e^{at}f(t)\right)\right](s) = F(s - a).$$

4. *Translation:* $[\mathcal{L}\,(f(t - a)H(t - a))]\,(s) = e^{-as}F(s)$ for $a > 0$. This can be put in the form

$$[\mathcal{L}(f(t)H(t - a))]\,(s) = e^{-as}\,[\mathcal{L}(f(t + a))]\,(s),$$

where H is the Heaviside function. Examples:

(a) If

$$g(t) = \begin{cases} 0, & 0 \le t \le a, \\ (t - a)^\nu, & a \le t, \end{cases}$$

then $g(t) = f(t - a)H(t - a)$ where $f(t) = t^\nu\,(\mathrm{Re}\,\nu > -1)$. Since $\mathcal{L}(t^\nu) = \Gamma(\nu + 1)/s^{\nu+1}$, it follows that $(\mathcal{L}g)(s) = e^{-as}\Gamma(\nu + 1)/s^{\nu+1}$, for Re $s > 0$.

(b) If

$$g(t) = \begin{cases} t, & 0 \le t \le a, \\ 0, & a < t, \end{cases}$$

we may write $g(t) = t\,[H(t) - H(t - a)] = tH(t) - (t - a)H(t - a) - aH(t - a)$. Thus by properties (1) and (4),

$$G(s) = \frac{1}{s^2} - \frac{1}{s^2}e^{-as} - \frac{a}{s}e^{-as}.$$

5. *Differentiation of the transformed function*: If f is a differentiable function of exponential type, $\lim_{t \to 0+} f(t) = f(0+)$ exists, and f' is locally integrable on $[0, \infty)$, then the Laplace transform of f' exists, and

$$(\mathcal{L}f')(s) = sF(s) - f(0). \tag{6.28.5}$$

Note that although f is assumed to be of exponential type, f' need not be. For example, $f(t) = \sin e^{t^2}$, but $f'(t) = 2te^{t^2} \cos e^{t^2}$.

6. *Differentiation of higher orders*: Let f be an n times differentiable function so that $f^{(k)}$ (for $k = 0, 1, \ldots, n-1$) are of exponential type with the additional assumption that $\lim_{t \to 0+} f^{(k)}(t) = f^{(k)}(0+)$ exists. If $f^{(n)}$ is locally integrable on $[0, \infty)$, then its Laplace transform exists, and

$$\left[\mathcal{L}\left(f^{(n)}\right)\right](s) = s^n F(s) - s^{n-1} f(0) - s^{n-2} f'(0) - \ldots - f^{(n-1)}(0). \tag{6.28.6}$$

7. *Integration*: If $g(t) = \int_0^t f(x)\,dx$, then (if the transforms exist) $G(s) = F(s)/s$. Repeated applications of this rule result in

$$\left[\mathcal{L}\left(f^{(-n)}\right)\right](s) = \frac{1}{s^n} F(s), \tag{6.28.7}$$

where $f^{(-n)}$ is the nth anti-derivative of f defined by $f^{(-n)}(t) = \int_0^t dt_n \int_0^{t_n} dt_{n-1} \cdots \int_0^{t_2} f(t_1)\,dt_1$. Section 6.28.1 shows that the Laplace transform is an analytic function in a half-plane. Hence it has derivatives of all orders at any point in that half-plane. The next property shows that we can evaluate these derivatives by direct calculation.

8. *Multiplication by powers of t*: Let f be a locally integrable function whose Laplace integral converges absolutely and uniformly for $\text{Re}\ s > \sigma$. Then F is analytic in $\text{Re}\ s > \sigma$, and (for $n = 0, 1, 2, \ldots$, with $\text{Re}\ s > \sigma$)

$$[\mathcal{L}(t^n f(t))](s) = \left(-\frac{d}{ds}\right)^n F(s),$$
$$\left[\mathcal{L}\left(\left(t\frac{d}{dt}\right)^n f(t)\right)\right](s) = \left(-\frac{d}{ds}s\right)^n F(s), \tag{6.28.8}$$

where $\left(t\frac{d}{dt}\right)^n$ is the operator $\left(t\frac{d}{dt}\right)$ applied n times.

9. *Division by powers of t*: If f is a locally integrable function of exponential type such that $f(t)/t$ is a Laplace transformable function, then

$$\left[\mathcal{L}\left(\frac{f(t)}{t}\right)\right](s) = \int_s^\infty F(u)\,du, \tag{6.28.9}$$

or, more generally,

$$\left[\mathcal{L}\left(\frac{f(t)}{t^n}\right)\right](s) = \int_s^\infty \cdots \int_{s_3}^\infty \int_{s_2}^\infty F(s_1)\,ds_1\,ds_2 \cdots ds_n \tag{6.28.10}$$

is the nth repeated integral. It follows from properties (7) and (9) that

$$\left[\mathcal{L}\left(\int_0^t \frac{f(x)}{x}\,dx \right) \right](s) = \frac{1}{s}\int_s^\infty F(u)\,du. \tag{6.28.11}$$

10. *Periodic functions:* Let f be a locally integrable function that is periodic with period T. Then

$$[\mathcal{L}(f)](s) = \frac{1}{(1 - e^{-Ts})}\int_0^T f(t)e^{-st}\,dt. \tag{6.28.12}$$

11. *Hardy's theorem*: If $f(t) = \sum_{n=0}^\infty c_n t^n$ for $t \geq 0$ and $\sum_{n=0}^\infty \frac{c_n n!}{s_0^n}$ converges for some $s_0 > 0$, then $[\mathcal{L}(f)](s) = \sum_{n=0}^\infty \frac{c_n n!}{s^n}$ for Re $s > s_0$.

6.28.3 INVERSION FORMULAE

6.28.3.1   Inversion by integration

If $f(t)$ is a locally integrable function on $[0, \infty)$ such that

1. f is of bounded variation in a neighborhood of a point $t_0 \geq 0$ (a right-hand neighborhood if $t_0 = 0$),

2. The Laplace integral of f converges absolutely on the line Re $s = c$, then

$$\lim_{T \to \infty} \frac{1}{2\pi i}\int_{c-iT}^{c+iT} F(s)e^{st_0}\,ds = \begin{cases} 0, & \text{if } t_0 < 0, \\ f(0+)/2 & \text{if } t_0 = 0, \\ [f(t_0+) + f(t_0-)]/2 & \text{if } t_0 > 0. \end{cases}$$

In particular, if f is differentiable on $(0, \infty)$ and satisfies the above conditions, then

$$\lim_{T \to \infty} \frac{1}{2\pi i}\int_{c-iT}^{c+iT} F(s)e^{st}\,ds = f(t), \quad 0 < t < \infty. \tag{6.28.13}$$

The integral here is taken to be a Cauchy principal value since, in general, this integral may be divergent. For example, if $f(t) = 1$, then $F(s) = 1/s$ and, for $c = 1$ and $t = 0$, the integral $\int_{1-i\infty}^{1+i\infty} \frac{1}{s}\,ds$ diverges.

6.28.3.2 Inversion by partial fractions

Suppose that F is a rational function $F(s) = P(s)/Q(s)$ in which the degree of the denominator Q is greater than that of the numerator P. For instance, let F be represented in its most reduced form where P and Q have no common zeros, and assume that Q has only simple zeros at $a_1, \ldots, a_n$, then

$$f(t) = \mathcal{L}^{-1}\left(F(s)\right)(t) = \mathcal{L}^{-1}\left(\frac{P(s)}{Q(s)}\right)(t) = \sum_{k=1}^{n} \frac{P(a_k)}{Q'(a_k)} e^{a_k t}. \qquad (6.28.14)$$

EXAMPLE If $P(s) = s - 5$ and $Q(s) = s^2 + 6s + 13$, then $a_1 = -3 + 2i$, $a_2 = -3 - 2i$, and it follows that

$$f(t) = \mathcal{L}^{-1}\left(\frac{s-5}{s^2 + 6s + 13}\right) = \frac{(2i-8)}{4i} e^{(-3+2i)t} + \frac{(2i+8)}{4i} e^{(-3-2i)t}$$

$$= e^{-3t}(\cos 2t - 4\sin 2t).$$

6.28.4 CONVOLUTION

Let $f(t)$ and $g(t)$ be locally integrable functions on $[0, \infty)$, and assume that their Laplace integrals converge absolutely in some half-plane Re $s > \alpha$. Then the convolution operation, $\star$, associated with the Laplace transform, is defined by

$$h(t) = (f \star g)(t) = \int_0^t f(x)g(t-x)\, dx. \qquad (6.28.15)$$

The convolution of f and g is a locally integrable function on $[0, \infty)$ that is continuous if either f or g is continuous. Additionally, it has a Laplace transform given by

$$H(s) = (\mathcal{L}h)(s) = F(s)G(s), \qquad (6.28.16)$$

where $(\mathcal{L}f)(s) = F(s)$ and $(\mathcal{L}g)(s) = G(s)$.

6.29 HANKEL TRANSFORM

The *Hankel transform* of order ν of a real-valued function $f(x)$ is defined as

$$\mathcal{H}_\nu(f)(y) = F_\nu(y) = \int_0^\infty f(x)\sqrt{xy}\, J_\nu(yx)\, dx, \qquad (6.29.1)$$

for $y > 0$ and $\nu > -1/2$, where $J_\nu(z)$ is the Bessel function of the first kind of order ν.

The Hankel transforms of order $1/2$ and $-1/2$ are equal to the Fourier sine and cosine transforms, respectively, because

$$J_{1/2}(x) = \sqrt{\frac{2}{\pi x}} \sin x, \qquad J_{-1/2}(x) = \sqrt{\frac{2}{\pi x}} \cos x. \qquad (6.29.2)$$

As with the Fourier transform, there are many variations on the definition of the Hankel transform. Some authors define it as

$$G_\nu(y) = \int_0^\infty x g(x) J_\nu(yx) dx; \tag{6.29.3}$$

however, the two definitions are equivalent; we only need to replace $f(x)$ by $\sqrt{x}g(x)$ and $F_\nu(y)$ by $\sqrt{y}G_\nu(y)$.

6.29.1 PROPERTIES

1. *Existence*: Since $\sqrt{x}J_\nu(x)$ is bounded on the positive real axis, the Hankel transform of f exists if $f \in L^1(0, \infty)$.

2. *Multiplication by x^m*:

$$\mathcal{H}_\nu\left(x^m f(x)\right)(y) = y^{1/2-\nu}\left(\frac{1}{y}\frac{d}{dy}\right)^m \left[y^{\nu+m-1/2}F_{\nu+m}(y)\right].$$

3. *Division by x*:

$$\mathcal{H}_\nu\left(\frac{2\nu}{x}f(x)\right)(y) = y\left[F_{\nu-1}(y) + F_{\nu+1}(y)\right],$$

   and also

$$\mathcal{H}_\nu\left(\frac{f(x)}{x}\right)(y) = y^{1/2-\nu}\int_0^y t^{\nu-1/2}F_{\nu-1}(t)\,dt.$$

4. *Differentiation*:

$$\mathcal{H}_\nu\left(2\nu f'(x)\right)(y) = (\nu-1/2)y F_{\nu+1}(y) - (\nu+1/2)y F_{\nu-1}(y).$$

5. *Differentiation and multiplication by powers of x*:

$$\mathcal{H}_\nu\left[x^{1/2-\nu}\left(\frac{1}{x}\frac{d}{dx}\right)^m\left(x^{\nu+m-1/2}f(x)\right)\right](y) = y^m F_{\nu+m}(y).$$

6. *Parseval's relation*: Let F_ν and G_ν denote the Hankel transforms of order ν of f and g, respectively. Then

$$\int_0^\infty F_\nu(y)G_\nu(y)dy = \int_0^\infty f(x)g(x)dx. \tag{6.29.4}$$

   In particular,

$$\int_0^\infty |F_\nu(y)|^2\,dy = \int_0^\infty |f(x)|^2 dx. \tag{6.29.5}$$

7. *Inversion formula*: If f is absolutely integrable on $(0, \infty)$ and of bounded variation in a neighborhood of point x, then

$$\int_0^\infty F_\nu(y)\sqrt{xy}J_\nu(yx)dy = \frac{f(x+) + f(x-)}{2}, \tag{6.29.6}$$

   whenever the expression on the right-hand side of the equation has a meaning; the integral converges to $f(x)$ whenever f is continuous at x.

6.30 HARTLEY TRANSFORM

Define the function $\operatorname{cas} x = \cos x + \sin x$. The *Hartley transform* of the real function $g(t)$ is

$$(Hg)(\omega) = \int_{-\infty}^{\infty} g(t) \operatorname{cas}(2\pi\omega t)\, dt. \tag{6.30.1}$$

Let $(Eg)(\omega)$ and $(Og)(\omega)$ be the even and odd parts of $(Hg)(\omega)$,

$$\begin{aligned}
(Eg)(\omega) &= \frac{1}{2}\left((Hg)(\omega) + (Hg)(-\omega)\right), \\
(Og)(\omega) &= \frac{1}{2}\left((Hg)(\omega) - (Hg)(-\omega)\right),
\end{aligned} \tag{6.30.2}$$

so that $(Hg)(\omega) = (Eg)(\omega) + (Og)(\omega)$. The Fourier transform of g (using the kernel $e^{2\pi i\omega t}$) can then be written in terms of $(Eg)(\omega)$ and $(Og)(\omega)$ as

$$\int_{-\infty}^{\infty} g(t)e^{2\pi i\omega t}\, dt = (Eg)(\omega) + i(Og)(\omega). \tag{6.30.3}$$

Note that the Hartley transform, applied twice in succession, returns the original function.

6.31 HILBERT TRANSFORM

The *Hilbert transform* of f is defined as

$$(\mathcal{H}f)(x) = \tilde{f}(x) = \frac{1}{\pi}\int_{-\infty}^{\infty} \frac{f(t)}{t-x}\, dt = \frac{1}{\pi}\int_{-\infty}^{\infty} \frac{f(x+t)}{t}\, dt \tag{6.31.1}$$

where the integral is a Cauchy principal value. A table is on page 612.

Since the definition is given in terms of a singular integral, it is sometimes impractical to use. An alternative definition is given below. First, let f be an integrable function, and define $a(t)$ and $b(t)$ by

$$a(t) = \frac{1}{\pi}\int_{-\infty}^{\infty} f(x)\cos tx\, dx, \qquad b(t) = \frac{1}{\pi}\int_{-\infty}^{\infty} f(x)\sin tx\, dx. \tag{6.31.2}$$

Consider the function $F(z)$, defined by the integral

$$F(z) = \int_{0}^{\infty} (a(t) - ib(t))e^{izt}\, dt = U(z) + i\tilde{U}(z), \tag{6.31.3}$$

where $z = x + iy$. The real and imaginary parts of F are

$$\begin{aligned}
U(z) &= \int_{0}^{\infty} (a(t)\cos xt + b(t)\sin xt)e^{-yt}\, dt, \quad \text{and} \\
\tilde{U}(z) &= \int_{0}^{\infty} (a(t)\sin xt - b(t)\cos xt)e^{-yt}\, dt.
\end{aligned} \tag{6.31.4}$$

Formally,

$$\lim_{y \to 0} U(z) = f(x) = \int_0^\infty (a(t) \cos xt + b(t) \sin xt)\, dt, \qquad (6.31.5)$$

and

$$\lim_{y \to 0} \tilde{U}(z) = -\tilde{f}(x) = \int_0^\infty (a(t) \sin xt - b(t) \cos xt)\, dt. \qquad (6.31.6)$$

The Hilbert transform of a function f, given by Equation (6.31.5), is defined as the function $\tilde{f}$ given by Equation (6.31.6).

6.31.1 EXISTENCE

If $f \in L^1(\mathbb{R})$, then its Hilbert transform $(\mathcal{H}f)(x)$ exists for almost all x. For $f \in L^p(\mathbb{R}), p > 1$, there is the following stronger result:

THEOREM 6.31.1

Let $f \in L^p(\mathbb{R})$ for $1 < p < \infty$. Then $(\mathcal{H}f)(x)$ exists for almost all x and defines a function that also belongs to $L^p(\mathbb{R})$ with

$$\int_{-\infty}^\infty |(\mathcal{H}f)(x)|^p\, dx \le C_p \int_{-\infty}^\infty |f(x)|^p\, dx. \qquad (6.31.7)$$

In the special case of $p = 2$, we have

$$\int_{-\infty}^\infty |(\mathcal{H}f)(x)|^2\, dx = \int_{-\infty}^\infty |f(x)|^2\, dx. \qquad (6.31.8)$$

The theorem is not valid if $p = 1$ because, although it is true that $(\mathcal{H}f)(x)$ is defined almost everywhere, it is not necessarily in $L^1(\mathbb{R})$. The function $f(t) = (t \log^2 t)^{-1} H(t)$ provides a counterexample.

6.31.2 PROPERTIES

1. *Translation*: The Hilbert transform commutes with the translation operator $(\mathcal{H}f)(x + a) = \mathcal{H}(f(t + a))(x)$.

2. *Dilation*: The Hilbert transformation also commutes with the dilation operator

$$(\mathcal{H}f)(ax) = \mathcal{H}(f(at))(x) \quad a > 0,$$

   but

$$(\mathcal{H}f)(ax) = -\mathcal{H}(f(at))(x) \quad \text{for } a < 0.$$

3. *Multiplication by t*: $\mathcal{H}(tf(t))(x) = x(\mathcal{H}f)(x) + \dfrac{1}{\pi} \displaystyle\int_{-\infty}^\infty f(t)\, dt.$

4. *Differentiation*: $\mathcal{H}(f'(t))(x) = (\mathcal{H}f)'(x)$, provided that $f(t) = O(t)$ as $|t| \to \infty$.

5. *Orthogonality*: The Hilbert transform of $f \in L^2(\mathbb{R})$ is orthogonal to f in the sense $\int_{-\infty}^{\infty} f(x)(\mathcal{H}f)(x)dx = 0$.

6. *Parity*: The Hilbert transform of an even function is odd and that of an odd function is even.

7. *Inversion formula*: If $(\mathcal{H}f)(x) = \dfrac{1}{\pi} \displaystyle\int_{-\infty}^{\infty} \dfrac{f(t)}{t-x} dt$, then

$$f(t) = -\frac{1}{\pi} \int_{-\infty}^{\infty} \frac{(\mathcal{H}f)(x)}{x-t} dx \text{ or, symbolically,}$$

$$\mathcal{H}\left(\mathcal{H}f\right)(x) = -f(x), \tag{6.31.9}$$

that is, applying the Hilbert transform twice returns the negative of the original function. Moreover, if $f \in L^1(\mathbb{R})$ has a bounded derivative, then the allied integral (see Equation 6.24.6) equals $(\mathcal{H}f)(x)$.

8. *Meromorphic invariance*:

$$(\mathcal{H}f)(u(x)) = \tilde{f}(u(x)) = \frac{1}{\pi} \int_{-\infty}^{\infty} \frac{f(t)}{t-u(x)} dt = \frac{1}{\pi} \int_{-\infty}^{\infty} \frac{f[u(t)]}{t-x} dt$$

where

$$u(t) = t - \sum_{n=1}^{\infty} \frac{a_n}{t-b_n},$$

for arbitrary $a_n \geq 0$ and b_n real.

6.31.3 RELATIONSHIP WITH THE FOURIER TRANSFORM

From Equations (6.31.3)–(6.31.6), we obtain

$$\lim_{y \to 0} F(z) = F(x) = \int_0^{\infty} (a(t) - ib(t))e^{ixt} dt = f(x) - i(\mathcal{H}f)(x),$$

where $a(t)$ and $b(t)$ are given by Equation (6.31.2).

Let g be a real-valued integrable function and consider its Fourier transform $\hat{g}(x) = \frac{1}{\sqrt{2\pi}} \int_{-\infty}^{\infty} g(t)e^{ixt} dt$. If we denote the real and imaginary parts of $\hat{g}$ by f and $\tilde{f}$, respectively, then

$$f(x) = \frac{1}{\sqrt{2\pi}} \int_{-\infty}^{\infty} g(t) \cos xt \, dt, \quad \text{and}$$

$$\tilde{f}(x) = \frac{1}{\sqrt{2\pi}} \int_{-\infty}^{\infty} g(t) \sin xt \, dt.$$

Splitting g into its even and odd parts, g_e and g_o, respectively, we obtain

$$g_e(t) = \frac{g(t) + g(-t)}{2} \quad \text{and} \quad g_o(t) = \frac{g(t) - g(-t)}{2};$$

hence

$$f(x) = \sqrt{\frac{2}{\pi}} \int_0^\infty g_e(t) \cos xt \, dt, \quad \text{and} \quad \tilde{f}(x) = \sqrt{\frac{2}{\pi}} \int_0^\infty g_o(t) \sin xt \, dt,$$

$$(6.31.10)$$

or

$$f(x) = \frac{1}{\sqrt{2\pi}} \int_{-\infty}^\infty g_e(t) e^{ixt} \, dt, \quad \text{and} \quad \tilde{f}(x) = \frac{-i}{\sqrt{2\pi}} \int_{-\infty}^\infty g_o(t) e^{ixt} \, dt.$$

$$(6.31.11)$$

This shows that, if the Fourier transform of the even part of a real-valued function represents a function $f(x)$, then the Fourier transform of the odd part represents the Hilbert transform of f (up to multiplication by i).

THEOREM 6.31.2

Let $f \in L^1(\mathbb{R})$ and assume that $\mathcal{H}f$ is also in $L^1(\mathbb{R})$. Then

$$\mathcal{F}(\mathcal{H}f)(\omega) = -i \, \mathrm{sgn}(\omega) \mathcal{F}(f)(\omega), \tag{6.31.12}$$

where $\mathcal{F}$ denotes the Fourier transformation. Similarly, if $f \in L^2(\mathbb{R})$, then $(\mathcal{H}f) \in L^2(\mathbb{R})$, and Equation (6.31.12) remains valid.

6.32   Z-TRANSFORM

The *Z-transform* of a sequence $\{f(n)\}_{-\infty}^\infty$ is defined by

$$\mathcal{Z}[f(n)] = F(z) = \sum_{n=-\infty}^\infty f(n) z^{-n}, \tag{6.32.1}$$

for all complex numbers z for which the series converges.

The series converges at least in a ring of the form $0 \le r_1 < |z| < r_2 \le \infty$, whose radii, r_1 and r_2, depend on the behavior of $f(n)$ at $\pm\infty$:

$$r_1 = \limsup_{n \to \infty} \sqrt[n]{|f(n)|}, \qquad r_2 = \liminf_{n \to \infty} \frac{1}{\sqrt[n]{|f(-n)|}}. \tag{6.32.2}$$

If there is more than one sequence involved, we may denote r_1 and r_2 by $r_1(f)$ and $r_2(f)$ respectively. It may happen that $r_1 > r_2$, so that the function is nowhere defined. The function $F(z)$ is analytic in this ring, but it may be possible to continue it analytically beyond the boundaries of the ring. If $f(n) = 0$ for $n < 0$, then $r_2 = \infty$, and if $f(n) = 0$ for $n \geq 0$, then $r_1 = 0$.

Let $z = re^{i\theta}$. Then the Z-transform evaluated at $r = 1$ is the Fourier transform of the sequence $\{f(n)\}_{-\infty}^{\infty}$,

$$\sum_{n=-\infty}^{\infty} f(n)e^{-in\theta}. \tag{6.32.3}$$

6.32.1 EXAMPLES

1. Let a be a complex number and define $f(n) = a^n$, for $n \geq 0$, and zero otherwise, then

$$\mathcal{Z}[f(n)] = \sum_{n=0}^{\infty} a^n z^{-n} = \frac{z}{z-a} \quad , \quad |z| > |a|. \tag{6.32.4}$$

Special case: unit step function. If $a = 1$ then $f(n) = u(n) = \begin{cases} 1, & n \geq 0, \\ 0, & n < 0, \end{cases}$

and $\mathcal{Z}[u(n)] = \dfrac{z}{z-1}$.

2. If $f(n) = na^n$, for $n \geq 0$, and zero otherwise, then

$$\mathcal{Z}[f(n)] = \sum_{n=0}^{\infty} na^n z^{-n} = \frac{az}{(z-a)^2} \quad , \quad |z| > |a|.$$

3. Let $\delta(n) = \begin{cases} 1, & n = 0, \\ 0, & \text{otherwise,} \end{cases}$ then $\mathcal{Z}[\delta(n-k)] = z^{-k}$ for $k = 0, \pm 1, \pm 2, \ldots$.

6.32.2 PROPERTIES

Let the region of convergence of the Z-transform of the sequence $\{f(n)\}$ be denoted by D_f.

1. *Linearity*:

$$\mathcal{Z}[af(n)+bg(n)] = a\mathcal{Z}[f(n)]+b\mathcal{Z}[g(n)] = aF(z)+bG(z), \quad z \in D_f \cap D_g.$$

The region $D_f \cap D_g$ contains the ring $r_1 < |z| < r_2$, where $r_1 = \text{maximum } \{r_1(f), r_1(g)\}$ and $r_2 = \text{minimum } \{r_2(f), r_2(g)\}$.

2. *Translation*: $\mathcal{Z}[f(n-k)] = z^{-k}F(z)$.

3. *Multiplication by exponentials*: $\mathcal{Z}[(a^n f(n))] = F(z/a)$ when $|a|r_1 < |z| < |a|r_2$.

4. *Multiplication by powers of n:* For $k = 0, 1, 2, \ldots$ and $z \in D_f$,

$$\mathcal{Z}[(n^k f(n))] = (-1)^k \left(z \frac{d}{dz} \right)^k F(z). \tag{6.32.5}$$

5. *Complex conjugation:* $\mathcal{Z}[\overline{f}(-n)] = \overline{F} \left(\dfrac{1}{z} \right).$

6. *Initial and final values:* If $f(n) = 0$ for $n < 0$, then $\lim_{z \to \infty} F(z) = f(0)$ and, conversely, if $F(z)$ is defined for $r_1 < |z|$ and for some integer m, $\lim_{z \to \infty} z^m F(z) = A$ (with $A \neq \pm\infty$), then $f(m) = A$ and $f(n) = 0$, for $n < m$.

7. *Parseval's relation:* Let $F, G \in L^2(-\pi, \pi)$, and let $F(z)$ and $G(z)$ be the Z-transforms of $\{f(n)\}$ and $\{g(n)\}$, respectively. Then

$$\sum_{n=-\infty}^{\infty} f(n)\overline{g}(n) = \frac{1}{2\pi} \int_{-\pi}^{\pi} F(e^{i\omega})\overline{G}(e^{i\omega}) \, d\omega. \tag{6.32.6}$$

In particular,

$$\sum_{n=-\infty}^{\infty} |f(n)|^2 = \frac{1}{2\pi} \int_{-\pi}^{\pi} |F(e^{i\omega})|^2 \, d\omega. \tag{6.32.7}$$

6.32.3   INVERSION FORMULA

Consider the sequences

$$f(n) = u(n) = \begin{cases} 1, & n \geq 0, \\ 0, & n < 0, \end{cases} \quad \text{and} \quad g(n) = -u(-n-1) = \begin{cases} -1, & n < 0, \\ 0, & n \geq 0. \end{cases}$$

Note that $F(z) = \dfrac{z}{z-1}$ for $|z| > 1$, and $G(z) = \dfrac{z}{z-1}$ for $|z| < 1$. Hence, the inverse Z-transform of the function $z/(z-1)$ is not unique. In general, the inverse Z-transform is not unique, unless its region of convergence is specified.

1. *Inversion by using series representation:*

   If $F(z)$ is given by its series

   $$F(z) = \sum_{n=-\infty}^{\infty} a_n z^{-n}, \quad r_1 < |z| < r_2,$$

   then its inverse Z-transform is unique and is given by $f(n) = a_n$ for all n.

2. *Inversion by using complex integration*:

   If $F(z)$ is given in a closed form as an algebraic expression and its domain of analyticity is known, then its inverse Z-transform can be obtained by using the relationship

   $$f(n) = \frac{1}{2\pi i} \oint_\gamma F(z) z^{n-1} dz, \tag{6.32.8}$$

   where γ is a closed contour surrounding the origin once in the positive (counter-clockwise) direction in the domain of analyticity of $F(z)$.

3. *Inversion by using Fourier series*:

   If the domain of analyticity of F contains the unit circle, $|r| = 1$, and if F is single valued therein, then $F(e^{i\theta})$ is a periodic function with period 2π, and, consequently, it can be expanded in a Fourier series. The coefficients of the series form the inverse Z-transform of F and they are given explicitly by

   $$f(n) = \frac{1}{2\pi} \int_{-\pi}^{\pi} F(e^{i\theta}) e^{in\theta} d\theta. \tag{6.32.9}$$

   (This is a special case of (2) with $\gamma(\theta) = e^{i\theta}$.)

4. *Inversion by using partial fractions*:

   Dividing Equation (6.32.4) by z and differentiating both sides with respect to z results in

   $$Z^{-1} \left[(z-a)^{-k} \right] = \binom{n-1}{n-k} a^{n-k} u(n-k), \tag{6.32.10}$$

   for $k = 1, 2, \ldots$ and $|z| > |a| > 0$. Moreover,

   $$Z^{-1} \left[z^{-k} \right] = \delta(n-k). \tag{6.32.11}$$

   Let $F(z)$ be a rational function of the form

   $$F(z) = \frac{P(z)}{Q(z)} = \frac{a_N z^N + \ldots + a_1 z + a_0}{b_M z^M + \ldots + b_1 z + b_0}$$

   with $a_N \neq 0$ and $b_M \neq 0$.

   (a) Consider the case $N < M$. The denominator $Q(z)$ can be factored over the field of complex numbers as $Q(z) = c(z - z_1)^{k_1} \cdots (z - z_m)^{k_m}$, where c is a constant and $k_1, \ldots, k_m$ are positive integers satisfying $k_1 + \ldots + k_m = M$. Hence, F can be written in the form

   $$F(z) = \sum_{i=1}^{m} \sum_{j=1}^{k_i} \frac{A_{i,j}}{(z - z_i)^j}, \tag{6.32.12}$$

   where

   $$A_{i,j} = \frac{1}{(k_i - j)!} \lim_{z \to z_i} \frac{d^{k_i - j}}{dz^{k_i - j}} (z - z_i)^{k_i} F(z). \tag{6.32.13}$$

The inverse Z-transform of the decomposition in Equation (6.32.12) in the region that is exterior to the smallest circle containing all the zeros of $Q(z)$ can be obtained by using Equation (6.32.10).

(b) Consider the case $N \geq M$. We must divide until F can be reduced to the form

$$F(z) = H(z) + \frac{R(z)}{Q(z)},$$

where the remainder polynomial, $R(z)$, has degree less than or equal to $M - 1$, and the quotient, $H(z)$, is a polynomial of degree, at most, $N - M$. The inverse Z-transform of the quotient polynomial can be obtained by using Equation (6.32.11) and that of $R(z)/Q(z)$ can be obtained as in the case $N < M$.

EXAMPLE   To find the inverse Z-transform of the function,

$$F(z) = \frac{z^4 + 5}{(z - 1)^2(z - 2)}, \qquad |z| > 2, \tag{6.32.14}$$

the partial fraction expansion,

$$\frac{z^4 + 5}{(z - 1)^2(z - 2)} = z + 4 - \frac{6}{(z - 1)^2} - \frac{10}{(z - 1)} + \frac{21}{(z - 2)}, \tag{6.32.15}$$

is computed. With the aid of Equation (6.32.10) and Equation (6.32.11),

$$Z^{-1}[F(z)] = \delta(n + 1) + 4\delta(n) - 6(n - 1)u(n - 2)$$
$$- 10u(n - 1) + 21 \cdot 2^{n-1}u(n - 1), \tag{6.32.16}$$

or $f(n) = -6n - 4 + 21 \cdot 2^{n-1}$, for $n \geq 2$, with the initial values $f(-1) = 1$, $f(0) = 4$, and $f(1) = 11$.

6.32.4   CONVOLUTION AND PRODUCT

The convolution of the two sequences, $\{f(n)\}_{-\infty}^{\infty}$ and $\{g(n)\}_{-\infty}^{\infty}$ is the sequence $\{h(n)\}_{-\infty}^{\infty}$ defined by $h(n) = \sum_{k=-\infty}^{\infty} f(k)g(n - k)$. The Z-transform of the convolution of two sequences is the product of their Z-transforms,

$$Z[h(n)] = Z[f(n)]Z[g(n)], \tag{6.32.17}$$

for $z \in D_f \cap D_g$, or $H(z) = F(z)G(z)$.

The Z-transform of the product of two sequences is given by

$$Z[f(n)g(n)] = \frac{1}{2\pi i} \oint_\gamma F(\omega)G\left(\frac{z}{\omega}\right)\frac{d\omega}{\omega}, \tag{6.32.18}$$

where γ is a closed contour surrounding the origin in the positive direction in the domain of convergence of $F(\omega)$ and $G(z/\omega)$.

6.33   TABLES OF TRANSFORMS

Finite sine transforms

$$f_s(n) = \int_0^\pi F(x) \sin nx \, dx, \text{ for } n = 1, 2, \dots.$$

No.	$f_s(n)$	$F(x)$
1	$(-1)^{n+1} f_s(n)$	$F(\pi - x)$
2	$1/n$	$\pi - x/\pi$
3	$(-1)^{n+1}/n$	x/π
4	$1 - (-1)^n/n$	1
5	$\dfrac{2}{n^2} \sin \dfrac{n\pi}{2}$	$\begin{cases} x & \text{when } 0 < x < \pi/2 \\ \pi - x & \text{when } \pi/2 < x < \pi \end{cases}$
6	$(-1)^{n+1}/n^3$	$x(\pi^2 - x^2)/6\pi$
7	$1 - (-1)^n/n^3$	$x(\pi - x)/2$
8	$\dfrac{\pi^2 (-1)^{n-1}}{n} - \dfrac{2[1 - (-1)^n]}{n^3}$	x^2
9	$\pi(-1)^n \left(\dfrac{6}{n^3} - \dfrac{\pi^2}{n} \right)$	x^3
10	$\dfrac{n}{n^2 + c^2} [1 - (-1)^n e^{c\pi}]$	e^{cx}
11	$\dfrac{n}{n^2 + c^2}$	$\dfrac{\sinh c(\pi - x)}{\sinh c\pi}$
12	$\dfrac{n}{n^2 - k^2}$ with $k \neq 0, 1, 2, \dots$	$\dfrac{\sin k(\pi - x)}{\sin k\pi}$
13	$\begin{cases} \pi/2 & \text{when } n = m \\ 0 & \text{when } n \neq m, m = 1, 2, \dots \end{cases}$	$\sin mx$
14	$\dfrac{n}{n^2 - k^2} [1 - (-1)^n \cos k\pi]$ with $k \neq 1, 2, \dots$ (0 if $n = k$)	$\cos kx$
15	$\dfrac{n}{(n^2 - k^2)^2}$ with $k \neq 0, 1, 2, \dots$	$\dfrac{\pi \sin kx}{2k \sin^2 k\pi} - \dfrac{x \cos k(\pi - x)}{2k \sin k\pi}$
16	$\dfrac{b^n}{n}$ with $|b| \leq 1$	$\dfrac{2}{\pi} \tan^{-1} \dfrac{b \sin x}{1 - b \cos x}$
17	$\dfrac{1 - (-1)^n}{n} b^n$ with $|b| \leq 1$	$\dfrac{2}{\pi} \tan^{-1} \dfrac{2b \sin x}{1 - b^2}$

Finite cosine transforms

$$f_c(n) = \int_0^\pi F(x) \cos nx \, dx, \text{ for } n = 0, 1, 2, \dots.$$

No.	$f_c(n)$	$F(x)$
1	$(-1)^n f_c(n)$	$F(\pi - x)$
2	$\begin{cases} \pi & n = 0 \\ 0 & n = 1, 2, \dots \end{cases}$	1
3	$\begin{cases} 0 & n = 0 \\ \dfrac{2}{n}\sin\dfrac{n\pi}{2} & n = 1, 2, \dots \end{cases}$	$\begin{cases} 1 & \text{for } 0 < x < \pi/2 \\ -1 & \text{for } \pi/2 < x < \pi \end{cases}$
4	$\begin{cases} \dfrac{\pi^2}{2} & n = 0 \\ (-1)^n - 1/n^2 & n = 1, 2, \dots \end{cases}$	x
5	$\begin{cases} \dfrac{\pi^2}{6} & n = 0 \\ (-1)^n/n^2 & n = 1, 2, \dots \end{cases}$	$\dfrac{x^2}{2\pi}$
6	$\begin{cases} 0 & n = 0 \\ 1/n^2 & n = 1, 2, \dots \end{cases}$	$\dfrac{(x-\pi)^2}{2\pi} - \dfrac{\pi}{6}$
7	$\begin{cases} \dfrac{\pi^4}{4} & n = 0 \\ 3\pi^2\dfrac{(-1)^n}{n^2} - 6\dfrac{1-(-1)^n}{n^4} & \\ & n = 1, 2, \dots \end{cases}$	x^3
8	$\dfrac{(-1)^n e^c \pi - 1}{n^2 + c^2}$	$\frac{1}{c} e^{cx}$
9	$\dfrac{1}{n^2 + c^2}$	$\dfrac{\cosh c(\pi - x)}{c \sinh c\pi}$
10	$\dfrac{k}{n^2 - k^2}[(-1)^n \cos \pi k - 1]$ with $k \neq 0, 1, 2, \dots$	$\sin kx$
11	$\begin{cases} 0 & m = 1, 2, \dots \\ \dfrac{(-1)^{n+m} - 1}{n^2 - m^2} & m \neq 1, 2, \dots \end{cases}$	$\dfrac{1}{m}\sin mx$
12	$\dfrac{1}{n^2 - k^2}$ with $k \neq 0, 1, 2, \dots$	$-\dfrac{\cos k(\pi - x)}{k \sin k\pi}$
13	$\begin{cases} \pi/2 & \text{when } n = m \\ 0 & \text{when } n \neq m \end{cases}$	$\cos mx \qquad (m = 1, 2, \dots)$

Fourier sine transforms

$$F(\omega) = F_s(f)(\omega) = \sqrt{\frac{2}{\pi}} \int_0^\infty f(x) \sin(\omega x)\, dx, \quad \omega > 0.$$

No.	$f(x)$	$F(\omega)$
1	$\begin{cases} 1 & 0 < x < a \\ 0 & x > a \end{cases}$	$\sqrt{\dfrac{2}{\pi}} \left(\dfrac{1 - \cos \omega a}{\omega} \right)$
2	$x^{p-1} \qquad (0 < p < 1)$	$\sqrt{\dfrac{2}{\pi}} \dfrac{\Gamma(p)}{\omega^p} \sin \dfrac{p\pi}{2}$
3	$\begin{cases} \sin x & 0 < x < a \\ 0 & x > a \end{cases}$	$\dfrac{1}{\sqrt{2\pi}} \left(\dfrac{\sin[a(1-\omega)]}{1-\omega} - \dfrac{\sin[a(1+\omega)]}{1+\omega} \right)$
4	e^{-x}	$\sqrt{\dfrac{2}{\pi}} \dfrac{\omega}{1+\omega^2}$
5	$xe^{-x^2/2}$	$\omega e^{-\omega^2/2}$
6	$\cos \dfrac{x^2}{2}$	$\sqrt{2}\left[\sin \dfrac{\omega^2}{2} C\left(\dfrac{\omega^2}{2} \right) - \cos \dfrac{\omega^2}{2} S\left(\dfrac{\omega^2}{2} \right) \right]$
7	$\sin \dfrac{x^2}{2}$	$\sqrt{2}\left[\cos \dfrac{\omega^2}{2} C\left(\dfrac{\omega^2}{2} \right) + \sin \dfrac{\omega^2}{2} S\left(\dfrac{\omega^2}{2} \right) \right]$

Fourier cosine transforms

$$F(\omega) = F_c(f)(\omega) = \sqrt{\frac{2}{\pi}} \int_0^\infty f(x) \cos(\omega x)\, dx, \quad \omega > 0.$$

No.	$f(x)$	$F(\omega)$
1	$\begin{cases} 1 & 0 < x < a \\ 0 & x > a \end{cases}$	$\sqrt{\dfrac{2}{\pi}} \dfrac{\sin a\omega}{\omega}$
2	$x^{p-1} \qquad (0 < p < 1)$	$\sqrt{\dfrac{2}{\pi}} \dfrac{\Gamma(p)}{\omega^p} \cos \dfrac{p\pi}{2}$
3	$\begin{cases} \cos x & 0 < x < a \\ 0 & x > a \end{cases}$	$\dfrac{1}{\sqrt{2\pi}} \left(\dfrac{\sin[a(1-\omega)]}{1-\omega} + \dfrac{\sin[a(1+\omega)]}{1+\omega} \right)$
4	e^{-x}	$\sqrt{\dfrac{2}{\pi}} \dfrac{1}{1+\omega^2}$
5	$e^{-x^2/2}$	$e^{-\omega^2/2}$
6	$\cos \dfrac{x^2}{2}$	$\cos \left(\dfrac{\omega^2}{2} - \dfrac{\pi}{4} \right)$
7	$\sin \dfrac{x^2}{2}$	$\cos \left(\dfrac{\omega^2}{2} + \dfrac{\pi}{4} \right)$

Fourier transforms: functional relations

$$F(\omega) = \mathcal{F}(f)(\omega) = \frac{1}{\sqrt{2\pi}} \int_{-\infty}^{\infty} f(x) e^{i\omega x}\, dx$$

No.	$f(x)$	$F(\omega)$
1	$ag(x) + bh(x)$	$aG(\omega) + bH(\omega)$
2	$f(ax)$ $\qquad a \neq 0, \operatorname{Im} a = 0$	$\frac{1}{|a|} F\left(\frac{\omega}{a}\right)$
3	$f(-x)$	$F(-\omega)$
4	$\overline{f(x)}$	$\overline{F(-\omega)}$
5	$f(x - \tau)$ $\qquad \operatorname{Im} \tau = 0$	$e^{i\omega\tau} F(\omega)$
6	$e^{i\Omega x} f(x)$ $\qquad \operatorname{Im} \Omega = 0$	$F(\omega + \Omega)$
7	$F(x)$	$f(-\omega)$
8	$\frac{d^n}{dx^n} f(x)$	$(-i\omega)^n F(\omega)$
9	$(ix)^n f(x)$	$\frac{d^n}{d\omega^n} F(\omega)$
10	$\frac{\partial}{\partial a} f(x, a)$	$\frac{\partial}{\partial a} F(\omega, a)$

Fourier transforms

$$F(\omega) = \mathcal{F}(f)(\omega) = \frac{1}{\sqrt{2\pi}} \int_{-\infty}^{\infty} f(x) e^{i\omega x}\, dx$$

No.	$f(x)$	$F(\omega)$
1	$\delta(x)$	$1/\sqrt{2\pi}$
2	$\delta(x - \tau)$	$e^{i\omega\tau}/\sqrt{2\pi}$
3	$\delta^{(n)}(x)$	$(-i\omega)^n/\sqrt{2\pi}$
4	$H(x) = \begin{cases} 1 & x > 0 \\ 0 & x < 0 \end{cases}$	$-\dfrac{1}{i\omega\sqrt{2\pi}} + \sqrt{\dfrac{\pi}{2}}\delta(\omega)$
5	$\operatorname{sgn}(x) = \begin{cases} 1 & x > 0 \\ -1 & x < 0 \end{cases}$	$-\sqrt{\dfrac{2}{\pi}}\dfrac{1}{i\omega}$
6	$\begin{cases} 1 & |x| < a \\ -1 & |x| > a \end{cases}$	$\sqrt{\dfrac{2}{\pi}}\dfrac{\sin a\omega}{\omega}$
7	$\begin{cases} e^{i\Omega t} & |x| < a \\ 0 & |x| > a \end{cases}$	$\sqrt{\dfrac{2}{\pi}}\dfrac{\sin a(\Omega + \omega)}{\Omega + \omega}$
8	$e^{-a|x|}$ $\qquad a > 0$	$-\sqrt{\dfrac{2}{\pi}}\dfrac{a}{a^2+\omega^2}$

Fourier transforms

$$F(\omega) = \mathcal{F}(f)(\omega) = \frac{1}{\sqrt{2\pi}} \int_{-\infty}^{\infty} f(x)e^{i\omega x} \, dx$$

No.	$f(x)$	$F(\omega)$
9	$\frac{\sin \Omega x}{x}$	$\sqrt{\frac{\pi}{2}} \left[H(\Omega - \omega) - H(-\Omega - \omega) \right]$
10	$\sin ax/x$	$\begin{cases} \sqrt{\frac{\pi}{2}} & \|\omega\| < a \\ 0 & \|\omega\| > a \end{cases}$
11	$\begin{cases} e^{iax} & p < x < q \\ 0 & x < p, x > q \end{cases}$	$\frac{i}{\sqrt{2\pi}} e^{ip(\omega+a)} - e^{iq(\omega+a)} / \omega + a$
12	$\begin{cases} e^{-cx+iax} & x > 0 \\ 0 & x < 0 \end{cases}$ $(c > 0)$	$\frac{i}{\sqrt{2\pi}(\omega + a + ic)}$
13	e^{-px^2} $\quad$ Re $p > 0$	$\frac{1}{\sqrt{2p}} e^{-\omega^2/4p}$
14	$\cos px^2$	$\frac{1}{\sqrt{2p}} \cos\left(\frac{\omega^2}{4p} - \frac{\pi}{4} \right)$
15	$\sin px^2$	$\frac{1}{\sqrt{2p}} \cos\left(\frac{\omega^2}{4p} + \frac{\pi}{4} \right)$
16	$\|x\|^{-p}$ $\quad (0 < p < 1)$	$\sqrt{\frac{2}{\pi}} \frac{\Gamma(1-p) \sin \frac{p\pi}{2}}{\|\omega\|^{1-p}}$
17	$e^{-a\|x\|}/\sqrt{\|x\|}$	$\frac{\sqrt{\sqrt{a^2+\omega^2}+a}}{\sqrt{\omega^2+a^2}}$
18	$\frac{\cosh ax}{\cosh \pi x}$ $\quad (-\pi < a < \pi)$	$\sqrt{\frac{2}{\pi}} \frac{\cos \frac{a}{2} \cosh \frac{\omega}{2}}{\cos a + \cosh \omega}$
19	$\frac{\sinh ax}{\sinh \pi x}$ $\quad (-\pi < a < \pi)$	$\frac{1}{\sqrt{2\pi}} \frac{\sin a}{\cos a + \cosh \omega}$
20	$\begin{cases} \frac{1}{\sqrt{a^2-x^2}} & \|x\| < a \\ 0 & \|x\| > a \end{cases}$	$\sqrt{\frac{\pi}{2}} J_0(a\omega)$
21	$\frac{\sin[b\sqrt{a^2+x^2}]}{\sqrt{a^2+x^2}}$	$\begin{cases} 0 & \|\omega\| > b \\ \sqrt{\frac{\pi}{2}} J_0(a\sqrt{b^2 - \omega^2}) & \|\omega\| < b \end{cases}$
22	$\begin{cases} P_n(x) & \|x\| < 1 \\ 0 & \|x\| > 1 \end{cases}$	$\frac{i^n}{\sqrt{\omega}} J_{n+1/2}(\omega)$
23	$\begin{cases} \frac{\cos[b\sqrt{a^2-x^2}]}{\sqrt{a^2-x^2}} & \|x\| < a \\ 0 & \|x\| > a \end{cases}$	$\sqrt{\frac{\pi}{2}} J_0(a\sqrt{\omega^2 + b^2})$
24	$\begin{cases} \frac{\cosh[b\sqrt{a^2-x^2}]}{\sqrt{a^2-x^2}} & \|x\| < a \\ 0 & \|x\| > a \end{cases}$	$\sqrt{\frac{\pi}{2}} J_0(a\sqrt{\omega^2 - b^2})$

Multidimensional Fourier transforms

$$F(\mathbf{u}) = (2\pi)^{-n/2} \int \cdots \int_{\mathbb{R}^n} f(\mathbf{x}) e^{i(\mathbf{x}\cdot\mathbf{u})} \, d\mathbf{x}$$

No.	$f(\mathbf{x})$		$F(\mathbf{u})$
	In n-dimensions		
1	$f(a\mathbf{x})$	$\mathrm{Im}\, a = 0$	$\lvert a \rvert^{-n} F(a^{-1}\mathbf{u})$
2	$f(\mathbf{x} - \mathbf{a})$		$e^{-i\mathbf{a}\cdot\mathbf{u}} F(\mathbf{u})$
3	$e^{i\mathbf{a}\cdot\mathbf{x}} f(\mathbf{x})$		$F(\mathbf{u} + \mathbf{a})$
4	$F(\mathbf{x})$		$(2\pi)^n f(-\mathbf{u})$
	Two dimensions: let $\mathbf{x} = (x, y)$ and $\mathbf{u} = (u, v)$.		
5	$f(ax, by)$		$\dfrac{1}{\lvert ab \rvert} F\!\left(\dfrac{u}{a}, \dfrac{v}{b}\right)$
6	$f(x - a, y - b)$		$e^{i(au+bv)} F(u, v)$
7	$e^{i(ax+by)} f(x, y)$		$F(u + a, v + b)$
8	$F(x, y)$		$(2\pi)^2 F(-u, -v)$
9	$\delta(x - a)\delta(y - b)$		$\dfrac{1}{2\pi} e^{-i(au+bv)}$
10	$e^{-x^2/4a - y^2/4b} \qquad a, b > 0$		$2\sqrt{ab}\, e^{-au^2 - bv^2}$
11	$\begin{cases} 1 & \lvert x \rvert < a, \lvert y \rvert < b \\ 0 & \text{otherwise} \end{cases}$ (rectangle)		$\dfrac{2\sin au \sin bv}{\pi uv}$
12	$\begin{cases} 1 & \lvert x \rvert < a \\ 0 & \text{otherwise} \end{cases}$ (strip)		$\dfrac{2\sin au}{\pi uv}\delta(v)$
13	$\begin{cases} 1 & x^2 + y^2 < a^2 \\ 0 & \text{otherwise} \end{cases}$ (circle)		$\dfrac{aJ_1(a\sqrt{u^2 + v^2})}{\sqrt{u^2 + v^2}}$
	Three dimensions: let $\mathbf{x} = (x, y, z)$ and $\mathbf{u} = (u, v, w)$.		
14	$\delta(x - a)\delta(y - b)\delta(z - c)$		$\dfrac{1}{(2\pi)^{3/2}} e^{-i(au+bv+cw)}$
15	$e^{-x^2/4a - y^2/4b - z^2/4c} \quad a, b, c > 0$		$2^{3/2}\sqrt{abc}\, e^{-au^2 - bv^2 - cw^2}$
16	$\begin{cases} 1 & \lvert x \rvert < a, \lvert y \rvert < b, \lvert z \rvert < c \\ 0 & \text{otherwise} \qquad\text{(box)} \end{cases}$		$\left(\dfrac{2}{\pi}\right)^{3/2} \dfrac{\sin au \sin bv \sin cw}{uvw}$
17	$\begin{cases} 1 & x^2 + y^2 + z^2 < a^2 \\ 0 & \text{otherwise} \end{cases}$ (ball)		$\dfrac{\sin a\rho - a\rho\cos a\rho}{\sqrt{2\pi}\rho^3}$ $\rho^2 = u^2 + v^2 + w^2$

Laplace transforms: functional relations

$$F(s) = \mathcal{L}(f)(s) = \int_0^\infty f(t)e^{-st}\, dt.$$

No.	$f(t)$	$F(s)$
1	$af(t) + bg(t)$	$aF(s) + bG(s)$
2	$f'(t)$	$sF(s) - F(0+)$
3	$f''(t)$	$s^2 F(s) - sF(0+) - F'(0+)$
4	$f^{(n)}(t)$	$s^n F(s) - \displaystyle\sum_{k=0}^{n-1} s^{n-1-k} F^{(k)}(0+)$
5	$\int_0^t f(\tau)\, d\tau$	$\frac{1}{s} F(s)$
6	$\int_0^t \int_0^\tau f(u)\, du\, d\tau$	$\frac{1}{s^2} F(s)$
7	$\int_0^t f_1(t - \tau) f_2(\tau)\, d\tau = f_1 * f_2$	$F_1(s) F_2(s)$
8	$t f(t)$	$-F'(s)$
9	$t^n f(t)$	$(-1)^n F^{(n)}(s)$
10	$\frac{1}{t} f(t)$	$\int_s^\infty F(z)\, dz$
11	$e^{at} f(t)$	$F(s - a)$
12	$f(t - b)$ with $f(t) = 0$ for $t < 0$	$e^{-bs} F(s)$
13	$\frac{1}{c} f\left(\frac{t}{c}\right)$	$F(cs)$
14	$\frac{1}{c} e^{bt/c} f\left(\frac{t}{c}\right)$	$F(cs - b)$
15	$f(t + a) = f(t)$	$\int_0^a e^{-st} f(t)\, dt / 1 - e^{-as}$
16	$f(t + a) = -f(t)$	$\int_0^a e^{-st} f(t)\, dt / 1 + e^{-as}$
17	$\sum_{k=1}^n \frac{p(a_k)}{q'(a_k)} e^{a_k t}$ with $q(t) = (t - a_1)\cdots(t - a_n)$	$\dfrac{p(s)}{q(s)}$
18	$e^{at} \displaystyle\sum_{k=1}^n \frac{\phi^{(n-k)}(a)}{(n-k)!} \frac{t^{k-1}}{(k-1)!}$	$\dfrac{\phi(s)}{(s - a)^n}$

Laplace transforms

$$F(s) = \mathcal{L}(f)(s) = \int_0^\infty f(t)e^{-st}\,dt.$$

No.	$f(t)$		$F(s)$	
1	$\delta(t)$,	delta function	1	
2	$H(t)$,	unit step function or Heaviside function	$1/s$	
3	t		$1/s^2$	
4	$\frac{t^{n-1}}{(n-1)!}$		$1/s^n$	$(n = 1, 2, \ldots)$
5	$1/\sqrt{\pi t}$		$1/\sqrt{s}$	
6	$2\sqrt{t/\pi}$		$s^{-3/2}$	
7	$\frac{2^n t^{n-1/2}}{\sqrt{\pi}(2n-1)!!}$		$s^{-(n+1/2)}$	$(n = 1, 2, \ldots)$
8	t^{k-1}		$\frac{\Gamma(k)}{s^k}$	$(k > 0)$
9	e^{at}		$\frac{1}{s-a}$	
10	te^{at}		$\frac{1}{(s-a)^2}$	
11	$\frac{1}{(n-1)!}t^{n-1}e^{at}$		$\frac{1}{(s-a)^n}$	$(n = 1, 2, \ldots)$
12	$t^{k-1}e^{at}$		$\frac{\Gamma(k)}{(s-a)^k}$	$(k > 0)$
13	$\frac{1}{a-b}\left(e^{at} - e^{bt}\right)$		$\frac{1}{(s-a)(s-b)}$	$(a \neq b)$
14	$\frac{1}{a-b}\left(ae^{at} - be^{bt}\right)$		$\frac{s}{(s-a)(s-b)}$	$(a \neq b)$
15	$-\frac{(b-c)e^{at}+(c-a)e^{bt}+(a-b)e^{ct}}{(a-b)(b-c)(c-a)}$		$\frac{1}{(s-a)(s-b)(s-c)}$	$(a, b, c \text{ distinct})$
16	$\frac{1}{a}\sin at$		$\frac{1}{s^2+a^2}$	
17	$\cos at$		$\frac{s}{s^2+a^2}$	
18	$\frac{1}{a}\sinh at$		$\frac{1}{s^2-a^2}$	
19	$\cosh at$		$\frac{s}{s^2-a^2}$	
20	$\frac{1}{a^2}(1 - \cos at)$		$\frac{1}{s(s^2+a^2)}$	
21	$\frac{1}{a^3}(at - \sin at)$		$\frac{1}{s^2(s^2+a^2)}$	
22	$\frac{1}{2a^3}(\sin at - at\cos at)$		$\frac{1}{(s^2+a^2)^2}$	
23	$\frac{t}{2a}\sin at$		$\frac{s}{(s^2+a^2)^2}$	

Laplace transforms

$$F(s) = \mathcal{L}(f)(s) = \int_0^\infty f(t)e^{-st}\,dt.$$

No.	$f(t)$	$F(s)$	
24	$\frac{1}{2a}\left(\sin at + at\cos at\right)$	$\frac{s^2}{(s^2+a^2)^2}$	
25	$t\cos at$	$\frac{s^2-a^2}{(s^2+a^2)^2}$	
26	$\frac{\cos at - \cos bt}{b^2-a^2}$	$\frac{s}{(s^2+a^2)(s^2+b^2)}$	$(a^2 \neq b^2)$
27	$\frac{1}{b}e^{at}\sin bt$	$\frac{1}{(s-a)^2+b^2}$	
28	$e^{at}\cos bt$	$\frac{s-a}{(s-a)^2+b^2}$	
29	$-\frac{e^{-at}}{4^{n-1}b^{2n}}\sum_{k=1}^{n}\binom{2n-k-1}{n-1}$ $\times(-2t)^{k-1}\frac{d^k}{dt^k}[\cos bt]$	$\frac{1}{[(s-a)^2+b^2]^n}$	
30	$\frac{e^{-at}}{4^{n-1}b^{2n}}\left\{\sum_{k=1}^{n}\binom{2n-k-1}{n-1}\frac{(-2t)^{k-1}}{(k-1)!}\right.$ $\times\frac{d^k}{dt^k}[a\cos bt + b\sin at]$ $-2b\sum_{k=1}^{n-1}\binom{2n-k-2}{n-1}\frac{(-2t)^{k-1}}{(k-1)!}$ $\left.\times\frac{d^k}{dt^k}[\sin bt]\right\}$	$\frac{s}{[(s-a)^2+b^2]^n}$	
31	e^{-at} $-e^{at/2}\left(\cos\frac{at\sqrt{3}}{2}-\sqrt{3}\sin\frac{at\sqrt{3}}{2}\right)$	$\frac{3a^2}{s^3+a^3}$	
32	$\sin at\cosh at - \cos at\sinh at$	$\frac{4a^3}{s^4+4a^4}$	
33	$\frac{1}{2a^2}\sin at\sinh at$	$\frac{s}{s^4+4a^4}$	
34	$\frac{1}{2a^3}\left(\sinh at - \sin at\right)$	$\frac{1}{s^4-a^4}$	
35	$\frac{1}{2a^2}\left(\cosh at - \cos at\right)$	$\frac{s}{s^4-a^4}$	
36	$(1+a^2t^2)\sin at - \cos at$	$\frac{8a^3s^2}{(s^2+a^2)^3}$	
37	$\frac{e^t}{n!}\frac{d^n}{dt^n}\left(t^n e^{-t}\right)$	$\frac{1}{s}\left(\frac{s-1}{s}\right)^n$	
38	$\frac{1}{\sqrt{\pi t}}e^{at}(1+2at)$	$\frac{s}{(s-a)^{3/2}}$	
39	$\frac{1}{2\sqrt{\pi t^3}}\left(e^{bt}-e^{at}\right)$	$\sqrt{s-a}-\sqrt{s-b}$	
40	$\frac{1}{\sqrt{\pi t}}-ae^{a^2t}\operatorname{erfc}(a\sqrt{t})$	$\frac{1}{\sqrt{s}+a}$	
41	$\frac{1}{\sqrt{\pi t}}+ae^{a^2t}\operatorname{erf}(a\sqrt{t})$	$\frac{\sqrt{s}}{s-a^2}$	

Laplace transforms

$$F(s) = \mathcal{L}(f)(s) = \int_0^\infty f(t)e^{-st}\,dt.$$

No.	$f(t)$	$F(s)$	
42	$\dfrac{1}{\sqrt{\pi t}} - \dfrac{2a}{\sqrt{\pi}}e^{-a^2t}\int_0^{a\sqrt{t}}e^{\tau^2}\,d\tau$	$\dfrac{\sqrt{s}}{s+a^2}$	
43	$\dfrac{1}{a}e^{a^2t}\operatorname{erf}(a\sqrt{t})$	$\dfrac{1}{\sqrt{s}(s-a^2)}$	
44	$\dfrac{2}{a\sqrt{\pi}}e^{-a^2t}\int_0^{a\sqrt{t}}e^{\tau^2}\,d\tau$	$\dfrac{1}{\sqrt{s}(s+a^2)}$	
45	$\begin{aligned}&e^{a^2t}[b - a\operatorname{erf}(a\sqrt{t})]\\&\quad -be^{b^2t}\operatorname{erfc}(b\sqrt{t})\end{aligned}$	$\dfrac{b^2 - a^2}{(s - a^2)(b + \sqrt{s})}$	
46	$e^{a^2t}\operatorname{erfc}(a\sqrt{t})$	$\dfrac{1}{\sqrt{s}(\sqrt{s}+a)}$	
47	$\dfrac{1}{\sqrt{b-a}}e^{-at}\operatorname{erf}(\sqrt{b-a}\sqrt{t})$	$\dfrac{1}{(s+a)\sqrt{s+b}}$	
48	$\begin{aligned}&e^{a^2t}\left[\tfrac{b}{a}\operatorname{erf}(a\sqrt{t}) - 1\right]\\&\quad +e^{b^2t}\operatorname{erfc}(b\sqrt{t})\end{aligned}$	$\dfrac{b^2-a^2}{\sqrt{s}(s-a^2)(\sqrt{s}+b)}$	
49	$\dfrac{n!}{(2n)!\sqrt{\pi t}}H_{2n}(\sqrt{t})$	$\dfrac{(1-s)^n}{s^{n+1/2}}$	
50	$-\dfrac{n!}{(2n+1)!\sqrt{\pi}}H_{2n+1}(\sqrt{t})$	$\dfrac{(1-s)^n}{s^{n+3/2}}$	
51	$ae^{-at}\left[I_1(at) + I_0(at)\right]$	$\dfrac{\sqrt{s+2a}}{\sqrt{s}} - 1$	
52	$e^{-(a+b)t/2}I_0\left(\tfrac{a-b}{2}t\right)$	$\dfrac{1}{\sqrt{s+a}\sqrt{s+b}}$	
53	$\begin{aligned}&\sqrt{\pi}\left(\tfrac{t}{a-b}\right)^{k-1/2}e^{-(a+b)t/2}\\&\quad \times I_{k-1/2}\left(\tfrac{a-b}{2}t\right)\end{aligned}$	$\dfrac{\Gamma(k)}{(s+a)^k(s+b)^k}$	$(k \geq 0)$
54	$\begin{aligned}&te^{-(a+b)t/2}\left[I_0\left(\tfrac{a-b}{2}t\right)\right.\\&\quad \left.+ I_1\left(\tfrac{a-b}{2}t\right)\right]\end{aligned}$	$\dfrac{1}{\sqrt{s+a}(s+b)^{3/2}}$	
55	$\dfrac{1}{t}e^{-at}I_1(at)$	$\dfrac{\sqrt{s+2a}-\sqrt{s}}{\sqrt{s+2a}+\sqrt{s}}$	
56	$J_0(at)$	$\dfrac{1}{\sqrt{s^2+a^2}}$	
57	$a^kJ_k(at)$	$\dfrac{(\sqrt{s^2+a^2}-s)^k}{\sqrt{s^2+a^2}}$	$(k > -1)$
58	$\dfrac{\sqrt{\pi}}{\Gamma(k)}\left(\tfrac{t}{2a}\right)^{k-1/2}J_{k-1/2}(at)$	$\dfrac{1}{(s^2+a^2)^k}$	$(k > 0)$
59	$\dfrac{ka^k}{t}J_k(at)$	$(\sqrt{s^2+a^2}-s)^k$	$(k > 0)$
60	$a^kI_k(at)$	$\dfrac{(s-\sqrt{s^2-a^2})^k}{\sqrt{s^2-a^2}}$	$(k > -1)$
61	$\dfrac{\sqrt{\pi}}{\Gamma(k)}\left(\tfrac{t}{2a}\right)^{k-1/2}I_{k-1/2}(at)$	$\dfrac{1}{(s^2-a^2)^k}$	$(k > 0)$

Laplace transforms

$$F(s) = \mathcal{L}(f)(s) = \int_0^\infty f(t)e^{-st}\,dt.$$

No.	$f(t)$	$F(s)$	
62	$\begin{cases} 0 & \text{when } 0 < t < k \\ 1 & \text{when } t > k \end{cases}$	$\dfrac{e^{-ks}}{s}$	
63	$\begin{cases} 0 & \text{when } 0 < t < k \\ t-k & \text{when } t > k \end{cases}$	$\dfrac{e^{-ks}}{s^2}$	
64	$\begin{cases} 0 & \text{when } 0 < t < k \\ \dfrac{(t-k)^{p-1}}{\Gamma(p)} & \text{when } t > k \end{cases}$	$\dfrac{e^{-ks}}{s^p}$	$(p > 0)$
65	$\begin{cases} 1 & \text{when } 0 < t < k \\ 0 & \text{when } t > k \end{cases}$	$\dfrac{1 - e^{-ks}}{s}$	
66	$\lvert \sin at \rvert$	$\dfrac{a}{s^2+a^2} \coth \dfrac{\pi s}{2a}$	
67	$J_0(2\sqrt{at})$	$\dfrac{1}{s}e^{-a/s}$	
68	$\dfrac{1}{\sqrt{\pi t}} \cos 2\sqrt{at}$	$\dfrac{1}{\sqrt{s}}e^{-a/s}$	
69	$\dfrac{1}{\sqrt{\pi t}} \cosh 2\sqrt{at}$	$\dfrac{1}{\sqrt{s}}e^{a/s}$	
70	$\dfrac{1}{\sqrt{\pi a}} \sin 2\sqrt{at}$	$\dfrac{1}{s^{3/2}}e^{-a/s}$	
71	$\dfrac{1}{\sqrt{\pi a}} \sinh 2\sqrt{at}$	$\dfrac{1}{s^{3/2}}e^{a/s}$	
72	$\left(\dfrac{t}{a}\right)^{(k-1)/2} J_{k-1}(2\sqrt{at})$	$\dfrac{1}{s^k}e^{-a/s}$	$(k > 0)$
73	$\left(\dfrac{t}{a}\right)^{(k-1)/2} I_{k-1}(2\sqrt{at})$	$\dfrac{1}{s^k}e^{a/s}$	$(k > 0)$
74	$\dfrac{a}{2\sqrt{\pi t^3}}e^{-a^2/4t}$	$e^{-a\sqrt{s}}$	$(a > 0)$
75	$\operatorname{erfc}\left(\dfrac{a}{2\sqrt{t}}\right)$	$\dfrac{1}{s}e^{-a\sqrt{s}}$	$(a \geq 0)$
76	$\dfrac{1}{\sqrt{\pi t}}e^{-a^2/4t}$	$\dfrac{1}{\sqrt{s}}e^{-a\sqrt{s}}$	$(a \geq 0)$
77	$2\sqrt{\dfrac{t}{\pi}}e^{-a^2/4t} - a\operatorname{erfc}\left(\dfrac{a}{2\sqrt{t}}\right)$	$s^{-3/2}e^{-a\sqrt{s}}$	$(a \geq 0)$
78	$e^{ak+a^2t}\operatorname{erfc}\left(a\sqrt{t} + \dfrac{k}{2\sqrt{t}}\right)$	$\dfrac{e^{-k\sqrt{s}}}{\sqrt{s}(a+\sqrt{s})}$	$(k \geq 0)$
79	$J_0(a\sqrt{t^2 + 2kt})$	$\dfrac{e^{-k(\sqrt{s^2+a^2}-s)}}{\sqrt{s^2+a^2}}$	$(k \geq 0)$
80	$\Gamma'(1) - \log t$	$\dfrac{1}{s}\log s$	
81	$t^{k-1}\left[\dfrac{\Gamma'(k)}{\lvert\Gamma(k)\rvert^2} - \dfrac{\log t}{\Gamma(k)}\right]$	$\dfrac{1}{s^k}\log s$	$(k > 0)$
82	$e^{at}[\log a - \operatorname{Ei}(-at)]$	$\dfrac{\log s}{s-a}$	$(a > 0)$

Laplace transforms

$$F(s) = \mathcal{L}(f)(s) = \int_0^\infty f(t)e^{-st}\,dt.$$

No.	$f(t)$	$F(s)$	
83	$\cos t\,\mathrm{Si}(t) - \sin t\,\mathrm{Ci}(t)$	$\frac{\log s}{s^2+1}$	
84	$-\sin t\,\mathrm{Si}(t) - \cos t\,\mathrm{Ci}(t)$	$\frac{t\log s}{s^2+1}$	
85	$-\mathrm{Ei}\left(-\frac{t}{a}\right)$	$\frac{1}{s}\log(1+as)$	$(a>0)$
86	$\frac{1}{t}\left(e^{bt} - e^{at}\right)$	$\log\frac{s-a}{s-b}$	
87	$-2\,\mathrm{Ci}\left(-\frac{t}{a}\right)$	$\frac{1}{s}\log(1+a^2s^2)$	
88	$2\log a - 2\,\mathrm{Ci}(at)$	$\frac{1}{s}\log(s^2+a^2)$	$(a>0)$
89	$\frac{2}{a}[at\log a + \sin at - at\,\mathrm{Ci}(at)]$	$\frac{1}{s^2}\log(s^2+a^2)$	$(a>0)$
90	$\frac{2}{t}(1-\cos at)$	$\log\frac{s^2+a^2}{s^2}$	
91	$\frac{2}{t}(1-\cosh at)$	$\log\frac{s^2-a^2}{s^2}$	
92	$\frac{1}{t}\sin at$	$\tan^{-1}\frac{a}{s}$	
93	$\frac{1}{a\sqrt{\pi}}e^{-t^2/4a^2}$	$e^{a^2s^2}\mathrm{erfc}(as)$	$(a>0)$
94	$\mathrm{erf}\left(\frac{t}{2a}\right)$	$\frac{1}{s}e^{a^2s^2}\mathrm{erfc}(as)$	$(a>0)$
95	$\frac{\sqrt{a}}{\pi\sqrt{t}(t+a)}$	$e^{as}\mathrm{erfc}(\sqrt{as})$	$(a>0)$
96	$\frac{1}{\sqrt{\pi}(t+a)}$	$\frac{1}{\sqrt{s}}e^{as}\mathrm{erfc}(\sqrt{as})$	$(a>0)$
97	$\frac{1}{\pi t}\sin(2a\sqrt{t})$	$\mathrm{erf}\left(\frac{a}{\sqrt{s}}\right)$	
98	$\frac{1}{t+a}$	$-e^{as}\,\mathrm{Ei}(-as)$	$(a>0)$
99	$\frac{1}{(t+a)^2}$	$\frac{1}{a} + se^{as}\,\mathrm{Ei}(-as)$	$(a>0)$
100	$\frac{1}{t^2+1}$	$\left[\frac{\pi}{2}-\mathrm{Si}(s)\right]\cos s + \mathrm{Ci}(s)\sin s$	
101	$\begin{cases} 0 & \text{when } 0<t<a \\ (t^2-a^2)^{-1/2} & \text{when } t>a \end{cases}$	$K_0(as)$	

Hankel transforms

$$\mathcal{H}_\nu(f)(y) = F_\nu(y) = \int_0^\infty f(x)\sqrt{xy}\,J_\nu(yx)\,dx, \quad y > 0.$$

No.	$f(x)$	$F_\nu(y)$
1	$\begin{cases} x^{\nu+1/2}, & 0 < x < 1 \\ 0, & 1 < x \end{cases}$ $\operatorname{Re}\nu > -1$	$y^{-1/2}J_{\nu+1}(y)$
2	$\begin{cases} x^{\nu+1/2}(a^2-x^2)^\mu, & 0 < x < a \\ 0, & a < x \end{cases}$ $\operatorname{Re}\nu, \quad \operatorname{Re}\mu > -1$	$2^\mu\Gamma(\mu+1)a^{\nu+\mu+1}y^{-\mu-1/2}$ $\times J_{\nu+\mu+1}(ay)$
3	$x^{\nu+1/2}(x^2+a^2)^{-\nu-1/2},$ $\operatorname{Re}a > 0, \quad \operatorname{Re}\nu > -1/2$	$\sqrt{\pi}y^{\nu-1/2}2^{-\nu}e^{-ay}$ $\times [\Gamma(\nu+1/2)]^{-1}$
4	$x^{\nu+1/2}e^{-ax},$ $\operatorname{Re}a > 0, \quad \operatorname{Re}\nu > -1$	$a(\pi)^{-1/2}2^{\nu+1}y^{\nu+1/2}\Gamma(\nu+3/2)$ $\times (a^2+y^2)^{-\nu-3/2}$
5	$x^{\nu+1/2}e^{-ax^2},$ $\operatorname{Re}a > 0, \quad \operatorname{Re}\nu > -1$	$y^{\nu+1/2}(2a)^{-\nu-1}\exp\left(-y^2/4a\right)$
6	$e^{-ax}/\sqrt{x},$ $\operatorname{Re}a > 0, \quad \operatorname{Re}\nu > -1$	$y^{-\nu+1/2}\left[\sqrt{(a^2+y^2)}-a\right]^\nu$ $\times (a^2+y^2)^{-1/2}$
7	$x^{-\nu-1/2}\cos(ax),$ $a > 0, \quad \operatorname{Re}\nu > -1/2$	$\sqrt{\pi}2^{-\nu}y^{-\nu+1/2}[\Gamma(\nu+1/2)]^{-1}$ $\times (y^2-a^2)^{\nu-1/2}H(y-a)$
8	$x^{1/2-\nu}\sin(ax),$ $a > 0, \quad \operatorname{Re}\nu > 1/2$	$a2^{1-\nu}\sqrt{\pi}y^{\nu+1/2}[\Gamma(\nu-1/2)]^{-1}$ $\times (y^2-a^2)^{\nu-3/2}H(y-a)$
9	$x^{-1/2}J_{\nu-1}(ax),$ $a > 0, \quad \operatorname{Re}\nu > -1$	$a^{\nu-1}y^{-\nu+1/2}H(y-a)$
10	$x^{-1/2}J_{\nu+1}(ax),$ $a > 0, \quad \operatorname{Re}\nu > -3/2$	$\begin{cases} a^{-\nu-1}y^{\nu+1/2}, & 0 < y < a \\ 0, & a < y \end{cases}$

Hilbert transforms

$$\mathcal{H}(f)(y) = F(y) = \frac{1}{\pi} \int_{-\infty}^{\infty} \frac{f(x)}{x-y} \, dx.$$

No.	$f(x)$	$F(y)$
1	1	0
2	$\begin{cases} 0, & -\infty < x < a \\ 1, & a < x < b \\ 0, & b < x < \infty \end{cases}$	$\frac{1}{\pi} \log \left| (b-y)(a-y)^{-1} \right|$
3	$\begin{cases} 0, & -\infty < x < a \\ x^{-1}, & a < x < \infty \end{cases}$	$(\pi y)^{-1} \log \left| a(a-y)^{-1} \right|,$ $\qquad 0 \neq y \neq a, \quad a > 0$
4	$(x+a)^{-1} \qquad\qquad \text{Im } a > 0$	$i(y+a)^{-1}$
5	$\dfrac{1}{1+x^2}$	$-\dfrac{y}{1+y^2}$
6	$\dfrac{1}{1+x^4}$	$-\dfrac{y(1+y^2)}{\sqrt{2}(1+y^4)}$
7	$\sin(ax), \qquad\qquad\quad a > 0$	$\cos(ay)$
8	$\dfrac{\sin(ax)}{x}, \qquad\qquad a > 0$	$\dfrac{\cos(ay) - 1}{y}$
9	$\dfrac{\sin x}{1+x^2}$	$\dfrac{\cos y - e^{-1}}{1+y^2}$
10	$\cos(ax), \qquad\qquad\quad a > 0$	$-\sin(ay)$
11	$\dfrac{1 - \cos(ax)}{x}, \qquad a > 0$	$\dfrac{\sin(ay)}{y}$
12	$\text{sgn}(x)\sin(a|x|^{1/2}) \qquad a > 0$	$\cos(a|y|^{1/2}) + \exp(-a|y|^{1/2})$
13	$e^{iax} \qquad\qquad\qquad\quad a > 0$	ie^{iay}

Mellin transforms

$$f^*(s) = \mathcal{M}[f(x); s] = \int_0^\infty f(x)x^{s-1}\,dx.$$

No.	$f(x)$	$f^*(s)$
1	$ag(x) + bh(x)$	$ag^*(s) + bh^*(s)$
2	$f^{(n)}(x)$†	$(-1)^n \frac{\Gamma(s)}{\Gamma(s-n)} f^*(s-n)$
3	$x^n f^{(n)}(x)$†	$(-1)^n \frac{\Gamma(s+n)}{\Gamma(s)} f^*(s)$
4	$I_n f(x)$‡	$(-1)^n \frac{\Gamma(s)}{\Gamma(s+n)} f^*(s+n)$
5	e^{-x}	$\Gamma(s)$ Re $s > 0$
6	e^{-x^2}	$\frac{1}{2}\Gamma(\frac{1}{2}s)$ Re $s > 0$
7	$\cos x$	$\Gamma(s)\cos(\frac{1}{2}\pi s)$ $0 < $ Re $s < 1$
8	$\sin x$	$\Gamma(s)\sin(\frac{1}{2}\pi s)$ $0 < $ Re $s < 1$
9	$(1-x)^{-1}$	$\pi\cot(\pi s)$ $0 < $ Re $s < 1$
10	$(1+x)^{-1}$	$\pi\operatorname{cosec}(\pi s)$ $0 < $ Re $s < 1$
11	$(1+x^a)^{-b}$	$\frac{\Gamma(s/a)\Gamma(b-s/a)}{a\Gamma(b)}$ $0 < $ Re $s < ab$
12	$\log(1+ax)$ $\|\arg a\| < \pi$	$\pi s^{-1}a^{-s}\operatorname{cosec}(\pi s)$ $-1 < $ Re $s < 0$
13	$\tan^{-1} x$	$-\frac{1}{2}\pi s^{-1}\sec(\frac{1}{2}\pi s)$ $-1 < $ Re $s < 0$
14	$\cot^{-1} x$	$\frac{1}{2}\pi s^{-1}\sec(\frac{1}{2}\pi s)$ $0 < $ Re $s < 1$
15	$\operatorname{csch} ax$ Re $a > 0$	$2(1-2^{-s})a^{-s}\Gamma(s)\zeta(s)$ Re $s > 1$
16	$\operatorname{sech}^2 ax$ Re $a > 0$	$4(2a)^{-s}\Gamma(s)\zeta(s-1)$ Re $s > 2$
17	$\operatorname{csch}^2 ax$ Re $a > 0$	$4(2a)^{-s}\Gamma(s)\zeta(s-1)$ Re $s > 2$
18	$K_\nu(ax)$	$a^{-s}2^{s-2}\Gamma((s-\nu)/2)$ $\times\, \Gamma((s+\nu)/2)$ Re $s > \|$Re $\nu\|$

†Assuming that $\lim_{x\to 0+} x^{s-r-1} f^{(r)}(x) = 0$ for $r = 0, 1, \ldots, n-1$.

‡Where I_n denotes the nth repeated integral of $f(x)$: $I_0 f(x) = f(x)$, $I_n f(x) = \int_0^x I_{n-1}(t)\,dt$.

Chapter 7

Probability and Statistics

7.1 PROBABILITY THEORY **617**
- 7.1.1 Introduction 617
- 7.1.2 Multivariate distributions 621
- 7.1.3 Random sums of random variables 622
- 7.1.4 Transforming variables 623
- 7.1.5 Central limit theorem 623
- 7.1.6 Inequalities 623
- 7.1.7 Averages over vectors 625
- 7.1.8 Geometric probability 625

7.2 CLASSICAL PROBABILITY PROBLEMS **627**
- 7.2.1 Raisin cookie problem 627
- 7.2.2 Gambler's ruin problem 627
- 7.2.3 Card games 628
- 7.2.4 Distribution of dice sums 629
- 7.2.5 Birthday problem 629

7.3 PROBABILITY DISTRIBUTIONS **630**
- 7.3.1 Discrete distributions 630
- 7.3.2 Continuous distributions 633

7.4 QUEUING THEORY **637**
- 7.4.1 Variables 637
- 7.4.2 Theorems 638

7.5 MARKOV CHAINS **640**
- 7.5.1 Transition function and matrix 640
- 7.5.2 Recurrence 641
- 7.5.3 Stationary distributions 641
- 7.5.4 Random walks 643
- 7.5.5 Ehrenfest chain 643

7.6 RANDOM NUMBER GENERATION **644**
- 7.6.1 Methods of pseudorandom number generation 644
- 7.6.2 Generating non-uniform random variables 647

1-58488-291-3/02/$0.00+$1.50
© 2003 CRC Press, Inc.

7.7   CONTROL CHARTS AND RELIABILITY **650**
   7.7.1   *Control charts* *650*
   7.7.2   *Acceptance sampling* *652*
   7.7.3   *Reliability* *653*
   7.7.4   *Failure time distributions* *655*

7.8   RISK ANALYSIS AND DECISION RULES **656**

7.9   STATISTICS **658**
   7.9.1   *Descriptive statistics* *658*
   7.9.2   *Statistical estimators* *661*
   7.9.3   *Cramer–Rao bound* *664*
   7.9.4   *Order statistics* *664*
   7.9.5   *Classic statistics problems* *665*

7.10   CONFIDENCE INTERVALS **666**
   7.10.1 *Confidence interval: sample from one population* *666*
   7.10.2 *Confidence interval: samples from two populations* *667*

7.11   TESTS OF HYPOTHESES **669**
   7.11.1 *Hypothesis tests: parameter from one population* *670*
   7.11.2 *Hypothesis tests: parameters from two populations* *673*
   7.11.3 *Hypothesis tests: distribution of a population* *677*
   7.11.4 *Hypothesis tests: distributions of two populations* *680*
   7.11.5 *Sequential probability ratio tests* *681*

7.12   LINEAR REGRESSION **682**
   7.12.1 *Linear model $y_i = \beta_0 + \beta_1 x_i + \epsilon$* *683*
   7.12.2 *General model $y = \beta_0 + \beta_1 x_1 + \beta_2 x_2 + \cdots + \beta_n x_n + \epsilon$* *685*

7.13   ANALYSIS OF VARIANCE (ANOVA) **686**
   7.13.1 *One-factor ANOVA* *687*
   7.13.2 *Unreplicated two-factor ANOVA* *689*
   7.13.3 *Replicated two-factor ANOVA* *692*

7.14   PROBABILITY TABLES **695**
   7.14.1 *Critical values* *695*
   7.14.2 *Table of the normal distribution* *696*
   7.14.3 *Percentage points, Student's t-distribution* *702*
   7.14.4 *Percentage points, chi-square distribution* *703*
   7.14.5 *Percentage points, F-distribution* *704*
   7.14.6 *Cumulative terms, binomial distribution* *710*
   7.14.7 *Cumulative terms, Poisson distribution* *712*
   7.14.8 *Critical values, Kolmogorov–Smirnov test* *717*
   7.14.9 *Critical values, two sample Kolmogorov–Smirnov test* *717*
   7.14.10 *Critical values, Spearman's rank correlation* *718*

7.15   SIGNAL PROCESSING **718**
   7.15.1 *Estimation* *718*
   7.15.2 *Kalman filters* *719*
   7.15.3 *Matched filtering (Wiener filter)* *721*
   7.15.4 *Walsh functions* *722*
   7.15.5 *Wavelets* *723*

7.1 PROBABILITY THEORY

7.1.1 INTRODUCTION

A *sample space* S associated with an experiment is a set S of elements such that any outcome of the experiment corresponds to a unique element of the set. An *event* E is a subset of a sample space S. An element in a sample space is called a *sample point* or a *simple event*.

7.1.1.1 Definition of probability

If an experiment can occur in n mutually exclusive and equally likely ways, and if exactly m of these ways correspond to an event E, then the *probability* of E is given by

$$P(E) = \frac{m}{n}. \tag{7.1.1}$$

If E is a subset of S, and if to each element subset of S, a non-negative number, called the probability, is assigned, and if E is the union of two or more different simple events, then the probability of E, denoted $P(E)$, is the sum of the probabilities of those simple events whose union is E.

7.1.1.2 Marginal and conditional probability

Suppose a sample space S is partitioned into rs disjoint subsets where the general subset is denoted $E_i \cap F_j$ (with $i = 1, 2, \ldots, r$ and $j = 1, 2, \ldots, s$). Then the *marginal probability* of E_i is defined as

$$P(E_i) = \sum_{j=1}^{s} P(E_i \cap F_j), \tag{7.1.2}$$

and the marginal probability of F_j is defined as

$$P(F_j) = \sum_{i=1}^{r} P(E_i \cap F_j). \tag{7.1.3}$$

The *conditional probability* of E_i, given that F_j has occurred, is defined as

$$P(E_i \mid F_j) = \frac{P(E_i \cap F_j)}{P(F_j)}, \qquad \text{when } P(F_j) \neq 0 \tag{7.1.4}$$

and that of F_j, given that E_i has occurred, is defined as

$$P(F_j \mid E_i) = \frac{P(E_i \cap F_j)}{P(E_i)}, \qquad \text{when } P(E_i) \neq 0. \tag{7.1.5}$$

7.1.1.3 Probability theorems

1. If $\emptyset$ is the null set, then $P(\emptyset) = 0$.

2. If S is the sample space, then $P(S) = 1$.

3. If E and F are two events, then

$$P(E \cup F) = P(E) + P(F) - P(E \cap F). \qquad (7.1.6)$$

4. If E and F are *mutually exclusive events*, then

$$P(E \cup F) = P(E) + P(F). \qquad (7.1.7)$$

5. If E and E' are *complementary events*, then

$$P(E) = 1 - P(E'). \qquad (7.1.8)$$

6. Two events are said to be *independent* if and only if

$$P(E \cap F) = P(E)\, P(F). \qquad (7.1.9)$$

The event E is said to be *statistically independent* of the event F if $P(E \mid F) = P(E)$ and $P(F \mid E) = P(F)$.

7. The events $\{E_1, \ldots, E_n\}$ are called *mutually independent* for all combinations if and only if every combination of these events taken any number of times is independent.

8. *Bayes' rule*: If $\{E_1, \ldots, E_n\}$ are n mutually exclusive events whose union is the sample space S, and if E is any arbitrary event of S such that $P(E) \neq 0$, then

$$P(E_k \mid E) = \frac{P(E_k)\, P(E \mid E_k)}{P(E)} = \frac{P(E_k)\, P(E \mid E_k)}{\sum_{j=1}^{n} P(E_j)\, P(E \mid E_j)}. \qquad (7.1.10)$$

9. For a uniform probability distribution,

$$P(A) = \frac{\text{Number of outcomes in event } A}{\text{Total number of outcomes}}. \qquad (7.1.11)$$

7.1.1.4 Terminology

1. A function whose domain is a sample space S and whose range is some set of real numbers is called a *random variable*. This random variable is called *discrete* if it assumes only a finite or denumerable number of values. It is called *continuous* if it assumes a continuum of values.

2. Random variables are usually represented by capital letters.

3. "iid" or "i.i.d." is often used for the phrase "independent and identically distributed".

4. Many probability distributions have special representations:

   (a) χ_n^2: chi-square random variable with n degrees of freedom
   (b) $E(\lambda)$: exponential distribution with parameter λ
   (c) $N(\mu, \sigma)$: normal random variable with mean μ and standard deviation σ
   (d) $P(\lambda)$: Poisson distribution with parameter λ
   (e) $U[a, b]$: uniform random variable on the interval $[a, b]$

7.1.1.5 Characterizing random variables

The *density function* is defined as follows:

1. When X is a continuous random variable, let $f(x)\, dx$ denote the probability that X lies in the region $[x, x + dx]$; $f(x)$ is called the *probability density function*. (We require $f(x) \geq 0$ and $\int f(x)\, dx = 1$.) For any event E,

$$P(E) = P(X \text{ is in } E) = \int_E f(x)\, dx. \qquad (7.1.12)$$

2. When X is a discrete random variable, let p_k for $k = 0, 1, \ldots$ be the probability that $X = x_k$ (with $p_k \geq 0$ and $\sum_k p_k = 1$). Mathematically, for any event E,

$$P(E) = P(X \text{ is in } E) = \sum_{x_k \in E} p_k. \qquad (7.1.13)$$

A discrete random variable can be written with the continuous density $f(x) = \sum_k p_k \delta(x - x_k)$.

The *cumulative distribution function*, or simply the *distribution function*, is defined by

$$F(x) = P(X \leq x) = \begin{cases} \sum_{x_k \leq x} p_k, & \text{in the discrete case,} \\ \int_{-\infty}^x f(t)\, dt, & \text{in the continuous case.} \end{cases} \qquad (7.1.14)$$

Note that $F(-\infty) = 0$ and $F(\infty) = 1$. The probability that X is between a and b is

$$P(a < X \leq b) = P(X \leq b) - P(X \leq a) = F(b) - F(a). \qquad (7.1.15)$$

Let $g(X)$ be a function of X. The *expected value* (or *expectation*) of $g(X)$, denoted by $\mathrm{E}\,[g(X)]$, is defined by

$$\mathrm{E}\,[g(X)] = \begin{cases} \sum_k p_k g(x_k), & \text{in the discrete case,} \\ \int_{\mathbb{R}} g(t) f(t)\, dt, & \text{in the continuous case.} \end{cases} \qquad (7.1.16)$$

1. $E[aX + bY] = aE[X] + bE[Y]$.
2. $E[XY] = E[X]E[Y]$ if X and Y are statistically independent.

The *moments* of X are defined by $\mu'_k = E[X^k]$. The first moment, μ'_1, is called the *mean* of X; it is usually denoted by $\mu = \mu'_1 = E[X]$. The *centered moments* of X are defined by $\mu_k = E[(X - \mu)^k]$. The second centered moment is called the *variance* and is denoted by $\sigma^2 = \mu_2 = E[(X - \mu)^2]$. Here, σ is called the *standard deviation*. The *skewness* is $\gamma_1 = \mu_3/\sigma^3$, and the *excess* or *kurtosis* is $\gamma_2 = (\mu_4/\sigma^4) - 3$.

Using σ_Z^2 to denote the variance for the random variable Z, we have

1. $\sigma_{cX}^2 = c^2\sigma_X^2$.
2. $\sigma_{c+X}^2 = \sigma_X^2$.
3. $\sigma_{aX+b}^2 = a^2\sigma_X^2$.

7.1.1.6 Generating and characteristic functions

In the case of a discrete distribution, the *generating function* corresponding to X (when it exists) is given by $G(s) = G_X(s) = E[s^X] = \sum_{k=0}^{\infty} p_k s^{x_k}$. From this function, the moments may be found from

$$\mu'_n = \left(s\frac{\partial}{\partial s}\right)^n G(s)\bigg|_{s=1}. \qquad (7.1.17)$$

1. If c is a constant, then the generating function of $c + X$ is $s^c G(s)$.

2. If c is a constant, then the generating function of cX is $G(s^c)$.

3. If $Z = X + Y$ where X and Y are independent discrete random variables, then $G_Z(s) = G_X(s)G_Y(s)$.

4. If $Y = \sum_{i=1}^{n} X_i$, the $\{X_i\}$ are independent, and each X_i has the common generating function $G_X(s)$, then the generating function of Y is $[G_X(s)]^n$.

In the case of a continuous distribution, the *characteristic function* corresponding to X is given by $\phi(t) = E[e^{itX}] = \int_{-\infty}^{\infty} e^{itx} f(x)\,dx$, the Fourier transform of $f(x)$. From this function, the moments may be found: $\mu'_n = i^{-n}\phi^{(n)}(0)$. If $Z = X + Y$ where X and Y are independent continuous random variables, then $\phi_Z(t) = \phi_X(t)\phi_Y(t)$.

The *cumulant function* is defined as the logarithm of the characteristic function. The nth *cumulant*, κ_n, is defined as a coefficient in the Taylor series of the cumulant function,

$$\log \phi(t) = \sum_{n=0}^{\infty} \kappa_n \frac{(it)^n}{n!}. \qquad (7.1.18)$$

Note that $\kappa_1 = \mu$, $\kappa_2 = \sigma^2$, $\kappa_3 = \mu_3$, and $\kappa_4 = \mu_4 - 3\mu_2^2$. For a normal probability distribution, $\kappa_n = 0$ for $n \geq 3$. The centered moments in terms of cumulants are

$$\mu_2 = \kappa_2,$$
$$\mu_3 = \kappa_3,$$
$$\mu_4 = \kappa_4 + 3\kappa_2^2, \tag{7.1.19}$$
$$\mu_5 = \kappa_5 + 10\kappa_3\kappa_2,$$
$$\mu_6 = \kappa_6 + 15\kappa_4\kappa_2 + 10\kappa_3^2 + 15\kappa_2^3.$$

7.1.2 MULTIVARIATE DISTRIBUTIONS

7.1.2.1 Discrete case

The k-dimensional random variable $(X_1, \ldots, X_k)$ is a k-dimensional discrete random variable if it assumes values only at a finite or denumerable number of points $(x_1, \ldots, x_k)$. Define

$$P(X_1 = x_1, X_2 = x_2, \ldots, X_k = x_k) = f(x_1, x_2, \ldots, x_k) \tag{7.1.20}$$

for every value that the random variable can assume. The function $f(x_1, \ldots, x_k)$ is called the *joint density* of the k-dimensional random variable. If E is any subset of the set of values that the random variable can assume, then

$$P(E) = P\left[(X_1, \ldots, X_k) \text{ is in } E\right] = \sum_E f(x_1, \ldots, x_k) \tag{7.1.21}$$

where the sum is over all the points $(x_1, \ldots, x_k)$ in E. The cumulative distribution function is defined as

$$F(x_1, x_2, \ldots, x_k) = \sum_{z_1 \leq x_1} \sum_{z_2 \leq x_2} \cdots \sum_{z_k \leq x_k} f(z_1, z_2, \ldots, z_k). \tag{7.1.22}$$

7.1.2.2 Continuous case

The k random variables $(X_1, \ldots, X_k)$ are said to be jointly distributed if a function f exists so that $f(x_1, \ldots, x_k) \geq 0$ for all $-\infty < x_i < \infty$ $(i = 1, \ldots, k)$ and so that, for any given event E,

$$P(E) = P[(X_1, X_2, \ldots, X_k) \text{ is in } E]$$
$$= \int_E \cdots \int f(x_1, x_2, \ldots, x_k) \, dx_1 \, dx_2 \cdots dx_k. \tag{7.1.23}$$

The function $f(x_1, \ldots, x_k)$ is called the *joint density* of the random variables X_1, $X_2, \ldots, X_k$. The cumulative distribution function is defined as

$$F(x_1, x_2, \ldots, x_k) = \int_{-\infty}^{x_1} \int_{-\infty}^{x_2} \cdots \int_{-\infty}^{x_k} f(z_1, z_2, \ldots, z_k) \, dz_k \cdots dz_2 \, dz_1. \tag{7.1.24}$$

Given the cumulative distribution function, the probability density may be found from

$$f(x_1, x_2, \ldots, x_k) = \frac{\partial}{\partial x_1} \frac{\partial}{\partial x_2} \cdots \frac{\partial}{\partial x_k} F(x_1, x_2, \ldots, x_k). \tag{7.1.25}$$

7.1.2.3   Moments

The rth moment of X_i is defined as

$$E\left[X_i^r\right] = \begin{cases} \sum_{x_1} \cdots \sum_{x_k} x_i^r f(x_1, \ldots, x_k), & \text{in the discrete case,} \\ \int\limits_{-\infty}^{\infty} \cdots \int\limits_{-\infty}^{\infty} x_i^r f(x_1, \ldots, x_k) \, dx_1 \cdots dx_k & \text{in the continuous case.} \end{cases}$$

(7.1.26)

Joint moments about the origin are defined as $E\left[X_1^{r_1} X_2^{r_2} \cdots X_k^{r_k}\right]$ where $r_1 + r_2 + \cdots + r_k$ is the order of the moment. Joint moments about the mean are defined as $E\left[(X_1 - \mu_1)^{r_1}(X_2 - \mu_2)^{r_2} \cdots (X_k - \mu_k)^{r_k}\right]$, where $\mu_k = E\left[X_k\right]$.

7.1.2.4   Marginal and conditional distributions

If the random variables $X_1, \ldots, X_k$ have the joint density function $f(x_1, \ldots, x_k)$, then the *marginal distribution* of the subset of the random variables, say, $X_1, \ldots, X_p$ (with $p < k$), is given by

$$g(x_1, x_2, \ldots, x_p) =$$
$$\begin{cases} \sum_{x_{p+1}} \sum_{x_{p+2}} \cdots \sum_{x_k} f(x_1, x_2, \ldots, x_k), & \text{in the discrete case,} \\ \int\limits_{-\infty}^{\infty} \cdots \int\limits_{-\infty}^{\infty} f(x_1, \ldots, x_k) \, dx_{p+1} \cdots dx_k, & \text{in the continuous case.} \end{cases}$$

(7.1.27)

The *conditional distribution* of a certain subset of the random variables is the joint distribution of this subset under the condition that the remaining variables are given certain values. The conditional distribution of $X_1, X_2, \ldots, X_p$, given $X_{p+1}, X_{p+2}, \ldots, X_k$, is

$$h(x_1, \ldots, x_p \mid x_{p+1}, \ldots, x_k) = \frac{f(x_1, x_2, \ldots, x_k)}{g(x_{p+1}, x_{p+2}, \ldots, x_k)} \tag{7.1.28}$$

if $g(x_{p+1}, x_{p+2}, \ldots, x_k) \neq 0$.

The *variance* σ_{ii} of X_i and the *covariance* σ_{ij} of X_i and X_j are given by

$$\sigma_{ii}^2 = \sigma_i^2 = E\left[(X_i - \mu_i)^2\right], \tag{7.1.29}$$
$$\sigma_{ij}^2 = \rho_{ij}\sigma_i\sigma_j = E\left[(X_i - \mu_i)(X_j - \mu_j)\right],$$

where ρ_{ij} is the *correlation coefficient*, and σ_i and σ_j are the standard deviations of X_i and X_j.

7.1.3   RANDOM SUMS OF RANDOM VARIABLES

If $T = \sum_{i=1}^{N} X_i$, and if N is an integer-valued random variable with generating function $G_N(s)$, and if the $\{X_i\}$ are discrete independent and identically distributed random variables with generating function $G_X(s)$, and the $\{X_i\}$ are independent of N, then the generating function for T is $G_T(s) = G_N(G_X(s))$. (If the $\{X_i\}$ are continuous random variables, then $\phi_T(t) = G_N(\phi_X(t))$.) Hence,

1. $\mu_T = \mu_N \mu_X$.
2. $\sigma_T^2 = \mu_N \sigma_X^2 + \mu_X \sigma_N^2$.

7.1.4 TRANSFORMING VARIABLES

1. Suppose that the random variable X has the probability density function $f_X(x)$ and the random variable Y is defined by $Y = g(X)$. If g is measurable and one-to-one, then

$$f_Y(y) = f_X(h(y)) \left| \frac{dh}{dy} \right| \qquad (7.1.30)$$

where $h(y) = g^{-1}(y)$.

2. If the random variables X and Y are independent and if their densities f_X and f_Y, respectively, exist almost everywhere, then the probability density of their sum, $Z = X + Y$, is given by the convolution

$$f_Z(z) = \int_{-\infty}^{\infty} f_X(x) f_Y(z - x)\, dx. \qquad (7.1.31)$$

3. If the random variables X and Y are independent and if their densities f_X and f_Y, respectively, exist almost everywhere, then the probability density of their product, $Z = XY$, is given by the formula,

$$f_Z(z) = \int_{-\infty}^{\infty} \frac{1}{|x|} f_X(x) f_Y\left(\frac{z}{x}\right) dx. \qquad (7.1.32)$$

7.1.5 CENTRAL LIMIT THEOREM

If $\{X_i\}$ are independent and identically distributed random variables with mean μ and finite variance σ^2, then the random variable

$$Z = \frac{(X_1 + X_2 + \cdots + X_n) - n\mu}{\sqrt{n}\sigma} \qquad (7.1.33)$$

tends (as $n \to \infty$) to a normal random variable with mean zero and variance one.

7.1.6 INEQUALITIES

1. *Markov's Inequality:* If X is a random variable which takes only non-negative values, then for any $a > 0$,

$$P(X \geq a) \leq \frac{\mathrm{E}[X]}{a}. \qquad (7.1.34)$$

2. *Cauchy–Schwartz Inequality:* Let X and Y be random variables for which $\mathrm{E}[X^2]$ and $\mathrm{E}[Y^2]$ exist, then

$$(\mathrm{E}[XY])^2 \leq \mathrm{E}[X^2]\,\mathrm{E}[Y^2]. \qquad (7.1.35)$$

3. *One-Sided Chebyshev Inequality:* Let X be a random variable with zero mean (i.e., $E[X] = 0$) and variance σ^2. Then, for any positive a,

$$P(X > a) \leq \frac{\sigma^2}{\sigma^2 + a^2}. \qquad (7.1.36)$$

4. *Chebyshev's Inequality:* Let c be any real number and let X be a random variable for which $E\left[(X - c)^2\right]$ is finite. Then, for every $\epsilon > 0$ the following holds:

$$P(|X - c| \geq \epsilon) \leq \frac{1}{\epsilon^2} E\left[(X - c)^2\right]. \qquad (7.1.37)$$

5. *Bienaymé–Chebyshev's Inequality:* If $E[|X|^r] < \infty$ for all $r > 0$ (r not necessarily an integer) then, for every $a > 0$,

$$P(|X| \geq a) \leq \frac{E[|X|^r]}{a^r}. \qquad (7.1.38)$$

6. *Generalized Bienaymé–Chebyshev's Inequality:* Let $g(x)$ be a non-decreasing non-negative function defined on $(0, \infty)$. Then, for $a \geq 0$,

$$P(|X| \geq a) \leq \frac{E[g(|X|)]}{g(a)}. \qquad (7.1.39)$$

7. *Chernoff bound:* This bound is useful for sums of random variables. Let $Y_n = \sum_{i=1}^{n} X_i$ where the $\{X_i\}$ are iid. Let $M(t) = E_x[e^{tX}]$ be the same moment generating function for each of the $\{X_i\}$, and define $g(t) = \log M(t)$. Then (the prime in this formula denotes a derivative),

$$\begin{aligned} P(Y_n \geq ng'(t)) &\leq e^{-n[tg'(t) - g(t)]}, &\text{if } t \geq 0, \\ P(Y_n \leq ng'(t)) &\leq e^{-n[tg'(t) - g(t)]}, &\text{if } t \leq 0. \end{aligned}$$

8. *Kolmogorov's Inequality:* Let $X_1, X_2, \ldots, X_n$ be n independent random variables such that $E[X_i] = 0$ and $\text{Var}(X_i) = \sigma_{X_i}^2$ is finite. Then, for all $a > 0$,

$$P\left(\max_{i=1,\ldots,n} |X_1 + X_2 + \cdots + X_i| > a\right) \leq \sum_{i=1}^{n} \frac{\sigma_i^2}{a^2}. \qquad (7.1.40)$$

9. *Jensen's Inequality:* If $E[X]$ exists, and if $f(x)$ is a convex $\cup$ ("convex cup") function, then
$$E[f(X)] \geq f(E[X]). \qquad (7.1.41)$$

7.1.7 AVERAGES OVER VECTORS

Let $\overline{f(\mathbf{n})}$ denote the expectation of the function f as the unit vector $\mathbf{n}$ varies uniformly in all directions in three dimensions. If $\mathbf{a}, \mathbf{b}, \mathbf{c}$, and $\mathbf{d}$ are constant vectors, then

$$\overline{|\mathbf{a} \cdot \mathbf{n}|^2} = |\mathbf{a}|^2/3,$$
$$\overline{(\mathbf{a} \cdot \mathbf{n})(\mathbf{b} \cdot \mathbf{n})} = (\mathbf{a} \cdot \mathbf{b})/3,$$
$$\overline{(\mathbf{a} \cdot \mathbf{n})\mathbf{n}} = \mathbf{a}/3,$$
$$\overline{|\mathbf{a} \times \mathbf{n}|^2} = 2|\mathbf{a}|^2/3, \tag{7.1.42}$$
$$\overline{(\mathbf{a} \times \mathbf{n}) \cdot (\mathbf{b} \times \mathbf{n})} = 2\mathbf{a} \cdot \mathbf{b}/3,$$
$$\overline{(\mathbf{a} \cdot \mathbf{n})(\mathbf{b} \cdot \mathbf{n})(\mathbf{c} \cdot \mathbf{n})(\mathbf{d} \cdot \mathbf{n})} = [(\mathbf{a} \cdot \mathbf{b})(\mathbf{c} \cdot \mathbf{d}) + (\mathbf{a} \cdot \mathbf{c})(\mathbf{b} \cdot \mathbf{d}) + (\mathbf{a} \cdot \mathbf{d})(\mathbf{b} \cdot \mathbf{c})]/15.$$

Now let $\overline{f(\mathbf{n})}$ denote the average of the function f as the unit vector $\mathbf{n}$ varies uniformly in all directions in two dimensions. If $\mathbf{a}$ and $\mathbf{b}$ are constant vectors, then

$$\overline{|\mathbf{a} \cdot \mathbf{n}|^2} = |\mathbf{a}|^2/2,$$
$$\overline{(\mathbf{a} \cdot \mathbf{n})(\mathbf{b} \cdot \mathbf{n})} = (\mathbf{a} \cdot \mathbf{b})/2, \tag{7.1.43}$$
$$\overline{(\mathbf{a} \cdot \mathbf{n})\mathbf{n}} = \mathbf{a}/2.$$

7.1.8 GEOMETRIC PROBABILITY

1. Points in a line segment:

   If A and B are uniformly and independently chosen from the interval $[0, 1)$, and X is the distance between A and B (that is, $X = |A - B|$) then the probability density of X is $f_X(x) = 2(1 - x)$.

2. Many points in a line segment:

   Uniformly and independently choose $n-1$ random values in the interval $[0, 1)$. This creates n intervals.

   $P_k(x)$ = Probability (exactly k intervals have length larger than x)

   $$= \binom{n}{k}\left\{[1 - kx]^{n-1} - \binom{n-k}{1}[1 - (k+1)x]^{n-1} + \right. \tag{7.1.44}$$
   $$\left. \cdots + (-1)^s \binom{n-k}{s}[1 - (k+s)x]^{n-1}\right\},$$

   where $s = \left\lfloor \dfrac{1}{x} - k \right\rfloor$. From this, the probability that the largest interval length exceeds x is

   $$1 - P_0(x) = \binom{n}{1}(1 - x)^{n-1} - \binom{n}{2}(1 - 2x)^{n-1} + \ldots. \tag{7.1.45}$$

3. Points in the plane:

   Assume that the number of points in any region A of the plane is a Poisson variate with mean λA (λ is the "density" of the points). Given a fixed point P define $R_1, R_2, \ldots$, to be the distance to the point nearest to P, second nearest to P, etc. Then

   $$f_{R_s}(r) = \frac{2(\lambda\pi)^s}{(s-1)!} r^{2s-1} e^{-\lambda\pi r^2}. \tag{7.1.46}$$

4. Points in three-dimensional space:

   Assume that the number of points in any volume V is a Poisson variate with mean λV (λ is the "density" of the points). Given a fixed point P define R_1, $R_2, \ldots$, to be the distance to the point nearest to P, second nearest to P, etc. Then

   $$f_{R_s}(r) = \frac{3\left(\frac{4}{3}\lambda\pi\right)^s}{\Gamma(s)} r^{3s-1} e^{-\frac{4}{3}\lambda\pi r^3}. \tag{7.1.47}$$

5. Points on a checkerboard:

   Consider the unit squares on a checkerboard and select one point uniformly and independently in each square. The following results concern the average distance between points:

   (a) For adjacent squares (a black and white square with a common side) the mean distance between points is 1.088.

   (b) For diagonal squares (two white squares with a point in common) the mean between points is 1.473.

6. Points in a cube:

   Choose two points uniformly and independently within a unit cube. The distance between these points has mean 0.66171 and standard deviation 0.06214.

7. Points in an n-dimensional cube:

   Select two points uniformly and independently within a unit n-dimensional cube. The expected distance between the points, $\Delta(n)$, is

   $$\Delta(1) = \tfrac{1}{3} \qquad\qquad\qquad \Delta(5) \approx 0.87852$$
   $$\Delta(2) \approx 0.54141 \qquad\qquad \Delta(6) \approx 0.96895$$
   $$\Delta(3) \approx 0.66171 \qquad\qquad \Delta(7) \approx 1.05159$$
   $$\Delta(4) \approx 0.77766 \qquad\qquad \Delta(8) \approx 1.12817$$

8. Points on a circle:

   Select three points uniformly and independently on a unit circle. These points determine a triangle with area A. The mean and variance of this area are:

   $$\mu_A = \frac{3}{2\pi} \approx 0.4775$$
   $$\sigma_A^2 = \frac{3\left(\pi^2 - 6\right)}{8\pi^2} \approx 0.1470 \tag{7.1.48}$$

9. Buffon's needle problem:

A needle of length L is placed at random on a plane on which are ruled parallel lines a distance D apart. If $\frac{L}{D} < 1$ then only one intersection is possible. The probability P that the needle intersects a line is

$$
P = \begin{cases}
\dfrac{2L}{\pi D} & \text{if } 0 < L \le D, \\[2em]
\dfrac{2L}{\pi D}\left(1 - \sqrt{1 - \left(\dfrac{D}{L}\right)^2}\right) + \left(1 - \dfrac{2}{\pi}\sin^{-1}\dfrac{D}{L}\right) & \text{if } 0 < D \le L.
\end{cases}
$$

$$(7.1.49)$$

7.2 CLASSICAL PROBABILITY PROBLEMS

7.2.1 RAISIN COOKIE PROBLEM

A baker creates enough cookie dough for $C = 1000$ raisin cookies. The number of raisins to be added to the dough, R, is to be determined.

1. If you want to be 99% certain that the *first* cookie will have at least one raisin, then $1 - \left(\frac{C-1}{C}\right)^R = 1 - \left(\frac{999}{1000}\right)^R \ge 0.99$, or $R \ge 4603$.

2. If you want to be 99% certain that *every* cookie will have at least one raisin, then $P(C, R) \ge 0.99$, where $P(C, R) = C^{-R} \sum_{i=0}^{C} \binom{C}{i}(-1)^i (C - i)^R$. Hence $R \ge 11508$.

7.2.2 GAMBLER'S RUIN PROBLEM

A gambler starts with z dollars. For each turn, with probability p he wins one dollar, with probability q he loses one dollar (with $p + q = 1$). Gambling stops when he has either last z dollars ("is ruined", the gambler holds zero dollars), or won $a - z$ dollars ("gambler's success", the gambler holds a dollars).

If q_z denotes the probability of stopping with zero dollars ("is ruined") then

$$
q_z = \begin{cases}
\dfrac{(q/p)^a - (q/p)^z}{(q/p)^a - 1} & \text{if } p \ne q, \\[2em]
1 - \dfrac{z}{a} & \text{if } p = q = \frac{1}{2}.
\end{cases}
$$

$$(7.2.1)$$

For example:

	p	q	z	a	q_z
fair	0.5	0.5	9	10	.900
game	0.5	0.5	90	100	.900
	0.5	0.5	900	1000	.900
	0.5	0.5	9000	10000	.900
biased	0.4	0.6	90	100	.017
game	0.4	0.6	90	99	.667

7.2.3 CARD GAMES

If the odds are $a{:}b$ against, the probability of the event is $p = \frac{b}{a+b}$; If the odds are $a{:}b$ for, the probability of the event is $p = \frac{a}{a+b}$.

1. Poker hands

   The number of distinct 5-card poker hands is $\binom{52}{5} = 2{,}598{,}960$.

Hand	Probability	Odds against
royal flush	$1.54 \times 10^{-6} = 4/\binom{52}{5}$	649,739:1
straight flush	$1.39 \times 10^{-5} = 36/\binom{52}{5}$	72,192:1
four of a kind	$2.40 \times 10^{-4} = 624/\binom{52}{5}$	4,164:1
full house	$1.44 \times 10^{-3} = 3744/\binom{52}{5}$	693:1
flush	1.97×10^{-3}	508:1
straight	3.92×10^{-3}	254:1
three of a kind	0.0211	46:1
two pair	0.0475	20:1
one pair	$0.423 = \frac{352}{833}$	1.37:1

2. Bridge hands

   The number of distinct 13-card bridge hands is $\binom{52}{13} = 635{,}013{,}559{,}600$.

   In bridge, the *honors* are the ten, jack, queen, king, and ace of each of the four suits. Obtaining the three top cards (ace, king, and queen) of three suits and the ace, king, queen, and jack of the remaining suit is called *13 top honors*. Obtaining all cards of the same suit is called a *13-card suit*. Obtaining 12 cards of the same suit with ace high and the 13th card not an ace is called a *12-card suit, ace high*. Obtaining no honors is called a *Yarborough*.

Hand	Probability	Odds against
13 top honors	$6.30 \times 10^{-12} = 4/\binom{52}{13}$	158,753,389,899:1
13-card suit	$6.30 \times 10^{-12} = 4/\binom{52}{13}$	158,753,389,899:1
12-card suit, ace high	2.72×10^{-9}	367,484,698:1
Yarborough	5.47×10^{-4}	1,827:1
four aces	2.64×10^{-3}	378:1
nine honors	9.51×10^{-3}	104:1

7.2.4 DISTRIBUTION OF DICE SUMS

A common die is a cube with six faces; the faces are numbered one through six. It is usually unbiased, all faces are equally likely.

When rolling two dice, the probability distribution of the sum is

$$\text{Prob (sum of } s) = \frac{6 - |s - 7|}{36} \qquad \text{for } 2 \leq s \leq 12. \qquad (7.2.2)$$

When rolling three dice, the probability distribution of the sum is

$$\text{Prob (sum of } s) = \frac{1}{216} \begin{cases} \frac{1}{2}(s-1)(s-2) & \text{for } 3 \leq s \leq 8 \\ -s^2 + 21s - 83 & \text{for } 9 \leq s \leq 14 \\ \frac{1}{2}(19-s)(20-s) & \text{for } 15 \leq s \leq 18 \end{cases} \qquad (7.2.3)$$

For two dice, the most common roll is a 7 (probability $\frac{1}{6}$). For three dice, the most common rolls are 10 and 11 (probability $\frac{1}{8}$ each). For four dice, the most common roll is a 14 (probability $\frac{73}{648}$).

7.2.5 BIRTHDAY PROBLEM

The probability that n people have different birthdays (neglecting February 29 th) is

$$q_n = \left(\frac{364}{365}\right) \cdot \left(\frac{363}{365}\right) \cdots \left(\frac{366 - n}{365}\right). \qquad (7.2.4)$$

Let $p_n = 1 - q_n$. For 23 independent people the probability of at least two people having the same birthday is more than half ($p_{23} = 1 - q_{23} > 1/2$).

n	10	20	23	30	40	50
p_n	0.117	0.411	0.507	0.706	0.891	0.970

That is, the number of people needed to have a 50% chance of two people having the same birthday is 23. The number of people needed to have a 50% chance of three people having the same birthday is 88. For four, five, and six people having the same birthday the number of people necessary is 187, 313, and 460.

The number of people needed so that there is a 50% chance that two people have a birthday within one day of each other is 14. In general, in an n-day year the probability that p people all have birthdays at least k days apart (so $k = 1$ is the original birthday problem) is

$$\text{probability} = \binom{n - p(k-1) - 1}{p - 1} \frac{(p-1)!}{n^{p-1}}. \qquad (7.2.5)$$

7.3 PROBABILITY DISTRIBUTIONS

7.3.1 DISCRETE DISTRIBUTIONS

1. *Discrete uniform distribution*: If the random variable X has a probability density function given by

$$P(X = x) = f(x) = \frac{1}{n}, \qquad \text{for } x = x_1, x_2, \ldots, x_n, \qquad (7.3.1)$$

   then the variable X is said to possess a discrete uniform probability distribution.

   Properties: When $x_i = i$ for $i = 1, 2, \ldots, n$ then

$$\text{Mean} = \mu = \frac{n+1}{2},$$

$$\text{Variance} = \sigma^2 = \frac{n^2 - 1}{12},$$

$$\text{Standard deviation} = \sigma = \sqrt{\frac{n^2 - 1}{12}}, \qquad (7.3.2)$$

$$\text{Moment generating function} = G(t) = \frac{e^t(1 - e^{nt})}{n(1 - e^t)}.$$

2. *Binomial distribution*: If the random variable X has a probability density function given by

$$P(X = x) = f(x) = \binom{n}{x} \theta^x (1 - \theta)^{n-x}, \qquad \text{for } x = 0, 1, \ldots, n, \qquad (7.3.3)$$

   then the variable X is said to possess a binomial distribution. Note that $f(x)$ is the general term in the expansion of $[\theta + (1 - \theta)]^n$.

   Properties:

$$\text{Mean} = \mu = n\theta,$$

$$\text{Variance} = \sigma^2 = n\theta(1 - \theta),$$

$$\text{Standard deviation} = \sigma = \sqrt{n\theta(1 - \theta)}, \qquad (7.3.4)$$

$$\text{Moment generating function} = G(t) = [\theta e^t + (1 - \theta)]^n.$$

   As $n \to \infty$ the binomial distribution approximates a normal distribution with a mean of $n\theta$ and variance of $n\theta(1 - \theta)$; see Figure 7.1.

3. *Geometric distribution*: If the random variable X has a probability density function given by

$$P(X = x) = f(x) = \theta(1 - \theta)^{x-1} \qquad \text{for } x = 1, 2, 3, \ldots, \qquad (7.3.5)$$

FIGURE 7.1

Comparison of $P(x)$ for a binomial distribution and the approximating normal distribution. Left figure is for $(n = 8, \theta = 0.2)$, right figure is for $(n = 8, \theta = 0.4)$; horizontal axis is x.

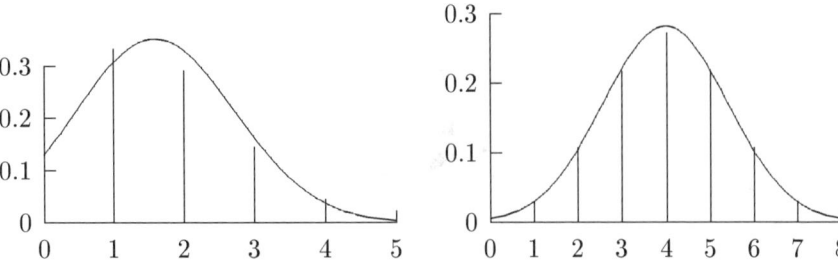

then the variable X is said to possess a geometric distribution.

Properties:

$$
\begin{aligned}
\text{Mean} &= \mu = \frac{1}{\theta}, \\
\text{Variance} &= \sigma^2 = \frac{1 - \theta}{\theta^2}, \\
\text{Standard deviation} &= \sigma = \sqrt{\frac{1 - \theta}{\theta^2}}, \\
\text{Moment generating function} &= G(t) = \frac{\theta e^t}{1 - e^t(1 - \theta)}.
\end{aligned}
\tag{7.3.6}
$$

4. *Hypergeometric distribution*: If the random variable X has a probability density function given by

$$
P(X = x) = f(x) = \frac{\binom{k}{x}\binom{N-k}{n-x}}{\binom{N}{n}} \quad \text{for } x = 1, 2, 3, \ldots, \min(n, k) \tag{7.3.7}
$$

then the variable X is said to possess a hypergeometric distribution.

Properties:

$$
\begin{aligned}
\text{Mean} &= \mu = \frac{kn}{N}, \\
\text{Variance} &= \sigma^2 = \frac{k(N - k)n(N - n)}{N^2(N - 1)}, \\
\text{Standard deviation} &= \sigma = \sqrt{\frac{k(N - k)n(N - n)}{N^2(N - 1)}}.
\end{aligned}
\tag{7.3.8}
$$

5. *Negative binomial distribution*: If the random variable X has a probability density function given by

$$P(X = x) = f(x) = \binom{x + r - 1}{r - 1} \theta^r (1 - \theta)^x \qquad \text{for } x = 0, 1, 2, \ldots,$$

(7.3.9)

then the variable X is said to possess a negative binomial distribution (also known as a *Pascal* or *Polya distribution*).

Properties:

$$\text{Mean} = \mu = \frac{r}{\theta} - r,$$

$$\text{Variance} = \sigma^2 = \frac{r}{\theta} \left(\frac{1}{\theta} - 1 \right) = \frac{r(1 - \theta)}{\theta^2},$$

$$\text{Standard deviation} = \sqrt{\frac{r}{\theta} \left(\frac{1}{\theta} - 1 \right)} = \sqrt{\frac{r(1 - \theta)}{\theta^2}},$$

(7.3.10)

$$\text{Moment generating function} = G(t) = \theta^r [1 - (1 - \theta)e^t]^{-r}.$$

6. *Poisson distribution*: If the random variable X has a probability density function given by

$$P(X = x) = f(x) = \frac{e^{-\lambda} \lambda^x}{x!} \qquad \text{for } x = 0, 1, 2, \ldots,$$

(7.3.11)

with $\lambda > 0$, then the variable X is said to possess a Poisson distribution.

Properties:

$$\text{Mean} = \mu = \lambda,$$

$$\text{Variance} = \sigma^2 = \lambda,$$

$$\text{Standard deviation} = \sigma = \sqrt{\lambda},$$

(7.3.12)

$$\text{Moment generating function} = G(t) = e^{\lambda(e^t - 1)}.$$

7. *Multinomial distribution*: If a set of random variables $X_1, X_2, \ldots, X_n$ has a probability function given by

$$P(X_1 = x_1, X_2 = x_2, \ldots, X_n = x_n) = f(x_1, x_2, \ldots, x_n)$$

$$= N! \prod_{i=1}^{n} \frac{\theta_i^{x_i}}{x_i!}$$

(7.3.13)

where the $\{x_i\}$ are positive integers, each $\theta_i > 0$, and

$$\sum_{i=1}^{n} \theta_i = 1 \quad \text{and} \quad \sum_{i=1}^{n} x_i = N,$$

(7.3.14)

then the joint distribution of $X_1, X_2, \ldots, X_n$ is called the multinomial distribution. Note that $f(x_1, x_2, \ldots, x_n)$ is a term in the expansion of $(\theta_1 + \theta_2 + \cdots + \theta_n)^N$.

Properties:

$$\text{Mean of } X_i = \mu_i = N\theta_i,$$
$$\text{Variance of } X_i = \sigma_i^2 = N\theta_i(1 - \theta_i),$$
$$\text{Covariance of } X_i \text{ and } X_j = \sigma_{ij}^2 = -N\theta_i\theta_j,$$
$$\text{Joint moment generating function} = (\theta_1 e^{t_1} + \cdots + \theta_n e^{t_n})^N.$$

(7.3.15)

7.3.2 CONTINUOUS DISTRIBUTIONS

1. *Uniform distribution*: If the random variable X has a density function of the form

$$f(x) = \frac{1}{\beta - \alpha}, \qquad \text{for } \alpha < x < \beta, \tag{7.3.16}$$

then the variable X is said to possess a uniform distribution.

Properties:

$$\text{Mean} = \mu = \frac{\alpha + \beta}{2},$$
$$\text{Variance} = \sigma^2 = \frac{(\beta - \alpha)^2}{12},$$
$$\text{Standard deviation} = \sigma = \sqrt{\frac{(\beta - \alpha)^2}{12}},$$
$$\text{Moment generating function} = G(t) = \frac{e^{\beta t} - e^{\alpha t}}{(\beta - \alpha)t}$$
$$= \frac{2}{(\beta - \alpha)t} \sinh\left[\frac{(\beta - \alpha)t}{2}\right] e^{(\alpha + \beta)t/2}.$$

(7.3.17)

2. *Normal distribution*: If the random variable X has a density function of the form

$$f(x) = \frac{1}{\sqrt{2\pi}\sigma} \exp\left(-\frac{(x - \mu)^2}{2\sigma^2}\right), \qquad \text{for } -\infty < x < \infty, \tag{7.3.18}$$

then the variable X is said to possess a normal distribution.

Properties:

$$\text{Mean} = \mu,$$
$$\text{Variance} = \sigma^2,$$
$$\text{Standard deviation} = \sigma,$$
$$\text{Moment generating function} = G(t) = \exp\left(\mu t + \frac{\sigma^2 t^2}{2}\right).$$

(7.3.19)

(a) Set $y = \frac{x-\mu}{\sigma}$ to obtain a standard normal distribution.

(b) The cumulative distribution function is

$$F(x) = \Phi(x) = \frac{1}{\sqrt{2\pi}\sigma} \int_{-\infty}^{x} \exp\left(-\frac{(t-\mu)^2}{2\sigma^2}\right) dt.$$

3. *Multi-dimensional normal distribution*:

The random vector $\mathbf{X}$ is said to be a *multivariate normal* (or a multi-dimensional normal) if and only if the linear combination $\mathbf{a}^{\mathrm{T}}\mathbf{X}$ is normal for all vectors $\mathbf{a}$. If the mean of $\mathbf{X}$ is μ, and if the second moment matrix $R = \mathrm{E}\left[(\mathbf{X} - \mu)(\mathbf{X} - \mu)^{\mathrm{T}}\right]$ is non-singular, the density function of $\mathbf{X}$ is

$$f(\mathbf{x}) = \frac{1}{(2\pi)^{n/2}\sqrt{\det R}} \exp\left[-\frac{1}{2}(\mathbf{x} - \mu)^{\mathrm{T}}R^{-1}(\mathbf{x} - \mu)\right]. \qquad (7.3.20)$$

Sometimes integrals of the form $I_k = \int_{-\infty}^{\infty} \cdots \int_{-\infty}^{\infty} (\mathbf{x}^{\mathrm{T}}M\mathbf{x})^k f(\mathbf{x})\, d\mathbf{x}$ are desired. Defining $a_k = \mathrm{tr}(MR)^k$, we find:

$$\begin{aligned}
I_0 &= 1, \\
I_1 &= a_1, \\
I_2 &= a_1^2 + 2a_2, \\
I_3 &= a_1^3 + 6a_1a_2 + 8a_3, \\
I_4 &= a_1^4 + 12a_1^2a_2 + 32a_1a_3 + 12a_2^2 + 48a_4.
\end{aligned} \qquad (7.3.21)$$

4. *Gamma distribution*: If the random variable X has a density function of the form

$$f(x) = \frac{1}{\Gamma(1+\alpha)\beta^{1+\alpha}} x^\alpha e^{-x/\beta}, \qquad \text{for } 0 < x < \infty, \qquad (7.3.22)$$

with $\alpha > -1$ and $\beta > 0$, then the variable X is said to possess a gamma distribution.

Properties:

$$\begin{aligned}
\text{Mean} &= \mu = \beta(1 + \alpha), \\
\text{Variance} &= \sigma^2 = \beta^2(1 + \alpha), \\
\text{Standard deviation} &= \sigma = \beta\sqrt{1 + \alpha}
\end{aligned} \qquad (7.3.23)$$

Moment generating function $= G(t) = (1 - \beta t)^{-1-\alpha}$, for $t < \beta^{-1}$.

5. *Exponential distribution*: If the random variable X has a density function of the form

$$f(x) = \frac{e^{-x/\theta}}{\theta}, \qquad \text{for } 0 < x < \infty, \qquad (7.3.24)$$

where $\theta > 0$, then the variable X is said to possess an exponential distribution.

Properties:

$$\text{Mean} = \mu = \theta,$$
$$\text{Variance} = \sigma^2 = \theta^2,$$
$$\text{Standard deviation} = \sigma = \theta,$$
$$\text{Moment generating function} = G(t) = (1 - \theta t)^{-1}. \tag{7.3.25}$$

6. *Beta distribution*: If the random variable X has a density function of the form

$$f(x) = B(1 + \alpha, 1 + \beta)x^\alpha(1 - x)^\beta = \frac{\Gamma(\alpha + \beta + 2)}{\Gamma(1 + \alpha)\Gamma(1 + \beta)}x^\alpha(1 - x)^\beta, \tag{7.3.26}$$

for $0 < x < 1$, where $\alpha > -1$ and $\beta > -1$, then the variable X is said to possess a beta distribution.

Properties:

$$\text{Mean} = \mu = \frac{1 + \alpha}{2 + \alpha + \beta}, \tag{7.3.27}$$

$$\text{Variance} = \sigma^2 = \frac{(1 + \alpha)(1 + \beta)}{(2 + \alpha + \beta)^2(3 + \alpha + \beta)},$$

$$r^{\text{th}} \text{ moment about the origin} = \nu_r = \frac{\Gamma(2 + \alpha + \beta)\Gamma(1 + \alpha + r)}{\Gamma(2 + \alpha + \beta + r)\Gamma(1 + \alpha)}.$$

7. *Chi-square distribution*: If the random variable X has a density function of the form

$$f(x) = \frac{x^{(n-2)/2}e^{-x/2}}{2^{n/2}\Gamma(n/2)} \qquad \text{for } 0 < x < \infty \tag{7.3.28}$$

then the variable X is said to possess a chi-square (χ^2) distribution with n degrees of freedom.

Properties:

$$\text{Mean} = \mu = n,$$
$$\text{Variance} = \sigma^2 = 2n, \tag{7.3.29}$$
$$\text{Standard deviation} = \sigma = \sqrt{2n}.$$

(a) If $Y_1, Y_2, \ldots, Y_n$ are independent and identically distributed normal random variables with a mean of 0 and a variance of 1, then $\chi^2 = \sum_{i=1}^{n} Y_i^2$ is distributed as chi-square with n degrees of freedom.

(b) If $\chi_1^2, \chi_2^2, \ldots, \chi_k^2$, are independent random variables and have chi-square distributions with $n_1, n_2, \ldots, n_k$ degrees of freedom, then $\sum_{i=1}^{k} \chi_i^2$ has a chi-squared distribution with $n = \sum_{i=1}^{k} n_i$ degrees of freedom.

8. *Snedecor's F-distribution*: If the random variable X has a density function of the form

$$f(x) = \frac{\Gamma\left(\frac{n+m}{2}\right)\left(\frac{m}{n}\right)^{m/2} x^{(m-2)/2}}{\Gamma\left(\frac{m}{2}\right)\Gamma\left(\frac{n}{2}\right)\left(1 + \frac{m}{n}x\right)^{(n+m)/2}}, \quad \text{for } 0 < x < \infty, \qquad (7.3.30)$$

then the variable X is said to possess a F-distribution with m and n degrees of freedom.

Properties:

$$\text{Mean} = \mu = \frac{n}{n-2}, \qquad \text{for } n > 2,$$

$$\text{Variance} = \frac{2n^2(m+n-2)}{m(n-2)^2(n-4)}, \qquad \text{for } n > 4. \qquad (7.3.31)$$

(a) The transformation $w = \dfrac{mx/n}{1 + \frac{mx}{n}}$ transforms the F-density to the beta density.

(b) If the random variable X has a χ^2-distribution with m degrees of freedom, the random variable Y has a χ^2-distribution with n degrees of freedom, and X and Y are independent, then $F = \dfrac{X/m}{Y/n}$ is distributed as an F-distribution with m and n degrees of freedom.

9. *Student's t-distribution*: If the random variable X has a density function of the form

$$f(x) = \frac{\Gamma\left(\frac{n+1}{2}\right)}{\sqrt{n\pi}\,\Gamma\left(\frac{n}{2}\right)\left(1 + \frac{x^2}{n}\right)^{(n+1)/2}}, \quad \text{for } -\infty < x < \infty. \qquad (7.3.32)$$

then the variable X is said to possess a t-distribution with n degrees of freedom.

Properties:

$$\text{Mean} = \mu = 0,$$

$$\text{Variance} = \sigma^2 = \frac{n}{n-2}, \qquad \text{for } n > 2. \qquad (7.3.33)$$

(a) If the random variable X is normally distributed with mean 0 and variance σ^2, and if Y^2/σ^2 has a χ^2 distribution with n degrees of freedom, and if X and Y are independent, then $t = \dfrac{X\sqrt{n}}{Y}$ is distributed as a t-distribution with n degrees of freedom.

FIGURE 7.2
Conceptual layout of a queue.

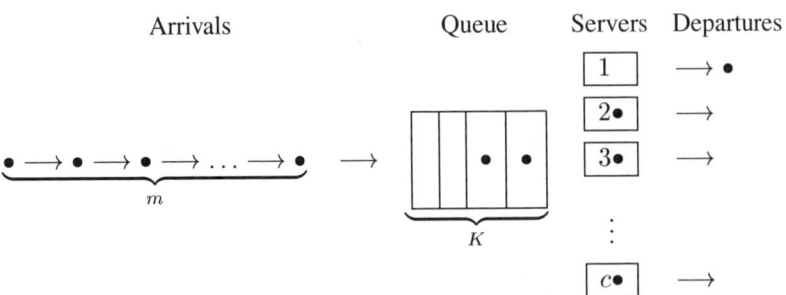

7.4 QUEUING THEORY

A queue is represented as $A/B/c/K/m/Z$ where (see Figure 7.2):

1. A and B represent the interarrival times and service times:

   GI general independent interarrival time,
   G general service time distribution,
   H_k k-stage hyperexponential interarrival or service time distribution,
   E_k Erlang-k interarrival or service time distribution,
   M exponential interarrival or service time distribution,
   D deterministic (constant) interarrival or service time distribution.

2. c is the number of identical servers.
3. K is the system capacity.
4. m is the number in the source.
5. Z is the queue discipline:

   FCFS first come, first served (also known as FIFO: "first in, first out"),
   LIFO last in, first out,
   RSS random,
   PRI priority service.

When not all variables are present, the trailing ones have the default values, $K = \infty$, $m = \infty$, and Z is FIFO. Note that the m and Z are rarely used.

7.4.1 VARIABLES

1. Proportions

   (a) a_n: proportion of customers that find n customers already in the system when they arrive.
   (b) d_n: proportion of customers leaving behind n customers in the system.

(c) p_n: proportion of time the system contains n customers.

2. Intrinsic queue parameters

(a) λ: average arrival rate of customers to the system (number per unit time).
(b) μ: average service rate per server (number per unit time), $\mu = 1/\mathrm{E}[T_s]$.
(c) u: traffic intensity, $u = \lambda/\mu$.
(d) ρ: server utilization, the probability that any particular server is busy,
$\rho = u/c = (\lambda/\mu)/c$.

3. Derived queue parameters

(a) L: average number of customers in the system.
(b) L_Q: average number of customers in the queue.
(c) N: number in system.
(d) W: average time for customer in system.
(e) W_Q: average time for customer in the queue.
(f) T_s: service time.

4. Probability functions

(a) $f_s(x)$: probability density function of customer's service time.
(b) $f_w(x)$: probability density function of customer's time in system.
(c) $\pi(z)$: probability generating function of p_n: $\pi_n = \sum_{n=0}^{\infty} p_n z^n$.
(d) $\pi_Q(z)$: probability generating function of the number in the queue.
(e) $\alpha(s)$: Laplace Transform of $f_w(x)$: $\alpha(s) = \int_0^{\infty} f_w(x)e^{-xs}\,dx$.
(f) $\alpha_T(s)$: Laplace Transform of the service time.

7.4.2 THEOREMS

1. *Little's law:* $L = \lambda W$ and $L_Q = \lambda W_Q$.

2. If the arrivals have a Poisson distribution: $p_n = a_n$.

3. If customers arrive one at a time and are served one at a time: $a_n = d_n$.

4. For an $M/M/1$ queue with $\rho < 1$,

(a) $p_n = (1-\rho)\rho^n$,
(b) $L = \rho/(1-\rho)$,
(c) $L_Q = \rho^2/(1-\rho)$,
(d) $W = 1/(\mu - \lambda)$,
(e) $W_Q = \rho/(\mu - \lambda)$,
(f) $\pi(z) = (1-\rho)/(1-z\rho)$,
(g) $\alpha(s) = (\lambda - \mu)/(\lambda - \mu - s)$.

5. For an $M/M/c$ queue with $\rho < 1$ (so that $\mu_n = n\mu$ for $n = 1, 2, \ldots, c$ and $\mu_n = c\mu$ for $n > c$),

   (a) $p_0 = \left[\dfrac{u^c}{c!(1-\rho)} + \displaystyle\sum_{n=0}^{c-1} \dfrac{u^n}{n!} \right]^{-1}$,

   (b) $p_n = \begin{cases} p_0 u^n / n! & \text{for } n = 0, 1, \ldots, c, \\ p_0 u^n / c! c^{n-c} & \text{for } n > c, \end{cases}$

   (c) $L_Q = p_0 u^c \rho / c! (1-\rho)^2$,

   (d) $W_Q = L_Q / \lambda$,

   (e) $W = W_Q + 1/\mu$,

   (f) $L = \lambda W$.

6. *Pollaczek–Khintchine formula*: For an $M/G/1$ queue with $\rho < 1$ and $\mathrm{E}\left[T_s^2\right] < \infty$

   (a) $L = L_Q + \rho$,

   (b) $L = \lambda W$,

   (c) $L_Q = \dfrac{\lambda^2 \mathrm{E}\left[T_s^2\right]}{2(1-\rho)}$,

   (d) $L_Q = \lambda W_Q$,

   (e) $\pi(z) = \pi_Q(z)\alpha_T(\lambda - \lambda z)$,

   (f) $\pi_Q(z) = (1-\rho)(1-z)/(\alpha(\lambda - \lambda z) - z)$.

7. For an $M/G/\infty$ queue

   (a) $p_n = e^{-u} u^n / n!$,

   (b) $\pi(z) = \exp\left(-(1-z)u\right)$.

8. *Erlang B formula*: For an $M/G/c/c$ queue, $p_c = \dfrac{u^c}{c!}\left(\displaystyle\sum_{k=0}^{c} \dfrac{u^k}{k!}\right)^{-1}$.

9. *Distributional form of Little's law*: For any single server system for which: (i) Arrivals are Poisson at rate λ, (ii) all arriving customers enter the system and remain in the system until served (i.e., there is no balking or reneging), (iii) the customers leave the system one at a time in order of arrival, (iv) for any time t, the arrival process after time t and the time in the system for any customer arriving before t are independent, then (here L and W do not denote averages)

   (a) $\pi(z) = \alpha(\lambda(1-z))$,

   (b) $\mathrm{E}[L^n] = \sum_{k=1}^{n} S(n,k)\mathrm{E}\left[(\lambda W)^k\right]$
       where $S(n, k)$ is Stirling number of the second kind. For example:

       i. $\mathrm{E}[L] = \lambda \mathrm{E}[W]$ (*Little's law*),

       ii. $\mathrm{E}\left[L^2\right] = \mathrm{E}\left[(\lambda W)^2\right] + \mathrm{E}[\lambda W]$.

7.5 MARKOV CHAINS

A *discrete parameter stochastic process* is a collection of random variables $\{X(t),\ t = 0,1,2,\ldots\}$. The values of $X(t)$ are called the *states* of the process. The collection of states is called the *state space*. The values of t usually represent points in time. The number of states is either finite or countably infinite. A discrete parameter stochastic process is called a *Markov chain* if, for any set of n time points $t_1 < t_2 < \cdots < t_n$, the conditional distribution of $X(t_n)$ given values for $X(t_1)$, $X(t_2), \ldots, X(t_{n-1})$ depends only on $X(t_{n-1})$. It is expressed by

$$P\left[X(t_n) \le x_n \mid X(t_1) = x_1, \ldots, X(t_{n-1}) = x_{n-1}\right]$$
$$= P\left[X(t_n) \le x_n \mid X(t_{n-1}) = x_{n-1}\right]. \quad (7.5.1)$$

A Markov chain is said to be *stationary* if the value of the conditional probability $P\left[X(t_{n+1}) = x_{n+1} | X(t_n) = x_n\right]$ is independent of n. This discussion will be restricted to stationary Markov chains.

7.5.1 TRANSITION FUNCTION AND MATRIX

7.5.1.1 Transition function

Let x and y be states and let $\{t_n\}$ be time points in $T = \{0, 1, 2, \ldots\}$. The *transition function*, $P(x, y)$, is defined by

$$P(x,y) = P_{n,n+1}(x,y) = P\left[X(t_{n+1}) = y \mid X(t_n) = x\right], \qquad t_n, t_{n+1} \in T. \quad (7.5.2)$$

$P(x,y)$ is the probability that a Markov chain in state x at time t_n will be in state y at time t_{n+1}. Some properties of the transition function are that $P(x,y) \ge 0$ and $\sum_y P(x,y) = 1$. The values of $P(x,y)$ are commonly called the *one-step transition probabilities*.

The function $\pi_0(x) = P(X(0) = x)$, with $\pi_0(x) \ge 0$ and $\sum_x \pi_0(x) = 1$, is called the *initial distribution* of the Markov chain. It is the probability distribution when the chain is started. Thus,

$$P\left[X(0) = x_0, X(1) = x_1, \ldots, X(n) = x_n\right]$$
$$= \pi_0(x_0)P_{0,1}(x_0, x_1)P_{1,2}(x_1, x_2) \cdots P_{n-1,n}(x_{n-1}, x_n). \quad (7.5.3)$$

7.5.1.2 Transition matrix

A convenient way to summarize the transition function of a Markov chain is by using the *one-step transition matrix*. It is defined as

$$\mathbf{P} = \begin{bmatrix} P(0,0) & P(0,1) & \cdots & P(0,n) & \cdots \\ P(1,0) & P(1,1) & \cdots & P(1,n) & \cdots \\ \vdots & \vdots & \ddots & \vdots & \\ P(n,0) & P(n,1) & \cdots & P(n,n) & \cdots \\ \vdots & \vdots & & \vdots & \end{bmatrix}. \quad (7.5.4)$$

Define the *n–step transition matrix* by $\mathbf{P}^{(n)}$ as the matrix with entries

$$P^{(n)}(x,y) = P\left[X(t_{m+n}) = y \mid X(t_m) = x\right]. \tag{7.5.5}$$

This can be written in terms of the one-step transition matrix as $\mathbf{P}^{(n)} = \mathbf{P}^n$.

Suppose the state space is finite. The one-step transition matrix is said to be *regular* if, for some positive power m, all of the elements of $\mathbf{P}^m$ are strictly positive.

THEOREM 7.5.1 *(Chapman–Kolmogorov equation)*

Let $P(x,y)$ be the one-step transition function of a Markov chain and define $P^{(0)}(x,y) = 1$, if $x = y$, and 0, otherwise. Then, for any pair of non-negative integers, s and t, such that $s + t = n$,

$$P^{(n)}(x,y) = \sum_z P^{(s)}(x,z)P^{(t)}(z,y). \tag{7.5.6}$$

7.5.2 RECURRENCE

Define the probability that a Markov chain starting in state x returns to state x for the first time after n steps by

$$f^n(x,x) = P\left[X(t_n) = x, X(t_{n-1}) \neq x, \ldots, X(t_1) \neq x \mid X(t_0) = x\right]. \tag{7.5.7}$$

It follows that $P^n(x,x) = \sum_{k=1}^n f^k(x,x)P^{n-k}(x,x)$. A state x is said to be *recurrent* if $\sum_{n=1}^\infty f^n(x,x) = 1$. This means that a state x is recurrent if, after starting in x, the probability of returning to it after some finite length of time is one. A state which is not recurrent is said to be *transient*.

THEOREM 7.5.2

A state x of a Markov chain is recurrent if and only if $\sum_{n=1}^\infty P^n(x,x) = \infty$.

Two states, x and y, are said to *communicate* if, for some $n > 0$, $P^n(x,y) > 0$. This theorem implies that, if x is a recurrent state and x communicates with y, y is also a recurrent state. A Markov chain is said to be *irreducible* if every state communicates with every other state and with itself.

Let x be a recurrent state and define T_x the *(return time)* as the number of stages for a Markov chain to return to state x, having begun there. A recurrent state x is said to be *null recurrent* if $\mathrm{E}\left[T_x\right] = \infty$. A recurrent state that is not null recurrent is said to be *positive recurrent*.

7.5.3 STATIONARY DISTRIBUTIONS

Let $\{X(t), t = 0, 1, 2, \ldots\}$ be a Markov chain having a one-step transition function $P(x,y)$. A function $\pi(x)$ where each $\pi(x)$ is non-negative, $\sum_x \pi(x)P(x,y) = \pi(y)$, and $\sum_y \pi(y) = 1$, is called a *stationary distribution*. If a Markov chain has a stationary distribution and $\lim_{n\to\infty} P^n(x, y) = \pi(y)$ for every x, then, regardless of the initial distribution, $\pi_0(x)$, the distribution of $X(t_n)$ approaches $\pi(x)$ as n tends to infinity. When this happens, $\pi(x)$ is often referred to as the *steady state distribution*. The following categorizes those Markov chains with stationary distributions.

THEOREM 7.5.3

Let X_P denote the set of positive recurrent states of a Markov chain.

1. *If X_P is empty, the chain has no stationary distribution.*
2. *If X_P is a non-empty irreducible set, the chain has a unique stationary distribution.*
3. *If X_P is non-empty but not irreducible, the chain has an infinite number of distinct stationary distributions.*

The *period* of a state x is denoted by $d(x)$ and is defined as the greatest common divisor of all integers, $n \geq 1$, for which $P^n(x, x) > 0$. If $P^n(x, x) = 0$ for all $n \geq 1$, then define $d(x) = 0$. If each state of a Markov chain has $d(x) = 1$, the chain is said to be *aperiodic*. If each state has period $d > 1$, the chain is said to be *periodic* with period d. The vast majority of Markov chains encountered in practice are aperiodic. An irreducible, positive recurrent, aperiodic Markov chain always possesses a steady-state distribution. An important special case occurs when the state space is finite. Suppose that $X = \{1, 2, \ldots, K\}$. Let $\pi_0 = \{\pi_0(1), \pi_0(2), \ldots, \pi_0(K)\}$.

THEOREM 7.5.4

Let $\mathbf{P}$ be a regular one-step transition matrix and π_0 be an arbitrary vector of initial probabilities. Then $\lim_{n\to\infty} \pi_0(x)\mathbf{P}^n = \mathbf{y}$, where $\mathbf{y}\mathbf{P} = \mathbf{y}$, and $\sum_{i=1}^{K} \pi_0(y_i) = 1$.

7.5.3.1   Example: A simple three-state Markov chain

A Markov chain having three states $\{0, 1, 2\}$ with a one-step transition matrix of

$$P = \begin{bmatrix} \frac{1}{2} & 0 & \frac{1}{2} \\ \frac{1}{4} & \frac{3}{4} & 0 \\ 0 & \frac{3}{4} & \frac{1}{4} \end{bmatrix}$$ is diagrammed below.

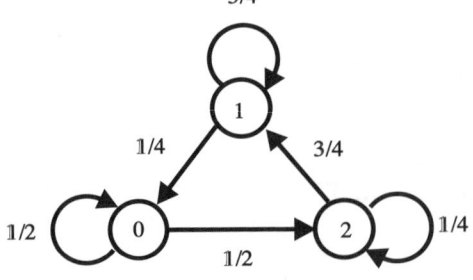

The one-step transition matrix gives a two–step transition matrix

$$P^{(2)} = P^2 = \begin{bmatrix} \frac{1}{4} & \frac{3}{8} & \frac{3}{8} \\ \frac{5}{16} & \frac{9}{16} & \frac{1}{8} \\ \frac{3}{16} & \frac{3}{4} & \frac{1}{16} \end{bmatrix}. \tag{7.5.8}$$

The one-step transition matrix is regular. This Markov chain is irreducible, and all three states are recurrent. In addition, all three states are positive recurrent. Since all states have period 1, the chain is aperiodic. The unique steady state distribution is $\pi(0) = 3/11$, $\pi(1) = 6/11$, and $\pi(2) = 2/11$.

7.5.4 RANDOM WALKS

Let $\eta(t_1), \eta(t_2), \ldots$ be independent random variables having a common density $f(x)$, and let $t_1, t_2, \ldots$ be integers. Let X_0 be an integer-valued random variable that is independent of $\eta(t_1), \eta(t_2), \ldots$, and $X(t_n) = X_0 + \sum_{i=1}^{n} \eta(t_i)$. The sequence $\{X(t_i), i = 0, 1, \ldots\}$ is called a *random walk*. An important special case is a *simple random walk*. It is defined by

$$P(x, y) = \begin{cases} p, & \text{if } y = x - 1, \\ r, & \text{if } y = x, \\ q, & \text{if } y = x + 1, \end{cases} \quad \text{where } p + q + r = 1, \text{ and } P(0,0) = p + r. \tag{7.5.9}$$

Here, an object begins at a certain point in a lattice and at each step either stays at that point or moves to a neighboring lattice point.

This one-dimensional random walk can be extended to higher-dimensional lattices. A common case is that an object can only transition to an adjacent lattice point, and all such transitions are equally likely. In this case, with a one- or two-dimensional lattice, if a random walk begins at a lattice point x, then it will return to that lattice point with probability 1. In this case, with a three-dimensional lattice, the probability that it will return to its starting point is only about 0.3405.

7.5.5 EHRENFEST CHAIN

A simple model of gas exchange between two isolated bodies is as follows. Suppose that there are two boxes, Box I and Box II, where Box I contains K molecules numbered $1, 2, \ldots, K$ and Box II contains $N - K$ molecules numbered $K + 1, K + 2, \ldots, N$. A number is chosen at random from $\{1, 2, \ldots, N\}$, and the molecule with that number is transferred from its box to the other one. Let $X(t_n)$ be the number of molecules in Box I after n trials. Then the sequence $\{X(t_n), n = 0, 1, \ldots\}$ is a Markov chain with one-stage transition function of

$$P(x, y) = \begin{cases} \frac{x}{N}, & y = x - 1, \\ 1 - \frac{x}{N}, & y = x + 1, \\ 0, & \text{otherwise}. \end{cases} \tag{7.5.10}$$

7.6   RANDOM NUMBER GENERATION

7.6.1   METHODS OF PSEUDORANDOM NUMBER GENERATION

In Monte Carlo applications, and other computational situations where randomness is required, one must appeal to random numbers for assistance. While it has been argued that numbers measured from a physical process known to be random should be used, it has been infinitely more practical to use simple recursions that produce numbers that behave as random in applications and with respect to statistical tests of randomness. These are so-called *pseudorandom numbers* and are produced by a pseudorandom number generator (PRNG). Depending on the application, either integers in some range or floating point numbers in $[0, 1)$ are the desired output from a PRNG. Since most PRNGs use integer recursions, a conversion into integers in a desired range or into a floating point number in $[0, 1)$ is required. If x_n is an integer produced by some PRNG in the range $0 \leq x_n \leq M - 1$, then an integer in the range $0 \leq x_n \leq N - 1$, with $N \leq M$, is given by $y_n = \lfloor \frac{N}{M} x_n \rfloor$. If $N \ll M$, then $y_n = x_n \pmod{N}$ may be used. Alternately, if a floating point value in $[0, 1)$ is desired, let $y_n = x_n / M$.

7.6.1.1   Linear congruential generators

Perhaps the oldest generator still in use is the *linear congruential generator* (LCG). The underlying integer recursion for LCGs is

$$x_n = ax_{n-1} + b \pmod{M}. \tag{7.6.1}$$

Equation (7.6.1) defines a periodic sequence of integers modulo M starting with x_0, the initial seed. The constants of the recursion are referred to as the *modulus M*, *multiplier a*, and *additive constant b*. If $M = 2^m$, a very efficient implementation is possible. Alternately, there are theoretical reasons why choosing M prime is optimal. Hence, the only moduli that are used in practical implementations are $M = 2^m$ or the prime $M = 2^p - 1$ (i.e., M is a Mersenne prime). With a Mersenne prime or any modulus "close to" 2^p, modular multiplication can be implemented at about twice the computational cost of multiplication modulo 2^p.

Equation (7.6.1) yields a sequence $\{x_n\}$ whose period, denoted $\mathrm{Per}(x_n)$, depends on M, a, and b. The values of the maximal period for the three most common cases used and the conditions required to obtain them are

a	b	M	$\mathrm{Per}(x_n)$
Primitive root of M	Anything	Prime	$M - 1$
3 or 5 $\pmod 8$	0	2^m	2^{m-2}
1 $\pmod 4$	1 $\pmod 2$	2^m	2^m

A major shortcoming of LCGs modulo a power-of-two compared with prime modulus LCGs derives from the following theorem for LCGs:

THEOREM 7.6.1

Define the following LCG sequence: $x_n = ax_{n-1} + b \pmod{M_1}$. *If M_2 divides M_1 then $y_n = x_n \pmod{M_2}$ satisfies $y_n = ay_{n-1} + b \pmod{M_2}$.*

Theorem 7.6.1 implies that the k least-significant bits of any power-of-two modulus LCG with $\mathrm{Per}(x_n) = 2^m = M$ has $\mathrm{Per}(y_n) = 2^k, 0 < k \leq m$. Since a long period is crucial in PRNGs, when these types of LCGs are employed in a manner that makes use of only a few least-significant-bits, their quality may be compromised. When M is prime, no such problem arises.

Since LCGs are in such common usage, here is a list of parameter values mentioned in the literature. The Park–Miller LCG is widely considered a minimally acceptable PRNG. Using any values other than those in the following table may result in a "weaker" LCG.

a	b	M	Source
7^5	0	$2^{31} - 1$	Park–Miller
131	0	2^{35}	Neave
16333	25887	2^{15}	Oakenfull
3432	6789	9973	Oakenfull
171	0	30269	Wichman–Hill

7.6.1.2 Shift-register generators

Another popular method of generating pseudorandom numbers is using binary shift-register sequences to produce pseudorandom bits. A *binary shift-register sequence* (SRS) is defined by a binary recursion of the type,

$$x_n = x_{n-j_1} \oplus x_{n-j_2} \oplus \cdots \oplus x_{n-j_k}, \qquad j_1 < j_2 < \cdots < j_k = \ell, \qquad (7.6.2)$$

where $\oplus$ is the exclusive "or" operation. Note that $x \oplus y \equiv x + y \pmod{2}$. Thus the new bit, x_n, is produced by adding k previously computed bits together modulo 2. The implementation of this recurrence requires keeping the last ℓ bits from the sequence in a shift register, hence the name. The longest possible period is equal to the number of non-zero ℓ-dimensional binary vectors, namely $2^\ell - 1$.

A sufficient condition for achieving $\mathrm{Per}(x_n) = 2^\ell - 1$ is that the characteristic polynomial, corresponding to Equation (7.6.2), be primitive modulo 2. Since primitive trinomials of nearly all degrees of interest have been found, SRSs are usually implemented using two-term recursions of the form,

$$x_n = x_{n-k} \oplus x_{n-\ell}, \qquad 0 < k < \ell. \qquad (7.6.3)$$

In these two-term recursions, k is the *lag* and ℓ is the register length. Proper choice of the pair (ℓ, k) leads to SRSs with $\mathrm{Per}(x_n) = 2^\ell - 1$. Here is a list with suitable (ℓ, k) pairs:

Primitive trinomial exponents					
(5,2)	(7,1)	(7,3)	(17,3)	(17,5)	(17,6)
(31,3)	(31,6)	(31,7)	(31,13)	(127,1)	(521,32)

7.6.1.3 Lagged-Fibonacci generators

Another way of producing pseudorandom numbers uses lagged-Fibonacci generators. The term "lagged-Fibonacci" refers to two-term recurrences of the form,

$$x_n = x_{n-k} \diamond x_{n-\ell}, \qquad 0 < k < \ell, \tag{7.6.4}$$

where $\diamond$ refers to one of the three common methods of combination: (1) addition modulo 2^m, (2) multiplication modulo 2^m, or (3) bitwise exclusive 'OR'ing of m-long bit vectors. Combination method (3) can be thought of as a special implementation of a two-term shift-register sequence.

Using combination method (1) leads to *additive lagged-Fibonacci sequences* (ALFSs). If x_n is given by

$$x_n = x_{n-k} + x_{n-\ell} \pmod{2^m}, \qquad 0 < k < \ell, \tag{7.6.5}$$

then the maximal period is $\mathrm{Per}(x_n) = (2^\ell - 1)2^{m-1}$.

ALFSs are especially suitable for producing floating point deviates using the real-valued recursion $y_n = y_{n-k} + y_{n-\ell} \pmod 1$. This circumvents the need to convert from integers to floating point values and allows floating point hardware to be used. One caution with ALFSs is that Theorem 7.6.1 holds, and so the low-order bits have periods that are shorter than the maximal period. However, this is not nearly the problem as in the LCG case. With ALFSs, the j least-significant bits will have period $(2^\ell - 1)2^{j-1}$, so, if ℓ is large, there really is no problem. Note that one can use the table of primitive trinomial exponents to find (ℓ, k) pairs that give maximal period ALFSs.

7.6.1.4 Non-linear generators

A recent development among PRNGs are non-linear integer recurrences. For example, if in Equation (7.6.4) "$\diamond$" referred to multiplication modulo 2^m, then this recurrence would be a *multiplicative lagged-Fibonacci generator* (MLFG), a non-linear generator. The mathematical structure of non-linear generators is qualitatively different than that of linear generators. Thus, their defects and deficiencies are thought to be complementary to their linear counterparts.

The maximal period of a MLFG is $\mathrm{Per}(x_n) = (2^\ell - 1)2^{m-3}$, a factor of 4 shorter than the corresponding ALFS. However, there are benefits to using multiplication as the combining function due to the bit mixing achieved. Because of this, the perceived quality of the MLFG is considered superior to an ALFS with the same lag, ℓ.

We conclude by defining two non-linear generators, the *inversive congruential generators* (ICGs), which were designed as non-linear analogs of the LCG.

1. The *implicit ICG* is defined by the following recurrence that is almost that of an LCG

$$x_n = a\overline{x_{n-1}} + b \pmod{M}. \tag{7.6.6}$$

   The difference is that we must also take the multiplicative inverse of x_{n-1}, which is defined by $\overline{x_{n-1}}\, x_{n-1} \equiv 1 \pmod{M}$, and $\overline{0} = 0$. This recurrence is indeed non-linear, and avoids some of the problems inherent in linear recurrences, such as the fact that linear tuples must lie on hyperplanes.

2. The *explicit ICG* is

$$x_n = \overline{an + b} \quad (\text{mod } M). \tag{7.6.7}$$

One drawback of ICGs is the cost of inversion, which is $O(\log_2 M)$ times the cost of multiplication modulo M.

7.6.2 GENERATING NON-UNIFORM RANDOM VARIABLES

Suppose we want deviates from a distribution with probability density function $f(x)$ and distribution function $F(x) = \int_{-\infty}^{x} f(u)\, du$. In the following "$y$ is $U[0,1)$" means y is uniformly distributed on $[0,1)$.

Two general techniques for converting uniform random variables into those from other distributions are as follows:

1. The *inverse transform method*:

    If y is $U[0,1)$, then the random variable $F^{-1}(y)$ will have its density equal to $f(x)$. (Note that $F^{-1}(y)$ exists since $0 \le F(x) \le 1$.)

2. The *acceptance-rejection method*:

    Suppose the density can be written as $f(x) = Ch(x)g(x)$ where $h(x)$ is the density of a computable random variable, the function g satisfies $0 < g(x) \le 1$, and $C^{-1} = \int_{-\infty}^{\infty} h(u)g(u)\, du$ is a normalization constant. If x is $U[0,1)$, y has density $h(x)$, and if $x < g(y)$, then x has density $f(x)$. Thus one generates $\{x, y\}$ pairs, rejecting both if $x \ge g(y)$ and returning x if $x < g(y)$.

Examples of the inverse transform method:

1. *Exponential distribution*: The exponential distribution with rate λ has $f(x) = \lambda e^{-\lambda x}$ (for $x \ge 0$) and $F(x) = 1 - e^{-\lambda x}$. Thus $u = F(x)$ can be solved to give $x = F^{-1}(u) = -\lambda^{-1} \ln(1 - u)$. If u is $U[0,1)$, then so is $1 - u$. Hence $x = -\lambda^{-1} \ln u$ is exponentially distributed with rate λ.

2. *Normal distribution*: Suppose the z_i's are normally distributed with density function $f(z) = \frac{1}{\sqrt{2\pi}} e^{-z^2/2}$. The polar transformation then gives random variables $r = \sqrt{z_1^2 + z_2^2}$ (exponentially distributed with $\lambda = 2$) and $\theta = \tan^{-1}(z_2/z_1)$ (uniformly distributed on $\left[-\frac{\pi}{2}, \frac{\pi}{2}\right]$). Inverting these relationships results in $z_1 = \sqrt{-2 \ln x_1} \cos 2\pi x_2$ and $z_2 = \sqrt{-2 \ln x_1} \sin 2\pi x_2$; each is normally distributed when x_1 and x_2 are $U[0,1)$. (This is the *Box–Muller technique*.)

Examples of the rejection method:

1. *Exponential distribution with $\lambda = 1$*:

    (a) Generate random numbers $\{U_i\}_{i=1}^{N}$ uniformly in $[0, 1]$, stopping at $N = \min\{n \mid U_1 \ge U_2 \ge U_{n-1} < U_n\}$.

    (b) If N is even, accept that run, and go to step (c). If N is odd reject the run, and return to step (a).

(c) Set X equal to the number of failed runs plus U_1 (the first random number in the successful run).

2. *Normal distribution*:

    (a) Select two random variables (V_1, V_2) from $U[0, 1)$. Form $R = V_1^2 + V_2^2$.
    (b) If $R > 1$, then reject the (V_1, V_2) pair, and select another pair.
    (c) If $R < 1$, then $x = V_1 \sqrt{-2 \dfrac{\ln R}{R}}$ has a $N(0, 1)$ distribution.

3. *Normal distribution*:

    (a) Select two exponentially distributed random variables with rate $1 : (V_1, V_2)$.
    (b) If $V_2 \geq (V_1 - 1)^2 / 2$, then reject the (V_1, V_2) pair, and select another pair.
    (c) Otherwise, V_1 has a $N(0, 1)$ distribution.

4. *Cauchy distribution*: To generate values of X from $f(x) = \frac{1}{\pi(1+x^2)}$ on $-\infty < x < \infty$,

    (a) Generate random numbers U_1, U_2 (uniform on $[0, 1)$), and set
    $Y_1 = U_1 - \frac{1}{2}, Y_2 = U_2 - \frac{1}{2}$.
    (b) If $Y_1^2 + Y_2^2 \leq \frac{1}{4}$, then return $X = Y_1/Y_2$. Otherwise return to step (a).

    To generate values of X from a Cauchy distribution with parameters β and θ,
    $$f(x) = \frac{\beta}{\pi\left[\beta^2 + (x - \theta)^2\right]}, \text{ for } -\infty < x < \infty, \text{ construct } X \text{ as above, and}$$
    then use $\beta X + \theta$.

7.6.2.1   Discrete random variables

The density function of a discrete random variable that attains finitely many values can be represented as a vector $\mathbf{p} = (p_0, p_1, \ldots, p_{n-1}, p_n)$ by defining the probabilities $P(x = j) = p_j$ (for $j = 0, \ldots, n$). The distribution function can be defined by the vector $\mathbf{c} = (c_0, c_1, \ldots, c_{n-1}, 1)$, where $c_j = \sum_{i=0}^{j} p_i$. Given this representation of $F(x)$, we can apply the inverse transform by computing x to be $U[0, 1)$, and then finding the index j so that $c_j \leq x < c_{j+1}$. In this case event j will have occurred. Examples:

1. (Binomial distribution) The binomial distribution with n trials of mean p has $p_j = \binom{n}{j} p^j (1 - p)^{n-j}$, for $j = 0, \ldots, n$.

    (a) As an example, consider the result of flipping a fair coin. In 2 flips, the probability of obtaining $(0, 1, 2)$ heads is $\mathbf{p} = (\frac{1}{4}, \frac{1}{2}, \frac{1}{4})$. Hence $\mathbf{c} = (\frac{1}{4}, \frac{3}{4}, 1)$. If x (chosen from $U[0, 1)$) turns out to be say, 0.4, then "1 head" is returned (since $\frac{1}{4} \leq 0.4 < \frac{3}{4}$).

    (b) Note that, when n is large, it is costly to compute the density and distribution vectors. When n is large and relatively few binomially distributed pseudorandom numbers are desired, an alternative is to use the normal approximation to the binomial.

(c) Alternately, one can form the sum $\sum_{i=1}^{n} \lfloor u_i + p \rfloor$, where each u_i is $U[0, 1)$.

2. (Geometric distribution) To simulate a value from $P(X = i) = p(1 - p)^{i-1}$ for $i \geq 1$, use $X = 1 + \left\lceil \dfrac{\log U}{\log(1 - p)} \right\rceil$.

3. (Poisson distribution) The Poisson distribution with mean λ has $p_j = \lambda^j e^{-\lambda} / j!$ for $j \geq 0$. The Poisson distribution counts the number of events in a unit time interval if the times are exponentially distributed with rate λ. Thus if the times t_i are exponentially distributed with rate λ, then j will be Poisson distributed with mean λ when $\sum_{i=0}^{j} t_i \leq 1 \leq \sum_{i=0}^{j+1} t_i$. Since $t_i = -\lambda^{-1} \ln u_i$, where u_i is $U[0, 1)$, the previous equation may be written as $\prod_{i=0}^{j} u_i \geq e^{-\lambda} \geq \prod_{i=0}^{j+1} u_i$. This allows us to compute Poisson random variables by iteratively computing $P_j = \prod_{i=0}^{j} u_i$ until $P_j < e^{-\lambda}$. The first such j that makes this inequality true will have the desired distribution.

Random variables can be simulated using the following table (each U and U_i is uniform on the interval $[0, 1)$):

Distribution	Density	Formula for deviate
Binomial	$p_j = \dbinom{n}{j} p^j (1 - p)^{n-j}$	$\sum_{i=1}^{n} \lfloor U_i + p \rfloor$
Cauchy	$f(x) = \dfrac{\sigma}{\pi(x^2 + \sigma^2)}$	$\sigma \tan(\pi U)$
Exponential	$f(x) = \lambda e^{-\lambda x}$	$-\lambda^{-1} \ln U$
Pareto	$f(x) = ab^a / x^{a+1}$	$b / U^{1/a}$
Rayleigh	$f(x) = x/\sigma e^{-x^2/2\sigma^2}$	$\sigma \sqrt{-\ln U}$

7.6.2.2 Testing pseudorandom numbers

The prudent way to check a complicated computation that makes use of pseudorandom numbers is to run it several times with different types of pseudorandom number generators and see if the results appear consistent across the generators. The fact that this is not always possible or practical has led researchers to develop statistical tests of randomness that should be passed by general purpose pseudorandom number generators. Some common tests are the spectral test, the equidistribution test, the serial test, the runs test, the coupon collector test, and the birthday spacing test.

7.7 CONTROL CHARTS AND RELIABILITY

7.7.1 CONTROL CHARTS

Control charts are graphical tools used to assess and maintain the stability of a process. They are used to separate random variation from specific causes. Data measurements are plotted versus time along with upper and lower control limits and a center line. If the process is in control and the underlying distribution is normal, then the control limits represent three standard deviations from the center line (mean).

If all of the data points are contained within the control limits, the process is considered stable and the mean and standard deviations can be reliably calculated. The variations between data points occur from random causes. Data outside the control limits or forming abnormal patterns point to unstable, out-of-control processes.

In the tables, k denotes the number of samples taken, i is an index for the samples ($i = 1 \dots k$), n is the sample size (number of elements in each sample), and R is the range of the values in a sample (maximum element value minus minimum element value). The mean is μ and the standard deviation is σ. Control chart upper and lower control limits are denoted UCL and LCL.

Types of control charts, their statistics, and uses

Chart	Statistics	Statistical quantity
	Applications	
$\bar{x} - R$	Gaussian	Average value and range
	Charts continuous measurable quantities. Measurements taken on small sample sets.	
$\tilde{x} - R$	Gaussian	Median value and range
	Similar to $\bar{x} - R$ chart but fewer calculations needed for plotting.	
$x - Rs$	Gaussian	Individual measured values
	Similar to $\bar{x} - R$ chart but single measurements are made. Used when measurements are expensive or dispersion of measured values is small. $Rs = |x_i - x_{i-1}|$.	
pn	Binomial	Number of defective units
	Charts number of defective units in sets of fixed size.	
p	Binomial	Percent defective
	Charts number of defective units in sets of varying size.	
c	Poisson	Number of defects
	Charts number of flaws in a product of fixed size.	
u	Poisson	Defect density (defects per quantity unit)
	Charts the defect density on a product of varying size.	

Types of control charts and limits ("P" stands for parameter)

Chart	(μ, σ) known?	P	Centerline	UCL	LCL
$\overline{x} - R$	No	$\overline{x}$	$\overline{\overline{x}} = \frac{\sum \overline{x}}{k}$	$\overline{\overline{x}} + A_2\overline{R}$	$\overline{\overline{x}} - A_2\overline{R}$
$\overline{x} - R$	No	R	$\overline{R} = \frac{\sum R}{k}$	$D_4\overline{R}$	$D_3\overline{R}$
$\overline{x} - R$	Yes	$\overline{x}$	$\overline{\overline{x}} = \mu$	$\mu + \frac{3\sigma}{\sqrt{n}}$	$\mu - \frac{3\sigma}{\sqrt{n}}$
$\overline{x} - R$	Yes	R	$\overline{R} = d_2\sigma$	$D_2\sigma$	$D_1\sigma$
$\tilde{x} - R$	No	$\tilde{x}$	$\overline{\tilde{x}} = \frac{\sum \tilde{x}}{k}$	$\tilde{x} + m_3 A_2$	$\tilde{x} - m_3 A_2$
$\tilde{x} - R$	No	R	$\overline{R} = \frac{\sum R}{k}$	$D_4\overline{R}$	$D_3\overline{R}$
$x - Rs$	No	x	$\overline{x} = \frac{\sum x}{k}$	$\overline{x} + 2.66\overline{R}s$	$\overline{x} - 2.66\overline{R}s$
$x - Rs$	No	Rs	$\overline{R}s = \frac{\sum \overline{R}s}{k}$	$3.27\overline{R}s$	—
pn	No	pn	$\overline{p}n = \frac{\sum pn}{k}$	$\overline{p}n + \sqrt{\overline{p}n(1 - \overline{p})}$	$\overline{p}n - \sqrt{\overline{p}n(1 - \overline{p})}$
p	No	p	$\overline{p} = \frac{\sum pn}{\sum n}$	$\overline{p}n + 3\sqrt{\frac{\overline{p}(1-\overline{p})}{n}}$	$\overline{p}n - 3\sqrt{\frac{\overline{p}(1-\overline{p})}{n}}$
c	No	c	$\overline{c} = \frac{\sum c}{k}$	$\overline{c} + 3\sqrt{\overline{c}}$	$\overline{c} - 3\sqrt{\overline{c}}$
u	No	u	$\overline{u} = \frac{\sum c}{\sum n}$	$\overline{u} + 3\sqrt{\frac{\overline{u}}{n}}$	$\overline{u} - 3\sqrt{\frac{\overline{u}}{n}}$

Sample size n	A_2	d_2	D_1	D_2	D_3	D_4	m_3	$m_3 A_2$
2	1.880	1.128	0	3.686	–	3.267	1.000	1.880
3	1.023	1.693	0	4.358	–	2.575	1.160	1.187
4	0.729	2.059	0	4.698	–	2.282	1.092	0.796
5	0.577	2.326	0	4.918	–	2.115	1.198	0.691
6	0.483	2.534	0	5.078	–	2.004	1.135	0.549
7	0.419	2.704	0.205	5.203	0.076	1.924	1.214	0.509
8	0.373	2.847	0.387	5.307	0.136	1.864	1.160	0.432
9	0.337	2.970	0.546	5.394	0.184	1.816	1.223	0.412
10	0.308	3.078	0.687	5.469	0.223	1.777	1.176	0.363
11	0.285	3.173	0.812	5.534	0.256	1.744		
12	0.266	3.258	0.924	5.592	0.284	1.716		
13	0.249	3.336	1.026	5.646	0.308	1.692		
14	0.235	3.407	1.121	5.693	0.329	1.671		
15	0.223	3.472	1.207	5.737	0.348	1.652		
16	0.212	3.532	1.285	5.779	0.364	1.636		
17	0.203	3.588	1.359	5.817	0.379	1.621		
18	0.194	3.640	1.426	5.854	0.392	1.608		
19	0.187	3.689	1.490	5.888	0.404	1.596		
20	0.180	3.735	1.548	5.922	0.414	1.586		
21	0.173	3.778	1.605	5.951	0.425	1.575		
22	0.167	3.819	1.659	5.979	0.434	1.566		
23	0.162	3.858	1.710	6.006	0.443	1.557		
24	0.157	3.895	1.759	6.031	0.452	1.548		
25	0.153	3.931	1.806	6.056	0.459	1.541		

Abnormal Distributions of Points in Control Charts

Abnormality	Description
Sequence	Seven or more consecutive points on one side of the center line. Denotes the average value has shifted.
Bias	Fewer than seven consecutive points on one side of the center line, but most of the points are on that side. • 10 of 11 consecutive points • 12 or more of 14 consecutive points • 14 or more of 17 consecutive points • 16 or more of 20 consecutive points
Trend	Seven or more consecutive rising or falling points.
Approaching the limit	Two out of three or three or more out of seven consecutive points are more than two-thirds the distance from the center line to a control limit.
Periodicity	The data points vary in a regular periodic pattern.

7.7.2 ACCEPTANCE SAMPLING

Expression	Meaning
AQL	acceptable quality level
AOQ	average outgoing quality
AOQL	average outgoing quality limit (maximum value of AOQ for varying incoming quality)
LTPD	lot tolerance percent defective
producer's risk	Type I error (percentage of "good" lots rejected)
consumer's risk	Type II error (percentage of "bad" lots accepted)

Military standard 105 D is a widely used sampling plan. There are three general levels of inspection corresponding to different consumer's risks. (Inspection level II is usually chosen; level I uses smaller sample sizes and level III uses larger sample sizes.) There are also three types of inspections: normal, tightened, and reduced. Tables are available for single, double, and multiple sampling.

To use MIL-STD-105 D for single sampling, determine the sample size code letter from Figure 7.3. Using this sample size code letter find the sample size and the acceptance and rejection numbers from the table on page 654.

EXAMPLE   Suppose that MIL-STD-105 D is to be used with incoming lots of 1,000 items, inspection level II is to be used in conjunction with normal inspection, and an AQL of 2.5 percent is desired. How should the inspections be carried out?

1. From Figure 7.3 the sample size code letter is J.

2. From page 654, for column J, the lot size is 80. Using the row labeled 2.5 the acceptance number is 5 and the rejection number is 6.

3. Thus, if a single sample of size 80 (selected randomly from each lot of 1,000 items) contains 5 or fewer defectives then the lot is to be accepted. If it contains 6 or more

FIGURE 7.3
Sample size code letters for MIL-STD-105 D.

Lot or batch size			general inspection levels		
			I	II	III
2	to	8	A	A	B
9	to	15	A	B	C
16	to	25	B	C	D
26	to	50	C	D	E
51	to	90	C	E	F
91	to	150	D	F	G
151	to	280	E	G	H
281	to	500	F	H	J
501	to	1,200	G	J	K
1,201	to	3,200	H	K	L
3,201	to	10,000	J	L	M
10,001	to	35,000	K	M	N
35,001	to	150,000	L	N	P
150,001	to	500,000	M	P	Q
500,001	and	over	N	Q	R

defectives, then the lot is to be rejected.

7.7.3 RELIABILITY

1. The *reliability* of a product is the probability that the product will function within specified limits for at least a specified period of time.
2. A *series system* is one in which the entire system will fail if any of its components fail.
3. A *parallel system* is one in which the entire system will fail only if all of its components fail.
4. Let R_i denote the reliability of the ith component.
5. Let R_s denote the reliability of a series system.
6. Let R_p denote the reliability of a parallel system.

The *product law of reliabilities* states

$$R_s = \prod_{i=1}^{n} R_i. \tag{7.7.1}$$

The *product law of unreliabilities* states

$$R_p = 1 - \prod_{i=1}^{n} (1 - R_i). \tag{7.7.2}$$

FIGURE 7.4

Master table for single sampling inspection (normal inspection) MIL-STD-105 D.

Sample size code letter and sample size. Cell values are given as Ac/Re (Accept number / Reject number). ↑ and ↓ are the routing arrows described in the legend.

AQL	A	B	C	D	E	F	G	H	J	K	L	M	N	P	Q	R
sample size →	2	3	5	8	13	20	32	50	80	125	200	315	500	800	1250	2000
0.010	↑	↑	↑	↑	↑	↑	↑	↑	↑	↑	↑	↑	↑	↑	0/1	↓
0.015	↑	↑	↑	↑	↑	↑	↑	↑	↑	↑	↑	↑	↑	0/1	↓	↑
0.025	↑	↑	↑	↑	↑	↑	↑	↑	↑	↑	↑	↑	0/1	↓	↑	1/2
0.040	↑	↑	↑	↑	↑	↑	↑	↑	↑	↑	↑	0/1	↓	↑	1/2	2/3
0.065	↑	↑	↑	↑	↑	↑	↑	↑	↑	↑	0/1	↓	↑	1/2	2/3	3/4
0.10	↑	↑	↑	↑	↑	↑	↑	↑	↑	0/1	↓	↑	1/2	2/3	3/4	5/6
0.15	↑	↑	↑	↑	↑	↑	↑	↑	0/1	↓	↑	1/2	2/3	3/4	5/6	7/8
0.25	↑	↑	↑	↑	↑	↑	↑	0/1	↓	↑	1/2	2/3	3/4	5/6	7/8	10/11
0.40	↑	↑	↑	↑	↑	↑	0/1	↓	↑	1/2	2/3	3/4	5/6	7/8	10/11	14/15
0.65	↑	↑	↑	↑	↑	0/1	↓	↑	1/2	2/3	3/4	5/6	7/8	10/11	14/15	21/22
1.0	↑	↑	↑	↑	0/1	↓	↑	1/2	2/3	3/4	5/6	7/8	10/11	14/15	21/22	30/31
1.5	↑	↑	↑	0/1	↓	↑	1/2	2/3	3/4	5/6	7/8	10/11	14/15	21/22	30/31	44/45
2.5	↑	↑	0/1	↓	↑	1/2	2/3	3/4	5/6	7/8	10/11	14/15	21/22	30/31	44/45	↓
4.0	↑	0/1	↓	↑	1/2	2/3	3/4	5/6	7/8	10/11	14/15	21/22	30/31	44/45	↓	↓
6.5	0/1	↓	↑	1/2	2/3	3/4	5/6	7/8	10/11	14/15	21/22	30/31	44/45	↓	↓	↓
10	↓	↑	1/2	2/3	3/4	5/6	7/8	10/11	14/15	21/22	30/31	44/45	↓	↓	↓	↓
15	↑	1/2	2/3	3/4	5/6	7/8	10/11	14/15	21/22	30/31	44/45	↓	↓	↓	↓	↓
25	1/2	2/3	3/4	5/6	7/8	10/11	14/15	21/22	30/31	44/45	↓	↓	↓	↓	↓	↓
40	2/3	3/4	5/6	7/8	10/11	14/15	21/22	30/31	44/45	↓	↓	↓	↓	↓	↓	↓
65	3/4	5/6	7/8	10/11	14/15	21/22	30/31	44/45	↓	↓	↓	↓	↓	↓	↓	↓
100	5/6	7/8	10/11	14/15	21/22	30/31	44/45	↓	↓	↓	↓	↓	↓	↓	↓	↓
150	7/8	10/11	14/15	21/22	30/31	44/45	↓	↓	↓	↓	↓	↓	↓	↓	↓	↓
250	10/11	14/15	21/22	30/31	44/45	↓	↓	↓	↓	↓	↓	↓	↓	↓	↓	↓
400	14/15	21/22	30/31	44/45	↓	↓	↓	↓	↓	↓	↓	↓	↓	↓	↓	↓
650	21/22	30/31	44/45	↓	↓	↓	↓	↓	↓	↓	↓	↓	↓	↓	↓	↓
1000	30/31	44/45	↓	↓	↓	↓	↓	↓	↓	↓	↓	↓	↓	↓	↓	↓

AQL   Acceptable quality level (normal inspection).

Ac|Re   Accept if Ac or fewer are found, reject if Re or more are found.

←   Use first sampling procedure to left.

→   Use first sampling procedure to right. If sample size equals, or exceeds, lot or batch size, do 100 percent inspection.

7.7.4 FAILURE TIME DISTRIBUTIONS

1. Let the probability of an item failing between times t and $t + \Delta t$ be $f(t)\Delta t + o(\Delta t)$ as $\Delta t \to 0$.

2. The probability that an item will fail in the interval from 0 to t is

$$F(t) = \int_0^t f(x)\, dx. \tag{7.7.3}$$

3. The *reliability function* is the probability that an item survives to time t

$$R(t) = 1 - F(t). \tag{7.7.4}$$

4. The *instantaneous hazard rate*, $Z(t)$, is approximately the probability of failure in the interval from t to $t + \Delta t$, given that the item survived to time t

$$Z(t) = \frac{f(t)}{R(t)} = \frac{f(t)}{1 - F(t)}. \tag{7.7.5}$$

Note the relationships:

$$R(t) = e^{-\int_0^t Z(x)\, dx} \qquad\qquad f(t) = Z(t)e^{-\int_0^t Z(x)\, dx} \tag{7.7.6}$$

EXAMPLE If $f(t) = \alpha\beta t^{\beta-1}e^{\alpha t^\beta}$ with $\alpha > 0$ and $\beta > 0$, the probability distribution function for a Weibull random variable, then the failure rate is $Z(t) = \alpha\beta t^{\beta-1}$ and $R(t) = e^{-\alpha t^\beta}$. Note that failure rate decreases with time if $\beta < 1$ and increases with time if $\beta > 1$.

7.7.4.1 Use of the exponential distribution

If the hazard rate is a constant $Z(t) = \alpha$ (with $\alpha > 0$) then $f(t) = \alpha e^{-\alpha t}$ (for $t > 0$) which is the probability density function for an exponential random variable. If a failed item is replaced with another having the same constant hazard rate α, then the sequence of occurrence of failures is a Poisson process. The constant $1/\alpha$ is called the *mean time between failures* (MTBF). The reliability function is $R(t) = e^{-\alpha t}$.

If a series system has n components, each with constant hazard rate $\{\alpha_i\}$, then

$$R_s(t) = \exp\left(-\sum_{i=1}^n \alpha_i\right). \tag{7.7.7}$$

The MTBF for the series system is μ_s

$$\mu_s = \frac{1}{\frac{1}{\mu_1} + \frac{1}{\mu_2} + \cdots + \frac{1}{\mu_n}}. \tag{7.7.8}$$

If a parallel system has n components, each with identical constant hazard rate α, then the MTBF for the parallel system is μ_p

$$\mu_p = \frac{1}{\alpha}\left(1 + \frac{1}{2} + \cdots + \frac{1}{n}\right). \tag{7.7.9}$$

7.8   RISK ANALYSIS AND DECISION RULES

Suppose knowledge of a specific state of a system is desired, and those states can be delineated as $\{\theta_1, \theta_2, \dots\}$. (In a weather application the states might be *rain* and *no rain*.) Decision rules are actions that may be taken based on the state of a system. For example, in making a decision about a trip, there are the decision rules: *stay home*, *go with an umbrella*, and *go without an umbrella*.

A *loss function* is a function that depends on a specific state and a decision rule. For example, consider the following loss function $\ell(\theta, a)$:

Loss function data

Possible actions		System state	
		θ_1 (rain)	θ_2 (no rain)
Stay home	a_1	4	4
Go without an umbrella	a_2	5	0
Go with an umbrella	a_3	2	5

It is possible to determine the "best" decision, under different models, even without obtaining any data.

1. **Minimax principle**

   With this principle one should expect and prepare for the worst. That is, for each action it is possible to determine the minimum possible loss that may be incurred. This loss is assigned to each action; the action with the smallest (or minimum) maximum loss is the action chosen.

   For the given loss function data the maximum loss is 4 for action a_1 and 5 for either of the actions a_2 or a_3. Under a minimax principle, the chosen action would be a_1 and the minimax loss would be 4.

2. **Minimax principle for mixed actions**

   It is possible to minimize the maximum loss when the action taken is a statistical distribution, **p**, of actions. Assume that action a_i is taken with probability p_i (with $p_1 + p_2 + p_3 = 1$). Then the expected loss $L(\theta_i)$ is given by $L(\theta_i) = E_a[\ell(\theta_i, a)] = p_1\ell(\theta_i, a_1) + p_2\ell(\theta_i, a_2) + p_3\ell(\theta_i, a_3)$. The given loss function data results in the following expected losses:

$$\begin{bmatrix} L(\theta_1) \\ L(\theta_2) \end{bmatrix} = p_1 \begin{bmatrix} 4 \\ 4 \end{bmatrix} + p_2 \begin{bmatrix} 5 \\ 0 \end{bmatrix} + p_3 \begin{bmatrix} 2 \\ 5 \end{bmatrix}. \tag{7.8.1}$$

   It can be shown that the minimax point of this mixed action case has to satisfy $L(\theta_1) = L(\theta_2)$. Solving equation (7.8.1) with this constraint leads to $5p_2 = 3p_3$. Using this and $p_1 + p_2 + p_3 = 1$ in equation (7.8.1) results in $L(\theta_1) = L(\theta_2) = 4 - 7p_3/5$. This indicates that p_3 should be as large as possible. Hence, the maximum value is obtained by the mixed distribution $\mathbf{p} = (\frac{0}{8}, \frac{3}{8}, \frac{5}{8})$.

Hence, if action a_2 is chosen 3/8's of the time, and action a_3 is chosen 5/8's of the time, then the minimax loss is equal to $L = 25/8$. This is a smaller loss than using a pure strategy of only choosing a single action.

3. **Bayes actions**

   If the probability distribution of the states $\{\theta_1, \theta_2, \dots\}$ is given by the density function $g(\theta_i)$, then the loss has a known distribution with an expectation of $B(a) = E_i[\ell(\theta_i, a)] = \sum_i g(\theta_i)\ell(\theta_i, a)$. This quantity is known as the *Bayes loss* for action a. A *Bayes action* is an action that minimizes the Bayes loss.

   For example, assuming that the prior distribution is given by $g(\theta_1) = 0.4$ and $g(\theta_2) = 0.6$, then $B(a_1) = 4$, $B(a_2) = 2$, and $B(a_3) = 3.8$. This leads to the choice of action a_2.

A course of action can also be based on data about the states of interest. For example, a weather report Z will give data for the predictions of *rain* and *no rain*. Continuing the example, assume that the correctness of these predictions is given as follows:

		θ_1 (rain)	θ_2 (no rain)
Predict rain	z_1	0.8	0.1
Predict no rain	z_2	0.2	0.9

That is, when it will rain, then the prediction is correct 80% of the time.

A *decision function* is an assignment of data to actions. Since there are finitely many possible actions and finitely many possible values of Z, the number of decision functions is finite. For this example there are $3^2 = 9$ possible decision functions, $\{d_1, d_2, \dots, d_9\}$; they are defined to be:

	Decision functions								
	d_1	d_2	d_3	d_4	d_5	d_6	d_7	d_8	d_9
Predict z_1, take action	a_1	a_2	a_3	a_1	a_2	a_1	a_3	a_2	a_3
Predict z_2, take action	a_1	a_2	a_3	a_2	a_1	a_3	a_1	a_3	a_2

The *risk function* $R(\theta, d_i)$ is the expected value of the loss when a specific decision function is being used: $R(\theta, d_i) = E_Z[\ell(\theta, d_i(Z))]$. It is straightforward to compute the risk function for all values of $\{d_i\}$ and $\{a_j\}$. This results in the following values:

Risk function evaluation		
Decision Function	θ_1 (rain)	θ_2 (no rain)
d_1	4	4
d_2	5	0
d_3	2	5
d_4	4.2	0.4
d_5	4.8	3.6
d_6	3.6	4.9
d_7	2.4	4.1
d_8	4.4	4.5
d_9	2.6	0.5

This array can now be treated as though it gave the loss function in a no–data problem. The minimax principle for mixed action results in the "best" solution being rule d_3 for $\frac{7}{17}$'s of the time and rule d_9 for $\frac{10}{17}$'s of the time. This leads to a minimax loss of $\frac{40}{17}$. Before the data Z is received, the minimax loss was $\frac{25}{8}$. Hence, the data Z is "worth" $\frac{25}{8} - \frac{40}{17} = \frac{105}{136}$ in using the minimax approach.

The *regret function* (also called the *opportunity loss function*) $r(\theta, a)$ is the loss, $\ell(\theta, a)$, minus the minimum loss for the given θ: $r(\theta, a) = \ell(\theta, a) - \min_b \ell(\theta, b)$. For each state, the least loss is determined if that state were known to be true. This is the contribution to loss that even a good decision cannot avoid. The quantity $r(\theta, a)$ represents the loss that could have been avoided had the state been known—hence the term regret.

For the given loss function data, the minimum loss for $\theta = \theta_1$ is 2, and the minimum loss for $\theta = \theta_2$ is 0. Hence, the regret function is

	θ_1 (rain)	θ_2 (no rain)
a_1	2	4
a_2	3	0
a_3	0	5

Most of the computations performed for a loss function could also be performed with the risk function. If the minimax principle is used to determine the "best" action, then, in this example, the "best" action is a_2.

7.9 STATISTICS

7.9.1 DESCRIPTIVE STATISTICS

1. Sample distribution and density functions

    (a) Sample distribution function:

$$\widehat{F}(x) = \frac{1}{n} \sum_{i=1}^{n} u(x - x_i) \qquad (7.9.1)$$

   where $u(x)$ is the unit step function (or Heaviside function) defined by $u(x) = 0$ for $x \le 0$ and $u(x) = 1$ for $x > 0$.

    (b) Sample density function or histogram:

$$\hat{f}(x) = \frac{\widehat{F}(x_0 + (i+1)w) - \widehat{F}(x_0 + iw)}{w} \qquad (7.9.2)$$

   for $x \in [x_0 + iw, x_0 + (i+1)w)$. The interval $[x_0 + iw, x_0 + (i+1)w)$ is called the ith bin, w is the bin width, and $f_i = \widehat{F}(x_0 + (i+1)w) - \widehat{F}(x_0 + iw)$ is the *bin frequency*.

2. Order statistics and quantiles:

   (a) Order statistics are obtained by arranging the sample values $\{x_1, \ldots, x_n\}$ in increasing order, denoted by

   $$x_{(1)} \leq x_{(2)} \leq \cdots \leq x_{(n)}. \qquad (7.9.3)$$

   i. $x_{(1)}$ and $x_{(n)}$ are the minimum and maximum data values, respectively.

   ii. For $i = 1, \ldots, n$, $x_{(i)}$ is called the ith *order statistic*.

   (b) *Quantiles*: If $0 < p < 1$, then the *quantile of order p*, ξ_p, is defined as the $p(n+1)^{\text{th}}$ order statistic. It may be necessary to interpolate between successive values.

   i. If $p = j/4$ for $j = 1, 2$, or 3, then $\xi_{\frac{j}{4}}$ is called the jth *quartile*.

   ii. If $p = j/10$ for $j = 1, 2, \ldots, 9$, then $\xi_{\frac{j}{10}}$ is called the jth *decile*.

   iii. If $p = j/100$ for $j = 1, 2, \ldots, 99$, then $\xi_{\frac{j}{100}}$ is called the jth *percentile*.

3. Measures of central tendency

   (a) *Arithmetic mean*:

   $$\overline{x} = \frac{1}{n} \sum_{i=1}^{n} x_i = \frac{x_1 + x_2 + \cdots + x_n}{n}. \qquad (7.9.4)$$

   (b) *α-trimmed mean*:

   $$\overline{x}_\alpha = \frac{1}{n(1 - 2\alpha)} \left((1 - r)\left(x_{(k+1)} + x_{(n-k)}\right) + \sum_{i=k+2}^{n-k-1} x_{(i)} \right), \qquad (7.9.5)$$

   where $k = \lfloor \alpha n \rfloor$ is the greatest integer less than or equal to αn, and $r = \alpha n - k$. If $\alpha = 0$ then $\overline{x}_\alpha = \overline{x}$.

   (c) *Weighted mean*: If to each x_i is associated a weight $w_i \geq 0$ so that

   $$\sum_{i=1}^{n} w_i = 1, \quad \text{then} \quad \overline{x}_{\text{w}} = \sum_{i=1}^{n} w_i x_i. \qquad (7.9.6)$$

   (d) *Geometric mean*:

   $$\text{G.M.} = \left(\prod_{i=1}^{n} x_i \right)^{\frac{1}{n}} = (x_1 x_2 \cdots x_n)^{\frac{1}{n}}. \qquad (7.9.7)$$

(e) *Harmonic mean*:

$$\text{H.M.} = \frac{n}{\sum_{i=1}^{n} \frac{1}{x_i}} = \frac{n}{\frac{1}{x_1} + \frac{1}{x_2} + \cdots + \frac{1}{x_n}}. \tag{7.9.8}$$

(f) Relationship between arithmetic, geometric, and harmonic means:

$$\text{H.M.} \leq \text{G.M.} \leq \overline{x} \tag{7.9.9}$$

with equality holding only when all sample values are equal.

(g) The *mode* is the data value that occurs with the greatest frequency. Note that the mode may not be unique.

(h) *Median*:

   i. If n is odd and $n = 2k + 1$, then $M = x_{(k+1)}$.
   ii. If n is even and $n = 2k$, then $M = (x_{(k)} + x_{(k+1)})/2$.

(i) *Midrange*:

$$\text{mid} = \frac{x_{(1)} + x_{(n)}}{2}. \tag{7.9.10}$$

4. Measures of dispersion

   (a) *Mean deviation* or *absolute deviation*:

$$\text{M.D.} = \frac{1}{n} \sum_{i=1}^{n} |x_i - \overline{x}|, \quad \text{or} \quad \text{A.D.} = \frac{1}{n} \sum_{i=1}^{n} |x_i - M|. \tag{7.9.11}$$

   (b) *Sample standard deviation*:

$$s = \sqrt{\frac{1}{n-1} \sum_{i=1}^{n} (x_i - \overline{x})^2} = \sqrt{\frac{\sum_{i=1}^{n} x_i^2 - n\overline{x}^2}{n-1}}. \tag{7.9.12}$$

   (c) The *sample variance* is the square of the sample standard deviation.

   (d) *Root mean square*: $\text{R.M.S.} = \sqrt{\dfrac{1}{n} \sum_{i=1}^{n} x_i^2}.$

   (e) *Sample range*: $x_{(n)} - x_{(1)}$.

   (f) *Interquartile range*: $\xi_{\frac{3}{4}} - \xi_{\frac{1}{4}}$.

   (g) The quartile deviation or semi-interquartile range is one half the interquartile range.

5. Higher-order statistics

   (a) Sample moments: $m_k = \dfrac{1}{n} \sum_{i=1}^{n} x_i^k.$

   (b) Sample central moments, or sample moments about the mean:

$$\mu_k = \frac{1}{n} \sum_{i=1}^{n} (x_i - \overline{x})^k. \tag{7.9.13}$$

7.9.2 STATISTICAL ESTIMATORS

7.9.2.1 Definitions

1. A function of a set of random variables is a *statistic*. It is a function of observable random variables that does not contain any unknown parameters. A statistic is itself an observable random variable.

2. Let θ be a parameter appearing in the density function for the random variable X. Suppose that we know a formula for computing an approximate value $\widehat{\theta}$ of θ from a given sample $\{x_1, \ldots, x_n\}$ (call such a function g). Then $\widehat{\theta} = g(x_1, x_2, \ldots, x_n)$ can be considered as a single observation of the random variable $\widehat{\Theta} = g(X_1, X_2, \ldots, X_n)$. The random variable $\widehat{\Theta}$ is an *estimator* for the parameter θ.

3. A *hypothesis* is an assumption about the distribution of a random variable X. This may usually be cast into the form $\theta \in \Theta_0$. We use H_0 to denote the *null hypothesis* and H_1 to denote an *alternative hypothesis*.

4. In *significance testing*, a *test statistic* $T = T(X_1, \ldots, X_n)$ is used to *reject* H_0, or to *not reject* H_0. Generally, if $T \in C$, where C is a *critical region*, then H_0 is rejected.

5. A *type I error*, denoted α, is to reject H_0 when it should not be rejected. A *type II error*, denoted β, is to not reject H_0 when it should be rejected.

6. The power of a test is $\eta = 1 - \beta$.

	Unknown truth	
	H_0	H_1
Do not reject H_0	True decision. Probability is $1 - \alpha$	Type II error. Probability is β
Reject H_0	Type I error. Probability is α	True decision. Probability is $\eta = 1 - \beta$

7.9.2.2 Consistent estimators

Let $\widehat{\Theta} = g(X_1, X_2, \ldots, X_n)$ be an estimator for the parameter θ, and suppose that g is defined for arbitrarily large values of n. If the estimator has the property, $\mathrm{E}[(\widehat{\Theta} - \theta)^2] \to 0$, as $n \to \infty$, then the estimator is called a *consistent estimator*.

1. A consistent estimator is not unique.
2. A consistent estimator may be meaningless.
3. A consistent estimator is not necessarily unbiased.

7.9.2.3   Efficient estimators

An unbiased estimator $\widehat{\Theta} = g(X_1, X_2, \ldots, X_n)$ for a parameter θ is said to be *efficient* if it has finite variance ($\mathrm{E}\left[(\widehat{\Theta} - \Theta)^2\right] < \infty$) and if there does not exist another estimator $\widehat{\Theta}^* = g^*(X_1, X_2, \ldots, X_n)$ for θ, whose variance is smaller than that of $\widehat{\Theta}$. The *efficiency* of an unbiased estimator is the ratio,

$$\frac{\text{Cramer–Rao lower bound}}{\text{Actual variance}}.$$

The *relative efficiency* of two unbiased estimators is the ratio of their variances.

7.9.2.4   Maximum likelihood estimators (MLE)

Suppose X is a random variable whose density function is $f(x; \theta)$, where $\theta = (\theta_1, \ldots, \theta_r)$. If the independent sample values $x_1, \ldots, x_n$ are obtained, then define the likelihood function as $L = \prod_{i=1}^{n} f(x_i; \theta)$. The MLE estimate for θ is the solution of the simultaneous equations, $\frac{\partial L}{\partial \theta_i} = 0$, for $i = 1, \ldots, r$.

1. A MLE need not be consistent.
2. A MLE may not be unbiased.
3. A MLE need not be unique.
4. If a single sufficient statistic T exists for the parameter θ, the MLE of θ must be a function of T.
5. Let $\widehat{\Theta}$ be a MLE of θ. If $\tau(\cdot)$ is a function with a single-valued inverse, then a MLE of $\tau(\theta)$ is $\tau(\widehat{\Theta})$.

Define $\overline{x} = \sum_{i=1}^{n} X_i/n$ and $S^2 = \sum_{i=1}^{n} (X_i - \overline{x})^2/n$ (note that $S \neq s$). Then:

Distribution	Estimated parameter	MLE estimate of parameter
Exponential $E(\lambda)$	$1/\lambda$	$1/\overline{x}$
Exponential $E(\lambda)$	$\lambda^2 = \sigma^2$	$\overline{x}^2$
Normal $N(\mu, \sigma)$	μ	$\overline{x}$
Normal $N(\mu, \sigma)$	σ^2	S^2
Poisson $P(\lambda)$	λ	$\overline{x}$
Uniform $U(0, \theta)$	θ	$X_{\max}$

7.9.2.5   Method of moments (MOM)

Let $\{X_i\}$ be independent and identically distributed random variables with density $f(x; \theta)$. Let $\mu'_r(\theta) = E[X^r]$ be the rth moment (if it exists). Let $m'_r = \frac{1}{n} \sum_{i=1}^{n} x_i^r$ be the rth sample moment. Form the k equations, $\mu'_r = m'_r$, and solve to obtain an estimate of θ.

1. MOM estimators are not necessarily uniquely defined.
2. MOM estimators may not be functions of sufficient or complete statistics.

7.9.2.6 Sufficient statistics

A statistic $G = g(X_1, \ldots, X_n)$ is called a *sufficient statistic* if, and only if, the conditional distribution of H, given G, does not depend on θ for any statistic $H = h(X_1, \ldots, X_n)$.

Let $\{X_i\}$ be independent and identically distributed random variables, with density $f(x; \theta)$. The statistics $\{G_1, \ldots, G_r\}$ are said to be *jointly sufficient statistics* if, and only if, the conditional distribution of $X_1, X_2, \ldots, X_n$ given $G_1 = g_1, G_1 = g_2,$ $\ldots, G_r = g_r$ does not depend on θ.

1. A single sufficient statistic may not exist.

7.9.2.7 UMVU estimators

A *uniformly minimum variance unbiased* estimator, called a UMVU estimator, is unbiased and has the minimum variance among all unbiased estimators.

Define, as usual, $\overline{x} = \sum_{i=1}^{n} X_i / n$ and $s^2 = \sum_{i=1}^{n} (X_i - \overline{x})^2 / (n-1)$. Then:

Distribution	Estimated parameter	UMVU estimate of parameter	Variance of estimator
Exponential $E(\lambda)$	λ	$\dfrac{n-1}{s}$	$\dfrac{\lambda^2}{n-2}$
Exponential $E(\lambda)$	$\dfrac{1}{\lambda}$	$\overline{x}$	$\dfrac{1}{n\lambda^2}$
Normal $N(\mu, \sigma)$	μ	$\overline{x}$	$\dfrac{\sigma^2}{n}$
Normal $N(\mu, \sigma)$	σ^2	s^2	$\dfrac{2\sigma^4}{n-1}$
Poisson $P(\lambda)$	λ	$\overline{x}$	$\dfrac{\lambda^2}{n}$
Uniform $U(0, \theta)$	θ	$\dfrac{n+1}{n} X_{\max}$	$\dfrac{\theta^2}{n(n+2)}$

7.9.2.8 Unbiased estimators

An estimator $g(X_1, X_2, \ldots, X_n)$ for a parameter θ is said to be *unbiased* if

$$E[g(X_1, X_2, \ldots, X_n)] = \theta. \tag{7.9.14}$$

1. An unbiased estimator may not exist.
2. An unbiased estimator is not unique.
3. An unbiased estimator may be meaningless.
4. An unbiased estimator is not necessarily consistent.

7.9.3 CRAMER–RAO BOUND

The *Cramer–Rao bound* gives a lower bound on the variance of an unknown un-biased statistical parameter, when n samples are taken. When the single unknown parameter is θ,

$$\sigma^2(\theta) \geq \frac{1}{-n\mathrm{E}\left[\frac{\partial^2}{\partial\theta^2}\log f(x;\theta)\right]} = \frac{1}{n\mathrm{E}\left[\left(\frac{\partial}{\partial\theta}\log f(x;\theta)\right)^2\right]}. \tag{7.9.15}$$

EXAMPLES

1. For a normal random variable with unknown mean θ and known variance σ^2, the density is $f(x;\theta) = \frac{1}{\sqrt{2\pi}\sigma}\exp\left(-\frac{(x-\theta)^2}{2\sigma^2}\right)$. Hence, $\frac{\partial}{\partial\theta}\log f(x;\theta) = (x-\theta)/\sigma^2$. The computation

$$\mathrm{E}\left[\frac{(x-\theta)^2}{\sigma^4}\right] = \int_{-\infty}^{\infty}\frac{(x-\theta)^2}{\sigma^4}\frac{1}{\sqrt{2\pi}\sigma}e^{-(x-\theta)^2/2\sigma^2}\,dx = \frac{1}{\sigma^2}$$

   results in $\sigma^2(\theta) \geq \frac{\sigma^2}{n}$.

2. For a normal random variable with known mean μ and unknown variance $\theta = \sigma^2$, the density is $f(x;\theta) = \frac{1}{\sqrt{2\pi\theta}}\exp\left(-\frac{(x-\mu)^2}{2\theta^2}\right)$. Hence, $\frac{\partial}{\partial\theta}\log f(x;\theta) = ((x-\mu)^2 - 2\theta)/(2\theta)^2$. The computation $\mathrm{E}\left[\frac{(x-\mu)^2-2\theta}{(2\theta)^2}\right] = \frac{1}{2\theta^2} = \frac{1}{2\sigma^4}$ results in $\sigma^2(\theta) \geq 2\sigma^4/n$.

3. For a Poisson random variable with unknown mean θ, the density is $f(x;\theta) = \theta^x e^{-\theta}/x!$. Hence, $\frac{\partial}{\partial\theta}\log f(x;\theta) = x/\theta - 1$. The computation

$$\mathrm{E}\left[\left(\frac{x}{\theta}-1\right)^2\right] = \sum_{x=0}^{\infty}\left(\frac{x}{\theta}-1\right)^2\frac{\theta^x e^{-\theta}}{x!} = \frac{1}{\theta}\text{ results in }\sigma^2(\theta) \geq \theta/n.$$

7.9.4 ORDER STATISTICS

When $\{X_i\}$ are n independent and identically distributed random variables with the common distribution function $F_X(x)$, let Z_m be the mth largest of the values $(m = 1, 2, \ldots, n)$. Hence Z_1 is the maximum of the n values and Z_n is the minimum of the n values. Then

$$F_{Z_m}(x) = \sum_{i=m}^{n}\binom{n}{i}[F_X(x)]^i[1 - F_X(x)]^{n-i}. \tag{7.9.16}$$

Hence

$$F_{\max}(z) = [F_X(z)]^n, \qquad f_{\max}(z) = n[F_X(z)]^{n-1}f_X(z), \tag{7.9.17}$$

$$F_{\min}(z) = 1 - [1 - F_X(z)]^n, \qquad f_{\min}(z) = n[1 - F_X(z)]^{n-1}f_X(z). \tag{7.9.18}$$

The expected value of the ith order statistic is given by

$$\mathrm{E}\left[x_{(i)}\right] = \frac{n!}{(i-1)!(n-i)!}\int_{-\infty}^{\infty}xf(x)F^{i-1}(x)[1 - F(x)]^{n-i}\,dx. \tag{7.9.19}$$

7.9.4.1 Uniform distribution:

If X is uniformly distributed on the interval $[0, 1)$ then

$$E\left[x_{(i)}\right] = \frac{n!}{(i-1)!(n-i)!} \int_0^1 x^i (1-x)^{n-i}\, dx. \tag{7.9.20}$$

The expected value of the largest of n samples is $\frac{n}{n+1}$; the expected value of the least of n samples is $\frac{1}{n+1}$.

7.9.4.2 Normal distribution:

The following table gives values of $E\left[x_{(i)}\right]$ for a standard normal distribution. Missing values (indicated by a dash) may be obtained from $E\left[x_{(i)}\right] = -E\left[x_{(n-i+1)}\right]$.

i	$n=2$	3	4	5	6	7	8	10
1	0.5642	0.8463	1.0294	1.1630	1.2672	1.3522	1.4236	1.5388
2	—	0.0000	0.2970	0.4950	0.6418	0.7574	0.8522	1.0014
3		—	—	0.0000	0.2016	0.3527	0.4728	0.6561
4			—	—	—	0.0000	0.1522	0.3756
5				—	—	—	—	0.1226
6					—	—	—	—

EXAMPLE   If a person of average intelligence takes five intelligence tests (each test having a normal distribution with a mean of 100 and a standard deviation of 20), then the expected value of the largest score is $100 + (1.1630)(20) \approx 123$.

7.9.5 CLASSIC STATISTICS PROBLEMS

7.9.5.1 Sample size problem

Suppose that a Bernoulli random variable is to be estimated from a sample. What sample size n is required so that, with 99% certainty, the error is no more than $e = 5$ percentage points (i.e., $\mathrm{Prob}(|\hat{p} - p| < 0.05) > 0.99$)?

If an *a priori* estimate of p is available, then the minimum sample size is $n_p = z_{\alpha/2}^2 p(1-p)/e^2$. If no *a priori* estimate is available, then $n_n = z_{\alpha/2}^2/4e^2 \geq n_p$. For the numbers above, $n \geq n_n = 664$.

7.9.5.2 Large scale testing with infrequent success

Suppose that a disease occurs in one person out of every 1000. Suppose that a test for this disease has a type I and a type II error of 1% (that is, $\alpha = \beta = 0.01$). Imagine that 100,000 people are tested. Of the 100 people who have the disease, 99 will be diagnosed as having it. Of the 99,900 people who do not have the disease, 999 will be diagnosed as having it. Hence, only $\frac{99}{1098} \approx 9\%$ of the people who test positive for the disease actually have it.

7.10 CONFIDENCE INTERVALS

A probability distribution may have one or more unknown parameters. A *confidence interval* is an assertion that an unknown parameter lies in a computed range, with a specified probability. Before constructing a confidence interval, first select a confidence coefficient, denoted $1 - \alpha$. Typically, $1 - \alpha = 0.95, 0.99$, or the like. The definitions of z_α, t_α, and χ^2_α are in Section 7.14.1 on page 695.

7.10.1 CONFIDENCE INTERVAL: SAMPLE FROM ONE POPULATION

The following confidence intervals assume a random sample of size n, given by $\{x_1, x_2, \ldots, x_n\}$.

1. Find mean μ of the normal distribution with known variance σ^2.

   (a) Determine the critical value $z_{\alpha/2}$ such that $\Phi\left(z_{\alpha/2}\right) = 1 - \alpha/2$, where $\Phi(z)$ is the standard normal distribution function.
   (b) Compute the mean $\bar{x}$ of the sample.
   (c) Compute $k = z_{\alpha/2}\sigma/\sqrt{n}$.
   (d) The $100(1-\alpha)$ percent confidence interval for μ is given by $[\bar{x} - k, \bar{x} + k]$.

2. Find mean μ of the normal distribution with unknown variance σ^2.

   (a) Determine the critical value $t_{\alpha/2}$ such that $F\left(t_{\alpha/2}\right) = 1 - \alpha/2$, where $F(\cdot)$ is the t-distribution with $n - 1$ degrees of freedom.
   (b) Compute the mean $\bar{x}$ and standard deviation s of the sample.
   (c) Compute $k = t_{\alpha/2}s/\sqrt{n}$.
   (d) The $100(1-\alpha)$ percent confidence interval for μ is given by $[\bar{x} - k, \bar{x} + k]$.

3. Find the probability of success p for Bernoulli trials with large sample size.

   (a) Determine the critical value $z_{\alpha/2}$ such that $\Phi\left(z_{\alpha/2}\right) = 1 - \alpha/2$, where $\Phi(z)$ is the standard normal distribution function.
   (b) Compute the proportion $\hat{p}$ of "successes" out of n trials.
   (c) Compute $k = z_{\alpha/2}\sqrt{\dfrac{\hat{p}(1 - \hat{p})}{n}}$.
   (d) The $100(1-\alpha)$ percent confidence interval for p is given by $[\hat{p} - k, \hat{p} + k]$.

4. Find variance σ^2 of the normal distribution.

   (a) Determine the critical values $\chi^2_{\alpha/2}$ and $\chi^2_{1-\alpha/2}$ such that $F\left(\chi^2_{\alpha/2}\right) = 1 - \alpha/2$ and $F\left(\chi^2_{1-\alpha/2}\right) = \alpha/2$, where $F(z)$ is the chi-square distribution function with $n - 1$ degrees of freedom.
   (b) Compute the standard deviation s.

(c) Compute $k_1 = \dfrac{(n-1)s^2}{\chi^2_{\alpha/2}}$ and $k_2 = \dfrac{(n-1)s^2}{\chi^2_{1-\alpha/2}}$.

(d) The $100(1-\alpha)$ percent confidence interval for σ^2 is given by $[k_1, k_2]$.

(e) The $100(1-\alpha)$ percent confidence interval for the standard deviation σ is given by $\left[\sqrt{k_1}, \sqrt{k_2}\right]$.

5. Find quantile ξ_p of order p for large sample sizes.

(a) Determine the critical value $z_{\alpha/2}$ such that $\Phi\left(z_{\alpha/2}\right) = 1 - \alpha/2$, where $\Phi(z)$ is the standard normal distribution function.

(b) Compute the order statistics $x_{(1)}, x_{(2)}, \ldots, x_{(n)}$.

(c) Compute $k_1 = \left\lfloor np - z_{\alpha/2}\sqrt{np(1-p)} \right\rfloor$ and

$k_2 = \left\lceil np + z_{\alpha/2}\sqrt{np(1-p)} \right\rceil$.

(d) The $100(1-\alpha)$ percent confidence interval for ξ_p is given by $\left[x_{(k_1)}, x_{(k_2)}\right]$.

6. Find median M based on the Wilcoxon one-sample statistic for a large sample.

(a) Determine the critical value $z_{\alpha/2}$ such that $\Phi\left(z_{\alpha/2}\right) = 1 - \alpha/2$, where $\Phi(z)$ is the standard normal distribution function.

(b) Compute the order statistics $w_{(1)}, w_{(2)}, \ldots, w_{(N)}$ of the $N = n(n-1)/2$ averages $(x_i + x_j)/2$, for $1 \le i < j \le n$.

(c) Compute $k_1 = \left\lfloor \dfrac{N}{2} - \dfrac{z_{\alpha/2}N}{\sqrt{3n}} \right\rfloor$ and $k_2 = \left\lceil \dfrac{N}{2} + \dfrac{z_{\alpha/2}N}{\sqrt{3n}} \right\rceil$.

(d) The $100(1-\alpha)$ percent confidence interval for M is given by $\left[w_{(k_1)}, w_{(k_2)}\right]$.

7.10.2 CONFIDENCE INTERVAL: SAMPLES FROM TWO POPULATIONS

The following confidence intervals assume random samples from two large populations: one sample of size n, given by $\{x_1, x_2, \ldots, x_n\}$, and one sample of size m, given by $\{y_1, y_2, \ldots, y_m\}$.

1. Find the difference in population means μ_x and μ_y from independent samples with known variances σ_x^2 and σ_y^2.

(a) Determine the critical value $z_{\alpha/2}$ such that $\Phi\left(z_{\alpha/2}\right) = 1 - \alpha/2$, where $\Phi(z)$ is the standard normal distribution function.

(b) Compute the means $\bar{x}$ and $\bar{y}$.

(c) Compute $k = z_{\alpha/2}\sqrt{\dfrac{\sigma_x^2}{n} + \dfrac{\sigma_y^2}{m}}$.

(d) The $100(1-\alpha)$ percent confidence interval for $\mu_x - \mu_y$ is given by $[(\bar{x} - \bar{y}) - k, (\bar{x} - \bar{y}) + k]$.

2. Find the difference in population means μ_x and μ_y from independent samples with unknown variances σ_x^2 and σ_y^2.

   (a) Determine the critical value $z_{\alpha/2}$ such that $\Phi\left(z_{\alpha/2}\right) = 1 - \alpha/2$, where $\Phi\left(z\right)$ is the standard normal distribution function.

   (b) Compute the means $\bar{x}$ and $\bar{y}$, and the standard deviations s_x and s_y.

   (c) Compute $k = z_{\alpha/2}\sqrt{\dfrac{s_x^2}{n} + \dfrac{s_y^2}{m}}$.

   (d) The $100(1 - \alpha)$ percent confidence interval for $\mu_x - \mu_y$ is given by $[(\bar{x} - \bar{y}) - k, (\bar{x} - \bar{y}) + k]$.

3. Find the difference in population means μ_x and μ_y from independent samples with unknown but equal variances $\sigma_x^2 = \sigma_y^2$.

   (a) Determine the critical value $t_{\alpha/2}$ such that $F\left(t_{\alpha/2}\right) = 1 - \alpha/2$, where $F(\cdot)$ is the t-distribution with $n + m - 2$ degrees of freedom.

   (b) Compute the means $\bar{x}$ and $\bar{y}$, the standard deviations s_x and s_y, and the pooled standard deviation estimate,

   $$s = \sqrt{\frac{(n-1)s_x^2 + (m-1)s_y^2}{n+m-2}}. \qquad (7.10.1)$$

   (c) Compute $k = t_{\alpha/2}\, s\sqrt{\dfrac{1}{n} + \dfrac{1}{m}}$.

   (d) The $100(1 - \alpha)$ percent confidence interval for $\mu_x - \mu_y$ is given by $[(\bar{x} - \bar{y}) - k, (\bar{x} - \bar{y}) + k]$.

4. Find the difference in population means μ_x and μ_y for paired samples with unknown but equal variances $\sigma_x^2 = \sigma_y^2$.

   (a) Determine the critical value $t_{\alpha/2}$ such that $F\left(t_{\alpha/2}\right) = 1 - \alpha/2$, where $F(\cdot)$ is the t-distribution with $n - 1$ degrees of freedom.

   (b) Compute the mean $\bar{\mu}_d$ and standard deviation s_d of the paired differences $x_1 - y_1, x_2 - y_2, \ldots, x_n - y_n$.

   (c) Compute $k = t_{\alpha/2}\, s_d/\sqrt{n}$.

   (d) The $100(1 - \alpha)$ percent confidence interval for $\mu_d = \mu_x - \mu_y$ is given by $[\bar{\mu}_d - k, \bar{\mu}_d + k]$.

5. Find the difference in Bernoulli trial success rates, $p_x - p_y$, for large, independent samples.

   (a) Determine the critical value $z_{\alpha/2}$ such that $\Phi\left(z_{\alpha/2}\right) = 1 - \alpha/2$, where $\Phi\left(z\right)$ is the standard normal distribution function.

   (b) Compute the proportions $\hat{p}_x$ and $\hat{p}_y$ of "successes" for the samples.

   (c) Compute $k = z_{\alpha/2}\sqrt{\dfrac{\hat{p}_x\left(1 - \hat{p}_x\right)}{n} + \dfrac{\hat{p}_y\left(1 - \hat{p}_y\right)}{m}}$.

(d) The $100(1 - \alpha)$ percent confidence interval for $p_x - p_y$ is given by $[(\hat{p}_x - \hat{p}_y) - k, (\hat{p}_x - \hat{p}_y) + k]$.

6. Find the difference in medians $M_x - M_y$ based on the Mann–Whitney–Wilcoxon procedure.

    (a) Determine the critical value $z_{\alpha/2}$ such that $\Phi\left(z_{\alpha/2}\right) = 1 - \alpha/2$, where $\Phi\left(z\right)$ is the standard normal distribution function.

    (b) Compute the order statistics $w_{(1)}, w_{(2)}, \ldots, w_{(N)}$ of the $N = nm$ differences $x_i - y_j$, for $1 \le i \le n$ and $1 \le j \le m$.

    (c) Compute

$$k_1 \;=\; \frac{nm}{2} + \left\lfloor 0.5 - z_{\alpha/2}\sqrt{\frac{nm\left(n + m + 1\right)}{12}} \right\rfloor$$

and

$$k_2 \;=\; \left\lceil \frac{nm}{2} - 0.5 + z_{\alpha/2}\sqrt{\frac{nm\left(n + m + 1\right)}{12}} \right\rceil .$$

    (d) The $100(1 - \alpha)$ percent confidence interval for $M_x - M_y$ is given by $\left[w_{(k_1)}, w_{(k_2)}\right]$.

7. Find the ratio of variances σ_x^2/σ_y^2, for independent samples.

    (a) Determine the critical values $F_{\alpha/2}$ and $F_{1-\alpha/2}$ such that $F\left(F_{\alpha/2}\right) = 1 - \alpha/2$ and $F\left(F_{1-\alpha/2}\right) = \alpha/2$, where $F(\cdot)$ is the F-distribution with $m - 1$ and $n - 1$ degrees of freedom.

    (b) Compute the standard deviations s_x and s_y of the samples.

    (c) Compute $k_1 = F_{1-\alpha/2}$ and $k_2 = F_{\alpha/2}$.

    (d) The $100(1 - \alpha)$ percent confidence interval for σ_x^2/σ_y^2 is given by

$$\left[\frac{s_x^2}{s_y^2}k_1, \frac{s_x^2}{s_y^2}k_2 \right] .$$

7.11 TESTS OF HYPOTHESES

A statistical hypothesis is a statement about the distribution of a random variable. A statistical test of a hypothesis is a procedure in which a sample is used to determine whether we should "reject" or "not reject" the hypothesis. Before employing a hypothesis test, first select a *significance level* α. Typically, $\alpha = 0.05, 0.01$, or the like.

7.11.1 HYPOTHESIS TESTS: PARAMETER FROM ONE POPULATION

The following hypothesis tests assume a random sample of size n, given by $\{x_1, x_2, \ldots, x_n\}$.

1. Test of the hypothesis $\mu = \mu_0$ against the alternative $\mu \neq \mu_0$ of the mean of a normal distribution with known variance σ^2:

   (a) Determine the critical value $z_{\alpha/2}$ such that $\Phi\left(z_{\alpha/2}\right) = 1 - \alpha/2$, where $\Phi(z)$ is the standard normal distribution function.

   (b) Compute the mean $\bar{x}$ of the sample.

   (c) Compute the test statistic $z = \dfrac{(\bar{x} - \mu_0)\sqrt{n}}{\sigma}$.

   (d) If $|z| > z_{\alpha/2}$, then reject the hypothesis. If $|z| \leq z_{\alpha/2}$, then do not reject the hypothesis.

2. Test of the hypothesis $\mu = \mu_0$ against the alternative $\mu > \mu_0$ (or $\mu < \mu_0$) of the mean of a normal distribution with known variance σ^2:

   (a) Determine the critical value z_α such that $\Phi(z_\alpha) = 1 - \alpha$, where $\Phi(z)$ is the standard normal distribution function.

   (b) Compute the mean $\bar{x}$ of the sample.

   (c) Compute the test statistic $z = \dfrac{(\bar{x} - \mu_0)\sqrt{n}}{\sigma}$. (For the alternative $\mu < \mu_0$, multiply z by -1.)

   (d) If $z > z_\alpha$, then reject the hypothesis. If $z \leq z_\alpha$, then do not reject the hypothesis.

3. Test of the hypothesis $\mu = \mu_0$ against the alternative $\mu \neq \mu_0$ of the mean of a normal distribution with unknown variance σ^2:

   (a) Determine the critical value $t_{\alpha/2}$ such that $F\left(t_{\alpha/2}\right) = 1 - \alpha/2$, where $F(\cdot)$ is the t-distribution with $n - 1$ degrees of freedom.

   (b) Compute the mean $\bar{x}$ and standard deviation s of the sample.

   (c) Compute the test statistic $t = \dfrac{(\bar{x} - \mu_0)\sqrt{n}}{s}$.

   (d) If $|t| > t_{\alpha/2}$, then reject the hypothesis. If $|t| \leq t_{\alpha/2}$, then do not reject the hypothesis.

4. Test of the hypothesis $\mu = \mu_0$ against the alternative $\mu > \mu_0$ (or $\mu < \mu_0$) of the mean of a normal distribution with unknown variance σ^2:

   (a) Determine the critical value t_α such that $F(t_\alpha) = 1 - \alpha$, where $F(\cdot)$ is the t-distribution with $n - 1$ degrees of freedom.

   (b) Compute the mean $\bar{x}$ and standard deviation s of the sample.

   (c) Compute the test statistic $t = \dfrac{(\bar{x} - \mu_0)\sqrt{n}}{s}$. (For the alternative $\mu < \mu_0$, multiply t by -1.)

(d) If $t > t_\alpha$, then reject the hypothesis. If $t \leq t_\alpha$, then do not reject the hypothesis.

5. Test of the hypothesis $p = p_0$ against the alternative $p \neq p_0$ of the probability of success for a binomial distribution, large sample:

   (a) Determine the critical value $z_{\alpha/2}$ such that $\Phi\left(z_{\alpha/2}\right) = 1 - \alpha/2$, where $\Phi\left(z\right)$ is the standard normal distribution function.
   (b) Compute the proportion $\hat{p}$ of "successes" for the sample.
   (c) Compute the test statistic $z = \dfrac{\hat{p} - p_0}{\sqrt{\dfrac{p_0(1-p_0)}{n}}}$.
   (d) If $|z| > z_{\alpha/2}$, then reject the hypothesis. If $|z| \leq z_{\alpha/2}$, then do not reject the hypothesis.

6. Test of the hypothesis $p = p_0$ against the alternative $p > p_0$ (or $p < p_0$) of the probability of success for a binomial distribution, large sample:

   (a) Determine the critical value z_α such that $\Phi\left(z_\alpha\right) = 1 - \alpha$, where $\Phi\left(z\right)$ is the standard normal distribution function.
   (b) Compute the proportion $\hat{p}$ of "successes" for the sample.
   (c) Compute the test statistic $z = \dfrac{\hat{p} - p_0}{\sqrt{\dfrac{p_0(1-p_0)}{n}}}$. (For the alternative $p < p_0$, multiply z by -1.)
   (d) If $z > z_\alpha$, then reject the hypothesis. If $z \leq z_\alpha$, then do not reject the hypothesis.

7. Wilcoxon signed rank test of the hypothesis $M = M_0$ against the alternative $M \neq M_0$ of the median of a population, large sample:

   (a) Determine the critical value $z_{\alpha/2}$ such that $\Phi\left(z_{\alpha/2}\right) = 1 - \alpha/2$, where $\Phi\left(z\right)$ is the standard normal distribution.
   (b) Compute the quantities $|x_i - M_0|$, and keep track of the sign of $x_i - M_0$. If $|x_i - M_0| = 0$, then remove it from the list and reduce n by one.
   (c) Order the $|x_i - M_0|$ from smallest to largest, assigning rank 1 to the smallest and rank n to the largest; $|x_i - M_0|$ has rank r_i if it is the r_i^{th} entry in the ordered list. In case of ties (i.e., $|x_i - M_0| = |x_j - M_0|$ for 2 or more values) assign each the average of their ranks.
   (d) Compute the sum of the signed ranks $R = \displaystyle\sum_{i=1}^{n} \text{sign}\left(x_i - M_0\right) r_i$.
   (e) Compute the test statistic $z = \dfrac{R}{\sqrt{\dfrac{n(n+1)(2n+1)}{6}}}$.
   (f) If $|z| > z_{\alpha/2}$, then reject the hypothesis. If $|z| \leq z_{\alpha/2}$, then do not reject the hypothesis.

8. Wilcoxon signed rank test of the hypothesis $M = M_0$ against the alternative $M > M_0$ (or $M < M_0$) of the median of a population, large sample:

   (a) Determine the critical value z_α such that $\Phi(z_\alpha) = 1 - \alpha$, where $\Phi(z)$ is the standard normal distribution.
   (b) Compute the quantities $|x_i - M_0|$, and keep track of the sign of $x_i - M_0$. If $|x_i - M_0| = 0$, then remove it from the list and reduce n by one.
   (c) Order the $|x_i - M_0|$ from smallest to largest, assigning rank 1 to the smallest and rank n to the largest; $|x_i - M_0|$ has rank r_i if it is the r_i^{th} entry in the ordered list. If $|x_i - M_0| = |x_j - M_0|$, then assign each the average of their ranks.
   (d) Compute the sum of the signed ranks $R = \displaystyle\sum_{i=1}^{n} \text{sign}(x_i - M_0)\, r_i$.
   (e) Compute the test statistic $z = \dfrac{R}{\sqrt{\frac{n(n+1)(2n+1)}{6}}}$.   (For the alternative $M < M_0$, multiply the test statistic by -1.)
   (f) If $z > z_\alpha$, then reject the hypothesis. If $z \le z_\alpha$, then do not reject the hypothesis.

9. Test of the hypothesis $\sigma^2 = \sigma_0^2$ against the alternative $\sigma^2 \ne \sigma_0^2$ of the variance of a normal distribution:

   (a) Determine the critical values $\chi^2_{\alpha/2}$ and $\chi^2_{1-\alpha/2}$ such that $F\left(\chi^2_{\alpha/2}\right) = 1 - \alpha/2$ and $F\left(\chi^2_{1-\alpha/2}\right) = \alpha/2$, where $F(\cdot)$ is the chi-square distribution function with $n - 1$ degrees of freedom.
   (b) Compute the standard deviation s of the sample.
   (c) Compute the test statistic $\chi^2 = \dfrac{(n-1)s^2}{\sigma_0^2}$.
   (d) If $\chi^2 < \chi^2_{1-\alpha/2}$ or $\chi^2 > \chi^2_{\alpha/2}$, then reject the hypothesis.
   (e) If $\chi^2_{1-\alpha/2} \le \chi^2 \le \chi^2_{\alpha/2}$, then do not reject the hypothesis.

10. Test of the hypothesis $\sigma^2 = \sigma_0^2$ against the alternative $\sigma^2 > \sigma_0^2$ (or $\sigma^2 < \sigma_0^2$) of the variance of a normal distribution:

   (a) Determine the critical value χ^2_α ($\chi^2_{1-\alpha}$ for the alternative $\sigma^2 < \sigma_0^2$) such that $F\left(\chi^2_\alpha\right) = 1 - \alpha$ ($F\left(\chi^2_{1-\alpha}\right) = \alpha$), where $F(\cdot)$ is the chi-square distribution function with $n - 1$ degrees of freedom.
   (b) Compute the standard deviation s of the sample.
   (c) Compute the test statistic $\chi^2 = \dfrac{(n-1)s^2}{\sigma_0^2}$.
   (d) If $\chi^2 > \chi^2_\alpha$ ($\chi^2 < \chi^2_{1-\alpha}$), then reject the hypothesis.
   (e) If $\chi^2 \le \chi^2_\alpha$ ($\chi^2_{1-\alpha} \le \chi^2$), then do not reject the hypothesis.

7.11.2 HYPOTHESIS TESTS: PARAMETERS FROM TWO POPULATIONS

The following hypothesis tests assume a random sample of size n, given by $\{x_1, x_2, \ldots, x_n\}$, and a random sample of size m, given by $\{y_1, y_2, \ldots, y_m\}$.

1. Test of the hypothesis $\mu_x = \mu_y$ against the alternative $\mu_x \neq \mu_y$ of the means of independent normal distributions with known variances σ_x^2 and σ_y^2:

    (a) Determine the critical value $z_{\alpha/2}$ such that $\Phi\left(z_{\alpha/2}\right) = 1 - \alpha/2$, where $\Phi(z)$ is the standard normal distribution function.
    (b) Compute the means, $\overline{x}$ and $\overline{y}$, of the samples.
    (c) Compute the test statistic $z = \dfrac{\overline{x} - \overline{y}}{\sqrt{\dfrac{\sigma_x^2}{n} + \dfrac{\sigma_y^2}{m}}}$.
    (d) If $|z| > z_{\alpha/2}$, then reject the hypothesis. If $|z| \leq z_{\alpha/2}$, then do not reject the hypothesis.

2. Test of the hypothesis $\mu_x = \mu_y$ against the alternative $\mu_x > \mu_y$ (or $\mu_x < \mu_y$) of the means of independent normal distributions with known variances σ_x^2 and σ_y^2:

    (a) Determine the critical value z_α such that $\Phi(z_\alpha) = 1 - \alpha$, where $\Phi(z)$ is the standard normal distribution function.
    (b) Compute the means $\overline{x}$ and $\overline{y}$ of the samples.
    (c) Compute the test statistic $z = \dfrac{\overline{x} - \overline{y}}{\sqrt{\dfrac{\sigma_x^2}{n} + \dfrac{\sigma_y^2}{m}}}$. (For the alternative $\mu_x < \mu_y$, multiply z by -1.)
    (d) If $z > z_\alpha$, then reject the hypothesis. If $z \leq z_\alpha$, then do not reject the hypothesis.

3. Test of the hypothesis $\mu_x = \mu_y$ against the alternative $\mu_x \neq \mu_y$ of the means of independent normal distributions with unknown variances σ_x^2 and σ_y^2, large sample:

    (a) Determine the critical value $z_{\alpha/2}$ such that $\Phi\left(z_{\alpha/2}\right) = 1 - \alpha/2$, where $\Phi(z)$ is the standard normal distribution.
    (b) Compute the means, $\overline{x}$ and $\overline{y}$, and standard deviations, s_x^2 and s_y^2, of the samples.
    (c) Compute the test statistic $z = \dfrac{\overline{x} - \overline{y}}{\sqrt{\dfrac{s_x^2}{n} + \dfrac{s_y^2}{m}}}$.
    (d) If $|z| > z_{\alpha/2}$, then reject the hypothesis. If $|z| \leq z_{\alpha/2}$, then do not reject the hypothesis.

4. Test of the hypothesis $\mu_x = \mu_y$ against the alternative $\mu_x > \mu_y$ (or $\mu_x < \mu_y$) of the means of independent normal distributions with unknown variances, σ_x^2 and σ_y^2, large sample:

(a) Determine the critical value z_α such that $\Phi\left(z_\alpha\right) = 1 - \alpha$, where $\Phi\left(z\right)$ is the standard normal distribution function.

(b) Compute the means, $\overline{x}$ and $\overline{y}$, and standard deviations, s_x^2 and s_y^2, of the samples.

(c) Compute the test statistic $z = \dfrac{\overline{x} - \overline{y}}{\sqrt{\frac{s_x^2}{n} + \frac{s_y^2}{m}}}$. (For the alternative $\mu_x < \mu_y$, multiply z by -1.)

(d) If $z > z_\alpha$, then reject the hypothesis. If $z \leq z_\alpha$, then do not reject the hypothesis.

5. Test of the hypothesis $\mu_x = \mu_y$ against the alternative $\mu_x \neq \mu_y$ of the means of independent normal distributions with unknown variances $\sigma_x^2 = \sigma_y^2$:

(a) Determine the critical value $t_{\alpha/2}$ such that $F\left(t_{\alpha/2}\right) = 1 - \alpha/2$, where $F(\cdot)$ is the t-distribution with $n + m - 2$ degrees of freedom.

(b) Compute the means, $\overline{x}$ and $\overline{y}$, and standard deviations, s_x^2 and s_y^2, of the samples.

(c) Compute the test statistic $t = \dfrac{\overline{x} - \overline{y}}{\sqrt{\frac{(n-1)s_x^2 + (m-1)s_y^2}{n+m-2}}\sqrt{\frac{1}{n} + \frac{1}{m}}}$.

(d) If $|t| > t_{\alpha/2}$, then reject the hypothesis. If $|t| \leq t_{\alpha/2}$, then do not reject the hypothesis.

6. Test of the hypothesis $\mu_x = \mu_y$ against the alternative $\mu_x > \mu_y$ (or $\mu_x < \mu_y$) of the means of independent normal distributions with unknown variances $\sigma_x^2 = \sigma_y^2$:

(a) Determine the critical value t_α such that $F\left(t_\alpha\right) = 1 - \alpha$, where $F(\cdot)$ is the t-distribution with $n + m - 2$ degrees of freedom.

(b) Compute the means, $\overline{x}$ and $\overline{y}$, and standard deviations, s_x^2 and s_y^2, of the samples.

(c) Compute the test statistic $t = \dfrac{\overline{x} - \overline{y}}{\sqrt{\frac{(n-1)s_x^2 + (m-1)s_y^2}{n+m-2}}\sqrt{\frac{1}{n} + \frac{1}{m}}}$. (For the alternative $\mu_x < \mu_y$, multiply t by -1.)

(d) If $t > t_\alpha$, then reject the hypothesis. If $t \leq t_\alpha$, then do not reject the hypothesis.

7. Test of the hypothesis $\mu_x = \mu_y$ against the alternative $\mu_x \neq \mu_y$ of the means of paired normal samples:

(a) Determine the critical value $t_{\alpha/2}$ so that $F\left(t_{\alpha/2}\right) = 1 - \alpha/2$, where $F(\cdot)$ is the t-distribution with $n - 1$ degrees of freedom.

(b) Compute the mean, $\hat{\mu}_d$, and standard deviation, s_d, of the differences $x_1 - y_1, x_2 - y_2, \ldots, x_n - y_n$.

(c) Compute the test statistic $t = \dfrac{\hat{\mu}_d \sqrt{n}}{s_d}$.

(d) If $|t| > t_{\alpha/2}$, then reject the hypothesis. If $|t| \le t_{\alpha/2}$, then do not reject the hypothesis.

8. Test of the hypothesis $\mu_x = \mu_y$ against the alternative $\mu_x > \mu_y$ (or $\mu_x < \mu_y$) of the means of paired normal samples:

   (a) Determine the critical value t_α so that $F(t_\alpha) = 1 - \alpha$, where $F(\cdot)$ is the t-distribution with $n - 1$ degrees of freedom.

   (b) Compute the mean, $\hat\mu_d$, and standard deviation, s_d, of the differences $x_1 - y_1, x_2 - y_2, \ldots, x_n - y_n$.

   (c) Compute the test statistic $t = \dfrac{\hat\mu_d \sqrt{n}}{s_d}$. (For the alternative $\mu_x < \mu_y$, multiply t by -1.)

   (d) If $t > t_\alpha$, then reject the hypothesis. If $t \le t_\alpha$, then do not reject the hypothesis.

9. Test of the hypothesis $p_x = p_y$ against the alternative $p_x \ne p_y$ of the probability of success for a binomial distribution, large sample:

   (a) Determine the critical value $z_{\alpha/2}$ such that $\Phi(z_{\alpha/2}) = 1 - \alpha/2$, where $\Phi(z)$ is the standard normal distribution function.

   (b) Compute the proportions, $\hat p_x$ and $\hat p_y$, of "successes" for the samples.

   (c) Compute the test statistic $z = \dfrac{\hat p_x - \hat p_y}{\sqrt{\dfrac{\hat p_x(1-p_x)}{n} + \dfrac{\hat p_y(1-p_y)}{m}}}$.

   (d) If $|z| > z_{\alpha/2}$, then reject the hypothesis. If $|z| \le z_{\alpha/2}$, then do not reject the hypothesis.

10. Test of the hypothesis $p_x = p_y$ against the alternative $p_x > p_y$ (or $p_x < p_y$) of the probability of success for a binomial distribution, large sample:

   (a) Determine the critical value z_α such that $\Phi(z_\alpha) = 1 - \alpha$, where $\Phi(z)$ is the standard normal distribution function.

   (b) Compute the proportions, $\hat p_x$ and $\hat p_y$, of "successes" for the samples.

   (c) Compute the test statistic $z = \dfrac{\hat p_x - \hat p_y}{\sqrt{\dfrac{\hat p_x(1-p_x)}{n} + \dfrac{\hat p_y(1-p_y)}{m}}}$. (Multiply it by -1 for the alternative $p_x < p_y$.)

   (d) If $z > z_\alpha$, then reject the hypothesis. If $z \le z_\alpha$, then do not reject the hypothesis.

11. Mann–Whitney–Wilcoxon test of the hypothesis $M_x = M_y$ against the alternative $M_x \ne M_y$ of the medians of independent samples, large sample:

   (a) Determine the critical value $z_{\alpha/2}$ such that $\Phi(z_{\alpha/2}) = 1 - \alpha/2$, where $\Phi(z)$ is the standard normal distribution.

   (b) Pool the $N = m + n$ observations, but keep track of which sample the observation was drawn from.

(c) Order the pooled observations from smallest to largest, assigning rank 1 to the smallest and rank N to the largest; an observation has rank r_i if it is the r_i^{th} entry in the ordered list. If two observations are equal, then assign each the average of their ranks.

(d) Compute the sum of the ranks from the first sample T_x.

(e) Compute the test statistic $z = \dfrac{T_x - \frac{m(N+1)}{2}}{\sqrt{\frac{mn(N+1)}{12}}}$.

(f) If $|z| > z_{\alpha/2}$, then reject the hypothesis. If $|z| \le z_{\alpha/2}$, then do not reject the hypothesis.

12. Mann–Whitney–Wilcoxon test of the hypothesis $M_x = M_y$ against the alternative $M_x > M_y$ (or $M_x < M_y$) of the medians of independent samples, large sample:

(a) Determine the critical value z_α such that $\Phi(z_\alpha) = 1 - \alpha$, where $\Phi(z)$ is the standard normal distribution.

(b) Pool the $N = m + n$ observations, but keep track of which sample the observation was drawn from.

(c) Order the pooled observations from smallest to largest, assigning rank 1 to the smallest and rank N to the largest; an observation has rank r_i if it is the r_i^{th} entry in the ordered list. If two observations are equal, then assign each the average of their ranks.

(d) Compute the sum of the ranks from the first sample T_x.

(e) Compute the test statistic $z = \dfrac{T_x - \frac{m(N+1)}{2}}{\sqrt{\frac{mn(N+1)}{12}}}$. (For the alternative $M_x < M_y$, multiply the test statistic by -1.)

(f) If $|z| > z_\alpha$, then reject the hypothesis. If $|z| \le z_\alpha$, then do not reject the hypothesis.

13. Wilcoxon signed rank test of the hypothesis $M_x = M_y$ against the alternative $M_x \ne M_y$ of the medians of paired samples, large sample:

(a) Determine the critical value $z_{\alpha/2}$ such that $\Phi(z_{\alpha/2}) = 1 - \alpha/2$, where $\Phi(z)$ is the standard normal distribution.

(b) Compute the paired differences $d_i = x_i - y_i$, for $i = 1, 2, \ldots, n$.

(c) Compute the quantities $|d_i|$ and keep track of the sign of d_i. If $d_i = 0$, then remove it from the list and reduce n by one.

(d) Order the $|d_i|$ from smallest to largest, assigning rank 1 to the smallest and rank n to the largest; $|d_i|$ has rank r_i if it is the r_i^{th} entry in the ordered list. In case of ties (i.e., $|d_i| = |d_j|$ for 2 or more values) assign each the average of their ranks.

(e) Compute the sum of the signed ranks $R = \displaystyle\sum_{i=1}^{n} \text{sign}(d_i)\, r_i$.

(f) Compute the test statistic $z = \dfrac{R}{\sqrt{\frac{n(n+1)(2n+1)}{6}}}$.

(g) If $|z| > z_{\alpha/2}$, then reject the hypothesis. If $|z| \leq z_{\alpha/2}$, then do not reject the hypothesis.

14. Wilcoxon signed rank test of the hypothesis $M_x = M_y$ against the alternative $M_x > M_y$ (or $M_x < M_y$) of the medians of paired samples, large sample:

(a) Determine the critical value z_α such that $\Phi(z_\alpha) = 1 - \alpha$, where $\Phi(z)$ is the standard normal distribution.

(b) Compute the paired differences $d_i = x_i - y_i$, for $i = 1, 2, \ldots, n$.

(c) Compute the quantities $|d_i|$ and keep track of the sign of d_i. If $d_i = 0$, then remove it from the list and reduce n by one.

(d) Order the $|d_i|$ from smallest to largest, assigning rank 1 to the smallest and rank n to the largest; $|d_i|$ has rank r_i if it is the r_i^{th} entry in the ordered list. In case of ties (i.e., $|d_i| = |d_j|$ for 2 or more values) assign each the average of their ranks.

(e) Compute the sum of the signed ranks $R = \sum_{i=1}^{n} \text{sign}(d_i) r_i$.

(f) Compute the test statistic $z = \dfrac{R}{\sqrt{\dfrac{n(n+1)(2n+1)}{6}}}$. (For the alternative $M_x < M_y$, multiply the test statistic by -1.)

(g) If $z > z_\alpha$, then reject the hypothesis. If $z \leq z_\alpha$, then do not reject the hypothesis.

15. Test of the hypothesis $\sigma_x^2 = \sigma_y^2$ against the alternative $\sigma_x^2 \neq \sigma_y^2$ (or $\sigma_x^2 > \sigma_y^2$) of the variances of independent normal samples:

(a) Determine the critical value $F_{\alpha/2}$ (F_α for the alternative $\sigma_x^2 > \sigma_y^2$) such that $F(F_{\alpha/2}) = 1 - \alpha/2$ ($F(F_\alpha) = 1 - \alpha$), where $F(\cdot)$ is the F-distribution function with $n - 1$ and $m - 1$ degrees of freedom.

(b) Compute the standard deviations s_x and s_y of the samples.

(c) Compute the test statistic $F = \dfrac{s_x^2}{s_y^2}$. (For the two-sided test, put the larger value in the numerator.)

(d) If $F > F_{\alpha/2}$ ($F > F_\alpha$), then reject the hypothesis. If $F \leq F_{\alpha/2}$ ($F \leq F_\alpha$), then do not reject the hypothesis.

7.11.3 HYPOTHESIS TESTS: DISTRIBUTION OF A POPULATION

The following hypothesis tests assume a random sample of size n, given by $\{x_1, x_2, \ldots, x_n\}$.

1. Run test for randomness of a sample of binary values, large sample:

(a) Determine the critical value $z_{\alpha/2}$ such that $\Phi\left(z_{\alpha/2}\right) = 1 - \alpha/2$, where $\Phi(z)$ is the standard normal distribution function.

(b) Since the data are binary, denote the possible values of x_i by 0 and 1. Count the total number of zeros, and call this n_1; count the total number of ones, and call this n_2. Group the data into maximal sub-sequences of consecutive zeros and ones, and call each such sub-sequence a *run*. Let R be the number of runs in the sample.

(c) Compute $\mu_R = \dfrac{2n_1n_2}{n_1 + n_2} + 1$, and $\sigma_R^2 = \dfrac{(\mu_R-1)(\mu_R-2)}{n_1+n_2-1}$.

(d) Compute the test statistic $z = \dfrac{R - \mu_R}{\sigma_R}$.

(e) If $|z| > z_{\alpha/2}$, then reject the hypothesis. If $|z| \le z_{\alpha/2}$, then do not reject the hypothesis.

2. Run test for randomness against an alternative that a trend is present in a sample of binary values, large sample:

(a) Determine the critical value z_α such that $\Phi(z_\alpha) = 1 - \alpha$, where $\Phi(z)$ is the standard normal distribution function.

(b) Since the data are binary, denote the possible values of x_i by 0 and 1. Count the total number of zeros, and call this n_1; count the total number of ones, and call this n_2. Group the data into maximal sub-sequences of consecutive zeros and ones, and call each such sub-sequence a run. Let R be the number of runs in the sample.

(c) Compute $\mu_R = \dfrac{2n_1n_2}{n_1 + n_2} + 1$, and $\sigma_R^2 = \dfrac{(\mu_R-1)(\mu_R-2)}{n_1+n_2-1}$.

(d) Compute the test statistic $z = \dfrac{R - \mu_R}{\sigma_R}$.

(e) If $z < -z_\alpha$, then reject the hypothesis (this suggests the presence of a trend in the data). If $z \ge -z_\alpha$, then do not reject the hypothesis.

3. Run test for randomness against an alternative that the data are periodic for a sample of binary values, large sample:

(a) Determine the critical value z_α such that $\Phi(z_\alpha) = 1 - \alpha$, where $\Phi(z)$ is the standard normal distribution function.

(b) Since the data are binary, denote the possible values of x_i by 0 and 1. Count the total number of zeros, and call this n_1; count the total number of ones, and call this n_2. Group the data into maximal sub-sequences of consecutive zeros and ones, and call each such sub-sequence a run. Let R be the number of runs in the sample.

(c) Compute $\mu_R = \dfrac{2n_1n_2}{n_1 + n_2} + 1$, and $\sigma_R^2 = \dfrac{(\mu_R-1)(\mu_R-2)}{n_1+n_2-1}$.

(d) Compute the test statistic $z = \dfrac{R - \mu_R}{\sigma_R}$.

(e) If $z > z_\alpha$, then reject the hypothesis (this suggests the data are periodic). If $z \le z_\alpha$, then do not reject the hypothesis.

4. Chi-square test that the data are drawn from a specific k-parameter multinomial distribution, large sample:

   (a) Determine the critical value χ_α^2 such that $F\left(\chi_\alpha^2\right) = 1 - \alpha$, where $F(x)$ is the chi-square distribution with $k - 1$ degrees of freedom.

   (b) The k-parameter multinomial has k possible outcomes $A_1, A_2, \ldots, A_k$ with probabilities $p_1, p_2, \ldots, p_k$. For $i = 1, 2, \ldots, k$, compute n_i, the number of x_j's corresponding to A_i.

   (c) For $i = 1, 2, \ldots, k$, compute the sample multinomial parameters $\hat{p}_i = n_i/n$.

   (d) Compute the test statistic $\chi^2 = \sum_{i=1}^{k} \dfrac{(n_i - np_i)^2}{np_i}$.

   (e) If $\chi^2 > \chi_\alpha^2$, then reject the hypothesis. If $\chi^2 \leq \chi_\alpha^2$, then do not reject the hypothesis.

5. Chi-square test for independence of attributes A and B having possible outcomes $A_1, A_2, \ldots, A_k$ and $B_1, B_2, \ldots, B_m$:

   (a) Determine the critical value χ_α^2 such that $F\left(\chi_\alpha^2\right) = 1 - \alpha$, where $F(\cdot)$ is the chi-square distribution with $(k - 1)(m - 1)$ degrees of freedom.

   (b) For $i = 1, 2, \ldots, k$ and $j = 1, 2, \ldots, m$, define o_{ij} to be the number of observations having attributes A_i and B_j, and define $o_{i \cdot} = \sum_{j=1}^{m} o_{ij}$ and $o_{\cdot j} = \sum_{i=1}^{k} o_{ij}$.

   (c) The variables defined above are often collected into a table, called a *contingency table*:

Attribute	B_1	B_2	$\cdots$	B_m	Totals
A_1	o_{11}	o_{12}	$\cdots$	o_{1m}	$o_{1 \cdot}$
A_2	o_{21}	o_{22}	$\cdots$	o_{2m}	$o_{2 \cdot}$
$\vdots$	$\vdots$	$\vdots$	$\ddots$	$\vdots$	$\vdots$
A_k	o_{k1}	o_{k2}	$\cdots$	o_{km}	$o_{k \cdot}$
Totals	$o_{\cdot 1}$	$o_{\cdot 2}$	$\cdots$	$o_{\cdot m}$	n

   (d) For $i = 1, 2, \ldots, k$ and $j = 1, 2, \ldots, m$, compute the sample mean number of observations in the ij^{th} cell of the contingency table $e_{ij} = \dfrac{o_{i \cdot} o_{\cdot j}}{n}$.

   (e) Compute the test statistic, $\chi^2 = \sum_{i=1}^{k} \sum_{j=1}^{m} \dfrac{(o_{ij} - e_{ij})^2}{e_{ij}}$.

   (f) If $\chi^2 > \chi_\alpha^2$, then reject the hypothesis (that is, conclude that the attributes are not independent). If $\chi^2 \leq \chi_\alpha^2$, then do not reject the hypothesis.

6. Kolmogorov–Smirnov test that $F_0(x)$ is the distribution of the population from which the sample was drawn:

   (a) Determine the critical value D_α such that $Q(D_\alpha) = 1 - \alpha$, where $Q(D)$ is the distribution function for the Kolmogorov–Smirnov test statistic D.

(b) Compute the sample distribution function $\widehat{F}(x)$.

(c) Compute the test statistic, given the maximum deviation of the sample and target distribution functions $D = \max \left| \widehat{F}(x) - F_0(x) \right|$.

(d) If $D > D_\alpha$, then reject the hypothesis (that is, conclude that the data are not drawn from $F_0(x)$). If $D \leq D_\alpha$, then do not reject the hypothesis.

7.11.4 HYPOTHESIS TESTS: DISTRIBUTIONS OF TWO POPULATIONS

The following hypothesis tests assume a random sample of size n, given by $\{x_1, x_2, \ldots, x_n\}$, and a random sample of size m, given by $\{y_1, y_2, \ldots, y_m\}$.

1. Chi-square test that two k-parameter multinomial distributions are equal, large sample:

   (a) Determine the critical value χ_α^2 such that $F\left(\chi_\alpha^2\right) = 1 - \alpha$, where $F(\cdot)$ is the chi-square distribution with $k - 1$ degrees of freedom.

   (b) The k-parameter multinomials have k possible outcomes $A_1, A_2, \ldots,$ A_k. For $i = 1, 2, \ldots, k$, compute n_i, the number of x_j's corresponding to A_i, and compute m_i, the number of y_j's corresponding to A_i.

   (c) For $i = 1, 2, \ldots, k$, compute the sample multinomial parameters $\hat{p}_i = \dfrac{n_i + m_i}{n + m}$.

   (d) Compute the test statistic,

   $$\chi^2 = \sum_{i=1}^{k} \frac{(n_i - n\hat{p}_i)^2}{n\hat{p}_i} + \sum_{i=1}^{k} \frac{(m_i - m\hat{p}_i)^2}{m\hat{p}_i}. \qquad (7.11.1)$$

   (e) If $\chi^2 > \chi_\alpha^2$, then reject the hypothesis. If $\chi^2 \leq \chi_\alpha^2$, then do not reject the hypothesis.

2. Mann–Whitney–Wilcoxon test for equality of independent continuous distributions, large sample:

   (a) Determine the critical value $z_{\alpha/2}$ such that $\Phi\left(z_{\alpha/2}\right) = 1 - \alpha/2$, where $\Phi(z)$ is the normal distribution function.

   (b) For $i = 1, 2, \ldots, n$ and $j = 1, 2, \ldots, m$, define $S_{ij} = 1$ if $x_i < y_j$ and $S_{ij} = 0$ if $x_i > y_j$.

   (c) Compute $U = \displaystyle\sum_{i=1}^{n} \sum_{j=1}^{m} S_{ij}$.

   (d) Compute the test statistic $z = \dfrac{U - \frac{mn}{2}}{\sqrt{\frac{mn(m+n+1)}{12}}}$.

   (e) If $|z| > z_{\alpha/2}$, then reject the hypothesis. If $|z| \leq z_{\alpha/2}$, then do not reject the hypothesis.

3. Spearman rank correlation coefficient for independence of paired samples, large sample:

   (a) Determine the critical value $R_{\alpha/2}$ such that $F\left(R_{\alpha/2}\right) = 1 - \alpha/2$, where $F(R)$ is the distribution function for the Spearman rank correlation coefficient.

   (b) The samples are ordered, with the smallest x_i assigned the rank r_1 and the largest assigned the rank r_n; for $i = 1, 2, \ldots, n$, x_i is assigned rank r_i if it occupies the ith position in the ordered list. Similarly the y_i's are assigned ranks s_i. In case of a tie within a sample, the ranks are averaged.

   (c) Compute the test statistic

$$R = \frac{n \sum_{i=1}^{n} r_i s_i - \left(\sum_{i=1}^{n} r_i\right)\left(\sum_{i=1}^{n} s_i\right)}{\sqrt{\left(n \sum_{i=1}^{n} r_i^2 - \left(\sum_{i=1}^{n} r_i\right)^2\right)\left(n \sum_{i=1}^{n} s_i^2 - \left(\sum_{i=1}^{n} s_i\right)^2\right)}}.$$

   (d) If $|R| > R_{\alpha/2}$, then reject the hypothesis. If $|R| \leq R_{\alpha/2}$, then do not reject the hypothesis.

7.11.5 SEQUENTIAL PROBABILITY RATIO TESTS

Given two simple hypotheses and m observations, compute:

1. $P_{0m} = $ Prob (observations $\mid H_0$).
2. $P_{1m} = $ Prob (observations $\mid H_1$).
3. $v_m = P_{1m}/P_{0m}$.

Then make one of the following decisions:

1. If $v_m \geq \dfrac{1 - \beta}{\alpha}$ then reject H_0.
2. If $v_m \leq \dfrac{\beta}{1 - \alpha}$ then reject H_1.
3. If $\dfrac{\beta}{1 - \alpha} < v_m < \dfrac{1 - \beta}{\alpha}$ then make another observation.

Hence, the number of samples taken is not fixed *a priori*, but determined as sampling occurs.

EXAMPLES

1. Let θ denote the fraction of defective items. Two simple hypotheses are H_0: $\theta = \theta_0 = 0.05$ and H_1: $\theta = \theta_1 = 0.15$. Choose $\alpha = 5\%$ and $\beta = 10\%$ (i.e., reject lot with

$\theta = \theta_0$ about 5% of the time; accept lot with $\theta = \theta_1$ about 10% of the time). If, after m observations, there are d defective items, then

$$P_{im} = \binom{m}{d} \theta_i^d (1 - \theta_i)^{m-d} \qquad \text{and} \qquad v_m = \left(\frac{\theta_1}{\theta_0}\right)^d \left(\frac{1 - \theta_1}{1 - \theta_0}\right)^{m-d} \qquad (7.11.2)$$

or $v_m = 3^d (0.895)^{m-d}$, using the above numbers. The critical values are $\frac{\beta}{1-\alpha} = 0.105$ and $\frac{1-\beta}{\alpha} = 18$. The decision to perform another observation depends on whether or not

$$0.105 \le 3^d (0.895)^{m-d} \le 18. \qquad (7.11.3)$$

Taking logarithms, a $(m - d, d)$ control chart can be drawn with the following lines: $d = 0.101(m - d) - 2.049$ and $d = 0.101(m - d) + 2.63$. On the figure below, a sample path leading to rejection of H_0 has been indicated:

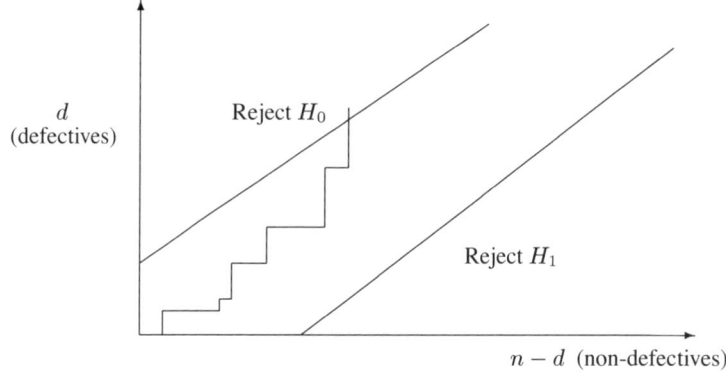

2. Let X be normally distributed with unknown mean μ and known standard deviation σ. Consider the two simple hypotheses, $H_0 : \mu = \mu_0$ and $H_1 : \mu = \mu_1$. If Y is the sum of the first m observations of X, then a (Y, m) control chart is constructed with the two lines:

$$\begin{aligned} Y &= \frac{\mu_0 + \mu_1}{2} m + \frac{\sigma^2}{\mu_1 - \mu_0} \log \frac{\beta}{1 - \alpha}, \\ Y &= \frac{\mu_0 + \mu_1}{2} m + \frac{\sigma^2}{\mu_1 - \mu_0} \log \frac{1 - \beta}{\alpha}. \end{aligned} \qquad (7.11.4)$$

7.12   LINEAR REGRESSION

1. The general linear statistical model assumes that the observed data values $\{y_1, y_2, \ldots, y_m\}$ are of the form

$$y_i = \beta_0 + \beta_1 x_{i1} + \beta_2 x_{i2} + \cdots + \beta_n x_{in} + \epsilon_i,$$

for $i = 1, 2, \ldots, m$.

2. For $i = 1, 2, \ldots, m$ and $j = 1, 2, \ldots, n$, the independent variables x_{ij} are known (nonrandom).

3. $\{\beta_0, \beta_1, \beta_2, \cdots, \beta_n\}$ are unknown parameters.

4. For each i, ϵ_i is a zero-mean normal random variable with unknown variance σ^2.

7.12.1 LINEAR MODEL $y_i = \beta_0 + \beta_1 x_i + \epsilon$

1. Point estimate of β_1:

$$\widehat{\beta}_1 = \frac{m \sum_{i=1}^{m} x_i y_i - \left(\sum_{i=1}^{m} x_i \right) \left(\sum_{i=1}^{m} y_i \right)}{m \left(\sum_{i=1}^{m} x_i^2 \right) - \left(\sum_{i=1}^{m} x_i \right)^2}.$$

2. Point estimate of β_0:

$$\widehat{\beta}_0 = \bar{y} - \widehat{\beta}_1 \bar{x}.$$

3. Point estimate of the correlation coefficient:

$$r = \widehat{\rho} = \frac{m \sum_{i=1}^{m} x_i y_i - \left(\sum_{i=1}^{m} x_i \right) \left(\sum_{i=1}^{m} y_i \right)}{\sqrt{m \left(\sum_{i=1}^{m} x_i^2 \right) - \left(\sum_{i=1}^{m} x_i \right)^2} \sqrt{m \left(\sum_{i=1}^{m} y_i^2 \right) - \left(\sum_{i=1}^{m} y_i \right)^2}}.$$

4. Point estimate of error variance σ^2: $\widehat{\sigma^2} = \dfrac{\sum_{i=1}^{m} \left(y_i - \widehat{\beta}_0 - \widehat{\beta}_1 x_i \right)^2}{m-2}.$

5. The standard error of the estimate is defined as $s_e = \sqrt{\widehat{\sigma^2}}$.

6. Least-squares regression line: $\quad \widehat{y} = \widehat{\beta}_0 + \widehat{\beta}_1 x.$

7. Confidence interval for β_0:

   (a) Determine the critical value $t_{\alpha/2}$ such that $F\left(t_{\alpha/2}\right) = 1 - \alpha/2$, where $F\left(\cdot\right)$ is the cumulative distribution function for the t-distribution with $m - 2$ degrees of freedom.

   (b) Compute the point estimate $\widehat{\beta}_0$.

   (c) Compute $k = t_{\alpha/2} s_e \sqrt{\dfrac{1}{m} + \dfrac{\bar{x}^2}{\sum_{i=1}^{m} (x_i - \bar{x})^2}}.$

(d) The $100(1 - \alpha)$ percent confidence interval for β_0 is given by
$$\left[\widehat{\beta}_0 - k, \widehat{\beta}_0 + k \right].$$

8. Confidence interval for β_1:

   (a) Determine the critical value $t_{\alpha/2}$ such that $F\left(t_{\alpha/2}\right) = 1 - \alpha/2$, where $F\left(\cdot\right)$ is the cumulative distribution function for the t-distribution with $m - 2$ degrees of freedom.

   (b) Compute the point estimate $\widehat{\beta}_1$.

   (c) Compute $k = t_{\alpha/2} \dfrac{s_e}{\sqrt{\displaystyle\sum_{i=1}^{m} (x_i - \overline{x})^2}}$.

   (d) The $100(1 - \alpha)$ percent confidence interval for β_1 is given by
   $$\left[\widehat{\beta}_1 - k, \widehat{\beta}_1 + k \right].$$

9. Confidence interval for σ^2:

   (a) Determine the critical values $\chi^2_{\alpha/2}$ and $\chi^2_{1-\alpha/2}$ such that $F\left(\chi^2_{\alpha/2}\right) = 1 - \alpha/2$ and $F\left(\chi^2_{1-\alpha/2}\right) = \alpha/2$, where $F\left(\cdot\right)$ is the cumulative distribution function for the chi-square distribution function with $m - 2$ degrees of freedom.

   (b) Compute the point estimate $\widehat{\sigma^2}$.

   (c) Compute $k_1 = \dfrac{(n - 2)\widehat{\sigma^2}}{\chi^2_{\alpha/2}}$ and $k_2 = \dfrac{(n - 2)\widehat{\sigma^2}}{\chi^2_{1-\alpha/2}}$.

   (d) The $100(1 - \alpha)$ percent confidence interval for σ^2 is given by $[k_1, k_2]$.

10. Confidence interval (predictive interval) for y, given x_0:

    (a) Determine the critical value $t_{\alpha/2}$ such that $F\left(t_{\alpha/2}\right) = 1 - \alpha/2$, where $F\left(\cdot\right)$ is the cumulative distribution function for the t-distribution with $m - 2$ degrees of freedom.

    (b) Compute the point estimates $\widehat{\beta}_0$, $\widehat{\beta}_1$, and s_e.

    (c) Compute $k = t_{\alpha/2} s_e \sqrt{\dfrac{1}{m} + \dfrac{(x_0 - \overline{x})^2}{\displaystyle\sum_{i=1}^{m} (x_i - \overline{x})^2}}$ and $\widehat{y} = \widehat{\beta}_0 + \widehat{\beta}_1 x_0$.

    (d) The $100(1 - \alpha)$ percent confidence interval for β_1 is given by $[\widehat{y} - k, \widehat{y} + k]$.

11. Test of the hypothesis $\beta_1 = 0$ against the alternative $\beta_1 \neq 0$:

    (a) Determine the critical value $t_{\alpha/2}$ such that $F\left(t_{\alpha/2}\right) = 1 - \alpha/2$, where $F\left(\cdot\right)$ is the cumulative distribution function for the t-distribution with $m - 2$ degrees of freedom.

(b) Compute the point estimates $\widehat{\beta}_1$ and s_e.

(c) Compute the test statistic $t = \dfrac{\widehat{\beta}_1}{s_e} \sqrt{\displaystyle\sum_{i=1}^{m} (x_i - \overline{x})^2}$.

(d) If $|t| > t_{\alpha/2}$, then reject the hypothesis. If $|t| \leq t_{\alpha/2}$, then do not reject the hypothesis.

7.12.2 GENERAL MODEL $y = \beta_0 + \beta_1 x_1 + \beta_2 x_2 + \cdots + \beta_n x_n + \epsilon$

1. The m equations ($i = 1, 2, \ldots, m$)

$$y_i = \beta_0 + \beta_1 x_{i1} + \beta_2 x_{i2} + \cdots + \beta_n x_{in} + \epsilon_i \qquad (7.12.1)$$

can be written in matrix notation as $\mathbf{y} = \mathbf{X}\beta + \epsilon$ where

$$\mathbf{y} = \begin{bmatrix} y_1 \\ y_2 \\ \vdots \\ y_m \end{bmatrix}, \qquad \beta = \begin{bmatrix} \beta_0 \\ \beta_1 \\ \vdots \\ \beta_n \end{bmatrix}, \qquad \epsilon = \begin{bmatrix} \epsilon_1 \\ \epsilon_2 \\ \vdots \\ \epsilon_m \end{bmatrix}, \qquad (7.12.2)$$

and

$$\mathbf{X} = \begin{bmatrix} 1 & x_{11} & x_{12} & \cdots & x_{1n} \\ 1 & x_{21} & x_{22} & \cdots & x_{2n} \\ \vdots & \vdots & \vdots & \ddots & \vdots \\ 1 & x_{m1} & x_{m2} & \cdots & x_{mn} \end{bmatrix}. \qquad (7.12.3)$$

2. Throughout the remainder of the section, we assume $\mathbf{X}$ has full column rank.

3. The least-squares estimate $\widehat{\beta}$ satisfies the *normal equations* $\mathbf{X}^{\mathrm{T}}\mathbf{X}\widehat{\beta} = \mathbf{X}^{\mathrm{T}}\mathbf{y}$. That is, $\widehat{\beta} = \left(\mathbf{X}^{\mathrm{T}}\mathbf{X}\right)^{-1} \mathbf{X}^{\mathrm{T}}\mathbf{y}$.

4. Point estimate of σ^2:

$$\widehat{\sigma^2} = \frac{1}{m - n - 1} \left(\mathbf{y}^{\mathrm{T}}\mathbf{y} - \widehat{\beta}^{\mathrm{T}}(\mathbf{X}^{\mathrm{T}}\mathbf{y})\right).$$

5. The standard error of the estimate is defined as $s_e = \sqrt{\widehat{\sigma^2}}$.

6. Least-squares regression line: $\widehat{y} = \mathbf{x}^{\mathrm{T}}\widehat{\beta}$.

7. In the following, let c_{ij} denote the $(i,j)^{\text{th}}$ entry in the matrix $\left(\mathbf{X}^{\mathrm{T}}\mathbf{X}\right)^{-1}$.

8. Confidence interval for β_i:

(a) Determine the critical value $t_{\alpha/2}$ such that $F\left(t_{\alpha/2}\right) = 1 - \alpha/2$, where $F\left(\cdot\right)$ is the cumulative distribution function for the t-distribution with $m - n - 1$ degrees of freedom.

(b) Compute the point estimate $\widehat{\beta}_i$ by solving the normal equations, and compute s_e.

(c) Compute $k_i = t_{\alpha/2} s_e \sqrt{c_{ii}}$.

(d) The $100(1 - \alpha)$ percent confidence interval for β_i is given by $\left[\widehat{\beta}_i - k_i, \widehat{\beta}_i + k_i \right]$.

9. Confidence interval for σ^2:

(a) Determine the critical values $\chi^2_{\alpha/2}$ and $\chi^2_{1-\alpha/2}$ such that $F\left(\chi^2_{\alpha/2} \right) = 1 - \alpha/2$ and $F\left(\chi^2_{1-\alpha/2} \right) = \alpha/2$, where $F\left(\cdot \right)$ is the cumulative distribution function for the chi-square distribution function with $m - n - 1$ degrees of freedom.

(b) Compute the point estimate $\widehat{\sigma^2}$.

(c) Compute $k_1 = \dfrac{(m - n - 1)\widehat{\sigma^2}}{\chi^2_{\alpha/2}}$ and $k_2 = \dfrac{(m - n - 1)\widehat{\sigma^2}}{\chi^2_{1-\alpha/2}}$.

(d) The $100(1 - \alpha)$ percent confidence interval for σ^2 is given by $[k_1, k_2]$.

10. Confidence interval (predictive interval) for y, given $\mathbf{x}_0$:

(a) Determine the critical value $t_{\alpha/2}$ such that $F\left(t_{\alpha/2} \right) = 1 - \alpha/2$, where $F\left(\cdot \right)$ is the cumulative distribution function for the t-distribution with $n - m - 1$ degrees of freedom.

(b) Compute the point estimate $\widehat{\beta}_i$ by solving the normal equations, and compute s_e.

(c) Compute $k = t_{\alpha/2}\, s_e \sqrt{1 + \mathbf{x}_0^T \left(\mathbf{X}^T \mathbf{X} \right)^{-1} \mathbf{x}_0}$ and $\widehat{y} = \mathbf{x}_0^T \widehat{\beta}$.

(d) The $100(1-\alpha)$ percent confidence interval for β_1 is given by $[\widehat{y} - k, \widehat{y} + k]$.

11. Test of the hypothesis $\beta_i = 0$ against the alternative $\beta_i \neq 0$:

(a) Determine the critical value $t_{\alpha/2}$ such that $F\left(t_{\alpha/2} \right) = 1 - \alpha/2$, where $F\left(\cdot \right)$ is the cumulative distribution function for the t-distribution with $m - n - 1$ degrees of freedom.

(b) Compute the point estimates $\widehat{\beta}_i$ and s_e by solving the normal equations.

(c) Compute the test statistic $t = \dfrac{\widehat{\beta}_i}{s_e \sqrt{c_{ii}}}$.

(d) If $|t| > t_{\alpha/2}$, then reject the hypothesis. If $|t| \leq t_{\alpha/2}$, then do not reject the hypothesis.

7.13 ANALYSIS OF VARIANCE (ANOVA)

Analysis of variance (ANOVA) is a statistical methodology for determining information about means. The analysis uses variances both between and within samples.

7.13.1 ONE-FACTOR ANOVA

1. Suppose we have k samples from k populations, with the jth population consisting of n_j observations,

$$y_{11}, y_{21}, \ldots, y_{n_1 1}$$
$$y_{12}, y_{22}, \ldots, y_{n_2 2}$$
$$\vdots$$
$$y_{1k}, y_{2k}, \ldots, y_{n_k k}.$$

2. One-factor model:

   (a) The one-factor ANOVA assumes that the ith observation from the jth sample is of the form $y_{ij} = \mu + \tau_j + e_{ij}$.

   (b) For $j = 1, 2, \ldots, k$, the parameter $\mu_j = \mu + \tau_j$ is the unknown mean of the jth population, and $\sum_{j=1}^{k} \tau_j = 0$.

   (c) For $i = 1, 2, \ldots, k$ and $j = 1, 2, \ldots, n_j$, the random variables e_{ij} are independent and normally distributed with mean zero and variance σ^2.

   (d) The total number of observations is $n = n_1 + n_2 + \cdots + n_k$.

3. Point estimates of means:

   (a) Total sample mean $\widehat{y} = \dfrac{1}{n} \displaystyle\sum_{j=1}^{k} \sum_{i=1}^{n_j} y_{ij}$.

   (b) Sample mean of jth sample $\widehat{y}_j = \dfrac{1}{n_j} \displaystyle\sum_{i=1}^{n_j} y_{ij}$.

4. Sums of squares:

   (a) Sum of squares between samples $SS_b = \displaystyle\sum_{j=1}^{k} n_j \left(\widehat{y}_j - \widehat{y} \right)^2$.

   (b) Sum of squares within samples $SS_w = \displaystyle\sum_{j=1}^{k} \sum_{i=1}^{n_j} \left(y_{ij} - \widehat{y}_j \right)^2$.

   (c) Total sum of squares Total $SS = \displaystyle\sum_{j=1}^{k} \sum_{i=1}^{n_j} \left(y_{ij} - \widehat{y} \right)^2$.

   (d) Partition of total sum of squares Total $SS = SS_b + SS_w$.

5. Degrees of freedom:

   (a) Between samples, $k - 1$.

   (b) Within samples, $n - k$.

   (c) Total, $n - 1$.

6. Mean squares:

   (a) Obtained by dividing sums of squares by their respective degrees of freedom.

   (b) Between samples, $\text{MS}_b = \dfrac{\text{SS}_b}{k - 1}$.

   (c) Within samples (also called the *residual mean square*), $\text{MS}_w = \dfrac{\text{SS}_w}{n - k}$.

7. Test of the hypothesis $\mu_1 = \mu_2 = \cdots = \mu_k$ against the alternative $\mu_i \neq \mu_j$ for some i and j; equivalently, test the null hypothesis $\tau_1 = \tau_2 = \cdots = \tau_k = 0$ against the hypothesis $\tau_j \neq 0$ for some j:

   (a) Determine the critical value F_α such that $F(F_\alpha) = 1 - \alpha$, where $F(\cdot)$ is the cumulative distribution function for the F-distribution with $k - 1$ and $n - k$ degrees of freedom.

   (b) Compute the point estimates $\widehat{y}$ and $\widehat{y}_j$ for $j = 1, 2, \ldots, k$.

   (c) Compute the sums of squares SS_b and SS_w.

   (d) Compute the mean squares MS_b and MS_w.

   (e) Compute the test statistic $F = \dfrac{\text{MS}_b}{\text{MS}_w}$.

   (f) If $F > F_\alpha$, then reject the hypothesis. If $F \leq F_\alpha$, then do not reject the hypothesis.

   (g) The above computations are often organized into an ANOVA table:

Source	SS	D.O.F.	MS	F Ratio
Between samples	SS_b	$k - 1$	MS_b	$F = \frac{\text{MS}_b}{\text{MS}_w}$
Within samples	SS_w	$n - k$	MS_w	
Total	Total SS	$n - 1$		

8. Confidence interval for $\mu_i - \mu_j$, for $i \neq j$:

   (a) Determine the critical value $t_{\alpha/2}$ such that $F(t_{\alpha/2}) = 1 - \alpha/2$, where $F(\cdot)$ is the cumulative distribution function for the t-distribution with $n - k$ degrees of freedom.

   (b) Compute the point estimates $\widehat{y}_i$ and $\widehat{y}_j$.

   (c) Compute the residual mean square MS_w.

   (d) Compute $k = t_{\alpha/2} \sqrt{\text{MS}_w \left(\dfrac{1}{n_i} + \dfrac{1}{n_j} \right)}$.

   (e) The $100(1 - \alpha)$ percent confidence interval for $\mu_i - \mu_j$ is given by $[(\widehat{y}_i - y_j) - k, (\widehat{y}_i - y_j) + k]$.

9. Confidence interval for contrast in the means, defined by $C = c_1 \mu_1 + c_2 \mu_2 + \cdots + c_k \mu_k$, where $c_1 + c_2 + \cdots + c_k = 0$:

   (a) Determine the critical value F_α such that $F(F_\alpha) = 1 - \alpha$, where $F(\cdot)$ is the cumulative distribution function for the F-distribution with $k - 1$ and $n - k$ degrees of freedom.

(b) Compute the point estimates $\widehat{y}_j$ for $j = 1, 2, \ldots, k$.

(c) Compute the residual mean square MS_w.

(d) Compute $k = \sqrt{F_\alpha \text{MS}_w \left(\dfrac{k-1}{n} \displaystyle\sum_{j=1}^{k} c_j^2 \right)}$.

(e) The $100(1 - \alpha)$ percent confidence interval for the contrast C is given

by $\left[\displaystyle\sum_{j=1}^{k} c_j \widehat{y}_j - k, \displaystyle\sum_{j=1}^{k} c_j \widehat{y}_j + k \right]$.

7.13.2 UNREPLICATED TWO-FACTOR ANOVA

1. Suppose we have a sample of observations y_{ij} indexed by two factors $i = 1, 2, \ldots, m$ and $j = 1, 2, \ldots, n$.

2. Unreplicated two-factor model:

   (a) The unreplicated two-factor ANOVA assumes that the ij^{th} observation is of the form $y_{ij} = \mu + \beta_i + \tau_j + e_{ij}$.

   (b) μ is the overall mean, β_i is the ith differential effect of factor one, τ_j is the j differential effect of factor two, and

$$\sum_{i=1}^{m} \beta_i = \sum_{j=1}^{n} \tau_j = 0.$$

   (c) For $i = 1, 2, \ldots, m$ and $j = 1, 2, \ldots, n$, the random variables e_{ij} are independent and normally distributed with mean zero and variance σ^2.

   (d) Total number of observations is mn.

3. Point estimates of means:

   (a) Total sample mean $\widehat{y} = \dfrac{1}{mn} \displaystyle\sum_{i=1}^{m} \sum_{j=1}^{n} y_{ij}$.

   (b) i^{th} factor-one sample mean $\widehat{y}_{i\cdot} = \dfrac{1}{n} \displaystyle\sum_{j=1}^{n} y_{ij}$.

   (c) j^{th} factor-two sample mean $\widehat{y}_{\cdot j} = \dfrac{1}{m} \displaystyle\sum_{i=1}^{m} y_{ij}$.

4. Sums of squares:

   (a) Factor-one sum of squares $\text{SS}_1 = n \displaystyle\sum_{i=1}^{m} \left(\widehat{y}_{i\cdot} - \widehat{y} \right)^2$.

   (b) Factor-two sum of squares $\text{SS}_2 = m \displaystyle\sum_{j=1}^{n} \left(\widehat{y}_{\cdot j} - \widehat{y} \right)^2$.

(c) Residual sum of squares $\text{SS}_r = \sum\limits_{i=1}^{m} \sum\limits_{j=1}^{n} (y_{ij} - \widehat{y}_{i\cdot} - \widehat{y}_{\cdot j} + \widehat{y})^2$.

(d) Total sum of squares $\text{Total SS} = \sum\limits_{j=1}^{n} \sum\limits_{i=1}^{m} (y_{ij} - \widehat{y})^2$.

(e) Partition of total sum of squares $\text{Total SS} = \text{SS}_1 + \text{SS}_2 + \text{SS}_r$.

5. Degrees of freedom:

   (a) Factor one, $m - 1$.
   (b) Factor two, $n - 1$.
   (c) Residual, $(m - 1)(n - 1)$.
   (d) Total, $mn - 1$.

6. Mean squares:

   (a) Obtained by dividing sums of squares by their respective degrees of freedom.
   (b) Factor-one mean square $\text{MS}_1 = \dfrac{\text{SS}_1}{m - 1}$.
   (c) Factor-two mean square $\text{MS}_2 = \dfrac{\text{SS}_2}{n - 1}$.
   (d) Residual mean square $\text{MS}_r = \dfrac{\text{SS}_r}{(m - 1)(n - 1)}$.

7. Test of the null hypothesis $\beta_1 = \beta_2 = \cdots = \beta_m = 0$ (no factor-one effects) against the alternative hypothesis $\beta_i \neq 0$ for some i:

   (a) Determine the critical value F_α such that $F(F_\alpha) = 1 - \alpha$, where $F(\cdot)$ is the cumulative distribution function for the F-distribution with $m - 1$ and $(m - 1)(n - 1)$ degrees of freedom.
   (b) Compute the point estimates $\widehat{y}$ and $\widehat{y}_{i\cdot}$ for $i = 1, 2, \ldots, m$.
   (c) Compute the sums of squares SS_1 and SS_r.
   (d) Compute the mean squares MS_1 and MS_r.
   (e) Compute the test statistic $F = \dfrac{\text{MS}_1}{\text{MS}_r}$.
   (f) If $F > F_\alpha$, then reject the hypothesis. If $F \leq F_\alpha$, then do not reject the hypothesis.

8. Test of the null hypothesis $\tau_1 = \tau_2 = \cdots = \tau_n = 0$ (no factor-two effects) against the alternative hypothesis $\tau_j \neq 0$ for some j:

   (a) Determine the critical value F_α such that $F(F_\alpha) = 1 - \alpha$, where $F(\cdot)$ is the cumulative distribution function for the F-distribution with $n - 1$ and $(m - 1)(n - 1)$ degrees of freedom.
   (b) Compute the point estimates $\widehat{y}$ and $\widehat{y}_{\cdot j}$ for $j = 1, 2, \ldots, n$.
   (c) Compute the sums of squares SS_2 and SS_r.

(d) Compute the mean squares MS_2 and MS_r.

(e) Compute the test statistic $F = \dfrac{MS_2}{MS_r}$.

(f) If $F > F_\alpha$, then reject the hypothesis. If $F \le F_\alpha$, then do not reject the hypothesis.

(g) The above computations are often organized into an ANOVA table:

Source	SS	D.O.F.	MS	F Ratio
Factor one	SS_1	$m-1$	MS_1	$F = \frac{MS_1}{MS_r}$
Factor two	SS_2	$n-1$	MS_2	$F = \frac{MS_2}{MS_r}$
Residual	SS_r	$(m-1)(n-1)$	MS_r	
Total	Total SS	$mn-1$		

9. Confidence interval for contrast in the factor-one means, defined by $C = c_1\beta_1 + c_2\beta_2 + \cdots + c_m\beta_m$, where $c_1 + c_2 + \cdots + c_m = 0$:

   (a) Determine the critical value F_α such that $F(F_\alpha) = 1 - \alpha$, where $F(\cdot)$ is the cumulative distribution function for the F-distribution with $m-1$ and $(m-1)(n-1)$ degrees of freedom.

   (b) Compute the point estimates $\widehat{y}_{i\cdot}$ for $i = 1, 2, \ldots, m$.

   (c) Compute the residual mean square MS_r.

   (d) Compute $k = \sqrt{F_\alpha MS_r \left(\dfrac{m-1}{n} \sum_{i=1}^{m} c_i^2 \right)}$.

   (e) The $100(1 - \alpha)$ percent confidence interval for the contrast C is given by
   $$\left[\sum_{i=1}^{m} c_i\widehat{y}_{i\cdot} - k, \sum_{i=1}^{m} c_i\widehat{y}_{i\cdot} + k \right].$$

10. Confidence interval for contrast in the factor-two means, defined by $C = c_1\tau_1 + c_2\tau_2 + \cdots + c_n\tau_n$, where $c_1 + c_2 + \cdots + c_n = 0$:

   (a) Determine the critical value F_α such that $F(F_\alpha) = 1 - \alpha$, where $F(\cdot)$ is the cumulative distribution function for the F-distribution with $n-1$ and $(m-1)(n-1)$ degrees of freedom.

   (b) Compute the point estimates $\widehat{y}_{\cdot j}$ for $j = 1, 2, \ldots, n$.

   (c) Compute the residual mean square MS_r.

   (d) Compute $k = \sqrt{F_\alpha MS_r \left(\dfrac{n-1}{m} \sum_{j=1}^{n} c_j^2 \right)}$.

   (e) The $100(1 - \alpha)$ percent confidence interval for the contrast C is given by
   $$\left[\sum_{j=1}^{n} c_j\widehat{y}_{\cdot j} - k, \sum_{j=1}^{n} c_j\widehat{y}_{\cdot j} + k \right].$$

7.13.3 REPLICATED TWO-FACTOR ANOVA

1. Suppose we have a sample of observations y_{ijk} indexed by two factors $i = 1, 2, \ldots, m$ and $j = 1, 2, \ldots, n$. Moreover, there are p observations per factor pair (i, j), indexed by $k = 1, 2, \ldots, p$.

2. Replicated two-factor model:

   (a) The replicated two-factor ANOVA assumes that the ijk^{th} observation is of the form $y_{ijk} = \mu + \beta_i + \tau_j + \gamma_{ij} + e_{ijk}$.

   (b) μ is the overall mean, β_i is the ith differential effect of factor one, τ_j is the j differential effect of factor two, and

$$\sum_{i=1}^{m} \beta_i = \sum_{j=1}^{n} \tau_j = 0.$$

   (c) For $i = 1, 2, \ldots, m$ and $j = 1, 2, \ldots, n$, γ_{ij} is the ij^{th} interaction effect of factors one and two.

   (d) For $i = 1, 2, \ldots, m$, $j = 1, 2, \ldots, n$, and $k = 1, 2, \ldots, p$, the random variables e_{ijk} are independent and normally distributed with mean zero and variance σ^2.

   (e) Total number of observations is mnp.

3. Point estimates of means:

   (a) Total sample mean $\widehat{y} = \dfrac{1}{mnp} \sum_{i=1}^{m} \sum_{j=1}^{n} \sum_{k=1}^{p} y_{ijk}$.

   (b) i^{th} factor-one sample mean $\widehat{y}_{i\cdot\cdot} = \dfrac{1}{np} \sum_{j=1}^{n} \sum_{k=1}^{p} y_{ijk}$.

   (c) j^{th} factor-two sample mean $\widehat{y}_{\cdot j\cdot} = \dfrac{1}{mp} \sum_{i=1}^{m} \sum_{k=1}^{p} y_{ijk}$.

   (d) ij^{th} interaction mean $\widehat{y}_{ij\cdot} = \dfrac{1}{p} \sum_{k=1}^{p} y_{ijk}$.

4. Sums of squares:

   (a) Factor-one sum of squares $SS_1 = np \sum_{i=1}^{m} (\widehat{y}_{i\cdot\cdot} - \widehat{y})^2$.

   (b) Factor-two sum of squares $SS_2 = mp \sum_{j=1}^{n} (\widehat{y}_{\cdot j\cdot} - \widehat{y})^2$.

   (c) Interaction sum of squares $SS_{12} = p \sum_{i=1}^{m} \sum_{j=1}^{n} (\widehat{y}_{ij\cdot} - \widehat{y})^2$.

(d) Residual sum of squares $\mathrm{SS_r} = \sum\limits_{i=1}^{m}\sum\limits_{j=1}^{n}\sum\limits_{k=1}^{p}\left(y_{ijk} - \widehat{y}_{i\cdot\cdot} - \widehat{y}_{\cdot j\cdot} + \widehat{y}\right)^2$.

(e) Total sum of squares Total SS $= \sum\limits_{i=1}^{m}\sum\limits_{j=1}^{n}\sum\limits_{k=1}^{p}\left(y_{ijk} - \widehat{y}\right)^2$.

(f) Partition of total sum of squares Total SS $= \mathrm{SS_1} + \mathrm{SS_2} + \mathrm{SS_{12}} + \mathrm{SS_r}$.

5. Degrees of freedom:

   (a) Factor one, $m - 1$.
   (b) Factor two, $n - 1$.
   (c) Interaction, $(m - 1)(n - 1)$.
   (d) Residual, $mn(p - 1)$.
   (e) Total, $mnp - 1$.

6. Mean squares:

   (a) Obtained by dividing sums of squares by their respective degrees of freedom.
   (b) Factor-one mean square $\mathrm{MS_1} = \dfrac{\mathrm{SS_1}}{m - 1}$.
   (c) Factor-two mean square $\mathrm{MS_2} = \dfrac{\mathrm{SS_2}}{n - 1}$.
   (d) Interaction mean square $\mathrm{MS_{12}} = \dfrac{\mathrm{SS_{12}}}{(m - 1)(n - 1)}$.
   (e) Residual mean square $\mathrm{MS_r} = \dfrac{\mathrm{SS_r}}{mn(p - 1)}$.

7. Test of the null hypothesis $\beta_1 = \beta_2 = \cdots = \beta_m = 0$ (no factor-one effects) against the alternative hypothesis $\beta_i \neq 0$ for some i:

   (a) Determine the critical value F_α such that $F(F_\alpha) = 1 - \alpha$, where $F(\cdot)$ is the cumulative distribution function for the F-distribution with $m - 1$ and $mn(p - 1)$ degrees of freedom.
   (b) Compute the point estimates $\widehat{y}$ and $\widehat{y}_{i\cdot\cdot}$ for $i = 1, 2, \ldots, m$.
   (c) Compute the sums of squares $\mathrm{SS_1}$ and $\mathrm{SS_r}$.
   (d) Compute the mean squares $\mathrm{MS_1}$ and $\mathrm{MS_r}$.
   (e) Compute the test statistic $F = \dfrac{\mathrm{MS_1}}{\mathrm{MS_r}}$.
   (f) If $F > F_\alpha$, then reject the hypothesis. If $F \leq F_\alpha$, then do not reject the hypothesis.

8. Test of the null hypothesis $\tau_1 = \tau_2 = \cdots = \tau_n = 0$ (no factor-two effects) against the alternative hypothesis $\tau_j \neq 0$ for some j:

   (a) Determine the critical value F_α such that $F(F_\alpha) = 1 - \alpha$, where $F(\cdot)$ is the cumulative distribution function for the F-distribution with $n - 1$ and $mn(p - 1)$ degrees of freedom.

(b) Compute the point estimates $\widehat{y}$ and $\widehat{y}_{.j}$ for $j = 1, 2, \ldots, n$.

(c) Compute the sums of squares SS_2 and SS_r.

(d) Compute the mean squares MS_2 and MS_r.

(e) Compute the test statistic $F = \dfrac{MS_2}{MS_r}$.

(f) If $F > F_\alpha$, then reject the hypothesis. If $F \leq F_\alpha$, then do not reject the hypothesis.

9. Test of the null hypothesis $\gamma_{ij} = 0$ for $i = 1, 2, \ldots, m$ and $j = 1, 2, \ldots, n$ (no factor-one effects) against the alternative hypothesis $\gamma_{ij} \neq 0$ for some i and j:

(a) Determine the critical value F_α such that $F(F_\alpha) = 1 - \alpha$, where $F(\cdot)$ is the cumulative distribution function for the F-distribution with $(m - 1)(n - 1)$ and $mn(p - 1)$ degrees of freedom.

(b) Compute the point estimates $\widehat{y}, \widehat{y}_{i..}, \widehat{y}_{.j.}$, and $\widehat{y}_{ij.}$ for $i = 1, 2, \ldots, m$ and $j = 1, 2, \ldots, n$.

(c) Compute the sums of squares SS_{12} and SS_r.

(d) Compute the mean squares MS_{12} and MS_r.

(e) Compute the test statistic $F = \dfrac{MS_{12}}{MS_r}$.

(f) If $F > F_\alpha$, then reject the hypothesis. If $F \leq F_\alpha$, then do not reject the hypothesis.

(g) The above computations are often organized into an ANOVA table:

Source	SS	D.O.F.	MS	F Ratio
Factor one	SS_1	$m - 1$	MS_1	$F = MS_1/MS_r$
Factor two	SS_2	$n - 1$	MS_2	$F = MS_2/MS_r$
Interaction	SS_{12}	$(m - 1)(n - 1)$	MS_{12}	$F = MS_{12}/MS_r$
Residual	SS_r	$mn(p - 1)$	MS_r	
Total	Total SS	$mnp - 1$		

10. Confidence interval for contrast in the factor-one means, defined by $C = c_1\beta_1 + c_2\beta_2 + \cdots + c_m\beta_m$, where $c_1 + c_2 + \cdots + c_m = 0$:

(a) Determine the critical value F_α such that $F(F_\alpha) = 1 - \alpha$, where $F(\cdot)$ is the cumulative distribution function for the F-distribution with $m - 1$ and $mn(p - 1)$ degrees of freedom.

(b) Compute the point estimates $\widehat{y}_{i..}$ for $i = 1, 2, \ldots, m$.

(c) Compute the residual mean square MS_r.

(d) Compute $k = \sqrt{F_\alpha MS_r \left(\dfrac{m - 1}{np} \displaystyle\sum_{i=1}^{m} c_i^2 \right)}$.

(e) The $100(1 - \alpha)$ percent confidence interval for the contrast C is given by

$$\left[\sum_{i=1}^{m} c_i \widehat{y}_{i..} - k, \sum_{i=1}^{m} c_i \widehat{y}_{i..} + k \right].$$

11. Confidence interval for contrast in the factor-two means, defined by $C = c_1\tau_1 + c_2\tau_2 + \cdots + c_n\tau_n$, where $c_1 + c_2 + \cdots + c_n = 0$:

   (a) Determine the critical value F_α such that $F(F_\alpha) = 1 - \alpha$, where $F(\cdot)$ is the cumulative distribution function for the F-distribution with $n - 1$ and $mn(p - 1)$ degrees of freedom.
   (b) Compute the point estimates $\widehat{y}_{\cdot j\cdot}$ for $j = 1, 2, \ldots, n$.
   (c) Compute the residual mean square $\mathrm{MS_r}$.

   (d) Compute $k = \sqrt{F_\alpha \mathrm{MS_r}\left(\dfrac{n-1}{mp}\sum_{j=1}^{n} c_j^2\right)}$.

   (e) The $100(1 - \alpha)$ percent confidence interval for the contrast C is given by

   $$\left[\sum_{j=1}^{n} c_j\widehat{y}_{\cdot j\cdot} - k, \sum_{j=1}^{n} c_j\widehat{y}_{\cdot j\cdot} + k\right].$$

FIGURE 7.5
The shaded region is defined by $X \geq z_\alpha$ and has area α (here X is $N(0,1)$).

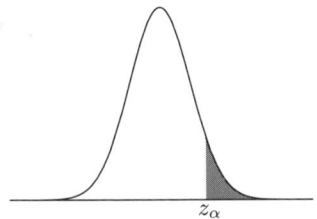

z_α

7.14 PROBABILITY TABLES

7.14.1 CRITICAL VALUES

1. The critical value z_α satisfies $\Phi(z_\alpha) = 1 - \alpha$ (where, as usual, $\Phi(z)$ is the distribution function for the standard normal). See Figure 7.5.

$$\Phi(z) = \frac{1}{\sqrt{2\pi}}\int_{-\infty}^{\infty} e^{-t^2/2}\, dt = \frac{1}{2}\left(1 + \mathrm{erf}\left(\frac{z}{\sqrt{2}}\right)\right) \qquad (7.14.1)$$

2. The critical value t_α satisfies $F(t_\alpha) = 1 - \alpha$ where $F(\cdot)$ is the distribution function for the t-distribution (for a specified number of degrees of freedom).

FIGURE 7.6

Illustration of σ and 2σ regions of a normal distribution.

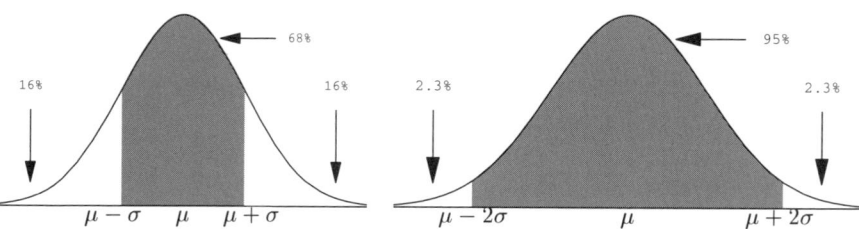

3. The critical value χ_α^2 satisfies $F(\chi_\alpha^2) = 1 - \alpha$ where $F(\cdot)$ is the distribution function for the χ^2-distribution (for a specified number of degrees of freedom).

4. The critical value F_α satisfies $F(F_\alpha) = 1 - \alpha$ where $F(\cdot)$ is the distribution function for the F-distribution (for a specified number of degrees of freedom).

7.14.2 TABLE OF THE NORMAL DISTRIBUTION

For a standard normal random variable (see Figure 7.6):

Limits		Proportion of the total area (%)	Remaining area (%)
$\mu - \lambda\sigma$	$\mu + \lambda\sigma$	(%)	(%)
$\mu - \sigma$	$\mu + \sigma$	68.27	31.73
$\mu - 1.65\sigma$	$\mu + 1.65\sigma$	90	10
$\mu - 1.96\sigma$	$\mu + 1.96\sigma$	95	5
$\mu - 2\sigma$	$\mu + 2\sigma$	95.45	4.55
$\mu - 2.58\sigma$	$\mu + 2.58\sigma$	99.0	0.99
$\mu - 3\sigma$	$\mu + 3\sigma$	99.73	0.27
$\mu - 3.09\sigma$	$\mu + 3.09\sigma$	99.8	0.2
$\mu - 3.29\sigma$	$\mu + 3.29\sigma$	99.9	0.1

x	1.282	1.645	1.960	2.326	2.576	3.090
$\Phi(x)$	0.90	0.95	0.975	0.99	0.995	0.999
$2[1 - \Phi(x)]$	0.20	0.10	0.05	0.02	0.01	0.002

x	3.09	3.72	4.26	4.75	5.20	5.61	6.00	6.36
$1 - \Phi(x)$	10^{-3}	10^{-4}	10^{-5}	10^{-6}	10^{-7}	10^{-8}	10^{-9}	10^{-10}

For large values of x:

$$\left[\frac{e^{-x^2/2}}{\sqrt{2\pi}} \left(\frac{1}{x} - \frac{1}{x^3} \right) \right] < 1 - \Phi(x) < \left[\frac{e^{-x^2/2}}{\sqrt{2\pi}} \left(\frac{1}{x} \right) \right] \qquad (7.14.2)$$

x	$F(x)$	$1 - F(x)$	$f(x)$	x	$F(x)$	$1 - F(x)$	$f(x)$
0.01	0.50399	0.49601	0.39892	0.02	0.50798	0.49202	0.39886
0.03	0.51197	0.48803	0.39876	0.04	0.51595	0.48405	0.39862
0.05	0.51994	0.48006	0.39844	0.06	0.52392	0.47608	0.39822
0.07	0.52790	0.47210	0.39797	0.08	0.53188	0.46812	0.39767
0.09	0.53586	0.46414	0.39733	0.10	0.53983	0.46017	0.39695
0.11	0.54380	0.45621	0.39654	0.12	0.54776	0.45224	0.39608
0.13	0.55172	0.44828	0.39559	0.14	0.55567	0.44433	0.39505
0.15	0.55962	0.44038	0.39448	0.16	0.56356	0.43644	0.39387
0.17	0.56749	0.43250	0.39322	0.18	0.57142	0.42858	0.39253
0.19	0.57534	0.42466	0.39181	0.20	0.57926	0.42074	0.39104
0.21	0.58317	0.41683	0.39024	0.22	0.58706	0.41294	0.38940
0.23	0.59095	0.40905	0.38853	0.24	0.59484	0.40516	0.38762
0.25	0.59871	0.40129	0.38667	0.26	0.60257	0.39743	0.38568
0.27	0.60642	0.39358	0.38466	0.28	0.61026	0.38974	0.38361
0.29	0.61409	0.38591	0.38251	0.30	0.61791	0.38209	0.38139
0.31	0.62172	0.37828	0.38023	0.32	0.62552	0.37448	0.37903
0.33	0.62930	0.37070	0.37780	0.34	0.63307	0.36693	0.37654
0.35	0.63683	0.36317	0.37524	0.36	0.64058	0.35942	0.37391
0.37	0.64431	0.35569	0.37255	0.38	0.64803	0.35197	0.37115
0.39	0.65173	0.34827	0.36973	0.40	0.65542	0.34458	0.36827
0.41	0.65910	0.34090	0.36678	0.42	0.66276	0.33724	0.36526
0.43	0.66640	0.33360	0.36371	0.44	0.67003	0.32997	0.36213
0.45	0.67365	0.32636	0.36053	0.46	0.67724	0.32276	0.35889
0.47	0.68082	0.31918	0.35723	0.48	0.68439	0.31561	0.35553
0.49	0.68793	0.31207	0.35381	0.50	0.69146	0.30854	0.35207
0.51	0.69497	0.30503	0.35029	0.52	0.69847	0.30153	0.34849
0.53	0.70194	0.29806	0.34667	0.54	0.70540	0.29460	0.34482
0.55	0.70884	0.29116	0.34294	0.56	0.71226	0.28774	0.34105
0.57	0.71566	0.28434	0.33912	0.58	0.71904	0.28096	0.33718
0.59	0.72240	0.27759	0.33521	0.60	0.72575	0.27425	0.33322
0.61	0.72907	0.27093	0.33121	0.62	0.73237	0.26763	0.32918
0.63	0.73565	0.26435	0.32713	0.64	0.73891	0.26109	0.32506
0.65	0.74215	0.25785	0.32297	0.66	0.74537	0.25463	0.32086
0.67	0.74857	0.25143	0.31874	0.68	0.75175	0.24825	0.31659
0.69	0.75490	0.24510	0.31443	0.70	0.75804	0.24196	0.31225
0.71	0.76115	0.23885	0.31006	0.72	0.76424	0.23576	0.30785
0.73	0.76731	0.23270	0.30563	0.74	0.77035	0.22965	0.30339
0.75	0.77337	0.22663	0.30114	0.76	0.77637	0.22363	0.29887
0.77	0.77935	0.22065	0.29659	0.78	0.78231	0.21769	0.29430
0.79	0.78524	0.21476	0.29200	0.80	0.78814	0.21185	0.28969

x	$F(x)$	$1 - F(x)$	$f(x)$	x	$F(x)$	$1 - F(x)$	$f(x)$
0.81	0.79103	0.20897	0.28737	0.82	0.79389	0.20611	0.28504
0.83	0.79673	0.20327	0.28269	0.84	0.79955	0.20045	0.28034
0.85	0.80234	0.19766	0.27798	0.86	0.80510	0.19490	0.27562
0.87	0.80785	0.19215	0.27324	0.88	0.81057	0.18943	0.27086
0.89	0.81327	0.18673	0.26848	0.90	0.81594	0.18406	0.26609
0.91	0.81859	0.18141	0.26369	0.92	0.82121	0.17879	0.26129
0.93	0.82381	0.17619	0.25888	0.94	0.82639	0.17361	0.25647
0.95	0.82894	0.17106	0.25406	0.96	0.83147	0.16853	0.25164
0.97	0.83398	0.16602	0.24923	0.98	0.83646	0.16354	0.24681
0.99	0.83891	0.16109	0.24439	1.00	0.84135	0.15865	0.24197
1.01	0.84375	0.15625	0.23955	1.02	0.84614	0.15386	0.23713
1.03	0.84849	0.15151	0.23471	1.04	0.85083	0.14917	0.23230
1.05	0.85314	0.14686	0.22988	1.06	0.85543	0.14457	0.22747
1.07	0.85769	0.14231	0.22506	1.08	0.85993	0.14007	0.22265
1.09	0.86214	0.13786	0.22025	1.10	0.86433	0.13567	0.21785
1.11	0.86650	0.13350	0.21546	1.12	0.86864	0.13136	0.21307
1.13	0.87076	0.12924	0.21069	1.14	0.87286	0.12714	0.20831
1.15	0.87493	0.12507	0.20594	1.16	0.87698	0.12302	0.20357
1.17	0.87900	0.12100	0.20121	1.18	0.88100	0.11900	0.19886
1.19	0.88298	0.11702	0.19652	1.20	0.88493	0.11507	0.19419
1.21	0.88686	0.11314	0.19186	1.22	0.88877	0.11123	0.18954
1.23	0.89065	0.10935	0.18724	1.24	0.89251	0.10749	0.18494
1.25	0.89435	0.10565	0.18265	1.26	0.89616	0.10383	0.18037
1.27	0.89796	0.10204	0.17810	1.28	0.89973	0.10027	0.17585
1.29	0.90148	0.09853	0.17360	1.30	0.90320	0.09680	0.17137
1.31	0.90490	0.09510	0.16915	1.32	0.90658	0.09342	0.16694
1.33	0.90824	0.09176	0.16474	1.34	0.90988	0.09012	0.16256
1.35	0.91149	0.08851	0.16038	1.36	0.91309	0.08692	0.15823
1.37	0.91466	0.08534	0.15608	1.38	0.91621	0.08379	0.15395
1.39	0.91774	0.08226	0.15183	1.40	0.91924	0.08076	0.14973
1.41	0.92073	0.07927	0.14764	1.42	0.92220	0.07780	0.14556
1.43	0.92364	0.07636	0.14350	1.44	0.92507	0.07493	0.14146
1.45	0.92647	0.07353	0.13943	1.46	0.92785	0.07215	0.13742
1.47	0.92922	0.07078	0.13542	1.48	0.93056	0.06944	0.13343
1.49	0.93189	0.06811	0.13147	1.50	0.93319	0.06681	0.12952
1.51	0.93448	0.06552	0.12758	1.52	0.93575	0.06426	0.12566
1.53	0.93699	0.06301	0.12376	1.54	0.93822	0.06178	0.12188
1.55	0.93943	0.06057	0.12001	1.56	0.94062	0.05938	0.11816
1.57	0.94179	0.05821	0.11632	1.58	0.94295	0.05705	0.11450
1.59	0.94408	0.05592	0.11270	1.60	0.94520	0.05480	0.11092

x	$F(x)$	$1 - F(x)$	$f(x)$	x	$F(x)$	$1 - F(x)$	$f(x)$
1.61	0.94630	0.05370	0.10916	1.62	0.94738	0.05262	0.10741
1.63	0.94845	0.05155	0.10568	1.64	0.94950	0.05050	0.10396
1.65	0.95053	0.04947	0.10226	1.66	0.95154	0.04846	0.10059
1.67	0.95254	0.04746	0.09892	1.68	0.95352	0.04648	0.09728
1.69	0.95449	0.04551	0.09566	1.70	0.95544	0.04457	0.09405
1.71	0.95637	0.04363	0.09246	1.72	0.95728	0.04272	0.09089
1.73	0.95818	0.04181	0.08933	1.74	0.95907	0.04093	0.08780
1.75	0.95994	0.04006	0.08628	1.76	0.96080	0.03920	0.08478
1.77	0.96164	0.03836	0.08329	1.78	0.96246	0.03754	0.08183
1.79	0.96327	0.03673	0.08038	1.80	0.96407	0.03593	0.07895
1.81	0.96485	0.03515	0.07754	1.82	0.96562	0.03438	0.07614
1.83	0.96637	0.03363	0.07477	1.84	0.96712	0.03288	0.07341
1.85	0.96784	0.03216	0.07207	1.86	0.96856	0.03144	0.07074
1.87	0.96926	0.03074	0.06943	1.88	0.96995	0.03005	0.06814
1.89	0.97062	0.02938	0.06687	1.90	0.97128	0.02872	0.06562
1.91	0.97193	0.02807	0.06438	1.92	0.97257	0.02743	0.06316
1.93	0.97320	0.02680	0.06195	1.94	0.97381	0.02619	0.06076
1.95	0.97441	0.02559	0.05960	1.96	0.97500	0.02500	0.05844
1.97	0.97558	0.02442	0.05730	1.98	0.97615	0.02385	0.05618
1.99	0.97671	0.02329	0.05508	2.00	0.97725	0.02275	0.05399
2.01	0.97778	0.02222	0.05292	2.02	0.97831	0.02169	0.05186
2.03	0.97882	0.02118	0.05082	2.04	0.97933	0.02067	0.04980
2.05	0.97982	0.02018	0.04879	2.06	0.98030	0.01970	0.04780
2.07	0.98077	0.01923	0.04682	2.08	0.98124	0.01876	0.04586
2.09	0.98169	0.01831	0.04491	2.10	0.98214	0.01786	0.04398
2.11	0.98257	0.01743	0.04307	2.12	0.98300	0.01700	0.04217
2.13	0.98341	0.01659	0.04128	2.14	0.98382	0.01618	0.04041
2.15	0.98422	0.01578	0.03955	2.16	0.98461	0.01539	0.03871
2.17	0.98500	0.01500	0.03788	2.18	0.98537	0.01463	0.03706
2.19	0.98574	0.01426	0.03626	2.20	0.98610	0.01390	0.03547
2.21	0.98645	0.01355	0.03470	2.22	0.98679	0.01321	0.03394
2.23	0.98713	0.01287	0.03319	2.24	0.98745	0.01255	0.03246
2.25	0.98778	0.01222	0.03174	2.26	0.98809	0.01191	0.03103
2.27	0.98840	0.01160	0.03034	2.28	0.98870	0.01130	0.02966
2.29	0.98899	0.01101	0.02899	2.30	0.98928	0.01072	0.02833
2.31	0.98956	0.01044	0.02768	2.32	0.98983	0.01017	0.02705
2.33	0.99010	0.00990	0.02643	2.34	0.99036	0.00964	0.02582
2.35	0.99061	0.00939	0.02522	2.36	0.99086	0.00914	0.02463
2.37	0.99111	0.00889	0.02406	2.38	0.99134	0.00866	0.02349
2.39	0.99158	0.00842	0.02294	2.40	0.99180	0.00820	0.02240

x	$F(x)$	$1 - F(x)$	$f(x)$	x	$F(x)$	$1 - F(x)$	$f(x)$
2.41	0.99202	0.00798	0.02186	2.42	0.99224	0.00776	0.02134
2.43	0.99245	0.00755	0.02083	2.44	0.99266	0.00734	0.02033
2.45	0.99286	0.00714	0.01984	2.46	0.99305	0.00695	0.01936
2.47	0.99324	0.00676	0.01888	2.48	0.99343	0.00657	0.01842
2.49	0.99361	0.00639	0.01797	2.50	0.99379	0.00621	0.01753
2.51	0.99396	0.00604	0.01709	2.52	0.99413	0.00587	0.01667
2.53	0.99430	0.00570	0.01625	2.54	0.99446	0.00554	0.01585
2.55	0.99461	0.00539	0.01545	2.56	0.99477	0.00523	0.01506
2.57	0.99491	0.00509	0.01468	2.58	0.99506	0.00494	0.01431
2.59	0.99520	0.00480	0.01394	2.60	0.99534	0.00466	0.01358
2.61	0.99547	0.00453	0.01323	2.62	0.99560	0.00440	0.01289
2.63	0.99573	0.00427	0.01256	2.64	0.99586	0.00415	0.01223
2.65	0.99598	0.00402	0.01191	2.66	0.99609	0.00391	0.01160
2.67	0.99621	0.00379	0.01129	2.68	0.99632	0.00368	0.01100
2.69	0.99643	0.00357	0.01071	2.70	0.99653	0.00347	0.01042
2.71	0.99664	0.00336	0.01014	2.72	0.99674	0.00326	0.00987
2.73	0.99683	0.00317	0.00961	2.74	0.99693	0.00307	0.00935
2.75	0.99702	0.00298	0.00909	2.76	0.99711	0.00289	0.00885
2.77	0.99720	0.00280	0.00860	2.78	0.99728	0.00272	0.00837
2.79	0.99736	0.00264	0.00814	2.80	0.99745	0.00255	0.00792
2.81	0.99752	0.00248	0.00770	2.82	0.99760	0.00240	0.00748
2.83	0.99767	0.00233	0.00727	2.84	0.99774	0.00226	0.00707
2.85	0.99781	0.00219	0.00687	2.86	0.99788	0.00212	0.00668
2.87	0.99795	0.00205	0.00649	2.88	0.99801	0.00199	0.00631
2.89	0.99807	0.00193	0.00613	2.90	0.99813	0.00187	0.00595
2.91	0.99819	0.00181	0.00578	2.92	0.99825	0.00175	0.00562
2.93	0.99830	0.00169	0.00545	2.94	0.99836	0.00164	0.00530
2.95	0.99841	0.00159	0.00514	2.96	0.99846	0.00154	0.00499
2.97	0.99851	0.00149	0.00485	2.98	0.99856	0.00144	0.00470
2.99	0.99860	0.00139	0.00457	3.00	0.99865	0.00135	0.00443
3.01	0.99869	0.00131	0.00430	3.02	0.99874	0.00126	0.00417
3.03	0.99878	0.00122	0.00405	3.04	0.99882	0.00118	0.00393
3.05	0.99886	0.00114	0.00381	3.06	0.99889	0.00111	0.00369
3.07	0.99893	0.00107	0.00358	3.08	0.99896	0.00103	0.00347
3.09	0.99900	0.00100	0.00337	3.10	0.99903	0.00097	0.00327
3.11	0.99906	0.00093	0.00317	3.12	0.99910	0.00090	0.00307
3.13	0.99913	0.00087	0.00298	3.14	0.99916	0.00085	0.00288
3.15	0.99918	0.00082	0.00279	3.16	0.99921	0.00079	0.00271
3.17	0.99924	0.00076	0.00262	3.18	0.99926	0.00074	0.00254
3.19	0.99929	0.00071	0.00246	3.20	0.99931	0.00069	0.00238

x	$F(x)$	$1 - F(x)$	$f(x)$	x	$F(x)$	$1 - F(x)$	$f(x)$
3.21	0.99934	0.00066	0.00231	3.22	0.99936	0.00064	0.00224
3.23	0.99938	0.00062	0.00216	3.24	0.99940	0.00060	0.00210
3.25	0.99942	0.00058	0.00203	3.26	0.99944	0.00056	0.00196
3.27	0.99946	0.00054	0.00190	3.28	0.99948	0.00052	0.00184
3.29	0.99950	0.00050	0.00178	3.30	0.99952	0.00048	0.00172
3.31	0.99953	0.00047	0.00167	3.32	0.99955	0.00045	0.00161
3.33	0.99957	0.00043	0.00156	3.34	0.99958	0.00042	0.00151
3.35	0.99960	0.00040	0.00146	3.36	0.99961	0.00039	0.00141
3.37	0.99962	0.00038	0.00136	3.38	0.99964	0.00036	0.00132
3.39	0.99965	0.00035	0.00128	3.40	0.99966	0.00034	0.00123
3.41	0.99967	0.00032	0.00119	3.42	0.99969	0.00031	0.00115
3.43	0.99970	0.00030	0.00111	3.44	0.99971	0.00029	0.00108
3.45	0.99972	0.00028	0.00104	3.46	0.99973	0.00027	0.00100
3.47	0.99974	0.00026	0.00097	3.48	0.99975	0.00025	0.00094
3.49	0.99976	0.00024	0.00090	3.50	0.99977	0.00023	0.00087
3.51	0.99978	0.00022	0.00084	3.52	0.99979	0.00021	0.00081
3.53	0.99979	0.00021	0.00078	3.54	0.99980	0.00020	0.00076
3.55	0.99981	0.00019	0.00073	3.56	0.99982	0.00018	0.00071
3.57	0.99982	0.00018	0.00068	3.58	0.99983	0.00017	0.00066
3.59	0.99984	0.00016	0.00063	3.60	0.99984	0.00016	0.00061
3.61	0.99985	0.00015	0.00059	3.62	0.99985	0.00015	0.00057
3.63	0.99986	0.00014	0.00055	3.64	0.99986	0.00014	0.00053
3.65	0.99987	0.00013	0.00051	3.66	0.99987	0.00013	0.00049
3.67	0.99988	0.00012	0.00047	3.68	0.99988	0.00012	0.00046
3.69	0.99989	0.00011	0.00044	3.70	0.99989	0.00011	0.00042
3.71	0.99990	0.00010	0.00041	3.72	0.99990	0.00010	0.00039
3.73	0.99990	0.00010	0.00038	3.74	0.99991	0.00009	0.00037
3.75	0.99991	0.00009	0.00035	3.76	0.99991	0.00009	0.00034
3.77	0.99992	0.00008	0.00033	3.78	0.99992	0.00008	0.00032
3.79	0.99992	0.00007	0.00030	3.80	0.99993	0.00007	0.00029
3.81	0.99993	0.00007	0.00028	3.82	0.99993	0.00007	0.00027
3.83	0.99994	0.00006	0.00026	3.84	0.99994	0.00006	0.00025
3.85	0.99994	0.00006	0.00024	3.86	0.99994	0.00006	0.00023
3.87	0.99995	0.00005	0.00022	3.88	0.99995	0.00005	0.00021
3.89	0.99995	0.00005	0.00021	3.90	0.99995	0.00005	0.00020
3.91	0.99995	0.00005	0.00019	3.92	0.99996	0.00004	0.00018
3.93	0.99996	0.00004	0.00018	3.94	0.99996	0.00004	0.00017
3.95	0.99996	0.00004	0.00016	3.96	0.99996	0.00004	0.00016
3.97	0.99996	0.00004	0.00015	3.98	0.99997	0.00003	0.00015
3.99	0.99997	0.00003	0.00014	4.00	0.99997	0.00003	0.00013

7.14.3 PERCENTAGE POINTS, STUDENT'S t-DISTRIBUTION

For a given value of n and α this table gives the value of $t_{\alpha,n}$ such that

$$F(t_{\alpha,n}) = \int_{-\infty}^{t_{\alpha,n}} \frac{\Gamma((n+1)/2)}{\sqrt{n\pi}\Gamma(n/2)} \left(1 + \frac{x^2}{n}\right)^{-(n+1)/2} dx = 1 - \alpha. \qquad (7.14.3)$$

The t-distribution is symmetrical, so that $F(-t) = 1 - F(t)$.

EXAMPLE   The table gives $t_{\alpha=0.60, n=2} = 0.325$.
Hence, when $n = 2$, $F(0.325) = 0.4$.

n	\multicolumn{9}{c}{$F(t) =$}
	0.6000	0.7500	0.9000	0.9500	0.9750	0.9900	0.9950	0.9990	0.9995
1	0.325	1.000	3.078	6.314	12.706	31.821	63.657	318.309	636.619
2	0.289	0.816	1.886	2.920	4.303	6.965	9.925	22.327	31.599
3	0.277	0.765	1.638	2.353	3.182	4.541	5.841	10.215	12.924
4	0.271	0.741	1.533	2.132	2.776	3.747	4.604	7.173	8.610
5	0.267	0.727	1.476	2.015	2.571	3.365	4.032	5.893	6.869
6	0.265	0.718	1.440	1.943	2.447	3.143	3.707	5.208	5.959
7	0.263	0.711	1.415	1.895	2.365	2.998	3.499	4.785	5.408
8	0.262	0.706	1.397	1.860	2.306	2.896	3.355	4.501	5.041
9	0.261	0.703	1.383	1.833	2.262	2.821	3.250	4.297	4.781
10	0.260	0.700	1.372	1.812	2.228	2.764	3.169	4.144	4.587
11	0.260	0.697	1.363	1.796	2.201	2.718	3.106	4.025	4.437
12	0.259	0.695	1.356	1.782	2.179	2.681	3.055	3.930	4.318
13	0.259	0.694	1.350	1.771	2.160	2.650	3.012	3.852	4.221
14	0.258	0.692	1.345	1.761	2.145	2.624	2.977	3.787	4.140
15	0.258	0.691	1.341	1.753	2.131	2.602	2.947	3.733	4.073
16	0.258	0.690	1.337	1.746	2.120	2.583	2.921	3.686	4.015
17	0.257	0.689	1.333	1.740	2.110	2.567	2.898	3.646	3.965
18	0.257	0.688	1.330	1.734	2.101	2.552	2.878	3.610	3.922
19	0.257	0.688	1.328	1.729	2.093	2.539	2.861	3.579	3.883
20	0.257	0.687	1.325	1.725	2.086	2.528	2.845	3.552	3.850
25	0.256	0.684	1.316	1.708	2.060	2.485	2.787	3.450	3.725
50	0.255	0.679	1.299	1.676	2.009	2.403	2.678	3.261	3.496
100	0.254	0.677	1.290	1.660	1.984	2.364	2.626	3.174	3.390
∞	0.253	0.674	1.282	1.645	1.960	2.326	2.576	3.091	3.291

7.14.4 PERCENTAGE POINTS, CHI-SQUARE DISTRIBUTION

For a given value of n this table gives the value of χ^2 such that

$$F(\chi^2) = \int_0^{\chi^2} \frac{x^{(n-2)/2}e^{-x/2}}{2^{n/2}\Gamma(n/2)}\, dx \qquad (7.14.4)$$

is a specified number.

n	$F=$ 0.995	0.990	0.975	0.950	0.900	0.750	0.500	0.250	0.100	0.050	0.025	0.010	0.005
1	7.88	6.63	5.02	3.84	2.71	1.32	0.455	0.102	0.0158	0.00393	0.0009821	0.0001571	0.0000393
2	10.6	9.21	7.38	5.99	4.61	2.77	1.39	0.575	0.211	0.103	0.0506	0.0201	0.0100
3	12.8	11.3	9.35	7.81	6.25	4.11	2.37	1.21	0.584	0.352	0.216	0.115	0.0717
4	14.9	13.3	11.1	9.49	7.78	5.39	3.36	1.92	1.06	0.711	0.484	0.297	0.207
5	16.7	15.1	12.8	11.1	9.24	6.63	4.35	2.67	1.61	1.15	0.831	0.554	0.412
6	18.5	16.8	14.4	12.6	10.6	7.84	5.35	3.45	2.20	1.64	1.24	0.872	0.676
7	20.3	18.5	16.0	14.1	12.0	9.04	6.35	4.25	2.83	2.17	1.69	1.24	0.989
8	22.0	20.1	17.5	15.5	13.4	10.2	7.34	5.07	3.49	2.73	2.18	1.65	1.34
9	23.6	21.7	19.0	16.9	14.7	11.4	8.34	5.90	4.17	3.33	2.70	2.09	1.73
10	25.2	23.2	20.5	18.3	16.0	12.5	9.34	6.74	4.87	3.94	3.25	2.56	2.16
11	26.8	24.7	21.9	19.7	17.3	13.7	10.3	7.58	5.58	4.57	3.82	3.05	2.60
12	28.3	26.2	23.3	21.0	18.5	14.8	11.3	8.44	6.30	5.23	4.40	3.57	3.07
13	29.8	27.7	24.7	22.4	19.8	16.0	12.3	9.30	7.04	5.89	5.01	4.11	3.57
14	31.3	29.1	26.1	23.7	21.1	17.1	13.3	10.2	7.79	6.57	5.63	4.66	4.07
15	32.8	30.6	27.5	25.0	22.3	18.2	14.3	11.0	8.55	7.26	6.26	5.23	4.60
16	34.3	32.0	28.8	26.3	23.5	19.4	15.3	11.9	9.31	7.96	6.91	5.81	5.14
17	35.7	33.4	30.2	27.6	24.8	20.5	16.3	12.8	10.1	8.67	7.56	6.41	5.70
18	37.2	34.8	31.5	28.9	26.0	21.6	17.3	13.7	10.9	9.39	8.23	7.01	6.26
19	38.6	36.2	32.9	30.1	27.2	22.7	18.3	14.6	11.7	10.1	8.91	7.63	6.84
20	40.0	37.6	34.2	31.4	28.4	23.8	19.3	15.5	12.4	10.9	9.59	8.26	7.43
21	41.4	38.9	35.5	32.7	29.6	24.9	20.3	16.3	13.2	11.6	10.3	8.90	8.03
22	42.8	40.3	36.8	33.9	30.8	26.0	21.3	17.2	14.0	12.3	11.0	9.54	8.64
23	44.2	41.6	38.1	35.2	32.0	27.1	22.3	18.1	14.8	13.1	11.7	10.2	9.26
24	45.6	43.0	39.4	36.4	33.2	28.2	23.3	19.0	15.7	13.8	12.4	10.9	9.89
25	46.9	44.3	40.6	37.7	34.4	29.3	24.3	19.9	16.5	14.6	13.1	11.5	10.5
30	53.7	50.9	47.0	43.8	40.3	34.8	29.3	24.5	20.6	18.5	16.8	15.0	13.8
35	60.3	57.3	53.2	49.8	46.1	40.2	34.3	29.1	24.8	22.5	20.6	18.5	17.2
50	79.5	76.2	71.4	67.5	63.2	56.3	43.9	42.9	37.7	34.8	32.4	29.7	28.0

7.14.5 PERCENTAGE POINTS, F-DISTRIBUTION

Given n and m this gives the value of f such that

$$F(f) = \int_0^f \frac{\Gamma((n+m)/2)}{\Gamma(m/2)\Gamma(n/2)} m^{m/2} n^{n/2} x^{m/2-1}(n+mx)^{-(m+n)/2}\, dx = \mathbf{0.9}.$$

n	$m=1$	2	3	4	5	6	7	8	9	10	50	100	∞
1	39.86	49.50	53.59	55.83	57.24	58.20	58.91	59.44	59.86	60.19	62.69	63.01	63.33
2	8.53	9.00	9.16	9.24	9.29	9.33	9.35	9.37	9.38	9.39	9.47	9.48	9.49
3	5.54	5.46	5.39	5.34	5.31	5.28	5.27	5.25	5.24	5.23	5.15	5.14	5.13
4	4.54	4.32	4.19	4.11	4.05	4.01	3.98	3.95	3.94	3.92	3.80	3.78	3.76
5	4.06	3.78	3.62	3.52	3.45	3.40	3.37	3.34	3.32	3.30	3.15	3.13	3.10
6	3.78	3.46	3.29	3.18	3.11	3.05	3.01	2.98	2.96	2.94	2.77	2.75	2.72
7	3.59	3.26	3.07	2.96	2.88	2.83	2.78	2.75	2.72	2.70	2.52	2.50	2.47
8	3.46	3.11	2.92	2.81	2.73	2.67	2.62	2.59	2.56	2.54	2.35	2.32	2.29
9	3.36	3.01	2.81	2.69	2.61	2.55	2.51	2.47	2.44	2.42	2.22	2.19	2.16
10	3.29	2.92	2.73	2.61	2.52	2.46	2.41	2.38	2.35	2.32	2.12	2.09	2.06
11	3.23	2.86	2.66	2.54	2.45	2.39	2.34	2.30	2.27	2.25	2.04	2.01	1.97
12	3.18	2.81	2.61	2.48	2.39	2.33	2.28	2.24	2.21	2.19	1.97	1.94	1.90
13	3.14	2.76	2.56	2.43	2.35	2.28	2.23	2.20	2.16	2.14	1.92	1.88	1.85
14	3.10	2.73	2.52	2.39	2.31	2.24	2.19	2.15	2.12	2.10	1.87	1.83	1.80
15	3.07	2.70	2.49	2.36	2.27	2.21	2.16	2.12	2.09	2.06	1.83	1.79	1.76
16	3.05	2.67	2.46	2.33	2.24	2.18	2.13	2.09	2.06	2.03	1.79	1.76	1.72
17	3.03	2.64	2.44	2.31	2.22	2.15	2.10	2.06	2.03	2.00	1.76	1.73	1.69
18	3.01	2.62	2.42	2.29	2.20	2.13	2.08	2.04	2.00	1.98	1.74	1.70	1.66
19	2.99	2.61	2.40	2.27	2.18	2.11	2.06	2.02	1.98	1.96	1.71	1.67	1.63
20	2.97	2.59	2.38	2.25	2.16	2.09	2.04	2.00	1.96	1.94	1.69	1.65	1.61
25	2.92	2.53	2.32	2.18	2.09	2.02	1.97	1.93	1.89	1.87	1.61	1.56	1.52
50	2.81	2.41	2.20	2.06	1.97	1.90	1.84	1.80	1.76	1.73	1.44	1.39	1.34
100	2.76	2.36	2.14	2.00	1.91	1.83	1.78	1.73	1.69	1.66	1.35	1.29	1.20
∞	2.71	2.30	2.08	1.94	1.85	1.77	1.72	1.67	1.63	1.60	1.24	1.17	1.00

Given n and m this gives the value of f such that

$$F(f) = \int_0^f \frac{\Gamma((n+m)/2)}{\Gamma(m/2)\Gamma(n/2)} m^{m/2} n^{n/2} x^{m/2-1} (n+mx)^{-(m+n)/2}\, dx = \mathbf{0.95}.$$

n							$m=$							
	1	2	3	4	5	6	7	8	9	10	50	100	∞
1	161.4	199.5	215.7	224.6	230.2	234.0	236.8	238.9	240.5	241.9	251.8	253.0	254.3
2	18.51	19.00	19.16	19.25	19.30	19.33	19.35	19.37	19.38	19.40	19.48	19.49	19.50
3	10.13	9.55	9.28	9.12	9.01	8.94	8.89	8.85	8.81	8.79	8.58	8.55	8.53
4	7.71	6.94	6.59	6.39	6.26	6.16	6.09	6.04	6.00	5.96	5.70	5.66	5.63
5	6.61	5.79	5.41	5.19	5.05	4.95	4.88	4.82	4.77	4.74	4.44	4.41	4.36
6	5.99	5.14	4.76	4.53	4.39	4.28	4.21	4.15	4.10	4.06	3.75	3.71	3.67
7	5.59	4.74	4.35	4.12	3.97	3.87	3.79	3.73	3.68	3.64	3.32	3.27	3.23
8	5.32	4.46	4.07	3.84	3.69	3.58	3.50	3.44	3.39	3.35	3.02	2.97	2.93
9	5.12	4.26	3.86	3.63	3.48	3.37	3.29	3.23	3.18	3.14	2.80	2.76	2.71
10	4.96	4.10	3.71	3.48	3.33	3.22	3.14	3.07	3.02	2.98	2.64	2.59	2.54
11	4.84	3.98	3.59	3.36	3.20	3.09	3.01	2.95	2.90	2.85	2.51	2.46	2.40
12	4.75	3.89	3.49	3.26	3.11	3.00	2.91	2.85	2.80	2.75	2.40	2.35	2.30
13	4.67	3.81	3.41	3.18	3.03	2.92	2.83	2.77	2.71	2.67	2.31	2.26	2.21
14	4.60	3.74	3.34	3.11	2.96	2.85	2.76	2.70	2.65	2.60	2.24	2.19	2.13
15	4.54	3.68	3.29	3.06	2.90	2.79	2.71	2.64	2.59	2.54	2.18	2.12	2.07
16	4.49	3.63	3.24	3.01	2.85	2.74	2.66	2.59	2.54	2.49	2.12	2.07	2.01
17	4.45	3.59	3.20	2.96	2.81	2.70	2.61	2.55	2.49	2.45	2.08	2.02	1.96
18	4.41	3.55	3.16	2.93	2.77	2.66	2.58	2.51	2.46	2.41	2.04	1.98	1.92
19	4.38	3.52	3.13	2.90	2.74	2.63	2.54	2.48	2.42	2.38	2.00	1.94	1.88
20	4.35	3.49	3.10	2.87	2.71	2.60	2.51	2.45	2.39	2.35	1.97	1.91	1.84
25	4.24	3.39	2.99	2.76	2.60	2.49	2.40	2.34	2.28	2.24	1.84	1.78	1.71
50	4.03	3.18	2.79	2.56	2.40	2.29	2.20	2.13	2.07	2.03	1.60	1.52	1.45
100	3.94	3.09	2.70	2.46	2.31	2.19	2.10	2.03	1.97	1.93	1.48	1.39	1.28
∞	3.84	3.00	2.60	2.37	2.21	2.10	2.01	1.94	1.88	1.83	1.35	1.25	1.00

Given n and m this gives the value of f such that

$$F(f) = \int_0^f \frac{\Gamma((n+m)/2)}{\Gamma(m/2)\Gamma(n/2)} m^{m/2} n^{n/2} x^{m/2-1}(n+mx)^{-(m+n)/2}\, dx = \mathbf{0.975}.$$

n						$m =$								
	1	2	3	4	5	6	7	8	9	10	50	100	∞
1	647.8	799.5	864.2	899.6	921.8	937.1	948.2	956.7	963.3	968.6	1008	1013	1018
2	38.51	39.00	39.17	39.25	39.30	39.33	39.36	39.37	39.39	39.40	39.48	39.49	39.50
3	17.44	16.04	15.44	15.10	14.88	14.73	14.62	14.54	14.47	14.42	14.01	13.96	13.90
4	12.22	10.65	9.98	9.60	9.36	9.20	9.07	8.98	8.90	8.84	8.38	8.32	8.26
5	10.01	8.43	7.76	7.39	7.15	6.98	6.85	6.76	6.68	6.62	6.14	6.08	6.02
6	8.81	7.26	6.60	6.23	5.99	5.82	5.70	5.60	5.52	5.46	4.98	4.92	4.85
7	8.07	6.54	5.89	5.52	5.29	5.12	4.99	4.90	4.82	4.76	4.28	4.21	4.14
8	7.57	6.06	5.42	5.05	4.82	4.65	4.53	4.43	4.36	4.30	3.81	3.74	3.67
9	7.21	5.71	5.08	4.72	4.48	4.32	4.20	4.10	4.03	3.96	3.47	3.40	3.33
10	6.94	5.46	4.83	4.47	4.24	4.07	3.95	3.85	3.78	3.72	3.22	3.15	3.08
11	6.72	5.26	4.63	4.28	4.04	3.88	3.76	3.66	3.59	3.53	3.03	2.96	2.88
12	6.55	5.10	4.47	4.12	3.89	3.73	3.61	3.51	3.44	3.37	2.87	2.80	2.72
13	6.41	4.97	4.35	4.00	3.77	3.60	3.48	3.39	3.31	3.25	2.74	2.67	2.60
14	6.30	4.86	4.24	3.89	3.66	3.50	3.38	3.29	3.21	3.15	2.64	2.56	2.49
15	6.20	4.77	4.15	3.80	3.58	3.41	3.29	3.20	3.12	3.06	2.55	2.47	2.40
16	6.12	4.69	4.08	3.73	3.50	3.34	3.22	3.12	3.05	2.99	2.47	2.40	2.32
17	6.04	4.62	4.01	3.66	3.44	3.28	3.16	3.06	2.98	2.92	2.41	2.33	2.25
18	5.98	4.56	3.95	3.61	3.38	3.22	3.10	3.01	2.93	2.87	2.35	2.27	2.19
19	5.92	4.51	3.90	3.56	3.33	3.17	3.05	2.96	2.88	2.82	2.30	2.22	2.13
20	5.87	4.46	3.86	3.51	3.29	3.13	3.01	2.91	2.84	2.77	2.25	2.17	2.09
25	5.69	4.29	3.69	3.35	3.13	2.97	2.85	2.75	2.68	2.61	2.08	2.00	1.91
50	5.34	3.97	3.39	3.05	2.83	2.67	2.55	2.46	2.38	2.32	1.75	1.66	1.54
100	5.18	3.83	3.25	2.92	2.70	2.54	2.42	2.32	2.24	2.18	1.59	1.48	1.37
∞	5.02	3.69	3.12	2.79	2.57	2.41	2.29	2.19	2.11	2.05	1.43	1.27	1.00

Given n and m this gives the value of f such that

$$F(f) = \int_0^f \frac{\Gamma((n+m)/2)}{\Gamma(m/2)\Gamma(n/2)} m^{m/2} n^{n/2} x^{m/2-1} (n+mx)^{-(m+n)/2}\, dx = \mathbf{0.99}.$$

n	$m=1$	2	3	4	5	6	7	8	9	10	50	100	∞
1	4052	5000	5403	5625	5764	5859	5928	5981	6022	6056	6303	6334	6336
2	98.50	99.00	99.17	99.25	99.30	99.33	99.36	99.37	99.39	99.40	99.48	99.49	99.50
3	34.12	30.82	29.46	28.71	28.24	27.91	27.67	27.49	27.35	27.23	26.35	26.24	26.13
4	21.20	18.00	16.69	15.98	15.52	15.21	14.98	14.80	14.66	14.55	13.69	13.58	13.46
5	16.26	13.27	12.06	11.39	10.97	10.67	10.46	10.29	10.16	10.05	9.24	9.13	9.02
6	13.75	10.92	9.78	9.15	8.75	8.47	8.26	8.10	7.98	7.87	7.09	6.99	6.88
7	12.25	9.55	8.45	7.85	7.46	7.19	6.99	6.84	6.72	6.62	5.86	5.75	5.65
8	11.26	8.65	7.59	7.01	6.63	6.37	6.18	6.03	5.91	5.81	5.07	4.96	4.86
9	10.56	8.02	6.99	6.42	6.06	5.80	5.61	5.47	5.35	5.26	4.52	4.41	4.31
10	10.04	7.56	6.55	5.99	5.64	5.39	5.20	5.06	4.94	4.85	4.12	4.01	3.91
11	9.65	7.21	6.22	5.67	5.32	5.07	4.89	4.74	4.63	4.54	3.81	3.71	3.60
12	9.33	6.93	5.95	5.41	5.06	4.82	4.64	4.50	4.39	4.30	3.57	3.47	3.36
13	9.07	6.70	5.74	5.21	4.86	4.62	4.44	4.30	4.19	4.10	3.38	3.27	3.17
14	8.86	6.51	5.56	5.04	4.69	4.46	4.28	4.14	4.03	3.94	3.22	3.11	3.00
15	8.68	6.36	5.42	4.89	4.56	4.32	4.14	4.00	3.89	3.80	3.08	2.98	2.87
16	8.53	6.23	5.29	4.77	4.44	4.20	4.03	3.89	3.78	3.69	2.97	2.86	2.75
17	8.40	6.11	5.18	4.67	4.34	4.10	3.93	3.79	3.68	3.59	2.87	2.76	2.65
18	8.29	6.01	5.09	4.58	4.25	4.01	3.84	3.71	3.60	3.51	2.78	2.68	2.57
19	8.18	5.93	5.01	4.50	4.17	3.94	3.77	3.63	3.52	3.43	2.71	2.60	2.49
20	8.10	5.85	4.94	4.43	4.10	3.87	3.70	3.56	3.46	3.37	2.64	2.54	2.42
25	7.77	5.57	4.68	4.18	3.85	3.63	3.46	3.32	3.22	3.13	2.40	2.29	2.17
50	7.17	5.06	4.20	3.72	3.41	3.19	3.02	2.89	2.78	2.70	1.95	1.82	1.70
100	6.90	4.82	3.98	3.51	3.21	2.99	2.82	2.69	2.59	2.50	1.74	1.60	1.45
∞	6.63	4.61	3.78	3.32	3.02	2.80	2.64	2.51	2.41	2.32	1.53	1.32	1.00

Given n and m this gives the value of f such that

$$F(f) = \int_0^f \frac{\Gamma((n+m)/2)}{\Gamma(m/2)\Gamma(n/2)} m^{m/2} n^{n/2} x^{m/2-1} (n+mx)^{-(m+n)/2} \, dx = \mathbf{0.995}.$$

$m =$

n	1	2	3	4	5	6	7	8	9	10	50	100	∞
1	16211	20000	21615	22500	23056	23437	23715	23925	24091	24224	25211	25337	25465
2	198.5	199.0	199.2	199.2	199.3	199.3	199.4	199.4	199.4	199.4	199.5	199.5	199.5
3	55.55	49.80	47.47	46.19	45.39	44.84	44.43	44.13	43.88	43.69	42.21	42.02	41.83
4	31.33	26.28	24.26	23.15	22.46	21.97	21.62	21.35	21.14	20.97	19.67	19.50	19.32
5	22.78	18.31	16.53	15.56	14.94	14.51	14.20	13.96	13.77	13.62	12.45	12.30	12.14
6	18.63	14.54	12.92	12.03	11.46	11.07	10.79	10.57	10.39	10.25	9.17	9.03	8.88
7	16.24	12.40	10.88	10.05	9.52	9.16	8.89	8.68	8.51	8.38	7.35	7.22	7.08
8	14.69	11.04	9.60	8.81	8.30	7.95	7.69	7.50	7.34	7.21	6.22	6.09	5.95
9	13.61	10.11	8.72	7.96	7.47	7.13	6.88	6.69	6.54	6.42	5.45	5.32	5.19
10	12.83	9.43	8.08	7.34	6.87	6.54	6.30	6.12	5.97	5.85	4.90	4.77	4.64
11	12.23	8.91	7.60	6.88	6.42	6.10	5.86	5.68	5.54	5.42	4.49	4.36	4.23
12	11.75	8.51	7.23	6.52	6.07	5.76	5.52	5.35	5.20	5.09	4.17	4.04	3.90
13	11.37	8.19	6.93	6.23	5.79	5.48	5.25	5.08	4.94	4.82	3.91	3.78	3.65
14	11.06	7.92	6.68	6.00	5.56	5.26	5.03	4.86	4.72	4.60	3.70	3.57	3.44
15	10.80	7.70	6.48	5.80	5.37	5.07	4.85	4.67	4.54	4.42	3.52	3.39	3.26
16	10.58	7.51	6.30	5.64	5.21	4.91	4.69	4.52	4.38	4.27	3.37	3.25	3.11
17	10.38	7.35	6.16	5.50	5.07	4.78	4.56	4.39	4.25	4.14	3.25	3.12	2.98
18	10.22	7.21	6.03	5.37	4.96	4.66	4.44	4.28	4.14	4.03	3.14	3.01	2.87
19	10.07	7.09	5.92	5.27	4.85	4.56	4.34	4.18	4.04	3.93	3.04	2.91	2.78
20	9.94	6.99	5.82	5.17	4.76	4.47	4.26	4.09	3.96	3.85	2.96	2.83	2.69
25	9.48	6.60	5.46	4.84	4.43	4.15	3.94	3.78	3.64	3.54	2.65	2.52	2.38
50	8.63	5.90	4.83	4.23	3.85	3.58	3.38	3.22	3.09	2.99	2.10	1.95	1.81
100	8.24	5.59	4.54	3.96	3.59	3.33	3.13	2.97	2.85	2.74	1.84	1.68	1.51
∞	7.88	5.30	4.28	3.72	3.35	3.09	2.90	2.74	2.62	2.52	1.60	1.36	1.00

Given n and m this gives the value of f such that

$$F(f) = \int_0^f \frac{\Gamma((n+m)/2)}{\Gamma(m/2)\Gamma(n/2)} m^{m/2} n^{n/2} x^{m/2-1} (n+mx)^{-(m+n)/2}\, dx = \mathbf{0.999}.$$

n	$m=$ 1	2	3	4	5	6	7	8	9	10	50	100	∞
2	998.5	999.0	999.2	999.2	999.3	999.3	999.4	999.4	999.4	999.4	999.5	999.5	999.5
3	167.0	148.5	141.1	137.1	134.6	132.8	131.6	130.6	129.9	129.2	124.7	124.1	123.5
4	74.14	61.25	56.18	53.44	51.71	50.53	49.66	49.00	48.47	48.05	44.88	44.47	44.05
5	47.18	37.12	33.20	31.09	29.75	28.83	28.16	27.65	27.24	26.92	24.44	24.12	23.79
6	35.51	27.00	23.70	21.92	20.80	20.03	19.46	19.03	18.69	18.41	16.31	16.03	15.75
7	29.25	21.69	18.77	17.20	16.21	15.52	15.02	14.63	14.33	14.08	12.20	11.95	11.70
8	25.41	18.49	15.83	14.39	13.48	12.86	12.40	12.05	11.77	11.54	9.80	9.57	9.33
9	22.86	16.39	13.90	12.56	11.71	11.13	10.70	10.37	10.11	9.89	8.26	8.04	7.81
10	21.04	14.91	12.55	11.28	10.48	9.93	9.52	9.20	8.96	8.75	7.19	6.98	6.76
11	19.69	13.81	11.56	10.35	9.58	9.05	8.66	8.35	8.12	7.92	6.42	6.21	6.00
12	18.64	12.97	10.80	9.63	8.89	8.38	8.00	7.71	7.48	7.29	5.83	5.63	5.42
13	17.82	12.31	10.21	9.07	8.35	7.86	7.49	7.21	6.98	6.80	5.37	5.17	4.97
14	17.14	11.78	9.73	8.62	7.92	7.44	7.08	6.80	6.58	6.40	5.00	4.81	4.60
15	16.59	11.34	9.34	8.25	7.57	7.09	6.74	6.47	6.26	6.08	4.70	4.51	4.31
16	16.12	10.97	9.01	7.94	7.27	6.80	6.46	6.19	5.98	5.81	4.45	4.26	4.06
17	15.72	10.66	8.73	7.68	7.02	6.56	6.22	5.96	5.75	5.58	4.24	4.05	3.85
18	15.38	10.39	8.49	7.46	6.81	6.35	6.02	5.76	5.56	5.39	4.06	3.87	3.67
19	15.08	10.16	8.28	7.27	6.62	6.18	5.85	5.59	5.39	5.22	3.90	3.71	3.51
20	14.82	9.95	8.10	7.10	6.46	6.02	5.69	5.44	5.24	5.08	3.77	3.58	3.38
25	13.88	9.22	7.45	6.49	5.89	5.46	5.15	4.91	4.71	4.56	3.28	3.09	2.89
50	12.22	7.96	6.34	5.46	4.90	4.51	4.22	4.00	3.82	3.67	2.44	2.25	2.06
100	11.50	7.41	5.86	5.02	4.48	4.11	3.83	3.61	3.44	3.30	2.08	1.87	1.65
∞	10.83	6.91	5.42	4.62	4.10	3.74	3.47	3.27	3.10	2.96	1.75	1.45	1.00

7.14.6 CUMULATIVE TERMS, BINOMIAL DISTRIBUTION

$$B(n, x; p) = \sum_{k=0}^{x} \binom{n}{k} p^k (1 - p)^{n-k}.$$

Note that $B(n, x; p) = B(n, n - x; 1 - p)$.

If p is the probability of success, then $B(n, x; p)$ is the probability of x or fewer successes in n independent trials. For example, if a biased coin has a probability $p = 0.4$ of being a head, and the coin is independently flipped 5 times, then there is a 68% chance that there will be 2 or fewer heads (since $B(5, 2; 0.4) = 0.6826$).

n	x	0.05	0.10	0.15	0.20	0.25	0.30	0.40	0.50
2	0	0.9025	0.8100	0.7225	0.6400	0.5625	0.4900	0.3600	0.2500
	1	0.9975	0.9900	0.9775	0.9600	0.9375	0.9100	0.8400	0.7500
3	0	0.8574	0.7290	0.6141	0.5120	0.4219	0.3430	0.2160	0.1250
	1	0.9928	0.9720	0.9393	0.8960	0.8438	0.7840	0.6480	0.5000
	2	0.9999	0.9990	0.9966	0.9920	0.9844	0.9730	0.9360	0.8750
4	0	0.8145	0.6561	0.5220	0.4096	0.3164	0.2401	0.1296	0.0625
	1	0.9860	0.9477	0.8905	0.8192	0.7383	0.6517	0.4752	0.3125
	2	0.9995	0.9963	0.9880	0.9728	0.9492	0.9163	0.8208	0.6875
	3	1.0000	0.9999	0.9995	0.9984	0.9961	0.9919	0.9744	0.9375
5	0	0.7738	0.5905	0.4437	0.3277	0.2373	0.1681	0.0778	0.0312
	1	0.9774	0.9185	0.8352	0.7373	0.6328	0.5282	0.3370	0.1875
	2	0.9988	0.9914	0.9734	0.9421	0.8965	0.8369	0.6826	0.5000
	3	1.0000	0.9995	0.9978	0.9933	0.9844	0.9692	0.9130	0.8125
	4	1.0000	1.0000	0.9999	0.9997	0.9990	0.9976	0.9898	0.9688
6	0	0.7351	0.5314	0.3771	0.2621	0.1780	0.1177	0.0467	0.0156
	1	0.9672	0.8857	0.7765	0.6554	0.5339	0.4202	0.2333	0.1094
	2	0.9978	0.9841	0.9527	0.9011	0.8306	0.7443	0.5443	0.3438
	3	0.9999	0.9987	0.9941	0.9830	0.9624	0.9295	0.8208	0.6562
	4	1.0000	1.0000	0.9996	0.9984	0.9954	0.9891	0.9590	0.8906
	5	1.0000	1.0000	1.0000	0.9999	0.9998	0.9993	0.9959	0.9844
7	0	0.6983	0.4783	0.3206	0.2097	0.1335	0.0824	0.0280	0.0078
	1	0.9556	0.8503	0.7166	0.5767	0.4450	0.3294	0.1586	0.0625
	2	0.9962	0.9743	0.9262	0.8520	0.7564	0.6471	0.4199	0.2266
	3	0.9998	0.9973	0.9879	0.9667	0.9294	0.8740	0.7102	0.5000
	4	1.0000	0.9998	0.9988	0.9953	0.9871	0.9712	0.9037	0.7734
	5	1.0000	1.0000	0.9999	0.9996	0.9987	0.9962	0.9812	0.9375
	6	1.0000	1.0000	1.0000	1.0000	0.9999	0.9998	0.9984	0.9922

		p							
n	x	0.05	0.10	0.15	0.20	0.25	0.30	0.40	0.50
8	0	0.6634	0.4305	0.2725	0.1678	0.1001	0.0576	0.0168	0.0039
	1	0.9428	0.8131	0.6572	0.5033	0.3671	0.2553	0.1064	0.0352
	2	0.9942	0.9619	0.8948	0.7969	0.6785	0.5518	0.3154	0.1445
	3	0.9996	0.9950	0.9787	0.9437	0.8862	0.8059	0.5941	0.3633
	4	1.0000	0.9996	0.9971	0.9896	0.9727	0.9420	0.8263	0.6367
	5	1.0000	1.0000	0.9998	0.9988	0.9958	0.9887	0.9502	0.8555
	6	1.0000	1.0000	1.0000	0.9999	0.9996	0.9987	0.9915	0.9648
	7	1.0000	1.0000	1.0000	1.0000	1.0000	0.9999	0.9993	0.9961
9	0	0.6302	0.3874	0.2316	0.1342	0.0751	0.0403	0.0101	0.0019
	1	0.9288	0.7748	0.5995	0.4362	0.3003	0.1960	0.0705	0.0195
	2	0.9916	0.9470	0.8591	0.7382	0.6007	0.4628	0.2318	0.0898
	3	0.9994	0.9917	0.9661	0.9144	0.8343	0.7297	0.4826	0.2539
	4	1.0000	0.9991	0.9944	0.9804	0.9511	0.9012	0.7334	0.5000
	5	1.0000	0.9999	0.9994	0.9969	0.9900	0.9747	0.9006	0.7461
	6	1.0000	1.0000	1.0000	0.9997	0.9987	0.9957	0.9750	0.9102
	7	1.0000	1.0000	1.0000	1.0000	0.9999	0.9996	0.9962	0.9805
	8	1.0000	1.0000	1.0000	1.0000	1.0000	1.0000	0.9997	0.9980
10	0	0.5987	0.3487	0.1969	0.1074	0.0563	0.0283	0.0060	0.0010
	1	0.9139	0.7361	0.5443	0.3758	0.2440	0.1493	0.0464	0.0107
	2	0.9885	0.9298	0.8202	0.6778	0.5256	0.3828	0.1673	0.0547
	3	0.9990	0.9872	0.9500	0.8791	0.7759	0.6496	0.3823	0.1719
	4	0.9999	0.9984	0.9901	0.9672	0.9219	0.8497	0.6331	0.3770
	5	1.0000	0.9999	0.9986	0.9936	0.9803	0.9526	0.8338	0.6230
	6	1.0000	1.0000	0.9999	0.9991	0.9965	0.9894	0.9452	0.8281
	7	1.0000	1.0000	1.0000	0.9999	0.9996	0.9984	0.9877	0.9453
	8	1.0000	1.0000	1.0000	1.0000	1.0000	0.9999	0.9983	0.9893
	9	1.0000	1.0000	1.0000	1.0000	1.0000	1.0000	0.9999	0.9990
11	0	0.5688	0.3138	0.1673	0.0859	0.0422	0.0198	0.0036	0.0005
	1	0.8981	0.6974	0.4922	0.3221	0.1971	0.1130	0.0302	0.0059
	2	0.9848	0.9104	0.7788	0.6174	0.4552	0.3127	0.1189	0.0327
	3	0.9984	0.9815	0.9306	0.8389	0.7133	0.5696	0.2963	0.1133
	4	0.9999	0.9972	0.9841	0.9496	0.8854	0.7897	0.5328	0.2744
	5	1.0000	0.9997	0.9973	0.9883	0.9657	0.9218	0.7535	0.5000
	6	1.0000	1.0000	0.9997	0.9980	0.9924	0.9784	0.9006	0.7256
	7	1.0000	1.0000	1.0000	0.9998	0.9988	0.9957	0.9707	0.8867
	8	1.0000	1.0000	1.0000	1.0000	0.9999	0.9994	0.9941	0.9673
	9	1.0000	1.0000	1.0000	1.0000	1.0000	1.0000	0.9993	0.9941
	10	1.0000	1.0000	1.0000	1.0000	1.0000	1.0000	1.0000	0.9995

n	x	0.05	0.10	0.15	0.20	0.25	0.30	0.40	0.50
12	0	0.5404	0.2824	0.1422	0.0687	0.0317	0.0138	0.0022	0.0002
	1	0.8816	0.6590	0.4435	0.2749	0.1584	0.0850	0.0196	0.0032
	2	0.9804	0.8891	0.7358	0.5584	0.3907	0.2528	0.0834	0.0193
	3	0.9978	0.9744	0.9078	0.7946	0.6488	0.4925	0.2253	0.0730
	4	0.9998	0.9957	0.9761	0.9274	0.8424	0.7237	0.4382	0.1938
	5	1.0000	0.9995	0.9954	0.9806	0.9456	0.8821	0.6652	0.3872
	6	1.0000	1.0000	0.9993	0.9961	0.9858	0.9614	0.8418	0.6128
	7	1.0000	1.0000	0.9999	0.9994	0.9972	0.9905	0.9427	0.8062
	8	1.0000	1.0000	1.0000	0.9999	0.9996	0.9983	0.9847	0.9270
	9	1.0000	1.0000	1.0000	1.0000	1.0000	0.9998	0.9972	0.9807
	10	1.0000	1.0000	1.0000	1.0000	1.0000	1.0000	0.9997	0.9968
	11	1.0000	1.0000	1.0000	1.0000	1.0000	1.0000	1.0000	0.9998
13	0	0.5133	0.2542	0.1209	0.0550	0.0238	0.0097	0.0013	0.0001
	1	0.8646	0.6213	0.3983	0.2336	0.1267	0.0637	0.0126	0.0017
	2	0.9755	0.8661	0.6920	0.5017	0.3326	0.2025	0.0579	0.0112
	3	0.9969	0.9658	0.8820	0.7473	0.5843	0.4206	0.1686	0.0461
	4	0.9997	0.9935	0.9658	0.9009	0.7940	0.6543	0.3530	0.1334
	5	1.0000	0.9991	0.9925	0.9700	0.9198	0.8346	0.5744	0.2905
	6	1.0000	0.9999	0.9987	0.9930	0.9757	0.9376	0.7712	0.5000
	7	1.0000	1.0000	0.9998	0.9988	0.9943	0.9818	0.9023	0.7095
	8	1.0000	1.0000	1.0000	0.9998	0.9990	0.9960	0.9679	0.8666
	9	1.0000	1.0000	1.0000	1.0000	0.9999	0.9993	0.9922	0.9539
	10	1.0000	1.0000	1.0000	1.0000	1.0000	0.9999	0.9987	0.9888
	11	1.0000	1.0000	1.0000	1.0000	1.0000	1.0000	0.9999	0.9983
	12	1.0000	1.0000	1.0000	1.0000	1.0000	1.0000	1.0000	0.9999

7.14.7 CUMULATIVE TERMS, POISSON DISTRIBUTION

$$F(x;\lambda) = \sum_{k=0}^{x} e^{-\lambda}\frac{\lambda^k}{k!}.$$

If λ is the rate of Poisson arrivals, then $F(x;\lambda)$ is the probability of x or fewer arrivals occurring in a unit of time. For example, if customers arrive at the rate of $\lambda = 0.5$ customers per hour, then the probability of having no customers in any specified hour is 0.61 (the probability of one or fewer customers is 0.91).

λ	0	1	2	3	4	5	6	7	8	9
0.02	0.980	1.000								
0.04	0.961	0.999	1.000							
0.06	0.942	0.998	1.000							
0.08	0.923	0.997	1.000							
0.10	0.905	0.995	1.000							
0.15	0.861	0.990	1.000	1.000						
0.20	0.819	0.983	0.999	1.000						
0.25	0.779	0.974	0.998	1.000						
0.30	0.741	0.963	0.996	1.000						

λ	x									
	0	1	2	3	4	5	6	7	8	9
0.35	0.705	0.951	0.995	1.000						
0.40	0.670	0.938	0.992	0.999	1.000					
0.45	0.638	0.925	0.989	0.999	1.000					
0.50	0.607	0.910	0.986	0.998	1.000					
0.55	0.577	0.894	0.982	0.998	1.000					
0.60	0.549	0.878	0.977	0.997	1.000					
0.65	0.522	0.861	0.972	0.996	0.999	1.000				
0.70	0.497	0.844	0.966	0.994	0.999	1.000				
0.75	0.472	0.827	0.960	0.993	0.999	1.000				
0.80	0.449	0.809	0.953	0.991	0.999	1.000				
0.85	0.427	0.791	0.945	0.989	0.998	1.000				
0.90	0.407	0.772	0.937	0.987	0.998	1.000				
0.95	0.387	0.754	0.929	0.984	0.997	1.000				
1.00	0.368	0.736	0.920	0.981	0.996	0.999	1.000			
1.1	0.333	0.699	0.900	0.974	0.995	0.999	1.000			
1.2	0.301	0.663	0.879	0.966	0.992	0.999	1.000			
1.3	0.273	0.627	0.857	0.957	0.989	0.998	1.000			
1.4	0.247	0.592	0.834	0.946	0.986	0.997	0.999	1.000		
1.5	0.223	0.558	0.809	0.934	0.981	0.996	0.999	1.000		
1.6	0.202	0.525	0.783	0.921	0.976	0.994	0.999	1.000		
1.7	0.183	0.493	0.757	0.907	0.970	0.992	0.998	1.000		
1.8	0.165	0.463	0.731	0.891	0.964	0.990	0.997	0.999	1.000	
1.9	0.150	0.434	0.704	0.875	0.956	0.987	0.997	0.999	1.000	
2.0	0.135	0.406	0.677	0.857	0.947	0.983	0.996	0.999	1.000	
2.2	0.111	0.355	0.623	0.819	0.927	0.975	0.993	0.998	1.000	
2.4	0.091	0.308	0.570	0.779	0.904	0.964	0.988	0.997	0.999	1.000
2.6	0.074	0.267	0.518	0.736	0.877	0.951	0.983	0.995	0.999	1.000
2.8	0.061	0.231	0.469	0.692	0.848	0.935	0.976	0.992	0.998	0.999
3.0	0.050	0.199	0.423	0.647	0.815	0.916	0.967	0.988	0.996	0.999
3.2	0.041	0.171	0.380	0.603	0.781	0.895	0.955	0.983	0.994	0.998
3.4	0.033	0.147	0.340	0.558	0.744	0.871	0.942	0.977	0.992	0.997
3.6	0.027	0.126	0.303	0.515	0.706	0.844	0.927	0.969	0.988	0.996
3.8	0.022	0.107	0.269	0.473	0.668	0.816	0.909	0.960	0.984	0.994
4.0	0.018	0.092	0.238	0.433	0.629	0.785	0.889	0.949	0.979	0.992
4.2	0.015	0.078	0.210	0.395	0.590	0.753	0.868	0.936	0.972	0.989
4.4	0.012	0.066	0.185	0.359	0.551	0.720	0.844	0.921	0.964	0.985
4.6	0.010	0.056	0.163	0.326	0.513	0.686	0.818	0.905	0.955	0.981
4.8	0.008	0.048	0.142	0.294	0.476	0.651	0.791	0.887	0.944	0.975
5.0	0.007	0.040	0.125	0.265	0.441	0.616	0.762	0.867	0.932	0.968
5.2	0.005	0.034	0.109	0.238	0.406	0.581	0.732	0.845	0.918	0.960
5.4	0.004	0.029	0.095	0.213	0.373	0.546	0.702	0.822	0.903	0.951

λ	0	1	2	3	4	5	6	7	8	9
5.6	0.004	0.024	0.082	0.191	0.342	0.512	0.670	0.797	0.886	0.941
5.8	0.003	0.021	0.071	0.170	0.313	0.478	0.638	0.771	0.867	0.929
6.0	0.003	0.017	0.062	0.151	0.285	0.446	0.606	0.744	0.847	0.916
6.2	0.002	0.015	0.054	0.134	0.259	0.414	0.574	0.716	0.826	0.902
6.4	0.002	0.012	0.046	0.119	0.235	0.384	0.542	0.687	0.803	0.886
6.6	0.001	0.010	0.040	0.105	0.213	0.355	0.511	0.658	0.780	0.869
6.8	0.001	0.009	0.034	0.093	0.192	0.327	0.480	0.628	0.755	0.850
7.0	0.001	0.007	0.030	0.082	0.173	0.301	0.450	0.599	0.729	0.831
7.2	0.001	0.006	0.025	0.072	0.155	0.276	0.420	0.569	0.703	0.810
7.4	0.001	0.005	0.022	0.063	0.140	0.253	0.392	0.539	0.676	0.788
7.6	0.001	0.004	0.019	0.055	0.125	0.231	0.365	0.510	0.648	0.765
7.8	0.000	0.004	0.016	0.049	0.112	0.210	0.338	0.481	0.620	0.741
8.0	0.000	0.003	0.014	0.042	0.100	0.191	0.313	0.453	0.593	0.717
8.5	0.000	0.002	0.009	0.030	0.074	0.150	0.256	0.386	0.523	0.653
9.0	0.000	0.001	0.006	0.021	0.055	0.116	0.207	0.324	0.456	0.587
9.5	0.000	0.001	0.004	0.015	0.040	0.088	0.165	0.269	0.392	0.522
10.0	0.000	0.001	0.003	0.010	0.029	0.067	0.130	0.220	0.333	0.458
10.5	0.000	0.000	0.002	0.007	0.021	0.050	0.102	0.178	0.279	0.397
11.0	0.000	0.000	0.001	0.005	0.015	0.037	0.079	0.143	0.232	0.341
11.5	0.000	0.000	0.001	0.003	0.011	0.028	0.060	0.114	0.191	0.289
12.0	0.000	0.000	0.001	0.002	0.008	0.020	0.046	0.089	0.155	0.242
12.5	0.000	0.000	0.000	0.002	0.005	0.015	0.035	0.070	0.125	0.201
13.0	0.000	0.000	0.000	0.001	0.004	0.011	0.026	0.054	0.100	0.166
13.5	0.000	0.000	0.000	0.001	0.003	0.008	0.019	0.042	0.079	0.135
14.0	0.000	0.000	0.000	0.001	0.002	0.005	0.014	0.032	0.062	0.109
14.5	0.000	0.000	0.000	0.000	0.001	0.004	0.011	0.024	0.048	0.088
15.0	0.000	0.000	0.000	0.000	0.001	0.003	0.008	0.018	0.037	0.070

λ	10	11	12	13	14	15	16	17	18	19
2.8	1.000									
3.0	1.000									
3.2	1.000									
3.4	0.999	1.000								
3.6	0.999	1.000								
3.8	0.998	0.999	1.000							
4.0	0.997	0.999	1.000							
4.2	0.996	0.999	1.000							
4.4	0.994	0.998	0.999	1.000						
4.6	0.992	0.997	0.999	1.000						
4.8	0.990	0.996	0.999	1.000						
5.0	0.986	0.995	0.998	0.999	1.000					
5.2	0.982	0.993	0.997	0.999	1.000					
5.4	0.978	0.990	0.996	0.999	1.000					
5.6	0.972	0.988	0.995	0.998	0.999	1.000				

λ	x									
	10	11	12	13	14	15	16	17	18	19
5.8	0.965	0.984	0.993	0.997	0.999	1.000				
6.0	0.957	0.980	0.991	0.996	0.999	1.000	1.000			
6.2	0.949	0.975	0.989	0.995	0.998	0.999	1.000			
6.4	0.939	0.969	0.986	0.994	0.997	0.999	1.000			
6.6	0.927	0.963	0.982	0.992	0.997	0.999	1.000	1.000		
6.8	0.915	0.955	0.978	0.990	0.996	0.998	0.999	1.000		
7.0	0.901	0.947	0.973	0.987	0.994	0.998	0.999	1.000		
7.2	0.887	0.937	0.967	0.984	0.993	0.997	0.999	1.000		
7.4	0.871	0.926	0.961	0.981	0.991	0.996	0.998	0.999	1.000	
7.6	0.854	0.915	0.954	0.976	0.989	0.995	0.998	0.999	1.000	
7.8	0.835	0.902	0.945	0.971	0.986	0.993	0.997	0.999	1.000	
8.0	0.816	0.888	0.936	0.966	0.983	0.992	0.996	0.998	0.999	1.000
8.5	0.763	0.849	0.909	0.949	0.973	0.986	0.993	0.997	0.999	1.000
9.0	0.706	0.803	0.876	0.926	0.959	0.978	0.989	0.995	0.998	0.999
9.5	0.645	0.752	0.836	0.898	0.940	0.967	0.982	0.991	0.996	0.998
10.0	0.583	0.697	0.792	0.865	0.916	0.951	0.973	0.986	0.993	0.997
10.5	0.521	0.639	0.742	0.825	0.888	0.932	0.960	0.978	0.989	0.994
11.0	0.460	0.579	0.689	0.781	0.854	0.907	0.944	0.968	0.982	0.991
11.5	0.402	0.520	0.633	0.733	0.815	0.878	0.924	0.954	0.974	0.986
12.0	0.347	0.462	0.576	0.681	0.772	0.844	0.899	0.937	0.963	0.979
12.5	0.297	0.406	0.519	0.628	0.725	0.806	0.869	0.916	0.948	0.969
13.0	0.252	0.353	0.463	0.573	0.675	0.764	0.836	0.890	0.930	0.957
13.5	0.211	0.304	0.409	0.518	0.623	0.718	0.797	0.861	0.908	0.942
14.0	0.176	0.260	0.358	0.464	0.570	0.669	0.756	0.827	0.883	0.923
14.5	0.145	0.220	0.311	0.412	0.518	0.619	0.711	0.790	0.853	0.901
15.0	0.118	0.185	0.268	0.363	0.466	0.568	0.664	0.749	0.820	0.875

λ	x									
	20	21	22	23	24	25	26	27	28	29
8.5	1.000									
9.0	1.000									
9.5	0.999	1.000								
10.0	0.998	0.999	1.000							
10.5	0.997	0.999	0.999	1.000						
11.0	0.995	0.998	0.999	1.000						
11.5	0.993	0.996	0.998	0.999	1.000					
12.0	0.988	0.994	0.997	0.999	0.999	1.000				
12.5	0.983	0.991	0.995	0.998	0.999	0.999	1.000			
13.0	0.975	0.986	0.992	0.996	0.998	0.999	1.000			
13.5	0.965	0.980	0.989	0.994	0.997	0.998	0.999	1.000		
14.0	0.952	0.971	0.983	0.991	0.995	0.997	0.999	0.999	1.000	
14.5	0.936	0.960	0.976	0.986	0.992	0.996	0.998	0.999	1.000	1.000
15.0	0.917	0.947	0.967	0.981	0.989	0.994	0.997	0.998	0.999	1.000

					x					
λ	5	6	7	8	9	10	11	12	13	14
16	0.001	0.004	0.010	0.022	0.043	0.077	0.127	0.193	0.275	0.367
17	0.001	0.002	0.005	0.013	0.026	0.049	0.085	0.135	0.201	0.281
18	0.000	0.001	0.003	0.007	0.015	0.030	0.055	0.092	0.143	0.208
19	0.000	0.001	0.002	0.004	0.009	0.018	0.035	0.061	0.098	0.150
20	0.000	0.000	0.001	0.002	0.005	0.011	0.021	0.039	0.066	0.105
21	0.000	0.000	0.000	0.001	0.003	0.006	0.013	0.025	0.043	0.072
22	0.000	0.000	0.000	0.001	0.002	0.004	0.008	0.015	0.028	0.048
23	0.000	0.000	0.000	0.000	0.001	0.002	0.004	0.009	0.017	0.031
24	0.000	0.000	0.000	0.000	0.000	0.001	0.003	0.005	0.011	0.020
25	0.000	0.000	0.000	0.000	0.000	0.001	0.001	0.003	0.006	0.012
26	0.000	0.000	0.000	0.000	0.000	0.000	0.001	0.002	0.004	0.008
27	0.000	0.000	0.000	0.000	0.000	0.000	0.000	0.001	0.002	0.005
28	0.000	0.000	0.000	0.000	0.000	0.000	0.000	0.001	0.001	0.003
29	0.000	0.000	0.000	0.000	0.000	0.000	0.000	0.000	0.001	0.002
30	0.000	0.000	0.000	0.000	0.000	0.000	0.000	0.000	0.000	0.001

					x					
λ	15	16	17	18	19	20	21	22	23	24
16	0.467	0.566	0.659	0.742	0.812	0.868	0.911	0.942	0.963	0.978
17	0.371	0.468	0.564	0.655	0.736	0.805	0.862	0.905	0.937	0.959
18	0.287	0.375	0.469	0.562	0.651	0.731	0.799	0.855	0.899	0.932
19	0.215	0.292	0.378	0.469	0.561	0.647	0.726	0.793	0.849	0.893
20	0.157	0.221	0.297	0.381	0.470	0.559	0.644	0.721	0.787	0.843
21	0.111	0.163	0.227	0.302	0.384	0.471	0.558	0.640	0.716	0.782
22	0.077	0.117	0.169	0.233	0.306	0.387	0.472	0.556	0.637	0.712
23	0.052	0.082	0.123	0.175	0.238	0.310	0.389	0.472	0.555	0.635
24	0.034	0.056	0.087	0.128	0.180	0.243	0.314	0.392	0.473	0.554
25	0.022	0.038	0.060	0.092	0.134	0.185	0.247	0.318	0.394	0.473
26	0.014	0.025	0.041	0.065	0.097	0.139	0.191	0.252	0.321	0.396
27	0.009	0.016	0.027	0.044	0.069	0.102	0.144	0.195	0.256	0.324
28	0.005	0.010	0.018	0.030	0.048	0.073	0.106	0.148	0.200	0.260
29	0.003	0.006	0.011	0.020	0.033	0.051	0.077	0.110	0.153	0.204
30	0.002	0.004	0.007	0.013	0.022	0.035	0.054	0.081	0.115	0.157

					x					
λ	25	26	27	28	29	30	31	32	33	34
16	0.987	0.993	0.996	0.998	0.999	0.999	1.000			
17	0.975	0.985	0.991	0.995	0.997	0.999	0.999	1.000		
18	0.955	0.972	0.983	0.990	0.994	0.997	0.998	0.999	1.000	
19	0.927	0.951	0.969	0.981	0.988	0.993	0.996	0.998	0.999	0.999
20	0.888	0.922	0.948	0.966	0.978	0.987	0.992	0.995	0.997	0.999
21	0.838	0.883	0.917	0.944	0.963	0.976	0.985	0.991	0.995	0.997
22	0.777	0.832	0.877	0.913	0.940	0.960	0.974	0.983	0.990	0.994
23	0.708	0.772	0.827	0.873	0.908	0.936	0.956	0.971	0.981	0.988
24	0.632	0.704	0.768	0.823	0.868	0.904	0.932	0.953	0.969	0.979

7.14.8 CRITICAL VALUES, KOLMOGOROV–SMIRNOV TEST

One-sided test	$p = 0.90$	0.95	0.975	0.99	0.995
Two-sided test	$p = 0.80$	0.90	0.95	0.98	0.99
$n = 1$	0.900	0.950	0.975	0.990	0.995
2	0.684	0.776	0.842	0.900	0.929
3	0.565	0.636	0.708	0.785	0.829
4	0.493	0.565	0.624	0.689	0.734
5	0.447	0.509	0.563	0.627	0.669
6	0.410	0.468	0.519	0.577	0.617
7	0.381	0.436	0.483	0.538	0.576
8	0.358	0.410	0.454	0.507	0.542
9	0.339	0.387	0.430	0.480	0.513
10	0.323	0.369	0.409	0.457	0.489
11	0.308	0.352	0.391	0.437	0.468
12	0.296	0.338	0.375	0.419	0.449
13	0.285	0.325	0.361	0.404	0.432
14	0.275	0.314	0.349	0.390	0.418
15	0.266	0.304	0.338	0.377	0.404
20	0.232	0.265	0.294	0.329	0.352
25	0.208	0.238	0.264	0.295	0.317
30	0.190	0.218	0.242	0.270	0.290
35	0.177	0.202	0.224	0.251	0.269
40	0.165	0.189	0.210	0.235	0.252
Approximation for $n > 40$:	$\dfrac{1.07}{\sqrt{n}}$	$\dfrac{1.22}{\sqrt{n}}$	$\dfrac{1.36}{\sqrt{n}}$	$\dfrac{1.52}{\sqrt{n}}$	$\dfrac{1.63}{\sqrt{n}}$

7.14.9 CRITICAL VALUES, TWO SAMPLE KOLMOGOROV–SMIRNOV TEST

The value of D listed below is so large that the hypothesis H_0, the two distributions are the same, is to be rejected at the indicated level of significance. Here, n_1 and n_2 are assumed to be large, and $D = \max |F_{n_1}(x) - F_{n_2}(x)|$.

Level of significance	Value of D
$\alpha = 0.10$	$1.22\sqrt{\dfrac{n_1+n_2}{n_1 n_2}}$
$\alpha = 0.05$	$1.36\sqrt{\dfrac{n_1+n_2}{n_1 n_2}}$
$\alpha = 0.025$	$1.48\sqrt{\dfrac{n_1+n_2}{n_1 n_2}}$
$\alpha = 0.01$	$1.63\sqrt{\dfrac{n_1+n_2}{n_1 n_2}}$
$\alpha = 0.005$	$1.73\sqrt{\dfrac{n_1+n_2}{n_1 n_2}}$
$\alpha = 0.001$	$1.95\sqrt{\dfrac{n_1+n_2}{n_1 n_2}}$

7.14.10 CRITICAL VALUES, SPEARMAN'S RANK CORRELATION

Spearman's coefficient of rank correlation, ρ_s, measures the correspondence between two rankings. Let d_i be the difference between the ranks of the ith pair of a set of n pairs of elements. Then Spearman's rho is defined as

$$\rho_s = 1 - \frac{6 \sum_{i=1}^{n} d_i^2}{n^3 - n} = 1 - \frac{6 S_r}{n^3 - n}$$

where $S_r = \sum_{i=1}^{n} d_i^2$. The table below gives critical values for S_r when there is complete independence.

n	$p = 0.90$	$p = 0.95$	$p = 0.99$	$p = 0.999$
4	0.8000	0.8000	—	—
5	0.7000	0.8000	0.9000	—
6	0.6000	0.7714	0.8857	—
7	0.5357	0.6786	0.8571	0.9643
8	0.5000	0.6190	0.8095	0.9286
9	0.4667	0.5833	0.7667	0.9000
10	0.4424	0.5515	0.7333	0.8667
11	0.4182	0.5273	0.7000	0.8364
12	0.3986	0.4965	0.6713	0.8182
13	0.3791	0.4780	0.6429	0.7912
14	0.3626	0.4593	0.6220	0.7670
15	0.3500	0.4429	0.6000	0.7464
20	0.2977	0.3789	0.5203	0.6586
25	0.2646	0.3362	0.4654	0.5962
30	0.2400	0.3059	0.4251	0.5479

7.15 SIGNAL PROCESSING

7.15.1 ESTIMATION

Let $\{e_t\}$ be a white noise process (so that $\mathrm{E}\,[e_t] = \mu$, $\mathrm{Var}\,[e_t] = \sigma^2$, and $\mathrm{Cov}\,[e_t, e_s] = 0$ for $s \neq t$). Suppose that $\{X_t\}$ is a time series. A non-anticipating linear model presumes that $\sum_{u=0}^{\infty} h_u X_{t-u} = e_t$, where the $\{h_u\}$ are constants. This can be written $H(z)X_t = e_t$ where $H(z) = \sum_{u=0}^{\infty} h_u z^u$ and $z^n X_t = X_{t-n}$. Alternately, $X_t = H^{-1}(z)e_t$. In practice, several types of models are used:

1. AR(k), autoregressive model of order k: This assumes that $H(z) = 1 + a_1 z + \cdots + a_k z_k$ and so

$$X_t + a_1 X_{t-1} + \ldots a_k X_{t-k} = e_t. \qquad (7.15.1)$$

2. MA(l), moving average of order l: This assumes that
$H^{-1}(z) = 1 + b_1 z + \cdots + b_k z_k$ and so

$$X_t = e_t + b_1 e_{t-1} + \ldots b_l e_{t-l}. \tag{7.15.2}$$

3. ARMA(k, l), mixed autoregressive/moving average of order (k, l): This assumes that $H^{-1}(z) = \frac{1+b_1 z+\cdots+b_l z_l}{1+a_1 z+\cdots+a_k z_k}$ and so

$$X_t + a_1 X_{t-1} + \ldots a_k X_{t-k} = e_t + b_1 e_{t-1} + \cdots + b_l e_{t-l}. \tag{7.15.3}$$

7.15.2   KALMAN FILTERS

Kalman filtering is a linear least-squares recursive estimator. It is used when the state space has a higher dimension than the observation space. For example, in some airport radars the distance to aircraft is measured and the velocity of each aircraft is inferred.

1. $\mathbf{x}$ is the unknown state to be estimated
2. $\widehat{\mathbf{x}}$ is the estimate of $\mathbf{x}$
3. $\mathbf{z}$ is an observation
4. $\{\mathbf{w}, \mathbf{v}\}$ are noise terms
5. $\{Q, R\}$ are spectral density matrices
6. "$\mathbf{a} \sim N(\mathbf{b}, C)$" means that the random variable $\mathbf{a}$ has a normal distribution with a mean of $\mathbf{b}$ and a covariance matrix of C.
7. "Extended Kalman filter": State propagation is achieved through sequential linearizations of the system model and the measurement model.
8. "$(-)$" is the value before a new discrete observation and "$(+)$" is the value after a new discrete observation

7.15.2.1   Discrete Kalman filter

System model	$\mathbf{x}_k = \Phi_{k-1}\mathbf{x}_{k-1} + \mathbf{w}_{k-1};$	$\mathbf{w}_k \sim N(\mathbf{0}, Q_k)$
Measurement model	$\mathbf{z}_k = H_k\mathbf{x}_k + \mathbf{v}_k;$	$\mathbf{v}_k \sim N(\mathbf{0}, R_k)$
Initial conditions	$\mathrm{E}\left[\mathbf{x}(0)\right] = \widehat{\mathbf{x}}_0,\ \mathrm{E}\left[(\mathbf{x}(0) - \widehat{\mathbf{x}}_0)(\mathbf{x}(0) - \widehat{\mathbf{x}}_0)^{\mathrm{T}}\right] = P_0$	
Other assumptions	$\mathrm{E}\left[\mathbf{w}_k\mathbf{v}_j^{\mathrm{T}}\right] = 0$ for all j and k	
State estimate extrapolation	$\widehat{\mathbf{x}}_k(-) = \Phi_{k-1}\widehat{\mathbf{x}}_{k-1}(+)$	
Error covariance extrapolation	$P_k(-) = \Phi_{k-1}P_{k-1}(+)\Phi_{k-1}^{\mathrm{T}} + Q_{k-1}$	
State estimate update	$\widehat{\mathbf{x}}_k(+) = \widehat{\mathbf{x}}_k(-) + K_k[\mathbf{z}_k - H_k\widehat{\mathbf{x}}_k(-)]$	
Error covariance update	$P_k(+) = [I - K_k H_k]P_k(-)$	
Kalman gain matrix	$K_k = P_k(-)H_k^{\mathrm{T}}\left[H_k P_k(-)H_k^{\mathrm{T}} + R_k\right]^{-1}$	

7.15.2.2 Continuous Kalman filter

System model	$\dot{\mathbf{x}}(t) = F(t)\mathbf{x}(t) + G(t)\mathbf{w}(t); \qquad \mathbf{w}(t) \sim N(\mathbf{0}, Q(t))$
Measurement model	$\mathbf{z}(t) = H(t)\mathbf{x}(t) + \mathbf{v}(t); \qquad \mathbf{v}(t) \sim N(\mathbf{0}, R(t))$
Initial conditions	$\widehat{\mathbf{x}}(0) = \widehat{\mathbf{x}}_0, \quad P(0) = P_0,$
	$\mathrm{E}[\mathbf{x}(0)] = \widehat{\mathbf{x}}_0, \quad \mathrm{E}\left[(\mathbf{x}(0) - \widehat{\mathbf{x}}_0)(\mathbf{x}(0) - \widehat{\mathbf{x}}_0)^{\mathrm{T}}\right] = P_0$
Other assumptions	$R^{-1}(t)$ exists, $\quad \mathrm{E}\left[\mathbf{w}(t)\mathbf{v}^{\mathrm{T}}(\tau)\right] = C(t)\delta(t - \tau)$
State estimate propagation	$\dot{\widehat{\mathbf{x}}}(t) = F(t)\widehat{\mathbf{x}}(t) + K(t)\left[\mathbf{z}(t) - H(t)\widehat{\mathbf{x}}(t)\right]$
Error covariance propagation	$\dot{P}(t) = F(t)P(t) + P(t)F^{\mathrm{T}}(t)$
	$\qquad + G(t)Q(t)G^{\mathrm{T}}(t) - K(t)R(t)K^{\mathrm{T}}(t)$
Kalman gain matrix	$K(t) = \left[P(t)H^{\mathrm{T}}(t) + G(t)C(t)\right] R^{-1}(t)$

7.15.2.3 Continuous extended Kalman filter

System model	$\dot{\mathbf{x}}(t) = \mathbf{f}(\mathbf{x}(t), t) + \mathbf{w}(t); \qquad \mathbf{w}(t) \sim N(\mathbf{0}, Q(t))$
Measurement model	$\mathbf{z}(t) = \mathbf{h}(\mathbf{x}(t)) + \mathbf{v}(t); \qquad \mathbf{v}(t) \sim N(\mathbf{0}, R(t))$
Initial conditions	$\mathbf{x}(0) \sim N(\widehat{\mathbf{x}}_0, P_0)$
Other assumptions	$\mathrm{E}\left[\mathbf{w}(t)\mathbf{v}^{\mathrm{T}}(\tau)\right] = 0$ for all t and all τ
State estimate propagation	$\dot{\widehat{\mathbf{x}}}(t) = \mathbf{f}(\widehat{\mathbf{x}}(t), t) + K(t)\left[\mathbf{z}(t) - \mathbf{h}(\widehat{\mathbf{x}}(t), t)\right]$
Error covariance propagation	$\dot{P}(t) = F(\widehat{\mathbf{x}}(t), t)P(t) + P(t)F^{\mathrm{T}}(\widehat{\mathbf{x}}(t), t) + Q(t)$
	$\qquad - P(t)H^{\mathrm{T}}(\widehat{\mathbf{x}}(t), t)R^{-1}(t)H(\widehat{\mathbf{x}}(t), t)P(t)$
Gain equation	$K(t) = P(t)H^{\mathrm{T}}(\widehat{\mathbf{x}}(t), t)R^{-1}(t)$
Definitions	$F(\widehat{\mathbf{x}}(t), t) = \left. \dfrac{\partial \mathbf{f}(\mathbf{x}(t), t)}{\partial \mathbf{x}(t)} \right|_{\mathbf{x}(t) = \widehat{\mathbf{x}}(t)}$
	$H(\widehat{\mathbf{x}}(t), t) = \left. \dfrac{\partial \mathbf{h}(\mathbf{x}(t), t)}{\partial \mathbf{x}(t)} \right|_{\mathbf{x}(t) = \widehat{\mathbf{x}}(t)}$

7.15.2.4 Continuous-discrete extended Kalman filter

System model	$\dot{\mathbf{x}}(t) = \mathbf{f}(\mathbf{x}(t), t) + \mathbf{w}(t); \qquad \mathbf{w}(t) \sim N(\mathbf{0}, Q(t))$
Measurement model	$\mathbf{z}_k = \mathbf{h}_k(\mathbf{x}(t_k)) + \mathbf{v}_k; \qquad \mathbf{v}_k \sim N(\mathbf{0}, R_k)$
	$\qquad\qquad\qquad\qquad\qquad\qquad k = 1, 2, \ldots$
Initial conditions	$\mathbf{x}(0) \sim N(\widehat{\mathbf{x}}_0, P_0)$
Other assumptions	$\mathrm{E}\left[\mathbf{w}(t)\mathbf{v}_k^{\mathrm{T}}\right] = 0$ for all k and all t
State estimate propagation	$\dot{\widehat{\mathbf{x}}}(t) = \mathbf{f}(\widehat{\mathbf{x}}(t), t)$
Error covariance propagation	$\dot{P}(t) = F(\widehat{\mathbf{x}}(t), t)P(t) + P(t)F^{\mathrm{T}}(\widehat{\mathbf{x}}(t), t) + Q(t)$
State estimate update	$\widehat{\mathbf{x}}_k(+) = \widehat{\mathbf{x}}_k(-) + K_k\left[\mathbf{z}_k - \mathbf{h}_k(\widehat{\mathbf{x}}_k(-))\right]$
Error covariance update	$P_k(+) = \left[I - K_k H_k(\widehat{\mathbf{x}}_k(-))\right] P_k(-)$
Gain matrix	$K_k = P_k(-)H_k^{\mathrm{T}}(\widehat{\mathbf{x}}_k(-))$
	$\qquad \times \left[H_k(\widehat{\mathbf{x}}_k(-))P_k(-)H_k^{\mathrm{T}}(\widehat{\mathbf{x}}_k(-)) + R_k\right]^{-1}$
Definitions	$F(\widehat{\mathbf{x}}(t), t) = \left. \dfrac{\partial \mathbf{f}(\mathbf{x}(t), t)}{\partial \mathbf{x}(t)} \right|_{\mathbf{x}(t) = \widehat{\mathbf{x}}(t)}$
	$H_k(\widehat{\mathbf{x}}_k(-)) = \left. \dfrac{\partial \mathbf{h}_k(\mathbf{x}(t_k))}{\partial \mathbf{x}(t_k)} \right|_{\mathbf{x}(t_k) = \widehat{\mathbf{x}}_k(-)}$

7.15.3 MATCHED FILTERING (WIENER FILTER)

Let $X(t)$ represent a signal to be recovered, let $N(t)$ represent noise, and let $Y(t) = X(t) + N(t)$ represent the observable signal. A prediction of the signal is

$$X_p(t) = \int_0^\infty K(z)Y(t - z)\, dz, \qquad (7.15.4)$$

where $K(z)$ is a filter. The mean square error is $\mathrm{E}\left[(X(t) - X_p(t))^2\right]$; this is minimized by the optimal (*Wiener*) filter $K_{\mathrm{opt}}(z)$.

When X and Y are stationary, define their autocorrelation functions as $R_{XX}(t - s) = \mathrm{E}[X(t)X(s)]$ and $R_{YY}(t - s) = \mathrm{E}[Y(t)Y(s)]$. If $\mathcal{F}$ represents the Fourier transform (see page 576), then the optimal filter is given by

$$\mathcal{F}[K_{\mathrm{opt}}(t)] = \frac{1}{2\pi} \frac{\mathcal{F}[R_{XX}(t)]}{\mathcal{F}[R_{YY}(t)]}. \qquad (7.15.5)$$

For example, if X and N are uncorrelated, then

$$\mathcal{F}[K_{\mathrm{opt}}(t)] = \frac{1}{2\pi} \frac{\mathcal{F}[R_{XX}(t)]}{\mathcal{F}[R_{XX}(t)] + \mathcal{F}[R_{NN}(t)]}. \qquad (7.15.6)$$

In the case of no noise, $\mathcal{F}[K_{\mathrm{opt}}(t)] = \frac{1}{2\pi}$, $K_{\mathrm{opt}}(t) = \delta(t)$, and $S_p(t) = Y(t)$.

7.15.4   WALSH FUNCTIONS

The *Rademacher functions* are defined by $r_k(x) = \operatorname{sgn}\sin\left(2^{k+1}\pi x\right)$. If the binary expansion of n has the form $n = 2^{i_1} + 2^{i_2} + \cdots + 2^{i_m}$, then the *Walsh function* of order n is $W_n(x) = r_{i_1}(x)r_{i_2}(x)\ldots r_{i_m}(x)$.

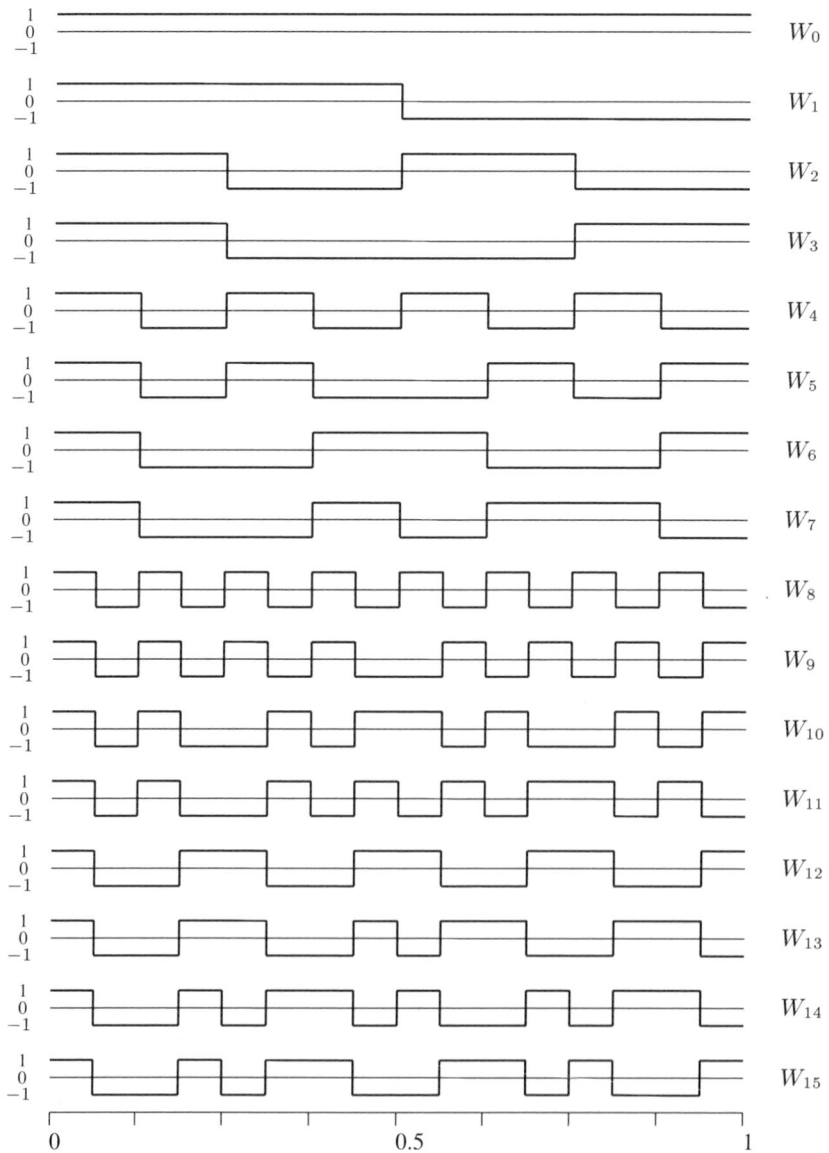

7.15.5 WAVELETS

The *Haar wavelet* is $H(x) = 1$ if $0 \le x < 1/2$, -1 if $1/2 \le x < 1$, and 0 otherwise. Define $H_{j,k}(x) = 2^{j/2}H(2^j x - k)$. Then the *Haar system*, $\{H_{j,k}\}_{j,k=-\infty}^{+\infty}$, forms an orthonormal basis for the Hilbert space, $L^2(\mathbb{R})$, consisting of functions f with finite energy, i.e., $\int_{-\infty}^{+\infty} |f(x)|^2 \, dx < \infty$.

7.15.5.1 Definitions

The construction of other wavelet orthonormal bases $\{\psi_{j,k}\}_{j,k=-\infty}^{+\infty}$ begins by choosing real coefficients $h_0, \ldots, h_n$ which satisfy the following conditions (we set $h_k = 0$ if $k < 0$ or $k > n$):

1. *Normalization*: $\sum_k h_k = \sqrt{2}$.
2. *Orthogonality*: $\sum_k h_k \, h_{k-2j} = 1$ if $j = 0$ and 0 if $j \ne 0$.
3. *Accuracy p*: $\sum_k (-1)^k k^j h_k = 0$ for $j = 0, \ldots, p-1$ with $p > 0$.
4. *Cohen–Lawton criterion*: A technical condition only rarely violated by coefficients which satisfy the normalization, orthogonality, and accuracy p conditions.

The terms *order of approximation* or *Strang–Fix conditions* are often used in place of "accuracy". The orthogonality condition implies that n is odd.

The four conditions above imply the existence of a solution $\varphi \in L^2(\mathbb{R})$, called the *scaling function*, to the following *refinement equation*:

$$\varphi(x) = \sqrt{2} \sum_{k=0}^{n} h_k \, \varphi(2x - k). \qquad (7.15.7)$$

The scaling function has a non-vanishing integral which we normalize: $\int \varphi(x) \, dx = 1$. Then $\varphi(x)$ is unique, and it vanishes outside of the interval $[0, n]$. The maximum possible accuracy is $p = (n+1)/2$. Thus, increasing the accuracy requires increasing the number of coefficients h_k. High accuracy is desirable, as it implies that each of the polynomials $1, x, \ldots, x^{p-1}$ can be written as an infinite linear combination of the integer translates $\varphi(x - k)$. In particular, $\sum_k \varphi(x - k) = 1$. Also, the smoothness of φ is limited by the accuracy; φ can have at most $n - 2$ derivatives, although in practice it usually has fewer.

For each fixed integer j, let V_j be the closed subspace of $L^2(\mathbb{R})$ spanned by the functions, $\{\varphi_{j,k}\}_{k=-\infty}^{+\infty}$, where $\varphi_{j,k}(x) = 2^{j/2}\varphi(2^j x - k)$. The sequence of subspaces $\{V_j\}_{j=-\infty}^{+\infty}$ forms a *multi-resolution analysis* for $L^2(\mathbb{R})$, meaning that:

1. The subspaces are nested: $\cdots \subset V_{-1} \subset V_0 \subset V_1 \cdots$.
2. They are obtained from each other by dilation by 2: $v(x) \in V_j \iff v(2x) \in V_{j+1}$.
3. V_0 is integer translation invariant: $v(x) \in V_0 \iff v(x + 1) \in V_0$.
4. The V_j increase to all of $L^2(\mathbb{R})$ and decrease to zero: $\cup V_j$ is dense in $L^2(\mathbb{R})$ and $\cap V_j = \{0\}$.
5. The set of integer translates $\{\varphi(x - k)\}_{k=-\infty}^{+\infty}$ forms an orthonormal basis for V_0.

The projection of $f(x)$ onto the subspace V_j is an approximation at resolution level 2^{-j}. It is given by $f_j(x) = \sum_k c_{j,k}\,\varphi_{j,k}(x)$ with

$$c_{j,k} = \langle f, \varphi_{j,k}\rangle = \int f(x)\,\varphi_{j,k}(x)\,dx. \qquad (7.15.8)$$

The *wavelet* ψ is derived from the scaling function φ by the formula,

$$\psi(x) = \sqrt{2}\sum_{k=0}^{n} g_k\,\varphi(2x - k), \qquad \text{where } g_k = (-1)^k\,h_{n-k}. \qquad (7.15.9)$$

The wavelet ψ has the same smoothness as φ, and the accuracy p condition implies vanishing moments for ψ: $\int x^j \psi(x)\,dx = 0$ for $j = 0, \ldots, p-1$. The functions $\psi_{j,k}(x) = 2^{j/2}\psi(2^j x - k)$ are orthonormal, and the entire collection $\{\psi_{j,k}\}_{j,k=-\infty}^{+\infty}$ forms an orthonormal basis for $L^2(\mathbb{R})$.

With j fixed, let W_j be the closed subspace of $L^2(\mathbb{R})$ spanned by $\psi(2^j x - k)$ for integer k. Then V_j and W_j are orthogonal subspaces whose direct sum is V_{j+1}. Let f be a function and let $p_j = \sum_k d_{j,k}\,\varphi_{j,k}(x)$ be its projection onto W_j, where $d_{j,k} = \langle f, \psi_{j,k}\rangle$. Then the approximation f_{j+1} with resolution $2^{-(j+1)}$ is $f_{j+1} = f_j + p_j$, the sum of the approximation f_j at resolution 2^{-j} and the additional fine details p_j needed to give the next higher resolution level.

The *discrete wavelet transform* is an algorithm for computing the coefficients $c_{j,k}$ and $d_{j,k}$ from the coefficients $c_{j+1,k}$. It can also be interpreted as an algorithm dealing directly with discrete data, dividing data $c_{j+1,k}$ into a *low-pass* part $c_{j,k}$ and a *high-pass* part $d_{j,k}$. Specifically,

$$c_{j,l} = \sum_k h_{k-2l}\,c_{j+1,k} \qquad \text{and} \qquad d_{j,l} = \sum_k g_{k-2l}\,c_{j+1,k}. \qquad (7.15.10)$$

The inverse transform is

$$c_{j+1,k} = \sum_l h_{k-2l}\,c_{j,l} + \sum_l g_{k-2l}\,d_{j,l}. \qquad (7.15.11)$$

The discrete wavelet transform is closely related to engineering techniques known as *sub-band coding* and *quadrature mirror filtering*.

EXAMPLE   *The Daubechies family.* For each *even* integer $N > 0$, there is a unique set of coefficients $h_0, \ldots, h_{N-1}$ which satisfy the normalization and orthogonality conditions with maximal accuracy $p = N/2$. The corresponding φ and ψ are the *Daubechies scaling function* D_N and *Daubechies wavelet* W_N. The Haar wavelet H is the same as the Daubechies wavelet W_2. For the Haar wavelet, the coefficients are $h_0 = h_1 = 1/\sqrt{2}$ and the subspace V_j consists of all functions which are piecewise constant on each interval $[k2^{-j}, (k+1)2^{-j})$. The coefficients for D_4 are: $h_0 = (1 + \sqrt{3})/(4\sqrt{2})$, $h_1 = (3 + \sqrt{3})/(4\sqrt{2})$, $h_2 = (3 - \sqrt{3})/(4\sqrt{2})$, and $h_3 = (1 - \sqrt{3})/(4\sqrt{2})$.

7.15.5.2 Generalizations

For a given number of coefficients, the Daubechies wavelet has the highest accuracy. Other wavelets reduce the accuracy in exchange for other properties. In the *Coiflet* family the scaling function and the wavelet possesses vanishing moments, leading to simple one-point quadrature formulae. The "least asymmetric" wavelets are close to being symmetric or antisymmetric (perfect symmetry is incompatible with orthogonality, except for the Haar wavelet).

Wavelet packets are libraries of basis functions defined recursively from the scaling functions ϕ and ψ. The *Walsh functions* are wavelet packets based on the Haar wavelet.

Allowing infinitely many coefficients results in wavelets supported on the entire real line. The *Meyer wavelet* is band-limited, possesses infinitely many derivatives, and has accuracy $p = \infty$. The *Battle–Lemarié* wavelets are piecewise splines and have exponential decay.

M-band wavelets replace the ubiquitous dilation factor 2 by another integer M. *Multiwavelets* replace the coefficients h_k by $r \times r$ matrices and the scaling function and wavelet by vector-valued functions $(\varphi_1, \ldots, \varphi_r)$ and $(\psi_1, \ldots, \psi_r)$, resulting in an orthonormal basis for $L^2(\mathbb{R})$ generated by the several wavelets $\psi_1, \ldots, \psi_r$.

For higher dimensions, a *separable* wavelet basis is constructed via a tensor product, with scaling function $\varphi(x)\varphi(y)$ and three wavelets $\varphi(x)\psi(y)$, $\psi(x)\varphi(y)$, $\psi(x)\psi(y)$. *Non-separable wavelets* replace the dilation factor 2 by a *dilation matrix*.

Biorthogonal wavelets allow greater flexibility of design by relaxing the requirement that the wavelet system $\{\psi_{j,k}\}$ form an orthonormal basis to requiring only that it form a Riesz basis. The *Cohen–Daubechies–Feauveau wavelets* are symmetric and their coefficients h_k are dyadic rationals. The *Chui–Wang–Aldroubi–Unser semiorthogonal wavelets* are splines with explicit analytic formulae.

The basis condition may be further relaxed by allowing $\{\psi_{j,k}\}$ to be a *frame*, an over-complete system with basis-like properties. Introducing further redundancy, the *continuous wavelet transform* uses all possible dilates and translates $a^{1/2}\psi(ax + b)$.

7.15.5.3 Wavelet coefficients and figures

Coefficients for Daubechies scaling functions[1]

D_2 :	h_0:	0.707106781187	D_6 :	h_0:	0.332670552950
	h_1:	0.707106781187		h_1:	0.806891509311
				h_2:	0.459877502118
D_4 :	h_0:	0.482962913145		h_3:	−0.135011020010
	h_1:	0.836516303738		h_4:	−0.085441273882
	h_2:	0.224143868042		h_5:	0.035226291886
	h_3:	−0.129409522551			

---

[1]The tables and figures shown below are reprinted with permission from Daubechies, I., *Ten Lectures on Wavelets.* Copyright ©1992 by the Society for Industrial and Applied Mathematics, Philadelphia, PA. All rights reserved.

$D_8:$

$h_0:$	0.230377813309
$h_1:$	0.714846570553
$h_2:$	0.630880767930
$h_3:$	-0.027983769417
$h_4:$	-0.187034811719
$h_5:$	0.030841381836
$h_6:$	0.032883011667
$h_7:$	-0.010597401785

$D_{10}:$

$h_0:$	0.160102397974
$h_1:$	0.603829269797
$h_2:$	0.724308528438
$h_3:$	0.138428145901
$h_4:$	-0.242294887066
$h_5:$	-0.032244869585
$h_6:$	0.077571493840
$h_7:$	-0.006241490213
$h_8:$	-0.012580751999
$h_9:$	0.003335725285

$D_{12}:$

$h_0:$	0.111540743350
$h_1:$	0.494623890398
$h_2:$	0.751133908021
$h_3:$	0.315250351709
$h_4:$	-0.226264693965
$h_5:$	-0.129766867567
$h_6:$	0.097501605587
$h_7:$	0.027522865530
$h_8:$	-0.031582039317
$h_9:$	0.000553842201
$h_{10}:$	0.004777257511
$h_{11}:$	-0.001077301085

D_4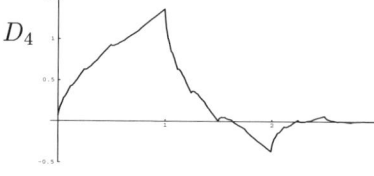

W_4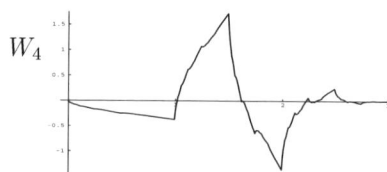

D_6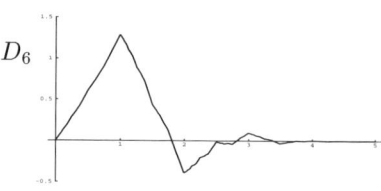

W_6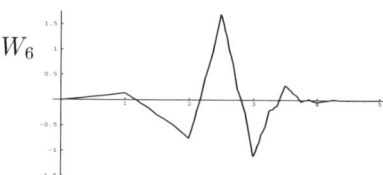

D_8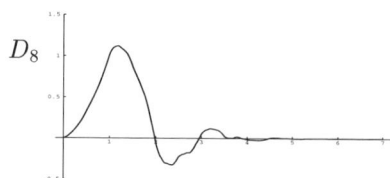

W_8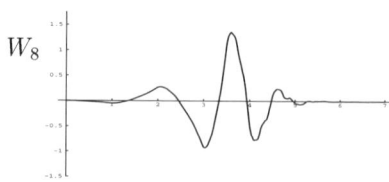

D_{10}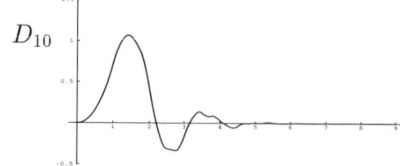

W_{10}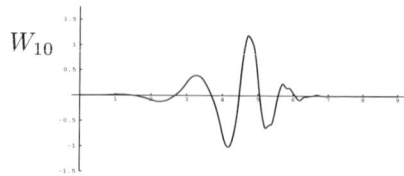

Chapter 8

Scientific Computing

8.1   BASIC NUMERICAL ANALYSIS *728*

   8.1.1   Approximations and errors   728

   8.1.2   Solution to algebraic equations   729

   8.1.3   Interpolation   733

   8.1.4   Fitting equations to data   737

8.2   NUMERICAL LINEAR ALGEBRA *740*

   8.2.1   Solving linear systems 740

   8.2.2   Gaussian elimination   740

   8.2.3   Gaussian elimination algorithm 741

   8.2.4   Pivoting   741

   8.2.5   Eigenvalue computation 742

   8.2.6   Householder's method   744

   8.2.7   QR algorithm 745

   8.2.8   Non-linear systems and numerical optimization   748

8.3   NUMERICAL INTEGRATION AND DIFFERENTIATION *750*

   8.3.1   Numerical integration 750

   8.3.2   Numerical differentiation   764

   8.3.3   Numerical summation 776

8.4   PROGRAMMING TECHNIQUES *777*

The text *Numerical Analysis*, Seventh Edition, Brooks/Cole, Pacific Grove, CA, 2001, by R. L. Burden and J. D. Faires, was the primary reference for most of the information presented in this chapter.

1-58488-291-3/02/$0.00+$1.50

© 2003 CRC Press, Inc.

8.1   BASIC NUMERICAL ANALYSIS

8.1.1   APPROXIMATIONS AND ERRORS

Numerical methods involve finding approximate solutions to mathematical problems. Errors of approximation can result from two sources: error inherent in the method or formula used and round-off error. *Round-off error* results when a calculator or computer is used to perform real-number calculations with a finite number of significant digits. All but the first specified number of digits are either *chopped* or *rounded* to that number of digits.

If p^* is an approximation to p, the *absolute error* is defined to be $|p - p^*|$ and the *relative error* is $|p - p^*|/|p|$, provided that $p \neq 0$.

Iterative techniques often generate sequences that (ideally) converge to an exact solution. It is sometimes desirable to describe the *rate of convergence*.

DEFINITION 8.1.1

Suppose $\lim \beta_n = 0$ *and* $\lim \alpha_n = \alpha$. *If a positive constant* K *exists with* $|\alpha_n - \alpha| < K|\beta_n|$ *for large* n, *then* $\{\alpha_n\}$ *is said to converge to* α *with a rate of convergence* $O(\beta_n)$. *This is read "big oh of* β_n*" and written* $\alpha_n = \alpha + O(\beta_n)$.

DEFINITION 8.1.2

Suppose $\{p_n\}$ *is a sequence that converges to* p, *with* $p_n \neq p$, *for all* n. *If positive constants* λ *and* α *exist with* $\lim_{n \to \infty} \dfrac{|p_{n+1} - p|}{|p_n - p|^{\alpha}} = \lambda$, *then* $\{p_n\}$ *converges to* p *of order* α, *with asymptotic error constant* λ.

In general, a higher order of convergence yields a more rapid rate of convergence. A sequence has *linear convergence* if $\alpha = 1$ and *quadratic convergence* if $\alpha = 2$.

8.1.1.1   Aitken's Δ^2 method

DEFINITION 8.1.3

Given $\{p_n\}_{n=0}^{\infty}$, *the forward difference* Δp_n *is defined by* $\Delta p_n = p_{n+1} - p_n$, *for* $n \geq 0$. *Higher powers* $\Delta^k p_n$ *are defined recursively by* $\Delta^k p_n = \Delta(\Delta^{k-1} p_n)$, *for* $k \geq 2$. *In particular,* $\Delta^2 p_n = \Delta(p_{n+1} - p_n) = p_{n+2} - 2p_{n+1} + p_n$.

If a sequence $\{p_n\}$ converges linearly to p and $(p_n - p)(p_{n-1} - p) > 0$, for sufficiently large n, then the new sequence $\{\widehat{p}_n\}$ generated by called *Aitken's* Δ^2

method,

$$\widehat{p}_n = p_n - \frac{(\triangle p_n)^2}{\triangle^2 p_n} \qquad (8.1.1)$$

for all $n \geq 0$, satisfies $\lim_{n \to \infty} \dfrac{\widehat{p}_n - p}{p_n - p} = 0$.

8.1.1.2 Richardson's extrapolation

Improved accuracy can be achieved by combining *extrapolation* with a low-order formula. Suppose the unknown value M is approximated by a formula $N(h)$ for which

$$M = N(h) + K_1 h + K_2 h^2 + K_3 h^3 + \dots \qquad (8.1.2)$$

for some unspecified constants $K_1, K_2, K_3, \dots$. To apply extrapolation, set $N_1(h) = N(h)$, and generate new approximations $N_j(h)$ by

$$N_j(h) = N_{j-1}\left(\frac{h}{2}\right) + \frac{N_{j-1}\left(\frac{h}{2}\right) - N_{j-1}(h)}{2^{j-1} - 1}. \qquad (8.1.3)$$

Then $M = N_j(h) + O(h^j)$. A table of the following form is generated, one row at a time:

$$
\begin{array}{llll}
N_1(h) & & & \\
N_1\left(h/2\right) & N_2(h) & & \\
N_1\left(h/4\right) & N_2\left(h/2\right) & N_3(h) & \\
N_1\left(h/8\right) & N_2\left(h/4\right) & N_3\left(h/2\right) & N_4(h).
\end{array}
$$

Extrapolation can be applied whenever the truncation error for a formula has the form $\sum_{j=1}^{m-1} K_j h^{\alpha_j} + O(h^{\alpha_m})$ for constants K_j and $\alpha_1 < \alpha_2 < \dots < \alpha_m$. In particular, if $\alpha_j = 2j$, the following computation can be used:

$$N_j(h) = N_{j-1}\left(\frac{h}{2}\right) + \frac{N_{j-1}\left(\frac{h}{2}\right) - N_{j-1}(h)}{4^{j-1} - 1}, \qquad (8.1.4)$$

and the entries in the jth column of the table have order $O(h^{2j})$.

8.1.2 SOLUTION TO ALGEBRAIC EQUATIONS

Iterative methods generate sequences $\{p_n\}$ that converge to a solution p of an equation.

DEFINITION 8.1.4

A solution p of $f(x) = 0$ is a zero of multiplicity m if $f(x)$ can be written as $f(x) = (x - p)^m q(x)$, for $x \neq p$, where $\lim_{x \to p} q(x) \neq 0$. A zero is called simple if $m = 1$.

8.1.2.1   Fixed point iteration

A *fixed point* p for a function g satisfies $g(p) = p$. Given p_0, generate $\{p_n\}$ by

$$p_{n+1} = g(p_n) \quad \text{for } n \geq 0. \tag{8.1.5}$$

If $\{p_n\}$ converges, then it will converge to a fixed point of g and the value p_n can be used as an approximation for p. The following theorem gives conditions that guarantee convergence.

THEOREM 8.1.1   *(Fixed point theorem)*

Let $g \in C[a, b]$ and suppose that $g(x) \in [a, b]$ for all x in $[a, b]$. Suppose also that g' exists on (a, b) with $|g'(x)| \leq k < 1$, for all $x \in (a, b)$. If p_0 is any number in $[a, b]$, then the sequence defined by Equation (8.1.5) converges to the (unique) fixed point p in $[a, b]$. Both of the error estimates $|p_n - p| \leq \frac{k^n}{1-k} |p_0 - p_1|$ and $|p_n - p| \leq k^n max\{p_0 - a, b - p_0\}$ hold, for all $n \geq 1$.

The iteration sometimes converges even if the conditions are not all satisfied.

THEOREM 8.1.2

Suppose g is a function that satisfies the conditions of Theorem 8.1.1 and g' is also continuous on (a, b). If $g'(p) \neq 0$, then for any number p_0 in $[a, b]$, the sequence generated by Equation (8.1.5) converges only linearly to the unique fixed point p in $[a, b]$.

THEOREM 8.1.3

Let p be a solution of the equation $x = g(x)$. Suppose that $g'(p) = 0$ and g'' are continuous and bounded by a constant on an open interval I containing p. Then there exists a $\delta > 0$ such that, for $p_0 \in [p - \delta, p + \delta]$, the sequence defined by Equation (8.1.5) converges at least quadratically to p.

8.1.2.2   Steffensen's method

For a linearly convergent fixed-point iteration, convergence can be accelerated by applying Aitken's $\triangle^2$ method. This is called *Steffensen's method*. Define $p_0^{(0)} = p_0$, compute $p_1^{(0)} = g(p_0^{(0)})$ and $p_2^{(0)} = g(p_1^{(0)})$. Set $p_0^{(1)} = \widehat{p}_0$ which is computed using Equation (8.1.1) applied to $p_0^{(0)}, p_1^{(0)}$, and $p_2^{(0)}$. Use fixed-point iteration to compute $p_1^{(1)}$ and $p_2^{(1)}$ and then Equation (8.1.1) to find $p_0^{(2)}$. Continuing, generate $\{p_0^{(n)}\}$.

THEOREM 8.1.4

Suppose that $x = g(x)$ has the solution p with $g'(p) \neq 1$. If there exists a $\delta > 0$ such that $g \in C^3[p - \delta, p + \delta]$, then Steffensen's method gives quadratic convergence for the sequence $\{p_0^{(n)}\}$ for any $p_0 \in [p - \delta, p + \delta]$.

FIGURE 8.1
Illustration of Newton's method.[1]

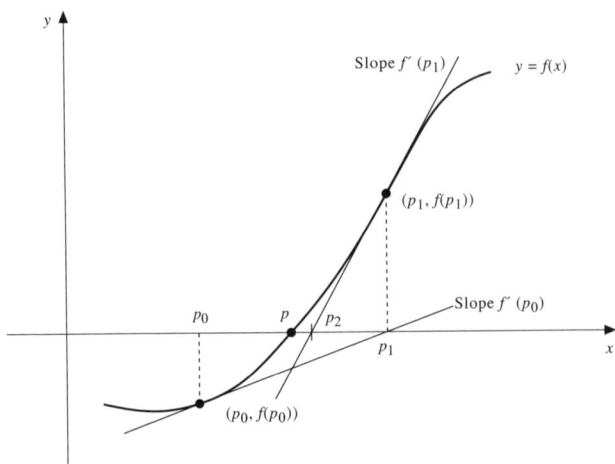

8.1.2.3 Newton–Raphson method (Newton's method)

To solve $f(x) = 0$, given an initial approximation p_0, generate $\{p_n\}$ using

$$p_{n+1} = p_n - \frac{f(p_n)}{f'(p_n)}, \quad \text{for } n \geq 0. \tag{8.1.6}$$

Figure 8.1 describes the method geometrically. Each value p_{n+1} represents the x-intercept of the tangent line to the graph of $f(x)$ at the point $[p_n, f(p_n)]$.

THEOREM 8.1.5

Let $f \in C^2[a, b]$. If $p \in [a, b]$ is such that $f(p) = 0$ and $f'(p) \neq 0$, then there exists a $\delta > 0$ such that Newton's method generates a sequence $\{p_n\}$ converging to p for any initial approximation $p_0 \in [p - \delta, p + \delta]$.

Note:

1. Generally the conditions of the theorem cannot be checked. Therefore one usually generates the sequence $\{p_n\}$ and observes whether or not it converges.
2. An obvious limitation is that the iteration terminates if $f'(p_n) = 0$.
3. For *simple* zeros of f, Theorem 8.1.5 implies that Newton's method converges quadratically. Otherwise, the convergence is much slower.

---

[1]From R.L. Burden and J.D. Faires, *Numerical Analysis*, 7th ed., Brooks/Cole, Pacific Grove, CA, 2001. With permission.

8.1.2.4 Modified Newton's method

Newton's method converges only linearly if p has multiplicity larger than one. However, the function $u(x) = \frac{f(x)}{f'(x)}$ has a simple zero at p. Hence, the Newton iteration formula applied to $u(x)$ yields quadratic convergence to a root of $f(x) = 0$. The iteration simplifies to

$$p_{n+1} = p_n - \frac{f(p_n)f'(p_n)}{[f'(p_n)]^2 - f(p_n)f''(p_n)}, \qquad \text{for } n \geq 0. \qquad (8.1.7)$$

8.1.2.5 Secant method

To solve $f(x) = 0$, the *secant method* uses the x-intercept of the secant line passing through $(p_n, f(p_n))$ and $(p_{n-1}, f(p_{n-1}))$. The derivative of f is not needed. Given p_0 and p_1, generate the sequence with

$$p_{n+1} = p_n - f(p_n)\frac{(p_n - p_{n-1})}{f(p_n) - f(p_{n-1})}, \qquad \text{for } n \geq 1. \qquad (8.1.8)$$

8.1.2.6 Root-bracketing methods

Suppose $f(x)$ is continuous on $[a, b]$ and $f(a)f(b) < 0$. The Intermediate Value Theorem guarantees a number $p \in (a, b)$ exists with $f(p) = 0$. A *root-bracketing method* constructs a sequence of nested intervals $[a_n, b_n]$, each containing a solution of $f(x) = 0$. At each step, compute $p_n \in [a_n, b_n]$ and proceed as follows:

If $f(p_n) = 0$, stop the iteration and $p = p_n$.

Else,

if $f(a_n)f(p_n) < 0$, then set $a_{n+1} = a_n$, $b_{n+1} = p_n$.
Else, set $a_{n+1} = p_n$, $b_{n+1} = b_n$.

8.1.2.7 Bisection method

This is a special case of the root-bracketing method. The values p_n are computed by

$$p_n = a_n + \frac{b_n - a_n}{2} = \frac{a_n + b_n}{2}, \qquad \text{for } n \geq 1. \qquad (8.1.9)$$

Clearly, $|p_n - p| \leq (b - a)/2^n$ for $n \geq 1$. The rate of convergence is $O(2^{-n})$. Although convergence is slow, the exact number of iterations for a specified accuracy ϵ can be determined. To guarantee that $|p_N - p| < \epsilon$, use

$$N > \log_2\left(\frac{b - a}{\epsilon}\right) = \frac{\ln(b - a) - \ln \epsilon}{\ln 2}. \qquad (8.1.10)$$

8.1.2.8 False position (*regula falsi*)

$$p_n = b_n - f(b_n)\frac{b_n - a_n}{f(b_n) - f(a_n)}, \qquad \text{for } n \geq 1. \qquad (8.1.11)$$

This root-bracketing method also converges if the initial criteria are satisfied.

8.1.2.9 Horner's method with deflation

If Newton's method is used to solve for roots of the polynomial $P(x) = 0$, then the polynomials P and P' are repeatedly evaluated. Horner's method efficiently evaluates a polynomial of degree n using only n multiplications and n additions.

8.1.2.10 Horner's algorithm

To evaluate $P(x) = a_n x^n + a_{n-1} x^{n-1} + \ldots + a_0$ and its derivative at x_0:

INPUT: degree n, coefficients $\{a_0, a_1, \ldots, a_n\}$; x_0,

OUTPUT: $y = P(x_0)$; $z = P'(x_0)$.

Algorithm:

1. Set $y = a_n$; $z = a_n$.
2. For $j = n - 1, n - 2, \ldots, 1$,
   set $y = x_0 y + a_j$; $z = x_0 z + y$.
3. Set $y = x_0 y + a_0$.
4. OUTPUT (y, z). STOP.

When satisfied with the approximation $\widehat{x}_1$ for a root x_1 of P, use synthetic division to compute $Q_1(x)$ so that $P(x) \approx (x - \widehat{x}_1) Q_1(x)$. Estimate a root of $Q_1(x)$ and write $P(x) \approx (x - \widehat{x}_1)(x - \widehat{x}_2) Q_2(x)$, and so on. Eventually, $Q_{n-2}(x)$ will be a quadratic, and the quadratic formula can be applied. This procedure, finding one root at a time, is called *deflation*.

Note: Care must be taken since $\widehat{x}_1$ is an *approximation* for x_1. Some inaccuracy occurs when computing the coefficients of $Q_1(x)$, etc. Although the estimate $\widehat{x}_2$ of a root of $Q_1(x)$ can be very accurate, it may not be as accurate when estimating a root of $P(x)$.

8.1.3 INTERPOLATION

Interpolation involves fitting a function to a set of data points (x_0, y_0), (x_1, y_1), $\ldots$, (x_n, y_n). The x_i are unique and the y_i may be regarded as the values of some function $f(x)$, that is, $y_i = f(x_i)$ for $i = 0, 1, \ldots, n$. The following are polynomial interpolation methods.

8.1.3.1 Lagrange interpolation

The *Lagrange interpolating polynomial*, denoted $P_n(x)$, is the unique polynomial of degree at most n for which $P_n(x_k) = f(x_k)$ for $k = 0, 1, \ldots, n$. It is given by

$$P(x) = \sum_{k=0}^{n} f(x_k) L_{n,k}(x) \tag{8.1.12}$$

where $\{x_0, \ldots, x_n\}$ are called *node points*, and

$$L_{n,k}(x) = \frac{(x - x_0)(x - x_1) \cdots (x - x_{k-1})(x - x_{k+1}) \cdots (x - x_n)}{(x_k - x_0)(x_k - x_1) \cdots (x_k - x_{k-1})(x_k - x_{k+1}) \cdots (x_k - x_n)}$$

$$= \prod_{i=0, i \neq k}^{n} \frac{(x - x_i)}{(x_k - x_i)}, \qquad \text{for } k = 0, 1, \ldots, n. \tag{8.1.13}$$

THEOREM 8.1.6   *(Error formula)*

If $x_0, x_1, \ldots, x_n$ are distinct numbers in $[a, b]$ and $f \in C^{n+1}[a, b]$, then, for each x in $[a, b]$, a number $\xi(x)$ in (a, b) exists with

$$f(x) = P(x) + \frac{f^{(n+1)}(\xi(x))}{(n + 1)!}(x - x_0)(x - x_1) \cdots (x - x_n), \tag{8.1.14}$$

where P is the interpolating polynomial given in Equation (8.1.12).

Although the Lagrange polynomial is unique, it can be expressed and evaluated in several ways. Equation (8.1.12) is tedious to evaluate, and including more nodes affects the *entire* expression. *Neville's method* evaluates the Lagrange polynomial at a single point *without* explicitly finding the polynomial and the method adapts easily when new nodes are included.

8.1.3.2   Neville's method

Let $P_{m_1, m_2, \ldots, m_k}$ denote the Lagrange polynomial using distinct nodes $\{x_{m_1}, x_{m_2}, \ldots, x_{m_k}\}$. If $P(x)$ denotes the Lagrange polynomial using nodes $\{x_0, x_1, \ldots, x_k\}$ and x_i and x_j are two distinct numbers in this set, then

$$P(x) = \frac{(x - x_j)P_{0,1,\ldots,j-1,j+1,\ldots,k}(x) - (x - x_i)P_{0,1,\ldots,i-1,i+1,\ldots,k}(x)}{(x_i - x_j)}.$$

$$\tag{8.1.15}$$

8.1.3.3   Neville's algorithm

Generate a table of entries $Q_{i,j}$ for $j \geq 0$ and $0 \leq i \leq j$ where the terms are $Q_{i,j} = P_{i-j,i-j+1,\ldots,i-1,i}$. Calculations use Equation (8.1.15) for a specific value of x as shown:

$$
\begin{array}{lllll}
x_0 & Q_{0,0} = P_0 & & & \\
x_1 & Q_{1,0} = P_1 & Q_{1,1} = P_{0,1} & & \\
x_2 & Q_{2,0} = P_2 & Q_{2,1} = P_{1,2} & Q_{2,2} = P_{0,1,2} & \\
x_3 & Q_{3,0} = P_3 & Q_{3,1} = P_{2,3} & Q_{3,2} = P_{1,2,3} & Q_{3,3} = P_{0,1,2,3}
\end{array}
$$

Note that $P_k = P_k(x) = f(x_k)$ and $\{Q_{i,i}\}$ represents successive estimates of $f(x)$ using Lagrange polynomials. Nodes may be added until $|Q_{i,i} - Q_{i-1,i-1}| < \epsilon$ as desired.

8.1.3.4 Divided differences

Some interpolation formulae involve *divided differences*. Given an ordered sequence of values, $\{x_i\}$, and the corresponding function values $f(x_i)$, the *zeroth divided difference* is $f[x_i] = f(x_i)$. The *first divided difference* is defined by

$$f[x_i, x_{i+1}] = \frac{f(x_{i+1}) - f(x_i)}{x_{i+1} - x_i} = \frac{f[x_{i+1}] - f[x_i]}{x_{i+1} - x_i}. \tag{8.1.16}$$

The kth *divided difference* is defined by

$$f[x_i, x_{i+1}, \dots, x_{i+k-1}, x_{i+k}]$$
$$= \frac{f[x_{i+1}, x_{i+2}, \dots, x_{i+k}] - f[x_i, x_{i+1}, \dots, x_{i+k-1}]}{x_{i+k} - x_i}. \tag{8.1.17}$$

Divided differences are usually computed by forming a triangular table.

x	$f(x)$	First divided differences	Second divided differences	Third divided differences
x_0	$f[x_0]$			
		$f[x_0, x_1]$		
x_1	$f[x_1]$		$f[x_0, x_1, x_2]$	
		$f[x_1, x_2]$		$f[x_0, x_1, x_2, x_3]$
x_2	$f[x_2]$		$f[x_1, x_2, x_3]$	
		$f[x_2, x_3]$		
x_3	$f[x_3]$			

8.1.3.5 Newton's interpolatory divided-difference formula

(also known as the Newton polynomial)

$$P_n(x) = f[x_0] + \sum_{k=1}^{n} f[x_0, x_1, \dots, x_k](x - x_0) \cdots (x - x_{k-1}). \tag{8.1.18}$$

Labeling the nodes as $\{x_n, x_{n-1}, \dots, x_0\}$, a formula similar to Equation (8.1.18) results in *Newton's backward divided-difference formula*,

$$P_n(x) = f[x_n] + f[x_n, x_{n-1}](x - x_n)$$
$$+ f[x_n, x_{n-1}, x_{n-2}](x - x_n)(x - x_{n-1}) \tag{8.1.19}$$
$$+ \dots + f[x_n, x_{n-1}, \dots, x_0](x - x_n) \cdots (x - x_1).$$

If the nodes are equally spaced (that is, $x_i - x_{i-1} = h$), define the parameter s by the equation $x = x_0 + sh$. The following formulae evaluate $P_n(x)$ at a single point:

1. Newton's interpolatory divided-difference formula,

$$P_n(x) = P_n(x_0 + sh) = \sum_{k=0}^{n} \binom{s}{k} k! h^k f[x_0, x_1, \dots, x_k]. \tag{8.1.20}$$

2. Newton's forward-difference formula (Newton–Gregory),

$$P_n(x) = P_n(x_0 + sh) = \sum_{k=0}^{n} \binom{s}{k} \Delta^k f(x_0). \qquad (8.1.21)$$

3. Newton–Gregory backward formula (fits nodes x_{-n} to x_0),

$$P_n(x) = f(x_0) + \binom{s}{1} \Delta f(x_{-1}) + \binom{s+1}{2} \Delta^2 f(x_{-2}) +$$

$$\ldots + \binom{s+n-1}{n} \Delta^n f(x_{-n}). \quad (8.1.22)$$

4. Newton's backward-difference formula,

$$P_n(x) = \sum_{k=0}^{n}(-1)^k \binom{-s}{k} \nabla^k f(x_n), \qquad (8.1.23)$$

where $\nabla^k f(x_n)$ is the kth *backward difference*, defined for a sequence $\{p_n\}$, by $\nabla p_n = p_n - p_{n-1}$ for $n \geq 1$. Higher powers are defined recursively by $\nabla^k p_n = \nabla(\nabla^{k-1} p_n)$ for $k \geq 2$. For notation, set $\nabla^0 p_n = p_n$.

5. Stirling's formula (for equally spaced nodes $x_{-m}, \ldots, x_{-1}, x_0, x_1, \ldots, x_m$),

$$P_n(x) = P_{2m+1}(x) = f[x_0] + \frac{sh}{2}(f[x_{-1}, x_0] + f[x_0, x_1]) +$$

$$s^2 h^2 f[x_{-1}, x_0, x_1] + \frac{s(s^2 - 1)h^3}{2}(f[x_{-2}, x_{-1}, x_0, x_1] + f[x_{-1}, x_0, x_1, x_2])$$

$$+ \ldots + s^2(s^2 - 1)(s^2 - 4) \cdots (s^2 - (m-1)^2)h^{2m} f[x_{-m}, \ldots, x_m]$$

$$+ \frac{s(s^2 - 1) \cdots (s^2 - m^2)h^{2m+1}}{2}(f[x_{-m-1}, \ldots, x_m] + f[x_{-m}, \ldots, x_{m+1}]).$$

Use the entire formula if $n = 2m + 1$ is odd, and omit the last term if $n = 2m$ is even. The following table identifies the desired divided differences used in Stirling's formula:

x	$f(x)$	First divided differences	Second divided differences	Third divided differences
x_{-2}	$f[x_{-2}]$			
		$f[x_{-2}, x_{-1}]$		
x_{-1}	$f[x_{-1}]$		$f[x_{-2}, x_{-1}, x_0]$	
		$f[x_{-1}, x_0]$		$f[x_{-2}, x_{-1}, x_0, x_1]$
x_0	$f[x_0]$		$f[x_{-1}, x_0, x_1]$	
		$f[x_0, x_1]$		$f[x_{-1}, x_0, x_1, x_2]$
x_1	$f[x_1]$		$f[x_0, x_1, x_2]$	
		$f[x_1, x_2]$		
x_2	$f[x_2]$			

8.1.3.6 Inverse interpolation

Any method of interpolation which does not require the nodes to be equally spaced may be applied by interchanging the nodes (x values) and the function values (y values).

8.1.3.7 Hermite interpolation

Given distinct numbers $\{x_0, x_1, \ldots, x_n\}$, the *Hermite interpolating polynomial* for a function f is the unique polynomial $H(x)$ of degree at most $2n + 1$ that satisfies $H(x_i) = f(x_i)$ and $H'(x_i) = f'(x_i)$ for each $i = 0, 1, \ldots, n$.

A technique and formula similar to Equation (8.1.18) can be used. For distinct nodes $\{x_0, x_1, \ldots, x_n\}$, define $\{z_0, z_1, \ldots, z_{2n+1}\}$ by $z_{2i} = z_{2i+1} = x_i$ for $i = 0, 1, \ldots, n$. Construct a divided difference table for the ordered pairs $(z_i, f(z_i))$ using $f'(x_i)$ in place of $f[z_{2i}, z_{2i+1}]$, which would be undefined. Denote the Hermite polynomial by $H_{2n+1}(x)$.

8.1.3.8 Hermite interpolating polynomial

$$
\begin{aligned}
H_{2n+1}(x) &= f[z_0] + \sum_{k=1}^{2n+1} f[z_0, z_1, \ldots, z_k](x - z_0) \cdots (x - z_{k-1}) \\
&= f[z_0] + f[z_0, z_1](x - x_0) + f[z_0, z_1, z_2](x - x_0)^2 \\
&\quad + f[z_0, z_1, z_2, z_3](x - x_0)^2(x - x_1) \\
&\quad + \ldots + f[z_0, \ldots, z_{2n+1}](x - x_0)^2 \cdots (x - x_{n-1})^2(x - x_n).
\end{aligned}
\tag{8.1.24}
$$

THEOREM 8.1.7 *(Error formula)*

If $f \in C^{2n+2}[a, b]$, *then*

$$
f(x) = H_{2n+1}(x) + \frac{f^{(2n+2)}(\xi(x))}{(2n+2)!}(x - x_0)^2 \cdots (x - x_n)^2
\tag{8.1.25}
$$

for some $\xi(x) \in (a, b)$ *and where* $x_i \in [a, b]$ *for each* $i = 0, 1, \ldots, n$.

8.1.4 FITTING EQUATIONS TO DATA

8.1.4.1 Piecewise polynomial approximation

An interpolating polynomial has large degree and tends to oscillate greatly for large data sets. *Piecewise polynomial approximation* divides the interval into a collection of subintervals and constructs an approximating polynomial on each subinterval. *Piecewise linear interpolation* consists of simply joining the data points with line segments. This collection is continuous but not differentiable at the node points. Hermite polynomials would require derivative values. *Cubic spline interpolation* is popular since no derivative information is needed.

DEFINITION 8.1.5

Given a function f defined on $[a, b]$ and a set of numbers $a = x_0 < x_1 < \ldots < x_n = b$, a cubic spline interpolant, S, for f is a function that satisfies

1. *S is a piecewise cubic polynomial, denoted S_j, on $[x_j, x_{j+1}]$ for each $j = 0, 1, \ldots, n - 1$.*

2. *$S(x_j) = f(x_j)$ for each $j = 0, 1, \ldots, n$.*

3. *$S_{j+1}(x_{j+1}) = S_j(x_{j+1})$ for each $j = 0, 1, \ldots, n - 2$.*

4. *$S'_{j+1}(x_{j+1}) = S'_j(x_{j+1})$ for each $j = 0, 1, \ldots, n - 2$.*

5. *$S''_{j+1}(x_{j+1}) = S''_j(x_{j+1})$ for each $j = 0, 1, \ldots, n - 2$.*

6. *One of the following sets of boundary conditions is satisfied:*

    (a) *$S''(x_0) = S''(x_n) = 0$ (free or natural boundary),*
    (b) *$S'(x_0) = f'(x_0)$ and $S'(x_n) = f'(x_n)$ (clamped boundary).*

If a function f is defined at all node points, then f has a unique natural spline interpolant. If, in addition, f is differentiable at a and b, then f has a unique clamped spline interpolant. To construct a cubic spline, set

$$S_j(x) = a_j + b_j(x - x_j) + c_j(x - x_j)^2 + d_j(x - x_j)^3$$

for each $j = 0, 1, \ldots, n - 1$. The constants $\{a_j, b_j, c_j, d_j\}$ are found by solving a tridiagonal system of linear equations, which is included in the following algorithms.

8.1.4.2   Algorithm for natural cubic splines

INPUT: n, $\{x_0, x_1, \ldots, x_n\}$,
   $a_0 = f(x_0), a_1 = f(x_1), \ldots, a_n = f(x_n)$.
OUTPUT: $\{a_j, b_j, c_j, d_j\}$ for $j = 0, 1, \ldots, n - 1$.
Algorithm:

1. For $i = 0, 1, \ldots, n - 1$, set $h_i = x_{i+1} - x_i$.

2. For $i = 1, 2, \ldots, n-1$, set $\alpha_i = \dfrac{3}{h_i}(a_{i+1} - a_i) - \dfrac{3}{h_{i-1}}(a_i - a_{i-1})$.

3. Set $\ell_0 = 1, \mu_0 = 0, z_0 = 0$.

4. For $i = 1, 2, \ldots, n - 1$,
      set $\ell_i = 2(x_{i+1} - x_{i-1}) - h_{i-1}\mu_{i-1}$;
      set $\mu_i = h_i/\ell_i$;
      set $z_i = (\alpha_i - h_{i-1}z_{i-1})/\ell_i$.

5. Set $\ell_n = 1, z_n = 0, c_n = 0$.

6. For $j = n - 1, n - 2, \ldots, 0$,
      set $c_j = z_j - \mu_j c_{j+1}$;
      set $b_j = (a_{j+1} - a_j)/h_j - h_j(c_{j+1} + 2c_j)/3$;
      set $d_j = (c_{j+1} - c_j)/(3h_j)$.

7. OUTPUT $(a_j, b_j, c_j, d_j$ for $j = 0, 1, \ldots, n - 1)$.   STOP.

8.1.4.3 Algorithm for clamped cubic splines

INPUT: n, $\{x_0, x_1, \ldots, x_n\}$,
  $a_0 = f(x_0)$, $a_1 = f(x_1)$, $\ldots$, $a_n = f(x_n)$,
  $F_0 = f'(x_0)$, $F_n = f'(x_n)$.
OUTPUT: $\{a_j, b_j, c_j, d_j\}$ for $j = 0, 1, \ldots, n-1$.
Algorithm

1. For $i = 0, 1, \ldots, n-1$, set $h_i = x_{i+1} - x_i$.

2. Set $\alpha_0 = 3(a_1 - a_0)/h_0 - 3F_0$, $\alpha_n = 3F_n - 3(a_n - a_{n-1})/h_{n-1}$.

3. For $i = 1, 2, \ldots, n-1$, set $\alpha_i = \dfrac{3}{h_i}(a_{i+1} - a_i) - \dfrac{3}{h_{i-1}}(a_i - a_{i-1})$.

4. Set $\ell_0 = 2h_0$, $\mu_0 = 0.5$, $z_0 = \alpha_0/\ell_0$.

5. For $i = 1, 2, \ldots, n-1$,
     set $\ell_i = 2(x_{i+1} - x_{i-1}) - h_{i-1}\mu_{i-1}$;
     set $\mu_i = h_i/\ell_i$;
     set $z_i = (\alpha_i - h_{i-1}z_{i-1})/\ell_i$.

6. Set $\ell_n = h_{n-1}(2 - \mu_{n-1})$, $z_n = (\alpha_n - h_{n-1}z_{n-1})/\ell_n$, $c_n = z_n$.

7. For $j = n-1, n-2, \ldots, 0$,
     set $c_j = z_j - \mu_j c_{j+1}$;
     set $b_j = (a_{j+1} - a_j)/h_j - h_j(c_{j+1} + 2c_j)/3$;
     set $d_j = (c_{j+1} - c_j)/(3h_j)$.

8. OUTPUT $(a_j, b_j, c_j, d_j$ for $j = 0, 1, \ldots, n-1)$. STOP.

8.1.4.4 Discrete approximation

Another approach to fit a function to a set of data points $\{(x_i, y_i) \mid i = 1, 2, \ldots, m\}$ is *approximation*. If a polynomial of degree n is used, the polynomial $P_n(x) = \sum_{k=0}^{n} a_k x^k$ is found that minimizes the *least-squares error* $E = \sum_{i=1}^{m}[y_i - P_n(x_i)]^2$.

To find $\{a_0, a_1, \ldots, a_n\}$, solve the linear system, called the *normal equations*, created by setting partial derivatives of E taken with respect to each a_k equal to zero. The coefficient of a_0 in the first equation is actually the number of data points, m.

8.1.4.5 Normal equations

$$a_0 \sum_{i=1}^{m} x_i^0 + a_1 \sum_{i=1}^{m} x_i^1 + a_2 \sum_{i=1}^{m} x_i^2 + \ldots + a_n \sum_{i=1}^{m} x_i^n = \sum_{i=1}^{m} y_i x_i^0,$$

$$a_0 \sum_{i=1}^{m} x_i^1 + a_1 \sum_{i=1}^{m} x_i^2 + a_2 \sum_{i=1}^{m} x_i^3 + \ldots + a_n \sum_{i=1}^{m} x_i^{n+1} = \sum_{i=1}^{m} y_i x_i^1, \quad (8.1.26)$$

$$\vdots$$

$$a_0 \sum_{i=1}^{m} x_i^n + a_1 \sum_{i=1}^{m} x_i^{n+1} + a_2 \sum_{i=1}^{m} x_i^{n+2} + \ldots + a_n \sum_{i=1}^{m} x_i^{2n} = \sum_{i=1}^{m} y_i x_i^n.$$

Note: $P_n(x)$ can be replaced by a function f of specified form. Unfortunately, to minimize E, the resulting system is generally *not* linear. Although these systems can be solved, one technique is to "linearize" the data. For example, if $y = f(x) = be^{ax}$, then $\ln y = \ln b + ax$. The method applied to the data points $(x_i, \ln y_i)$ produces a linear system. Note that this technique does *not* find the approximation for the original problem but, instead, minimizes the least-squares for the "linearized" data.

8.1.4.6 Best-fit line

Given the points $P_1 = (x_1, y_1)$, $P_2 = (x_2, y_2)$, ..., $P_n = (x_n, y_n)$, the *line of best-fit* is given by $y - \overline{y} = m(x - \overline{x})$ where

$$\overline{x} = \frac{1}{n} \sum_{i=1}^{n} x_i = \frac{(x_1 + x_2 + \ldots + x_n)}{n},$$

$$\overline{y} = \frac{1}{n} \sum_{i=1}^{n} y_i = \frac{(y_1 + y_2 + \ldots + y_n)}{n}, \tag{8.1.27}$$

$$m = \frac{(x_1 y_1 + x_2 y_2 + \ldots + x_n y_n) - n\overline{x}\,\overline{y}}{(x_1^2 + x_2^2 + \ldots + x_n^2) - n\overline{x}^2} = \frac{\overline{xy} - \overline{x}\,\overline{y}}{\overline{x^2} - \overline{x}^2}.$$

8.2 NUMERICAL LINEAR ALGEBRA

8.2.1 SOLVING LINEAR SYSTEMS

The solution of systems of linear equations using *Gaussian elimination* with backward substitution is described in Section 8.2.2. The algorithm is highly sensitive to round-off error. *Pivoting strategies* can reduce round-off error when solving an $n \times n$ system. For a linear system $\mathbf{Ax} = \mathbf{b}$, assume that the equivalent matrix equation $\mathbf{A}^{(k)}\mathbf{x} = \mathbf{b}^{(k)}$ has been constructed. Call the entry, $a_{kk}^{(k)}$, the *pivot element*.

8.2.2 GAUSSIAN ELIMINATION

To solve the system $A\mathbf{x} = \mathbf{b}$, Gaussian elimination creates the *augmented matrix*

$$A' = [A \,\vdots\, \mathbf{b}] = \begin{bmatrix} a_{11} & \cdots & a_{1n} & b_1 \\ \vdots & & \vdots & \vdots \\ a_{n1} & \cdots & a_{nn} & b_n \end{bmatrix}. \tag{8.2.1}$$

This matrix is turned into an upper-triangular matrix by a sequence of (1) row permutations, and (2) subtracting a multiple of one row from another. The result is a matrix of the form (the primes denote that the quantities have been modified)

$$\begin{bmatrix} a'_{11} & a'_{12} & \cdots & a'_{1n} & b'_1 \\ 0 & a'_{22} & \cdots & a'_{2n} & b'_2 \\ \vdots & \ddots & \ddots & \vdots & \vdots \\ 0 & \cdots & 0 & a'_{nn} & b'_n \end{bmatrix}.$$ (8.2.2)

This matrix represents a linear system that is equivalent to the original system. If the solution exists and is unique, then back substitution can be used to successively determine $\{x_n, x_{n-1}, \dots\}$.

8.2.3   GAUSSIAN ELIMINATION ALGORITHM

INPUT: number of unknowns and equations n, matrix A, and vector **b**.
OUTPUT: solution $\mathbf{x} = (x_1, \dots, x_n)^{\mathrm{T}}$ to the linear system $A\mathbf{x} = \mathbf{b}$,
      or message that the system does not have a unique solution.
Algorithm:

1. Construct the augmented matrix $A' = [A \vdots \mathbf{b}] = (a'_{ij})$

2. For $i = 1, 2, \dots, n-1$ do (a)–(c): *(Elimination process)*

   (a) Let p be the least integer with $i \le p \le n$ and $a'_{pi} \ne 0$
       If no integer can be found, then
       OUTPUT("no unique solution exists"). STOP.

   (b) If $p \ne i$ interchange rows p and i in A'. Call the new matrix A'.

   (c) For $j = i + 1, \dots, n$ do i–ii:
       i. Set $m_{ij} = a'_{ji}/a'_{ii}$.
       ii. Subtract from row j the quantity $(m_{ij}$ times row $i)$.
          Replace row j with this result.

3. If $a'_{nn} = 0$ then OUTPUT ("no unique solution exists"). STOP.

4. Set $x_n = a'_{n,n+1}/a'_{nn}$. *(Start backward substitution)*.

5. For $i = n - 1, \dots, 2, 1$ set $x_i = \left[a'_{i,n+1} - \sum_{j=i+1}^{n} a'_{ij} x_j \right]/a'_{ii}$.

6. OUTPUT $(x_1, \dots, x_n)$, *(Procedure completed successfully)*. STOP.

8.2.4   PIVOTING

8.2.4.1   Maximal column pivoting

Maximal column pivoting (often called *partial pivoting*) finds, at each step, the element in the same column as the pivot element that lies on or below the main diagonal having the largest magnitude and moves it to the pivot position. Determine the least $p \ge k$ such that $\left| a_{pk}^{(k)} \right| = \max_{k \le i \le n} \left| a_{ik}^{(k)} \right|$ and interchange the kth equation with the pth equation before performing the elimination step.

8.2.4.2   Scaled-column pivoting

Scaled-column pivoting sometimes produces better results, especially when the elements of A differ greatly in magnitude. The desired pivot element is chosen to have the largest magnitude *relative* to the other values in its row. For each row define a *scale factor* s_i by $s_i = \max_{1 \le j \le n} |a_{ij}|$. The desired pivot element at the kth step is determined by choosing the smallest integer p with $\left| a_{pk}^{(k)} \right| / s_p = \max_{k \le j \le n} \left| a_{jk}^{(k)} \right| / s_j$.

8.2.4.3   Maximal (or complete) pivoting

The desired pivot element at the kth step is the entry of largest magnitude among $\{a_{ij}\}$ with $i = k, k+1, \ldots, n$ and $j = k, k+1, \ldots, n$. Both row and column interchanges are necessary and additional comparisons are required, resulting in additional execution time.

8.2.5   EIGENVALUE COMPUTATION

8.2.5.1   Power method

Assume that the $n \times n$ matrix A has n eigenvalues $\{\lambda_1, \lambda_2, \ldots, \lambda_n\}$ with linearly independent eigenvectors $\{\mathbf{v}^{(1)}, \mathbf{v}^{(2)}, \ldots, \mathbf{v}^{(n)}\}$. Assume further that A has a unique dominant eigenvalue λ_1, that is $|\lambda_1| > |\lambda_2| \ge |\lambda_3| \ge \cdots \ge |\lambda_n|$. Note that for any $\mathbf{x} \in \mathbb{R}^n$, $\mathbf{x} = \sum_{j=1}^{n} \alpha_j \mathbf{v}^{(j)}$.

The algorithm is called the *power method* because powers of the input matrix are taken: $\lim_{k \to \infty} A^k \mathbf{x} = \lim_{k \to \infty} \lambda_1^k \alpha_1 \mathbf{v}^{(1)}$. However, this sequence converges to zero if $|\lambda_1| < 1$ and diverges if $|\lambda_1| > 1$, provided $\alpha_1 \ne 0$. Appropriate scaling of $A^k \mathbf{x}$ is necessary to obtain a meaningful limit. Begin by choosing a unit vector $\mathbf{x}^{(0)}$ having a component $x_{p_0}^{(0)}$ such that $x_{p_0}^{(0)} = 1 = \left\| \mathbf{x}^{(0)} \right\|_\infty$.

The algorithm inductively constructs sequences of vectors $\{\mathbf{x}^{(m)}\}_{m=0}^{\infty}$ and $\{\mathbf{y}^{(m)}\}_{m=0}^{\infty}$ and a sequence of scalars $\{\mu^{(m)}\}_{m=1}^{\infty}$ by

$$\mathbf{y}^{(m)} = A\mathbf{x}^{(m-1)}, \qquad \mu^{(m)} = y_{p_{m-1}}^{(m)}, \qquad \mathbf{x}^{(m)} = \frac{\mathbf{y}^{(m)}}{y_{p_m}^{(m)}}, \qquad (8.2.3)$$

where, at each step, p_m represents the least integer for which $\left| y_{p_m}^{(m)} \right| = \left\| \mathbf{y}^{(m)} \right\|_\infty$.

The sequence of scalars satisfies $\lim_{m \to \infty} \mu^{(m)} = \lambda_1$, provided $\alpha_1 \ne 0$, and the sequence of vectors $\{\mathbf{x}^{(m)}\}_{m=0}^{\infty}$ converges to an eigenvector associated with λ_1 that has l_∞ norm one.

8.2.5.2   Power method algorithm

    INPUT: dimension n, matrix A, vector $\mathbf{x}$, tolerance TOL, and
        maximum number of iterations N.
    OUTPUT: approximate eigenvalue μ,
        approximate eigenvector $\mathbf{x}$ (with $\|\mathbf{x}\|_\infty = 1$),

or a message that the maximum number of
iterations was exceeded.

Algorithm:

1. Set $k=1$.

2. Find the smallest integer p with $1 \le p \le n$ and $|x_p| = \|\mathbf{x}\|_\infty$.

3. Set $\mathbf{x} = \mathbf{x}/x_p$.

4. While $(k \le N)$ do (a)–(g):

   (a) Set $\mathbf{y} = A\mathbf{x}$.

   (b) Set $\mu = y_p$.

   (c) Find the smallest integer p with $1 \le p \le n$ and $|y_p| = \|\mathbf{y}\|_\infty$.

   (d) If $y_p = 0$ then OUTPUT ("Eigenvector", $\mathbf{x}$, "corresponds to
       eigenvalue 0. Select a new vector $\mathbf{x}$ and restart."); STOP.

   (e) Set ERR $= \|\mathbf{x} - \mathbf{y}/y_p\|_\infty$; $\mathbf{x} = \mathbf{y}/y_p$.

   (f) If ERR < TOL then OUTPUT $(\mu, \mathbf{x})$
       (procedure successful) STOP.

   (g) Set $k = k + 1$.

5. OUTPUT ("Maximum number of iterations exceeded"). STOP.

Notes:

1. The method does not really require that λ_1 be unique. If the multiplicity is
   greater than one, the eigenvector obtained depends on the choice of $\mathbf{x}^{(0)}$.

2. The sequence constructed converges linearly, so that Aitken's Δ^2 method
   (Equation (8.1.1)) can be applied to accelerate convergence.

8.2.5.3 Inverse power method

The *inverse power method* modifies the power method to yield faster convergence
by finding the eigenvalue of A that is closest to a specified number q. Assume that
A satisfies the conditions as before. If $q \ne \lambda_i$, for $i = 1, 2, \ldots, n$, the eigenvalues
of $(A - qI)^{-1}$ are $\frac{1}{\lambda_1 - q}, \frac{1}{\lambda_2 - q}, \ldots, \frac{1}{\lambda_n - q}$, with the same eigenvectors $\mathbf{v}^{(1)}, \ldots, \mathbf{v}^{(n)}$.
Apply the power method to $(A - qI)^{-1}$. At each step, $\mathbf{y}^{(m)} = (A - qI)^{-1}\mathbf{x}^{(m-1)}$.
Generally, $\mathbf{y}^{(m)}$ is found by solving $(A - qI)\mathbf{y}^{(m)} = \mathbf{x}^{(m-1)}$ using Gaussian elimination with pivoting. Choose the value q from an initial approximation to the eigen-
vector $\mathbf{x}^{(0)}$ by $q = \mathbf{x}^{(0)\mathrm{T}} A\mathbf{x}^{(0)} / \left(\mathbf{x}^{(0)\mathrm{T}} \mathbf{x}^{(0)}\right)$.

The only changes in the algorithm for the power method (see page 742) are to
set an initial value q as described (do this prior to step 1), determine $\mathbf{y}$ in step (4a)
by solving the linear system $(A - qI)\mathbf{y} = \mathbf{x}$ (if the system does not have a unique
solution, output a message that q is an eigenvalue and stop), delete step (4d), and
replace step (4f) with

$$\text{if} \quad \text{ERR} < \text{TOL then set } \mu = \frac{1}{\mu} + q; \qquad \text{OUTPUT}(\mu, \mathbf{x}); \qquad \text{STOP.}$$

8.2.5.4   Wielandt deflation

Once the dominant eigenvalue has been found, remaining eigenvalues can be found by using *deflation techniques.* A new matrix B is formed having the same eigenvalues as A, except that the dominant eigenvalue of A is replaced by 0. One method is *Wielandt deflation* which defines $\mathbf{x} = \frac{1}{\lambda_1 v_i^{(1)}}[a_{i1}\ a_{i2}\ \dots\ a_{in}]^{\mathrm{T}}$, where $v_i^{(1)}$ is a coordinate of $\mathbf{v}^{(1)}$ that is non-zero, and the values $\{a_{i1}, a_{i2}, \dots, a_{in}\}$ are the entries in the ith row of A. Then the matrix $B = A - \lambda_1\mathbf{v}^{(1)}\mathbf{x}^{\mathrm{T}}$ has eigenvalues $0, \lambda_2, \lambda_3, \dots, \lambda_n$ with associated eigenvectors $\{\mathbf{v}^{(1)}, \mathbf{w}^{(2)}, \mathbf{w}^{(3)}, \dots, \mathbf{w}^{(n)}\}$, where

$$\mathbf{v}^{(i)} = (\lambda_i - \lambda_1)\mathbf{w}^{(i)} + \lambda_1(\mathbf{x}^{\mathrm{T}}\mathbf{w}^{(i)})\mathbf{v}^{(1)} \qquad (8.2.4)$$

for $i = 2, 3, \dots, n$. The ith row of B consists entirely of zero entries and B may be replaced with an $(n-1) \times (n-1)$ matrix B' obtained by deleting the ith row and i^{th} column of B. The power method can be applied to B' to find *its* dominant eigenvalue and so on.

8.2.6   HOUSEHOLDER'S METHOD

DEFINITION 8.2.1

Two $n \times n$ matrices A and B are said to be similar if a non-singular matrix S exists with $A = S^{-1}BS$. (Note that if A is similar to B, then they have the same set of eigenvalues.)

Householder's method constructs a symmetric tridiagonal matrix B that is similar to a given symmetric matrix A. After applying this method, the *QR algorithm* can be used efficiently to approximate the eigenvalues of the resulting symmetric tridiagonal matrix.

8.2.6.1   Algorithm for Householder's method

To construct a symmetric tridiagonal matrix $A^{(n-1)}$ similar to the symmetric matrix $A = A^{(1)}$, construct matrices $A^{(2)}, A^{(3)}, \dots, A^{(n-1)}$, where $A^{(k)} = (a_{ij}^{(k)})$ for $k = 1, 2, \dots, n-1$.

    INPUT: dimension n, matrix A.
    OUTPUT: $A^{(n-1)}$.   *(At each step, A can be overwritten.)*
    Algorithm:

      1. For $k = 1, 2, \dots, n-2$ do (a)–(k).

        (a) Set $q = \sum_{j=k+1}^{n}(a_{jk}^{(k)})^2$.

        (b) If $a_{k+1,k}^{(k)} = 0$ then set $\alpha = -q^{\frac{1}{2}}$

           else set $\alpha = -q^{\frac{1}{2}}a_{k+1,k}^{(k)}/\left|a_{k+1,k}^{(k)}\right|$.

        (c) Set RSQ $= \alpha^2 - \alpha a_{k+1,k}^{(k)}$.

(d) Set $v_k = 0$. (Note: $v_1 = \cdots = v_{k-1} = 0$, but are not needed.)

set $v_{k+1} = a_{k+1,k}^{(k)} - \alpha$;

for $j = k + 2, \ldots, n$ set $v_j = a_{jk}^{(k)}$.

(e) For $j = k, k + 1, \ldots, n$ set $u_j = \left(\sum_{i=k+1}^{n} a_{ji}^{(k)} v_i \right) / \text{RSQ}$.

(f) Set $\text{PROD} = \sum_{i=k+1}^{n} v_i u_i$.

(g) For $j = k, k + 1, \ldots, n$ set $z_j = u_j - (\text{PROD}/2\,\text{RSQ}) v_j$.

(h) For $\ell = k + 1, k + 2, \ldots, n - 1$ do i–ii.

  i. For $j = \ell + 1, \ldots, n$ set $a_{j\ell}^{(k+1)} = a_{j\ell}^{(k)} - v_\ell z_j - v_j z_\ell$;

  $a_{\ell j}^{(k+1)} = a_{j\ell}^{(k+1)}$.

  ii. Set $a_{\ell\ell}^{(k+1)} = a_{\ell\ell}^{(k)} - 2 v_\ell z_\ell$.

(i) Set $a_{nn}^{(k+1)} = a_{nn}^{(k)} - 2 v_n z_n$.

(j) For $j = k + 2, \ldots, n$ set $a_{kj}^{(k+1)} = a_{jk}^{(k+1)} = 0$.

(k) Set $a_{k+1,k}^{(k+1)} = a_{k+1,k}^{(k)} - v_{k+1} z_k$; $a_{k,k+1}^{(k+1)} = a_{k+1,k}^{(k+1)}$.

(Note: The other elements of $A^{(k+1)}$ are the same as $A^{(k)}$.)

2. OUTPUT $(A^{(n-1)})$. STOP.

($A^{(n-1)}$ is symmetric, tridiagonal, and similar to A.)

8.2.7 QR ALGORITHM

The *QR algorithm* is generally used (instead of deflation) to determine all of the eigenvalues of a symmetric matrix. The matrix must be symmetric and tridiagonal. If necessary, first apply Householder's method. Suppose the matrix A has the form

$$A = \begin{bmatrix} a_1 & b_2 & 0 & \cdots & 0 & 0 & 0 \\ b_2 & a_2 & b_3 & & 0 & 0 & 0 \\ 0 & b_3 & a_3 & & 0 & 0 & 0 \\ \vdots & & & \ddots & & & \vdots \\ 0 & 0 & 0 & & a_{n-2} & b_{n-1} & 0 \\ 0 & 0 & 0 & & b_{n-1} & a_{n-1} & b_n \\ 0 & 0 & 0 & \cdots & 0 & b_n & a_n \end{bmatrix}. \tag{8.2.5}$$

If $b_2 = 0$ or $b_n = 0$, then A has the eigenvalue a_1 or a_n, respectively. If $b_j = 0$ for some j, $2 < j < n$, the problem is reduced to considering the smaller matrices

$$\begin{bmatrix} a_1 & b_2 & 0 & \cdots & 0 \\ b_2 & a_2 & b_3 & & 0 \\ 0 & b_3 & a_3 & & 0 \\ \vdots & & & & \\ 0 & 0 & & a_{j-2} & b_{j-1} \\ 0 & 0 & & b_{j-1} & a_{j-1} \end{bmatrix} \quad \text{and} \quad \begin{bmatrix} a_j & b_{j+1} & 0 & \cdots & 0 \\ b_{j+1} & a_{j+1} & b_{j+2} & & 0 \\ 0 & b_{j+2} & a_{j+2} & & 0 \\ \vdots & & & & \\ 0 & 0 & & a_{n-1} & b_n \\ 0 & 0 & & b_n & a_n \end{bmatrix}. \tag{8.2.6}$$

If no b_j equals zero, the algorithm constructs $A^{(1)}$, $A^{(2)}$, $A^{(3)}$, ... as follows:

1. $A^{(1)} = A$ is factored as $A^{(1)} = Q^{(1)}R^{(1)}$, with $Q^{(1)}$ orthogonal and $R^{(1)}$ upper-triangular.

2. $A^{(2)}$ is defined as $A^{(2)} = R^{(1)}Q^{(1)}$.

In general, $A^{(i+1)} = R^{(i)}Q^{(i)} = (Q^{(i)^T}A^{(i)})Q^{(i)} = Q^{(i)^T}A^{(i)}Q^{(i)}$. Each $A^{(i+1)}$ is symmetric and tridiagonal with the same eigenvalues as $A^{(i)}$ and, hence, has the same eigenvalues as A.

8.2.7.1 Algorithm for QR

To obtain eigenvalues of the symmetric, tridiagonal $n \times n$ matrix

$$A \equiv A_1 = \begin{bmatrix} a_1^{(1)} & b_2^{(1)} & 0 & \cdots & 0 & 0 & 0 \\ b_2^{(1)} & a_2^{(1)} & b_3^{(1)} & & 0 & 0 & 0 \\ 0 & b_3^{(1)} & a_3^{(1)} & & 0 & 0 & 0 \\ \vdots & & & \ddots & & & \vdots \\ 0 & 0 & 0 & & a_{n-2}^{(1)} & b_{n-1}^{(1)} & 0 \\ 0 & 0 & 0 & & b_{n-1}^{(1)} & a_{n-1}^{(1)} & b_n^{(1)} \\ 0 & 0 & 0 & \cdots & 0 & b_n^{(1)} & a_n^{(1)} \end{bmatrix}. \tag{8.2.7}$$

INPUT: n; $\{a_1^{(1)}, \ldots, a_n^{(1)}, b_2^{(1)}, \ldots, b_n^{(1)}\}$, tolerance TOL, and maximum number of iterations M.

OUTPUT: eigenvalues of A, or recommended splitting of A, or a message that the maximum number of iterations was exceeded.

Algorithm:

1. Set $k = 1$; SHIFT $= 0$. (Accumulated shift)

2. While $k \le M$, do steps 3–12.

3. Test for success:

   (a) If $\left| b_n^{(k)} \right| \le$ TOL, then set $\lambda = a_n^{(k)} +$ SHIFT;
   OUTPUT (λ); set $n = n - 1$.

   (b) If $\left| b_2^{(k)} \right| \le$ TOL then set $\lambda = a_1^{(k)} +$ SHIFT;
   OUTPUT (λ);
   set $n = n - 1$; $a_1^{(k)} = a_2^{(k)}$;
   for $j = 2, \ldots, n$ set $a_j^{(k)} = a_{j+1}^{(k)}$; $b_j^{(k)} = b_{j+1}^{(k)}$.

   (c) If $n = 0$ then STOP.

   (d) If $n = 1$ then set $\lambda = a_1^{(k)} +$ SHIFT;
   OUTPUT(λ); STOP.

   (e) For $j = 3, \ldots, n - 1$
   if $\left| b_j^{(k)} \right| \le$ TOL then
   OUTPUT ("split into", $\{a_1^{(k)}, \ldots, a_{j-1}^{(k)}, b_2^{(k)}, \ldots, b_{j-1}^{(k)}\}$,
   "and" $\{a_j^{(k)}, \ldots, a_n^{(k)}, b_{j+1}^{(k)}, \ldots, b_n^{(k)}\}$, SHIFT); STOP.

4. Compute shift:

   Set $b = -(a_{n-1}^{(k)} + a_n^{(k)})$; $c = a_n^{(k)} a_{n-1}^{(k)} - [b_n^{(k)}]^2$; $d = (b^2 - 4c)^{\frac{1}{2}}$.

5. If $b > 0$, then set $\mu_1 = -2c/(b+d)$; $\mu_2 = -(b+d)/2$;

   else set $\mu_1 = (d-b)/2$; $\mu_2 = 2c/(d-b)$.

6. If $n = 2$, then set $\lambda_1 = \mu_1 + \text{SHIFT}$; $\lambda_2 = \mu_2 + \text{SHIFT}$;

   OUTPUT (λ_1, λ_2); STOP.

7. Choose s so that $\left| s - a_n^{(k)} \right| = \min\left(\left| \mu_1 - a_n^{(k)} \right|, \left| \mu_2 - a_n^{(k)} \right| \right)$.

8. Accumulate shift: Set $\text{SHIFT} = \text{SHIFT} + s$.

9. Perform shift: For $j = 1, \ldots, n$ set $d_j = a_j^{(k)} - s$.

10. Compute $R^{(k)}$:

    (a) Set $x_1 = d_1$; $y_1 = b_2$.

    (b) For $j = 2, \ldots, n$

    set $z_{j-1} = (x_{j-1}^2 + [b_j^{(k)}]^2)^{\frac{1}{2}}$; $c_j = x_{j-1}/z_{j-1}$;

    set $s_j = b_j^{(k)}/z_{j-1}$; $q_{j-1} = c_j y_{j-1} + s_j d_j$;

    set $x_j = -s_j y_{j-1} + c_j d_j$.

    If $j \neq n$ then set $r_{j-1} = s_j b_{j+1}^{(k)}$; $y_j = c_j b_{j+1}^{(k)}$.

    (At this point, $A_j^{(k)} = P_j A_{j-1}^{(k)}$ has been computed (P_j is a rotation matrix) and $R^{(k)} = A_n^{(k)}$.)

11. Compute $A^{(k+1)}$.

    (a) Set $z_n = x_n$; $a_1^{(k+1)} = s_2 q_1 + c_2 z_1$; $b_2^{(k+1)} = s_2 z_2$.

    (b) For $j = 2, 3 \ldots, n - 1$,

    set $a_j^{(k+1)} = s_{j+1} q_j + c_j c_{j+1} z_j$;

    set $b_{j+1}^{(k+1)} = s_{j+1} z_{j+1}$.

    (c) Set $a_n^{(k+1)} = c_n z_n$.

12. Set $k = k + 1$.

13. OUTPUT ("Maximum number of iterations exceeded");

    (Procedure unsuccessful.) STOP.

8.2.8 NON-LINEAR SYSTEMS AND NUMERICAL OPTIMIZATION

8.2.8.1 Newton's method

Many iterative methods exist for solving systems of non-linear equations. *Newton's method* is a natural extension from solving a single equation in one variable. Convergence is generally quadratic but usually requires an initial approximation that is near the true solution. Assume $\mathbf{F}(\mathbf{x}) = \mathbf{0}$ where $\mathbf{x}$ is an n-dimensional vector, $\mathbf{F} : \mathbb{R}^n \to \mathbb{R}^n$, and $\mathbf{0}$ is the zero vector. That is,

$$\mathbf{F}(\mathbf{x}) = \mathbf{F}(x_1, x_2, \ldots, x_n) = [f_1(x_1, x_2, \ldots, x_n), \ldots, f_n(x_1, x_2, \ldots, x_n)]^{\mathrm{T}}.$$
(8.2.8)

A fixed-point iteration is performed on $\mathbf{G}(\mathbf{x}) = \mathbf{x} - (J(\mathbf{x}))^{-1}\mathbf{F}(\mathbf{x})$ where $J(\mathbf{x})$ is the *Jacobian matrix*,

$$J(\mathbf{x}) = \begin{bmatrix} \frac{\partial f_1(\mathbf{x})}{\partial x_1} & \frac{\partial f_1(\mathbf{x})}{\partial x_2} & \cdots & \frac{\partial f_1(\mathbf{x})}{\partial x_n} \\ \frac{\partial f_2(\mathbf{x})}{\partial x_1} & \frac{\partial f_2(\mathbf{x})}{\partial x_2} & \cdots & \frac{\partial f_2(\mathbf{x})}{\partial x_n} \\ \vdots & & & \vdots \\ \frac{\partial f_n(\mathbf{x})}{\partial x_1} & \frac{\partial f_n(\mathbf{x})}{\partial x_2} & \cdots & \frac{\partial f_n(\mathbf{x})}{\partial x_n} \end{bmatrix}.$$
(8.2.9)

The iteration is given by

$$\mathbf{x}^{(k)} = \mathbf{G}(\mathbf{x}^{(k-1)}) = \mathbf{x}^{(k-1)} - \left[J(\mathbf{x}^{(k-1)})\right]^{-1} \mathbf{F}(\mathbf{x}^{(k-1)}).$$
(8.2.10)

The algorithm avoids calculating $(J(\mathbf{x}))^{-1}$ at each step. Instead, it finds a vector $\mathbf{y}$ so that $J(\mathbf{x}^{(k-1)})\mathbf{y} = -\mathbf{F}(\mathbf{x}^{(k-1)})$, and then sets $\mathbf{x}^{(k)} = \mathbf{x}^{(k-1)} + \mathbf{y}$.

For the special case of a two-dimensional system (the equations $f(x, y) = 0$ and $g(x, y) = 0$ are to be satisfied), Newton's iteration becomes:

$$\begin{aligned} x_{n+1} &= x_n - \left.\frac{fg_y - f_yg}{f_xg_y - f_yg_x}\right|_{x=x_n, y=y_n}, \\ y_{n+1} &= y_n - \left.\frac{f_xg - fg_x}{f_xg_y - f_yg_x}\right|_{x=x_n, y=y_n}. \end{aligned}$$
(8.2.11)

8.2.8.2 Method of steepest-descent

The method of *steepest-descent* determines the *local minimum* for a function of the form $g : \mathbb{R}^n \to \mathbb{R}$. It can also be used to solve a system $\{f_i\}$ of non-linear equations. The system has a solution $\mathbf{x} = (x_1, x_2, \ldots, x_n)^{\mathrm{T}}$ when the function

$$g(x_1, x_2, \ldots, x_n) = \sum_{i=1}^{n} [f_i(x_1, x_2, \ldots, x_n)]^2$$
(8.2.12)

has the minimal value zero.

This method converges only linearly to the solution but it usually converges even for poor initial approximations. It can be used to locate initial approximations that are close enough so that Newton's method will converge. Intuitively, a local minimum for a function $g : \mathbb{R}^n \to \mathbb{R}$ can be found as follows:

1. Evaluate g at an initial approximation $\mathbf{x}^{(0)} = (x_1^{(0)}, \ldots, x_n^{(0)})^{\mathrm{T}}$.
2. Determine a direction from $\mathbf{x}^{(0)}$ that results in a decrease in the value of g.
3. Move an appropriate distance in this direction and call the new vector $\mathbf{x}^{(1)}$.
4. Repeat steps 1 through 3 with $\mathbf{x}^{(0)}$ replaced by $\mathbf{x}^{(1)}$.

The direction of greatest decrease in the value of g at $\mathbf{x}$ is the direction given by $-\nabla g(\mathbf{x})$ where $\nabla g(\mathbf{x})$ is the *gradient* of g.

DEFINITION 8.2.2

If $g : \mathbb{R}^n \to \mathbb{R}$, the gradient of g at $\mathbf{x} = (x_1, x_2, \ldots, x_n)^T$, denoted $\nabla g(\mathbf{x})$, is

$$\nabla g(\mathbf{x}) = \left(\frac{\partial g}{\partial x_1}(\mathbf{x}), \frac{\partial g}{\partial x_2}(\mathbf{x}), \ldots, \frac{\partial g}{\partial x_n}(\mathbf{x}) \right)^T. \qquad (8.2.13)$$

Thus, set $\mathbf{x}^{(1)} = \mathbf{x}^{(0)} - \alpha \nabla g(\mathbf{x}^{(0)})$ for some constant $\alpha > 0$. Ideally the value of α minimizes the function $h(\alpha) = g\left(\mathbf{x}^{(0)} - \alpha \nabla g(\mathbf{x}^{(0)})\right)$. Instead of tedious direct calculation, the method interpolates h with a quadratic polynomial using nodes α_1, α_2, and α_3 that are hopefully close to the minimum value of h.

8.2.8.3 Algorithm for steepest-descent

To approximate a solution to the minimization problem $\min\limits_{\mathbf{x} \in \mathbb{R}^n} g(\mathbf{x})$, given an initial approximation $\mathbf{x}$.

INPUT: number n of variables, initial approximation $\mathbf{x} = (x_1, x_2, \ldots, x_n)^{\mathrm{T}}$, tolerance TOL, and maximum number of iterations N.
OUTPUT: approximate solution $\mathbf{x} = (x_1, x_2, \ldots, x_n)^{\mathrm{T}}$
or a message of failure.
Algorithm:

1. Set $k = 1$.

2. While $(k \leq N)$, do steps (a)–(k).

    (a) Set: $g_1 = g(x_1, \ldots, x_n)$; (Note: $g_1 = g(\mathbf{x}^{(k)})$.)
        $\mathbf{z} = \nabla g(x_1, \ldots, x_n)$; (Note: $\mathbf{z} = \nabla g(\mathbf{x}^{(k)})$.)
        $z_0 = \|\mathbf{z}\|_2$.

    (b) If $z_0 = 0$ then OUTPUT ("Zero gradient");
        OUTPUT $(x_1, \ldots, x_n, g_1)$;
        (Procedure completed, may have a minimum.) STOP.

    (c) Set $\mathbf{z} = \mathbf{z}/z_0$. (Make $\mathbf{z}$ a unit vector.)
        Set $\alpha_1 = 0$; $\alpha_3 = 1$; $g_3 = g(\mathbf{x} - \alpha_3 \mathbf{z})$.

    (d) While $(g_3 \geq g_1)$, do steps i–ii.

        i. Set $\alpha_3 = \alpha_3/2$; $g_3 = g(\mathbf{x} - \alpha_3 \mathbf{z})$.

        ii. If $\alpha_3 < \text{TOL}/2$, then
            OUTPUT ("No likely improvement");

OUTPUT $(x_1, \ldots, x_n, g_1)$;
(Procedure completed, may have a minimum.)
STOP.

(e) Set $\alpha_2 = \alpha_3/2$; $g_2 = g(\mathbf{x} - \alpha_2 \mathbf{z})$.

(f) Set: $h_1 = (g_2 - g_1)/\alpha_2$; $h_2 = (g_3 - g_2)/(\alpha_3 - \alpha_2)$;
$h_3 = (h_2 - h_1)/\alpha_3$.

(g) Set: $\alpha_0 = (\alpha_2 - h_1/h_3)/2$ (critical point occurs at α_0.)
$g_0 = g(\mathbf{x} - \alpha_0 \mathbf{z})$.

(h) Find α from $\{\alpha_0, \alpha_3\}$ so that $g = g(\mathbf{x} - \alpha \mathbf{z}) = \min\{g_0, g_3\}$.

(i) Set $\mathbf{x} = \mathbf{x} - \alpha \mathbf{z}$.

(j) If $|g - g_1| < \text{TOL}$ then OUTPUT $(x_1, \ldots, x_n, g)$;
(Procedure completed successfully.) STOP.

(k) Set $k = k + 1$.

3. OUTPUT ("Maximum iterations exceeded");
(Procedure unsuccessful.) STOP.

8.3 NUMERICAL INTEGRATION AND DIFFERENTIATION

8.3.1 NUMERICAL INTEGRATION

Numerical quadrature involves estimating $\int_a^b f(x)\, dx$ using a formula of the form

$$\int_a^b f(x)\, dx \approx \sum_{i=0}^{n} c_i f(x_i). \tag{8.3.1}$$

8.3.1.1 Newton–Cotes formulae

A *closed Newton–Cotes formula* uses nodes $x_i = x_0 + ih$ for $i = 0, 1, \ldots, n$, where $h = (b - a)/n$. Note that $x_0 = a$ and $x_n = b$.

An *open Newton–Cotes formula* uses nodes $x_i = x_0 + ih$ for $i = 0, 1, \ldots, n$, where $h = (b - a)/(n + 2)$. Here $x_0 = a + h$ and $x_n = b - h$. Set $x_{-1} = a$ and $x_{n+1} = b$. The nodes actually used lie in the *open interval* (a, b).

In all formulae, ξ is a number for which $a < \xi < b$ and f_i denotes $f(x_i)$.

8.3.1.2 Closed Newton–Cotes formulae

1. ($n = 1$) Trapezoidal rule

$$\int_a^b f(x)\, dx = \frac{h}{2}[f(x_0) + f(x_1)] - \frac{h^3}{12} f''(\xi).$$

2. $(n = 2)$ Simpson's rule

$$\int_a^b f(x)\, dx = \frac{h}{3}[f(x_0) + 4f(x_1) + f(x_2)] - \frac{h^5}{90} f^{(4)}(\xi).$$

3. $(n = 3)$ Simpson's three-eighths rule

$$\int_a^b f(x)\, dx = \frac{3h}{8}[f(x_0) + 3f(x_1) + 3f(x_2) + f(x_3)] - \frac{3h^5}{80} f^{(4)}(\xi).$$

4. $(n = 4)$ Milne's rule (also called Boole's rule)

$$\int_a^b f(x)\, dx = \frac{2h}{45}[7f_0 + 32f_1 + 12f_2 + 32f_3 + 7f_4] - \frac{8h^7}{945} f^{(6)}(\xi).$$

5. $(n = 5)$

$$\int_a^b f(x)\, dx = \frac{5h}{288}[19f_0 + 75f_1 + 50f_2 + 50f_3 + 75f_4 + 19f_5] - \frac{275h^7}{12096} f^{(6)}(\xi).$$

6. $(n = 6)$ Weddle's rule

$$\int_a^b f(x)\, dx = \frac{h}{140}[41f_0 + 216f_1 + 27f_2 + 272f_3 + 27f_4 + 216f_5 + 41f_6]$$
$$- \frac{9h^9}{1400} f^{(8)}(\xi).$$

7. $(n = 7)$

$$\int_a^b f(x)\, dx = \frac{7h}{17280}[751f_0 + 3577f_1 + 1323f_2 + 2989f_3 + 2989f_4 + 1323f_5$$
$$+ 3577f_6 + 751f_7] - \frac{8183h^9}{518400} f^{(8)}(\xi).$$

8.3.1.3 Open Newton–Cotes formulae

1. $(n = 0)$ Midpoint rule

$$\int_a^b f(x)\, dx = 2hf(x_0) + \frac{h^3}{3} f''(\xi).$$

2. $(n = 1)$

$$\int_a^b f(x)\, dx = \frac{3h}{2}[f(x_0) + f(x_1)] + \frac{3h^3}{4} f''(\xi).$$

3. ($n = 2$)

$$\int_a^b f(x)\,dx = \frac{4h}{3}[2f(x_0) - f(x_1) + 2f(x_2)] + \frac{14h^5}{45}f^{(4)}(\xi).$$

4. ($n = 3$)

$$\int_a^b f(x)\,dx = \frac{5h}{24}[11f(x_0) + f(x_1) + f(x_2) + 11f(x_3)] + \frac{95h^5}{144}f^{(4)}(\xi).$$

5. ($n = 4$)

$$\int_a^b f(x)\,dx = \frac{3h}{10}[11f_0 - 14f_1 + 26f_2 - 14f_3 + 11f_4] + \frac{41h^7}{140}f^{(6)}(\xi).$$

6. ($n = 5$)

$$\int_a^b f(x)\,dx = \frac{7h}{1440}[611f_0 - 453f_1 + 562f_2 + 562f_3 - 453f_4 + 611f_5]$$
$$+ \frac{5257h^7}{8640}f^{(6)}(\xi).$$

8.3.1.4 Composite rules

Some Newton–Cotes formulae extend to *composite formulae*. This consists of dividing the interval into subintervals and using Newton–Cotes formulae on each subinterval. In the following, note that $a < \mu < b$.

1. Composite trapezoidal rule for n subintervals: If $f \in C^2[a, b]$, $h = (b - a)/n$, and $x_j = a + jh$, for $j = 0, 1, \ldots, n$, then

$$\int_a^b f(x)\,dx = \frac{h}{2}\left[f(a) + 2\sum_{j=1}^{n-1} f(x_j) + f(b)\right] - \frac{b - a}{12}h^2 f''(\mu).$$

2. Composite Simpson's rule for n subintervals: If $f \in C^4[a, b]$, n is even, $h = (b - a)/n$, and $x_j = a + jh$, for $j = 0, 1, \ldots, n$, then

$$\int_a^b f(x)\,dx = \frac{h}{3}\left[f(a) + 2\sum_{j=1}^{(n/2)-1} f(x_{2j}) + 4\sum_{j=1}^{n/2} f(x_{2j-1}) + f(b)\right]$$
$$- \frac{b - a}{180}h^4 f^{(4)}(\mu).$$

3. Composite midpoint rule for $n + 2$ subintervals: If $f \in C^2[a, b]$, n is even, $h = (b - a)/(n + 2)$, and $x_j = a + (j + 1)h$, for $j = -1, 0, 1, \ldots, n + 1$, then

$$\int_a^b f(x)\,dx = 2h\sum_{j=0}^{n/2} f(x_{2j}) + \frac{b - a}{6}h^2 f''(\mu).$$

8.3.1.5 Romberg integration

Romberg integration uses the composite trapezoidal rule beginning with $h_1 = b - a$ and $h_k = (b - a)/2^{k-1}$, for $k = 1, 2, \ldots$, to give preliminary estimates for $\int_a^b f(x) \, dx$ and improves these estimates using Richardson's extrapolation. Since many function evaluations would be repeated, the first column of the extrapolation table (with entries denoted $R_{i,j}$) can be more efficiently determined by the following recursion formula:

$$R_{1,1} = \frac{h_1}{2}[f(a) + f(b)] = \frac{b-a}{2}[f(a) + f(b)],$$

$$R_{k,1} = \frac{1}{2}\left[R_{k-1,1} + h_{k-1}\sum_{i=1}^{2^{k-2}} f(a + (2i - 1)h_k)\right], \tag{8.3.2}$$

for $k = 2, 3, \ldots$. Now apply Equation (8.1.4) to complete the extrapolation table.

8.3.1.6 Gregory's formula

Using f_j to represent $f(x_0 + jh)$,

$$\int_{x_0}^{x_0+nh} f(y) \, dy = h(\frac{1}{2}f_0 + f_1 + \ldots + f_{n-1} + \frac{1}{2}f_n)$$

$$+ \frac{h}{12}(\Delta f_0 - \Delta f_{n-1}) - \frac{h}{24}(\Delta^2 f_0 + \Delta^2 f_{n-2})$$

$$+ \frac{19h}{720}(\Delta^3 f_0 - \Delta^3 f_{n-3}) - \frac{3h}{160}(\Delta^4 f_0 + \Delta^4 f_{n-4}) + \ldots \tag{8.3.3}$$

where Δ's represent forward differences. The first expression on the right in Equation (8.3.3) is the composite trapezoidal rule, and additional terms provide improved approximations. Care must be taken not to carry this process too far because Gregory's formula is only asymptotically convergent in general and round-off error can be significant when computing higher differences.

FIGURE 8.2

Formulae for integration rules with various weight functions

Weight $w(x)$	Interval (α,β)	Abcissas are zeros of	x_i	w_i	K_n
1	$(-1,1)$	$P_n(x)$	See table on page 755	$\dfrac{-2}{(n+1)P_{n+1}(x_i)P'_n(x_i)}$	$\dfrac{2^{2n+1}(n!)^4}{(2n+1)[(2n)!]^3}$
e^{-x}	$(0,\infty)$	$L_n(x)$	See table on page 756	$\dfrac{(n!)^2 x_i}{(n+1)^2 L_{n+1}^2(x_i)}$	$\dfrac{(n!)^2}{(2n)!}$
e^{-x^2}	$(-\infty,\infty)$	$H_n(x)$	See table on page 756	$\dfrac{2^{n-1}n!\sqrt{\pi}}{n^2 H_{n-1}^2(x_i)}$	$\dfrac{n!\sqrt{\pi}}{2^n(2n)!}$
$\dfrac{1}{\sqrt{1-x^2}}$	$(-1,1)$	$T_n(x)$	$\cos\left(\dfrac{(2i-1)\pi}{2n}\right)$	$\dfrac{\pi}{n}$	$\dfrac{2\pi}{2^{2n}(2n)!}$
$\sqrt{1-x^2}$	$(-1,1)$	$U_n(x)$	$\cos\left(\dfrac{i\pi}{n+1}\right)$	$\dfrac{\pi}{n+1}\sin^2\left(\dfrac{i\pi}{n+1}\right)$	$\dfrac{\pi}{2^{2n+1}(2n)!}$
$\sqrt{\dfrac{x}{1-x}}$	$(0,1)$	$\dfrac{T_{2n+1}(\sqrt{x})}{\sqrt{x}}$	$\cos^2\left(\dfrac{(2i-1)\pi}{4n+2}\right)$	$\dfrac{2\pi}{2n+1}\cos^2\left(\dfrac{(2i-1)\pi}{4n+2}\right)$	$\dfrac{\pi}{2^{4n+1}(2n)!}$
$\sqrt{\dfrac{1-x}{1+x}}$	$(0,1)$	$J_n\left(x,\dfrac{1}{2},-\dfrac{1}{2}\right)$	$\cos\left(\dfrac{2i\pi}{2n+1}\right)$	$\dfrac{4\pi}{2n+1}\sin^2\left(\dfrac{i\pi}{2n+1}\right)$	$\dfrac{\pi}{2^{2n}(2n)!}$
$\dfrac{1}{\sqrt{x}}$	$(0,1)$	$P_{2n}(\sqrt{x})$	$(x_i^+)^2$	$2h_i$	$\dfrac{2^{4n+1}[(2n)!]^3}{(4n+1)[(4n)!]^2}$
$\sqrt{x}$	$(0,1)$	$\dfrac{P_{2n+1}(\sqrt{x})}{\sqrt{x}}$	$(x_i^+)^2$	$2h_i\,(x_i^+)^2$	$\dfrac{2^{4n+3}[(2n+1)!]^4}{(4n+3)[(4n+2)!]^2(2n)!}$

In this table, P_n, L_n, H_n, T_n, U_n, and J_n denote the nth Legendre, Laguerre, Hermite, Chebyshev (first kind T_n, second kind U_n), and Jacobi polynomials, respectively. Also, x_i^+ denotes the ith positive root of $P_{2n}(x)$ or $P_{2n+1}(x)$ of the previous column, and h_i denotes the corresponding weight for x_i^+ in the Gauss–Legendre formula $(w(x)=1)$.

8.3.1.7 Gaussian quadrature

A quadrature formula, whose nodes (abscissae) x_i and coefficients w_i are chosen to achieve a maximum order of accuracy, is called a *Gaussian quadrature formula*. The integrand usually involves a *weight function* w. An integral in t on an interval (a, b) must be converted into an integral in x over the interval (α, β) specified for the weight function involved. This can be accomplished by the transformation $x = \frac{(b\alpha - a\beta)}{(b-a)} + \frac{(\beta - \alpha)t}{(b-a)}$. Gaussian quadrature formulae generally take the form

$$\int_\alpha^\beta w(x)f(x)\,dx = \sum_i w_i f(x_i) + E_n \qquad (8.3.4)$$

where $E_n = K_n f^{(2n)}(\xi)$ for some $\alpha < \xi < \beta$ and K_n is a specified constant. Many popular weight functions and their associated intervals are summarized in the table on page 754.

The following tables give abscissae and weights for selected formulae. If some x_i are specified (such as one or both end points), then the formulae of Radau and Lobatto may be used.

8.3.1.8 Gauss–Legendre quadrature

Weight function is $w(x) = 1$. $\displaystyle\int_{-1}^1 f(x)\,dx \approx \sum_{i=1}^n w_i f(x_i).$

n	Nodes $\{\pm x_i\}$	Weights $\{w_i\}$	n	Nodes $\{\pm x_i\}$	Weights $\{w_i\}$
2	0.5773502692	1	8	0.1834346425	0.3626837834
				0.5255324099	0.3137066459
3	0	0.8888888889		0.7966664774	0.2223810345
	0.7745966692	0.5555555556		0.9602898565	0.1012285363
4	0.3399810436	0.6521451549	9	0	0.3302393550
	0.8611363116	0.3478548451		0.3242534234	0.3123470770
				0.6133714327	0.2606106964
5	0	0.5688888889		0.8360311073	0.1806481607
	0.5384693101	0.4786286705		0.9681602395	0.0812743883
	0.9061798459	0.2369268851			
			10	0.1488743390	0.2955242247
6	0.2386191861	0.4679139346		0.4333953941	0.2692667193
	0.6612093865	0.3607615730		0.6794095683	0.2190863625
	0.9324695142	0.1713244924		0.8650633667	0.1494513492
				0.9739065285	0.0666713443
7	0	0.4179591837			
	0.4058451514	0.3818300505			
	0.7415311856	0.2797053915			
	0.9491079123	0.1294849662			

8.3.1.9   Gauss–Laguerre quadrature

Weight function is $w(x) = e^{-x}$.

$$\int_0^\infty e^{-x} f(x)\, dx \approx \sum_{i=1}^n w_i f(x_i).$$

n	Nodes $\{x_i\}$	Weights $\{w_i\}$	n	Nodes $\{x_i\}$	Weights $\{w_i\}$
2	0.5857864376	0.8535533905	6	0.2228466041	0.4589646739
	3.4142135623	0.1464466094		1.1889321016	0.4170008307
				2.9927363260	0.1133733820
3	0.4157745567	0.7110930099		5.7751435691	0.0103991974
	2.2942803602	0.2785177335		9.8374674183	0.0002610172
	6.2899450829	0.0103892565		15.9828739806	0.0000008985
4	0.3225476896	0.6031541043	7	0.1930436765	0.4093189517
	1.7457611011	0.3574186924		1.0266648953	0.4218312778
	4.5366202969	0.0388879085		2.5678767449	0.1471263486
	9.3950709123	0.0005392947		4.9003530845	0.0206335144
5	0.2635603197	0.5217556105		8.1821534445	0.0010740101
	1.4134030591	0.3986668110		12.7341802917	0.0000158654
	3.5964257710	0.0759424496		19.3957278622	0.0000000317
	7.0858100058	0.0036117586			
	12.6408008442	0.0000233699			

8.3.1.10   Gauss–Hermite quadrature

Weight function is $w(x) = e^{-x^2}$.

$$\int_{-\infty}^\infty e^{-x^2} f(x)\, dx \approx \sum_{i=1}^n w_i f(x_i).$$

n	Nodes $\{\pm x_i\}$	Weights $\{w_i\}$	n	Nodes $\{\pm x_i\}$	Weights $\{w_i\}$
2	0.7071067811	0.8862269254	7	0	0.8102646175
				0.8162878828	0.4256072526
3	0	1.1816359006		1.6735516287	0.0545155828
	1.2247448713	0.2954089751		2.6519613568	0.0009717812
4	0.5246476232	0.8049140900	8	0.3811869902	0.6611470125
	1.6506801238	0.0813128354		1.1571937124	0.2078023258
				1.9816567566	0.0170779830
5	0	0.9453087204		2.9306374202	0.0001996040
	0.9585724646	0.3936193231			
	2.0201828704	0.0199532420	9	0	0.7202352156
				0.7235510187	0.4326515590
6	0.4360774119	0.7246295952		1.4685532892	0.0884745273
	1.3358490740	0.1570673203		2.2665805845	0.0049436242
	2.3506049736	0.0045300099		3.1909932017	0.0000396069

8.3.1.11 Radau quadrature

$$\int_{-1}^{1} f(x)\, dx = w_1 f(-1) + \sum_{i=2}^{n} w_i f(x_i) + \frac{2^{2n-1}[n(n-1)!]^4}{[(2n-1)!]^3} f^{(2n-1)}(\xi),$$

where each free node x_i is the ith root of $\frac{P_{n-1}(x)+P_n(x)}{x+1}$ and $w_i = \frac{1-x_i}{n^2[P_{n-1}(x_i)]^2}$ for $i = 2, \ldots, n$; see the following table. Note that $x_1 = -1$ and $w_1 = 2/n^2$.

n	Nodes	Weights $\{w_i\}$	n	Nodes	Weights $\{w_i\}$
3	−0.2898979485	1.0249716523	8	−0.8874748789	0.1853581548
	0.6898979485	0.7528061254		−0.6395186165	0.3041306206
				−0.2947505657	0.3765175453
4	−0.5753189235	0.6576886399		0.0943072526	0.3915721674
	0.1810662711	0.7763869376		0.4684203544	0.3470147956
	0.8228240809	0.4409244223		0.7706418936	0.2496479013
				0.9550412271	0.1145088147
5	−0.7204802713	0.4462078021			
	−0.1671808647	0.6236530459	9	−0.9107320894	0.1476540190
	0.4463139727	0.5627120302		−0.7112674859	0.2471893782
	0.8857916077	0.2874271215		−0.4263504857	0.3168437756
				−0.0903733696	0.3482730027
6	−0.8029298284	0.3196407532		0.2561356708	0.3376939669
	−0.3909285467	0.4853871884		0.5713830412	0.2863866963
	0.1240503795	0.5209267831		0.8173527842	0.2005532980
	0.6039731642	0.4169013343		0.9644401697	0.0907145049
	0.9203802858	0.2015883852			
			10	−0.9274843742	0.1202966705
7	−0.8538913426	0.2392274892		−0.7638420424	0.2042701318
	−0.5384677240	0.3809498736		−0.5256460303	0.2681948378
	−0.1173430375	0.4471098290		−0.2362344693	0.3058592877
	0.3260306194	0.4247037790		0.0760591978	0.3135824572
	0.7038428006	0.3182042314		0.3806648401	0.2906101648
	0.9413671456	0.1489884711		0.6477666876	0.2391934317
				0.8512252205	0.1643760127
				0.9711751807	0.0736170054

8.3.1.12 Lobatto quadrature

$$\int_{-1}^{1} f(x)\, dx = w_1 f(-1) + w_n f(1)$$

$$+ \sum_{i=2}^{n-1} w_i f(x_i) - \frac{n(n-1)^3 2^{2n-1}[(n-2)!]^4}{(2n-1)[(2n-2)!]^3} f^{(2n-2)}(\xi)$$

where x_i is the $(i-1)^{\text{st}}$ root of $P'_{n-1}(x)$ and $w_i = \frac{2}{n(n-1)[P_{n-1}(x_i)]^2}$ for $i = 2, \ldots, n-1$. Note that $x_1 = -1$, $x_n = 1$, and $w_1 = w_n = 2/(n(n-1))$. The table lists all x_i and w_i for $n < 13$.

n	Nodes $\{\pm x_i\}$	Weights $\{w_i\}$
3	0	1.3333333333
	1	0.3333333333
4	0.4472135954	0.8333333333
	1	0.1666666666
5	0	0.7111111111
	0.6546536707	0.5444444444
	1	0.1000000000
6	0.2852315164	0.5548583770
	0.7650553239	0.3784749562
	1	0.0666666666
7	0	0.4876190476
	0.4688487934	0.4317453812
	0.8302238962	0.2768260473
	1	0.0476190476
8	0.2092992179	0.4124587946
	0.5917001814	0.3411226924
	0.8717401485	0.2107042271
	1	0.0357142857
9	0	0.3715192743
	0.3631174638	0.3464285109
	0.6771862795	0.2745387125
	0.8997579954	0.1654953615
	1	0.0277777777

n	Nodes $\{\pm x_i\}$	Weights $\{w_i\}$
10	0.1652789576	0.3275397611
	0.4779249498	0.2920426836
	0.7387738651	0.2248893420
	0.9195339081	0.1333059908
	1	0.0222222222
11	0	0.3002175954
	0.2957581355	0.2868791247
	0.5652353269	0.2480481042
	0.7844834736	0.1871698817
	0.9340014304	0.1096122732
	1	0.0181818181
12	0.1365529328	0.2714052409
	0.3995309409	0.2512756031
	0.6328761530	0.2125084177
	0.8192793216	0.1579747055
	0.9448992722	0.0916845174
	1	0.0151515151
13	0	0.2519308493
	0.2492869301	0.2440157903
	0.4829098210	0.2207677935
	0.6861884690	0.1836468652
	0.8463475646	0.1349819266
	0.9533098466	0.0778016867
	1	0.0128205128

8.3.1.13   Chebyshev quadrature

$$\int_{-1}^{1} f(x)\,dx \approx \frac{2}{n} \sum_{i=1}^{n} f(x_i).$$

n	Nodes $\{\pm x_i\}$
2	0.5773502691
3	0
	0.7071067811
4	0.1875924740
	0.7946544722

n	Nodes $\{\pm x_i\}$
5	0
	0.3745414095
	0.8324974870
6	0.2666354015
	0.4225186537
	0.8662468181

n	Nodes $\{\pm x_i\}$
7	0
	0.3239118105
	0.5296567752
	0.8838617007

8.3.1.14   Multiple integrals

Quadrature methods can be extended to multiple integrals. The general idea, using a double integral as an example, involves writing the double integral in the form of an iterated integral, applying the quadrature method to the "inner integral" and then applying the method to the "outer integral".

8.3.1.15   Simpson's double integral over a rectangle

To integrate a function $f(x, y)$ over the rectangular region $R = \{(x, y) \mid a \le x \le b, c \le y \le d\}$ using the composite Simpson's Rule produces an approximating formula given below. Intervals $[a, b]$ and $[c, d]$ must be partitioned using even integers n and m to identify evenly-spaced mesh points $x_0, x_1, \ldots, x_n$ and $y_0, y_1, \ldots, y_m$, respectively.

$$\int\int_R f(x, y)\, dy dx = \int_a^b \int_c^d f(x, y)\, dy dx = \frac{hk}{9} \sum_{i=0}^n \sum_{j=0}^m c_{i,j} f(x_i, y_j) + E \tag{8.3.5}$$

where the error term E is given by

$$E = -\frac{(d - c)(b - a)}{180} \left[h^4 \frac{\partial^4 f}{\partial x^4}(\bar{\eta}, \bar{\mu}) + k^4 \frac{\partial^4 f}{\partial y^4}(\hat{\eta}, \hat{\mu}) \right] \tag{8.3.6}$$

for some $(\bar{\eta}, \bar{\mu})$ and $(\hat{\eta}, \hat{\mu})$ in R with h and k determined by $h = (b - a)/n$ and $k = (d - c)/m$, and the coefficients $c_{i,j}$ are the entries in the following table.

m	1	4	2	4	2	4	$\ldots$	2	4	1
$m - 1$	4	16	8	16	8	16	$\ldots$	8	16	4
$m - 2$	2	8	4	8	4	8	$\ldots$	4	8	2
$m - 3$	4	16	8	16	8	16	$\ldots$	8	16	4
$\vdots$	$\vdots$	$\vdots$	$\vdots$	$\vdots$	$\vdots$	$\vdots$		$\vdots$	$\vdots$	$\vdots$
2	2	8	4	8	4	8	$\ldots$	4	8	2
1	4	16	8	16	8	16	$\ldots$	8	16	4
0	1	4	2	4	2	4	$\ldots$	2	4	1
j										
i	0	1	2	3	4	5	$\ldots$	$n - 2$	$n - 1$	n

Similarly, Simpson's Rule can be extended for regions that are not rectangular. It is simpler to give the following algorithm than to state a general formula.

8.3.1.16   Simpson's double integral algorithm

To approximate the integral $I = \int_a^b \int_{c(x)}^{d(x)} f(x, y)\, dy\, dx$:

INPUT   endpoints a, b; even positive integers m, n;
    functions $c(x)$, $d(x)$, and $f(x, y)$
OUTPUT   approximation J to I.

Algorithm:

1. Set $h = (b - a)/n$; $J_1 = 0$; (End terms.)
   $J_2 = 0$; (Even terms.) $J_3 = 0$. (Odd terms.)

2. For $i = 0, 1, ..., n$ do (a)–(d).

   (a) Set $x = a + ih$;
       (Composite Simpson's method for x.)
       $HX = (d(x) - c(x))/m$;
       $K_1 = f(x, c(x)) + f(x, d(x))$;   (End terms.)
       $K_2 = 0$; (Even terms.)
       $K_3 = 0$.   (Odd terms.)

   (b) For $j = 1, 2, ..., m - 1$ do i–ii.
       i. Set $y = c(x) + jHX$; $Q = f(x, y)$.
       ii. If j is even then set $K_2 = K_2 + Q$ else set $K_3 = K_3 + Q$.

   (c) Set $L = (K_1 + 2K_2 + 4K_3)HX/3$.
       $$\left(L \approx \int_{c(x_i)}^{d(x_i)} f(x_i, y) \, dy \quad \text{by composite Simpson's method.} \right)$$

   (d) If $i = 0$ or $i = n$ then set $J_1 = J_1 + L$;
       else if i is even then set $J_2 = J_2 + L$;
       else set $J_3 = J_3 + L$.

3. Set $J = h(J_1 + 2J_2 + 4J_3)/3$.

4. OUTPUT(J);
   STOP.

8.3.1.17   Gaussian double integral

To apply Gaussian quadrature to $I = \int_a^b \int_{c(x)}^{d(x)} f(x, y) \, dy \, dx$ first requires transforming, for each x in $[a, b]$, the interval $[c(x), d(x)]$ to $[-1, 1]$ and then applying Gaussian quadrature. This is performed in the following algorithm.

8.3.1.18   Gauss–Legendre double integral

To approximate the integral $I = \int_a^b \int_{c(x)}^{d(x)} f(x, y) \, dy \, dx$:

INPUT   endpoints a, b; positive integers m, n.
   (The roots $r_{i,j}$ and coefficients $c_{i,j}$ are found in 8.3.1.8 for
   $i = \max\{m, n\}$ and for $1 \le j \le i$.)
OUTPUT   approximation J to I.
Algorithm:

1. Set $h_1 = (b - a)/2$; $h_2 = (b + a)/2$; $J = 0$.

2. For $i = 1, 2, ..., m$ do (a)–(c).

(a) Set $JX = 0$;

$x = h_1 r_{m,i} + h_2$;

$d_1 = d(x)$; $c_1 = c(x)$;

$k_1 = (d_1 - c_1)/2$; $k_2 = (d_1 + c_1)/2$.

(b) For $j = 1, 2, ..., n$ do

set $y = k_1 r_{n,j} + k_2$;

$Q = f(x, y)$;

$JX = JX + c_{n,j} Q$.

(c) Set $J = J + c_{m,i} k_1 JX$.

3. Set $J = h_1 J$.

4. OUTPUT(J); STOP.

8.3.1.19 Double integrals of polynomials over polygons

If the vertices of the polygon A are $\{(x_1, y_1), (x_2, y_2), \ldots, (x_p, y_p)\}$, and we define $w_i = x_i y_{i+1} - x_{i+1} y_i$ (with $x_{p+1} = x_1$ and $y_{p+1} = y_1$) then

$$\int\int_A x^m y^n \, dA = \tag{8.3.7}$$

$$\frac{m! n!}{(m+n+2)!} \sum_{i=1}^{p} w_i \sum_{j=0}^{m} \sum_{k=0}^{n} \binom{j+k}{j} \binom{m+n-j-k}{n-k} x_i^{m-j} x_{i+1}^{j} y_i^{n-k} y_{i+1}^{k}.$$

8.3.1.20 Monte–Carlo methods

Monte–Carlo methods, in general, involve the generation of random numbers (actually *pseudorandom* when computer-generated) to represent independent, uniform random variables over $[0, 1]$. Section 7.6 describes random number generation. Such a simulation can provide insight into the solutions of very complex problems.

Monte–Carlo methods are generally not competitive with other numerical methods of this section. However, if the function fails to have continuous derivatives of moderate order, those methods may not be applicable. One advantage of Monte–Carlo methods is that they extend to multidimensional integrals quite easily, although here only a few techniques for one-dimensional integrals $I = \int_a^b g(x) \, dx$ are given.

8.3.1.21 Hit or miss method

Suppose $0 \leq g(x) \leq c$, $a \leq x \leq b$, and $\Omega = \{(x, y) \mid a \leq x \leq b, 0 \leq y \leq c\}$. If (X, Y) is a random vector which is uniformly distributed over Ω, then the probability p that (X, Y) lies in S (see Figure 8.3) is $p = I/(c(b-a))$.

If N independent random vectors $\{(X_i, Y_i)\}_{i=1}^{N}$ are generated, the parameter p can be estimated by $\hat{p} = N_H/N$ where N_H is the number of times $Y_i \leq g(X_i)$, $i = 1, 2, \ldots, N$, called the number of *hits*. (Likewise $N - N_H$ is the number of *misses*.) The value of I is then estimated by the unbiased estimator $\theta_1 = c(b-a)N_H/N$.

FIGURE 8.3

Illustration of the Monte–Carlo method. The sample points are shown as circles. The solid circle is counted as a "hit", the empty circle is counted as a "miss".

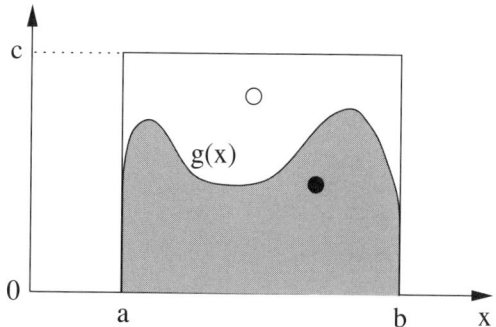

8.3.1.22   Hit or miss algorithm

1. Generate $\{U_j\}_{j=1}^{2N}$ of $2N$ random numbers, uniformly distributed in $[0, 1)$.
2. Arrange the sequence into N pairs $(U_1, U_1'), (U_2, U_2'), \ldots, (U_N, U_N')$, so that each U_j is used exactly once.
3. Compute $X_i = a + U_i(b - a)$ and $g(X_i)$ for $i = 1, 2, \ldots, N$.
4. Count the number of cases N_H for which $g(X_i) \geq cU_i'$.
5. Compute $\theta_1 = c(b - a)N_H/N$. (This is an estimate of I.)

The number of trials N necessary for $P(|\theta_1 - I| < \epsilon) \geq \alpha$ is given by

$$N \geq \frac{(1 - p)p[c(b - a)]^2}{(1 - \alpha)\epsilon^2}. \tag{8.3.8}$$

With the usual notation of z_α for the value of the standard normal random variable Z for which $P(Z > z_\alpha) = \alpha$ (see page 695), a confidence interval for I with confidence level $1 - \alpha$ is

$$\theta_1 \pm z_{\frac{\alpha}{2}} \frac{\sqrt{\hat{p}(1 - \hat{p})}(b - a)c}{\sqrt{N}}. \tag{8.3.9}$$

8.3.1.23   Sample-mean Monte–Carlo method

Write the integral $I = \int_a^b g(x)\, dx$ as $\int_a^b \frac{g(x)}{f_X(x)} f_X(x)\, dx$, where f is any probability density function for which $f_X(x) > 0$ when $g(x) \neq 0$. Then $I = \mathrm{E}\left[\frac{g(X)}{f_X(X)}\right]$ where the random variable X is distributed according to $f_X(x)$. Values from this distribution can be generated by the methods discussed in Section 7.6.2. For the case where $f_X(x)$ is the uniform distribution on $[0, 1]$, $I = (b - a)\mathrm{E}[g(X)]$. An unbiased estimator of I is its sample mean

$$\theta_2 = (b - a)\frac{1}{N}\sum_{i=1}^{N} g(X_i). \tag{8.3.10}$$

It follows that the variance of θ_2 is less than or equal to the variance of θ_1. In fact,

$$\text{var } \theta_1 = \frac{I}{N}[c(b-a) - I],$$

$$\text{var } \theta_2 = \frac{1}{N}\left[(b-a)\int_a^b g^2(x)\,dx - I^2\right]. \tag{8.3.11}$$

Note that to estimate I with θ_1 or θ_2, $g(x)$ is not needed explicitly. It is only necessary to evaluate $g(x)$ at any point x.

8.3.1.24 Sample-mean algorithm

1. Generate $\{U_i\}_{i=1}^N$ of N random numbers, uniformly distributed in $[0,1)$.
2. Compute $X_i = a + U_i(b-a)$, for $i = 1, 2, \ldots, N$.
3. Compute $g(X_i)$ for $i = 1, 2, \ldots, N$.
4. Compute θ_2 according to Equation (8.3.10). (This is an estimate of I.)

8.3.1.25 Integration in the presence of noise

Suppose $g(x)$ is measured with some error: $\tilde{g}(x_i) = g(x_i) + \epsilon_i$, for $i = 1, 2, \ldots, N$, where ϵ_i are independent identically distributed random variables with $\mathrm{E}\,[\epsilon_i] = 0$, $\text{var}(\epsilon_i) = \sigma^2$, and $|\epsilon_i| < k < \infty$.

If (X, Y) is uniformly distributed on the rectangle $a \le x \le b$, $0 \le y \le c_1$, where $c_1 \ge g(x) + k$, set $\tilde{\theta}_1 = c_1(b-a)N_H/N$ as in the hit or miss method. Similarly, set $\tilde{\theta}_2 = \frac{1}{N}(b-a)\sum_{i=1}^N \tilde{g}(X_i)$ as in the sample-mean method. Then both $\tilde{\theta}_1$ and $\tilde{\theta}_2$ are unbiased and converge almost surely to I. Again, var $\tilde{\theta}_2 \le$ var $\tilde{\theta}_1$.

8.3.1.26 Weighted Monte–Carlo integration

Estimate the integral $I = \int_0^1 g(x)\,dx$ according to the following algorithm:

1. Generate numbers $\{U_1, U_2, \ldots, U_N\}$ from the uniform distribution on $[0,1)$.

2. Arrange $U_1, U_2, \ldots, U_N$ in the increasing order $U_{(1)}, U_{(2)}, \ldots, U_{(N)}$.

3. Compute $\theta_3 = \dfrac{1}{2}\left[\displaystyle\sum_{i=0}^N (g(U_{(i)}) + g(U_{(i+1)}))(U_{(i+1)} - U_{(i)})\right]$, where $U_{(0)} \equiv 0$, $U_{(N+1)} \equiv 1$. This is an estimate of I.

If $g(x)$ has a continuous second derivative on $[0,1]$, then the estimator θ_3 satisfies var $\theta_3 = \mathrm{E}\,[(\theta_3 - I)^2] \le k/N^4$, where k is some positive constant.

8.3.2 NUMERICAL DIFFERENTIATION

8.3.2.1 Derivative estimates

Selected formulae to estimate the derivative of a function at a single point, with error terms, are given. Nodes are equally spaced with $x_i - x_{i-1} = h$; h may be positive or negative and, in the error formulae, ξ lies between the smallest and largest nodes. To shorten some of the formulae, f_j is used to denote $f(x_0 + jh)$ and some error formulae are expressed as $O(h^k)$.

1. Two-point formula for $f'(x_0)$

$$f'(x_0) = \frac{1}{h}(f(x_0 + h) - f(x_0)) - \frac{h}{2}f''(\xi). \qquad (8.3.12)$$

   This is called the *forward-difference formula* if $h > 0$ and the *backward-difference formula* if $h < 0$.

2. Three-point formulae for $f'(x_0)$

$$f'(x_0) = \frac{1}{2h}[-3f(x_0) + 4f(x_0 + h) - f(x_0 + 2h)] + \frac{h^2}{3}f^{(3)}(\xi)$$
$$= \frac{1}{2h}[f(x_0 + h) - f(x_0 - h)] - \frac{h^2}{6}f^{(3)}(\xi). \qquad (8.3.13)$$

3. Four-point formula (or five uniformly spaced points) for $f'(x_0)$

$$f'(x_0) = \frac{1}{12h}[f_{-2} - 8f_{-1} + 8f_1 - f_2] + \frac{h^4}{30}f^{(5)}(\xi). \qquad (8.3.14)$$

4. Five-point formula for $f'(x_0)$

$$f'(x_0) = \frac{1}{12h}[-25f_0 + 48f_1 - 36f_2 + 16f_3 - 3f_4] + \frac{h^4}{5}f^{(5)}(\xi). \quad (8.3.15)$$

5. Formulae for the second derivative

$$f''(x_0) = \frac{1}{h^2}[f_{-1} - 2f_0 + f_1] - \frac{h^2}{12}f^{(4)}(\xi)$$
$$= \frac{1}{h^2}[f_0 - 2f_1 + f_2] + \frac{h^2}{6}f^{(4)}(\xi_1) - hf^{(3)}(\xi_2). \qquad (8.3.16)$$

6. Formulae for the third derivative

$$f^{(3)}(x_0) = \frac{1}{h^3}[f_3 - 3f_2 + 3f_1 - f_0] + O(h)$$
$$= \frac{1}{2h^3}[f_2 - 2f_1 + 2f_{-1} - f_{-2}] + O(h^2). \qquad (8.3.17)$$

7. Formulae for the fourth derivative

$$f^{(4)}(x_0) = \frac{1}{h^4}[f_4 - 4f_3 + 6f_2 - 4f_1 + f_0] + O(h)$$

$$= \frac{1}{h^4}[f_2 - 4f_1 + 6f_0 - 4f_{-1} + f_{-2}] + O(h^2).$$

(8.3.18)

Richardson's extrapolation can be applied to improve estimates. The error term of the formula must satisfy Equation (8.1.2) and an extrapolation procedure must be developed. As a special case, however, Equation (8.1.4) may be used when first-column entries are generated by Equation (8.3.13).

8.3.2.2 Computational molecules

A computational molecule is a graphical depiction of an approximate partial derivative formula. The following computational molecules are for $h = \Delta x = \Delta y$:

(a) $\left.\dfrac{\partial u}{\partial x}\right|_{i,j} = \dfrac{1}{2h}(u_{1,0} - u_{-1,0}) + O(h^2) = \dfrac{1}{2h}\left\{ \underset{}{\boxed{-1}} \!-\! \underset{i,j}{\boxed{0}} \!-\! \boxed{1} \right\} + O(h^2)$

(b) $\left.\dfrac{\partial u}{\partial y}\right|_{i,j} = \dfrac{1}{2h}(u_{0,1} - u_{0,-1}) + O(h^2) = \dfrac{1}{2h}\left\{ \begin{array}{c} \boxed{1} \\ | \\ \underset{i,j}{\boxed{0}} \\ | \\ \boxed{-1} \end{array} \right\} + O(h^2)$

(c) $\left.\dfrac{\partial^2 u}{\partial x^2}\right|_{i,j} = \dfrac{1}{h^2}(u_{1,0} - 2u_{0,0} + u_{-1,0}) + O(h^2)$

$$= \dfrac{1}{h^2}\left\{ \boxed{1} \!-\! \underset{i,j}{\boxed{-2}} \!-\! \boxed{1} \right\} + O(h^2)$$

(d) $\left.\dfrac{\partial^2 u}{\partial x\,\partial y}\right|_{i,j} = \dfrac{1}{4h^2}(u_{1,1} - u_{1,-1} - u_{-1,1} + u_{-1,-1}) + O(h^2)$

$$= \dfrac{1}{4h^2}\left\{ \begin{array}{ccc} \boxed{-1} & \boxed{0} & \boxed{1} \\ \boxed{0} & \underset{i,j}{\boxed{0}} & \boxed{0} \\ \boxed{1} & \boxed{0} & \boxed{-1} \end{array} \right\} + O(h^2)$$

(e) $\left. \nabla^2 u \right|_{i,j} = \dfrac{1}{h^2}(u_{1,0} + u_{0,1} + u_{-1.0} + u_{0,-1} - 4u_{00}) + O(h^2)$

$$= \dfrac{1}{h^2}\left\{ \begin{array}{ccc} & \boxed{1} & \\ \boxed{1} & \underset{i,j}{\boxed{-4}} & \boxed{1} \\ & \boxed{1} & \end{array} \right\} + O(h^2)$$

8.3.2.3   Numerical solution of differential equations

Numerical methods to solve differential equations depend on whether small changes in the statement of the problem cause small changes in the solution.

DEFINITION 8.3.1

The initial-value problem,

$$\frac{dy}{dt} = f(t, y), \quad a \le t \le b, \quad y(a) = \alpha, \tag{8.3.19}$$

is said to be well posed if

1. *A unique solution, $y(t)$, to the problem exists.*
2. *For any $\epsilon > 0$, there exists a positive constant $k(\epsilon)$ with the property that, whenever $|\epsilon_0| < \epsilon$ and $\delta(t)$ is continuous with $|\delta(t)| < \epsilon$ on $[a, b]$, a unique solution, $z(t)$, to the problem,*

$$\frac{dz}{dt} = f(t, z) + \delta(t), \quad a \le t \le b, \quad z(a) = \alpha + \epsilon_0,$$

exists and saisfies $|z(t) - y(t)| < k(\epsilon)\epsilon$, for all $a \le t \le b$.

This is called the *perturbed problem* associated with the original problem. Although other criteria exist, the following result gives conditions that are easy to check to guarantee that a problem is well posed.

THEOREM 8.3.1   *(Well posed condition)*

Suppose that f and f_y (its first partial derivative with respect to y) are continuous for t in $[a, b]$. Then the initial-value problem given by Equation (8.3.19) is well posed.

Using Taylor's theorem, numerical methods for solving the well posed, first-order differential equation given by Equation (8.3.19) can be derived. Using equally-spaced *mesh points* $t_i = a + ih$ (for $i = 0, 1, 2, \ldots, N$) and w_i to denote an approximation to $y_i \equiv y(t_i)$, then methods generally use *difference equations* of the form

$$w_0 = \alpha, \qquad w_{i+1} = w_i + h\phi(t_i, w_i),$$

for each $i = 0, 1, 2, \ldots, N - 1$. Here ϕ is a function depending on f. The difference method has *local truncation error* given by

$$\tau_{i+1}(h) = \frac{y_{i+1} - y_i}{h} - \phi(t_i, y_i),$$

for each $i = 0, 1, 2, \ldots, N - 1$. The following formulae are called *Taylor methods*. Each has local truncation error

$$\frac{h^n}{(n+1)!} f^{(n)}(\xi_i, y(\xi_i)) = \frac{h^n}{(n+1)!} y^{(n+1)}(\xi_i, y(\xi_i))$$

for each $i = 0, 1, 2, \ldots, N - 1$, where $\xi_i \in (t_i, t_{i+1})$. Thus, if $y \in C^{n+1}[a, b]$ the local truncation error is $O(h^n)$.

1. Euler's method ($n = 1$):

$$w_{i+1} = w_i + hf(t_i, w_i). \tag{8.3.20}$$

2. Taylor method of order n:

$$w_{i+1} = w_i + hT^{(n)}(t_i, w_i), \tag{8.3.21}$$

where $T^{(n)}(t_i, w_i) = f(t_i, w_i) + \frac{h}{2}f'(t_i, w_i) + \ldots + \frac{h^{n-1}}{n!}f^{(n-1)}(t_i, w_i)$.

The *Runge–Kutta methods* below are derived from the nth degree Taylor polynomial in two variables.

3. Midpoint method:

$$w_{i+1} = w_i + h\left[f\left(t_i + \frac{h}{2}, w_i + \frac{h}{2}f(t_i, w_i)\right)\right]. \tag{8.3.22}$$

If all second-order partial derivatives of f are bounded, this method has local truncation error $O(h^2)$, as do the following two methods.

4. Modified Euler method:

$$w_{i+1} = w_i + \frac{h}{2}\left\{f(t_i, w_i) + f[t_{i+1}, w_i + hf(t_i, w_i)]\right\}. \tag{8.3.23}$$

5. Heun's method:

$$w_{i+1} = w_i + \frac{h}{4}\left\{f(t_i, w_i) + 3f\left[t_i + \frac{2}{3}h, w_i + \frac{2}{3}hf(t_i, w_i)\right]\right\}.$$

6. Runge–Kutta method of order four:

$$w_{i+1} = w_i + \frac{1}{6}(k_1 + 2k_2 + 2k_3 + k_4), \tag{8.3.24}$$

where

$$k_1 = hf(t_i, w_i),$$
$$k_2 = hf\left(t_i + \frac{h}{2}, w_i + \frac{1}{2}k_1\right),$$
$$k_3 = hf\left(t_i + \frac{h}{2}, w_i + \frac{1}{2}k_2\right),$$
$$k_4 = hf(t_{i+1}, w_i + k_3).$$

The local truncation error is $O(h^4)$ if the solution $y(t)$ has five continuous derivatives.

8.3.2.4   Multistep methods and predictor-corrector methods

A *multistep method* is a technique whose difference equation to compute w_{i+1} involves more prior values than just w_i. An *explicit method* is one in which the computation of w_{i+1} does not depend on $f(t_{i+1}, w_{i+1})$ whereas an *implicit method* does involve $f(t_{i+1}, w_{i+1})$. For each formula, $i = n - 1, n, \ldots, N - 1$.

8.3.2.5 Adams–Bashforth n-step (explicit) methods

1. ($n = 2$):

    $w_0 = \alpha, w_1 = \alpha_1, w_{i+1} = w_i + \frac{h}{2}[3f(t_i, w_i) - f(t_{i-1}, w_{i-1})].$

    Local truncation error is $\tau_{i+1}(h) = \frac{5}{12}y^{(3)}(\mu_i)h^2$, for some $\mu_i \in (t_{i-1}, t_{i+1})$.

2. ($n = 3$):

    $w_0 = \alpha, w_1 = \alpha_1, w_2 = \alpha_2, w_{i+1} = w_i + \frac{h}{12}[23f(t_i, w_i) - 16f(t_{i-1}, w_{i-1}) + 5f(t_{i-2}, w_{i-2})].$

    Local truncation error is $\tau_{i+1}(h) = \frac{3}{8}y^{(4)}(\mu_i)h^3$, for some $\mu_i \in (t_{i-2}, t_{i+1})$.

3. ($n = 4$):

    $w_0 = \alpha, w_1 = \alpha_1, w_2 = \alpha_2, w_3 = \alpha_3, w_{i+1} = w_i + \frac{h}{24}[55f(t_i, w_i) - 59f(t_{i-1}, w_{i-1}) + 37f(t_{i-2}, w_{i-2}) - 9f(t_{i-3}, w_{i-3})].$

    Local truncation error is $\tau_{i+1}(h) = \frac{251}{720}y^{(5)}(\mu_i)h^4$, for some $\mu_i \in (t_{i-3}, t_{i+1})$.

4. ($n = 5$):

    $w_0 = \alpha, \ w_1 = \alpha_1, w_2 = \alpha_2, w_3 = \alpha_3, w_4 = \alpha_4, w_{i+1} = w_i + \frac{h}{720}[1901f(t_i, w_i) - 2774f(t_{i-1}, w_{i-1}) + 2616f(t_{i-2}, w_{i-2}) - 1274f(t_{i-3}, w_{i-3}) + 251f(t_{i-4}, w_{i-4})].$

    Local truncation error is $\tau_{i+1}(h) = \frac{95}{288}y^{(6)}(\mu_i)h^5$, for some $\mu_i \in (t_{i-4}, t_{i+1})$.

8.3.2.6 Adams–Moulton n-step (implicit) methods

1. ($n = 2$):

    $w_0 = \alpha, w_1 = \alpha_1, w_{i+1} = w_i + \frac{h}{12}[5f(t_{i+1}, w_{i+1}) + 8f(t_i, w_i) - f(t_{i-1}, w_{i-1})].$

    Local truncation error is $\tau_{i+1}(h) = -\frac{1}{24}y^{(4)}(\mu_i)h^3$, for some $\mu_i \in (t_{i-1}, t_{i+1})$.

2. ($n = 3$):

    $w_0 = \alpha, w_1 = \alpha_1, w_2 = \alpha_2, w_{i+1} = w_i + \frac{h}{24}[9f(t_{i+1}, w_{i+1}) + 19f(t_i, w_i) - 5f(t_{i-1}, w_{i-1}) + f(t_{i-2}, w_{i-2})].$

    Local truncation error is $\tau_{i+1}(h) = -\frac{19}{720}y^{(5)}(\mu_i)h^4$,

    for some $\mu_i \in (t_{i-2}, t_{i+1})$.

3. ($n = 4$):

    $w_0 = \alpha, w_1 = \alpha_1, w_2 = \alpha_2, w_3 = \alpha_3, w_{i+1} = w_i + \frac{h}{720}[251f(t_{i+1}, w_{i+1}) + 646f(t_i, w_i) - 264f(t_{i-1}, w_{i-1}) + 106f(t_{i-2}, w_{i-2}) - 19f(t_{i-3}, w_{i-3})].$

    Local truncation error is $\tau_{i+1}(h) = -\frac{3}{160}y^{(6)}(\mu_i)h^5$,

    for some $\mu_i \in (t_{i-3}, t_{i+1})$.

In practice, implicit methods are not used by themselves. They are used to improve approximations obtained by explicit methods. An explicit method *predicts* an approximation and the implicit method *corrects* this prediction. The combination is called a *predictor-corrector method.* For example, the Adams–Bashforth method with $n = 4$ might be used with the Adams–Moulton method with $n = 3$ since both have comparable errors. Initial values may be computed, say, by the Runge–Kutta method of order four, Equation (8.3.24).

8.3.2.7 Higher-order differential equations and systems

A system of m first-order initial-value problems can be expressed in the form

$$\frac{du_1}{dt} = f_1(t, u_1, u_2, \ldots, u_m), \qquad u_1(a) = \alpha_1,$$

$$\frac{du_2}{dt} = f_2(t, u_1, u_2, \ldots, u_m), \qquad u_2(a) = \alpha_2,$$

$$\vdots$$

$$\frac{du_m}{dt} = f_m(t, u_1, u_2, \ldots, u_m), \qquad u_m(a) = \alpha_m. \tag{8.3.25}$$

Generalizations of methods for solving first-order equations can be used to solve such systems. An example here uses the Runge-Kutta method of order four.

Partition $[a, b]$ as before, and let $w_{i,j}$ denote the approximation to $u_i(t_j)$ for $j = 0, 1, \ldots, N$ and $i = 1, 2, \ldots, m$. For the initial conditions, set $w_{1,0} = \alpha_1$, $w_{2,0} = \alpha_2, \ldots, w_{m,0} = \alpha_m$. From the values $\{w_{1,j}, w_{2,j}, \ldots, w_{m,j}\}$ previously computed, obtain $\{w_{1,j+1}, w_{2,j+1}, \ldots, w_{m,j+1}\}$ from

$$k_{1,i} = hf_i\left(t_j, w_{1,j}, w_{2,j}, \ldots, w_{m,j}\right),$$

$$k_{2,i} = hf_i\left(t_j + \frac{h}{2}, w_{1,j} + \frac{1}{2}k_{1,1}, w_{2,j} + \frac{1}{2}k_{1,2}, \ldots, w_{m,j} + \frac{1}{2}k_{1,m}\right),$$

$$k_{3,i} = hf_i\left(t_j + \frac{h}{2}, w_{1,j} + \frac{1}{2}k_{2,1}, w_{2,j} + \frac{1}{2}k_{2,2}, \ldots, w_{m,j} + \frac{1}{2}k_{2,m}\right),$$

$$k_{4,i} = hf_i\left(t_j + h, w_{1,j} + k_{3,1}, w_{2,j} + k_{3,2}, \ldots, w_{m,j} + k_{3,m}\right),$$

$$w_{i,j+1} = w_{i,j} + \frac{1}{6}[k_{1,i} + 2k_{2,i} + 2k_{3,i} + k_{4,i}], \tag{8.3.26}$$

where $i = 1, 2, \ldots, m$ for each of the above.

A differential equation of high order can be converted into a *system* of first-order equations. Suppose that a single differential equation has the form

$$y^{(m)} = f(t, y, y', y'', \ldots, y^{(m-1)}), \quad a \le t \le b \tag{8.3.27}$$

with initial conditions $y(a) = \alpha_1, y'(a) = \alpha_2, \ldots, y^{(m-1)}(a) = \alpha_m$. All derivatives are with respect to t. That is, $y^{(k)} = \frac{d^k y}{dt^k}$. Define $u_1(t) = y(t), u_2(t) = y'(t), \ldots,$ $u_m(t) = y^{(m-1)}(t)$. This yields first-order equations

$$\frac{du_1}{dt} = u_2, \quad \frac{du_2}{dt} = u_3, \quad \cdots \quad \frac{du_{m-1}}{dt} = u_m, \quad \frac{du_m}{dt} = f(t, u_1, u_2, \ldots, u_m),$$

$$\tag{8.3.28}$$

with initial conditions $u_1(a) = \alpha_1, \ldots, u_m(a) = \alpha_m$.

8.3.2.8 Partial differential equations

To develop difference equations for partial differential equations, one needs to estimate the partial derivatives of a function, say, $u(x, y)$. For example,

$$\frac{\partial u}{\partial x}(x, y) = \frac{u(x + h, y) - u(x, y)}{h} - \frac{h}{2} \frac{\partial^2 u(\xi, y)}{\partial x^2} \quad \text{for } \xi \in (x, x + h), \quad (8.3.29)$$

$$\frac{\partial^2 u}{\partial x^2}(x, y) = \frac{1}{h^2}[u(x + h, y) - 2u(x, y) + u(x - h, y)] - \frac{h^2}{12} \frac{\partial^4 u(\xi, y)}{\partial x^4}, \quad (8.3.30)$$

$$\text{for } \xi \in (x - h, x + h).$$

Notes:

1. Equation (8.3.29) is simply Equation (8.3.12) applied to estimate the partial derivative. It is given here to emphasize its application for forming difference equations for partial differential equations. A similar formula applies for $\partial u / \partial y$, and others could follow from the formulae in Section 8.3.2.1.

2. An estimate of $\partial^2 u / \partial y^2$ is similar. A formula for $\partial^2 u / \partial x \partial y$ could be given. However, in practice, a change of variables is generally used to eliminate this mixed second partial derivative from the problem.

If a partial differential equation involves partial derivatives with respect to only one of the variables, the methods described for ordinary differential equations can be used. If, however, the equation involves partial derivatives with respect to both variables, the approximation of the partial derivatives requires increments in both variables. The corresponding difference equations form a *system* of linear equations that must be solved.

Three specific forms of partial differential equations with popular methods of solution are given. The domains are assumed to be rectangular. Otherwise, additional considerations must be made for the boundary conditions.

8.3.2.9 Poisson equation

The Poisson equation is an elliptic partial differential equation that has the form

$$\nabla^2 u(x, y) = \frac{\partial^2 u}{\partial x^2}(x, y) + \frac{\partial^2 u}{\partial y^2}(x, y) = f(x, y) \quad (8.3.31)$$

for $(x, y) \in R = \{(x, y) \mid a < x < b, c < y < d\}$, with $u(x, y) = g(x, y)$ for $(x, y) \in S$, where $S = \partial R$. When the function $f(x, y) = 0$ the equation is called *Laplace's equation*.

To begin, partition $[a, b]$ and $[c, d]$ by choosing integers n and m, define step sizes $h = (b - a)/n$ and $k = (d - c)/m$, and set $x_i = a + ih$ for $i = 0, 1, \ldots, n$ and $y_j = c + jk$ for $j = 0, 1, \ldots, m$. The lines $x = x_i$, $y = y_j$, are called *grid lines* and their intersections are called *mesh points*. Estimates $w_{i,j}$ for $u(x_i, y_j)$ can be generated using Equation (8.3.30) to estimate $\frac{\partial^2 u}{\partial x^2}$ and $\frac{\partial^2 u}{\partial y^2}$. The method described here is called the *finite-difference method*.

Start with the values

$$w_{0,j} = g(x_0, y_j), \quad w_{n,j} = g(x_n, y_j), \quad w_{i,0} = g(x_i, y_0), \quad w_{i,m} = g(x_i, y_m),$$
$$(8.3.32)$$

and then solve the resulting system of linear algebraic equations

$$2\left[\left(\frac{h}{k}\right)^2 + 1\right] w_{i,j} - (w_{i+1,j} + w_{i-1,j}) - \left(\frac{h}{k}\right)^2 (w_{i,j+1} + w_{i,j-1}) = -h^2 f(x_i, y_j),$$
$$(8.3.33)$$

for $i = 1, 2, \ldots, n-1$ and $j = 1, 2, \ldots, m-1$. The local truncation error is $O(h^2 + k^2)$.

If the interior mesh points are labeled $P_\ell = (x_i, y_j)$ and $w_\ell = w_{i,j}$ where $\ell = i + (m-1-j)(n-1)$, for $i = 1, 2, \ldots, n-1$, and $j = 1, 2, \ldots, m-1$, then the two-dimensional array of values becomes a one-dimensional array. This results in a banded linear system. The case $n = m = 4$ yields $\ell = (n-1)(m-1) = 9$. Using the relabeled grid points, $f_\ell = f(P_\ell)$, the equations at the points P_i are

$$
\begin{array}{rrcl}
P_1: & 4w_1 - w_2 - w_4 & = & w_{0,3} + w_{1,4} - h^2 f_1, \\
P_2: & 4w_2 - w_3 - w_1 - w_5 & = & w_{2,4} - h^2 f_2, \\
P_3: & 4w_3 - w_2 - w_6 & = & w_{4,3} + w_{3,4} - h^2 f_3, \\
P_4: & 4w_4 - w_5 - w_1 - w_7 & = & w_{0,2} - h^2 f_4, \\
P_5: & 4w_5 - w_6 - w_4 - w_2 - w_8 & = & 0 - h^2 f_5, \\
P_6: & 4w_6 - w_5 - w_3 - w_9 & = & w_{4,2} - h^2 f_6, \\
P_7: & 4w_7 - w_8 - w_4 & = & w_{0,1} + w_{1,0} - h^2 f_7, \\
P_8: & 4w_8 - w_9 - w_7 - w_5 & = & w_{2,0} - h^2 f_8, \\
P_9: & 4w_9 - w_8 - w_6 & = & w_{3,0} + w_{4,1} - h^2 f_9,
\end{array}
$$

where the right-hand sides of the equations are obtained from the boundary conditions.

The following algorithm can be used to solve the Poisson equation. Note that, for simplicity, the algorithm incorporates an iterative procedure called Gauss–Seidel for solving linear systems. Instead, Gaussian elimination is recommended (because stability with respect or round-off errors is assured) when the order is small (say, less than 100). For large systems, the SOR (Successive Over-Relaxation) method is recommended. The Gauss–Seidel and SOR methods can be found in Burden and Faires.

8.3.2.10   Poisson equation finite-difference algorithm

To approximate the solution to the Poisson equation

$$\nabla^2 u(x, y) = \frac{\partial^2 u}{\partial x^2}(x, y) + \frac{\partial^2 u}{\partial y^2}(x, y) = f(x, y),$$
$$(8.3.34)$$

for $a \leq x \leq b$ and $c \leq y \leq d$,

subject to $u(x, y) = g(x, y)$ if $x = a$ or $x = b$ and $c \leq y \leq d$ and $u(x, y) = g(x, y)$ if $y = c$ or $y = d$ and $a \leq x \leq b$:

INPUT endpoints a, b, c, d; integers $m \geq 3$, $n \geq 3$; tolerance TOL; maximum number of iterations N.

OUTPUT approximations $w_{i,j}$ to $u(x_i, y_j)$ for $i = 1, \ldots, n-1$ and $j = 1, \ldots, m-1$ or a message that N was exceeded.

Algorithm:

1. Set $h = (b-a)/n$; $k = (d-c)/m$.

2. For $i = 1, \ldots, n-1$ set $x_i = a + ih$.

3. For $j = 1, \ldots, m-1$ set $y_j = c + jk$.

4. For $i = 1, \ldots, n-1$

   for $j = 1, \ldots, m-1$ set $w_{i,j} = 0$.

5. Set $\lambda = h^2/k^2$; $\mu = 2(1+\lambda)$; $\ell = 1$.

6. While $\ell \leq N$ do (a)–(i)

   (a) Set
   $$z = \left(-h^2 f(x_1, y_{m-1}) + g(a, y_{m-1}) + \lambda g(x_1, d) \right.$$
   $$\left. +\lambda w_{1,m-2} + w_{2,m-1} \right)/\mu;$$
   NORM $= |z - w_{1,m-1}|$; $w_{1,m-1} = z$.

   (b) For $i = 2, \ldots, n-2$
   set $z = \left(-h^2 f(x_i, y_{m-1}) + \lambda g(x_i, d) + w_{i-1,m-1} \right.$
   $$\left. +w_{i+1,m-1} + \lambda w_{i,m-2} \right)/\mu;$$
   if $|w_{i,m-1} - z| >$ NORM then set NORM $= |w_{i,m-1} - z|$;
   set $w_{i,m-1} = z$.

   (c) Set $z = \left(-h^2 f(x_{n-1}, y_{m-1}) + g(b, y_{m-1}) + \lambda g(x_{n-1}, d) \right.$
   $$\left. +w_{n-2,m-1} + \lambda w_{n-1,m-2} \right)/\mu;$$
   if $|w_{n-1,m-1} - z| >$ NORM then set NORM $= |w_{n-1,m-1} - z|$;
   set $w_{n-1,m-1} = z$.

   (d) For $j = m-2, \ldots, 2$ do i–iii.

   i. Set $z = \left(-h^2 f(x_1, y_j) + g(a, y_j) + \lambda w_{1,j+1} + \lambda w_{1,j-1} \right.$
   $$\left. +w_{2,j} \right)/\mu;$$
   if $|w_{1,j} - z| >$ NORM then set NORM $= |w_{1,j} - z|$;
   set $w_{1,j} = z$.

   ii. For $i = 2, \ldots, n-2$
   set $z = \left(-h^2 f(x_i, y_j) + w_{i-1,j} + \lambda w_{i,j+1} + w_{i+1,j} \right.$
   $$\left. +\lambda w_{i,j-1} \right)/\mu;$$
   if $|w_{i,j} - z| >$ NORM then set NORM $= |w_{i,j} - z|$;
   set $w_{i,j} = z$.

   iii. Set $z = \left(-h^2 f(x_{n-1}, y_j) + g(b, y_j) + w_{n-2,j} + \lambda w_{n-1,j+1} \right.$
   $$\left. +\lambda w_{n-1,j-1} \right)/\mu;$$
   if $|w_{n-1,j} - z| >$ NORM then set NORM $= |w_{n-1,j} - z|$;
   set $w_{n-1,j} = z$.

   (e) Set $z = \left(-h^2 f(x_1, y_1) + g(a, y_1) + \lambda g(x_1, c) + \lambda w_{1,2} \right.$
   $$\left. +w_{2,1} \right)/\mu;$$
   if $|w_{1,1} - z| >$ NORM then set NORM $= |w_{1,1} - z|$;
   set $w_{1,1} = z$.

(f) For $i = 2, \ldots, n - 2$

set $z = \big(- h^2 f(x_i, y_1) + \lambda g(x_i, c) + w_{i-1,1} + \lambda w_{i,2}$

$\qquad\qquad + w_{i+1,1}\big)/\mu;$

if $|w_{i,1} - z| >$ NORM then set NORM $= |w_{i,1} - z|;$

set $w_{i,1} = z.$

(g) Set $z = \big(- h^2 f(x_{n-1}, y_1) + g(b, y_1) + \lambda g(x_{n-1}, c) + w_{n-2,1}$

$\qquad\qquad + \lambda w_{n-1,2}\big)/\mu;$

if $|w_{n-1,1} - z| >$ NORM then set NORM $= |w_{n-1,1} - z|;$

set $w_{n-1,1} = z.$

(h) If NORM $\leq$ TOL then do i–ii.

    i. For $i = 1, \ldots, n - 1$

        for $j = 1, \ldots, m - 1$ OUTPUT$(x_i, y_j, w_{i,j}).$

    ii. STOP. (Procedure successful.)

(i) Set $\ell = \ell + 1.$

7. OUTPUT('Maximum number of iterations exceeded.');

(Procedure unsuccessful.)

STOP.

8.3.2.11   Heat or diffusion equation

The heat, or diffusion, equation is a parabolic partial differential equation of the form

$$\frac{\partial u}{\partial t}(x, t) = \alpha^2 \frac{\partial^2 u}{\partial x^2}(x, t), \quad 0 < x < \ell, \quad t > 0, \qquad (8.3.35)$$

where $u(0, t) = 0 = u(\ell, t) = 0$, for $t > 0$, and $u(x, 0) = f(x)$, for $0 \leq x \leq \ell$. An efficient method for solving this type of equation is the *Crank–Nicolson method*.

To apply the method, select an integer $m > 0$, set $h = \ell/m$, and select a time-step size k. Here $x_i = ih$, $i = 0, \ldots, m$ and $t_j = jk$, $j = 0, \ldots.$ The difference equation is given by:

$$\frac{w_{i,j+1} - w_{i,j}}{k}$$

$$- \frac{\alpha^2}{2}\left[\frac{w_{i+1,j} - 2w_{i,j} + w_{i-1,j}}{h^2} + \frac{w_{i+1,j+1} - 2w_{i,j+1} + w_{i-1,j+1}}{h^2}\right] = 0$$

$$(8.3.36)$$

and has local truncation error $O(k^2 + h^2)$. The difference equations can be represented in the matrix form $A\mathbf{w}^{(j+1)} = B\mathbf{w}^{(j)}$, for each $j = 0, 1, 2, \ldots$, where

$\lambda = \alpha^2 k / h^2$, $\mathbf{w}^{(j)} = (w_{1,j}, w_{2,j}, \ldots, w_{m-1,j})^{\mathrm{T}}$, and the matrices A and B are

$$
A = \begin{bmatrix}
(1+\lambda) & -\lambda/2 & 0 & 0 & \cdots & 0 & 0 \\
-\lambda/2 & (1+\lambda) & -\lambda/2 & 0 & \cdots & 0 & 0 \\
0 & -\lambda/2 & (1+\lambda) & -\lambda/2 & & 0 & 0 \\
0 & 0 & -\lambda/2 & (1+\lambda) & & 0 & 0 \\
\vdots & \vdots & & & \ddots & & \\
0 & 0 & 0 & 0 & & (1+\lambda) & -\lambda/2 \\
0 & 0 & 0 & 0 & & -\lambda/2 & (1+\lambda)
\end{bmatrix}
$$

(8.3.37)

$$
B = \begin{bmatrix}
(1-\lambda) & \lambda/2 & 0 & 0 & \cdots & 0 & 0 \\
\lambda/2 & (1-\lambda) & \lambda/2 & 0 & \cdots & 0 & 0 \\
0 & \lambda/2 & (1-\lambda) & \lambda/2 & & 0 & 0 \\
0 & 0 & \lambda/2 & (1-\lambda) & & 0 & 0 \\
\vdots & \vdots & & & \ddots & & \\
0 & 0 & 0 & 0 & & (1-\lambda) & \lambda/2 \\
0 & 0 & 0 & 0 & & \lambda/2 & (1-\lambda)
\end{bmatrix}.
$$

8.3.2.12   Crank–Nicolson algorithm

To approximate the solution to the parabolic partial differential equation

$$
\frac{\partial u}{\partial t}(x,t) - \alpha^2 \frac{\partial^2 u}{\partial x^2}(x,t) = 0, \qquad 0 < x < \ell, \qquad 0 < t < T,
$$

subject to boundary conditions $u(0,t) = u(\ell,t) = 0$ for $0 < t < T$, and the initial conditions $u(x,0) = f(x)$ for $0 \le x \le \ell$.

INPUT endpoint ℓ; maximum time T; constant α;
      integers $m \ge 3$; $N \ge 1$.
OUTPUT approximations $w_{i,j}$ to $u(x_i, t_j)$ for $i = 1, \ldots, m-1$, and
      $j = 1, \ldots, N$.
Algorithm:

1. Set $h = \ell/m$; $k = T/N$; $\lambda = \alpha^2 k/h^2$; $w_m = 0$.

2. For $i = 1, \ldots, m-1$ set $w_i = f(ih)$.

3. Set $\ell_1 = 1 + \lambda$; $u_1 = -\lambda/(2\ell_1)$.

4. For $i = 2, \ldots, m-2$
      set $\ell_i = 1 + \lambda + \lambda u_{i-1}/2$; $u_i = -\lambda/(2\ell_i)$.

5. Set $\ell_{m-1} = 1 + \lambda + \lambda u_{m-2}/2$.

6. For $j = 1, \ldots, N$ do (a)–(e).

   (a) Set $t = jk$; $z_1 = \left[(1-\lambda)w_1 + \frac{\lambda}{2}w_2 \right]/\ell_1$.

   (b) For $i = 2, \ldots, m-1$
        set $z_i = \left[(1-\lambda)w_i + \frac{\lambda}{2}(w_{i+1} + w_{i-1} + z_{i-1}) \right]/\ell_i$.

   (c) Set $w_{m-1} = z_{m-1}$.

   (d) For $i = m-2, \ldots, 1$ set $w_i = z_i - u_i w_{i+1}$.

(e) OUTPUT(t);   (Note: $t = t_j$.)
　　For $i = 1, \ldots, m - 1$ set $x = ih$; OUTPUT(x, w_i).
　　(Note: $w_i = w_{i,j}$.)

7. STOP. (Procedure completed.)

8.3.2.13 Wave equation

The wave equation is an example of a hyperbolic partial differential equation and has the form

$$\frac{\partial^2 u}{\partial t^2}(x, t) - \alpha^2 \frac{\partial^2 u}{\partial x^2}(x, t) = 0, \quad 0 < x < \ell, \quad t > 0 \qquad (8.3.38)$$

(where α is a constant) subject to $u(0, t) = u(\ell, t) = 0$ for $t > 0$, and $u(x, 0) = f(x)$ and $\frac{\partial u}{\partial t}(x, 0) = g(x)$ for $0 \le x \le \ell$.

Select an integer $m > 0$, time-step size $k > 0$, and using $h = \ell/m$, mesh points (x_i, t_j) are defined by $x_i = ih$ and $t_j = jk$. Using $w_{i,j}$ to represent an approximation of $u(x_i, t_j)$ and $\lambda = \alpha k/h$, the difference equation becomes

$$w_{i,j+1} = 2(1 - \lambda^2)w_{i,j} + \lambda^2(w_{i+1,j} + w_{i-1}, j) - w_{i,j-1},$$

with $w_{0,j} = w_{m,j} = 0$ and $w_{i,0} = f(x_i)$, for $i = 1, \ldots, m - 1$ and $j = 1, 2, \ldots$. Also needed is an estimate for $w_{i,1}$, for each $i = 1, \ldots, m - 1$, which can be written

$$w_{i,1} = (1 - \lambda^2)f(x_i) + \frac{\lambda^2}{2}f(x_{i+1}) + \frac{\lambda^2}{2}f(x_{i-1}) + kg(x_i).$$

The local truncation error of the method is $O(h^2 + k^2)$ but the method is extremely accurate if the true solution is infinitely differentiable. For the method to be stable, it is necessary that $\lambda \le 1$. The following algorithm, applied with $\lambda \le 1$, is $O(h^2 + k^2)$ convergent if f and g are sufficiently differentiable.

8.3.2.14 Wave equation algorithm

To approximate the solution to the wave equation

$$\frac{\partial^2 u}{\partial t^2}(x, t) - \alpha^2 \frac{\partial^2 u}{\partial x^2}(x, t) = 0, \quad 0 < x < \ell, \quad 0 < t < T,$$

subject to $u(0, t) = u(\ell, t) = 0$ for $0 < t < T$, $u(x, 0) = f(x)$ for $0 \le x \le \ell$, and $\frac{\partial u}{\partial t}(x, 0) = g(x)$ for $0 \le x \le \ell$.

INPUT endpoint ℓ; maximum time T; constant α; integers $m \ge 2$; $N \ge 2$.
OUTPUT approximations $w_{i,j}$ to $u(x_i, t_j)$, $i = 0, \ldots, m, j = 0, \ldots, N$.
Algorithm:

1. Set $h = \ell/m$; $k = T/N$; $\lambda = k\alpha/h$.

2. For $j = 1, \ldots, N$ set $w_{0,j} = 0$; $w_{m,j} = 0$.

3. Set $w_{0,0} = f(0)$; $w_{m,0} = f(\ell)$.

4. For $i = 1, \ldots, m - 1$

   set $w_{i,0} = f(ih)$;

   $$w_{i,1} = (1 - \lambda^2)f(ih) + \frac{\lambda^2}{2}[f((i+1)h)$$
   $$+ f((i-1)h)] + kg(ih).$$

5. For $j = 1, \ldots, N - 1$ (Perform matrix multiplication.)

   for $i = 1, \ldots, m - 1$ set

   $$w_{i,j+1} = 2(1 - \lambda^2)w_{i,j} + \lambda^2(w_{i+1,j} + w_{i-1,j}) + w_{i,j-1}.$$

6. For $j = 1, \ldots, N$

   set $t = jk$;

   for $i = 0, \ldots, m$

   set $x = ih$; OUTPUT($x, t, w_{i,j}$).

7. STOP. (Procedure completed.)

8.3.3 NUMERICAL SUMMATION

A sum of the form $\sum_{j=0}^{n} f(x_0 + jh)$ (n may be infinite) can be approximated by the *Euler–MacLaurin sum formula*,

$$\sum_{j=0}^{n} f(x_0 + jh) = \frac{1}{h} \int_{x_0}^{x_0+nh} f(y)\, dy + \frac{1}{2}[f(x_0 + nh) + f(x_0)]$$

$$+ \sum_{k=1}^{m} \frac{B_{2k}}{(2k)!} h^{2k-1}[f^{(2k-1)}(x_0 + nh) - f^{(2k-1)}(x_0)] + E_m \quad (8.3.39)$$

where $E_m = \frac{nh^{2m+2}B_{2m+2}}{(2m+2)!} f^{(2m+2)}(\xi)$, with $x_0 < \xi < x_0 + nh$. The B_n here are *Bernoulli numbers* (see Section 1.2.7).

The above formula is useful even when n is infinite, although the error can no longer be expressed in this form. A useful error estimate (which also holds when n is finite) is that the error is less than the magnitude of the first neglected term in the summation on the second line of Equation (8.3.39) if $f^{(2m+2)}(x)$ and $f^{(2m+4)}(x)$ do not change sign and are of the same sign for $x_0 < x < x_0 + nh$. If just $f^{(2m+2)}(x)$ does not change sign in the interval, then the error is less than twice the first neglected term.

Quadrature formulae result from Equation (8.3.39) using estimates for the derivatives.

8.4 PROGRAMMING TECHNIQUES

Efficiency and accuracy are the ultimate goal when solving any problem. Listed here are several suggestions to consider when developing algorithms and computer programs.

1. *Every algorithm must have an effective stopping rule.* For example, popular stopping rules for iteration methods described in Section 8.1.2 are based on the estimate of the absolute error, relative error, or function value. One might choose to stop when a combination of the following conditions are satisfied:

$$|p_n - p_{n-1}| < \epsilon_1, \quad \frac{|p_n - p_{n-1}|}{|p_n|} < \epsilon_2, \quad |f(p_n)| < \epsilon_3,$$

where each ϵ_i represents a prescribed tolerance. However, since some iterations are not guaranteed to converge, or converge very slowly, it is *recommended* that an *upper bound*, say N, *is specified for the number of iterations* to be performed (see algorithm on page 742). This will avoid infinite loops.

2. *Avoid the use of arrays whenever possible.* Subscripted values often do not require the use of an array. For example, in Newton's method (see page 731) the calculations may be performed using $p = p_0 - \frac{f(p_0)}{f'(p_0)}$. Then check the stopping rule, say, if $|p - p_0| < \epsilon$, and update the current value by setting $p_0 = p$ before computing the next value of the sequence.

3. *Limit the use of arrays when forming tables.* A two-dimensional array can often be avoided. For example, a divided difference table can be formed and printed as a lower triangular matrix. The entries of any row depend only on the entries of the preceding row. Thus, one-dimensional arrays may be used to save the preceding row and the current row being calculated. It is important to note that usually the *entire array need not be saved.* For example, only special values in the table are needed for the coefficients of an interpolating polynomial.

4. *Avoid using formulae that may be highly susceptible to round off error.* Exercise caution when computing quotients of extremely small values as in Equation (8.3.12) with a very small value of h.

5. *Alter formulae* for iterations to obtain a "small correction" to an approximation. For example, writing $\frac{a+b}{2}$ as $a + \frac{b-a}{2}$ in the bisection method (see page 732) is recommended. Many of the iteration formulae in this chapter have this form.

6. *Pivoting strategies* are recommended when solving linear systems to reduce round-off error.

7. *Eliminate unnecessary steps* that may increase execution time or round-off error.

8. Some methods converge very rapidly, when they do converge, but rely on reasonably close initial approximations. A weaker, but reliable, method (such as the bisection method) to obtain such an approximation can be combined with a more powerful method (such as Newton's method). The weaker method might converge slowly and, by itself, is not very efficient. The powerful method might not converge at all. The combination, however, might remedy both difficulties.

Chapter 9

Financial Analysis

9.1 FINANCIAL FORMULAE 779
  9.1.1 Definition of financial terms 779
  9.1.2 Formulae connecting financial terms 780
  9.1.3 Examples 781

9.2 FINANCIAL TABLES 783
  9.2.1 Compound interest: find final value 783
  9.2.2 Compound interest: find interest rate 785
  9.2.3 Compound interest: find annuity 787

9.1 FINANCIAL FORMULAE

9.1.1 DEFINITION OF FINANCIAL TERMS

A amount that P is worth, after n time periods, with i percent interest per period
B total amount borrowed
P principal to be invested (equivalently, present value)
a future value multiplier after one time period
i percent interest per time period (expressed as a decimal)
m amount to be paid each time period
n number of time periods

Note that the units of A, B, P, and m must all be the same, for example, dollars.

1-58488-291-3/02/$0.00+$1.50
© 2003 CRC Press, Inc.

9.1.2 FORMULAE CONNECTING FINANCIAL TERMS

1. **Interest**: Let the principal amount P be invested at an interest rate of $i\%$ per time period (expressed as a decimal), for n time periods. Let A be the amount that this is worth after n time periods. Then

   (a) *Simple interest*:

   $$A = P(1 + ni) \quad \text{and} \quad P = \frac{A}{(1 + ni)} \quad \text{and} \quad i = \frac{1}{n}\left(\frac{A}{P} - 1\right).$$

   $$(9.1.1)$$

   (b) *Compound interest* (see the tables beginning on page 783 for A and the tables beginning on page 785 for i):

   $$A = P(1 + i)^n \quad \text{and} \quad P = \frac{A}{(1 + i)^n} \quad \text{and} \quad i = \left(\frac{A}{P}\right)^{1/n} - 1.$$

   $$(9.1.2)$$

   When interest is compounded q times per time period for n time periods, it is equivalent to an interest rate of $(i/q)\%$ per time period for nq time periods.

   $$A = P\left(1 + \frac{i}{q}\right)^{nq},$$

   $$P = A\left(1 + \frac{i}{q}\right)^{-nq}, \qquad (9.1.3)$$

   $$i = q\left[\left(\frac{A}{P}\right)^{1/nq} - 1\right].$$

   Continuous compounding occurs when the interest is compounded infinitely often in each time period (i.e., $q \to \infty$). In this case: $A = Pe^{in}$.

2. **Present value**: If A is to be received after n time periods of $i\%$ interest per time period, then the present value P of such an investment is given by (from Equation (9.1.2)) $P = A(1 + i)^{-n}$.

3. **Annuities**: Suppose that the amount B (in dollars) is borrowed, at a rate of $i\%$ per time period, to be repaid at a rate of m (in dollars) per time period, for a total of n time periods. Then (see the tables beginning on page 787):

   $$m = Bi\frac{(1 + i)^n}{(1 + i)^n - 1}, \qquad (9.1.4)$$

   $$B = \frac{m}{i}\left(1 - \frac{1}{(1 + i)^n}\right). \qquad (9.1.5)$$

   Using $a = (1 + i)$, these equations can be written more compactly as

   $$m = Bi\frac{a^n}{a^n - 1} \quad \text{and} \quad B = \frac{m}{i}\left(1 - \frac{1}{a^n}\right). \qquad (9.1.6)$$

9.1.3   EXAMPLES

1. **Question**: If $100 is invested at 5% per year, compounded annually for 10 years, what is the resulting amount?

   - **Analysis**: Using Equation (9.1.2), we identify
     - (a) Principal invested, $P = 100$ (the units are dollars)
     - (b) Time period, 1 year
     - (c) Interest rate per time period, $i = 5\% = 0.05$
     - (d) Number of time periods, $n = 10$

   - **Answer**: $A = P(1+i)^n$ or $A = 100(1+0.05)^{10} = \$162.89$.
     (Or, see tables starting on page 783.)

2. **Question**: If $100 is invested at 5% per year and the interest is compounded quarterly (4 times a year) for 10 years, what is the final amount?

   - **Analysis**: Using Equation (9.1.3) we identify
     - (a) Principal invested, $P = 100$ (the units are dollars)
     - (b) Time period, 1 year
     - (c) Interest rate per time period, $i = 5\% = 0.05$
     - (d) Number of time periods, $n = 10$
     - (e) Number of compounding time periods, $q = 4$

   - **Answer**: $A = P\left(1 + \frac{i}{q}\right)^{nq}$ or $A = 100(1 + \frac{0.05}{4})^{4 \cdot 10} = 100(1.0125)^{40}$
     $= \$164.36$.

   - **Alternate analysis**: Using Equation (9.1.2), we identify
     - (a) Principal invested, $P = 100$ (the units are dollars)
     - (b) Time period, quarter of a year
     - (c) Interest rate per time period, $i = \frac{5\%}{4} = \frac{0.05}{4} = 0.0125$
     - (d) Number of time periods, $n = 10 \cdot 4 = 40$

   - **Alternate answer**: $A = P(1+i)^n$ or $A = 100(1.0125)^{40} = \$164.36$.
     (Or, see tables starting on page 783.)

3. **Question**: If $100 is invested now, and we wish to have $200 at the end of 10 years, what yearly compound interest rate must we receive?

   - **Analysis**: Using Equation (9.1.2), we identify
     - (a) Principal invested, $P = 100$ (the units are dollars)
     - (b) Final amount, $A = 200$
     - (c) Time period, 1 year
     - (d) Number of time periods, $n = 10$

- **Answer**: $i = \left(\frac{A}{P}\right)^{1/n} - 1$ or $i = \left(\frac{200}{100}\right)^{1/10} - 1 = 0.0718$. (Or, see tables starting on page 785.) Hence, we must receive an annual interest rate of 7.2%.

4. **Question**: An investment returns $10,000 in 10 years time. If the interest rate will be 10% per year, what is the present value? (That is, how much money would have to be invested now to obtain this amount in ten years?)

   - **Analysis**: Using Equation (9.1.2), we identify
     - (a) Final amount, $A = 10,000$ (the units are dollars)
     - (b) Time period, 10 years
     - (c) Interest rate per time period, $i = 10\% = 0.1$
     - (d) Number of time periods, $n = 10$

   - **Answer**: $P = A(1+i)^{-n} = 10000(1.1)^{-10} = 3855.43$; the present value of this investment is $3,855.43$. (Or, the table on page 784 gives the value 2.5937; the present value of this investment is then $\frac{\$10,000}{2.5937} = 3,855.43$).

5. **Question**: A mortgage of $100,000 is obtained with which to buy a house. The mortgage will be repaid at an interest rate of 9% per year, compounded monthly, for 30 years. What is the monthly payment?

   - **Analysis**: Using Equation (9.1.6), we identify
     - (a) Amount borrowed, $B = 100,000$ (the units are dollars)
     - (b) Time period, 1 month
     - (c) Interest rate per time period, $i = 0.09/12 = 0.0075$
     - (d) Number of time periods, $n = 30 \cdot 12 = 360$

   - **Answer**: $a = 1+i = 1.0075$ and $m = Bi\frac{a^n}{a^n-1} = (100,000)(.0075)$
     $\times \frac{(1.0075)^{360}}{(1.0075)^{360}-1} = 804.62$. (Or, see tables starting on page 787.) The monthly payment is $804.62.

6. **Question**: Suppose that interest rates on 15-year mortgages are currently 6%, compounded monthly. By spending $800 per month, what is the largest mortgage obtainable?

   - **Analysis**: Using Equation (9.1.6), we identify
     - (a) Time period, 1 month
     - (b) Payment amount, $m = 800$ (the units are dollars)
     - (c) Interest rate per time period, $i = 0.06/12 = 0.005$
     - (d) Number of time periods, $n = 15 \cdot 12 = 180$

   - **Answer**: $a = 1+i = 1.005$ and $B = \frac{m}{i}\left(1 - \frac{1}{a^n}\right) = \frac{800}{0.005}\left(1 - \frac{1}{1.005^{180}}\right)$
     $= 94802.81$. (Or, see tables starting on page 787.) The largest mortgage amount obtainable is $94,802.81.

9.2 FINANCIAL TABLES

9.2.1 COMPOUND INTEREST: FIND FINAL VALUE

These tables use Equation (9.1.2) to determine the final value in dollars (A) when one dollar ($P = 1$) is invested at an interest rate of i per time period, the length of investment time being n time periods. For example, if $1 is invested at a return of 3% per time period, for 60 time periods, then the final value would be $5.89 (see the following table). Analogously, if $10 had been invested, then the final value would be $59.92.

n	Interest rate (i)					
	1.00%	1.50%	2.00%	2.50%	3.00%	3.50%
2	1.0201	1.0302	1.0404	1.0506	1.0609	1.0712
4	1.0406	1.0614	1.0824	1.1038	1.1255	1.1475
6	1.0615	1.0934	1.1262	1.1597	1.1941	1.2293
8	1.0829	1.1265	1.1717	1.2184	1.2668	1.3168
10	1.1046	1.1605	1.2190	1.2801	1.3439	1.4106
12	1.1268	1.1956	1.2682	1.3449	1.4258	1.5111
20	1.2202	1.3469	1.4860	1.6386	1.8061	1.9898
24	1.2697	1.4295	1.6084	1.8087	2.0328	2.2833
36	1.4308	1.7091	2.0399	2.4325	2.8983	3.4503
48	1.6122	2.0435	2.5871	3.2715	4.1322	5.2136
60	1.8167	2.4432	3.2810	4.3998	5.8916	7.8781
72	2.0471	2.9212	4.1611	5.9172	8.4000	11.9043

n	Interest rate (i)					
	4.00%	4.50%	5.00%	5.50%	6.00%	6.50%
2	1.0816	1.0920	1.1025	1.1130	1.1236	1.1342
4	1.1699	1.1925	1.2155	1.2388	1.2625	1.2865
6	1.2653	1.3023	1.3401	1.3788	1.4185	1.4591
8	1.3686	1.4221	1.4775	1.5347	1.5938	1.6550
10	1.4802	1.5530	1.6289	1.7081	1.7909	1.8771
12	1.6010	1.6959	1.7959	1.9012	2.0122	2.1291
20	2.1911	2.4117	2.6533	2.9178	3.2071	3.5236
24	2.5633	2.8760	3.2251	3.6146	4.0489	4.5331
36	4.1039	4.8774	5.7918	6.8721	8.1472	9.6513
48	6.5705	8.2715	10.4013	13.0653	16.3939	20.5485
60	10.5196	14.0274	18.6792	24.8398	32.9877	43.7498
72	16.8423	23.7888	33.5451	47.2256	66.3777	93.1476

n	Interest rate (i)					
	7.00%	7.50%	8.00%	8.50%	9.00%	9.50%
2	1.1449	1.1556	1.1664	1.1772	1.1881	1.1990
4	1.3108	1.3355	1.3605	1.3859	1.4116	1.4377
6	1.5007	1.5433	1.5869	1.6315	1.6771	1.7238
8	1.7182	1.7835	1.8509	1.9206	1.9926	2.0669
10	1.9671	2.0610	2.1589	2.2610	2.3674	2.4782
12	2.2522	2.3818	2.5182	2.6617	2.8127	2.9715
20	3.8697	4.2478	4.6610	5.1120	5.6044	6.1416
24	5.0724	5.6729	6.3412	7.0846	7.9111	8.8296
36	11.4239	13.5115	15.9682	18.8569	22.2512	26.2366
48	25.7289	32.1815	40.2106	50.1912	62.5852	77.9611
60	57.9464	76.6492	101.2571	133.5932	176.0313	231.6579
72	130.5065	182.5616	254.9825	355.5831	495.1170	688.3615

n	Interest rate (i)					
	10.00%	10.50%	11.00%	11.50%	12.00%	12.50%
2	1.2100	1.2210	1.2321	1.2432	1.2544	1.2656
4	1.4641	1.4909	1.5181	1.5456	1.5735	1.6018
6	1.7716	1.8204	1.8704	1.9215	1.9738	2.0273
8	2.1436	2.2228	2.3045	2.3889	2.4760	2.5658
10	2.5937	2.7141	2.8394	2.9699	3.1059	3.2473
12	3.1384	3.3140	3.4985	3.6923	3.8960	4.1099
20	6.7275	7.3662	8.0623	8.8206	9.6463	10.5451
24	9.8497	10.9823	12.2392	13.6332	15.1786	16.8912
36	30.9127	36.3950	42.8181	50.3379	59.1356	69.4210
48	97.0172	120.6117	149.7970	185.8633	230.3908	285.3127
60	304.4816	399.7023	524.0572	686.2653	897.5969	1172.6039
72	955.5938	1324.5978	1833.3884	2533.9057	3497.0161	4819.2740

n	Interest rate (i)					
	13.00%	13.50%	14.00%	14.50%	15.00%	15.50%
2	1.2769	1.2882	1.2996	1.3110	1.3225	1.3340
4	1.6305	1.6595	1.6890	1.7188	1.7490	1.7796
6	2.0819	2.1378	2.1950	2.2534	2.3131	2.3741
8	2.6584	2.7540	2.8526	2.9542	3.0590	3.1671
10	3.3946	3.5478	3.7072	3.8731	4.0456	4.2249
12	4.3345	4.5704	4.8179	5.0777	5.3502	5.6362
20	11.5231	12.5869	13.7435	15.0006	16.3665	17.8501
24	18.7881	20.8882	23.2122	25.7829	28.6252	31.7664
36	81.4374	95.4665	111.8342	130.9174	153.1518	179.0406
48	352.9923	436.3162	538.8066	664.7577	819.4007	1009.1024
60	1530.0535	1994.1218	2595.9187	3375.4307	4383.9987	5687.4691

9.2.2 COMPOUND INTEREST: FIND INTEREST RATE

These tables use Equation (9.1.2) to determine the compound interest rate i that must be obtained from an investment of one dollar ($P = 1$) to yield a final value of A (in dollars) when the initial amount is invested for n time periods. For example, if $1 is invested for 60 time periods, and the final amount obtained is $4.00, then the actual interest rate has been 2.34% per time period (see the following table). Analogously, if $100 had been invested, and the final amount was $400, then the interest rate would also be 2.34% per time period.

n	Annuity (A)					
	2.0	2.5	3.0	3.5	4.0	4.5
1	100.00	150.00	200.00	250.00	300.00	350.00
2	41.42	58.11	73.20	87.08	100.00	112.13
3	25.99	35.72	44.23	51.83	58.74	65.10
4	18.92	25.74	31.61	36.78	41.42	45.65
5	14.87	20.11	24.57	28.47	31.95	35.10
10	7.18	9.60	11.61	13.35	14.87	16.23
12	5.95	7.93	9.59	11.00	12.25	13.35
20	3.53	4.69	5.65	6.46	7.18	7.81
24	2.93	3.89	4.68	5.36	5.95	6.47
36	1.94	2.58	3.10	3.54	3.93	4.27
48	1.46	1.93	2.31	2.64	2.93	3.18
60	1.16	1.54	1.85	2.11	2.34	2.54
72	0.97	1.28	1.54	1.75	1.94	2.11

n	Annuity (A)					
	5.0	5.5	6.0	6.5	7.0	7.5
1	400.00	450.00	500.00	550.00	600.00	650.00
2	123.61	134.52	144.95	154.95	164.57	173.86
3	71.00	76.52	81.71	86.63	91.29	95.74
4	49.53	53.14	56.51	59.67	62.66	65.49
5	37.97	40.63	43.10	45.41	47.58	49.63
10	17.46	18.59	19.62	20.58	21.48	22.32
12	14.35	15.27	16.10	16.88	17.61	18.28
20	8.38	8.90	9.37	9.81	10.22	10.60
24	6.94	7.36	7.75	8.11	8.45	8.76
36	4.57	4.85	5.10	5.34	5.55	5.76
48	3.41	3.62	3.80	3.98	4.14	4.29
60	2.72	2.88	3.03	3.17	3.30	3.42
72	2.26	2.40	2.52	2.63	2.74	2.84

	Annuity (A)					
n	8.0	8.5	9.0	9.5	10.0	10.5
1	700.00	750.00	800.00	850.00	900.00	950.00
2	182.84	191.55	200.00	208.22	216.23	224.04
3	100.00	104.08	108.01	111.79	115.44	118.98
4	68.18	70.75	73.20	75.56	77.83	80.01
5	51.57	53.42	55.19	56.87	58.49	60.04
10	23.11	23.86	24.57	25.25	25.89	26.51
12	18.92	19.52	20.09	20.64	21.15	21.65
20	10.96	11.29	11.61	11.91	12.20	12.48
24	9.05	9.33	9.59	9.83	10.07	10.29
36	5.95	6.12	6.29	6.45	6.61	6.75
48	4.43	4.56	4.68	4.80	4.91	5.02
60	3.53	3.63	3.73	3.82	3.91	4.00
72	2.93	3.02	3.10	3.18	3.25	3.32

	Annuity (A)					
n	11.0	12.0	13.0	14.0	15.0	16.0
1	1000.00	1100.00	1200.00	1300.00	1400.00	1500.00
2	231.66	246.41	260.56	274.17	287.30	300.00
3	122.40	128.94	135.13	141.01	146.62	151.98
4	82.12	86.12	89.88	93.43	96.80	100.00
5	61.54	64.38	67.03	69.52	71.88	74.11
10	27.10	28.21	29.24	30.20	31.10	31.95
12	22.12	23.01	23.83	24.60	25.32	25.99
20	12.74	13.23	13.68	14.11	14.50	14.87
24	10.51	10.91	11.28	11.62	11.95	12.25
36	6.89	7.15	7.38	7.61	7.81	8.01
48	5.12	5.31	5.49	5.65	5.80	5.95
60	4.08	4.23	4.37	4.50	4.62	4.73
72	3.39	3.51	3.63	3.73	3.83	3.93

	Annuity (A)					
n	17.0	18.0	19.0	20.0	25.0	30.0
1	1600.00	1700.00	1800.00	1900.00	2400.00	2900.00
2	312.31	324.26	335.89	347.21	400.00	447.72
3	157.13	162.07	166.84	171.44	192.40	210.72
4	103.05	105.98	108.78	111.47	123.61	134.03
5	76.23	78.26	80.20	82.06	90.36	97.44
10	32.75	33.51	34.24	34.93	37.97	40.51
12	26.63	27.23	27.81	28.36	30.77	32.77
20	15.22	15.55	15.86	16.16	17.46	18.54
24	12.53	12.80	13.05	13.29	14.35	15.22
36	8.19	8.36	8.52	8.68	9.35	9.91
48	6.08	6.21	6.33	6.44	6.94	7.34
60	4.83	4.93	5.03	5.12	5.51	5.83
72	4.01	4.10	4.17	4.25	4.57	4.84

9.2.3 COMPOUND INTEREST: FIND ANNUITY

These tables use Equation (9.1.4) to determine the annuity (or mortgage) payment that must be paid each time period, for n time periods, at an interest rate of $i\%$ per time period, to pay off a loan of one dollar ($B = 1$). For example, if $1 is borrowed at 3% interest per time period, and the amount is to be paid back in equal amounts over 10 time periods, then the amount paid back per time period is $0.12 (see the following table). Analogously, if $100 had been borrowed, then the mortgage amount would be $11.72.

n	Interest rate (i)					
	2.00%	2.25%	2.50%	2.75%	3.00%	3.25%
1	1.0200	1.0225	1.0250	1.0275	1.0300	1.0325
2	0.5151	0.5169	0.5188	0.5207	0.5226	0.5245
3	0.3468	0.3484	0.3501	0.3518	0.3535	0.3552
4	0.2626	0.2642	0.2658	0.2674	0.2690	0.2706
5	0.2122	0.2137	0.2152	0.2168	0.2183	0.2199
6	0.1785	0.1800	0.1815	0.1831	0.1846	0.1861
7	0.1545	0.1560	0.1575	0.1590	0.1605	0.1620
8	0.1365	0.1380	0.1395	0.1410	0.1425	0.1440
9	0.1225	0.1240	0.1255	0.1269	0.1284	0.1299
10	0.1113	0.1128	0.1143	0.1157	0.1172	0.1187
12	0.0946	0.0960	0.0975	0.0990	0.1005	0.1020
20	0.0612	0.0626	0.0641	0.0657	0.0672	0.0688
24	0.0529	0.0544	0.0559	0.0575	0.0590	0.0607
36	0.0392	0.0408	0.0425	0.0441	0.0458	0.0475
72	0.0263	0.0282	0.0301	0.0320	0.0340	0.0361

n	Interest rate (i)					
	3.50%	3.75%	4.00%	4.25%	4.50%	4.75%
1	1.0350	1.0375	1.0400	1.0425	1.0450	1.0475
2	0.5264	0.5283	0.5302	0.5321	0.5340	0.5359
3	0.3569	0.3586	0.3604	0.3621	0.3638	0.3655
4	0.2722	0.2739	0.2755	0.2771	0.2787	0.2804
5	0.2215	0.2230	0.2246	0.2262	0.2278	0.2294
6	0.1877	0.1892	0.1908	0.1923	0.1939	0.1955
7	0.1635	0.1651	0.1666	0.1681	0.1697	0.1713
8	0.1455	0.1470	0.1485	0.1501	0.1516	0.1532
9	0.1315	0.1330	0.1345	0.1360	0.1376	0.1391
10	0.1202	0.1218	0.1233	0.1248	0.1264	0.1279
12	0.1035	0.1050	0.1066	0.1081	0.1097	0.1112
20	0.0704	0.0720	0.0736	0.0752	0.0769	0.0785
24	0.0623	0.0639	0.0656	0.0673	0.0690	0.0707
36	0.0493	0.0511	0.0529	0.0547	0.0566	0.0585
72	0.0382	0.0403	0.0425	0.0447	0.0470	0.0492

	Interest rate (i)					
n	5.00%	5.25%	5.50%	5.75%	6.00%	6.25%
1	1.0500	1.0525	1.0550	1.0575	1.0600	1.0625
2	0.5378	0.5397	0.5416	0.5435	0.5454	0.5474
3	0.3672	0.3689	0.3706	0.3724	0.3741	0.3758
4	0.2820	0.2837	0.2853	0.2869	0.2886	0.2903
5	0.2310	0.2326	0.2342	0.2358	0.2374	0.2390
6	0.1970	0.1986	0.2002	0.2018	0.2034	0.2050
7	0.1728	0.1744	0.1760	0.1776	0.1791	0.1807
8	0.1547	0.1563	0.1579	0.1595	0.1610	0.1626
9	0.1407	0.1423	0.1438	0.1454	0.1470	0.1486
10	0.1295	0.1311	0.1327	0.1343	0.1359	0.1375
12	0.1128	0.1144	0.1160	0.1177	0.1193	0.1209
20	0.0802	0.0819	0.0837	0.0854	0.0872	0.0890
24	0.0725	0.0742	0.0760	0.0779	0.0797	0.0815
36	0.0604	0.0624	0.0644	0.0664	0.0684	0.0704
72	0.0515	0.0539	0.0562	0.0585	0.0609	0.0633

	Interest rate (i)					
n	6.50%	6.75%	7.00%	7.25%	7.50%	7.75%
1	1.0650	1.0675	1.0700	1.0725	1.0750	1.0775
2	0.5493	0.5512	0.5531	0.5550	0.5569	0.5588
3	0.3776	0.3793	0.3810	0.3828	0.3845	0.3863
4	0.2919	0.2936	0.2952	0.2969	0.2986	0.3002
5	0.2406	0.2423	0.2439	0.2455	0.2472	0.2488
6	0.2066	0.2082	0.2098	0.2114	0.2130	0.2147
7	0.1823	0.1839	0.1855	0.1872	0.1888	0.1904
8	0.1642	0.1658	0.1675	0.1691	0.1707	0.1724
9	0.1502	0.1519	0.1535	0.1551	0.1568	0.1584
10	0.1391	0.1407	0.1424	0.1440	0.1457	0.1474
12	0.1226	0.1242	0.1259	0.1276	0.1293	0.1310
20	0.0908	0.0926	0.0944	0.0962	0.0981	0.1000
24	0.0834	0.0853	0.0872	0.0891	0.0911	0.0930
36	0.0725	0.0746	0.0767	0.0789	0.0810	0.0832
72	0.0657	0.0681	0.0705	0.0730	0.0754	0.0779

	Interest rate (i)					
n	8.00%	8.25%	8.50%	8.75%	9.00%	9.25%
1	1.0800	1.0825	1.0850	1.0875	1.0900	1.0925
2	0.5608	0.5627	0.5646	0.5665	0.5685	0.5704
3	0.3880	0.3898	0.3915	0.3933	0.3951	0.3968
4	0.3019	0.3036	0.3053	0.3070	0.3087	0.3104
5	0.2505	0.2521	0.2538	0.2554	0.2571	0.2588
6	0.2163	0.2180	0.2196	0.2213	0.2229	0.2246
7	0.1921	0.1937	0.1954	0.1970	0.1987	0.2004
8	0.1740	0.1757	0.1773	0.1790	0.1807	0.1824
9	0.1601	0.1618	0.1634	0.1651	0.1668	0.1685
10	0.1490	0.1507	0.1524	0.1541	0.1558	0.1575
12	0.1327	0.1344	0.1361	0.1379	0.1396	0.1414
20	0.1018	0.1037	0.1057	0.1076	0.1095	0.1115
24	0.0950	0.0970	0.0990	0.1010	0.1030	0.1051
36	0.0853	0.0875	0.0898	0.0920	0.0942	0.0965
72	0.0803	0.0828	0.0852	0.0877	0.0902	0.0927

	Interest rate (i)					
n	9.50%	10.00%	10.50%	11.00%	11.50%	12.00%
1	1.0950	1.1000	1.1050	1.1100	1.1150	1.1200
2	0.5723	0.5762	0.5801	0.5839	0.5878	0.5917
3	0.3986	0.4021	0.4057	0.4092	0.4128	0.4163
4	0.3121	0.3155	0.3189	0.3223	0.3258	0.3292
5	0.2604	0.2638	0.2672	0.2706	0.2740	0.2774
6	0.2263	0.2296	0.2330	0.2364	0.2398	0.2432
7	0.2020	0.2054	0.2088	0.2122	0.2157	0.2191
8	0.1840	0.1874	0.1909	0.1943	0.1978	0.2013
9	0.1702	0.1736	0.1771	0.1806	0.1841	0.1877
10	0.1593	0.1628	0.1663	0.1698	0.1734	0.1770
12	0.1432	0.1468	0.1504	0.1540	0.1577	0.1614
20	0.1135	0.1175	0.1215	0.1256	0.1297	0.1339
24	0.1071	0.1113	0.1155	0.1198	0.1241	0.1285
36	0.0988	0.1033	0.1080	0.1126	0.1173	0.1221
72	0.0951	0.1001	0.1051	0.1101	0.1150	0.1200

	Interest rate (i)					
n	12.50%	13.00%	13.50%	14.00%	14.50%	15.00%
1	1.1250	1.1300	1.1350	1.1400	1.1450	1.1500
2	0.5956	0.5995	0.6034	0.6073	0.6112	0.6151
3	0.4199	0.4235	0.4271	0.4307	0.4344	0.4380
4	0.3327	0.3362	0.3397	0.3432	0.3467	0.3503
5	0.2808	0.2843	0.2878	0.2913	0.2948	0.2983
6	0.2467	0.2501	0.2536	0.2572	0.2607	0.2642
7	0.2226	0.2261	0.2296	0.2332	0.2368	0.2404
8	0.2048	0.2084	0.2120	0.2156	0.2192	0.2228
9	0.1913	0.1949	0.1985	0.2022	0.2059	0.2096
10	0.1806	0.1843	0.1880	0.1917	0.1955	0.1993
12	0.1652	0.1690	0.1728	0.1767	0.1806	0.1845
20	0.1381	0.1424	0.1467	0.1510	0.1554	0.1598
24	0.1329	0.1373	0.1418	0.1463	0.1509	0.1554
36	0.1268	0.1316	0.1364	0.1413	0.1461	0.1510
72	0.1250	0.1300	0.1350	0.1400	0.1450	0.1500

	Interest rate (i)					
n	15.50%	16.00%	16.50%	17.00%	17.50%	18.00%
1	1.1550	1.1600	1.1650	1.1700	1.1750	1.1800
2	0.6190	0.6230	0.6269	0.6308	0.6348	0.6387
3	0.4416	0.4453	0.4489	0.4526	0.4562	0.4599
4	0.3538	0.3574	0.3609	0.3645	0.3681	0.3717
5	0.3019	0.3054	0.3090	0.3126	0.3162	0.3198
6	0.2678	0.2714	0.2750	0.2786	0.2823	0.2859
7	0.2440	0.2476	0.2513	0.2550	0.2586	0.2624
8	0.2265	0.2302	0.2339	0.2377	0.2415	0.2452
9	0.2133	0.2171	0.2209	0.2247	0.2285	0.2324
10	0.2031	0.2069	0.2108	0.2147	0.2186	0.2225
12	0.1884	0.1924	0.1964	0.2005	0.2045	0.2086
20	0.1642	0.1687	0.1732	0.1777	0.1822	0.1868
24	0.1600	0.1647	0.1693	0.1740	0.1787	0.1835
36	0.1559	0.1608	0.1657	0.1706	0.1755	0.1805
72	0.1550	0.1600	0.1650	0.1700	0.1750	0.1800

Miscellaneous

10.1   UNITS *792*
10.1.1   SI system of measurement *792*
10.1.2   United States customary system of weights and measures *793*
10.1.3   Physical constants *794*
10.1.4   Dimensional analysis/Buckingham pi *795*
10.1.5   Units of physical quantities *796*
10.1.6   Conversion: metric to English *796*
10.1.7   Conversion: English to metric *797*
10.1.8   Miscellaneous conversions *797*
10.1.9   Temperature conversion *798*

10.2   INTERPRETATIONS OF POWERS OF 10 . . . . *798*

10.3   CALENDAR COMPUTATIONS *799*
10.3.1   Leap years *799*
10.3.2   Day of week for any given day *799*
10.3.3   Number of each day of the year *800*

10.4   AMS CLASSIFICATION SCHEME *801*

10.5   FIELDS MEDALS *802*

10.6   GREEK ALPHABET *803*

10.7   COMPUTER LANGUAGES *803*
10.7.1   Software contact information *803*

10.8   PROFESSIONAL MATHEMATICAL ORGANIZATIONS *804*

10.9   ELECTRONIC MATHEMATICAL RESOURCES *807*

10.10 BIOGRAPHIES OF MATHEMATICIANS *810*

1-58488-291-3/02/$0.00+$1.50
© 2003 CRC Press, Inc.

10.1   UNITS

10.1.1   SI SYSTEM OF MEASUREMENT

SI, the abbreviation of the French words "Systeme Internationale d'Unites", is the accepted abbreviation for the International Metric System.

1. There are seven base units

Quantity measured	Unit	Symbol
Length	meter	m
Mass	kilogram	kg
Time	second	s
Amount of substance	mole	mol
Electric current	ampere	A
Luminous intensity	candela	cd
Thermodynamic temperature	kelvin	K

2. There are 22 derived units with special names and symbols

Quantity	SI Name	Symbol	Combination of other SI units (or base units)
Absorbed dose	gray	Gy	J/kg
Activity (radiation source)	becquerel	Bq	1/s
Capacitance	farad	F	C/V
Catalytic activity	katal	kat	s^{-1} mol
Celsius temperature	degree Celsius	°C	K
Conductance	siemen	S	A/V
Dose equivalent	sievert	Sv	J/kg
Electric charge	coulomb	C	A s
Electric potential	volt	V	W/A
Electric resistance	ohm	Ω	V/A
Energy	joule	J	N m
Force	newton	N	kg m/s^2
Frequency	hertz	Hz	1/s
Illuminance	lux	lx	lm/m^2
Inductance	henry	H	Wb/A
Luminous flux	lumen	lm	cd sr
Magnetic flux density	tesla	T	Wb/m^2
Magnetic flux	weber	Wb	V s
Plane angle	radian	rad	$m \cdot m^{-1}$ (unitless)
Power	watt	W	J/s
Pressure or stress	pascal	Pa	N/m^2
Solid angle	steradian	sr	$m^2 \cdot m^{-2}$ (unitless)

3. The following units are accepted for use with SI units.

Name	Symbol	Value in SI units
(angle) degree	°	$1° = (\pi/180)$ rad
(angle) minute	′	$1' = (1/60)° = (\pi/10800)$ rad
(angle) second	″	$1'' = (1/60)' = (\pi/648000)$ rad
(time) day	d	$1\ d = 24\ h = 86400\ s$
(time) hour	h	$1\ h = 60\ min = 3600\ s$
(time) minute	min	$1\ min = 60\ s$
astronomical unit	au	$1\ au \approx 1.49598 \times 10^{11}\ m$
bel	B	$1\ B = (1/2) \ln 10\ Np$
		(Note that $1\ dB = 0.1\ B$)
electronvolt	eV	$1\ eV \approx 1.6021764 \times 10^{-19}\ C$
liter	L	$1\ L = 1\ dm^3 = 10^{-3}\ m^3$
metric ton	t	$1\ t = 10^3\ kg$
neper	Np	$1\ Np = 1$ (unitless)
unified atomic mass unit	u	$1\ u \approx 1.66054 \times 10^{-27}\ kg$

4. The following units are currently accepted for use with SI units (subject to further review).

Name	Symbol	Value in SI units
angstrom	Å	$1Å = 0.1\ nm = 10^{-10}\ m$
are	a	$1\ a = 1\ dam^2 = 10^2\ m^2$
barn	b	$1\ b = 100\ fm^2 = 10^{-28}\ m^2$
bar	bar	$1\ bar = 0.1\ MPa = 100\ kPa = 1000\ hPa = 10^5\ Pa$
curie	Ci	$1\ Ci = 3.7 \times 10^{10}\ Bq$
hectare	ha	$1\ ha = 1\ hm^2 = 10^4\ m^2$
knot		1 nautical mile per hour $= (1852/3600)$ m/s
nautical mile		1 nautical mile $= 1852\ m$
rad	rad	$1\ rad = 1\ cGy = 10^{-2}\ Gy$
rem	rem	$1\ rem = 1\ cSv = 10^{-2}\ Sv$
roentgen	R	$1\ R = 2.58 \times 10^{-4}\ C/kg$

10.1.2   UNITED STATES CUSTOMARY SYSTEM OF WEIGHTS AND MEASURES

Linear measure		
1 mile	=	5280 feet or 320 rods
1 rod	=	16.5 feet or 5.5 yards
1 yard	=	3 feet
1 foot	=	12 inches

Linear measure: nautical		
1 fathom	=	6 feet
1° of latitude	≈	69 miles
1° of longitude at 40° latitude	≈	46 nautical miles ≈ 53 miles
1 nautical mile	≈	6076.1 feet ≈ 1.1508 statute miles

Square measure		
1 square mile	=	640 acres
1 acre	=	43,560 square feet
Volume measure		
1 cubic yard	=	27 cubic feet
1 cubic foot	=	1728 cubic inches
Dry measure		
1 bushel	=	4 pecks
1 peck	=	8 quarts
1 quart	=	2 pints
Liquid measure		
1 cubic foot	=	7.4805 gallons
1 gallon	=	4 quarts
1 quart	=	2 pints
1 pint	=	4 gills
Liquid measure: Apothecaries'		
1 pint	=	16 fluid ounces
1 fluid ounce	=	8 drams
1 fluid dram	=	60 minims
Weight: Avoirdupois		
1 ton	=	2000 pounds
1 pound	=	16 ounces or 7000 grains
1 ounce	=	16 drams or 437.5 grains
Weight: Troy		
1 pound	=	12 ounces
1 ounce	=	20 pennyweights
1 pennyweight	=	24 grains
Weight: Apothecaries'		
1 pound	=	12 ounces
1 ounce	=	8 drams
1 dram	=	3 scruples
1 scruple	=	20 grains

10.1.3 PHYSICAL CONSTANTS

c (speed of light) = 299,792,458 m/s   (exact value)

e (charge of electron) $\approx 1.6021764 \times 10^{-19}$ C

G (gravitational constant) $\approx (6.673 \pm 0.003) \times 10^{-8}$ cm^3/g s^2

$\hbar$ (Plank constant over 2π) $\approx 1.0545716 \times 10^{-34}$ J s

k (Boltzmann constant) $\approx 1.38065 \times 10^{-23}$ J/K

1 knot = 1 nautical mile/hour ≈ 1.6878 ft/s ≈ 1.1508 statute miles/hr

Acceleration, sea level, latitude $45° \approx 9.806194$ m/s$^2 \approx 32.1726$ ft/s^2

Avogadro's constant ≈ 6.022142 mol^{-1}

Density of mercury, at $0°$C ≈ 13.5951 g/mL

Density of water (maximum), at $3.98\,^{\circ}\text{C} \approx 0.99997496$ g/mL

Density of water, at $0\,^{\circ}\text{C} \approx 0.9998426$ g/mL

Density of dry air, at $0\,^{\circ}\text{C}$, 760 mm of Hg ≈ 1.2927 g/L

Earth: equatorial radius ≈ 6378.388 km ≈ 3963.34 statute miles

Earth: polar radius ≈ 6356.912 km ≈ 3949.99 statute miles

Earth: mean density ≈ 5.522 g/cm^3 ≈ 344.7 lb/ft^3

Heat of fusion of water, at $0\,^{\circ}\text{C} \approx 333.6$ J/g

Heat of vaporization of water, at $100\,^{\circ}\text{C} \approx 2256.8$ J/g

Mass of hydrogen atom $\approx 1.67353 \times 10^{-24}$ g

Velocity of sound, dry air, at $0\,^{\circ}\text{C} \approx 331.36$ m/s ≈ 1087.1 ft/s

Wavelength of orange-red line of krypton 86 ≈ 6057.802 Å

10.1.4 DIMENSIONAL ANALYSIS/BUCKINGHAM PI

The units of the parameters in a system constrain all the derivable quantities, regardless of the equations describing the system. In particular, all derived quantities are functions of dimensionless combinations of parameters. The number of dimensionless parameters and their forms are given by the *Buckingham pi theorem*.

In a system, the quantity $u = f(W_1, W_2, \ldots, W_n)$ is to be determined in terms of the n measurable variables and parameters $\{W_i\}$ where f is an unknown function. Let the quantities $\{u, W_i\}$ involve m fundamental dimensions labeled by $L_1, L_2, \ldots, L_m$ (such as length, mass, time, or charge). The dimensions of any of the $\{u, W_i\}$ are given by a product of powers of the fundamental dimensions. For example, the dimensions of W_i are $L_1^{b_{i1}} L_2^{b_{i2}} L_3^{b_{i3}} \cdots L_m^{b_{im}}$ where the $\{b_{ij}\}$ are real and called the *dimensional exponents*. A quantity is called dimensionless if all of its dimensional exponents are zero. Let $\mathbf{b}_i = \begin{bmatrix} b_{i1} & b_{i2} & \cdots & b_{im} \end{bmatrix}^{\mathrm{T}}$ be the dimension vector of W_i and let $B = \begin{bmatrix} \mathbf{b}_1 & \mathbf{b}_2 & \cdots & \mathbf{b}_n \end{bmatrix}$ be the $m \times n$ dimension matrix of the system. Let $\mathbf{a} = \begin{bmatrix} a_1 & a_2 & \cdots & a_m \end{bmatrix}^{\mathrm{T}}$ be the dimension vector of u and let $\mathbf{y} = \begin{bmatrix} y_1 & y_2 & \cdots & y_n \end{bmatrix}^{\mathrm{T}}$ represent a solution of $B\mathbf{y} = -\mathbf{a}$. Then,

1. The number of dimensionless quantities is $k + 1 = n + 1 - \text{rank}(B)$.

2. The measurable quantity u can be expressed in terms of dimensionless parameters as

$$u = W_1^{-y_1} W_2^{-y_2} \cdots W_n^{-y_n} g(\pi_1, \pi_2, \ldots, \pi_k) \qquad (10.1.1)$$

where g is an unknown function of its parameters and the $\{\pi_i\}$ are dimensionless quantities. Specifically, let $\mathbf{x}^{(i)} = \begin{bmatrix} x_{1i} & x_{2i} & \cdots & x_{ni} \end{bmatrix}^{\mathrm{T}}$ be one of $k = n - r(B)$ linearly independent solutions of the system $B\mathbf{x} = \mathbf{0}$ and define $\pi_i = W_1^{x_{1i}} W_2^{x_{2i}} \cdots W_n^{x_{ni}}$.

10.1.5 UNITS OF PHYSICAL QUANTITIES

In the following, read "kilograms" for the mass M, "meters" for the length L, "seconds" for the time T, and "degrees" for the temperature θ. For example, acceleration is measured in units of L/T^2, or meters per second squared.

Quantity	Dimensions	Quantity	Dimensions
Acceleration	L/T^2	Mass	M
Angular acceleration	$1/T^2$	Mass density	M/L^3
Angular frequency	$1/T$	Momentum	ML/T
Angular momentum	ML^2/T	Period	T
Angular velocity	$1/T$	Power	ML^2/T^3
Area	L^2	Pressure	M/LT^2
Displacement	L	Moment of inertia	ML^2
Energy or work	ML^2/T^2	Time	T
Energy, kinetic	ML^2/T^2	Torque	ML^2/T^2
Energy, potential	ML^2/T^2	Velocity	L/T
Energy, total	ML^2/T^2	Volume	L^3
Force	ML/T^2	Wavelength	L
Frequency	$1/T$	Work	ML^2/T^2
Gravitational field strength	ML/T^2	Entropy	$ML^2/T^2\theta$
Gravitational potential	ML^2/T^2	Internal energy	ML^2/T^2
Length	L	Heat	ML^2/T^2

10.1.6 CONVERSION: METRIC TO ENGLISH

Multiply	By	To obtain
centimeters	0.3937008	inches
cubic meters	1.307951	cubic yards
cubic meters	35.31467	cubic feet
grams	0.03527396	ounces
kilograms	2.204623	pounds
kilometers	0.6213712	miles
liters	0.2641721	gallons (US)
meters	1.093613	yards
meters	3.280840	feet
milliliters	0.03381402	fluid ounces
milliliters	0.06102374	cubic inches
square centimeters	0.1550003	square inches
square meters	1.195990	square yards
square meters	10.76391	square feet

10.1.7 CONVERSION: ENGLISH TO METRIC

Multiply	By	To obtain
cubic feet	0.02831685	cubic meters
cubic inches	16.38706	milliliters
cubic yards	0.7645549	cubic meters
feet	0.3048000	meters
fluid ounces	29.57353	milliliters
gallons (US)	3.785412	liters
inches	2.540000	centimeters
miles	1.609344	kilometers
mils	25.4	micrometers
ounces	28.34952	grams
pounds	0.4535924	kilograms
square feet	0.09290304	square meters
square inches	6.451600	square centimeters
square yards	0.8361274	square meters
yards	0.9144000	meters

10.1.8 MISCELLANEOUS CONVERSIONS

Multiply	By	To obtain
feet of water at 4°C	2.950×10^{-2}	atmospheres
inches of mercury at 4°C	3.342×10^{-2}	atmospheres
pounds per square inch	6.804×10^{-2}	atmospheres
foot-pounds	1.285×10^{-3}	BTU
joules	9.480×10^{-4}	BTU
cords	128	cubic feet
radian	57.29578	degree (angle)
foot-pounds	1.356×10^{7}	ergs
atmospheres	33.90	feet of water at 4°C
miles	5280	feet
horsepower	3.3×10^{4}	foot-pounds per minute
horsepower-hours	1.98×10^{6}	foot-pounds
kilowatt-hours	2.655×10^{6}	foot-pounds
foot-pounds per second	1.818×10^{-3}	horsepower
atmospheres	2.036	inches of mercury at 0°C
BTU	1.055060×10^{3}	joules
foot-pounds	1.35582	joules
BTU per minute	1.758×10^{-2}	kilowatts
foot-pounds per minute	2.26×10^{-5}	kilowatts
horsepower	0.7457	kilowatts
miles per hour	0.8689762	knots

Multiply	By	To obtain
feet	1.893939×10^{-4}	miles
miles	0.8689762	nautical miles
degrees	1.745329×10^{-2}	radians
acres	43560	square feet
BTU per minute	17.5796	watts

10.1.9 TEMPERATURE CONVERSION

If t_F is the temperature in degrees Fahrenheit and t_C is the temperature in degrees Celsius, then

$$t_C = \frac{5}{9}(t_F - 32) \quad \text{and} \quad t_F = \frac{9}{5}t_C + 32. \qquad (10.1.2)$$

$-40°C$	$0°C$	$10°C$	$20°C$	$37°C$	$100°C$
$-40°F$	$32°F$	$50°F$	$68°F$	$98.6°F$	$212°F$

If T_K is the temperature in kelvin and T_R is the temperature in degrees Rankine, then

$$T_R = t_F + 459.69 \quad \text{and} \quad T_K = t_C + 273.15 = \frac{5}{9}T_R. \qquad (10.1.3)$$

10.2 INTERPRETATIONS OF POWERS OF 10

10^{-15}	the radius of the hydrogen nucleus (a proton) in meters
10^{-11}	the likelihood of being dealt 13 top honors in bridge
10^{-10}	the radius of a hydrogen atom in meters
10^{-9}	the number of seconds it takes light to travel one foot
10^{-6}	the likelihood of being dealt a royal flush in poker
10^{0}	the density of water is 1 gram per milliliter
10^{1}	the number of fingers that people have
10^{2}	the number of stable elements in the periodic table
10^{5}	the number of hairs on a human scalp
10^{6}	the number of possible chess board positions after 4 moves
10^{7}	the number of seconds in a year
10^{8}	the speed of light in meters per second
10^{9}	the number of heartbeats in a lifetime for most mammals
10^{10}	the number of people on the earth
10^{15}	the surface area of the earth in square meters
10^{16}	the age of the universe in seconds
10^{18}	the volume of water in the earth's oceans in cubic meters
10^{19}	the number of possible positions of Rubik's cube
10^{21}	the volume of the earth in cubic meters

10^{24}	the number of grains of sand in the Sahara desert
10^{28}	the mass of the earth in grams
10^{33}	the mass of the solar system in grams
10^{50}	the number of atoms in the earth
10^{78}	the volume of the universe in cubic meters

(Note: these numbers have been rounded to the nearest power of ten.)

10.3 CALENDAR COMPUTATIONS

10.3.1 LEAP YEARS

If a year is divisible by 4, then it will be a leap year, unless the year is divisible by 100 (when it will not be a leap year), unless the year is divisible by 400 (when it will be a leap year). Hence the list of leap years includes 1896, 1904, 1908, 1992, 1996, 2000, 2004, 2008 and the list of non-leap years includes 1900, 1998, 1999, 2001.

10.3.2 DAY OF WEEK FOR ANY GIVEN DAY

The following formula gives the day of the week for the Gregorian calendar (i.e., for any date after 1582):

$$W \equiv \left(k + \lfloor 2.6m - 0.2 \rfloor - 2C + Y + \left\lfloor \frac{Y}{4} \right\rfloor + \left\lfloor \frac{C}{4} \right\rfloor \right) \pmod{7} \qquad (10.3.1)$$

where

- W is the day of the week ($0 =$ Sunday, $\ldots$, $6 =$ Saturday).
- k is the day of the month (1 to 31).
- m is the month ($1 =$ March, $\ldots$, $10 =$ December, $11 =$ January, $12 =$ February). (January and February are treated as months of the preceding year.)
- C is century minus one (1997 has $C = 19$, 2005 has $C = 20$).
- Y is the year (1997 has $Y = 97$ except $Y = 96$ for January and February).
- $\lfloor \cdot \rfloor$ denotes the integer floor function.
- The "mod" function returns a non-negative value.

In any given year the following days fall on the same day of the week: 4/4, 6/6, 8/8, 10/10, 12/12, 9/5, 5/9, 7/11, 11/7, and the last day of February.

EXAMPLE   Consider the date 16 March 1997 (for which $k = 16$, $m = 1$, $C = 19$, and $Y = 97$). From Equation (10.3.1), we compute $W \equiv 16 + \lfloor 2.4 \rfloor - 38 + 97 + \lfloor \frac{97}{4} \rfloor + \lfloor \frac{19}{4} \rfloor \pmod{7} \equiv 2 + 2 - 3 + 6 + 3 + 4 \pmod{7} \equiv 0 \pmod{7}$. So this date was a Sunday.

Because 7 does not divide 400, January 1 occurs more frequently on some days of the week than on others! In a cycle of 400 years, January 1 and March 1 occur on the following days with the following frequencies:

	Sun	Mon	Tue	Wed	Thu	Fri	Sat
January 1	58	56	58	57	57	58	56
March 1	58	56	58	56	58	57	57

10.3.3   NUMBER OF EACH DAY OF THE YEAR

Day	Jan	Feb	Mar	Apr	May	Jun	Jul	Aug	Sep	Oct	Nov	Dec
1	1	32	60	91	121	152	182	213	244	274	305	335
2	2	33	61	92	122	153	183	214	245	275	306	336
3	3	34	62	93	123	154	184	215	246	276	307	337
4	4	35	63	94	124	155	185	216	247	277	308	338
5	5	36	64	95	125	156	186	217	248	278	309	339
6	6	37	65	96	126	157	187	218	249	279	310	340
7	7	38	66	97	127	158	188	219	250	280	311	341
8	8	39	67	98	128	159	189	220	251	281	312	342
9	9	40	68	99	129	160	190	221	252	282	313	343
10	10	41	69	100	130	161	191	222	253	283	314	344
11	11	42	70	101	131	162	192	223	254	284	315	345
12	12	43	71	102	132	163	193	224	255	285	316	346
13	13	44	72	103	133	164	194	225	256	286	317	347
14	14	45	73	104	134	165	195	226	257	287	318	348
15	15	46	74	105	135	166	196	227	258	288	319	349
16	16	47	75	106	136	167	197	228	259	289	320	350
17	17	48	76	107	137	168	198	229	260	290	321	351
18	18	49	77	108	138	169	199	230	261	291	322	352
19	19	50	78	109	139	170	200	231	262	292	323	353
20	20	51	79	110	140	171	201	232	263	293	324	354
21	21	52	80	111	141	172	202	233	264	294	325	355
22	22	53	81	112	142	173	203	234	265	295	326	356
23	23	54	82	113	143	174	204	235	266	296	327	357
24	24	55	83	114	144	175	205	236	267	297	328	358
25	25	56	84	115	145	176	206	237	268	298	329	359
26	26	57	85	116	146	177	207	238	269	299	330	360
27	27	58	86	117	147	178	208	239	270	300	331	361
28	28	59	87	118	148	179	209	240	271	301	332	362
29	29	*	88	119	149	180	210	241	272	302	333	363
30	30		89	120	150	181	211	242	273	303	334	364
31	31		90		151		212	243		304		365

*In leap years, after February 28, add 1 to the tabulated number.

10.4 AMS CLASSIFICATION SCHEME

00	General	45	Integral equations
01	History and biography	46	Functional analysis
03	Mathematical logic and foundations	47	Operator theory
04	This section has been deleted	49	Calculus of variations and optimal control; optimization
05	Combinatorics	51	Geometry
06	Order, lattices, ordered algebraic structures	52	Convex and discrete geometry
08	General algebraic systems	53	Differential geometry
11	Number theory	54	General topology
12	Field theory and polynomials	55	Algebraic topology
13	Commutative rings and algebras	57	Manifolds and cell complexes
14	Algebraic geometry	58	Global analysis, analysis on manifolds
15	Linear and multilinear algebra; matrix theory	60	Probability theory and stochastic processes
16	Associative rings and algebras	62	Statistics
17	Non-associative rings and algebras	65	Numerical analysis
18	Category theory; homological algebra	68	Computer science
19	K-theory	70	Mechanics of particles and systems
20	Group theory and generalizations	73	This section has been deleted
22	Topological groups, Lie groups	74	Mechanics of deformable solids
26	Real functions	76	Fluid mechanics
28	Measure and integration	78	Optics, electromagnetic theory
30	Functions of a complex variable	80	Classical thermodynamics, heat transfer
31	Potential theory	81	Quantum theory
32	Several complex variables and analytic spaces	82	Statistical mechanics, structure of matter
33	Special functions	83	Relativity and gravitational theory
34	Ordinary differential equations	85	Astronomy and astrophysics
35	Partial differential equations	86	Geophysics
37	Dynamical systems and ergodic theory	90	Operations research, mathematical programming
39	Difference and functional equations	91	Game theory, economics, social and behavioral sciences
40	Sequences, series, summability	92	Biology and other natural sciences
41	Approximations and expansions	93	Systems theory; control
42	Fourier analysis	94	Information and communication, circuits
43	Abstract harmonic analysis	97	Mathematics education
44	Integral transforms, operational calculus		

See `www.ams.org/msc/` for details.

10.5   FIELDS MEDALS

The Fields medal is the most prestigious award that can be bestowed upon a mathematician. It is awarded to someone no more than 40 years of age.

(1)	1936	Ahlfors, Lars	29	Harvard University
(2)	1936	Douglas, Jesse	39	MIT
(3)	1950	Schwartz, Laurent	35	Universite de Nancy
(4)	1950	Selberg, Atle	33	Princeton/Inst. for Advanced Study
(5)	1954	Kodaira, Kunihiko	39	Princeton University
(6)	1954	Serre, Jean-Pierre	27	College de France
(7)	1958	Roth, Klaus	32	University of London
(8)	1958	Thom, Rene	35	University of Strasbourg
(9)	1962	Hormander, Lars	31	University of Stockholm
(10)	1962	Milnor, John	31	Princeton University
(11)	1966	Atiyah, Michael	37	Oxford University
(12)	1966	Cohen, Paul	32	Stanford University
(13)	1966	Grothendieck, Alexander	38	University of Paris
(14)	1966	Smale, Stephen	36	University of California at Berkeley
(15)	1970	Baker, Alan	31	Cambridge University
(16)	1970	Hironaka, Heisuke	39	Harvard University
(17)	1970	Novikov, Serge	32	Moscow University
(18)	1970	Thompson, John	37	University of Chicago
(19)	1974	Bombieri, Enrico	33	University of Pisa
(20)	1974	Mumford, David	37	Harvard University
(21)	1978	Deligne, Pierre	33	IHES
(22)	1978	Fefferman, Charles	29	Princeton University
(23)	1978	Margulis, Gregori	32	InstPrblmInfTrans
(24)	1978	Quillen, Daniel	38	MIT
(25)	1982	Connes, Alain	35	IHES
(26)	1982	Thurston, William	35	Princeton University
(27)	1982	Yau, Shing-Tung	33	IAS
(28)	1986	Donaldson, Simon	27	Oxford University
(29)	1986	Faltings, Gerd	32	Princeton University
(30)	1986	Freedman, Michael	35	University of California at San Diego
(31)	1990	Drinfeld, Vladimir	36	Phys. Inst. Kharkov
(32)	1990	Jones, Vaughan	38	University of California at Berkeley
(33)	1990	Mori, Shigefumi	39	University of Kyoto
(34)	1990	Witten, Edward	38	Princeton/Inst. for Advanced Study
(35)	1994	Bourgain, Jean	40	Princeton/Inst. for Advanced Study
(36)	1994	Lions, Pierre-Louis	38	Universite de Paris-Dauphine
(37)	1994	Yoccoz, Jean-Chrisophe	36	Universite de Paris-Sud
(38)	1994	Zelmanov, Efim	39	University of Wisconsin
(39)	1998	Borcherds, Richard E.	38	Cambridge University
(40)	1998	Gowers, William T.	34	Cambridge University
(41)	1998	Kontsevich, Maxim	34	Institut des Hautes Etudes Scientifiques and Rutgers University
(42)	1998	McMullen, Curtis T.	40	Harvard University

10.6   GREEK ALPHABET

For each Greek letter, we illustrate the form of the capital letter and the form of the
lower case letter. In some cases, there is a popular variation of the lower case letter.

Greek letter			Greek name	English equivalent	Greek letter			Greek name	English equivalent
A	α		Alpha	a	N	ν		Nu	n
B	β		Beta	b	Ξ	ξ		Xi	x
Γ	γ		Gamma	g	O	o		Omicron	o
Δ	δ		Delta	d	Π	π	ϖ	Pi	p
E	ϵ	ε	Epsilon	e	P	ρ	ϱ	Rho	r
Z	ζ		Zeta	z	Σ	σ	ς	Sigma	s
H	η		Eta	e	T	τ		Tau	t
Θ	θ	ϑ	Theta	th	Υ	υ		Upsilon	u
I	ι		Iota	i	Φ	ϕ	φ	Phi	ph
K	κ		Kappa	k	X	χ		Chi	ch
Λ	λ		Lambda	l	Ψ	ψ		Psi	ps
M	μ		Mu	m	Ω	ω		Omega	o

10.7   COMPUTER LANGUAGES

The following is a sampling of computer languages used by scientists and engineers:

1. Numerical languages

   - Matlab and Octave
   - C and C++
   - Fortran
   - Lisp

2. Statistical languages

   - SPSS
   - Minitab

3. Optimization languages

   - GAMS (AMPL)
   - MINOS
   - MINTO

4. Symbolic languages

   - Derive
   - Maple
   - Mathematica
   - Reduce

10.7.1   SOFTWARE CONTACT INFORMATION

1. Derive   http://www.derive.com
2. Fortran   http://www.fortran.com
3. Maple   http://www.maplesoft.com
4. MathCad   http://www.mathsoft.com
5. Matlab   http://www.mathworks.com
6. Mathematica   http://www.wolfram.com

10.8 PROFESSIONAL MATHEMATICAL ORGANIZATIONS

1. **American Mathematical Society (AMS)**
   201 Charles Street, Providence, RI 02904
   Telephone: 800/321-4AMS
   Electronic address: `www.ams.org`

2. **American Mathematical Association of Two-Year Colleges**
   Southwest Tennessee Community College
   5983 Macon Cove, Memphis, TN 38134
   Telephone: 901/333-4643
   Electronic address: `www.amatyc.org`

3. **American Statistical Association**
   1429 Duke Street, Alexandria, VA 22314
   Telephone: 703/684-1221
   Electronic address: `www.amstat.org`

4. **Association for Symbolic Logic**
   Box 742, Vassar College, 124 Raymond Avenue
   Poughkeepsie, New York 12604
   Telephone: 845/437-7080
   Electronic address: `www.aslonline.org`

5. **Association for Women in Mathematics**
   4114 Computer & Space Sciences Building, University of Maryland,
   College Park, MD 20742
   Telephone: 301/405-7892
   Electronic address: `www.awm-math.org`

6. **Canadian Applied Mathematics Society**
   Department of Mathematics and Statistics, Simon Fraser University,
   Burnaby, British Columbia, Canada V5A 1S6
   Telephone: 604/291-3337, 604/291-3332
   Electronic address: `www.caims.ca`

7. **Canadian Applied and Industrial Mathematics Society**
   577 King Edward, Suite 109, P. O. Box 450, Station A,
   Ottawa, Ontario, Canada K1N 6N5
   Telephone: 613/562-5702
   Electronic address: `www.cms.math.ca`

8. **Casualty Actuarial Society**
   1100 North Glebe Road, Suite 600, Arlington, VA 22201
   Telephone: 703/276-3100
   Electronic address: `www.casact.org`

9. **Conference Board of the Mathematical Sciences**
   1529 Eighteenth Street, N.W., Washington, DC 20036
   Telephone: 202/293-1170
   Electronic address: `www.cbmsweb.org`

10. **The Consortium for Mathematics and Its Applications (COMAP)**
    57 Bedford Street, Suite 210, Lexington, MA 02420
    Telephone: 800/77-COMAP
    Electronic address: `www.comap.com`

11. **Council on Undergraduate Research**
    Council on Undergraduate Research. 734 15th St. N.W., Suite 550, Washington, DC 20005
    Telephone: 202/783-4810
    Electronic address: `www.cur.org`

12. **The Fibonacci Association**
    Chase Building, Dalhousie University
    Halifax, Nova Scotia, Canada B3H 3J5
    Telephone: 902/494-2572
    Electronic address: `www.mscs.dal.ca/Fibonacci/`

13. **Institute for Operations Research and the Management Sciences (INFORMS)**
    940-A Elkridge Landing Road, Linthicum, MD 21090
    Telephone: 800/4IN-FORMS
    Electronic address: `www.informs.org`

14. **Institute of Mathematical Statistics**
    P.O. Box 22718
    Beachwood, OH 44122
    Telephone: 216/295-2340
    Electronic address: `www.imstat.org`

15. **International Mathematics Union (IMU)**
    Estrada Dona Castorina, 110, Jardim Botánico,
    Rio de Janeiro – RJ 22460 Brazil
    Telephone: 55-21-294 9032, 55-21-5111749
    Electronic address: `www.mathunion.org`

16. **Joint Policy Board for Mathematics**
    1 Oxford Street #325, Cambridge, MA 02138
    Electronic address: `www.jpbm.org`

17. **Kappa Mu Epsilon ($\kappa\mu\epsilon$)**
    Electronic address: `www.cst.cmich.edu/org/kme_nat`

18. **The Mathematical Association of America (MAA)**
    1529 Eighteenth Street, N.W., Washington, DC 20036
    Telephone: 202/387-5200
    Electronic address: `www.maa.org`

19. **Mathematical Programming Society**
    3600 University City Science Center, Philadelphia, PA 19104
    Telephone: 215/382-9800, x323
    Electronic address: `www.caam.rice.edu/~mathprog/`

20. **Mu Alpha Theta ($\mu\alpha\theta$)**
    University of Oklahoma, 610 Elm Avenue, Room 423
    Norman, OK 73019
    Telephone: 405/325-4489
    Electronic address: `www.mualphatheta.org`

21. **National Association of Mathematicians**
    Department of Mathematics, Morehouse College
    Atlanta, GA 30314
    Electronic address: `www.caam.rice.edu/~nated/orgs/nam/index.html`

22. **The National Council of Teachers of Mathematics**
    1906 Association Drive, Reston, VA 22091
    Telephone: 703/620-9840
    Electronic address: `www.nctm.org`

23. **ORSA (see *INFORMS*)**

24. **Pi Mu Epsilon ($\pi\mu\epsilon$)**
    Electronic address: `www.pme-math.org`

25. **Rocky Mountain Mathematics Consortium**
    Arizona State University, Box 871904, Tempe, AZ 85287
    Telephone: 602/965-3788
    Electronic address: `math.la.asu.edu/~rmmc`

26. **Society of Industrial and Applied Mathematics (SIAM)**
    3600 University City Science Center, Philadelphia, PA 19104
    Telephone: 215/382-9800
    Electronic address: `www.siam.org`

27. **The Society for Mathematical Biology**
    Electronic address: `www.smb.org`

28. **Society of Actuaries**
    475 North Martingale Road, Suite 800, Schaumburg, IL 60173
    Telephone: 847/706-3500
    Electronic address: `www.soa.org`

29. **Statistical Society of Canada**
    1485 Lapérrire St., Ottawa, Ontario K1Z 7S8
    Telephone: 613/725-2253
    Electronic address: `www.ssc.ca`

10.9 ELECTRONIC MATHEMATICAL RESOURCES

1. General web sites related to mathematics

    (a) `http://dir.yahoo.com/Science/mathematics/`

    A very large list of useful sites relating to mathematics. It is perhaps the best place to start researching an arbitrary mathematical question not covered elsewhere in this list.

    (b) `http://web.math.fsu.edu/Science/math.html`

    The mathematics WWW virtual library has a very comprehensive collection of links to other mathematics-related sites.

    (c) `http://mathworld.wolfram.com`

    A comprehensive on-line encyclopedia of mathematics with more than 10,000 entries, 4,000 figures, and 100 animated graphics.

    (d) `http://www.sosmath.com`

    A collection of sites of mathematical interest on the web.

    (e) `http://carbon.cudenver.edu/~hgreenbe/glossary/intro.html`

    A *mathematical programming glossary.*

    (f) `http://thesaurus.maths.org`

    A *mathematical thesaurus.*

    (g) `http://www.cs.unb.ca/~alopez-o/math-faq/math-faq.html`

    The *FAQ* (frequently asked questions) listing from the news group `sci.math`.

2. Web sites that respond to user input

    (a) `http://www-neos.mcs.anl.gov/`

    The *NEOS server for optimization* will run many different optimization packages on an input user problem.

    (b) `http://www.theory.csc.UVic.CA/~cos/`

    The *Combinatorial Object Server* creates combinatorial objects such as necklaces, permutations, combinations, etc.

    (c) `http://www.research.att.com/~njas/sequences/`

    The *On-Line Encyclopedia of Integer Sequences* allows the "next term" in a sequence to be determined. (See page 25.)

    (d) `http://www.cecm.sfu.ca/projects/ISC/`

    If a real number is input to the *Inverse Symbolic Calculator* it will determine where this number might have come from.

3. Societies

    (a) `http://www.ams.org`

    The American Mathematical Society with the Combined Membership List of the AMS and the Math Reviews subject classifications.

    (b) `http://www.maa.org/`

    The Mathematics Association of America.

(c) http://www.siam.org/

The Society for Industrial and Applied Mathematics.

(d) http://www.ima.umn.edu

Institute for Mathematics and its Applications at the University of Minnesota.

(e) http://www.comap.com

The Consortium for Mathematics and its Applications, with links appropriate for elementary, high school, and college undergraduates. They also sponsor contests in mathematics for college students and high school students.

4. Software

(a) http://gams.cam.nist.gov

The Guide to Available Mathematical Software.

(b) http://www.mathtools.net

Mathematical tools (programs) in many different computing languages.

(c) http://www.gnu.org/software/gsl/

The GNU scientific library is a freely available numerical library in C and C++.

(d) http://www.netlib.org

The master listing for Netlib, containing many standard programs, including linpack, eispack, hompack, SPARC packages, and ODEpack.

(e) http://www.nag.co.uk

The home page of the Numerical Algorithms Group.

5. Journals, pre-prints, and essays

(a) http://dlmf.nist.gov

The "Digital Library of Mathematical Functions" from the National Institute of Standards and Technology.

(b) http://arxiv.org/archive/math

A mathematics preprint server based at the Los Alamos National Laboratory.

(c) http://www.mathcad.com/library/Constants/index.htm

A large collection of essays devoted to constants arising in mathematics.

(d) http://ejde.math.swt.edu/

The Electronic Journal of Differential Equations.

(e) http://nyjm.albany.edu:8000/nyjm.html

The New York Journal of Mathematics, the first electronic journal devoted to general mathematics.

(f) http://www.wavelet.org

The Wavelet Digest contains questions and answers about wavelets, and announcements of papers, books, journals, software, and conferences.

(g) http://www.math.ohio-state.edu/JAT

The Journal of Approximation Theory.

(h) http://www.combinatorics.org/ejc-wce.html

The Electronic Journal of Combinatorics and *World Combinatorics Exchange.*

(i) `http://rattler.cameron.edu/swjpam/swjpam.html`
    The Southwest Journal of Pure and Applied Mathematics.

6. Miscelleneous mathematical web sites

   (a) `http://www.math.hmc.edu/codee/main.html`
       The Consortium of Ordinary Differential Equation Experiments.

   (b) `http://www.utm.edu/research/primes/largest.html`
       Information about primes, including largest known primes of various types.

   (c) `http://www.dartmouth.edu/~chance/index.html`
       Material about teaching a "quantitative literacy course".

   (d) `http://aleph0.clarku.edu/~djoyce/julia/explorer.html`
       Useful for exploring the Mandelbrot and Julia sets.

   (e) `ftp://megrez.math.u-bordeaux.fr/pub/numberfields/`
       The Computational Number Theory group in Bordeaux has made available (by anonymous ftp at the above URL) extensive tables of number fields (almost 550000 number fields). For the number fields belonging to tables of reasonable length, this site contains the signature, the Galois group of the Galois closure of the field, the discriminant of the number field, the class number, the structure of the class group as a product of cyclic groups, an ideal in the class for each class generating these cyclic groups, the regulator, the number of roots of unity in the field, a generator of the torsion part of the unit group, and a system of fundamental units.

   (f) `http://www.eccpage.com`
       The Error Correcting Codes (ECC) home page provides free software implementing several important error-correcting codes.

   (g) `http://www.georgehart.com/virtual-polyhedra/vp.html`
       An online "Encyclopedia of Polyehdra".

   (h) `http://www.earlham.edu/~peters/knotlink.htm`
       A collection of internet links related to knots.

   (i) `http://www.mathsoft.com/asolve/`
       A large collection of unsolved mathematical problems, and pointers to other collections.

   (j) `http://www.clarku.edu/~djoyce/wallpaper`
       Pictures and descriptions of the 17 crystallographic groups; see page 307.

   (k) `http://hcoonce.math.mankato.msus.edu/`
       The mathematics genealogy project; given the name of a PhD mathematician, this site will tell you who their thesis advisor was.

EXAMPLE   The editor-in-chief of this book has the ancestral sequence of advisors:
D. I. Zwillinger $\longrightarrow$ B. S. White $\longrightarrow$ G. C. Papanicolaou $\longrightarrow$ J. B. Keller
$\longrightarrow$ R. Courant $\longrightarrow$ D. Hilbert $\longrightarrow$ C. L. F. Lindemann $\longrightarrow$ C. F. Klein
$\longrightarrow$ R. O. S. Lipschitz and J. Plucker $\longrightarrow$ G. P. L. Dirichlet $\longrightarrow$ S. D. Poisson
$\longrightarrow$ J. L. Lagrange $\longrightarrow$ L. Euler $\longrightarrow$ J. Bernoulli

10.10 BIOGRAPHIES OF MATHEMATICIANS

In alphabetical order:

1. Agnesi, Maria (page 814)
2. Ah'mose (page 810)
3. al-Haytham, Abu Ali (page 811)
4. al-Khwarizmi, Muhammad (page 811)
5. al-Tusi, Nasir al-Din (page 811)
6. Archimedes (page 810)
7. Banneker, Benjamin (page 814)
8. Bernoulli, Johann (page 813)
9. Bhaskara (page 811)
10. Brahmagupta (page 811)
11. Cardano, Gerolamo (page 812)
12. Cauchy, Augustin-Louis (page 814)
13. Cayley, Arthur (page 815)
14. Dedekind, Richard (page 815)
15. Descartes, René (page 812)
16. Dickson, Leonard Eugene (page 816)
17. Euclid (page 810)
18. Euler, Leonhard (page 813)
19. Fermat, Pierre de (page 813)
20. Gauss, Carl Friedrich (page 814)

21. Gerson, Levi ben (page 812)
22. Hamilton, William Rowan (page 814)
23. Hilbert, David (page 816)
24. Hypatia (page 811)
25. Jiushao, Qin (page 812)
26. Kovalevskaya, Sofia (page 815)
27. Lagrange, Joseph (page 814)
28. Leibniz, Gottfried Wilhelm (page 813)
29. Leonardo of Pisa (page 811)
30. Napier, John (page 812)
31. Newton, Isaac (page 813)
32. Noether, Emmy (page 816)
33. Pascal, Blaise (page 813)
34. Poincaré, Henri (page 816)
35. Ptolemy (page 811)
36. Riemann, Georg Bernhard (page 815)
37. Stevin, Simon (page 812)
38. Turing, Alan (page 816)
39. Viète, François (page 812)
40. Weierstrass, Karl (page 815)

In chronological order:

Ah'mose *(c. 1650 B.C.E.)* was the scribe responsible for copying the *Rhind Papyrus*, the most detailed original document still extant on ancient Egyptian mathematics. The papyrus contains some 87 problems with solutions dealing with what we consider first-degree equations, arithmetic progressions, areas and volumes of rectangular and circular regions, proportions, and several other topics. It also contains a table of the results of the division of 2 by every odd number from 3 to 101.

Euclid *(c. 300 B.C.E.)* is responsible for the most famous mathematics text of all time, the *Elements*. Not only does this work deal with the standard results of plane geometry, but it also contains three chapters on number theory, one long chapter on irrational quantities, and three chapters on solid geometry, culminating with the construction of the five regular solids. The axiom-definition-theorem-proof style of Euclid's work has become the standard for formal mathematical writing up to the present day.

Archimedes *(287–212 B.C.E.)* not only wrote several works on mathematical topics more advanced than Euclid, but also was the first mathematician to derive quantitative results from the creation of mathematical models of physical problems on earth. In several of his books, he described the reasoning process by which he arrived at his results in addition to giving formal proofs. For example, he showed how to calculate the areas of a segment of a parabola and the region bounded by one turn of a spiral, and the volume of a paraboloid of revolution.

Ptolemy *(c. 100–178 C.E.)* is most famous for the *Almagest*, a work in thirteen books, which contains a complete mathematical description of the Greek model of the universe with parameters for the various motions of the sun, moon, and planets. The first book provides the strictly mathematical material detailing the plane and spherical trigonometry, all based solely on the chord function, necessary for astronomical computations.

Hypatia *(c. 370–415)*, the first woman mathematician on record, lived in Alexandria. She was given a very thorough education in mathematics and philosophy by her father Theon and was responsible for detailed commentaries on several important Greek works, including Ptolemy's *Almagest*, Apollonius's *Conics*, and Diophantus's *Arithmetica*.

Brahmagupta *(c. 598–670)*, from Rajasthan in India, is most famous for his *Brahmasphutasiddhanta (Correct Astronomical System of Brahma)*, an astronomical work which contains many chapters on mathematics. Among the mathematical problems he considered and gave solution algorithms for were systems of linear congruences, quadratic equations, and special cases of the Pell equation $Dx^2 \pm 1 = y^2$. He also gave the earliest detailed treatment of rules for operating with positive and negative numbers.

Muhammad al-Khwarizmi *(c. 780–850)*, originally from Khwarizm in what is now Uzbekistan, was one of the first scholars called to the House of Wisdom in Baghdad by the caliph al-Ma'mun. He is best known for his algebra text, in which he gave a careful treatment of solution methods for quadratic equations. This Arabic text, after being translated into Latin in the twelfth century, provided Europeans with an introduction to algebra, a subject not considered by the ancient Greeks. Al-Khwarizmi's book on arithmetic provided Europe with one of its earliest looks at the Hindu-Arabic number system.

Abu Ali ibn al-Haytham *(965–1039)*, who spent much of his life in Egypt, is most famous for his work on optics, a work read and commented on for many centuries in Europe. In pure mathematics, he developed an inductive procedure for calculating formulae for the sums of integral powers of the first n integers, and used the formula for fourth powers to calculate the volume of the solid formed by revolving a parabola about a line perpendicular to its axis.

Bhaskara *(1114–1185)*, the most famous of medieval Indian mathematicians, gave a complete algorithmic solution to the Pell equation. In addition, he dealt with techniques of solving systems of linear equations with more unknowns than equations and was familiar with the basic combinatorial formulae, giving many examples, though no proofs, of their use.

Leonardo of Pisa *(1170–1240)*, often known today as Fibonacci, is most famous for his *Liber Abbaci (Book of Calculation)*, which contains the earliest publication of the Fibonacci numbers in the problem of how many pairs of rabbits can be bred in one year from one pair. Many of the sources of the book are in the Islamic world, where Leonardo spent much of his early life. The work contains the rules for computing with the new Hindu–Arabic numerals, many practical problems in such topics as calculation of profits and currency conversions, and topics now standard in algebra texts such as motion problems, mixture problems, and quadratic equations.

Nasir al-Din al-Tusi *(1201–1274)* was the head of a large group of astronomers at the observatory in Maragha, in what is now Iran. He computed a new set of very accurate astronomical tables and developed some new ideas on planetary motion which may have influenced Copernicus in working out his heliocentric system. In pure mathematics, al-Tusi's attempted proof of the parallel postulate was modified by his son and later published in Rome, where it influenced European work on non-Euclidean geometry. Al-Tusi also wrote the first systematic work on plane and spherical trigonometry, independent of astronomy, and gave the earliest proof of the theorem of sines.

Qin Jiushao *(1202–1261)*, born in Sichuan, published a general procedure for solving systems of linear congruences—the Chinese remainder theorem—in his *Shushu jiuzhang (Mathematical Treatise in Nine Sections)* in 1247, a procedure which makes essential use of the Euclidean algorithm. He also gave a complete description of a method for solving numerically polynomial equations of any degree. Qin's method was developed in China over a period of a thousand years or more and is very similar to what is now called Horner's method of solution, published by William Horner in 1819.

Levi ben Gerson *(1288–1344)* was a French rabbi and also an astronomer, philosopher, biblical commentator, and mathematician. His most famous mathematical work is the *Maasei Hoshev (The Art of the Calculator)*, which contains detailed proofs of the standard combinatorial formulae, some of which use the principle of mathematical induction.

Gerolamo Cardano *(1501–1576)*, a physician and gambler as well as a mathematician, wrote one of the earliest works containing systematic probability calculations, not all of which were correct. He is most famous, however, for his *Ars Magna (The Great Art, 1545)*, an algebra text which contained the first publication of the rules for solving cubic equations algebraically. Some of the rules had been discovered earlier in the sixteenth century by Scipione del Ferro and Niccoló Tartaglia.

François Viète *(1540–1603)*, a lawyer and advisor to two kings of France, was one of the earliest cryptanalysts and successfully decoded intercepted messages for his patrons. Although a mathematician only by avocation, he made important contributions to the development of algebra. In particular, he introduced letters to stand for numerical constants, thus enabling him to break away from the style of verbal algorithms of his predecessors and treat general examples by formulae rather than by giving rules for specific problems.

Simon Stevin *(1548–1620)* spent much of his life in the service of Maurice of Nassau, the Stadhouder of Holland, as a military engineer, advisor in finance and navigation, and quartermaster general of the Dutch army. In his book *De Thiende (The Art of Tenths)*, Stevin introduced decimal fractions to Europe, although they had previously been used in the Islamic world. Stevin's notation is different from our own, but he had a clear understanding of the advantage of decimals and advocated their use in all forms of measurement.

John Napier *(1550–1617)* was a Scottish laird who worked for years on the idea of producing a table which would enable one to multiply any desired numbers together by performing additions. These tables of logarithms first appeared in his 1614 book *Mirifici Logarithmorum Canonis Descriptio (Description of the Wonderful Canon of Logarithms)*. Napier's logarithms are different from, but related to, natural logarithms. His ideas were soon adapted by Henry Briggs, who eventually created the first table of common logarithms by 1628.

René Descartes *(1596–1650)* published the *Geometry* in 1637 as a supplement to his philosophical work, the *Discourse on the Method for Rightly Directing One's Reason and Searching for Truth in the Sciences*. In it, he developed the principles of analytic geometry, showing how to derive algebraic equations which represented geometric curves. The *Geometry* also contained methods for solving polynomial equations, including the modern factor theorem and Descartes' rule of signs.

Pierre de Fermat *(1601–1665)* was a French lawyer who spent his spare time doing mathematics. Not only was he a coinventor of analytic geometry, although his methods were somewhat different from those of Descartes, but he also was instrumental in the early development of probability theory and made many contributions to the theory of numbers. He is most remembered for the statement of his so-called "last theorem", that the equation $x^n + y^n = z^n$ has no non-trivial integral solution if $n > 2$, a theorem whose proof was finally completed by Andrew Wiles in 1994.

Blaise Pascal *(1623–1662)* showed his mathematical precocity with his *Essay on Conics* of 1640 in which he stated his theorem that the opposite sides of a hexagon inscribed in a conic section always intersect in three collinear points. Pascal is better known, however, for his detailed study of what is now called Pascal's triangle of binomial coefficients, the basic facts of which had been known in the Islamic and Chinese worlds for centuries. He also introduced the differential triangle in his *Treatise on the Sines of a Quadrant of a Circle*, an idea adopted by Leibniz in his calculus.

Isaac Newton *(1642–1727)*, the central figure in the Scientific Revolution, is most famous for his *Philosophiae Naturalis Principia Mathematica (Mathematical Principles of Natural Philosophy, 1687)*, in which he derived his system of the world based on his laws of motion and his law of universal gravitation. Over 20 years earlier, however, Newton had consolidated and generalized all the material on tangents and areas worked out by his predecessors into the magnificent problem solving tool of the calculus. He also developed the power series as a method of investigating various transcendental functions, stated the general binomial theorem, and, although never establishing his methods with the rigor of Greek geometry, did demonstrate an understanding of the concept of limit quite sufficient for him to apply the calculus to solve many important mathematical and physical problems.

Gottfried Wilhelm Leibniz *(1646–1716)*, born in Leipzig, developed his version of the calculus some ten years after Isaac Newton, but published it much earlier. Leibniz based his calculus on the inverse relationship of sums and differences, generalized to infinitesimal quantities called differentials. By clever manipulation of differentials, based in part on the geometrical model of the differential triangle, Leibniz was able to derive all of the basic rules of the differential and integral calculus and apply them to solve physical problems expressible in terms of differential equations. Leibniz's d and $\int$ notation for differentials and integrals turned out to be much more flexible and useful than Newton's dot notation and remains the notation of calculus to the present day.

Johann Bernoulli *(1667–1748)*, one of a number of prominent mathematicians of his Swiss family, was one of the earliest proponents of Leibniz's differential and integral calculus. Bernoulli helped to stimulate the development of the new techniques by proposing challenge problems to mathematicians, the most important probably being that of describing the brachistochrone, the curve representing the path of descent of a body between two given points in the shortest possible time. Many of the problems he posed required the solution of differential equations, and Bernoulli developed many techniques useful toward this end, including the calculus of the logarithmic and exponential functions.

Leonhard Euler *(1707–1783)*, a student of Johann Bernoulli in Basel who became one of the earliest members of the St. Petersburg Academy of Sciences founded by Peter the Great of Russia, was the most prolific mathematician of all time. His series of analysis texts, *Introduction to Analysis of the Infinite, Methods of the Differential Calculus*, and *Methods of the Integral Calculus*, established many of the notations and methods still in use today. Among his numerous contributions to every area of mathematics and physics are his development of the calculus of the trigonometric functions, the establishment of the theory of surfaces in differential geometry, and the creation of the calculus of variations.

Maria Agnesi *(1718–1799)*, the eldest child of a professor of mathematics at the University of Bologna, in 1748 published the clearest text on calculus up to that point. Based on the work of Leibniz and his followers, the work explained concepts lucidly and provided numerous examples, including some that have become standard in calculus texts to this day. Curiously, her name is often attached to a small item in her book not even original with her, a curve whose equation was $y = \pm \frac{a\sqrt{a-x}}{\sqrt{x}}$. This curve was called *la versiera*, derived from the Latin meaning "to turn"; unfortunately the word also was the abbreviation of the Italian word meaning "wife of the devil" and so was translated into English as "witch". The curve has ever since been known as the "witch of Agnesi".

Benjamin Banneker *(1731–1806)*, the first American black to achieve distinction in science, taught himself sufficient mathematics and astronomy to publish a series of well-regarded almanacs in the 1790s. He also assisted Andrew Ellicott in the survey of the boundaries of the District of Columbia. He was fond of solving mathematical puzzles and problems and recorded many of these in his notebooks.

Joseph Lagrange *(1736–1813)*, was born in Turin, becoming at age 19 a professor of mathematics at the Royal Artillery School there. He is most famous for his *Analytical Mechanics* (1788), a work which extended the mechanics of Newton and Euler, and demonstrated how problems in mechanics can generally be reduced to solutions of ordinary or partial differential equations. In 1797 he published his *Theory of Analytic Functions*, which attempted to reduce the ideas of calculus to those of algebraic analysis by assuming that every function could be represented as a power series. Although his central idea was incorrect, many of the proofs of basic theorems of calculus in this work were subsequently adapted by Cauchy into the forms still in use today.

Carl Friedrich Gauss *(1777–1855)* published his important work on number theory, the *Disquisitiones Arithmeticae*, when he was only 24, a work containing not only an extensive discussion of the theory of congruences, culminating in the quadratic reciprocity theorem, but also a detailed treatment of cyclotomic equations in which he showed how to construct regular n-gons by Euclidean techniques whenever n is prime and $n - 1$ is a power of 2. Gauss also made fundamental contributions to the differential geometry of surfaces in his *General Investigations of Curved Surfaces* in 1827, as well as to complex analysis, astronomy, geodesy, and statistics during his long tenure as a professor at the University of Göttingen. Many ideas later published by others, including the basics of non-Euclidean geometry, were found in his notebooks after his death.

Augustin-Louis Cauchy *(1789–1857)*, the most prolific mathematician of the nineteenth century, wrote several textbooks in analysis for use at the École Polytechnique, textbooks which became the model for calculus texts for the next hundred years. In his texts, Cauchy based the calculus on the notion of limit, using, for the first time, a definition which could be applied arithmetically to give proofs of some of the important results. Among numerous other subjects to which he contributed important ideas were complex analysis, in which he gave the first proof of the Cauchy integral theorem, the theory of matrices, in which he demonstrated that every symmetric matrix can be diagonalized by use of an orthogonal substitution, and the theory of permutations, in which he was the earliest to consider these from a functional point of view.

William Rowan Hamilton *(1805–1865)* became the Astronomer Royal of Ireland in 1827 because of his original work in optics accomplished during his undergraduate years at Trinity College, Dublin. In 1837, he showed how to introduce complex numbers into algebra axiomatically by considering $a + ib$ as a pair (a, b) of real numbers with appropriate computational rules. After many years of seeking an appropriate definition for multiplication rules for triples of numbers which could be applied to vector analysis in three-space, he discovered that it was in fact necessary to consider quadruplets of numbers. It was out of the natural definition of multiplication of these quaternions that the modern notions of dot product and cross product of vectors evolved.

Karl Weierstrass *(1815–1897)* taught for many years at German gymnasia before producing a series of brilliant mathematical papers in the 1850s which resulted in his appointment to a professorship at the University of Berlin. It was in his lectures there that he insisted on defining every concept of analysis arithmetically, including such ideas as uniform convergence and uniform continuity, thus completing the transformation away from the use of terms such as "infinitely small". Since he himself never published many of these ideas, his primary influence was through the work of his numerous students.

Arthur Cayley *(1821–1895)*, although graduating from Trinity College, Cambridge, as Senior Wrangler, became a lawyer because there was no suitable mathematics position available in England. He produced nearly 300 mathematical papers during his 14 years as a lawyer, however, and finally secured a professorship at Cambridge in 1863. Among his numerous mathematical achievements are the earliest abstract definition of a group in 1854, out of which he was able to calculate all possible groups of order up to eight and the basic rules for operating with matrices, including a statement (without rigorous proof) of the Cayley–Hamilton theorem that every matrix satisfies its characteristic equation.

Georg Bernhard Riemann *(1826–1866)*, in his 1854 inaugural lecture at the University of Göttingen entitled "On the Hypotheses which Lie at the Foundation of Geometry", discussed the general notion of an n-dimensional manifold, developed the idea of a metric relation on such a manifold, and gave criteria which would determine whether a three-dimensional manifold is Euclidean, or "flat". This lecture had enormous influence on the development of geometry, including non-Euclidean geometry, as well as on the development of a new concept of our physical space ultimately necessary for the theory of general relativity. Among his other achievements, Riemann's work on complex functions and their associated Riemann surfaces became one of the foundations of combinatorial topology.

Richard Dedekind *(1831–1916)* solved the problem of the lack of unique factorization in rings of algebraic integers by introducing ideals and their arithmetic and demonstrating that every ideal is either prime or can be expressed uniquely as a product of prime ideals. During his teaching at Zurich in the late 1850s, he realized that, although differential calculus deals with continuous magnitudes, there was no satisfactory definition available of what it means for the set of real numbers to be continuous. He therefore worked out a definition of irrational numbers through his idea of what is now called a Dedekind cut in the set of rational numbers. Somewhat later Dedekind also considered the basic ideas of set theory and gave a set theoretic characterization of the natural numbers.

Sofia Kovalevskaya *(1850–1891)* was the first European woman since the Renaissance to earn a Ph.D. in mathematics (1874), a degree based on her many new results in the theory of partial differential equations. Because women were generally not permitted to study mathematics officially in European universities, Kovalevskaya had been forced to study privately with Weierstrass. Her mathematical talents eventually earned her a professorship at the University of Stockholm, an editorship of the journal *Acta Mathematica*, and the Prix Bordin of the French Academy of Sciences for her work on the revolution of a solid body about a fixed point. Unfortunately, her career was cut short by her untimely death from pneumonia at the age of 41.

Henri Poincaré *(1854–1912)*, one of the last of the universal mathematicians, contributed to virtually every area of mathematics, including physics and theoretical astronomy. Among his many contributions was the introduction of the idea of homology into topology, the creation of a model of Lobachevskian geometry which helped to convince mathematicians that this non-Euclidean geometry was as valid as Euclid's, and a detailed study of the non-linear partial differential equations governing planetary motion aimed at answering questions about the stability of the solar system. Toward the end of his life, Poincaré wrote several popular books emphasizing the importance of science and mathematics.

David Hilbert *(1862–1943)* is probably most famous for his lecture at the International Congress of Mathematicians in Paris in 1900 in which he presented a list of 23 problems which he felt would be of central importance for 20th century mathematics. Most of the problems have now been solved, while significant progress has been achieved in the remainder. Hilbert himself made notable contributions to the study of algebraic forms, algebraic number theory, the foundations of geometry, integral equations, theoretical physics, and the foundations of mathematics.

Leonard Eugene Dickson *(1874–1954)* was the first recipient of a doctorate in mathematics at the University of Chicago, where he ultimately spent most of his mathematical career. Dickson helped to develop the abstract approach to algebra by developing sets of axioms for such constructs as groups, fields, and algebras. Among his important books was his monumental three volume *History of the Theory of Numbers*, which traced the evolution of every important concept in that field.

Emmy Noether *(1882–1935)* received her doctorate from the University of Erlangen in 1908, a few years later moving to Göttingen to assist Hilbert in the study of general relativity. During her 18 years there, she was extremely influential in stimulating a new style of thinking in algebra by always emphasizing its structural rather than computational aspects. She is most famous for her work on what are now called Noetherian rings, but her inspiration of others is still evident in today's textbooks in abstract algebra.

Alan Turing *(1912–1954)* developed the concept of a "Turing machine" in 1936 to answer the questions of what a computation is and whether a given computation can in fact be carried out. This notion today lies at the basis of the modern all-purpose computer, a machine which can be programmed to do any desired computation. During World War II, Turing led the successful effort in England to crack the German "Enigma" code, an effort central to the defeat of Nazi Germany.

List of References

Chapter 1   *Analysis*

1. J. W. Brown and R. V. Churchill, *Complex variables and applications*, 6th edition, McGraw–Hill, New York, 1996.

2. L. B. W. Jolley, *Summation of Series*, Dover Publications, New York, 1961.

3. S. G. Krantz, *Real Analysis and Foundations,* CRC Press, Boca Raton, FL, 1991.

4. S. G. Krantz, *The Elements of Advanced Mathematics,* CRC Press, Boca Raton, FL, 1995.

5. J. P. Lambert, "Voting Games, Power Indices, and Presidential Elections", *The UMAP Journal*, Module 690, **9**, No. 3, pages 214–267, 1988.

6. L. D. Servi, "Nested Square Roots of 2", *American Mathematical Monthly*, to appear in 2003.

7. N. J. A. Sloane and S. Plouffe, *Encyclopedia of Integer Sequences*, Academic Press, New York, 1995.

Chapter 2   *Algebra*

1. C. Caldwell and Y. Gallot, "On the primality of $n! \pm 1$ and $2 \times 3 \times 5 \times \cdots \times p \pm 1$", *Mathematics of Computation*, 71:237, pages 441–448, 2002.

2. I. N. Herstein, *Topics in Algebra*, 2nd edition, John Wiley & Sons, New York, 1975.

3. P. Ribenboim, *The book of Prime Number Records*, Springer–Verlag, New York, 1988.

4. G. Strang, *Linear Algebra and Its Applications*, 3rd edition, International Thomson Publishing, 1988.

Chapter 3   *Discrete Mathematics*

1. B. Bollobás, *Graph Theory*, Springer–Verlag, Berlin, 1979.

1-58488-291-3/02/$0.00+$1.50
© 2003 CRC Press, Inc.

2. C. J. Colbourn and J. H. Dinitz, *Handbook of Combinatorial Designs*, CRC Press, Boca Raton, FL, 1996.

3. F. Glover, "Tabu Search: A Tutorial", *Interfaces*, 20(4), pages 74–94, 1990.

4. D. E. Goldberg, *Genetic Algorithms in Search, Optimization, and Machine Learning*, Addison–Wesley, Reading, MA, 1989.

5. J. Gross, *Handbook of Graph Theory & Applications*, CRC Press, Boca Raton, FL, 1999.

6. D. Luce and H. Raiffa, *Games and Decision Theory*, Wiley, 1957.

7. F. J. MacWilliams and N. J. A. Sloane, *The Theory of Error-Correcting Codes*, North–Holland, Amsterdam, 1977.

8. N. Metropolis, A. W. Rosenbluth, M. N. Rosenbluth, A. H. Teller and E. Teller, "Equation of State Calculations by Fast Computing Machines", *J. Chem. Phys.*, V 21, No. 6, pages 1087–1092, 1953.

9. K. H. Rosen, *Handbook of Discrete and Combinatorial Mathematics*, CRC Press, Boca Raton, FL, 2000.

10. J. O'Rourke and J. E. Goodman, *Handbook of Discrete and Computational Geometry*, CRC Press, Boca Raton, FL, 1997.

Chapter 4   *Geometry*

1. A. Gray, *Modern Differential Geometry of Curves and Surfaces,* CRC Press, Boca Raton, FL, 1993.

2. C. Livingston, *Knot Theory*, The Mathematical Association of America, Washington, D.C., 1993.

3. D. J. Struik, *Lectures in Classical Differential Geometry*, 2nd edition, Dover, New York, 1988.

Chapter 5   *Continuous Mathematics*

1. A. G. Butkovskiy, *Green's Functions and Transfer Functions Handbook*, Halstead Press, John Wiley & Sons, New York, 1982.

2. I. S. Gradshteyn and M. Ryzhik, *Tables of Integrals, Series, and Products*, edited by A. Jeffrey and D. Zwillinger, 6th edition, Academic Press, Orlando, Florida, 2000.

3. N. H. Ibragimov, Ed., *CRC Handbook of Lie Group Analysis of Differential Equations*, Volume 1, CRC Press, Boca Raton, FL, 1994.

4. A. J. Jerri, *Introduction to Integral Equations with Applications*, Marcel Dekker, New York, 1985.

5. P. Moon and D. E. Spencer, *Field Theory Handbook*, Springer-Verlag, Berlin, 1961.

6. A. D. Polyanin and V. F. Zaitsev, *Handbook of Exact Solution for Ordinary Differential Equations*, CRC Press, Boca Raton, FL, 1995.

7. J. A. Schouten, *Ricci-Calculus*, Springer–Verlag, Berlin, 1954.

8. J. L. Synge and A. Schild, *Tensor Calculus*, University of Toronto Press, Toronto, 1949.

9. D. Zwillinger, *Handbook of Differential Equations*, 3rd ed., Academic Press, New York, 1997.

10. D. Zwillinger, *Handbook of Integration*, A. K. Peters, Boston, 1992.

Chapter 6 *Special Functions*

1. Staff of the Bateman Manuscript Project, A. Erdélyi, Ed., *Tables of Integral Transforms*, in 3 volumes, McGraw–Hill, New York, 1954.

2. I. S. Gradshteyn and M. Ryzhik, *Tables of Integrals, Series, and Products*, edited by A. Jeffrey and D. Zwillinger, 6th edition, Academic Press, Orlando, Florida, 2000.

3. W. Magnus, F. Oberhettinger, and R. P. Soni, *Formulas and Theorems for the Special Functions of Mathematical Physics*, Springer–Verlag, New York, 1966.

4. N. I. A. Vilenkin, *Special Functions and the Theory of Group Representations*, American Mathematical Society, Providence, RI, 1968.

Chapter 7 *Probability and Statistics*

1. I. Daubechies, *Ten Lectures on Wavelets*, SIAM Press, Philadelphia, 1992.

2. W. Feller, *An Introduction to Probability Theory and Its Applications*, Volume 1, John Wiley & Sons, New York, 1968.

3. J. Keilson and L. D. Servi, "The Distributional Form of Little's Law and the Fuhrmann–Cooper Decomposition", *Operations Research Letters*, Volume 9, pages 237–247, 1990.

4. *Military Standard 105 D*, U.S. Government Printing Office, Washington, D.C., 1963.

5. S. K. Park and K. W. Miller, "Random number generators: good ones are hard to find", *Comm. ACM*, October 1988, 31, 10, pages 1192–1201.

6. G. Strang and T. Nguyen, *Wavelets and Filter Banks*, Wellesley–Cambridge Press, Wellesley, MA, 1995.

7. D. Zwillinger and S. Kokoska, *Standard Probability and Statistics Tables and Formulae*, Chapman & Hall/CRC, Boca Raton, Florida, 2000.

Chapter 8 *Scientific Computing*

1. R. L. Burden and J. D. Faires, *Numerical Analysis*, 7th edition, Brooks/Cole, Pacific Grove, CA, 2001.

2. G. H. Golub and C. F. Van Loan, *Matrix Computations*, 2nd ed., The Johns Hopkins Press, Baltimore, 1989.

3. W. H. Press, S. A. Teukolsky, W. T. Vetterling, and B. P. Flannery, *Numerical Recipes in C++: The Art of Scientific Computing*, 2nd edition, Cambridge University Press, New York, 2002.

4. A. Ralston and P. Rabinowitz, *A First Course in Numerical Analysis*, 2nd edition, McGraw–Hill, New York, 1978.

5. R. Rubinstein, *Simulation and the Monte Carlo Method*, Wiley, New York, 1981.

Chapter 10   *Miscellaneous*

1. American Mathematical Society, *Mathematical Sciences Professional Directory*, Providence, 1995.

2. E. T. Bell, *Men of Mathematics,* Dover, New York, 1945.

3. C. C. Gillispie, Ed., *Dictionary of Scientific Biography*, Scribners, New York, 1970–1990.

4. H. S. Tropp, "The Origins and History of the Fields Medal", *Historia Mathematica*, 3, pages 167–181, 1976.

5. E. W. Weisstein, *CRC Concise Encyclopedia of Mathematics*, CRC Press, Boca Raton, FL, 1999.

List of Figures

2.1   Depiction of right-hand rule 134

3.1   Hasse diagrams 205
3.2   Three graphs that are isomorphic 225
3.3   Examples of graphs with 6 or 7 vertices. 231
3.4   Trees with 7 or fewer vertices 242
3.5   Trees with 8 vertices 243
3.6   Julia sets 274
3.7   The Mandlebrot set 274
3.8   Directed network modeling a flow problem 282

4.1   Change of coordinates by a rotation 301
4.2   Cartesian coordinates: the 4 quadrants 302
4.3   Polar coordinates 302
4.4   Homogeneous coordinates 304
4.5   Oblique coordinates 304
4.6   A shear with factor $r = \frac{1}{2}$ 313
4.7   A perspective transformation 314
4.8   The normal form of a line 315
4.9   Simple polygons 317
4.10   Notation for a triangle 319
4.11   Triangles: isosceles and right 320
4.12   Ceva's theorem and Menelaus's theorem 321
4.13   Quadrilaterals 322
4.14   Conics: ellipse, parabola, and hyperbola 325
4.15   Conics as a function of eccentricity 326
4.16   Ellipse and components 326
4.17   Hyperbola and components 327
4.18   Arc of a circle 335
4.19   Angles within a circle 335
4.20   The general cubic parabola 336
4.21   Curves: semi-cubic parabola, cissoid of Diocles, witch of Agnesi 337
4.22   The folium of Descartes in two positions, and the strophoid . . . 338

4.23	Cassini's ovals	338
4.24	The conchoid of Nichomedes	339
4.25	The limaçon of Pascal	339
4.26	Cycloid and trochoids	340
4.27	Epicycloids: nephroid, and epicycloid	341
4.28	Hypocycloids: deltoid and astroid	342
4.29	Spirals: Bernoulli, Archimedes, and Cornu	343
4.30	Cartesian coordinates in space	346
4.31	Cylindrical coordinates	346
4.32	Spherical coordinates	347
4.33	Relations between Cartesian, cylindrical, and spherical coordinates	347
4.34	Euler angles	350
4.35	The Platonic solids	359
4.36	Cylinders: oblique and right circular	361
4.37	Right circular cone and frustram	362
4.38	A torus of revolution	363
4.39	The five nondegenerate real quadrics	366
4.40	Spherical cap, zone, and segment	367
4.41	Right spherical triangle and Napier's rule	369
5.1	Types of critical points	468
6.1	Notation for trigonometric functions	504
6.2	Definitions of angles.	504
6.3	Sine and cosine	505
6.4	Tangent and cotangent	506
6.5	Different triangles requiring solution	515
6.6	Graphs of $\Gamma(x)$ and $1/\Gamma(x)$	541
6.7	Cornu spiral	547
6.8	Sine and cosine integrals $\mathrm{Si}(x)$ and $\mathrm{Ci}(x)$	549
6.9	Legendre functions	555
6.10	Graphs of the Airy functions $\mathrm{Ai}(x)$ and $\mathrm{Bi}(x)$	566
7.1	Approximation to binomial distributions	631
7.2	Conceptual layout of a queue	637
7.3	Sample size code letters for MIL-STD-105 D	653
7.4	Master table for single sampling inspection (normal inspection) .	654
7.5	Area of a normal random variable	695
7.6	Illustration of σ and 2σ regions of a normal distribution	696
8.1	Illustration of Newton's method	731
8.2	Formulae for integration rules with various weight functions . .	754
8.3	Illustration of the Monte–Carlo method	762

$\Phi(x)$ normal distribution function ...634
Θ asymptotic function 75
$\Upsilon(G)$ graph arboricity 220
α
    $\alpha(G)$ graph independence number
        225
    $\alpha(k)$ function, related to zeta
        function 23
    one minus the confidence
        coefficient 666
    probability of type I error 661
β
    β probability of type II error .. 661
    $\beta(k)$ function, related to zeta
        function 23
χ
    $\chi'(G)$ chromatic index 221
    $\chi(G)$ chromatic number 221
    χ^2-distribution 703
    χ^2_α critical value 696
    χ^2_n chi-square distributed 619
δ
    $\delta(G)$ minimum vertex degree ..223
    $\delta(x)$ delta function 76
    δ_{ij} Kronecker delta 483
    designed distance 257
    Feigenbaum's constant 272
$\epsilon_{i_1 \cdots i_n}$ Levi–Civita symbol 489
η
    η power of a test 661
    $\eta(x, y)$ component of infinitesimal
        generator 466
γ
    γ Euler's constant
        definition 15
        in different bases 16
        value 16
    $\gamma(G)$ graph genus 224
    $\gamma(k)$ function, related to zeta
        function 23
    γ_1 skewness 620
    γ_2 excess 620
κ
    $\kappa(G)$ connectivity 222
    $\kappa(s)$ curvature 374
    κ_n cumulant 620
λ
    $\lambda(G)$ edge connectivity 223
    average arrival rate 638
    eigenvalue 152, 477, 478
    number of blocks 241

μ
    $\mu(n)$ Möbius function 102
    μ_k centered moments 620
    μ'_k moments 620
    μ_p MTBF for parallel system ..655
    μ_s MTBF for series system ... 655
    average service rate 638
    mean 620
ν
    $\overline{\nu}(G)$ rectilinear graph crossing
        number 222
    $\nu(G)$ graph crossing number .. 222
$\omega(G)$ size of the largest clique 221
ϕ
    $\phi(n)$ totient function 128, 169
    $\phi(t)$ characteristic function 620
    Euler constant 21
    golden ratio
        defined 16
        value 16
    incidence mapping 219
    zenith 346
π
    $\pi(x)$
        prime counting function 103
        probability distribution 640
    constants containing 14
    continued fraction 97
    distribution of digits 15
    identities 14
    number 13
        in different bases 16
    permutation 172
    sums involving 24
    $\psi(z)$ logarithmic derivative of the
        gamma function 543
ρ
    $\rho(A)$ spectral radius 154
    $\rho(s)$ radius of curvature 374
    ρ_{ij} correlation coefficient 622
    server utilization 638
σ
    σ standard deviation 620
    $\sigma(n)$ sum of divisors 128
    σ^2 variance 620
    σ_i singular value of a matrix .. 152
    $\sigma_k(n)$ sum of kth powers of divisors
        128
    σ_{ii} variance 622
    σ_{ij} covariance 622

τ

τ Ramanujan function 31
$\tau(n)$ number of divisors 128
$\tau(s)$ torsion 374

θ

$\theta(G)$ graph thickness 227
angle in polar coordinates 302
argument of a complex number . 53
azimuth 346

ξ

$\xi(x, y)$ component of infinitesimal
generator 466
ξ_p quantile of order p 659
$\zeta(k)$ Riemann zeta function 23

Numbers

$()^{-1}$ group inverse 161
$()^{-1}$ matrix inverse 138
0 null vector 137
1
1, group identity 161
1-form 395
10, powers of 6, 13, 798
105 D standard 652
16, powers of 12
17 crystallographic groups 307
2
2^A power set of A 203
2^*22 crystallographic group ... 310
2, negative powers of 10
2, powers of 6, 10, 27
2-$(v,3,1)$ Steiner triple system . 249
2-form 396
2-sphere 491
2-switch 227
22^* crystallographic group 309
$22^\times$ crystallographic group 309
2222 crystallographic group ... 310
230 crystallographic groups,
three-dimensional 307
3
3^*3 crystallographic group 311
3, powers of 29
3-design (Hadamard matrices) . 250
3-form 397
3-sphere 491
333 crystallographic group 311
360, degrees in a circle 503

4
4^*2 crystallographic group 310
4, powers of 30
442 crystallographic group 310
5
5, powers of 30
5-(12,6,1) table 244
5-design, Mathieu 244
632 crystallographic group 311

Roman Letters

A

A

A interarrival time 637
$A(n, d)$ number of codewords . 259
$\mathcal{A}(T)$ skew symmetric part of a
tensor 484
A ampere 792
$A_{()}$

A_4 alternating group on 4 elements
188
A_k radius of circumscribed circle
324
A_n alternating group 163, 172
$A/B/c/K/m/Z$ queue 637
$Ai(z)$ Airy function 465, 565
ALFS additive lagged-Fibonacci
sequence 646
AMS American Mathematical Society
801
ANOVA analysis of variance 686
AOQ average outgoing quality 652
AOQL average outgoing quality limit 652
AQL acceptable quality level 652
$AR(k)$ autoregressive model 718
$ARMA(k, l)$ mixed model 719
$Aut(G)$ graph automorphism group . 220
$a_{()}$

$\mathbf{a}_i$ unit vector 492
a_n Fourier coefficients 48
a_n proportion of customers 637
$a.e.$ almost everywhere 74
am amplitude 572
arg argument 53

B

B

B amount borrowed 779
B service time 637
$B(p,q)$ beta function 544
$\mathcal{B}$ set of blocks 241

$B_{()}$

B_n Bell number 211
B_n Bernoulli number 19
B_n a block 241
$B_n(x)$ Bernoulli polynomial ... 19
B.C.E (before the common era, B.C.) 810
BFS basic feasible solution 283
Bi(z) Airy function 465, 565
BIBD balanced incomplete block design 245
Bq becquerel 792
b unit binormal vector 374

C

C

C channel capacity 255
$C(n,r)$ r-combination ... 206, 215
$C(x)$ Fresnel integral 547
$C^R(n,r)$ combinations with replacement 206
$\mathbb{C}$ complex numbers 3, 167
$\mathbb{C}^n$ complex n element vectors 131
$\mathcal{C}$ integration contour 399, 404
C coulomb 792
C Roman numeral (100) 4

$C_{()}$

C_2 cyclic group of order 2 178
$C_2 \times C_2 \times C_2$ direct group product 181
C_3 cyclic group of order 3 178
$C_3 \times C_3$ direct group product . 184
C_4 cyclic group of order 4 178
$C_4 \times C_2$ direct group product . 181
C_5 cyclic group of order 5 179
C_6 cyclic group of order 6 179
C_7 cyclic group of order 7 180
C_8 cyclic group of order 8 180
C_9 cyclic group of order 9 184
C_n Catalan numbers 212
C_n cycle graph 229
C_n cyclic group 172
C_{10} cyclic group of order 10 .. 185
C.E. (common era, A.D.) 810
Ci(z) cosine integral 549

c

c cardinality of real numbers .. 204
c number of identical servers .. 637
c speed of light 794
cas combination of sin and cos 591
cd candela 792
cm crystallographic group 309
cmm crystallographic group 310
c_n Fourier coefficients 50
cn(u,k) elliptic function 572
cof$_{ij}(A)$ cofactor of matrix A 145
cond(A) condition number 148
cos trigonometric function 505
cosh hyperbolic function 524
cot trigonometric function 505
coth hyperbolic function 524
covers trigonometric function 505
csc trigonometric function 505
csch hyperbolic function 524
cyc number of cycles 172

D

D

D constant service time 637
D diagonal matrix 138
D differentiation operator 456, 466
D Roman numeral (500) 4

$D_{()}$

D_4 dihedral group of order 8 .. 182
D_5 dihedral group of order 10 . 185
D_6 dihedral group of order 12 . 186
D_f region of convergence 595
D_n derangement 210
D_n dihedral group 163, 172
DFT discrete Fourier transform 582
DLG$(n;a,b)$ double loop graph 230

d

$d(u,v)$ distance between vertices 223
derivative operator 386
exterior derivative 397
minimum distance 256

$d_{()}$

d_n proportion of customers 637
$d_H(\mathbf{u},\mathbf{v})$ Hamming distance ... 256
$dx_k(\mathbf{a})$ projection 395
det(A) determinant of matrix A 144
diam(G) graph diameter 223
div divergence 493
dn(u,k) elliptic function 572

dS differential surface area 405
dV differential volume 405
$d\mathbf{x}$ fundamental differential 377

E

E

E edge set 219
E event 617
$E(u, v)$ first fundamental metric
    coefficient 377
$\mathrm{E}[\,]$ expectation operator 619

$E_{(\,)}$

E_k Erlang-k service time 637
E_n Euler numbers 20
$E_n(x)$ Euler polynomial 20
$E_n(x)$ exponential integral 550
E_p identity group 172
E_{ij} elementary matrix 138
Ei exbi 6

e

e algebraic identity 161
e charge of electron 794
e constants containing 15
e continued fraction 97
e definition 15
e eccentricity 325
e in different bases 16
$e(u, v)$ second fundamental metric
    coefficient 377

$e_{(\,)}$

$\mathbf{e}$ vector of ones 137
$\mathbf{e}_i$ unit vector 137
$e_{i_1 \cdots i_n}$ permutation symbol ...489
ecc(v) eccentricity of a vertex 223
erf error function 545
erfc complementary error function ..545
exsec trigonometric function 505

F

F

$F(u, v)$ first fundamental metric
    coefficient 377
$F(x)$ Dawson's integral 546
$F(x)$ probability distribution
    function 619
$\mathcal{F}$ Fourier transform 576
$F(a, b; c; z)$ hypergeometric
    function 553
$\widehat{F}(x)$ sample distribution function
    658

F farad 792

$F_{(\,)}$

F_n Fibonacci numbers 21
F_α critical value 696
F_{p^n} Galois field 169
F_c Fourier cosine transform ...582
F_s Fourier sine transform 582
$\mathcal{F}_N$ discrete Fourier transform .582
$_pF_q$ hypergeometric function ...36
FCFS first come, first served 637
FFT fast Fourier transform 584
FIFO first in, first out 637

f

$\widehat{f}(x)$ sample density function ..658
$f(u, v)$ second fundamental metric
    coefficient 377
$f(x)$ probability density function
    619
$f_{X(t)}$ density function 467

G

G

G Green's function 471
G general service time distribution
    637
G generating matrix 256
G graph 219
G gravitational constant 794
G primitive root 195
$G(k)$ Waring's problem 100
$G(s)$ generating function 620
$G(u, v)$ first fundamental metric
    coefficient 377
$G(x; z)$ Green's function 463
$G[x_n]$ generating function 268
$\mathbf{G}$ Catalan constant 23
$\mathcal{G}_{n,m}$ isomorphism classes 241
GCD greatest common divisor 101
$GF(p^n)$ Galois field 169
GI general interarrival time 637
Gi gibi 6
$GL(n, \mathbb{C})$ matrix group 171
$GL(n, \mathbb{R})$ matrix group 171
G.M. geometric mean 659
Gy gray 792
gir($(\,)G$) graph girth 224

g

g determinant of the metric tensor
    489
g metric tensor 487

g primitive root 195
$g(k)$ Waring's problem 100
$g(x)$ generating polynomial ... 256
gd function 530
gd x Gudermannian function 530
g^{ij} covariant metric 486
g_{ij} contravariant metric 486
glb greatest lower bound 68

H

H

H mean curvature 377
H parity check matrix 256
$H(\mathbf{p}_X)$ entropy 253
$H(x)$ Haar wavelet 723
$H(x)$ Heaviside function .. 77, 408
$\mathcal{H}$ Hilbert transform 591
^{H} Hermitian conjugate 138
H henry 792

$H_{(\,)}$

H_0 null hypothesis 661
H_1 alternative hypothesis 661
$H_\nu^{(1)}$ Hankel function 559
$H_\nu^{(2)}$ Hankel function 559
H_k k-stage hyperexponential
    service time 637
H_n harmonic numbers 32
$H_n(x)$ Hermite polynomials .. 532
$\mathcal{H}_\nu$ Hankel transform 589
H.M. harmonic mean 660
Hz hertz 792
hav trigonometric function 372, 505
h_i metric coefficients 492

I

I

I first fundamental form 377
I identity matrix 138
$I(X,Y)$ mutual information .. 254
I Roman numeral (1) 4
ICG inversive congruential generator 646
II second fundamental form 377
Im imaginary part of a complex number
    53
I_n identity matrix 138
Inv number of invariant elements ... 172
IVP initial-value problem 265

i

$\mathbf{i}$ unit vector 494
$\hat{\mathbf{i}}$ unit vector 135
i imaginary unit 53
i interest rate 779
iid independent and identically
    distributed 619
inf greatest lower bound 68
infimum greatest lower bound 68

J

J

J Jordan form 154
J joule 792

j

$\mathbf{j}$ unit vector 494
$\hat{\mathbf{j}}$ unit vector 135

$J_{(\,)}$

$J_\nu(z)$ Bessel function 559
J_c Julia set 273

$j_{(\,)}$

$j_n(z)$ half order Bessel function
    563
$j_{\nu,k}$ zero of Bessel function ... 563

K

K

K Gaussian curvature 377
K system capacity 637
K Kelvin (degrees) 792

$K_{(\,)}$

K_n complete graph 229
$K_{m,n}$ complete bipartite graph 230
$K_{n_1,\ldots,n_k}$ complete multipartite
    graph 230
$\overline{K}_n$ empty graph 229
Ki kibi 6

k

$\mathbf{k}$ curvature vector 374
$\mathbf{k}$ unit vector 494
$\hat{\mathbf{k}}$ unit vector 135
k Boltzmann constant 794
k dimension of a code 258
$k(x,t)$ kernel 478

$k_{(\,)}$

$\mathbf{k}_g$ geodesic curvature 377
$\mathbf{k}_n$ normal curvature vector 377
k_j block size 241
kg kilogram 792

L

L

L average number of customers 638
L period 48
$L(\theta)$ expected loss function ... 656
$\mathcal{L}$ Laplace transform 585
L length 796
L Roman numeral (50) 4

$L_{(\,)}$

L_1 norm 133
L_2 norm 133
L_Q average number of customers 638
L_p norm 73
L_r Lie group 466
L_∞ space of measurable functions 73

LCG linear congruential generator .. 644
LCL lower control limit 650
LCM least common multiple 101
$\mathrm{Li}_1(z)$ logarithm 551
$\mathrm{Li}_2(z)$ dilogarithm 551
LIFO last in, first out 637
$\mathrm{Li}_n(z)$ polylogarithm 551
$\mathrm{li}(x)$ logarithmic integral 550
LP linear programming 280
LTPD lot tolerance percent defective 652
$\ell(\theta, a)$ loss function 656
lim limits 70, 385
liminf limit inferior 70
limsup limit superior 70
lm lumen 792
ln logarithmic function 522
log logarithmic function 522
$\log_b$ logarithm to base b 522
lub least upper bound 68
lux lux 792

M

M

M Mandelbrot set 273
M exponential service time ... 637
M number of codewords 258
$M(P)$ measure of a polynomial 93
$\mathcal{M}$ Mellin transform 612
M mass 796
M Roman numeral (1000) 4
$\mathrm{MA}(l)$ moving average 719
M.D. mean deviation 660

MFLG multiplicative lagged-Fibonacci generator 646
$M/G/1$ queue 639
$M/G/c/c$ queue 639
$M/G/\infty$ queue 639
Mi mebi 6
MLE maximum likelihood estimator 662
$M/M/1$ queue 638
$M/M/c$ queue 639
M_n Möbius ladder graph 229
MOLS mutually orthogonal Latin squares 251
MOM method of moments 662
MTBF mean time between failures ..655

m

m mortgage amount 779
m number in the source 637
m meter 792
mid midrange 660
mod modular arithmetic 94
mol mole 792

N

N

N number of zeros 58
$N(A)$ null space 149
$N(\mu, \sigma)$ normal random variable 619
N unit normal vector 378
$\mathcal{N}$ normal vector 377
$\mathbb{N}$ natural numbers 3
N newton 792
$N_q(n)$ number of monic irreducible polynomials 261

n

n principal normal unit vector . 374
$\hat{\mathbf{n}}$ unit normal vector 135
n code length 258
n number of time periods 779
n order of a plane 248

O

O asymptotic function 75
$O(n)$ matrix group 171
O_n odd graph 229
o asymptotic function 75

P

P

P number of poles 58

P principal 779

$P(B \mid A)$ conditional probability
617

$P(E)$ probability of event E .. 617

$P(\nu, z)$ auxiliary function 561

$P(m, r)$ r-permutation 215

$P(n, r)$ r-permutation 206

$P(x, y)$ Markov transition function
640

$P\{ \}$ Riemann P function 465

$P_{(\ )}$

$P_G(x)$ chromatic polynomial .. 221

P_n path (type of graph) 229

$P_n(x)$ Lagrange interpolating
polynomial 733

$P_n(x)$ Legendre function 465

$P_n(x)$ Legendre polynomials .. 534

$P_n^{(\alpha,\beta)}(x)$ Jacobi polynomials . 533

$P_\nu(z)$ Legendre function 554

$P_\nu^\mu(x)$ associated Legendre
functions 557

Pa pascal 792

Per(x_n) period of a sequence 644

Pi pebi 6

PID principal ideal domain 165

$P^n(x, y)$ n-step Markov transition
matrix 641

$P^R(n, r)$ permutations with replacement
206

PRI priority service 637

PRNG pseudorandom number generator
644

p

$p(n)$ partitions 210

$p\#$ product of prime numbers . 106

p1 crystallographic group 309, 311

p2 crystallographic group 310

p3 crystallographic group 311

p31m crystallographic group 311

p3m1 crystallographic group 311

p4 crystallographic group 310

p4g crystallographic group 310

p4m crystallographic group 310

p6 crystallographic group 311

p6m crystallographic group 311

per permanent 145

pg crystallographic group 309

pgg crystallographic group 309

pm crystallographic group 309

pmg crystallographic group 309

pmm crystallographic group 310

$p_{(\ )}$

$\mathbf{p}_{X \times Y}$ joint probability distribution
254

p_k discrete probability 619

$p_k(n)$ partitions 207

$p_m(n)$ restricted partitions 210

p_n proportion of time 638

Q

Q

Q quaternion group 182

$Q(\nu, z)$ auxiliary function 561

$\mathbb{Q}$ rational numbers 3, 167

$Q_{(\ )}$

Q_n cube (type of graph) 229

$Q_n(x)$ Legendre function 465

$Q_\nu(z)$ Legendre function 554

$Q_\nu^\mu(x)$ associated Legendre functions
557

q nome 574

R

R

R Ricci tensor 485, 488

R Riemann tensor 488

R curvature tensor 485

R radius (circumscribed circle) 319,
513

R range 650

R rate of a code 255

$R(A)$ range space 149

$R(\theta, d_i)$ risk function 657

$R(t)$ reliability function 655

$\mathbb{R}$ continuity in 71

$\mathbb{R}$ convergence in 70

$\mathbb{R}$ real numbers 3, 167

$R_{(\ )}$

R_i reliability of a component .. 653

R_p reliability of parallel system 653

R_s reliability of series system . 653

$R_\oplus$ radius of the earth 372

$\mathbb{R}^n$ real n element vectors 131

$\mathbb{R}^{n \times m}$ real $n \times m$ matrices 137

Re real part of a complex number 53

R.M.S. root mean square 660

RSS random service 637
r
    r distance in polar coordinates . 302
    r modulus of a complex number 53
    r radius (inscribed circle) . 318, 512
    r shearing factor 352
    $r(\theta, a)$ regret function 658
rad(()G) radius of graph 226
rad radian 792
r_i replication number 241
$r_k(x)$ Rademacher functions 722

S

S
    S sample space 617
    S torsion tensor 485
    $S(x)$ Fresnel integral 547
    S_n symmetric group 163
    $S(n, k)$ Stirling number second
        kind 213
    $\mathcal{S}(T)$ symmetric part of a tensor
        484
    S siemen 792
$S_{()}$
    S_3 symmetric group 180
    S_k area of inscribed polygon . . 324
    S_n star (type of graph) 229
    S_n symmetric group 172
    $S_n(r)$ surface area of a sphere . 368
SA simulated annealing 291
SI Systeme Internationale d'Unites . . 792
Si(z) sine integral 549
$SL(n, \mathbb{C})$ matrix group 171
$SL(n, \mathbb{R})$ matrix group 171
$SO(2)$ matrix group 172
$SO(n)$ matrix group 172
SPRT sequential probability ratio test 681
SRS shift-register sequence 645
STS Steiner triple system 249
$SU(n)$ matrix group 172
SVD singular value decomposition . . 156
s
    $s(n, k)$ Stirling number first kind
        213
    s arc length parameter 373
    s sample standard deviation . . . 660
    s semi-perimeter 512
    s second 792

$s_{()}$
    s_k area of circumscribed polygon
        324
    s_k elementary symmetric functions
        84
sec trigonometric function 505
sech hyperbolic function 524
sgn signum function 77, 144
sin trigonometric function 505
sinh hyperbolic function 524
sn(u, k) elliptic function 572
sr steradian 792
sup least upper bound 68
supremum least upper bound 68

T

T
    $^\mathrm{T}$ transpose 131
    T tesla 792
    T time interval 796
    transpose 138
$T_{()}$
    $T_n(x)$ Chebyshev polynomials 534
    $T_{n,m}$ isomorphism class of trees
        241
Ti tebi 6
TN(w, s) Toeplitz network 230
tr(A) trace of matrix A 150
t-(v, k, λ) design nomenclature 241
$t_{()}$
    t_α critical value 695
    $t_{x,y}$ transition probabilities 255
tan trigonometric function 505
tanh hyperbolic function 524
t unit tangent vector 374

U

U
    U universe 201
    $U(n)$ matrix group 172
    $U[a, b)$ uniform random variable
        619
    $U_n(x)$ Chebyshev polynomials 535
UCL upper control limit 650
UFD unique factorization domain . . . 165
UMVU type of estimator 663
URL Uniform Resource Locators . . . 803
u traffic intensity 638
$u(n)$ unit step function 595
u_i distance 492

V

V

V Klein four group 179
V vertex set 219
V Roman numeral (5) 4
V volt 792
Vec vector operation 158
$V_n(r)$ volume of a sphere 368
vers trigonometric function 505

W

W

W average time 638
$W(u, v)$ Wronskian 462
W watt 792
$W_{()}$
W_N root of unity 582
W_Q average time 638
W_n wheel (type of graph) 229
$W_n(x)$ Walsh functions 722
Wb weber 792

X

X

X infinitesimal generator 466
X set of points 241
X Roman numeral (10) 4
$X^{(1)}$ first prolongation 466
$X^{(2)}$ second prolongation 466
$x_{(i)}$ i^{th} order statistic 659
x_i rectangular coordinates 492

Y

$Y_\nu(z)$ Bessel function 559
$y_{()}$
$y_h(x)$ homogeneous solution .. 456
$y_n(z)$ half order Bessel function 563
$y_p(x)$ particular solution 456
$y_{\nu,k}$ zero of Bessel function ... 563

Z

Z

Z queue discipline 637
$Z(G)$ center of a graph 221
$Z(t)$ instantaneous hazard rate . 655
$\mathbb{Z}$ integers 3, 167
$\mathcal{Z}$ Z-transform 594
$Z_{()}$
$Z_3 \rtimes Z_4$ semidirect group product 187
$\mathbb{Z}_n$ integers modulo n 167
$\mathbb{Z}_n^*$ a group 163
$\mathbb{Z}_p$ integers modulo p 167
z complex number 53
z_α critical value 695

Index

A

Abel
    integral equation 480
    summation 41
Abelian
    binary operation 161
    groups .... *see* groups, abelian
        characters 191
        number of 25, 177
abscissa of convergence 35
absolute
    convergence of series 32
    covariant derivative, defined 485
    deviation 660
    error 728
    tensor 483
    tensor, Kronecker deltas as 489
abstract algebra 160
abundant numbers 26
acceleration
    sea level 794
    units 796
acceptable quality level 652
acceptance
    -rejection method 647
    sampling 652
accumulation point, defined 69
acres 794, 798
actions, Bayes 657
acute angle, defined 503
acute triangle 318
acyclic graph, defined 220
Adams–Bashforth methods 768
Adams–Moulton methods 768

addition
    of complex numbers 54
    of fractions 3
    of hexadecimal numbers 8
    of integers 81
    of matrices 142
    of powers of integers 22
    of tensors, defined 483
    of vectors 132
    tables, fields 190
additive
    constant 644
    group 161
    lagged-Fibonacci sequences 646
    shift register ....*see* shift register
adjacency
    matrix 220
    matrix, collapsed 226
    of graph edges 220
affine
    planes
        and mutually orthogonal Latin
            squares 251
        defined 247
        order 248
    transformations in space 352
    transformations, defined 313
affinely-parameterized geodesic 485
    equations 488
Agnesi
    curve 337
    Maria 814
Ah'mose 810

835

Airy
  equation (ODE) 465
  functions 565
    figure 566
Aitken's $\triangle^2$ method 728
  and fixed point iteration 730
al-Haytham, Abu Ali ibn 811
al-Khwarizmi, Muhammad 811
al-Tusi, Nasir al-Din 811
aleph null ($\aleph_0$) 204
algebra
  abstract 160
  basic 83
  fundamental theorem 84
  Lie 467
  vector 131, 132
algebraic
  curves 336
  equations 83
    numerical solution 729
  extension 166
  functions, series for 42
  identities, polynomials 85
  identity element 161
  integer, defined 167
  number field 168
  numbers 3
  structure
    defined 160
    order 160
algorithm
  assignment 289
  augmenting path 288
  branch and bound 287
  classifying a conic 328
  continued fractions 97
  Crank–Nicolson 774
  cubic splines
    clamped 739
    natural 738
  design principles 777
  Dijkstra's (shortest-path) 290
  dynamic programming 289
  Euclidean, greatest common divisor
    101
  fast Fourier transform 584
  Gauss–Legendre double integral . 760
  Gaussian elimination for solving
    linear equations 741
  genetic (optimization) 293
  GWZ 38

algorithm (*continued*)
  hit or miss, Monte–Carlo quadrature
    762
  Horner's (polynomial evaluation) 733
  Householder (similar tridiagonal
    matrices) 744
  Hungarian (assignment problem) 288
  inverse power method 743
  Karmarkar's method 285
  linear programming 283
  matrix multiplication, Strassen ... 143
  maximum flow 288
  Monte–Carlo integration 763
  Neville, evaluating Lagrange
    interpolating polynomial 734
  Poisson equation, finite-difference
    algorithm 771
  power method (eigenvalues) 742
  probabilistic primality test 104
  QR 745, 746
  sample mean (quadrature) .. 762, 763
  shortest path, Dijkstra's 290
  simplex method 283
  Simpson's double integral 759
  steepest-descent, for solving
    non-linear equations 749
  Strassen (matrix multiplication) .. 143
  wave equation 775
allied Fourier integral, defined 577
allowable
  change of parameter, class of curves
    373
  coordinate system 482
  parameter transformation, surfaces
    376
Almagest 811
almost everywhere 74
alphabet
  and Shannon's coding theorem .. 255
  code, English text 260
  Greek 803
  sequences over 24
alternating
  group 188
    defined 163
    defined as permutation 172
  harmonic series 32, 33
  path in a graph 225
  prime knots, number of 27
  series test 34

alternative
    Fredholm (integral equation) 479
    hypothesis 661
altitude, of a triangle, defined 318
American Mathematical Society 807
ampere 792
amplitude, elliptic functions 572
analysis
    complex 53
    interval 65
    of variance 686
    real 66
    tensor 482
analytic function 54, 159
Analytical Mechanics 814
Analytical Theory of Heat 576
"and", in propositional calculus 199, 200
angles *503*
    acute 503
    between lines 316
    between planes 355
    between tangents to surface 378
    conversion between degrees and
        radians 381
    convex 503
    difference relations
        circular functions 509
        inverse circular functions 520
    direction 353
    Euler (rotation in space) 349
    exterior, polygon 318
    figure 504
    formulae
        double- 509
        half- 510, 514
        multiple- 510
    half- formulae 514
    in polar coordinates 302
    interior, polygon 318
    interior, sum in right triangle 512
    lines in space 356
    obtuse 503
    right 503
    special, circular functions 506
    sum relations 509
    trigonometric functions 505
    trisection 339, 345
angular
    acceleration, units 796
    momentum, units 796
    velocity, units 796

annuity
    defined 780
    example 782
    tables 787
ANOVA 686
    one-factor 687
    table 688, 691, 694
    two-factor 692
anti-chain in posets, defined 205
antiderivatives .... *see* integrals
antisymmetric relation in sets 204
Apothecaries' weight 794
apothem of regular polygon 324
Appel–Haken four-color theorem ... 235
approximations
    and errors 728
    asymptotic .... *see* asymptotic
        approximations
    least-squares 739
    theorem 72
arboricity
    graphs 220
    theorem 232
arc cos .... *see* inverse circular functions
arc cot .... *see* inverse circular functions
arc csc .... *see* inverse circular functions
arc length
    and metric tensor 486
    as integral 409
    curve on surface 378
    ellipse 330
    hyperbola 332
    of a circle 334
    of regular curve 373
    parabola 333
    parameter 373
arc sec .... *see* inverse circular functions
arc sin .... *see* inverse circular functions
arc tan .... *see* inverse circular functions
    computing π using 13
Archimedes 810
    property of real numbers 68
    spiral 342
area
    circle 13, 334, 368
    circular torus 363
    cone 362
    cylinder 361
    differential element of 396
    ellipse 330, 331
    ellipsoid 364

area (*continued*)

    enclosed by a curve 409, 410

    frustum 363

    polygon 318

       polar coordinates 318

       regular 324

    quadrilateral 322

    random

       triangle 626

    sphere 367

       n-dimensional 368

    spherical

       cap 367

       polygon 367

       segment 368

       zone 367

    surface of revolution 363

    surface, as integral 409

    triangle 319, 513

    units 796

argument

    of a complex number 53

    principle 58

arithmetic

    interval, properties 65

    mean 659

    mean–geometric mean inequality . 33, 73, 83

    modular 94

    on infinite cardinals 204

    progression 86

       primes in 103

    series 38

arrangements 210

    preferential 29

arranging distinct objects 210

array, Costas 252

arrow notation 4

Ars Magna 812

Arzela, Ascoli– theorem 72

Ascoli–Arzela theorem 72

assignment problem 288

associate, in quadratic field 168

associated Legendre equation 557

associated Legendre function 557

associative laws 200

associativity

    in a ring 164

    of binary operation 161

    of interval arithmetic 65

    of matrix multiplication 143

astroid, defined and figure 342

asymptote, hyperbola 327

asymptotic

    approximations, of a function 75

    direction, coordinate patch 377

    equivalence, of functions 75

    expansions, of a function 75

    expansions, of integrals 408

    line, coordinate patch 377

    relations 75

    series of functions 75

    stability of ODE 466

Atkin–Morain certificate of primality 104

atmospheres 797

atomic statement, in tautologies 199

atto, as prefix 6

augmented matrix, Gaussian elimination 740

augmenting path

    algorithm 288

    in a graph 225

autocorrelation 721

    binary sequences, table 264

    Costas array 252

    in periodic sequence 263

automorphism

    graph 226

       defined 220

       group table 238

    group

       of graphs 220

       of graphs, theorems 238

    of a structure, defined 170

autonomous equation (ODE) 461

autoregressive model 718, 719

average

    outgoing quality 652

    over random vectors 625

Avogadro's constant 794

Avoirdupois weight 794

axes

    Cartesian coordinates in space . . . 345

    Cartesian coordinates, defined . . . 301

    conjugate, hyperbola 327

    major, ellipse 326

    minor, ellipse 326

    of rotation, matrix representation 141

    transverse, hyperbola 327

axioms

    completeness 68

    of order 67

azimuth (spherical coordinates) 346

B

Bachet, equation 98
backward
    difference for interpolation 736
    difference formula, numerical
        differentiation 764
Baker transformation (chaos) 272
balanced incomplete block design ... 245
    and Hadamard matrices 250
    existence table 246
ball 220
balls into cells 206
Banach space 72
band, hearing 478
band-limited function, and Shannon
        sampling theorem 581
bandwidth (matrix) 141
bang-bang control 498
Banneker, Benjamin 814
Banzhaf power index 279
Barker sequences 263
bases
    constants in different 16
    conversion 5
    representation 5
Bashforth, Adams– methods 768
basic
    2-forms 396
    feasible solution (linear
        programming) 283
basis, and Galois field 170
Bayes
    actions 657
    loss 657
    rule (probability) 618
BCH codes .... *see* code, BCH
becquerel 792
Beijing, distance to New York 373
Bell numbers 28, 211
    counting equivalence relations 66
Bellman's equations (dynamic
        programming) 290
Bernoulli
    equation (ODE) 461, 465
    Johann 809, 813
    lemniscate of 338
    numbers 19, 40, 44
        in gamma function 542
    polynomials 19
    spiral 342

Bessel
    equation (ODE) 465
    Fourier– series 34
    functions
        defined and properties ... 559–564
        figure 559
        generating function 39
        integrals 448
        relations 564
        relations with Airy 566
        table 567
    inequality 74
    series 34
best
    -fit line 740
    decision 656
beta
    function (table) 545
    function, defined and properties .. 544
    probability distribution 635
        and F-distribution 636
Bhaskara 811
Bianchi's identity (torsion tensor) ... 487
Bienaymé–Chebyshev inequality ... 624
bifurcation, in dynamical systems ... 272
biharmonic equation (PDE) 469
bijective function, defined 66
bilinear
    form (matrix) 138
    transformations 58
billion 6
bin 658
binary
    BCH codes 257, 259
    codes, table 259
    Golay code 258
    Hamming codes 257
    m-sequence 263, 264
    notation, number of 1's 25
    operations
        on a graph 228
        on a set 160
    partitions 27
    polynomials
        irreducible, table 261
        primitive, table 262
    representation 5
    sequences 263
        table 264

binomial
   coefficients 36, 208
      central, list of 26, 28
      fractional, table of 218
      generating functions for 208
      relations 209
      sums of 42
   distribution 710
   probability distribution 630
   series 37
binormal line, regular curve 374
binormal vector, regular curve 374
biographies 810
biorthogonal wavelets 725
bipartite graph
   complete 230
   cube as a 229
   defined 226
   in assignment problem 289
   star as a 229
biquadratic residues (difference sets) 247
birthday coincidence, probability ... 629
bisection method, numerical solution of
        algebraic equations 732
bits
   defined 253
   non-zero in binary expansions 25
block
   codes 256
      bounds 258
      Hamming bound 258
      Plotkin bound 258
      Singleton bound 258
      Varsharmov–Gilbert bound ...258
   combinatorial design 241
   design 245
   Jordan (matrix) 154
   of a graph 220
Bôcher equation 465
Boltzmann constant 794
Bolzano–Weierstrass theorem 70
Boole's rule (quadrature) 751
Boolean functions, number of 26
Borel, Heine– theorem 70
borrowed amount 779
Bose (BCH) codes 257
boundary
   conditions, PDE 470
   operator, for a graph, defined 221
   value problem, equivalence to integral
      equation 479

bounded
   functions 54
   variation, defined 576
bounds
   BCH (coding) 256
   Chernoff (probability) 624
   Cramer–Rao (probability) 664
   greatest lower bound 68
   Hamming (coding) 258
   least upper 68
   lower 68
   Moore (graphs) 235
   Nordhaus–Gaddum (graphs) 234
   on the variance 664
   Plotkin (coding) 258
   upper 68
Box–Muller technique 647
Braess paradox 278
Brahmagupta 811
   formula 323
Brahmasphutasiddhanta 811
branch and bound algorithm 287
bridge
   card game 798
   card game (probabilities) 628
   in a graph 221
Brooks' theorem (graphs) 234
Brownian motion, stochastic differential
      equation 467
Bruck–Ryser–Chowla theorem 245
Bruijn
   de sequences 24
   Moser–de sequence 30
Brun constant (twin primes) 103
BTU 797
Buckingham pi 795
Buffon's needle problem 627
Burgers' equation (PDE) 469
Burnside
   formula (for $n!$) 17
   lemma (permutation group) 173
bushel 794
byte, definition of 6

C

cactus 221
cage 221
calculus
   differential 385
   fundamental theorem 388, 400
   of finite differences 265

calculus (*continued*)
    predicate 201
    propositional ...... *see* propositional
        calculus
    vector 390
calendars 799
cancellation
    interval arithmetic 65
    law (group) 162
    law (ring) 164
candela 792
Cantor
    dust, capacity dimension 344
    Georg 204
    set, capacity dimension 344
cap, spherical 367
capacity
    dimension, defined 344
    of a channel, defined 255
Capell, Narayana–Zidek– numbers .. 26
card games (probability) 628
Cardano
    formula (cubic polynomial) 89
    Gerolamo 812
Cardinal series 581
cardinality
    of a set, defined 67
    of infinite sets 204
cardioid 341
    defined 339
    figure 341
Carleman's inequality 74
Carmichael numbers 31, 94
carpet, Sierpiński, dimension 344
Cartesian
    coordinates
        changing to polar 396
        defined 301
        in space 345
        in space, figure 346
        relation to polar 303
        symmetries 349
        symmetry formulae 305
    form of complex number 53
    tensor, defined and properties 489
Casorati–Weierstrass theorem 57
Casoratian of sequences, defined 266
Cassini's oval, properties and figure . 337
Catalan
    constant 23
    numbers 28, 212

catenary 340
Cauchy
    –Goursat integral theorem 406
    –Riemann equations 55
    –Schwartz inequality 73, 623
    Augustin-Louis 814
    conditions, PDEs 470
    inequality 56
    integral equation 480
    integral formula 55, 406
    integral theorem 55
    kernel, integral equation 478
    principal value integral 399, 588, 591
    root test 33
    sequence 70
        convergence in metric space ... 70
caustic 341
Cayley
    –Hamilton theorem (matrix) 153
    Arthur 815
    formula (counting trees) 239
    graph 230
    theorem (group isomorphism) ... 171
ceiling function 520
cells, balls into 206
cells, in crystallographic groups 308
Celsius 798
center
    critical point (ODE) 468
    of a circle 333
    of a graph 221
    of curvature, plane curve 375
    sphere, defined 365
centered moments 620
centered square numbers 30
centering transformation for linear
        programming 285
centi, as prefix 6
centimeters 796, 797
central
    angle, of circle 334
    binomial coefficients, list of ... 26, 28
    conics 329
    limit theorem (random variables) 623
    moments 660
    quadrics 364
    tendency 659
    vertex of a graph 221
certificate of primality 104
Cesaro summation 41, 49
Ceva's theorem (triangle) 321

chain
  Ehrenfest 643
  poset, defined 205
  rule 388
chair 280
changing variables
  in PDEs 469
  random 623
channel capacity, defined 255
chaos
  Baker transformation 272
  definition 272
  logistic map 272
  maps, examples 272
Chapman–Kolmogorov equation 641
character, of a group 191
  defined 170
  table 191
characteristic
  equation
    damping, linear ODE 460
    of a difference equation 267
    of a matrix 152
    paraboloid of revolution 380
    surface 378
  function, random variable 620
  of a field, defined 167
  polynomial
    of a graph 221
    of a matrix 152
    of a shift register 645
    spectrum of a graph 227
  roots 267
    of matrix 152
characterization of real numbers 68
charts, control 650
Chaudhuri (BCH) codes 257
Chebyshev
  Bienaymé– inequality 624
  inequalities (random variable) . . . 624
  polynomials 754
    first kind 534
    generating function 39
    second kind 535
    table 536
  quadrature 758
  theorem (prime number theorem) 103
checkerboard, distance between points
  626
Chernoff bound (probability) 624
chess positions 798

chi-square
  probability distribution 635
    notation 619
    tables 703
  test 679
Chinese remainder theorem 96, 812
Cholesky factorization (matrix) 157
choosing objects 208
chord
  number of 27
  of a circle 334
Chowla, Bruck–Ryser– theorem 245
Christoffel symbols
  first kind
    2-sphere 491
    3-sphere 492
    defined 487
    orthogonal coordinates 490
    polar coordinates 492
  second kind
    2-sphere 491
    3-sphere 492
    defined 487
    orthogonal coordinates 490
chromatic
  index, of a graph 221
  numbers
    for given genus 30
    of a graph 221
    table 234
  polynomial
    form of 235
    graph 221
    table 234
circles
  arc length 334
  area 13, 368
  center 333
  circumference 13, 368
  circumscribed, triangle 319, 513
  equation 333
    oblique coordinates 335
  Gerschgorin theorem (eigenvalues)
  158
  inscribed, triangle 318, 512
  mapping 58
  osculating 375
  points in, probability 626
  polar equation for 335
  polynomials 537
  properties 333

circles (*continued*)
    radius 333
    squaring 345
    ways to join points on 27
circuit
    Eulerian 224
    graph 221
    in a graph 233
circulant
    graph 221, 230
    graph, cycle as a 229
    matrix 149, 221
    matrix, defined 141
circular
    cylindrical coordinates 494
    functions *505*
       and exponentials 508
       angle difference relations 509
       angle sum relations 509
       double-angle formulae 509
       figures 505, 506
       half-angle formulae 510
       identities among 509
       inverse 518, 519
       multiple-angle formulae 510
       powers 511
       products 511
       relations among (table) 508
       relations with Gudermannian . 531
       relations with hyperbolic 525, 529,
         530
       special angles (tables) 506
       sum 511
       symmetries 507
       table of values 516
    helix 375
    parts, Napier's rules 368
circumference
    circle 13, 334, 368
    ellipse 330
    graph 221
circumscribable quadrilateral 323
circumscribed
    circle, regular polygon 324
    circle, triangle 319, 513
    sphere, tetrahedron 358
cissoid of Diocles 336
Civita, Levi–
    connection 487
    symbol 489
    tensor 489

Clairaut equation (ODE) 461
clamped cubic spline 739
class
    equivalence 66
    number 809
    of curves 373
    parallel, containing lines 248
classical constructions, geometric . . . 345
classical problems 665
classification
    AMS 801
    conics 328, 329
    crystallographic groups 312
    groups 162
    integral equation kernels 478
    integral equations 478
    ODEs 461
    PDEs 468
    quadrics 365
Clebsch–Gordan coefficients 574
    and spherical harmonics 539
    table 575
clique 233
    graph 221
    number 221
closed
    Newton–Cotes formulae 750
    sets 69
clothoid 343
cluster point 70
coalition, in voting game 278
coboundary operator 221
cocycle vector 222
Codazzi, Gauss–Mainardi– equations
    379
code
    alphabetical, English text 260
    BCH 257
       table 259
    best binary, table 259
    binary, table 259
    block . . . . *see* block codes
    cyclic 256
       BCH bound 256
    defined 256
    diagram 256
    dimension 256
       binary BCH code 257
    dual 257
    error correction 809
       block coding 256

code (*continued*)

    error detection, block coding 256

    German "Enigma" 816

    Golay 244

        as perfect 257

        binary 258

        ternary 258

    Gray 261

    Hamming 257

        as perfect 257

    Huffman, English text 260

    linear 256

    MDS ....*see* MDS codes

    Morse, table 260

    perfect 257

    rate 255

    Reed–Solomon 257

    self-dual 257

    theorem 255

codewords

    and Shannon's theorem 255

    number in best binary code 259

    number of in block codes 258

coefficients

    Clebsch–Gordan 574

    connection (tensor analysis) 484

    polynomial, relation to roots 84

cofactor of a matrix, defined 145

coincident planes, equation 364

collapsed adjacency matrix of a graph
    226

collinear points 317

color classes of a graph 226

coloring

    a graph *222*, 234

    a necklace 173

    a set (permutation group) 173

    the corners of a polygon 174, 175

    the corners of a square 173

column

    of a matrix 137

    rank of a matrix 147

    vector *131*, 137

COMAP 808

combinations

    generating 807

    table of 215

    with replacement 206

combinatorial design

    Fisher's inequality 244

    theory 241

combinatorial sums 38

combinatorics 206, 808

committee 280

common

    divisor, greatest 101

    fractions 7

    multiple, least 101

communication theory 253

commutative

    division ring, field as 167

    interval arithmetic 65

    laws 200

    ring *164*

    ring, table 167

commutativity

    binary operation 161

    matrix multiplication 143

commutator 467

    of matrices 155

compact

    sets 69

    subsets, Heine–Borel theorem 70

comparable elements, in posets 205

comparison test 33

compass, in geometric constructions 345

compatible norm, matrix 147

complement

    of a graph 226

        defined 222

    of a set 69, 203

complementary

    design 244

    error function 545

    events 618

    integrals (elliptic) 569

    modulus (elliptic integrals) 570

    slackness 285

    slackness theorem 286

complete

    bipartite graph 230

    elliptic integral 569

    graph 222

    graph, description 229

    multipartite graph 230

    orthonormal set, defined 74

    pivoting, solving linear equations 742

    set 74

    space 70

completeness

    axiom 68

    relation 539

complex
   analysis 53
   conjugate 54
   Fourier series 50
   functions 54
      analytic 55
      transformations and mappings . 58
   numbers *3*
      arithmetic operations 54
      Cartesian form of 53
      defined 53
      polar form 53, 87
      powers 54
      properties 3
      roots 54
      roots of -1 87
component
   functions
      class of curves 373
      coordinate patch on surface .. 376
   of a graph, defined 222
   of a vector, defined 131
   series 32
composite
   midpoint rule 752
   numbers, in arithmetic sequence . 103
   rules for numerical integration ...752
   Simpson rule 752
   trapezoidal rule 752
composition
   of functions 67
   operation on graphs 228
compound
   event, connection to set theory ...203
   interest
      defined 780
      example 781
      tables of annuity 787
      tables of final value 783
      tables of interest rate 785
   statement, in tautologies 199
compounding 780
computational molecules 765
computer
   languages 803
   programming techniques 777
   speed 76
conchoid of Nichomedes 338
conclusion 200
concurrent lines 317
concurrent planes 355

condition number
   and determinants 149
   of matrix 148–149
      table 149
conditional
   convergence of series 32
   distributions, multivariate 622
   expectation 468
conditional probability 617
conditions
   boundary (PDE) 470
   Cauchy (PDE) 470
   Dini's (Fourier transform) 580
   Dirichlet (PDE) 470
   Moore–Penrose (matrix
      pseudo-inverse) 151
   Neumann (PDE) 470
   radiation (PDE) 471
   separability (PDE) 474
cone
   defined and properties 361
   moment of inertia 411
   quadric, equation 364
   surface area 362
   volume 362
confidence
   coefficient 666
   level 762
confidence interval .. 666, 667, 683–686,
      688, 694, 695
   for mean 666
      difference 667, 668
   for median 667, 669
   for one population 666
   for quantile 667
   for success 666
      difference 668
   for two populations 667
   for variance 666
conformability of matrices 142
conformal mapping 58
   table 60
congruences of integers 94
   properties 95
congruential generator 644
conic
   central 329
   classifying 328
   degenerate 328, 329
   rotating and scaling 328
   section 325–336

conical coordinates 496
conjecture, Goldbach (sum of primes)
     103
conjugate
     axis, hyperbola 327
     complex 54
     Hermitian, matrix 138
     hyperbola, defined 331
     of an algebraic integer 168
conjunction, operation on graphs 228
connected
     graph 222
     graphs, number of 28
connection
     coefficients (tensor analysis) 484
     Levi–Civita 487
     linear (tensor analysis) 484
     pseudo-Riemannian 487
connectives
     in propositional calculus 199
     in truth tables 200
connectivity
     in graphs 222
     in graphs, theorems 231
     line 223
consistency, of matrix norms 147
consistent
     estimator 661
     system of linear equations 148
constant
     -coefficient ODE 461
     failure rate 655
     of integration 412
constants
     Avogadro 794
     Boltzmann 794
     Brun (twin primes 103
     essays on 808
     Feigenbaum 272
     in different bases 16
     mathematical 3–10
     physical 794
     Plank 794
     section 3
     special 13
constraints on resource 281
constructions, classical geometric ... 345
consumer's risk 652
contests 808
contingency table 679

continued fractions .... *see* fractions,
     continued
continuity in $\mathbb{R}$ 71
continuous
     multivariate distributions 621
     at a point 71
     compounding 780
     distributions 633
     evaluation of an integral 408
     extended Kalman filter 720, 721
     Kalman filter 720
     random variables 618
     wavelet transform 725
continuum hypothesis 204
contour
     integral 399, 406
     of complex function 57
     of integration 399
contraction
     graph 222
     mapping theorem 71
     of a function 71
     of a tensor 484
contrapositive law 200
contravariant
     indices, mixed tensor 483
     Levi–Civita tensor 489
     metric
          tensor field 486
     metric tensor
          2-sphere 491
          3-sphere 491
          orthogonal coordinates 490
     permutation symbol, as tensor ... 489
     valence, mixed tensor 483
     vector field 485
          Ricci identity 487
contributors vii
control
     charts 650
     theory 497
controllable system 497
convective operator in orthogonal
          coordinates 493
convergence
     abscissa 35
     improving 40
     in $\mathbb{R}$ 70
     infinite products 47
     interval 36
     linear, defined 728

convergence (*continued*)
  of functions, in the mean 73
  pointwise 71
  quadratic, defined 728
  rate of 728
  series
    absolute 32
    conditional 32
    definition 32
    slow, examples of 42
    tests 33
      integrals 407
    uniform 71
convergents to a continued fraction ... 96
conversion
  English to metric 797
  fractions 7, 9
  from one radix to another 5
  metric to English 796
convex
  angle 503
  polyhedra 357
convolution
  discrete Fourier transform 583
  Fourier transform, defined 579
  Laplace transform 589
  of probability densities 623
  of sequences 583, 598
  Vandermonde 208
Conway notation, crystallographic
      groups 307
cookie problem (probability) 627
cooling schedule (neighborhood search)
      291
coordinate system
  allowable 482
  Cartesian .. *see* Cartesian coordinates
    defined 301
  circular cylindrical 494
  conical 496
  cylindrical 346, 492
  ellipsoidal 497
  elliptic cylindrical 495
  homogeneous 306, 348, 351
  in space 345
  manifold 482
  normal (tensor analysis) 488
  oblate spheroidal 496
  oblique, defined 303

coordinate system (*continued*)
  orthogonal 492
    metric tensor 490
    table 494
  parabolic 474, 496
    cylindrical 495
  paraboloidal 497
  plane, conventions 299
  polar .... *see* polar coordinates
  projective 303
  prolate spheroidal 496
  rectangular 494
    and angles 504
    defined 301
    in space 345
  spherical 346, 495, 539
    polar 77
  transformation 349, 482
coordinates
  curvilinear, orthogonal 490
  oblique planar
    rotation 304
    translation 305
  patch on surface 376
  planar
    and transformations 300
    change of, defined 299
    homogeneous, defined 303
    rotation 301
    translation 301
  polar, planar, defined 302
  representation (tensor) 483
coplanar
  lines in space 357
  points 355
cords 798
Cornu spiral 343
  defined and figure 547
correlation coefficient 683
  multivariate 622
cosecant
  derivative 386
  function 505
  table of values 516
coset of subgroup, defined 162
coset, defined 164
cosine
  derivative 386
  function 505
  integral 549, 551
  table of values 516

cosine (*continued*)
   transform 35
      and Hankel transform 589
      Fourier *582*
      table 601
cosines
   direction 353
   law of (triangle) 319
   laws of spherical 370
cost function, minimization 498
Costas array 252
cotangent
   derivative 386
   function 505
   table of values 516
Cotes, Newton– formulae 750
coulomb 792
countably infinite set, defined 67
Courant–Fischer minimax theorem
      (eigenvalues) 157
covariance, multivariate, defined 622
covariant
   derivative 484
      operator, properties 487
      raising or lowering indices ... 488
   differentiation, tensor, symbol ... 484
   indices, mixed tensor 483
   metric tensor
      2-sphere 491
      3-sphere 491
      field 486
   skew-symmetric tensor, dual 489
   valence, mixed tensor 483
cover 222
covering a set, ways to 30
covering sequence 94
coversed sine, defined 505
Cramer's rule (linear equations) 158
Cramer–Rao bound 664
Crank–Nicolson method 773, 774
critical
   damping, linear ODE 460
   line 24
   points, ODE 468
   region 661
   values 695
cross
   product, vectors 135
   ratio 58
crosscap number 222

crossing
   in a graph 222
   number (graph) 222
      rectilinear 222
   number of knots 27
crossover operator (genetic algorithms)
      293
crystallographic group 307–312, 809
cubes
   coloring 174
   constructing 360
   cycle index and pattern inventory 176
   defined 357
   doubling 345
   graph type, described 229
   list of 31
   points in, probability 626
   properties and figure 359
   Rubik 798
   volume 345, 357
cubic
   feet 794, 796–798
   graph 222
   inches 794, 796, 797
   meters 796, 797
   parabola, defined and figure 336
   polynomial
      discriminant 89
      roots 89
      trigonometric solution 89
   resolvent, quartic polynomials 90
   splines 737, 738
      clamped 739
   yards 794, 796, 797
Cullen
   numbers 29
   primes, table 106
cumulants and cumulant function ... 620
cumulative
   distribution function 619
      continuous multivariate 621
      discrete multivariate 621
   terms
      binomial distribution 710
      Poisson distribution 712
curl
   of a gradient 398
   operation in orthogonal coordinates
      493
curtate cycloid, defined 340

curvature
  Gaussian
    coordinate patch 377, 379
    paraboloid of revolution 380
  geodesic, coordinate patch 377
  invariant (tensor) 488
  line, coordinate patch 377
  line, paraboloid of revolution 380
  mean, coordinate patch 377, 379
  mean, paraboloid of revolution .. 380
  normal, coordinate patch 377
  principal
    coordinate patch 377–379
    paraboloid of revolution 380
  radius of, regular curve 374
  regular curve 374
  tensor
    defined 485
    identities 487
    Riemann 379
  vector, normal, coordinate patch . 377
  vector, regular curve 374
curves
  algebraic 336
  differential geometry of 373
  fractal 343
  in polar coordinates 342
  parameterized (tensor) 483
  Peano 343
  quartic 90
  rational 337
  regular, of class 373
  spirograph 340
  surface, arc length 378
curvilinear coordinates 490
customary weights and measures 793
cut
  of a graph 222
  space of a graph 223
  vector for a graph 223
  vertex of a graph 223
cutset (linear programming) 288
cycle
  graph 223
  graph type 229, 233
  Hamiltonian 224
    in double loop graph 230
  in permutations 163, 212

cycle (*continued*)
  index
    examples 175
    of group, defined 172
    permutation group 173
  space 223
cyclic
  codes
    BCH bound 256
    defined 256
  difference sets
    and binary sequence 263
    defined 246
    table 247
  group 178–181, 184, 185, 187
    defined 162
    defined as permutation 172
  quadrilateral 323
  subgroup, defined 162
cyclical identity (curvature tensor) .. 487
cycloid and trochoid, figures 340
cycloid, defined 340
cyclotomic polynomials
  defined 91
  table 92
cylinder
  defined 361
  elliptic, equation 364
  figure 361
  hyperbolic, equation 364
  moment of inertia 411
  parabolic, equation 364
cylindrical coordinates
  in space 346
  metric coefficients 492

D

damping, linear ODE 460
data-fitting by equations 737
Daubechies wavelet 724
Dawson's integral 546
day of week or year 799, 800
de Bruijn sequences 24
De Morgan's laws 200
De Thiende 812
de Bruijn, Moser– sequence 30
deadlock 280
decagon, area and apothem 324
decay, linear ODE 460
deci, as prefix 6
decile 659

decimal
   equivalents of fractions 7
   fractions to hexadecimal 9
   multiples 6
decimation
   in periodic sequence 263
   in time, FFT 584
decision
   best 656
   function 657
   point of a process 289
   problems in networks 288
   rules 656
   variables (linear programming) . . 281
decomposition
   Schur (matrix) 154
   spectral (matrix diagonalization) . 154
Dedekind
   numbers 26
   Richard 815
deductions, in propositional calculus 201
defective items 681
deficient improper node (ODE) 468
definite integrals 399
   properties 400
   table 448–455
deflation
   for finding eigenvalues 744
   for finding roots 733
   Wielandt method 744
degenerate conic 329
degree
   of a graph vertex 223
   sequence
      graph 223
      of graphs, theorems on 237
degrees
   angle *503*, 798
   Celsius 798
   conversion to radians 381
   Fahrenheit 798
   Kelvin 792, 798
   of freedom 636, 687
      chi-square distribution 635
      F-distribution 636
   of latitude 794
   Rankine 798
deka, as prefix 6
delta
   Kronecker 483
   sequence, defined 77

delta function
   approximations 77
   definition and properties 76
deltoid, defined and figure 341
DeMoivre's theorem 54, 87
dense
   property of rational numbers 68
   set, defined 70
density
   function, continuous multivariate 621
   function, random variable 619
   function, sample 658
   of mercury 795
   of water 795
denumerable set, defined 67
dependent variable missing (ODE) . . 461
depth, and projective transformation . 353
derangements 28, 210
derivative operator, covariant 487
derivatives
   common functions 386
   covariant (tensor) 484
   defined 386
   determinant 394
   exterior 397
   formulae 387, 394
   matrix 391
   of Kronecker product 160
   partial, defined 386
   theorems 388
   vector 391
Derive 803
derived design 244
Descartes
   folium of, defined and figure 337
   René 812
descent, steepest- (non-linear equations)
      748
descriptive statistics 658
design
   2-$(v,3,1)$, as Steiner triple system 249
   5-(12,6,1), table 244
   balanced incomplete block 245
   complementary 244
   derived 244
   distance, binary BCH code 257
   residual 244

design (*continued*)
    symmetric
        Bruck–Ryser–Chowla theorem
           245
        defined 245
        properties 245
        theory 241
determinants
    and condition numbers 149
    and eigenvalues 153
    definition 144
    in wedge products 395
    of matrix exponentials 155
    of metric tensor 489
    of orthogonal matrix 139
    of sequence values 266
    properties 144–145
deviation
    absolute 660
    mean 660
    standard
        random variable 620
        sample 660
DFT 582
diagonal
    component, matrix 137
    matrix *138*
diagonalization
    of a matrix 154
    of matrix exponentials 155
diagram
    Hasse (posets) 205
    of trees 241
    Venn 203
diameter
    graph 223
    of a circle 334
diamond (parallelogram), properties . 323
dice, probability distribution 629
Dickson, Leonard Eugene 816
dictator, in voting game 278
difference
    backward for interpolation 736
    formulae, numerical differentiation
        764
    forward, in Aitken's $\triangle^2$ method . 728
    kernel, integral equation 478
    of hyperbolic function arguments 526
    of hyperbolic functions 526
    of inverse hyperbolic functions . . 529
    of vectors 132

difference (*continued*)
    sets
        and (v, k, λ)-design 246
        cyclic, table 247
        defined 246
        properties of families 246
difference equations
    approximating ODEs 766
    characteristic equation and roots . 267
    constant coefficients 267
    eigenvalues of 267
    fundamental system 266
    homogeneous 265, 267
    initial-value problem 265
    linear 265
        independence of solutions 266
    logistic 271, 272
    named 269
    non-homogeneous 265, 268
    Riccati 270
    section 265–271
    solution
        defined 265
        using generating functions . . . 268
differences
    divided, in interpolation 735
    finite, calculus 265
differential
    1-form 395
    2-form 396
    3-form 397
    calculus 385
    forms 395
    fundamental, coordinate patch . . . 377
    geometry 373
    operator, eigenvalues 477
    total 397
differential equations
    Airy (ODE) 465
    associated Legendre (ODE) 557
    autonomous 461
    Bôcher (ODE) 465
    Bernoulli (ODE) 461, 465
    Bessel (ODE) 465, 559
    biharmonic (PDE) 469
    boundary-value problem, equivalence
        to integral equation 479
    Burgers' (PDE) 469
    Cauchy–Riemann (PDE) 55
    Chebyshev polynomials 534, 535
    Clairaut (ODE) 461

differential equations (*continued*)
  constant-coefficient ODE ... 460, 461
  conversion of higher-order to
      first-order system 769
  diffusion (PDE) 469
  Duffing's (ODE) 465
  elliptic (PDE) 469
  Emden–Fowler (ODE) 465
  Euler (ODE) 461
  exact 461
  first-order linear 462
  forward Kolmogorov (PDE) 467
  Hamilton–Jacobi (PDE) 469
  heat (PDE) 469
  Helmholtz (PDE) 469, 474
  Helmholtz, separability 474
  Hermite polynomials (ODE) 532
  homogeneous 461
  hyperbolic (PDE) 469
  hypergeometric (ODE) 465
  initial-value problem, equivalence to
      integral equation 479
  integrating factor 462
  invariant 467
  Jacobi polynomials (ODE) 533
  Korteweg de Vries (PDE) 469
  Laplace (PDE) 469, 474, 475
  Legendre (ODE) 465, 554
  Legendre polynomials (ODE) ... 534
  linear 456
  Mathieu (ODE) 465
  named, ODE 465
  named, PDE 469
  Navier–Stokes (PDE) 469
  numerical solutions 766, 770
  ordinary .... *see* ordinary differential
      equations
  Painlevé transcendent (ODE) 465
  parabolic (PDE) 469
  parabolic cylinder (ODE) 466
  partial .... *see* partial differential
      equations
  Poisson (PDE) 469
  quasi-linear differential 472
  reducible to homogeneous 462
  reducible to separable 462
  Riccati (ODE) 466
  Schrödinger (PDE) 469
  second-order, constant coefficients
      460
  Sine–Gordon (PDE) 469

differential equations (*continued*)
  solution techniques 461
  stochastic 467
  system 466
  systems and higher-order 769
  telegraph (PDE) 469
  Tricomi (PDE) 469
  variation of parameters 462
  wave (PDE) 469
  wave, solutions 475
  web 808, 809
differentiation
  exterior ..... *see* derivatives, exterior
  numerical 764
  of power series 32, 36
  of tensors 484
  operator 456
diffusion equation 773
  in a circle 473
  n-dimensional 471
  PDE 469
digital math library 808
digraphs *223*
digraphs, number of, table 239
dihedral group 178, 182, 185, 186
  characters 191
  defined 163
  defined as permutations 172
dihedral symmetry, in crystallographic
      groups 307
Dijkstra's algorithm, shortest-path .. 290
dilation
  Fourier transform 578
  Hilbert transform 592
  Laplace transform 586
  matrix *725*
dilemma, prisoners' 277
dilogarithm
  defined and properties 551
  function, and integration 408
dimension, capacity, defined 344
dimensional analysis 795
dimensional exponents 795
Dini's condition (Fourier transform) . 580
Diocles, cissoid of 336
Diophantine equation 98
Dirac
  delta function 76
  theorem (graphs) 233
direct product of groups 181, 184
  defined 163

directed graphs
   number of 29
   transitive, number of 30
direction
   angles 353
   asymptotic, coordinate patch 377
   cosines 353
   principal
      coordinate patch 377, 379
      defined 379
      paraboloid of revolution 380
directrix
   cone 361
   cylinder 361
   ellipse 326
   of a conic, defined 325
Dirichlet
   conditions
      PDEs 470
      required for Fourier series 48
   problem 477
   series 35
   theorem (prime numbers) 103
discontinuous evaluation, integral ...408
discrete
   multivariate distributions 621
   binomial probability distribution . 630
   dynamical systems and chaos 272
   Fourier transform 582
   Kalman filter 719, 721
   least-squares approximation 739
   probability distributions 630
   random variables 618, 648
   uniform probability distribution ..630
discriminant
   cubic polynomial 89
   of a polynomial 84
   quadratic polynomial 89
diseases 665
disjoint
   internally 225
   union, of graphs 229
dispersion
   function, plasma 546
   measures 660
displacement, units 796
Disquisitiones Arithmeticae 814
distance
   between point and line 82

distance (*continued*)
   between points 316
      in a cube, probability 626
      in a plane, probability 626
      in an n-cube, probability 626
      in space, probability 626
      on a circle, probability 626
      on checkerboard, probability . 626
   designed, binary BCH code 257
   from point to line 316
   function, on a set 69
   graph property 223
   Hamming, defined 256
   in graphs 223, 235
   measure, in vector space 133
   minimum Hamming 256
   on earth 372
   points in space 356
distinct objects 208
distinguishable objects, sampling ... 206
distribution
   beta 635
   binomial 630
   chi-square 635
   conditional, multivariate 622
   exponential 634, 647, 655
   F ....*see* F-distribution
   failure time 655
   function 619
      continuous multivariate 621
      discrete multivariate 621
      sample 658
   gamma 634
   geometric 630
   hypergeometric 631
   marginal, multivariate 622
   multidimensional 254
   multinomial 632
   negative binomial 632
   normal 633, 665, 696
      multidimensional 254
   normal, and binomial 630
   Pascal 632
   Poisson 632
   Polya 632
   probability, entropy 253
   Snedecor's F 636
   Student's t 636
   student's t ....*see* t-distribution
   t ....*see* t-distribution

distribution (*continued*)
  test 679, 680
  uniform, discrete 630
  Weibull 655
distributive law 65
distributivity
  in a ring 164
  matrices 143
divergence
  of a curl 398
  of a vector 488, 493
  tests 33
    integrals 407
  theorem 405
divergent series 32
  summability methods 41
divided differences, in interpolation . 735
divisibility of integers 93
division
  of complex numbers 54
  ring
    defined 165
    field as 167
    normed 165
    of quaternions 166
    table 167
  synthetic 733
divisors
  greatest common divisor 101
  of n 35
    number of 25
  of zero (ring) 164
  sum of (totient function) 28, 128
Dobinski's formula 211
dodecagon, area and apothem 324
dodecahedron
  constructing 360
  properties and figure 359
domain
  Euclidean (ring), table 167
  factorization domain 165
  in crystallographic groups 308
  integral (ring), defined 164
  integral (ring), table 167
  of a function 66
  principal ideal (ring), defined 165
  principal ideal (ring), table 167
  simply-connected, Riemann's
    mapping theorem 58
dominant strategy 275
dot product 133

double
  -angle formulae 509
  factorial 17
    list of 28, 29
  loop graph 230
  root of polynomial 83
doubling, cube 345
drams 794
drawing graphs 223, 236
drum, hearing 477
dry measure 794
dual
  code, defined 257
  problem (linear programming) ... 285
  tensor 489
  variable (linear programming) ... 285
duality theorem 286
Duffing's equation (ODE) 465
dummy, in voting game 278
duodecimal 5
Dupin's indicatrix, coordinate patch . 377
dust, Cantor, capacity dimension 344
dynamic programming 289
dynamical systems and chaos 272

E

e .... *see* notation index
earth
  desnity 795
  distance between points on 372
  mass 799
  radius 795
  surface area 798
  volume 798
eccentricity
  ellipse 326, 330
  graph 223
  hyperbola 327
  of a conic, defined 325
edge
  adjacency in a graph 220
  chromatic numbers, table 234
  connectivity of a graph 223
  cover of a graph 222
  difference, graph operation 228
  Euler formula (convex polyhedra) 357
  in a graph 219
  in a polygon 317
  set in a graph 219
  similarity in a graph 226
  space of a graph 224

edge (*continued*)

    sum, graph operation 228

    symmetric graph 227

editors iii

EEC, voting game 279

efficiency, relative 662

efficient estimator 662

Ehrenfest chain 643

eigenfrequencies 477

eigenstructure of matrix 152

eigenvalues

    and central conics 330

    and ellipse 330

    and quadratic forms 155

    computation 742, 746

    difference equation 267

    distinct 153

    matrix

        defined 152

        Hermitian 153

        idempotent 153

        nilpotent 140

        similar 150

        sum is trace 151

        symmetric 153

        triangular 153

        unitary 140

    of a differential operator 477

    of an integral equation 478

    of graph spectrum 237

    of Kronecker product 159

    of Kronecker sum 160

    of matrix functions, table 153

    of specific matrices, table 153

    product, and determinant 153

    signs 158

eigenvectors

    and quadratic forms 155

    in diagonalization of a matrix 154

    linearly independent 153

    of a matrix, computation 742

    of Kronecker product 159

    of matrix, defined 153

    of similar matrices 150

Einstein

    summation convention 482

    tensor, 3-sphere 492

    tensor, defined and properties 488

electric current 792

electron charge 794

electronic journal 808

elementary

    event, connection to set theory . . .203

    matrix 138

    symmetric functions 84

Elements 810

elimination

    Gaussian

        elementary matrices in 139

        solving linear equations 740

    of cases, in propositional calculus 200

ellipse

    area 330, 331

    axes 331

    circumference 330

    defined 325

    eccentricity 330

    equation 325

    figure 325

    imaginary 328

    polar equation for 331

    properties 325, 330

ellipsoid

    and positive definite matrix 156

    equation 364

    moment of inertia 411

ellipsoidal coordinates 497

elliptic

    coordinates, Hamiltonian 473

    cylinder coordinates 495

    cylinder, equation 364

    function, and integration 409

    functions, defined and properties .572

    integrals

        defined and properties 568

        figure 571

        table 570

    partial differential equations 469, 770

        boundary conditions, table . . . 470

    points, coordinate patch 377

    points, paraboloid of revolution . .380

embedding (graphs) 224, 236

Emden–Fowler equation (ODE) 465

empty graph, defined 229

empty set 202

encyclopedia

    of integer sequences 31, 807

    of mathematics 807

    of polyhedra 809

end vertex 224

energy, units 796

English
    text, coding 260
    units 796, 797
entire functions 47, 54
    defined 54
entropy
    and uniform random variable 253
    continuous, defined 254
    of a distribution, defined 253
    relative 254
enumeration of graphs 239
epicycloid, defined 341
epicycloid, figure 341
epitrochoid, defined 341
equations
    algebraic, numerical solution 729
    Bachet 98
    Bellman's (dynamic programming)
      290
    Chapman–Kolmogorov 641
    characteristic ...... *see* characteristic
      equations
    difference ... *see* difference equations
    differential . *see* differential equations
    Diophantine (integer solutions) ... 98
    elliptic partial differential 770
    Euler–Lagrange 407
    fitting to data 737
    Gauss, surfaces 379
    Gauss–Mainardi–Codazzi 379
    integral .... *see* integral equations
    Lamé, orthogonal coordinates ... 491
    linear, systems 148
    logistic, difference 272
    named, integral 480
    normal (least squares) 685
    normal (least-squares) 739
    Pell 99
    Poisson, finite-difference algorithm
      771
    Poisson, partial differential 770
    quadratic, conic 328
    Riccati 270
    Serret–Frenet (regular curve) 375
    solving systems of linear 740
    Weingarten, surfaces 379
equatorial plane 346
equicontinuity, of a function
    Ascoli–Arzela theorem 72
    defined 71
equilateral triangle 320

equilibrium
    mixed zero-sum game 276
    Nash, non-zero-sum game 277
    pure zero-sum game 275
    solution, difference equation 265
equivalence
    asymptotic .... *see* asymptotic
      equivalence
    class 66
    class (permutation group) 173
    of a parametric representation ... 373
    relations (permutation group) 173
    relations, defined 66
equivalent functions, defined 74
erf .... *see* error function
ergs 798
Erlang formula 639
Erlang-k service time 637
errata, website v
error
    absolute 728
    correction, block coding 256
    detection, block coding 256
    formula, interpolating polynomial
      Hermite 737
      Lagrange 734
    function, defined and properties ..545
    functions, table 548
    least-squares 739
    mean square, defined 74
    minimization by least-squares ... 739
    relative 728
    round-off 728
    standard 683
    types I and II 661
escape set (chaos) 273
essential singularity 57
estimation 718
estimator 661
    consistent 661
    efficient 662
    maximum likelihood 662
    unbiased 663
Etherington, Wedderburn– numbers ..26
Euclid 810
Euclidean
    algorithm for greatest common divisor
      101
    field 168
    norm of a vector 133
    norm, metric on $\mathbb{R}^2$ 72

Euler
  –Lagrange equation 407
  –MacLaurin sum formula 40, 776
  angles (rotation in space) 349
  constant 15
  equation (ODE) 461
  Fermat– , theorem (congruences) . 95
  formula (polyhedra) 357
  graphs, number of 27
  Leonhard 809, 813
  method 767
  method, modified 767
  numbers 20, 28, 44
    list of 30
  polynomials 20
  theorem 233
  totient function 25, 128
    and primitive polynomials 169
    and primitive roots 195
Eulerian
  circuit in a graph 224
  graph 233
    defined 224
    table 233
  trail in a graph 224
even
  function, integral of 400
  graph 224
  permutations, in symmetric group 163
  sequence 584
  series 50
event
  compound 203
  elementary 203
  in sample space 617
  probability, connection to set theory
    203
everywhere, almost 74
evolution, genetic algorithms for 293
exa, as prefix 6
exact
  covering sequence 94
  differential equation 461
examples
  balls into cells 207, 213
  colorings 174
  difference sets 246
  exterior calculus 396, 397
  fields 167
    quadratic 169
  Gray code 261

examples (*continued*)
  groups 163
  integration 400, 401, 403
  l'Hôpital's rule 389
  Lagrange multipliers 390
  limits 386
  pairing terms 212
  partitioning a set 211
  permutations 212
  pseudo-inverse 152
  rings 165
  selecting a sample 206, 209
  subsets of a given size 208
  tensors 489
exbi, definition of 6
excess, random variable 620
excluded middle, law of 200
exclusion rule for set properties 204
exclusive or 645
existence 577, 585, 590, 592
  solution of PDE 470
  solution to difference equation ... 265
  table for BIBDs 246
  theorem 375
existential quantifier 201
expansion
  asymptotic *see* asymptotic expansions
  of a number as continued fraction . 97
expectation
  Bienaymé–Chebyshev inequality 624
  Cauchy–Schwartz inequality 623
  Chebyshev inequalities 624
  conditional 468
  connection to set theory 203
  defined 619
  Jensen's inequality 624
  random vectors 625
expected value, defined 619
explicit method (numerical differential
    equations) 767, 768
exponential
  derivative 386
  distribution 647, 655
  Fourier series 50
  function 520, 521
    and trigonometric functions .. 508
    definition 15
    integrals 443
    series for 43
    table 531
  integral, defined and properties .. 550

exponential (*continued*)

    integral, table 551

    matrix 155

    numbers 28

    probability distribution 634

        notation 619

    relation to logarithm 523

    service time 637

exponentiation 520

exponents 261

    algebraic 86

    dimensional 795

    laws of 520

exsecant, defined 505

extended Kalman filter 719

extension

    field, defined 167

    of reals by $+\infty$ 68

exterior

    angle, polygon 318

    derivative 397

    product of 1-forms 395

extrapolation, Richardson 729

extremum of function 389

F

F-distribution

    probability 636

    tables 704–709

faces, Euler formula 357

factor

    group, defined 162

    integrating, differential equation . 462

    of a graph 224

factored graph 224

factorial

    definition 17

    double, list of 28, 29

    falling 17

    list of 28

    non-integer arguments 17

    polynomial 213, 214

    primes, table 106

    rising 17

    shifted 17, 36

factorization domain 165

factorizations

    Cholesky (matrix) 157

    matrix 156

    of $2^m - 1$ 128

    of $x^n - 1$ mod 2 196

factorizations (*continued*)

    ordered, number of 25

    QR (matrix) 157

    table 125

    unique, in quadratic field 168

Fahrenheit 798

failure 655

    rate constant 655

    time distribution 655

falling factorial 17

false position method, numerical solution

        of algebraic equations ... 732

families of difference sets 246

FAQ 807

farad 792

Fary embedding (graphs) 236

fast Fourier transform 584

fathom 794

fathomed (branch and bound) 287

feasible solution, linear programming

    281

feet 797

    in other units 794, 796

    of water 798

Feigenbaum's constant 272

Feit, Thompson– theorem (groups) . 162

femto, as prefix 6

Fermat

    –Euler theorem (congruences) 95

    last theorem 98, 201

    Pierre de 813

    primes 106

        and polygons 345

    theorem (congruences) 95

    theorem (primality test) 104

FFT 584

Fibonacci 811

    numbers 21, 268

        list of 26

field

    addition and multiplication tables 190

    algebraic number 168

    characteristic, and extension 169

    characteristic, defined 167

    commutative ring as 165

    defined 167

    Euclidean 168

    examples 167

    extension, defined 167, 169

    finite, properties 169

    Galois, defined 169

field (*continued*)
  quadratic 167
    defined 167
    properties 168
  skew 165
  small finite, tables 189
  tensor 483
Fields medals 802
figures, list 821
filters
  Kalman 719
  matched 721
  optimal 721
  Wiener 721
financial
  analysis 779–790
  tables 783
  terms 779
finite
  -difference method for partial
    differential equations 770
  cosine transform 599
  differences, calculus 265
  extension field, defined 169
  fields 261
    properties 169
  geometry 247
  integration range 402
  set, defined 67
  sine transform 599
first
  come, first served 637
  fundamental form 380
    defined 377
  fundamental metric coefficients . . 380
    coordinate patch 377
  in, first out 637
  kind integral equation 478
Fischer
  Courant– minimax theorem
    (eigenvalues) 157
  Riesz– theorems 73, 74
Fisher
  inequality (combinatorial design) 244
  inequality (symmetric design) . . . 245
fitting equations to data 737
fixed point
  defined 730
  in logistic map 272
  iteration 730
  theorem 71, 730

flat pseudo-Riemannian manifold . . . 488
floor function 520
flows
  maximum network 282
  solution by linear programming . . 288
fluid
  dram 794
  flow, Navier–Stokes equations . . . 469
  ounces 794, 796, 797
focus
  critical point (ODE) 468
  ellipse 326
  hyperbola 327
  of a conic, defined 325
folding stamps, number of ways 27
folium of Descartes 337
foot 794
foot-pounds 797
force, units 796
forest 224
forest, spanning, of graphs 220
forms, differential 395
formula
  Brahmagupta 323
  Burnside (for $n!$) 17
  Cardano (cubic polynomial) 89
  Cauchy integral 55, 406
  Cayley 239
  Dobinski 211
  Erlang 639
  Euler (polyhedra) 357
  Euler–MacLaurin (sums) 40, 776
  Gauss (divergence theorem) 405
  Gauss (spherical triangles) 371
  Gregory (integration) 753
  Heron (area of triangle) 319, 513
  inversion (Z-transform) 596
  inversion (DFT) 582
  inversion (Fourier transform) 580
  inversion (Hankel transform) 590
  inversion (Hilbert transform) 593
  inversion (Laplace transform) . . . 588
  Möbius inversion 102
  Newton–Cotes 750
  Plana 40
  Poisson summation 40, 580
  Pollaczek–Khintchine 639
  Ptolemy 323, 335
  quadratic 89
  Rodrigues . . . . *see* Rodrigues formula
  surface 379

formula (*continued*)
    Stirling (for $n!$) 17, 542
    Stirling (interpolation) 736
    Viete's 84
Fortran 803
forward
    difference 265
       formula, numerical differentiation
       764
       in Aitken's $\triangle^2$ method 728
    Kolmogorov equation 467
four-color theorem 235, 236
Fourier
    –Bessel series 34
    allied integral, defined 577
    coefficients 48
       generalized, defined 74
       generalized, Parseval's identity 75
       Riesz–Fischer theorems 74
    cosine
       coefficients 577
       transform 582, 601
       transform, and Hankel transform
       589
    expansion, Bessel function 561
    repeated integral, defined 577
    series 35, 48, 76
       complex 50
       convergence in the mean 74
       example of Bessel's inequality . 74
       exponential 50
       generalized, defined 74
       in inversion of Z-transform . . 597
       sine 76
    sine
       coefficients 577
       series 76
       transform 582, 600
       transform, and Hankel transform
       589
    spectrum 50
Fourier transform 76, 576
    and Z-transform 595
    and Hartley transform 591
    and Hilbert transform 593
    and probability distribution 620
    convolution, defined 579
    discrete 582
    fast 584
    for linear ODE 464
    multidimensional 585

Fourier transform (*continued*)
    multidimensional, table 604
    pair, defined 577
    table 602, 603
Fowler, Emden– equation (ODE) . . . 465
fractal 343, 344
    defined 273
fractional binomial coefficients, table 218
fractions
    addition and multiplication 3
    common 7
    comparison 3
    continued 96–98
       Pell's equation 99
       periodic 97
    decimal equivalents 7
    defined 3
    hexadecimal conversion to decimal . 9
    partial 87–88
Fredholm
    alternative (integral equation) 479
    integral equation 478
       equivalence to boundary-value
          problem 479
       Neumann series 482
       second kind, resolvent kernel . 481
       separable kernel 480
       symmetric kernel 481
Frenet, Serret– equations (curve) . . . 375
frequency 796
Fresnel integral
    defined and properties 547
    table 548
Frobenius
    –Perron theorem (eigenvalues) . . . 157
    norm of matrix 146
Frucht's theorem (graphs) 238
frustum
    area 363
    cone 362
    volume 363
full rank 147
functions
    Airy 565
    algebraic, series for 42
    analytic 159
       defined 54
    asymptotic . . *see* asymptotic function
    Bessel and properties 559–564
    beta, defined and properties 544
    bijective, defined 66

functions (*continued*)

   Boolean, number of monotone 26

   bounded 54

   ceiling 520

   circular *505*

   component, class of curves 373

   composition 67

   continuity of in ℝ 71

   contraction of 71

   convergence of a sequence of 71

   decision 657

   derivatives, table 386

   Dirac delta .. *see* Dirac delta function

   domain, defined 66

   doubly periodic, elliptic 573

   elliptic, defined and properties ... 572

   entire 54

   equicontinuity of, defined 71

   equivalent, defined 74

   error, defined and properties 545

   Euler totient 25, 128

   exponential 520

     defined and properties 521

     definition 15

     series for 43

   floor 520

   gamma 540–544

   generalized 76

   generating .. *see* generating functions

   greatest integer 520

   Green's, in solution of ODE 463

   Gudermannian 530

   Hankel, defined 559

   Heaviside .... *see* Heaviside function

   hyperbolic

     inverse, series for 46

     series for 45

   hypergeometric 36, 552

   injective, defined 66

   inverse 67

     hyperbolic, series for 46

   inverse trigonometric ..... *see* inverse

     functions

   Jacobian elliptic 572

   least integer 520

   Legendre 554

   limit, defined 71

   logarithmic 522

     series for 43

   loss 656

   mappings 58

functions (*continued*)

   maxima 389

   measurable, equivalence of 74

   minima 389

   Möbius 25, 35

   modified Bessel 564

   modular, coefficients of 31

   Neumann 560

   normal probability, and error function 545

   of a complex variable 54

   one-to-one, defined 66

   onto, defined 66

   opportunity loss 658

   periodic

     and Laplace transform 588

     expansion of 51

     representation by Fourier series 48

   plasma dispersion 546

   Ramanujan τ 31

   range, defined 66

   regret 658

   represented as continued fraction .. 97

   Riemann zeta 552

   risk 657

   signum .... *see* signum function

   simple, defined 58

   sum of divisors 26

   surjective, defined 66

   transformations 58

   trigonometric . *505, see* trigonometric

     functions

   types 66

   uniformly continuous 71

   univalent, defined 66

fundamental

   differential, coordinate patch 377

   domains 308

   existence and uniqueness theorem 375

   form

     coordinate patch 377

     defined 377

     first 380

     for paraboloid of revolution .. 380

   metric, coordinate patch 377

   system, difference equations 266

   theorem

     of algebra 84

     of calculus 388, 400

     of surface theory 379

future value, financial 779

G

Gaddum, Nordhaus– bounds (graphs)
234
Galileo 81
gallons 794, 796, 797
Galois field, defined 169
Galois group 809
gambler's ruin 627
game
    matrix 278
    mixed zero-sum 276
    non-zero-sum 277
    prisoners' dilemma 277
    theory section 274–278
    two-person non-cooperative 274
    voting 278
    zero-sum 275
gamma
    function 540–544
        and Bessel functions 560
        figure 541
        relation to beta 544
        table 543
    probability distribution 634
Gauss
    –Hermite quadrature 756
    –Laguerre quadrature 756
    –Legendre double 760
    –Legendre quadrature 755
    –Mainardi–Codazzi equations ... 379
    Carl Friedrich 814
    equations, surfaces 379
    formulae (divergence theorem) .. 405
    formulae (spherical triangles) 371
    hypergeometric function 553
    law of quadratic reciprocity 95
    test 34
    theorem, totient function 128
    theorema egregium (surfaces) ... 379
Gaussian
    curvature
        coordinate patch 377, 379
        paraboloid of revolution 380
    double integral 760
    elimination
        elementary matrices in 139
        solving linear equations 740
    integers 169
    integers, as ring 166
    probability distribution 467

Gaussian (*continued*)
    quadrature 755, 756
    random variable ..*see* normal random
        variable
genealogy project 809
genera 224
general triangles, properties 512
generalized
    functions 76
    inverse of a matrix 151
    Riemann–Lebesgue lemma 578
generating
    matrix, linear codes 256
    polynomial, and cyclic code 256
    random variables 647
generating functions 39
    Bell numbers 211
    Bernoulli numbers 19
    Bernoulli polynomials 19
    binomial coefficients 208
    Chebyshev polynomials 535
    Euler numbers 20
    Euler polynomials 20
    Hermite polynomials 532
    Jacobi polynomials 533
    joint moment 633
    Laguerre polynomials 533
    Legendre polynomials 534
    moments
        binomial distribution 630
        discrete random variable 630
        exponential distribution 635
        gamma distribution 634
        geometric distribution 631
        hypergeometric distribution .. 631
        negative binomial distribution 632
        normal distribution 633
        Poisson distribution 632
        uniform distribution 633
    Möbius function 102
    partitions 211
    polylogarithm 552
    random variables 620
    solving difference equations 268
    totient function 129
    trees 239
generator
    cylinder 361
    non-linear 646
genetic algorithms (optimization) ... 293
Genocchi numbers 30

genus of a graph 224
geodesic
    affinely-parameterized 485
    affinely-parameterized, equations 488
    coordinate patch 377
    curvature, coordinate patch 377
geometric
    arithmetic mean– mean inequality 33,
    73
    mean 659
        defined 87
    probability distribution 630
    probability problems 625
    progression 86
    series 38, 82
geometry
    differential 373
    finite 247
    spherical 368
Gerschgorin circle theorem (eigenvalues)
    158
Gerson, Levi ben 812
gibi, definition of 6
giga, as prefix 6
gills 794
girth 224
Givens rotation, defined 141
glide-reflection
    formula 306
    in a plane 351
    isometry of space 349
    plane isometry 305
GNU 808
Gödel incompleteness theorem 202
Golay code
    as perfect 257
    binary 258
    in Mathieu 5-design 244
    ternary 258
Goldbach conjecture (sum of primes) 103
golden ratio 16
    relation to Fibonacci numbers 21
Golomb ruler 29
googol 6
googolplex 42
    defined 6
Gordan, Clebsch– coefficients 574
    and spherical harmonics 539
Gordon, sine– equation (PDE) 469
Goursat, Cauchy– integral theorem . 406
grad .... *see* gradient

gradient 390, 749
    in orthogonal coordinates 493
    of a scalar invariant, as tensor 489
grains 794
grams 796, 797
graphs
    acyclic, defined 220
    algebraic methods 237
    Appel–Haken four-color theorem 235
    arboricity
        defined 220
        Nash–Williams theorem 232
    automorphism 220, 226
        groups of, table 238
    bipartite 226, 229
        assignment problem 289
        defined 226
    block 220
    bridge 221
    Brooks' theorem 234
    cactus 221
    cage 221
    Cayley 230
    center 221
    characteristic polynomial ... 221, 227
    chromatic
        index 221
        numbers 221
        numbers, table 234
        polynomial 221
        polynomial, form of 235
        polynomials, table 234
    circuit 221
    circulant 221, 230
    circulant, cycle as a 229
    circumference 221
    clique 221
    clique number 221
    coboundary operator 221
    cocycle vector 222
    color classes 226
    coloring 226
        defined 222
        theorems 234
    complement 222, 226
    complete
        bipartite 230
        bipartite, star as a 229
        defined 222
        description 229
        multipartite 230

graphs (*continued*)

graphs (*continued*)

component 222

connected 222

   number of 28

connectivity

   defined 222

   theorems 231

contraction 222

cover 222

crosscap number 222

crossing number 222

cube, described 229

cubic 222

cut

   defined 222

   space 223

   vector 223

   vertex 223

cycle 223

cycle space 223

cycle, described 229

defined 219

degree

   sequence 223

   sequence, theorems on 237

degree, defined 223

diameter 223

Dirac's theorem 233

directed 223

   number of 29

distance, defined 223

distance, theorems on 235

double loop 230

drawing 223

drawing, theorems on 236

eccentricity 223

edge

   -chromatic numbers, table 234

   connectivity 223

   cover 222

   defined 219

   space 224

   symmetric 227

eigenvalues in spectrum 237

embedding 224

embedding, theorems on 236

empty, defined 229

end vertex 224

enumeration of, theorems 239

Euler, number of 27

Eulerian 233

   circuit 224

   defined 224

   table 233

   trail 224

even 224

examples with 6 or 7 vertices 231

factor 224

Fary embedding 236

forest 224

four-color

   theorem and colorability 236

   theorem for planar 235

genera 224

genus 224

girth 224

Hall's theorem 238

Hamiltonian

   4-connected planar 233

   complete graph as 229

   components of subgraph 233

   cycle 224

   cycle as 229

   defined 224

   Dirac's theorem 233

   Ore's theorem 233

   path 224

   table 233

   Toeplitz graph as 230

handshaking lemma 237

Havel theorem on degree sequence
237

Heawood map coloring theorem . 236

highly regular 226

homeomorphic 224

identification of vertices 224

incidence

   defined 224

   matrix 224

independence number 225

independent set of vertices, defined
225

internally disjoint paths 225

intersection 231

interval 231

interval, as an intersection graph . 231

isolated vertex 225

isomorphism 225

graphs (*continued*)

   isomorphism classes

      descriptions 241

      digraphs, table 240

      table 239, 240

   k-coloring 222

   k-partite 226

   Kirchhoff matrix-tree theorem ... 232

   Kneser, defined 229

   König's theorem 238

   Kuratowski's theorem 236

   labeled 225

   labeled, number of, table 240

   ladder, defined 229

   loop 220, 230

   matching 225

   matching, theorems on 238

   metric space, theorems on 235

   monotone property 226

   Moore, defined 235

   multigraph 220

   multipartite

      complete 230

      defined 226

   Nash–Williams arboricity theorem 232

   neighbor 226

   neighborhood 226

   Nordhaus–Gaddum bounds 234

   notation 219

   number of

      table 239

      transitive and directed 29

   odd, defined 229

   operations 228

      composition 228

      conjunction 228

      edge difference 228

      edge sum 228, 229

      join 228

      power 228

      product 229

      square 228

      union 228, 229

   order 226

   Ore's theorem 233

   partitions, theorems 234

   path 229

      defined 226

      theorems 231

   perfect 226

graphs (*continued*)

   Petersen, defined 229

   planar 226

   planar, four-color theorem 235

   planarity, theorems on 236

   prism, defined 230

   pseudograph 220

   radius 226

   Ramsey's theorem 233

   rectilinear crossing number 222

   regular 226

   regular, Toeplitz graph as 230

   rooted 226

   saturated matching 225

   section 219–241

   self-complementary 226

   similar edges 226

   similar vertices 226

   simple 220

   size 227

   small, descriptions 241

   spanning tree 227

   spectrum 227

   sphere 227

   star 229

   subdivision 227

   subgraph 227

   supergraph 227

   switch 227

   symmetric 227

   Szekeres–Wilf theorem 234

   thickness 227

   thickness, theorems on 236

   Toeplitz 230

   toroidal mesh 230

   trail 227

   transitive, number of 30

   trees .... *see* trees

   triangle 227

   trivial 227

   Turán 230

   Turán's theorem 233

   unicyclic 227

   unlabeled 225

   unlabeled nodes, number of 27

   vertex *219*

      cover, condition for 233

      degree, theorems on 237

      space 227

      symmetric 227

   Vizing's theorem 234

graphs (*continued*)

    walks 228

        theorems 231

    wheel 229

gravitational

    constant 794

    field strength, units 796

    potential, units 796

gray 792

Gray code 261

great circle distance on earth 372

greatest

    common divisor 101

    integer function 520

    lower bound of a subset 68

Greek alphabet 803

Green's

    function (ODE) 463

        table 463

    function (PDE) 471

    theorem (determining area) 409

    theorem (line integral) 404

    theorem (surface integral) 405

Gregory

    formula for approximating integrals 753

    Newton– formulae for interpolation 736

    series 23

grouping terms, ways to 212

groups *161*

    abelian 162

        number of 25, 177

    additive 161

    alternating 188

        defined 163

        defined as permutation 172

    automorphism

        of a graph 220

        of graphs, table 238

        of graphs, theorems 238

    character, defined 170

    characters 191

    classification 162

        of crystallographic 312

    crystallographic 307–312, 809

        classification 312

    cyclic .. 162, 178–181, 184, 185, 187

        and finite field 169

        defined as permutation 172

        homomorphism 171

groups (*continued*)

    dihedral 182, 185, 186

        defined 163

        defined as permutation 172

    direct product, defined 163

    examples 163

    factor, defined 162

    finite simple 162

    fundamental theorem, isomorphism 171

    Galois 809

    homomorphism

        defined 170

        theorem 171

    identity, defined as permutation .. 172

    isomorphism, defined 170

    kernel of homomorphism 170

    Klein four group 179

    Lie 162, 466

        and second-order ODE 467

    linear character, defined 191

    matrices 163

    multiplication modulo n 163

    multiplicative 161

    non-isomorphic, number of 25

    number of abelian, table 177

    number of, table 176

    of matrices 171

    one-parameter 466

    order n, number of 25

    orders of 26

    orthogonal, defined 171

    permutation

        creating 172

        defined 163

        table 172

    prime order 162

    product 181, 184, 187

        defined 163

        of cyclic groups, isomorphism 171

    properties 162

    quaternion 182

    quotient, defined 162

    rotations 172

    simple ...... 162, *see* simple, groups

        defined 162

        orders of 26

        sporadic 162

        sporadic, table 191

    small order 178

    tables 178

groups (*continued*)
 solvable 162
 special linear 171
 sporadic 162
  table 191
 symmetric 163, 180
  defined as permutation 172
 symmetry, crystallographic 307
 unimodular 171
 units of a ring 165
 wallpaper 307, 809
Gudermannian
 function 530
 table 531
guide to mathematical software 808
Guldinus, Pappus– theorem (surface of
   revolution) 363

H

Haar wavelet 723
Hadamard matrix
 3-design 250
 balanced incomplete block design 250
 defined 249
 symmetric design 250
Haken, Appel- four-color theorem .. 235
half
 -angle formulae 510
 -angle formulae, spherical 370
 -side formulae, spherical 371
Hall's theorem (graphs) 238
Hamilton
 –Jacobi equation (PDE) 469
 Cayley– theorem (matrix) 153
 William Rowan 814
Hamiltonian
 complete graph 229
 cycle in a graph 229, 230
  defined 224
 graph
  4-connected planar 233
  components of subgraph 233
  defined 224
  Dirac's theorem 233
  Ore's theorem 233
  table 233
 in cost minimization 498
 path in a graph 224
 separability 473
 Toeplitz graph as 230

Hamming
 bound, block codes 258
 bound, perfect code 257
 code, as perfect 257
 codes 257
 distance, defined 256
handshaking lemma 237
Hankel
 function
  and Bessel function 562
  defined 559
  solution of potential equation . 471
 transform, defined and properties 589
 transform, table 610
Hardy's theorem (Laplace transform) 588
harmonic
 mean 660
 mean, defined 87
 numbers 32
 series 32
  alternating 32
harmonics, spherical 538
Hartley transform 591
Hasse diagram (posets) 205
Havel theorem (graphs) 237
haversine, defined 505
haversine, spherical triangles 372
hazard rate 655
hear 477, 478
heat
 equation 469
  in a circle 473
 of fusion 795
 of vaporization 795
 partial differential equation 773
Heaviside function 77
 and Laplace transform 586
 in integral evaluation 408
 in solution to wave equation 472
Heawood map coloring theorem 236
hecto, as prefix 6
Heine–Borel theorem 70
helix, circular 375
Helmholtz equation
 defined 469
 separability 474
 separated components in orthogonal
  coordinates 494
henry 792
heptagon, area and apothem 324
heptillion 6

Hermite
  Gauss– quadrature 756
  interpolating polynomial 737
  interpolation 737
  polynomials 532, 754
    generating function 39
    table 535
Hermitian
  kernel, integral equation 478
  matrix
    and multiplication 143
    conjugate 138
    defined 139
    form 138
    skew, defined 139
Heron formula (area of triangle) 319, 513
hertz 792
Hessian 392
Heun's method (for DEs) 767
heuristic search 290
hexacode, self-dual MDS code 257
hexadecimal 5
  addition and subtraction 8
  fraction conversion to decimal 9
  multiplication 8
  notation 5
  scale 12, 13
hexagonal numbers 31
hexagons
  area and apothem 324
  coloring the corners 175
hexillion 6
high school 808
high-pass 724
highly regular graph 226
Hilbert
  –Schmidt kernel, integral equation
    478
  –Schmidt norm of matrix 146
  David 809, 816
  matrix 149
  space 74
  transform 591
  transform, table 611
histogram 658
hit or miss method, Monte–Carlo
  quadrature 761
Hocquenghem (BCH) codes 257
Hölder inequality 33, 73
  integrals 406
holomorphic function 54

homeomorphic graph 224
homogeneity, scalar 72
homogeneous
  coordinates 306
    in space 348
    planar, defined 303
  coordinates in space 351
  difference equations 267
  integral equation 478
  ODE 461, 462
  solution, linear ODE 456
homomorphism
  group, defined 170
  properties 171
  ring 170
  theorem for groups 171
  theorem for rings 171
homothety 352
  in similarity 313
honors in bridge (card game) 628
Horner's method (polynomial evaluation)
    733
horsepower-hours 797
l'Hôspital's rule (limits) 389, 564
Householder
  method for constructing similar
    tridiagonal matrices 744
  transformation 141
Huffman code, English text 260
Hungarian method (linear programming)
    288
Hurwitz–Radon numbers 25
Hypatia 811
hyperbola
  conjugate, defined 331
  defined 325
  equation 326
  figure 325
  polar equation for 332
  properties 326, 331
  rectangular, defined 327
hyperbolic
  cylinder, equation 364
  partial differential equations 469, 775
    boundary conditions, table ... 470
  point, coordinate patch 377
hyperbolic functions 523
  defined 524
  derivatives 386, 527
  graph 523
  half-argument formulae 527

hyperbolic functions (*continued*)
   integrals 445
   inverse
      defined 527
      properties 527
      relation with circular 519
      relations 528
      relations with circular 529
      relations with logarithmic 529
      series for 46
      sum and difference 529
   multiple argument relations 526
   product 527
   ranges 524
   relations 524, 525
      with circular functions .. 525, 530
      with Gudermannian 531
   series expansions 525
   series for 45
   sum and difference 526
   sum and difference of arguments . 526
   symmetry relations 525
   table 531
hyperboloid, equation 364
hyperexponential service time 637
hypergeometric
   equation (ODE) 465
   function 36, 552
      and elliptic integrals 569
      solution of polynomials 89
   function, and Legendre function . 554
   probability distribution 631
   series 36
hypocycloids, defined 341
hypotenuse, of a right triangle 320
hypothesis 200
   alternative 661
   continuum 204
   null 661
   Riemann 24
   testing 669
hypotrochoid, defined 341

I

icosahedron
   constructing 360
   properties and figure 359
ideal
   defined 164
   maximal 164
   prime in 164

ideal (*continued*)
   principal, defined 164
   proper, defined 164
idempotent
   laws 200
   matrix 150
      defined 140
      eigenvalues 153
      generalized inverse 151
identification of graph vertices 224
identities
   algebraic, polynomials 85
   among circular functions 509
   Bianchi (torsion tensor) 487
   cyclical (curvature tensor) 487
   index-cycling (octonions) 166
   index-doubling (octonions) 166
   inverse circular functions 519
   Jacobi 48
   Parseval's 75
   Pfaff–Saalschutz 38
   quintuple product 48
   Ricci (contravariant vector field) . 487
   Rogers–Ramanujan 47
   triple product 48
identity
   binary operation 161
   group, defined as permutation ... 172
   in interval arithmetic 65
   isometry of space 349
   matrix, defined 138
   plane isometries 305
if and only if (propositional calculus) 199
ill-conditioned matrix, defined 149
image
   inverse, of a set 66
   of a set, defined 66
imaginary
   ellipse 328
   part of complex number 53
   quadratic field, defined 167
   unit, defined 53
implication, transitivity of 200
implicit methods (numerical DEs) .. 767, 768
implies, in propositional calculus ... 199
improper integral 399
improper node (critical point, ODE) . 468
impulse function 76
   Z-transform 595

inches
  in other units 794, 796
  of mercury 798
incidence
  graph 224
  mapping, graph definition 220
  mapping, graph notation 219
  matrix
    combinatorial design, defined .241
    graph 224
  of vertices 224
  structure (combinatorial design) . 241
inclusion rule for set properties 204
incomplete block design 245
inconsistent linear equations 148
incumbent solution (linear programming)
    287
indefinite
  integrals 398
    properties 399
    table 412–448
  metric 486
independence number of a graph 225
independent
  discrete variables, generating function
    620
  events, defined 618
  set 233
    of vertices of a graph 225
index
  -cycling identity (octonions) 166
  -doubling identity (octonions) ... 166
  chromatic 221
indicatrix, Dupin's, coordinate patch 377
indices
  contravariant, mixed tensor 483
  covariant, mixed tensor 483
  lowering, and covariant derivative 488
  lowering, tensor 486
  of $\mathbb{Z}_n^*$, table 192
  raising, and covariant derivative ..488
  raising, tensor 486
induced subgraph 227
inequalities
  arithmetic mean–geometric mean .33,
    73, 83
  Bessel 74
  Bienaymé–Chebyshev 624
  Carleman 74
  Cauchy 56
  Cauchy–Schwartz 73, 623

inequalities (*continued*)
  Chebyshev (random variable) 624
  Fisher's 244, 245
  Hölder 33, 73, 406
  in real analysis 73
  integral 406
  Jensen 93, 624
  Kantorovich 33
  Kolmogorov (random variables) . 624
  Markov (random variable) 623
  Minkowski 33, 73, 406
  probabilistic 623
  real numbers, properties 68
  Schwartz 32, 73, 406
  triangle 69, 72, 133, 321
inertia
  moments of, table 410
  Sylvester's law (eigenvalues) 158
inference, rule of 200
infinite
  integration range 402
  order, group element 161
  products 47
  series 42, 47
  sets 204
infinitesimal generator 466
infinity, defined 68
information
  mutual 254
  symbols, linear codes 256
  theory 253
initial
  ray
    in polar coordinates 302
    in space 346
    spherical coordinates 346
  value problem 265
    differential equation 766
    equivalence to integral equation
      479
    systems of differential equations
      769
injective function, defined 66
inner product 133
  and metric tensor 486
  of orthogonal vectors 132
inscribed circle
  regular polygon 324
  triangle 318, 512
inspection levels 652
instantaneous hazard rate 655

Institute for Mathematics 808
integer
    algebraic, defined 167
    defined 3
    divisibility 93
    expressed as sum of powers (Waring's
        problem) 100
    Gaussian 169
    positive 3
    powers 21
    programming, linear 287
    properties 93
    sequences 25–31, 817
    series 25
integral
    asymptotic expansions of 408
    contour, in complex plane 57
    convergence 33
    derivative of 388
    domain (ring), table 167
    domain, defined 164
    equations 478
        classification 478
        named 480
    formula, Cauchy 55
    inequalities 406
    line 404
    principal value 399, 588, 591
    repeated 588
    surface 404
    table 411–455
    test 33
    theorem, Cauchy 55
    transforms 576
        tables 599–613
integrals
    definite, properties 400
    discontinuities 408
    elementary 412
    evaluation by partial fractions . . . 401
    evaluation by substitution 400
    miscellaneous forms 428
    transformations of 402
    with $\sqrt{2ax - x^2}$ 428
    with $\sqrt{a + bx}$ 420, 421
    with $\sqrt{a + bx + cx^2}$ 426
    with $\sqrt{a^2 - x^2}$ 424
    with $\sqrt{x^2 \pm a^2}$ 422
    with $a + bx + cx^2$ 419
    with $a + bx$ 413
    with $a + bx$ and $c + dx$ 415

integrals (*continued*)
    with $a + bx^n$ 416
    with $c^2 \pm x^2$ 415
    with $c^3 \pm x^3$ 418
    with $c^4 \pm x^4$ 418
    with Bessel functions 448
    with exponentials 443
    with hyperbolic functions 445
    with inverse trigonometric functions
        439
    with logarithms 441
    with trigonometric functions 430
integrand, defined 398
integrating factor (ODE) 462
integration 398
    applications 409
    by parts 402
        defined 399
        example 465
        extended 404
    contour 399, 404
    numerical . . . . 750, *see* quadrature
    of power series 32, 36
interarrival time 637
intercept, of plane 354
interest
    compound 780, 781
    compound, tables of annuity 787
    compound, tables of final value . . 783
    compound, tables of interest rate . 785
    financial 779
    rate, tables for compound 785
    simple 780
interior
    angle, polygon 318
    point method (linear programming)
        284
intermediate value theorem 388
internally disjoint paths in a graph . . 225
international nautical mile 794
interpolating polynomial
    Hermite 737
    Lagrange 733
interpolation 737
    Hermite 737
    inverse 737
    Lagrange 733
    of a function 733
interquartile range 660
intersecting planes, equation 364

intersection
   graph 231
   of elements 69
   of plane 354
   of sets, defined 203
interval
   analysis 65
   arithmetic properties 65
   arithmetic rules 65
   graph 231
   graph, as an intersection graph ...231
   of convergence 36
   thin 65
invariant
   arc length of regular curve 374
   curvature (tensor) 488
   graphs 219
   scalar (tensor) 483
inverse
   circular functions 518, 519
      defined by integral 518
      identities 519
      negative arguments 519
      principal value 518
   function 67
   hyperbolic functions 519
      defined 527
      derivatives 386
      properties 527
      relations 528
      relations with circular 529
      relations with logarithmic 529
      series for 46
      sum and difference 529
   image, of a set 66
   in algebra 161
   interpolation 737
   matrix 138
      generalized 151
   power method (eigenvalues) 743
   symbolic calculator 807
   transform method 647
   trigonometric functions 518
      derivatives 386
      in integrals 412
      series for 45
inversion
   discrete Fourier transform 582
   Fourier transform 580
   Hankel transform 590
   Hilbert transform 593
   Laplace transform 588

inversion (*continued*)
   of power series 36
   theorem (integral transforms) 580
   Z-transform 596
inversive, congruential generator 646
invertible matrix 138
irreducible
   element (integral domain) 164
   polynomials
      binary, table 261
      defined 166
      number of 25
      number of monic 261
      tables 194
irrotational vector field, defined 390
isolated
   singularity 56
   vertex in a graph 225
isometries
   of space 348
   of the plane 305
isomorphism
   classes of trees 241
   classes, of graphs, descriptions .. 241
   graph 225
   of a group, defined 170
   properties 171
isosceles triangle 320
iteration, fixed point 730

J

Jacobi
   Hamilton– equation (PDE) 469
   polynomials 533, 754
      table 537
   symbol (quadratic residue) 94
   triple product identity 48
Jacobian 77, 392
   elliptic functions 572
   matrix (coordinate transformation)
      483
   matrix (in Newton's method) 748
Jensen's inequality
   polynomials 93
   random variables 624
join, operation on graphs 228
joint
   density 621
   moment generating function 633
   moments 622
      multivariate distributions 622

jointly sufficient statistics 663
Jordan blocks (matrix) 154
joules 792, 798
Julia set (chaos) 273, 809
jump discontinuities 48

K

k-coloring of a graph 222
k-partite graph, defined 226
Kalman filter 719
Kantorovich inequality 33
Kapteyn series 34
Karmarkar's method (linear
          programming) 285
katal 792
Kelvin 798
    degrees 792
Kendall–Mann numbers 28
Kent vi
kernel
    integral equation, classification .. 478
    of a group/ring homomorphism .. 170
    of an integral equation 478
    resolvent (integral equation) 481
kibi, definition of 6
kibibyte, definition of 6
kilo, as prefix 6
kilograms 792, 796, 797
kilometers 796, 797
kilowatt-hours 797
kilowatts 797
Kirchhoff matrix-tree theorem 232
Kirkman schoolgirl problem 249
Klein bottle, crystallographic groups 308
Klein four group 179
Kneser graph, defined 229
knots 809
    number of 27
    prime 27
    speed 794, 798
    table and figures 382
Koch snowflake 344
Kolmogorov
    –Smirnov
       table 717
       test 679
    equation, forward 467
Kolmogorov's inequality 624
König's theorem (graphs) 238
Korteweg de Vries equation (PDE) .. 469
Kovalevskaya, Sofia 815

Kronecker
    delta 483
    delta, as tensor 489
    product (matrices) 159
    sum (matrices) 160
Kuratowski's theorem (graphs) 236
kurtosis, random variable 620
Kutta, Runge– method 767

L

l'Hôspital's rule 389, 564
labeled graph 225
ladder graph, defined 229
lag 645
lagged-Fibonacci sequences 646
Lagrange
    Euler– equation 407
    interpolation 733
    Joseph 809, 814
    multipliers 389
    remainder 37
    series 39
    theorem (group order) 162
    theorem (sum of squares) 100
Laguerre
    Gauss– quadrature 756
    polynomials 533, 754
       generalized 533
       generating function 39
       table 536
Lahire's theorem (ellipse) 331
Lamé equations 491
Laplace
    development 145
    equation 469
       separability 474
       separated components in
          orthogonal coordinates 494
       solutions 474
    integral, defined 585
    method (integral evaluation) 408
    partial differential equation 770
    transform 35, 585, 638
       convolution 589
       for linear ODE 464
       in control theory 497
    transform, table 605
Laplacian 474, 475
    in orthogonal coordinates 493
    of scalar invariant (tensor analysis)
      488

large numbers, arrow notation 4
latera recta 326, 327
Latin square, defined 251
latitude 794
latus rectum 327
Laurent series 56
laws
    associative 200
    cancellation 162, 164
    commutative 200
    contrapositive 200
    De Morgan's 200
    idempotent 200
    Little's . . . . 639, *see* Little's law
    of cosines, spherical 370
    of cosines, triangle 319, 513
    of exponents 86, 520
    of inertia (eigenvalues) 158
    of quadratic reciprocity 95
    of reliabilities 653
    of sines, spherical 370
    of sines, triangle 319, 513
    of tangents, spherical 370
    of tangents, triangle 513
    of the excluded middle 200
    of unreliabilities 653
leap years 799
least
    common multiple 101
    integer function 520
    norm solution of linear equations 148
    squares
        approximation 739
        estimator 719
        problem 150, 151
        solution of linear equations . . . 148
    upper bound 68
    upper bound axiom 68
Lebesgue: Riemann– lemma 578
left
    -handed circular helix 376
    cancellation law, groups 162
    coset of subgroup, defined 162
    distributive law, ring 164
    divisor of zero (ring) 164
    hand limits 49
    limit 576
    matrix inverse 151
    shift, in periodic sequence 263

Legendre
    equation (ODE) 465
    functions 554
        associated 557
        figures 554
    Gauss– double integral 760
    Gauss– quadrature 755
    polynomials 534, 754
        and Legendre function 556
        generating function 39
        series 557
        table 536
    relation (elliptic integrals) 570
    symbol (quadratic residue) 94
Lehmer, Lucas– primality test 104
Leibniz
    Gottfried Wilhelm 813
    rule (derivative of integral) 388
lemma
    Burnside's (permutation group) . . 173
    handshaking 237
    Riemann–Lebesgue 578
lemniscate of Jakob Bernoulli 338
length 792
    arc, and metric tensor 486
    arc, as integral 409
    arc, regular curve 373
    units 796
Levi–Civita
    connection 487
    symbol 489
Liapunov method (ODE) 466
Lie algebra 467
Lie groups 162, 466
    and second-order ODE 467
light speed 794
likelihood function 662
limaçon of Pascal, defined and figure 338
liminf . . . . *see* limit inferior
limit
    inferior (liminf) 70
    left-hand 49
    of a function, at infinity 71
    of a function, defined 71
    of a sequence, defined 70
    of an infinite series 32
    of integration 398
    point
        Bolzano–Weierstrass theorem . 70
        defined 69, 70
    properties 385

limit (*continued*)

    right-hand 49

    superior (limsup) 70

    test 33

    theorem, central 623

limsup .... *see* limit superior

line

    binormal, regular curve 374

    element 486

        polar coordinates (tensor) 492

    integral 404

    segment

        points in 625

linear

    algebra 137

    characters of groups 191

    code, defined 256

    congruential generator 644

    connection (tensor analysis) 484

    convergence, defined 728

    differential equations 456, 462

    equations

        perturbed 149

        solution by Cramer's rule 158

        solving 740

        systems 148

    fractional transformation 58

    generating function property 268

    independence, sequences, defined 266

    integral equation 478

    interpolation, piecewise 737

    mappings 149

    measure 794

    ODE, constant coefficients,

        homogeneous solution 456

    programming 280, 283

        integer 287

        Karmarkar's method 285

        optimization of mixed strategy 276

    regression 682

    spaces 149

    spiral 342

lines 314

    angle between 316, 356

    asymptotic, coordinate patch 377

    at infinity 249

    best-fit 740

    combinatorial design 241

    concurrent 317

    critical 24

    distance from point 316

lines (*continued*)

    in space 355

    length of segment 316

    mapping 58

    normal, coordinate patch 377

    of curvature

        coordinate patch 377

        paraboloid of revolution 380

    parallel, defined 316

    perpendicular, defined 316

    principal normal, regular curve .. 374

    slope 314

    straight, equation (Cartesian) 314

    straight, equation (polar) 316

    tangent, regular curve 374

    with prescribed properties 315

    x-intercept 314

    y-intercept 314

Liouville theorem 54

Lipschitz function 71

liquid measure 794

list

    of figures 821

    of notation 823

    of references 817

liters 796, 797

Little's law 638, 639

Lobatto

    quadrature 757

    weight functions (quadrature) ... 755

local

    integrability, defined 576

    minimum, in steepest-descent method

        748

    property of coordinate patch 376

    truncation error (differential

        equations) 766

logarithmic

    derivative, of gamma function ... 543

    functions 522

        integrals 441

        relations with hyperbolic 529

        series for 43

        table 531

    integral, defined and properties .. 550

    integral, table 551

    spiral 342

logarithms

    conversion of base 522

    derivative 386, 523

    identities 523

logarithms (*continued*)
    integration 523
    natural, base 15
    of 2, in different bases 16
    relation to exponential 523
    series expansions 523
    special values 522
logical implication, in propositional
        calculus 199
logistic
    difference equation 271, 272
    equation, capacity dimension 344
    map (chaos) 272
longitude 794
loop 230
loop, in a graph, defined 220
Lorentzian manifold 486
Los Alamos 808
loss function 656
low-pass 724
lower
    -triangular matrix 139, 157
    bandwidth of a matrix 141
    bound of a subset 68
    control limit 650
lowering indices of tensor 486
lowering indices, and covariant derivative
    488
Lucas numbers 29
Lucas–Lehmer primality test 104
lucky numbers 29
lumen 792
luminous intensity 792
lune (spherical segment) 368
Lux 792

M

m-sequences 263
Maasei Hoshev 812
MacLaurin
    Euler– sum formula 40, 776
    series 37
MacMahon's numbers 27
magic squares 101
Mahler measure of a polynomial 93
main diagonal, matrix 137
Mainardi, Gauss– –Codazzi equations
    379
major axis, ellipse 326
majority 280
majority rule (voting) 280

Mandelbrot set 273, 809
manifold
    Lorentzian 486
    n-dimensional coordinate 482
    pseudo-Riemannian
        and metric tensor field 487
        defined 486
        flat 488
    Riemannian 486
Mann
    –Whitney–Wilcoxon procedure .. 669
    –Whitney–Wilcoxon test ... 675, 676,
    680
    Kendall– numbers 28
Maple 803
mappings
    among points, number of 29
    conformal 58
        table 60
    of complex functions 58
    types 66
marginal distributions, multivariate .. 622
marginal player (voting power) 279
marginal probability 617
Markov
    chains 640
    inequality (random variable) 623
    matrix, defined 141
mass
    base unit 792
    earth 799
    hydrogen atom 795
    units 796
matched filtering 721
matching of graphs 225, 238
MathCad 803
Mathematica 803
mathematical library 808
Mathematical Society 807
mathematical software 808
mathematicians 802, 810
    genealogy 809
Mathematics Association of America 807
Mathieu
    5-design 244
    equation (ODE) 465
Matlab 803
matrix
    addition 142
    adjacency 220
    algebra 137

matrix (*continued*)

    augmented, Gaussian elimination 740
    bandwidth, defined 141
    Cholesky factorization 157
    circulant 141, 221
        condition number 149
    classes that are groups 171
    cofactor, defined 145
    collapsed adjacency, graph 226
    column 139
    condition number, defined 148
    defined 137
    derivatives 391
    determinant 144
    diagonal
        bandwidth 141
        defined 138
        inverse 147
    diagonalization of 154
    dilation 725
    eigenvalues
        computation 746
        specific matrices, table 153
    elementary 138
    exponentials 155
        of Kronecker sum 160
    factorization 156, 157
        Cholesky 157
        QR 157
    Frobenius–Perron theorem 157
    games 274, 278
    generalized inverse 151
    generating, linear codes 256
    group 163
    Hadamard 249
    Hermitian
        conjugate 138
        defined 139
        skew, defined 139
    Hessenberg, defined 140
    Hilbert 149
    Householder transformation 141
    idempotent 140, 150, 151
        eigenvalues 153
    identity, defined 138
    incidence 224
        combinatorial design, defined .241
    inverse 138
        left 151
        pseudo-inverse 151
        right 151

matrix (*continued*)

    invertible 138, *144*
    Jacobian
        in Newton's method 748
        of coordinate transformation . 483
    Kronecker product 159
    Kronecker sum 160
    lower-triangular, bandwidth 141
    Markov, defined 141
    multiplication *142*
        by scalar 142
        fast 143
        figure 142
        properties 142
        sequential 143
    nilpotent 140, 153
    non-singular *144*, 147
    norm, defined 146
    normal, defined 140
    null 138
    null space, defined 149
    orthogonal, condition number ... *139*, 149
    parity check, linear codes 256
    permanent 145
    permutation 140
        and permanents 145
        and vector operation 158
        in matrix factorization 157
    positive definite 155
    positive semi-definite 156
    principal sub-matrix, defined 140
    projection *140*, 150
    pseudo-inverse 151, 152
        and condition number 149
    quadratic forms 155
    range space, defined 149
    rank 147, 685
    rectangular 137
    regular 138
    representation 183
    rotation *139*, 141
        Givens 141
    row 139
    scaling 143
    similar 744
    similarity 150
    singular *144*, 147
    skew symmetric 139
    square 137
    square, types of 139

matrix (*continued*)
  Stäckel 473
  subtraction 142
  symmetric *139*
    eigenvalue computation 746
    QR algorithm for eigenvalues . 745
  Toeplitz, defined 140
  trace 150
  transition 640
    and channel capacity 255
  transpose 138
  tree theorem, Kirchhoff 232
  triangular
    eigenvalue computation 746
    eigenvalues of 153
    lower- 139
    upper, in matrix factorization . 157
    upper- 139
  tridiagonal 744
    bandwidth 141
  unitary, as group 172
  unitary, defined 140
  upper-triangular, bandwidth 141
  Vandermonde, defined 140
  vector representation 158
  zero 138
maxima of functions 389
maximal
  chain in posets, defined 205
  column pivoting (solving linear
    equations) 741
  ideal 164
  pivoting, in solving linear equations
    742
maximin strategy, zero-sum game ... 275
maximum
  distance separable code .... see MDS
    code
  flow (linear programming) 288
  in a sample 664
  likelihood estimator 662
  network flow 282
MDS code 257
  cyclic 257
  hexacode 257
  self-dual 257
mean
  arithmetic 87, 659
  confidence interval 666–668

mean (*continued*)
  curvature
    coordinate patch 377, 379
    paraboloid of revolution 380
  density of the earth 795
  deviation 660
  geometric 87, 659
  harmonic 87, 660
  random variable 620
  square error, defined 74
  test 670, 673–675
  time between failures 655
  trimmed 659
  value theorem 388
  weighted 659
meandric numbers 27
  closed, list of 28
measurable functions, equivalence of . 74
measure
  dry 794
  linear 794
  liquid 794
  nautical 794
  of a polynomial 93
  of a set 203
mebi, definition of 6
medals, Fields 802
median
  confidence interval 667, 669
  of a triangle, defined 318
  property 83
  test 671, 672, 675–677
mega, as prefix 6
Mellin transform, table 612
Menelaus's theorem (triangle) 321
Menger sponge, capacity dimension . 344
Menger's theorem 232
meromorphic invariance (Hilbert
    transform) 593
Mersenne prime .. see primes, Mersenne
mesh
  planar 230
  points, in difference equations ... 766
  toroidal 230
message, in codes 256
meters 792, 796, 797
method of moments 662
metric
  coefficients, in orthogonal coordinate
    system 492
  contravariant, tensor field 486

metric (*continued*)
    covariant, tensor field 486
    indefinite 486
    Lorentzian 486
    on a set 69
    signature, defined 486
    space . . . . *see* space
        defined 69
        in graphs, theorems on 235
        separable 70
        vector space 133
    tensor 486
    units 792, 796, 797
metric tensor
    determinant 489
    polar coordinates 492
micro, as prefix 6
micrometers 797
midpoint
    method (numerical differential
        equations) 767
    of a line segment 316
    rule (quadrature) 751
midrange 660
MIL-STD-105 D 652
miles 794, 796, 797
    per hour 798
military standard 652
milli, as prefix 6
milliliters 796, 797
million 6
Milne rule for (quadrature) 751
mils 797
minima of functions 389
minimax
    principle 656
    strategy, zero-sum game 275
    theorem (eigenvalues) 157
minimization of error, by least-squares
        approximation 739
minims 794
minimum
    by heuristic search 291
    distance, Hamming 256
    in a sample 664
    local, in steepest-descent method   748
Minkowski
    inequality 33, 73
        integrals 406
    sausage (dimension) 344
minor axis, ellipse 326

minus (sets), defined 203
minutes, angle, conversion 381
mixed
    actions 656
    product rule (Kronecker, matrices)
        159
    strategy 275
    tensor, defined 483
    zero-sum game 276
mixture problem (linear programming)
        281
Möbius
    function 25, 35, 102
        and irreducible polynomials . 170,
        261
    inversion formula 102
    ladder 229
    strip, crystallographic groups 308
    transformation 58
mode 660
modeling, linear programming 281
modification, rank-one, identity matrix
        139
modified
    Bessel functions 564
    Euler's method (numerical differential
        equations) 767
    Newton's method (numerical solution
        of equations) 732
modular arithmetic 94
modular function, coefficients of 31
modulation
    discrete Fourier transform 583
    Fourier transform 578
modulus 644
    complementary (elliptic integrals) 570
    elliptic integrals 570
    of a complex number 53
modus ponens 200
modus tollens 200
mole 792
molecules, computational 765
Mollweide's formulae (triangle) 513
moment generating function
    binomial distribution 630
    discrete random variable 630
    exponential distribution 635
    gamma distribution 634
    geometric distribution 631
    hypergeometric distribution 631
    negative binomial distribution . . . 632

moment generating function (*continued*)
    normal distribution 633
    Poisson distribution 632
    uniform distribution 633
moments
    beta distribution 635
    centered, random variable 620
    joint 622
    multivariate distributions 622
    of inertia 796
       table 410
    random variable 620
    sample 660
momentum, units 796
Monge patch
    paraboloid of revolution 380
    surfaces 376
monic polynomials 166
    primitive, table 194
monoid
    defined 161
    examples 161
monotone property of a graph 226
Monte–Carlo methods for quadrature
    761, 763
Moore
    –Penrose conditions (matrix
       pseudo-inverse) 151
    bound (graphs) 235
    graph, defined 235
Morain, Atkin– certificate of primality
    104
Morse
    code, table 260
    Thue– sequence 25
mortgage payments, calculation of .. 782
mortgage, tables of payments 787
Moser–de Bruijn sequence 30
motion, Brownian 467
Motzkin numbers 27
Moulton, Adams– methods (numerical
    differential equations) ... 768
moving trihedron, regular curve 374
Muller, Box– technique 647
multi-resolution analysis 723
multidimensional Fourier transforms 585
    table 604
multigraph 220
    Euler's theorem 233

multinomial
    coefficients 207, 209, 210
    probability distribution 632
multipartite graph 226
    complete 230
multiple
    -angle formulae 510
    argument relations, hyperbolic
       functions 526
    integrals, numerical estimation .. 759
    least common multiple 101
    series 39
multiplication
    hexadecimal 8
    of complex numbers 54
    of fractions 3
    of matrices 142
    of tensors 483
    tables, fields 190
multiplicative
    group 161
    lagged-Fibonacci sequences 646
    shift register ....*see* shift register
multiplicity, of root of polynomial 83
multiplier 644
    Lagrange 389
multistep methods 767
multivariate distributions 621
    normal 634
multiwavelets 725
mutation operator (genetic algorithms)
    295
mutual information (random variables)
    254
mutually
    exclusive events 618
    independent events 618
    orthogonal Latin squares 251

N

n-tuple 131
    in de Bruijn sequences 24
named
    difference equations 269
    integral equations 480
    ordinary differential equations ... 465
    partial differential equations 469
nano, as prefix 6

Napier
    analogs (spherical triangles) 371
    John 812
    rules (right spherical triangle) ... 368
Napierian logarithm *see* natural logarithm
Narayana–Zidek–Capell numbers 26
Nash equilibrium (game theory) 277
Nash–Williams arboricity theorem .. 232
natural
    cubic spline 738
    logarithms 522
        base 15
    numbers, defined 3
    representation of class of curves . 374
nautical measure 794
nautical miles 794, 798
Navier–Stokes equations (PDE) 469
nearest neighbor
    in a plane, probability 626
    in space, probability 626
necklaces 25, 807
    coloring 173
needle problem, Buffon's 627
negative binomial distribution 632
neighbor
    graph 226
    nearest
        in a plane, probability 626
        in space, probability 626
neighborhood 226
    graph 226
    of a point in a metric space, defined 69
    search methods 291
nephroid, defined 341
nephroid, figure 341
Netlib 808
nets, polyhedra 360
network flow 282
    solution by linear programming .. 288
Neumann
    conditions, PDEs 470
    function 560
    series 34, 482
Neville's method for evaluating Lagrange
        interpolating polynomial . 734
New York, distance to Beijing 373
Newton
    –Cotes formulae 750, 752
    –Gregory formulae 736
    –Raphson method 731
    backward divided-differences 735

Newton (*continued*)
    backward-differences 736
    formulae (triangle) 514
    forward-difference formula 736
    interpolatory divided-differences . 735
    Isaac 813
    method
        example 71
        for non-linear equations . 731, 748
        modified 732
        polynomial 735
        unit of force 792
Nichomedes, conchoid of 338
Nicolson, Crank– method 773, 774
nilpotent matrix *140*
    eigenvalues 153
NIST 808
Noether, Emmy 816
noise, in Monte–Carlo quadrature ... 763
nome 574
non-cooperative games 274
non-isomorphic groups, number of ... 25
non-linear
    generators 646
    system of differential equations .. 466
    systems of equations 748
non-proportional scaling transformation
        313
    in space 352
non-repeating sequence 25
non-singular matrix 138, 147
non-zero-sum game 277
nonagon, area and apothem 324
nonary 5
Nordhaus–Gaddum bounds (graphs) 234
norm
    L_1 .... 133, *146*
    L_2 .... *146*
    L_p 73, *146*
    L_∞ 73, 133, *146*
    comparable, of polynomials 91
    compatible (matrix) 147
    Euclidean, of a vector 133
    Frobenius, of matrix 146
    Hilbert–Schmidt, of matrix 146
    in quadratic field 168
    matrix, defined 146
    of an algebraic integer 168
    on a vector space 72
    polynomial, defined 91
    vector 133

normal
    basis, and Galois field 170
    coordinate system (tensor analysis)
       488
    curvature vector, coordinate patch 377
    curvature, coordinate patch 377
    direction, to plane 354
    distribution 254, 665
        multi-dimensional 634
        table 696–701
    equations 685
        least-squares 739
    form, equation for line 315
    form, equation for plane 354
    line, coordinate patch 377
    line, principal, regular curve 374
    matrix, defined 140
    plane, regular curve 374
    polynomials, and Galois field 170
    probability distribution 633
        and binomial 630
        error function 545
        notation 619
    random variables
        chi-square distribution 635
        from central limit 623
    random vector 254
    subgroup 162
        alternating group 163
        defined 162
        homomorphism 171
    vector
        and vector product 135
        coordinate patch 377, 378
        paraboloid of revolution 380
        principal unit, regular curve .. 374
normed division rings 165
not, in propositional calculus 199
notation 652
    Conway, crystallographic groups .307
    graphs 219
    index 823, 835
    list 823
n^{th} power function 521
n^{th} repeated integral 588
n^{th} root function, defined 521
null
    hypothesis 661
    matrix 138
    set 202
    space of matrix, defined 149

null (*continued*)
    vector, and metric tensor 486
    vector, defined 137
number
    clique, of a graph 221
    complex, polar form 87
    crosscap 222
    crossing (of a graph) 222
    field, algebraic 168
    independence, of a graph 225
    replication, combinatorial design .241
    theory 93
        website 809
number of
    atoms in earth 799
    day of year 800
    digraphs 239
        isomorphisms classes 240
    graphs 239
        isomorphism classes 239
        isomorphisms classes 240
        labeled 240
    trees 239
        labeled 239
        rooted 239
numbers
    abundant 26
    algebraic 3
    Bell 28, 211
        counting equivalence relations .66
    Bernoulli 19, 40, 44–46
    Carmichael 31, 94
    Catalan 28, 212
    centered square 30
    chromatic 221, *see* chromatic number
    complex ...... *see* complex numbers
    composite ... *see* composite numbers
    Cullen 29
    Dedekind 26
    double factorial 28, 29
    edge-chromatic, table 234
    Euler 20, 28, 30, 44–46
    exponential 28
    factorial 28
    Fibonacci 21
    Genocchi 30
    harmonic 32
    hexagonal 31
    highly abundant 26
    Hurwitz–Radon 25
    in different bases 16

numbers (*continued*)

Kendall–Mann 28
large 798
Lucas 29
lucky 29
MacMahon's 27
meandric 27, 28
Motzkin 27
Narayana–Zidek–Capell 26
natural .... *see* natural numbers
octagonal 31
of possible relations 30
partitions of 210
Pell 27
pentagonal 30
perfect 31, 129
prime .... *see* prime numbers
not .... *see* composite numbers
pronic 28
pyramidal 30
rational .... *see* rational numbers
real .... *see* real numbers
rencontres 210
representation 5
Robbins 28
Roman 4
Sarrus 31
segmented 27
small 798
special 10–24
Stirling .... *see* Stirling numbers
tangent 28
tetrahedral 30
transcendental 3
π 13
e 15
triangular 29, 30
types 3
Ulam 26
Wedderburn–Etherington 26

numerical

analysis 728
differential equations, solution ...766
differentiation 764
integration 750
linear algebra 740
optimization 748
quadrature 750
solution of algebraic equations ...729
summation 776

O

oblate

spheroid 364
spheroidal coordinates 496

oblique

coordinates, defined 303
coordinates, relations between ...304
spherical triangles 369
solution (table) 372

observable system 498
observable vector, control theory 497
obtuse angle 503
obtuse triangle 318
occupancy problems 206
ocean, volume of 798
octagon, area and apothem 324
octagonal numbers 31

octahedron

constructing 360
properties and figure 359

octal notation 5

octonions

division ring 165
example of ring 166

odd

function, integral of 399
graph, defined 229
permutations, in symmetric group 163
sequence 584
series 50

odds, in card games 628
ODE .. *see* ordinary differential equation
off-diagonal component, matrix 137
ohm 792
one-factor ANOVA 687
one-form (differential) 395

one-step

transition matrix 640
transition probabilities 640

one-to-one correspondence, defined .. 67
one-to-one function, defined 66
onto function, defined 66

open

Newton–Cotes formulae 751
sets 69

operations

binary, on a set 160
on graphs 228
per second 76
research 280

opportunity loss function 658
optimal
    control 498
    filter 721
    Pareto 275
optimality principle (dynamic
        programming) 290
optimization of non-linear equations 748
or
    exclusive 645
    in propositional calculus 199
order
    affine plane 248
    axioms of 67
    infinite, group element 161
    of a curve 336
    of a graph 226
    of a zero 56
    of an algebraic structure 160
    of approximation 723
    partial 204
    statistics 659, 664
ordered
    factorizations, number of 25
    pairs 66
ordering of sets 204
ordinary differential equations . 456–468
    named 465
    numerical solutions 766
    solution techniques 461
Ore's theorem (graphs) 233
oriented Cartesian tensor 489
origin
    Cartesian coordinates, defined ... 301
    in space 345
orthogonal
    coordinates 492
        table 494
    curvilinear coordinates, metric tensor
    490
    functions, defined 74
    group, defined 171
    Latin squares 251
        examples 251
    matrix 149
    matrix, defined 139
    polynomials 532
        tables 535
    vectors, defined 132
orthogonality, Hilbert transform 593

orthonormal
    basis 723
    functions, defined 74
    rows and columns, matrix 139
    set
        defined 74
        Riesz–Fischer theorems 74
    vectors, defined 132
oscillation, linear ODE 460
osculating
    circle, plane curve 375
    sphere 374
osculating plane
    regular curve 374
ounces 794, 796, 797
outer product, tensors 483
oval, Cassini's 337
overdamping, linear ODE 460
overdetermined (PDE) 470

P

Painlevé transcendent (ODE) 465
pairing terms, ways to 212
Paley difference set 246
Pappus–Guldinus theorem 363
parabola 325
    equation 327
    figure 325
    polar equation for 333
    properties 327, 333
parabolic
    coordinates 474, 496
        Hamiltonian 473
    cylinder equation (ODE) 466
    cylinder, equation 364
    cylindrical coordinates 495
    partial differential equations 469, 773
        boundary conditions, table ... 470
    point, coordinate patch 377
    reflector 333
paraboloid of revolution 380
paraboloid, equation 364
paraboloidal coordinates 497
paradox
    Braess 278
    of set theory 203
    Russell's 203
parallel
    classes, lines of affine plane 248

parallel (*continued*)
    lines *316*
        Buffon's needle problem 627
        in space 356, 357
    planes 355
        equation 364
    postulate 248
    system 653
    vector fields 488
    vectors, defined 131
parallelepiped *357*
    defined by matrix 145
    determined by vectors 136, 397
    moment of inertia 411
    rectangular, defined 357
    volume 357
parallelogram determined by vectors 135
parallelogram, properties 322
parameterized
    curve (tensor) 483
    geodesic 485
    geodesic, equations 488
parametric representation
    ellipse 330
    hyperbola 332
Pareto optimal outcome
    defined 275
    zero-sum game 275
parity
    and Pythagorean triples 99
    check matrix, linear codes 256
    Hilbert transform 593
Parseval relation
    Bessel function 561
    Fourier transform 579
        discrete 583
        multidimensional 585
    Hankel transform 590
    Z-transform 596
Parseval's identity 75
partial
    derivative, defined 386
    differentiation, tensor, symbol ... 484
    fractions 589
        in evaluation of integrals 401
        inversion using 597
        of polynomials 87
    order 204
    pivoting, solving linear equations 741
    quotients, in continued fraction ... 97
    sum, definition 31

partial differential equations ... 468–477
    diffusion 773
    heat 773
    hyperbolic 775
    numerical solutions 770
    parabolic 773
    transformations 469
    wave 775
partially ordered sets 204
particular solutions (ODE) 456, 457
particular solutions (PDE) 477
partitioning
    for matrix inversion 147
    integers into blocks 213
    of determinants 145
partitions
    binary 27
    generating function for 211
    in combinatorics 207
    number of 26
    of n, number of 25
    of a graph 234
    of a number 25, 210
    of a set 211
    planar, number of 29
parts, integration by 402
Pascal
    Blaise 813
    distribution 632
    limaçon, defined and figure 338
    triangle 208
pascal 792
patch, coordinate, on surface 376
path
    alternating, graph 225
    augmenting, graph 225
    graph 226, 229
    Hamiltonian 224
    shortest 290
    theorems 231
pattern inventory 173
PDE .... *see* partial differential equation
Peano curve 343
pebi, definition of 6
peck 794
Pell
    equation 99
    numbers 27
pennyweights 794
Penrose, Moore– conditions 151
pentaflake, capacity dimension 344

pentagonal numbers 30
pentagons
    area and apothem 324
    coloring the corners 175
    cycle index and pattern inventory 175
percentage points 702–709
percentile 659
perfect
    codes, defined 257
    cubes, list of 31
    graph 226
    numbers 31, 129
    partitions, number of 25
    squares, list of 30
perfectly-conditioned matrix, defined 149
perimeter, semi- triangle 319, 512
period of function 48
periodic
    autocorrelation, binary sequences 264
    autocorrelation, in sequence 263
    continued fraction 97
    functions
        and Laplace transform 588
        doubly elliptic 573
        expansion of 51
        representation by Fourier series 48
    sequences, defined 263
periodicity of trigonometric functions
    507
periods, of time, financial 779
permanent of matrix, defined 145
permutations
    Costas array 252
    cycle decomposition 172
    cycles in 212
    derangements 28, 210
    even, in symmetric group 163
    generating 807
    group *163*
        creating 172
        table 172
    in determinant 144
    matrix, and permanents 145
    matrix, and vector operation 158
    matrix, defined 140
    matrix, in matrix factorization ... 157
    odd, in symmetric group 163
    of indices, tensor 484
    symbol, as tensor 489
    table of 215
    with replacement 206

perpendicular
    lines 316
    lines in space 356
    planes 355
Perron, Frobenius– theorem
    (eigenvalues) 157
perspective transformation, figure ...314
persymmetry, in Toeplitz matrices ...140
perturbations of an integral 407
perturbed initial-value problem 766
peta, as prefix 6
Petersen graph, defined 229
Pfaff–Saalschutz identity 38
physical constants 794
π .... *see* notation index
pico, as prefix 6
piecewise polynomial approximation 737
pint 794
Pisa, Leonardo of 811
pivotal player, and power index 278
pivoting (linear equations) 740–742
Plana formula 40
planar
    curves 336
    graph *226*, 235, 236
        mesh 230
    partitions, number of 29
    point, coordinate patch 377
    transformations, defined 300
    triangles, figure 512
plane
    curve 375
    normal, regular curve 374
    osculating, regular curve 374
    rectifying, regular curve 374
    tangent, coordinate patch 378
planes *354*
    affine, defined 247
    angle between 355
    coincident, equation 364
    coordinate systems, conventions . 299
    equation of 354
    equatorial (spherical coordinates) 346
    graph embedding in 226
    intersecting, equation 364
    isometries 305
    order of affine 248
    parallel 355
        equation 364
    perpendicular 355
    points in, probability 626

planes (*continued*)
    projective *248*
        construction 249
    properties 354
    symmetries 305
    transformations 312
Plank constant 794
plasma dispersion function 546
Platonic solids, properties and figure 358
Plotkin bound (block codes) 258
Pochhammer symbol 17
Poincaré, Henri 816
point
    distance to plane 354
    estimate 683, 687, 689, 692
    in combinatorial design 241
    method (interior) 284
    power with respect to circle 336
    set, and coordinate manifold 482
points
    accumulation, defined 69
    cluster 70
    collinear 317
    critical, ODE 468
    density in a plane 626
    density in space 626
    distance between, in space 356
    distance on a sphere 372
    elliptic, coordinate patch 377
    fixed, in logistic map 272
    hyperbolic, coordinate patch 377
    in a line segment, probability 625
    in a plane, probability 626
    in space, probability 626
    limit 69, *69*
    nearest in a plane, probability 626
    nearest in space, probability 626
    number of mappings among 29
    of closure, defined 69
    parabolic, coordinate patch 377
    percentage 702–709
    planar, coordinate patch 377
    umbilical *378*
        coordinate patch 379
pointwise convergence 71
Poisson
    arrivals 638
    distribution, table 712
    equation 469, 770
        finite-difference algorithm ... 771

Poisson (*continued*)
    probability distribution 632
        notation 619
    process 655
    summation (Fourier transform) .. 580
    summation formula 40
poker 798
    probabilities 628
polar
    axis (spherical coordinates) 346
    coordinates
        area of polygon 318
        changing from Cartesian 396
        curves in 342
        Hamiltonian 473
        planar, defined 302
        relation to Cartesian 303
        symmetries 307
        tensor quantities 492
    form of complex number 53
    triangle 369
pole
    coordinate origin 302
    of a cycloid 340
    of a roulette 340
    of complex functions 57
Pollaczek–Khintchine formula 639
Polya
    distribution 632
    theorem (permutation group) 173
    theory 172
        tables 174
polygons
    area 318
    defined 317
    integrating over 761
    regular *317*, 323
        and dihedral group 163
        coloring the corners 175
    simple, defined 317
    spherical 367
    ways to cut into triangles 29
polyhedra
    convex 357
        regular 358
    nets 360
    online encyclopedia 809
polylogarithm, defined and properties 551
polynomials *83*
    approximation theorem 72
    Bernoulli 19

polynomials (*continued*)

    characteristic ...... *see* characteristic polynomial

    Chebyshev 754

        first kind 534

        second kind 535

        table 536

    chromatic . *see* chromatic polynomial

    circle 537

    cubic

        discriminant 89

        roots 89

        trigonometric solution 89

    cyclotomic

        defined 91

        table 92

    degree *166*

    Euler 20

    evaluation, Horner's method 733

    factorial 213, 214

    finite field 170

    generating, and cyclic code 256

    Hermite 532, 737, 754

        table 535

    interpolating 737

    irreducible

        binary, table 261

        defined 166

        number of 25

        number of monic 261

        table 194

    Jacobi 533, 754

        table 537

    Laguerre 533, 754

        generalized 533

        table 536

    Legendre 534, 754

        and Legendre function 556

        table 536

    measure 93

    monic 194

    monic, defined 166

    Newton, interpolation 735

    norm, defined 91

    normal, and Galois field 170

    orthogonal *532*

        tables 535

    piecewise approximation 737

    prime-representing 103

polynomials (*continued*)

    primitive 166, 194

        and Galois field 170

        binary, table 262

        monic, table 194

    properties 93

    quadratic

        discriminant 89

        roots 89

    quartic, roots and resolvent 90

    quintic, solvable by radicals 90

    quotient, and partial fractions 87

    radial 537

    resultant 85

    ring, defined 166

    roots, properties 84

    solution by radicals 89

    symmetric functions 84

    symmetric representation 93

    tables ... 92, 194, 234, 261, 262, 535, 537

    valuation 93

    Zernike 537

polyominoes 27

posets *204*

    chain and anti-chain 205

    examples 205

    Hasse diagram 205

positive

    definite 69

        matrix 155

        metric 486

        norm, metric, or distance 72

    integers 3

    numbers, axiom of order 67

    semi-definite matrix 156

postulate, parallel 248

potential equation (PDE) 471

pounds 796, 797

    in other units 794

    per square inch 798

power

    index (voting power) 278

        Banzhaf 279

    method (eigenvalues) 742

        inverse 743

    of a test 661

    of complex numbers 54

    of point with respect to circle 336

    operation on graphs 228

power (*continued*)
    residues of $\mathbb{Z}_n^*$, table 192
    residues of $\mathbb{Z}_p$, table 193
    series .... 32, 36, *see* series
        integration and differentiation . 32
        inversion 36
    set, defined 203
    units 796
    voting .... *see* voting power
powers of
    $2^m - 1$, factorizations of 128
    2 .... 6, 10, 27
        partitions of $2n$ into 27
    3 29
    4 30
    5 30
    10 .... 13, 798
        names 6
    16, list of 12
    circular functions 511
    integers 21
        sums of 22
    integers, negative 23
powers, sums of 19
    alternating 20
Pratt's certificate of primality 104
predicate calculus 201
prediction 657
predictor-corrector methods (numerical
        differential equations) ... 767
preface v
preferential arrangements 29
prefix symbol 6
prefixes 6
preprints 808
present value 779, *780*
    example 782
pressure, units 796
primality certificate 104
prime
    factors, table 125
    in quadratic field 168
    knots, number of 27
    number theorem 103
primes
    Cullen, table 106
    definition 103
    factorial, table 106
    Fermat 106
        and polygons 345
    formulae for 103

primes (*continued*)
    in an ideal 164
    in an integral domain 165
    in arithmetic progressions .. 103, 106
    in quadratic field 168
    large 809
    largest known 105
    less than a given number
        asymptotic formula 103
        table 103
    MacMahon's 27
    Mersenne 104, 644
        and perfect numbers 129
        list of 26
        table 106
    number dividing n 25
    powers 26
    primorial, table 106
    probable, defined 104
    products of 106
    proof of primality 104
    quadratic residue 95
    relative, Euler totient function ... 128
    relative, probability of 101
    Sophie Germain 105
    special forms 105–108
    sum of (Goldbach conjecture) ... 103
    table 25, 26
        less than 100,000 108–125
    testing, Lucas–Lehmer 104
    testing, probabilistic 104
    twin, defined 103
    twin, largest known 105
    Wieferich 106
    Wilson 106
    Woodall, table 106
primitive
    element, and Galois field 170
    element, of finite field 169
    normal basis, and Galois field ... 170
    polynomials 166, 194
        and Galois field 170
        binary, table 262
        monic, table 194
        normal, table 170
    roots 195
        modulo n 163
        table 195
        used for primality test 104
    trinomial exponents 645
primorial primes, table 106

principal
  curvatures
    coordinate patch 377–379
    paraboloid of revolution 380
  diagonal, matrix 137
  directions
    coordinate patch 377, 379
    defined 379
    paraboloid of revolution 380
  ideal *164*
    domain 165
    domain, table 167
  normal line, regular curve 374
  normal unit vector, regular curve .374
  root, of a complex number 54
  sub-matrix *140*, 156
  to be invested 779
  value
    inverse circular functions 518
    of integral, Cauchy 399
principle
  minimax 656
  of optimality (dynamic programming)
    290
  of the argument 58
  Raleigh's (eigenvalues) 157
priority service 637
prism (graph) 230
prism (polyhedra) 357
prisoner set (chaos) 273
prisoners' dilemma 277
probabilistic algorithm, primality test 104
probability *617*
  and sets 203
  classical problems 627
  density function 467, 619
    continuous multivariate 621
  geometric problems 625
  in card games 628
  of relative primes 101
  ratio tests 681
  tables 695
  theorems 618
probability distribution *630*
  beta 635
  binomial 630
  chi-square 635
  discrete 630
  discrete, uniform 630
  entropy 253

probability distribution (*continued*)
  exponential 634
  F ....*see F*-distribution
  function 619
  gamma 634
  geometric 630
  hypergeometric 631
  multi-dimensional normal 634
  multinomial 632
  negative binomial 632
  normal 633
  normal, and binomial 630
  normal, and error function 545
  notation 619
  Pascal 632
  Poisson 632
  Polya 632
  Snedecor's F .....*see F*-distribution
  student's t ....*see t*-distribution
  t ....*see t*-distribution
  uniform, continuous 633
probable prime *104*
  strong 105
problem
  assignment 288
  least-squares 151
  schoolgirl 249
  traveling salesman 292, 295
producer's risk 652
product 31
  circular functions 509
  exterior, 1-forms 395
  group, defined 163
  infinite 47
    convergence of 47
  Kronecker (matrices) 159
  law of reliabilities 653
  law of unreliabilities 653
  mix problem (linear programming)
    281
  of cyclic groups, isomorphism ...171
  of forms 395
  of graphs 229
  of random variables 623
  of sets 66
  of sine and cosine 511
  outer, tensors 483
  pseudoscalar 467
  rule (Kronecker, matrices) 159
  wedge, 1-forms 395
professional math organizations 804

programming
    dynamic 289
    integer, linear 287
    linear .... *see* linear programming
    techniques 777
programs 808
progression
    arithmetic 86
    geometric 86
projecting a scene 353
projection matrix *140*, 150
projective
    coordinates 303
    matrix, and homogeneous coordinates
      351
    plane *248*
        construction 249
    transformations 314
        in space 353
prolate
    cycloid, defined 340
    spheroid 364
    spheroidal coordinates 496
prolongation 466
pronic numbers 28
proof
    by cases 200
    of primality 104
    without words 81
proper ideal, defined 164
proper node (critical point, ODE) ... 468
properties
    indefinite integrals 399
    local, coordinate patch 376
    median 83
    of sets, inclusion/exclusion 204
proportion, definition and properties .. 86
proportional scaling transformation . 313,
    352
propositional calculus 199
    inference, rule of 200
    relation to predicate calculus 201
provability (Gödel incompleteness) . 202
pseudo-inverse
    and condition number 149
    of a matrix 151
pseudo-Riemannian
    connection 487
    manifold *486*
        and metric tensor field 487
pseudograph 220

pseudorandom number generator ... 644
pseudoscalar product 467
Ptolemy 811
    formula 323, 335
pure quadratic form 155
pyramid
    defined 357
    truncated, defined 357
pyramidal numbers 30
Pythagorean
    theorem
        circular functions 509
        figure 81
        right triangle 81, 320, 512
        triple 98

Q

Qin Jiushao 812
QR algorithm (eigenvalues 745
QR algorithm for eigenvalues of a
        symmetric matrix 746
QR factorization of matrix 157
quadrangles, defined 317
quadrant *503*
    Cartesian coordinates, figure 302
    determination
        right spherical triangle 369
        spherical triangle 371
    signs of circular functions 505
quadratic
    convergence, defined 728
    equation of a conic 328
    field 167
        defined 167
        examples 169
        properties 168
    form (matrix) 138, 155
    formula 89
    polynomial
        discriminant 89
        roots 89
    reciprocity 95
    residue 94, 95
quadrature
    Chebyshev 758
    Gauss–Hermite 756
    Gauss–Laguerre 756
    Gauss–Legendre 755
    Gaussian 755
    Lobatto 757
    Monte–Carlo methods 761

quadrature (*continued*)
    multiple dimensions 759
    over a polygon 761
    over a rectangle 759
    Radau 757
quadrics
    cone, equation 364
    figures 366
    surfaces in space 364
quadrilaterals
    area 322
    cyclic 323
    defined 317
    figures 317, 322
quadrillion 6
quality level 652
quantifier
    existential, defined 201
    universal, defined 201
quantiles 659
quartic curves 90
quartic polynomials
    resolvent 90
    roots 90
quartile 659
    confidence interval 667
    deviation 660
quarts 794
quasi-linear differential equations . . . 472
quaternary 5
quaternion
    division ring 165, 166
    group 178, 182
        characters 191
queue discipline 637
queuing theory 637–639
quinary 5
quintic polynomial (solvable) 90
quintillion 6
quintuple product identity 48
quota, in voting game 278
quotient
    circular functions 509
    group, defined 162
    partial, in continued fraction 97
    ring 164
    space 308

R

r-combination 206
r-permutation 206

Radau quadrature 757
Radau weight function 755
Rademacher function 722
radial polynomials 537
radian *503*
    conversion to degrees 381
radiation condition (PDE) 471
radicals, in polynomial solution 89
radius
    circumscribed circle, triangle 513
    inscribed circle, triangle 512
    of a circle 333
    of a graph 226
    of a nucleus 798
    of an atom 798
    of curvature, regular curve 374
    of the earth 795
    spectral (matrix eigenvalues) 154
    sphere, defined 365
radix 5
Radon, Hurwitz– numbers 25
raisin cookie problem (probability) . . 627
raising indices
    and covariant derivative 488
    of tensor 486
Raleigh's principle (eigenvalues) 157
Ramanujan
    Rogers– identities 47
    τ function 31
Ramsey's theorem (graphs) 233
random
    number generation 644
    service 637
    strategy 275
    triangles 626
    variables 203
        and entropy 253
        continuous, defined 618
        discrete, defined 618
        normal 254
        properties 619
        sums 622, 623
        transformation 623
    vector, normal 254
    vectors, average 625
    walks 627, 643
range 660
    interquartile 660
    of a function 66
    space of matrix, defined 149

rank
    -changing perturbations 152
    -one modification, identity matrix 139
    of matrix 147
Rankine 798
Raphson, Newton– method (equations)
        731
rate of a code 255
rate of convergence 728
ratio
    cross 58
    golden 16
    test 33
rational
    curve 337
    numbers *3*
        contained in reals 68
        dense property of 68
        equivalence relation on 66
real
    analysis 66
    line subset 65
    numbers *3*
        characterization of 68
        inequality properties 68
        separability 70
        sets of 67
    part of complex number 53
    quadratic field, defined 167
reciprocal relations among circular
        functions 509
reciprocal system (vectors) 136
rectangle, integrating over 759
rectangle, properties 322
rectangular
    coordinates *301*, 494
        in space 345
    hyperbola, defined 327
    matrix 137
    parallelepiped, defined 357
rectifying plane, regular curve 374
rectilinear crossing number, graph .. 222
recurrence relation
    Bessel function 560
    Catalan numbers 212
    Chebyshev polynomials 535
    Fibonacci numbers 21
    gamma function 540
    Hermite polynomials 532
    hypergeometric function 554
    Jacobi polynomials 533

recurrence relation (*continued*)
    Laguerre polynomials 533
    Legendre function 556
    Legendre polynomials 534
recurrence, Markov chain 641
reductio ad absurdum 200
Reed–Solomon codes 257
references 817
refinement equation 723
reflection
    formula (plane Cartesian) ...305, 306
    formula (plane polar) 307
    Householder 141
    in space 350
    isometry of space 349
    plane isometry 305
reflector, parabolic 333
reflexive
    relation 66
    relation in sets 204
register .... *see* shift register
regression, linear 682
regret function 658
regula falsi (non-linear equations) ...732
regular
    curve of class 373
    graph 226
    matrix 138
    parametric representation, curves 373
    polygons 317, 323
    polyhedra 358
    Toeplitz graph as 230
rejecting a hypothesis 661
relations 66
    equivalence, defined 66
    in sets 204
    Legendre (elliptic integrals) 570
    number of possible 30
    reflexive 66
    symmetric 66
    transitive 66
relative
    efficiency 662
    entropy 254
    error 728
    primes, probability of 101
    tensor 483
relaxation (branch and bound) 287
reliabilities, law of 653
reliability 653
    function 655

remainder theorem, Chinese 96
remainder, power series, Lagrange's .. 37
removable singularity
    defined 57
    theorem 57
rencontres numbers 210
repeated Fourier integral, defined ... 577
repeated integral 588
replication number 241
repunits (primes) 107
residual design 244
residual mean square 688
residue
    biquadratic, and difference sets .. 247
    of complex functions 57, 399
    quadratic 94
    theorem (contour integration) 406
resolvable Steiner triple system 249
resolvent
    kernel (integral equation) 481
    quartic polynomials 90
resource constraints (linear
            programming) 281
resultant of polynomials 85
return time 641
revolution . *see* surface of revolution, *see*
        volume of revolution
    paraboloid 380
Rhind Papyrus 810
rhombus, properties 323
Riccati equation
    difference 270
    differential 466
Ricci
    identity (vector field) 487
    tensor *485*
        2-sphere 491
        3-sphere 492
        properties 488
Richardson extrapolation 729, 753
    numerical differentiation 765
Riemann
    –Lebesgue lemma 578
    Cauchy– equations 55
    curvature tensor 379
    Georg Bernhard 815
    hypothesis 24
    mapping theorem 58
    P function 465
    removable singularity theorem 57
    tensor *488*, 491

Riemann (*continued*)
    zeta function 23, 35
        and Möbius function 102
        and polylogarithm 552
Riemannian manifold 486
    and orthogonal coordinates 490
Riesz–Fischer theorems 73, 74
right
    -hand rule, vectors 134, 135
    -handed axes in space 345
    -handed circular helix 376
    angle 503
    cancellation law, groups 162
    circular cone 362
    circular cylinder 361
    coset of subgroup, defined 162
    cylinder 361
    distributive law, ring 164
    divisor of zero (ring) 164
    hand limits 49
    limit 576
    matrix inverse 151
    spherical triangles 368
    triangle *318*, 320
        and circular functions 512
        Pythagorean triple 98
ring *164*
    commutative *164*
        table 167
    division, defined 165
    division, table 167
    examples 165–167
    field as commutative division ...167
    fundamental theorem, homomorphism
        171
    homomorphism 170
    isomorphism 170
    kernel of homomorphism 170
    moment of inertia 411
    normed division 165
    polynomial 166
    properties 165
    quotient 164
    unit in 164
    with unity 164
risk
    analysis 656
    consumer's 652
    function 657
    producer's 652
Robbins numbers 28

rod, moment of inertia 410
Rodrigues formula
     Chebyshev polynomials 535
     Hermite polynomials 532
     Jacobi polynomials 533
     Laguerre polynomials 533
     Legendre polynomials 534
     surface 379
rods 794
Rogers–Ramanujan identities 47
Rolle's theorem (derivatives) 388
Roman numerals 4
Romberg integration 753
Room squares, defined 252
root
     bracketing-methods 732
     complex, of -1 87
     mean square 660
     minimal primitive of $\mathbb{Z}_p$, table . . . 193
     multiplicity of 47
     of complex numbers 54
         principal 54
     of polynomial *83*
         properties 84
     primitive 195
         table 195
     square . . . . 16, *see* square roots
     test 33
     type of vertex of a graph 226
rooted graph 226
rotation
     -reflection (isometry of space) . . . 349
     change of planar coordinates 301, 304
     formula (plane Cartesian) . . . 305, 306
     formula (plane polar) 307
     Givens, defined 141
     in space 349
     isometry of space 349
     matrix representation 139, 141
     of polygons, and coloring 174
     plane isometry 305
roulette, defined 340
round-off error 728
round-robin tournament 27
     and Room square 252
     number of scores 27
row
     of a matrix 137
     rank of a matrix 147
     vector *131, 137*
Rubik's cube 798

ruin problem (probability) 627
rule
     chain 388
     composite (numerical integration) 752
     Cramer's (linear equations) 158
     l'Hôspital's (limits) 389
     Leibniz (derivative of integral) . . . 388
     of inference 200
ruler, Golomb 29
run test 677, 678
Runge–Kutta method 767
Russell's paradox 203
Ryser, Bruck– –Chowla theorem 245

S

Saalschutz, Pfaff– identity 38
saddle (critical point, ODE) 468
salesman problem
     genetic algorithms for 295
     solution by tabu search 292
sample
     central moments 660
     density function 658
     distribution function 658
     from distinguishable objects 206
     mean algorithm (quadrature) 762, 763
     moments 660
     point 617
     range 660
     size 665
     space 617
         connection to set theory 203
     standard deviation 660
     variance 660
sampling
     acceptance 652
     plan 652
     theorem, Shannon 581
Sarrus numbers 31
saturated matching in a graph 225
sausage, Minkowski (dimension) . . . 344
scalar *131*
     Helmholtz equation, separability . 474
     homogeneity 72
     invariant *483*
         Laplacian of 488
     multiplication of vector 132
     product 133
     triple product 136
scale factor (linear equations) 742
scaled-column pivoting 742

scalene triangle 320
scaling
    Fourier transform 578
    function 723
    transformation, in similarity 313
Schlömilch series 34
Schmidt
    Hilbert– kernel, integral equation 478
    Hilbert– norm 146
schoolgirl problem, Kirkman 249
Schrödinger equation (PDE) 469
Schroeder's problems 29
Schur decomposition (matrix) 154
Schwartz
    Cauchy– inequality 73, 623
    inequality 32, 73
    inequality (integral) 406
scores in round-robin tournament 27
screen coordinates, and projective
            transformation 353
screw motion 350
    isometry of space 349
scruples 794
search (optimization) 290
secant
    derivative 386
    function 505
    method, numerical solution of
        algebraic equations 732
    table of values 516
second
    -order ODE, Lie groups 467
    -order ODE, constant coefficients 460
    angle, conversion 381
    fundamental form 380
        coordinate patch 377
    fundamental metric coefficients . 378,
        380
    kind integral equation 478
    unit of measurement 792
sector of a circle 334
security council, UN (voting) 280
seed 644
segment
    of a circle 334
    spherical 368
segmented numbers 27
self-complementary graph 226
self-dual code, defined 257
semi
    -axes 364
    -cubic parabola 336

semi (*continued*)
    -integrals 455
    -interquartile range 660
    -major axis 331
    -perimeter, triangle 319, 512
semidirect product of groups 187
semigroup *161*
    examples 161
senary 5
separability
    condition 474
    of Hamiltonian 473
separable
    kernel, integral equation 478
    ODE 462
    space 70
    wavelet basis 725
separation of variables (ODE) 462
separation of variables (PDE) 472
septenary 5
sequences *67*
    Barker 263
    binary 263
        table 264
    bounded, limits 70
    Casoratian 266
    Cauchy .... *see* Cauchy sequence
    convergence *70*
        to limsup and liminf 70
    convolution of 598
    de Bruijn 24
    degree 223
    delta .... *see* delta sequence
    even 584
    geometric 86
    identifier 31, 807
    limit
        defined 70
        inferior 70
        point, defined 70
        superior 70
    linear independence, defined 266
    m-sequences 263
    Moser–de Bruijn 30
    non-repeating 25
    odd 584
    of integers 25–31
    periodic, defined 263
    Stern's 26
    Thue–Morse 25
sequential probability ratio tests 681

series 31, 482, 525
  algebraic functions 42
  alternating 32
  arithmetic 38
  arithmetic–geometric 38
  asymptotic ..... *see* asymptotic series
  Bessel 34, 560
  binomial 37
  Cardinal 581
  component 32
  convergence 32
  Dirichlet 35
  elliptic functions 573
  error function 546
  even 50
  exponential functions 43
  Fourier .... *see* Fourier series
  geometric 38, 82
  Gregory's 23
  harmonic 32, 33
  hyperbolic functions 45
  hypergeometric 36
  infinite 32, 42, 47
  integration and differentiation 32
  inverse hyperbolic functions 46
  inverse trigonometric functions ... 45
  inversion 36
  Kapteyn 34
  Lagrange 39
  Laurent 56
  Legendre polynomials 557
  logarithms 43, 523
  MacLaurin 37
  miscellaneous 41
  multiple 39
  Neumann 34, 482
  odd 50
  of integers 25
  operations 41
  polylogarithm 552
  power 32, 36
  Schlömilch 34
  system 653
  Taylor 36, *55*
  telescoping 37
  theta 39
  trigonometric functions 44
    and Fourier transform 576
    inverse 45
  types 34, 38
  useful 50

Serret–Frenet equations (curve) 375
service time 637
sets *202*
  cardinality 67
  closed 69
  combinatorial design theory 241
  compact 69
  complement 69, 203
  complete 74, *74*
  countably infinite 67
  dense, defined 70
  denumerable 67
  difference, defined 246
  distance function on 69
  edge, graph notation 219
  empty 202
  escape (chaos) 273
  families of subsets 30
  finite 67
  image 66
  independent, of a graph 225
  infinite 204
  intersection of 203
  inverse image 66
  Julia (chaos) 273
  Mandelbrot (chaos) 273
  measure of 203
  metric on 69
  minus 203
  null 202
  of distinguishable objects 206
  open 69
  operations on 203
  orthonormal 74
  partially ordered 204
  power 203
  prisoner (chaos) 273
  probability, connection to 203
  product 66
  relations 202
  symmetric difference 203
  theory 202
  union 203
  vertex, graph notation 219
  ways to cover 30
sexagesimal 5
shadow price (linear programming) . 285
Shanks transformation 40
Shannon
  coding theorem 255
  sampling theorem 581

Shapley–Shubik power index 278
shear (affine transformation) 313
shear in space .352
sheet, moment of inertia 410, 411
shell, spherical, moment of inertia . . 411
shift
    in periodic sequence 263
    register 264
        length 645
        sequence 645
shifted factorial 17, 36
shortest-path problem 290
Shubik, Shapley– power index 278
Shushu jiuzhang 812
SI system 792
siemen 792
Sierpiński
    carpet (capacity dimension) 344
    sieve (capacity dimension) 344
sievert 792
signal processing 718
signature of the metric 486
signum function 77, 144
similar
    edges of a graph 226
    matrices *150*, 744
        Kronecker product 160
    vertices of a graph 226
similarities 352
    plane transformations 312
simple
    event 617
    function 58
    graph 220
    groups
        defined .162
        orders of 26
        sporadic 162
        sporadic, table 191
    interest 780
    polygon 317
    root of polynomial 83
    zero 729
simplex method 283
Simpson
    double integral 759
    rule (quadrature) 751, 752
    three-eighths rule (quadrature) . . .751
simulated annealing (search) 291

sine
    –Gordon equation (PDE) 469
    derivative 386
    function 505
    integral 549, 551
    series . . . . *see* Fourier sine series
    table of values 516
    transform
        and Hankel transform 589
        Fourier *582*, 600
sines
    law of (spherical triangle) 370
    law of (triangle) 319
Singer difference set 247
singleton bound
    block codes .258
    maximum distance separable code
        257
singular
    -value decomposition 152, 156
    integral equation 478
        difference kernel 480
    kernel, integral equation 478
    matrix 147
    values of a matrix 156
singularity
    essential 57
    isolated 56
    of complex functions 56
    of matrix 147
    removable 57
size of a graph 227
skew
    -Hermitian matrix, defined 139
    field 165
    symmetric
        matrix 139
        part of tensor 484
        tensor 484
skewness, random variable 620
slackness theorem (linear programming)
        286
slope of a line 314
slot notation, in PDEs 470
small order groups 178
Snedecor's *F* . . . . *see F*-distribution
snowflake, Koch 344
Society for Industrial and Applied
        Mathematics 808
software, mathematical 808
solenoidal vector field, defined 390

solids
     Platonic, properties and figure ... 358
     volume, as integral 409
Solomon, Reed– codes 257
solution
     feasible, linear programming 281
     homogeneous (ODE) 456
     particular (ODE) 456, 457
     particular (PDE) 477
solvable group, defined 162
solving linear equations 740
Sophie Germain prime 105
sound velocity 795
source coding, English text 260
space
     Banach 72
     complete 70
     coordinate system in 345
     cycle 223
     edge 224
     Hilbert 74
     metric 69
          vector space 133
     points in, probability 626
     sample 203
     separable 70
     topological 69
span property, binary sequences 263
spanning
     forest, of graphs 220
     subgraph 227
     tree of a graph 227
Spearman rank correlation coefficient
          681, 718
special
     constants 13
     curves 336
     linear group 171
     numbers 10
spectral
     decomposition 154
     radius (matrix eigenvalues) 154
spectrum of a graph 227
spectrum, Fourier 50
speed of light 794
sphere *365*
     -packing bound (block codes) ... 258
     circumscribed, tetrahedron 358
     distance between points 372
     graph 227
     moment of inertia 411

sphere (*continued*)
     n-dimensional
          area 368
          surface area 368, 475
          volume 368
     osculating 374
     surface area 13
     volume 13
spherical
     cap 367
     coordinates 346, 495, 539
     excess (triangle) 369
     geometry 368
     half-angle formulae 370, 371
     harmonics 538
          Clebsch–Gordan coefficients . 574
          table 540
     mirror 341
     polar coordinates 77
     polygon 367
     segment 368
     shell, moment of inertia 411
     triangles 368
          oblique 369
          solution (table) 372
     trigonometry 368
     volume 368
     zone 367
spheroid of revolution (inertia) 411
spiral *342*
     Archimedean 342
     Bernoulli 342
     Cornu 343
     critical point, ODE 468
     linear 342
     logarithmic 342
spirograph curves 340
splines
     clamped 739
     cubic, algorithms 738
     interpolation 737
sponge, Menger, capacity dimension 344
sporadic finite simple groups, table .. 191
square *323*
     centimeters 796, 797
     feet 794, 796–798
     graph type, defined 229
     inches 796, 797
     Latin, defined 251
     matrix 137
     measure 794

square (*continued*)
    meters 796, 797
    mile 794
    numbers, centered 30
    operation on graphs 228
    Room, defined 252
    roots 16
        nested 14, 16
    yards 796, 797
squares 25
    area and apothem 324
    coloring the corners 175
    coloring the corners of 173
    cycle index and pattern inventory 175
    list of 30
    magic 101
    on a checkerboard 626
    sums of 2 27
squaring the circle 345
Stäckel matrix 473
stability, solution of PDE 470
stamps, folding, number of ways 27
standard
    deviation
        random variable 620
        sample 660
    error 683
    form 282
    normal distribution .... *see* normal
        distribution
Stanton–Sprott difference set 247
star (critical point, ODE) 468
star (type of graph) 229
state variables (dynamic programming)
    289
state vector, control theory 497
statements .... *see* propositional calculus
stationary 640, 721
    distributions 641
    phase (integral evaluation) 408
statistically independent events 618
statistics 661, 809
    descriptive 658
    estimators 661
    jointly sufficient 663
    sufficient 663
statute miles 794, 795
steepest-descent (non-linear equations)
    748
Steffensen's method 730
Steiner triple system 249

step function 77
    in integral evaluation 408
    in solution to wave equation 472
    Z-transform 595
steradian 792
Stern's sequence 26
Stevin, Simon 812
Stirling
    cycle numbers 29, 212
        table 213
    first kind number 213
    formula (for $n!$) 17, 542
    formula, interpolation 736
    second kind number ... 213, 239, 639
    subset numbers 31, 207, 211, 213
        table 214
stochastic differential equations 467
Stokes
    Navier– equations (PDE) 469
    theorem (line integral) 405
straight line
    equation (Cartesian) 314
    equation (polar) 316
    in space 355
straightedge, in geometric constructions
    345
strain tensor, as Cartesian tensor 490
Strang–Fix conditions 723
Strassen algorithm (matrix
        multiplication) 143
strategy (game theory) 275, 277
stress tensor, as Cartesian tensor 490
stripe, of circulant matrix 141
strong duality theorem 286
strong probable prime 105
strophoid, equations and figure 337
sub-cancellation, interval arithmetic .. 65
sub-distributivity, interval arithmetic . 65
sub-factorials 210
sub-matrix 140
subdivision of a graph 227
subgraph 227
subgroup *161*
    cyclic 162
    normal 162, *162*
        alternating group as 163
        homomorphism 171
subring 164
subsets
    families of 30
    Heine–Borel theorem 70
    of real line 65

substitution
    in evaluation of integrals 400
    planar coordinates, defined 299
subtraction
    of complex numbers 54
    of hexadecimal numbers 8
    of matrices 142
    of vectors 132
sufficient statistic 663
summability methods 41
summation
    Abel 41
    Cesaro 41, 49
    convention, Einstein 482
    formula 40
    Poisson formula 40
        Fourier transform 580
sums 509
    by parts 32
    combinatorial 38
    Euler–MacLaurin 776
    integers, figure 81
    interior angles in right triangle ... 512
    inverse circular functions 520
    involving π 24
    miscellaneous 41, 50
    numerical 776
    of 2 squares 27
    of circular functions 511
    of divisors function 26
    of divisors of n 28
    of hyperbolic function arguments 526
    of hyperbolic functions 526
    of inverse hyperbolic functions .. 529
    of negative powers 23
    of parts .... *see* partitions
    of powers (Waring's problem) ... 100
    of powers of integers 22
    of random variables 622, 623
    of squares 138
        Lagrange's theorem 100
    of tensors 483
    of vectors 132
    partial, definition 31
supergraph of a graph 227
superposition principle (difference
        equations) 267
supremum of a subset 68
surface
    differential geometry of 376
    integral 404

surface (*continued*)
    of revolution 363
    sphere, distance between points .. 372
    theory, fundamental theorem 379
surface area
    as integral 409
    earth 798
    sphere 13
        n-dimensional 368, 475
surjective function, defined 66
swing player (voting power) 279
switch 227
Sylvester's law of inertia 158
symbol
    Levi–Civita 489
    Pochhammer 17
    prefix 6
symbolic calculator 807
symbolic logic 199
symmetric
    channel 255
    designs *245*
        and Hadamard matrices 250
        Bruck–Ryser–Chowla theorem
        245
        properties 245
    difference 203
    functions, polynomial roots and
        coefficients 84
    graph 227
    group 180
        characters 191
        defined 163
        defined as permutation 172
    kernel, integral equation 478
    matrix *139*
        QR algorithm for eigenvalues 745,
        746
    part of tensor 484
    relation 66
    representation of polynomials 93
    tensor 484
symmetries
    formula, Cartesian coordinates ... 305
    of a metric or distance function ... 69
    of regular polygon (dihedral group)
        163
    of space 348
    of the plane 305
        crystallographic 307
    of trigonometric functions 507

symmetrization of tensor 487
syndrome 256
   linear codes 256
synthetic division 733
systems
   capacity 637
   controllable 497
   non-linear equations 748
   observable 498
   of congruences 96
   of differential equations 466, 769
   of linear equations 148
   reciprocal (vectors) 136
   Steiner triple, resolvable 249
Szekeres–Wilf theorem (graphs) 234

T

t-distribution 636, 702
t-designs 241
tables
   χ^2-distribution 703
   ANOVA 694
   binomial coefficients
      fractional 218
   characters of a group 191
   combinations 215
   conformal mappings 60
   contingency 679
   coordinate systems operations ... 494
   cosine transform 599
   cyclotomic polynomials 92
   definite integrals 448–455
   eigenvalues of specific matrices .. 153
   Eulerian graphs 233
   exponential function 531
   F-distribution 704–709
   factorizations of $x^n - 1$ mod 2 .. 196
   fields
      addition and multiplication ... 190
      small finite 189
   Fourier
      cosine transform 601
      sine transform 600
      transform 601–604
   Green's function (ODE) 463
   groups
      names 178
      non-isomorphic 176
      non-isomorphic, abelian 177
      representations 178
      sporadic simple 191

tables (*continued*)
   Gudermannian 531
   Hamiltonian graphs 233
   Hankel transform 610
   Hilbert transform 611
   hyperbolic functions 531
   indefinite integrals 412–448
   indices of $\mathbb{Z}_n^*$ 192
   integral transforms 599–613
   integrals 411–455
   Laplace transform 604–610
   logarithmic functions 531
   Mellin transform 612
   minimal primitive root of $\mathbb{Z}_p$ 193
   of transformations 59
   permutations 215
   Polya theory 174
   polynomials
      irreducible 194
      primitive monic 194
   power residues 192, 193
   primitive roots 195
   probability 695
   sine transform 599
   Stirling
      cycle numbers 213
      subset numbers 214
   t-distribution 702
   trigonometric
      functions 516
tabu search 292
tangent
   derivative 386
   function 505
   line, regular curve 374
   numbers 28
   plane, coordinate patch 378
   table of values 516
   vector
      as tensor 489
      parallel 485
      regular curve 374
   vectors, angle between 486
tangents
   laws of spherical 370
   to surface, angle between 378
tautologies, defined 199
Taylor
   Janet vi
   methods 766, 767
   series 36, 37, 55

tebi, definition of 6
telegraph equation (PDE) 469
telescoping series 37
temperature 792
    conversion 798
tensor
    absolute 483
    addition 483, 487
    analysis 482
    Cartesian, defined and properties . 489
    contraction 484, 487
    curvature, defined 485
    curvature, identities 487
    differentiation 484
    dual 489
    Einstein
        3-sphere 492
        defined and properties 488
    examples 489
    field 483
    fundamental metric coefficients .. 378
    Levi–Civita 489
    metric 486
    mixed, defined 483
    multiplication 483, 487
    operations 483
    product (matrices) 159
    relative 483
    Ricci *485*, 488
        2-sphere 491
        3-sphere 492
    Riemann
        curvature 379
        defined and properties 488
        orthogonal coordinates 491
    skew symmetric 484
    symmetric part 484
    symmetric, defined 484
    torsion *485*
        identities 487
    torsion-free 487
    weight 483
tent map 272
tera, as prefix 6
ternary 5
    Golay code 258
tesla 792
test
    alternating series 34
    chi-square 679
    comparison 33

test (*continued*)
    convergence and divergence 33
    distribution 679, 680
    Gauss 34
    integral 33
    Kolmogorov–Smirnov 679
    limit 33
    Mann–Whitney–Wilcoxon . 675, 676, 680
    mean 670, 673–675
    median 671, 672, 675–677
    power 661
    probability ratio 681
    ratio 33
    root 33
    runs 677, 678
    statistic 661
    variance 672, 677
tetrahedral numbers 30
tetrahedron
    coloring 174
    constructing 360
    defined 357
    properties and figure 359
    volume 358
tetrix, capacity dimension 344
text coding 260
theorems
    approximation 72
    Ascoli–Arzela 72
    automorphism group 238
    Bolzano–Weierstrass 70
    Brooks' (graphs) 234
    Bruck–Ryser–Chowla 245
    Buckingham pi 795
    Casorati–Weierstrass 57
    Cauchy–Goursat integral 406
    Cayley (group isomorphism) 171
    Cayley–Hamilton (matrix) 153
    central limit (random variables) .. 623
    Ceva (triangle) 321
    Chebyshev (prime number) 103
    Chinese remainder 96, 812
    coding 255
    contraction mapping 71
    Courant–Fischer minimax 157
    DeMoivre 54, 87
    derivatives 388
    Dirac (graphs) 233
    Dirichlet (primes) 103
    divergence 405

theorems (*continued*)

 duality (linear programming) 286
 Euler (multigraphs) 233
 existence 375
 Fermat
   –Euler (congruences) 95
   congruences 95
   last 98, 201
   primality test 104
 fixed point 71, 730
 four-color, and graph colorability 236
 four-color, and planar graphs 235
 Frobenius–Perron (eigenvalues) . . 157
 Frucht's (graphs) 238
 fundamental
   algebra 84
   calculus 388, 400
   finite abelian groups 171
   finitely-generated abelian groups
     171
   homomorphism for groups . . . 171
   homomorphism for rings 171
   surface theory 379
 Gauss
   theorema egregium, surface . . 379
   totient function 128
 Gerschgorin circle (eigenvalues) . 158
 Gödel incompleteness 202
 Green's (line integral) 404
 Green's (surface integral) 405
 Gödel incompleteness 202
 Hall's (graphs) 238
 Hardy (Laplace transform) 588
 Havel (graphs) 237
 Heawood map coloring 236
 Heine–Borel 70
 incompleteness 202
 intermediate value 388
 inversion (integral transform) 580
 Kirchhoff matrix-tree 232
 König's (graphs) 238
 Kuratowski's (graphs) 236
 Lagrange (group order) 162
 Lagrange (sum of squares) 100
 Lahire's (ellipse) 331
 Liouville 54
 Lucas–Lehmer (primes) 104
 mean value 388
 Menelaus (triangle) 321
 Menger 232
 minimax (eigenvalues) 157

theorems (*continued*)

 Nash–Williams arboricity 232
 Ore's (graphs) 233
 Pappus–Guldinus (surface of
   revolution) 363
 Polya (permutation group) 173
 polynomial approximation 72
 prime number (number of primes) 103
 primes in arithmetic progressions 103
 probability 618
 Pythagorean
   circular functions 509
   figure 81
   right triangle 512
 queueing theory 638
 Ramsey's (graphs) 233
 removable singularity 57
 residue (contour integration) 406
 Riemann (mapping) 58
 Riemann (singularity) 57
 Riesz–Fischer 73, 74
 Rolle's (derivatives) 388
 Shannon (coding) 255
 Shannon (sampling) 581
 slackness (linear programming) . . 286
 Stokes (line integral) 405
 Szekeres–Wilf (graphs) 234
 Thompson–Feit (groups) 162
 Turán (graphs) 233
 uniqueness 375
 unnamed 70, 72, 75
 Vizing's (graphs) 234
 Weierstrass 47
   polynomial approximation 72
 Wilson (congruences) 95
theta series 39
thickness of graphs 227, 236
thin interval 65
third kind integral equation 478
Thompson–Feit theorem (groups) . . . 162
three-dimensional Cartesian coordinates
   345
three-form (differential) 397
three-sphere, tensor quantities 491
Thue–Morse sequence 25
time 792
 between failures 655
 periods, financial 779
 separability 289
 to failure 655
 units 796

Toeplitz
  graph 230
  matrix, defined 140
tolerance percent defective 652
ton 794
topological space 69
topology *69*
  number of 29
toroidal mesh (graph) 230
torque, units 796
torsion
  -free tensor 487
  regular curve 374
  tensor *485*
    identities 487
torus
  circular 363
  of revolution, and Cassini oval ...338
  surface of revolution 363
total differential 397
totient function 25, 128
  and primitive roots 195
  in Fermat–Euler theorem 95
tournament, round-robin 27
  and Room square 252
trace
  function, binary sequences 263
  of a matrix 150
    derivative 393
traffic intensity 638
trail
  Eulerian 224
  in a graph 227
transcendental numbers 3
transfer function, control theory 497
transform
  cosine 35
    table 599
  discrete Fourier 582
  discrete wavelet 724
  fast Fourier 584
  finite cosine 599
  finite sine 599
  Fourier ....*see* Fourier transform
    cosine *582*, 601
    for linear ODE 464
    multidimensional 585
    multidimensional, table 604
    sine *582*, 600
    table 602, 603

transform (*continued*)
  Hankel 589
    table 610
  Hartley 591
  Hilbert 591
    table 611
  integral 576
  Laplace 35, 585, 638
    for linear ODE 464
    table 605
  Mellin, table 612
  multidimensional Fourier 585
    table 604
  sine, table 599
  tables .... *see* tables
  wavelet
    continuous 725
    discrete 724
  Z 594
transformations
  affine *313*
    in space 352
  Baker (chaos) 272
  bilinear 58
  centering 285
  coordinate, and manifold 482
  glide-reflection in space 351
  Householder 141
  in changes of coordinates 300
  linear fractional 58
  Möbius 58
  non-proportional scaling 313
    in space 352
  of complex functions 58
  of integrals 402
  of the plane, defined 300
  PDEs 469
  plane isometries (Cartesian) 306
  plane isometries (polar) 307
  projective 314
    in space 353
  proportional scaling 313, 352
  random variables 623
  reflection in space 350
  rotation in space 349
  screw motion in space 350
  Shanks 40
  table 59
  translation in space 349
  Tschirnhaus 91
transforms, integral, tables 599–613

transition
    function 640
    matrix 255, 640
    probabilities 255
transitive
    directed graphs 29, 30
    relation 66
    relation in sets 204
transitivity
    in propositional calculus 200
    of implication 200
translation
    change of planar coordinates 301, 305
    discrete Fourier transform 583
    formula (plane Cartesian) . . . 305, 306
    Fourier transform 578
    Hilbert transform 592
    in space 349
    invariance of generating function 268
    isometry of space 349
    Laplace transform 586
    plane isometry 305
    Z-transform 595
transpose
    defined 131
    matrix 138
        and multiplication 143
transpositions, in permutation group . 163
transverse axis, hyperbola 327
trapezoid, defined 323
trapezoidal rule (quadrature) 750
traveling salesman problem
    genetic algorithms for 295
    solution by tabu search 292
trees *227*
    diagrams 241
    graph 229
    isomorphism classes of 241
    number of 26, 29
        labeled 239
        table 239
    properties 232
    rooted, number of 27
    spanning 227
triangles *317*
    acute 318
    altitude, defined 318
    angle-side-angle given 515
    area 319, 513
    Cholesky (matrix factorization) . . 157
    circumscribed circle 513

triangles (*continued*)
    coloring the corners 175
    cut into polygon 29
    cycle index and pattern inventory 175
    equilateral 320
    figure 319, 320
    general, properties 512
    graph type 227, 229
    inequality 321
        of a metric or distance function 69
        of a norm on a vector space 72
        vector norm 133
    inscribed circle 512
    isosceles 320
    law of cosines 513
    law of sines 513
    law of tangents 513
    median, defined 318
    Mollweide's formulae 513
    Newton's formulae 514
    obtuse 318
    Pascal 208
    planar, figure 512
    point inside, testing for 321
    polar 369
    random 626
    right *318*, 320
        and circular functions 512
        Pythagorean triples 98
    scalene 320
    side-angle-side given 515
    side-side-angle given 515
    sides from other components 513
    solution of 514
    spherical 367–369
    three sides given 514
triangular
    matrix 139
    numbers 29, 30
trichotomy, axiom of order 67
Tricomi equation (PDE) 469
tridiagonal matrices 744
    bandwidth 141
trigonometric
    series
        and Fourier transform 576
        example of Bessel's inequality . 74
        solution of cubic polynomials 89
trigonometric functions . . . *505*, 503–515
    and exponentials 508
    angle difference relations 509

trigonometric functions (*continued*)
 angle sum relations 509
 derivatives 386
 double-angle formulae 509
 figures 505, 506
 half-angle formulae 510
 identities among 509
 integrals 430
 inverse 518
  series for 45
 multiple-angle formulae 510
 powers 511
 products 511
 relations among (table) 508
 series for 44
 special angles (tables) 506
 sum 511
 symmetries 507
 table of values 516
trigonometry, spherical 368
trihedron, moving, regular curve 374
trillion 6
trimmed mean 659
triple
 products identity 48
 products of vectors 136
 Pythagorean 98
 root of polynomial 83
trisecting an angle 339, 345
trivial graph 227
trochoid, defined 340
Troy weight 794
truncated pyramid, defined 357
truth tables, as functions 200
truth tables, in propositional calculus 199
Tschebyscheff .... *see* Chebyshev
Tschirnhaus' transformation 91
Turán graph 230
Turán theorem 233
Turing, Alan 816
twin primes .... *see* primes, twin
two-
 factor ANOVA 689, 692
 form (differential) 396
 person games 274
 sample Kolmogorov–Smirnov ...717
 valued autocorrelation 264
two-sphere, tensor quantities 491
types
 of numbers 3
 of series 34

U

u-parameter curves, surfaces ...376, 379
Ulam numbers 26
umbilical point
 coordinate patch 379
 definition 378
 paraboloid of revolution 380
UN security council (voting) 280
unbiased estimator 663
uncertainty principle (Fourier transform) 581
undecagon, area and apothem 324
undenary 5
underdamping, linear ODE 460
undetermined linear equations 148
unicyclic graph 227
uniform
 convergence 71
 probability distribution
  continuous 633
  discrete 630
  notation 619
 random variable, and entropy 253
 resource locator .... *see* URL
uniform probability distribution 618
uniformly
 continuous function 71
 minimum variance unbiased estimator 663
unimodular group 171
unimodularity property 288
union
 of elements 69
 of sets 203
 operation on graphs 228, 229
unique
 factorization domain 165
 factorization property, in quadratic
  field 168
uniqueness
 generating function 268
 solution of PDE 470
 solution to difference equation ...265
 theorem 375
unit
 binormal vector, regular curve ...374
 cells, in crystallographic groups . 308
 impulse function 76
  Z-transform 595
 in a ring 164

unit (*continued*)
  in quadratic field 168
  normal vector 135
    coordinate patch 378
    paraboloid of revolution 380
  step function 77, 658
    Z-transform 595
  tangent vector, regular curve 374
  vector *133*
    normal, regular curve 374
    reciprocal system 136
unitary matrix, defined 140
United States 793
units
  English 796, 797
  entropy 253
  metric 796, 797
  of physical quantities 796
  United States 793
unity (ring property), table 167
unity, in a ring 164
univalent function, defined 66
universal quantifier 201
universal set 203
universe 202
unlabeled graph 225
unreliabilities, law of 653
unsolved problems 809
upper
  -triangular matrix 139
    in matrix factorization 157
  bandwidth, matrix 141
  bound of a subset 68
  control limit 650
  Hessenberg matrix 140
URL 31, 801
  contact information 803
  errata v
  math organizations 804–806
  mathematical resources 807–809

V

v-parameter curves, surfaces ...376, 379
valence
  contravariant, mixed tensor 483
  covariant, mixed tensor 483
value
  critical 695
  future 779
  present 779
Vandermonde convolution 208

Vandermonde matrix 140
variables, decision 281
variables, random .. *see* random variables
variance
  analysis 686
  bounds 664
  confidence interval 666
  multivariate, defined 622
  random variable, defined 620
  sample 660
  test 672, 677
variation of parameters
  difference equations 268
  differential equations 462
variational principles 407
Varsharmov–Gilbert bound (block codes)
  258
vector fields
  contravariant 485
    Ricci identity 487
  parallel 485, 488
vectors *131*
  addition 132
  algebra 131, 132
  average over random 625
  calculus of 390
  cocycle 222
  column *131*, *137*, 139
  cross product 135
  curvature, regular curve 374
  cut 223
  derivatives 391
  direction cosines 353
  divergence (tensor analysis) 488
  dot product 133
  equality 131
  Householder 141
  inner product 133
    and metric tensor 486
  Laplacian in orthogonal coordinates
    494
  moving trihedron, regular curve ..375
  multiplication by scalar 132, 142
  negative 131
  norm 133
  normal
    and vector product 135
    coordinate patch 377
    curvature, coordinate patch ...377
    unit, coordinate patch 378
    unit, paraboloid of revolution . 380

null *137*
   and metric tensor 486
observable, control theory 497
operation 158
orthogonal, defined 132
orthonormal, defined 132
parallel, defined 131
physical, examples 131
principal normal unit, regular curve
   374
product 135
   derivatives 391
representation 456
row *131, 137*, 139
scalar product 133
space
   and Galois field 170
   norm defined 72
state, control theory 497
tangent, angle between 486
tangent, as tensor 489
triple product 136
unit *133*
   binomial, regular curve 374
   tangent, regular curve 374
velocity
   of light 794
   of sound 795
   units 796
Venn diagrams 203
versed cosine, defined 505
versed sine, defined 505
versine, defined 505
vertex *317*
   central 221
   cone 361
   cover, condition for 233
   cut 223
   degree 223
     sequence 223
     theorems on (graphs) 237
   end 224
   Euler formula (convex polyhedra) 357
   graph 219
   identification 224
   isolated 225
   set, graph notation 219
   similarity 226
   space of a graph 227
   symmetric graph 227
veto power, in voting game 278

Viete formulae, polynomial roots and
   coefficients 84
Viete, François 812
vigesimal 5
Vizing's theorem (graphs) 234
volt 792
Volterra integral equation 478
   difference kernel 480
   equivalence to initial-value problem
   479
   second kind 481
volume
   circular torus 363
   cone 362
   cubes 357
     doubling 345
   cylinder 361
   earth 798
   ellipsoid 364
   enclosed by a surface 410
   frustum 362, 363
   measure 794
   of revolution, as integral 409
   of the oceans 798
   of the universe 799
   prism 357
   pyramid 357
   rectangular parallelepiped 357
   regular polyhedra 360
   sphere 13, 367
     n-dimensional 368
   spherical cap 367
   spherical segment 368
   spherical zone 367
   surface of revolution 363
   tetrahedron 358
   truncated pyramid 357
   units 796
voting power 278–280, 817
   examples 279

W

walk
   and connectivity 231
   in a graph 228
wallpaper groups 307, 809
Walsh function 722, 725
Waring's problem 100
watts 792, 798

wave equation
  algorithm for solution 775
  PDE 469, 471
  section 775
  solutions 475
wavelength 795
wavelets 723–726, 808
  Battle–Lemarié 725
  biorthogonal 725
  coefficients 725
  Daubechies 724
  figures 726
  Haar 723
  Meyer 725
  packets 725
weak duality theorem 286
weather example 656
web sites ....*see* URL
weber 792
Wedderburn–Etherington numbers ...26
Weddle's rule (quadrature) 751
wedge product of 1-forms 395
Weibull distribution 655
Weierstrass
  Bolzano– theorem 70
  Casorati– theorem 57
  Karl 815
  theorem (infinite product) 47
  theorem (polynomial approximation)
     72
weight 794
  block coding 256
  function, in orthogonal polynomials
     532
  of a mixed tensor 483
weighted
  mean 659
  Monte–Carlo integration 763
  sum of squares, defined 138
  voting game 278
Weingarten equation, surfaces 379
well posed initial-value problem 766
well-conditioned matrix, defined 149
well-posedness, PDEs 470
wheel (type of graph) 229
white noise 718
Whitney . *see* Mann–Whitney–Wilcoxon
Wieferich prime 106
Wielandt deflation (eigenvalues) 744
Wiener filter 721
Wigner coefficients 574

Wilcoxon *see* Mann–Whitney–Wilcoxon
Wilf, Szekeres– theorem (graphs) ..234
Williams, Nash– arboricity theorem 232
Wilson prime 106
Wilson theorem 95
witch of Agnesi 337, 814
Woodall primes, table 106
work, units 796
Wronskian
  in variation of parameters for ODE
     462, 463
  relation, Airy functions 566
WWW ....*see* URL

X

x-intercept of a line 314

Y

y-intercept of a line 314
Yarborough, in bridge (card game) ..628
yards 797
  in other units 794, 796
year
  day of 800
  leap 799
yocto, as prefix 6
yotta, as prefix 6

Z

Z-transform 594
  sequence 269
zenith (spherical coordinates) 346
zepto, as prefix 6
Zernike polynomials 537
zero
  -sum game 275
    mixed 276
  matrix 138
  of Bessel functions 563
  of complex function 56
  simple 729
zeta function 23, 35
  in Möbius function 102
  in terms of polylogarithm 552
  Riemann hypothesis 24
  table of values 24
zetta, as prefix 6
Zidek, Narayana––Capell numbers ..26
zone, spherical 367